The Greek Alphabet

Alpha	A	α	Iota	I	ι	Rho	P	ρ
Beta	B	β	Kappa	K	κ	Sigma	Σ	σ
Gamma	Γ	γ	Lambda	Λ	λ	Tau	T	τ
Delta	Δ	δ	Mu	M	μ	Upsilon	Υ	υ
Epsilon	E	ϵ	Nu	N	ν	Phi	Φ	ϕ
Zeta	Z	ζ	Xi	Ξ	ξ	Chi	X	χ
Eta	H	η	Omicron	O	o	Psi	Ψ	ψ
Theta	Θ	θ	Pi	Π	π	Omega	Ω	ω

Abbreviations for Units

A	ampere	ℓ	liter
Å	angstrom (10^{-10} m)	m	meter
atm	atmosphere	MeV	megaelectron volts
C	coulomb	mi	mile
°C	degree Celsius	min	minute
cal	calorie	mm	millimeter
cm	centimeter	msec	millisecond
eV	electron volt	N	newton
°F	degree Fahrenheit	nm	nanometer (10^{-9} m)
fm	femtometer, fermi (10^{-15} m)	pt	pint
ft	foot	qt	quart
G	gauss	rev	revolution
gm	gram	sec	second
H	henry	T	tesla
h	hour	u	unified mass unit
Hz	hertz	V	volt
in	inch	W	watt
J	joule	y	year
K	kelvin	yd	yard
kg	kilogram	μm	micrometer (10^{-6} m)
km	kilometer	μsec	microsecond
keV	kiloelectron volts	μC	microcoulomb
lb	pound	Ω	ohm

PHYSICS

Physics

Paul A. Tipler

Oakland University
Rochester, Michigan

WORTH PUBLISHERS, INC.

Physics

Paul A. Tipler

Printed in the United States of America

Library of Congress Catalog Card No. 74-82693

ISBN: 0-87901-041-X

First Printing, January 1976

Designed by Malcolm Grear Designers

Composition by Progressive Typographers, Inc.

Illustrated by Felix Cooper

Picture research by Susan Haggerty

Worth Publishers, Inc.

444 Park Avenue South

New York, NY 10016

For Libor Velinsky

Preface

This is a textbook for the standard two-semester or three-semester elementary physics course for engineering and science majors. It is assumed that the students have taken, or are concurrently taking, calculus. The usual topics in classical physics are presented: Mechanics (Chapters 1–16), Thermodynamics (Chapters 17–19), Waves (Chapters 20–27), and Electromagnetism (Chapters 29–41). In addition, special relativity is discussed in Chapter 28 and quantization in Chapter 42.

Although I have deviated only slightly from the traditional order of topics, I have tried to emphasize the unity of physics and the economy of physical theory by relating many phenomena through a common theoretical description. Many topics are introduced as examples illustrating more general physical principles. For example, there is no separate chapter on fluid mechanics. Instead, Archimedes' principle is discussed as an example of the application of Newton's laws in Chapter 5, and Bernoulli's equation is derived as an application of the work-energy theorem in Chapter 9. Similarly, the relationship between gas pressure and molecular kinetic energy is obtained in Chapter 11 as an application of the impulse-momentum equation. This immediate application of general principles to the description of interesting phenomena makes the theory less abstract for the beginning student.

The teaching of elementary physics is complicated by the wide range of student backgrounds and abilities in a typical class. Some have had good physics courses and a calculus course in high school; others have had no high school physics and are just beginning to struggle with their first calculus course. I have tried to write this book so that the least prepared students will be able to read it, enjoy it, and gain confidence because they can work most of the exercises. At the same time, optional sections and some difficult problems are included to challenge the better-prepared students. Because of these varied mathematical backgrounds, I have included a review of trigonometry in the first chapter and discussed differentiation and integration while introducing the kinematic concepts of velocity and acceleration in Chapter 2.

Although there is beauty and elegance in the mathematical derivation of specific results from a general theory, most students learn more

easily when new concepts are introduced in an intuitive, nonmathe-
matical way, proceeding from a specific example to the more general
relation. I have followed this approach in many cases, supplying the
general or formal derivation in an optional paragraph at the end of the
section or chapter. Whenever possible, concepts are introduced in one
dimension and later generalized to three dimensions.

Derivations are given without calculus whenever possible, with
calculus-based derivations included for comparison. Simple differentiation and integration are illustrated in worked examples and in
problems throughout the book. Material that is more difficult mathematically, or that can be omitted with no loss in continuity, is labeled
"optional" and is set in the right column of the page, bounded on
either side by colored rules. Many derivations are relegated to this
optional column.

Frequently, to show the interrelatedness of phenomena, a concept is
mentioned briefly even though it will be treated in detail later. For
example, Newton's law of gravitation and Coulomb's law are introduced as examples of basic forces (Chapter 6) and later treated in detail
(in Chapters 16 and 29). Similarly, gravitational potential energy is
mentioned in terms of recovery of work (at the end of Chapter 7), then
discussed more fully when the idea of conservative forces is introduced.

Material dealing with applications and techniques is often placed in
"how-to-do-it" sections and chapters. This allows the use of many
examples, graded in difficulty, without disturbing the continuity of the
discussion of the concepts. For example, Newton's laws are discussed
in Chapter 4, while their application to problems is shown by examples in Chapter 5. The electric field and its properties are discussed
in Chapter 29, while calculation of the electric field for various charge
distributions is illustrated in Chapter 30.

A minor deviation from the traditional order of topics is the brief
discussion of the four basic interactions presented early in classical
mechanics (Chapter 6) along with a discussion of molecular forces and
such empirical contact forces as support forces, friction, and the forces
exerted by strings and springs. The mention of gravitational forces and
electrostatic forces here allows for more varied examples of potential-
energy functions in Chapter 8 than just those associated with constant
forces or spring forces. It also permits discussions of such interesting
phenomena as escape velocity, binding energy, and molecular dissociation as examples of the use of energy conservation in Chapter 9.

A more significant change from the traditional order is the unified
treatment of waves. Reflection, refraction, interference, diffraction, dispersion, etc., are discussed as general wave phenomena with applications to waves on strings, sound, and light. This integration of various
wave phenomena illustrates the wide applicability of physical theory
while making it easier for students to understand the important properties of sound and light. In particular, the general discussion of diffraction and the validity of the ray approximation in Chapter 25 lays
the foundation for the understanding of the wave-particle duality of
nature (Chapter 42), and for the student's future study of quantum
theory. I have placed this material before electricity and magnetism for
two reasons: 1. the study of simple harmonic motion is then relatively
fresh in the students' minds, and 2. most students find the material on
waves, with its many possibilities for classroom demonstrations, easier
and more intuitive than the more abstract ideas in electricity and
magnetism. Instructors who prefer to present electricity and magnetism

first can easily do so by skipping to Chapter 29 after completing Chapter 19 and later returning to Chapter 20.

Within each chapter are questions placed at the ends of various sections. Some are routine and can easily be answered from the material in the preceding section. Others are open-ended and can serve as a basis for classroom discussion. At the end of each chapter is a review section. This section generally consists of a checklist of words and phrases which students are asked to "define, explain, or otherwise identify" and a set of true-false questions. The page number where the definition of the term is given appears after each item in the checklist, and answers to all the true-false questions are given in the back of the book. Following the review section, there is an extensive set of exercises (usually about 30) organized by the sections within the chapter. These exercises are not difficult, and each involves material in that section of the chapter only. The exercises are followed by a set of problems, which tend to be somewhat more difficult than the exercises, involve more mathematics, or require the student to use material from more than one section and perhaps supply reasonable data from his or her experience. Answers (and sometimes hints) to all the odd-numbered exercises and problems are given at the back of the book. I suggest the assignment of many more exercises than problems. It is my experience that self-confidence is an important factor in learning and can be easily destroyed by too many challenging problems. The exercises are designed to build the student's self-confidence.

Fifteen essays have been written especially for this book. Three are biographical (on Isaac Newton, Albert Einstein, and Benjamin Franklin), many are on applications (e.g., xerography, transistors, and radar astronomy), and others are on topics of current interest (e.g., thermal pollution, black holes, and energy resources). These brief essays are for the enjoyment and enlightenment of students and instructors. There are no exercises or problems connected with the essays.

Although the international system of units (SI) is used predominately, the British engineering system is introduced and used often in the mechanics chapters, and commonly used units such as the calorie, eV, atmosphere, etc., are also used where convenient.

For the sake of completeness and flexibility, the book contains more material than can be covered in two semesters. There should be enough for a three-semester course or, perhaps with some supplement in modern physics, a four-semester course. To use this book in a two-semester course such as I teach, some sections and chapters in addition to those labeled "optional" must be omitted or at least skimmed over. One method for broadening a course (while sacrificing some of the depth) is to assign a chapter or section to be read, without assigning the associated exercises or problems. I would recommend this to those who are tempted to omit Chapter 15, because the important ideas of resonance should be at least presented to students as early as possible. The inclusion or omission of certain topics naturally depends on the judgment of the instructor and the particular needs of the students. For a two-semester course, I would suggest, in addition to the optional sections, the omission of some of the following: Sections 5-4 (Motion with a Retarding Force Proportional to the Speed), 9-3 (Bernoulli's Equation), 10-6 (Using Integration to Find the Center of Mass), 11-7 (Elastic Collisions in Three Dimensions), 11-9 (Reaction Threshold), 11-10 (Coefficient of Restitution), 12-6 (Static Equilibrium of a Rigid Body), 13-6 (Motion of a Gyroscope), 14-8 (Combinations of Two Simple Harmonic Motions), 21-7 (Energy and Intensity of Harmonic Sound Waves), 23-3 (Harmonic Analysis and Synthesis), 23-4 (Wave Packets), 23-5 (Dispersion), 25-8 (Bragg Scattering), 26-8 (Diffraction Gratings), 29-8 (Electric Dipole in Electric Fields), 34-6 (Conservation of Charge and Approach to Electrostatic Equilibrium), 37-3 (Special Relativity and the Magnetic Field), and Chapters 28 (Special Relativity), 40 (Alternating-Current Circuits), 41 (Maxwell's Equations and Electromagnetic Waves), and 42 (Quantization).

Acknowledgments

Many people have contributed to this book. I would like to thank all those who have reviewed the various drafts and preliminary edition and have offered many helpful suggestions. In particular, James Gerhart was a constant source of information, inspiration, and encouragement through all the early drafts and preliminary edition. His generous hospitality during my pleasant and productive summer at the University of Washington and his help with writing problems, drawing figures, and criticizing the manuscript was largely responsible for getting the project off the ground. John McKinley's line-by-line review of the entire final manuscript, his checking of the final figures, and his overall help with all phases of the final production, including the reading of the galleys, was invaluable. Jerry Griggs suggested many of the interesting exercises and problems, and worked all the odd-numbered exercises and problems to check the answer section. Stanley Williams reviewed all the exercises and problems, supervised the working of them, and suggested exercises and problems for the latter chapters. Others who reviewed various parts of the manuscript include Reuben E. Alley, Jr., U.S. Naval Academy; Wallace Arthur, Fairleigh Dickinson University; Leo L. Baggerly, California State College, Bakersfield; Bill Beam, Rose-Hulman Institute of Technology; James R. Benbrook, University of Houston; Robert A. Boyer, Muhlenberg College; Morton K. Brussel, University of Illinois; Charles Buchanan, UCLA; Robert Folk, Lehigh University; Kenneth N. Geller, Drexel University; Sigmund P. Harris, Los Angeles Pierce College; Alvin M. Hudson, Occidental College; Granvil C. Kyker, Rose-Hulman Institute of Technology; David A. Nordling, U.S. Naval Academy; Jack Prince, Bronx Community College; Edward A. Saunders, U.S. Military Academy; D. J. Schlueter, Purdue University; Charles S. Shapiro, California State University, San Francisco; Stanley J. Shepherd, Pennsylvania State University; Dan Sober, UCLA; Martin Tiersten, CUNY; Miro M. Todorovich, Bronx Community College; and Al Verone, Oakland Community College.

I would like to thank the students at Oakland University and at Rose-Hulman Institute of Technology, who used the preliminary edition and various early drafts and who offered the kind of criticism only students can give. The preliminary edition was skillfully typed by La Verna Cole and the entire final manuscript, as well as several early drafts, by Sue Nast. I am grateful to my wife Sue for her continuing support and for her expert work on the index. Finally, I received much help and encouragement from Worth Publishers.

PAUL A. TIPLER

Supplements

Instructor's Manual   This comprehensive manual provides many aids and suggestions for teaching the course, including classroom demonstrations, additional applications and topics, suggestions for outside reading, and a guide to films that can be rented or purchased, keyed to the chapters in the text.

Student Study Guide   Available for student purchase, this includes the following for each chapter in the text:

A chapter summary including the fundamental equations.

A list of key items for review and self-testing.

True-false questions.

Essay questions to allow students to apply what they have learned.

Worked-out examples to supplement those in the text. Guidance in solving given problems is provided in three steps so students can take just as much help as they need.

Answers (and explanations) to all true-false and essay questions.

Contents in Brief

Contents

PHYSICS

CHAPTER 1   Introduction

Man has always been curious about the world around him. Since the beginnings of his recorded thought he has sought ways to impose order on the bewildering diversity of observed events. This search for order takes a variety of forms. One is religion, another is art, a third is science. The word science has its origins in a Latin verb meaning "to know," but science has come to mean not merely knowledge but knowledge specifically of the natural world and, most importantly, a body of knowledge organized in a specific and rational way.

Although the roots of science are as deep as those of religion or of art, its traditions are much more modern. Only in the last few centuries have there been methods for studying nature systematically. They include techniques of observation, rules for reasoning and prediction, the idea of planned experimentation, and ways for communicating experimental and theoretical results—all loosely referred to as the *scientific method*. An essential part of the advance of our understanding of nature is the open communication of experimental results, theoretical calculations, speculations, and summaries of knowledge. A textbook is one of these forms. An elementary textbook such as this has two purposes: It is designed, first, to introduce the newcomer in the field of science to material which is already widely known in the scientific and technical community and which will form the basis of his or her more advanced studies of this knowledge. It may also serve to acquaint a student not majoring in science with information and with a way of thinking that is having a cumulative effect upon our way of life.

This book concentrates on the subjects of *classical physics,* a term that usually refers to mechanics, light, heat, sound, electricity, and magnetism, subjects well understood in the late nineteenth century before the advent of relativity and quantum theory, which were developed in the early years of the twentieth century. The application of special relativity, and particularly quantum theory, to the description of such microscopic systems as atoms, molecules, and nuclei and to a detailed understanding of solids, liquids, and gases is often referred to as *modern physics*. Although modern physics has made many important contributions to technology, the great bulk of technical knowledge and skill is still based squarely on classical physics. While concentrating on classical physics, we shall often discuss modifications based on quantum theory and special relativity, and compare the predictions of the classical and modern theories with each other and with the results of experiments.

Except for the interior of the atom and for motion at speeds near the speed of light, classical physics correctly and precisely describes the behavior of the physical world. It is through applications of classical physics that we have been able to exploit natural resources successfully, and it is largely through applications of classical physics that we will find the technical means necessary to preserve our environment for successful and controlled future use. Moreover, modern physics builds on the concepts of classical physics and cannot be understood without them. Consider, for a moment, the modern descriptions of the atom. In the first successful theory of the atom, the Bohr theory of 1913, the atom was pictured as a miniature solar system with the electrons revolving around the nucleus in circular or elliptical orbits. The attraction of the electron to the nucleus was described by the classical theory of electricity. The fact that the orbits had to be circular or elliptical followed directly from Newton's laws of mechanics. In the Bohr theory, the classical principles of *conservation of mechanical energy* and of *angular momentum* were fundamental assumptions. In the modern quantum-mechanical theory of the atom, the electron is represented by a *wave packet*. The energy of the atom is *quantized* by postulating that the *electron wave function* obeys a *standing-wave condition*. As with the Bohr theory, *energy conservation* determines the possible *energies of radiation* when the atom changes *energy states*. When the atom is in a *magnetic field,* the *radiation spectrum* of the atom is slightly different. This difference is related to the frequency with which the *magnetic moment* precesses about the *magnetic field* due to the *torque* exerted by the *field* on the atom. This *precession* is similar to that of a spinning top or gyroscope.

Many of the terms and concepts mentioned in the above paragraph are probably unfamiliar to you now. Don't let this bother you. All of these concepts—angular momentum, energy, potential energy, torque, precession, wave packet, wave function, standing-wave condition, conservation of energy, magnetic field—are concepts of classical physics which we discuss in this book. With a thorough understanding of these concepts and familiarity with the role they play in classical physics, you will be well equipped to study the modern theory of the atom or any other topic in modern physics.

1-1   Mechanical Models

Classical mechanics, fascinating in its own right, is a particularly good subject to start with. Since so much of our everyday experience is directly applicable, classical mechanics is less abstract than subjects like electric and magnetic fields. Many abstract concepts in physics, e.g., potential energy, are easiest to learn in the framework of classical mechanics. But classical mechanics is important for other reasons. Many of the modern microscopic theories of matter are stated in terms

Marshall Lockman/Black Star

Mechanical model of a DNA molecule.

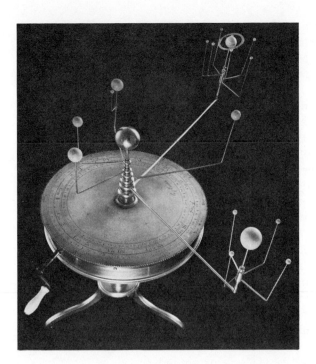

Eighteenth-century mechanical model of the solar system. (*Courtesy of Harvard University.*)

of mechanical models, like the Bohr model of the atom, already mentioned. We say, for example, that the magnetic moment of the electron precesses like the precession of a spinning top. In a useful mechanical model of solid matter, atoms are pictured as being connected to each other by springs, and the vibrations of the atoms are coupled to each other like those of coupled mechanical oscillators. The classical mechanical theory of oscillation is therefore central to the modern theory of solids.

There is great economy in the methods of physics. Consider the motion of a mass on a spring, which we shall look at in detail when we study simple harmonic motion. Later we shall consider different types of oscillation and resonance. What you learn from these studies is directly applicable to many other phenomena; e.g., the behavior of electric circuits with capacitance, inductance, and resistance is described by exactly the same kind of mathematical equations as a mass oscillating on a spring. The exchange between the electric energy in the capacitor and magnetic energy in the inductor and the dissipation of energy in the resistor are analogous to the exchange between potential and kinetic energy of a mass on a spring and the dissipation of energy by friction. This same study is applicable to wave motion, whether it is the simple harmonic motion of a string segment for standing waves on a string, the oscillations of air particles during the passage of sound waves through air, or the oscillations of the electric and magnetic field vectors in a light wave. In modern theory a thorough knowledge of the classical oscillator is necessary in many quantum-mechanical problems.

1-2 Mathematics and Physics

The laws of physics are generalizations from observations and experimental results. For example, Newton's law of universal gravitation was based on a variety of observations: the paths of planets in their motion about the sun, the acceleration of objects near the earth, the acceleration of the moon in its orbit, the daily and seasonal variation in the tides, etc. Physical laws are usually expressed as mathematical equations.[1] The law of gravitation states that any two objects attract each other with a force which is proportional to the mass of each object and inversely proportional to the square of the distance between them. It is written $F = Gm_1m_2/r^2$. The equation can then be used to make predictions about other phenomena and to test the range of validity of the law. For example, together Newton's law of gravitation and Newton's laws of motion can be used to predict the orbits of planets, comets, and satellites. Understanding such predictions requires knowledge of elementary calculus and the ability to manipulate and solve simple differential equations, for the laws of motion are expressed as second-order differential equations. The mathematical equations governing wave motion are partial differential equations. Thus to understand physics on any level beyond a qualitative description, a considerable amount of mathematics is needed. It is usually easiest to learn the physics and the necessary mathematics at about the same time since the immediate application of mathematics to a physical situation helps you understand both the physics and the mathematics. In this book it is assumed that you have taken or are now taking a course in calculus. Much of the calculus needed to understand physics is presented in outline form. Since calculus is a new and unfamiliar tool for most students beginning physics, it is used here as simply as possible. As this book is about physics and not mathematics, mathematical details and rigor are not stressed. In the more difficult mathematical problems, plausibility arguments are often substituted for rigorous derivations in order to appeal to your intuition.

1-3   How to Learn Physics

A textbook is only one tool for learning physics. A good teacher, lecture demonstrations, films, and experimental work in the laboratory are indispensible. Outside reading is highly recommended. While you concentrate on obtaining a rigorous understanding of classical physics in your introductory course, you should be broadening your familiarity with contemporary physics by reading widely in the many excellent popular and semipopular accounts of modern science, e.g., those in *Scientific American.*

At the end of each chapter are a review, exercises, and problems. The importance of solving problems to learn physics cannot be overemphasized. Only in working problems can you find out whether you have really grasped the text material. Many details can be brought out in problems that cannot be treated in any other way. You should do as many problems as possible, whether assigned or not. One way to gain practice and experience in problem solving is to use the examples in the text as problems. Read the statement or question in the example and then attempt to answer the question or work through the example without looking at the text. When you finish or if you get stuck, look at the worked example. This approach will demonstrate how well you understand the material and what mistakes you may be making.

---

[1] Some laws are expressed as mathematical inequalities; e.g., the second law of thermodynamics can be stated: the entropy change of the universe in any period of time is always greater than or equal to zero.

(Don't be discouraged if you are often stuck. Although many examples are rather direct applications of material discussed in the text, some introduce a new method of solution or approach. A few examples cover famous results and are really an extension of the text. In these cases, you cannot expect to work through the examples on your first try without difficulty.)

Scattered throughout the text are questions. Some are very easy and are answered in the previous discussion. If you have understood the discussion, you should be able to answer them with little trouble. Other questions are meant to extend or apply the discussion in the text. Some have no simple answers but are food for thought. Whether or not you are taking a simultaneous laboratory course, you should be alert to simple experiments or observations relating the concepts in the book to your experience in the real world. The questions should stimulate your ability to relate your study to everyday experience.

It is possible to state the laws of physics in concise statements and equations and use them to deduce the behavior of many systems under various conditions. Although there is a certain aesthetic appeal in this deductive approach, it is not an easy way to learn physics or to develop an understanding of the workings of nature. What we now know of nature results from the efforts of many different people and many years of experimentation, theoretical proposals, and debate. Sometimes discussing the history of an idea helps us learn the presently accepted view. Considering rather special cases can be a helpful preliminary to the general discussion of a physical law. For example, some familiarity with many new concepts can be obtained by treating one-dimensional problems. The extension to three dimensions is then much easier than starting with the general three-dimensional problem would be. We shall consider one-dimensional motion first to introduce some of the concepts of velocity and acceleration before generalizing to two- and three-dimensional motion. Similarly, the ideas of work, potential energy, conservative forces, and many of the properties of waves are first introduced in one dimension before the general three-dimensional situation is treated.

1-4   Units

The laws of physics express relationships between physical quantities such as length, time, force, energy, and temperature. Measuring such a quantity involves comparison with some unit value of the quantity. The most elementary measurement is probably that of distance. In order to measure the distance between two points we need a standard unit, e.g., a meter stick or a ruler. The statement that a certain distance is 25 meters means that it is 25 times the length of the unit meter; i.e., a standard meter stick fits into that distance 25 times. It is important to include the unit meters along with the number 25 in expressing a distance because there are other units of distance in common use. To say that a distance is 25 is meaningless.

The units of all physical quantities can be expressed in terms of a small number of fundamental units. For example, a unit of speed, e.g., meters per second or miles per hour, is expressed in terms of a unit of length and a unit of time. Similarly, any unit of energy can be expressed in terms of the units of length, time, and mass. In fact, all quantities occurring in the study of mechanics can be expressed in

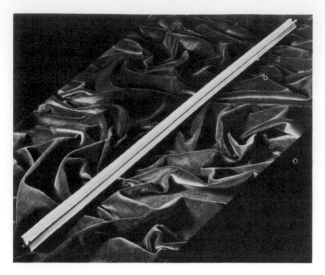

Standard meter bar, kept at the National Bureau of Standards, Washington, D.C., the United States national standard from 1893 to 1960. The bar, made of a platinum-iridium alloy, is an exact duplicate of the international prototype meter housed at Sèvres, France. (*Courtesy of the National Bureau of Standards.*)

Figure 1-1
The standard of length, the meter, was chosen so that the distance from the equator to the North Pole along the meridian through Paris would be 10^7 m. The circumference of the earth is therefore 4×10^7 m, and the radius is $(2 \times 10^7)/\pi$ m $= 6.37 \times 10^6$ m.

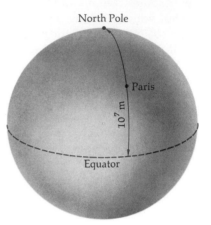

terms of these three fundamental units. The choice of standard units for these fundamental quantities determines a system of units for all mechanical quantities. In the system used universally in the scientific community, which we shall use most often, the standard length is the meter, the standard time is the second, and the standard mass is the kilogram. This system of units is called the mks system (after the meter, kilogram, and second). The standard of length, the meter (abbreviated m), was originally indicated by two scratches on a platinum-iridium alloy bar kept at the International Bureau of Weights and Measures in Sèvres, France. This length was chosen so that the distance between the equator and the North Pole along the meridian through Paris would be 10 million meters (Figure 1-1). (After construction of the standard meter bar, it was found that this distance differs by a few hundredths of a percent from 10^7 m.) The standard meter was used to construct secondary standards, which are used to calibrate measuring rods throughout the world. The standard meter is now defined in terms of the wavelength of a particular spectral line of an isotope of krypton. It is 1,650,763.73 wavelengths of this light. This change made comparison of lengths throughout the world easier and more accurate.

The unit of time, the second (sec), was originally defined in terms of the rotation of the earth to be $\frac{1}{60}(\frac{1}{60})(\frac{1}{24})$ of the mean solar day. The second is now defined in terms of a characteristic frequency associated with the cesium atom.

The unit of mass, the kilogram (kg), equal to 1000 grams (gm), is defined to be the mass of a particular standard body, also kept at Sèvres. We shall discuss the concept of mass in detail in Chapter 4.

The units of other quantities in mechanics, such as speed and momentum, are derived from the three fundamental units of length, time, and mass. The choice of these three as the fundamental quantities is somewhat arbitrary. For example, we could replace the choice of a standard length with a standard speed, e.g., the speed of light in vacuum, and give this unit a name. The unit of length would then be derived from the product of this speed unit and the unit of time.

In studying thermodynamics and electricity, we shall need two more fundamental physical units, the unit of temperature, the kelvin (formerly the degree kelvin), and the unit of current, the ampere. These five fundamental units, the meter (m), the second (sec), the kilogram (kg), the kelvin (K), and the ampere (A), constitute the international system of units (SI).[1] The unit of every physical quantity can be expressed in terms of these fundamental SI units. Some important combinations are given special names. For example, the SI unit of force, kg-m/sec², is called a newton (N). Similarly, the SI unit of power, kg-m²/sec³ = N-m/sec, is called a watt (W). Prefixes for common multiples and submultiples of SI units are listed in Table 1-1. These multiples are all powers of 10. Such a system is called a *decimal system;* the decimal system based on the meter is called the *metric system.* The prefixes can be applied to any SI unit; for example, 10^{-2} meter is 1 centimeter (cm); 10^{-3} watt is 1 milliwatt (mW).

In another system of units used in the United States, the *British engineering system,* the unit of force, the pound, is chosen to be a fundamental unit instead of mass. (We shall discuss the relation between force and mass in detail in Chapter 4.) The unit of force is defined in terms of the gravitational attraction of the earth at a particular place for a standard body. The other fundamental units in this system are the foot and the second. The second is defined as in the international system. The foot is defined as exactly one-third of a yard, which is now legally defined in terms of the meter

$$1 \text{ yd} = 0.9144 \text{ m}$$

$$1 \text{ ft} = \tfrac{1}{3} \text{ yd} = 0.3048 \text{ m}$$

making the inch exactly 2.54 cm. This system is not a decimal system. It is less convenient than the SI or other decimal systems because common multiples of the unit are not powers of 10. For example, 1 ft = $\frac{1}{3}$ yd, and 1 in = $\frac{1}{12}$ ft. We shall see in Chapter 4 that mass is a better choice of a fundamental unit than force because mass can be defined without reference to the gravitational attraction of the earth. Relations between the British engineering system and SI units are given in Appendix C.

Questions

1. What are the advantages and disadvantages of using the length of a person's arm for a standard of length?

2. A certain clock is consistently 10 percent fast when compared with the standard cesium clock. A second clock varies in a random way by 1 percent. Which clock would make a more useful secondary standard for a laboratory? Why?

1-5   Conversion of Units

We have said that the magnitude of a physical quantity must include both a number and a unit. When such quantities are added, sub-

Table 1-1
Prefixes for powers of 10

Multiple	Prefix	Abbreviation
10^{12}	tera	T
10^{9}	giga	G
10^{6}	mega	M
10^{3}	kilo	k
10^{-1}	deci	d
10^{-2}	centi	c
10^{-3}	milli	m
10^{-6}	micro	μ
10^{-9}	nano	n
10^{-12}	pico	p
10^{-15}	femto	f
10^{-18}	atto	a

Definition of yard and foot

Roman-Greek tablet showing the length of the standard foot about A.D. 300–500. (*Courtesy of the Science Museum, London.*)

[1] SI stands for *Système International.* The sixth SI unit, the candela, a unit of luminous intensity, we have no occasion to use in this book.

tracted, multiplied, or divided in an algebraic equation, the unit can be treated like any other algebraic quantity. For example, suppose you wish to find the distance traveled in 3 hours (h) by a car moving at a constant rate of 60 miles/hour (mi/h). The distance x is just the speed v times the time t:

$$x = vt = 60 \,\frac{\text{mi}}{\cancel{\text{h}}} \times 3 \,\cancel{\text{h}} = 180 \text{ mi}$$

We cancel the unit of time, the hour, just as we would any algebraic quantity to obtain the distance in the proper unit of length, the mile. This method makes it easy to convert from one unit of distance to another. Suppose we want to convert our answer of 180 mi into feet. We use the fact that

$$5280 \text{ ft} = 1 \text{ mi}$$

If we divide each side of this equation by 1 mi, we obtain

$$\frac{5280 \text{ ft}}{1 \text{ mi}} = 1$$

We can now change 180 mi to feet by multiplying by the factor 5280 ft/mi:

$$180 \text{ mi} = 180 \text{ mi} \times \frac{5280 \text{ ft}}{1 \text{ mi}} = 9.50 \times 10^5 \text{ ft}$$

We have rounded the result to three significant figures, consistent with the fact that the distance 180 mi is given to only three significant figures. The factor 5280 ft/mi is called a *conversion factor*. It has the value 1 and is used to convert a quantity expressed in one unit of measure into the equivalent in another unit of measure. By writing out the units explicitly, we need not think about whether we multiply by 5280 or divide by 5280 to change miles to feet or feet to miles because the units tell us if we have chosen the correct or incorrect factor.

Example 1-1 What is the equivalent of 60 mi/h in feet per second?
   We use the above conversion factor plus the facts that 60 sec = 1 min and 60 min = 1 h. Then

$$60 \,\frac{\cancel{\text{mi}}}{\cancel{\text{h}}} \times \frac{5280 \text{ ft}}{1 \,\cancel{\text{mi}}} \times \frac{1 \,\cancel{\text{h}}}{60 \,\cancel{\text{min}}} \times \frac{1 \,\cancel{\text{min}}}{60 \text{ sec}} = 88 \,\frac{\text{ft}}{\text{sec}}$$

The quantity 60 mi/h is multiplied by a set of conversion factors each having the value 1, so that the value of the speed is not changed.

1-6   Dimensions of Physical Quantities

The area of a plane figure is found by multiplying one length by another. For example, the area of a rectangle of sides 2 and 3 in is $A = (2 \text{ in})(3 \text{ in}) = 6 \text{ in}^2$. The units of this area are square inches. Because area is the product of two lengths, it is said to have the *dimensions* of length times length, or length squared, often written L^2. The idea of dimensions is easily extended to other nongeometric quantities. For example, speed is said to have the dimensions of length divided by time, or L/T. The dimensions of any quantity in mechanics are written in terms of the fundamental quantities length, time, and mass. Adding two physical quantities makes sense only if the quantities have the same dimensions. For example, we cannot add an area to a speed to obtain a meaningful sum. If we have an equation like

$$A = B + C$$

the quantities A, B, and C must all have the same dimensions. Addition of B and C requires that these quantities be in the same units. For example, if B is an area of 500 in^2 and C is 4 ft^2, we must either convert B into square feet or C into square inches to find the sum of the two areas.

We can often find mistakes in a calculation by checking the dimensions or units of the quantities in the calculation. Suppose, for example, we are using a formula for distance x given by

$$x = x_0 + vt + \tfrac{1}{2}at$$

where t is the time, x_0 is the initial distance at time $t = 0$, v is the speed, and a is the acceleration, which (as we shall see in the next chapter) has dimensions of L/T^2 and SI units of meters per second per second (m/sec^2). We can see immediately that this formula cannot be correct: since x has dimensions of length, each term on the right side of the equation must have dimensions of length. Both x_0 and vt have dimensions of length, but the dimensions of $\tfrac{1}{2}at$ are $(L/T^2)T = L/T$. Since the last term does not have the correct dimensions, an error has been made somewhere in obtaining the formula. Dimensional consistency is a necessary condition for an equation to be correct. It is, of course, not sufficient. An equation can have the correct dimensions in each term without describing any physical situation.

Our knowledge of the dimensions of quantities often enables us to recall a formula or sometimes even guess at it. Suppose we remember that there is a simple formula relating the velocity of a wave v (dimensions L/T), the wavelength λ[1] (dimensions L), and frequency f (dimensions T^{-1}). There is only one dimensionally consistent equation expressing one of these quantities as the product of the other two; it is

$$v = f\lambda$$

This equation happens to be correct. If we were to write down $f = v\lambda$, we could see immediately by checking the dimensions that it is incorrect.

Questions

3. The speed v of an object dropped from rest depends on the time t and on the acceleration of gravity g, which is a constant and has dimensions LT^{-2}. From dimensional considerations only, guess at a possible relation between v, g, and t.

4. The distance d an object falls when dropped from rest also depends on the time t and the acceleration of gravity g. Guess at a relation among d, g, and t.

---

[1] λ is the Greek letter lambda. A list of the Greek alphabet will be found inside the front cover.

1-7   Review of Trigonometry

Since the elements of trigonometry are used in analyzing almost all physical problems, you must be able to use trigonometry easily, quickly, and correctly. With practice, the physical scientist or engineer can handle simple trigonometry almost as easily as algebra or arithmetic. This section gives the main results of trigonometry used in physics. You should read it through to be sure you are familiar with these results. If you encounter unfamiliar material, study it carefully. This section may also be a useful reference as you read further or work on problems.

The angle made by two intersecting straight lines is defined as follows. The two lines are connected with the arc of a circle whose center is at the intersection of the lines (see Figure 1-2). The angle made by the two lines is defined to be the ratio of the arc length and the radius. This ratio is the same no matter where the arc is drawn. If r is the radius of the arc and s the distance along it, i.e., the arc length, the angle θ is given by

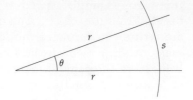

Figure 1-2
The angle θ is defined to be the ratio s/r, where s is the arc length intercepted on a circle of radius r. The angle so defined is measured in radians.

$$\theta = \frac{s}{r} \qquad\qquad\qquad 1\text{-}1 \qquad \textit{Definition of angle}$$

The angle expressed this way is measured in *radians* (rad). Since the angle is the ratio of two lengths, it is dimensionless. The radian is therefore a dimensionless unit. In dimensional analysis it can be ignored. A more familiar unit of angular measurement is the degree, defined as the angle for which the arc length s is $\frac{1}{360}$ times the circumference of the circle. Since the circumference of a circle has an arc length of $2\pi r$, the angle subtended by a full circle in radians is $2\pi r/r = 2\pi$. The degree and radian are related by

$$360° = 2\pi \text{ rad}$$

or

$$1 \text{ rad} = \frac{360°}{2\pi} = 57.3° \qquad\qquad 1\text{-}2$$

Figure 1-3 shows a right triangle formed by drawing the line BC perpendicular to AC. The lengths of the sides are labeled a, b, and c. The trigonometric functions $\sin \theta$, $\cos \theta$, and $\tan \theta$ are defined for an acute angle θ by

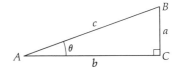

Figure 1-3
Right triangle used to define the trigonometric functions $\sin \theta$, $\cos \theta$, and $\tan \theta$.

$$\sin \theta = \frac{a}{c} = \frac{\text{opposite side}}{\text{hypotenuse}}$$

$$\cos \theta = \frac{b}{c} = \frac{\text{adjacent side}}{\text{hypotenuse}} \qquad 1\text{-}3 \qquad \textit{Definition of trigonometric functions}$$

$$\tan \theta = \frac{a}{b} = \frac{\sin \theta}{\cos \theta} = \frac{\text{opposite side}}{\text{adjacent side}}$$

Three other trigonometric functions are defined by the inverse of these functions:

$$\sec \theta = \frac{c}{b} = (\cos \theta)^{-1} \qquad \csc \theta = \frac{c}{a} = (\sin \theta)^{-1}$$

$$\cot \theta = \frac{b}{a} = (\tan \theta)^{-1} = \frac{\cos \theta}{\sin \theta}$$

which are not often used in physics. The pythagorean theorem gives some useful identities:

$$a^2 + b^2 = c^2 \qquad\qquad\qquad 1\text{-}4$$

If we divide each term in this equation by c^2, we obtain

$$\frac{a^2}{c^2} + \frac{b^2}{c^2} = 1 \qquad\qquad 1\text{-}5$$

or from the definitions of $\sin\theta$ and $\cos\theta$,

$$\sin^2\theta + \cos^2\theta = 1 \qquad\qquad 1\text{-}6$$

Similarly we can divide each term in Equation 1-4 by a^2 or b^2 and obtain

$$1 + \cot^2\theta = \csc^2\theta \qquad\qquad 1\text{-}7$$

and

$$1 + \tan^2\theta = \sec^2\theta \qquad\qquad 1\text{-}8$$

Example 1-2 Use the isosceles right triangle shown in Figure 1-4 to find the sine, cosine, and tangent of 45°.

It is clear from the figure that the two acute angles of this triangle are equal. Since the sum of the three angles in a triangle must equal 180° and the right angle is 90°, each acute angle must be 45°. If we multiply each side of any triangle by a common factor, we obtain a similar triangle with the same angles as before. Since the trigonometric functions involve only ratios of two sides of a right triangle, we can choose any convenient length for one side. Let the equal sides of this triangle have the length 1 unit. The length of the hypotenuse is then found from the pythagorean theorem (Equation 1-4):

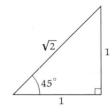

Figure 1-4
Isoceles right triangle for Example 1-2. This triangle is also called a 45-45-90 triangle.

$$c = \sqrt{a^2 + b^2} = \sqrt{1^2 + 1^2} = \sqrt{2} \text{ units}$$

The trigonometric functions for the angle 45° are then given by Equation 1-3:

$$\sin 45° = \frac{1}{\sqrt{2}} = 0.707$$

$$\cos 45° = \frac{1}{\sqrt{2}} = 0.707$$

$$\tan 45° = \frac{1}{1} = 1.00$$

Example 1-3 The sine of 30° is exactly $\frac{1}{2}$. Find the ratios of the sides of a 30-60 right triangle.

This common triangle is shown in Figure 1-5. We choose the length 1 unit for the side opposite the 30° angle. The hypotenuse is then obtained from the fact that $\sin 30° = 0.5$:

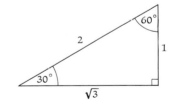

Figure 1-5
A 30-60-90 triangle for Example 1-3.

$$c = \frac{a}{\sin 30°} = \frac{1}{0.5} = 2 \text{ units}$$

The length of the side opposite the 60° angle is then found from the pythagorean theorem:

$$b = \sqrt{c^2 - a^2} = \sqrt{2^2 - 1} = \sqrt{3}$$

From these results we can obtain the other trigonometric functions for the angles 30 and 60°:

$$\cos 30° = \frac{b}{c} = \frac{\sqrt{3}}{2} = 0.866$$

$$\tan 30° = \frac{a}{b} = \frac{1}{\sqrt{3}} = 0.577$$

$$\sin 60° = \frac{b}{c} = \cos 30° = \frac{\sqrt{3}}{2} = 0.866$$

$$\cos 60° = \frac{a}{c} = \sin 30° = \tfrac{1}{2} = 0.500$$

$$\tan 60° = \frac{b}{a} = \frac{\sqrt{3}}{1} = 1.732$$

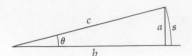

Figure 1-6
For small angles, $\theta = s/c$,
$\sin \theta = a/c$, and $\tan \theta = a/b$
are approximately equal.

The calculation of the trigonometric functions for another common right triangle, the 3-4-5 right triangle, is left as an exercise.

For small angles, the length a is nearly equal to the arc length s, as can be seen from Figure 1-6. Also, the lengths c and b are nearly equal. The angle $\theta = s/c$ is therefore nearly equal to $\sin \theta = a/c$:

$$\sin \theta \approx \theta \qquad \text{small } \theta \qquad\qquad\qquad 1\text{-}9$$

Small-angle approximations

Similarly, $\tan \theta = a/b$ is nearly equal to both θ and $\sin \theta$ for small θ:

$$\tan \theta \approx \sin \theta \approx \theta \qquad \text{small } \theta \qquad\qquad\qquad 1\text{-}10$$

Since $\cos \theta = b/c$ and these lengths are nearly equal for small θ, we have

$$\cos \theta \approx 1 \qquad \text{small } \theta \qquad\qquad\qquad 1\text{-}11$$

Equations 1-9 and 1-10 hold only if θ is measured in radians.

Example 1-4 By how much do $\sin \theta$, $\tan \theta$, and θ differ when $\theta = 15°$? This angle in radians is

$$\theta = 15° \frac{2\pi \text{ rad}}{360°} = 0.262 \text{ rad}$$

From a table of trigonometric functions,

$$\sin 15° = 0.259 \qquad \tan 15° = 0.268$$

Thus $\sin \theta$ and θ (in radians) differ by 0.003, or about 1 percent, and $\tan \theta$ and θ differ by 0.006, or about 2 percent. For smaller angles the approximation $\theta \approx \sin \theta \approx \tan \theta$ is even more accurate.

This example shows that if accuracy of a few percent is needed, these approximations can be used only for angles of about 15° or smaller. Figure 1-7 graphs θ, $\sin \theta$, and $\tan \theta$ versus θ for small θ.

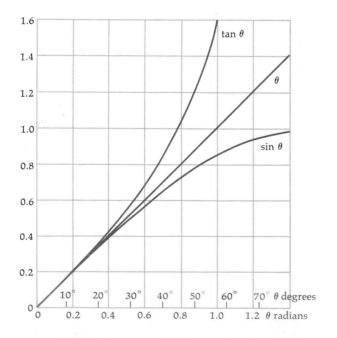

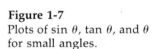

Figure 1-7
Plots of $\sin \theta$, $\tan \theta$, and θ
for small angles.

Other trigonometric identities which we shall use occasionally follow directly from the definitions given:

$$\sin (\theta_1 + \theta_2) = \sin \theta_1 \cos \theta_2 + \cos \theta_1 \sin \theta_2$$

$$\cos (\theta_1 + \theta_2) = \cos \theta_1 \cos \theta_2 - \sin \theta_1 \sin \theta_2$$

$$\sin \theta_1 + \sin \theta_2 = 2 \sin \tfrac{1}{2}(\theta_1 + \theta_2) \cos \tfrac{1}{2}(\theta_1 - \theta_2)$$

$$\cos \theta_1 + \cos \theta_2 = 2 \cos \tfrac{1}{2}(\theta_1 + \theta_2) \cos \tfrac{1}{2}(\theta_1 - \theta_2)$$

1-12

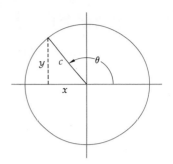

Figure 1-8
Diagram defining trigonometric functions for an obtuse angle.

Figure 1-8 shows an obtuse angle with vertex at the origin and one side along the x axis. The trigonometric functions are defined for a general angle by

$$\sin \theta = \frac{y}{c} \qquad \cos \theta = \frac{x}{c} \qquad \tan \theta = \frac{y}{x}$$

1-13

Figure 1-9 plots these functions versus θ. Some useful relations which can be obtained from these figures are

$$\sin (\pi - \theta) = \sin \theta \qquad \cos (\pi - \theta) = -\cos \theta$$

$$\sin (\tfrac{1}{2}\pi - \theta) = \cos \theta \qquad \cos (\tfrac{1}{2}\pi - \theta) = \sin \theta$$

1-14

All trigonometric functions have a period of 2π. That is, when an angle changes by 2π or $360°$, the original value of the function is obtained again. Thus $\sin (\theta + 2\pi) = \sin \theta$, etc.

Figure 1-9
The trigonometric functions sin θ, cos θ, and tan θ versus θ.

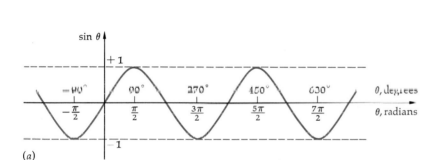

(a)

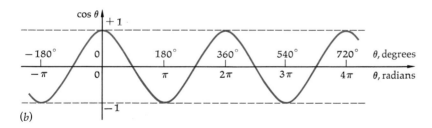

(b)

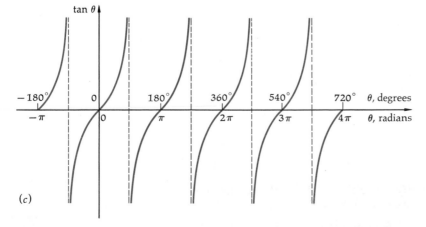

(c)

1-8   The Binomial Expansion

The binomial theorem is very useful for making approximations. One form of this theorem is

$$(1 + x)^n = 1 + nx + \frac{n(n-1)}{2} x^2 + \frac{n(n-1)(n-2)}{(3)(2)} x^3$$

$$+ \frac{n(n-1)(n-2)(n-3)}{(4)(3)(2)} x^4 + \cdots \qquad \text{1-15}$$

If n is a positive integer, there are just $n + 1$ terms in this series. If n is a real number other than a positive integer, there are an infinite number of terms. The series converges (is valid) for any value of n if x^2 is less than 1. It also converges for $x^2 = 1$ if n is positive. The series is particularly useful if $|x|$ is much less than 1. Then each term in Equation 1-15 is much smaller than the previous term and we can drop all but the first two or three terms in the equation. If $|x|$ is much less than 1, we have

$$(1 + x)^n \approx 1 + nx \qquad |x| << 1 \qquad \qquad \text{1-16} \qquad \textit{Binomial approximation}$$

Example 1-5 Use Equation 1-16 to find an approximate value for the square root of 101.

First state the problem to give an expression of the form $(1 + x)^n$ with x much less than 1:

$$(101)^{1/2} = (100 + 1)^{1/2} = (100)^{1/2}(1 + 0.01)^{1/2} = 10(1 + 0.01)^{1/2}$$

Now we can use Equation 1-16 with $n = \frac{1}{2}$ and $x = 0.01$:

$$(1 + 0.01)^{1/2} \approx 1 + \tfrac{1}{2}(0.01) = 1.005$$

$$(101)^{1/2} = 10(1 + 0.01)^{1/2} \approx 10(1.005) = 10.05$$

We can get an idea of the accuracy of this approximation by looking at the first term in Equation 1-15 that is neglected. This term is

$$\frac{n(n-1)}{2} x^2$$

Since x in this case is 1 percent of 1, the quantity x^2 is 0.01 percent of 1. By dropping this term, we are making an error of the order of

$$\frac{n(n-1)}{2} x^2 = \tfrac{1}{8}(0.01)^2 \approx 0.001\%$$

We thus expect our answer to be correct to within about 0.001 percent. The actual value of $(101)^{1/2}$ is 10.049876. The difference between this value and our approximation is 0.000124, which is about 0.001 percent of 10.05.

Example 1-6 Using $\sin \theta \approx \theta$ and $\cos \theta = (1 - \sin^2 \theta)^{1/2}$, find an approximation for $\cos \theta$ for small θ.

We have

$$\cos \theta = (1 - \sin^2 \theta)^{1/2} \approx (1 - \theta^2)^{1/2} \approx 1 + \tfrac{1}{2}(-\theta^2) = 1 - \tfrac{1}{2}\theta^2$$

This approximation is often useful as a correction to the more drastic approximation $\cos \theta \approx 1$ given by Equation 1-11.

Review

A. Define, explain, or otherwise identify the following (the numbers give the page on which the term is discussed or defined):

Units, 5   Fundamental units, 5

B. True or false. State whether each statement is true or false and explain your reasoning. If you can, give an example supporting a true statement and a counterexample contradicting a false statement.

1. Two quantities which are to be added must have the same dimensions.

2. Two quantities which are to be multiplied must have the same dimensions.

3. All conversion factors have the value 1.

4. If the angle θ is very small, sin θ and tan θ are approximately equal.

5. If the angle θ is very small, cos θ and θ are approximately equal.

Exercises

Section 1-4, Units; Section 1-5, Conversion of Units; and Section 1-6, Dimensions of Physical Quantities

1. In the following equations, the distance x is in meters and the time t is in seconds. What are the SI units of the constants C_1, C_2, and C_3?

(a) $x = C_1 + C_2 t + C_3 t^2$

(b) $x = C_1 \cos C_2 t$

(c) $x = C_1 \sin C_2 t$

Hint: The arguments of trigonometric functions and exponentials must be dimensionless.

2. What are the dimensions of the constants in Exercise 1?

3. An equation relating the speed v to the distance x is

$$v^2 = C_1 x$$

(a) What are the dimensions of the constant C_1? (b) If the speed is in meters per second and the distance in meters, what are the SI units of C_1?

4. If x is in feet and t in seconds in the equations in Exercise 1, what are the units of the constants in that exercise in the British engineering system?

5. The speed of a particle as a function of time is given by $v = C_1 \cos C_2 t$. Find the dimensions and the SI units of the constants C_1 and C_2.

6. From the original definition of the meter in terms of the distance from the equator to the North Pole, find in meters (a) the circumference and (b) the radius of the earth.

7. Find the circumference and radius of the earth in miles by converting your answers in Exercise 6 from meters to miles.

8. Find the conversion factor to convert miles per hour to kilometers per hour.

9. (a) Find the number of seconds in a year. (b) If one could count $1 per second, how many years would it take to count 1 billion dollars (1 billion $= 10^9$)? (c) If one could count one molecule per second, how many years would it take to count the molecules in a mole? (The number of molecules in a mole is Avogadro's number, $N_A = 6 \times 10^{23}$.)

10. The SI unit of force, the kilogram-meter per second per second (kg-m/sec²), is called the newton (N). Find the dimensions and the SI units of the constant G in Newton's law of gravitation $F = Gm_1 m_2/r^2$.

11. Sometimes a conversion factor can be derived from the knowledge of a constant in two different systems. (a) The speed of light in vacuum is 186,000 mi/sec = 3×10^8 m/sec. Use this fact to find the number of kilometers in a mile. (b) The weight of 1 ft³ of water is 62.4 lb. Use this and the fact that 1 cm³ of water has a mass of 1 gm to find the weight in pounds of a 1-kg mass.

Section 1-7, Review of Trigonometry

12. Find the measure of a right angle in radians directly from the definition $\theta = s/r$.

13. Find: (a) $\sin \frac{1}{2}\pi$; (b) $\cos \frac{1}{2}\pi$; (c) $\tan \pi$; (d) $\tan (\pi/4)$; (e) $\sin (\pi/4)$; (f) $\cos (3\pi/4)$.

14. Convert from radians to degrees: (a) $\pi/4$; (b) $3\pi/2$; (c) $\pi/2$; (d) $\pi/6$; (e) $5\pi/6$.

15. Convert from degrees to radians: (a) $60°$; (b) $90°$; (c) $30°$; (d) $45°$; (e) $180°$; (f) $37°$; (g) $720°$.

16. Find conversion factors for revolutions per minute to radians per second and degrees per second.

17. A record rotates at a rate of 33.3 revolutions per minute (rev/min). Find the angle in radians and in degrees through which the record rotates in 1 sec.

18. The diameter of a record is 12 in, and the record rotates at 33.3 rev/min. (a) What is the distance traveled by a point on the rim (measured along the circular arc) during 1 rev? (b) At what rate in inches per second does this point travel?

19. A right triangle has sides 3, 4, and 5 m. If θ_1 is the smaller acute angle and θ_2 the larger acute angle of the triangle, find the sine, cosine, and tangent of each of these angles. Find the measures of these angles in degrees using Appendix F.

20. A right triangle has sides $a = 2$ in, $b = 8$ in, and $c = ?$ (a) Find the length of the hypotenuse c. (b) Find the sine, cosine, and tangent of the angles of the triangle. (c) From Appendix F find the angles.

21. Use the small-angle approximations to find approximate values for (a) $\sin 8°$ and (b) $\tan 5°$. *Hint:* First convert these angles to radian measure.

22. The angle subtended by the moon's diameter at a point on the earth is about $0.524°$. Use this and the fact that the moon is about 2.36×10^5 mi away to find the diameter of the moon.

Section 1-8, The Binomial Expansion

23. Write out the complete binomial expansion (Equation 1-15) for $(1 + x)^3$ and check by finding $(1 + x)^3$ directly by multiplication.

24. The quantity $1/(1 + x)^2$ can be written as an infinite power series $1/(1 + x)^2 = 1 + a_1 x + a_2 x^2 + a_3 x^3 + \cdots + \cdots + a_n x^n + \cdots$, where a_n are constants. Find these constants using the binomial equation 1-15.

25. Use the approximation $(1 + x)^n \approx 1 + nx$, $|x| \ll 1$, to find approximate values for: (a) $\sqrt{99}$; (b) $1/1.01$; (c) $124^{1/3}$.

26. Use the approximation $(1 + x)^n \approx 1 + nx$ to find approximate values for: (a) 9.8^2; (b) 0.999^5; (c) 1.002^{-2}.

Problems

1. The accompanying table gives experimental results for a measurement of the period of motion T of an object of mass m suspended on a spring versus the mass of the object.

Mass m, kg	0.10	0.20	0.40	0.50	0.75	1.00	1.50
Period T, sec	0.56	0.83	1.05	1.28	1.55	1.75	2.22

These data are consistent with a simple equation expressing T as a function of m of the form $T = Cm^n$, where C and n are constants (n is not necessarily an integer). (a) Find n and C. (There are several ways to do this. One is to guess the value of n and check by plotting T versus m^n on ordinary graph paper. If your guess is right, the plot will be a straight line. Another is to plot T versus m on log log paper. The slope of the straight line on this paper is n.) (b) Which data points deviate the most from a straight-line plot of T versus m^n?

2. The table below gives the period T and orbit radius r for the motions of four satellites orbiting a dense, heavy asteroid.

Period T, y	0.44	1.61	3.88	7.89
Radius r, 10^5 km	0.88	2.08	3.74	6.00

(a) These data can be fitted by the formula $T = Cr^n$. Find C and n. (b) A fifth satellite is discovered to have a period of 6.20 y. Find the radius for the orbit of this satellite, which fits the same formula.

3. The period T of a simple pendulum depends on the length L of the pendulum and the acceleration of gravity g (dimensions L/T^2). (a) Find a simple combination of L and g which has the dimensions of time. (b) Check the dependence of the period T on the length L by measuring the period (time for a complete to-and-fro swing) of a pendulum for two different values of L. (c) The correct formula relating T to L and g involves a constant which is a multiple of π and cannot be obtained by the dimensional analysis of part (a). It can be found by experiment as in part (b) if g is known. Using the value $g = 9.8$ m/sec^2 and your experimental results from part (b), find the formula relating T to L and g.

4. A projectile fired at an angle of 45° travels a total distance R, called the range, which depends only on the initial speed v and the acceleration of gravity g (dimensions L/T^2). Using dimensional analysis, find how R depends on the speed and on g.

5. A ball thrown horizontally from a height H with speed v travels a total horizontal distance R. (a) Do you expect R to increase or decrease with increasing H? With increasing v? (b) From dimensional analysis, find a possible dependence of R on H, v, and g.

6. Show that $\sin 2\theta = 2 \sin \theta \cos \theta$. Hint: Use Equation 1-12.

7. From Equation 1-12, and the identity $\tan \theta = (\sin \theta)/(\cos \theta)$, find an expression for $\tan (\theta_1 + \theta_2)$ in terms of $\tan \theta_1$ and $\tan \theta_2$.

8. A man begins at an intersection of a north-south road with an east-west road and walks 10 mi in the direction north of east along a line which makes an angle of 30° with the easterly direction. (a) How far is he from the north-south road? From the east-west road? (b) He now walks along a line perpendicular to his original direction until he reaches the north-south road. How far must he walk?

9. For each case in Exercise 25, compute the next term $\frac{1}{2}n(n-1)x^2$ in the binomial expansion used and express this term as a percentage of your answer found in that exercise. (b) Find the difference between your approximate answer for each case in that exercise and the exact calculation (from tables or from any exact method of calculation) and express this difference as a percentage of the exact answer.

10. Repeat Problem 9 using the data of Exercise 26.

CHAPTER 2   Motion in One Dimension

To simplify our discussion of motion, we start with objects whose position can be described by locating one point. Such an object is called a *particle*. One tends to think of a particle as a very small object, e.g., a piece of shot, but actually no size limit is implied by the word particle. If we are not interested in the extension of an object or in its rotational motion, any object can be considered as a particle. For example, it is sometimes convenient to consider the earth as a particle moving around the sun in a nearly circular path. In this case we are interested only in the motion of the center of the earth and are ignoring the size of the earth and its rotation. In some astronomical problems, the solar system or even a whole galaxy is treated as a particle.

When we are interested in the internal motion or internal structure of an object, it can no longer be treated as a particle. However, our study of particle motion is useful even in these cases because any complex object can be treated as a system of particles. Even so small a thing as an atomic nucleus, with a diameter of only about 10^{-15} m (1 fm), turns out to be a rather complicated system of particles when its structure is examined in detail. In the first nine chapters of this book we shall concentrate on the motion of a single particle (or, at times, two particles) while we develop the important concepts and methods of classical mechanics. In Chapter 10 we extend our discussion to systems of particles.

To describe the motion of a particle, we need the concepts of *displacement, velocity,* and *acceleration*. In the general motion of a particle in three dimensions, these quantities are vectors, which have direction as well as magnitude. We study vectors and the general motion of a particle in Chapter 3; in this chapter, we confine our discussion to motion along a straight line, i.e., motion in one dimension. For such restricted motion, there are only two possible directions. They are distinguished by designating one positive and the other negative. A simple example of one-dimensional motion is a car moving along a flat, straight, narrow road. We can choose any convenient point on the

Size and structure are unimportant for particles

The simplest motion is one-dimensional

car for the location of this "particle." Since we are interested for the moment only in one-dimensional motion, we can neglect the width of the road and consider it a line.

Questions

1. Give several examples of the motion of large objects where treating the object as a particle is an adequate approximation. Give other examples where it is not.

2. In many respects the motion of a body along any curved line is similar to straight-line motion. Can you suggest reasons why these two cases require separate treatment?

2-1 Displacement and Average Velocity

Let us set up a coordinate system by choosing some reference point on a line for the origin O. To every other point on the line we assign a number x which indicates how far the point is from the origin. The value of x depends on the unit chosen as the measure of distance. The sign of x depends on its position relative to the origin O; if it is to the right, it is positive; to the left, negative.

Suppose that our particle is at position x_1 at some time t_1 and at point x_2 at time t_2. We determine its position at several other times along the way. Figure 2-1 shows a graph of the function $x(t)$. The measured points have been connected with a smooth curve. How accurately this curve represents the motion of the particle depends upon the number of points used and on the complexity of the motion. The change in the position of the particle, $x_2 - x_1$, is called the *displacement* of the particle. It is customary to use the Greek letter Δ (capital delta) to indicate the change in a quantity. Thus the change in x is written Δx. The notation Δx (read "delta x") stands for a single quantity, the change in x (it is not a product of Δ and x any more than cos x is a product of cos and x):

$$\Delta x = x_2 - x_1 \qquad\qquad 2\text{-}1$$

Displacement defined

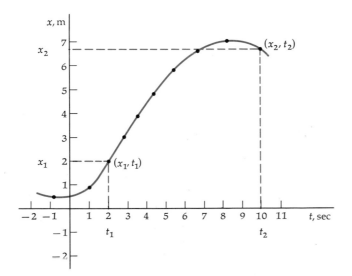

Figure 2-1
Plot of x versus t for a particle moving in one dimension. Measured points are connected with a smooth curve.

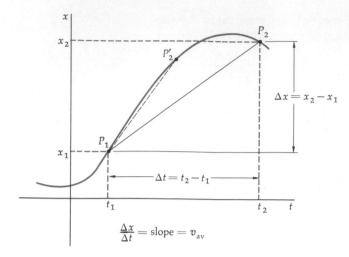

$$\frac{\Delta x}{\Delta t} = \text{slope} = v_{av}$$

Figure 2-2
A plot of x versus t for a particle moving in one dimension, as in Figure 2-1. Here, a straight line is drawn from position P_1 to position P_2. The average velocity for the time interval $t_2 - t_1$ is defined to be the slope of this line $v_{av} = \Delta x/\Delta t$. The average velocity depends on the time interval, as indicated by the fact that the line from P_1 to P'_2 has a greater slope than that from P_1 to P_2.

The average velocity of the particle is defined to be the ratio of the displacement Δx and the time interval $\Delta t = t_2 - t_1$:

$$v_{av} = \frac{\Delta x}{\Delta t} = \frac{x_2 - x_1}{t_2 - t_1} \qquad \text{2-2} \qquad \textit{Average velocity defined}$$

Note that the displacement and the average velocity may be either positive or negative, depending on whether x_2 is greater or less than x_1 (assuming positive Δt). A positive value indicates motion to the right; a negative value indicates motion to the left.

In Figure 2-2 a straight line has been drawn from point (x_1,t_1) to point (x_2,t_2). This line is the hypotenuse of a triangle with sides Δt and Δx. The ratio $\Delta x/\Delta t$ is called the *slope* of this straight line. In geometric terms, it is a measure of the steepness of the straight line in the graph. *Slope: a measure of steepness* For a given interval Δt, the steeper the line, the greater the value of $\Delta x/\Delta t$. Since the slope of this line is just the average velocity for the time interval Δt, we have a geometric interpretation of average velocity. It is the slope of the straight line connecting the points (x_1,t_1) and (x_2,t_2). In general, the average velocity depends on the time interval chosen. For example, in Figure 2-2, if we chose a smaller time interval by choosing a time t'_2 closer to t_1, the average velocity would be greater, as indicated by the greater steepness of the line connecting points P_1 and P'_2.

Example 2-1 If $x_1 = 18$ m at $t_1 = 2$ sec and $x_2 = 3$ m at $t_2 = 7$ sec, find the displacement and the average velocity for this time interval.

By the definition, the displacement is

$$\Delta x = x_2 - x_1 = 3 \text{ m} - 18 \text{ m} = -15 \text{ m}$$

and the average velocity is

$$v_{av} = \frac{\Delta x}{\Delta t} = \frac{x_2 - x_1}{t_2 - t_1} = \frac{3 \text{ m} - 18 \text{ m}}{7 \text{ sec} - 2 \text{ sec}} = \frac{-15 \text{ m}}{5 \text{ sec}} = -3 \text{ m/sec}$$

The displacement and average velocity are negative, indicating that the particle moved to the left, toward decreasing values of x.

Note that the unit meters per second is included as part of the answer for the average velocity found in Example 2-1. Since there are many other possible choices for units of length and time (feet, inches, miles, light-years, etc., for length and hours, days, years, etc., for time) it is essential to include the unit with a numerical answer. The statement "the average velocity of a particle is -3" is meaningless.

Example 2-2 How far does a car go in 5 min if its average velocity is 60 mi/h?

In this example we are interested in the displacement during a time interval of 5 min. From Equation 2-2 the displacement Δx is given by

$$\Delta x = v_{av}\, \Delta t$$

Thus

$$\Delta x = 60 \text{ mi/h} \times 5 \text{ min} = 300\, \frac{\text{mi-}\cancel{\text{min}}}{\cancel{\text{h}}}\, \frac{1\, \cancel{\text{h}}}{60\, \cancel{\text{min}}} = 5 \text{ mi}$$

We use the conversion factor $1 \text{ h} = 60 \text{ min}$ to change the unit of displacement from mi-min/h to miles.

Questions

3. What sense, if any, does the following statement make? "The average velocity of the car at 9 A.M. was 60 mi/h."

4. Is it possible that the average velocity for some interval may be zero although the average velocity for a shorter interval included in the first interval is not zero? Explain.

2-2 Instantaneous Velocity

At first glance, it might seem impossible to define the velocity of a particle at a single instant, i.e., at a specific time. At a time t_1, the particle is at a single point x_1. If it is at a single point, how can it be moving? On the other hand, if it is not moving, shouldn't it stay at the same point? This is an age-old paradox, which can be resolved when we realize that to observe motion and thus define it, we must look at the position of the object at more than one time. It is then possible to define the velocity at an instant by a limiting process.

Figure 2-3 is the same x-versus-t curve as Figure 2-2, showing a sequence of time intervals indicated, Δt_1, Δt_2, Δt_3 . . . , each smaller

Velocity at a point

Figure 2-3
Plot of $x(t)$ from Figures 2-1 and 2-2. As the time interval beginning at t_1 is decreased, the average velocity for that interval approaches the slope of the line tangent to the curve at time t_1. The instantaneous velocity at time t_1 is defined to be the slope of the line tangent to the curve at t_1.

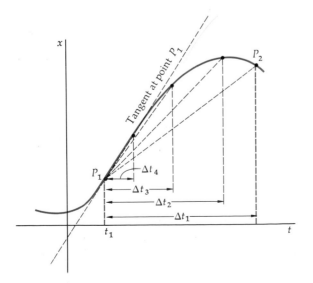

A. J. Wyatt

Ken Regan/Camera 5

Motion depicted by an artist in Duchamp's "Nude Descending a Staircase." (*Marcel Duchamp, No. 2, 1912, Louise and Walter Arensberg Collection, Philadelphia Museum of Art.*) What features in the photograph of a runner imply motion? Can we tell with certainty from a single high-speed photograph that the athlete is moving?

than the previous one. For each time interval Δt, the average velocity is the slope of the dashed line appropriate for that interval. This figure shows that as the time interval becomes smaller, the dashed lines get steeper but never incline more than the line tangent to the curve at point t_1. We define the slope of this tangent line to be the *instantaneous velocity* at the time t_1. The instantaneous velocity is the limit of the ratio $\Delta x/\Delta t$ as Δt approaches zero:

$$v(t) = \lim_{\Delta t \to 0} \frac{\Delta x}{\Delta t} = \text{slope of line tangent to } x(t) \text{ curve} \qquad \text{2-3}$$

Instantaneous velocity defined

This limit is called the *derivative of x with respect to t* at the point t_1. In the usual calculus notation[1] the derivative is written dx/dt:

$$v(t) = \lim_{\Delta t \to 0} \frac{\Delta x}{\Delta t} = \frac{dx}{dt} \qquad \text{2-4}$$

Derivative defined

It is important to realize that the displacement Δx depends on the time interval Δt. As Δt approaches zero, Δx does also (as can be seen from Figure 2-3) and the ratio $\Delta x/\Delta t$ approaches the slope of the line tangent to the curve. In computing this limit, we do not set Δt equal to zero since then Δx would be zero also and the ratio $\Delta x/\Delta t$ would not be defined.

This concept of the limit of a function can be understood only with careful study. We omit further discussion here, assuming that it is covered in your calculus course.

[1] Strictly speaking, the notation dx/dt stands for the single quantity, the derivative of x with respect to t, which is the slope of the line tangent to the $x(t)$ curve. Later, we shall define the quantities dx and dt (called *differentials*) so that their ratio is in fact this slope. If Δx and Δt are chosen small enough, the ratio $\Delta x/\Delta t$ is *nearly* equal to the slope of the tangent line. Therefore it is convenient, though not mathematically rigorous, to think of dt as a very small change in t, and dx as the corresponding very small change in x.

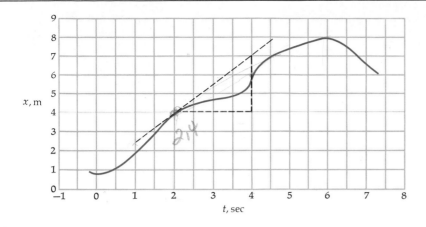

Figure 2-4
Plot of x versus t for Example 2-3. The instantaneous velocity at time $t = 2$ sec can be found by measuring the slope of the line tangent to the curve at that time.

Example 2-3 The position of a particle is given by the function shown in Figure 2-4. Find the instantaneous velocity at the time $t = 2$ sec. When is the velocity greatest? When is it zero? Is it ever negative?

In the figure we have sketched the line tangent to the curve at time $t = 2$ sec. The slope of this line is measured from the figure to be (3 m)/(2 sec) = 1.5 m/sec. Thus $v = 1.5$ m/sec at time $t = 2$ sec. According to the figure, the slope is greatest at about $t = 4$ sec. The velocity is zero at times $t = 0$ and $t = 6$ sec, as indicated by the fact that the tangent lines at these times are horizontal with zero slope. After $t = 6$ sec, the curve has a negative slope, indicating that the velocity is negative. (The slope of the line tangent to a curve is often referred to merely as the *slope of the curve*.)

Example 2-4 A particle is at point $x = 5$ ft at time $t = 0$ and moves with a constant velocity of 10 ft/sec. Sketch its position x as a function of time.

A constant velocity means that the slope of the x-versus-t curve is constant. The curve which has a constant slope is a straight line. The general expression for a straight-line curve is

$$x = mt + b$$

where m is the slope and b is the value of x at $t = 0$, called the *intercept*. Thus $m = 10$ ft/sec, $b = 5$ ft, and the equation for the curve is

$$x = 5 \text{ ft} + (10 \text{ ft/sec})t$$

This curve is sketched in Figure 2-5. In general, if a particle is at posi-

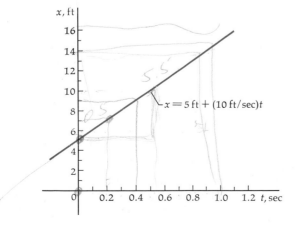

$$x = 5 \text{ ft} + (10 \text{ ft/sec})t$$

Figure 2-5
Plot of the position function $x = 5$ ft + (10 ft/sec)t from Example 2-4.

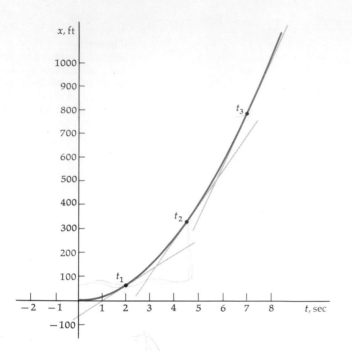

Figure 2-6
Plot of the function $x(t) = 16t^2$. Tangent lines are drawn at times t_1, t_2, and t_3. The slope of these tangent lines increases steadily from t_1 to t_2 to t_3, indicating that the instantaneous velocity increases steadily with time.

tion x_0 at time $t = 0$ and has a constant velocity v, its position is given by

$$x = x_0 + vt$$

Example 2-5 The position of a stone dropped from rest from a cliff is given by $x = 16t^2$, where x is in feet measured downward from the original position at $t = 0$ and t is in seconds. Find the velocity at any time. (We omit explicit indication of the units to simplify the notation.)

This function is shown in Figure 2-6, where tangent lines are drawn for three different times, t_1, t_2, and t_3. The slopes of these lines differ. As time passes, the slope of the curve increases, indicating that the instantaneous velocity is increasing with time. We can compute the velocity at some time t_1 by computing the derivative dx/dt directly from the definition in Equation 2-3. At time t_1 the position is

$$x_1 = 16t_1{}^2$$

At a later time $t_2 = t_1 + \Delta t$ the position is x_2, given by

$$\begin{aligned} x_2 &= 16t_2{}^2 = 16(t_1 + \Delta t)^2 \\ &= 16\,[t_1{}^2 + 2t_1\,\Delta t + (\Delta t)^2] \\ &= 16t_1{}^2 + 32t_1\,\Delta t + 16(\Delta t)^2 \end{aligned}$$

The displacement for this time interval is thus

$$\Delta x = x_2 - x_1 = 32t_1\,\Delta t + 16(\Delta t)^2$$

The average velocity for this time interval is

$$v_{av} = \frac{\Delta x}{\Delta t} = 32\,t_1 + 16\Delta t$$

As we consider shorter and shorter time intervals, Δt approaches zero and the second term $16\Delta t$ approaches zero, though the first is unaltered. The instantaneous velocity at time t_1 is thus

$$v = \lim_{\Delta t \to 0} \frac{\Delta x}{\Delta t} = 32t_1$$

This function, the instantaneous velocity, is proportional to the time. For any general time t, the instantaneous velocity is $v(t) = 32t$.

It is instructive to examine the limiting process numerically by computing the average velocity for smaller and smaller time intervals. Table 2-1 gives the average velocity for Example 2-5 computed for $t_1 = 2$ sec and various time intervals Δt, each smaller than the previous one. The table shows that for very small time intervals, the average velocity is very nearly equal to the instantaneous velocity 64 ft/sec. The difference between $\Delta x / \Delta t$ and $\lim_{\Delta t \to 0} (\Delta x / \Delta t)$ can be made arbitrarily small by choosing Δt sufficiently small.

It is important to distinguish carefully between average and instantaneous velocity. By custom, however, the word velocity alone is assumed to mean instantaneous velocity. The magnitude of the instantaneous velocity is called the *speed*.

Speed defined

Questions

5. If the instantaneous velocity does not change from instant to instant, will the average velocities for different intervals differ?

6. If $v_{av} = 0$ for some time interval Δt, must the instantaneous velocity v be zero at some time in the interval? Support your answer by a sketch of a possible x-versus-t curve which has $\Delta x = 0$ for some interval Δt.

2-3 Acceleration

When the instantaneous velocity of a particle is changing with time, as in Example 2-5, the particle is said to be *accelerating*. The average acceleration for a particular time interval $\Delta t = t_2 - t_1$ is defined as the ratio $\Delta v / \Delta t$, where $\Delta v = v_2 - v_1$ is the change in instantaneous velocity for that time interval:

$$a_{av} = \frac{\Delta v}{\Delta t} \qquad \text{2-5}$$

Average acceleration defined

The dimensions of acceleration are length divided by time². Convenient units are meters per second per second (m/sec²) or feet per second per second (ft/sec²). For automobile accelerations miles per hour per second (mi/h-sec) is often used. *Instantaneous acceleration* is the limit of this ratio as Δt approaches zero. If we plot the velocity versus time, the instantaneous acceleration is defined to be the slope of the tangent line to this curve:

$$a = \lim_{\Delta t \to 0} \frac{\Delta v}{\Delta t} = \text{slope of } v\text{-versus-}t \text{ curve} \qquad \text{2-6}$$

Instantaneous acceleration defined

The acceleration is thus the derivative of the velocity with respect to time. The calculus notation for this derivative is dv/dt. Since the velocity is also the derivative of the position x with respect to t, the acceleration is the *second derivative* of x with respect to t. The common nota-

Table 2-1
Displacement and average velocity for various time intervals Δt beginning at $t = 2$ sec for the function $x = 16t^2$

Δt, sec	Δx, ft	$\Delta x / \Delta t$, ft/sec
1.00	80	80
0.50	36	72
0.20	13.44	67.2
0.10	6.56	65.6
0.05	3.24	64.8
0.01	0.6416	64.16
0.005	0.3204	64.08
0.001	0.064016	64.016
0.0001	0.00640016	64.0016

tion for the second derivative of x with respect to t is d^2x/dt^2. We can see the origin of this notation by writing the acceleration as dv/dt and replacing v with dx/dt:

$$a = \frac{dv}{dt} = \frac{d(dx/dt)}{dt} = \frac{d^2x}{dt^2}$$
2-7   *Notation for second derivative*

If the velocity is constant, as in Example 2-4, the acceleration is zero since $\Delta v = 0$ for all time intervals. In this case the slope of the x-versus-t curve does not change. In Example 2-5 we found that for the position function $x = (16 \text{ ft/sec}^2)t^2$ the velocity increases linearly with time according to $v = (32 \text{ ft/sec}^2)t$. In this case the acceleration is constant and equal to 32 ft/sec^2, the constant slope of the v-versus-t curve. We can get at least a qualitative idea of the acceleration from the x-versus-t curve. From Figure 2-6 we can see that the velocity is increasing with time, as indicated by the increasing steepness of the tangent lines; i.e., this curve shows that the acceleration is positive.

Example 2-6 From the x-versus-t curve in Figure 2-7, discuss qualitatively the velocity and acceleration at the point $t = 5$ sec.

At $t = 5$ sec the tangent line to the curve is horizontal with zero slope. Thus the instantaneous velocity at this point is zero. However, the fact that the velocity is zero does *not* imply that the acceleration must be zero also. The acceleration is the rate of change of the instantaneous velocity. To find the acceleration at some time we must know how the velocity is changing at that time. For this example, to find the acceleration at $t = 5$ sec we must know the velocity at other times in the neighborhood of $t = 5$ sec. From the figure we see that the slope of the curve is positive just before $t = 5$ sec and negative just after. Thus the velocity changes from a positive value, to zero, to a negative value, and acceleration at this time is negative. In Figure 2-8 we have sketched the velocity-versus-t curve corresponding to the position function given in Figure 2-7. Note that at $t = 5$ sec, the velocity is zero; nevertheless the slope of the v-versus-t curve is not zero but negative.

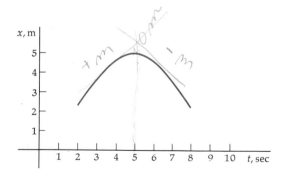

Figure 2-7
x-versus-t curve for Example 2-6. The slope of the tangent line at $t = 5$ sec is zero, indicating that the instantaneous velocity is zero at this time. Just before $t = 5$ sec, the slope is positive; just after $t = 5$ sec, the slope is negative. Since the slope is decreasing at $t = 5$ sec, the acceleration at this time is negative.

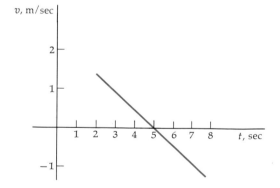

Figure 2-8
Velocity-versus-time curve corresponding to the position curve in Figure 2-7. At $t = 5$ sec, the velocity is zero and decreasing.

Example 2-7 The position of a particle is given by $x = Ct^3$, where C is a constant with the dimensions of m/sec^3. Find the velocity and acceleration as functions of time.

The graph of x versus t (not shown) resembles the graph in Example 2-5 except that the increase of x with time is faster. The slope of the curve increases with time, indicating an increasing velocity. As in that example, we can compute the velocity by computing the derivative dx/dt directly from its definition (Equation 2-3). At some time t_1 the position is $x_1 = Ct_1^3$. At a later time $t_2 = t_1 + \Delta t$ the position is

$$x_2 = Ct_2^3 = C(t_1 + \Delta t)^3 = Ct_1^3 + 3Ct_1^2\,\Delta t + 3Ct(\Delta t)^2 + C(\Delta t)^3$$

The displacement is thus

$$\Delta x = x_2 - x_1 = 3Ct_1^2\,\Delta t + 3Ct_1(\Delta t)^2 + C(\Delta t)^3$$

The average velocity for this time interval is

$$v_{\text{av}} = \frac{\Delta x}{\Delta t} = 3Ct_1^2 + 3Ct\,\Delta t + C(\Delta t)^2$$

As $\Delta t \to 0$, the two right-hand terms also approach zero and the term $3Ct_1^2$ remains unchanged. The instantaneous velocity at time t is thus

$$v = 3Ct^2$$

(Having computed the velocity at time t_1, we can drop the subscript 1 on the time and write the velocity for a general time t.) We find the acceleration by repeating the process and thus finding the derivative of v with respect to t, which is the second derivative of x with respect to t. We omit some of the algebra in this derivation because it is similar to that shown above and in Example 2-5. The change in velocity for the time interval from t to $t + \Delta t$ is

$$\Delta v = 3C(t + \Delta t)^2 - 3Ct^2 = 6Ct\,\Delta t + 3C(\Delta t)^2$$

We divide this expression by Δt to obtain the average acceleration for this interval:

$$a_{\text{av}} = \frac{\Delta v}{\Delta t} = 6Ct + 3C\,\Delta t$$

The instantaneous acceleration is thus

$$a = \lim_{\Delta t \to 0} \frac{\Delta v}{\Delta t} = 6Ct$$

In this example the acceleration is not constant but increases with time.

In the examples so far we have computed derivatives directly from the definition by taking the appropriate limit explicitly. It is useful to examine various properties of the derivative and to develop rules permitting us to calculate the derivatives of many functions quickly without applying the definition each time. Table E-1 in Appendix E contains a list of such rules, followed by a brief discussion of their origin. We use these rules without comment and leave their detailed study to your calculus course.

Rule 7 in Table E-1 is used so often we repeat it here. If x is a simple power function of t, such as

$$x = Ct^n$$

where C and n are any constants, the derivative of x with respect to t is given by

$$\frac{dx}{dt} = Cnt^{n-1}$$

We have already seen applications of this rule for $n = 1, 2$, and 3 in the examples; e.g., for the position function of Example 2-7, $x = Ct^3$, we found $v = dx/dt = 3Ct^2$ and $a = dv/dt = 6Ct$.

Questions

7. Give an example of a motion for which the velocity is negative but the acceleration is positive; e.g., sketch a graph of v versus t.

8. Give an example of a motion for which both the acceleration and velocity are negative.

9. Is it possible for a body to have zero velocity and nonzero acceleration?

2-4   Finding $x(t)$ from $v(t)$: The Initial-Value Problem

We have learned how to obtain the velocity and acceleration functions from a given position function by differentiation. The inverse problem is to find the position function x given the velocity v or the acceleration a. To solve this problem of finding a function which has a given derivative we make use of our knowledge of the derivatives of various types of functions, e.g., those listed in Appendix Table E-1. Consider, for example, constant velocity $v = v_0$. We have already found the position function x corresponding to this given velocity in Example 2-4. We make use of our knowledge that the curve x versus t which has a constant slope is a straight line. Thus x must vary linearly with t. One such function is

Finding the position from the velocity

$$x = v_0 t$$

However, this is not a general solution to the problem.

We can add any constant to the right-hand side of this equation without changing the derivative since the derivative of a constant is zero. Calling this constant x_0, we have for the general position function of a particle with constant velocity v_0

$$x = x_0 + v_0 t$$

The constant x_0 is the position at time $t = 0$, as can be seen by substituting $t = 0$ into this equation.

Let us consider a slightly more difficult example, a particle whose velocity is proportional to time:

$$v = a_0 t$$

The proportionality constant a_0 is just the acceleration, as can be seen by computing $dv/dt = a_0$. We now ask: What function $x(t)$ has the derivative $a_0 t$? We note from rule 7 in Table E-1 that the derivative of t^n is nt^{n-1}. Our position function must therefore be proportional to t^2, and we can verify that the correct function is $\frac{1}{2}a_0 t^2$. Again, we can add any constant to this function without changing its derivative. The general position function having the derivative $a_0 t$ is

$$x = x_0 + \tfrac{1}{2}a_0 t^2$$

where the additive constant x_0 is the value of x when $t = 0$. If the value of x is given for any particular time, we can determine the value of x_0.

Finding the function from its given derivative is called *antidifferentiation*. The function determined is the *antiderivative* (often called the indefinite integral) of the function. It always contains an additive constant which is related to the choice of origin. If the derivative of a function is given everywhere and the value of the function is known at one point, the function is uniquely determined because the actual value of the additive constant can be determined from the given information.

Antiderivatives

Let us now consider an example in which the acceleration is given.

Example 2-8 The acceleration of a particle is constant, $a(t) = a_0$. Find the position function $x(t)$.

We must first find the velocity $v(t)$ and use it to find $x(t)$. We thus go through the process of antidifferentiation twice. Each antiderivative contains an additive constant. Consequently there will be two constants in the function $x(t)$ which are as yet unspecified. These constants are usually determined from the velocity at some given time and from the position at some given (and perhaps different) time. We are given $dv/dt = a_0$. We did this type of problem earlier in the example at the beginning of this section. The general function which has a constant derivative is $v(t) = a_0 t + v_0$. Here the additive constant v_0 is the value of $v(t)$ at $t = 0$. We now need to find the function $x(t)$ which has the derivative $dx/dt = a_0 t + v_0$. On the basis of rule 2 in Appendix Table E-1 (the derivative of a sum of terms is the sum of the derivatives of the terms), we can treat each term separately. The function $\tfrac{1}{2}a_0 t^2$ has the derivative $a_0 t$, and the function $v_0 t$ has the derivative v_0. The additive constant in the antiderivatives can be combined into one constant. Calling this constant x_0, we have for the complete solution

$$x(t) = x_0 + v_0 t + \tfrac{1}{2}a_0 t^2$$

In Example 2-8 we saw that when $x(t)$ is found from the acceleration $a(t)$, two constants appear. We were able to relate these constants to the initial position x_0 and the initial velocity v_0. The problem "given $a(t)$, find $x(t)$" is called the *initial-value problem*. The solution depends on the form of the function $a(t)$, on the value of v at some particular time, and on the value of x at some particular time; these are often taken as initial values, i.e., values at time $t = 0$. The problem is particularly important in physics because the acceleration of a particle is determined by the forces acting on it. Thus if we know the forces acting on a particle and its position and velocity at some particular time, we can find its position uniquely at all other times.

Finding position from acceleration

2-5 The Antiderivative and Integration

The problem of finding the antiderivative is related to the problem of *integration,* or finding the area under a curve. This fundamental

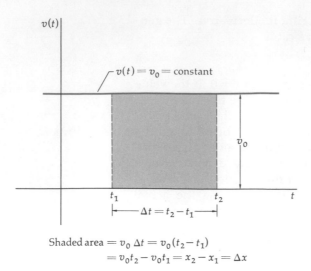

Shaded area $= v_0 \, \Delta t = v_0 (t_2 - t_1)$
$= v_0 t_2 - v_0 t_1 = x_2 - x_1 = \Delta x$

Figure 2-9
The displacement for the interval Δt equals the area
under the velocity-versus-
time curve for that interval.
For $v(t) = v_0 =$ constant, the
displacement equals the area
of the rectangle shown.

problem in calculus will be discussed only briefly.

Consider the case of constant velocity v_0. The displacement during some time interval Δt is just the velocity times the time interval:

$$\Delta x = v_0 \, \Delta t$$

This is the area under the v-versus-t curve, as can be seen in Figure 2-9. The geometric interpretation of displacement as the area under the v-versus-t curve is quite general. Figure 2-10 is the v-versus-t curve when the velocity is proportional to time $v = a_0 t$. In the previous section, we found the position function to be $x = x_0 + \frac{1}{2} a_0 t^2$. The displacement during the time interval $t = 0$ to $t = t_1$ is

$$\Delta x = x_1 - x_0 = \tfrac{1}{2} a_0 t_1{}^2$$

As can be seen from the figure, again this equals the area under the curve for this time interval.

Figure 2-11 shows a general v-versus-t curve. If we know the algebraic function $v(t)$ for this curve, we can always compute the antiderivative and determine the position function $x(t)$ except for the initial value x_0, which must be found from additional information, e.g., the value of x at some specified time. We then can find the displacement during any interval $\Delta t = t_2 - t_1$ by subtracting $x(t_1)$ from $x(t_2)$ (this gives us Δx even if x_0 is not known since it subtracts out).

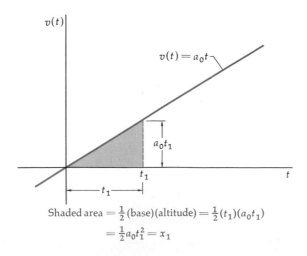

Shaded area $= \frac{1}{2}$(base)(altitude) $= \frac{1}{2}(t_1)(a_0 t_1)$
$= \frac{1}{2} a_0 t_1^2 = x_1$

Figure 2-10
For $v(t) = a_0 t$, the displacement during the interval from
$t = 0$ to $t = t_1$ equals the triangular area indicated. Since the
base of the triangle is t_1 and
the height is $a_0 t_1$, the area is
$\frac{1}{2}(a_0 t_1)(t_1) = \frac{1}{2} a_0 t_1^2$.

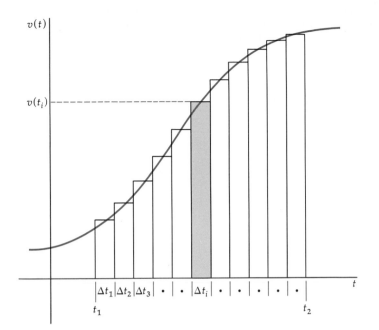

Figure 2-11
Plot of general $v(t)$ versus t.
The displacement for the interval Δt_i is approximately
$v(t_i)\,\Delta t_i$, indicated by the
shaded rectangular area. The
total displacement from t_1 to
t_2 is the area under the curve
for this interval, which can be
approximated by summing
the areas of the rectangles.

However, even if we do not know the algebraic function for $v(t)$, we
can obtain the displacement Δx for some time interval $t_2 - t_1$ directly
from the curve $v(t)$ versus t by finding the area under the curve for this
time interval.

In Figure 2-11 the interval $t_2 - t_1$ is divided into a number of small
intervals Δt_i. At time t_i, which is any time in the interval Δt_i, the
instantaneous velocity is

$$v(t_i) = \lim_{\Delta t_i \to 0} \frac{\Delta x_i}{\Delta t_i}$$

If Δt_i is very small, the instantaneous velocity is approximately the
average velocity:

$$v(t_i) \approx \frac{\Delta x_i}{\Delta t_i}$$

The displacement during the small interval Δt_i is thus

$$\Delta x_i \approx v(t_i)\,\Delta t_i$$

The approximation improves if we make Δt_i smaller. From the figure
we see that Δx_i is the area of the rectangle indicated. This area is
approximately the area under the curve for small Δt_i. The total displacement from t_1 to t_2 is then the sum of all the Δx_i:

$$\Delta x = \sum_{t_1}^{t_2} \Delta x_i \approx \sum_{t_1}^{t_2} v(t_i)\,\Delta t_i$$

where the sum is taken from the first rectangle at t_1 to the last one at t_2.
The number of terms in the sum depends on the number of subin-
tervals chosen. As we take smaller and smaller intervals Δt_i, the
number of terms increases and the sum becomes a better approximation of the area under the curve and displacement Δx. Thus

$$\Delta x = x(t_2) - x(t_1) = \lim_{\Delta t \to 0} \sum_{t_1}^{t_2} v(t_i)\,\Delta t_i$$

$$= \text{area under } v(t)\text{-versus-}t \text{ curve from } t_1 \text{ to } t_2 \qquad 2\text{-}8$$

Equation 2-8 is actually the definition of the area under a curve. The area under a curve can be measured in various ways. One is to plot the curve on graph paper containing many small squares of known area, count the squares, and estimate the fractional squares cut by the curve. Another is to cut the paper out along the curve, weigh it, and compare the weight with that of a standard rectangle of known area.

The definite integral is the area under a curve

The limit of the sum in Equation 2-8 is called a *definite integral*, usually written

$$\lim_{\Delta t_i \to 0} \sum_{t_1}^{t_2} v(t_i)\, \Delta t_i = \int_{t_1}^{t_2} v(t)\, dt = x(t_2) - x(t_1) \qquad 2\text{-}9$$

(The integral sign $\int$ is an elongated S, denoting a sum.) Equation 2-9 is also useful in finding the area under a curve for which the algebraic function is given. Since this area is just $x(t_2) - x(t_1)$, where $x(t)$ is the antiderivative, we need only compute the antiderivative at t_1 and t_2 and subtract. The antiderivative is also called the indefinite integral and written without limits on the integral sign:

$$x(t) = \int v(t)\, dt$$

Example 2-9 Find the area under the curve $v = Ct^2$ from $t = 0$ to $t = 4$, where $C = 1$ m/sec³.

Using rule 7, the antiderivative, or indefinite integral, of Ct^2 is $Ct^3/3$ plus any constant. (This can be checked, of course, by differentiating $Ct^3/3$.) Then

$$x(4) - x(0) = \int_0^4 Ct^2\, dt = \frac{4^3 C}{3} - \frac{0^3 C}{3} = \frac{64}{3} C = 21\tfrac{1}{3}\, C$$

You should check this by plotting the curve $v = Ct^2$ versus t and measuring the area by counting the squares.

The average velocity has a simple geometric interpretation in terms of area under a curve. Consider the v-versus-t curve in Figure 2-12. The displacement during the time interval $\Delta t = t_2 - t_1$ is indicated by the shaded area. By definition of the average velocity for this interval (Equation 2-2), the displacement is also the product of v_{av} and Δt:

$$\Delta x = v_{av}\, \Delta t$$

The average velocity is indicated in Figure 2-12 by the horizontal line, drawn so that the area under it from t_1 to t_2 equals the area under the actual v-versus-t curve:

$$v_{av}\, \Delta t = \Delta x = \int_{t_1}^{t_2} v\, dt \qquad 2\text{-}10$$

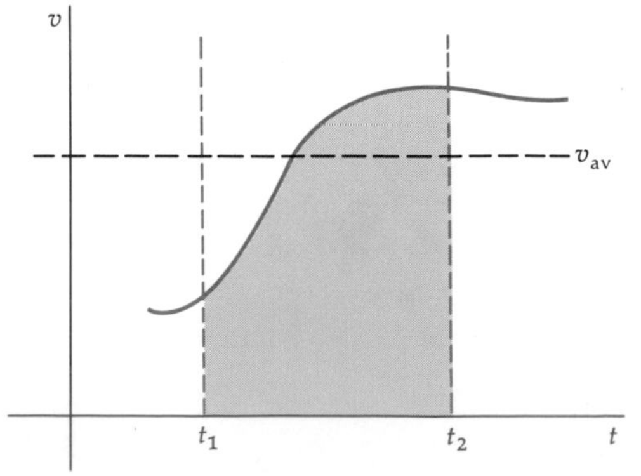

Figure 2-12
Geometric interpretation of average velocity. By definition, $v_{av} = \Delta x/\Delta t$. Thus the rectangular area $v_{av}(t_2 - t_1)$ must equal the displacement in the time interval $t_2 - t_1$. It follows that the rectangular area $v_{av}(t_2 - t_1)$ and the shaded area under the curve must be equal.

2-6 Motion with Constant Acceleration

In Example 2-8 we found the expressions for $v(t)$ and $x(t)$ for motion with constant acceleration. This type of motion is common in nature. Near the surface of the earth, all objects fall vertically with the constant acceleration of gravity $g = 32$ ft/sec² $- 9.8$ m/sec² if air resistance can be neglected and if there are no forces on the objects other than the pull of gravity. The examples in this section will increase your familiarity with the concepts of velocity and acceleration.

Free fall has constant acceleration

The equations found in Example 2-8 for constant acceleration a_0 are

$$v(t) = v_0 + a_0 t \qquad\qquad\qquad 2\text{-}11$$

Equations for constant acceleration

$$x(t) = x_0 + v_0 t + \tfrac{1}{2}a_0 t^2 \qquad\qquad 2\text{-}12$$

A third equation which is sometimes useful can be obtained by eliminating the variable t between these two equations. Using $t = (v - v_0)/a_0$ from Equation 2-11 in Equation 2-12, we have

$$x = x_0 + v_0 \frac{v - v_0}{a_0} + \tfrac{1}{2}a_0 \left(\frac{v - v_0}{a_0}\right)^2$$

This can be simplified and rearranged to yield

$$v^2 = v_0{}^2 + 2a_0(x - x_0) \qquad\qquad 2\text{-}13$$

Constant acceleration with time eliminated

In Figure 2-13 the velocity is plotted versus time. The average velocity for some time interval $\Delta t = t_2 - t_1$ is indicated in the figure. The average velocity is just the arithmetic mean of the value v_1 at the beginning of the time interval and v_2 at the end of the interval:

$$v_{\text{av}} = \tfrac{1}{2}(v_1 + v_2) \qquad \text{for constant acceleration only}$$

This simple result for average velocity holds only when the instantaneous velocity varies linearly with time, i.e., when the acceleration is constant. When the acceleration is not constant, v_{av} is not generally the arithmetic mean of v_1 and v_2, as can be seen from Figure 2-12.

Since Equation 2-13 was obtained from Equations 2-11 and 2-12, these three equations are not independent. Only two of them (sometimes only one) are needed to solve any problem in which the acceleration is constant. Which equation is chosen depends on the problem.

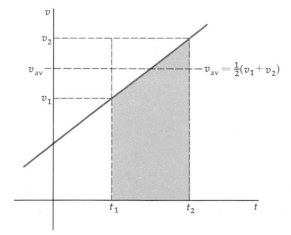

Figure 2-13
For constant acceleration, the average velocity for some time interval is just the mean of the initial and final velocities for that interval.

Example 2-10 A ball is dropped from rest from the top of a 100-ft building and has a constant acceleration downward of 32 ft/sec². (*a*) How fast is it moving just before it hits the ground? (*b*) How long does it take to reach the ground?

Let us choose our coordinate system with the origin at the top of the building with the positive x axis *down*. Then $x_0 = 0$, $v_0 = 0$, and $a_0 = +32$ ft/sec². Equations 2-11 to 2-13 for this problem become

(1) $v(t) = (32 \text{ ft/sec}^2)t$

(2) $x(t) = (16 \text{ ft/sec}^2)t^2$

(3) $v^2 = (64 \text{ ft/sec}^2)x$

(*a*) One way to find the speed at the bottom is to find the time it takes to fall from Equation (2) and then find $v(t)$ for this time from Equation (1). This amounts to doing part (*b*) first. If we are interested only in the final velocity, we can use Equation (3), which relates v to x without involving the time. For $x = 100$ ft, we have

$$v^2 = (64 \text{ ft/sec}^2)(100 \text{ ft}) = 6400 \text{ ft}^2/\text{sec}^2$$
$$v = 80 \text{ ft/sec}$$

Note how the units are treated as algebraic quantities in this calculation.

(*b*) Having found the velocity v at the bottom, we can find the time from Equation (1):

$$80 \text{ ft/sec} = (32 \text{ ft/sec}^2)t$$

$$t = \frac{80 \text{ ft/sec}}{32 \text{ ft/sec}^2} = 2.5 \text{ sec}$$

We could also find the time from $\Delta t = \Delta x/v_{av}$. Since the initial velocity is zero and the final velocity is 80 ft/sec, $v_{av} = \frac{1}{2}(0 + 80 \text{ ft/sec}) = 40$ ft/sec. Then $\Delta t = (100 \text{ ft})/(40 \text{ ft/sec}) = 2.5$ sec. If we had not found v in part (*a*), we could find the time from Equation (2), which relates distance to time, 100 ft = (16 ft/sec²)t^2:

$$t^2 = \frac{100 \text{ ft}}{16 \text{ ft/sec}^2} = (\tfrac{10}{4})^2 \text{ sec}^2 \qquad \text{or} \qquad t = \tfrac{10}{4} \text{ sec} = 2.5 \text{ sec}$$

Figure 2-14 shows a plot of $x(t)$ versus t for this problem. The slope of

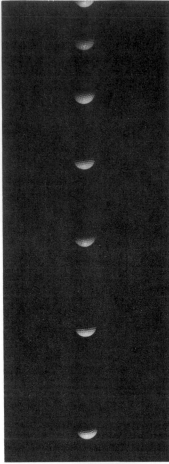

Fundamental Photographs

Multiflash photograph of a golf ball falling with constant acceleration. The position of the ball is recorded at 1/25-sec intervals using a strobe light which flickers 25 times per second. The increase in spacing between successive positions indicates that the speed is increasing.

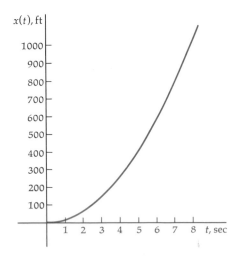

Figure 2-14
Plot of $x(t) = (16 \text{ ft/sec}^2)t^2$ for Example 2-10. The slope is zero at $t = 0$ and increases with t, indicating that the velocity is initially zero and increases with t.

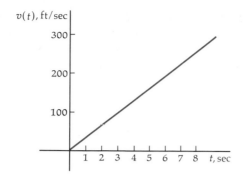

Figure 2-15
Plot of $v(t) = (32 \text{ ft/sec}^2)t$ for
Example 2-10. Note that $v(t)$
versus t is a straight line, indicating that the acceleration
in this example is a constant.

this curve is zero at the origin (indicating that the initial velocity is
zero), and the slope increases with increasing t. Figure 2-15 is a plot of
$v(t)$ versus t. This curve has a constant slope equal to 32 ft/sec².

Example 2-11 A ball thrown vertically upward with an initial velocity of 96 ft/sec has a constant acceleration downward of 32 ft/sec². (*a*)
How high will the ball go? (*b*) How long is the ball in the air? (*c*) At
what times is the ball 80 ft above the ground?

Let us choose our coordinate system with origin at the initial position of the ball and with x positive *upward*. Then $x_0 = 0$, $v_0 = 96$ ft/sec,
and $a_0 = -32$ ft/sec². For these conditions the constant-acceleration
equations 2-11 to 2-13 become

(1) $v = 96 - 32t$

(2) $x = 96t - 16t^2$

(3) $v^2 = 96^2 - 64x$

In this example, we omit the units in the equations except when giving
the results. In the above equations it is to be understood that x is in feet,
t in seconds, and v in feet per second. Figure 2-16 shows a sketch of
x versus t [Equation (2)]. We see that x increases to a maximum value
and then decreases to zero and that the time to reach maximum equals
the time to fall back to $x = 0$. The slope of this curve is 96 ft/sec at $t = 0$

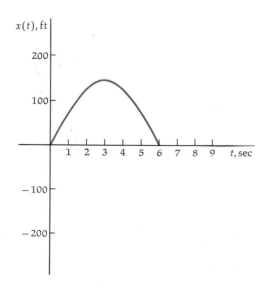

Figure 2-16
Plot of $x = 96t - 16t^2$ for Example 2-12. The slope is positive at $t = 0$, decreases to 0 at
$t = 3$ sec, and continues to decrease with increasing t. The
acceleration has the constant
negative value of -32 ft/sec².

(the original velocity) and decreases to zero at the highest point. After that the slope *continues to decrease,* becoming more and more negative. The symmetry of the figure means that just before the ball hits the ground, the velocity must be −96 ft/sec. Figure 2-17 shows a plot of $v(t)$ versus t from Equation (1). As observed from the slope of the $x(t)$ curve, the velocity begins at 96 ft/sec and always decreases, with a constant negative slope which is the constant acceleration −32 ft/sec. Note that when $v = 0$, the acceleration is *not zero* (despite the common misconception). When $v = 0$, the velocity is decreasing at the same rate as at any other time.

(*a*) Since $v = 0$ when the ball reaches the maximum height, the easiest way to find the maximum height is to set $v = 0$ in Equation (3) and find x:

$$0 = 96^2 - 64x$$

$$x = \frac{96^2}{64} = 144 \text{ ft}$$

(*b*) We can find the total time in the air by finding the times for which $x = 0$ from Equation (2):

$$0 = 96t - 16t^2 = 16t(6 - t)$$

Hence, $t = 0$ and $t = 6$ sec are the two times for which $x = 0$. The first is the time when the ball leaves the ground. The second is the time when it returns to the ground. Another way would be to recognize the symmetry from the figure, find the time to reach the highest point, and multiply by 2. Since $v = 0$ at the highest point, we set $v = 0$ in Equation (1) and obtain $96 - 32t = 0$, or $t = 3$ sec. The total time to reach the ground again is thus 2×3 sec $= 6$ sec.

(*c*) We can see from the sketch of $x(t)$ versus t that there will be two times when $x = 80$ ft. We can find them from Equation (2):

$$80 = 96t - 16t^2 \qquad t^2 - 6t + 5 = 0$$

$$(t - 1)(t - 5) = 0 \qquad \text{or} \qquad t = 1 \text{ sec and } t = 5 \text{ sec}$$

At $t = 1$ sec, the ball passes the height 80 ft on its way up. At $t = 5$ sec, the ball passes the same point on the way down.

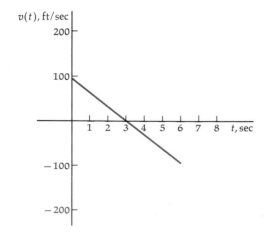

Figure 2-17
Plot of $v(t) = 96 - 32t$ for Example 2-11. The slope is the constant acceleration of −32 ft/sec².

Questions

10. Two boys standing on a bridge throw rocks straight down into the water below. They throw two rocks at the same time, but one hits the water before the other. How can this be if the rocks have the same acceleration g?

11. A ball is thrown straight up. What is its velocity at the top of its flight? What is its acceleration at that point?

Optional

2-7   The Differential

An important quantity related to the derivative and useful in making approximations is the differential. Consider the position function we have studied several times,

$$x = 16t^2$$

The distance traveled between times t_1 and $t_2 = t_1 + \Delta t$ is

$$\Delta x = x(t_1 + \Delta t) - x(t_1) = 16(t_1 + \Delta t)^2 - 16t_1^2$$
$$= 32t_1 \Delta t + 16(\Delta t)^2 \qquad\qquad\qquad 2\text{-}14$$

The first term in the equation is Δt times the derivative of x with respect to t. This quantity is called the *differential of x*, written dx,

Definition of differential

$$dx = \left(\lim_{\Delta t \to 0} \frac{\Delta x}{\Delta t}\right) \Delta t \qquad\qquad\qquad 2\text{-}15$$

Equation 2-14 is then

$$\Delta x = dx + 16(\Delta t)^2$$

If the time interval Δt is very small, the second term in this equation is small compared with the first term because $(\Delta t)^2$ will be small compared to Δt. Thus, to a good approximation,

$$\Delta x \approx dx$$

This approximation is useful because it is often easier to calculate dx than Δx when Δx is wanted.

Figure 2-18 illustrates the difference between the actual change Δx in x, and its approximation, the differential dx. This approximation is essentially that of replacing the curve $x(t)$ by the straight line tangent

Figure 2-18
Comparison of the differential approximation dx with the actual change Δx for the interval t_1 to t_2. The differential dx is the change in x which would result if the slope of $x(t)$ versus t did not change in the interval t_1 to t_2. The approximation is good if the time interval is small.

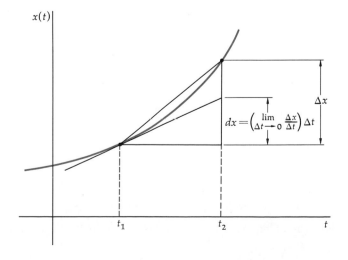

to the curve at t_1. The validity of the approximation obviously depends on how small Δt is and on how much the curve $x(t)$ deviates from a straight line in the range t_1 to $t_1 + \Delta t$. For the curve $x(t) = Ct$, the differential $dx = C\,\Delta t$ equals Δx for all Δt.

It is customary to write dt for the change Δt in the independent variable t. Then the differential dx is just the derivative of x with respect to t times dt. This is the origin of the traditional notation for the derivative we have been using

$$\frac{dx}{dt} = \lim_{\Delta t \to 0} \frac{\Delta x}{\Delta t}$$

We may regard dx/dt as the single quantity, the derivative of x with respect to t, as we have been doing up to now, or we may regard it as the ratio of the differentials dx and dt.

Example 2-12 For a falling body whose position is given by $x = (16$ ft/sec$^2)t^2$, find the distance fallen during the 0.5-sec time interval between $t_1 = 10$ sec and $t_2 = 10.5$ sec.

The derivative of this function is just the velocity:

$$v = \frac{dx}{dt} = (32 \text{ ft/sec}^2)t$$

and the differential dx is

$$dx = (32 \text{ ft/sec}^2)t\, dt$$

Substituting $t = 10$ sec and $dt = 0.5$ sec gives

$$dx = (32 \text{ ft/sec}^2)(10 \text{ sec})(0.5 \text{ sec}) = 160 \text{ ft}$$

This is the distance the particle would fall during 0.5 sec if its velocity were equal to 320 ft/sec, the value of the instantaneous velocity at $t = 10$ sec. Thus the differential approximation here neglects the fact that the velocity increases slightly during the 0.5-sec interval from $t = 10$ sec to $t = 10.5$ sec. Since the velocity at $t = 10.5$ is $32 \times 10.5 = 336$ ft/sec, the average velocity during this interval is 328 ft/sec. The actual distance the body falls during this interval is 164 ft. This result is greater by 4 ft than that found using the differential approximation, an error of about $2\frac{1}{2}$ percent.

Review

A. Define, explain, or otherwise identify:

Particle, 18
Displacement, 19
Average velocity, 20
Instantaneous velocity, 21
Derivative, 22
Slope, 23
Speed, 25

Average acceleration, 25
Instantaneous acceleration, 25
Initial-value problem, 29
Antiderivative, 29
Definite integral, 32
Indefinite integral, 32
Differential, 37

B. True or false: If the statement is true, explain; if it is false, give a counterexample, i.e., a known example which contradicts the statement.

1. The equation $x = x_0 + v_0 t + \frac{1}{2}a_0 t^2$ holds for all motion in one dimension.

2. If the acceleration is zero, the particle cannot be moving.

3. The displacement equals the area under the velocity-versus-time curve.

4. The average velocity always equals the mean value of the initial and final velocities.

Exercises

Unless instructed otherwise, use the approximate value $g = 32$ ft/sec$^2 = 9.8$ m/sec^2 for the acceleration of gravity in Exercises and Problems.

Section 2-1, Displacement and Average Velocity

1. A particle is at $x = +5$ m at $t = 0$, $x = -7$ m at $t = 6$ sec, and $x = +2$ m at $t = 10$ sec. Find the average velocity of the particle during the intervals (*a*) $t = 0$ to $t = 6$ sec; (*b*) $t = 6$ sec to $t = 10$ sec; (*c*) $t = 0$ to $t = 10$ sec.

2. A driver begins a 300-mi trip at noon. (*a*) He drives nonstop and arrives at his destination at 5:30 P.M. Calculate his average velocity for the trip. (*b*) He drives for 3 h, rests for $\frac{1}{2}$h, and continues driving, arriving at 5:30 P.M. Calculate his average velocity. (*c*) After resting 2 h, he drives back home, taking 6 h for the return trip. What is his average velocity for the total round trip? (*d*) What is his displacement?

3. A car travels in a straight line with average velocity of 60 mi/h for 2.5 h and then with average velocity of 30 mi/h for 1.5 h. (*a*) What is the total displacement for the 4-h trip? (*b*) What is the average velocity for the total trip?

4. Figure 2-19 shows the position of a particle versus time. Find the average velocity for the time intervals *a*, *b*, *c*, and *d* indicated.

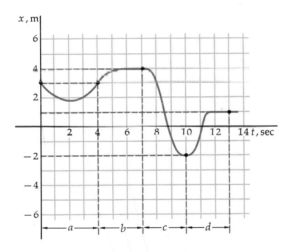

Figure 2-19
x versus *t* for Exercise 4.

Section 2-2, Instantaneous Velocity

5. For each of the four graphs of *x* versus *t* in Figure 2-20, indicate whether the velocity at time t_2 is greater than, less than, or equal to the velocity at time t_1. Compare also the speeds at these times.

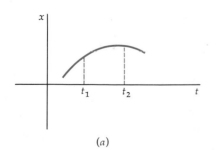

(*a*)

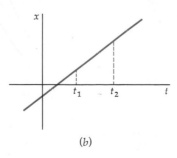

(*b*)

Figure 2-20
x-versus-*t* curves for Exercise 5. (Continued on next page)

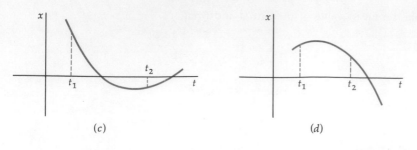

(c)   (d)

Figure 2-20
(Continued)

6. In the graph of x versus t in Figure 2-21, (a) find the average velocity between the times $t = 0$ and $t = 2$ sec. (b) Find the instantaneous velocity at $t = 2$ sec by measuring the slope of the tangent line indicated.

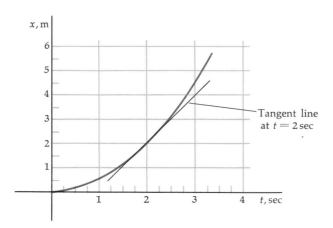

Tangent line
at $t = 2$ sec

Figure 2-21
x versus t for Exercise 6 with tangent line drawn at $t = 2$ sec.

7. For the graph of x versus t in Figure 2-22 find the average velocity for the time intervals $\Delta t = t_2 - 0.75$ sec when t_2 is 1.75, 1.5, 1.25, and 1.0 sec. What is the instantaneous velocity at $t = 0.75$ sec?

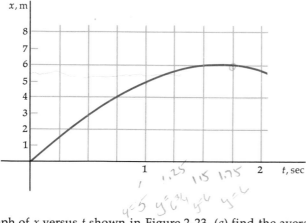

Figure 2-22
x-versus-t curve for Exercise 7.

8. For the graph of x versus t shown in Figure 2-23, (a) find the average velocity for the interval $t = 1$ sec to $t = 5$ sec. (b) Find the instantaneous velocity at $t = 4$ sec. (c) At what time is the velocity of the particle zero?

9. The position of a particle depends on the time according to $x = (1 \text{ m/sec}^2)t^2 - (5 \text{ m/sec})t + 1$ m. (a) Find the displacement and average velocity for the interval $t = 3$ sec to $t = 4$ sec. (b) Find a general formula for the displacement for the time interval from t to $t + \Delta t$. (c) Find the instantaneous velocity for any time t.

10. The height of a certain projectile is related to time by $y = -5(t - 5)^2 + 125$, where y is in meters and t in seconds. (a) Sketch y versus t for $t = 0$ to $t = 10$ sec. (b) Find the average velocity for each of the 1-sec time intervals between

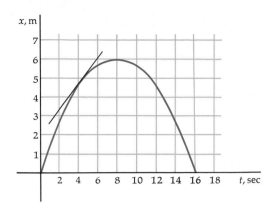

Figure 2-23
x-versus-t curve for Exercise 8 with tangent line drawn at $t = 4$ sec.

integral time values from $t = 0$ to $t = 10$ sec. Sketch v_{av} versus t. (c) Find the instantaneous velocity as a function of time.

Section 2-3, Acceleration

11. A fast car can accelerate from 0 to 60 mi/h in 5 sec. What is the average acceleration during this interval? What is the ratio of this acceleration to the free-fall acceleration of gravity?

12. A car is traveling at 45 mi/h at time $t = 0$. It accelerates at a constant rate of 10 mi/h-sec. (a) How fast is the car going at $t = 1$ sec? At $t = 2$ sec? (b) What is its speed at a general time t?

13. At $t = 5$ sec an object is traveling at 5 m/sec. At $t = 8$ sec its velocity is -1 m/sec. Find the average acceleration for this interval.

14. A particle moves with velocity given by $v = 8t - 7$, where v is in meters per second and t in seconds. (a) Find the average acceleration for the 1-sec intervals beginning at $t = 3$ sec and $t = 4$ sec. (b) Sketch v versus t. What is the instantaneous acceleration at any time?

15. State whether the acceleration is positive, negative, or zero for each of the *position* functions $x(t)$ in Figure 2-24.

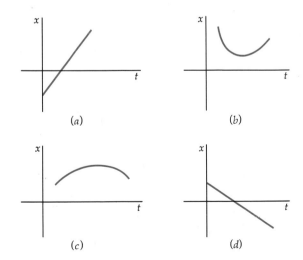

Figure 2-24
x-versus-t curves for Exercise 15.

16. The position of a particle versus time is given by

t, sec	0	1	2	3	4	5	6	7	8	9	10	11
x, m	0	5	15	45	65	70	60	-30	-50	-50	-55	-55

Plot x versus t and draw a smooth curve $x(t)$. Indicate the times or time intervals for which (a) the velocity is greatest; (b) the velocity is least; (c) the velocity is zero; (d) the velocity is constant; (e) the acceleration is positive; (f) the acceleration is negative.

17. The position of an object is related to time by $x = 4t^2 - 8t + 6$, where x is in feet and t in seconds. Find the instantaneous velocity and acceleration as functions of time.

18. A mass oscillating on a spring has a position function given by $x = A \sin \omega t$, where A and ω (the Greek letter omega) are constants. Find the instantaneous velocity and acceleration functions. (Rules 3 and 8 in Appendix Table E-1 may be helpful in finding the derivatives.) Show that the acceleration is proportional to the displacement x and that the velocity is greatest when the acceleration is zero.

19. The velocity of a particle under the influence of viscous forces (such as air resistance) is given by $v = Ae^{-bt}$, where A and b are constants. (a) What is the physical significance of the constant A? What are the dimensions of b? (b) Show that the acceleration is proportional to the velocity.

Section 2-4, Finding $x(t)$ from $v(t)$: The Initial-Value Problem

20. The velocity of a particle is given by $v = 6$ m/sec for all time t. Find the most general position function $x(t)$.

21. The velocity of a particle in meters per second is given by $v = 7t + 5$, where t is in seconds. Find the most general position function $x(t)$.

22. An object is dropped from rest at $t = 0$ from a height of 100 m. Its acceleration is $a = -9.8$ m/sec². (a) Find the velocity v at any time t. (b) Find the position x at any time t.

23. The velocity of an oscillating particle is given by $v = v_0 \cos \omega t$, where v_0 and ω are constants. Find the most general position function $x(t)$.

24. The acceleration of a certain rocket is given by $a = Ct$, where C is a constant. (a) Find the most general position function $x(t)$. (b) Find the position and velocity at $t = 5$ sec if $x = 0$ and $v = 0$ at $t = 0$ and $C = 3$ m/sec³.

25. A particle moves with a constant acceleration of 3 m/sec². At $t = 4$ sec it is at $x = 100$ m. At $t = 6$ sec it has a velocity $v = 15$ m/sec. Find its position at $t = 6$ sec.

Section 2-5, The Antiderivative and Integration

26. The velocity of a particle is given by $v = 6t + 3$, where t is in seconds and v in meters per second. (a) Sketch $v(t)$ versus t and find the area under the curve for the interval $t = 0$ to $t = 5$ sec. (b) Find the most general position function $x(t)$. Use this to calculate the displacement during the interval $t = 0$ to $t = 5$ sec.

27. The velocity of a particle in meters per second is given by $v = 7 - 4t$, where t is in seconds. (a) Sketch $v(t)$ versus t and find the area between the curve and the t axis from $t = 2$ sec to $t = 6$ sec. (b) Find the position function $x(t)$ by integration and use it to find the displacement during the interval $t = 2$ sec to $t = 6$ sec. (c) What is the average velocity for this interval?

28. Figure 2-25 shows the velocity of a particle versus time. (a) What is the magnitude in meters of the area of the rectangle indicated? (b) Find the displacement of the particle for the 1-sec intervals beginning at $t = 1$ sec and $t = 2$ sec. (c) What is the average velocity for the interval from $t = 1$ sec to $t = 3$ sec?

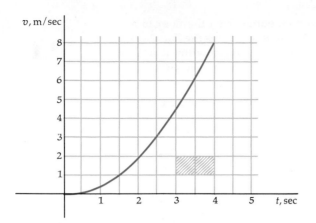

Figure 2-25
v-versus-t curve for Exercise 28.

29. Figure 2-26 shows the acceleration of a particle versus time. (*a*) What is the magnitude of the area of the rectangle indicated? (*b*) The particle starts from rest at $t = 0$. Find the velocity at the times $t = 1$, 2, and 3 sec by counting the squares under the curve. (*c*) Sketch the curve $v(t)$ versus t from your results of part (*b*) and estimate how far the particle traveled in the interval $t = 0$ to $t = 3$ sec.

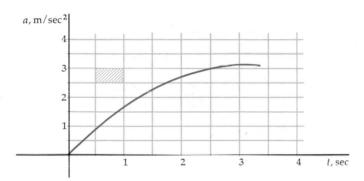

Figure 2-26
a-versus-t curve for Exercise 29.

30. The equation of the curve shown in Figure 2-25 is $v = (0.5 \text{ m/sec})t^2$. Find the displacement of the particle for the interval $t = 1$ to $t = 3$ sec by integration and compare with your answer for Exercise 28. Is the average velocity equal to the mean of the initial and final velocities for this case?

Section 2-6, Motion with Constant Acceleration

31. A car accelerates from rest at a constant rate of 8 m/sec². (*a*) How fast is it going after 10 sec? (*b*) How far has it gone after 10 sec? (*c*) What is its average velocity for the interval $t = 0$ to $t = 10$ sec?

32. An object with initial velocity of 5 m/sec has a constant acceleration of 2 m/sec. When its speed is 15 m/sec, how far has it traveled?

33. An object with constant acceleration has velocity $v = 10$ m/sec when it is at $x = 6$ m and $v = 15$ m/sec when it is at $x = 10$ m. What is its acceleration?

34. An object has constant acceleration $a = 4$ m/sec². Its velocity is 1 m/sec at $t = 0$, at which time it is at $x = 7$ m. How fast is it moving when it is at $x = 8$ m? At what time is this?

35. How long does it take for a particle to travel 100 m if it begins from rest and accelerates at 10 m/sec²? What is the velocity when it has traveled 100 m? What is the average velocity for this time?

36. A ball is thrown upward with initial velocity of 80 ft/sec. Its acceleration is constant, equal to 32 ft/sec² downward. (*a*) How long is the ball in the air? (*b*) What is the greatest height reached by the ball? (*c*) When is the ball 96 ft above the ground?

37. A ball is dropped from a height of 3 ft and rebounds from the floor to a height of 2 ft. (a) What is the velocity of the ball just as it reaches the floor? (b) What is its velocity just as it leaves the floor? (c) If it is in contact with the floor for 0.02 sec, what are the magnitude and direction of its average acceleration during this interval?

38. The minimum distance for a controlled stop with no wheels locked for a certain car is 170 ft for level braking from 60 mi/h. Find the acceleration, assuming it to be constant, and express your answer as a fraction of the free-fall acceleration of gravity. How long does it take to stop?

Section 2-7, The Differential

39. The distance a body falls from rest in a time t is $x = (16 \text{ ft/sec}^2)t^2$. (a) Find the exact distance the body falls from time $t_1 = 20$ sec to time $t_2 = 20.3$ sec. (b) Find the velocity at time $t = 20$ sec and use the differential approximation to find the distance fallen during this interval.

40. The area of a circle of radius r is $A = \pi r^2$. Use the differential approximation to find the change in area ΔA for a small change in the radius Δr. Interpret the quantity dA geometrically.

41. Show that the fractional change in area of a circle is approximately related to the fractional change in radius by $\Delta A/A = 2\Delta r/r$. If the radius changes by 1 percent, what is the percentage change in the area?

42. The volume of a sphere of radius r is $V = (4\pi/3)r^3$. (a) Find the approximate small change in the volume dV for a small change in radius dr and interpret dV geometrically. (b) Show that the fractional changes are related by $dV/V = 3dr/r$. (c) The volume of a sphere is determined by measuring its radius. If a 2 percent error is made in this measurement, what is the resulting percentage error in the volume?

Problems

1. A car traveling at a constant speed of 20 m/sec passes an intersection at time $t = 0$, and 5 sec later a second car passes the same intersection traveling 30 m/sec in the same direction. (a) Sketch the position functions $x_1(t)$ and $x_2(t)$ for the two cars. (b) Find when the second car overtakes the first. (c) How far have the cars traveled when this happens?

2. Hare and Tortoise begin a 10-km race at time $t = 0$. Hare runs at 4 m/sec and quickly outdistances Tortoise, who runs at 1 m/sec (about 10 times faster than a tortoise can actually run but convenient for this problem). After running 5 min, Hare stops and falls asleep. His nap lasts 135 min. He awakes and again runs at 4 m/sec but loses the race. Plot an x-versus-t graph for each on the same axes. At what time does Tortoise pass Hare? How far behind is Hare when Tortoise crosses the finish line? How long can Hare nap and still win the race?

3. According to a mean-value theorem in calculus, in any time interval Δt there is at least one time t' at which the instantaneous velocity equals the average velocity for the whole interval. Sketch any continuous smooth curve x versus t and choose two times t_1 and t_2. Draw the line whose slope is the average velocity for the time interval $\Delta t = t_2 - t_1$ and find a time t' in your interval at which the tangent line to your curve has the same slope.

4. Figure 2-27 shows the position of a car plotted as a function of time. At which of the times t_0 to t_7 is (a) the velocity negative, (b) the velocity positive, (c) the velocity zero, (d) the acceleration negative, (e) the acceleration positive, (f) the acceleration zero?

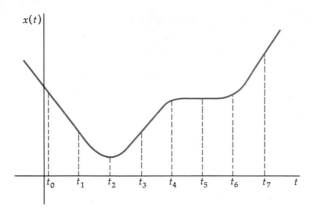

Figure 2-27
Position of a car versus time
for Problem 4.

5. Sketch a single v-versus-t curve in which there are points or segments for which (a) the acceleration is zero and constant while the velocity is not zero; (b) the acceleration is zero but not constant; (c) the velocity and acceleration are both positive; (d) the velocity and acceleration are both negative; (e) the velocity is positive and the acceleration negative; (f) the velocity is negative and the acceleration positive; (g) the velocity is zero, but the acceleration is not.

6. For each of the graphs in Figure 2-28 indicate (a) at what times the acceleration of the object is positive, negative, and zero. (b) At what times is the acceleration constant? (c) At what times is the instantaneous velocity zero?

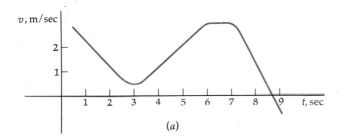

(a)

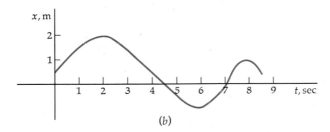

(b)

7. A car is traveling 50 mi/h in a school zone. A police car starts from rest just as the speeder passes it and accelerates at a constant rate of 5 mi/h-sec. (a) Make a sketch of $x(t)$ for both cars. (b) When does the police car catch the speeding car? (c) How fast is the police car traveling when it finally catches the speeder?

8. The police car in Problem 7 chases a speeder traveling 80 mi/h. The maximum speed of the police car is 120 mi/h. The police car travels from rest with constant acceleration, 5 mi/h-sec, until its speed reaches 120 mi/h and then moves at constant speed. (a) When does it catch the speeder if it starts just as the speeder passes? (b) How far has each car traveled? (c) Sketch $x(t)$ for each car.

9. When the police car of Problem 8 traveling at 120 mi/h is 100 yd behind the speeder (traveling 80 mi/h), the speeder sees the car and slams on his brakes, locking the wheels. (a) Assuming that each car can brake at 20 ft/sec² and that the driver of the police car brakes as soon as he sees the brake lights of the

speeder, i.e., with no reaction time, show that the cars collide. (b) What is the speed of the police car relative to the speeder when they collide? At what time do they collide after the brakes are applied? (c) The time interval between the policeman's seeing the brake lights and putting on his own brakes is called his *reaction time T*. Estimate or measure your reaction time and discuss how it affects this problem.

10. When a car traveling at speed v_1 rounds a corner, the driver sees another car traveling at a slower speed v_2 a distance d ahead of him. (a) If the maximum acceleration his brakes can provide is a, show that the distance d must be greater than $(v_1 - v_2)^2/2a$ if a collision is to be avoided. (b) Evaluate this distance for $v_1 = 60$ mi/h, $v_2 = 30$ mi/h, and $a = 20$ ft/sec². (c) Estimate or measure your reaction time and calculate its effect on the distance found in part (b).

11. Figure 2-29 shows a plot of x versus t for a body moving along a straight line. Sketch rough graphs of v versus t and a versus t for this motion.

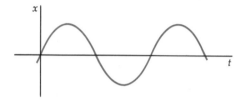

Figure 2-29
x-versus-t curve for Problem 11.

12. Figure 2-30 is a graph of v versus t for a particle moving along a straight line. The position of the particle at time $t = 0$ is $x_0 = 5$ m. (a) Find x for various times t by counting squares and sketch x versus t. (b) Sketch the acceleration a versus t.

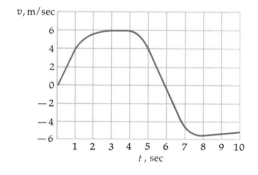

Figure 2-30
v-versus-t curve for Problem 12.

13. A passenger is running at his maximum velocity of 10 ft/sec to catch a train. When he is a distance d from the nearest entry to the train, the train starts from rest with constant acceleration $a = 1.0$ ft/sec² away from the passenger. (a) If $d = 40$ ft and he keeps running, will he be able to jump onto the train? (b) Sketch the position function $x(t)$ for the train, choosing $x = 0$ at $t = 0$. On the same graph sketch $x(t)$ for the passenger for various values of initial separation distances d including $d = 40$ ft and the critical value d_c such that he just catches the train. (c) For the critical separation distance d_c, what is the speed of the train when the passenger catches it? What is its average speed for the time $t = 0$ until he catches it? What is the value of d_c?

14. A load of bricks is being lifted by a crane at the steady velocity of 16 ft/sec, but 20 ft above the ground one brick falls off. Describe the motion of the free brick by sketching $x(t)$. (a) What is the greatest height the brick reaches above the ground? (b) How long does it take to reach the ground? (c) What is its speed just before it hits the ground?

15. The position of a body oscillating on a spring is given by $x = A \sin \omega t$, where A and ω are constants and have the values $A = 5$ cm and

$\omega = 10°/\text{sec} = 0.175$ rad/sec. Sketch x versus t for times from $t = 0$ to $t = 36$ sec. (a) Measure the slope of your graph at $t = 0$ to find the velocity at this time. (b) Using sine tables, calculate the average velocity for a series of intervals beginning at $t = 0$ and ending at $t = 6, 3, 2, 1, 0, 0.5,$ and 0.25 sec. (c) Compute dx/dt and find the velocity at time $t = 0$.

16. The kinetic energy of a particle is $E_k = \frac{1}{2}mv^2$, where m is the mass of the particle and v is its speed. Show that if the kinetic energy is plotted as a function of x the slope of this curve is proportional to the acceleration dv/dt. Hint: Use the chain rule

$$\frac{dv}{dt} = \frac{dv}{dx}\frac{dx}{dt}$$

17. Table 2-2 lists some world track records for short sprints. A simple model of sprinting assumes that a sprinter starts from rest, accelerates with constant acceleration a for a short time T, and then runs with constant speed $v_0 = aT$. According to this model, for times t greater than T, the distance x varies linearly with time. (a) Make a graph of distance x versus time t from the data in the table. (b) Set up an equation for x versus t according to the simple model described and show that for $t > T$, x can be written $x = v_0(t - \frac{1}{2}T)$. (c) Connect the points on your graph with a straight line and determine the slope and the intercept of the line with the time axis. From the fact that the slope is v_0 and the intercept is $\frac{1}{2}T$, compute the acceleration a. (d) The record for $x = 200$ m is 19.5 sec. Discuss the applicability of this simple model to races of 200 m or more.

Table 2-2		
x		
yd	m	t, sec
50		5.1
	50	5.5
60		5.9
	60	6.5
100		9.1
	100	9.9

18. The acceleration of a particle falling under the influence of gravity and a resistive force such as air resistance is given by

$$a = \frac{dv}{dt} = g - Bv$$

where g is the free-fall acceleration of gravity and B is a constant which depends on the mass and shape of the particle and on the medium. Suppose the particle begins with zero velocity at time $t = 0$. (a) Discuss qualitatively how the speed v varies with time from your knowledge of the rate of change dv/dt given by this equation. What is the value of the velocity when the acceleration is zero? This is called the *terminal velocity*. (b) Sketch the solution $v(t)$ versus t without solving the equation. This can be done as follows. At $t = 0$, v is zero and the slope is g. Sketch a straight-line segment, neglecting any change in slope for a short time interval. At the end of the interval the velocity is not zero, and so the slope is less than g. Sketch another straight-line segment with a smaller slope. Continue until the slope is zero and the velocity equals the terminal velocity.

19. (a) Show that the equation in Problem 18 can be written $df/f = -B\, dt$, where $f = g - Bv$. (b) Find the indefinite integral of each side of this equation and solve for v, using the fact that exp $(\ln f) = f$. Find the constant of integration in terms of g and B from the initial condition $v = 0$ at $t = 0$. (c) Write your solution in the form $v = v_t(1 - e^{-t/T})$, where v_t is the terminal velocity and $T = 1/B$. (d) Plot the points for $t = 0, 1, 3, 6, 10,$ and 20 sec for $B = 0.1$ sec^{-1}.

20. Show that if $y = Cx^n$, where c and n are any constants, the differential dy is related to dx by

$$\frac{dy}{y} = n\frac{dx}{x}$$

CHAPTER 3   Motion in Two and Three Dimensions

We now extend our description of the motion of a particle to the more general cases of motion in two and three dimensions. For this more general motion, displacement, velocity, and acceleration are vectors, which have direction in space as well as magnitude. In this chapter, we investigate the properties of vectors in general and of these three vectors in particular. We also discuss two important special cases of motion in a plane, projectile motion and circular motion.

When the motion of a particle is confined to a plane, its position can be described by two numbers. For example, we might choose the distance x from the y axis and the distance y from the x axis, where the x and y axes are perpendicular axes which intersect at origin O, as in Figure 3-1. Alternatively, we can specify the same position by giving the distance r from the origin, and the angle θ made by the x axis and the line from the origin to the point. There are of course other choices, but in each case two numbers are needed to specify the location of a point on a plane.

Position on a plane

If the particle is not confined to a plane but moves in three dimensions, three numbers are needed to specify its position. A convenient method is to use the three coordinates x, y, and z. An alternative is to use the spherical coordinates r, θ, and ϕ. The relations between spher-

Position in space

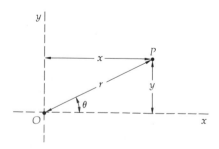

Figure 3-1
Coordinates of a point P in a plane. The rectangular coordinates (x,y) and polar coordinates (r,θ) are related by $x = r \cos \theta$, $y = r \sin \theta$, $r = \sqrt{x^2 + y^2}$, $\tan \theta = y/x$.

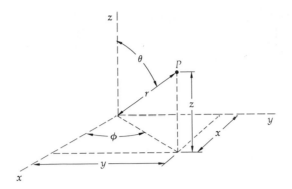

Figure 3-2
Coordinates of a point P in three dimensions. The rectangular coordinates (x,y,z) and spherical coordinates (r,θ,ϕ) are related by
$x = r \sin \theta \cos \phi$
$y = r \sin \theta \sin \phi$
$z = r \cos \theta$
$r = \sqrt{x^2 + y^2 + z^2}$
$\tan \theta = \sqrt{x^2 + y^2}/z$
$\tan \phi = y/x$

ical and rectangular coordinates are shown in Figure 3-2. Spherical coordinates are convenient when there is spherical symmetry. For example, if a particle moves on the surface of a sphere, these coordinates are useful because r is constant and only θ and ϕ vary.

Most of the interesting features of motion in more than one dimension can be brought out in two-dimensional motion, which is easier to illustrate on paper or a blackboard. Most of our examples are therefore limited to two dimensions, with an indication of the extension to three-dimensional motion.

3-1 The Displacement Vector

Consider a particle which moves from some point P_1 to another P_2. As in one-dimensional motion, we call the change in position of a particle its *displacement*. In two or three dimensions, we indicate the displacement by an arrow from the original position P_1 to the new position P_2, as in Figure 3-3. We see that the displacement has direction as well as magnitude. We should note also that the displacement as we have defined it does not depend on the path taken by the particle as it moves from P_1 to P_2 but only on the end points P_1 and P_2. A second displacement is indicated in Figure 3-3 by an arrow from point P_2 to P_3. The arrow from the original point P_1 to P_3 indicates the *resultant* displacement from the original position P_1 to the last position P_3; it is called the sum of the two successive displacements. Quantities which have magnitude and direction *and* which obey the addition law illus-

Displacement in space

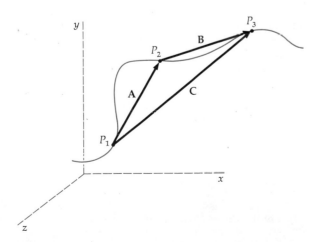

Figure 3-3
The addition of vectors. The displacement **C** is equivalent to the successive displacements **A** and **B**; that is, **C** = **A** + **B**.

trated by these displacements are called *vectors*. Many quantities in physics besides displacements are vectors, e.g., velocity, acceleration, force, momentum, and electric field. We denote vector quantities by boldface type, as in **A**, **B**, and **C** for the three displacements in Figure 3-3. (In handwriting, we indicate a vector by drawing an arrow over the symbol.) The addition law for vectors illustrated in this figure is simply

$$\mathbf{C} = \mathbf{A} + \mathbf{B} \qquad\qquad\qquad 3\text{-}1$$

The *magnitude*, or length, of the vector **A** is written as $|\mathbf{A}|$ or A. The magnitude of a vector ordinarily has physical units, like speed or acceleration, discussed in Chapter 2. A displacement vector, for example, has a magnitude which can be expressed in feet or meters or any other measure of distance. Note that the sum of the magnitudes of **A** and **B** does not equal the magnitude of $\mathbf{C} = \mathbf{A} + \mathbf{B}$ unless **A** and **B** are in the same direction:

$$C \neq A + B \qquad \text{unless } \mathbf{A} \text{ and } \mathbf{B} \text{ are in the same direction}$$

Quantities which are completely specified by only a number (with units), e.g., distance, mass, or temperature, are called *scalars*.[1]

Questions

1. Can the displacement of a particle in any time interval have a magnitude which is less than the distance traveled by the particle along its path in that interval? Can its magnitude be more than the distance traveled? Explain.

2. Give an example in which the distance traveled is a significant amount yet the corresponding displacement is zero.

3. Is displacement always in the direction of motion?

4. If $A = 7$ m and $B = 3$ m, what are the greatest and least values possible for C if $\mathbf{C} = \mathbf{A} + \mathbf{B}$?

3-2 Components of a Vector

The projection of a vector on a line is called a *component* of the vector in the direction of that line. An important example of vector components is the projections of the vector on the axes of a rectangular coordinate system. These projections, called the *rectangular* components of the vector, are denoted by A_x, A_y, and A_z for the vector **A**. Figure 3-4 illustrates the rectangular components of the displacements **A**, **B**, and **C** of Figure 3-3. We denote the coordinates of P_1 by (x_1, y_1) and use similar designations for P_2 and P_3; then the components of these vectors are

$A_x = x_2 - x_1$	$A_y = y_2 - y_1$	
$B_x = x_3 - x_2$	$B_y = y_3 - y_2$	3-2
$C_x = x_3 - x_1$	$C_y = y_3 - y_1$	

[1] Strictly speaking, only quantities which are independent of the choice of coordinate system are scalars. Thus distance, mass, and temperature are scalars, but, for example, the x coordinate of a point is not strictly a scalar because it depends on the coordinate system. This distinction will not be important for our work.

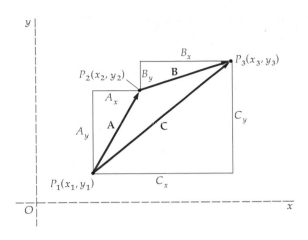

Figure 3-4
The x and y components of vectors $\mathbf{A}$, $\mathbf{B}$, and $\mathbf{C}$ from Figure 3-3. $A_x = x_2 - x_1$, $A_y = y_2 - y_1$, $B_x = x_3 - x_2$, $B_y = y_3 - y_2$, $C_x = x_3 - x_1 = A_x + B_x$, $C_y = y_3 - y_1 = A_y + B_y$.

Note that the *components* of a vector are not vectors. They are positive or negative numbers with units.[1] It should be clear from the figure that

$$\mathbf{C} = \mathbf{A} + \mathbf{B} \text{ implies both } C_x = A_x + B_x \text{ and } C_y = A_y + B_y \qquad 3\text{-}3$$

The magnitude of a vector can be written in terms of the rectangular components of the vector. As shown in Figure 3-4, the rectangular components A_x and A_y form the two legs of a right triangle whose hypotenuse is the magnitude A. Hence, from the theorem of Pythagoras,

$$A = \sqrt{A_x{}^2 + A_y{}^2} \qquad 3\text{-}4$$

We can also specify the direction in terms of the rectangular components. If θ is the angle between the vector and a line parallel to the x axis,

$$\tan \theta = \frac{A_y}{A_x} \qquad 3\text{-}5$$

Example 3-1 A man walks 4.00 mi northeast and then 3.00 mi southeast. If the origin of coordinates is taken at his starting point and the x axis points straight east, (*a*) what are the rectangular components of his two displacements? (*b*) What are the rectangular components of his overall displacement? (*c*) What are his final distance and direction from the starting point?

(*a*) The successive displacements are shown in Figure 3-5. Displace-

[1] The component of a vector is not strictly a scalar because it depends on the choice of coordinate system. (see previous footnote).

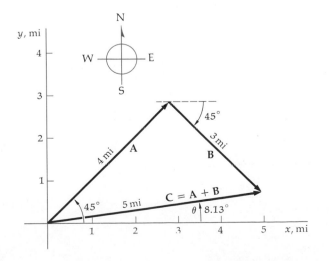

Figure 3-5
A plot of the displacements from Example 3-1. North is plotted in the $+y$ direction; east in the $+x$ direction.

ment **A** has magnitude 4.00 mi and is 45° north of east. Displacement **B** has magnitude 3.00 mi and is 45° south of east. The components of **A** are

$$A_x = A \cos 45° = (4.00 \text{ mi})(0.707) = 2.83 \text{ mi}$$

$$A_y = A \sin 45° = (4.00 \text{ mi})(0.707) = 2.83 \text{ mi}$$

The components of **B** are

$$B_x = B \cos 45° = (3.00 \text{ mi})(0.707) = 2.12 \text{ mi}$$

$$B_y = -B \sin 45° = -(3.00 \text{ mi})(0.707) = -2.12 \text{ mi}$$

The component B_y is negative, indicating that it is in the direction of decreasing, or negative, values of y rather than of increasing, or positive, values of y.

(b) The components of the resultant displacement **C** = **A** + **B** are calculated as follows:

$$C_x = A_x + B_x = 2.83 \text{ mi} + 2.12 \text{ mi} = 4.95 \text{ mi}$$

$$C_y = A_y + B_y = 2.83 \text{ mi} - 2.12 \text{ mi} = 0.71 \text{ mi}$$

(c) The man's final distance from his starting point is the magnitude of **C**:

$$C = \sqrt{C_x{}^2 + C_y{}^2} = \sqrt{4.95^2 + 0.71^2} = 5.0 \text{ mi}$$

The direction of the man's final position can be specified by giving the angle θ:

$$\tan \theta = \frac{C_y}{C_x} = \frac{0.71}{4.95} = 0.143$$

or

$$\theta = 8.13°$$

That is, the direction of the net displacement is 8.13° north of east.

We can write a vector conveniently in terms of its rectangular components by making use of *unit vectors*. A unit vector is a dimensionless vector which has the magnitude of 1 and points in any convenient direction. For example, let **i**, **j**, and **k** be unit vectors which point in the x, y, and z directions, respectively. A general vector **A** can then be written as the sum of three vectors each parallel to a coordinate:

$$\mathbf{A} = A_x \mathbf{i} + A_y \mathbf{j} + A_z \mathbf{k} \qquad 3\text{-}6 \qquad \textit{Unit vectors}$$

The vector $A_x\mathbf{i}$ is the product of the component A_x and the unit vector **i**. It is a vector parallel to the x axis with magnitude A_x. The vector sum indicated in Equation 3-6 is illustrated in Figure 3-6. The addition of

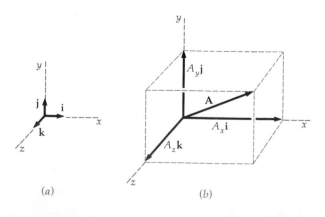

(a)   (b)

Figure 3-6
(a) The unit vectors **i**, **j**, and **k** in a rectangular coordinate system. (b) The vector **A** can be written in terms of the unit vectors **i**, **j**, **k**; that is, $\mathbf{A} = A_x \mathbf{i} + A_y \mathbf{j} + A_z \mathbf{k}$.

two vectors **A** and **B** can be written in terms of these unit vectors as

$$\mathbf{A} + \mathbf{B} = (A_x\mathbf{i} + A_y\mathbf{j} + A_z\mathbf{k}) + (B_x\mathbf{i} + B_y\mathbf{j} + B_z\mathbf{k})$$
$$= (A_x + B_x)\mathbf{i} + (A_y + B_y)\mathbf{j} + (A_z + B_z)\mathbf{k} \qquad 3\text{-}7$$

For example, the resultant displacement found in Example 3-1 is written

$$\mathbf{C} = 4.95\mathbf{i} + 0.71\mathbf{j}$$

where **C** is in miles.

We shall frequently want to use components of vectors along lines other than the rectangular axes. For these cases, discussed later, we define other unit vectors pointing in directions convenient for the particular problem.

Questions

5. Can a component of a vector have a magnitude greater than the magnitude of the vector? Under what circumstances can a component of a vector have a magnitude equal to the magnitude of the vector?

6. Can a vector be equal to zero and still have one or more components not equal to zero?

7. Are the components of $\mathbf{C} = \mathbf{A} + \mathbf{B}$ necessarily larger than the corresponding components of either **A** or **B**?

3-3 Properties of Vectors

This section summarizes some important properties of vectors.

Equality

Two vectors **A** and **B** are defined to be equal if they have the same magnitude *and* the same direction:

$$\mathbf{A} = \mathbf{B} \qquad \text{if } A = B \text{ } and \text{ their directions are the same} \qquad 3\text{-}8$$

Note that we do not require that the end points of the vectors be the same for them to be equal. All the vectors in Figure 3-7 are equal. Thus a vector is completely specified by its magnitude and direction. Since these properties can be written in terms of the rectangular components of the vector, we can also specify the vector completely by giving its

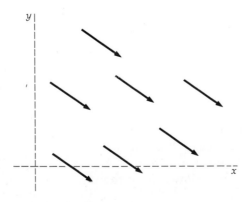

Figure 3-7
Vectors are equal if their magnitudes and directions are the same. The vectors in this figure are all equal.

rectangular components. Two vectors are equal if their corresponding rectangular components are equal:

$$\mathbf{A} = \mathbf{B} \quad \text{if } A_x = B_x,\ A_y = B_y, \text{ and } A_z = B_z \tag{3-9}$$

Figure 3-8
Addition of two vectors. The vector $\mathbf{C} = \mathbf{A} + \mathbf{B}$ is drawn from the tail of **A** to the head of **B** when the tail of **B** is placed at the head of **A**.

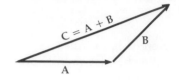

Vector Addition

The addition of two vectors **A** and **B** is defined geometrically as in Figure 3-8. Here, the tail of vector **B** is placed at the head of vector **A**. The sum $\mathbf{C} = \mathbf{A} + \mathbf{B}$ is the vector from the tail of **A** to the head of **B**, as shown.

An alternative construction for finding the sum of two vectors is to put the tails of the vectors together and construct a parallelogram, as in Figure 3-9. The resultant vector is along the diagonal of the parallelogram. The vector-addition law is sometimes referred to as the *parallelogram law of addition.* Figure 3-9 shows that this construction gives the same result as Figure 3-8, i.e., putting the tail of **B** at the head of **A** and drawing vector $\mathbf{C} = \mathbf{A} + \mathbf{B}$ from the tail of **A** to the head of **B**.

Figure 3-9
Parallelogram method of addition of vectors. The vector $\mathbf{C} = \mathbf{A} + \mathbf{B}$ is along the diagonal of the parallelogram formed by **A** and **B**. The result is the same as that in Figure 3-8. It can be seen from this figure that $\mathbf{A} + \mathbf{B}$ is the same as $\mathbf{B} + \mathbf{A}$.

By geometric construction, we can see that the sum of the two vectors is independent of the order of the vectors. This is the *commutative law of addition:*

$$\mathbf{A} + \mathbf{B} = \mathbf{B} + \mathbf{A} \tag{3-10}$$

The sum of three or more vectors is independent of the order in which they are added. This *associative law of addition,*

$$(\mathbf{A} + \mathbf{B}) + \mathbf{C} = \mathbf{A} + (\mathbf{B} + \mathbf{C}) \tag{3-11}$$

is illustrated for three vectors in Figure 3-10.

Figure 3-10
Vector addition is associative. $(\mathbf{A} + \mathbf{B}) + \mathbf{C} = \mathbf{A} + (\mathbf{B} + \mathbf{C})$.

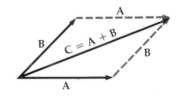

Negative of a Vector

The vector $-\mathbf{A}$ is defined to be the vector which when added to **A** gives zero:

$$\mathbf{A} + -\mathbf{A} = 0$$

The vector $-\mathbf{A}$ has the same magnitude as **A** but points in the opposite direction (sometimes stated by saying that $-\mathbf{A}$ is antiparallel to **A**). The components of $-\mathbf{A}$ are each the negative of the components of **A**. That is, if **A** has components (A_x, A_y, A_z), then $-\mathbf{A}$ has components $(-A_x, -A_y, -A_z)$.

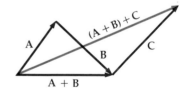

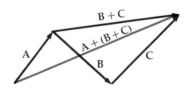

Subtraction of Vectors

The definition of the negative of a vector allows us to define vector subtraction in a simple way. **B** subtracted from **A** is defined to be $-\mathbf{B}$ added to **A**:

$$\mathbf{A} - \mathbf{B} = \mathbf{A} + (-\mathbf{B}) \tag{3-12}$$

Vector subtraction is illustrated in Figure 3-11. It is often useful to think of $\mathbf{A} - \mathbf{B}$ as the vector which is added to vector **B** to obtain vector **A**. This idea is illustrated in Figure 3-11*b*, where the vectors **A** and **B** are drawn tail to tail. Then the vector $\mathbf{A} - \mathbf{B}$ is drawn from the tip of **B** to the tip of **A**.

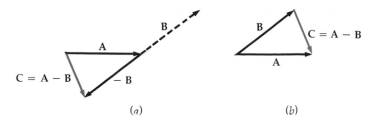

Figure 3-11
Subtraction of vectors. (*a*) **C** = **A** − **B** is found by adding −**B** to **A**. (*b*) An alternative method of finding **C** = **A** − **B**. Here **C** is the vector which when added to **B** gives the vector **A**.

Multiplication of a Vector by a Scalar

If we multiply a vector **A** by a positive scalar c, we obtain the vector $c\mathbf{A}$, which has the same direction as **A** and magnitude cA.[1] We used this definition when we expressed a vector in terms of its rectangular components and the unit vectors **i**, **j**, and **k**. Note that this definition is consistent with the result that when we add two equal vectors **A** + **A**, we obtain a vector parallel to **A** with twice the magnitude, which is $2A$.

If c is a negative scalar, the vector $c\mathbf{A}$ has the magnitude $|c|A$ and is antiparallel to **A**.

Questions

8. Can two vectors of unequal magnitudes be added to give zero?

9. Are the magnitudes of **A** and −**A** equal, or do they have opposite signs? What about the components of **A** and −**A**?

10. Does the magnitude of a vector change if the units in which it is expressed change? Does the magnitude of $c\mathbf{A}$ have the same units as **A**?

11. If **A**, **B**, and **C** are three successive displacement vectors, the total displacement resulting from all three can be obtained by adding the three vectorially in any order; but is the path described by displacement **B** followed by **C** followed by **A** the same as the path corresponding to displacements **A**, **B**, and **C** taken in that order?

3-4 The Velocity Vector

We are now ready to extend our concepts of velocity and acceleration to the general motion of a particle. Consider a particle moving along a curve in two dimensions, as in Figure 3-12. At some time t_1 it is at

[1] This result holds whether c is a strict scalar or any numerical quantity.

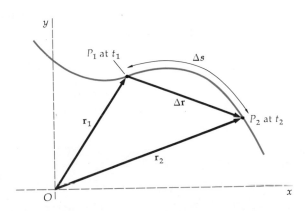

Figure 3-12
The position, or radius, vector **r** is the vector from the origin to the position of the particle. The displacement vector Δ**r** for the time interval from t_1 to t_2 is then the difference in the position vectors $\Delta\mathbf{r} = \mathbf{r}_2 - \mathbf{r}_1$.

point P_1, and at a later time t_2 it is at point P_2. We can describe the position of the particle by a vector $\mathbf{r}$ from the origin to the position of the particle. This *position vector*, or *radius vector* as it is sometimes called, is special in that we specify in its definition the location of its tail. From the figure we see that the displacement vector is the difference of the two position vectors $\mathbf{r}_2 - \mathbf{r}_1$. By analogy with our one-dimensional notation, in which we wrote Δr for a displacement in two (or three) dimensions,

$$\Delta \mathbf{r} = \mathbf{r}_2 - \mathbf{r}_1 \qquad\qquad 3\text{-}13$$

Radius vector

Displacement vector

The new position vector $\mathbf{r}_2$ is thus the sum of the original position vector $\mathbf{r}_1$ and the displacement vector $\Delta \mathbf{r}$, as can be seen from the figure:

$$\mathbf{r}_2 = \mathbf{r}_1 + \Delta \mathbf{r}$$

We define the average velocity vector as the ratio of the displacement vector $\Delta \mathbf{r}$ and the time interval for this displacement $\Delta t = t_2 - t_1$:

$$\mathbf{v}_{av} = \frac{\Delta \mathbf{r}}{\Delta t} \qquad\qquad 3\text{-}14$$

Average-velocity vector

The average velocity is the product of a scalar $(1/\Delta t)$ and a vector $\Delta \mathbf{r}$ and is therefore a vector. We note from Figure 3-12 that the magnitude of the displacement vector is *not* equal to the distance of travel Δs as measured along the curve. It is, in fact, less than this distance (unless the particle travels in a straight line between points P_1 and P_2). However, if we consider smaller and smaller time intervals, as indicated in Figure 3-13, the magnitude of the displacement approaches the distance traveled by the particle along the curve, and the direction of $\Delta \mathbf{r}$ approaches the direction of the line tangent to the curve at point P_1. We define the instantaneous-velocity vector as the limit of the average velocity as the time interval Δt approaches zero:

$$\mathbf{v} = \lim_{\Delta t \to 0} \frac{\Delta \mathbf{r}}{\Delta t} = \frac{d\mathbf{r}}{dt} \qquad\qquad 3\text{-}15$$

Instantaneous-velocity vector

The instantaneous velocity is thus the derivative of the position vector with respect to time. The direction of the instantaneous velocity is along the line tangent to the curve traveled by the particle in space. The magnitude of the instantaneous velocity is called the *speed ds/dt*, where s is the distance measured along the curve. The position vector

Speed is the magnitude of the velocity vector

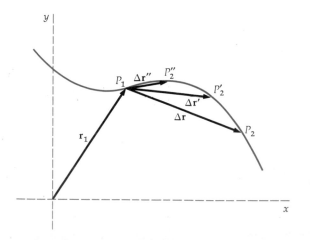

Figure 3-13
As smaller and smaller time intervals are considered, the magnitude of the displacement vector approaches the distance traveled along the curve and the direction of the displacement vector approaches the direction of the line tangent to the curve at point P_1.

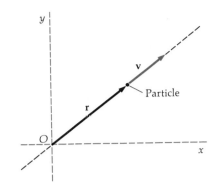

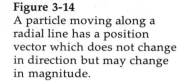

Figure 3-14
A particle moving along a radial line has a position vector which does not change in direction but may change in magnitude.

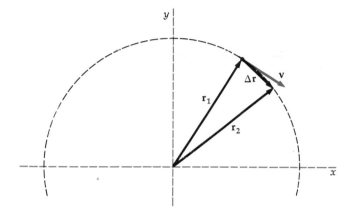

Figure 3-15
A particle moving in a circle has a position vector which does not change in magnitude but does change in direction. The velocity vector is tangent to the circle and perpendicular to the radius vector.

r can change in magnitude, direction, or both. If the particle moves along a radial line through the origin, as in Figure 3-14, r changes in magnitude only and the velocity vector is parallel to r. Our discussion of one-dimensional motion in Chapter 2 applies for this special case.

Figure 3-15 shows a particle moving around the origin in a circle of radius r. In this case, the position vector has a constant magnitude but is changing in direction. The velocity vector for this motion is tangent to the circle and therefore perpendicular to the radius vector. In the general case illustrated in Figure 3-12 the radius vector is changing in both magnitude and direction.

Example 3-2 A sailboat with initial coordinates $(x_1, y_1) = (100$ ft, 200 ft$)$ has the coordinates 1.00 min later $(x_2, y_2) = (120$ ft, 210 ft$)$. What are the components, magnitude, and direction of its average velocity for this 1-min interval?

$$v_{x,\text{av}} = \frac{x_2 - x_1}{\Delta t} = \frac{120 - 100 \text{ ft}}{1.00 \text{ min}} = 20.0 \text{ ft/min}$$

$$v_{y,\text{av}} = \frac{y_2 - y_1}{\Delta t} = \frac{210 - 200 \text{ ft}}{1.00 \text{ min}} = 10.0 \text{ ft/min}$$

$$v_{\text{av}} = \sqrt{(v_{x,\text{av}})^2 + (v_{y,\text{av}})^2}$$
$$= \sqrt{20.0^2 + 10.0^2} = \sqrt{500} = 22.4 \text{ ft/min}$$

$$\tan \theta = \frac{v_{y,\text{av}}}{v_{x,\text{av}}} = \frac{10 \text{ ft/min}}{20 \text{ ft/min}} = 0.500$$

or

$$\theta = 26.6°$$

Questions

12. For an arbitrary motion of a given particle does the direction of the velocity vector have any particular relationship to the direction of the position vector?

13. Give examples in which the directions of the velocity and position vectors are (*a*) opposite, (*b*) the same, (*c*) mutually perpendicular.

14. Can the velocity vector change direction without changing magnitude? If so, give an example.

3-5   The Acceleration Vector

The *average-acceleration vector* is defined as the ratio of the change in the instantaneous-velocity vector $\Delta \mathbf{v}$ and the time interval Δt:

$$\mathbf{a}_{av} = \frac{\Delta \mathbf{v}}{\Delta t} = \frac{\mathbf{v}_2 - \mathbf{v}_1}{\Delta t} \qquad \text{3-16}$$

Average-acceleration vector

The *instantaneous-acceleration vector* is defined as the derivative of the velocity vector with respect to time:

$$\mathbf{a} = \lim_{\Delta t \to 0} \frac{\Delta \mathbf{v}}{\Delta t} = \frac{d\mathbf{v}}{dt} \qquad \text{3-17}$$

Instantaneous-acceleration vector

It is particularly important to note that the velocity vector may be changing in magnitude, direction, or both. If the velocity vector is changing in any way, the particle is accelerating according to our definition. We are perhaps most familiar with acceleration in which the velocity vector changes in magnitude. In this case the speed changes (since the speed is just the magnitude of the velocity vector). However, a particle can be traveling with constant speed and still be accelerating if the direction of the velocity vector is changing. (A particularly important example of this is circular motion, discussed in Section 3-7.) This acceleration is just as real as when the speed is changing. We shall see in the next chapter that a force is necessary to produce an acceleration of a particle. The force required to produce a given acceleration (say 1 m/sec^2 downward) on a given particle is the same whether this acceleration is associated with a change in the magnitude of the velocity vector, a change in its direction, or a combination of both.

Changing direction involves acceleration

Questions

15. How is it possible for a particle moving at constant speed to be accelerating? Can a particle with constant velocity be accelerating at the same time?

16. Is it possible for a particle to round a curve without accelerating?

17. Show that the acceleration of a particle always points toward the concave side of its path.

18. Describe a circumstance in which a particle's velocity is horizontal and its acceleration is vertical.

3-6 Motion with Constant Acceleration: Projectile Motion

An important special case of motion in two or three dimensions occurs when the acceleration is constant in both magnitude and direction. An example of motion with constant acceleration is that of a projectile near the surface of the earth if air resistance can be neglected.

Let $\mathbf{a}_0$ be the instantaneous acceleration vector. Since $\mathbf{a}_0$ is constant, it is equal to the average acceleration for any time interval Δt. Equation 3-16 is then

$$\Delta \mathbf{v} = \mathbf{a}_0 \, \Delta t$$

If $\mathbf{v}_0$ is the velocity at time $t = 0$ and $\mathbf{v}$ is the instantaneous velocity at a general time t, we have

$$\mathbf{v} - \mathbf{v}_0 = \mathbf{a}_0 \, (t - 0)$$

or

$$\mathbf{v} = \mathbf{v}_0 + \mathbf{a}_0 t \qquad\qquad 3\text{-}18$$

The relation between the vectors $\mathbf{v}$, $\mathbf{v}_0$, and $\mathbf{a}_0$ is shown in Figure 3-16a.

The position vector $\mathbf{r}$ which has the derivative given by Equation 3-16 is

$$\mathbf{r} = \mathbf{r}_0 + \mathbf{v}_0 t + \tfrac{1}{2} \mathbf{a}_0 t^2 \qquad\qquad 3\text{-}19$$

where $\mathbf{r}_0$ is the position vector at $t = 0$. The validity of Equation 3-19 can be verified by computing the derivative $d\mathbf{r}/dt$ directly. Let $\mathbf{r}_1$ and $\mathbf{r}_2$ be the position vectors at times t and $t + \Delta t$, respectively. Then

$$\begin{aligned}
\mathbf{r}_2 &= \mathbf{r}_0 + \mathbf{v}_0 (t + \Delta t) + \tfrac{1}{2} \mathbf{a}_0 (t + \Delta t)^2 \\
&= \mathbf{r}_0 + \mathbf{v}_0 t + \mathbf{v}_0 \, \Delta t + \tfrac{1}{2} \mathbf{a}_0 t^2 + \mathbf{a}_0 t \, \Delta t + \tfrac{1}{2} \mathbf{a}_0 (\Delta t)^2 \\
&= \mathbf{r}_1 + \mathbf{v}_0 \, \Delta t + \mathbf{a}_0 t \, \Delta t + \tfrac{1}{2} \mathbf{a}_0 (\Delta t)^2
\end{aligned}$$

The average velocity for this time interval is

$$\mathbf{v}_{av} = \frac{\Delta \mathbf{r}}{\Delta t} = \frac{\mathbf{r}_2 - \mathbf{r}_1}{\Delta t} = \frac{\mathbf{v}_0 \, \Delta t + \mathbf{a}_0 t \, \Delta t + \tfrac{1}{2} \mathbf{a}_0 (\Delta t)^2}{\Delta t}$$

$$= \mathbf{v}_0 + \mathbf{a}_0 t + \tfrac{1}{2} \mathbf{a}_0 \, \Delta t$$

The limit of $\mathbf{v}_{av}$ as Δt approaches zero is the instantaneous velocity $\mathbf{v} = \mathbf{v}_0 + \mathbf{a}_0 t$.

Equations 3-18 and 3-19 are the three-dimensional generalizations of the one-dimensional constant-acceleration equations 2-11 and 2-12. Figure 3-16b shows the relation between the displacement vector $\mathbf{r} - \mathbf{r}_0$, the initial velocity $\mathbf{v}_0$, and the acceleration $\mathbf{a}_0$. The displacement at any time t is in the plane formed by $\mathbf{v}_0$ and $\mathbf{a}_0$. The motion is therefore two-dimensional.

Let us choose the z axis to be perpendicular to the plane of $\mathbf{v}_0$ and $\mathbf{a}_0$ and assume that the initial position of the particle is in the xy plane (Figure 3-17). The motion is then in the xy plane. The x and y components of Equations 3-18 and 3-19 are

$$v_x = v_{0x} + a_{0x} t \qquad\qquad 3\text{-}20$$

$$x = x_0 + v_{0x} t + \tfrac{1}{2} a_{0x} t^2$$

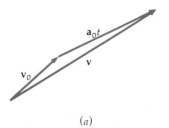

(a)

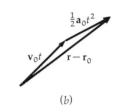

(b)

Figure 3-16
(a) The velocity vector $\mathbf{v}$ is the sum of the initial velocity $\mathbf{v}_0$ and the vector $\mathbf{a}_0 t$. (b) The displacement vector $\mathbf{r} - \mathbf{r}_0$ is the sum of $\mathbf{v}_0 t$ and $\tfrac{1}{2} \mathbf{a}_0 t^2$. The vector $\mathbf{v}_0 t$ would be the displacement if there were no acceleration, whereas the vector $\tfrac{1}{2} \mathbf{a}_0 t^2$ would be the displacement with constant acceleration if the initial velocity were zero. The displacement vector is in the plane of $\mathbf{v}_0$ and $\mathbf{a}_0$, indicating that motion with constant acceleration is confined to this plane.

Motion with constant acceleration is in a plane

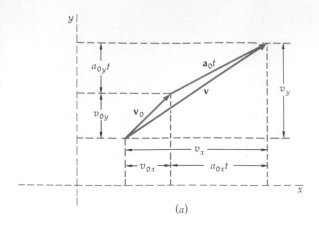

(a)

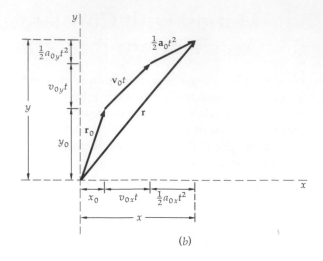

(b)

and

$$v_y = v_{0y} + a_{0y}t \qquad\qquad 3\text{-}21$$

$$y = y_0 + v_{0y}t + \tfrac{1}{2}a_{0y}t^2$$

Figure 3-17
The rectangular components
of (a) the velocity vector and
(b) the position vector **r**
assuming that the motion is
in the *xy* plane.

(If the particle is not assumed to be initially in the *xy* plane, the motion is parallel to the *xy* plane with the additional equations $v_z = 0$ and $z = z_0$.)

When the acceleration is constant, the motion takes place in a plane and the *x* and *y* motions can be described separately by equations identical to those for motion in one dimension with constant acceleration.

We can apply these results to the motion of a projectile, i.e., any body launched into the air and allowed to move freely. If air resistance is neglected, the acceleration of a projectile is the acceleration of gravity **g**. (We shall neglect the effects of air resistance for simplicity although such effects are important in the motion of real projectiles.) If the motion is near the surface of the earth and we can neglect the curvature of the earth, the acceleration **g** is a constant vector directed downward with magnitude of about 32 ft/sec² ≈ 9.8 m/sec². We choose the *xy* plane to be the plane of motion with the *y* axis vertical and the *x* axis parallel to the surface of the earth. If the positive *y* direction is chosen to be up, the acceleration of the projectile is

Acceleration of gravity

$$\mathbf{a} = -g\mathbf{j}$$

Equations 3-20 and 3-21 apply with $a_{0x} = 0$ and $a_{0y} = -g$. We have then

$$v_x = v_{0x} \qquad\qquad 3\text{-}22$$

$$x = x_0 + v_{0x}t \qquad\qquad 3\text{-}23$$

and

Projectile motion

$$v_y = v_{0y} - gt \qquad\qquad 3\text{-}24$$

$$y = y_0 + v_{0y}t - \tfrac{1}{2}gt^2 \qquad\qquad 3\text{-}25$$

(These equations can be slightly simplified by choosing the origin at the point of projection so that $x_0 = y_0 = 0$.)

The *x* and *y* components of the initial velocity are related to the initial speed v_0 and the angle of projection θ_0 (the angle between the initial velocity and the horizontal), as shown in Figure 3-18:

$$v_{0x} = v_0 \cos \theta_0 \qquad v_{0y} = v_0 \sin \theta_0$$

Example 3-3 A ball is thrown into the air with initial velocity of 100 ft/sec at 53° to the horizontal. Find the total time the ball is in the air and the total horizontal distance traveled.

The rectangular components of the initial velocity vector are

$$v_{0x} = (100 \text{ ft/sec}) \cos 53° = 60 \text{ ft/sec}$$

$$v_{0y} = (100 \text{ ft/sec}) \sin 53° = 80 \text{ ft/sec}$$

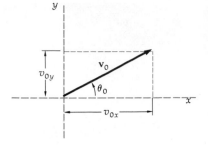

Figure 3-18
The components of the initial velocity of a projectile are $v_{0x} = v_0 \cos \theta_0$ and $v_{0y} = v_0 \sin \theta_0$, where θ_0 is the angle made by $\mathbf{v}_0$ and the horizontal x axis.

Figure 3-19 shows the height y versus t for this case. This curve is of course the same as that in Example 2-11 except for different initial velocity. Since for each second the projectile moves 60 ft in the x direction, we can also interpret this sketch as that of y versus x if we change the scale of the horizontal axis from a time scale to a distance scale by multiplying the time values by 60 ft/sec. Thus the path of the ball $y(x)$ is a parabola.

The time for the ball to reach the top of its path can be calculated from Equation 3-24, just as for one dimension in Example 2-11. Here

$$v_y = 80 \text{ ft/sec} - (32 \text{ ft/sec}^2)t$$

$$v_y = 0 \quad \text{at } t = \tfrac{80}{32} \text{ sec} = 2.5 \text{ sec}$$

The total time the ball is in the air is twice this time, or 5 sec. The total horizontal distance traveled, called the *range*, is $(60 \text{ ft/sec})(5 \text{ sec}) = 300$ ft.

We can apply these methods to find the range R for a general v_0 and θ_0. The time to reach the highest point is found by setting $v_y = 0$ in Equation 3-24:

$$t = \frac{v_{0y}}{g}$$

The range is then the horizontal distance traveled in twice this time:

$$R = v_{0x} \frac{2v_{0y}}{g} = \frac{2v_{0x}v_{0y}}{g} \qquad\qquad 3\text{-}26a$$

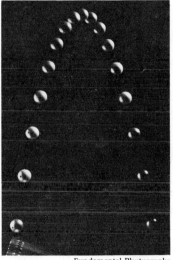

Fundamental Photographs

Multiflash photograph of a ball thrown into the air. The position of the ball is recorded at approximately 0.43-sec intervals.

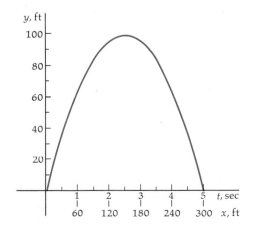

Figure 3-19
Plot of $y(t)$ and $y(x)$ for Example 3-3. Here $y = 80t - 16t^2$, where y is in feet and t in seconds. The scale can be converted to a horizontal distance scale by multiplying each time by 60 ft/sec because x (in feet) is related to t (in seconds) by $x = 60t$.

Illuminated fountains, St. Louis, Missouri, showing parabolic paths of water.

Ralph Krubner/Black Star

In terms of the initial speed v_0 and angle θ_0, the range is

Range of a projectile

$$R = \frac{2(v_0 \cos \theta_0)(v_0 \sin \theta_0)}{g} = \frac{v_0^2}{g}(2 \sin \theta_0 \cos \theta_0)$$

$$= \frac{v_0^2}{g} \sin 2\theta_0 \qquad \qquad \text{3-26}b$$

where we have used the trigonometric identity $2 \cos \theta_0 \sin \theta_0 = \sin 2\theta_0$.

Since the maximum value of $\sin 2\theta_0$ is 1 when $2\theta_0 = 90°$ or $\theta_0 = 45°$, the range is a maximum v_0^2/g when $\theta_0 = 45°$. (This calculation and the formulas used assume that there is no air resistance acting on the projectile. This often is a poor approximation, so that the correct launch angle θ_0 for maximum range is generally somewhat less than 45°.)

The general equation for the path $y(x)$ can be obtained from Equations 3-23 and 3-25 by eliminating the variable t between these equations. Choosing $x_0 = y_0 = 0$ and using $t = x/v_{0x}$ in Equation 3-25, we obtain

$$y = v_{0y} \frac{x}{v_{0x}} - \frac{1}{2} g \left(\frac{x}{v_{0x}} \right)^2$$

or

$$y = \frac{v_{0y}}{v_{0x}} x - \frac{1}{2} \frac{g}{v_{0x}^2} x^2 \qquad \qquad \text{3-27}$$

Equation of projectile's path

This equation is of the same form as $y(t)$ in Equation 3-25. Figure 3-20 shows the path of a projectile with the velocity vector and its components indicated at several points.

If we view the motion of the projectile in a second coordinate system parallel to the first but moving with speed v_{0x} relative to it, the motion is in one dimension. Let us call the origin of our second coordinate system O' and the axes x' and y'. If we let the two coordinate systems be coincident at $t = 0$, the coordinates are related by

$$y' = y \qquad x' = x - v_{0x}t \qquad \qquad \text{3-28}$$

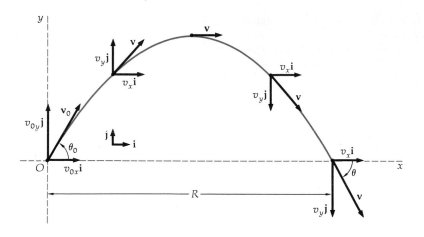

Figure 3-20
Path of a projectile with the velocity vector and its rectangular components indicated at several points. The total horizontal distance traveled is the range R.

The equations for the projectile in this coordinate system are found from Equations 3-23 and 3-25. Choosing x_0 and y_0 to be zero, we have

$$y' = v_{0y}t - \tfrac{1}{2}gt^2 \qquad\qquad\qquad 3\text{-}29$$

$$x' = 0 \qquad\qquad\qquad\qquad\qquad 3\text{-}30$$

In this coordinate system, the projectile moves in one dimension along the y' axis.

According to our analysis of projectile motion, an object dropped from a height h above the ground will hit the ground in the same time as one projected horizontally from the same height. In each case, the distance the object *falls* is given by $d = \tfrac{1}{2}gt^2$ (measuring d downward from the original height). This remarkable fact can easily be demonstrated.

It was first commented upon during the Renaissance. Galileo Galilei (1564–1642), the first to give the modern, quantitative description of projectile motion we have discussed, used this observation to illustrate the validity of treating the horizontal and vertical components of a projectile's motion as independent motions. Galileo wrote:[1]

> Let the ship be motionless and the fall of the stone from the mast take two pulse beats. Then cause the ship to move, and drop the same stone from the same place; from what has been said, it will take two

[1] Galileo Galilei, *Dialogue Concerning Two Chief World Systems—Ptolemaic and Copernican,* pp. 154–155, trans. Stillman Drake, University of California Press, Berkeley, 1953; reprinted by permission of The Regents of the University of California.

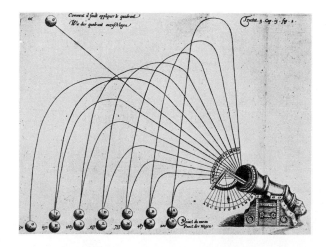

Path of cannonballs for various angles of projection, drawn by Diego Ufano in 1621. Which features of the motion are correct, and which are incorrect? (*Courtesy of the New York Public Library, Rare Book Division, Astor, Lenox, and Tilden Foundations.*)

pulse beats to arrive at the deck. In these two pulse beats, the ship will have gone, say, twenty yards so that the natural motion of the stone will have been a diagonal line much longer than the first straight and perpendicular one, which was merely the length of the mast; nevertheless, it will have traversed this distance in the same time. Now, assuming the ship to be speeded up still more, so that the stone in falling must follow a diagonal line very much longer still than the other, eventually the velocity of the ship may be increased by any amount while the falling rock will describe always longer and longer diagonals, and still pass over them in the same two pulse beats. Similarly, if a perfectly level cannon on a tower were fired parallel to the horizon, it would not matter whether a small charge or a great one was put in, so that the ball would fall a thousand yards away, four thousand, or six thousand, or ten thousand or more; all these shots would require equal times and each time would be equal to that which the ball would have taken in going from the mouth of the cannon to the ground if it were allowed to fall straight down without any other impulse.

Portrait of Galileo by Ottavio Leoni. (*Courtesy of Cabinet de Dessins, Musée du Louvre, Paris.*)

Example 3-4 A hunter with a blowgun wishes to shoot a monkey hanging from a branch. The hunter aims right at the monkey, not realizing that the dart will follow a parabolic path and thus fall below the monkey. The monkey, however, seeing the dart leave the gun, lets go of the branch and drops out of the tree, expecting to avoid the dart. Show that the monkey will be hit no matter what the initial velocity of the dart is so long as it is great enough to travel the horizontal distance to the tree before hitting the ground.

This problem is often demonstrated using a target suspended by an electromagnet. When the dart leaves the gun, the circuit is broken and the target falls. Let the horizontal distance to the tree be x and the original height of the monkey be H, as shown in Figure 3-21. Then the dart will be projected at an angle given by $\tan \theta = H/x$. If there were no gravity, the dart would reach the height H in the time t taken for it to travel the horizontal distance x:

$$y = v_{0y}t = H \quad \text{in time } t = \frac{x}{v_{0x}} \quad \text{with no gravity}$$

However, because of the gravity, the dart has an acceleration vertically

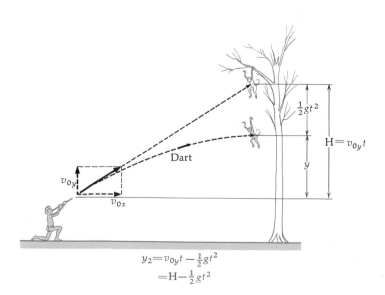

Figure 3-21
The monkey and the dart from Example 3-4. The height of the dart is given by $y_2 = v_{0y}t - \frac{1}{2}gt^2$. The height of the monkey is given by $y = H - \frac{1}{2}gt^2$. But since $H = v_{0y}t$, the height of the dart and the monkey will coincide if the monkey is released the instant the dart is fired.

down. In time $t = x/v_{0x}$, the dart reaches a height given by

$$y = v_{0y}t - \tfrac{1}{2}gt^2 = H - \tfrac{1}{2}gt^2$$

This is lower than H by $\tfrac{1}{2}gt^2$, which is just the amount the monkey falls in this time. In the usual lecture demonstration, the initial velocity of the dart is varied so that for large v_0 the target is hit very near its original height and for small v_0 it is hit just before it reaches the floor.

Questions

19. What is the acceleration of a projectile at the top of its flight?

20. Can the velocity of an object change direction while its acceleration is constant in both magnitude and direction? If so, cite an example.

21. A projectile is launched horizontally at the top of a cliff. Describe how the angle between its velocity and the acceleration varies as it falls.

3-7 Circular Motion with Constant Speed

A particle moving in a circular path illustrates many of the important features of velocity and acceleration vectors in two dimensions. In many natural phenomena the motion is circular or nearly circular, e.g., satellite motion or motion of the earth around the sun.

Consider first a particle moving in a circle of radius r with constant speed v. Though the magnitude of the velocity vector does not change, its direction does. Thus the particle is accelerating even though its speed is constant. Figure 3-22 shows the origin at the center of the circle, the position vectors $\mathbf{r}_1$ and $\mathbf{r}_2$ at t_1 and t_2, and the corresponding velocity vectors $\mathbf{v}_1$ and $\mathbf{v}_2$, which are tangent to the circle. The change in velocity in the time interval Δt is the difference of these two velocity vectors $\Delta\mathbf{v} = \mathbf{v}_2 - \mathbf{v}_1$, and the average acceleration is this change in velocity divided by Δt. We see from Figure 3-22 that for very small Δt, the velocity change (and therefore the average acceleration) is approxi-

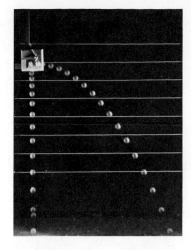

Comparison of a ball dropped with one projected horizontally. The vertical position is independent of the horizontal motion. (*From* PSSC Physics, *3d ed., p. 248, D. C. Heath and Company, Lexington, Mass., 1971.*)

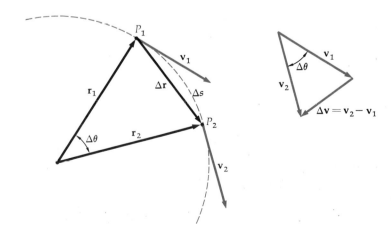

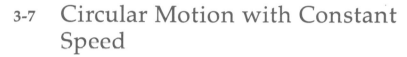

Figure 3-22
Calculation of the acceleration of a particle moving with constant speed in a circle of radius r. The position and velocity vectors are shown at times t_1 and t_2. Since $\mathbf{v}_1$ is perpendicular to $\mathbf{r}_1$ and $\mathbf{v}_2$ to $\mathbf{r}_2$, the angle between $\mathbf{v}_1$ and $\mathbf{v}_2$ is $\Delta\theta$, the same as between $\mathbf{r}_2$ and $\mathbf{r}_1$. For very short time intervals, the angle $\Delta\theta$ is very small and the change in velocity $\Delta\mathbf{v}$ is nearly perpendicular to $\mathbf{v}_1$ and points toward the center of the circle.

mately perpendicular to the velocity vectors in the direction pointing toward the center of the circle.

To calculate the magnitude of the acceleration, let Δs be the distance moved along the arc of the circle in Δt. The angular change measured in radians is

$$\Delta \theta = \frac{\Delta s}{r} \qquad\qquad 3\text{-}31$$

Thus

$$\Delta s = r \, \Delta \theta$$

The average speed equals the magnitude of the velocity since the speed is constant:

$$v = \frac{\Delta s}{\Delta t} = r \, \frac{\Delta \theta}{\Delta t}$$

From the similar triangles in Figure 3-22 we have

$$\frac{|\Delta \mathbf{v}|}{v} \approx \Delta \theta = \frac{\Delta s}{r} \qquad\qquad 3\text{-}32$$

or

$$|\Delta \mathbf{v}| \approx v \, \frac{\Delta s}{r} \qquad\qquad 3\text{-}33$$

The magnitude of the acceleration (assuming small Δt) is

$$a \approx \frac{|\Delta \mathbf{v}|}{\Delta t} \approx v \, \frac{\Delta s}{\Delta t} \frac{1}{r} = \frac{v^2}{r} \qquad\qquad 3\text{-}34$$

Acceleration for circular motion at constant speed

In the limit as Δt approaches zero, these approximate relations become exact. Thus the magnitude of the instantaneous acceleration is v^2/r, and the direction is radially in, toward the center.

As with projectile motion, the position, velocity, and acceleration vectors are not parallel. In projectile motion the acceleration vector points down, the velocity vector is tangent to the parabolic path, and the position vector points from the origin to the projectile. In circular motion, the velocity vector is tangent to the circle and perpendicular to the position vector, whereas the acceleration vector (for motion with constant speed) points toward the center of the circle, perpendicular to the velocity vector, and antiparallel to the position vector.

It is convenient to express the velocity and acceleration vectors in terms of radial and tangential components, i.e., components along lines which are parallel or perpendicular to the radius vector. We then have for the velocity

$$v_r = 0 \qquad v_t = v \qquad\qquad 3\text{-}35$$

and for the acceleration

$$a_r = -\frac{v^2}{r} \qquad\qquad 3\text{-}36$$

$$a_t = 0$$

Acceleration that points in toward the center of the circle is called *centripetal acceleration*. The minus sign in the expression for a_r indicates that this acceleration is toward the center and antiparallel to the radius vector.

Centripetal acceleration

Example 3-5 A satellite moves at constant speed in a circular orbit about the center of the earth and near the surface of the earth. Its acceleration has the magnitude 32.2 ft/sec². What is the speed, and how long does it take for one complete revolution?

This acceleration is the same as for any body falling freely near the surface of the earth. We take the radius of the earth, about 3960 mi, to be approximately the radius of the orbit. (For actual satellites put into orbit a few hundred miles above the earth's surface, the orbit radius is of course somewhat larger. The acceleration is slightly less than 32.2 ft/sec², due to the decrease in the gravitational attraction of the earth with distance from the center of the earth.) The speed of the satellite can be found from Equation 3-34:

$$v^2 = rg = (3960 \text{ mi})(32.2 \text{ ft/sec}^2)(5280 \text{ ft/mi})$$

$$= 6.73 \times 10^8 \text{ ft}^2/\text{sec}^2$$

$$v = 2.59 \times 10^4 \text{ ft/sec}$$

We can change this to miles per hour using the conversion factor 60 mi/h = 88 ft/sec:

$$v = 2.59 \times 10^4 \text{ ft/sec} \frac{60 \text{ mi/h}}{88 \text{ ft/sec}} = 1.77 \times 10^4 \text{ mi/h}$$

The time for a satellite of this speed to make one complete revolution is called the *period T*. Since it goes a distance of $2\pi r$ in this time T at speed v, the period is

$$T = \frac{2\pi r}{v} = \frac{2\pi (3960 \text{ mi})}{1.77 \times 10^4 \text{ mi/h}}$$

$$= 1.41 \text{ h} = 84.3 \text{ min}$$

People often wonder why a satellite does not fall toward the earth. The answer is that it does. However, because of its tangential velocity, it continually misses the earth. Consider Figure 3-23. If there were no force exerted on the satellite by the earth, the satellite would travel along a straight line as indicated. (This follows from Newton's laws as discussed in Chapter 4.) In a short time t it would move from point P_1 to point P_2. However, because it is accelerating toward the earth, it is actually at point P_2' after this time. If we take the time t small enough, points P_2 and P_2' are approximately on a radius, as shown in the figure. The distance fallen in this time is h, which is much smaller than the

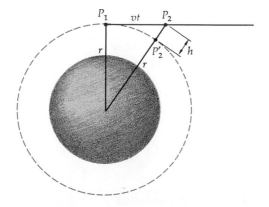

Figure 3-23
Satellite moving with speed v in a circular orbit of radius r about the earth. If the satellite did not accelerate toward the earth, it would move in a straight line from point P_1 to P_2 in time t. Because of its acceleration, it falls a distance h in this time. For small t, $h = \frac{1}{2}(v^2/r)t^2 = \frac{1}{2}at^2$.

orbit radius r for small t. We can calculate h from the right triangle of sides vt, r, and $r + h$. We have

$$(r + h)^2 = (vt)^2 + r^2$$

$$r^2 + 2hr + h^2 = v^2t^2 + r^2$$

If we cancel the r^2 terms and neglect h^2 compared with $2rh$, we obtain

$$2rh \approx v^2t^2$$

$$h = \frac{1}{2}\frac{v^2}{r}t^2$$

The distance the satellite falls in time t is thus $\frac{1}{2}at^2$, where the acceleration a is v^2/r. (The formula $\frac{1}{2}at^2$ applies to this case only if the time t is very small.)

Figure 3-24 is a drawing from Newton's *Principia* illustrating the connection between projectile motion and satellite motion. Newton's description reads:[1]

> That by means of centripetal forces the planets may be retained in certain . . . orbits, we may easily understand, if we consider the motions of projectiles; for a stone that is projected is by the pressure of its own weight forced out of the rectilinear path, which by the initial projection alone it should have pursued, and made to describe a curved line in the air; and through that crooked way is at last brought down to the ground; and the greater the velocity is with which it is projected, the farther it goes before it falls to the earth. We may therefore suppose the velocity to be so increased, that it would describe an arc of 1, 2, 5, 10, 100, 1000 miles before it arrived at the earth, till at last, exceeding the limits of the earth, it should pass into space without touching it.

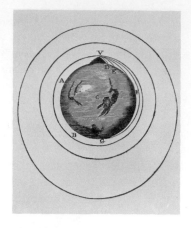

Figure 3-24
Drawing from Newton's *System of the World*, published in 1728, illustrating the connection between projectile motion and satellite motion. (*Courtesy of the Niels Bohr Library, American Institute of Physics.*)

3-8 Circular Motion with Varying Speed

Let us assume that the speed of the particle discussed in Section 3-7 is increasing as the particle moves around the circle instead of remaining constant. Then the velocity vector $\mathbf{v}_2$ will be greater than $\mathbf{v}_1$ in magnitude and different in direction, as shown in Fig. 3-25. The change in velocity is $\Delta\mathbf{v} = \mathbf{v}_2 - \mathbf{v}_1$. The change can be considered to consist of two parts:

$$\Delta\mathbf{v} = \Delta\mathbf{v}_r + \Delta\mathbf{v}_t$$

The first term, $\Delta\mathbf{v}_r$, is the change that would occur if the speed were constant. We computed this term in Section 3-7. For small Δt, this change in the velocity vector is approximately radial. The second term, $\Delta\mathbf{v}_t$, is due to the increase in magnitude of the velocity vector. For small time intervals Δt this term is approximately tangential to the circle.

The average acceleration is

$$\mathbf{a}_{av} = \frac{\Delta\mathbf{v}_r}{\Delta t} + \frac{\Delta\mathbf{v}_t}{\Delta t} \qquad\qquad 3\text{-}37$$

[1] *Philosophiae Naturalis Principia Mathematica*, 1686, trans. Andrew Motte, 1729, University of California Press, Berkeley, 1960; reprinted by permission of The Regents of the University of California.

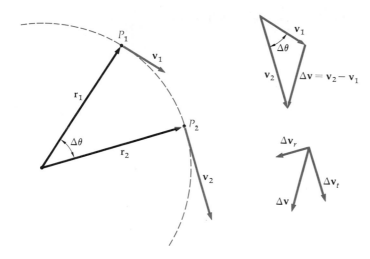

Figure 3-25
Calculation of the acceleration of a particle moving in a circle with variable speed. Here the velocity change $\Delta \mathbf{v}$ is due partly to the change in direction of motion and partly to the change in speed. For very small time intervals, $\Delta \theta$ is very small, and the change in $\mathbf{v}$ due to changing direction is approximately radial toward the center of the circle, while the change in $\mathbf{v}$ due to changing speed is approximately tangential to the motion. The instantaneous acceleration has an inward radial component of magnitude v^2/r and a tangential component of magnitude dv/dt.

In the limit as Δt approaches zero, the first term in Equation 3-24 has the magnitude v^2/r and points toward the center of the circle, as shown above. The second term has the magnitude dv/dt and is tangential to the circle and parallel to the velocity vector. Thus, in general, for circular motion of a particle with varying speed, the acceleration has both a radial and a tangential component:

$$a_r = -\frac{v^2}{r} \qquad a_t = \frac{dv}{dt} \qquad\qquad 3\text{-}38$$

Radial and tangential components of acceleration

This result, that *the radial component of the acceleration is associated with the change in direction of the velocity vector while the tangential component of acceleration is associated with the change in magnitude of the velocity,* is a general one not limited to circular motion. Consider a particle moving along any arbitrary curve in space. A small segment of the curve can be considered as the arc of a circle, as shown in Figure 3-26. The acceleration has a component dv/dt parallel to the velocity vector and a component v^2/r perpendicular to the velocity.

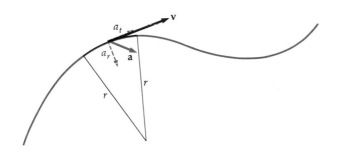

Figure 3-26
Particle moving along an arbitrary curve in space. A small segment of any curve in space can be considered to be the arc of a circle of radius r. The instantaneous-acceleration vector has a component a_t of magnitude dv/dt tangent to the curve and a component a_r of magnitude v^2/r toward the center of curvature of the arc.

3-9 Radial and Tangential Unit Vectors

We have seen that for circular motion it is convenient to use radial and tangential components of the velocity and acceleration vectors. We can write these vectors in terms of these components using radial and tangential unit vectors (Figure 3-27). In this figure, the angle θ and the distance s are measured from the x axis. The distance s is related to the polar coordinates r and θ by

$$\theta = \frac{s}{r} \qquad s = r\theta \qquad\qquad 3\text{-}39$$

Since r is constant, the speed of the particle is

$$v = \frac{ds}{dt} = r\,\frac{d\theta}{dt} \qquad\qquad 3\text{-}40$$

We define $\hat{\mathbf{r}}$ to be a unit vector parallel to the position vector $\mathbf{r}$ and $\hat{\boldsymbol{\theta}}$ to be a unit vector perpendicular to $\mathbf{r}$, tangent to the circle in the direction of increasing θ and s.

The position, velocity, and acceleration vectors for a particle moving in a circle of radius r are

$$\mathbf{r} = r\hat{\mathbf{r}} \qquad\qquad 3\text{-}41$$

$$\mathbf{v} = v\hat{\boldsymbol{\theta}} = \frac{ds}{dt}\,\hat{\boldsymbol{\theta}} = \frac{r\,d\theta}{dt}\,\hat{\boldsymbol{\theta}} \qquad\qquad 3\text{-}42$$

$$\mathbf{a} = -\frac{v^2}{r}\,\hat{\mathbf{r}} + \frac{dv}{dt}\,\hat{\boldsymbol{\theta}} = -\frac{(ds/dt)^2}{r}\,\hat{\mathbf{r}} + \frac{d^2 s}{dt^2}\,\hat{\boldsymbol{\theta}} \qquad\qquad 3\text{-}43$$

We can relate the unit vectors $\hat{\mathbf{r}}$ and $\hat{\boldsymbol{\theta}}$ to the rectangular unit vectors $\mathbf{i}$ and $\mathbf{j}$. From Figure 3-28 we have

$$\mathbf{r} = r\cos\theta\,\mathbf{i} + r\sin\theta\,\mathbf{j} \qquad\qquad 3\text{-}44$$

Therefore

$$\hat{\mathbf{r}} = \frac{\mathbf{r}}{r} = \cos\theta\,\mathbf{i} + \sin\theta\,\mathbf{j} \qquad\qquad 3\text{-}45$$

Similarly from Figure 3-28 we see that $\hat{\boldsymbol{\theta}}$ has rectangular components $\hat{\boldsymbol{\theta}}_x = -\sin\theta$ and $\hat{\boldsymbol{\theta}}_y = \cos\theta$; thus

$$\hat{\boldsymbol{\theta}} = -\sin\theta\,\mathbf{i} + \cos\theta\,\mathbf{j} \qquad\qquad 3\text{-}46$$

The unit vectors $\hat{\mathbf{r}}$ and $\hat{\boldsymbol{\theta}}$ differ from $\mathbf{i}$ and $\mathbf{j}$ in that their direction depends on r and θ, that is, on the location of the particle. Thus the unit vectors $\hat{\mathbf{r}}$ and $\hat{\boldsymbol{\theta}}$ change with time as the particle moves in a circle, while the unit vectors $\mathbf{i}$ and $\mathbf{j}$ are true constant vectors.

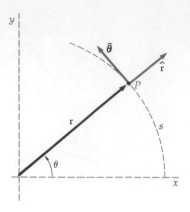

Figure 3-27
Definition of the unit vectors $\hat{\mathbf{r}}$ and $\hat{\boldsymbol{\theta}}$.

Unit vectors $\hat{\mathbf{r}}$ and $\hat{\boldsymbol{\theta}}$

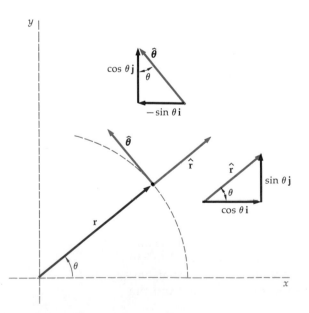

Figure 3-28
The unit vectors $\hat{\mathbf{r}}$ and $\hat{\boldsymbol{\theta}}$ are related to $\mathbf{i}$ and $\mathbf{j}$ by $\hat{\mathbf{r}} = \cos\theta\,\mathbf{i} + \sin\theta\,\mathbf{j}$ and $\hat{\boldsymbol{\theta}} = -\sin\theta\,\mathbf{i} + \cos\theta\,\mathbf{j}$.

We can compute the rate of change of the unit vectors $\hat{\mathbf{r}}$ and $\hat{\boldsymbol{\theta}}$ from Equations 3-45 and 3-46 and use them to compute the velocity and acceleration vectors for circular motion. We have

$$\frac{d\hat{\mathbf{r}}}{dt} = \frac{d}{dt}(\cos\theta\,\mathbf{i} + \sin\theta\,\mathbf{j})$$

$$= -\sin\theta\,\frac{d\theta}{dt}\,\mathbf{i} + \cos\theta\,\frac{d\theta}{dt}\,\mathbf{j}$$

$$= \frac{d\theta}{dt}(-\sin\theta\,\mathbf{i} + \cos\theta\,\mathbf{j}) = \frac{d\theta}{dt}\,\hat{\boldsymbol{\theta}} \qquad\qquad 3\text{-}47$$

$$\frac{d\hat{\boldsymbol{\theta}}}{dt} = \frac{d}{dt}(-\sin\theta\,\mathbf{i} + \cos\theta\,\mathbf{j})$$

$$= -\cos\theta\,\frac{d\theta}{dt}\,\mathbf{i} - \sin\theta\,\frac{d\theta}{dt}\,\mathbf{j}$$

$$= -\frac{d\theta}{dt}(\cos\theta\,\mathbf{i} + \sin\theta\,\mathbf{j}) = \frac{-d\theta}{dt}\,\hat{\mathbf{r}} \qquad\qquad 3\text{-}48$$

Let us now find the velocity and acceleration vectors by direct differentiation of the position vector for a particle moving in a circle

$$\mathbf{r} = r\mathbf{r} \qquad \mathbf{v} = \frac{d\mathbf{r}}{dt} = r\frac{d\hat{\mathbf{r}}}{dt} = r\frac{d\theta}{dt}\,\hat{\boldsymbol{\theta}}$$

Since

$$r\frac{d\theta}{dt} = \frac{ds}{dt} = v$$

the speed, we have

$$\mathbf{v} = v\hat{\boldsymbol{\theta}}$$

Then the acceleration is

$$\mathbf{a} = \frac{d\mathbf{v}}{dt} = v\frac{d\hat{\boldsymbol{\theta}}}{dt} + \frac{dv}{dt}\,\hat{\boldsymbol{\theta}}$$

where we have used the product rule for differentiating $\mathbf{v}$ when both quantities may depend on time. Using Equation 3-48 for $d\hat{\boldsymbol{\theta}}/dt$, we obtain

$$\mathbf{a} = v\frac{-d\theta}{dt}\,\hat{\mathbf{r}} + \frac{dv}{dt}\,\hat{\boldsymbol{\theta}}$$

This can be written in the usual form if we note that $d\theta/dt = v/r$:

$$\mathbf{a} = -\frac{v^2}{r}\,\hat{\mathbf{r}} + \frac{dv}{dt}\,\hat{\boldsymbol{\theta}}$$

Review

A. Define, explain, or otherwise identify:

Displacement, 49
Vector, 50
Scalar, 50
Component of a vector, 50
Unit vector, 52
Vector equality, 53

Radius vector, 56
Position vector, 56
Range, 61
Centripetal acceleration, 66
Tangential acceleration, 69

B. True or false:

1. The instantaneous velocity vector is always in the direction of motion.

2. The instantaneous acceleration vector is always in the direction of motion.

3. If the speed is constant, the acceleration must be zero.

4. If the acceleration is zero, the speed must be constant.

5. The component of a vector is a vector.

6. The magnitude of the sum of two vectors must be greater than the magnitude of either vector.

7. If a vector is zero, each of its rectangular components must be zero.

8. It is impossible to go around a curve without acceleration.

9. The time required for a bullet fired horizontally to reach the ground is the same as if it were dropped from rest from the same height.

Exercises

Unless otherwise instructed, use the approximate value $g = 32$ ft/sec^2 = 9.8 m/sec^2 for the acceleration of gravity in Exercises and Problems.

Section 3-1, The Displacement Vector

1. A man walks northeast for 10 m and then east for 10 m. Show each displacement graphically and find the resultant displacement vector.

2. (*a*) A man walks along a circular arc from the position $x = 5$ m, $y = 0$ to a final position $x = 0$, $y = 5$ m. What is his displacement? (*b*) A second man walks from the same initial position along the x axis to the origin and then along the y axis to $y = 5$ m and $x = 0$. What is his displacement?

3. A hiker sets off at 8 A.M. in level terrain. At 9 A.M. he is 2 mi due east of his starting point. At 10 A.M. he is 1 mi northwest of where he was at 9 A.M. At 11 A.M. he is 3 mi due north of where he was at 10 A.M. (*a*) Make a drawing showing these successive displacements as vectors, the tail of each being at the head of the previous one. What are the magnitudes and directions of these displacements? (Specify the direction of vectors by giving their angle with the eastward direction.) (*b*) What are the north and east components of these displacements? (*c*) How far is the hiker from his starting point at 11 A.M.? In what direction? (*d*) Add the three displacement vectors by drawing them to scale. Do these successive straight lines represent the actual path the hiker followed? Is the distance he walked the sum of the lengths of the three displacement vectors?

4. A circular path has a radius of 10 m. An xy coordinate system is established so that the center of the circle is on the y axis and the circle passes through the origin. A man starts at the origin and walks around the path at a steady speed, returning to the origin exactly 1 min after he started. (*a*) Find the magnitude and direction of his displacement from the origin 15, 30, 45, and 60 sec after he started. (*b*) Find the magnitude and direction of his displacement for each of the four successive 15-sec intervals of his walk. (*c*) How is his displacement for the first 15 sec related to that for the second 15 sec? (*d*) How is his displacement for the second 15-sec interval related to that for the last 15-sec interval?

Section 3-2, Components of a Vector

5. What are the rectangular components of the displacement vector for part (*a*) of Exercise 2? Write this vector in terms of the unit vectors **i** and **j**.

6. Find the rectangular components of the vectors which lie in the xy plane, have the magnitude A, and make an angle θ with the x axis, as shown in Figure 3-29 for the following values of A and θ: (*a*) $A = 10$ m, $\theta = 30°$; (*b*) $A = 5$ m,

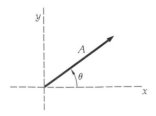

Figure 3-29
Exercise 6.

$\theta = 45°$; (c) $A = 7$ ft, $\theta = 60°$; (d) $A = 5$ ft, $\theta = 90°$; (e) $A = 15$ ft/sec, $\theta = 150°$; (f) $A = 10$ m/sec, $\theta = 240°$; (g) $A = 8$ m/sec^2, $\theta = 270°$.

7. Find the magnitude and direction of the following vectors: (a) $\mathbf{A} = 5\mathbf{i} + 3\mathbf{j}$; (b) $\mathbf{B} = 10\mathbf{i} - 7\mathbf{j}$; (c) $\mathbf{C} = -2\mathbf{i} - 3\mathbf{j} + 4\mathbf{k}$.

8. A cube of side 2 m has its faces parallel to the coordinate planes with one corner at the origin. A fly begins at the origin and walks along the three edges until it is at the far corner. Write the displacement vector of the fly using the unit vectors $\mathbf{i}$, $\mathbf{j}$, and $\mathbf{k}$ and find the magnitude of this displacement.

9. A plane is inclined at an angle of 30° with the horizontal. Choose the x axis parallel to the plane pointing down the slope and the y axis perpendicular to the plane pointing away from the plane. Find the components of the acceleration of gravity which has the magnitude 9.8 m/sec^2 and points vertically down.

10. Find the magnitude and direction of $\mathbf{A}$, $\mathbf{B}$, and $\mathbf{A} + \mathbf{B}$ for (a) $\mathbf{A} = -4\mathbf{i} - 7\mathbf{j}$, $\mathbf{B} = 3\mathbf{i} - 2\mathbf{j}$; (b) $\mathbf{A} = 1\mathbf{i} - 4\mathbf{j}$, $\mathbf{B} = 2\mathbf{i} + 6\mathbf{j}$.

11. Describe the following vectors by using the unit vectors $\mathbf{i}$ and $\mathbf{j}$: (a) a velocity of 10 m/sec at an angle of elevation of 60°; (b) a vector $\mathbf{A}$ of magnitude $A = 5$ m and $\theta = 225°$; (c) a displacement from the origin to the point $x = 14$ m, $y = -6$ m.

Section 3-3, Properties of Vectors

12. The displacement vectors $\mathbf{A}$ and $\mathbf{B}$ shown in Figure 3-30 both have magnitude 2 m. Find their x and y components. Find the components, magnitude, and direction of the sum $\mathbf{A} + \mathbf{B}$. Find the components, magnitude, and direction of the difference $\mathbf{A} - \mathbf{B}$.

13. For the two vectors $\mathbf{A}$ and $\mathbf{B}$ shown in Figure 3-30 find graphically (a) $\mathbf{A} + \mathbf{B}$; (b) $\mathbf{A} - \mathbf{B}$; (c) $2\mathbf{A} + \mathbf{B}$; (d) $\mathbf{B} - \mathbf{A}$; (e) $2\mathbf{B} - \mathbf{A}$.

14. For the vector $\mathbf{A} = 3\mathbf{i} + 4\mathbf{j}$ find any three other vectors $\mathbf{B}$ which also lie in the xy plane and have the property that $A = B$ but $\mathbf{A} \neq \mathbf{B}$. Write these vectors in terms of their components and show them graphically.

15. If $\mathbf{A} = 2\mathbf{i} - 6\mathbf{j}$, find $5\mathbf{A}$ and $-7\mathbf{A}$.

16. Two vectors $\mathbf{A}$ and $\mathbf{B}$ lie in the xy plane. Under what conditions does the ratio A/B equal A_x/B_x?

17. If $\mathbf{A} = 5\mathbf{i} - 4\mathbf{j}$ and $\mathbf{B} = -7.5\mathbf{i} + 6\mathbf{j}$, write an equation relating $\mathbf{A}$ to $\mathbf{B}$.

18. Draw any three nonparallel vectors $\mathbf{A}$, $\mathbf{B}$, and $\mathbf{C}$ which lie in a plane and show that the sum $(\mathbf{A} + \mathbf{B}) + \mathbf{C}$ equals $(\mathbf{A} + \mathbf{C}) + \mathbf{B}$.

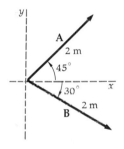

Figure 3-30
Exercises 12 and 13.

Section 3-4, The Velocity Vector

19. A vector $\mathbf{A}(t)$ has constant magnitude but is changing direction in a uniform way. Draw the vectors $\mathbf{A}(t + \Delta t)$ and $\mathbf{A}(t)$ for a small time interval Δt and find the difference $\Delta\mathbf{A} = \mathbf{A}(t + \Delta t) - \mathbf{A}(t)$ graphically. How is the direction of $\Delta\mathbf{A}$ related to $\mathbf{A}$ for small time intervals?

20. A stationary radar operator determines that a ship is 10 mi south of him. An hour later the same ship is 20 mi southeast of him. If the ship moved at constant speed always in the same direction, what was its velocity during this time?

21. A particle's position coordinates (x,y) are (2 m, 3 m) at $t = 0$; (6 m, 7 m) at $t = 2$ sec; and (13 m, 14 m) at $t = 5$ sec. (a) Find $\mathbf{v}_{av}$ from $t = 0$ to $t = 2$ sec. (b) Find $\mathbf{v}_{av}$ from $t = 0$ to $t = 5$ sec.

22. A particle travels with constant speed in a circular path around the origin of radius 5 m. It begins at $t = 0$ at $x = 5$ m, $y = 0$ and takes 100 sec for a complete revolution. (a) What is the speed of the particle? (b) Give the magnitude and direction of the position vector $\mathbf{r}$ at the times $t = 50$ sec, $t = 25$ sec, $t = 10$ sec, and $t = 0$. (c) Find the magnitude and indicate graphically the direction of $\mathbf{v}_{av}$ for each of the following time intervals: $t = 0$ to $t = 50$ sec; $t = 0$ to $t = 25$ sec; $t = 0$ to $t = 10$ sec. (d) How does $\mathbf{v}_{av}$ for the interval $t = 0$ to $t = 10$ sec compare with the instantaneous velocity at $t = 0$?

23. The position vector of a particle is given by $\mathbf{r}(t) = 5t\mathbf{i} + 10t\mathbf{j}$, where t is in seconds and $\mathbf{r}$ in meters. (a) Draw the path of the particle in the xy plane. (b) Find $v(t)$ in component form and find its magnitude.

Section 3-5, The Acceleration Vector

24. A ball is thrown directly upward. Consider the 2-sec time interval $\Delta t = t_2 - t_1$, where t_1 is 1 sec before the ball reaches its highest point and t_2 is 1 sec after it reaches its highest point. Find the (a) change in speed, (b) change in velocity, and (c) the average acceleration for this time interval.

25. Figure 3-31 shows the path of an automobile, made up of segments of straight lines and arcs of circles. The automobile starts from rest at point A. After it reaches point B, it travels at constant speed until it reaches point E. It comes to rest at point F. (a) At the middle of each segment (AB, BC, CD, DE, and EF) what is the direction of the velocity vector? (b) At which of these points does the automobile have an acceleration? In those cases, what is the direction of the acceleration? (c) How do the magnitudes of the acceleration compare for segments BC and DE?

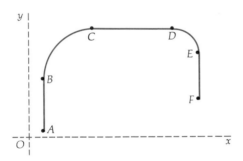

Figure 3-31
Exercise 25.

26. Initially a particle is moving due west with a speed of 40 ft/sec, and 5 sec later it is moving north with a speed of 30 ft/sec. (a) What was the change in the magnitude of the particle's velocity during this time? (b) What was the change in direction of the velocity? (c) What are the magnitude and direction of $\Delta\mathbf{v}$ for this interval? (d) What are the magnitude and direction of $\mathbf{a}_{av}$ for this interval?

27. At $t = 0$ a particle located at the origin has a speed of 40 m/sec at $\theta = 45°$. At $t = 3$ sec the particle is at $x = 100$ m, $y = 80$ m with speed of 30 m/sec at $\theta = 50°$. Calculate (a) the average velocity and (b) the average acceleration of the particle during this interval.

28. A particle has a position vector given by $\mathbf{r} = 30t\mathbf{i} + (40t - 5t^2)\mathbf{j}$, where $\mathbf{r}$ is in meters and t in seconds. Find the instantaneous velocity and acceleration vectors as functions of time t.

Section 3-6, Motion with Constant Acceleration: Projectile Motion

29. A bullet is fired horizontally with an initial velocity of 800 ft/sec. The gun is 4 ft above the ground. How long is the bullet in the air?

30. A supersonic transport is flying horizontally at an altitude of 10 mi and with speed of 1500 mi/h when an engine falls off. (*a*) How long does it take the engine to hit the ground? (*b*) How far horizontally is the engine from where it fell off when it hits the ground? (*c*) How far is the engine from the aircraft (assuming it continues to fly as if nothing had happened) when the engine hits the ground? Neglect air resistance.

31. A cannon is elevated at an angle of 45°. It fires a ball with a speed of 300 m/sec. (*a*) What height does the ball reach? (*b*) How long is the ball in the air? (*c*) What is the horizontal range?

32. Sketch the trajectory of a projectile and indicate the velocity and acceleration vectors at several points. (*a*) What is the magnitude of the acceleration at the top of the path? (*b*) Does the magnitude or direction of the acceleration change from point to point? (*c*) How does the angle between $\mathbf{a}$ and $\mathbf{v}$ change during the motion?

33. A projectile is launched with speed v_0 and at an angle θ_0 with the horizontal. Find an expression for the maximum height it reaches above its starting point in terms of v_0, θ_0, and g.

34. A projectile is fired with initial velocity of 100 ft/sec at 60° to the horizontal. At the highest point what is the velocity? The acceleration?

Section 3-7, Circular Motion with Constant Speed

35. A particle travels in a circular path of radius 5 m with a constant speed of 15 m/sec. What is the magnitude of its acceleration?

36. An object on the equator has an acceleration toward the center of the earth because of the earth's rotation and an acceleration toward the sun because of the earth's motion along the orbit. Calculate the magnitude of both accelerations and express them as fractions of the free-fall acceleration of gravity g.

37. A particle moves in a circle of radius 4 in. It takes 8 sec to make a complete round trip. Draw the path of the particle to scale and indicate the positions at 1-sec intervals. Draw the displacement vectors for these 1-sec intervals. These vectors also indicate the average velocity vectors for these intervals. Find graphically the change in the average velocity $\Delta\mathbf{v}$ for two consecutive 1-sec intervals. Compare $\Delta\mathbf{v}/\Delta t$ measured in this way with the instantaneous acceleration computed from $a_r = v^2/r$.

38. A boy whirls a ball on a string in a horizontal circle of radius 1 m. How many revolutions per minute must the ball make if its acceleration toward the center of the circle is to have the same magnitude as the acceleration of gravity?

39. An airplane pilot pulls out of a dive by following an arc of a circle whose radius is 1000 ft. At the bottom of the circle, where his speed is 300 mi/h, what are the direction and magnitude of his acceleration?

40. An object travels with a constant speed v in a circular path of radius r. (*a*) If v is doubled, how is the acceleration a affected? (*b*) If r is doubled, how is a affected? (*c*) Why is it impossible for an object to travel around a perfectly sharp angular turn?

Section 3-8, Circular Motion with Varying Speed

41. In parts (*a*) to (*c*) of Figure 3-32 particles are traveling in circular paths with varying speed. The velocity vectors are indicated. Find the magnitude of the average-acceleration vector between the two given positions in each case.

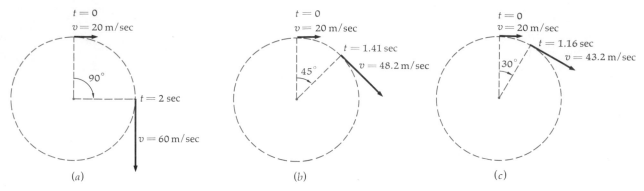

(*a*)   (*b*)   (*c*)

Figure 3-32
Exercise 41.

42. A particle moves clockwise in a circle of radius 1 m with center at $(x,y) = (1\ m, 0)$. It starts at rest at the origin at time $t = 0$. Its *speed* increases at the constant rate of $\pi/2$ m/sec². (*a*) How long does it take to travel halfway around the circle? (*b*) What is its speed at that time? (*c*) What is the direction of its velocity at that time? (*d*) What is its radial acceleration then? Its tangential acceleration? (*e*) What are the magnitude and direction of the total acceleration halfway around the circle?

43. In Figure 3-33 particles are traveling counterclockwise in a circle of radius 5 m with speeds which may be varying. The acceleration vectors are indicated at certain times. Find the values of v and dv/dt for each of these times.

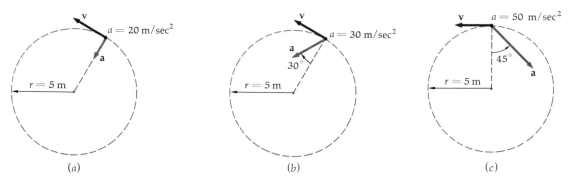

(*a*)   (*b*)   (*c*)

Figure 3-33
Exercise 43.

Section 3-9, Radial and Tangential Unit Vectors

44. For an object traveling in a circular path is it possible for $\hat{\mathbf{r}}$ at time t_1 to be equal to $\hat{\boldsymbol{\theta}}$ at another time t_2? Give examples.

45. A particle travels in a plane along a path described by $\mathbf{r} = 10\hat{\mathbf{r}}$ and $\theta = 2\pi t$, where $\mathbf{r}$ is in meters, θ in radians, and t in seconds. (*a*) Describe the motion. (*b*) Find the velocity vector $\mathbf{v} = d\mathbf{r}/dt$ by direct differentiation of $\mathbf{r}$. (*c*) Since the distance along the path is $s = r\theta$, find the speed by finding ds/dt. Is this the same as the magnitude of $\mathbf{v}$ found in part (*b*)? (*d*) Find the acceleration vector $\mathbf{a}$ in terms of the unit vectors $\hat{\mathbf{r}}$ and $\hat{\boldsymbol{\theta}}$.

Problems

1. Find the angle of projection such that the maximum height of the projectile is equal to the horizontal range.

2. Galileo showed that if air resistance is neglected, the ranges are equal for projectiles whose angles of projection exceed or fall short of 45° by the same amount. Prove this.

3. A baseball is struck by a bat, and 3 sec later it is caught 96 ft away. (a) If it was 3 ft above the ground when struck and caught, what was the greatest height it reached above the ground? (b) What were its horizontal and vertical components of velocity when it was struck? (c) What was its speed when it was caught? (d) At what angle with the horizontal did it leave the bat? (Neglect air resistance.)

4. A gun shoots bullets that leave the muzzle at 800 ft/sec. If the bullet is to hit a target 100 yd away at the level of the muzzle, the gun must be aimed at a point above the target. How far above the target is this point? (Neglect air resistance.)

5. A baseball is thrown toward a player with an initial speed of 20 m/sec and 45° with the horizontal. At the moment the ball is thrown, the player is 50 m from the thrower. At what speed and in what direction must he run to catch the ball at the same height at which it was released?

6. A freight train is moving at a constant speed of 20 mi/h. A man standing on a flatcar throws a ball into the air and catches it as it falls. Relative to the flatcar the initial velocity of the ball is 30 mi/h straight up. (a) What are the magnitude and direction of the initial velocity of the ball as seen by a second man standing next to the track? (b) How long is the ball in the air according to the man on the train? According to the man on the ground? (c) What horizontal distance has the ball traveled by the time it is caught according to the man on the train? According to the man on the ground? (d) What is the minimum speed of the ball during its flight according to the man on the train? According to the man on the ground? (e) What is the acceleration of the ball according to the man on the train? According to the man on the ground?

7. A car is traveling down a highway at 80 ft/sec. Just as the car crosses a per-pendicularly intersecting crossroad, the passenger throws out a beer can at 45° angle of elevation in a plane perpendicular to the motion of the car. The initial speed of the can relative to the car is 30 ft/sec. It is released at a height of 4 ft above the road. (a) Write the initial velocity of the beer can (relative to the road) in terms of the unit vectors $\mathbf{i}$, $\mathbf{j}$, and $\mathbf{k}$. (b) Where does the can land?

8. For short time intervals any path can be considered an arc of a circle. How can the radius of curvature of a path segment be determined from the instantaneous velocity and acceleration? Consider a projectile at the top of its path. Indicate the velocity vector just before and just after this point. Is the speed changing? What is the radius of curvature of the path segment at this point?

9. The position of a particle as a function of time is

$$\mathbf{R} = 4 \sin 2\pi t \, \mathbf{i} + 4 \cos 2\pi t \, \mathbf{j}$$

where $\mathbf{R}$ is in meters and t is in seconds. (a) Show that the path of this particle is a circle of radius 4 m with its center at the origin. (b) Compute the velocity vector. Show that $v_x/v_y = -y/x$. (c) Compute the acceleration vector and show that it is in the radial direction and has the magnitude v^2/R.

10. A particle has constant acceleration $\mathbf{a} = 6\mathbf{i} + 4\mathbf{j}$ m/sec². At time $t = 0$ the velocity is zero and the position vector is $\mathbf{r}_0 = (10 \text{ m})\mathbf{i}$. (a) Find the velocity and position vectors at any time t. (b) Find the equation of the path in the xy plane and sketch the path.

11. The position of a particle is given by

$$\mathbf{R} = 3 \sin 2\pi t \, \mathbf{i} + 2 \cos 2\pi t \, \mathbf{j}$$

where $\mathbf{R}$ is in meters and t is in seconds. (a) Plot the path of the particle in the xy plane. (b) Find the velocity vector. (c) Find the acceleration vector and show that its direction is along $\mathbf{R}$; that is, it is radial. (d) Find the times for which the speed is a maximum or minimum.

12. A large boulder rests on a cliff 400 m above a small village in such a position that if it should roll off, it would leave with a speed of 50 m/sec (Figure 3-34). There is a pond, diameter 200 m, with its edge 100 m from the base of

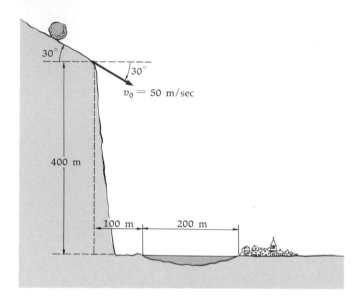

Figure 3-34
Problem 12.

the cliff, as shown. The village houses are at the edge of the pond. (a) A physics student says that the boulder will land in the pond. Is he right? (b) How fast will it be going when it hits? What will the horizontal component of its velocity be when it hits? (c) How long will the boulder be in the air?

13. A projectile is fired into the air from the top of a 600-ft cliff above a valley (Figure 3-35). Its initial velocity is 200 ft/sec at 60° to the horizontal. Neglecting air resistance, where does the projectile land?

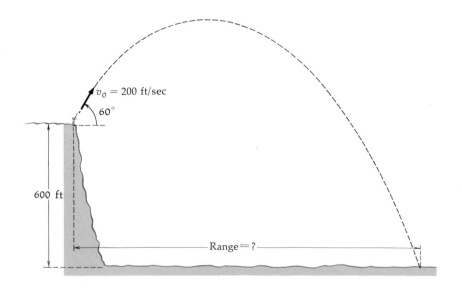

Figure 3-35
Problem 13.

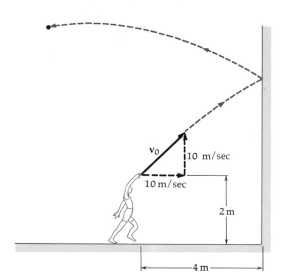

Figure 3-36
Problem 14.

14. A boy stands 4 m from a vertical wall and throws a ball (Figure 3-36). The ball leaves the boy's hand at 2 m above the ground with initial velocity $\mathbf{v} = 10\mathbf{i} + 10\mathbf{j}$ m/sec. When the ball hits the wall, its horizontal component of velocity is reversed and its vertical component remains unchanged. Where does the ball hit the ground?

15. A baseball just clears a 10-ft wall 400 ft from home plate. If it leaves the bat at 45° and 4 ft above the ground, what must its initial velocity be (again making the unreal assumption that air resistance can be ignored)?

16. An acceleration of 31g was withstood for 5 sec by R. F. Gray in 1959. How many revolutions per minute does a centrifuge have to undergo to produce an acceleration of 31g on somebody positioned in the arm at a radius of 5 m?

17. The position of a particle moving in the xy plane described by the polar coordinates r and θ and the unit vectors $\hat{\mathbf{r}}$ and $\hat{\boldsymbol{\theta}}$ is $\mathbf{r} = r\hat{\mathbf{r}}$. Show by direct differentiation, using Equations 3-47 and 3-48, that the velocity and acceleration vectors are given by

$$\mathbf{v} = \frac{dr}{dt}\,\hat{\mathbf{r}} + r\,\frac{d\theta}{dt}\,\hat{\boldsymbol{\theta}}$$

$$\mathbf{a} = \left[\frac{d^2r}{dt^2} - r\left(\frac{d\theta}{dt}\right)^2\right]\hat{\mathbf{r}} + \left(r\,\frac{d^2\theta}{dt^2} + 2\,\frac{dr}{dt}\,\frac{d\theta}{dt}\right)\hat{\boldsymbol{\theta}}$$

CHAPTER 4 Newton's Laws

Classical, or newtonian, mechanics is a theory of motion based on the ideas of mass and force and the laws connecting these physical concepts to the kinematic quantities—position, velocity, and acceleration—discussed in the preceding chapters. The fundamental relationships of classical mechanics are contained in Newton's laws of motion. We begin by stating Newton's laws. We then discuss in some detail the concepts of force, mass, and momentum and the significance and limitations of Newton's laws. In Chapter 5 we illustrate how these laws are applied in solving problems in mechanics.

A modern version of Newton's laws of motion follows:

1. If there are no external forces acting on it, a body (particle) in an inertial reference frame remains at rest or moves with constant velocity.

Newton's laws of motion

2. The time rate of change of the momentum of a body is equal to the resultant external force acting on the body:

$$\Sigma \mathbf{F} = \frac{d\mathbf{p}}{dt} \qquad 4\text{-}1$$

where the momentum $\mathbf{p}$ is defined as the product of the mass m and the velocity $\mathbf{v}$:

$$\mathbf{p} = m\mathbf{v} \qquad 4\text{-}2$$

If the mass of a particle is taken as constant, this equation can also be written

$$\Sigma \mathbf{F} = m\frac{d\mathbf{v}}{dt} = m\mathbf{a} \qquad 4\text{-}3$$

where $\mathbf{a} = d\mathbf{v}/dt$ is the acceleration.

3. Forces always occur in pairs. If body A exerts a force $\mathbf{F}_{AB}$ on body B, an equal but opposite force

$$\mathbf{F}_{BA} = -\mathbf{F}_{AB} \qquad 4\text{-}4$$

is exerted by body B on A.

SECTION 4-1 Force   81

4. Forces obey the parallelogram law of addition; i.e., forces are vectors.

It is interesting to read Newton's version of the laws of motion:[1]

> Law I. Every body continues in its state of rest, or in uniform motion in a right line unless it is compelled to change that state by forces impressed upon it.
>
> Law II. The change of motion is proportional to the motive force impressed; and is made in the direction of the right line in which that force is impressed.
>
> Law III. To every action there is always opposed an equal reaction; or, the mutual actions of two bodies upon each other are always directed to contrary parts.
>
> Corollary I. A body, acted on by two forces simultaneously, will describe the diagonal of a parallelogram in the same time as it would describe the sides by those forces separately.

Portrait of Newton by Johan Vanderbank. (*Courtesy of the National Portrait Gallery, London.*)

Newton's laws relate the acceleration of an object to its mass and the forces acting on it. We have intuitive ideas about the words force and mass. We think of a force as being a push or a pull, like that exerted by our muscles. We visualize a massive body as something large or heavy. These intuitive notions are all right for everyday conversation but not for the applications of Newton's laws to problems in physics or even for a precise statement of the laws. To understand Newton's laws fully and be able to apply them we must define these words carefully. This we do by outlining methods for their measurement, in what is called an *operational definition*. We shall find that Newton's second law follows directly from the definitions of force and mass.

Operational definition

4-1   Force

Consider an object, say a block of wood or metal, resting on a smooth horizontal surface, e.g., a table. We observe that if the body is at rest (relative to the table), it remains at rest unless we push on it or pull on it. If we project the body along the table, it slides along for a way, but eventually the speed decreases and the object comes to rest. We attribute the decrease in speed to the *frictional force* exerted on the body by the table because neither the table nor the body is perfectly smooth. If we polish the surface of the table and body, the body slides farther and its decrease in velocity in a given time is smaller. If we support the body by a thin cushion of air (this is possible with an air table or with a glider on an air track), the body will glide for a considerable time and distance with almost no perceptible change in its velocity. We extrapolate this experience to an *ideal smooth* surface which in no way impedes the movement of a body and state that on such a surface the velocity of a body will not change. We thus *define* the situation in which there are *no* (horizontal) forces acting on the body. If there are no forces acting on the body, the velocity of the body remains constant. This is Newton's first law, the *law of inertia*. (This extrapolation from the real world to an ideal situation is typical of

[1] *Philosophiae Naturalis Principia Mathematica,* 1686, trans. Andrew Motte, 1729, University of California Press, Berkeley, 1960.

physical theory. A perfectly smooth surface is a *model* with which we can make calculations. Real surfaces approximate our model of a smooth surface with varying degrees of accuracy, depending on the care taken in preparing the real surface. This extrapolation was made by Newton and others before him.)

The significance of the first law, or law of inertia, is that it defines, by an operational means, what we mean by saying there is no net or resultant force acting upon an object. It says that we can determine whether there is a net external force acting on a particle by observing the motion of the particle. If that motion has constant velocity, we conclude that there is no resultant external force. In other words, there is a net force acting only if there is acceleration, a time rate of change of velocity. A state of rest is but a special case of constant-velocity motion. Recall that constant-speed motion and constant-velocity motion are not the same. A body is accelerated if its speed is changing or if its direction of motion is changing or if both are changing. In all these cases, according to the law of inertia, a net external force must act to effect the change in velocity.

Motion with constant velocity defines the absence of a net force

If the velocity of a body is not constant, we conclude that a net force is acting on the body. Our next problem is to develop a quantitative measure of force. We do this by defining the magnitude and direction of a given force in terms of the acceleration it produces on a particular object, which we call our standard body. A convenient agent for exerting forces on bodies is a spring. It takes a push or a pull to compress or extend a spring from its natural length, and the greater the push or pull the greater the compression or extension. Consider a particular spring A attached to our standard body on a smooth horizontal surface, as in Figure 4-1. If the spring is extended from its natu-

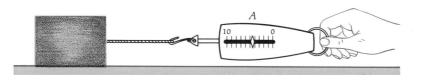

Figure 4-1
A horizontal force is applied to a body by the extended spring. The body accelerates in the direction of the resultant force, which is to the right in this case.

ral length, the body accelerates and (up to a certain limit, depending on the spring) the greater the extension the greater the acceleration. The extension that produces an acceleration of 1 m/sec² we shall call Δx_{A1}. We *define* the force exerted by the spring at this extension to be 1 unit. Similarly, if the extension Δx_{A2} produces an acceleration of 2 m/sec², we define the force exerted to be 2 units. Thus, by noting the extensions of the spring which correspond to different accelerations of this body, we can define a scale of force. In effect, we have provided a calibration of the spring in units of force. Figure 4-2 shows such a calibration curve. For common springs, the force defined in this way is proportional to the extension Δx for small extensions, an observation known as *Hooke's law*. However, this behavior of springs is not necessary in our definition of a force scale. We define the direction of a force acting on a body to be the direction of the acceleration it produces.

Force is defined in terms of acceleration

Now that we have a device for measuring force, we can calibrate other springs by attaching them to this same body and measuring the extension necessary to produce a given acceleration. Consider two calibrated springs attached to the body and extended so that they exert forces in different directions, as in Figure 4-3. We observe experi-

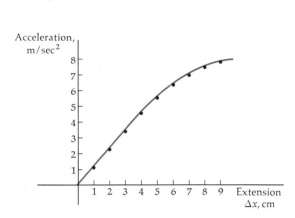

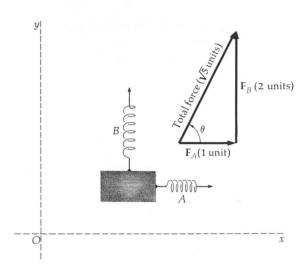

Figure 4-2
Calibration curve for a spring
scale like that in Figure 4-1.
The acceleration of a given
body is plotted as a function
of the extension of the spring
Δx from its equilibrium position.

Figure 4-3
Forces obey the vector-addi-
tion rule. The forces $\mathbf{F}_A$ and $\mathbf{F}_B$
produce accelerations of 1 and
2 m/sec², respectively, when
each acts alone on the standard body. Acting together at
right angles to each other, the
forces produce an acceleration
of $\sqrt{5}$ m/sec² on the standard
body in the direction found
using the vector-addition
rule.

mentally that the forces add as vectors; i.e., the resultant acceleration is
found by adding the vector accelerations each force would produce if
it were acting alone. For example, if spring A exerts a force of 1 unit
along the x axis and spring B exerts a force of 2 units along the y axis,
as in Figure 4-3, the acceleration observed has the magnitude
$a = \sqrt{1^2 + 2^2}$ m/sec² $= \sqrt{5}$ m/sec² and makes an angle θ with the x
axis given by $\tan \theta = 2$. This observation is known as the *parallelogram
law of addition of forces*; forces are vectors. We have listed this experimental result as Newton's fourth law. In particular, if two equal forces
act in opposite directions on the body, the acceleration is zero.

We can use our calibrated springs to measure other types of forces.
For example, the force exerted by the gravitational attraction of the
earth for an object is called its *weight*. When the weight of a body is
the only force acting on it, the acceleration of the body is 9.8 m/sec²
toward the earth. The weight of our standard body is thus 9.8 units
(defined by our spring measurements). If we hang our standard body
by one of our calibrated springs, as in Figure 4-4, the spring is
stretched by an amount corresponding to 9.8 units of force to balance
the downward gravitational force exerted by the earth.

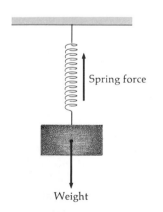

Example 4-1 A box rests on the front seat of a car next to the driver.
What forces must act to keep it from slipping on the seat as the car
travels at a steady velocity? As it speeds up? As it rounds a curve?

According to Newton's first law, no net force acts on a body unless
it accelerates. The measure of the net force is the acceleration produced
by the net force. In this example, since the box has no vertical component of acceleration in any of the three cases, we must conclude that

Figure 4-4
Forces on a body suspended
from a spring. The upward
force due to the spring balances the downward force
due to the gravitational attraction of the earth (the
weight of the body).

there is no resultant vertical force on the box at any time. This does not mean, however, that no vertical forces are acting. If the car seat were not there, the box would accelerate downward, indicating the presence of the downward force we call weight. We do not suppose, of course, that the mere presence of the seat removes the force of gravity and thus accounts for the lack of vertical acceleration of the box. So we must conclude that the seat exerts an upward force which exactly balances the weight of the box. These two forces must be present in all the cases mentioned, and they must always be equal though opposing.

When the car moves with steady velocity, there is no horizontal acceleration and hence no net horizontal force. Since there is no obvious horizontal force acting (like the weight in the case of the vertical forces), we can conclude that there are no horizontal forces at all.

When the car speeds up, it and the box must have forward acceleration. There must therefore be a forward horizontal force on the box. The greater the acceleration the greater the force on the box. This force must be exerted by the car seat, since it is the only physical thing that directly affects the box. We know this force by the name of friction. It is capable of acting on the box if neither the seat nor the box is smooth. It is also limited. If the acceleration of the car is too great, the friction force will not be enough to make the box accelerate with the car. In that event, the box slips back on the car seat—a very familiar occurrence.

When the car rounds a corner, even if its speed is constant, it has an acceleration toward the concave side of the curve. There must also be a horizontal force on the box in this same direction. Again the force must be exerted by the car seat. The force is a friction force, as before. If the seat is too smooth, the corner too sharp, or the speed too great, there may not be enough friction to make the box accelerate with the car. Then, since the acceleration of the box is less than that of the car, the box will slide relative to the car seat in the direction opposite to that of the car's acceleration.

Example 4-2 A heavy block rests on the smooth surface of an ice-skating rink. If either of two men pulls separately on the block in a horizontal direction, he can give it an acceleration of 1 m/sec². If both pull together at 90° to each other, what will be the acceleration of the block?

We assume negligible friction between the block and the ice so that the acceleration of the block is wholly determined by the forces the two men exert upon it. The forces are of equal magnitude because each applied separately produces the same acceleration. Their net effect when they are applied together is determined by adding them vectorially. The two forces act at 90°, and their vector sum is calculated in Figure 4-5. The magnitude F of the resultant force is $\sqrt{2}$ times the magnitude of the force one man can exert. Hence, the acceleration produced by this resultant force is $\sqrt{2}$ times the acceleration either man can produce alone, or 1.4 m/sec².

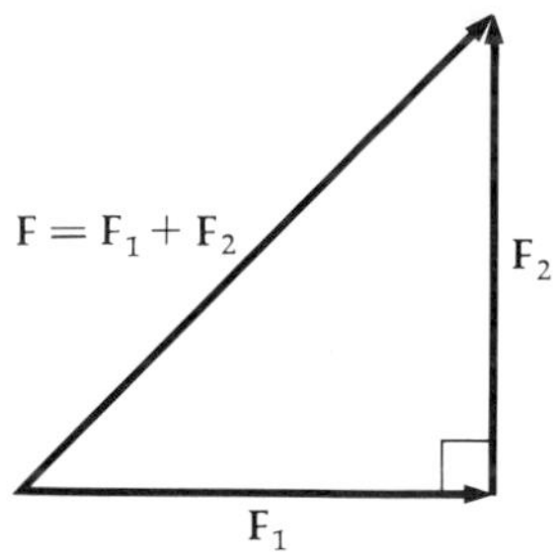

Figure 4-5
Two perpendicular forces of equal magnitude are applied to a single object. The resultant force is $\sqrt{2}$ times the magnitude of the individual forces.

Questions

1. How can you decide whether there is a net, or resultant, force acting on a body? Can you determine by this means the number of separate forces which act?

2. If a body has no acceleration, can you conclude that no forces act on it?

3. If only a single force acts on a body, must it be accelerated? Can it ever have zero velocity?

4. Is there a net force acting (*a*) when a body moves at constant speed along a circle, (*b*) when a body that is moving in a straight line slows down, (*c*) when a body moves at constant speed in a straight line?

5. Is it possible for an object to round any curve without a force being impressed upon it?

6. When a body rounds a curve, can the net force on it be toward the convex side of the curve?

7. If a single known force acts on a body, can you tell *in which direction* the body will *move* from this information alone?

8. If several forces of different magnitudes and directions are applied to a body initially at rest, how can you predict the direction in which it will move?

9. A string is just barely capable of supporting a 10-lb weight suspended at rest from it. What happens if the weight is set swinging? Explain.

10. The maximum tension a rope can withstand without breaking is 90 lb. Could the rope be used to lower a 100-lb weight? How? Is it possible that the rope might break in raising an 80-lb weight? Under what conditions?

4-2   Mass

So far we have discussed how to compare forces quantitatively by comparing the accelerations they produce in some particular single object. We now investigate the effect of a given force on different objects. Let us use one of our calibrated springs to produce a given force on a different body. We note that the acceleration produced for a given spring extension is not the same, in general, as it was for our first body. If our second body is "more massive" (according to our everyday use of this term), the acceleration produced by a force of 1 unit is observed to be less than 1 m/sec². If it is less massive, the acceleration produced is greater. For example, if we connect two identical bodies, the acceleration produced by a given force is exactly half that produced by the same force acting on just one of the bodies. This suggests that we quantify the concept of mass by considering the acceleration a given force will produce in different bodies.

Specifically, we define the ratio of the mass of one body to that of another body to be the *inverse* ratio of the accelerations produced in those two bodies by the *same force*. If a given force produces an acceleration a_1 when it acts on body 1 and an acceleration a_2 on body 2, the ratio of the masses of the two bodies is defined to be

$$\frac{m_2}{m_1} = \frac{a_1}{a_2}$$

4-5   *Definition of mass*

We find the ratio of the masses of any two bodies by applying the same force to each and comparing their accelerations. Having defined the ratio of masses for any two bodies, we can set up a mass scale by choosing one particular body to be a standard body and arbitrarily calling its mass 1 unit. The SI unit of mass is the *kilogram*. The international standard body is a cylinder of platinum carefully kept at the International Bureau of Weights and Measures at Sèvres, France. Its mass is defined to be one kilogram. (This was originally intended to be equal to the mass of 1000 cm³ of water, but the relationship proved inexact by a very small amount.) Theoretically, the mass of any other body can be compared to the standard mass, as described above, and assigned an appropriate number. For example, if a certain force produces an acceleration of $a_1 = 1$ m/sec² on the standard body and the same force produces an acceleration $a_2 = 2$ m/sec² on a second body, the mass of the second body is $m_2 = (a_1/a_2)m_1 = 0.5$ kg.

The ratio of accelerations a_2/a_1 produced by the same force on two bodies is independent of the magnitude and direction of the force. It is also independent of the kind of force used, i.e., whether the force is due to springs, the pull of gravity, electric or magnetic attraction or repulsion, etc. We note also that if mass m_2 is found to be twice the standard mass by direct comparison and mass m_3 is found to be 4 times the standard mass, m_3 is found to be twice m_2 when compared with it directly. Thus mass is an *intrinsic* property of the body which does *not* depend on the surroundings of the body, any external agent, or on the type of force used to measure it. It is a scalar quantity which obeys the ordinary rules of arithmetic and algebra.

The concept of mass, as we have defined it, is associated with the size of the acceleration produced by a given force. The larger the mass the smaller the acceleration produced by a given force. Thus mass is a quantitative measure of the property of *inertia*, or resistance to acceleration, which all objects have.

The standard kilogram, kept at the International Bureau of Weights and Measures, Sèvres, France. (*Courtesy of the Science Museum, London.*)

Mass is an intrinsic property

Mass is a quantitative measure of inertia

Questions

11. Can you judge the mass of an object by its size? If A is twice as big as B, does that mean that $m_A = 2m_B$?

12. Could a body have a different mass for electric forces than for gravitational forces?

13. Can the mass of a body be negative?

14. How do we know that we can add masses as scalars? That is, if one body has a mass of 2 kg and another of 3 kg, how do we know that the two bodies taken together have a mass of 5 kg?

15. Mass is sometimes said to be the measure of the quantity of matter in a body. How does this unscientific definition compare with the definition discussed above?

4-3   Newton's Second Law

The definitions we have given for force and mass are summarized in Newton's second law, which states that the net, or resultant, force $\Sigma \mathbf{F}$ acting on a body is proportional to its mass and its acceleration. In symbols

$$\Sigma \mathbf{F} = Km\mathbf{a}$$

where K represents a constant we are still free to determine. (The actual operational definitions we have used are not convenient for practical work. Their value is in clarifying our quantitative concepts, not in their direct application. More practical definitions, however, can be shown to be equivalent to ours though they are much more difficult to phrase.) It is convenient to define the unit of force so that the constant of proportionality is unity. Then a force of 1 unit acting on a mass of 1 kg produces an acceleration of 1 m/sec², or 1 unit of force = 1 kg-m/sec². This combination of units, 1 kg-m/sec², is called a newton (N):

$$1 \text{ N} = 1 \text{ kg-m/sec}^2 \qquad \text{4-6}$$

Definition of the newton

The unit of force is so defined that Newton's second law holds in the form

$$\Sigma \mathbf{F} = m\mathbf{a} \qquad \text{4-7}$$

Newton's second law

Our definitions of force and mass agree with our intuitive notions of the meaning of these words. These definitions are useful because they allow us to describe a wide variety of physical phenomena using just a few relatively simple force laws. For example, with the addition of Newton's law of gravitational attraction between two bodies we can calculate and explain such phenomena as the motion of the moon, the orbits of all the planets around the sun, the orbits of artificial satellites, the variation in the acceleration of gravity g with latitude due to the rotation of the earth, the variations in the acceleration of gravity due to the presence of mineral deposits, the paths of ballistic missiles, and many other motions.

Note that Newton's statement of the second law does not refer to mass and acceleration but to the change in "motion." What Newton called motion is now called *momentum*. The momentum of a particle is a dynamic quantity defined as the product of the mass and the velocity of a particle. Writing $\mathbf{a} = d\mathbf{v}/dt$ for the acceleration and assuming the mass of the particle to be constant, we have

$$\Sigma \mathbf{F} = m\frac{d\mathbf{v}}{dt} = \frac{d(m\mathbf{v})}{dt}$$

or

$$\Sigma \mathbf{F} = \frac{d\mathbf{p}}{dt} \qquad \text{4-8}$$

where

$$\mathbf{p} = m\mathbf{v} \qquad \text{4-9}$$

Momentum defined

is the momentum of the particle. Momentum is a vector parallel to the velocity vector. A body of mass 0.5 kg moving at 2000 m/sec has the same momentum as a body of mass 4 kg moving in the same direction at 250 m/sec. A force of 1 N will change the momentum of any body by 1 kg-m/sec in 1 sec.

In classical mechanics the mass of a particle is always constant, and Equations 4-7 and 4-8 are equivalent. However, when a particle moves with a speed near the speed of light (about 3×10^8 m/sec), the ratio of the force to the acceleration depends on the speed. As we shall see in Chapter 28, for high-speed particles, classical mechanics must be modified according to Einstein's theory of special relativity. In this theory, Equation 4-8 holds if the expression for momentum is taken to be

$$\mathbf{p} = \frac{m\mathbf{v}}{\sqrt{1 - v^2/c^2}} \qquad\qquad 4\text{-}10$$

where c is the speed of light in vacuum. When the speed of the particle is much less than that of light, the quantity $\sqrt{1 - v^2/c^2}$ is very nearly equal to 1 and the relativistic and classical expressions for momentum (Equations 4-9 and 4-10) are approximately equal.

4-4 Weight

The force most common in our everyday experience is the force of attraction of the earth for various bodies. This force is called the *weight* of the body. We can find the weight of a 1-kg mass by measuring its acceleration when the mass is falling freely so that the only force on it is its weight. The resulting acceleration is about 9.81 m/sec^2 downward. The weight w of a 1-kg mass thus has the magnitude

$$w = ma = 1 \text{ kg} \times 9.81 \text{ m/sec}^2 = 9.81 \text{ N}$$

The free-fall acceleration of any body (about 9.81 m/sec^2 at sea level) is independent of the mass of the body as long as air resistance can be neglected. Since the acceleration is the ratio of the resultant force (the weight in this case) and the mass, the fact that it is independent of the mass implies that the weight of a body is proportional to its mass,

$$\mathbf{w} = m\mathbf{g} \qquad\qquad 4\text{-}11$$

Weight is the gravitational force on a body

where $\mathbf{g}$ is the free-fall acceleration of gravity. Careful measurements of the free-fall acceleration of a body at various places show that this acceleration is not the same everywhere. Thus weight, unlike mass, is not an intrinsic property of the body. The force of attraction of the earth for a body varies with location. In particular, for points above the surface of the earth, this force varies inversely with the square of the distance of the body from the center of the earth. Thus a body weighs less at very high altitude than it does at sea level. The weight of a body also varies slightly with latitude because the earth is not exactly spherical but is flattened at the poles.[1]

Near the surface of the moon, the gravitational attraction of the moon is much stronger than that of the earth. The force exerted on the body by the moon is usually called the weight of the body when it is near the moon. Note again that the mass of a body is the same whether it is on the earth, on the moon, or somewhere in space. Mass is a property of the body itself, whereas weight depends on the nature and distance of other objects which exert gravitational forces on the body.

Since at any particular location the weight of a body is proportional to its mass, we can conveniently compare the mass of one body with that of another by comparing their weights as long as we determine the weights at the same place.

---

[1] If the free-fall acceleration of a body is measured relative to a point on the surface of the earth, there is a variation with latitude because the reference point has an acceleration due to the rotation of the earth. This variation, discussed in Section 4-7, is not a variation in $\mathbf{g}$, which we define as the gravitational force on the body divided by its mass.

Edward H. White II in a space walk during the third orbit of the Gemini-Titan 4 flight. He experiences weightlessness whether he is inside or outside the space capsule because the only force acting on him is gravity. He is in free fall accelerating toward the earth with the acceleration of gravity.

NASA

Our sensation of our own weight usually comes from other forces which balance it. For example, sitting on a chair, we feel the force exerted by the chair which balances our weight. When we stand on a spring scale, our feet feel the force exerted on us by the scale. The scale is calibrated to read the force it must exert (by compression of its springs) to balance our weight. The force which balances our weight is called our *apparent weight*. It is the apparent weight that is given by a spring scale. If there is no force to balance your weight, as in free fall, your apparent weight is zero. This condition, called *weightlessness*, is experienced by astronauts in orbiting satellites. Consider a satellite in a circular orbit near the surface of the earth with a centripetal acceleration v^2/r, where r is the orbit radius and v is the speed. The only force acting on the satellite is its weight. Thus it is in free fall with the acceleration of gravity. The astronaut is also in free fall. The only force on him is his weight, which produces the acceleration $g = v^2/r$. Since there is no force balancing the force of gravity, the astronaut's apparent weight is zero.

Apparent weight defined

Questions

16. From our definitions of mass and weight, would it be conceivable to use the same units for both?

17. Suppose an object were sent far out in space away from galaxies, stars, or other bodies. How would its mass change? Its weight?

18. How would an astronaut in a condition of weightlessness be aware of his mass?

19. We commonly compare the masses of objects by comparing their weights. Would this be possible if g depended on the kind of material an object is made of?

20. Under what circumstances would your apparent weight be greater than your true weight?

4-5   Units of Force and Mass

The definition of the unit of mass, the kilogram, as the mass of a particular standard body completes our definition of the three fundamental units of mechanics in the mks system. In this system, the unit of force, the newton (that force which produces an acceleration of 1 m/sec^2 on a 1-kg mass) is a derived unit; i.e., it can be expressed in terms of the three fundamental units, as in Equation 4-6. As mentioned in Chapter 1, the mks system of mechanics units is a subset of the international system of units (SI), which also includes units of temperature, electric current, and luminous intensity. These units are used almost exclusively throughout the world except in the United States, where they will also eventually become standard. Although we generally use SI units in this book, we need to know about two other systems, the *cgs system,* a metric system based on the centimeter, gram, and second, which is closely related to the mks system and used by many scientists, and the British engineering system, based on the foot, the second, and a force unit (the pound), which is still used today in the United States.

The unit of time, the second, is common to all three systems of units. In the cgs system, the unit of length is the centimeter (cm), now defined to be one-hundredth the length of the meter:

$$1 \text{ cm} = 10^{-2} \text{ m} \qquad\qquad 4\text{-}12$$

The unit of mass, the gram (gm), is now defined to be exactly one-thousandth the mass of the standard kilogram:

Definition of centimeter and gram

$$1 \text{ gm} = 10^{-3} \text{ kg} \qquad\qquad 4\text{-}13$$

The gram was originally chosen to be the mass of one cubic centimeter of water at standard pressure and temperature. The unit of force in the cgs system, called the dyne, is the force which applied to a one-gram mass produces an acceleration of one centimeter per second per second:

$$1 \text{ dyne} = 1 \text{ gm-cm/sec}^2 \qquad\qquad 4\text{-}14$$

Because the units are so small, the cgs system is less convenient than the mks system for practical work. For example, the mass of a penny is about 3 gm. Since the free-fall acceleration of gravity is 981 cm/sec^2, the weight of a penny in the cgs system is about

$$w = mg = (3 \text{ gm})(981 \text{ cm/sec}^2) = 2.94 \times 10^3 \text{ dynes}$$

The dyne is a very small unit of force. The relation between the dyne and the newton is

$$1 \text{ dyne} = \frac{1 \text{ gm-cm}}{\text{sec}^2} \frac{1 \text{ kg}}{10^3 \text{ gm}} \frac{1 \text{ m}}{10^2 \text{ cm}} = 10^{-5} \text{ kg-m/sec}^2$$

or

$$1 \text{ dyne} = 10^{-5} \text{ N} \qquad\qquad 4\text{-}15$$

The British engineering system differs from both the mks and cgs systems in that a unit of force is chosen as a fundamental unit rather than a unit of mass. (Another difference is that it is not a decimal system.) The pound was originally defined as the weight of a particular standard body at point where the acceleration of gravity is exactly 9.80665 m/sec^2 = 32.1740 ft/sec^2. It is now defined in terms of the standard kilogram. One pound is defined to be the weight of a body of

Definition of pound

mass 0.45359237 kg at a point where the acceleration of gravity has the value given above. The relation between the standard pound and the newton is thus

$$1 \text{ pound} = (0.45359237 \text{ kg})(9.80665 \text{ m/sec}^2) = 4.448222 \text{ N}$$

or

$$1 \text{ lb} \approx 4.45 \text{ N} \tag{4-16}$$

Since 1 kg weighs 9.81 N, its weight in pounds is

$$9.81 \text{ N} \times \frac{1 \text{ lb}}{4.45 \text{ N}} = 2.20 \text{ lb} \tag{4-17}$$

Weight of 1 kg in pounds

The unit of mass in the British system is that mass which will be given an acceleration of one foot per second per second when a force of one pound is applied to it. This unit is called a *slug*. From its definition

Definition of slug

$$1 \text{ slug} = \frac{1 \text{ lb}}{1 \text{ ft/sec}^2} = 1 \text{ lb-sec}^2/\text{ft} \tag{4-18}$$

The weight of a slug near sea level is about

$$w = mg = (1 \text{ lb-sec}^2/\text{ft})(32.2 \text{ ft/sec}^2) = 32.2 \text{ lb} \tag{4-19}$$

In practice, the slug is seldom used. It is more convenient to work problems in this system by writing the mass as w/g, where w is the weight and g is the acceleration of gravity.

Example 4-3 The net force acting on a 10.0-lb body is 3.00 lb. What is its acceleration?

The acceleration is the force divided by the mass:

$$a = \frac{F}{m} = \frac{F}{w/g} = \frac{3.00 \text{ lb}}{(10.0 \text{ lb})/(32.2 \text{ ft/sec}^2)} = 9.66 \text{ ft/sec}^2$$

Although the weight of an object varies from place to place because of changes in the acceleration of gravity, this variation is too small to be noticed in most practical applications. Thus, in our everyday experience, the weight of a body appears to be as much a constant characteristic of the body as its mass.[1]

Questions

21. What is your weight in newtons?

22. What is your mass in kilograms? In slugs?

23. What would your weight be in pounds on the moon, where objects fall freely with acceleration of about $5\frac{1}{3}$ ft/sec²?

24. What is the weight at sea level of a 10-kg mass? Give your answer in newtons and in pounds. What would be the weight of this same mass on the moon, where the acceleration of gravity is about one-sixth what it is at sea level? What is the mass of this object on the moon?

---

[1] This fact has led to everyday use of two other units, which often is confusing. One is a unit of force, the kilogram force, which is the weight of a 1-kg mass. A kilogram force is equal to 9.81 N or 2.20 lb. A second practical unit is the pound mass, which is the mass of a body which weighs 1 lb. Since by definition, a pound force is the weight of 0.454 kg, a pound mass is equivalent to 0.454 kg. We shall not use these practical but confusing units.

4-6 Newton's Third Law and Conservation of Momentum

Newton's third law describes an important property of forces: they always occur in pairs. For each force exerted on some body A there must be some external agent, say another body B, exerting the force. The third law states that body A exerts an equal but opposite force on the agent B. For example, earth exerts a gravitational force on a projectile, causing it to accelerate toward the earth with acceleration $\mathbf{g} = \mathbf{F}/m = \mathbf{w}/m$, where $\mathbf{w}$ is the weight of the projectile and m its mass. According to the third law, the projectile in turn exerts a force on the earth equal in magnitude and opposite in direction. Thus the projectile exerts a force $\mathbf{w}'$ on the earth toward the projectile. The earth, in response to the force exerted by the projectile, must accelerate. Because of the great mass of the earth, this contribution to its total acceleration is negligible and unobserved.

In discussions of Newton's third law the words action and reaction are frequently used. If the force exerted on body A is called the *action* of B upon A, then the force body A exerts back on body B is called the *reaction* of A upon B. It does not matter which force in such a pair is called the action and which the reaction. The important point is that forces always occur in action-reaction pairs and that the reaction force is equal in magnitude and opposite in direction to the action force.

Action and reaction

Note that the action and reaction forces can never balance each other because they act on different objects. This is illustrated in Figure 4-6, which shows two action-reaction pairs of forces for a block resting on a table. The force acting downward on the block is the weight $\mathbf{w}$ due to the attraction of the earth. An equal and opposite force $\mathbf{w}'$ is exerted by the block *on the earth*. These are an action-reaction pair. If they were the only forces acting, the block would accelerate down because it would have only a single force acting on it. However, the table in contact with the block exerts a force $\mathbf{N}$ upward on it. This force balances the weight of the block. The block also exerts a force $\mathbf{N}'$ downward on the table. The forces $\mathbf{N}$ and $\mathbf{N}'$ are also an action-reaction pair.

Action-reaction forces can never balance

There is a simple but important consequence of the third law for two objects isolated from their surroundings so that the only forces on them are the ones they exert on each other. *The sum of the momenta of the two objects remains constant in time.* This can be seen from the form of Newton's second law in Equation 4-8. The rate of change of the momentum of each object equals the force acting on it. Each object has a single force acting on it, and the two forces are equal but opposite.

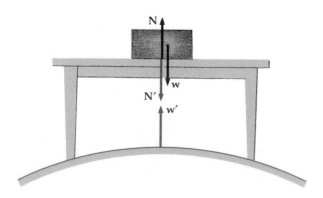

Figure 4-6
Action-reaction forces. The weight $\mathbf{w}$ is the force exerted on the block by the earth. The equal and opposite reaction force is $\mathbf{w}'$, exerted on the earth by the block. Similarly, the table exerts a force $\mathbf{N}$ on the block, and the block exerts an equal and opposite force $\mathbf{N}'$ on the table. Action-reaction forces are exerted on different objects.

Thus the change of momentum of one object equals the negative of the change of momentum of the other, and the sum of the changes is zero. Let $\mathbf{F}_1$ be the force on object 1 and $\mathbf{p}_1 = m_1 \mathbf{v}_1$ be its momentum. Then Newton's second law for this object is

$$\mathbf{F}_1 = \frac{d\mathbf{p}_1}{dt}$$

Similarly, if $\mathbf{F}_2$ is the force acting on object 2 and $\mathbf{p}_2 = m_2 \mathbf{v}_2$ is its momentum, we have

$$\mathbf{F}_2 = \frac{d\mathbf{p}_2}{dt}$$

But $\mathbf{F}_2 = -\mathbf{F}_1$, and so

$$\frac{d\mathbf{p}_2}{dt} = -\frac{d\mathbf{p}_1}{dt}$$

or

$$\frac{d\mathbf{p}_1}{dt} + \frac{d\mathbf{p}_2}{dt} = \frac{d}{dt}(\mathbf{p}_1 + \mathbf{p}_2) = 0$$

Thus the rate of change of the sum of the momenta $\mathbf{p}_1 + \mathbf{p}_2$ is zero. Then

$$\mathbf{p}_1 + \mathbf{p}_2 = \text{constant} \qquad\qquad 4\text{-}20$$

Conservation of momentum

We can take Equation 4-20 as an equivalent statement of Newton's third law. In fact, Newton seems to have arrived at his statement of action-reaction by studying the momentum of two bodies before and after collisions. When two bodies collide, they exert very large forces on each other during the short time they are in contact. Even if there are other forces on the bodies, they are usually much smaller than these contact forces and can be neglected. From the careful measurements made by his predecessors Newton knew that no matter what kind of collision occurs, the sum of the momenta of the two colliding bodies is the same after the collision as before. Starting from Equation 4-20, which describes this result, he arrived at his statement that action equals reaction. We need only differentiate Equation 4-20 and substitute $\mathbf{F}_1$ for $d\mathbf{p}_1/dt$ and $\mathbf{F}_2$ for $d\mathbf{p}_2/dt$. This result is known as the *law of conservation of momentum*. We have shown that it applies to two bodies each of which experiences only a force exerted by the other. That is, the total momentum of the two bodies is constant (conserved) if they are isolated from influences from other sources. By a straightforward generalization, the same law can be shown to apply to any isolated system of bodies, no matter how great their number.

The extrapolation of the action-reaction principle for forces exerted by bodies in contact to bodies far apart presents conceptual difficulties of which Newton was well aware. We have seen that the statement that action equals reaction is equivalent to the statement that the rate at which one body gains momentum equals the rate at which the second body loses momentum. This is easily imagined if the bodies are in contact, but if they are widely separated, it implies that momentum is instantaneously transmitted from one to the other across the intervening space. This concept, called *action at a distance*, is difficult to accept. For example, applied to the earth-sun system, it suggests that the momentum lost by one travels instantly across the 93 million miles between them to be taken up by the other (Figure 4-7). Newton justified his extension of the third law to action-at-a-distance forces

Action at a distance

Figure 4-7
Action-reaction forces for
widely separated bodies.
Modern theory treats the
action-at-a-distance problem
by introducing the concept of
a field.

because the assumption enabled him to calculate the orbits of the
planets correctly from the law of gravitation. He perceived action at a
distance as a flaw in his theory but avoided giving any other hypothesis. In 1692 Newton made a famous comment[1] about the concept
of action at a distance:

> It is inconceivable that inanimate, brute matter should, without the
> mediation of something else, which is not material, operate upon,
> and affect other matter without mutual contact, as it must be if gravi-
> tation, in the sense of Epicurus, be essential and inherent in it. And
> this is one reason why I desired you would not ascribe innate gravity
> to me. That gravity should be innate, inherent, and essential to
> matter, so that one body may act upon another at a distance through a
> vacuum, without the mediation of anything else, by and through
> which their action and force may be conveyed from one to another, is
> to me so great an absurdity that I believe no man who has in philo-
> sophical matters a competent faculty of thinking can ever fall into it.

Today, we treat the problem of action at a distance by introducing
the concept of a field. For example, we consider the attraction of the
earth by the sun in two steps. The sun creates a condition in space
which we call the *gravitational field*. This field produces a force on the
earth. The field is thus the intermediary agent. Similarly, the earth
produces a gravitational field which exerts a force on the sun. If the
earth suddenly moves to a new position, the field of the earth is
changed. This change is not propagated through space instantly but
with the velocity $c = 3 \times 10^8$ m/sec $= 1.86 \times 10^5$ mi/sec, which is also
the velocity of light. If we can neglect the time it takes for propagation
of the field, we can ignore this intermediary agent and treat the forces
as if they were exerted by the sun and the earth directly on each other.
For example, during the 8 min it takes for propagation of the gravitational field from the earth to the sun, the earth moves only a
small fraction of its total orbit around the sun. (The angular displacement of the earth after 8 min is only 9.6×10^{-5} rad $= 5.5 \times 10^{-3}$ deg.)

The third law is only an approximate law for two separated bodies.
It holds if we can neglect the time of propagation of momentum
between the interacting bodies. The law of conservation of momentum
for *two bodies* is also only approximate: it takes time for the momentum to be transferred from one body to another. However, the

The field concept

[1] Isaac Newton, Third Letter to Bentley (Feb. 25, 1692), R. and J. Dodsley, London, 1756.

conservation of momentum can be rephrased as an exact law by introducing the idea that the field itself can have momentum. Then during the time of transit, the momentum lost by the two bodies is carried by the field. It can be demonstrated in the analogous case of the electromagnetic force between two separated charges that the electromagnetic field can indeed transport momentum. It is more difficult to demonstrate this for the gravitational field.

Newton's extension of the action-reaction law to separated bodies generated the conceptual difficulties we have discussed, but the difficulties were indeed conceptual and not practical. The action-reaction law is generally an exceptionally good approximation and is very useful in practical problems.

Questions

25. When a ball is bounced from the ground, the earth exerts the force on the ball necessary to reverse its velocity. Is there a simultaneous force exerted on the earth? If so, why don't we perceive any acceleration of the earth?

26. When an object absorbs or reflects light, it experiences a very small but measurable force. Consider the following case. A supernova (stellar explosion) occurs in which the star becomes millions of times brighter than normal for a few weeks. Light from the star reaches the earth centuries after the explosion and causes the deflection of a delicately balanced mirror. Does the action-reaction law apply to the interaction of the star and the mirror?

27. In a tug of war the forces each team exerts on the other are equal in magnitude and opposite in direction. How can either team win?

4-7 Reference Frames

It might appear that the first law is merely a special case of the second when $\Sigma \mathbf{F} = 0$, since then $\mathbf{a} = 0$ and $\mathbf{v} = $ constant. However, we shall interpret the first law as a definition of an *inertial reference frame*. If the net force on a body is zero and it maintains a constant velocity relative to some reference frame, the reference frame is called an inertial frame. Before examining this concept of an inertial reference frame more fully, we must discuss what is meant by a reference frame.

To measure the velocity and acceleration of a particle, we need a coordinate system, as illustrated in Figure 4-8. Consider another coordinate system $x'y'z'$ *at rest* relative to xyz. For convenience, we have

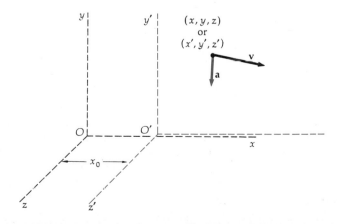

Figure 4-8
Two coordinate systems at rest relative to each other. A particle with position (x,y,z) relative to O has position $x' = x - x_0$, $y' = y$, $z' = z$ relative to O'. The velocity $\mathbf{v}$ and acceleration $\mathbf{a}$ of the particle are the same in both systems. These coordinate systems are said to be in the same reference frame.

assumed the axes parallel and the origin O' to be on the x axis at a distance x_0 from O. A particle at x with velocity $\mathbf{v}$ and acceleration $\mathbf{a}$ relative to O will have the same velocity $\mathbf{v}' = \mathbf{v}$ and acceleration $\mathbf{a}' = \mathbf{a}$ relative to O', but its coordinate will be different, $x' = x - x_0$. *The set of coordinate systems at rest relative to a given system is called a reference frame.* Clearly, any one coordinate system in a reference frame is as good as any other for describing the motion of a particle, for $\mathbf{v}$ and $\mathbf{a}$ are the same in all such coordinate frames.

Reference frame defined

Now consider the case in which two coordinate systems are in motion relative to each other. Let axes x', y', z' move with *constant velocity* $\mathbf{v}_0$ relative to axes x, y, z. (For example, the axes x', y', z' might be fixed to a train moving at constant velocity relative to the track.) For convenience, let the origins coincide at $t = 0$, and take $\mathbf{v}_0$ along the x or x' axes, as in Figure 4-9. We now have two different reference frames, which we shall call S and S'. S' moves with velocity $v_x = v_0$ relative to S whereas S moves with velocity $v'_x = -v_0$ relative to S'. A particle at rest in one frame is not at rest in the other. The coordinates are still related by

$$x' = x - x_0 \qquad\qquad 4\text{-}21$$

but now x_0 is not a constant. It is given by $x_0 = v_0 t$; thus

$$x' = x - v_0 t \qquad\qquad 4\text{-}22$$

The velocities of the particle measured in the two reference frames are related by

$$v'_x = \frac{dx'}{dt} = \frac{d}{dt}\,(x - v_0 t) = \frac{dx}{dt} - v_0 = v_x - v_0 \qquad\qquad 4\text{-}23$$

Differentiating again, we find that the accelerations in the two frames are equal, since the relative velocity v_0 was assumed to be constant.

$$a'_x = \frac{dv'_x}{dt} = \frac{d}{dt}(v_x - v_0) = \frac{dv_x}{dt} = a_x \qquad\qquad 4\text{-}24$$

Suppose we have a body of mass m attached to a spring which is extended by an amount Δx and thus exerts a force F_x in reference frame S. The acceleration will be $a_x = F_x/m$ assuming that Newton's law holds in this frame.

In frame S', the position x' and velocity v'_x of the body are not the same, but the acceleration a'_x is the same as a_x. Furthermore, the extension of the spring is the same (see Exercise 34). Thus the force is the same, and if $\Sigma \mathbf{F} = m\mathbf{a}$ holds in S, it holds in S' also.

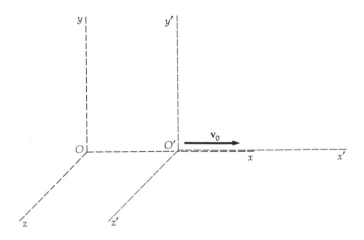

Figure 4-9
Two coordinate systems moving relative to each other are in different reference frames. The reference frame S' containing coordinate system $x'y'z'$ moves with velocity $\mathbf{v}_0$ relative to reference frame S containing coordinate system xyz. The position and velocity of a particle are different in the two reference frames, but the acceleration of a particle is the same if $\mathbf{v}_0$ is constant.

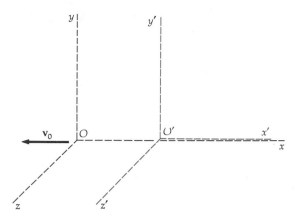

Figure 4-10
The two coordinate systems
of Figure 4-9 from the point
of view of reference frame S'.
In this frame, the coordinate
system $x'y'z'$ is at rest, and
the system xyz moves to the
left with speed v_0. Only the
relative speed v_0 between the
reference frames is of significance.

We could just as well have made the drawing from the point of
view of S' and had S moving with speed v_0 to the left (Figure 4-10). Is
there any way to determine which frame is really at rest and which
frame is moving? That is, is it possible to determine absolute velocity?
Since only acceleration and not velocity appears in Newton's laws,
there are no mechanics experiments available for distinguishing absolute motion. This *principle of relativity* was discussed in the fourteenth
century, but it became well known only with the work of Galileo early
in the seventeenth century. It is interesting to read Galileo's remarks
on the impossibility of observing absolute motion:[1]

Principle of relativity

> Shut yourself up with some friend in the main cabin below decks on
> some large ship, and have with you there some flies, butterflies, and
> other small flying animals. Have a large bowl of water with some fish
> in it; hang up a bottle that empties drop by drop into a wide vessel
> beneath it. With the ship standing still, observe carefully how the
> little animals fly with equal speed to all sides of the cabin. The fish
> swim indifferently in all directions; the drops fall into the vessel
> beneath; and, in throwing something to your friend, you need throw
> it no more strongly in one direction than another, the distances being
> equal; jumping with your feet together, you pass equal spaces in
> every direction. When you have observed all these things carefully
> (though there is no doubt that when the ship is standing still every-
> thing must happen in this way), have the ship proceed with any
> speed you like, so long as the motion is uniform and not fluctuating
> this way and that. You will discover not the least change in all the ef-
> fects named, nor could you tell from any of them whether the ship
> was moving or standing still. In jumping you will pass on the floor
> the same spaces as before, nor will you make larger jumps toward the
> stern than toward the prow even though the ship is moving quite
> rapidly, despite the fact that during the time you are in the air the
> floor under you will be going in a direction opposite to your jump. In
> throwing something to your companion, you will need no more force
> to get it to him whether he is in the direction of the bow or the stern,
> with yourself situated opposite. The droplets will fall as before into
> the vessel beneath without dropping toward the stern, although
> while the droplets are in the air the ship runs many spans. The fish in
> their water will swim toward the front of their bowl with no more ef-
> fort than toward the back, and will go with equal ease toward bait
> placed anywhere around the edges of the bowl. Finally the butterflies

---

[1] Galileo Galilei, *Dialogue Concerning Two Chief World Systems—Ptolemaic and Copernican,*
pp. 186–187, trans. Stillman Drake, University of California Press, Berkeley, 1953; reprinted by permission of The Regents of the University of California.

DPI

Midair refueling. Although the velocity of each plane is very great in a reference frame attached to the earth, the velocity of one plane is approximately zero in a reference frame attached to the other plane.

and flies will continue their flights indifferently toward every side, nor will it ever happen that they are concentrated toward the stern, as if tired out from keeping up with the course of the ship, from which they have been separated during long intervals by keeping themselves in the air. And if smoke is made by burning some incense, it will be seen going up in the form of a little cloud, remaining still and moving no more toward one side than the other.

In the late nineteenth century it was thought that a careful measurement of the velocity of light relative to the earth would reveal absolute motion of the earth through space. Such an observation would contradict the principle of relativity, which was not yet accepted as a fundamental law of nature. Careful experiments by Michelson and Morley, however, gave a null result for the absolute velocity of the earth: they could not detect motion of the earth by measuring the velocity of light. The principle of relativity was again enunciated as a universal law by Poincaré and Einstein: absolute uniform motion cannot be detected by any experiment. The consequences of this principle, along with the assumption, based on experiment, that the speed of light is the same for all observers independent of their motion relative to the light source, were investigated by Einstein in 1905. His revolutionary results are known as the *special theory of relativity*. We shall study Einstein's theory in detail in Chapter 28.

A reference frame in which Newton's laws hold is called an inertial reference frame. All reference frames moving with constant velocity relative to an inertial frame are also inertial reference frames. Consider now the case in which S is an inertial reference frame and S' is *accelerating* relative to S. S' might be, for example, a coordinate frame at rest in a railroad car which is accelerating with acceleration $\mathbf{a}_0$ relative to the tracks. We find that Newton's laws do not hold in the railroad car. For example, a body on a smooth table does not remain at rest or move with constant velocity relative to the car; instead it has an acceleration $-\mathbf{a}_0$. Figure 4-11 shows a body suspended by a cord from the ceiling of the car. There is an unbalanced force to the right, but there is no acceleration (relative to the car). Since Newton's laws do not hold in S', this is not an inertial reference frame.

It is important to note that in order to classify a reference frame as inertial or not, we need to know all the possible forces acting on a

Accelerated reference frames

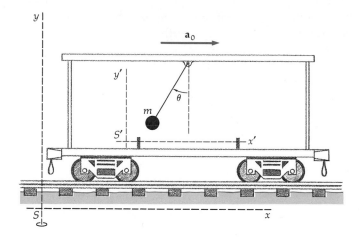

Figure 4-11
Body suspended by a cord
from the ceiling of an accelerated car. The horizontal component of the tension in the
cord is unbalanced, causing
the body to accelerate to the
right relative to the inertial
reference frame S attached to
the tracks. Reference frame S'
attached to the car is not an
inertial reference frame because Newton's laws do not
hold in this frame. Although
there is an unbalanced force
to the right, the body remains
at rest and does not accelerate
in frame S'.

body. We can identify the possible forces by finding the agent, i.e., the other body, responsible for each force. If we suspect the existence of a force because of an observed acceleration of the body but cannot find any agent responsible for it, we must conclude that the observed acceleration results because our reference frame is a noninertial frame.

Consider a body dropped in the accelerating railroad car. Its acceleration (relative to the car) is

$$a_y = -g \qquad a_x = -a_0$$

We describe the y component of acceleration as being due to the pull of gravity of the earth. Might not the x component be due to a similar cause, i.e., a large mass to the left of the car? To determine whether the car is in an inertial frame we must look outside for possible forces. Because there is no apparent agent (such as a large massive object to the left of the car) responsible for the observed acceleration to the left, we label the reference frame in the car as noninertial.

Two questions now arise. Can absolute acceleration be detected? Is there any natural inertial reference frame? Newton gave the following demonstration of the absoluteness of acceleration. Consider a pail of water suspended by a rope. The surface of the water is level. Now give the pail a twist. At first the pail rotates, but the water does not. Because of viscosity, the water begins to rotate, and the surface is no longer level but assumes a parabolic shape. Eventually the water and pail have the same angular velocity, and thus no relative velocity or acceleration. If the pail is now suddenly stopped, the water continues to rotate. The shape of the water surface does not depend on the relative motion of the water and pail but only on the absolute rotation of the water, or the rotation relative to the "fixed stars." Suppose, however, that we could rotate *all* the surroundings of the water including the fixed stars while leaving the water at "rest." What shape would the surface of the water take? If it remained flat, we would have to believe in absolute acceleration. If instead it assumed the parabolic shape as when it was rotating, we would conclude that only relative acceleration is observable.

We can state this question in terms of our accelerating box car. Suppose we leave the box car at rest and give all the stars in the universe an acceleration $-\mathbf{a}_0$ relative to the box car. Would a body dropped in the box car accelerate toward the earth, i.e., obey Newton's laws, or would it accelerate also to the left, as when we accelerated only the box car? These questions cannot be answered with certainty, for no such experiment or its equivalent has been performed. We may, with Newton, believe in absolute acceleration.

A reference frame attached to the surface of the earth is certainly useful in describing the motion of everyday objects, but it is not an inertial frame. Because of the earth's rotation, such a frame has an acceleration directed toward the axis of rotation. At the equator, this acceleration is

$$a = \frac{v^2}{R_E} = \frac{(2\pi R_E/T)^2}{R_E} = \frac{4\pi^2 R_E}{T^2}$$

$$\approx 3.4 \times 10^{-2} \text{ m/sec}^2 \qquad\qquad 4\text{-}25$$

using $T \approx 86{,}400$ sec for the period of rotation, 1 day (d), and $R_E = 6.4 \times 10^6$ m for the radius of the earth.[1] When this acceleration is very small compared to others, it can be neglected and the surface of the earth taken as approximately an inertial frame. Alternatively we can compensate for the earth's rotation by choosing a frame which rotates *relative to the earth* with a period of 1 d. Even this frame is accelerating, however, because of the revolution of the earth about the sun. Since $T = 1$ y $\approx 3.16 \times 10^7$ sec for the period and $r \approx 1.49 \times 10^{11}$ m for the radius of the orbit, this acceleration is

$$a_{\text{rev}} \approx 5.9 \times 10^{-3} \text{ m/sec}^2 \qquad\qquad 4\text{-}26$$

Suppose we choose a reference frame at rest relative to the sun and other fixed stars. To Newton, such a choice seemed a likely possibility for an inertial reference frame. Today we know more about the motion of the fixed stars. The acceleration of the sun corresponding to the rotation of the galaxy is about 10^{-10} m/sec^2, which is small enough to be neglected for most purposes. There is apparently no simple physical system such as the earth or the sun that is exactly at rest in an inertial reference frame. We can of course set up an inertial reference frame without dependence on this result. We need merely to take a first approximation, such as a rectangular coordinate system at rest relative to the sun, project masses along the axis, and measure any deviation from constant velocity. We can then correct for any observed acceleration by choosing a second coordinate system accelerating relative to the first. In practice, we need only find a frame in which $\Sigma \mathbf{F} = m\mathbf{a}$ holds within the accuracy of our measuring apparatus.

Questions

28. Suppose you are inside a closed railroad car (no sound or light from outsides reaches you). How can you decide whether the car is moving? How can you tell whether it is speeding up? Slowing down? Rounding a curve?

29. According to the principle of relativity you cannot detect your own constant velocity. Yet even a deaf and blind man riding at constant velocity along a freeway can tell that the car is not at rest. How?

30. Should it be possible to tell that the earth itself is moving by any experiment conducted in a closed room? Can you think of any possible experiments?

[1] The free-fall acceleration of a body *relative to the surface of the earth* is thus less than g by 3.4 cm/sec^2 at the equator.

Isaac Newton (1642–1727)

I. Bernard Cohen
Harvard University

When Isaac Newton was once asked how he had made his great discoveries, he replied, "By always thinking unto them." He is also reported to have said, "I keep the subject constantly before me and wait till the first dawnings open little by little into the full light." This ability to concentrate is a particular quality of Newton's genius, and it fits in well with his character and personality. For he was a solitary man, without close and intimate friends or confidants. He never married, and spent his early boyhood deprived of father (who died before young Isaac was born on Christmas day in 1642) and of mother (who remarried within 2 years and left Isaac to be reared by an aged grandmother).

A lonesome man, he developed "unusual powers of continuous concentrated introspection." In these words, his biographer Lord Keynes, the economist, epitomized Newton's "power of holding continuously in his mind a purely mental problem . . . for hours and days and weeks until it surrendered to him its secret." And then, in keeping with this character of inwardness, he was satisfied to keep his discoveries to himself, neither rushing into print, as his contemporary fellow scientists were wont to do, nor even communicating to his associates what he had accomplished. Accordingly, it has been said that every discovery of Newton's had two phases; Newton made the discovery, and then others had to find out that he had done so.

In 1684, Edmund Halley (the astronomer, after whom Halley's comet is named) went from London to Cambridge, where Newton was a professor, to ask him about a fundamental problem that had baffled the major scientists of the Royal Society (the world's oldest existing scientific organization), including Robert Hooke (of Hooke's law) and Christopher Wren (architect as well as scientist). What "curve would be described by the Planets," Halley asked Newton, "supposing the force of attraction towards the Sun to be reciprocal to the square of their distance from it?" Newton "replied immediately that it would be an Ellipsis," i.e., an ellipse. Halley, "struck with joy and amazement asked him how he knew it. Why, saith he, I have calculated it." In these final four words, Newton revealed that he had solved the major scientific problem of the century: to find the law of force that holds the solar system together, that causes the planets to move around the sun according to Kepler's laws, and that equally produces the same keplerian motion in satellites moving around planets.

Once Christiaan Huygens, in 1673, had published the law of "centrifugal force" (as he named it), one might easily compute that in uniform *circular motion* the central force is inversely proportional to the square of the distance from the center; and so, in fact, it was no great feat to *guess,* as Halley, Hooke, and Wren had done, that the same inverse-square law might apply to the *elliptical* orbits of planets and their satellites. However Newton had done far more. He had not only proved rigorously that an inverse-square force produces a keplerian elliptical orbit, he had also found the dynamical significance of Kepler's two other laws. He then went on to show that it is one and the same universal gravitational force that keeps the planets in their observed orbits around the sun and the satellites in their orbits around the planets, that (by the pull of the moon) produces tides in the oceans, and that causes bodies to fall to earth with the observed acceleration. And he computed the law of this force, directly proportional to the product of the masses of any gravitating bodies and inversely proportional to the distance between them. His actual reply to Halley ("I have calculated it") reveals the character of the man, since he had been satisfied merely to have made the calculations and had not rushed to make known the primary scientific discovery of the age: the law of universal gravitation.

The seeds of Newton's great achievements in science go back to a period of some 18 months after graduation from college (1665–1667), when fear of the plague had caused the university to be shut and Newton returned to the family farm in Woolsthorpe, Lincolnshire, where he had been born. Here he worked out his most fundamental contributions to mathematics, the methods of the differential and integral calculus (an innovation for which he must share the credit with the German philosopher and mathematician Leibniz, an independent codiscoverer). During this same time, he found the law of the inverse square and tested it by a rough calculation of the moon's motion. In this work he independently discovered the law of centrifugal force: that in uniform circular motion along a circle of radius r at speed v this force or acceleration must be proportional to v^2/r. He knew that for any system of bodies moving about a central body (such as the planets going around the sun or the set of moons around Jupiter) there holds a form of Kepler's third law: that in each such system r^3/T^2 is a constant k, where T is the period of revolution. Thus, considering the orbits to be circular (an approximation, but a rather good one at a first stage), the speed v is $2\pi r/T$, and so $v^2/r = 4\pi^2 r^2/T^2 r$. A clever man would see at once that multiplying by r/r gives

$$\frac{v^2}{r} = \frac{4\pi^2 r^3}{T^2 r^2} = \frac{4\pi^2}{r^2}\frac{r^3}{T^2} = \frac{4\pi^2 k}{r^2}$$

That is, v^2/r is equal to a constant multiplied by $1/r^2$; or the force is inversely proportional to the square of the distance. But is it? Newton's test was to note that the moon is about 60 times as far from the center of the earth as an apple. Accordingly, the "force" or acceleration of the moon, as it "falls" toward the earth and constantly "falls" away from straight-line inertial motion, should be just $1/60^2$, or $1/3600$ of the acceleration of free fall of an apple. The calculations, Newton said, agreed "pretty nearly." In his own words, "I thereby compared the force requisite to keep the Moon in her orb with the force of gravity at the surface of the Earth, and found them answer pretty nearly."

Later, Newton had to show that this inverse-square law also holds in elliptical orbits, and he had to take account of the modification required of Kepler's laws in an earth-moon system, because he proved that each of these two bodies (and any other pair) will move about their common center of mass. And he had to use his mathematical prowess to prove the fundamental theorem, that a uniform sphere (or a sphere made up of concentric uniform shells) will attract gravitationally as if all its mass were concentrated at its geometric center. Eventually, Newton cast out the misleading concept of centrifugal force, and introduced a new kind of force which he named *centripetal*.

A third set of discoveries made by Newton during his period of retreat during the plague years was in the field of optics. Newton is celebrated not only for his work in pure mathematics and for having founded the science of celestial mechanics; he is also known as an experimental scientist, primarily for his discoveries relating to light and color and that class of effects known today as *interference phenomena*. He bought a triangular glass prism, he tells us,

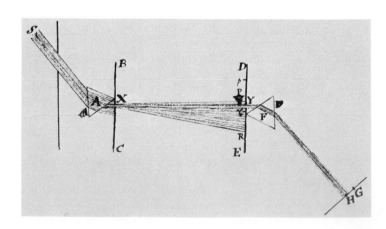

Manuscript drawing in Newton's papers, showing the "crucial experiment." Sunlight enters a darkened room at the left-hand side from S, producing a spectrum on passing through the first prism *A*. A hole at *Y* in the board *DE* permits light of a single color to pass through the second prism *F*, with the result of a deviated beam of light, without any further alteration in color. (*Courtesy of Cambridge University Library*.)

to try "the celebrated phenomenon of colours." But where others had played around with the colored spectrum produced by such a prism, Newton proceeded to analyze the phenomena of dispersion and the composition of white light and was led to devise and construct a new form of telescope, which in the succeeding three centuries has been astronomy's most powerful tool.

Newton's experiments on color showed that "white" light, or sunlight, is composite, a mixture of light of all the different colors. He demonstrated that the prism produces a spectrum, or separates out the different colors, because each color has its own index of refraction or degree of being turned or deviated from the original path. The violet light he found to be deviated the most, the red light the least. In order to prove that this is so, he devised what he called a "crucial experiment." He allowed a narrow beam of sunlight to enter a darkened room through a small hole in a window shutter. This beam then passed through a prism and produced a spectrum. Using an opaque board with a small hole in it, Newton could separate out of this spectrum a beam of light of a single color—red, orange, green, blue—and allow this monochromatic light to pass through a second prism. If prisms produce a spectrum by "altering" the incident light in some way, this effect should be apparent in the way the second prism would alter or affect an incident beam of monochromatic light. When monochromatic light entered the second prism, it emerged with no change of color; the second prism had produced only a further deviation, or change of path by refraction.

These investigations not only led Newton to an understanding of why objects appear to have the colors they display, in terms of the colors of the light they absorb or transmit or reflect, but also showed him that telescopes with simple or single lenses have definite limitations. Since a biconvex lens can be thought of as a pair of prisms base to base (Figure 1), we can see at once that there will be a different focus for each color, or a poor image due to *chromatic aberration.* Newton accordingly proceeded to invent a wholly new kind of telescope, a reflector rather than a refractor, which by using the front surface of a concave mirror to form the image avoided chromatic aberration.

A most notable set of optical investigations centered on the effects produced by thin plates, or films, as in the familiar colored patterns seen on films of oil. Newton studied the alternating fringed rings of darkness and light that are produced by the thin layer of air between a flat glass surface and the convex side of a plano-convex lens of large focal length, a phenomenon known today as *Newton's rings.* His measurements were so precise that they were used a hundred years later by Thomas Young to compute the wavelength of light.

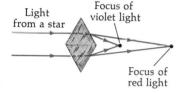

Figure 1
Two prisms, one on top of the other, act like a lens.

The alternating circles of brightness and darkness in monochromatic light, known as *Newton's rings. (From Newton's* Opticks, *1704.)*

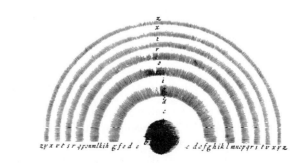

Newton's explanation of the phenomenon of Newton's rings, produced by the thin "film" of air trapped between a plane glass surface *AB* and the convex side of the plano-convex lens *CED.* This diagram shows Newton's conception of the alternation of transmission and reflection of the light from the curved surface. *(From Newton's* Opticks, *1704.)*

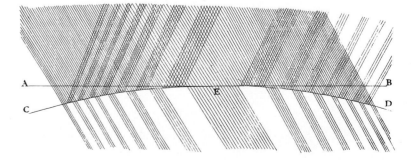

This is a remarkable set of discoveries for any lifetime, and we can only contemplate in amazement how in so short a time a young man barely out of college should have made such fundamental contributions to pure mathematics, to theoretical celestial mechanics and dynamics, and to experimental physics. As Newton himself later said, "All this was in the two plague years of 1665 and 1666, for in those days I was in the prime of my age for invention, and minded mathematics and philosophy[1] more than at any time since."

The remainder of Newton's scientific life was devoted to elaboration of the fundamental discoveries we have described. He became Lucasian Professor of Mathematics at Cambridge, and in 1672 his paper on light and color, together with a description of the newly invented telescope, was read to the Royal Society. After some aspects of his work had been misunderstood and criticized, he decided that he would publish no more. His mathematical innovations were distributed only in manuscript until later in his life when some were printed. Then, in 1684, after Halley's visit—and with Halley's urging—he wrote up his discoveries in dynamics and celestial mechanics in the *Philosophiae Naturalis Principia Mathematica* (London, 1687), which may be translated as *The Mathematical Principles of Celestial Mechanics*. Here he not only set forth and applied the law of universal gravitation but produced the three famous laws of motion that bear his name and that were the "axioms" on which his system of dynamics was constructed. In this work, too, he set forth for the first time a clear concept of mass and momentum and set up a series of beautiful experiments to show that mass (as measured by inertia) is proportional to weight.

In the 1690s, Newton became bored or disenchanted with the cloistered life of a university professor and moved to London, where he became director of the Mint, a post he held until his death in 1727. He became president of the Royal Society and ruled British science with a firm control. During the London years, he brought out two revised editions of his *Principia* (1713, 1726), published his book of optics (1704 and later editions) and various tracts on mathematics. He died in 1727 and was buried with state honors in Westminster Abbey.

From early manhood, Newton devoted only a small fraction of his creative intellectual life to orthodox scientific pursuits: pure and applied mathematics, astronomy and celestial mechanics, dynamics, experimental physics, and optics. During all these years he was an ardent student of theology, reading and taking innumerable notes, and writing tracts and even whole books on religious subjects. Some of these dealt with fundamental questions of interpretation of theological doctrines, others with the meaning of the prophetic books of the Bible, and yet others with the problem of unraveling Church history. He also developed a whole new system of world chronology, in part based on astronomy, that proved to be of little value. And his chief subject of study was alchemy—reading extensively, copying out whole sections of books, making experiments, all to what end we do not know.

Through his readings of mystical philosophers, theologians, and speculative alchemists, Newton may have gained a vision of a unified system of knowledge that would embrace both physical and divine science, revealing both the laws of the created world and the plan of its Creator. Various hints of this aspect of Newton's thought appear in his writings, e.g., "And thus to discourse about God from phenomena does belong to experimental philosophy." It is this vision of a knowledge which he never attained that may have caused him to devalue his monumental scientific achievements. For shortly before his death he said, "I do not know what I may appear to the world; but to myself I seem to have been only like a boy, playing on the sea-shore, and diverting myself in now and then finding a smoother pebble or a prettier shell than ordinary, whilst the great ocean of truth lay all undiscovered before me."

---

[1] The word "philosophy" as used at that time meant "natural philosophy," or science.

Review

A. Define, explain, or otherwise identify:

Force, 81
Mass, 85
Kilogram, 86
Inertia, 86
Momentum, 87
Weight, 88
Apparent weight, 89
Gram, 90

Dyne, 90
Newton, 90
Pound, 90
Slug, 91
Inertial reference frame, 95
Reference frame, 96
Principle of relativity, 97

B. True or false:

1. If there are no forces acting on a body, the body will not accelerate.

2. If a body is not accelerating, there must be no forces acting on it.

3. The motion of a body is always in the direction of the resultant force.

4. Action-reaction forces never act on the same body.

5. The mass of a body depends on its location.

6. The weight of a body depends on its location.

7. Action equals reaction only if the bodies are not accelerating.

8. Newton's laws hold only in inertial reference frames.

Exercises

Unless otherwise instructed, use the approximate value $g = 32$ ft/sec² = 9.8 m/sec² for the acceleration of gravity in Exercises and Problems.

Section 4-1, Force, and Section 4-2, Mass

1. An object experiences an acceleration of 4 m/sec² when a certain force F_0 acts on it. (*a*) What is its acceleration when the force is doubled? (*b*) A second object experiences an acceleration of 8 m/sec² under the influence of the force F_0. What is the ratio of the masses of the two objects? (*c*) If the two objects are tied together, what acceleration will the force F_0 produce?

2. An electric field gives a charged object an acceleration of 6×10^6 m/sec². Another field gives the same object an acceleration of 9×10^6 m/sec². (*a*) If the force exerted by the first field is F_0, what is the force exerted by the second field? What is the acceleration of the object (*b*) if the two fields act together on the same object in the same direction and (*c*) if the two fields act in opposite directions on the object?

3. Figure 4-12 shows the path taken by an automobile. It consists of straight lines and arcs of circles. The automobile starts from rest at point *A* and accelerates until it reaches point *B*. It then proceeds at constant speed until it reaches point *E*. From point *E* on it slows down, coming to rest at point *F*. What is the direction of the net force, if any, on the automobile at the midpoint of each section of the path?

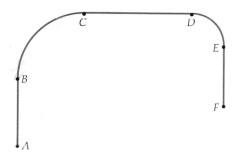

Figure 4-12
Path taken by automobile in Exercise 3.

4. The graph in Figure 4-13 shows a plot of v_x versus t for an object of mass 10 kg moving along a straight line. Make a plot of the net force on the object as a function of time.

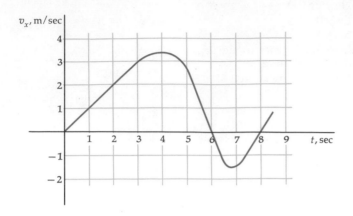

Figure 4-13
Graph of velocity v_x versus time t for Exercise 4.

5. Figure 4-14 shows the position versus time of a particle moving in one dimension. During what periods of time is there a net force acting on the particle? Give the direction (+ or −) of the net force for these times.

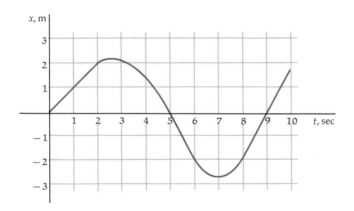

Figure 4-14
Graph of position x versus time t for Exercise 5.

6. In each part of Figure 4-15 illustrate the forces acting on the designated object: (a) a golf ball in midflight; (b) a box sliding down a rough-surfaced ramp at constant speed; (c) a test tube in a centrifuge on a table top, (d) a pendulum bob swinging in vacuum.

Figure 4-15
(a) Golf ball in flight; (b) box sliding down rough ramp; (c) test tube in centrifuge; and (d) pendulum in vacuum for Exercise 6.

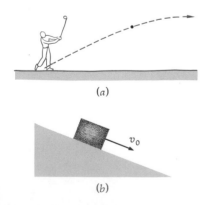

(a)

(b)

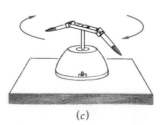

(c)

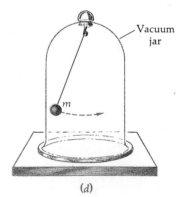

(d)

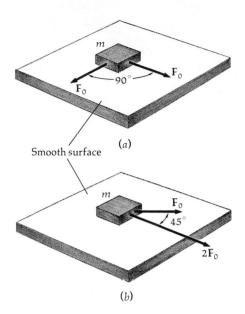

Figure 4-16
Forces acting on object for Exercise 7.

7. A force F_0 causes an acceleration of 5 m/sec² when acting on an object of mass m. Find the acceleration of the same mass when acted on by the forces shown in Figure 4-16a and b.

8. A certain force applied to mass m_1 gives it an acceleration of 20 ft/sec². The same force applied to m_2 gives it an acceleration of 30 ft/sec². If the two masses are tied together and the same force applied to the combination, find the acceleration.

9. A mass is pulled in a straight line along a level, frictionless surface with a constant force. The increase in its speed in a 10-sec interval is 5 mi/h. When a second constant force is applied in the same direction in addition to the first force, the speed increases by 15 mi/h in a 10-sec interval. How do the magnitudes of the two forces compare?

Section 4-3, Newton's Second Law

10. A 12-kg mass undergoes an acceleration of 4 m/sec². What is the magnitude of the resultant force acting on it?

11. A force of 15 N is applied to a mass m. The mass moves in a straight line with its speed increasing by 10 m/sec every 2 sec. Find the mass m.

12. A force of $F = 6i - 3j$ N acts on a mass of 2 kg. Find the acceleration a. What is the magnitude a?

13. A single force of 10 N acts on a mass m. The mass starts from rest and travels in a straight line a distance of 18 m in 6 sec. Find the mass.

14. A 10-kg block slides down a straight incline. When first observed, its speed is 3 m/sec. One second later its speed is 5 m/sec. After two seconds its speed is 7 m/sec. After three seconds it is 9 m/sec. What is the magnitude and direction of the resultant force on the block?

15. A 3-kg mass is acted on by a single force F_0 perpendicular to the velocity of the mass. The mass travels in a circle of radius 2 m. It makes one complete revolution in each 3 sec. (a) What is the magnitude of the acceleration? (b) What is the magnitude of F_0?

16. In order to drag a 100-kg log along the ground at constant velocity you have to pull on it with a force of 300 N (horizontally). (a) What is the resistive force exerted by the ground? (b) What force must you exert if you want to give the log an acceleration of 2 m/sec²?

17. Show that a newton-second is a unit of momentum.

18. A 3-kg mass moves with speed 4 m/sec in the x direction. (*a*) What is its momentum? (*b*) If a 3-N force also in the x direction acts on the mass, what will its momentum be after 2 sec? (*c*) If this force acts in the negative x direction, what will the momentum of the mass be after 2 sec? After 5 sec? (Assume that it is moving at 4 m/sec initially in the positive x direction.)

19. For each case the momentum of a certain particle is given in kilogram-meters/second as a function of time in seconds; find the force $\mathbf{F}$ as a function of time: (*a*) $\mathbf{p} = (6t + 7)\mathbf{i}$, (*b*) $\mathbf{p} = 2t\mathbf{i} + 3t\mathbf{j}$, (*c*) $\mathbf{p} = 3t^2\mathbf{i} + 3t\mathbf{j}$, (*d*) $\mathbf{p} = 6e^{-3t}\mathbf{i}$.

Section 4-4, Weight, and Section 4-5, Units of Force and Mass

20. Find the weight of a 50-kg girl in newtons and pounds.

21. Find the mass of a 175-lb man (*a*) in slugs, (*b*) in kilograms, and (*c*) in grams.

22. A newspaper reports that 3 tons of marijuana was found in a ship docked at Los Angeles. What is its mass in kilograms?

23. Find the weight of a 50-gm mass in dynes and newtons.

24. An 140-lb girl weighs herself by standing on a scale in an elevator. What does the scale read when (*a*) the elevator is descending at a constant rate of 10 ft/sec; (*b*) the elevator is accelerating downward at 4 ft/sec²; (*c*) the elevator is ascending at 10 ft/sec but the speed is decreasing by 4 ft/sec in each second?

25. The acceleration of gravity can be written $a = gR_E^2/(R_E + h)^2$, where R_E is the radius of the earth, h is the height above the earth, and $g = 32$ ft/sec². (*a*) Find the weight of an 160-lb man at a height of 500 mi above the earth. (*b*) What is the mass of the man at this altitude? (Take the radius of the earth to be 4000 mi.)

Section 4-6, Newton's Third Law and Conservation of Momentum

26.   A 2-kg mass hangs at rest from a string attached to the ceiling. (*a*) Draw a diagram showing the forces acting on the mass and indicate each reaction force. (*b*) Do the same for the forces acting on the string.

27. A hand pushes two masses on a smooth horizontal surface, as shown in Figure 4-17. One mass is 2 kg, and the other is 1 kg. The hand exerts a force of 5 N on the 2-kg mass. (*a*) What is the acceleration of the system? (*b*) What is the acceleration of the 1-kg mass? What force is exerted on it? What is the origin of this force? (*c*) Show all the forces acting on the 2-kg mass. What is the net force acting on this mass?

Figure 4-17
Exercise 27

5 N

2 kg

1 kg

28. A box slides down a rough inclined plane. Draw a diagram showing the forces acting on the box. For each force in your diagram, indicate the reaction force.

29. A 1-kg mass and a 2-kg mass are moving toward each other with equal speed. They collide head on and rebound, moving away from each other with unequal speed. Compare the change in (*a*) momentum and (*b*) velocity of the 1-kg mass with that of the 2-kg mass.

30. A 10-ton freight car rolls along a horizontal track at 5 ft/sec with negligible friction. It hits a second stationary car weighing 15 tons. They couple together and roll with speed v. (a) Find the initial momentum of the 10-ton car (1 ton = 2000 lb). (b) Find the speed v.

31. A 2000-lb car traveling north at 60 mi/h collides with a 5000-lb car traveling east at 40 mi/h. The cars stick together after the collision. Find the velocity (magnitude and direction) of the cars just after the collision.

Section 4-7, Reference Frames

32. Two coordinate frames S and S' are coincident at $t = 0$. The origin O' is moving in the x direction with speed 3 m/sec relative to O. A 2-kg mass is moving along the x axis with velocity v'_x relative to the origin O'. (a) What is the velocity relative to O? (b) A force of 5 N in the x direction acts on the mass. If it starts from rest in frame S' at $t = 0$, find its velocity v'_x at times $t = 1$ sec, $t = 2$ sec, and $t = 3$ sec. (c) Use your result from part (a) to find the velocity v_x in S at these times. (d) What is the acceleration of the mass in frame S'? In frame S?

33. Water in a river flows in the x direction at 4 m/sec. A boat crosses the river with speed 10 m/sec relative to the water. Set up a reference frame with its origin fixed relative to the water and another with its origin fixed relative to the shore. Write expressions for the velocity of the boat in each reference frame.

34. A spring parallel to the x axis has extension $\Delta x = x_2 - x_1$ measured in reference frame S. Show that the extension is the same when measured in frame S' if the coordinates x and x' are related by Equation 4-22.

Problems

1. The acceleration versus spring length observed when a 0.5-kg mass is pulled along a frictionless table by a single spring is

L, cm	4	5	6	7	8	9	10	11	12	13	14
a, m/sec²	0	2.0	3.8	5.6	7.4	9.2	11.2	12.8	14.01	14.6	14.6

(a) Make a plot of the force exerted by the spring versus length L. (b) If the spring is extended to 12.5 cm, what force does it exert? (c) By how much is the spring extended when the mass is at rest suspended from it near sea level, where $g = 9.8$ m/sec²?

2. Explain how you could measure the mass of an object if you were in a spaceship moving with constant velocity far from the earth or any other planet or large gravitational mass.

3. A 4000-lb car traveling at 60 mi/h can be brought to rest in 242 ft by a constant braking force. (a) How long does it take to stop? (b) What is the force necessary to stop the car in this distance? (c) If the same force were applied, how far would the car travel before stopping if it were initially traveling 30 mi/h? If it were traveling 80 mi/h?

4. A 2-ton load is being moved by a crane. (a) As the load is lifted off the ground, its upward acceleration is 4 ft/sec². What is the tension in the cable that supports it? (b) After a brief period of acceleration, the load is lifted at constant speed. What is the tension then? (c) When the load has nearly reached the desired height, it accelerates downward at 4 ft/sec². What is the tension in the cable?

5. A platform scale calibrated in newtons is placed on the bed of a truck driven at a constant speed of 14 m/sec. A box weighing 500 N is placed on the

scale. (*a*) When the truck passes over the crest of a hill with radius of curvature 100 m, what is the reading on the scale? (*b*) When the truck passes through the bottom of a dip with radius of curvature 80 m, what is the scale reading?

6. Two men pull with equal force on the ends of a rope. A weight *w* is suspended at the center of the rope, as shown in Figure 4-18. (*a*) For any given angle θ made by the rope with the horizontal, find the tension in the rope. (*b*) If the mass of each man is *m*, what force must be exerted by the ground on the man to keep him stationary? (*c*) Evaluate parts (*a*) and (*b*) for $\theta = 5°$, $w = 10$ lb, and $mg = 160$ lb. (*d*) Is it possible for the men to hold the weight suspended with the rope horizontal? Is it possible for them to pull in such a way that the rope is horizontal momentarily? How?

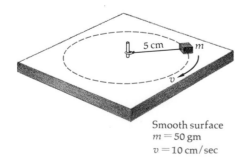

Figure 4-18
Two 160-lb men supporting a 10-lb weight on a rope (Problem 6).

7. A 100-kg mass is pulled along a frictionless surface by a force **F** so that its acceleration is 6 m/sec² (see Figure 4-19). A 20-kg mass slides along the top of the 100-kg mass and has an acceleration of 4 m/sec². (It thus slides back relative to the 100-kg mass.) (*a*) What is the frictional force exerted by the 100-kg mass on the 20-kg mass? (*b*) What is the net force on the 100-kg mass? What is the force **F**? (*c*) After the 20-kg mass falls off the 100-kg mass, what is the acceleration of the 100-kg mass?

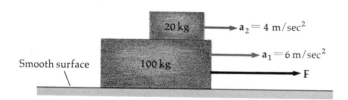

Figure 4-19
Problem 7.

8. A 50-gm mass which slides freely on a horizontal frictionless surface is attached by a light string 5 cm long to a pivot fixed on the surface (Figure 4-20). The mass is circling the pivot at a constant speed of 10 cm/sec. (*a*) What is the acceleration of the mass? (*b*) What is the force exerted by the string? (*c*) If the speed is doubled, what force must be exerted by the string? (*d*) If the mass moves at 10 cm/sec but the length of the string is doubled, what is the force? (*e*) If the string is 5 cm but the mass makes twice the number of revolutions per minute as originally, what is the force?

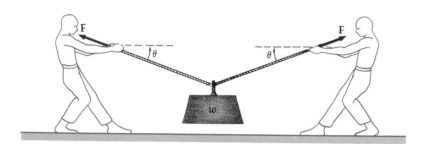

Figure 4-20
Mass moving in horizontal circle for Problem 8.

9. The velocity of a particle is 1 percent that of the velocity of light. (*a*) What is the velocity in meters per second? In miles per hour? (*b*) Expand $(1 - v^2/c^2)^{-1/2}$ using the binomial theorem. What percentage error is made if the relativistic correction to the momentum of the particle is ignored?

10. A small rocket with a mass of 10 kg is initially moving horizontally near the earth's surface with a speed of 10 m/sec. (*a*) What is the initial momentum of the rocket? (*b*) The rocket engine produces a force of 300 N on the rocket at an angle of 30° above the horizontal. (The force is in the plane determined by **g** and the initial momentum.) What is the resultant force on the rocket (use $g = 10$ m/sec^2)? (*c*) Find the speed and momentum of the rocket 10 sec after the engine is turned on (assume the mass of the rocket is constant).

11. (*a*) Show that a point on the surface of the earth at latitude θ has an acceleration relative to the center of the earth of magnitude 3.4 cos θ cm/sec^2. What is the direction of this acceleration? (*b*) Discuss the effect of this acceleration on the apparent weight of an object near the surface of the earth. (*c*) The free-fall acceleration of an object at sea level measured *relative to the earth's surface* has the value 978 cm/sec^2 at the equator and 981 cm/sec^2 at latitude $\theta = 45°$. What are the values of the acceleration of gravity g at these points?

12. A 100-kg mass is near the surface of the earth at the equator. It is at rest relative to the surface of the earth. What is its acceleration relative to the center of the earth when (*a*) it has zero acceleration relative to the surface of the earth and (*b*) it has acceleration 9.81 m/sec^2 downward relative to the surface of the earth? (*c*) What is its weight mg? (*d*) What is its apparent weight read on a scale on which it rests?

13. A simple accelerometer (a device for measuring acceleration) can be made by suspending a small mass from a string attached to a fixed point in the accelerating object, e.g., from the ceiling of a passenger car. When there is an acceleration, the mass will deflect and the string will make some angle with the vertical. (*a*) How is the direction in which the suspended mass deflects related to the direction of the acceleration? (*b*) Show that the acceleration *a* is related to the angle θ the string makes with the vertical by $a = g \tan \theta$. (*c*) Suppose the accelerometer is attached to the ceiling of an automobile which brakes to rest from 30 mi/h in a distance of 200 ft. What angle will the accelerator make? Will the mass swing forward or back?

14. An accelerometer of the type described in Problem 13 is suspended in a railroad car moving at a constant speed of 50 mi/h. The angle of the string with the vertical is 10°. The deflection is toward the right-hand side of the car when the observer faces the direction of motion. Which way is the car turning? What is the radius of the curve?

CHAPTER 5   Applications of Newton's Laws

Newton's laws are applied in two ways: (1) to determine the acceleration, velocity, and position of a particle as functions of time given all the forces acting on the particle and (2) to determine the forces acting on a particle given the acceleration, velocity, or position of the particle as a function of time. If all the forces acting on a particle are known, the acceleration is found from $\mathbf{a} = \Sigma\mathbf{F}/m$, and the velocity and position of the particle can be found (if their initial values are known) by integration. On the other hand, if the position or velocity of a particle is known as a function of time, the acceleration can be determined by differentiation and the *resultant* force from $\Sigma\mathbf{F} = m\mathbf{a}$. In this chapter we shall treat rather simple problems carefully in order to illustrate the general methods of solution. Careful study of these simple examples will make you aware of the content of newtonian mechanics and how it is applied. Practical problems are generally more complex than these examples, but the means of solving them are natural extensions of the methods illustrated here.

We begin in Section 5-1 by looking at examples in which the forces are constant and each object moves in a straight line. After illustrating a general method of attack which is useful in the solution of a wide variety of mechanics problems, we apply this method to problems involving circular motion in Section 5-2 and to Archimedes' principle in Section 5-3. Finally, we examine the motion of a particle under the influence of a velocity-dependent retarding force, the situation when a particle moves through a fluid such as air or water.

5-1 Linear Motion with Constant Forces

Example 5-1 A block resting on a smooth horizontal table is pulled horizontally by a light string. Find the acceleration of the block and the force exerted by the table.

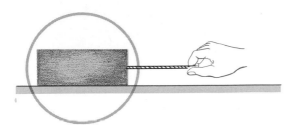

Figure 5-1
A block on a smooth surface with a horizontal force exerted upon it through a string. The first step in solving the problem is to isolate the system to be analyzed. In this case, the gray circle isolates the block from its surroundings.

To find the motion of the block, we need to find the resultant force acting on it. A first step in applying the newtonian laws is to choose the object whose acceleration is to be determined and upon which the forces to be considered act. Although this seems simple enough, many students' errors can be traced to failure to identify clearly what mass is being considered and what all the forces acting upon it are. In Figure 5-1 a circle is drawn around the block to help isolate it mentally from its surroundings. Forces acting on the block are due to agents external to the block. Our problem is to determine what those forces are. As a general rule, external forces may be exerted on a body at any point where the body touches something else. Such forces are called *contact forces*. For the block in this simple example, we would expect a contact force to be exerted on the block where it touches the surface supporting it. We also know that a force is exerted at the point where the string is attached to the block. We anticipate that forces will be exerted where the air touches the block. These are forces of air resistance, which for low speeds and heavy objects are usually negligible. We shall neglect them in this example.

Choose the object

Find what forces act on it

Contact forces

In addition to contact forces, there usually are *action-at-a-distance forces*, i.e., forces which act over a space between the mass whose motion is being analyzed and the body which exerts the force. The force of gravity, or weight, of the mass is the most familiar example. Electric and magnetic forces are also of this kind. To determine whether such forces act we must know whether there is any body nearby which can exert a significant gravitational force or whether there are electric charges nearby which can account for electric or magnetic forces.

Action-at-a-distance forces

Three significant external forces act on the block in this example. They are indicated in a *free-body diagram* in Figure 5-2:

Free-body diagram

w the weight, or force of gravity, pulling down on the block. This is an action-at-a-distance force.

N the contact force exerted by the table. In general, a surface like the table exerts a force which has both a tangential component, called the *force of friction*, and a perpendicular component, called the *normal* force

Friction
Normal force

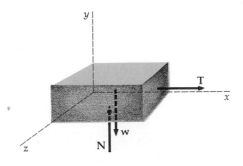

Figure 5-2
A *free-body diagram* for the block of Figure 5-1. The three significant forces acting on the block are its weight **w**, the normal force exerted by the table **N**, and the force exerted by the tension in the string **T**.

(the word normal means perpendicular). The phrase "smooth table" means that the frictional force is negligible compared with other forces in the problem. Thus the force exerted by a smooth surface is perpendicular, or normal, to the surface.

T the contact force exerted by the string, called the *tension* in the string.

Tension

A convenient coordinate system is also indicated in Figure 5-2. Note that **N** and **w** are drawn with equal magnitude. If this were not the case, there would be a net vertical force and either the block would crash through the table $(N < w)$ or the block would accelerate up off the table $(N > w)$. If the table is strong enough to hold the block and has no unusual properties that would make the block jump up, there will be no acceleration in the y direction; thus $\mathbf{N} + \mathbf{w} = 0$. Here we have used a known fact about the motion, namely $a_y = 0$, in combination with Newton's second law, $\Sigma F_y = ma_y = 0$, to infer something about the force exerted by the table. For the x component of $\Sigma \mathbf{F} = m\mathbf{a}$, we have, since **T** is the only horizontal force,

$$T = ma_x \quad \text{or} \quad a_x = \frac{T}{m} \qquad\qquad 5\text{-}1$$

Equation 5-1 thus gives a_x in terms of T and m assumed to be known. If T is *constant*, a_x is also constant and we can immediately write down the position x of the block by using the constant-acceleration equations from Chapter 2:

$$x(t) = x_0 + v_0 t + \frac{1}{2}\frac{T}{m} t^2 \qquad\qquad 5\text{-}2$$

where x_0 is the position at $t = 0$ and v_0 is the velocity at $t = 0$. The quantities x_0 and v_0 must be given in order to specify $x(t)$ completely, as discussed in Chapter 2.

Even in this simple example, both kinds of applications of Newton's laws were used: the horizontal acceleration, velocity, and position were found in terms of the given horizontal force **T**, and the vertical force exerted by the table **N** was found from the fact that the block remains on the table and thus $a_y = 0$. This second type of information, which limits the kind of motion possible for the block, is called a *constraint*.

Constraints

According to Newton's third law, forces always act in pairs. In Figure 5-2 we have only three forces. What are the action-reaction pairs in this example? Figure 5-3 shows three forces not shown in Figure 5-2, the gravitational force **w'** exerted *by the block on the earth*, the force **N'** exerted *by the block on the table*, and the force **T'** exerted *by the block on the string*. Note that none of these forces is exerted *on* the block. Therefore they have nothing to do with the motion of the block and must be omitted in the application of the second law to the motion of the block.

Example 5-1 illustrates a general method of attack for problems using Newton's laws. This method consists of the following steps.

1. Draw a neat diagram.

General method of attack

2. Isolate the body (particle) of interest and draw a free-body diagram

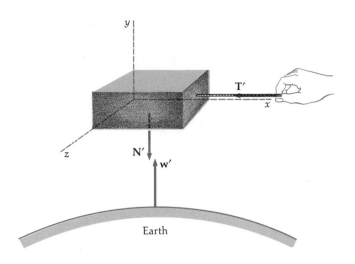

Figure 5-3
The reaction forces corresponding to the three forces
shown in Figure 5-2. Note
that these forces *do not* act on
the block; **T′** acts on the
man's hand, **N′** acts on the
table, and **w′** acts on the
earth.

indicating every external force acting on the body.[1] Do this for each
body if there is more than one in the problem, drawing a separate
diagram for each.

3. Choose a convenient coordinate system for each body and apply
Newton's law $\Sigma \mathbf{F} = m\mathbf{a}$ in component form.

4. Solve the resulting equations for the unknowns using whatever additional information is available, e.g., constraints. The unknowns generally will include the components of both the acceleration and some
of the forces.

5. Finally, inspect the results carefully, checking whether they correspond to reasonable expectations. Particularly valuable is to determine
what your solution predicts when variables in the solution are assigned extreme values. In this way you can check your work for errors.

Example 5-2 Find the acceleration of a block of mass m which moves
on a smooth, fixed surface inclined to the horizontal at an angle θ.

There are only two forces acting on the block, the weight **w** and the
force **N** exerted by the incline (see Figure 5-4). We neglect air resistance, and we are instructed that there is no friction at the contact with

---

[1] We assume that all conceivable forces that act on a body can be conveniently separated
into two classes: (1) a small number of dominant forces which can be enumerated and
(2) an unspecified number of other forces, e.g., air resistance in Example 5-1, the gravitational attraction of the moon or other bodies, etc., which are too small to affect the motion of the body and can reasonably be neglected. Only forces in the first class are indicated in the free-body diagram.

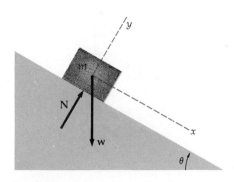

Figure 5-4
Forces acting on mass m on a
smooth incline. It is convenient to choose the x axis parallel to the incline.

the incline. Since the two forces are not in the same direction, they cannot add to zero and the block must therefore accelerate. As in Example 5-1, we have a constraint: the acceleration is along the incline. It is convenient for this problem to choose a coordinate frame with one axis parallel to the incline and the other perpendicular, as shown in Figure 5-4. Then the acceleration has only one component, a_x. For this choice, $\mathbf{N}$ is in the y direction, and the weight $\mathbf{w}$ has the components

$$w_x = w \sin \theta = mg \sin \theta$$
$$w_y = -w \cos \theta = -mg \cos \theta$$

5-3

where m is the mass and g is the acceleration of gravity (Figure 5-5). The resultant force in the y direction is $N - mg \cos \theta$. From Newton's second law and the fact that $a_y = 0$

$$\Sigma F_y = ma_y = N - mg \cos \theta = 0$$

and thus

$$N = mg \cos \theta$$

5-4

Similarly, for the x components

$$\Sigma F_x = ma_x = mg \sin \theta$$
$$a_x = g \sin \theta$$

5-5

The acceleration down the incline is constant and equal to $g \sin \theta$. It is useful to check our results at the extreme values of inclination, $\theta = 0$ and $\theta = 90°$. At $\theta = 0$, the surface is horizontal. The weight has only a y' component, which is balanced by the normal force $N =$

When friction is neglected, the body slides down the incline with acceleration $g \sin \theta$. (*Courtesy of the Museum of Modern Art/Film Archives, New York.*)

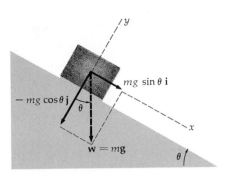

Figure 5-5
The weight of the block in Example 5-2 can be replaced by its components, $mg \cos \theta$ in the $-y$ direction and $mg \sin \theta$ in the $+x$ direction. Note that the normal force $\mathbf{N}$ exerted on the block by the incline has been omitted. Since there is no acceleration in the y direction, $\mathbf{N} = mg \cos \theta \, \mathbf{j}$.

$mg \cos 0° = mg$. The acceleration is of course zero; $a_x = g \sin 0° = 0$. At the opposite extreme, $\theta = 90°$, the incline is vertical. There the weight has only an x component along the incline, and the normal force is zero; $N = mg \cos 90° = 0$. The acceleration is $a_x = g \sin 90° = g$ since the block is in free fall.

Example 5-3 A 2-lb picture is supported by two wires of tension $\mathbf{T}_1$ and $\mathbf{T}_2$, as shown in Figure 5-6a. Find the tension in the wires.

This is a problem in static equilibrium. Since the picture does not accelerate, the net force acting on it must be zero. The three forces acting on the picture, its weight $m\mathbf{g}$, the tension $\mathbf{T}_1$, and the tension $\mathbf{T}_2$, must therefore sum to zero. Since the weight has only a vertical component mg downward, the horizontal components of the tensions $\mathbf{T}_1$ and $\mathbf{T}_2$ must be equal in magnitude and the vertical components of the tensions must balance the weight.

$$\Sigma F_x = T_1 \cos 30° - T_2 \cos 60° = 0$$

$$\Sigma F_y = T_1 \sin 30° + T_2 \sin 60° - mg = 0$$

Using $\cos 30° = \sqrt{3}/2 = \sin 60°$ and $\sin 30° = \frac{1}{2} = \cos 60°$ and solving for the tensions, we obtain

$$T_1 = \tfrac{1}{2}mg = 1 \text{ lb}$$

$$T_2 = \sqrt{3} \, T_1 = \frac{\sqrt{3}}{2} mg = 1.73 \text{ lb}$$

Example 5-4 Two blocks on a smooth table are connected by a light string (Figure 5-7), and one is drawn along by a second horizontal string. Find the acceleration of each block.

We assume that the given quantities in this problem are the masses m_1 and m_2 and the tension $\mathbf{T}_3$. We choose the x axis to be along the table, as in Figure 5-7.

Figure 5-6
(a) Picture supported by two wires in Example 5-3. (b) Choice of coordinate system and resolution of forces into x and y components. Since the picture is not accelerating,
$$F_x = 0 = T_{1x} - T_{2x}$$
$$F_y = 0 = T_{2y} + T_{1y} - w$$

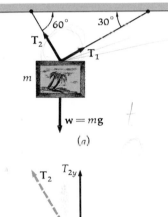

$\mathbf{w} = m\mathbf{g}$

(a)

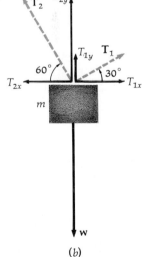

(b)

Figure 5-7
The two connected blocks of Example 5-4. The gray circles remind us to isolate each block separately.

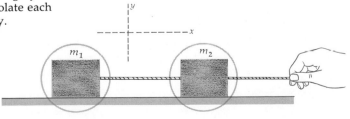

The free-body diagrams for these blocks are shown in Figure 5-8. There are three forces acting on m_1, its weight $\mathbf{w}_1 = m_1\mathbf{g}$, the normal force $\mathbf{N}_1$ exerted by the table, and the force $\mathbf{T}_1$ exerted by the connecting string. The x component of Newton's second law for this body is

$$T_1 = m_1 a_{x1} \qquad\qquad 5\text{-}6$$

The block of mass m_2 experiences four forces: its weight $\mathbf{w}_2 = m_2\mathbf{g}$, the normal supporting force $\mathbf{N}_2$ exerted by the table, the force $\mathbf{T}_2$ exerted by the connecting string, and the force $\mathbf{T}_3$ exerted by the second string. The x component of Newton's second law for this body is

$$T_3 - T_2 = m_2 a_{x2} \qquad\qquad 5\text{-}7$$

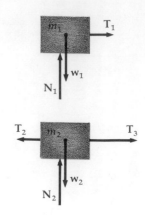

Figure 5-8
The free-body diagrams for the two blocks of Example 5-4. $\mathbf{T}_1$ and $\mathbf{T}_2$ are an action-reaction pair and therefore equal in magnitude.

For each of the bodies we have neglected frictional forces and forces of air resistance. As in our earlier examples, we have the constraint imposed by the table that there is no vertical acceleration (the blocks neither leap up from the table nor sink into it). Thus the weight of each block is supported fully by the corresponding normal force of the table; $N_1 = w_1 = m_1 g$ and $N_2 = w_2 = m_2 g$. The connecting string provides an additional constraint. If the string does not break or become slack, the distance between the two blocks remains constant. Thus the accelerations a_{x1} and a_{x2} must be equal. If we write a for this common acceleration, Equations 5-6 and 5-7 become

$$T_1 = m_1 a \qquad\qquad 5\text{-}8$$

and

$$T_3 - T_2 = m_2 a \qquad\qquad 5\text{-}9$$

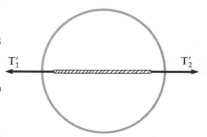

Figure 5-9
Forces acting on the connecting string in Figure 5-7. According to Newton's second law, $T_2' - T_1' = m_s a$, where m_s is the mass of the string. If the string is very light, T_2' and T_1' are approximately equal in magnitude.

We still have three unknowns, a, T_1, and T_2, and only two equations. (We are considering T_3 to be given.) We can relate the force T_1 and T_2 by applying Newton's laws to the connecting string. The free-body diagram for the connecting string (Figure 5-9) neglects the weight of the string. The forces $\mathbf{T}_1'$ and $\mathbf{T}_2'$ are the contact forces exerted *on the string* by the blocks at either end. According to Newton's third law, $\mathbf{T}_1'$ is equal and opposite to $\mathbf{T}_1$, and $\mathbf{T}_2'$ is equal and opposite to $\mathbf{T}_2$. (If the weight of the string is not negligible, the string will sag and the tensions will have vertical components. The fact that there is little sag in the string in practice indicates that we can neglect its weight.) Newton's second law applied to the connecting string is

$$T_2' - T_1' = m_s a \qquad\qquad 5\text{-}10a$$

where m_s is the mass of the string and a is the acceleration, which is the same as that of the blocks. Since the magnitudes T_1' and T_1 are equal, as are T_2 and T_2', we can write

$$T_2 - T_1 = m_s a \qquad\qquad 5\text{-}10b$$

Equation 5-10b gives us a third relation between the three unknowns T_1, T_2, and a. In most cases, the mass of the string is very small and can be neglected. Then the tensions T_1 and T_2 are equal; i.e., the tension in the connecting string is the same throughout the string. We can then simplify our notation by dropping the subscripts 1 and 2 from the forces $\mathbf{T}_1$ and $\mathbf{T}_2$. Equations 5-8 and 5-9 become

$$T = m_1 a \qquad\qquad 5\text{-}11$$

$$T_3 - T = m_2 a \qquad\qquad 5\text{-}12$$

These two equations are easily solved for the two unknowns T and a. (For example, simply adding them eliminates T.) The results are

$$a = \frac{T_3}{m_1 + m_2} \qquad\qquad 5\text{-}13$$

and

$$T = \frac{m_1}{m_1 + m_2} T_3 \qquad\qquad 5\text{-}14$$

The acceleration is the same as if there were only one body of mass $m = m_1 + m_2$ because the acceleration of each mass is the same in magnitude and direction. That is, they could be considered to be rigidly connected or one combined mass (Figure 5-10).

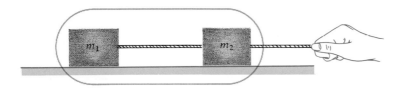

Figure 5-10
The two connected blocks of Example 5-4 treated as a single body. The forces exerted by the connecting string ($\mathbf{T}_1$ and $\mathbf{T}_2$ in Figure 5-8) are now forces internal to the system and have no effect on the acceleration of the system.

The results obtained in Equations 5-13 and 5-14 can be checked as follows. One extreme case is $m_1 = 0$; that is, this block is absent. Then our result for a should be the same as in Example 5-1 when a single block was being pulled by a horizontal string. In fact, putting $m_1 = 0$ in Equation 5-13 gives $a = T_3/m_2$, the expected result. In this same extreme case we would expect the tension in the connecting string to be zero (since it has nothing to pull). Putting $m_1 = 0$ in Equation 5-14 gives $T = 0$, as expected. The reader should check for himself that Equations 5-13 and 5-14 give the expected results in other extreme cases, such as $m_2 = 0$, $m_1 = \infty$, and $m_2 = \infty$.

The analysis in Example 5-3 of the forces associated with the connecting string is complete and rigorously correct. It is a lengthy argument, however, for a very common situation. The result is just that *a light string connecting two points has a tension which has constant magnitude throughout and which acts in the direction of the string at any point.* This approximation is valid if the mass of the string is negligible, so that there is no need for a difference between the forces at the ends to account for the acceleration of the string. It also assumes, of course, that no forces are applied to the string at intermediate points. In the next example we shall show that this statement is also true when a string passes over a smooth peg. The general conditions for the validity of this approximation are that (1) the mass of the string is negligible and (2) there are no forces on the string between the two points with component tangential to the string.

You use this approximation without going through the elaborate analysis as long as you recognize that it is only an approximation valid under the above conditions. In some practical applications these conditions do not hold because the string is massive or there is a tangential frictional force. In these situations, a more complete analysis must be used.

Example 5-5 A block hangs by a string which passes over a smooth peg and is connected to another block on a smooth table. Find the acceleration of each block and the tension in the string.

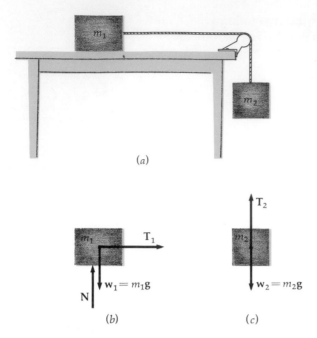

(a)

(b)　　　　　(c)

Figure 5-11
(a) The two blocks of Example
5-5. The free-body diagrams
(b) for m_1 and (c) for m_2.

Figure 5-11 shows the important elements of this problem. For block 1 on the table, the vertical forces, N and w_1, have equal magnitudes because of the constraint that the vertical acceleration is zero for m_1. Newton's second law applied to the horizontal components gives

$$T_1 = m_1 a_1 \qquad\qquad 5\text{-}15$$

where a_1 is the acceleration of m_1 along the horizontal surface. If we take the downward direction to be positive for the acceleration a_2 of block 2, the equation of motion for m_2 is

$$m_2 g - T_2 = m_2 a_2 \qquad\qquad 5\text{-}16$$

We can simplify these equations by noting that if the connecting string does not stretch, the accelerations a_1 and a_2 are both positive and equal in magnitude (but not in direction). Let us call this magnitude a. We shall now show that the tensions T_1 and T_2 are equal in magnitude.

Figure 5-12 shows a small segment of the string subtending an angle $\Delta\theta$ in contact with the peg. The tangent to the segment at the midpoint is shown as a dashed line. Since the peg is smooth (frictionless), the force it exerts on the segment is perpendicular to this tangent at the midpoint and bisects $\Delta\theta$. The other forces on the segment are the tensions T_A and T_B at the ends of the segment. If we assume the string to be massless, the three forces on the segment must add vectorially to zero. From the symmetry of the figure we see that this implies that the tensions T_A and T_B are equal in magnitude. We can extend this argument to other segments of the string. If there is no tangential force exerted on a string between two points and the mass of the string is negligible, the tension has the same magnitude at the two points.

Again we simplify our notation by dropping the subscripts and calling the magnitude of the tension T. Equations 5-15 and 5-16 then become

$$T = m_1 a \qquad\qquad 5\text{-}17$$

$$m_2 g - T = m_2 a \qquad\qquad 5\text{-}18$$

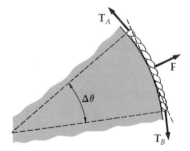

Figure 5-12
Forces on a small segment of string in contact with a smooth peg. The force F exerted by the peg is perpendicular to the peg at the midpoint and bisects the angle $\Delta\theta$ subtended by the segment. The forces T_A and T_B are due to the tension in the string. If the string is massless, these three forces must balance. Then the magnitudes T_A and T_B are equal.

Solution of these equations gives

$$a = \frac{m_2}{m_1 + m_2}\, g \qquad\qquad\qquad 5\text{-}19$$

$$T = \frac{m_1 m_2}{m_1 + m_2}\, g \qquad\qquad\qquad 5\text{-}20$$

Note that although the result for a appears to be the same as that for a mass $m = m_1 + m_2$ acted on by a force $m_2 g$, this is not the case because the magnitude of the acceleration of each mass is the same but the direction is not.

Example 5-6 Apparent Weight in an Accelerating Elevator A block rests on a scale in an elevator, as in Figure 5-13. What is the scale reading when the elevator is accelerating (*a*) up and (*b*) down?

(*a*) Since the block is at rest relative to the elevator, it is also accelerating up. The forces on the block are **N**, exerted by the scale platform on which it rests, and **w**, the force of gravity. If we call the upward acceleration **a**, Newton's second law gives

$$\mathbf{N} + \mathbf{w} = m\mathbf{a}$$

or, in terms of the upward components,

$$N - w = ma$$

$$N = w + ma = mg + ma \qquad\qquad 5\text{-}21$$

The force **N**′ exerted by the block on the scale determines the reading on the scale, the apparent weight. Since **N**′ and **N** are an action-reaction pair, they are equal in magnitude. Thus when the elevator accelerates up, the apparent weight of the block is greater than its true weight by the amount ma.

(*b*) Let us call the acceleration **a**′. Again Newton's second law gives

$$\mathbf{N} + \mathbf{w} = m\mathbf{a}'$$

but in this case the acceleration is downward. In terms of the magnitudes, we have

$$w - N = ma'$$

or

$$N = w - ma' = mg - ma' \qquad\qquad 5\text{-}22$$

Again, the scale reading, or apparent weight, equals N. In this case, the apparent weight is less than mg. If $a' = g$, as it would if the elevator were in free fall, the block would be apparently weightless. What if the acceleration of the elevator is greater than g? Assuming that the surface of the scale is not sticky, the scale cannot exert a force down on the block. (We assume that the scale is fastened to the floor of the elevator.) Since the downward force on the block cannot be greater than w, the scale will soon leave the block. The block will have the acceleration g, which is less than that of the elevator; so relative to the elevator, the block will accelerate up until it hits the ceiling. (As described in an inertial reference frame outside the elevator, the ceiling accelerates down faster than the block until it hits the block.) Then if the ceiling is strong enough, it can provide the force downward necessary to give the block the acceleration a'.

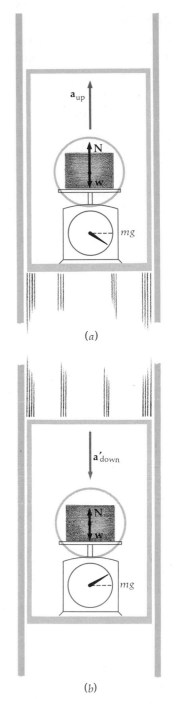

(*a*)

(*b*)

Figure 5-13
The block on a scale in an accelerating elevator for Example 5-6. The scale indicates the *apparent weight*, i.e., the upward force exerted on the block by the scale. The apparent weight is greater than mg when the acceleration is upward and is less than mg when the acceleration is downward.

Questions

1. A picture is supported by two wires as in Example 5-3. Do you expect the tension to be greater or less in the wire which is more nearly vertical?

2. A weight is hung on a wire which is originally horizontal. Can the wire remain horizontal? Explain.

3. Give an example in which the tension in a string does not have the same magnitude throughout.

4. If the two blocks in Example 5-4 were connected by a very light rod, would the analysis change?

5. What effect does the velocity of the elevator have on the apparent weight of the block in Example 5-6?

5-2   Circular Motion

If a particle moves with speed v in a circle of radius r, it has an acceleration of magnitude v^2/r toward the center of the circle. This centripetal acceleration is related to the change in the direction of the velocity of the particle, as discussed in Chapter 3. As with any acceleration, there must be a resultant force in the direction of the acceleration to produce it. If the speed of the particle is changing, there is also a component of the acceleration tangential to the circle of magnitude dv/dt. In this case, the resultant force has both radial and tangential components.

Example 5-7 A particle of mass m is suspended from a string of length L and travels at constant speed in a horizontal circle of radius r. The string makes an angle θ given by $\sin \theta = r/L$. Find the tension in the string and the speed of the particle.

This is known as a *conical pendulum* (Figure 5-14). The two forces acting on the particle are its weight $\mathbf{w} = m\mathbf{g}$, acting down, and the tension $\mathbf{T}$, which acts along the string. In this problem we know that the

This racetrack is banked so that the normal force exerted by the track on the car has a horizontal component toward the center of the circle to provide the centripetal force.

Ken Regan/Camera 5

acceleration is horizontal, toward the center of the circle, and of magnitude v^2/r. Thus the vertical component of the tension must balance the weight. The horizontal component of the tension is the resultant force. The vertical and horizontal components of $\Sigma \mathbf{F} = m\mathbf{a}$ therefore give

$$T \cos \theta - mg = 0 \qquad\qquad 5\text{-}23$$

$$T \sin \theta = ma = m \frac{v^2}{r} \qquad\qquad 5\text{-}24$$

The tension is found directly from Equation 5-23 since θ is given. We can find the speed v in terms of the known quantities r and θ by dividing one equation by the other to eliminate T:

$$\tan \theta = \frac{v^2}{rg} \qquad \text{or} \qquad v = \sqrt{rg \tan \theta}$$

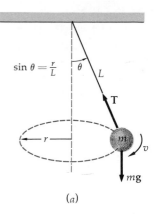

$\sin \theta = \frac{r}{L}$

(a)

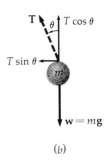

$T \cos \theta$

$T \sin \theta$

$w = mg$

(b)

Figure 5-14
(a) The conical pendulum of Example 5-7 and (b) its free-body diagram.

Example 5-8 A ball on a string moves in a vertical circle with speed v_t at the top and speed v_b at the bottom. What is the tension in the string (a) at the top and (b) at the bottom of the circle?

(a) The forces on the ball at the top of the circle are shown in Figure 5-15a. At this point both the contact force $\mathbf{T}$ exerted by the string and the force of gravity $\mathbf{w} = m\mathbf{g}$ point down. The acceleration is also down (toward the center of the circle) with magnitude v_t^2/r. Newton's second law gives

$$T_t + mg = m \frac{v_t^2}{r}$$

or

$$T_t = m \frac{v_t^2}{r} - mg \qquad\qquad 5\text{-}25$$

(b) At the bottom of the circle, the tension points up and the force of gravity down. The acceleration is still toward the center of the circle (which at this point is up) with magnitude v_b^2/r. Newton's second law gives

$$T_b - mg = m \frac{v_b^2}{r}$$

or

$$T_b = m \frac{v_b^2}{r} + mg \qquad\qquad 5\text{-}26$$

Figure 5-15
The forces on a ball moving in a vertical circle, as in Example 5-8. (The string is not shown.) (a) At the top of the circle, both the weight and the tension in the string (**w** and **T**) act downward and toward the center of the circle. (b) At the bottom of the circle, **w** and **T** act in opposite directions.

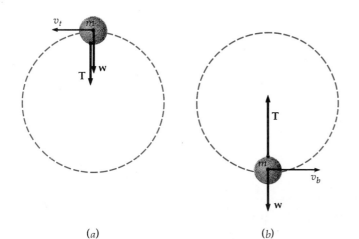

(a)   (b)

At the bottom of the circle, the tension exceeds the weight mg by the amount mv_b^2/r. If the ball were simply suspended at rest from the string, the tension in the string would have to just equal the weight. But if the ball is moving, it has a net upward acceleration. Thus, according to the second law, there must be a net upward force acting on it. In other words, the upward tension must exceed the downward weight in magnitude.

At the top of the circle, the tension in the string must be positive (otherwise the string would fall slack). For this to be true, the first term in Equation 5-25 must be greater than the second. That is,

$$\frac{mv_t^2}{r} > mg \qquad \text{or} \qquad v_t > \sqrt{rg}$$

If the speed of the ball becomes less than this value at any point in its swing, the string will become slack and the ball will begin to move in a parabola, since it then will be in free fall.

Example 5-9 A block on a string moves in a vertical circle. What is the tension when the string makes an angle θ with the vertical, as shown in Figure 5-16, and the speed is v?

From the diagram (Figure 5-16) we see that the tension and weight are not collinear. It is convenient to use the radial and tangential directions for the application of $\Sigma \mathbf{F} = m\mathbf{a}$. The only force in the tangential

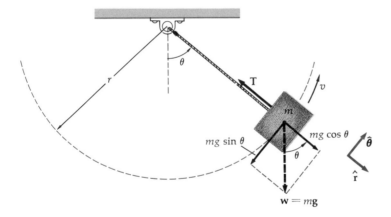

Figure 5-16
The forces on a block moving in a vertical circle (Example 5-9). The weight $\mathbf{w} = m\mathbf{g}$ is resolved into radial and tangential components. The tangential component of the weight is the only tangential force acting, and so the block has a tangential acceleration related to a change in its speed. The net radial force in the inward direction, $T - mg \cos \theta$, provides the centripetal acceleration v^2/r.

direction is the component of the weight, of magnitude $mg \sin \theta$. If we choose the direction of increasing θ for the positive tangential direction, this component is negative. The tangential component of Newton's second law thus gives

$$-mg \sin \theta = ma_t \qquad\qquad 5\text{-}27$$

or

$$a_t = -g \sin \theta \qquad\qquad 5\text{-}28$$

The tangential acceleration varies with the angle θ. The rate of change of the speed is thus

$$\frac{dv}{dt} = a_t = -g \sin \theta$$

This rate of change of speed is the same as that for a mass sliding down a smooth incline tangent to the circle at this point. Taking the outward direction for the positive radial direction, we have for the radial components of $\Sigma \mathbf{F} = m\mathbf{a}$,

$$mg \cos \theta - T = ma_r = \frac{-mv^2}{r}$$

or

$$T = \frac{mv^2}{r} + mg \cos \theta \qquad\qquad 5\text{-}29$$

This result includes both results of Example 5-8. At the bottom of the circle, $\theta = 0$, $\cos \theta = 1$, and $T = mv_b{}^2/r + mg$. At the top of the circle, $\theta = 180°$, $\cos \theta = -1$, and $T = mv_t{}^2/r - mg$.

Questions

6. Explain why the following statement is incorrect: In the circular motion of a ball on a string, the ball is in equilibrium because the outward force of the ball due to its motion is balanced by the tension in the string.

7. Could we simplify Example 5-8 by assuming that the ball on the string moves in a vertical circle with constant speed? Explain.

8. How would Example 5-9 be changed if the ball were attached to a very light rod instead of a flexible string?

5-3   Buoyant Forces; Archimedes' Principle

If a heavy object submerged in water is "weighed" by suspending it from a scale (Figure 5-17a), the scale reads less than when the object is weighed in air. Evidently the water exerts an upward force partially supporting the weight of the object. This force is even more evident when we submerge a piece of cork: it accelerates up toward the surface, where it floats partially submerged. The submerged cork experiences an upward force from the water greater than its weight.

   The force exerted by a fluid on a body submerged in it, called the *buoyant force*, depends on the density of the fluid and the volume of the body but not on the composition or shape of the body. It is equal in magnitude to the weight of the fluid displaced by the body. This result is known as *Archimedes' principle*:

A body wholly or partially submerged in a fluid is buoyed up by a force equal to the weight of the displaced fluid.

Archimedes' principle is a direct consequence of Newton's laws applied to a fluid. In deriving this result we shall first assume that the fluid is not accelerating relative to an inertial reference frame. We shall then investigate an accelerating fluid and show that the principle still holds if we interpret weight as meaning apparent weight. For example, if the fluid is in free fall, the apparent weight of the displaced fluid is zero and the buoyant force is zero.

   Figure 5-17b shows the vertical forces acting on an object being weighed while submerged, i.e., the weight **w** down, which is the attraction of the earth; the force of the spring balance $\mathbf{F}_s$ acting up; a

Figure 5-17
(a) Weighing a mass m submerged in a fluid. (b) The corresponding free-body diagram. $F_2 > F_1$ because of the greater pressure on the bottom surface of the object. The difference $F_2 - F_1$ is the buoyant force on the object.

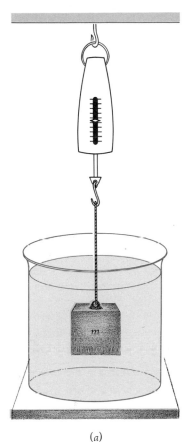

(a)

(b)

force $\mathbf{F}_1$ acting down because of the fluid pressing on the top surface of the object, and a force $\mathbf{F}_2$ acting up because of the fluid pressing on the bottom surface of the object. Since the spring balance reads a force less than the weight, the force $\mathbf{F}_2$ must be greater in magnitude than the force $\mathbf{F}_1$. The difference in magnitude of these two forces is the buoyant force $B = F_2 - F_1$. The buoyant force occurs because the pressure of the fluid at the bottom of the object is greater than that at the top.

Pressure is defined as follows. Consider a small surface of area A submersed in a fluid. Let F be the force normal to the surface exerted by the fluid on one side of the surface on the fluid on the other side. The pressure P is then

$$P = \frac{F}{A} \qquad\qquad 5\text{-}30 \qquad \textit{Pressure of a fluid}$$

The pressure so defined is independent of the size of A (assuming A to be small) and is independent of the orientation of the surface.

The force exerted by a fluid on a small area element of a submerged object is perpendicular to the area element and has the magnitude of the pressure times the area of the element. The vector sum of these forces over the total area of the object is the resultant force exerted by the fluid on the object which is the buoyant force. It is independent of the composition of the object.

In Figure 5-18 we have replaced the submerged object by an equal volume of fluid (indicated by the dotted lines) and eliminated the spring. As we have said, the buoyant force $B = F_2 - F_1$ acting on this volume of fluid is the same as that acting on our original object. Since this volume of fluid is in equilibrium, the resulting force acting on it must be zero. The buoyant force equals the weight of the fluid in this volume:

$$B = w_f \qquad\qquad 5\text{-}31$$

We have derived Archimedes' principle.

Let ρ_f be the mass per unit volume, called the *density*, of the fluid. A volume V of fluid then has mass $\rho_f V$ and weight $\rho_f g V$. If we submerge an object of volume V, the buoyant force on the object is $\rho_f g V$. The weight of the object can be written $\rho g V$, where ρ is the density of the object. If the density of the object is greater than that of the fluid, the weight will be greater than the buoyant force and the object will sink if not supported. If ρ is less than ρ_f, the buoyant force will be greater than the weight and the object will accelerate up to the top of the fluid unless held down. It will float in equilibrium with a fraction of its volume submerged so that the weight of the displaced fluid equals the weight of the object. Densities of various materials are listed in Appendix B, Table 4.

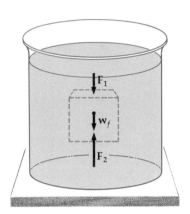

Figure 5-18
The situation of Figure 5-17 with the volume of the mass m replaced by fluid. The forces F_1 and F_2 are identical to the corresponding forces on the mass in Figure 5-17b.

Density defined

Example 5-10 A 1-lb block of aluminum (density $\rho = 2.7$ gm/cm³) is suspended by a spring scale and submerged in water ($\rho_w = 1.00$ gm/cm³), as in Figure 5-17a. What is the reading on the spring scale? If the block is weighed in air of density $\rho_a = 10^{-3}$ gm/cm³, what is the reading of the scale?

The forces acting on the block are its weight down, the force $\mathbf{F}_s$ up exerted by the spring and the buoyant force $\mathbf{B}$, also up. Since the block is not accelerating, Newton's second law gives

$$\mathbf{F}_s + \mathbf{B} + \mathbf{w} = 0 \qquad \text{or} \qquad F_s + B - w = 0$$

We can write the weight and buoyant force in terms of the volume of the block V: $B = \rho_w g V$ and $w = \rho g V$. Then the spring force is

$$F_s = w - B = \rho g V - \rho_w g V = \rho g V \left(1 - \frac{\rho_w}{\rho}\right) = w \left(1 - \frac{\rho_w}{\rho}\right) \qquad 5\text{-}32$$

But

$$\frac{\rho_w}{\rho} = \frac{1 \text{ gm/cm}^3}{2.7 \text{ gm/cm}^3} = 0.37$$

and so

$$F_s = (1 - 0.37)w = (0.63)(1 \text{ lb}) = 0.63 \text{ lb}$$

If we replace the density of water with that of air in Equation 5-32, we obtain the force exerted by the spring scale when the block is weighed in air. For this case,

$$\frac{\rho_a}{\rho} = \frac{10^{-3}}{2.7} = 3.7 \times 10^{-4}$$

and

$$1 - \frac{\rho_a}{\rho} = 1 - 0.00037 \approx 1$$

showing that we can usually neglect the buoyant force of air.

Ira Rosenberg/Detroit Free Press

The buoyant force of air is not always negligible. Here it equals the weight of the balloon plus passengers, so that the balloon floats in the air. When the air in the balloon is heated more, it expands and the buoyant force is greater, causing the balloon to rise.

Example 5-11 A cork has a density of $\rho = 0.2$ gm/cm^3. (a) Find what fraction of the volume of the cork is submerged when the cork floats in water. (b) Find the net force on the cork if it is completely submerged and then released.

(a) Let V be the volume of the cork and V' be the volume that is submerged when it floats on the water. The weight of the cork is $\rho g V$, and the buoyant force is $\rho_w g V'$. Since the cork is in equilibrium, the buoyant force equals the weight. Thus

$$B = w \qquad \rho_w g V' = \rho g V$$

$$\frac{V'}{V} = \frac{\rho}{\rho_w} = \frac{0.2}{1} = \frac{1}{5} \qquad 5\text{-}33$$

Thus one-fifth of the cork is submerged.

(b) When the cork is held completely submerged, the buoyant force is $\rho_w g V$. This is 5 times the weight of the cork since the density of water is 5 times that of the cork. The net force up is then

$$F = B - w = 5w - w = 4w$$

The net force upward is 4 times the weight of the cork. The upward acceleration of the cork is significantly less than $4g$ because as the cork moves upward an amount of water (which depends on the shape of the cork) must be accelerated to make room for the cork. The detailed calculation of the acceleration of the cork is a complicated problem in hydrodynamics.

We can use Archimedes' principle to find the pressure of a fluid as a function of depth. Consider a slab of fluid of area A and thickness Δh, where h is the depth measured from the top of the water (see Figure 5-19). If P_2 is the pressure at the bottom of the slab and P_1 that at the top, the buoyant force on the slab is $(P_2 - P_1)A = \Delta P\, A$, where ΔP

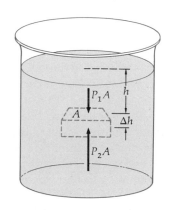

Figure 5-19
A slab of fluid of area A and thickness Δh. The difference between the forces on the upper and lower surfaces of the slab, $P_2 A - P_1 A$, must equal the weight of the slab.

is the pressure change in the depth change Δh. Setting this equal to the weight of the volume of fluid $A \, \Delta h$, we obtain

$$\Delta P \, A = mg = \rho A \, \Delta h \, g$$

$$\frac{\Delta P}{\Delta h} = g\rho$$

If we let Δh approach zero, the ratio $\Delta P / \Delta h$ becomes the derivative dP/dh

$$\frac{dP}{dh} = \rho g \qquad\qquad 5\text{-}34$$

Equation 5-34 holds for any fluid. To solve this equation we must know how the density of the fluid depends on the pressure. For water, the density is approximately independent of the pressure; i.e., water is nearly incompressible. Then according to Equation 5-34, $dP/dh =$ constant and the pressure varies linearly with h:

$$P = P_0 + \rho g h \qquad\qquad 5\text{-}35$$

where P_0 is the pressure at $h = 0$. We can also use this type of analysis to find an expression for the variation of atmospheric air pressure with height y above the ground. Since it is convenient to measure y upward from the ground rather than h downward, Equation 5-34 is replaced by

$$\frac{dP}{dy} = -\rho g \qquad\qquad 5\text{-}34a$$

However, the density of air is not independent of pressure. Air is compressible; the greater the pressure the greater the density. It is left as a problem to show that if the density is proportional to the pressure, the pressure decreases exponentially with height above the ground (see Problem 20).

Let us now consider an accelerating fluid. In Figure 5-20a a container of water is falling freely with the acceleration of gravity g. The resultant force on any part of the water must equal mg because the water is accelerating with $a = g$. Since this force equals the weight, there can be no buoyant force in this case. This result can be demonstrated by placing a cork at the bottom of a long tube of water and throwing the tube into the air. (Remember that whether the tube is moving up or down, its acceleration is g downward; i.e., it is a projectile.) The cork does not float to the top of the tube but remains where it was when the tube was thrown.

If we submerge a cork by suspending it from the bottom of a container of water with a string and accelerate the container in a horizontal direction (Figure 5-20b), the somewhat surprising result is that the string inclines forward toward the direction of acceleration. Let us first consider an element of water of volume V and mass $m_w = \rho_w V$. Since the water is accelerating to the right, the pressure on the left part of the water must be greater than that on the right, giving a net force to the right on the element equal to $m_w a$, where a is the acceleration. If we now replace this element of water with a cork, the cork will experience the same force to the right, but since the mass of the cork is less than m_w, the acceleration of the cork will be greater than that of the water and the container. The cork will thus accelerate to the right relative to the container. If the cork is attached by a string, as in Figure

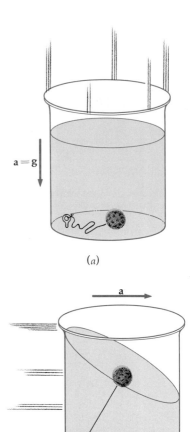

$a = g$

(a)

a

(b)

Figure 5-20
A cork is tied to the bottom of a container of water. (a) When the container is in free fall, $\mathbf{a} = \mathbf{g}$ and there is no buoyant force on the cork. (b) When the container accelerates to the right, the cork is deflected in the direction of the acceleration with the string perpendicular to the water surface.

Buoyant force on a submerged ball in an accelerated frame. The pendulum and water container are on a rotating platform and are therefore accelerating toward the center of rotation, which is toward the left in this photograph. Note that both strings are perpendicular to the water surface. (*Courtesy of Larry Langrill, Oakland University.*)

5-20*b*, the string provides a force to the left, so that the acceleration of the cork and the container are the same. How does this buoyant force to the right arise? If we look at the surface of the water, we see that it is not horizontal. The water piles up at the left, so that at a given distance from the bottom the water is deeper at the left of the cork than at the right. Note that the string holding the cork is perpendicular to the surface of the water.

This interesting behavior can be demonstrated by placing the container of water at the edge of a turntable and securing it so that it cannot fall off. When the table rotates, the acceleration of the container is toward the center and the string holding the cork inclines in that direction.

Questions

9. How might you estimate your average density at a swimming pool?

10. Smoke which usually rises from a smokestack is observed to sink on a very humid day. What can be concluded about the relative densities of humid and dry air?

11. Why is it easier to float in seawater than in fresh water?

12. A certain object has mass density nearly equal to that of water so that it floats almost completely submerged. However, it is much more compressible than water. What happens if the floating object is given a slight push to submerge it?

13. Fish can adjust their volume by varying the amount of oxygen and nitrogen gas (obtained from the blood) in a thin-walled sac under their spinal column called a swim bladder. Explain how this helps them swim.

5-4   Motion with a Retarding Force Proportional to the Speed

When an object moves through a fluid, in addition to the buoyant force there is a retarding force which depends on the velocity of the object. In general the greater the velocity the greater the retarding force. In some cases this retarding force is approximately proportional to the velocity

$$\mathbf{F}_r = -b\mathbf{v} \qquad\qquad 5\text{-}36$$

The proportionality constant b generally depends on the shape of the object and on the fluid. The negative sign in Equation 5-36 indicates that the retarding force is always in the direction opposite the velocity.

Example 5-12 A particle is dropped from rest and falls under the influence of its own weight and a retarding force given by Equation 5-36 (Figure 5-21). Describe its motion.

For simplicity we shall neglect the buoyant force (we can always include it by replacing the weight by w-B, as given by Equation 5-32). Since the particle falls down due to its weight, the retarding force is up. When the downward direction is taken to be positive, Newton's second law $\Sigma \mathbf{F} = m\mathbf{a}$ gives

$$mg - bv = ma = m\frac{dv}{dt}$$

or

$$\frac{dv}{dt} = g - \frac{b}{m}v \qquad\qquad 5\text{-}37$$

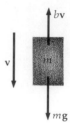

Figure 5-21
The free-body diagram for a particle falling through a resistive fluid. In this case, the buoyant force is considered to be negligible. This corresponds to the situation of a comparatively heavy mass falling through air.

When the retarding force equals his weight, the sky diver is in equilibrium and falls with constant velocity. The terminal velocity (without parachute) is approximately 120 mi/h.

Andy Keech

Equation 5-37 is a differential equation relating the speed v and its time rate of change dv/dt. It can be solved by standard methods of calculus, but since these methods may not be familiar to you yet, we shall consider the solution of this equation qualitatively. We can sketch the resultant v-versus-t curve from Equation 5-37 knowing only that dv/dt is the slope of this curve. We are given that the initial speed is zero. Thus initially there is no retarding force, and the acceleration is g. As the speed increases, the term bv increases and the acceleration decreases. Thus as time goes on, the speed increases (dv/dt is positive) but the rate of increase dv/dt gets smaller. Eventually, the retarding force bv equals the weight mg, and the acceleration is zero. From this time onward, the particle moves with constant velocity called its *terminal velocity*. An expression relating the terminal velocity to the constant b is obtained by setting $dv/dt = 0$ in Equation 5-37. The result is

$$mg - bv_t = 0$$

$$v_t = \frac{mg}{b} \qquad\qquad 5\text{-}38$$

A sketch of v versus t is shown in Figure 5-22. The function satisfying Equation 5-37 with $v = 0$ at $t = 0$ is

$$v = \frac{mg}{b}\,(1 - e^{-bt/m}) = v_t\,(1 - e^{-t/t_c}) \qquad\qquad 5\text{-}39$$

The time $t_c = m/b$ is the time for the particle to reach 63 percent of its terminal velocity.

The larger the constant b the smaller the terminal velocity and the sooner an appreciable fraction of the terminal velocity is reached. We have mentioned that the constant b depends on the shape of the object. For example, a parachute is designed to make b large so that the terminal velocity is small. On the other hand, cars are designed with b small to reduce the effect of wind resistance.

Example 5-13 A particle with initial velocity v_0 moves under the influence of a single force given by Equation 5-36. Describe its motion.
Newton's second law for this example is

$$-bv = m\,\frac{dv}{dt}$$

$$\frac{dv}{dt} = -\frac{b}{m}\,v \qquad\qquad 5\text{-}40$$

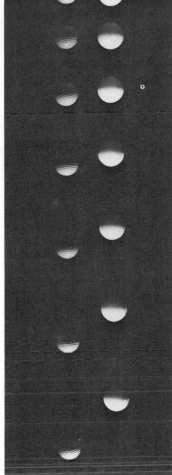

A golf ball and a styrofoam ball falling in air. The air resistance is negligible for the heavier golf ball, which falls with essentially constant acceleration. The styrofoam ball reaches terminal velocity quickly, as indicated by the nearly equal spacing at the bottom.

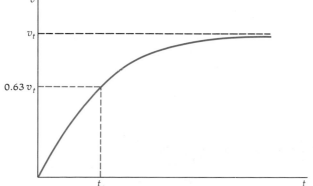

Figure 5-22
A plot of velocity versus time for an object falling through a resistive fluid; v_t is the maximum, or terminal, velocity attained by the object; t_c, usually called the time constant, is the time required for the object to attain 63 percent of its terminal velocity.

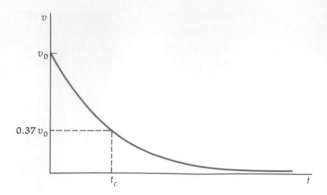

Figure 5-23
A plot of velocity versus time
for the particle of Example
5-13. Here, the particle has an
initial velocity v_0 and is subject to a single force given by
$\mathbf{F} = -b\mathbf{v}$. The velocity decreases exponentially with
time, reaching $e^{-1}v_0 = 0.37v_0$
after time t_c.

In this case the acceleration is always negative. The initial speed is v_0.
As time goes on, the speed decreases. The slope of the v-versus-t curve
is always negative. As v approaches zero, its slope dv/dt also approaches zero. A sketch of v versus t is shown in Figure 5-23. The
mathematical function describing this curve is

$$v = v_0 e^{-bt/m} \qquad\qquad\qquad 5\text{-}41$$

We can check that Equation 5-41 is a solution to Equation 5-40 by
direct differentiation:

$$\frac{dv}{dt} = \frac{d}{dt}\left(v_0 e^{-bt/m}\right) = v_0 \left(\frac{-b}{m}\right) e^{-bt/m} = -\frac{b}{m} v_0 e^{-bt/m} = -\frac{b}{m} v$$

Questions

14. How does a sky diver vary the value of b?

15. How would you expect the value of b for air resistance to depend
on the density of the air?

Review

A. Define, explain, or otherwise identify:

Contact force, 113	Constraint, 114
Action-at-a-distance force, 113	Buoyant force, 125
Free-body diagram, 113	Archimedes' principle, 125
Normal force, 113	Pressure, 126
Smooth surface, 114	Density, 126
Tension, 114	Terminal velocity, 131

B. Write out the steps involved in the general method of attack for solving
problems using Newton's laws of motion.

C. True or false:

1. Action-at-a-distance forces can be neglected when determining the acceleration of an object.

2. At any given time, the tension in a string must have the same magnitude at
any two points.

3. When the bob of a pendulum swings through its lowest point, the tension
in the string has the magnitude mg.

4. The terminal velocity of an object depends on its shape.

5. When an object floats in a fluid, the fraction submerged depends only on the
ratio of the densities of the object and fluid.

Exercises

Unless instructed otherwise, use the approximate value g = 32 ft/sec² = 9.8 m/sec²
for the acceleration of gravity in Exercises and Problems.

Section 5-1, Linear Motion with Constant Forces

1. A 10-kg object is subjected to the two forces $\mathbf{F}_1$ and $\mathbf{F}_2$, as shown in Figure 5-24. (*a*) Find the acceleration **a** of the object. (*b*) A third force $\mathbf{F}_3$ is applied so that the object is in static equilibrium. Find $\mathbf{F}_3$.

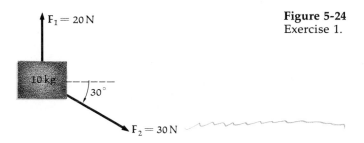

Figure 5-24
Exercise 1.

2. A 5-kg object is pulled along a frictionless horizontal surface by a horizontal force of 10 N. (*a*) If the object is at rest at $t = 0$, how fast is it moving after 3 sec? (*b*) How far does it travel from $t = 0$ to $t = 3$ sec? (*c*) Indicate on a diagram all the forces acting on the object.

3. A 4-kg object is subjected to two forces, $\mathbf{F}_1 = 2\mathbf{i} - 3\mathbf{j}$ N and $\mathbf{F}_2 = 4\mathbf{i} + 11\mathbf{j}$ N. The object is at rest at the origin at time $t = 0$. (*a*) What is the object's acceleration? (*b*) What is its velocity at time $t = 3$ sec? (*c*) Where is the object at time $t = 3$ sec?

4. A 3-lb object is suspended 4 ft from the ceiling by two 6-ft wires, as shown in Figure 5-25. Assuming the wires to have negligible mass, find the tension in the wires.

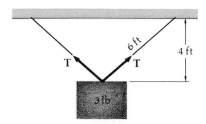

Figure 5-25
Exercise 4.

5. A 2-lb picture is hung by two wires of equal length which make an angle θ with the horizontal, as shown in Figure 5-26. (*a*) If $\theta = 30°$, find the tension in the wires. (*b*) Find the tension for general θ and weight w of the picture. For what angle θ is T the least? The greatest?

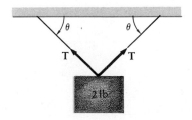

Figure 5-26
Exercise 5.

6. In Figure 5-27 the masses are attached to spring balances calibrated in newtons. Give the reading of the balances in each case. (The strings are massless, and the incline is smooth.)

Figure 5-27
Exercise 6.

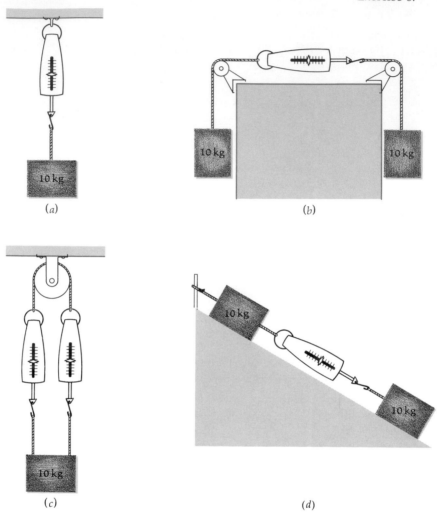

7. One of the first steps in the solution of a problem is drawing a correct diagram. In Figure 5-28a a mass is suspended from a horizontal cable. In Figure 5-28b a mass is suspended from a guy wire. (a) Correct the diagrams shown in this figure. (b) *Prove* that the diagrams are incorrect.

Figure 5-28
Exercise 7.

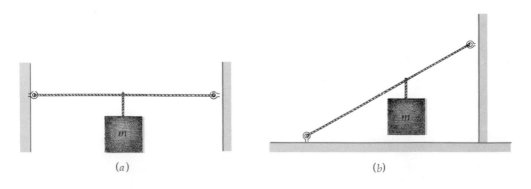

8. A mass is held in position by a cable along a smooth incline (Figure 5-29). (*a*) If $\theta = 60°$ and $m = 50$ kg, find the tension in the cable and the normal force exerted by the incline. (*b*) Find the tension as a function of θ and m and check your result for $\theta = 0$ and $\theta = 90°$.

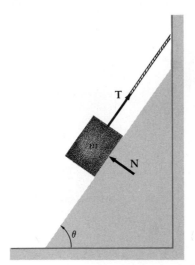

Figure 5-29
Mass held on smooth incline for Exercise 8.

9. A vertical force T is exerted on a 5-kg mass near the surface of the earth as shown in Figure 5-30. Find the acceleration of the mass if (*a*) $T = 5$ N; (*b*) $T = 10$ N; (*c*) $T = 100$ N.

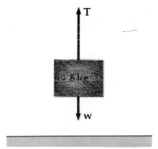

Figure 5-30
Exercise 9.

10. Two 5-kg masses are connected by a light string as shown in Figure 5-31. The table is smooth, and the string passes over a smooth peg. Find the acceleration of the masses and the tension in the string.

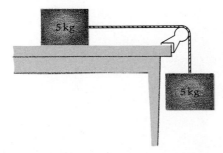

Figure 5-31
Exercise 10.

11. Two masses are connected by a light string as shown in Figure 5-32. The incline and peg are smooth. Find the acceleration of the masses and the tension in the string for $\theta = 30°$ and $m_1 = m_2 = 5$ kg.

Figure 5-32
Exercise 11.

12. Two masses are connected by a light string as shown in Figure 5-33. The string passes over a smooth peg. Find the acceleration of each mass and the tension in the string for $m_1 = 200$ gm and $m_2 = 100$ gm.

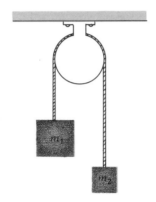

Figure 5-33
Exercise 12 and Problem 12.
Atwood's machine.

13. Two masses m_1 and m_2 rest on a smooth horizontal table. A force **F** is applied to m_1 (Figure 5-34). For $m_1 = 2$ kg, $m_2 = 4$ kg, and $F = 3$ N, find (*a*) the acceleration of the masses, (*b*) the *resultant* force on m_1, (*c*) the resultant force on m_2, and (*d*) the magnitude of the contact force exerted by one mass on the other.

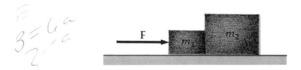

Figure 5-34
Exercise 13 and Problem 9.

14. A mass m_1 rests on a mass m_2, which rests on a smooth table. A force **F** is exerted on m_2, as shown in Figure 5-35. Consider $m_1 = 2$ kg, $m_2 = 4$ kg, and $F = 3$ N. (*a*) If the surface between m_1 and m_2 is perfectly smooth, find the acceleration of each mass. (*b*) If the surface between m_1 and m_2 is rough enough to ensure that m_2 does not slide on m_1, find the acceleration of the two masses. (*c*) In case (*b*) what are the resultant forces acting on each mass? What is the magnitude and direction of the contact force exerted by mass m_1 on m_2?

Figure 5-35
Exercise 14 and Problem 10.
The table is assumed to be smooth, but the surface between m_1 and m_2 is rough.

15. A 2-kg mass hangs from a spring balance (calibrated in newtons) attached to the ceiling of an elevator (Figure 5-36). What does the balance read (a) when the elevator is moving up with constant velocity of 30 m/sec, (b) when the elevator is moving down with constant velocity of 30 m/sec, (c) when the elevator is accelerating upward at 10 m/sec²? (d) From $t = 0$ to $t = 2$ sec the elevator moves upward at 10 m/sec. Its velocity is then reduced uniformly to zero in the next 2 sec so that it is at rest at $t = 4$ sec. Describe the reading on the balance during the time $t = 0$ to $t = 4$ sec.

16. A man holding a 10-kg mass on a cord made to withstand 150 N steps into an elevator. When the elevator starts up, the line breaks. What was the minimum acceleration of the elevator?

17. A 2-kg mass rests on a smooth wedge which has an inclination of 60° (Figure 5-37), and an acceleration a to the right such that the mass remains stationary relative to the wedge. Find a. What would happen if the wedge were given a greater acceleration?

Figure 5-37
A 2-kg mass resting on a smooth wedge which is accelerating so that the mass does not move relative to the wedge (Exercise 17).

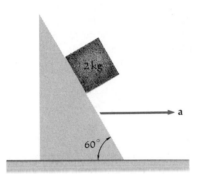

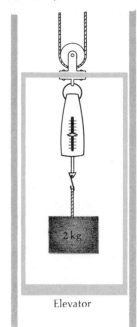

Figure 5-36
A 2-kg mass hanging from a spring balance in an elevator (Exercise 15).

Elevator

18. In a tug of war, two boys pull on a rope, each trying to pull the other over a line midway between their original positions. According to Newton's third law, the forces exerted by each boy on the other (transmitted by the rope) are equal and opposite. Draw a diagram to show how it is possible for one boy to win despite Newton's third law.

Section 5-2, Circular Motion

19. A 3-lb stone attached to a string is whirled in a horizontal circle of radius 2 ft (as in Figure 5-14). The string makes an angle $\theta = 30°$ with the vertical. (a) Find the magnitude and direction of the resultant force on the stone. (b) What is the tension in the string? (c) What is the speed of the stone?

20. A model airplane of mass 500 gm flies in a circle of radius 6 m attached to a horizontal string. The plane makes 1 rev every 4 sec. What is the tension in the string? (Assume that the only inward force on the plane is the tension in the string.)

21. A 12-in-diameter record rotates at 33.3 rev/min on a turntable. A penny of mass 3 gm rests on the outer edge of the record. What is the frictional force on the penny if it does not slide? By what factor must this force increase if the rotation of the record is increased to 45 rev/min?

22. A 2000-lb car rounds an unbanked curve which has a radius of curvature of 100 ft. The maximum frictional force the road can exert on the car is 800 lb. At what maximum speed (in miles per hour) can the car travel around the curve without skidding?

23. A curve of radius 100 ft is banked as shown in Figure 5-38 so that a car can round the curve at 30 mi/h even if the road is smooth and there is no friction. Show in a force diagram that a component of the normal force exerted by the road on the car can provide the centripetal force necessary if the angle θ is great enough. Find the banking angle θ for these conditions.

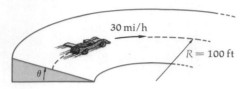

Figure 5-38
Car on banked road (Exercise 23).

24. A pail of water is swung in a vertical circle of radius 3 ft. What is the minimum speed at the top of the circle for the water to stay in the pail?

25. A pilot comes out of a vertical dive in a circular arc such that his upward acceleration is 9g. (*a*) If he weighs 175 lb, what is the magnitude of the force exerted by the airplane seat on him at the bottom of the arc? (*b*) If the speed of the plane is 200 mi/h, what is the radius of the circular arc?

26. A man stands on a scale at the equator, which rotates once a day in a circle of radius about 4000 mi. (*a*) By what fraction is the scale reading less than his true weight? (*b*) If the rotation of the earth were increased such that the scale reading were one-half his true weight, what would be the length of the day (in terms of our present hours)? (Assume that the earth maintains its present shape.)

Section 5-3, Buoyant Forces; Archimedes' Principle

27. A 1-kg piece of copper is submerged in water and suspended from a spring balance (Figure 5-39). The density of copper is 6 times that of water. (*a*) What force does the balance read? (*b*) The whole system is dropped out of the window. Describe the initial motion of the copper relative to the jar.

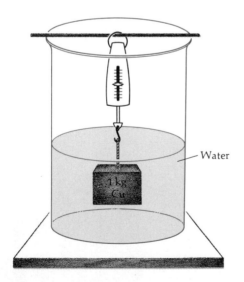

Figure 5-39
A 1-kg mass submerged in water and suspended by a spring balance (Exercise 27).

28. It is sometimes said that only the tip of an iceberg shows and that 90 percent lies beneath the surface. (*a*) Find the density of ice if this is true. (*b*) What fraction of an ice cube is beneath the surface in a glass of water on the moon, where the acceleration of gravity is about one-sixth that on earth?

29. A piece of copper (density about 6 gm/cm³) floats in mercury (density 13.6 gm/cm³). What fraction of the copper is submerged?

30. The pressure on the surface of a lake is atmospheric pressure $P_a = 1.01 \times 10^5$ N/m². (a) At what depth is the pressure twice atmospheric pressure? (b) If the pressure at the top of a deep pool of mercury is P_a, at what depth is the pressure $2P_a$?

31. A balloon filled with helium floats in air in a car. A string tied to the balloon is held down by a weight resting on the car seat. The car makes a sharp right turn, throwing the passengers to the left (relative to the car). Which way does the balloon move relative to the car? Why?

32. A 10 by 10 ft raft is 4 in thick and made of wood with density 0.6 gm/cm³. How many 150-lb people can stand on the raft without getting their feet wet when the water is calm (1 ft³ of water weighs 62.4 lb)?

Section 5-4, Motion with a Retarding Force Proportional to the Speed

33. If the retarding force is written $F_r = -bv$, what are the dimensions and mks units of the constant b?

34. In Example 5-12 show that the critical time t_c is just the time it would take to reach the terminal velocity v_t if the acceleration were constant and equal to g. Interpret this result on the graph of v versus t (see Figure 5-22).

35. A sky diver can slow himself to a constant speed of 60 mi/h by adjusting his form. If the diver weighs 150 lb, what is the retarding force, and what is the value of the constant b?

36. List some examples of an object moving with terminal velocity.

Problems

1. Your car is stuck in a mudhole. You are alone, but you have a long, strong rope. Having studied physics, you tie the rope taut to a tree and pull on it sideways, as in Figure 5-40. Find the force exerted by the rope on the car when the angle θ is 3° and you are pulling with a force of 100 lb but the car does not move. How strong must the rope be if it takes a force of 150 lb at an angle of $\theta = 3°$ to move the car?

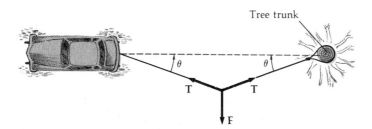

Tree trunk

Figure 5-40
Problem 1.

2. A box of mass m_1 is pulled by a force **F** exerted at the end of a rope which has a much smaller mass m_2, as shown in Figure 5-41. The box slides along a smooth horizontal surface. (a) Find the acceleration of the rope and box. (b) Find the tension in the rope at the point where it is attached to the box. (c) The diagram in Figure 5-41 with the rope horizontal is not quite correct for this situation. Correct the diagram and state how this correction affects your solution.

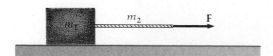

Figure 5-41
Force **F** applied to rope attached to block on smooth table (Problem 2).

3. Two 100-kg boxes are dragged along a smooth surface with a constant acceleration of 1 m/sec². Each of the ropes used has a mass of 1 kg (Figure 5-42). Find the force **F** and the tensions in the ropes at points A, B, and C.

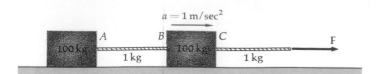

Figure 5-42
Problem 3.

Useful information about a situation can often be obtained by assuming a constant force even though it actually varies greatly with time. In this problem and the next problem, assume that the forces are constant.

4. A rifle bullet of mass 9 gm starts from rest and moves 0.6 m in the rifle barrel. The speed of the bullet as it leaves the barrel is 1200 m/sec. Assuming the force to be constant on the bullet while it is in the barrel, find (*a*) the average speed of the bullet in the barrel and the time it spends in the barrel and (*b*) the change in momentum of the bullet and the force exerted on the bullet.

5. A car traveling 60 mi/h comes to a sudden stop to avoid an accident. Fortunately the driver is wearing a seat belt. Using reasonable values for the mass of the driver and the time it takes to come to a stop, estimate the force (assumed constant) exerted on the driver by the seat belt.

6. Two boxes slide on smooth inclines, as shown in Figure 5-43. (*a*) Find the acceleration and the tension in the rope. (*b*) The two masses are replaced by two other masses m_1 and m_2 such that there is no acceleration. Find whatever information you can about the two masses.

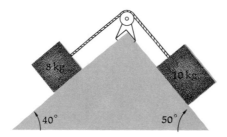

Figure 5-43
Two boxes on inclines
(Problem 6).

7. (*a*) Find the mass m in Figure 5-44 such that there is no acceleration. (*b*) Find the mass m such that m has a downward acceleration of $\frac{1}{2}g$. (*c*) Find m such that it has an upward acceleration $\frac{1}{4}g$. (*d*) What is the maximum upward acceleration possible for any mass m?

Figure 5-44
Problem 7.

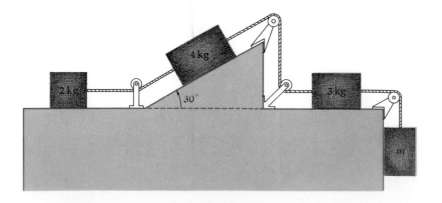

8. A 128-lb girl stands on a 32-lb aluminum platform to paint a house. A rope attached to the platform and passing over an overhead pulley allows the painter to raise herself and the platform (Figure 5-45). (*a*) To get started, she accelerates herself and the platform at a rate of 2 ft/sec². With what force must she pull on the rope? (*b*) After 1 sec she pulls so that she and the platform go up at a constant speed of 2 ft/sec. What force must she exert on the rope? (Ignore the mass of the rope.)

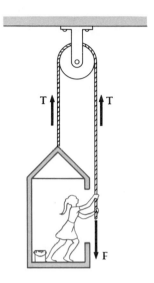

Figure 5-45
Painter on platform for Problem 8.

9. In Figure 5-34 (*a*) find the acceleration *a* and the contact force F_c for general masses m_1 and m_2 and applied force *F*. (*b*) If $m_2 = nm_1$, show that $F_c = nF/(n + 1)$.

10. For Figure 5-35 find the maximum force *F* that can be exerted without any slipping of m_1 if the maximum frictional contact force between the masses is $\mu m_1 g$, where μ is a fraction less than 1.

11. In Figure 5-46 find the maximum mass m_3 such that m_1 will not slip on m_2 if the maximum frictional contact force between m_1 and m_2 is $\mu m_1 g$.

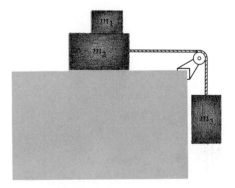

Figure 5-46
Problem 11.

12. The arrangement shown in Figure 5-33, called Atwood's machine, is used to measure the acceleration of gravity *g* by measuring the acceleration of the masses. (*a*) For general masses m_1 and m_2, find the acceleration *a* and the tension in the string *T*. Check your answers for the special cases $m_1 = m_2$ and for one mass much larger than the other. (*b*) The acceleration is measured by measuring the time *t* for the larger mass to fall a distance *L* from rest. Find an equation relating *g* to m_1, m_2, *L*, and *t*. (*c*) Show that if the time *t* changes by a small amount *dt*, there is a corresponding change in the acceleration *a* of

amount da, which is related to dt by $da/a = -2dt/t$. Show also that an error in the determination of a of the amount da results in an error dg in g related by $dg/g = da/a$. Find two masses m_1 and m_2 each less than 1 kg such that g can be measured to 5 percent with a watch that is accurate to 0.1 sec if $L = 3$ m. Assume that the only important uncertainty in the measurement is in the time of fall.

13. A constant force F is exerted on a smooth peg of mass m_1. Two masses m_2 and m_3 are connected to a string which goes around the peg as shown in Figure 5-47. Assuming that F is greater than $2T$, (a) find the acceleration of each of the masses and the tension in the string if $m_1 = m_2 = m_3$ and (b) the acceleration of each mass if $m_2 = m_1$ and $m_3 = 2m_1$.

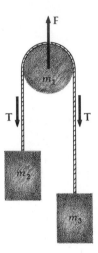

Figure 5-47
Atwood's machine for
Problem 13.

14. A 120-lb girl weighs herself by standing on a scale mounted on a skate board which rolls down an incline (Figure 5-48). (The force exerted by the incline on the board is perpendicular to the incline.) What is the reading on the scale if $\theta = 30°$?

Figure 5-48
Girl on scale·mounted on
skate board rolling down incline (Problem 14).

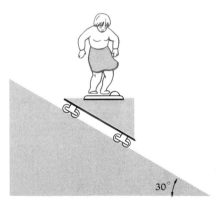

15. A small block of mass m slides on a smooth circular track in a vertical circle of radius R as shown in Figure 5-49. (a) Show that the speed of the block cannot be constant. (b) If the speed of the block at the top of the track is v, find the force exerted on the block by the track and the resultant force on the block. (c) What is the minimum value of v for the block to stay on the track? If the speed is less than this value, describe the path of the block.

16. A mass m_1 is attached to a cord of length L_1, which is fixed at one end. The mass moves in a horizontal circle supported by a smooth table. A second mass m_2 is attached to the first by a cord of length L_2 and also moves in a circle, as

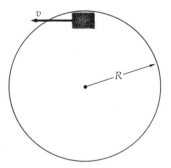

Figure 5-49
Small block sliding on circular
track (Problem 15).

shown in Figure 5-50. If the masses make f rev/sec, find the tension in each of the cords.

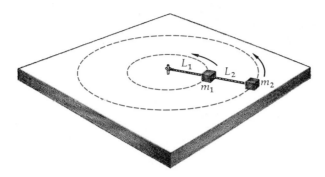

Figure 5-50
Two masses connected by a light cord moving in a horizontal circle (Problem 16).

17. A block of gold-aluminum alloy weighs 5 lb. When the block is submerged in water, it weighs 4 lb. How much gold is in the block? (The density of aluminum is 2.7 gm/cm³, and that of gold is 19.3 gm/cm³.)

18. When you weigh yourself in air, the scale reading is less than your weight because of the buoyant force of the air. Estimate the number of ounces you should add to correct for this.

19. A beaker which weighs 1 lb contains 2 lb of water and rests on a scale. A 2-lb block of aluminum (density 2.7 gm/cm³) suspended from a spring scale is submerged in the water. Find the readings of both scales (Figure 5-51).

Figure 5-51
Problem 19.

20. Consider a slab of air of area A and thickness dy, where y is the height of the air above the ground. Show that the difference dP in pressure between the bottom and top of the slab is related to dy by $dP = -\rho g\, dy$, where ρ is the density of air. This equation is equivalent to Equation 5-34 except that y is measured from the ground up (compared with measuring h from the top of a liquid down). However, the density of a gas is not independent of the pressure. For many gases, the density is proportional to the pressure $\rho = KP$, where K is a constant (assuming constant temperature). Show that P and y then obey the differential equation

$$\frac{dP}{P} = -Kg\, dy$$

Find the indefinite integral of both sides of the equation and show that the pressure can be written

$$P = e^C e^{-Kgy} = P_0 e^{-Kgy}$$

where C is the constant of integration and $P_0 = e^C$ is the pressure at $y = 0$.

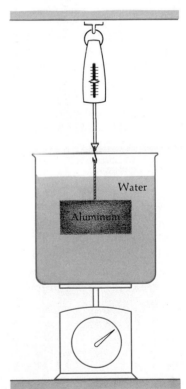

Water

Aluminum

21. (a) Show using the result of Problem 20 that the density of air varies with height y as $\rho = \rho_0 e^{-Kgy}$, where ρ_0 is the density at sea level $y = 0$. (b) Find the mass of a slab of air of area A and thickness dy at a height y above sea level. (c) Integrate your result to find the total mass of an air column of area A. Show that the total weight of the air column is $P_0 A$, where P_0 is the pressure at sea level.

22. A rectangular dam 100 ft wide supports a body of water to a height of 80 ft. (a) Neglecting atmospheric pressure, find the total force due to water pressure acting on a thin strip of height dy located at depth y. (b) Integrate the result in part (a) to find the total horizontal force due to the water acting on the dam. Why is it reasonable to neglect atmospheric pressure?

23. A water tank near the earth's surface is accelerated upward at a constant rate of $5g$. (a) Find the "buoyant force" acting on a 1-kg element of the water in the tank. (b) How does this force compare with the weight of a piece of aluminum which has the same volume as 1 kg of water? Will aluminum float in this tank while the tank is accelerating up? Explain.

CHAPTER 6   Forces in Nature

All the different forces observed in nature can be explained in terms of four basic interactions occurring among elementary atomic particles:

1. Gravitational forces

2. Electromagnetic forces

3. Strong nuclear forces

4. Weak nuclear forces

Most of the everyday forces we observe between macroscopic objects, e.g., contact forces, friction, and the forces exerted by springs and strings, are complicated manifestations of the basic electromagnetic interaction. In this chapter we take a brief look at some of the characteristics of the basic interactions and some empirical descriptions of the common, everyday forces that result from them. We shall study the gravitational and electromagnetic interactions in more detail later.

6-1   Gravity

All particles exert upon each other a gravitational force of attraction. The magnitude of the gravitational force exerted by a particle of mass m_1 on another particle of mass m_2 a distance r_{12} away is given by Newton's law of gravitation:

$$F_{12} = \frac{Gm_1m_2}{r_{12}{}^2}$$

Law of gravity

where G is a universal constant which has been determined experimentally. The direction of the force on m_2 is toward m_1. We can write this force in vector notation by defining a unit vector $\hat{r}_{12}$. If $\mathbf{r}_{12}$ is the

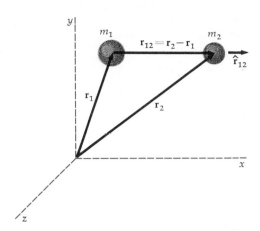

Figure 6-1
Definition of the unit vector, $\hat{\mathbf{r}}_{12}$; $\hat{\mathbf{r}}_{12}$ is 1 unit in length and has the direction of $\mathbf{r}_{12} = \mathbf{r}_2 - \mathbf{r}_1$.

vector displacement from m_1 to m_2, the unit vector $\hat{\mathbf{r}}_{12}$ (see Figure 6-1) is defined to be

$$\hat{\mathbf{r}}_{12} = \frac{\mathbf{r}_{12}}{r_{12}} \qquad\qquad 6\text{-}1$$

In terms of this unit vector, Newton's law of gravitation is written

$$\mathbf{F}_{12} = -\frac{Gm_1 m_2}{r_{12}{}^2} \hat{\mathbf{r}}_{12} \qquad\qquad 6\text{-}2$$

The minus sign in Equation 6-2 indicates that the force is an attractive force acting toward m_1. The force is inversely proportional to the square of the distance between the attracting particles and therefore decreases rapidly as they are separated. The force is proportional to the mass of each particle. For this reason, the attraction of a very large object like the earth is noticeable (as weight) even though the value of G is so small that the gravitational attractions between everyday objects are too small to be observed except with great difficulty.

We can apply Equation 6-2 to calculate the force exerted *on* m_1 *by* m_2 by interchanging the subscripts 1 and 2. The result obtained has the same magnitude as the force exerted on m_2 by m_1 but is in the opposite direction. (The unit vector $\hat{\mathbf{r}}_{21}$ is the negative of $\hat{\mathbf{r}}_{12}$.) Thus Newton's law of gravitation meets the requirements of Newton's third law.

The inverse-square nature of the gravitational force was suspected by Hooke and others during Newton's time. Using his laws of motion and Kepler's laws, the empirical laws describing planetary motion, Newton deduced the full law. As a check on its validity Newton compared the acceleration of the moon in its orbit with the acceleration of objects near the surface of the earth (such as the legendary apple), assuming that both accelerations are due to the same cause, the gravitational attraction of the earth. Let us consider this calculation in some detail.

Assuming moon and earth to be approximately point masses, the force exerted by the earth on the moon has the magnitude $GM_E m_m / r_m{}^2$, where M_E and m_m are the masses of the earth and moon, respectively, and $r_m = 2.39 \times 10^5$ mi $= 3.84 \times 10^8$ m is the distance from the center of the earth to the center of the moon. The direction of this force is toward the center of the earth. The acceleration of the moon, according to Newton's second law of motion, is

$$a_m = \frac{F}{m_m} = \frac{GM_E}{r_m{}^2} \qquad \text{toward center of earth} \qquad\qquad 6\text{-}3$$

Treating the earth and moon as point masses is a reasonable approximation because the distance between them is great compared with their sizes. However, the situation is different for the earth and a body near its surface. After a considerable effort, Newton was able to prove that if his law of gravity is correct, the force the whole earth exerts on a body on or outside its surface is the same as if all the earth's mass were concentrated at its center. The proof involves integral calculus, a tool developed by Newton for this problem, among others. We shall assume the result for now and consider the proof later. The force on a body of mass m near the surface of the earth is then directed toward the center of the earth and has magnitude

$$F = \frac{GM_E m}{R_E{}^2}$$

where R_E is the radius of the earth. The acceleration of gravity g (in an inertial reference frame) is then

$$g = \frac{F}{m} = \frac{GM_E}{R_E{}^2} \qquad\qquad 6\text{-}4$$

According to Equations 6-3 and 6-4, if both g and the acceleration of the moon are due to the earth's gravitational attraction, their ratio is just the ratio of the squares of the distance to the moon and the radius of the earth. Dividing Equation 6-4 by Equation 6-3, we obtain

$$\frac{g}{a_m} = \frac{GM_E/R_E{}^2}{GM_E/r_m{}^2} = \frac{r_m{}^2}{R_E{}^2} = \frac{(3.84 \times 10^8 \text{ m})^2}{(6.38 \times 10^6 \text{ m})^2} = 3.62 \times 10^3 \qquad 6\text{-}5$$

where we have taken $R_E = 6.38 \times 10^6$ m for the radius of the earth.[1] The centripetal acceleration of the moon is easily computed from its period $T = 27.3$ d $= 2.36 \times 10^6$ sec and the radius of its orbit r_m. Since it goes a distance $2\pi r_m$ in time T, its speed is $v = 2\pi r_m/T$ and its acceleration is

$$a_m = \frac{v^2}{r_m} = \frac{(2\pi r_m/T)^2}{r_m} = \frac{4\pi^2 r_m}{T^2}$$

$$= \frac{4\pi^2 (3.84 \times 10^8 \text{ m})}{(2.36 \times 10^6 \text{ sec})^2} = 2.72 \times 10^{-3} \text{ m/sec}^2$$

Taking 9.81 m/sec² for g at the equator, we obtain for the ratio of g and the acceleration of the moon

$$\frac{g}{a_m} = \frac{9.81}{2.72 \times 10^{-3}} = 3.61 \times 10^3$$

We see that g/a_m is indeed very nearly equal to $r_m{}^2/R_E{}^2$.

The constant G, called the *universal gravitational constant*, can be determined from either Equation 6-4 or 6-5 if the mass of the earth is known. Newton estimated the value of G from an estimate of the earth's density. The constant was first determined accurately by Cavendish in 1798 by directly measuring the force between two small spheres of known mass and separation. (Details of this measurement are given in Chapter 16.) The value of G is

$$G = 6.67 \times 10^{-11} \text{ N-m}^2/\text{kg}^2 \qquad\qquad 6\text{-}6$$

---

[1] The earth is not quite spherical. The radius 6.38×10^6 m is the distance from the center of the earth to the equator. The corresponding distance at the poles is 6.36×10^6 m. The radius of the sphere of volume equal to that of the earth, 6.37×10^6 m, is often taken as the average radius of the earth.

We can use this value to compute the gravitational attraction between two ordinary objects.

Example 6-1 What is the force of attraction between two balls each of mass 1 kg when their centers are 10 cm apart?

We can treat spheres as if the total mass were at a point at the center. The force exerted on either one by the other is then

$$F = \frac{(6.67 \times 10^{-11})(1)(1)}{(10^{-1})^2} = 6.67 \times 10^{-9} \text{ N}$$

This example demonstrates that the gravitational force exerted by one object of ordinary size on another such object is extremely small. For example, the weight of a 1-gm mass is 9.8×10^{-3} N, more than a million times the force computed in Example 6-1. We can usually neglect the gravitational force between two objects compared with other forces acting on them. The gravitational attraction can be noticed only if the objects are extremely massive, e.g., mountains, or if extreme care is taken to eliminate all other forces on the objects, as in the Cavendish experiment to measure G.

According to Equation 6-2, the weight of an object (and hence its free-fall acceleration) decreases as the object's distance from the center of the earth increases. Consider an object at a height h above the surface of the earth. The distance between the object and the center of the earth is then

$$r = R_E + h$$

and the free-fall acceleration of the object is

$$g(r) = \frac{F}{m} = \frac{GM_E}{r^2} = \frac{GM_E}{(R_E + h)^2} \qquad \text{6-7}$$

If h is small compared with the radius of the earth, we can obtain a useful approximation for this acceleration using the binomial expansion (Equations 1-15 and 1-16). We have

$$(R_E + h)^{-2} = R_E^{-2}\left(1 + \frac{h}{R_E}\right)^{-2} = R_E^{-2}\left(1 - 2\frac{h}{R_E} + \cdots\right)$$

Then

$$g(r) \approx \frac{GM_E}{R_E^2}\left(1 - 2\frac{h}{R_E}\right) = g\left(1 - 2\frac{h}{R_E}\right) \qquad \text{6-8}$$

where $g = GM_E/R_E^2 = 9.8$ m/sec² is the acceleration at the surface of the earth. The change in the acceleration of gravity for a given change h in the distance from the center of the earth is thus approximately   *Variation of g with altitude*

$$\Delta g = g(r) - g = -2g\frac{h}{R_E}$$

(The minus sign indicates that g decreases as r increases.) If h is only a few miles or less, we can usually neglect the variation in g since R_E is about 4000 mi. For larger heights with h still much less than R_E, we have the useful approximation for the fractional decrease in $g(r)$:

$$\frac{\Delta g}{g} \approx -2\frac{h}{R_E} \qquad \text{6-9}$$

Equation 6-9, found by using the binomial expansion, can also be obtained by using the differential dg to approximate Δg. The rate of change of $g(r)$ with respect to r is

$$\frac{dg}{dr} = \frac{d}{dr}\,(GM_E r^{-2}) = -2GM_E r^{-3} = \frac{-2}{r}\,\frac{GM_E}{r^2} = \frac{-2g}{r}$$

Thus when r changes by a small amount dr, the change in g is approximately dg, given by

$$dg = -2g\,\frac{dr}{r}$$

and the fractional change is

$$\frac{dg}{g} = -2\,\frac{dr}{r} \qquad\qquad\qquad 6\text{-}10$$

Equations 6-9 and 6-10 are the same except for the symbols. The fractional change in $g(r)$ is just twice the fractional change in distance if this change is small.

Example 6-2 What is the acceleration of a satellite in a circular orbit 100 mi above the earth's surface?

Taking 4000 mi for the earth's radius gives a fractional change in r of about

$$\frac{dr}{r} = \frac{h}{R_E} = \frac{100\text{ mi}}{4000\text{ mi}} = \frac{1}{40} = 2\tfrac{1}{2}\%$$

Then

$$\frac{dg}{g} = -2\,\frac{h}{R_E} = -5\%$$

$$dg = -(5\%)g = -\tfrac{5}{100}\,(9.8\text{ m/sec}^2) = -0.49\text{ m/sec}^2$$

$$g(r) \approx 9.8\text{ m/sec}^2 - 0.49\text{ m/sec}^2 = 9.3\text{ m/sec}^2$$

Questions

1. The greater the mass of an object the greater the force of gravity acting on it. Why, then, does a heavy body fall no faster than a light one? Why in some cases, e.g., a brick and a paper box of the same shape, does the heavy one fall faster?

2. An astronaut moves from the earth toward the moon. Describe qualitatively how his weight changes during the trip. Describe how his mass changes.

6-2   Electromagnetic Forces

Forces exerted by one particle on another because of the electric charges of the particles are called *electromagnetic forces*. Their description is considerably more complicated than that of the gravitational force. For one thing, the electromagnetic force can be either attractive or repulsive whereas the gravitational force is always attractive. A more serious complication arises because when charges are in motion, part of the electromagnetic force, the magnetic force, depends on both the magnitudes and directions of the velocities of the particles and is not generally along the line joining them. We avoid this complication

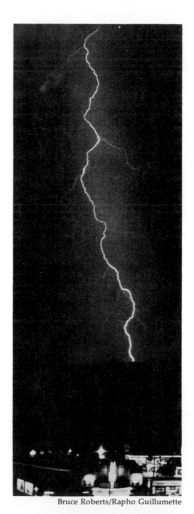

Bruce Roberts/Rapho Guillumette

A lightning discharge is a spectacular effect resulting from electromagnetic forces.

here by postponing any discussion of magnetism until Chapter 36. In our present brief survey of forces, we shall consider only the electrostatic force, i.e., the force between electric charges at rest.

There are several important properties of electric charge:

1. There are two kinds of electric charge. Like charges repel each other; unlike charges attract each other. The two kinds are labeled positive and negative, following the suggestion of Benjamin Franklin.

2. Charge is quantized. An electric charge always has the magnitude Ne, where N is an integer and e is a fundamental unit of charge equal in magnitude to the charge of an electron or a proton. All known elementary particles have a charge of $+e$, $-e$, or zero.

3. Charge is conserved. In certain nuclear reactions it is possible to create electrons and other particles. Whenever a particle with charge $-e$ (an electron) is created, a particle with charge $+e$ is created simultaneously or another particle with charge $-e$ is destroyed. The algebraic sum of the charges in an isolated system does not change.

A more detailed discussion of electric forces and charge is postponed until Chapter 29. The law which gives the electrostatic force between two point charges, Coulomb's law, is an empirical inverse-square law similar to Newton's law of gravitation. If there are two point charges q_1 and q_2 separated by a distance r_{12}, the force exerted on charge q_2 by charge q_1 is given by

$$\mathbf{F}_{12} = k \frac{q_1 q_2}{r_{12}{}^2} \hat{\mathbf{r}}_{12} \qquad \text{6-11}$$

Coulomb's law of electric force

where $\hat{\mathbf{r}}_{12}$ is a unit vector pointing from q_1 to q_2 and k is a constant, known as the coulomb constant. In the international system of units, the unit of charge is the coulomb (C), which is the amount of charge flowing past a point in a wire in one second when the current in the wire is one ampere. The unit of current, the ampere (A), defined in terms of magnetic-force measurement (Chapter 37), is the practical unit of current used in everyday electrical work.

The fact that like charges repel and unlike charges attract is included in Equation 6-11. If both charges have the same sign, the product $q_1 q_2$ is positive and the force on q_2 is directed away from q_1 parallel to $\hat{\mathbf{r}}_{12}$. If the charges have opposite signs, the product $q_1 q_2$ is negative and the force on q_2 is toward q_1 antiparallel to $\hat{\mathbf{r}}_{12}$. The value of k is

$$k = 8.99 \times 10^9 \ \text{N-m}^2/\text{C}^2 \approx 9 \times 10^9 \ \text{N-m}^2/\text{C}^2 \qquad \text{6-12}$$

The fundamental unit of charge is

$$e = 1.602 \times 10^{-19} \ \text{C} \qquad \text{6-13}$$

It takes a large number of electrons or protons to produce a charge with a magnitude of 1 C.

Since atoms normally contain an equal number of positively charged protons and negatively charged electrons (and no other charged particles), atoms are electrically neutral, i.e., have no net electric charge. An atom can sometimes lose one of its electrons or gain an additional electron to become a charged particle called an *ion*. Macroscopic objects usually contain a large number of neutral atoms and are therefore electrically neutral. Electrons, however, can often be transferred from one macroscopic object to another, leaving one with an excess positive charge and the other with an excess negative charge. For example, when a glass rod is rubbed with a silk cloth, electrons are transferred from the rod to the cloth. With no knowledge of the process of transfer of charge, Franklin arbitrarily labeled the charge of the glass rod positive and that of the cloth negative. It is for this reason that the electron charge is now taken to be negative and the proton charge positive rather than the other way around. Excess charges of the order of several microcoulombs can be obtained by putting certain objects in intimate contact, often by simply rubbing the surfaces together. Such charges produce relatively large forces, as we shall see in Example 6-3.

Example 6-3 What is the force between two point charges of 1 μC separated by a distance of 1.0 cm?

From Coulomb's law, the magnitude of the force is

$$F = \frac{(9 \times 10^9 \text{ N-m}^2/\text{C}^2)(10^{-6}\text{ C})(10^{-6}\text{ C})}{(10^{-2}\text{ m})^2} = 90 \text{ N} \approx 20 \text{ lb}$$

In metals, some electrons are so weakly attached to their atoms that they are essentially free to move about the whole volume of the metal. Such materials are called *electric conductors*. For example, in copper, about one electron per atom is free. In other materials, e.g., wood or glass, the electrons are tightly bound (by electric forces) to their molecules. These materials are called *insulators*. Even though the electrons in insulators do not leave their molecules, the presence of an external charged object will cause a slight separation of the positive and negative charge in the nearby molecules. For example, suppose that a balloon charged by rubbing against a piece of cloth acquires an excess of electrons and is thus negatively charged. When the balloon is held near an insulating wall, the negative charge in the molecules near the surface of the wall is repelled and the positive charge attracted, causing the slight separation illustrated schematically in Figure 6-2.

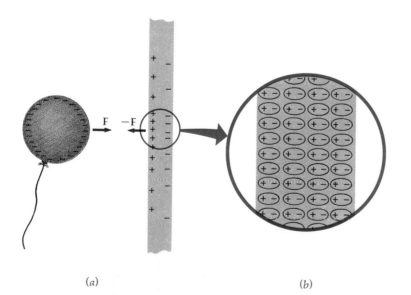

(a)   (b)

Figure 6-2
(a) The negative charge on the balloon attracts an excess of positive charge to the surface of the wall. The result is an attractive force between the balloon and the wall, even though the total charge of the wall is neutral. (b) An enlargement of the circled section of the wall from (a) showing schematically how the molecular charges in the wall are polarized by the proximity of the negatively charged balloon.

The molecules in the insulator are said to be *polarized*. Since the positive charge of these molecules is slightly closer to the negatively charged balloon than the negative charge, the balloon is attracted to the wall. It should be clear that if the balloon were positively charged, it would polarize the atoms in the wall in the opposite sense and again be attracted to the wall. Thus both positively and negatively charged objects are attracted by neutral insulators. In the same way, electrically charged objects attract small neutral objects, the most familiar example being small bits of paper or wood shavings attracted to a comb charged by running it through the hair. When the bits of paper touch the comb, they sometimes acquire some of its excess charge, and then fly off due to the repulsion of like charges.

Because both the electrostatic force and the gravitational force between two particles vary inversely with the square of the separation, the ratio of these forces is independent of the separation of the particles. We can thus compare the relative strengths of these forces for elementary particles, such as two protons, two electrons, or an electron and a proton. Consider the case of two protons. The mass of a proton is 1.67×10^{-27} kg, and its charge is 1.6×10^{-19} C. The ratio of the electrostatic force to the gravitational force between two protons at any separation is

$$\frac{F_e}{F_g} = \frac{kq_1q_2}{Gm_1m_2} = \frac{(9 \times 10^9)(1.6 \times 10^{-19})^2}{(6.67 \times 10^{-11})(1.67 \times 10^{-27})^2} \approx 10^{36} \qquad 6\text{-}14$$

Since electrons have a mass about $\frac{1}{1836}$ times that of protons, the ratio for two electrons is even greater.

The gravitational force between two elementary particles is so much smaller than the electrostatic force that it can be neglected in describing their interaction. It is only because such large masses as the earth contain almost exactly equal numbers of positive and negative charges that the gravitational force is important at all. If the positive and negative charges in bodies did not almost exactly cancel each other, the electric forces between them would be much greater than the gravitational attraction.

Questions

3. How can you tell whether the force acting on a body is gravitational or electric?

4. If all atoms are made up of electrically charged particles, and if electric forces are so much stronger than gravitational forces, why are we so aware of gravity in everyday life and so unfamiliar with electric forces?

5. A charged balloon is also attracted to an uncharged conductor regardless of the charge on the balloon. Explain.

6-3 Nuclear Forces

Besides the gravitational and electromagnetic forces the two other basic forces are the strong nuclear force and the weak nuclear force. It is clear from the stability of atomic nuclei that there must be some strong attractive forces besides electromagnetic and gravitational forces. Consider, for example, the helium nucleus, ^{4}He, which consists of two neutrons and two protons, a total of four particles contained in a volume with a radius of about 10^{-15} m. The repulsive electric force between protons this close together is quite large, and we have seen that by comparison the attractive gravitational force is negligible. Yet ^{4}He is very stable. The force which binds the neutrons and protons together in a nucleus is called the strong nuclear force. It acts between two protons, two neutrons, or a proton and neutron but only if the particles are very close together. When two nucleons[1] are within about 10^{-15} m of each other, the strong nuclear attractive force is about 10 times stronger than the repulsive electric force of two protons at this separation. However, this nuclear force has a short range; i.e., its strength decreases very rapidly with separation, much more rapidly than the inverse-square decrease in the electric force. At a separation greater than about 1.5×10^{-14} m the nuclear force is much less than the electric force and can be neglected.

Thus there is no macroscopic experiment which can be performed to investigate the properties of this force. Instead, information is obtained by bombarding neutrons, protons, and nuclei with neutrons, protons, and other particles and observing the results. Such *scattering experiments* have yielded much information about the nuclear force but have not produced a unique or simple formula like Coulomb's law or Newton's law of gravitation. Some of the known characteristics of the strong nuclear force are:

1. Charge independence: the nuclear force between two neutrons is the same as between two protons or between a neutron and a proton.

Characteristics of the strong nuclear force

2. Saturation: in a nucleus containing many nucleons, each nucleon bonds to only a few of the other nucleons. Thus the force needed to tear a neutron from a nucleus is approximately the same whether the nucleus contains five or six nucleons or one hundred nucleons.

3. Short range: at a distance of about 10^{-15} m, the strong nuclear force is attractive and about 10 times the electric force between two protons. The force decreases rapidly with increasing distance, becoming completely negligible at about 15 times this separation.

4. Repulsive at very short range: since at distances of about 0.4×10^{-15} m the nuclear force is repulsive, nuclei do not collapse.

A large class of particles, called *hadrons,* interact with one another via the strong nuclear force. Hadrons include *mesons* of many types and *baryons,* which in turn include nucleons and other particles. Particles in a much smaller class, the *leptons,* do not participate in the strong nuclear force at all. Leptons include electrons, positrons, muons, and neutrinos.

The *weak nuclear force* acts between all leptons and hadrons. It is also a short-range force. Its effect is to produce a degree of instability in nuclei and elementary particles. It is responsible for the type of radioactivity known as *beta decay,* and it explains, for example, why ^{12}C, a carbon nucleus with six protons and six neutrons, is stable while ^{11}C, a carbon nucleus with six protons and five neutrons, is not.

---

[1] A nucleon is either a proton or a neutron.

6-4 Molecular Forces

The forces which bind two or more atoms together to make a molecule are electromagnetic in origin, but the details of this bonding are extremely complicated. Consider one of the simpler cases, the NaCl molecule. When neutral sodium and chlorine atoms come close to each other, the sodium atom has a tendency to give up one of its electrons to the chlorine atom because of the structure of these atoms. We then have two ions, Na$^+$ with a net charge of $+e$ and Cl$^-$ with a net charge of $-e$. These ions attract each other (much like point charges), but at a certain separation the force of attraction becomes zero, and at smaller separations the ions repel each other. This repulsion is a complicated phenomenon which cannot be understood even approximately in terms of classical mechanics or electromagnetism but only in terms of quantum mechanics.[1] We can, however, give a qualitative sketch of the force between the ions as a function of separation, as in Figure 6-3, where repulsive forces are indicated as positive and attractive forces as negative. At the distance $r = r_0$ in this figure, the force is zero. This is called the *equilibrium separation* for the atoms, and for NaCl it is of the order of 10^{-10} m, the same order as the distance from the center of the atom to its outermost electrons. For small displacements from equilibrium (between points A and B in the figure) the force curve is linear

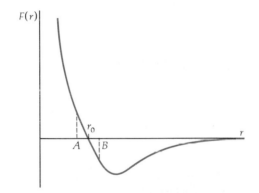

Figure 6-3
Force exerted by one ion on another as a function of separation r. At $r = r_0$ the force is zero. At greater separation the force is negative, indicating attraction. At separation less than r_0 the force is positive, indicating repulsion. In the range from A to B near the equilibrium separation r_0 the curve $F(r)$ versus r is approximately linear, similar to that for a spring (see Figure 6-5).

and resembles the force exerted by a spring when compressed or extended (see Figure 6-5). In its solid state, NaCl exists in a regular array of Na$^+$ and Cl$^-$ ions, each ion bound by ions of the other type into a lattice, as illustrated in Figure 6-4. Such a structure, called a *crystal*, is quite rigid. When we try to deform the solid by pushing or pulling on it, in an effort to change the separation distances between the ions, the strong forces between the ions resist compression or extension.

A more common type of molecular bonding, called *covalent bonding*, occurs, for example, between two hydrogen atoms in the H$_2$ molecule, between the carbon and oxygen atoms in CO, or between two carbon atoms in many organic compounds. In this type of bonding, the atoms do not give up or accept electrons to form ions; instead, some of the outer atomic electrons are shared by the atoms. In covalent bonding,

---

[1] Early in this century it became clear that newtonian mechanics is inadequate for treating the motions of particles within atoms. By 1926 a new and more general theory of mechanics, known as *quantum mechanics*, was developed which successfully describes the details of subatomic motions. Quantum mechanics applied on a macroscopic scale reduces to newtonian mechanics.

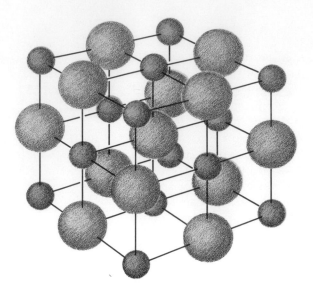

Figure 6-4
Arrangement of Na⁺ and Cl⁻ ions in an NaCl crystal.

both the attraction and the repulsion of the atoms are complicated quantum-mechanical phenomena. As with ionic bonding, the force-versus-separation curve looks like Figure 6-3.

There are other kinds of molecular bonds, e.g., metallic bonds, responsible, say, for the bonding of copper atoms in a solid piece of copper. In general, all forces between atoms and molecules are complex manifestations of the basic electromagnetic interaction, the details of which can be understood only within the framework of quantum mechanics. Nevertheless, we can often find empirical descriptions of these forces which are quite useful in determining the motion of macroscopic objects. At this level we must be content with such empirical descriptions.

6-5   Springs and Strings

A coil spring is made by winding a piece of stiff wire into a helix. If a spring is compressed or extended and released, it returns to its original, or natural, length provided the displacement is not too great. There is a limit to such displacements, beyond which the spring does not return to its original length but remains permanently deformed. If we allow only displacements below this limit, we can calibrate the extension or compression Δx in terms of the force needed to produce this extension or compression, as outlined in the beginning of Chapter 4. A typical calibration curve is shown in Figure 6-5. The sudden change in the slope of this curve at large compressions is at the point where the coils are completely collapsed so that adjacent turns are touching. Beyond this point it takes a very large force to produce any noticeable compression. We see from this curve that for small Δx the force exerted by the spring is approximately proportional to Δx. This result, known as *Hooke's law,* can be written

$$F_x = -k(x - x_0) = -k\,\Delta x \qquad\qquad 6\text{-}15$$ *Hooke's law*

where the empirically determined constant k is called the *force constant* of the spring. It is equal to the negative of the slope of the straight-line portion of the curve in Figure 6-5. The distance x is the coordinate of the free end of the spring or of any object attached to that end of the

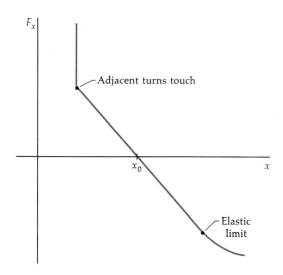

Figure 6-5
The force F_x exerted by the spring plotted versus the length x of the spring. x_0 is the natural or unextended length. The change in length is $\Delta x = x - x_0$, and $F_x \approx -k\,\Delta x$ in the linear portion of the curve near $x = x_0$. For compression, Δx is negative and F_x is positive (to the right); for extension, Δx is positive and F_x is negative (to the left).

spring. The constant x_0 is the value of this coordinate when the spring is unstretched from its equilibrium position.

If we attach our spring to a mass m on a smooth table, as in Figure 6-6, and extend the spring, there will be a resultant force on the mass to the left whether Equation 6-15 holds for this extension or not. If we release the mass from rest, it will accelerate to the left until $x = x_0$, when the spring has its natural length. At this point the force, and therefore the acceleration, is zero, but the mass is moving to the left and therefore continues. The spring then becomes compressed and exerts a force on the mass to the right tending to reduce its speed. Eventually the speed reaches zero at some point $x < x_0$, and the resultant force in the positive x direction now causes the speed to increase in the positive direction. The mass again overshoots the equilibrium position $x = x_0$, this time going to the right, and comes momentarily to rest at its original position. The motion then repeats. The mass oscillates back and forth.

The curve in Figure 6-5 is similar to the one we sketched for the force between two atoms in a molecule (Figure 6-4). It is often useful to visualize the atoms in a molecule as being connected by springs. For example, if we slightly increased the separation of the atoms in the molecule and released them, we would expect the atoms to oscillate back and forth as if they were two masses connected by a spring.

If we pull on a flexible string, the string stretches slightly and pulls back with an equal but opposite force (unless the string breaks). We can think of a string as a spring with such a large force constant that the extension of the string is negligible. If the string is flexible, however, we cannot exert a force of compression on it. When we push on a string, it merely flexes or bends.

Figure 6-6
(a) A mass m on a smooth table subject to the force of a spring. (b) When the displacement is to the right (positive), the force is to the left (negative). When the displacement is to the left (negative), the force is to the right (positive). Hence, the mass oscillates about x_0.

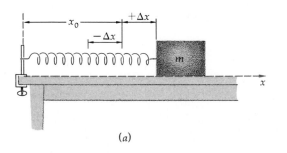

(a)

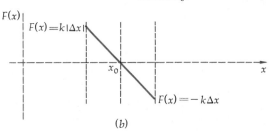

(b)

6-6   Contact Forces and Friction

Consider a block resting on a horizontal table. The weight of the block pulls the block downward, pressing it against the table. Because the molecules in the table have a great resistance to compression, as discussed previously, the table exerts a force upward on the block perpendicular, or normal, to the surface with no noticeable compression of the table. (Careful measurement would show that the supporting surface always does bend slightly in response to a load.) Similarly, the block exerts a downward force on the table.

Let us now exert a horizontal force on the block. The block does not move if the horizontal force is not too large. The table evidently exerts a horizontal force equal and opposite to the force we exert as long as the force is small enough. This force parallel to the surface of the table is called a *frictional force.* The block of course exerts an equal and opposite frictional force on the table, tending to drag it in the direction of the horizontal force we exert. This frictional force is due to the bonding of the molecules of the block and the table at the places where the surfaces are in very close contact.

We might expect the maximum force of friction to be proportional to the area of contact between the two surfaces. However, experimentally, to a good approximation it is independent of this area and simply proportional to the normal force exerted by one surface on the other. Figure 6-7, an enlarged picture of the contact between the block and the table, shows that the actual microscopic area of contact at which the molecules can bond together is just a small fraction of the macroscopic area. A possible model of static friction consistent with our intuition and with the empirical results is the following. The maximum force of friction *is* proportional to this microscopic area of contact, as expected, but this microscopic area is proportional to the total area and to the force per unit area exerted between the surfaces. Thus it does not depend on the total area of contact. Consider, for example, a 1-lb block with a side area of 30 in² and an end area of 10 in². When it is on its side on the table, a small fraction of the total 30 in² is actually in microscopic contact. When it is placed on end, the fraction of the

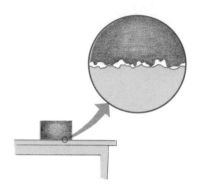

Figure 6-7
The microscopic area of contact between block and table. The microscopic contact is only a small fraction of the macroscopic area of the block. As the normal force on the block is increased, there is an approximately proportionate increase in the microscopic contact.

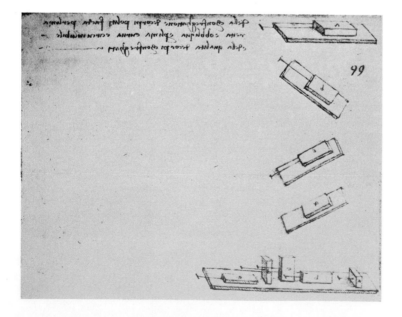

Leonardo da Vinci understood that the frictional force is independent of the macroscopic area of contact, as this drawing from one of his notebooks shows. (*Courtesy of the Deutsches Museum, Munich.*)

Cross-country skiers use a special ski wax to maximize static friction (so that they can walk up an incline) and minimize kinetic friction.

total area actually in microscopic contact is increased by a factor of 3 because the force per unit area is 3 times as great. However, since the area of the end is one-third that of the side, the actual microscopic area of contact is unchanged. The maximum force of static friction $f_{s,\max}$ is thus just proportional to the normal force between the surfaces,

$$f_{s,\max} = \mu_s N \qquad\qquad 6\text{-}16$$

where μ_s, called the *coefficient of static friction,* depends on the nature of the surfaces of the block and table. If we exert a smaller horizontal force on the block, the frictional force will just balance this horizontal force. In general we can write

$$f_s \lesssim \mu_s N \qquad\qquad 6\text{-}17 \qquad \textit{Static friction}$$

If we push the block hard enough, the static-friction force cannot prevent its motion. Then, as the block slides along the surface of the table, molecular bonds are continually made and broken and small pieces of the surfaces are broken off. The result is a force of sliding, or kinetic, friction which opposes the motion. Kinetic friction, like static friction, is a complicated phenomenon which even today is not completely understood. The coefficient of kinetic friction μ_k is defined as the ratio of the magnitudes of the frictional force f_k and the normal force N. Then

$$f_k = \mu_k N \qquad\qquad 6\text{-}18 \qquad \textit{Kinetic friction}$$

Experimentally it is found that:

1. μ_k is less than μ_s.

2. μ_k depends on the relative speed of the surfaces, but for speeds in the range from about 1 cm/sec to several meters per second μ_k is approximately constant.

3. μ_k depends on the nature of the surfaces but is independent of the macroscopic area of contact.

We shall neglect any variation in μ_k with speed and assume that it is a constant which depends only on the nature of the surfaces.

We can measure μ_s and μ_k for two surfaces simply by placing a block on a plane surface and inclining the plane until the block begins

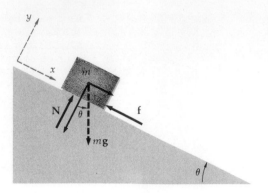

Figure 6-8
Block on a rough inclined plane. The component of the weight down the plane is $mg \cdot \sin \theta$. In equilibrium this is balanced by the force of static friction up the plane. As θ is increased, the force of static friction increases until it reaches its limiting value $\mu_s N$, where $N = mg \cos \theta$ is the normal force. Then $\mu_s = \tan \theta_c$. If θ is increased beyond this critical angle, the block slides down the plane and $f = \mu_k N$, which is less than $mg \sin \theta$.

to slide. Let θ_c be the critical angle at which sliding starts. For angles of inclination less than this, the block is in static equilibrium under the influence of its weight mg, the normal force **N**, and the force of static friction $\mathbf{f}_s$ (see Figure 6-8). Choosing the x axis parallel to the plane and the y axis perpendicular to the plane, we have

$$\Sigma F_y = N - mg \cos \theta = 0$$

and

$$\Sigma F_x = mg \sin \theta - f_s = 0$$

Eliminating the weight mg from these two equations gives

$$f_s = mg \sin \theta = \frac{N}{\cos \theta} \sin \theta = N \tan \theta$$

At the critical angle θ_c, the static friction is limiting, and we can replace f_s by $\mu_s N$. Then

$$\mu_s = \tan \theta_c$$

The coefficient of static friction equals the tangent of the angle of inclination at which the block just begins to slide.

At angles greater than θ_c, the block slides down the incline with acceleration a_x. In this case the frictional force is $\mu_k N$, and we have

$$F_x = mg \sin \theta - \mu_k N = ma_x$$

Substituting $mg \cos \theta$ for N, we obtain for the acceleration

$$a_x = g (\sin \theta - \mu_k \cos \theta)$$

A measurement of the acceleration a_x will then determine μ_k for these surfaces.

Questions

6. The friction between two surfaces is at first reduced by polishing both, but if the surfaces are polished until they are extremely smooth and flat, friction increases again. Explain.

7. Various objects lie on the floor on a truck. If the truck accelerates, what force acts on these objects to cause them to accelerate? Why will they slip if the acceleration of the truck is too great?

8. Any object resting on the floor of a truck will slip if the acceleration of the truck is too great. How does the critical acceleration at which a small box slips compare with that at which a much heavier object slips?

Optional

6-7   Pseudo Forces

Newton's laws hold only in inertial reference frames. When the acceleration of an object is measured relative to a reference frame which itself is accelerated relative to an inertial frame, the resultant force on the object does not equal the mass of the object times its acceleration. In some cases an object will be at rest relative to the noninertial frame even though there is obviously an unbalanced force acting on it. In other cases, the object has no forces acting on it but is accelerated relative to that frame. However, even in such an accelerated reference frame, we may use Newton's law $\Sigma F = ma$ if we introduce *fictitious,* or *pseudo, forces* which depend on the acceleration of the reference frame. These forces are not exerted by any agent. They are merely fictions introduced to make $\Sigma F = ma$ work when the acceleration a is measured relative to the noninertial frame. To observers in the noninertial frame, the pseudo forces appear as real as other forces because of our strong intuitive belief in Newton's second law. The most familiar pseudo force is the centrifugal force encountered in rotating reference frames. Unfortunately, the concept of centrifugal force is probably used incorrectly more often than correctly.

Let us first consider a railroad car moving in a straight line along a horizontal track with constant acceleration a relative to the track, which we assume to be in an inertial reference frame. If we drop an object in the car, it does not fall straight down but falls toward the back of the car. Relative to the car it has vertical acceleration g and horizontal acceleration $-a$ (Figure 6-9). If an object is placed on a smooth table so that the resulting force is zero, it accelerates toward the back of the car. Of course, from the point of view of an observer in an inertial frame on the tracks, the object does not accelerate. Instead, the car and table accelerate under the object. We can use Newton's second law in the reference frame of the car if we introduce a pseudo force $-ma$ acting on any object of mass m. Consider, for example, a lamp hanging by a cord from the ceiling of the car. The description of the acceleration of the lamp and forces acting on it from the inertial and noninertial frames is shown in Figure 6-10. The vertical compo-

Figure 6-9
An object is dropped inside a railway car which is initially at rest but has acceleration a to the right. (*a*) An observer on the ground in an inertial reference frame sees the ball fall straight down with the acceleration of gravity. (*b*) An observer in the car sees the ball accelerate downward and to the left relative to the car. Since the only force on the ball is its weight downward, Newton's laws do not hold in the accelerated reference frame. Newton's laws can be used in this frame if a pseudo force $-ma$ is introduced.

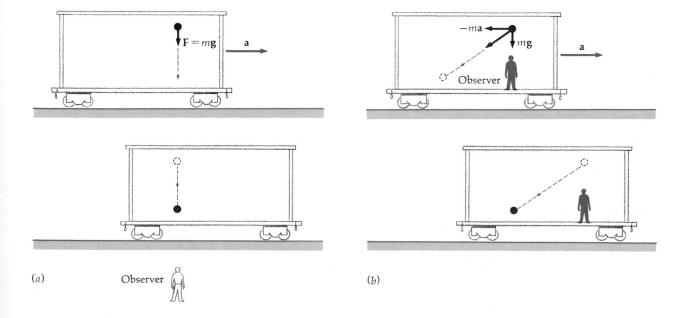

(*a*)   Observer   (*b*)

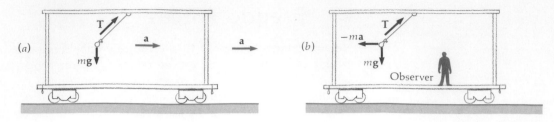

(a)   (b)

Observer

Figure 6-10
A lamp is hung by a cord from the roof of an accelerating car. (*a*) According to the inertial observer, the vertical component of **T** balances the weight *mg* and the horizontal component of **T** is the force which provides the observed acceleration. (*b*) According to the noninertial observer in the car, the lamp is at rest and not accelerating. He must introduce the pseudo force −*m***a** to balance the horizontal component of **T**.

nent of the tension of the cord equals the weight of the lamp according to each observer. In the inertial frame of the track, the lamp is accelerating. This acceleration is provided by the resultant force due to the horizontal component of the tension in the cord. In the frame of the car, the lamp is at rest and therefore has no acceleration. This is explained by the fact that the horizontal component of the tension balances the pseudo force −*m***a** observed on all objects in the car.

Figure 6-11 shows another noninertial frame, a rotating platform. Since each point on the platform is moving in a circle, it has centripetal acceleration. Thus a frame attached to the platform is a noninertial frame. In Figure 6-11 a block at rest relative to the platform is attached to the center post by a string. According to observers in an inertial frame, the block is moving in a circle with speed v. It is accelerating toward the center of the circle. This centripetal acceleration v^2/r is provided by the unbalanced force due to the string tension **T**. However, according to an observer on the platform, the block is at rest and not accelerating. In order to use $\Sigma\mathbf{F} = m\mathbf{a}$, he must introduce a pseudo force of magnitude mv^2/r acting radially outward to balance the string tension. This fictitious outward force, called the *centrifugal force*, appears quite real to the observer on the platform. If he wants to stand at "rest" on the platform an inward force of this magnitude must be exerted on him (by the floor) to "balance" the outward centrifugal force. We have occasion to use this pseudo force *only* in a rotating frame. Consider a satellite near the surface of the earth and observed in an inertial frame attached to the earth (we neglect the earth's rotation here). People often say that the satellite does not fall because the gravitational attraction of the earth "is balanced by the centrifugal force." This is incorrect. Pseudo forces such as the centrifugal force appear only in accelerated reference frames. In the earth frame the satellite does fall toward the earth with acceleration v^2/r produced by the single unbalanced force of gravity acting on it. However, an observer *in the satellite* who considers the satellite to be at rest can use $\Sigma\mathbf{F} = m\mathbf{a}$ only if he introduces an outward centrifugal force to balance gravity.

Centrifugal force

Figure 6-11
A block is tied to the center post of a rotating platform by a string. (*a*) An inertial observer sees the block accelerating toward the center of the platform at v^2/r due to the unbalanced force exerted by the tension in the string. (*b*) According to the noninertial observer on the platform, the block is not accelerating. If he is to apply Newton's laws, he must introduce the pseudo force mv^2/r away from the center of the platform to balance the tension in the string.

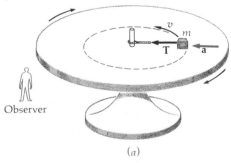
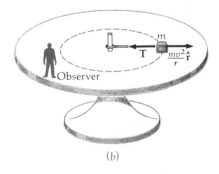

Observer

(a)   (b)

Figure 6-12
A boy at the center of a rotating platform throws a ball toward a friend on the edge of the platform. (*a*) According to the inertial observer, the ball travels in a straight line toward the initial position of the receiver and the receiver moves to the left of this point with the platform. (*b*) According to the noninertial observer on the platform, the receiver is stationary and the ball deflects to the right. The pseudo force which deflects the ball from a straight line in this frame is called the coriolis force.

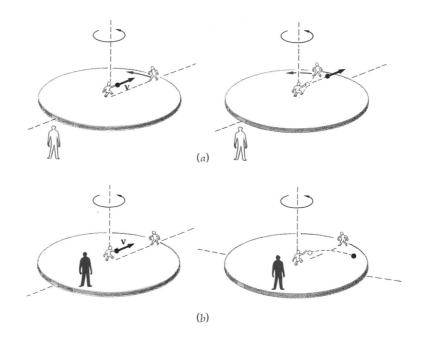

(*a*)

(*b*)

Coriolis force

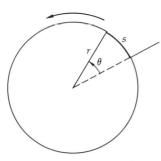

Figure 6-13
Diagram for calculation of coriolis force. The ball moves a radial distance $r = vt$, where v is its speed. The point on the platform at distance r moves a distance $s = r\omega t$ in this time where $\omega = d\theta/dt$. This can be written $s = \frac{1}{2}a_c t^2$, where $a_c = 2v\omega$ is the coriolis acceleration.

A second pseudo force which depends on the velocity of a particle must be introduced in a rotating frame in order to use $\Sigma\mathbf{F} = m\mathbf{a}$ in that frame. Called the *coriolis force*, it is perpendicular to the velocity of a particle (relative to the rotating frame) and causes a sideways deflection. Consider two observers standing along a radial line on a rotating platform and playing catch. If the ball is thrown outward along the radial line, they will see it deflect to the right and miss the receiver (Figure 6-12). In an inertial frame, the ball travels in a straight line after leaving the thrower and misses because the receiver is moving. The path of the ball relative to the rotating platform is the curved line shown in the figure. The ball must be thrown to the left of the receiver to take into account this sideways deflection.

We can calculate the magnitude of the coriolis force from a simple special case considered in an inertial frame. Let θ be the angle between a radial line fixed on the platform and a fixed line in space shown in Figure 6-13. This angle changes with time because the platform is rotating. The rate of change of θ with respect to time is called the *angular velocity* ω of the platform

$$\omega = \frac{d\theta}{dt}$$

The angular velocity ω describes the rotation of the platform. Any point fixed on the platform moves in a circular arc. The distance s moved is related to the angle θ and to the radial distance r of the point by

$$s = r\theta$$

Thus the speed of a point fixed on the platform is

$$\frac{ds}{dt} = r\frac{d\theta}{dt} = r\omega$$

Now consider a ball thrown from the origin with speed v along a radial line. After a time t the ball has moved a radial distance $r = vt$.

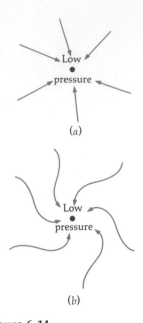

(a)

(b)

Figure 6-14
(a) Except for the coriolis force, winds would blow radially inward toward a low-pressure center. (b) Because of the deflection due to coriolis forces, winds form a counterclockwise pattern in the Northern Hemisphere (as seen from above). In the Southern Hemisphere the pattern is clockwise.

Hurricane Ginger, September 13, 1971, showing the counterclockwise rotation observed in the Northern Hemisphere.

But the point on the platform at a distance r has moved along a circular arc a distance s given by

$$s = r\omega t = (vt)\omega t$$

The distance s is just the distance the ball is deflected by the pseudo force, the coriolis force, in the rotating reference frame. In the rotating frame the deflection is to the right (looking along the velocity vector) and has the magnitude

$$s = \omega v t^2 = \tfrac{1}{2}(2\omega v)t^2 = \tfrac{1}{2}a_c t^2$$

where

$$a_c = 2\omega v \qquad\qquad\qquad 6\text{-}19$$

is called the coriolis acceleration. The coriolis force is

$$F_c = ma_c = 2m\omega v \qquad\qquad\qquad 6\text{-}20$$

This result, i.e., that there is a pseudo force perpendicular to the velocity vector with magnitude $2m\omega v$, is not limited to motions in the radial direction but applies to motion in any direction (perpendicular to the axis of rotation). The direction of the force is to the right looking along the velocity vector for counterclockwise rotation as seen from above.

These two pseudo forces, the centrifugal force and the coriolis forces for a rotating reference frame, have direct application to reference frames attached to the earth because of the earth's rotation. In particular coriolis forces are important for understanding weather. For example, these forces are responsible for the fact that cyclones (viewed from above) are counterclockwise in the Northern Hemisphere and clockwise in the Southern Hemisphere (see Figure 6-14).

NASA

Global Winds

Richard Goody
Harvard University

The question of forces acting on the air in a large meteorological system demonstrates the use of Newton's laws of motion in accelerated reference frames.

First, we must accept the idea of an identifiable parcel of air moving with the wind. This is an abstract concept since we have no convenient way of labeling a piece of air, but it is a concept we can reasonably accept.

This parcel will be subjected to many different forces. In the vertical direction there are gravitational forces and pressure forces, i.e., those due to the difference in pressure at different heights. If the pressure varies in the horizontal direction, there are horizontal pressure forces acting from high to low pressure. If the parcel moves relative to its surroundings or boundaries, there will be frictional forces. Finally, in the reference frame of the earth's surface, there are pseudo forces due to the earth's rotation: centrifugal forces and coriolis forces.

Obviously, it is a complicated problem to solve the equations for three-dimensional motion under the influence of all these forces. Fortunately, not all the forces are of equal importance under all circumstances. Let us make the following assumptions:

1. The flow is horizontal, so that we need not consider the vertical forces, which we take to balance.

2. The flow is slow enough to permit frictional forces to be neglected.

3. Acceleration does not take place along the trajectory, so that the horizontal forces balance.

We then have only the horizontal pressure forces, the coriolis forces, and possible horizontal centrifugal forces (in the frame moving with the parcel) if the air parcel is following a curved trajectory. The horizontal centrifugal force is of the order mv^2/r, where v is the speed, m the mass of the parcel, and r the radius of the trajectory. The ratio of this force to the coriolis force ($2m\omega v$) is $v/2\omega r$. For systems that are large enough and slow enough the horizontal centrifugal forces are negligible compared with the coriolis forces.

Cyclones and anticyclones and the average east or west winds encircling the globe are examples of such large, slow systems; in them there must be approximate balance between pressure and coriolis forces. Meteorologists call a balanced wind of this nature a *geostrophic wind*.

Lines of constant pressure are represented on a weather map by *isobars*. The pressure force acts at right angles to the isobars in the direction from high pressure to low pressure. The coriolis force must then also act at right angles to the isobars in such a direction as to balance the pressure forces. Since the coriolis force acts perpendicular to the wind velocity, the wind velocity must be parallel to the isobars. In counterclockwise rotating systems corresponding to the earth's Northern Hemisphere, the coriolis force is perpendicular to the velocity and to the right (looking along the velocity vector). It follows that a geostrophic wind blows parallel to the isobars and in such a direction that the high pressure is to the right in the Northern Hemisphere (Figure 1). This behavior of the winds was observed empirically before the physics was understood and is known as *Buys Ballot's law*. It is on the strength of such observations that we may conclude that the concept of geostrophic balance is a useful one. Let us use it to explain the average westerly winds (*toward* the east) which encircle the globe in midlatitudes at a level where the pressure is about 200 millibars (the average ground-level pressure is 1013 millibars).

Figure 1
Geostrophic balance in the Northern Hemisphere. The lines represent isobars (lines of constant pressure). The balanced wind blows along the isobars. (*From Richard Goody and James C. G. Walker,* Atmospheres, *Prentice-Hall Inc., Englewood Cliffs, New Jersey, 1972. Reprinted by permission of the publisher.*)

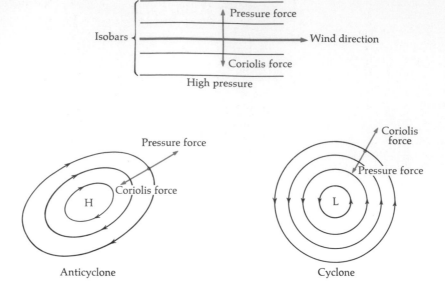

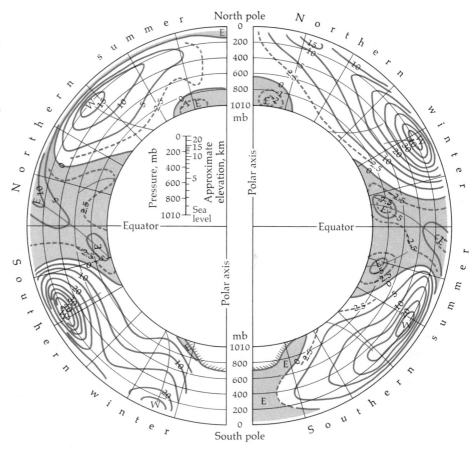

Figure 2
Mean zonal wind (east-west) averaged over latitude circles for northern summer (left) and northern winter (right). Winds are in m/sec. Easterly winds are shaded. The radial scale is pressure rather than height.

Average global winds for the two hemispheres and for the summer and winter half years are shown in Figure 2. This representation does not include average north-south winds, which are considerably smaller than the zonal winds (east-west) in middle latitudes. The jet stream is shown as a high-speed core at about 200 millibars and between latitudes 30 and 45 north or south.

In order to understand the horizontal pressure forces, we must discuss the vertical balance of the atmosphere. Since pressure decreases from about 1 atm at the surface to zero in space, there is a large upwardly directed pressure force which is approximately balanced by the large gravitational force.

Figure 3
Balance of forces on a horizontal atmospheric slab of unit area.

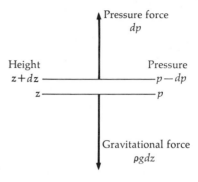

Consider the atmospheric slab shown in Figure 3 contained between the levels z and $z + dz$ (pressures p and $p - dp$). Since pressure is expressed as force per unit area, the upward force is dp. The downward force is mass $\rho\ dz$, where ρ is the density, multiplied by g. Since pressure must decrease as height increases, the balance of forces is expressed by

$$dp = -\rho g\ dz \qquad\qquad 1$$

But, according to the gas laws (Chapter 17),

$$p = \frac{k}{m}\rho T$$

where k is Boltzmann's constant, m the mass of a molecule of "air," and T the temperature in kelvins. If we eliminate ρ, we find

$$\frac{dp}{p} = -\frac{dz}{H} \qquad\qquad 2$$

where $H = kT/mg$ is the *scale height* of the atmosphere. For the earth's atmosphere at 288 K, the scale height is 8.4 km.

To clarify the issues, consider the idealized case of constant scale height. Equation 2 can then be integrated from the ground ($p = p_0$, $z = 0$) upward:

$$p = p_0 e^{-z/H} \qquad\qquad 3$$

The pressure at the surface p_0 is approximately the same over the globe. The pressure at a height z then depends mainly on a ratio z/H: the pressure will be high when H is large, i.e., when the temperature is high. Thus, in hot regions of the globe the pressure a few kilometers above the ground will exceed that in cooler regions. Since, on the average, the temperature decreases from equator to poles, at high altitudes there must be pressure forces directed from the equator toward the poles. At these altitudes, both hemispheres should behave like huge cyclones, and the geostrophic wind will be toward the east (Figure 1). We therefore anticipate a westerly wind circling the globe with its greatest intensity away from the ground, corresponding to the observations.

Review

A. List the four basic interactions.

B. Assign a letter to each basic interaction in your list in part A. After each term below, write the letter of the basic interaction (if any). If there is no associated basic interaction, explain.

1. Molecular forces
2. Centrifugal force
3. Coriolis force
4. Force holding the nucleus together
5. Beta decay

6. Friction
7. Force of a spring
8. Force holding a planet in orbit
9. Your weight

C. List three properties of electric charge.

D. List three properties of the strong nuclear force.

E. True or false:

1. The force of gravity is always attractive.

2. A proton and an electron have charges of equal magnitude.

3. Two protons always repel each other.

4. An example of the strong nuclear force is the force between the two oxygen nuclei in the O_2 molecule.

5. Contact forces and friction are related to molecular forces.

6. Pseudo forces appear only in accelerated reference frames.

Exercises

Unless instructed otherwise, use the approximate value $g = 32$ ft/sec² $= 9.8$ m/sec² for the acceleration of gravity in Exercises and Problems.

Section 6-1, Gravity

1. Using the values $G = 6.67 \times 10^{-11}$ N-m²/kg², $g = 9.8$ m/sec², and the radius of the earth $R_E = 6.37 \times 10^6$ m, calculate the mass of the earth.

2. From the known acceleration of the earth, calculate the mass of the sun.

3. Find the gravitational force which attracts a 140-lb boy to a 120-lb girl when they are 2 ft apart. (Assume that they are point masses.)

4. A particle is dropped from a height of 4000 mi above sea level. What is its initial acceleration?

5. Three 100-kg masses lie at the vertices of an equilateral triangle with 20-cm sides. Find the resultant gravitational force acting on one of the masses. (Indicate the direction of the force on your diagram.)

6. At what altitude above the earth's surface is the acceleration of gravity half its value at sea level?

7. Use Equation 6-9 to find the percentage decrease in the acceleration of gravity on top of a 14,000-ft mountain compared with that at sea level.

Section 6-2, Electromagnetic Forces

8. Two small spheres with charge of -1 μC each are 20 cm apart. (a) What is the minimum number of electrons on each sphere? (b) Find the magnitude of the electric force exerted by one sphere on the other.

9. Two small electrified objects separated by 5 cm attract each other with a force of 4 N. If their separation is increased to 7 cm, what force does one exert on the other?

10. Two small spheres each of mass $m = 10$ gm are suspended from a common point by threads of length $L = 50$ cm. When each carries a charge q, they come to equilibrium, each thread making an angle $\theta = 10°$ with the vertical, as shown in Figure 6-15. Find q.

11. Four equal charges $q = 5 \times 10^{-8}$ C are situated at the corners of a square of length $L = 4$ cm. Find the resultant electric force exerted on any one of the charges by the other three charges. (Indicate the direction of the force in your diagram.)

12. The most probable distance between the proton and the electron in the hydrogen atom is 0.53×10^{-10} m. Find the electric force of attraction between the electron and proton at this distance.

13. A penny has a mass of about 3 gm; 64 gm of copper contains Avogadro's number of atoms ($N_A = 6.02 \times 10^{23}$). Each copper atom contains 29 electrons. (a) Find the number of atoms, the number of electrons, and the total amount of negative charge in coulombs of the electrons in a penny. (b) How long would it take for this much charge to flow through a wire at a rate of 1 C/sec? (c) If 10 percent of the negative charge on one penny could be transferred to another penny, calculate the electrostatic force exerted by one penny on the other if they were separated by 20 cm. (Assume them to be point charges.)

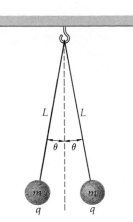

Figure 6-15
Exercise 10.

Section 6-3, Nuclear Forces, and Section 6-4, Molecular Forces

14. Calculate the magnitude of the strong nuclear force between two protons separated by a distance of 10^{-15} m assuming that this force is 10 times the electric force between the protons. Compare this force with that between the electron and proton in the hydrogen atom.

15. The equilibrium separation of the Na^+ and Cl^- ions in NaCl is about 2.4×10^{-10} m. Find the electric force of attraction between the ions at this separation assuming that they are point charges. Compare your result with the results of Exercise 12 for atomic forces and those of Exercise 14 for nuclear forces.

Section 6-5, Springs and Strings

16. A spring has a force constant $k = 2.0$ N/cm. A 5-kg mass is suspended motionless from the spring. Find (a) the numerical values of all the forces acting on the mass and (b) the extension of the spring from its equilibrium position.

17. A 100-kg mass is pulled with acceleration 0.5 m/sec² along a smooth table by a cable. The cable stretches 0.3 cm. Assuming that the cable is a spring that obeys Hooke's law, find its force constant.

18. A 2-kg mass rests on a smooth incline of angle $\theta = 30°$ supported by a spring (Figure 6-16). The spring stretches 3 cm. (a) Find the force constant of the spring. (b) If the mass is pulled down the incline 3 cm from its equilibrium position and released, what will its initial acceleration be?

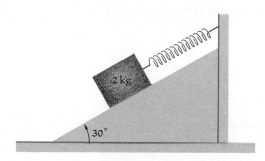

Figure 6-16
Exercise 18.

19. A mass m is attached to two springs along a line as shown in Figure 6-17. Each spring is stretched from its equilibrium position. The force constants of the springs are k_1 and k_2. (a) Find the ratio of the amount of stretching of the springs. (b) Show that if the mass is displaced a small distance x from equilibrium, the net restoring force is the same as if the mass were attached to a single spring of force constant $k = k_1 + k_2$.

Figure 6-17
Exercise 19.

Section 6-6, Contact Forces and Friction

20. The coefficient of friction between the tires of a car and the road is 0.5. (a) If the resultant force on the car is the frictional force, what is the acceleration of the car? (b) What is the least distance in which the car can stop if it is initially traveling at 88 ft/sec?

21. The force which accelerates a car along a flat road is the frictional force between the road and the car tires. (a) Explain why the acceleration is greater when the tires do not spin. (b) If a car is to accelerate from 0 to 60 mi/h in 12 sec at constant acceleration, what is the minimum coefficient of friction between the tires and the road?

22. The coefficient of static friction between a car's tires and the road is 0.6. What is the greatest speed with which the car can turn a corner of radius 200 ft if the road is flat and the car does not skid?

23. The coefficient of static friction between the floor of a truck and a box resting on it is 0.3. The truck is traveling at 50 mi/h. What is the least distance in which the truck can stop if the box is not to slide?

24. A chair slides across a polished floor. Its initial speed is 4 ft/sec. It comes to rest after sliding 4 ft. What is the coefficient of kinetic friction between the floor and the chair?

25. A 100-lb box must be moved across a level floor. The coefficient of static friction between the box and the floor is 0.6. One method is to push down on the box at an angle θ with the horizontal. Another method is to pull up on the box at an angle θ with the horizontal. (a) Explain why one method is better than the other. (b) Calculate the force necessary to move the box by each method if $\theta = 30°$ and compare these results with that for $\theta = 0°$.

26. The coefficient of friction between box A and the cart in Figure 6-18 is 0.6. The box has a mass of 2 kg. (a) Find the minimum acceleration a of the cart and box such that the box will not fall. (b) What is the magnitude of the frictional force in this case? (c) If the acceleration is greater than this minimum, will the frictional force be greater than in part (b)? Explain. (d) Show in general that for a box of any mass, the box will not fall if the acceleration is $a \geq g/\mu_s$, where μ_s is the coefficient of static friction.

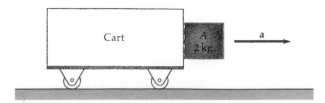

Figure 6-18
Exercise 26.

Section 6-7, Pseudo Forces

In Exercises 27 to 31, the situations described take place in a box car which has initial velocity $\mathbf{v} = 0$ *but an acceleration* $\mathbf{a} = (5 \text{ m/sec}^2)\mathbf{i}$ *(Figure 6-19). Work the exercises in the frame of the box car using pseudo forces and in an inertial frame using only real forces. Assume* $g = 10 \text{ m/sec}^2$.

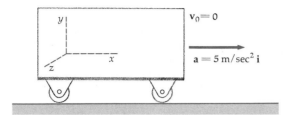

Figure 6-19
Box car initially at rest with acceleration 5 m/sec² to the right for Exercises 27 to 31.

27. A 2-kg object is slid along the smooth floor with initial velocity of (10 m/sec)$\mathbf{i}$. (*a*) Describe the motion of the object. (*b*) When does the object reach its original position relative to the box car?

28. A 2-kg object is slid along the smooth floor with initial transverse velocity of (10 m/sec)$\mathbf{k}$. Describe the motion.

29. A 2-kg object is slid along a rough floor (coefficient of sliding friction 0.3) with initial velocity (10 m/sec)$\mathbf{i}$. Describe the motion of the object. (Assume that the coefficient of static friction is greater than 0.5.)

30. A 2-kg object is suspended from the ceiling by a massless unstretchable string. (*a*) What angle does the string make with the vertical? (*b*) Indicate all the forces acting on the object in each frame.

31. A 6-kg mass is attached to the ceiling by a (massless) spring of unstretched length 0.5 m. The spring constant is 10 N/cm. By how much is the spring stretched?

Problems

1. Both the sun and the moon exert gravitational forces on the oceans of the earth, causing tides. (*a*) Show that the ratio of the force exerted by the sun to that exerted by the moon is $M_s R_m^2 / M_m R_s^2$, where M_s and M_m are the mass of the sun and moon and R_s and R_m are the distance from the earth to the sun and the moon. Evaluate this ratio using the values given in Appendix B. (*b*) Even though the sun exerts a much greater force on the ocean than the moon, the moon has a greater effect on the tides because it is the *change* in the force when the distance changes (due to rotation of the earth) that is important. Differentiate the expression $F = G m_1 m_2 / r^2$ to calculate the change in F due to a small change in r. Show that $dF/F = -2 dr/r$. (*c*) The largest change in distance due to rotation is twice the radius of the earth. Show that for a given small change in distance the change in the force exerted by the sun is related to the change in the force exerted by the moon by

$$\frac{\Delta F_s}{\Delta F_m} \approx \frac{M_s R_m^3}{M_m R_s^3}$$

and calculate this ratio.

2. Use Newton's law of gravitation along with the second law of motion to show that the period T of a planet moving in a circular orbit is related to the radius of the orbit by Kepler's law $T^2 = C r^3$ and find the constant C.

3. Kepler's law $T^2 = C r^3$ also applies to satellites in circular orbits around the earth. (*a*) Show that for this case, the constant C can be determined from the radius of the earth and the acceleration of gravity g without knowing the mass

of the earth or the constant G. (b) Calculate the period of a satellite orbiting near the surface of the earth. (c) Use your result in part (b) and Kepler's law to find the height above the earth for a satellite which appears to be stationary, i.e., which has a period of 24 h.

4. The planet Mars has a satellite orbiting at a distance $r = 9.4 \times 10^6$ m from the center of Mars with a period of 460 min. Use these data to calculate the mass of Mars.

5. A plumb bob near a large mountain is slightly deflected from the vertical by the gravitational attraction of the mountain. Estimate the order of magnitude of the angle of deflection using any assumptions you like. (Assuming the mountain to be a sphere of diameter equal to its height would allow the mountain to be treated as a point mass a distance r from the plumb bob.)

6. Show with a force diagram how a motorcycle can ride in a circle on a vertical wall. Assume reasonable parameters (coefficient of friction, radius of the circle, mass of the motorcycle, or whatever required) and calculate the minimum speed needed.

7. In Figure 6-20 the mass $m_2 = 10$ kg slides on a smooth table. The coefficients of static and kinetic friction between m_2 and the mass $m_1 = 5$ kg are $\mu_s = 0.6$ and $\mu_k = 0.4$. (a) What is the maximum acceleration of m_1? (b) What is the maximum value of the mass m_3 if m_1 moves with m_2 without slipping? (c) If $m_3 = 30$ kg, find the acceleration of each mass and the tension in the string.

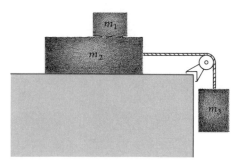

Figure 6-20
There is no friction between m_2 and the table, but there is friction between m_1 and m_2 (Problem 7).

8. A 100-lb box rests on a horizontal table. The coefficient of static friction is 0.6. The box is pulled by a massless rope with a force **F** at an angle θ, as shown in Figure 6-21. The minimum value of the force needed to move the box depends on the angle θ. Discuss qualitatively how you would expect this force to depend on θ. Compute the force for the angles $\theta = 0, 10, 20, 30, 40, 50$, and $60°$ and make a plot of F versus θ. From your plot, at what angle is it most efficient to apply the force to move the box?

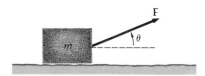

Figure 6-21
Box pulled along rough horizontal surface by force **F** making angle θ with horizontal (Problem 8).

9. A box of mass m rests on a horizontal table. The coefficient of friction is μ. A force **F** is applied at an angle θ, as in Figure 6-21. (a) Find the force F needed to move the box as a function of angle θ. (b) At the angle θ for which this force is minimum, the slope $dF/d\theta$ of the curve F versus θ is zero. Compute $dF/d\theta$ and show that this derivative is zero at the angle θ which obeys $\tan \theta = \mu$. Compare this general result with that obtained in Problem 8.

10. A mass m is attached to a spring of force constant k_1, which in turn is connected to a second spring of force constant k_2, as shown in Figure 6-22. When the springs are unstretched, the mass is at point x_0. Show that when the mass is at point x, the force on the mass is $F_x = -k(x - x_0)$, where k is related to k_1 and k_2 by $1/k = 1/k_1 + 1/k_2$.

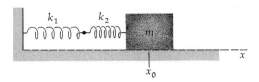

Figure 6-22
Problem 10.

11. A block of metal of mass m and density ρ is submerged in water and suspended by a spring of force constant k in an elevator. The water not only provides a buoyant force but also damps out oscillations of the metal so that it quickly comes to rest at its equilibrium position. The elevator accelerates upward with acceleration a for a short time, moves up at constant speed for a time, and then decelerates again with magnitude a until it stops at the top floor. The length of the spring is x_0 when the elevator is at rest. Describe the equilibrium position of the block of metal during the various parts of the ascent in terms of the given variables.

12. A 10-kg mass is suspended from the middle of a spring of natural length 30 cm and force constant 100 N/cm (Figure 6-23). By how much does the spring sag?

13. A student wishing to find the density of a block of wood submerges it in water and attaches a spring (of negligible volume) to the wood and to the bottom of the tank to hold the block under. He observes that the water rises 2 cm in his 20 by 40 cm tank and that his spring of force constant 4.9 N/cm stretches 1.6 cm. Find the density of the wood.

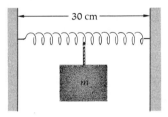

Figure 6-23
Problem 12.

14. A 2-kg mass is attached to a spring having a natural length of 30 cm and a force constant of 2000 N/m. The mass rotates in a circle (supported by a smooth horizontal surface) at $v = 4$ m/sec (Figure 6-24). (a) Write an exact equation for the amount x that the spring stretches in terms of the mass m, the force constant k, the natural length x_0, and the speed v. (b) Solve this quadratic equation for x for the values given. (c) Show that if the change in the radius of the circle because of the stretching of the spring is neglected, the stretching is given approximately by $x = mv^2/kx_0$. Find x for the values given in this approximation, and compare your result with the exact value found in part (b). (d) A better approximate value can be found by correcting the radius of the circle using the first approximation for x. Compute $x = mv^2/k(x_0 + x_1)$, where x_1 is the value found in part (c), and compare your result with the exact value.

Figure 6-24
Problem 14.

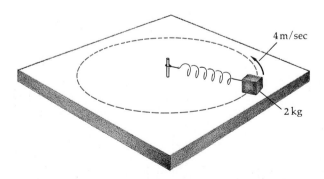

15. A block is on an incline whose angle can be varied. The angle is gradually increased from 0°. At 30°, the block starts to slide down the incline. It slides 3 m in 2 sec. Calculate the coefficients of static and kinetic friction between the block and incline.

16. Atoms are electrically neutral because they contain an equal number of positively charged protons and negatively charged electrons and because the magnitudes of the electron charge and proton charge are exactly equal. Suppose that each atom contained an equal number Z of electrons and protons but that the charge of the electron were slightly greater in magnitude than that of the proton so that the atom has a net charge $-fZe$, where f is the fractional charge difference. Estimate the order of magnitude of the fractional difference in charge that would cause the electric repulsion between the earth and the sun to equal the gravitational attraction, so that there would be no net force exerted on the earth by the sun. [This is not as difficult as it may seem at first because the mass and charge of either body is just proportional to the number of atoms. Let the mass of a typical atom be $2Zm_p$, where m_p is the mass of the proton (Appendix B) and the factor 2 is used because of the equal number of neutrons with approximately equal mass. Then find f so that the electrical repulsion of the charges fZe equals the gravitational attraction of the masses for two such atoms.]

17. A 2-kg block sits on a 4-kg block which rests on a smooth table (Figure 6-25). The coefficient of friction (both static and kinetic) between the blocks is 0.2. A force F is applied to the 4-kg block as shown. (a) What is the maximum F that can be applied if the 2-kg block is not to slide on the 4-kg block? (b) If F is twice this maximum, find the acceleration of each block. (c) If F is half this maximum, find the acceleration of each block and the friction force acting on each block.

Figure 6-25
Problem 17.

18. A road is banked so that a car traveling 50 mi/h can round a curve of radius 100 ft even if the road is so icy that the coefficient of friction is approximately zero. Find the range of speeds at which a car can travel around this curve without skidding if the coefficient of friction between the road and the tires is 0.3.

19. In an amusement-park ride participants are held against the walls of a spinning cylinder as the floor falls away. If the radius of the cylinder is 10 ft, find the minimum number of revolutions per minute necessary if the coefficient of friction between a rider and the wall is 0.4.

20. A 4-kg mass rests on a 30° incline attached to a cord which passes over a smooth peg and is attached to a second mass m, as shown in Figure 6-26. The coefficient of static friction between the mass and the incline is 0.4. (a) Find the range of possible values of m such that the system will be in static equilibrium. (b) If $m = 1$ kg, the system will be in static equilibrium. What is the frictional force on the 4-kg mass in this case?

Figure 6-26
Problem 20.

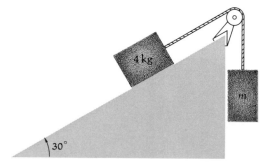

21. A smooth platform rotates counterclockwise (as seen from above) with angular velocity ω. A small object of mass m rests on the platform at a distance r from the center. The object is at rest relative to an inertial frame. (The smooth platform rotates under the mass.) (a) Describe the motion of the object relative to the platform. What are the magnitude and direction of the resultant pseudo force acting on the mass in the frame of the platform? (b) Show that in this case the coriolis force has twice the magnitude of the centrifugal force and is oppositely directed.

22. A space station has two compartments, as shown in Figure 6-27. The station rotates at B rev/min. (*a*) A mass m rests on the floor of one of the compartments a distance r from the center of rotation as shown. What is the normal force exerted by the floor on the mass? (*b*) The mass is now dropped from the ceiling of the compartment. Describe its motion relative to the compartment. What forces (including pseudo forces) act on the mass while it is falling? (*c*) Explain qualitatively why the mass falls to the floor from the point of view of an inertial reference frame in which there are no forces acting on the mass.

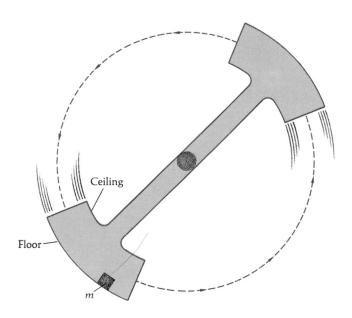

Figure 6-27
Rotating space platform
(Problem 22).

CHAPTER 7   Work and Energy

In most of our illustrations applying the laws of motion, we chose situations in which the forces are constant in magnitude in order to simplify the analysis. For one-dimensional motion with a constant force, for example, the position and velocity functions are determined by the constant-acceleration formulas 2-11 and 2-12, once the acceleration is determined by applying Newton's second law. If the force on a particle is a known function of time $F(t)$, determining the velocity and position functions is not much more difficult. Since $dv/dt = a(t) = F(t)/m$, the velocity function can be found by direct integration. In most physical situations, however, the force is neither constant nor a known function of time. It is more common to know the force on a particle as a function of the particle's position. For example, both the law of gravitation and Coulomb's law give the force exerted by one particle on another as an inverse-square function of the separation of the particles. Thus in the planetary problem we know the force on the planet exerted by the sun as a function of the distance from the planet to the sun. This distance varies as the planet moves along its elliptical orbit, and the force varies in both magnitude and direction. Similarly, in the empirical description of complex forces, we often find the force on a particle given as a function of position, such as $F = -k(x - x_0)$ for the force on a body attached to a spring which obeys Hooke's law.

In this chapter, we develop general methods for the solution of problems in which the force depends on the position of the particle. In the process we introduce the concept of the *work* done by a force as the particle on which it acts moves from one place to another. We shall see that this concept is closely related to the concept of *energy*, which plays a central role in our lives and in the description of the physical universe. In succeeding chapters we shall investigate the concept of energy more fully and see how to apply the law of conservation of energy to various problems.

7-1   Work in One Dimension

We define the work done on a particle by a force F_x as the particle moves from x_1 to x_2 to be the area under the curve F_x versus x between x_1 and x_2. That is, the work done by the force F_x is

$$W = \int_{x_1}^{x_2} F_x \, dx = \text{area under } F_x\text{-versus-}x \text{ curve} \qquad \text{7-1}$$

Definition of work in one dimension

This definition is equivalent to saying that the work dW done over a very short interval dx is

$$dW = F_x \, dx \qquad \text{7-1a}$$

The work dW is positive if the motion is in the same direction as the force and negative if it is in the opposite direction. The total work done over a finite displacement, calculated using Equation 7-1, may be either positive or negative, depending on the direction of the force during that interval, i.e., predominantly with the motion or against it.

The dimensions of work are those of force times distance. In the international system, the unit of work is the *joule* (J):

$$1 \text{ J} = 1 \text{ N-m} = 1 \text{ kg-m}^2/\text{sec}^2 \qquad \text{7-2}$$

The joule defined

where we have used the fact that $1 \text{ N} = 1 \text{ kg-m/sec}^2$. In the British engineering system the unit of work is the foot-pound. The relation between these units is easily found using the relations between pounds and newtons and between meters and feet:

$$1 \text{ J} = 0.738 \text{ ft-lb} \qquad \text{7-3}$$

Example 7-1 A block resting on a smooth table (Figure 7-1a) is attached to a horizontal spring which obeys Hooke's law and exerts a force $F_x = -kx$, where x is measured from the equilibrium length of the spring and the force constant is $k = 400$ N/m. The spring is compressed to

Jane Latta/Photo Researchers

Work is being done by these horses because they exert a force through a distance as they pull the harrow. Despite the fact that the Amish farm boy shown in this photograph is working according to the everyday use of the word, he does very little work according to the scientific definition because he exerts little force on the harrow; rather he guides its direction by use of the reins.

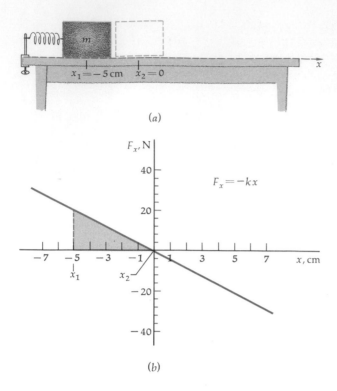

(a)

(b)

Figure 7-1
(*a*) Mass on a spring for Example 7-1. The spring is compressed 5 cm from equilibrium and released. (*b*) F_x versus x for mass on spring. The work done by the spring as the mass moves from $x_1 = -5$ cm to $x_2 = 0$ is the shaded area.

$x_1 = -5$ cm. Find the work done by the spring as the block moves from $x_1 = -5$ cm to its equilibrium position $x_2 = 0$.

Figure 7-1*b* is a sketch of this force versus distance. The work done as the block moves from x_1 to x_2 is by definition equal to the area under this curve between these limits, as indicated in the figure. This area is one-half the base times the height of the triangle. The base is 5 cm = 0.05 m, and the height is the value of the force at x_1, which is $F_x = -(400 \text{ N/m})(-0.05 \text{ m}) = +20$ N. The work done is thus $W = \frac{1}{2}(0.05 \text{ m})(20 \text{ N}) = 0.500$ N-m = 0.500 J. This work is positive because the force is in the direction of motion. This is indicated in the figure by the fact that the area is above the x axis.

7-2 Work Done by the Resultant Force; Kinetic Energy

If several forces are acting on a particle, we can compute the work done by each from the definition in Equation 7-1. The total work done on the particle by all forces acting on it is just the algebraic sum of the work done by each of the individual forces. (Work is a scalar quantity.) This total work is just the work done by the resultant force. For example, if there are three forces F_{x1}, F_{x2}, F_{x3}, the total work done when the particle moves a small distance dx is

$$dW_{\text{net}} = F_{x1} \, dx + F_{x2} \, dx + F_{x3} \, dx = (F_{x1} + F_{x2} + F_{x3}) \, dx$$
$$= \Sigma F_x \, dx \qquad \qquad \text{7-4}$$

The net work done as the particle moves from point x_1 to x_2 is

$$W_{\text{net}} = \int_{x_1}^{x_2} \Sigma F_x \, dx \qquad \qquad \text{7-5}$$

There is an important relation between the net work done on a particle and the speed of the particle at the original and final positions. This relation is obtained by using Newton's second law relating the resultant force to the acceleration of the particle $\Sigma F_x = ma$. We shall first consider the case in which the resultant force is constant and then show that the relation obtained is true for any force.

If the resultant force is constant, as in Figure 7-2, the area under the force-versus-x curve is just the force times the distance. The net work is thus

$$W_{\text{net}} = \int_{x_1}^{x_2} \Sigma F_x \, dx = \Sigma F_x \,(x_2 - x_1) = \Sigma F_x \, \Delta x = ma \, \Delta x \qquad 7\text{-}6$$

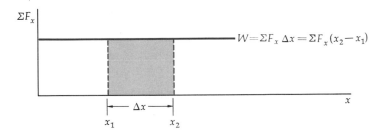

Figure 7-2
Constant resultant force ΣF_x versus x. The work done as the particle moves from x_1 to x_2 is indicated by the shaded area $W = \Sigma F_x(x_2 - x_1)$

Since the acceleration is constant, we can relate the distance moved to the initial and final speeds by the constant-acceleration formula 2-13:

$$v_2{}^2 - v_1{}^2 = 2a \, \Delta x$$

Thus

$$W_{\text{net}} = ma \, \Delta x = \tfrac{1}{2}mv_2{}^2 - \tfrac{1}{2}mv_1{}^2 \qquad 7\text{-}7$$

The quantity $\tfrac{1}{2}mv^2$, called the *kinetic energy* E_k of the particle, is a scalar quantity which depends on the particle's mass and speed. We can also write the kinetic energy in terms of the momentum of the particle, $p = mv$:

$$E_k = \tfrac{1}{2}mv^2 = \frac{(mv)^2}{2m} = \frac{p^2}{2m} \qquad 7\text{-}8 \qquad \textit{Kinetic energy defined}$$

The quantity on the right side of Equation 7-7 is the change in the kinetic energy of the particle, i.e., the kinetic energy $\tfrac{1}{2}mv_2{}^2$ at the end of the interval when the particle is at x_2 minus its kinetic energy $\tfrac{1}{2}mv_1{}^2$ at the beginning of the interval when the particle was at x_1. The work done by the resultant force equals the change in the kinetic energy of the particle:

$$W_{\text{net}} = \tfrac{1}{2}mv_2{}^2 - \tfrac{1}{2}mv_1{}^2 = \Delta E_k \qquad 7\text{-}9 \qquad \textit{Work-energy theorem}$$

We shall now show that this result, derived for the special case of a constant resultant force, is true for a general force which may vary with position. Consider a resultant force which varies with x as shown in Figure 7-3. The work done by this force as the particle moves from position x_1 to position x_2 is the area under this curve, i.e., the shaded region in the figure. In Figure 7-4 we have approximated this area by a series of rectangles. By choosing a large number of narrow rectangles, we can approximate the area to any desired accuracy. (In fact, the area under the original curve is defined to be the area of the rectangles in the limit as the number of the rectangles goes to infinity and the width

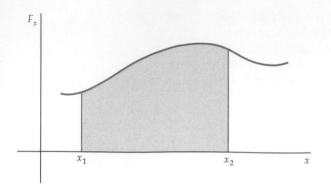

Figure 7-3
A general force F_x versus x. The work done as the particle moves from x_1 to x_2 is the area under the curve indicated.

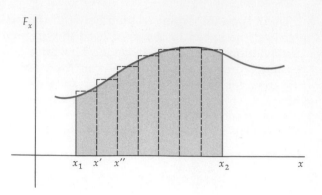

Figure 7-4
The shaded area from Figure 7-3 is broken up into a series of rectangles. Their sum approximates the area under the curve. The approximation is improved as the number of rectangles is increased and the width of each is decreased.

of each goes to zero.) Since for each rectangle the force is constant, we can apply Equation 7-9. The area of each rectangle equals the change in kinetic energy for that interval. For example, the area of the first rectangle for the interval $\Delta x_1 = x' - x_1$ equals $\frac{1}{2}mv'^2 - \frac{1}{2}mv_1^2$, where v' is the speed of the particle at position x'. Similarly, the area of the second rectangle for the interval $\Delta x_2 = x'' - x'$ equals $\frac{1}{2}mv''^2 - \frac{1}{2}mv'^2$. The sum of these two areas is $\frac{1}{2}mv''^2 - \frac{1}{2}mv_1^2$, the sum of the changes in kinetic energy, which is just the net change for the interval from x_1 to x''. The total area under the curve equals the sum of the changes in kinetic energy for each subinterval, or the net change in the kinetic energy $\frac{1}{2}mv_2^2 - \frac{1}{2}mv_1^2$.

An alternative proof of the validity of Equation 7-9 for a general force in one dimension makes use of the chain rule for derivatives (see Appendix Table E-1, rule 3). Combining Newton's second law with Equation 7-5 gives

$$W_{\text{net}} = \int_{x_1}^{x_2} \Sigma F_x \, dx = \int_{x_1}^{x_2} m \frac{dv}{dt} \, dx \qquad 7\text{-}10$$

Since the force is a function of x, the acceleration and speed are also functions of x. Then

$$\frac{dv}{dt} = \frac{dv}{dx} \frac{dx}{dt} = v \frac{dv}{dx}$$

since $dx/dt = v$. Substituting this into Equation 7-10, we have

$$W_{\text{net}} = \int_{x_1}^{x_2} mv \frac{dv}{dx} \, dx = \int_{v_1}^{v_2} mv \, dv = \frac{1}{2}mv_2^2 - \frac{1}{2}mv_1^2$$

In one dimension, the work done by the resultant force equals the change in kinetic energy whether the force is constant or not.

Example 7-2 A 3-kg block initially at rest is pulled with a constant horizontal force of 6 N for a distance of 5 m along a smooth horizontal surface (Figure 7-5). Find the work done and the final speed of the block.

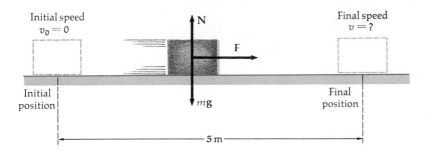

Figure 7-5
Example 7-2. The 3-kg block is accelerated from rest by a constant horizontal force of 6 N. Since the vertical forces cancel, this is the resultant force. The work done by the resultant force, 6 N $\times$ 5 m = 30 J, equals the change in kinetic energy of the block.

Since the weight of the block is balanced by the normal force of the table and there is no friction, the resultant force is just the applied force $F = 6$ N. The work done by this constant force is

$$W = F\,\Delta x = (6\text{ N})(5\text{ m}) = 30\text{ N-m} = 30\text{ J}$$

The initial kinetic energy is zero since the initial speed is zero. The final kinetic energy is $\frac{1}{2}mv^2$. Thus

$$W = \Delta E_k = \tfrac{1}{2}mv^2 - 0 = 30\text{ J}$$

$$v^2 = \frac{60\text{ J}}{3\text{ kg}} = 20\text{ m}^2/\text{sec}^2 \qquad v = 4.47\text{ m/sec}$$

Example 7-3 A 4-kg block is attached to the spring of Example 7-1. The spring is compressed 5 cm from equilibrium and released from rest. Find the speed of the block when the spring is at its equilibrium position.

The work done by the spring from $x_1 = -5$ cm to $x_2 = 0$ is 0.500 J, as computed in Example 7-1. Since the force exerted by the spring is the resultant force, this work equals the change in kinetic energy from 0 to $\frac{1}{2}mv^2$. Therefore

$$W = 0.500\text{ J} = \tfrac{1}{2}mv^2 = \tfrac{1}{2}(4.00)v^2$$

$$v^2 = 0.250\text{ m}^2/\text{sec}^2$$

$$v = 0.50\text{ m/sec} = 50\text{ cm/sec}$$

Example 7-4 Find the speed of the block in Example 7-3 when the spring is at its equilibrium position if the coefficient of kinetic friction between the table and block is 0.200.

In this case the work done by the spring is not the net work done on the block because the frictional force also does (negative) work. Since the frictional force is constant, the work done by it is just the force times the distance. This work is negative because the force $f_x = -\mu_k N = -\mu_k mg$ is opposite the direction of motion. The work done by the frictional force is

$$W_f = -\mu_k mgx = -(0.200)(4.00)(9.81)(0.050) = -0.392\text{ J}$$

The net work done on the block is then

$$W_{\text{net}} = 0.500\text{ J} - 0.392\text{ J} = 0.108\text{ J}$$

The net work equals the change in kinetic energy, which is $\frac{1}{2}mv^2$:

$$W_{\text{net}} = 0.108\text{ J} = \tfrac{1}{2}mv^2 = \tfrac{1}{2}(4.00)v^2$$

$$v^2 = 0.054\text{ m}^2/\text{sec}^2$$

$$v = 0.232\text{ m/sec} = 23.2\text{ cm/sec}$$

Questions

1. A block is pulled up an inclined plane a certain distance. Then it is pulled down the same distance. How does the work done by the friction as the block moves up compare with that done by friction as the block moves down?

2. How does the work done by gravity as the block in Question 1 moves up compare with the work done by gravity as it moves down?

3. Is it possible to exert a force which does work on a body without increasing its kinetic energy? If so, give an example.

4. How does the kinetic energy of a car change when its speed is doubled?

7-3 Work and Energy in Three Dimensions

The definition of work in one dimension given by Equation 7-1 is useful because the net work done by all the forces acting on a particle equals the change in kinetic energy of the particle. This result can be carried over to the motion of a particle in three dimensions if we modify the definition of work slightly to recognize that in general the resultant force is not parallel to the direction of motion of the particle, as indicated in Figure 7-6b. In general, the component of the force perpendicular to the velocity of the particle does not affect the speed of the particle. Instead, this component changes the direction of the velocity. (For example, if there is no component of the force along the instantaneous line of motion, i.e., if the resultant force is perpendicular to the velocity, the particle moves in a circle with constant speed. In this case the acceleration produced by the force is the centripetal acceleration, which is associated completely with the change in direction of the velocity.) On the other hand, the component of the force tangential to the curve (parallel or antiparallel to the velocity) produces a tangential acceleration which is just the time rate of change of the speed. It is this component which affects the kinetic energy of the particle.

Consider a particle moving along a curve as in Figure 7-6a. Let s be the distance traveled by the particle as measured along the curve. The work done by a force $\mathbf{F}$ acting on the particle as it moves a small distance ds is defined to be the product of the component of the force along a line tangent to the curve F_s and the distance ds:

Work in three dimensions defined

$$dW = F_s \, ds \qquad\qquad 7\text{-}11$$

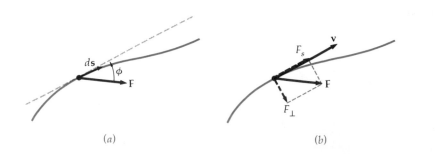

(a)   (b)

Figure 7-6
(a) Particle moving along a curve in space. (b) The component of the force perpendicular to the motion $F_\perp = F \sin \phi$ affects the direction of motion but not the speed. The component parallel to the motion $F_s = F \cos \phi$ equals the mass times the tangential acceleration, which is dv/dt. The work done by this force during the displacement ds is defined by $dW = F_s \, ds$.

The work done by the rope equals the force times the distance raised. (*Courtesy of the Museum of Modern Art/Film Archives, New York.*)

To find the total work done by the force as the particle moves from point 1 to point 2 we compute the product $F_s \, ds$ for each element of the path and sum:

$$W = \int F_s \, ds \qquad\qquad 7\text{-}12$$

As with motion in one dimension, the net work done by the resultant force equals the change in kinetic energy. This is most easily seen by noting that the component of the resulting force tangential to the curve equals the mass times the tangential acceleration, which is the rate of change of the speed

$$\Sigma F_s = ma_t = m \frac{dv}{dt}$$

We can now think of the speed as a function of the distance s measured along the curve and apply the chain rule for derivatives:

$$\frac{dv}{dt} = \frac{dv}{ds}\frac{ds}{dt} = v\frac{dv}{ds}$$

where we have used the fact that ds/dt is just the speed v. Thus the work done by the resultant force is

$$W_{\text{net}} = \int_{s_1}^{s_2} \Sigma F_s \, ds = \int_{s_1}^{s_2} m\frac{dv}{dt} \, ds$$

$$= \int_{s_1}^{s_2} mv\frac{dv}{ds} \, ds = \int_{v_1}^{v_2} mv \, dv = \tfrac{1}{2}mv_2{}^2 - \tfrac{1}{2}mv_1{}^2 \qquad 7\text{-}13$$

Work-energy theorem in three dimensions

Equation 7-13 (along with its one-dimensional counterpart, Equation 7-9) is known as the *work-energy theorem:*

The net work done by the resultant force equals the change in the kinetic energy of the particle.

The theorem follows directly from the definition of work ($dW = F_s\, ds$) and kinetic energy ($E_k = \frac{1}{2}mv^2$) and from Newton's second law of motion ($\Sigma \mathbf{F} = m\, d\mathbf{v}/dt$).

If ϕ is the angle between the force $\mathbf{F}$ and the element of displacement $d\mathbf{s}$, the component of the force parallel to $d\mathbf{s}$ is $F_s = F \cos \phi$, as can be seen from Figure 7-6. Therefore Equation 7-12 for the work done by a force $\mathbf{F}$ can be written

$$W = \int F_s\, ds = \int F \cos \phi \, ds \qquad\qquad 7\text{-}14$$

In general both F and ϕ vary as the particle moves along a curve in three dimensions. Note that $\cos \phi \, ds$ is the component of the displacement ds on a line parallel to the force. Thus we may think of the element of work $dW = F \cos \phi \, ds$ either as the product of the displacement ds and the parallel component of the force $F \cos \phi$ or as the product of the force F and the parallel component of the displacement $ds \cos \phi$. The second interpretation is sometimes useful when the force is constant in direction.

Example 7-5 A particle of mass m starts from rest and slides down a smooth plane inclined at angle θ from the horizontal. Find the work done by all the forces and the speed of the particle after it slides a distance s measured along the plane.

The forces acting on the particle are the weight $m\mathbf{g}$ and the contact force $\mathbf{N}$ exerted by the plane, which is perpendicular to the plane because the plane is smooth. These are indicated in Figure 7-7. Since the normal force $\mathbf{N}$ is perpendicular to the motion, it does no work. The only force that does work is the weight $m\mathbf{g}$, which has the component $mg \sin \theta$ in the direction of motion. (Note that the angle ϕ between the weight and the direction of motion is the complement of the angle of inclination θ. Thus the element of work done by the weight is $dW = mg \cos \phi \, ds = mg \sin \theta \, ds$.) Since the weight is constant, the work done over the distance s is merely $mg \sin \theta \, s$. This is the net work done by all the forces, and therefore it equals the change in kinetic energy, which is just $\frac{1}{2}mv^2$ because the particle begins from rest. Thus the work-energy theorem gives

$$W_{\text{net}} = mg \sin \theta \, s = \tfrac{1}{2}mv^2$$

or

$$v^2 = 2s \sin \theta \, g = 2gh$$

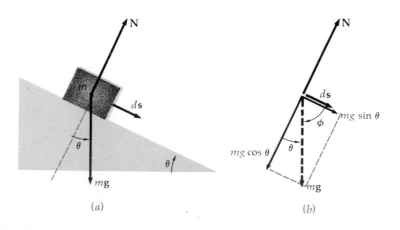

(a)   (b)

Figure 7-7
(a) Block on an inclined plane for Example 7-5. (b) The free-body diagram for the block. The resultant force is $mg \sin \theta$ down the incline. The work done by the resultant force is $mg \sin \theta \, s = mgh$, where s is the total distance traveled and $h = s \sin \theta$ is the vertical distance the block descends.

where $h = s \sin \theta$ is the total vertical height through which the particle descended. The work done by the earth on the particle is mgh, independent of the angle of the incline. If the angle θ were increased, the particle would travel a smaller distance s to drop the same vertical distance h but the component of the weight parallel to the motion $mg \cdot \sin \theta$ would be greater, making the work done $mg \sin \theta$ s the same.

The results of Example 7-5 can be generalized. Consider a particle sliding down a curve of any shape under the influence of gravity. Figure 7-8 shows a small displacement ds parallel to the curve. The work done by the earth during this displacement is $mg \cos \phi \, ds$, where ϕ is the angle between the displacement and the downward force of gravity. The quantity $ds \cos \phi$ is just dh, the vertical distance dropped. As the particle slides down the curve, the angle ϕ varies, but for each displacement ds, the downward component of displacement parallel to the weight is $ds \cos \phi = dh$ and the work done by the earth is $mg \, ds \cdot \cos \phi = mg \, dh$. Thus the total work done by the earth is mgh, where h is the total vertical distance the particle descends. If the curve is smooth (frictionless), the weight is the only force that does work. In this case the speed of the particle after descending a vertical distance h is obtained from $\frac{1}{2}mv^2 - \frac{1}{2}mv_0^2 = mgh$, where v_0 is the initial speed. If the curve is not smooth, the frictional force will do work (this work will be negative because the frictional force is in the direction opposite the motion). The work done by the friction force depends on the length and shape of the curve and on the coefficient of friction.

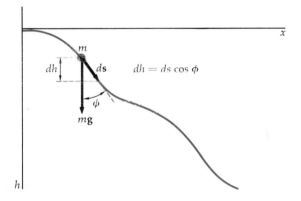

Figure 7-8
Particle of mass m sliding down a smooth curve of arbitrary shape. The work done by gravity for a displacement ds is $mg \cos \phi \, ds = mg \, dh$, where $dh = \cos \phi \, ds$ is the vertical component of the displacement. The total work done by gravity is mgh, where h is the total vertical distance descended.

Questions

5. A body moves in a circle at constant speed. Does the force that accounts for its acceleration do work on it? Explain.

6. When a body moves along some surface, what work does the normal force exerted by the surface do?

7. A heavy box is dragged across the floor from point A to point B. Will the work done by friction differ if the box is pulled along a zigzag path instead of along the shortest path from A to B? How?

8. The net force acting on a body does no work. Can the body be moving along a straight line?

9. How can a man who can exert a force of 100 lb raise a 200-lb box from the ground to a shelf a height 4 ft above the ground? Could a boy who can exert only 50 lb raise the box to the shelf? How much work would each have to do?

7-4   Potential Energy

Often the work done by a force applied to an object produces no increase in the kinetic energy of the object because other forces do an equal amount of negative work. For example, consider a block being pulled slowly up a smooth incline with constant velocity by an applied force F_{app} which just balances the component of the weight of the block parallel to the incline, $F_{app} = mg \sin \theta$.

The work done by the applied force in moving the block a distance s is

$$W = F_{app}s = mg \sin \theta \, s = mgh$$

where $h = s \sin \theta$ is the height the block has been raised above its initial level. In this case there is no increase in kinetic energy of the block because the weight does an equal but negative amount of work:

$$W_{earth} = -mg \sin \theta \, s = -mgh$$

However, we can convert the work done by the applied force into a change in kinetic energy by merely releasing the block and letting it slide back down the incline. The weight will then do a positive amount of work $mg \sin \theta \, s = mgh$, which equals the increase in kinetic energy because it is the only work done on the block.

We can use the gravitational attraction of the earth for the block to store the work we do on the block for later use in giving the block kinetic energy. We say that the block at the height h has *potential energy mgh* relative to the original position. The work done by the applied force increases the potential energy of the block. When the block slides back down under the influence of its weight alone, the work done by the earth decreases the potential energy while increasing the kinetic energy by an equal amount. Here potential energy is converted into kinetic energy. Since the loss of potential energy equals the gain in kinetic energy for each part of the downward motion, the sum of potential energy and kinetic energy is constant as the block slides down the incline. This is an example of the *conservation of energy*.

If we project an object up a smooth incline with initial velocity v_0, it will travel until the negative work done by the earth decreases the kinetic energy to zero. Since this negative work equals the increase in potential energy mgh, the maximum height attained by the object is given by $mgh = \frac{1}{2}mv_0^2$ (Figure 7-9). (Again, this height is independent of the angle of inclination of the plane.) The object stops momentarily at this height and then slides back down, gaining kinetic energy and losing potential energy. When it reaches its starting point its kinetic energy is again $\frac{1}{2}mv_0^2$.

Jen and Des Bartlett/Photo Researchers

These climbers do work in increasing their gravitational potential energy.

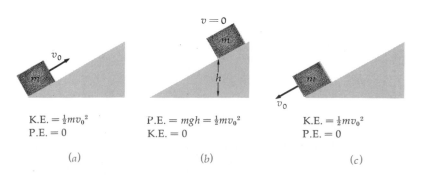

K.E. $= \frac{1}{2}mv_0^2$
P.E. $= 0$

(a)

P.E. $= mgh = \frac{1}{2}mv_0^2$
K.E. $= 0$

(b)

K.E. $= \frac{1}{2}mv_0^2$
P.E. $= 0$

(c)

Figure 7-9
A particle projected up a smooth inclined plane with initial speed v_0. (a) Initially its kinetic energy is $\frac{1}{2}mv_0^2$, and its potential energy is zero. (b) When the particle comes momentarily to rest, it has potential energy $mgh = \frac{1}{2}mv_0^2$ and zero kinetic energy. (c) It then slides back down the incline and has its original kinetic energy when it reaches its starting point.

There are many other kinds of potential energy. Suppose we have a mass on a spring at its equilibrium position. If we extend the spring by applying a force equal to that exerted by the spring (after we give the mass a slight start), the spring does negative work equal in magnitude to that done by the applied force and the kinetic energy does not change. In the extended position, the spring has potential energy equal to the magnitude of the (negative) work done by the spring. When the mass is released, the spring does positive work on the mass. This increases the kinetic energy of the mass and decreases the potential energy. We shall investigate this situation and other examples of potential energy in detail in the next chapter.

Atlas. There is no work done without motion.

7-5 Work and Energy

Let us compare the scientific definition of work with our everyday use of the word. Consider a person holding a weight a distance h off the floor, as in Figure 7-10. In everyday usage, we say that it takes work to do this: we have to exert a force, which is tiring to our muscles; but in our scientific definition, no work is done by a force acting on a stationary object. We could eliminate the effort of holding the weight by merely tying the string to some fixed object and the block would be supported with no help from us. It is useful to think of work, as we have defined it, as a quantity which has to be paid for in some way. The payment is in some kind of energy loss by whatever exerts the force. For example, we cannot lift the weight in Figure 7-10 to a greater height unless we supply energy in some form or another. If we attach a slightly greater weight to the string and lift our original weight by letting the second weight fall, we pay in the loss of potential energy of our second weight. If we merely lift the weight with our muscles, we pay for the work done in the loss of internal chemical energy of our body. In order to continue to do work with our muscles, we must eventually replenish this energy by taking in nourishment. If we attach the string to an electric motor to raise the weight, we spend electric energy to perform this work.

Figure 7-10
No work is done by the man holding the weight at a fixed height. The same task could be accomplished by tying the rope to a fixed point.

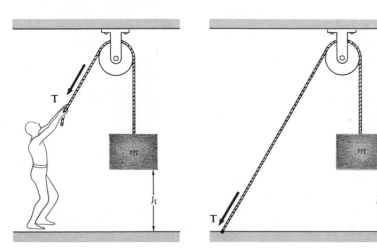

Energy Resources

Laurent Hodges
Iowa State University

Directly or indirectly, energy is essential for everything that exists or is done in modern society: coal in extracting metals from their ores, petroleum fuels for ground and air transportation, natural gas or fuel oil for heating buildings, electricity for industrial machinery and home appliances.

Primary energy resources, i.e., resources from the environment which are the ultimate sources of our energy, include the fossil fuels (coal, oil, and gas) and three types of electricity: hydroelectricity generated by falling water, electricity from nuclear power plants (ultimately from the heat of fission of ^{235}U), and electricity generated by geothermal power (the heat of steam and hot water from beneath the earth). These are really only the so-called *commercial* energy sources. Although such energy sources as wood, peat, and cow or camel dung are absent from international trade and account for perhaps only 3 percent of world energy consumption, they may represent up to about half the energy consumption in certain developing countries. Wood was the leading energy source in the United States throughout most of the nineteenth century, but coal displaced it in the 1880s and the fossil fuels have been predominant ever since.

Figure 1 shows the consumption of commercial energy by the world and the United States in the years 1950 and 1970. During that period coal declined in relative importance while crude oil and natural gas grew in popularity because of their greater convenience in transportation and use and their lesser environmental problems. Nuclear power was first used during that period but still made only a small contribution in the early 1970s. Nuclear energy and natural gas are relatively sophisticated fuels used mainly in the developed countries.

Figure 1
Commercial energy consumption in the United States and the world in 1950 and 1970.

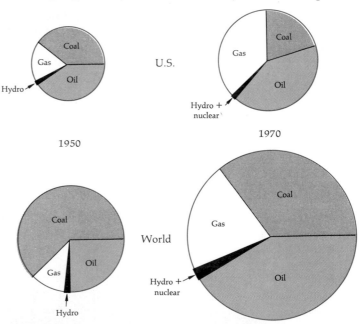

The energy actually used is often in a modified form, referred to as a *secondary energy resource*. For example, about 80 percent of the electricity used in the United States is generated by the combustion of fossil fuels; in this case the fossil fuel is the primary energy resource and the electricity is a secondary resource. Similarly the hydrogen fuel of any future "hydrogen economy" will be a secondary resource obtained from one or more primary resources via the electrolysis of water.

Figure 2 shows the per capita energy consumption in the world and different parts of the world. United States per capita consumption is about 2½ times that in the other developed countries (Canada, Europe, Australia, New Zealand, Japan, Israel, Soviet Russia) and 25 to 30 times that in the developing countries. It is often stated that the United States cannot cut its energy consumption significantly without returning to some unspecified "dark age." Actually a drastic 60 percent cut would only put United States per capita energy consumption at the level of that in the other developed countries, where the standard of living is certainly decent and civilized. A less drastic but supposedly intolerable cut of 30 percent would return us to consumption levels of about 10 years ago, which few of us remember as a dark age.

Annual increases in United States per capita consumption in recent years have about equaled the total per capita consumption in developing countries.

Figure 2
Per capita energy consumption in 1970.

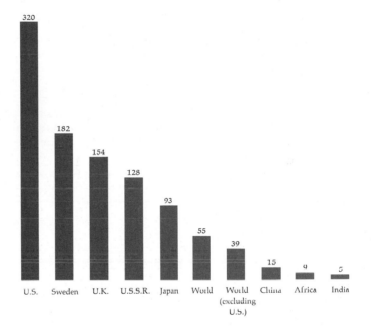

This is an unfortunate trend. It would be more equitable for the consumption increases to come in the developing countries, especially since they are providing most of the world's energy production increases. In the early 1950s both the United States and Great Britain were producing approximately as much energy as they consumed, but by the early 1970s they were importing about 15 and 50 percent, respectively, of the energy they consumed, most of the imports being produced in developing countries.

About 40 percent of United States energy consumption is by the industrial sector, 25 percent is by ground, air, and water transportation, 20 percent by residences, and 15 percent by commercial establishments (stores, offices, restaurants, etc.). Figure 3 shows the breakdown into end uses. All the small electrical appliances in use—clocks, radios, televisions, power tools, hair dryers—account for much less than 1 percent of total energy consumption in the United States.

The fossil fuels that constitute the bulk of our energy sources are a finite nonrenewable resource and cannot meet our energy needs forever. Although the recoverable reserves of these fuels can be estimated only imperfectly and future energy demands are uncertain, it seems clear that oil and gas can continue to be a major resource for only a few more decades at most and coal only a few centuries. Nuclear energy in the form of ^{238}U or ^{232}Th used in breeder reactors or deuterium or other light isotopes used in controlled thermonuclear fusion reactors (should they prove feasible) could supply projected energy needs for much longer—thousands or even millions of years assuming a leveling off of the growth of energy consumption.

Figure 3
Approximate energy use in
the United States as of the
early 1970s.

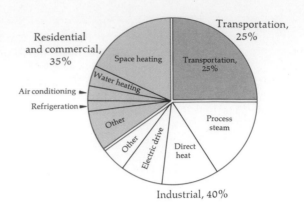

Other renewable or continuously available energy resources which are of negligible importance today may make significant contributions to our energy supply in the future: e.g., solar energy, wind power, organic wastes, tidal energy. Solar energy is especially abundant, though its low power density may make necessary the collection of solar radiation over large land areas.

Extensive use of renewable energy resources may be required if society is to have abundant and environmentally tolerable energy. Society has become increasingly concerned with the environmental costs of energy production and use. Some of the major environmental costs are land devastation by the strip mining of coal, high acid concentrations in the streams of coal mining areas, particulate matter and sulfur oxide air pollution from coal combustion, water pollution from oil-field operations, air pollution from petroleum refining, sulfur oxide pollution from petroleum combustion, carbon monoxide and hydrocarbon air pollution from the use of gasoline in motor vehicles, nitrogen oxide air pollution from any high-temperature combustion process, and radiation releases from uranium mining and from nuclear power plants.

Unfortunately much of the energy used today is wasted. For this reason energy conservation has become an important policy to implement. If we can use energy more efficiently, if we can devise new products with lower energy requirements (think of pocket calculators!), if we can change our life styles to reduce energy consumption, we can slow the rate of depletion of our energy resources and simultaneously decrease the adverse environmental effects of energy production and consumption. It will be especially advantageous to increase the use of renewable resources and thereby conserve fossil fuels for their other valuable uses, such as the manufacture of chemicals and drugs.

7-6   Power

The rate at which a force does work is called the *power input P* of the force. Consider a particle with instantaneous velocity **v**. In a short time interval dt, the particle has a displacement $d\mathbf{s} = \mathbf{v}\,dt$. The work done by a force **F** acting on the particle during this time interval is

$$dW = F \cos \phi \; ds = F \cos \phi \; v \; dt$$

where ϕ is the angle between **F** and **v**. The power input is

$$P = \frac{dW}{dt} = F \cos \phi \; v \qquad\qquad 7\text{-}15 \qquad \textit{Power defined}$$

The SI unit of power, one joule per second, is called a watt (W):

$$1 \text{ W} = 1 \text{ J/sec} \qquad\qquad 7\text{-}16 \qquad \textit{Watt defined}$$

When you buy energy from a power company, you are usually charged by the kilowatt-hour (kWh). A kilowatt-hour of energy is

$$1 \text{ kWh} = 10^3 \text{ W} \times 3600 \text{ sec} = 3.6 \times 10^6 \text{ W-sec} = 3.6 \times 10^6 \text{ J} \qquad 7\text{-}17$$

In the British engineering system, the unit for energy is the foot-pound and for power is the foot-pound per second. A more common multiple of this unit is called a horsepower:

$$1 \text{ horsepower} = 550 \text{ ft-lb/sec} = 746 \text{ W} \qquad 7\text{-}18$$

Example 7-6 A 1-ton automobile travels at a steady speed of 60 mi/h (= 88 ft/sec) along a road which gains altitude at a constant rate of 100 ft in every mile traveled. What power must be expended by the motor to do this?

The automobile is moving along a plane inclined at an angle θ given by

$$\sin \theta = \frac{100 \text{ ft}}{5280 \text{ ft}} = 0.0189$$

Since the car moves at constant speed, its motor must supply a force up along the incline which just balances the component of the car's weight down the incline. This applied force must be

$$F_{\text{app}} = mg \sin \theta = (2000 \text{ lb})(0.0189) = 37.9 \text{ lb}$$

Since $\mathbf{F}_{\text{app}}$ and $\mathbf{v}$ are parallel, the power required is

$$P = F_{\text{app}}v = (37.9 \text{ lb})(88 \text{ ft/sec}) = 3.3 \times 10^3 \text{ ft-lb/sec}$$
$$= 6.06 \text{ horsepower}$$

Actually additional power is needed to do work against friction in the moving parts of the car and against a considerable air resistance.

7-7   Scalar, or Dot, Product of Two Vectors

The scalar quantities work and power, defined by Equations 7-14 and 7-15, each involve the product of the magnitudes of two vectors and the cosine of the angle between them; the vectors are the force $\mathbf{F}$ and the displacement $d\mathbf{s}$ for work and the force $\mathbf{F}$ and the velocity $\mathbf{v}$ for power. A scalar function of two vectors occurs often in physics and is called the *scalar product*. The scalar product of two vectors $\mathbf{A}$ and $\mathbf{B}$ is written $\mathbf{A} \cdot \mathbf{B}$ and defined

$$\mathbf{A} \cdot \mathbf{B} = AB \cos \phi \qquad 7\text{-}19$$

Scalar, or dot, product defined

where ϕ is the angle between $\mathbf{A}$ and $\mathbf{B}$. Because of this notation, the scalar product is also called the *dot product*.[1] With this notation, the element of work dW is written

$$dW = \mathbf{F} \cdot d\mathbf{s} \qquad 7\text{-}20$$

---

[1] A vector function of two vectors, the vector product or cross product, written $\mathbf{A} \times \mathbf{B}$, will be defined in Chapter 13.

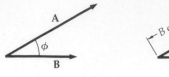
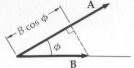
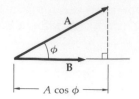

Figure 7-11
The geometric interpretation of the dot product $\mathbf{A} \cdot \mathbf{B}$. It can be thought of as the magnitude of $\mathbf{A}$ times the component of $\mathbf{B}$ in the direction of $\mathbf{A}$, or the magnitude of $\mathbf{B}$ times the component of $\mathbf{A}$ in the direction of $\mathbf{B}$; that is, $\mathbf{A} \cdot \mathbf{B} = A(B \cos \phi) = B(A \cos \phi)$.

and the power is written

$$P = \mathbf{F} \cdot \mathbf{v} \hspace{4cm} \text{7-21}$$

The dot product $\mathbf{A} \cdot \mathbf{B}$ can be thought of as the product of A and the component $B \cos \phi$ in the direction of $\mathbf{A}$ or as the product of B and the component $A \cos \phi$ in the direction of $\mathbf{B}$ (Figure 7-11). We have already discussed these interpretations applied to the dot product of $\mathbf{F}$ and $d\mathbf{s}$.

If $\mathbf{A}$ and $\mathbf{B}$ are perpendicular, their dot product is zero because $\phi = 90°$. Conversely, if $\mathbf{A} \cdot \mathbf{B} = 0$, either $\mathbf{A} = 0$, or $\mathbf{B} = 0$, or $\mathbf{A}$ and $\mathbf{B}$ are mutually perpendicular. If $\mathbf{A}$ and $\mathbf{B}$ are parallel vectors, the dot product is just the product of their magnitudes. The dot product of a vector with itself is the square of the magnitude of the vector:

$$\mathbf{A} \cdot \mathbf{A} = A^2$$

It follows from the definition that the scalar product is independent of the order of multiplication $\mathbf{A} \cdot \mathbf{B} = \mathbf{B} \cdot \mathbf{A}$. This is known as the *commutative rule of multiplication*. The scalar product also obeys the distributive rule of multiplication:

$$(\mathbf{A} + \mathbf{B}) \cdot \mathbf{C} = \mathbf{A} \cdot \mathbf{C} + \mathbf{B} \cdot \mathbf{C}$$

(The proof of this rule is given in Section 7-8.) We can use this result to write the dot product of two vectors in terms of their rectangular components:

$$
\begin{aligned}
\mathbf{A} \cdot \mathbf{B} = {}& (A_x\mathbf{i} + A_y\mathbf{j} + A_z\mathbf{k}) \cdot (B_x\mathbf{i} + B_y\mathbf{j} + B_z\mathbf{k}) \\
= {}& A_xB_x\mathbf{i} \cdot \mathbf{i} + A_xB_y\mathbf{i} \cdot \mathbf{j} + A_xB_z\mathbf{i} \cdot \mathbf{k} \\
& + A_yB_x\mathbf{j} \cdot \mathbf{i} + A_yB_y\mathbf{j} \cdot \mathbf{j} + A_yB_z\mathbf{j} \cdot \mathbf{k} \\
& + A_zB_x\mathbf{k} \cdot \mathbf{i} + A_zB_y\mathbf{k} \cdot \mathbf{j} + A_zB_z\mathbf{k} \cdot \mathbf{k}
\end{aligned}
$$

Since the unit vectors $\mathbf{i}$, $\mathbf{j}$, and $\mathbf{k}$ are mutually perpendicular, all the dot products of two different unit vectors in the expression above are zero. Thus six of the nine terms vanish. Each of the remaining terms contains the product of a unit vector with itself. Since each of these vectors has magnitude 1, these remaining dot products have just the value 1. The result for $\mathbf{A} \cdot \mathbf{B}$ thus reduces to

$$\mathbf{A} \cdot \mathbf{B} = A_xB_x + A_yB_y + A_zB_z \hspace{3cm} \text{7-22}$$

If the vectors $\mathbf{A}$ and $\mathbf{B}$ are given in terms of their rectangular components, we can use Equation 7-22 to find the angle between them.

Example 7-7 Find the angle between the vectors $\mathbf{A} = 2\mathbf{i} + 2\mathbf{j} + 4\mathbf{k}$ and $\mathbf{B} = -2\mathbf{i} + 5\mathbf{j} + 5\mathbf{k}$.

We have that $\mathbf{A} \cdot \mathbf{B} = AB \cos \phi = A_xB_x + A_yB_y + A_zB_z = (2)(-2) + (2)(5) + (4)(5) = -4 + 10 + 20 = 26$. The magnitudes of the vectors are

$$A = \sqrt{A_x^2 + A_y^2 + A_z^2} = \sqrt{4 + 4 + 16} = \sqrt{24}$$

$$B = \sqrt{B_x^2 + B_y^2 + B_z^2} = \sqrt{4 + 25 + 25} = \sqrt{54}$$

Thus

$$\cos \phi = \frac{\mathbf{A} \cdot \mathbf{B}}{AB} = \frac{26}{\sqrt{24} \ \sqrt{54}} = \frac{26}{36} = 0.72$$

$$\phi = 44°$$

If the vectors $\mathbf{A}$ and $\mathbf{B}$ are functions of time, we compute the derivative of $\mathbf{A} \cdot \mathbf{B}$ by applying the product rule for derivatives:

$$\frac{d\,(\mathbf{A} \cdot \mathbf{B})}{dt} = \mathbf{A} \cdot \frac{d\mathbf{B}}{dt} + \mathbf{B} \cdot \frac{d\mathbf{A}}{dt} \qquad \text{7-23}$$

This result can be obtained by taking the derivative of the right side of Equation 7-22. Since A_x and B_x are scalar functions of time, the product rule applies. Thus $d\,(A_x B_x)/dt = A_x\, dB_x/dt + B_x\, dA_x/dt$, with similar results for the other products. The six terms thus obtained are the same as the six terms obtained by writing out the right side of Equation 7-23 in terms of the components.

Example 7-8 Show that if the magnitude of a vector is constant in time, the rate of change of the vector is perpendicular to the vector.

We apply Equation 7-23 to the dot product of $\mathbf{A}$ with itself:

$$\mathbf{A} \cdot \mathbf{A} = A^2 = \text{constant}$$

$$\frac{d\,(\mathbf{A} \cdot \mathbf{A})}{dt} = \mathbf{A} \cdot \frac{d\mathbf{A}}{dt} + \mathbf{A} \cdot \frac{d\mathbf{A}}{dt} = 2\mathbf{A} \cdot \frac{d\mathbf{A}}{dt} = 0$$

Since the dot product of $\mathbf{A}$ and $d\mathbf{A}/dt$ is zero, they must be perpendicular (unless either $\mathbf{A}$ or $d\mathbf{A}/dt$ is zero). We have seen examples of this result in circular motion. Since the position vector $\mathbf{r}$ is constant in magnitude, the velocity $\mathbf{v} = d\mathbf{r}/dt$ is perpendicular to $\mathbf{r}$. If the motion is at constant speed, the acceleration $\mathbf{a} = d\mathbf{v}/dt$ is perpendicular to $\mathbf{v}$.

If we write the force and displacement in terms of their rectangular components, the element of work becomes

$$dW = \mathbf{F} \cdot d\mathbf{s} = (F_x\mathbf{i} + F_y\mathbf{j} + F_z\mathbf{k}) \cdot (dx\,\mathbf{i} + dy\,\mathbf{j} + dz\,\mathbf{k})$$
$$= F_x\, dx + F_y\, dy + F_z\, dz$$

The net work done by the resultant force can then be written

$$W_{\text{net}} = \int \Sigma F_x\, dx + \int \Sigma F_y\, dy + \int \Sigma F_z\, dz$$

The work-energy theorem in three dimensions can be derived from this equation by considering each term on the right side separately as a one-dimensional problem and using Newton's second law in component form ($\Sigma F_x = ma_x$). Then the first term can be written as the change in the quantity $\frac{1}{2}mv_x^2$, with similar expressions for the other two terms. The net work is thus the change in the quantity

$$\tfrac{1}{2}mv_x^2 + \tfrac{1}{2}mv_y^2 + \tfrac{1}{2}mv_z^2 = \tfrac{1}{2}m(v_x^2 + v_y^2 + v_z^2) = \tfrac{1}{2}m\mathbf{v} \cdot \mathbf{v} = \tfrac{1}{2}mv^2 = E_k$$

Questions

10. Can the scalar product of two vectors be negative? If so, how?

11. If the scalar product of two vectors is zero, does it follow that one of the vectors is zero?

Optional

7-8   Proof of the Distributive Law for the Dot Product

Figure 7-12 shows the vectors **A**, **B**, and **A** + **B** along with their projections on vector **C**. ϕ_A is the angle between **A** and **C**, ϕ_B the angle between **B** and **C**, and ϕ_{AB} the angle between **A** + **B** and **C**. From the figure it is clear that

$$|\mathbf{A} + \mathbf{B}| \cos \phi_{AB} = A \cos \phi_A + B \cos \phi_B$$

This multiplied by the magnitude of **C** is

$$|\mathbf{A} + \mathbf{B}|C \cos \phi_{AB} = AC \cos \phi_A + BC \cos \phi_B$$

But, by definition, the quantity on the left is (**A** + **B**) · **C**, and the two terms on the right are **A** · **C** and **B** · **C**. Hence, the last equation is the same as

$$(\mathbf{A} + \mathbf{B}) \cdot \mathbf{C} = \mathbf{A} \cdot \mathbf{C} + \mathbf{B} \cdot \mathbf{C}$$

Figure 7-12
Diagram to show that
$(\mathbf{A} + \mathbf{B}) \cdot \mathbf{C} = \mathbf{A} \cdot \mathbf{C} + \mathbf{B} \cdot \mathbf{C}$.

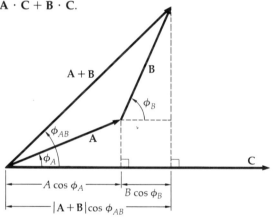

Review

A. Define, explain, or otherwise identify:

Work, 175
Joule, 175
Kinetic energy, 177
Work-energy theorem, 177
Power, 188

Watt, 188
Horsepower, 189
Scalar product, 189
Dot product, 189

B. True or false:

1. Work is the area under the force-versus-time curve.

2. Only the resultant force acting on an object can do work.

3. No work is done on a particle which remains at rest.

4. A force which is always perpendicular to the velocity of a particle does no work on the particle.

5. A kilowatt-hour is a unit of power.

6. If **A** and **B** are in opposite directions, **A** · **B** is zero.

7. No net work is done in carrying a heavy box at constant speed across a horizontal surface.

Exercises

Unless instructed otherwise, use the approximate value $g = 32$ ft/sec$^2 = 9.8$ m/sec^2 for the acceleration of gravity in Exercises and Problems.

Section 7-1, Work in One Dimension

1. A 5-kg mass is raised at constant velocity a distance of 10 m by a force F_0. Find the work done on the mass (a) by the force F_0 and (b) by the earth.

2. A force F_x varies with x as shown in Figure 7-13. Find the work done by the force acting on a particle as the particle moves from $x = 1$ m to $x = 4$ m.

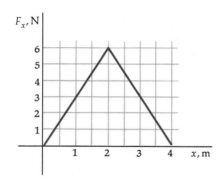

Figure 7-13
F_x versus x for Exercises 2 and 15.

3. A force F_x varies as shown in Figure 7-14. (a) Find the work done by the force acting on the particle as the particle moves from $x = 0$ to $x = 3$ m. (b) Find the work done by F_x as the particle moves from $x = 3$ m to $x = 6$ m. (c) Find the total work done by F_x over the distance $x = 0$ to $x = 6$ m.

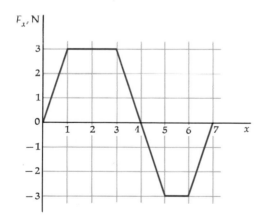

Figure 7-14
F_x versus x for Exercises 3 and 16.

4. A 10-kg box rests on a horizontal table. The coefficient of friction between the box and table is 0.4. A force F_x pulls the box at constant velocity a distance of 5 m. Find the work done (a) by F_x and (b) by friction.

5. The box described in Exercise 4 is pulled by a force F_x so that the box has an acceleration of 2 m/sec^2. (a) Find the work done by F_x while the box is pulled a distance of 5 m. (b) What is the work done by the frictional force in this case ($\mu = 0.4$)?

6. From the conversion factors for newtons to pounds and meters to feet derive the relation between the joule and the foot-pound expressed in Equation 7-3.

7. The unit of force in the cgs system is the dyne, defined as the force which produces an acceleration of 1 cm/sec^2 when acting on a 1-gm mass. The cgs

unit of work is the dyne-centimeter, called an erg. Find the number of ergs in a joule.

8. A force F_x is related to position by the formula $F_x = Cx^2$, where C is a constant. Find the work done by this force acting on a particle when the particle moves from $x = 0$ to $x = 3$ m.

Section 7-2, Work Done by the Resultant Force; Kinetic Energy

9. A 10-gm bullet has a velocity of 1000 m/sec. (a) What is its kinetic energy in joules? (b) If the speed is halved, what is its kinetic energy?

10. A 2-lb block moves with speed 10 ft/sec. (a) What is its kinetic energy in foot-pounds? (b) What is its kinetic energy in joules?

11. A 2-kg mass and a 200-kg mass have equal momenta of 40 kg-m/sec. Find the kinetic energy of each mass.

12. A 5-kg box initially at rest on a smooth table is pulled by a constant horizontal force of 10 N for a distance of 6 m. Find (a) the final kinetic energy of the box and (b) the final speed of the box.

13. A 5-kg mass is raised a distance of 4 m by a vertical force of 80 N. Find (a) the work done by the force and the work done by the earth and (b) the final kinetic energy of the mass if it was originally at rest.

14. A 2-kg box is initially at rest on a horizontal table. The coefficient of friction between the box and table is 0.4. The box is pulled along the table a distance of 3 m by a horizontal applied force of 10 N. (a) Find the work done by the applied force. (b) Find the work done by friction. (c) Find the change in kinetic energy of the box. (d) Find the speed of the box after it has traveled 3 m.

15. A particle of mass 2 kg is subjected to a single force F_x, which varies with x as shown in Figure 7-13. When the particle is at position $x = 1$ m, it has a velocity of 3 m/sec. (a) What is its kinetic energy at $x = 1$ m? (b) What is its kinetic energy at $x = 4$ m? (See Exercise 2.) (c) What is the speed of the particle when it is at $x = 4$ m?

16. A particle of mass 4 kg is subjected to a single force F_x shown in Figure 7-14. The particle is initially at rest at $x = 0$. Find the kinetic energy and speed of the particle (a) when it is at $x = 3$ m and (b) when it is at $x = 6$ m.

17. A 5-kg object is lifted by a rope exerting a force T so that it accelerates upward at 5 m/sec². The object is lifted 3 m. Find (a) the tension T, (b) the work done by T, (c) the work done by the earth, (d) the final kinetic energy of the object after being lifted 3 m if it is originally at rest.

18. A 4000-lb car moves with initial speed of 80 ft/sec. It is stopped in 200 ft by a constant frictional force. (a) What is the initial kinetic energy of the car? (b) How much work is done by the frictional force in stopping the car? (c) What is the coefficient of friction?

Section 7-3, Work and Energy in Three Dimensions

19. A 5-kg box is pulled a distance of 6 m along a smooth horizontal surface by a 20-N force which makes an angle of 30° with the horizontal. Find (a) the work done by the force and (b) the change in kinetic energy of the box and the final speed if the box is initially at rest.

20. A block of mass 6 kg slides down a rough incline starting from rest. The coefficient of friction is 0.2, and the angle of the incline is 60°. (a) List all the forces acting on the block and find the work done by each force when the block slides 2 m (measured along the incline). (b) What is the net work done on the block? (c) What is the velocity of the block after it has slid 2 m?

21. A 2-kg mass attached to a string moves on a smooth horizontal surface in a circle of radius 3 m. The speed of the mass is 1.5 m/sec. (a) Find the tension in the string. (b) List the forces acting on the mass and find the work done by each force during 1 rev.

22. A 200-lb cart is lifted up a 2-ft step by rolling the cart up an incline formed when a plank of length L is laid from the lower level to the top of the step. (Assume the rolling is equivalent to sliding without friction.) (a) Find the force parallel to the incline needed to push the cart up without acceleration for values of L of 3, 4, and 5 ft. (b) Calculate directly from Equation 7-12 the work needed to push the cart up the incline for each of these values of L. (c) Since the work found in (b) is the same for each value of L, what advantage if any is there in choosing one length over another?

23. A 2-kg box slides down a rough incline of length 10 m and angle of inclination 30°. The coefficient of friction is 0.3. (a) List all the forces acting on the box and find the work done by each. (b) Find the kinetic energy of the box at the bottom of the incline assuming that it began at the top with speed $v = 2$ m/sec.

24. A 2-kg mass is released from rest at point A on a smooth circular track of radius 50 cm (Figure 7-15). (a) Using the generalization of Example 7-4, find the work done by the earth while the mass slides from point A to the bottom of the track. (b) What is the velocity of the mass at the bottom of the track? (c) The mass slides past the bottom and up the right side of the track until the negative work done by the earth equals the kinetic energy of the mass at the bottom. At what point on the track will the mass come to rest momentarily? (d) Describe the subsequent motion of the mass on the track.

Figure 7-15
Particle on a smooth vertical circular track for Exercise 24.

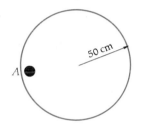

Section 7-4, Potential Energy

25. A man of mass 80 kg climbs up a stairway of height 5 m. (a) What is the increase in potential energy of the man? (b) What is the least amount of work the man must do to climb the stairway? (c) Explain why the work done by the man may be more than this minimum amount.

26. List some situations in which work done by some applied force is not stored; i.e., it cannot be recovered and changed into kinetic energy.

27. A man raises a 50-lb weight from the ground overhead to a resting position 7.5 ft above the ground. (a) Find the work done by the man and the work done by the earth. (b) What is the net work done by all the forces acting on the weight? (c) The weight is now released and allowed to fall to the ground. What is the kinetic energy of the weight just as it hits the ground? (d) Find the total work done by the man and the total work done by the earth during the complete cycle of moving the weight from the ground up to the height of 7.5 ft and back to the ground.

28. A 2-kg mass is held at a height of 20 m above the ground and released at time $t = 0$. (a) What is the original potential energy of the mass relative to the ground? (b) From Newton's laws, find the distance the mass falls after 1 sec and the velocity of the mass at this time. (c) Find the potential energy and kinetic energy of the mass at time $t = 1$ sec. What is the sum of the potential energy and kinetic energy at this time? (d) At time $t = 2$ sec, find the height of the mass, its potential energy, and its kinetic energy. What is the sum of kinetic and potential energy at this time?

29. A 2-kg box slides along a long smooth incline of angle 30°. It starts from rest at time $t = 0$ at the top of the incline a height of 20 m above the ground. (a) What is the initial potential energy of the mass? Find (b) the acceleration of the mass and (c) its position and speed after 2 sec. (d) At time $t = 2$ sec, find the potential energy of the mass, its kinetic energy, and the sum of the potential and kinetic energy. (e) What is the kinetic energy of the mass when it reaches the bottom of the incline?

Section 7-5, Work and Energy

There are no exercises for this section.

Section 7-6, Power

30. A 4-kg mass is lifted by a force F_0 equal to the weight of the mass. The mass moves upward at a constant velocity of 2 m/sec. (a) What is the power input of the force F_0? (b) What is the power input of the force exerted by the earth? (c) How much work is done by the force F_0 in 2 sec?

31. A constant horizontal force $F_0 = 3$ N drags a box along a rough horizontal surface at a constant velocity v_0. The force does work at the rate of 5 W. (a) What is the velocity v_0? (b) How much work is done by F_0 in 3 sec?

32. A single force of 5 N acts in the x direction on a 10-kg mass. (a) If the mass starts at rest at position $x = 0$ at time $t = 0$, find the velocity v and position x as a function of time t. (b) What is the power input of the force at time $t = 2$ sec? (c) Write an expression for the power input of the force as a function of time t.

33. A constant force of 4 N acts at an angle of $30°$ above the horizontal on a box of mass 2 kg resting on a rough horizontal surface. The box is dragged with a constant speed of 50 cm/sec. (a) Find the normal force exerted by the table on the box and the coefficient of friction. (b) What is the power input of the applied force? (c) How much work is done by the frictional force in 3 sec?

34. An engine is rated at 400 horsepower. How much work is done by it operating at full power for 1 min? Express your answer in foot-pounds and in kilowatt-hours.

35. A 4000-lb car drives up a road inclined at $2°$ to the horizontal at a rate of 60 mi/h (88 ft/sec). (a) What is the minimum force exerted by the road on the car tangent to the road? (b) At what rate (in horsepower) does the potential energy of the car increase? (c) Why must the car engine develop a greater horsepower than this?

Section 7-7, Scalar, or Dot, Product of Two Vectors

36. Two vectors **A** and **B** have magnitudes 6 ft and make an angle of $60°$ with each other. Find **A** $\cdot$ **B**.

37. The scalar product of two vectors, **A** and **B**, is 4 units. The magnitudes of the vectors are $A = 4$ and $B = 8$. Find the angle between the vectors.

38. Find **A** $\cdot$ **B** for the following vectors: (a) $\mathbf{A} = 2\mathbf{i} - 6\mathbf{j}$, $\mathbf{B} = -3\mathbf{i} + 2\mathbf{j}$; (b) $\mathbf{A} = 4\mathbf{i} + 4\mathbf{j}$, $\mathbf{B} = 2\mathbf{i} - 3\mathbf{j}$; (c) $\mathbf{A} = 3\mathbf{i} + 4\mathbf{j}$, $\mathbf{B} = 4\mathbf{i} - 3\mathbf{j}$.

39. Find the angle θ between the vectors **A** and **B** given in Exercise 38.

40. Find the power input of a force **F** acting on a particle which moves with velocity **v** for (a) $\mathbf{F} = 3\mathbf{i} + 4\mathbf{k}$ N, $\mathbf{v} = 2\mathbf{i}$ m/sec; (b) $\mathbf{F} = 5\mathbf{i} - 6\mathbf{j}$ N, $\mathbf{v} = -5\mathbf{i} + 4\mathbf{j}$ m/sec; (c) $\mathbf{F} = 2\mathbf{i} - 4\mathbf{j}$ N, $\mathbf{v} = 6\mathbf{i} + 3\mathbf{j}$ m/sec.

41. If **A** $\cdot$ **B** $= 0$ and $\mathbf{A} = 2\mathbf{i} - 3\mathbf{j}$, what can be said about **B**?

Problems

1. Derive the expression

$$W_{\text{net}} = \int \Sigma F_x \, dx = \int \frac{p}{m} \, dp$$

by using Newton's second law in the form $F_x = dp/dt$ and applying the chain rule to dp/dt. Consider the momentum p to be a function of x.

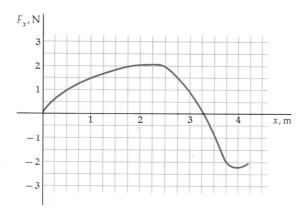

Figure 7-16
F_x versus x for Problem 2.

2. A 3-kg mass is moving with speed 1.00 m/sec in the x direction. When it passes the origin, it is acted on by a single force F_x, which varies with x as shown in Figure 7-16. (*a*) Find the work done by the force from $x = 0$ to $x = 2$ m. (*b*) What is the kinetic energy of the mass at $x = 2$ m? (*c*) What is the speed of the particle at $x = 2$ m? (*d*) Find the work done on the particle from $x = 0$ to $x = 4$ m. (*e*) What is the velocity of the particle at $x = 4$ m?

3. When you walk along a horizontal surface, you lift part of your weight a small amount in each step and therefore do work against the force of gravity. (*a*) Estimate the work done in each step and the total work you do in lifting yourself when you walk 1 mi along a horizontal surface. (*b*) If you walk 1 mi in 20 min, what is your horsepower output?

4. A 5-kg block is held against a spring of force constant 20 N/cm, compressing it 3 cm. The block is released, and the spring expands, pushing the block along a rough horizontal surface. The coefficient of friction between the surface and the block is 0.2. (*a*) Find the work done on the block by the spring as it extends from its compressed position to its equilibrium position. (*b*) Find the work done on the block by friction while it moves the 3 cm to the equilibrium position of the spring. (*c*) What is the speed of the block when the spring is at its equilibrium position? (*d*) If the block is not attached to the spring, how far will it slide along the rough surface before coming to rest? (*e*) Suppose the block is attached to the spring so that the spring is extended when the block slides past the equilibrium point. By how much will the spring be extended before the block comes momentarily to rest? Describe the subsequent motion.

5. Figure 7-17 shows a pendulum bob making an angle θ with the vertical. (*a*) Indicate all the forces acting on the bob at angle θ. Write the tangential component of the force as a function of θ. (*b*) Show that the work done by the earth when the angle increases by $d\theta$ is $-mgL \sin \theta \, d\theta$. Hint: Write the path length ds in terms of θ. (*c*) Find the total work done by the earth when the bob moves from $\theta = 0$ to $\theta = \theta_0$. Express your result in terms of the vertical distance H the bob is lifted from $\theta = 0$ to $\theta = \theta_0$.

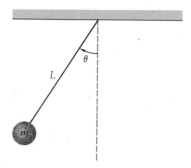

Figure 7-17
Problem 5.

6. A pendulum bob of mass m is raised to $\theta = 90°$ and released. (a) Find the work done by all forces acting on the bob while it swings from $\theta = 90°$ to $\theta = 0$. (b) Find the kinetic energy of the bob at $\theta = 0°$. (c) Show that the tension in the string at $\theta = 0°$ is 3 times the weight of the bob.

7. A particle of mass m moves in a horizontal circle of radius r on a rough table. It is attached to a string fixed at the center of the circle. The speed of the particle is initially v_0. After completing one full trip around the circle, the speed of the particle is $\frac{1}{2}v_0$. (a) Find the work done by friction during that 1 rev in terms of m, v_0, and r. (b) What is the coefficient of sliding friction? (c) How many more revolutions will the particle make before coming to rest?

8. A 2-kg box is projected with initial speed of 3 m/sec up a rough plane inclined at 60° to the horizontal. The coefficient of friction is 0.3. (a) List all the forces acting on the box and find the work done by each as the box slides up the plane. (b) How far does the box slide along the plane before it stops momentarily? (c) Find the work done by each force acting on the box as it slides back down the plane. (d) Find its speed when it reaches its initial position.

9. Vectors **A**, **B**, and **C** form a triangle, as shown in Figure 7-18. The angle between **A** and **B** is θ; the vectors are related by $\mathbf{C} = \mathbf{A} - \mathbf{B}$. Compute $\mathbf{C} \cdot \mathbf{C}$ in terms of A, B, and θ and derive the law of cosines $C^2 = A^2 + B^2 - 2AB \cos \theta$.

Figure 7-18
Problem 9.

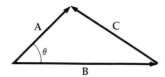

CHAPTER 8   # Potential Energy and Conservative Forces

We saw in Chapter 7 that in some cases work can be stored in the form of potential energy. For example, the work needed to lift an object of mass m a vertical distance h is mgh. At the height h, the object is said to have potential energy mgh relative to its original position $h = 0$. When the object moves back to its original position, the work done by the force of gravity is mgh no matter what path the object follows. In this chapter we investigate the concept of potential energy in detail. We shall see that potential energy is associated with the work done by a certain class of forces called *conservative forces*. Examples of conservative forces are the force of gravity, the force exerted by an elastic spring, and the electrostatic force. For each conservative force we define a potential-energy function U such that the work done by the force equals the decrease in the potential energy. If the conservative force is the only force acting on a particle or the only force that does work, the work done also equals the increase in the kinetic energy, as discussed in Chapter 7. Then the sum of the kinetic energy and the potential energy, called the *total mechanical energy*, remains constant because the decrease in potential energy is balanced by an increase in kinetic energy. (If the work done by the conservative force is negative, the potential energy increases and the kinetic energy decreases by an equal amount.) The total mechanical energy is said to be conserved.

8-1   Conservative Forces

The criterion for a force to be conservative can be simply stated. *A force is conservative if the total work it does on a particle is zero when the particle moves around any closed path returning to its initial position.* We have already seen that the force of gravity has this property. If we project an object up an inclined plane with some initial speed, the force of gravity does work $-mgh$ while the object moves up and $+mgh$ while it moves down, where h is the maximum vertical distance the

Conservative force defined

The waterfall in this 1961 lithograph by the Swiss artist M. C. Escher violates the law of the conservation of energy. When the water falls, part of its potential energy is converted into the kinetic energy of the waterwheel. How does the water get back to the top of the waterfall? (*Courtesy of the Escher Foundation, Haags Gemeentemuseum, The Hague.*)

object rises. If this conservative force is the only force acting, the object has the same kinetic energy when it arrives back at its starting point that it had originally.

An example of a nonconservative force is that of sliding friction. Since this force is always opposite to the direction of motion, it always does negative work. When an object makes a round trip ending at its original position, the total work done by sliding friction on it is negative. For example, if we project an object up a rough inclined plane with some initial speed, the frictional force does work $-\mu_k N s$ on the way up and $-\mu_k N s$ on the way down, where μ_k is the coefficient of friction, N is the normal force, and s is the distance traveled up the plane. When a nonconservative force such as friction acts on an object, the object does not have its original kinetic energy when it arrives back at its original position. In this example the kinetic energy is less by an amount $2\mu_k N s$.

An alternative statement of the criterion for a force to be conservative is that *the work done on a particle by a conservative force is independent of how the particle moves from one point to another*. Figure 8-1 shows several paths connecting points P_1 and P_2. If the work done by a conservative force is W when the particle moves from P_2 to P_1 along one of the paths, it must be $-W$ when the particle returns along any of the paths because the total round-trip work must be zero.

Alternative definition of conservative force

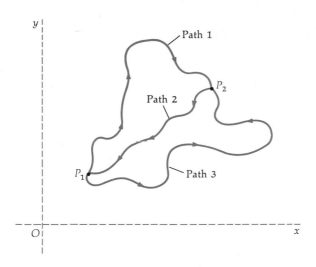

Figure 8-1
Several paths connecting points P_1 and P_2. By definition, the work done by a conservative force is zero for any round-trip path ending at its starting point. If a conservative force does work W on a particle moving from P_2 to P_1 along path 2, it must then do work $-W$ on the particle as it moves back along any return path such as path 1 or 3. It follows that the work done by a conservative force on a particle moving from point P_1 to point P_2 is the same for all paths connecting the points.

The work done on a particle by a force **F** as the particle moves from point P_1 to P_2 along some path can be written

$$W = \int_1^2 \mathbf{F} \cdot d\mathbf{s} \qquad \text{8-1}$$

The integral in this equation, called a *line integral,* is computed by taking the dot product of the force **F** and the displacement $d\mathbf{s}$ for each element along the path and summing. If the force is conservative, the work does not depend on which path is taken between the points or on the particle's speed. The work done by a conservative force depends only on the end points P_1 and P_2. It is therefore just a function of the positions P_1 and P_2. We define the potential-energy function U so that the work done equals the decrease in the potential energy:

$$W = \int_1^2 \mathbf{F} \cdot d\mathbf{s} = U_1 - U_2 = -\Delta U \qquad \text{8-2}u$$

where

$$\Delta U = U_2 - U_1 = -\int_1^2 \mathbf{F} \cdot d\mathbf{s} \qquad \text{8-2}b$$

For very small displacements $d\mathbf{s}$ we can write this definition of potential energy in differential form:

$$dU = -\mathbf{F} \cdot d\mathbf{s} \qquad \text{8-3}$$

Definition of potential energy

Since only the change in the potential energy is defined, the value of the function U at any point is not specified by the definition. We are free to choose the value of U at any one point arbitrarily. We usually choose U to be zero at some reference point. The potential energy at any other point is then the difference between the potential energy at that point and at the reference point.

The work done by a nonconservative force acting on a particle which moves from point P_1 to point P_2 can also be written as a line integral, as in Equation 8-1, but this work may depend on the path followed by the particle, on the speed of the particle, or on other quantities. Since this work is not just a function of the initial and final positions of the particle, there is no potential-energy function associated with a nonconservative force.

8-2   The Conservation of Mechanical Energy

The negative sign in the definition of potential energy in Equations 8-2 and 8-3 is introduced so that the work done by a conservative force equals the *decrease* in potential energy. If this force is the only force acting, or if it is the only force that does work on the particle, the work done by the force also equals the *increase* in *kinetic energy*, i.e.,

$$W = \int \mathbf{F} \cdot d\mathbf{s} = -\Delta U = +\Delta E_k$$

Hence

$$\Delta E_k + \Delta U = \Delta(E_k + U) = 0 \qquad \text{8-4}$$

Conservation of mechanical energy

This is an example of the conservation of energy, which we discuss more fully in the next chapter. We call the sum of the potential energy and kinetic energy the *total mechanical energy of the particle:*

$$E_{\text{total}} = E_k + U \qquad \text{8-5}$$

Equation 8-4 states that the total mechanical energy is constant if the only force that does work is a conservative force. (This is the origin of the expression conservative force.) When only such forces do work, the total mechanical energy is conserved.

8-3   Conservative Forces and Potential Energy in One Dimension

For one-dimensional motion, a necessary and sufficient condition for a force to be conservative is that the force be constant or depend only on the position of the particle. The work done by a force $\mathbf{F}$ acting on a particle moving along the x axis from point x_1 to point x_2 is

$$W = \int_{x_1}^{x_2} \mathbf{F} \cdot dx \, \mathbf{i} = \int_{x_1}^{x_2} F_x \, dx \qquad \text{8-6}$$

where F_x can be either positive or negative. In general, the force may depend on the position of the particle, the velocity of the particle, or other quantities. If the force is constant or depends only on the position x, the work done depends only on the initial and final positions x_1 and x_2 and the force is conservative. The potential-energy function $U(x)$ is defined by the one-dimensional equivalents of Equations 8-2 and 8-3,

$$\Delta U = U(x_2) - U(x_1) = -\int_{x_1}^{x_2} F_x \, dx \qquad \text{8-7}$$

or in differential form,

$$dU = -F_x \, dx \qquad \text{8-8}$$

An example of a conservative force in one dimension which we have already discussed is the force of gravity. (This force is also conservative for general three-dimensional motion, as we shall see later.) Let us consider motion near the surface of the earth along a vertical line which we take to be the y axis. The force of gravity on a particle of mass m is constant and directed downward. If we choose up for the positive direction of y, this force is

$$\mathbf{F} = F_y \mathbf{j} = -mg\mathbf{j}$$

The work done by gravity when the particle is displaced from point y_1 to point y_2 is

$$W = \int_{y_1}^{y_2} -mg\ (\mathbf{j} \cdot d\mathbf{s}) = \int_{y_1}^{y_2} -mg\ dy = -mg\,(y_2 - y_1)$$

Setting this work equal to the decrease in potential energy, we have

$$-mg\,(y_2 - y_1) = -\Delta U = U(y_1) - U(y_2)$$

or

$$\Delta U = U(y_2) - U(y_1) = mgy_2 - mgy_1$$

If we choose the potential energy equal to zero at the reference height $y = 0$, the potential energy at any height y relative to $y = 0$ is

$$U(y) = mgy \qquad\qquad 8\text{-}9$$

Gravitational potential energy near earth's surface

This potential-energy function can also be determined from the differential form, Equation 8-8 (using y instead of x since we are considering vertical motion),

$$dU = -F_y\ dy -- (-mg)\ dy = +mg\ dy$$

Integrating gives

$$U(y) = mgy + C$$

where C is an arbitrary constant of integration. This constant is related to the choice of zero potential energy. If we choose $U = 0$ at $y = 0$, the constant C is zero.

Example 8-1 A 2-kg mass is raised from height $y_1 = 0$ to $y_3 = 10$ m and then lowered to its final height $y_2 = 8$ m. Find the work done by the force of gravity and the change in potential energy.

The weight of the mass is $mg = (2\text{ kg})(9.81\text{ m/sec}^2) = 19.6$ N. The work done by gravity when the mass is raised is negative because the force is opposed to the motion:

$$W_1 = -mg\,(y_3 - y_1) = -(19.6\text{ N})(10\text{ m}) = -196\text{ J}$$

When the mass is lowered from 10 to 8 m, the force of gravity does positive work equal to

$$W_2 = (19.6\text{ N})(2\text{ m}) = 39.2\text{ J}$$

The total, or net, work done by the force of gravity is the sum of these:

$$W_{\text{total}} = W_1 + W_2 = -196\text{ J} + 39.2\text{ J} = -157\text{ J}$$

The negative of this work is $+157$ J, which is the change in the potential energy of the mass:

$$U = mg\,(y_2 - y_1) = (19.6\text{ N})(8\text{ m}) = 157\text{ J}$$

This work is the same as if the mass were just raised from $y = 0$ to $y = 8$ m. It does not depend on how the mass moves but only on its initial and final positions.

Joe Monroe/Photo Researchers

This boulder, located in the Arches National Park in Utah, has gravitational potential energy relative to ground level.

A slingshot is an example of elastic potential energy. The work done in stretching the rubber is stored as potential energy and then converted into kinetic energy of the missile.

Culver Pictures

Example 8-2 Find the potential-energy function associated with a spring which exerts a force $F_x = -kx$.

Here we have chosen the origin $x = 0$ at the equilibrium length of the spring. In this example, the force is not constant, but since it is a one-dimensional force which is a function of position only, it must be conservative. The potential-energy function $U(x)$ can be calculated from the differential form (Equation 8-8) or from Equation 8-7. Equation 8-8 gives

$$dU = -F_x \, dx = -(-kx) \, dx = +kx \, dx$$

Integrating gives

$$U = \tfrac{1}{2}kx^2 + C$$

where again the constant of integration C depends on our arbitrary choice of the position of zero potential energy. In this case, C is the value of the potential-energy function at the equilibrium length $x = 0$. If we choose the potential energy to be zero at this point, $C = 0$. The potential energy relative to $x = 0$ is then

$$U = \tfrac{1}{2}kx^2 \qquad\qquad\qquad 8\text{ }10$$

Potential energy of a spring

The potential-energy function can also be calculated from the integral form (Equation 8-7). If the spring is stretched from $x = 0$ to some distance x, the work done by the spring is

$$W = \int_0^x -kx \, dx = -\tfrac{1}{2}kx^2$$

This work is just the area under the F_x-versus-x curve in Figure 8-2.

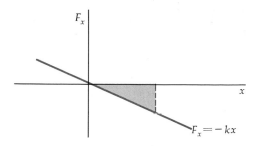

Figure 8-2
Work done by a spring when
it is stretched from its equilibrium length. The force F_x
exerted by the spring is
plotted versus extension x,
where $x = 0$ is chosen at the
equilibrium length. The work
indicated by the shaded area
is negative when the spring is
stretched because the force
opposes the motion.

The work is negative because the force opposes the motion. The negative of this work is the change in potential energy of the spring

$$\Delta U = U(x) - U(0) = +\tfrac{1}{2}kx^2$$

or

$$U(x) = \tfrac{1}{2}kx^2 + U(0)$$

This result is the same as Equation 8-10 when the potential energy at
$x = 0$ is chosen to be zero.

By our definition of potential energy, when a conservative force
does positive work, the potential energy decreases. The force is thus in
the direction of decreasing potential energy. A particle under the
influence of a single conservative force accelerates in the direction of
decreasing potential energy; with gravity the acceleration is downward; with a mass on a spring, the mass accelerates toward the equilibrium position $x = 0$.

If we apply a force equal and opposite to a conservative force so that
the particle moves (without acceleration) in the direction opposite the
conservative force, the conservative force does negative work and we
do positive work equal to the increase in potential energy. Thus to lift
a mass through a vertical distance y without acceleration the work we
do is mgh; to stretch a spring from its equilibrium length $x = 0$ to some
length x the work we do is $\tfrac{1}{2}kx^2$.

Questions

1. A mass moves along a vertical line. Must two observers (in the
same reference frame) studying its motion agree on the value of its
kinetic energy at any instant? Must they agree on the value of its potential energy?

2. A mass rests on a table 4 ft above the floor. Where would you
choose the reference point for potential energy to be so that the gravitational potential energy for the mass will be (a) positive, (b) zero, (c)
negative?

3. Estimate the amount by which your gravitational potential energy
changes when you walk up a flight of stairs.

4. The force exerted by a spring which obeys Hooke's law is a conservative force. Would the force of a spring which is stretched beyond its
elastic limit be conservative?

5. If a spring did not obey Hooke's law for large displacements but
did obey the law $F_x = -kx + bx^3$, where b is a constant, would the
force be conservative?

Linda Cronquist

Another example of elastic
potential energy. As the ball
is dropped, gravitational potential energy is converted
into kinetic energy, which is
greatest just before the ball
strikes the floor. Some of the
kinetic energy is then stored
momentarily as elastic potential energy of deformation.
This potential energy is
quickly converted into kinetic
energy, which is again
changed into gravitational
potential energy as the ball
rises.

8-4   Nonconservative Forces in One Dimension

It is important to realize that a potential-energy function $U(x)$ can be defined by Equations 8-7 and 8-8 only if the force **F** is conservative, i.e., if **F** is constant or a function of position only. The reason for this restriction can be seen by considering forces which depend on quantities other than position, e.g., the speed of the particle. In this case, the work done on a particle as it travels from point x_1 to x_2 depends on how fast the particle travels. Since we cannot write this work as a function of the points x_1 and x_2, we cannot define a potential energy $U(x)$ by Equation 8-7. An example of such a nonconservative force is viscous drag, which increases in magnitude with increasing speed and is in the direction opposite the velocity of the particle.

Another example of a nonconservative force, already mentioned, is sliding friction. Consider a box sliding on a rough plane from point x_1 to x_2. The frictional force depends on the *velocity* of the box and cannot be written as a function of the position of the box. For example, if the box is at rest with no applied force, the frictional force is zero. If the box moves to the right, the frictional force is to the left; if the box moves to the left, the frictional force is to the right. Even if the magnitude of the frictional force is approximately independent of speed, the direction of the force depends on the direction of the velocity. Therefore, we cannot write the work done by this force as a function of the initial and final positions, and we cannot define a potential-energy function associated with this force. If we move the box from $x_1 = 0$ to $x_2 = 5$ m, the work done by friction is negative, but we cannot get this work back when we release the box. If the box is moved from $x_1 = 0$ to $x_3 = 10$ m and then back to $x_2 = 5$ m, the frictional force does 3 times as much work as when we move it directly from 0 to 5 m (assuming constant magnitude of the frictional force). The work done by this force depends on how the object is moved from one point to another.

Another common example of a nonconservative force is a force applied by a human agent like a push or a pull. Suppose you decide to lift a weight through a vertical distance h. You may apply a vertical force equal to the weight, just balancing the force of gravity so that the weight does not accelerate. Or you may apply a force greater than the weight so that the weight is accelerated. The force you apply clearly does not depend merely on position and is thus nonconservative.

Questions

6. Is the work done by a nonconservative force always negative?

7. When you jump into the air, the forces exerted by your leg muscles do work on your body. Are these forces conservative?

8-5   Conservative Forces and Potential Energy in Three Dimensions

The problem of determining whether a given force is conservative is more complicated in three-dimensional motion. The work done by a force as a particle moves from one point to another is the line integral given in Equation 8-1. As in one dimension, if the force depends on velocity or any quantity other than position, the work done is not just a function of the initial and final positions and a potential-energy function of position cannot be defined by this work. Thus it is still necessary that the force not depend on any quantity other than position if it is to be conservative. However, in three-dimensional motion, even if a force is a function of position only, the work done by the force may depend on the path taken by the particle between the initial and final positions, in which case this force is nonconservative and a potential-energy function cannot be defined. We shall illustrate such a nonconservative force in Example 8-6.

If we can show that the work done by a force along two different paths connecting the same initial and final positions is different, we know immediately that the force is not conservative. Alternatively, we can show that a force is nonconservative if we can show that the work it does is not zero when a particle moves in a closed path ending at its initial position. To show that a given force is conservative, we must show that the work done is the same along *every* path connecting the initial and final positions. Surprisingly, this can be done rather easily for any given force using the properties of partial derivatives of a function of two or more variables. (We discuss this briefly in Section 8-10 for students familiar with this part of calculus.) Fortunately, for many forces of interest, we can show that the force is conservative by calculating the work done along a general path in space and showing that this work depends only on the initial and final positions.

The simplest example of a conservative force in three dimensions is a constant or uniform force. A force which has the same magnitude and direction at every point in space is called a *uniform force*. An example is the force of gravity near the surface of the earth. If we choose the y axis parallel to the direction of the force, such a force can be written

$$\mathbf{F} = F_0 \mathbf{j}$$

where F_0 is the magnitude of the force and $\mathbf{j}$ is the unit vector in the y direction. A general displacement in three dimensions can be written

$$d\mathbf{s} = dx\,\mathbf{i} + dy\,\mathbf{j} + dz\,\mathbf{k}$$

The work done by the force $\mathbf{F}$ for this displacement is

$$dW = \mathbf{F} \cdot d\mathbf{s} = (F_0\mathbf{j}) \cdot (dx\,\mathbf{i} + dy\,\mathbf{j} + dz\,\mathbf{k}) = F_0\,dy$$

The work done by the force when a particle moves from initial position (x_1, y_1, z_1) to final position (x_2, y_2, z_2) is

$$W = \int_{y_1}^{y_2} F_0\,dy = F_0(y_2 - y_1)$$

This work depends only on the initial and final positions, and so the force is conservative. If $\mathbf{F}$ is the force of gravity near the surface of the earth, $F_0 = -mg$ (assuming y to be positive upward). This result for W agrees with our result (discussed in Chapter 7) that the work done by the force of gravity when a particle moves from height y_1 to height y_2 is $-mg(y_2 - y_1) = mg(y_1 - y_2)$ no matter what path the particle follows (see Figure 7-8). The potential-energy function associated with the force of gravity near the surface of the earth is given by Equation 8-9.

In another important class of forces in three-dimensional motion are forces directed toward or away from the origin. Such a force is called a

William Vandivert

As the tree falls, potential energy is converted into kinetic energy. When it hits, the kinetic energy is lost as heat and sound.

central force. If the magnitude of the force depends only on the distance from the origin, the force is conservative. We can represent such a force in general by

$$\mathbf{F} = F_r \hat{\mathbf{r}} = f(r)\hat{\mathbf{r}} \qquad\qquad 8\text{-}11$$

where $f(r)$ is any function of r and $\hat{\mathbf{r}}$ is a unit vector pointing in the radial direction away from the origin. This general case includes the force of gravity, for which $f(r) = -Gm_1m_2/r^2$, and the electrostatic force, for which $f(r) = kq_1q_2/r^2$. Central forces of this type are conservative no matter what the function $f(r)$ may be.

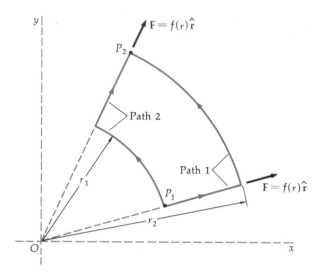

Figure 8-3
Two paths connecting points P_1 and P_2 in a radial force field $\mathbf{F} = f(r)\hat{\mathbf{r}}$. When a particle moves along either arc segment, the work done by the force is zero because the force is perpendicular to the motion. The work done by the force on a particle moving from r_1 to r_2 is the same along either radial segment because the magnitude of the force depends only on r. It follows that the work done by the force is the same for each of the paths connecting P_1 and P_2.

Consider the two simple paths between points P_1 and P_2 shown in Figure 8-3. Each path consists of a straight radial segment and a circular arc segment at a constant distance from the origin. The work done by the force $\mathbf{F}$ along a radial segment is

$$W = \int_{r_1}^{r_2} \mathbf{F} \cdot d\mathbf{s} = \int_{r_1}^{r_2} F_r\, dr = \int_{r_1}^{r_2} f(r)\, dr \qquad\qquad 8\text{-}12$$

This work is the same for the radial segments of each path because the magnitude of the force $F_r = f(r)$ depends only on r and not on θ. The work is in fact the same for all radial paths beginning at a distance r_1 and ending at a distance r_2. Along a circular arc segment, the displacement $d\mathbf{s}$ is always perpendicular to the force $\mathbf{F}$, so that the work done by the force is zero. Thus the work done by the force is the same along each of the paths shown between points P_1 and P_2.

We now consider a general path connecting these points shown in Figure 8-4. We can approximate this path by a series of radial segments and circular arc segments. The work done by the force along any radial segment is $dW = f(r)\, dr$. Along any circular arc segment the work is zero because $d\mathbf{s}$ is perpendicular to $\mathbf{F}$. Hence the total work done is just the sum of the work done along the radial segments, which is

$$W = \int_{r_1}^{r_2} f(r)\, dr$$

This is the same for all paths connecting P_1 and P_2. Thus the force given by Equation 8-11 is conservative. We shall calculate the potential-energy functions for the gravitational and electrostatic forces of this type in Examples 8-3 and 8-4.

Figure 8-4
Any path from P_1 to P_2 can be approximated by a series of radial segments and arc segments of constant radius. Since a radial force $\mathbf{F} = f(r)\hat{\mathbf{r}}$ does work only along the radial segments, the total work is the sum of the work along these segments. The total work is the same for all paths connecting P_1 and P_2.

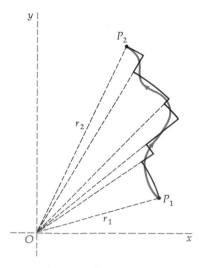

Example 8-3 Find the potential-energy function $U(r)$ associated with the gravitational force exerted by the earth on a particle which is not necessarily near the surface of the earth.

The gravitational force exerted by the earth on a particle of mass m at a distance r from the center of the earth is given by

$$\mathbf{F} = -\frac{GM_E m}{r^2}\,\hat{\mathbf{r}} \qquad\qquad 8\text{-}13$$

where M_E is the mass of the earth, r is the distance from the center of the earth to the mass, and $\hat{\mathbf{r}}$ is the unit vector pointing away from the center of the earth. This force is a conservative central force with $f(r) = -GM_E m/r^2$, as discussed above. The potential-energy function is most easily computed from the differential definition, Equation 8-3. For any displacement $d\mathbf{s}$, we have

$$dU = -\mathbf{F}\cdot d\mathbf{s} = -F_r\,dr = -\frac{-GM_E m}{r^2}\,dr = +\frac{GM_E m}{r^2}\,dr$$

Integrating gives

$$U = -\frac{GM_E m}{r} + C$$

where C is a constant of integration. The potential energy $U(r)$ increases as r increases to a maximum value $U = C$ at infinite distance. For this case the reference point for zero potential energy is customarily chosen to be at infinite distance so that the constant C is zero. Then

$$U = -\frac{GM_E m}{r} \qquad\qquad 8\text{-}14$$

Gravitational potential energy

With this choice of zero potential energy, the potential energy is always negative. Since only changes in U enter into its definition, the actual value at any one point is unimportant. This potential-energy function is shown in Figure 8-5. At $r = R_E$, the radius of the earth, the

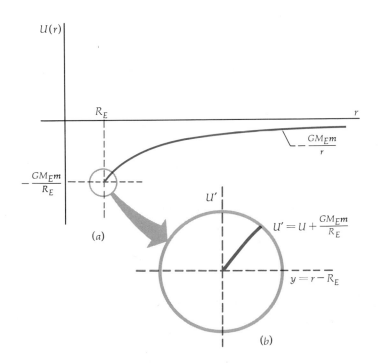

(a)

(b)

Figure 8-5
(a) Graph of the gravitational potential-energy function $U = -GM_E m/r$ versus r. At the surface of the earth ($r = R_E$) the potential energy has the value $-GM_E m/R_E$, and it increases to zero at infinite r. Near the surface of the earth the curve is approximately linear. (b) The same potential-energy function near the surface of the earth with the zero chosen at $r = R_E$. The function U' is related to U by $U' = U + GM_E m/R_E$. If $y = r - R_E$ is small compared with R_E, U' is approximately mgy.

potential energy is $U = -GM_E m/R_E$. At values of r less than R_E Equation 8-13 does not hold, and so the potential-energy function (Equation 8-14) does not hold either.

We can relate this potential-energy function to the one obtained previously for positions near the surface of the earth. Let U' be the potential-energy function which is defined to be zero at $r = R_E$. From Figure 8-5 we see that U' and U are related by

$$U' = U + \frac{GM_E m}{R_E} = -\frac{GM_E m}{r} + \frac{GM_E m}{R_E}$$

$$= \frac{GM_E m}{rR_E} (r - R_E)$$

But $r - R_E = y$, the height above the surface of the earth, and $GM_E/R_E^2 = g$, the free-fall acceleration of gravity. Then

$$U' = mgy \frac{R_E}{r} \qquad\qquad 8\text{-}15$$

Near the surface of the earth, r and R_E are approximately equal, and Equations 8-15 and 8-9 are the same.

Example 8-4 Electrostatic Potential Energy According to Coulomb's law, the force exerted by a charge q_1 at the origin on another charge q_2 at position **r** is

$$\mathbf{F} = \frac{kq_1 q_2}{r^2} \hat{\mathbf{r}}$$

where q_1 and q_2 are the magnitudes of the respective charges and k is the coulomb constant. This expression is similar to the gravitational force except that $GM_E m$ is replaced by $kq_1 q_2$ and there is no negative sign. That is, if q_1 and q_2 have the same sign, the force is repulsive, unlike the gravitational force, which is always attractive. The potential-energy function can be found as in the previous example. If q_2 is given displacement $d\mathbf{s}$, we have

$$dU = -\mathbf{F} \cdot d\mathbf{s} = -F_r \, dr = -\frac{kq_1 q_2}{r^2} \, dr$$

and

$$U = \frac{kq_1 q_2}{r} + C$$

Again, it is convenient to choose the potential energy to be zero at an infinite separation so that the constant C is zero. Then

$$U = +\frac{kq_1 q_2}{r} \qquad\qquad 8\text{-}16$$

Electrostatic potential energy

In this case the potential energy is positive if the charges have the same signs and negative if they have different signs. The magnitude decreases as r increases.

Example 8-5 Find the potential energy for an electron and proton at their most likely separation in the hydrogen atom, which is $r = 0.529 \times 10^{-10}$ m. The charges are

$$q_e = -e = 1.6 \times 10^{-19} \text{ C}$$

$$q_p = +e = 1.6 \times 10^{-19} \text{ C}$$

and the coulomb constant is $k = 9 \times 10^9$ N-m²/C². The potential energy is then

$$U = \frac{(9 \times 10^9 \text{ N-m}^2/\text{C}^2)(-1.6 \times 10^{-19} \text{ C})(+1.6 \times 10^{-19} \text{ C})}{0.529 \times 10^{-10} \text{ m}}$$

$$= -4.36 \times 10^{-18} \text{ N-m} = -4.36 \times 10^{-18} \text{ J}$$

Example 8-6 Show that the force $\mathbf{F} = f(r)\hat{\boldsymbol{\theta}}$ is nonconservative, where $f(r)$ is any function of r and $\hat{\boldsymbol{\theta}}$ is the unit vector perpendicular to $\mathbf{r}$.

At any distance r from the origin, this force is tangent to the circle of radius r and in the direction of increasing θ as in Figure 8-6*a*. This kind of force is exerted on a charge moving between the poles of an electromagnet when the magnetic field is changing. An example of its application is the betatron, a device for accelerating electrons to high energy for producing x-rays. We can show that the force is nonconservative by showing that the work done by the force when a particle makes a round trip is not zero.

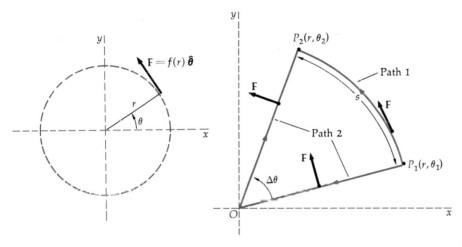

Figure 8-6
(*a*) The force $\mathbf{F} = f(r)\hat{\boldsymbol{\theta}}$ for Example 8-6. At distance r from the origin the force is tangent to the circle of radius r and in the direction of increasing θ. If a particle moves around the circle and returns to its original position, the work done by the force is not zero. The force is therefore not conservative. (*b*) Two paths connecting points P_1 and P_2. The work done by the force in (*a*) along path 1 is $f(r)s$, where s is the distance measured along the arc. Along path 2 the work done by the force is zero because it is perpendicular to the motion.

Consider a particle moving in a circle of radius r in the direction of the force. Since the force is constant and in the direction of motion, the work done is just the force $f(r)$ times the distance $s = 2\pi r$

$$W = f(r)2\pi r$$

Since this work is not zero when the particle has returned to its initial position, the force is not conservative. Alternatively, we can show that the work done by the force on a particle which moves from one point to another is different for different paths. Consider points P_1 and P_2 equidistant from the origin, as in Figure 8-6*b*. When the particle moves along the circular arc from point P_1 to point P_2, the work done is $f(r)s$, where s is the distance measured along the arc. If the particle first moves from point P_1 to the origin along a radial line and then back out to P_2 along another radial line, the force does no work because it is always perpendicular to the displacement for this path.

Questions

8. When you climb a mountain, is the work done on you by gravity different if you take a short steep trail instead of a long gentle trail? If not, why would you find one trail easier than the other?

9. What are the advantages or disadvantages in choosing the gravitational and electrostatic potential-energy functions to be zero at $r = \infty$?

8-6   The Electron Volt Energy Unit

The electrostatic potential energy of a particle divided by the charge of the particle is called the *electrostatic potential* of the particle. We shall study this concept in detail in Chapter 32; now we are interested only in the SI unit of potential, which is the joule per coulomb, called a *volt* (V):

$$1 \text{ V} = 1 \text{ J/C}$$

Energies in atomic physics are usually given in units of the fundamental electric charge e times 1 V. This unit, called the *electron volt* (eV), is related to the joule by

Electron volt defined

$$1 \text{ eV} = 1.6 \times 10^{-19} \text{ C-V} = 1.6 \times 10^{-19} \text{ J} \qquad \text{8-17}$$

For example, the potential energy found in Example 8-5, for the electron in the hydrogen atom, expressed in electron volts, is

$$U = -4.36 \times 10^{-18} \text{ J} \times \frac{1 \text{ eV}}{1.6 \times 10^{-19} \text{ J}} = -27.2 \text{ eV}$$

Common multiples are the kiloelectron volt (1 keV $= 10^3$ eV) and the megaelectron volt (1 MeV $= 10^6$ eV).

8-7   Force from Potential Energy

So far we have been finding the potential-energy function from a given force. We now consider the inverse problem, finding the force from a given potential-energy function. For most cases of interest to us (one-dimensional motion and central forces) the force has only one component, and the potential energy is a function of just one variable. In these cases the force is just the negative derivative of the potential energy. According to Equation 8-8, the force and potential-energy function for one dimension are related by

$$dU = -F_x \, dx$$

If we divide both sides by dx, we obtain

$$F_x = -\frac{dU}{dx} \qquad \text{8-18}$$

The force is thus the negative derivative of U with respect to x. For central forces, e.g., those of gravity and electrostatics, we found in Examples 8-3 and 8-4 that

$$dU = -F_r \, dr$$

In this case the force has only a radial component F_r, which is the negative derivative of U with respect to r:

$$F_r = -\frac{dU}{dr} \qquad \text{8-19}$$

Example 8-7 Given $U(x) = Cx^2$, where C is a constant, what is the force corresponding to this potential energy?

We have $dU/dx = 2Cx$; therefore, $F_x = -dU/dx = -2Cx$. We see that this is the same as for a spring of force constant $k = 2C$, with $x = 0$ at the equilibrium point.

Example 8-8 Given $U(x) = C(1/\sqrt{x^2 + a^2})$ with C and a constant, find F_x.

We have

$$\frac{dU}{dx} = -\tfrac{1}{2}C(x^2 + a^2)^{-3/2}2x = -Cx(x^2 + a^2)^{-3/2}$$

Thus the force is

$$F_x = -\frac{dU}{dx} = +\frac{Cx}{(x^2 + a^2)^{3/2}}$$

The general relation between force and potential energy is given by

$$dU = -\mathbf{F} \cdot d\mathbf{s}$$

Let us consider various displacements of magnitude ds but with different directions. If $d\mathbf{s}$ is perpendicular to $\mathbf{F}$, the potential-energy change is zero. The greatest change in U occurs when $d\mathbf{s}$ is parallel or antiparallel to the force. When $d\mathbf{s}$ is parallel to the force, we can write

$$F = -\frac{dU}{ds} \qquad d\mathbf{s} \parallel \mathbf{F} \qquad\qquad 8\text{-}20$$

A vector which points in the direction of the greatest change in a scalar function and whose magnitude equals the derivative of the function with respect to distance in that direction is called the *gradient* of the function. Thus in general, the force is the negative gradient of the potential energy.

8-8 Equilibrium and Potential Energy in One Dimension

Figure 8-7 shows a plot of potential energy versus x for a mass on a spring. The potential-energy function, from Equation 8-10, is

$$U(x) = \tfrac{1}{2}kx^2$$

The force is given by the negative slope of this curve,

$$F_x = -\frac{dU}{dx}$$

At the position where the slope is zero, the force is zero and the particle is in equilibrium. By equilibrium we mean that if F_x is the only force acting on it, the particle will remain at rest in this position if it is placed there at rest. This occurs at $x = 0$ when the spring is un-

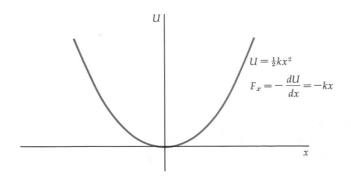

$$U = \tfrac{1}{2}kx^2$$

$$F_x = -\frac{dU}{dx} = -kx$$

Figure 8-7
Potential energy U versus displacement from equilibrium x for a mass on a spring. The force F_x is the negative slope $-dU/dx$ of this curve. For x positive, the slope is positive, indicating that F_x is negative. For x negative, the slope is negative, indicating that F_x is positive. The slope is zero at $x = 0$, indicating zero force. A minimum point in the potential-energy curve is a point of stable equilibrium because for displacements on either side of this point the force is toward the equilibrium position.

stretched. For x slightly greater than zero the slope is positive; thus F_x is negative and the particle would accelerate toward the equilibrium point $x = 0$. For x slightly less than zero the slope is negative; thus F_x is positive, and the particle again would accelerate toward $x = 0$. The equilibrium is thus stable; i.e., if a particle at rest at the equilibrium point $x = 0$ is displaced slightly, the force then accelerates it back toward the equilibrium point.

Stable equilibrium defined

Figure 8-8 shows the potential-energy curve from Example 8-8. At $x = 0$, the slope is zero. Thus $F_x = 0$ at $x = 0$, and the particle is in equilibrium. However, this equilibrium is unstable. For x slightly greater than zero, the slope is negative and $F_x = -dU/dx$ is positive. Thus if a particle in equilibrium is given a small positive displacement, it will experience a positive force F_x and will accelerate away from the equilibrium position. Similarly, for x slightly less than zero, the slope is positive, indicating that F_x is negative. If the particle is displaced slightly to the left of its equilibrium position, it will accelerate to the left, away from equilibrium. Both maximum and minimum points on a potential-energy curve are positions of equilibrium; at minimum points the equilibrium is stable, and at maximum points it is unstable. We also note that force always urges a particle toward lower potential energy.

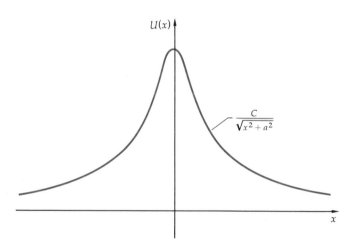

Figure 8-8
The potential energy $U(x) = C/\sqrt{x^2 + a^2}$ versus x for Example 8-8. The maximum at $x = 0$ is a position of equilibrium because the slope is zero and $F_x = -dU/dx = 0$. However, this equilibrium is unstable because for displacements in either direction the force is away from the equilibrium position.

Questions

10. You are shown a graph of $U(x)$ versus x. At some places the curve $U(x)$ is steep and at others not so steep. There are maximum and minimum points. At what points is the force associated with $U(x)$ large? Small? Zero?

11. For a certain range of x the potential energy is constant. What is the force in this range?

8-9   Potential Energy of a System of Particles

In our examples we have considered one particle in situations where there is assumed to be an external force which depends on position, e.g., the force due to gravity or a spring. We thus spoke of the potential energy of a particle in a given situation. This description is sufficient for many macroscopic problems in which we concentrate on one particle and treat the surroundings as merely an agent supplying a force F.

Let us now look at a more fundamental situation, a system consisting of two interacting particles. As an example, consider the special case of two charged particles, say protons. Let the particles be a distance r apart, as shown in Figure 8-9. The force on each is the electrostatic force of magnitude $F = kq_1q_2/r^2$ directed away from the other particle (assuming like charges and of course neglecting gravity). We cannot choose a coordinate system fixed on one particle and use Newton's laws because each particle accelerates. Let us instead choose an inertial frame with the origin on the line joining the particles. Let Δr_1 and Δr_2 be the distance moved by particles 1 and 2, respectively, in a short time interval Δt. Since the separation distance is $r = r_1 + r_2$, the change in separation is

$$\Delta r = \Delta r_1 + \Delta r_2$$

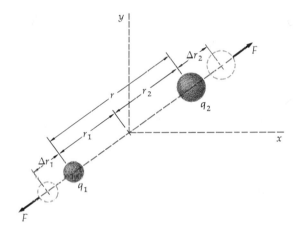

Figure 8-9
Two charges separated by a distance $r = r_1 + r_2$ are given displacements Δr_1 and Δr_2. The total work done on both charges by the electrostatic force of interaction equals the decrease in the electrostatic potential energy of the two-particle system.

If we choose Δt small enough, these distances will be so small that we can neglect the variation in the force during the motion. The work done on particle 1 is $F\,\Delta r_1$, and that done on particle 2 is $F\,\Delta r_2$. Since for each particle F is the resultant force, we can set this work done equal to the change in kinetic energy:

$$\Delta E_{k1} = F\,\Delta r_1 = \frac{kq_1q_2}{r^2}\,\Delta r_1$$

$$\Delta E_{k2} = F\,\Delta r_2 = \frac{kq_1q_2}{r^2}\,\Delta r_2$$

The total increase in kinetic energy of the system is

$$\Delta(E_{k1} + E_{k2}) = \frac{kq_1q_2}{r^2}\,(\Delta r_1 + \Delta r_2) = \frac{kq_1q_2}{r^2}\,\Delta r$$

We define the potential energy of the *system* by

$$U(r) = \frac{kq_1q_2}{r} \qquad\qquad \text{8-21}$$

For a small change in r, we have

$$\Delta U(r) \approx \frac{dU}{dr}\,\Delta r = \frac{d}{dr}\left(\frac{kq_1q_2}{r}\right)\Delta r = \frac{-kq_1q_2}{r^2}\,\Delta r \qquad\qquad \text{8-22}$$

Thus

$$\Delta(E_{k1} + E_{k2}) = -\Delta U \qquad\qquad 8\text{-}23$$

or

$$\Delta(E_{k1} + E_{k2} + U) = 0 \qquad\qquad 8\text{-}24$$

When we consider the total system, particle plus surroundings, the potential energy is a quantity associated with the configuration of the system rather than with a single particle. Equation 8-21 defines the electrostatic potential energy for a two-particle system. In general when two particles interact via any conservative force,

The potential energy of a two-particle system is the negative of the work done on both particles by the forces of interaction when the particles are brought from some reference separation to a given configuration.

We can think of this in a slightly different way. If we push the two protons together with zero acceleration, we must exert an applied force on each that is equal and opposite to the force of interaction. Thus the work done by the applied force is just the negative of the work done by the interaction forces. The potential energy of the system equals the work done by an applied force which brings the particles from infinite separation to their final configuration without acceleration.

Questions

12. Two particles interact through a conservative force. The potential energy of the system is U. Would you think it possible to divide U into two parts U_1 and U_2 ($U = U_1 + U_2$) which would be the potential energies of m_1 and m_2?

13. When you consider your own interaction with the earth, you ordinarily think of the potential energy as being yours rather than the earth's. Why is this possible?

14. The electrostatic force between an electron and a proton is attractive. Does their mutual potential energy increase or decrease as they move toward each other? Answer the same question for two protons, which repel each other.

Optional

8-10   General Test for Conservative Forces in Three Dimensions

In general the potential energy can be a function of all three coordinates x, y, and z, and the force has components F_x, F_y, and F_z. The potential-energy function $U(x,y,z)$ is related to a conservative force **F** by

$$dU = -\mathbf{F} \cdot d\mathbf{s} = -(F_x\, dx + F_y\, dy + F_z\, dz) \qquad\qquad 8\text{-}25$$

where we have written out the dot product in rectangular coordinates for a general displacement $d\mathbf{s} = dx\,\mathbf{i} + dy\,\mathbf{j} + dz\,\mathbf{k}$. We can relate any one of these components, say F_x, to the change in $U(x,y,z)$ when y and z are held fixed. Consider a displacement $d\mathbf{s} = dx\,\mathbf{i}$. Then

$$dU(x,y,z) = -\mathbf{F} \cdot d\mathbf{s} = -F_x\, dx \qquad y, z \text{ held fixed} \qquad 8\text{-}26$$

Thus F_x is the negative derivative of U with respect to x with y and z held constant. Such a derivative is called a partial derivative. It is written $\partial U/\partial x$. Thus in general, the components of the force are related to the potential energy by

$$F_x = -\frac{\partial U}{\partial x} \qquad F_y = -\frac{\partial U}{\partial y} \qquad F_z = -\frac{\partial U}{\partial z} \qquad \text{8-27}$$

If we take the derivative of a function with respect to x and then with respect to y, we get the same result as when we take the derivative first with respect to y and then x (assuming these second partial derivatives exist). Thus

$$\frac{\partial}{\partial y}\left(\frac{\partial U}{\partial x}\right) = \frac{\partial}{\partial x}\left(\frac{\partial U}{\partial y}\right)$$

Applying this to Equation 8-27, we have as a condition that $\mathbf{F}$ be conservative

$$\frac{\partial F_x}{\partial y} = \frac{\partial F_y}{\partial x} \qquad \text{8-28a}$$

Similarly

$$\frac{\partial F_y}{\partial z} = \frac{\partial F_z}{\partial y} \qquad \text{and} \qquad \frac{\partial F_x}{\partial z} = \frac{\partial F_z}{\partial x} \qquad \text{8-28b,c}$$

Equations 8-28 are a necessary and sufficient condition that the force $\mathbf{F}$ be conservative.

Example 8-9 A force is given by $\mathbf{F} = C(-y\mathbf{i} + x\mathbf{j})$, where C is a constant. Is this force conservative?

   Applying the test, we have

$$\frac{\partial F_x}{\partial y} = \frac{\partial(-Cy)}{\partial y} = -C$$

$$\frac{\partial F_y}{\partial x} = \frac{\partial(+Cx)}{\partial x} = +C \neq \frac{\partial F_y}{\partial x}$$

Thus this force is not conservative. It is in fact the type of force discussed in Example 8-6. If we change to polar coordinates, we have $x = r\cos\theta$, $y = r\sin\theta$; then

$$\mathbf{F} = C(-r\sin\theta\,\mathbf{i} + r\cos\theta\,\mathbf{j}) = rC(-\sin\theta\,\mathbf{i} + \cos\theta\,\mathbf{j}) = rC\hat{\boldsymbol{\theta}}$$

where $\hat{\boldsymbol{\theta}} = -\sin\theta\,\mathbf{i} + \cos\theta\,\mathbf{j}$ is a unit vector tangent to the circle around the origin.

Review

A. Define, explain, or otherwise identify:

Conservative force, 199
Potential energy, 201
Total mechanical energy, 202
Central force, 208

Electron volt, 212
Stable equilibrium, 214
Unstable equilibrium, 214

B. State the criterion for a force to be conservative.

C. Derive the law of conservation of mechanical energy.

D. True or false:

1. Only conservative forces do work.

2. There is a potential-energy function associated with every force.

3. If only conservative forces act, the kinetic energy of a particle does not change.

4. When a particle makes a round trip, the total work done by each conservative force is zero.

5. The work done by a conservative force decreases the potential energy associated with that force.

6. The total mechanical energy of a particle must be greater than zero.

Exercises

Section 8-1, Conservative Forces

1. A 2-kg ball moves from an initial position $x = 0$, $y = 4$ m to the final position $x = 4$ m, $y = 0$. (a) Calculate the work done by gravity along the segments AB and BC in Figure 8-10 and the total work done along this path (I in the figure). (b) Calculate the work done by gravity along the segments AD, DE, and EC and the total work done along path II. (c) Calculate directly the work done by gravity along the direct path AC (path III).

Figure 8-10
Exercise 1.

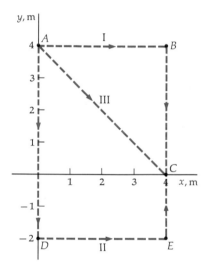

2. A particle is acted on by a force $\mathbf{F} = 4\mathbf{i} + 3\mathbf{j}$ N. The particle moves from the origin to point $x = 8$ m, $y = 6$ m. (a) Find the work done on the particle if it moves along the straight line from the origin to its final position (note that the force acts parallel to this line). (b) If the particle moves along the x axis to the point (8 m, 0) and then perpendicular to the x axis to its final position, find the work done by the force along each part of the path and the total work. (c) If the particle moves first along the y axis to the point (0, 6 m) and then parallel to the x axis to its final position, find the work done along each part of the path and the total work.

3. A particle is acted on by a force $\mathbf{F} = 4y\mathbf{i} + 5\mathbf{j}$, where $\mathbf{F}$ is in newtons and y is in meters. Show that this force is not conservative by computing the work done by the force in moving from the origin to the point $(x = 2$ m, $y = 2$ m) along two paths. (a) The first path is along the x axis to point (2 m, 0) and then parallel to the y axis along the line $x = 2$ m to the final point (2 m, 2 m). Find the component of the force parallel to the displacement for each of these parts of the path, the work done along each part of this path, and the total work. (b) The second path is first along the y axis to the point $x = 0$, $y = 2$ m and then parallel to the x axis. Find the work done along each part of this path and the total work.

Section 8-2, The Conservation of Mechanical Energy

4. List some situations in which the total mechanical energy of a particle is not conserved.

5. A particle is acted upon by a single conservative force which increases its kinetic energy by 10 J. (a) What is the change in the potential energy of the particle? (b) What is the change in the total energy of the particle?

6. A particle of mass 2 kg moves under the influence of a single conservative force which decreases its velocity from 20 to 10 m/sec. (a) Find the change in kinetic energy, the change in potential energy, and the change in total mechanical energy of the particle. (b) If the original potential energy of the particle was 100 J (when traveling at 20 m/sec), find the potential energy and the total energy when the speed is 10 m/sec.

7. A 0.5-kg ball is thrown straight up with initial speed of 10 m/sec. (a) Find the initial kinetic energy of the ball and total mechanical energy, assuming that the initial potential energy is zero. (b) Find the kinetic energy, potential energy, and total mechanical energy of the ball at the top of its flight. (c) Use the law of conservation of total mechanical energy to find the maximum height of the ball. (Take $g = 10$ m/sec² for convenience.) (d) Compare this method of

finding the maximum height of the ball with that in Chapter 2 using the constant-acceleration formulas. Can you find the time needed for the ball to reach its maximum height from the conservation of mechanical energy?

8. A particle moves under the influence of a single conservative force. Under what conditions (if any) is it possible for the potential energy to increase?

9. A conservative force does 25 J of work on a particle in moving it from point x_1 to x_2. (a) Find the change in potential energy of the particle when it moves from x_1 to x_2. (b) If the potential energy of the particle is zero at x_1, what is it at x_2? (c) If the mass of the particle is 5 gm and it starts from rest at x_1 and moves under the influence of this force only, what is its velocity at point x_2?

10. A particle moves under the influence of a single conservative force. In doing so the speed of the particle first decreases, then becomes zero momentarily, and then increases. (a) Describe the change in potential energy of the particle during this motion. (b) Suggest a physical situation corresponding to these conditions.

11. During its motion, the potential energy of a 10-kg mass increases by 200 J. (a) How much work was done on the mass by conservative forces? (b) If the particle travels at constant speed, how much work was done by nonconservative forces?

Section 8-3, Conservative Forces and Potential Energy in One Dimension

12. (a) Make a plot of the potential energy of a 10-lb weight versus its height above the ground for heights ranging from 0 to 50 ft. Use the ground as the reference level at which $U = 0$. (b) On the same graph plot the potential energy of the weight, using the height 25 ft as the point of zero potential energy.

13. How high must a mass be lifted to gain an amount of potential energy equal to the kinetic energy it has when moving at speed 60 mi/h?

14. A constant force acts in the negative x direction $F_x = -6$ N. Find the potential-energy function associated with this conservative force (a) if $U = 0$ at $x = 0$, (b) if $U = 0$ at $x = 5$ m, (c) if $U = 50$ J at $x = 0$.

15. A constant force is given by $F_x = 4$ N. (a) Find the potential-energy function associated with this force for arbitrary choice of zero potential energy. (b) Find $U(x)$ such that $U(x)$ is zero at $x = 6$ m. (c) Find $U(x)$ such that $U = 12$ J at $x = 6$ m.

16. A man places a 2-lb block against a horizontal spring of force constant $k = 20$ lb/ft and compresses it 9 in. (a) Find the work done by the man and the work done by the spring. (b) The spring is released from its compression of 9 in. and returns to its original position. Find the work done by the spring and the kinetic energy of the block at this point (assuming that only the spring does work on the block).

17. A spring obeys Hooke's law with force constant $k = 10^4$ N/m. How far must it be stretched for its potential energy to be (a) 100 J and (b) 50 J?

18. A spring is stretched so that its potential energy is 25 J and the force it exerts is 500 N. (a) By how much is the spring stretched from its natural length? (b) What is the force constant k?

Section 8-4, Nonconservative Forces in One Dimension

19. Which of the following forces are conservative, and which are nonconservative? (a) Frictional force exerted on a sliding box, (b) force exerted by a linear spring obeying Hooke's law, (c) force of gravity, (d) wind resistance on a moving car, (e) force exerted by a nonlinear spring obeying $F_x = -kx + Cx^2$, (f) force exerted by boy on ball when throwing the ball, (g) force exerted by floor on a dropped egg.

20. A man lifts a 100-lb weight from the floor to a shelf a height of 4 ft. The weight is at rest initially and finally. (a) What is the change in the potential energy of the weight? (b) List the forces acting on the weight as it is lifted and label them conservative or nonconservative.

Section 8-5, Conservative Forces and Potential Energy in Three Dimensions

21. Find the potential-energy function associated with the constant force $F = F_0\mathbf{k}$ ($\mathbf{k}$ is a unit vector in the z direction) such that the potential energy is zero at (a) $z = 0$, (b) $z = -15$ m, and (c) $z = 15$ m.

22. (a) Show that Equation 8-14 for the gravitational potential energy can be written

$$U = -\frac{mgR_E^2}{r}$$

where g is the free-fall acceleration of gravity at the earth's surface and R_E is the radius of the earth. (b) Use this result to calculate the potential energy in joules of a 1-kg mass at the positions $r = R_E$, $r = 2R_E$, and $r = 3R_E$. (Take R_E to be 6.37×10^6 m.)

23. A 1-kg mass is lifted at constant velocity by an applied force which just balances the gravitational force. (a) From your results of Exercise 22, calculate the work done by the applied force to raise the mass from $r = R_E$ to $r = 2R_E$. (b) Compare your result with that obtained assuming that the gravitational force had the constant value mg. (c) What are the minimum and maximum values of the magnitude of the applied force during this procedure?

24. A force similar to that discussed in Example 8-6 is given by

$$\mathbf{F} = (10 \text{ N})\hat{\theta}$$

Find the work done by the force on a particle which moves from the initial position $x = 5$ m, $y = 0$, $z = 0$ to the final position $x = -5$ m, $y = 0$, $z = 0$ if (a) the particle moves along a circular arc of radius 5 m in the direction of the force, and (b) the particle moves along the x axis. (c) Find a path between the initial and final positions such that the work done by the force is half that found in part (a).

Section 8-6, The Electron Volt Energy Unit

25. Convert the following energies in joules to electron volts, kiloelectron volts, or megaelectron volts: (a) 3×10^{-19} J; (b) 10^{-16} J; (c) 1.5×10^{-7} J; (d) 8.16×10^{-14} J.

26. A nucleus contains a positive electric charge of $50e$, where e is the magnitude of the electron charge. An electron is a distance of 2×10^{-10} m from the nucleus. Find the potential energy of the electron in electron volts and in joules.

27. The energy released in the fission of one uranium nucleus is about 200 MeV. (a) Convert this energy into joules. (b) Find the energy in joules released if 1 gm of uranium undergoes fission [1 gm contains about $(6.02 \times 10^{23})/235$ uranium nuclei].

Section 8-7, Force from Potential Energy, and Section 8-8, Equilibrium and Potential Energy in One Dimension

28. Figure 8-11 shows a potential-energy function $U(x)$ versus x. (a) At each point indicated, state whether the force F_x is positive, negative, or zero. (b) At which point does the force have the greatest magnitude? (c) Identify any equilibrium points and state whether the equilibrium is stable or unstable.

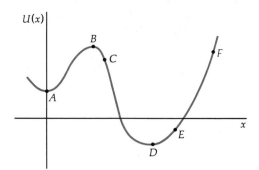

Figure 8-11
Exercise 28.

29. Repeat Exercise 28 for the potential-energy function $U(x)$ shown in Figure 8-12.

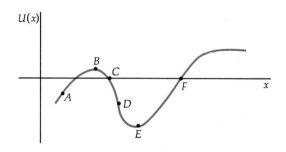

Figure 8-12
Exercise 29.

30. (a) Find the force F_x associated with the potential-energy function $U(x) = Ax^4$, where A is a constant. (b) At what point or points is the force zero?

31. A potential-energy function is given by $U(x) = U_0 \ln(x/x_0)$, where U_0 and x_0 are constants. (a) Find the force F_x. (b) Are there any points of equilibrium?

32. In the potential-energy curve $U(y)$ shown in Figure 8-13 the segments AB and CD are straight lines. Sketch the force F_y versus y.

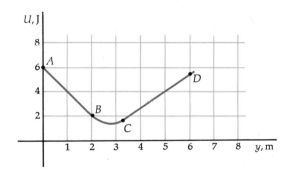

Figure 8-13
Exercise 32.

33. A potential-energy function is given by $U(r) = C/r$, where C is a positive constant. (a) Find the radial force F_r as a function of r. (b) Is this force toward the origin or away from it; i.e., is F_r negative or positive? (c) Does the potential energy increase or decrease as r increases? (d) Answer parts (b) and (c) if C is a negative constant.

Section 8-9, Potential Energy of a System of Particles

34. Calculate the electrostatic potential energy in electron volts of two protons separated by a distance of 10^{-15} m. This is a typical separation of protons in an atomic nucleus.

35. When ^{235}U fissions, the energy first appears as electrostatic potential energy of the two fission fragments. The fragments then fly apart as the potential energy is converted into kinetic energy of the fragments. Estimate the energy released in fission by assuming the $^{235}_{92}U$ nucleus is split into two equal nuclei (not usually true in practice), each carrying a charge of $+46e$ and separated by

15×10^{-15} m (this is about twice the radius of a nucleus of mass number 118), and calculating the electrostatic potential energy for these two particles. Express your answer in megaelectron volts.

36. Two bodies have masses 5 and 10 kg. They are 2 m apart and 2 m above the ground. (*a*) Calculate their potential energy because of their mutual gravitational attraction. (*b*) Compare this potential energy with the total potential energy relative to the ground level they have because of the gravitational attraction of the earth.

Section 8-10, General Test for Conservative Forces in Three Dimensions

37. Apply the general test (Equation 8-28) to determine whether the following forces are conservative: (*a*) $\mathbf{F} = y^3\mathbf{i} + 3xy^2\mathbf{j}$, (*b*) $\mathbf{F} = x\mathbf{i} + y\mathbf{j}$, (*c*) $\mathbf{F} = x^2y\mathbf{i} + x^2y\mathbf{j}$, (*d*) $\mathbf{F} = x^2\mathbf{i} + xy\mathbf{j}$.

38. A potential-energy function is given by $U(x,y,z) = C(1/\sqrt{x^2 + y^2 + z^2})$, where C is a constant. (*a*) Find the rectangular components of the force F_x, F_y, and F_z and show explicitly that Equation 8-28 is satisfied by these components. (*b*) Show that the force found in part (*a*) is a central force.

Problems

1. Figure 8-14 gives the force F_x on a particle as a function of its distance x from the origin. (*a*) Calculate from the graph the work done by the force when the particle moves from $x = 0$ to the following values of x: $-4, -3, -2, -1, 0, 1, 2, 3$, and 4 m. (*b*) How do you know that this force is conservative? (*c*) Plot the potential energy versus x for the range of x from -4 m to $+4$ m assuming that $U = 0$ at $x = 0$.

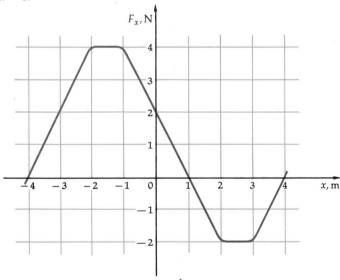

Figure 8-14
Problem 1.

2. Repeat Problem 1 for the force F_x shown in Figure 8-15.

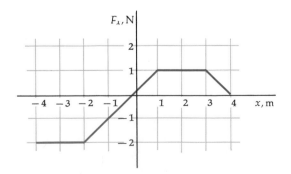

Figure 8-15
Problem 2.

3. The force on a particle moving along the x axis is given by $F_x = -ax^2$, where a is a constant. (a) Calculate the potential energy $U(x)$ relative to $U = 0$ at $x = 0$. (b) Sketch a graph of $U(x)$ versus x.

4. A force is given by $F_x = +A/x^3$, where $A = 8$ N-m³. (a) For positive x, does the potential energy associated with this force increase or decrease with increasing x? Convince yourself of the correct answer to this question by imagining a particle placed at rest at some point x and released. When it moves, its kinetic energy will increase and its potential energy therefore must decrease. (b) Find the potential-energy function $U(x)$ associated with this force such that U approaches zero as x approaches infinity. Sketch $U(x)$ versus x.

5. Figure 8-16 shows the potential-energy function $U(r)$ versus r for a diatomic molecule, where r is the separation of the atoms. (a) Indicate the equilibrium point on the graph and discuss its stability. (b) From this graph, sketch a graph of the force F_r exerted by one atom on the other versus separation distance r.

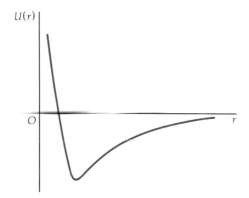

Figure 8-16
Problem 5.

6. Figure 8-17 shows the potential-energy function $U(r)$ versus r associated with the nuclear force between two protons, two neutrons, or a neutron and proton, where r is the separation distance. From this graph sketch the force F_r versus r.

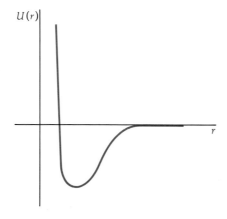

Figure 8-17
Problem 6.

7. A force acting in the xy plane is given by $\mathbf{F} = 10\mathbf{i} + 3x\mathbf{j}$, where $\mathbf{F}$ is in newtons and x is in meters. Suppose the force acts on a particle as it moves from initial position $x = 4$ m, $y = 1$ m to a final position $x = 4$ m, $y = 4$ m. (a) Show that this force is not conservative by computing the work done by the force for at least two different paths. (b) Find a path for which the work done by the force is zero.

8. A straight rod of negligible mass is mounted on a frictionless pivot, as shown in Figure 8-18. Masses m_1 and m_2 are suspended at distances l_1 and l_2 as shown. (a) Write the gravitational potential energy of the masses as a function of the angle θ made by the rod and the horizontal. (b) For what angle θ is the potential energy a minimum? Is the statement "systems tend to move toward minimum potential energy" consistent with your result? (c) Show that if $m_1 l_1 = m_2 l_2$, the potential energy is the same for all values of θ. (When this holds, the system will balance at any angle θ. This result is known as *Archimedes' law of the lever*.)

Figure 8-18
Problem 8.

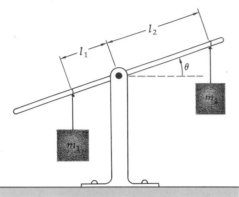

9. The force between two atoms in a diatomic molecule can be represented approximately by the potential-energy function

$$U(r) = U_0 \left[\left(\frac{r_0}{r} \right)^{12} - 2 \left(\frac{r_0}{r} \right)^6 \right]$$

where r_0 and U_0 are constants. (a) Find the force F_r. (b) At what value of r is the potential energy zero? At what value of r is the force zero? (c) At what value of r is the potential energy a minimum? What is the value of this minimum potential energy? (This minimum potential energy is approximately equal to the dissociation energy of the molecule.) (d) Sketch $U(r)$ versus r.

10. A theoretical form for the potential energy associated with the nuclear force between two protons, two neutrons, or a neutron and proton is the Yukawa potential

$$U = -U_0 \left(\frac{a}{r} \right) e^{-r/a}$$

where U_0 and a are constants. (a) Sketch U versus r using the parameters $U_0 = 23$ MeV and $a = 2.5 \times 10^{-15}$ m. (b) Find the force F_r. (c) Compare the magnitudes of the forces at the separations $r = a$, $r = 2a$, and $r = 5a$.

CHAPTER 9   Conservation of Energy

The potential energy of a particle is defined so that the work done on the particle by conservative forces equals the decrease in the potential energy. When only conservative forces do work, the total mechanical energy of the particle remains constant because the increase in kinetic energy equals the decrease in potential energy. This principle of conservation of energy is one of the most important in physics.

In the macroscopic world, nonconservative forces are always present, the most common being frictional forces. Another type of nonconservative force is involved in large deformations of bodies. For example, if a spring is stretched beyond its elastic limit, it becomes permanently deformed and the work done in the stretching is not recovered when the spring is released. When nonconservative forces do work on a particle, the sum of kinetic and potential energies of the particle is not constant. Because some kind of frictional force is nearly always present in the motion of a macroscopic body, the importance of energy and its conservation was not realized until the nineteenth century. Then it was discovered that the disappearance of *macroscopic mechanical energy* is accompanied by the appearance of *internal energy*, usually indicated by an increase in temperature. We now know that, on the microscopic scale, this internal energy of the surroundings consists of kinetic and potential energy of molecular motion. When the concept of energy is generalized to include this internal energy, the total energy of an object plus its surroundings does not change even when friction is present.

If we carefully define a system, e.g., several bodies and their local surroundings (as we did in solving force and acceleration problems in Chapter 5), we find that even when internal energy is included, the total energy of the system does not always remain constant. However, we are always able to account for the increase or decrease in the system energy by the appearance or disappearance of energy somewhere else. For example, energy of a system is often decreased because of some form of radiation, e.g., sound waves from a collision of two objects, water waves produced by a ship, or electromagnetic waves

The action of a pole vaulter demonstrates several kinds of energy. First he converts internal chemical energy into kinetic energy when he begins to run. Some of this energy is then converted into elastic potential energy of deformation of the pol᠉ and eventually into gravitational potential energy, which in turn changes into kinetic energy as he drops. The mechanical energy is finally lost as heat when he reaches the ground. (*Courtesy of Harold E. Edgerton.*)

produced by accelerated charges in a radio antenna. We can write a generalized statement of the conservation of energy in the following way. Let E_{sys} be the total energy of a given system and P_{in} be the power input, i.e., the rate at which energy is put into the system. Then

$$P_{in} = \frac{dE_{sys}}{dt} \qquad\qquad 9\text{-}1$$

where P_{in} is negative if energy is flowing out of the system. The significance of this statement is that if the energy concept is sufficiently generalized, changes in the total energy of the system are always precisely accounted for by energy flow into or out of the system. Energy is never created or destroyed in the system, though it may change from one form to another.

In this chapter, we consider only *mechanical energy,* the kinetic energy of a particle, $\frac{1}{2}mv^2$, and the various potential-energy functions associated with the forces acting on a particle. Consider a body acted on by a nonconservative force $\mathbf{F}_{nc}$ and several conservative forces $\mathbf{F}_i$ so that the resultant force is

$$\Sigma \mathbf{F} = \mathbf{F}_{nc} + \mathbf{F}_1 + \mathbf{F}_2 + \cdots$$

The work done by the resultant force, according to the general work-energy theorem, equals the increase in kinetic energy of the particle

$$W_{net} = \int \Sigma \mathbf{F} \cdot d\mathbf{s} = \int \mathbf{F}_{nc} \cdot d\mathbf{s} + \int \mathbf{F}_1 \cdot d\mathbf{s} + \int \mathbf{F}_2 \cdot d\mathbf{s} + \cdots = \Delta E_k$$

$$9\text{-}2$$

For each conservative force $\mathbf{F}_i$ we define a potential-energy function U_i in the usual way:

$$\int \mathbf{F}_i \cdot d\mathbf{s} = -\Delta U_i$$

Then Equation 9-2 can be written

$$\int \mathbf{F}_{nc} \cdot d\mathbf{s} - \Delta U_1 - \Delta U_2 + \cdots = \Delta E_k$$

or

$$\int \mathbf{F}_{nc} \cdot d\mathbf{s} = \Delta(E_k + U_1 + U_2 + \cdots) = \Delta E_{\text{total}} \qquad 9\text{-}3$$

Work-energy theorem modified to include potential energy

where the total mechanical energy of the particle is

$$E_{\text{total}} = E_k + U_1 + U_2 + \cdots \qquad 9\text{-}4$$

The work done by nonconservative forces equals the change in the total mechanical energy of the particle. This modified form of the work-energy theorem is the most convenient in many applications since explicit calculation of work done along the path is required only for the nonconservative forces. When the only forces that do work are conservative, the left side of Equation 9-3 is zero. Then there is no change in the total mechanical energy of the particle:

$$E_{\text{total}} = E_k + U_1 + U_2 + \cdots = \text{constant} \qquad 9\text{-}5$$

In this chapter we consider various examples of the use of the work-energy theorem and the conservation of energy expressed in Equations 9-3 and 9-5. Since we derived these equations from Newton's laws, we can never obtain more information about a mechanics problem than is contained in Newton's laws. Any problem that can be solved by using conservation of energy can also be solved directly from Newton's laws. In spite of this, energy conservation and the work-energy theorem are extremely useful tools for the analysis of problems. Since the kinetic energy is $\frac{1}{2}mv^2$ and the potential-energy functions are functions of position, Equation 9-5 relates the speed of a particle to its position. In any problem in which one of these quantities is to be found, the energy-conservation equation is generally simpler to use than a direct application of $\Sigma \mathbf{F} = m\mathbf{a}$. We shall illustrate this in Section 9-1. In many cases, examining the potential-energy function offers a qualitative understanding of the possible motion of a system without a detailed solution of the equations of motion.

9-1 Some Illustrative Examples

Example 9-1 Block on a Rough Incline Consider a block of mass m which is drawn up a rough incline by a constant applied force $\mathbf{T}$ parallel to the incline. The angle of the incline is θ. The block starts from rest at the bottom of the incline. Find its speed when it has moved a distance s up along the incline.

This problem is readily solved by the direct application of Newton's laws of motion, i.e., by analyzing the forces, finding the acceleration (which is constant), and then using the kinematic relationships for constant acceleration. We shall use the work-energy theorem instead to illustrate this method of analyzing a problem.

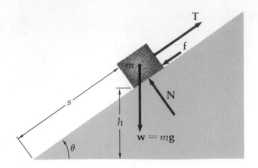

Figure 9-1
Block drawn up rough incline
by force **T** (Example 9-1).

There are four forces which act on the block as it moves up the incline (see Figure 9-1). The normal force **N**, which prevents the block from moving down into the incline, does no work during the motion because it is always perpendicular to the motion. The weight $m\mathbf{g}$ is a conservative force. As we have already seen, the increase in gravitational potential energy as the block ascends through the vertical distance $h = s \sin \theta$ is $U = mgh = mgs \sin \theta$.

The frictional force **f** is, of course, a nonconservative force. We must calculate the work it does as the block moves up the distance s along the incline. Since f is constant, and opposed to the motion, this work is just

$$W_f = -fs$$

Finally, the applied force **T** is a nonconservative force. (It depends on the action of whatever agency exerts it and not merely on the position of the block.) Hence we must compute its work. **T** is a constant force in the same direction as the motion; therefore the work is positive

$$W_T = +Ts$$

According to the work-energy theorem (Equation 9-3), the total work of the nonconservative forces equals the increase in total mechanical energy. The latter consists of the increase in gravitational potential energy plus the increase in the kinetic energy. Since the kinetic energy is zero at the bottom of the incline (the block starts from rest), the increase in kinetic energy is just the kinetic energy at the top, $\frac{1}{2}mv^2$.

Combining these facts gives

$$+Ts - fs = \tfrac{1}{2}mv^2 + mgs \sin \theta$$

The left side of this equation is the total work of the nonconservative forces. The right-side contains the increase in kinetic energy plus the increase in potential energy. Solving for the speed v gives

$$v = \sqrt{\frac{2s}{m} (T - f - mg \sin \theta)} \qquad\qquad 9\text{-}6$$

The quantity under the radical sign must be positive for its square root to have a meaningful value for the speed v. This requires that T be greater than $f + mg \sin \theta$. That, of course, means only that the applied force **T** must overbalance the sum of the opposing frictional force **f** and the component of the weight down the plane.

Equation 9-6 can also be obtained from the constant-acceleration formula $v^2 = 2as$, where the acceleration is the resultant force divided by the mass.

Example 9-2 The Simple Pendulum A mass m is suspended on a light string of length L. It is pulled aside so that the string makes an angle θ_0 with the vertical and is then released from rest. What is the speed v at the bottom of the swing, and what is the tension in the string there?

There are only two forces acting on m as it moves (neglecting air resistance). One is the tension **T** in the string, which is always perpendicular to the motion and therefore does no work. The other is the weight $m\mathbf{g}$, which is a conservative force. Here only conservative forces do work, and so the total mechanical energy of the mass is conserved.

Let us choose the gravitational potential energy to be zero at the bottom of the swing. At its original position a height h above the bottom the mass has potential energy mgh and no kinetic energy because it is at rest. As it swings down, the potential energy is converted into kinetic energy. Conservation of energy gives us for the speed v at the bottom,

$$\tfrac{1}{2}mv^2 = mgh$$

According to Figure 9-2, the distance h is related to the original angle θ_0 and the length of the pendulum L by

$$h = L - L \cos \theta_0 = L(1 - \cos \theta_0)$$

Thus

$$v^2 = 2gL(1 - \cos \theta_0) \qquad\qquad 9\text{-}7$$

The tension at the bottom is found as in Example 5-8. At this point the mass has centripetal acceleration $v^2/r = v^2/L$. The forces acting are

Multiflash photograph of a simple pendulum. As the bob descends, gravitational potential energy is converted into kinetic energy and the speed increases, as indicated by the increased spacing of the recorded positions. The speed decreases as the bob moves up, and the kinetic energy is changed into potential energy.

Figure 9-2
Simple pendulum for Example 9-2. The initial height above the bottom of the swing is related to the initial angle θ_0 by $h = L - L \cos \theta_0$. After the pendulum is released from rest, the speed of the bob at the bottom can be found from conservation of energy: $\tfrac{1}{2}mv^2 = mgh = mgL(1 - \cos \theta_0)$.

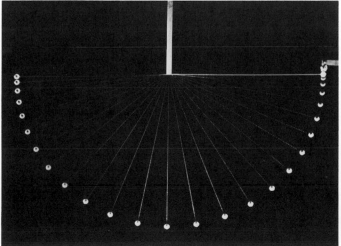

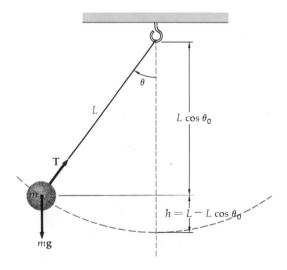

T up and mg down. Setting the resultant force equal to the mass times the acceleration gives

$$T - mg = \frac{mv^2}{r} = \frac{mv^2}{L} = 2mg(1 - \cos \theta_0)$$

$$T = mg + 2mg(1 - \cos \theta_0) \qquad\qquad 9\text{-}8$$

We now obtain Equation 9-7 directly from Newton's laws of motion to show how this method differs from the method of conservation of energy, just used. In Figure 9-3 the mass is at a point where the string makes some angle θ with the vertical. The mass has a velocity **v** tangent to the circular arc of motion. We can relate the speed v to the angle θ as follows. The distance s measured along the path length is related to θ by $s = L\theta$. Then the speed is

$$v = \frac{ds}{dt} = \frac{d(L\theta)}{dt} = L\frac{d\theta}{dt} \qquad\qquad 9\text{-}9$$

The tangential acceleration is just the rate of change of the speed:

$$a_t = \frac{dv}{dt} = L\frac{d^2\theta}{dt^2}$$

The only tangential force is the component of the weight $-mg \sin \theta$, where the minus sign indicates that this force component is in the direction of decreasing θ. The tangential component of $\Sigma \mathbf{F} = m\mathbf{a}$ is thus

$$-mg \sin \theta = ma_t = m\frac{dv}{dt} = mL\frac{d^2\theta}{dt^2} \qquad\qquad 9\text{-}10$$

or

$$L\frac{d^2\theta}{dt^2} = -g \sin \theta \qquad\qquad 9\text{-}11$$

A complete solution of Equation 9-11 for $\theta(t)$ is difficult. We shall consider an approximate solution (valid if θ is small) in Chapter 14, but we can obtain an exact solution for the speed $v = L \, d\theta/dt$. From Equation 9-10 we have

$$\frac{dv}{dt} = -g \sin \theta$$

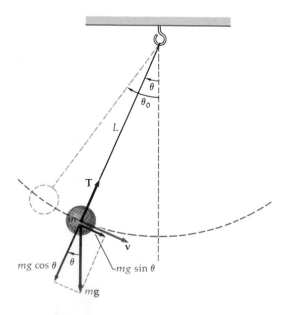

Figure 9-3
Forces on the pendulum bob when the string makes angle θ with the vertical and the bob moves with speed v. The resultant inward radial force $T - mg \cos \theta$ produces the centripetal acceleration v^2/L, while the resultant tangential force $mg \sin \theta$ produces the tangential acceleration dv/dt.

We can eliminate the time t by noting that since θ is a function of t, v must be a function of θ. Using the chain rule for derivatives, we have

$$\frac{dv}{dt} = \frac{dv}{d\theta}\frac{d\theta}{dt} = \frac{dv}{d\theta}\frac{1}{L}\left(L\frac{d\theta}{dt}\right) = \frac{v}{L}\frac{dv}{d\theta}$$

Then

$$\frac{v}{L}\frac{dv}{d\theta} = -g\sin\theta$$

or

$$v\,dv = -gL\sin\theta\,d\theta \qquad\qquad\qquad 9\text{-}12$$

Integrating θ from θ_0 to 0 and v from 0 to v gives

$$\int_0^v v\,dv = \int_{\theta_0}^0 -gL\sin\theta\,d\theta$$

$$\tfrac{1}{2}v^2\Big]_0^v = +gL\cos\theta\Big]_{\theta_0}^0$$

$$\tfrac{1}{2}v^2 - 0 = (gL\cos 0) - (gL\cos\theta_0)$$

$$v^2 = 2gL(1 - \cos\theta_0)$$

This is the same as the result we obtained using conservation of energy. The steps we have carried out here are essentially the same as those carried out in deriving the work-energy theorem. The integration on the right above is equivalent to calculating the work done by the component of the weight along the path. The integration on the left is essentially a calculation of the increase in kinetic energy similar to that carried out in our earlier derivation.

Example 9-3 A skier skis down a smooth hill of height H of arbitrary shape (see Figure 9-4). Find his speed at the bottom of the hill.

We have already considered this type of problem in Chapter 7. We repeat it here to contrast it with the pendulum problem. We have only one force that does work, i.e., gravity, which is conservative. The normal force exerted by the snow on the skis acts to constrain the skier to follow a certain path but does no work. Taking the potential energy to be zero at the bottom of the hill and assuming the initial speed at the top of the hill to be zero, we obtain from conservation of energy,

$$(E_k + U)_{\text{top}} = (E_k + U)_{\text{bottom}}$$

$$0 + mgH = \tfrac{1}{2}mv^2 + 0$$

$$v^2 = 2gH$$

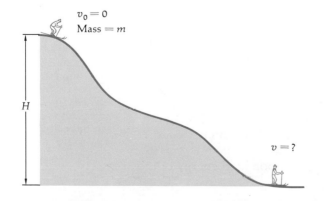

$v_0 = 0$
Mass $= m$

H

$v = ?$

Figure 9-4
Skier and smooth hill of arbitrary shape for Example 9-3. The total energy at the top of the hill is mgH assuming zero potential energy at the bottom. At the bottom of the hill the total energy is $\tfrac{1}{2}mv^2$. Conservation of energy gives $v^2 = 2gH$.

Helmut Gritscher/DPI

If there were no friction, the kinetic energy of the skier would equal the decrease in gravitational potential energy and his speed after descending through a given vertical distance would be the same as if he had fallen through that distance. In practice, frictional forces decrease the mechanical energy.

This example illustrates the great power of the principle of conservation of energy. To calculate v by direct application of Newton's second law would be impossible in this case because lacking a precise description of the curve along which the particle moves we cannot compute the tangential component of the weight at each point. Even if we had the equation of the curve, the direct application of $\Sigma \mathbf{F} = m\mathbf{a}$ would be very complex. On the other hand, a straightforward use of the conservation-of-energy principle enables us to obtain v quite easily.

Example 9-4 Atwood's Machine Figure 9-5 shows a device developed in the eighteenth century to make the first accurate determinations of the acceleration of gravity g. *Atwood's machine* consists of two masses m_1 and m_2 connected by a light string passing over a light pulley with negligible friction in its bearings. If the two masses are equal, they remain at rest at any position, each balancing the other. If m_2 is greater than m_1, m_2 accelerates downward and m_1 upward. The acceleration is constant, and if m_2 is only slightly greater than m_1, the acceleration is small and readily measured. From the measured acceleration it is possible to calculate g. We wish to find the equation to be used for this purpose. (This method was necessary because the acceleration g is too great to be measured directly with the instruments available 250 years ago. Direct measurement of g had to await the development of precise clocks capable of measuring times to accuracies of at least 0.01 sec.)

If friction in the pulley bearings is negligible and the pulley mass is so small that the work done to speed it up is negligible, the tension in the string is uniform. That is, the tension acting upward on m_1 equals the tension acting upward on m_2. The net work done by the tension on the masses is zero because when m_1 moves upward in the direction of T, m_2 moves downward the same distance in the direction opposite T. The only force that does net work on the two-mass system is gravity, and therefore the total kinetic plus gravitational potential energy of the two masses is conserved.

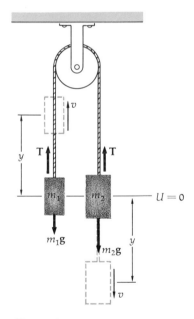

Figure 9-5
Atwood's machine (Example 9-4). When released from rest in the position shown, the larger mass m_2 falls and the smaller mass m_1 rises. After some time the masses are at the positions indicated by the dashed lines and moving with some speed v. The speed can be related to the distance moved by conservation of energy.

Let us choose the potential energy to be zero at the initial position of the masses. Since they are at rest at this position, the total energy is zero. Let v be the speed of m_1 after it has moved up a distance y. Then its kinetic energy is $\frac{1}{2}m_1v^2$, and its potential energy is m_1gy. Since the connecting string does not stretch, m_2 must move down the same distance y and acquire the same speed v. Then its kinetic energy is $\frac{1}{2}m_2v^2$, and its potential energy is $-m_2gy$. Conservation of energy gives

$$E = \tfrac{1}{2}m_1v^2 + m_1gy + \tfrac{1}{2}m_2v^2 - m_2gy = 0$$

or

$$\tfrac{1}{2}(m_1 + m_2)v^2 = (m_2 - m_1)gy \qquad\qquad 9\text{-}13$$

Solving this equation for v^2 gives

$$v^2 = \frac{2(m_2 - m_1)}{m_1 + m_2}\,gy \qquad\qquad 9\text{-}14$$

which relates the speed of either mass to the distance it moves. Comparing this with the constant-acceleration equation $v^2 = 2ay$, we see that the acceleration is

$$a = \frac{m_2 - m_1}{m_1 + m_2}\,g \qquad\qquad 9\text{-}15$$

In Example 9-4 we found the acceleration from the expression for v^2 using the constant-acceleration equation. In general, when a single conservative force acts, we can find the acceleration from the conservation-of-energy equation by differentiation. Consider such a one-dimensional problem in which a particle has initial energy E_0. At some time later it has kinetic energy $\frac{1}{2}mv^2$ and potential energy $U(x)$. Conservation of energy gives us

$$\tfrac{1}{2}mv^2 + U(x) = E_0 \qquad\qquad 9\text{-}16$$

We can obtain the acceleration from Equation 9-16 by differentiating with respect to time and using the chain rule:

$$\tfrac{1}{2}m2v\,\frac{dv}{dt} + \frac{dU}{dx}\frac{dx}{dt} = 0$$

Since dx/dt is just the speed v, we can divide out these terms. Then

$$m\frac{dv}{dt} = -\frac{dU}{dx} \qquad\qquad 9\text{-}17$$

This is just Newton's second law, since the negative derivative of U with respect to x is the force F_x. We could have obtained Equation 9-15 by differentiating Equation 9-14 (using $dy/dt = v$) without using the fact that the acceleration is constant.

Example 9-5 A Mass on a Horizontal Spring A mass m is placed on a smooth horizontal table and attached to a light spring of constant k. It is displaced a distance A from equilibrium and released (Figure 9-6a). Describe its motion.

The total mechanical energy is conserved since the only force that does work is the conservative force of the spring. If we choose $x = 0$ at the natural length of the spring, the potential energy of the spring is

$$U = \tfrac{1}{2}kx^2$$

Figure 9-6b shows a plot of $U(x)$ versus x. The minimum value of U occurs at $x = 0$, the equilibrium point where the slope $dU/dx = -F_x$ is

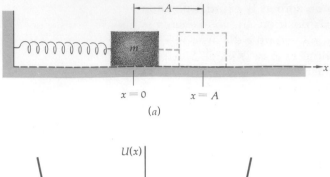

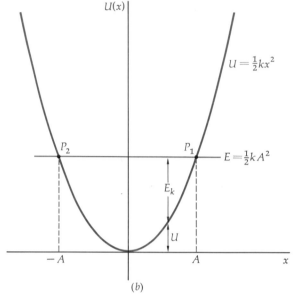

Figure 9-6
Mass on horizontal spring for
Example 9-5. (*a*) The mass is
displaced a distance *A* from
equilibrium and released. (*b*)
The potential-energy function
$U(x) = \frac{1}{2}kx^2$ for the mass on
the spring. The total energy
$E = \frac{1}{2}kA^2$ is indicated by the
horizontal line. The kinetic
energy $E_k = E - U(x)$ is represented by the vertical distance between the total energy line and $U(x)$. The points
P_1 and P_2 are the intersections
of E and $U(x)$. At these points
the kinetic energy is zero, and
the velocity of the particle reverses.

zero. Since the mass begins at $x = A$ with zero velocity, its total energy
at this point is potential energy:

$$E = \tfrac{1}{2}kA^2$$

This total energy, which is constant, is represented in Figure 9-6*b* by a
horizontal line. The speed v of the particle is related to its position x
through the conservation-of-energy equation,

$$\tfrac{1}{2}mv^2 + \tfrac{1}{2}kx^2 = E = \tfrac{1}{2}kA^2 \qquad\qquad 9\text{-}18$$

For any value of x, the kinetic energy $\frac{1}{2}mv^2 = E_k = E - U(x)$ is represented by the vertical distance between the total-energy line and the
potential-energy curve U, as shown in the figure. Since the kinetic
energy cannot be negative, the total-energy line must be above the potential-energy curve for physically possible motions. In Figure 9-6*b* the
curve $U(x)$ and the total-energy line intersect at $x = A$ (point P_1) and
$x = -A$ (point P_2). For this value of total energy, the motion is confined to the values of x between these points.

The points $x = +A$ and $x = -A$ are called *turning points* for the motion of the mass. At these points the velocity of the mass decreases to
zero and reverses. In this example, we start the mass from rest at
$x = +A$. The slope of $U(x)$ at this point is positive. The force $F_x =
-dU/dx$ is negative, and the particle accelerates to the left, decreasing
its potential energy and increasing its kinetic energy. When it reaches
the equilibrium point $x = 0$, its potential energy is zero and its kinetic
energy equals its total energy. Therefore its speed is maximum at this
point. After it passes the equilibrium point, the slope of $U(x)$ is negative, indicating a positive force. This force decreases the speed since it
is in the direction opposite the motion. The mass continues to the left

with its potential energy increasing and its kinetic energy decreasing until it reaches point P_2 at $x = -A$. Here it has no kinetic energy and is momentarily at rest. The force is still to the right, and so the particle turns and moves back toward the equilibrium point $x = 0$. Again, when it reaches this point, its kinetic energy is maximum and it continues moving to the right, until it comes momentarily to rest at the starting point P_1. After a full cycle of motion is completed, the mass begins a new identical cycle.

The symmetry of the potential-energy curve in Figure 9-6b makes the turning points equidistant from the equilibrium point. This distance A, called the *amplitude of the motion*, depends only on the total energy. It is also clear from the symmetry of $U(x)$ that the motion from $x = 0$ to $x = A$ and back to $x = 0$ is exactly like the motion from $x = 0$ to $x = -A$ and back except for reversal in direction; i.e., these two parts of the cycle of motion must take the same time.

We have been able to describe most of the features of the motion of a mass on a spring by using the energy diagram of Figure 9-6b. We shall use this type of analysis often in the next section. The length of time required for each oscillation (called the *period*) will be discussed in Chapter 14.

Example 9-6 A particle of mass m is dropped from a height h which is not necessarily small compared with the radius of the earth R_E. Find its speed when it reaches the surface of the earth assuming that air resistance can be neglected.

Gravity is the only force acting on the particle, and so its total mechanical energy is constant. Since the mass is not necessarily near the surface of the earth, we must use the general potential-energy function (Equation 8-14) rather than mgh. The initial kinetic energy of the particle is zero, and its initial potential energy is $-GM_Em/r_1$, where $r_1 = R_E + h$ is the initial distance from the center of the earth. When the particle reaches the surface of the earth, it has kinetic energy $\frac{1}{2}mv^2$ and potential energy $-GM_Em/R_E$. Conservation of energy gives

$$\tfrac{1}{2}mv^2 - \frac{GM_Em}{R_E} = -\frac{GM_Em}{r_1} \qquad\qquad 9\text{-}19$$

or

$$\tfrac{1}{2}mv^2 = \frac{GM_Em}{R_E} - \frac{GM_Em}{r_1} = \frac{GM_Em}{R_E r_1}(r_1 - R_E)$$

To simplify the calculation we write GM_E in terms of the free-fall acceleration near the surface of the earth, $g = GM_E/R_E^2$. Substituting gR_E^2 for GM_E gives

$$\tfrac{1}{2}mv^2 = \frac{mgR_E}{r_1}(r_1 - R_E)$$

or

$$v^2 = \frac{2gR_E}{r_1}(r_1 - R_E) = 2gh\,\frac{R_E}{r_1} = 2gh\,\frac{R_E}{R_E + h} \qquad\qquad 9\text{-}20$$

where we have written h for $r_1 - R_E$. The speed is smaller than the result $v^2 = 2gh$ obtained by neglecting the decrease in the force of gravity with distance.

In practice, the speed will be less than that given by Equation 9-20 because of air resistance.

Questions

1. Suggest two forces that may be important in skiing which are neglected in Example 9-3. Are these forces conservative? Is the work they do positive or negative? How would their inclusion affect the result $v^2 = 2gH$?

2. If the mass of the pulley in Atwood's machine is not negligible, the pulley has kinetic energy of rotation (to be studied in Chapter 12). Do you expect the inclusion of this kinetic energy to increase or decrease the value of the speed v for a given distance y?

3. Discuss the advantages and disadvantages of solving mechanics problems by energy methods compared with using Newton's laws.

9-2   Escape Velocity and Binding Energy

In the example of a mass oscillating on a spring we found that the mass is confined to a certain region of space bounded by turning points $x = \pm A$, where the amplitude A is related to the total energy by $A = \sqrt{2E/k}$. If we increase the total energy by initially pulling the mass out farther (and thus doing a greater amount of work on it), the turning points change. However, for this system, with potential energy given by Figure 9-6b, no matter how great the energy, the mass is confined because the total-energy line always intersects the potential energy at two values of x. Such a system is called a *bound system*. In this section we consider systems which are bound for values of the total energy below some critical value. For values of the total energy above this critical value the particles are not bound in any region of space.

Example 9-7 Escape Velocity Consider a particle of mass m a distance r from the center of the earth, where r is greater than the earth's radius R_E. The general gravitational potential-energy function is

$$U(r) = -\frac{GM_E m}{r} \qquad 9\text{-}21$$

As in Example 9-6, it simplifies calculation to write GM_E as gR_E^2, where g is the free-fall acceleration near the surface of the earth. The gravitational potential energy is then

$$U(r) = -\frac{mgR_E^2}{r} \qquad 9\text{-}22$$

At the surface of the earth $r = R_E$, and the gravitational potential energy is

$$U(R_E) = -\frac{mgR_E^2}{R_E} = -mgR_E \qquad 9\text{-}23$$

If we project the mass upward with some initial speed, it moves up with decreasing speed until the increase in potential energy equals the initial kinetic energy. It then stops and falls back toward the earth, losing potential energy and gaining kinetic energy. At an infinite distance from the earth, the potential energy is zero. For a particle initially at the surface of the earth with potential energy $-mgR_E$, the greatest possible increase in potential energy is mgR_E. This greatest possible increase in potential energy equals the greatest possible decrease in kinetic energy. If the initial kinetic energy of the particle at the surface of the earth is greater than mgR_E, the particle will never stop and fall back to the earth but escapes into outer space. The escape velocity is defined by

$$\tfrac{1}{2}mv_e{}^2 = mgR_E \qquad\qquad\qquad 9\text{-}24$$

or

$$v_e = \sqrt{2gR_E} \qquad\qquad\qquad 9\text{-}25$$

Using $g = 32.2$ ft/sec^2 = 32.2/5280 mi/sec^2 and $R_E = 3960$ mi, we obtain

$$v_e = \sqrt{2\,\frac{32.2}{5280}\,3960} = 6.95 \text{ mi/sec} = 2.50 \times 10^4 \text{ mi/h}$$

$$= 11.2 \text{ km/sec}$$

The potential-energy function $U(r)$ is sketched in Figure 9-7. The shaded region $r < R_E$ is inside the earth.

The total energy of a particle is the sum of the kinetic energy and the potential energy

$$E = E_k + U(r) = \tfrac{1}{2}mv^2 - \frac{mgR_E{}^2}{r} \qquad\qquad 9\text{-}26$$

If v_0 is the speed of the particle at the earth's surface, the total energy is

$$E = \tfrac{1}{2}mv_0{}^2 - \frac{mgR_E{}^2}{R_E} = \tfrac{1}{2}mv_0{}^2 - mgR_E \qquad\qquad 9\text{-}27$$

From Equation 9-24 we see that if v_0 equals the escape velocity v_e, the total energy is zero. If v_0 is greater than v_e, the total energy is positive; if v_0 is less than v_e, the total energy is negative. In terms of energy, the minimum kinetic energy needed by a particle to escape from the earth is that which makes the total energy zero.

In Figure 9-7, a horizontal line is drawn for total energy $E = E_1$ which is negative. The kinetic energy of a particle with total energy E_1 is $E_k = E_1 - U(r)$ represented by the vertical distance on the graph between the total energy and the potential energy. The kinetic energy

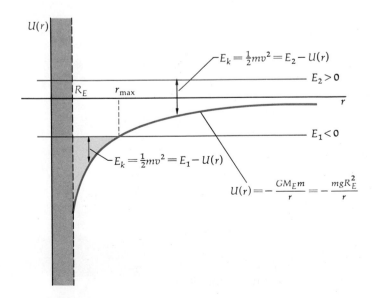

Figure 9-7
The potential-energy function $U(r) = -GM_E m/r = -mgR_E{}^2/r$ for a mass a distance r from the center of the earth, where r is greater than the radius of the earth R_E. If the particle has total energy E_1 less than zero, it is confined between R_E and r_{max}, where E_1 intersects $U(r)$. The kinetic energy $E_k = E_1 - U(r)$ is represented by the vertical distance between E_1 and $U(r)$. If the total energy is E_2, greater than zero, the particle is not confined. A positive total-energy line does not intersect the potential-energy curve.

decreases with increasing r and is zero at $r = r_{max}$, where the total-energy line E_1 intersects the potential-energy curve $U(r)$. The distance $r = r_{max}$ is the maximum distance from the center of the earth that can be reached by a particle with total energy E_1. The motion of a particle with total energy E_1 is therefore confined to the region between R_E and r_{max}, as indicated in the figure.

The horizontal line $E = E_2$ in Figure 9-7 indicates a total energy greater than zero. Again, the kinetic energy of a particle with this total energy is represented by the vertical distance between the total-energy line and the potential-energy curve. As r increases, the kinetic energy decreases, but it never reaches zero if E is positive. The positive total-energy line does not intersect the potential-energy curve. As r approaches infinity, the potential energy approaches zero and the kinetic energy approaches the total energy E_2. The motion of a particle with total energy greater or equal to zero is not confined; i.e., the particle is not bound to the earth. If a particle is projected away from the earth's surface with velocity greater than the escape velocity, it will escape whether its velocity is radial or at some angle, because the work done by gravity depends only on the radial distance traveled.

Example 9-8 Energy of an Orbiting Satellite Consider a particle of mass m orbiting the earth in a circular orbit above the surface of the earth. Its total energy is greater than $-GM_E m/r$ because it already has some kinetic energy. From Newton's law $\Sigma \mathbf{F} = m\mathbf{a}$ for circular motion we have

$$\frac{GM_E m}{r^2} = ma = \frac{mv^2}{r} \qquad\qquad 9\text{-}28$$

Thus the kinetic energy for a circular orbit is

$$\tfrac{1}{2}mv^2 = \frac{\tfrac{1}{2}GM_E m}{r} = \tfrac{1}{2}|U(r)| \qquad\qquad 9\text{-}29$$

This result (that the kinetic energy is half the magnitude of the potential energy) holds for any circular orbit in an inverse-square force field. On the energy diagram of Figure 9-7 the total energy of an orbiting satellite is thus halfway between $-GM_E m/r$ and zero. The minimum additional energy an orbiting satellite must be given to escape from the earth is just $\tfrac{1}{2}GM_E m/r$. The kinetic energy of the orbiting satellite must be doubled.

Example 9-9 Flight of Apollo 11 As a practical example of escape from the earth, consider the flight of Apollo 11, July 1969, during which man first set foot on the moon. The spacecraft was launched from the earth into a nearly circular orbit around the earth at an altitude of 119 mi and speed of 17,427 mi/h. From this orbit, the engine of the craft was fired to give it a speed of 24,245 mi/h. This speed, which nearly doubled the kinetic energy of the craft, is less than the escape velocity we calculated because the moon exerts a gravitational attraction which we have neglected. The spacecraft then traveled toward the moon, with its speed decreasing, to a point 23,800 mi from the moon, where the attraction of the moon equals the attraction of the earth. Beyond this point the craft accelerated toward the moon. As the craft passed the moon, its kinetic energy was greater than that needed to escape from the moon. On the far side of the moon, retarding rockets were fired to slow it down and put it into an orbit around the moon. (This

Blast-off of Saturn V rocket July 16, 1969, carrying Apollo 11 astronauts on man's first voyage to the moon.

NASA

was an elliptical orbit of maximum altitude of 195 mi and minimum altitude of 70 mi.) From this orbit, the lunar module Eagle, containing astronauts Armstrong and Aldrin, separated from the main craft and dropped into an orbit of about 50,000 ft altitude. From this orbit the module was flown to a landing on the surface of the moon.

Example 9-10 Ionization of the Hydrogen Atom In the Bohr model of the hydrogen atom, the electron follows a circular orbit around the proton. In the lowest energy state (called the *ground state*) the radius of the orbit is 0.529×10^{-10} m. We calculated the electrostatic potential energy for this distance in Example 8-5. The result is $U = -kq_1q_2/r = -4.36 \times 10^{-18}$ J $= -27.2$ eV. The energy diagram for this system has the same form as that for a satellite orbiting the earth. The kinetic energy of the electron in its circular orbit is

$$\tfrac{1}{2}mv^2 = \tfrac{1}{2}|U| = 13.6 \text{ eV}$$

The total energy of the electron is thus

$$E = \tfrac{1}{2}mv^2 + U = 13.6 \text{ eV} - 27.2 \text{ eV} = -13.6 \text{ eV} \qquad 9\text{-}30$$

To escape from the proton, the electron must have a total energy greater than or equal to zero. Thus to remove the electron, we must give it an additional energy of 13.6 eV, called the *binding energy* of the atom. Since removing an electron from an atom is called *ionization*, the binding energy is also called the *ionization energy*.

Example 9-11 Molecular Oscillations and Dissociation The force between two atoms or ions in a molecule, discussed briefly in Chapter 6 (see Figure 6-3), is similar to that exerted by a spring connecting the two atoms except that at great separation distances the force becomes zero. The potential energy U as a function of separation distance r is shown in Figure 9-8. This curve might apply to the two hydrogen atoms in the H_2 molecule or to the Na^+ and Cl^- ions in an NaCl molecule. At the point $r = r_0$, the slope of the potential-energy curve is zero, and hence the force is zero. This is a stable equilibrium point, since $U(r)$

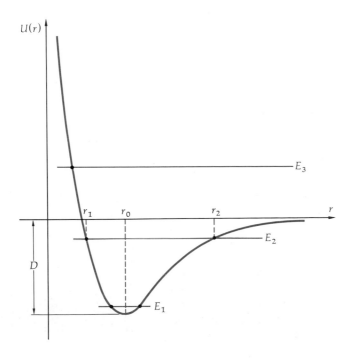

Figure 9-8
Potential energy versus separation for a diatomic molecule. E_1, E_2, and E_3 are successively greater total energies for the system. E_1 and E_2 are negative, and the system is bound between the maximum and minimum separations where the total-energy line intersects the potential-energy curve. E_3 is positive, and the system is not bound; i.e., the atoms or ions of the molecule are dissociated. The minimum energy for dissociation D is also called the binding energy.

increases on either side of this point. Very near this point the curve is similar to that for a mass on a spring. If the total energy is E_1, as shown in Figure 9-8, the atoms will oscillate about the equilibrium point like two masses joined by a spring. If the total energy is the larger value E_2, the motion will be an oscillation but not the simple, symmetric oscillation of two masses connected by a spring. The two turning points at r_1 and r_2 are different distances from the equilibrium point r_0. The two "halves" of the oscillation would take different times. Also, the average separation for this energy is slightly greater than r_0.

If the total energy is positive, e.g., the value E_3 in Figure 9-8, there is only one turning point. In this case the two particles are not bound. If initially the two particles were approaching, their separation would decrease until this single turning point was reached. Then they would move apart. Since there is no other turning point, they would continue to move apart.

The normal state of a molecule is for its parts to be nearly at rest relative to each other at the equilibrium separation r_0. If the molecule is given a greater energy, such as E_1, an oscillating motion ensues. This occurs, for example, when the body of which the molecule is a part is heated. If the heating is enough to increase the molecular energy to a value like E_2, the average separation of the particles is greater than r_0. In fact, the greater E_2 is, the greater the average separation. If the molecules are in a solid, this requires the solid to expand as E_2 is increased. In fact, this model accounts for the familiar phenomenon of the thermal expansion of solids.

We also can see from Figure 9-8 that there is a minimum energy that must be supplied to the molecule to make its parts separate completely. This energy, called the *binding* or *dissociation energy*, is evidently equal to the magnitude of the potential energy at the equilibrium separation r_0. (It is actually slightly less because the atoms are always oscillating with some total energy slightly greater than the minimum potential energy.)

Questions

4. If the gravitational potential energy of a mass m is chosen to be zero at the surface of the earth, what is its value when the mass is an infinite distance from the earth?

5. Two equal particles are separately projected with their escape velocities, but one is projected radially and the other tangentially to the surface of the earth. After 1 sec, how do their distances from the surface of the earth compare? (Neglect air resistance.) How do their speeds compare? When the particle projected tangentially is at a height $h = R_E$ above the earth's surface, how does its speed compare with that of the other particle at the same height?

6. In Example 9-9 it is stated that when Apollo 11 approached the moon, its kinetic energy was greater than that needed to escape from the moon. Explain why this must be true.

7. A comet enters the solar system, orbits around the sun, and leaves again. When it is near the sun, its gravitational potential energy (relative to infinity) plus its kinetic energy is found to be less than zero. Will the comet ever return?

Atmospheric Evaporation

Richard Goody
Harvard University

The expression for escape velocity of a particle near the earth's surface (Equation 9-25) can be generalized to apply to the escape of a particle from the surface of any planet or the moon. If R is the radius of the planet and g the acceleration of gravity at the surface, the escape velocity is given by

$$v_e = \sqrt{2gR}$$

This result is independent of the mass of the particle and therefore equally valid for a spacecraft, a projectile, or a molecule. The application to the escape of molecules from the surface of a planet leads to an important concept in planetary physics: using this equation and some results from the kinetic theory of gases, we can understand why some planets have atmospheres and others do not. We look to this mechanism to explain, for example, why the earth appears to have lost hydrogen to space following decomposition of water molecules, leaving an oxygen-rich atmosphere while, to a first approximation, Jupiter is a gigantic ball of hydrogen.

Several conditions must be met before a molecule (or atom) can escape from the surface of a planet. First, the magnitude of the molecule's velocity must be greater than the escape velocity v_e for the planet. Table 9-1 lists escape velocities for the moon and planets in the solar system. A second condition is that the direction of the velocity of the molecule must be away from the planet rather than toward it. Finally, the molecule must not collide with other molecules on its outward journey.

The chance that a molecule will collide with another in a particular region of space depends on the density of molecules in that region. The molecular density falls off rapidly with height above the surface of the planet (see Problems 20 and 21 in Chapter 5). There must therefore be a level, called the *escape level*, which varies with the planet and depends upon the atmospheric temperature, above which the collision probability is so small that a molecule heading away from a planet with velocity greater than v_e leaves the planet never to return. Figure 1 shows the trajectories of molecules above the escape level. Molecules with speeds greater than v_e escape, while those with speeds less than v_e fall back toward earth. These trajectories differ greatly from those

Table 9-1
Escape velocities for the moon and planets

Planet	Gravitational acceleration, m/sec²	Radius, km	Escape velocity v_e, km/sec
Mercury	3.76	2,439	4.3
Venus	8.88	6,049	10.3
Earth	9.81	6,371	11.2
Moon	1.62	1,738	2.3
Mars	3.73	3,390	5.0
Jupiter	26.2	69,500	60
Saturn	11.2	58,100	36
Uranus	9.75	24,500	22
Neptune	11.34	24,600	24

Figure 1
Molecules or atoms whose velocities exceed the escape velocity v_e can escape from the planet. These may form only a minute fraction of the total number at the critical level.

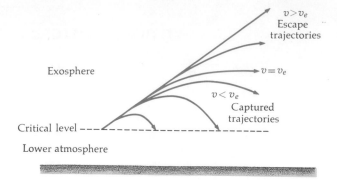

of gas molecules near the earth's surface, where a molecule can travel only a short distance before colliding with another. For earth, the escape level averages about 500 km above the ground with considerable diurnal and seasonal variation because of atmospheric temperature changes. For Venus, the escape level is about 200 km above the surface.

Although escape velocity does not depend on the mass of the molecule, the distribution of molecular speeds depends strongly on the mass of the molecule and on the temperature: in thermal equilibrium it is given by the Maxwell-Boltzmann distribution (Figure 2). In this figure, $f(v)\,dv$ is the fraction of all the molecules which have speeds in some interval dv. This fraction for an interval dv centered about some speed v is indicated by the shaded area in the figure. The most probable speed is related to the mass of the molecule m and the absolute temperature T by

$$v_0 = \sqrt{\frac{2kT}{m}}$$

where k is a constant called Boltzmann's constant.[1]

Some typical values of v_0 are shown in Table 9-2. Comparing Tables 9-1 and 9-2, we see that v_0 is greater than v_e only for hydrogen on the surface of the moon at temperatures greater than 300 K. Since lunar daytime temperatures exceed 300 K, any hydrogen would evaporate into space almost instantaneously.

Even when the escape speed v_e is much greater than the most probable speed v_0, there will be some molecules or atoms which have speeds greater than v_e and therefore have a chance of escaping. From Figure 1 we discern that most molecules have speeds not much different from v_0 but a few have speeds $2v_0$, $3v_0$, or even $100v_0$, though the fraction with speeds much greater than v_0 is very small and decreases rapidly with increasing v. Whether a certain type of molecule or atom is likely to be found in the atmosphere of a planet depends on the rate of escape of these molecules, which in turn depends crit-

Figure 2
The Maxwell-Boltzmann speed distribution. The fraction of molecules having speeds in some interval dv equals the area under the curve for that interval. The speed v_0 for which $f(v)$ is maximum is the most probable speed. The mean speed $\bar{v}$ and the root-mean-square speed v_{rms} differ slightly from the most probable speed.

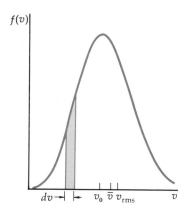

Table 9-2
Most probable velocities in kilometers per second

Atom	Atomic weight, μ	Temperature, K		
		300	600	900
H	1	2.24	3.16	3.87
He	4	1.12	1.58	1.94
O	16	0.56	0.79	0.97

[1] Because of the asymmetry of the curve for $f(v)$, the most probable speed v_0, the mean speed $\bar{v}$, and the root-mean-square speed v_{rms} are slightly different, though they all have the same dependence on the temperature and molecular mass. In Chapter 17, it will be shown that the root-mean-square speed is given by $v_{rms} = \sqrt{3kT/m}$.

Figure 3
The fraction of atoms or molecules with velocities greater than v_e in terms of the ratio v_e/v_0. The two parts of the figure differ only in scale.

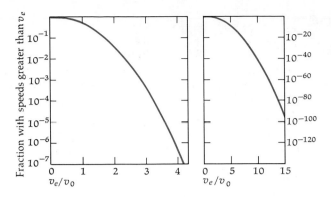

ically on the fraction of molecules or atoms above the escape level with speeds greater than v_e. This fraction is shown in Figure 3 as a function of the ratio v_e/v_0. When this ratio is greater than about 2, this fraction is given approximately by

$$f_{v>v_e} = \frac{4}{\pi} \frac{v_e}{v_0} e^{-(v_e/v_0)^2}$$

The temperature of the escape level on earth is close to 600 K, and v_0 is therefore 0.79 km/sec for atomic oxygen and about 3.16 km/sec for atomic hydrogen. Using $v_e = 11.2$ km/sec, we obtain for the fraction of hydrogen atoms with speeds greater than the escape speed:

$$f_{v>v_e}(\text{H}) = \frac{4}{\pi} \frac{11.2}{3.16} e^{-(11.2/3.16)^2} \approx 1.6 \times 10^{-5}$$

This result tells us that out of every million hydrogen atoms at the escape level about 16 will have speeds greater than v_e, and since half, on the average, will be moving away from the earth, 8 will escape. This fraction, seemingly small, is large enough to account for the almost complete absence of hydrogen from the earth's atmosphere now, approximately 4×10^9 years after formation of the earth.

For atomic oxygen, on the other hand, the fraction with speeds greater than v_e is

$$f_{v>v_e}(\text{O}) = \frac{4}{\pi} \frac{11.2}{0.79} e^{-(11.2/0.79)^2} \approx 9.3 \times 10^{-87}$$

This fraction for oxygen atoms is much smaller than for hydrogen atoms. If we want to evaluate the evaporation rates of hydrogen and oxygen numerically, our model of the escape process must be elaborated; nevertheless, the ratio of the numbers of hydrogen and oxygen atoms with velocities high enough for escape (more than 10^{81} to 1) is so large that the simultaneous retention of oxygen and escape of hydrogen is plausible. Because of the huge mass of Jupiter, its escape velocity of 60 km/sec enables it to retain hydrogen as easily as oxygen is retained on earth. We can therefore see in a general way why the two planets have such markedly different compositions.

9-3 Bernoulli's Equation

To illustrate the wide range of application of the work-energy theorem, we now apply it to the flow of fluids. Consider a fluid flowing steadily through a pipe that is not level. The pressure will change from point to point along the pipe, and work must be done to make the fluid flow up. If the cross-sectional area of the pipe varies, the speed of the fluid will vary and so will the pressure. The fundamental relationship which relates the pressure, flow speed, and height was first obtained in 1738 by Daniel Bernoulli and is named for him. It is in effect just the work-energy theorem applied to this situation. We shall derive Bernoulli's equation for the restricted case of an incompressible fluid (most liquids) which is nonviscous (no internal friction) and which flows steadily and without turbulence in a closed pipe. The result can be shown to hold in somewhat more general circumstances.

The Granger Collection

Daniel Bernoulli (1700–1782).

Figure 9-9a shows a sectional view of a length of pipe. We apply the work-energy theorem to the fluid contained initially between points 1 and 2. After a short time Δt, this fluid will have moved along the pipe and will be contained between points 1' and 2', as shown in Figure 9-9b.

If the flow is steady, the only change between Figure 9-9a and b is for the heavily shaded portions of the fluid mass. The fluid in the intermediate, lightly shaded region is different fluid, but the physical conditions of its location, pressure, speed, and energy are exactly the same as those of the fluid initially occupying this same position. The net change in time Δt is equivalent to lifting the fluid in the heavily shaded portion of Figure 9-9a to the position of the heavily shaded portion of Figure 9-9b and of changing its speed from v_1 at point 1 to v_2 at point 2.

Let Δm be the mass of this heavily shaded portion of the fluid. The mass can be written in terms of the density ρ and the volume ΔV

$$\Delta m = \rho \, \Delta V \qquad\qquad 9\text{-}31$$

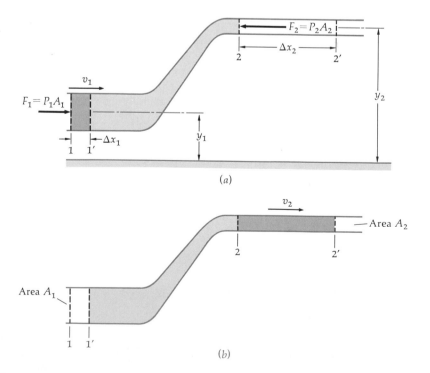

(a)

(b)

Figure 9-9
Fluid moving in a pipe for the derivation of Bernoulli's principle. The net work done by the forces P_1A_1 and P_2A_2 has the effect of raising the part of the fluid indicated by heavy shading from height y_1 to y_2 and changing its speed from v_1 to v_2. Bernoulli's equation is a statement of the work-energy theorem applied to the fluid.

If the fluid is incompressible, the density and volume of this part of the fluid will be the same at position 2 as at position 1. The volume at position 1 is $A_1 \Delta x_1$, where A_1 is the cross-sectional area of the pipe at this position. Similarly, the volume at position 2 is $A_2 \Delta x_2$. Thus for an incompressible fluid we have

$$A_2 \Delta x_2 = A_1 \Delta x_1 \qquad \text{9-32}$$

But $\Delta x_1 = v_1 \Delta t$, where v_1 is the speed at position 1, and $\Delta x_2 = v_2 \Delta t$. Substituting these into Equation 9-32 and canceling out the common factor Δt, we obtain a simple relation for the speeds of an incompressible fluid:

$$A_2 v_2 = A_1 v_1 \qquad \text{9-33}$$

Velocity-area product is constant for incompressible fluid

As the entire body of fluid we are considering moves in time Δt, the fluid following it in the pipe exerts a force on it of magnitude $F_1 = P_1 A_1$, where P_1 is the pressure at point 1. This force does work $W_1 = F_1 \Delta x_1 = P_1 A_1 \Delta x_1 = P_1 \Delta V$, where ΔV is just the volume of the heavily shaded mass Δm. At the same time, the fluid preceding that under consideration exerts a force $F_2 = P_2 A_2$ to the left in the figure. This force does negative work (since it opposes the motion) $W_2 = -F_2 \Delta x_2 = -P_2 A_2 \Delta x_2 = -P_2 \Delta V$. Thus the net work done by these forces is

$$W = P_1 \Delta V - P_2 \Delta V = (P_1 - P_2) \Delta V \qquad \text{9-34}$$

This work equals the change in the kinetic energy and gravitational potential energy of the fluid under consideration, i.e., just the change in energy of the portion of mass $\Delta m = \rho \Delta V$. The change in potential energy of this mass is

$$\Delta U = \Delta m\, g y_2 - \Delta m\, g y_1$$

The change in kinetic energy is

$$\Delta E_k = \tfrac{1}{2}(\Delta m) v_2^2 - \tfrac{1}{2}(\Delta m) v_1^2$$

Thus the work-energy theorem gives

$$(P_1 - P_2) \Delta V = \Delta m\, g y_2 - \Delta m\, g y_1 + \tfrac{1}{2}(\Delta m) v_2^2 - \tfrac{1}{2}(\Delta m) v_1^2$$

If we divide each term by ΔV and write the density $\rho = \Delta m / \Delta V$, we obtain

$$P_1 - P_2 = \tfrac{1}{2}\rho v_2^2 - \tfrac{1}{2}\rho v_1^2 + \rho g y_2 - \rho g y_1$$

When we collect all the quantities with subscript 2 on one side and those with subscript 1 on the other side, this equation becomes

$$P_1 + \rho g y_1 + \tfrac{1}{2}\rho v_1^2 = P_2 + \rho g y_2 + \tfrac{1}{2}\rho v_2^2 \qquad \text{9-35}$$

This result can be restated as

$$P + \rho g y + \tfrac{1}{2}\rho v^2 = \text{constant} \qquad \text{9-36}$$

Bernoulli's equation

meaning that this combination of quantities evaluated at any point along the pipe has the same value as at any other point. Equation 9-36 is known as *Bernoulli's equation* for steady, nonviscous flow of an incompressible fluid. To a considerable extent Bernoulli's equation also applies to compressible fluids like gases.

A special application of Bernoulli's equation occurs for a fluid at rest. Then $v_1 = v_2 = 0$, and we obtain

$$P_1 - P_2 = \rho g (y_2 - y_1) = \rho g h$$

where $h = y_2 - y_1$ is the difference in height between points 2 and 1.

This is just the result we obtained from Archimedes' principle, derived in Chapter 5 from Newton's laws. We now give some examples of the use of Bernoulli's equation in nonstatic situations.

Example 9-12 An incompressible fluid such as water flows through a horizontal pipe which has a constricted section as shown in Figure 9-10. Show that the pressure is *reduced* in the constriction.

Since both sections of the pipe are at the same elevation, we take $y_1 = y_2$ in Equation 9-35. We then have

$$P_1 + \tfrac{1}{2}\rho v_1{}^2 = P_2 + \tfrac{1}{2}\rho v_2{}^2 \qquad\qquad 9\text{-}37$$

But, from Equation 9-33,

$$A_1 v_1 = A_2 v_2 \qquad \text{or} \qquad v_2 = \frac{A_1}{A_2}\, v_1$$

Substituting this result in Equation 9-37 gives

$$P_1 + \tfrac{1}{2}\rho v_1{}^2 = P_2 + \tfrac{1}{2}\rho\, \frac{A_1{}^2 v_1{}^2}{A_2{}^2}$$

or

$$P_1 - P_2 = \tfrac{1}{2}\rho \left(\frac{A_1{}^2}{A_2{}^2} - 1\right) v_1{}^2 \qquad\qquad 9\text{-}38$$

Since A_1 is larger than A_2, the quantity on the right is positive. Hence P_2 is less than P_1. Note that the only circumstance in which P_2 and P_1 can be equal, according to Equation 9-38, is when $v_1 = 0$, that is, when the fluid is at rest in the pipe.

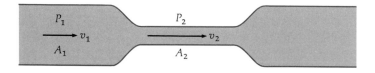

Figure 9-10
Constriction in a pipe carrying a moving fluid. The pressure is smaller in the narrow section of the pipe, in which the fluid is moving faster.

Example 9-13 A large water tower (Figure 9-11) is drained by a pipe of cross section A through a valve a distance h below the surface of the water in the tower. Show that the pressure in the pipe is reduced when the valve is opened.

In applying Equation 9-35 we take point 1 to be the top surface of the water in the tower, where the pressure is atmospheric pressure P_a. We also let $y_1 = h$. Point 2 is in the pipe just before the valve, and $y_2 = 0$. Because the surface area A_1 of the tower is so large compared to the area of the pipe, the speed v_1 with which the surface drops as the tower is drained can be neglected. Then, according to Equation 9-35,

$$P_a + \rho gh = P_2 + \tfrac{1}{2}\rho v_2{}^2$$

or

$$P_2 = P_a + \rho gh - \tfrac{1}{2}\rho v_2{}^2 \qquad\qquad 9\text{-}39$$

If the valve is closed, the water is at rest and $v_2 = 0$. Then the pressure in the pipe at the valve is, according to Equation 9-34, just $P_2 = P_a + \rho gh$. When the valve is opened, the water will move and v_2 will not be zero. Then, according to Equation 9-39, the pressure in the pipe will decrease by $\tfrac{1}{2}\rho v_2{}^2$.

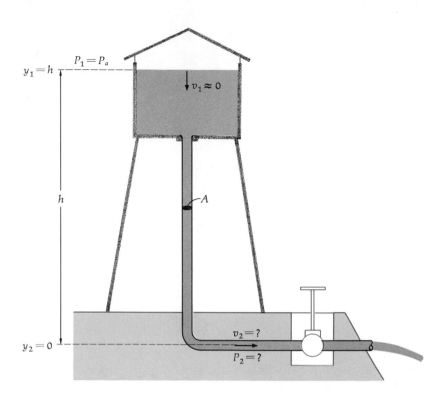

Figure 9-11
The water tower for Example 9-13. When the valve is opened, allowing water to flow out through the pipe, the pressure in the pipe is reduced.

Example 9-14 Calculate the pressure P_2 in Example 9-13 in pounds per square inch if $h = 40$ ft, $P_a = 14.6$ lb/in², and $v_2 = 48$ ft/sec with the valve open.

The density of water is 1 gm/cm³ = 1.94 slugs/ft³. In the British engineering system it is usually convenient to work with ρg in pounds per cubic foot rather than the density ρ. Using $g = 32.2$ ft/sec², we have for water,

$$\rho g = (1.94 \text{ slugs/ft}^3)(32.2 \text{ ft/sec}^2) = 62.5 \text{ lb/ft}^3$$

The quantity $\rho g h$ in Equation 9-39 is then

$$\rho g h = (62.5 \text{ lb/ft}^3)(40 \text{ ft}) = 2.50 \times 10^3 \text{ lb/ft}^2 = 17.4 \text{ lb/in}^2$$

The quantity $\frac{1}{2}\rho v_2^2$ is

$$\frac{1}{2}\rho v_2^2 = \frac{\frac{1}{2}\rho g v_2^2}{g} = \frac{\frac{1}{2}(62.5 \text{ lb/ft}^3)(48 \text{ ft/sec})^2}{32.2 \text{ ft/sec}^2}$$

$$= 2.24 \times 10^3 \text{ lb/ft}^2 = 15.5 \text{ lb/in}^2$$

Then with the valve closed and $v_2 = 0$ the pressure is

$$P_2 = P_a + \rho g h = 14.6 \text{ lb/in}^2 + 17.4 \text{ lb/in}^2 = 32.0 \text{ lb/in}^2$$

and with the valve open

$$P_2 = P_a + \rho g h - \frac{\frac{1}{2}\rho g v_2^2}{g}$$

$$= 14.6 \text{ lb/in}^2 + 17.4 \text{ lb/in}^2 - 15.5 \text{ lb/in}^2 = 16.5 \text{ lb/in}^2$$

Question

8. In Figure 9-10 the water entering the narrow part of the pipe is accelerated to a greater speed. What forces act on this water to produce this acceleration?

Review

A. Define, explain, or otherwise identify:

Work-energy theorem, 226   Ionization energy, 239
Escape velocity, 236   Dissociation energy, 240
Binding energy, 239   Bernoulli's equation, 245

B. True or false:

1. The total mechanical energy of a particle is always conserved.

2. When only conservative forces do work on a particle, its kinetic energy depends only on its position.

3. The time it takes a particle to move from one point to another can be found directly from the energy-conservation equation.

4. The kinetic energy needed to escape from the earth depends on the choice of reference point for the zero of potential energy.

5. Archimedes' principle can be derived from Bernoulli's equation.

Exercises

Section 9-1, Some Illustrative Examples

1. A 5-kg block is pushed along a horizontal rough surface by a constant horizontal force $F_0 = 25$ N. The coefficient of friction is 0.2. The block is pushed 3 m. (a) What is the work done on the block by the force F_0? (b) How much work is done on the block by friction? (c) What is the net increase in kinetic energy of the block? (d) What is the speed of the block after it has traveled 3 m?

2. A block is projected along a horizontal surface with initial velocity of 10 m/sec. The coefficient of friction between the block and the surface is 0.2. How far does the block slide before coming to rest?

3. A 2-kg block slides down a smooth curved ramp from rest at a height of 1 m (Figure 9-12). It slides 6 m on a rough horizontal surface before coming to rest. (a) What is the speed of the block at the bottom of the ramp? (b) How much work is done by friction on the block? (c) What is the coefficient of friction between the block and the horizontal surface?

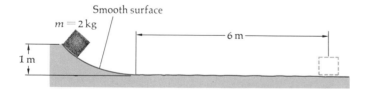

Figure 9-12
Block slides down smooth curve onto a rough surface for Exercise 3.

4. A 1-kg mass is released from rest at a height of 5 m on a curved smooth ramp. At the foot of the ramp is a spring of spring constant 400 N/m (Figure 9-13). The mass slides down the ramp and onto the spring, compressing it a distance x before coming momentarily to rest. (a) Find x. (b) What happens to the mass after it comes to rest?

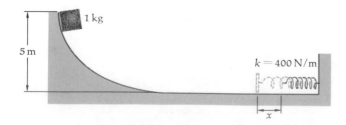

Figure 9-13
Exercise 4.

5. A 3-kg mass slides along a smooth horizontal surface with speed 5 m/sec and encounters a smooth ramp inclined at an angle of 30° to the horizontal. How far up the incline does the mass slide before coming momentarily to rest?

6. A 2-kg mass is pushed against a spring of force constant 500 N/m, compressing it 20 cm. It is released, and the spring projects the mass along a smooth horizontal surface and then up a smooth incline of angle 45°, as shown in Figure 9-14. (a) What is the speed of the mass when it leaves the spring? (b) How far up the incline does the mass travel?

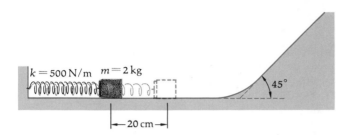

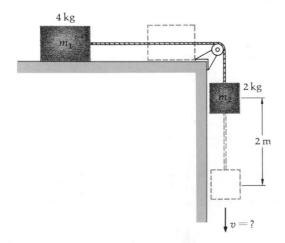

Figure 9-14
Exercise 6.

7. A 6-kg block slides down a 30° incline starting from rest. After sliding 10 m along the incline, its speed is 7 m/sec. (a) What work did gravity do on the block in this descent? (b) What was the increase in kinetic energy of the block? (c) By how much did the total energy of the block increase or decrease? (d) How much work was done by friction on the block?

8. For the Atwood's machine shown in Figure 9-15 find the velocity of each mass when all masses are at the same height. The system is at rest when the lower string is cut.

9. A 0.5-lb ball is thrown with initial speed of 80 ft/sec at an angle of 53° above the horizontal. (a) What is the total mechanical energy initially? (b) Write the initial kinetic energy in terms of the velocity components v_x and v_y. What is the minimum kinetic energy during the flight of the ball? At what point does the ball have this minimum kinetic energy? (c) What is the maximum potential energy of the ball during its flight? (d) Find the maximum height of the ball.

10. Two men stand at the edge of a 50-m cliff. Simultaneously they throw balls at initial speed of 10 m/sec, one straight up and the other straight down. (a) If air resistance is neglected, how do the speeds of the balls compare when they reach the bottom of the cliff? What are their speeds there? (b) How much later does one ball reach the bottom than the other? (c) Discuss qualitatively how your answers would differ if air resistance were taken in account.

11. In Figure 9-16, $m_1 = 4$ kg, $m_2 = 2$ kg, and there is no friction. Find the speed of the masses when m_2 has fallen 2 m.

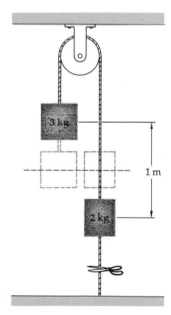

Figure 9-15
Exercise 8.

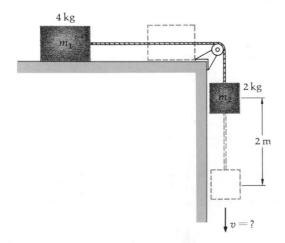

Figure 9-16
Exercise 11 and Problem 7.

12. A mass slides along the frictionless track shown in Figure 9-17. Initially it is at point P headed downhill with speed v_0. Describe the motion in as much detail as you can if (a) $v_0 = 7$ m/sec and (b) $v_0 = 12$ m/sec. (c) What is the least speed needed by the mass to get past point Q?

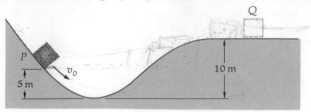

Figure 9-17
Exercise 12.

13. A pendulum of length L has a bob of mass m. It is released from some angle θ. The string hits a peg at a distance x directly below the pivot itself (Figure 9-18) and wraps itself around the peg, shortening the length of the pendulum. Show that the maximum height that the pendulum reaches equals its initial height. (Assume that initially the bob is below the height of the peg.)

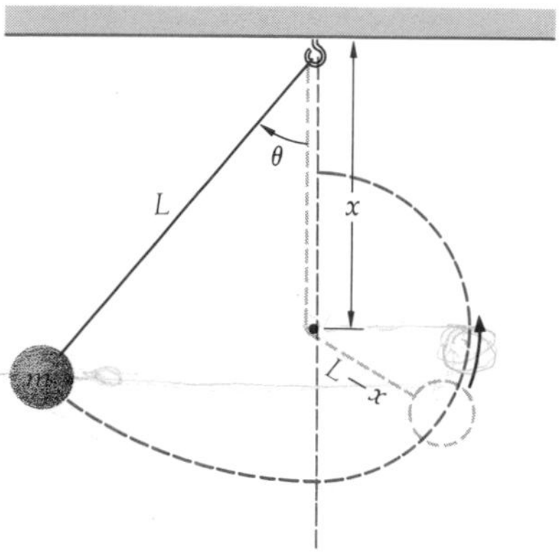

14. A mass m is attached to a spring of force constant k on a horizontal smooth surface. Its equilibrium position is $x = 0$. The mass is given a displacement $x = A$ and released. (a) What is the initial potential energy of the system? (b) Show that when the mass is at position x at any later time, its speed is given by $v = \sqrt{k/m} \ \sqrt{A^2 - x^2}$.

15. (a) Find the potential energy (relative to zero at infinity) of a 100-kg mass at the surface of the earth. (Take the radius of the earth to be $R_E = 6.4 \times 10^6$ m.) (b) Find the potential energy of the same mass at a height above the earth's surface equal to the earth's radius. (c) If the mass is dropped from this height, find its speed when it hits the surface of the earth. (Neglect air resistance; in practice, air resistance is of primary importance in this type of problem.)

Figure 9-18
Exercise 13 and Problem 3.

Section 9-2, Escape Velocity and Binding Energy

16. A particle is projected from the surface of the earth with velocity equal to twice the escape velocity. When the particle is very far from the earth, what is its speed?

17. Find the escape velocity for a rocket leaving the moon. (The acceleration of gravity on the moon is 0.166 times that on earth, and the moon's radius is $0.273 \ R_E$.)

18. The planet Saturn has mass of 95.2 times that of the earth and radius of 9.47 times that of the earth. Find the escape velocity for Saturn.

19. A space probe is to be sent from the earth so that it has a velocity of 50 km/sec when it is very far from the earth. What velocity is needed for the probe at the surface of the earth?

20. (a) Calculate the energy in joules necessary to send a 1-kg mass away from the earth with the escape velocity. (b) Convert this energy to kilowatt-hours. (c) If energy can be obtained at 3 cents per kilowatt-hour, what is the minimum cost of giving an 80-kg astronaut enough energy to escape the earth's gravitational field?

21. A satellite orbits the earth in a circular orbit. The orbit is then changed to another circular one of larger radius. Discuss the change in potential energy, total energy, and kinetic energy, i.e., whether each increases, decreases, or remains constant.

22. In the Bohr model of the hydrogen atom, the electron, with charge $-e$, moves in a circular orbit of radius r about the proton, with charge $+e$. Show that the kinetic energy of the electron is $E_k = \frac{1}{2}ke^2/r$ and the total energy is $E = -\frac{1}{2}ke^2/r$, where k is the coulomb constant.

Section 9-3, Bernoulli's Equation

23. Water flows through a 1-in-diameter hose with speed 2 ft/sec. The diameter of the nozzle is $\frac{1}{8}$ in. (a) At what speed does the water pass through the nozzle? (b) If the pump at one end and the nozzle at the other end are at the same height and the pressure at the nozzle is atmospheric, what is the pressure at the pump?

24. A large tank of water is tapped a distance h below the water surface by a small pipe, as shown in Figure 9-19. (a) Why is the pressure the same at points a and b? (b) Show that the speed of the water emerging at point b is $\sqrt{2gh}$ in the approximation that the speed at point a is negligible. (This is known as *Torricelli's law*.) Why is this a good approximation?

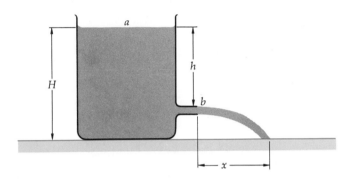

Figure 9-19
Exercise 24 and Problem 16.

25. Water is flowing at 3 m/sec in a horizontal pipe under a pressure of 2×10^5 N/m², which is about twice atmospheric pressure. The pipe narrows to half its original diameter. (a) What is the speed of flow in the narrow section? (b) What is the pressure in the narrower section of pipe? (c) How does the number of kilograms of water flowing through the narrow section each second compare with that flowing through the wider section?

26. The pressure in a section of 1-in-diameter horizontal pipe is 20.6 lb/in². Water flows through the pipe at 0.1 ft³/sec. What should the diameter of a constricted section of the pipe be for the pressure to be atmospheric pressure (14.6 lb/in²).

Problems

1. A small mass m slides without friction along the loop-the-loop track shown in Figure 9-20. The circular loop has radius R. The mass starts from rest at point P a distance h above the bottom of the loop. (a) What is the kinetic energy of m when it reaches the top of the loop? (b) What is its acceleration at the top of the loop assuming that it stays on the track? (c) What is the least value of h if m is to reach the top of the loop without leaving the track? (d) Assuming that h is greater than this least value, write an expression for the normal force exerted by the track on the mass.

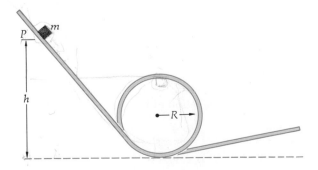

Figure 9-20
Problem 1.

2. The rate of energy loss of a certain system at a given time is directly proportional to the total energy of the system at that time. (*a*) Write a differential
equation relating the energy E to the rate of change of energy dE/dt expressing
the above property. Let the proportionality constant be C. (*b*) Solve your equation to obtain a general relationship for the energy of the system E as a function of time t. (*c*) Find the constant C in your result when 10 percent of the
energy is lost in 10 sec. (*d*) How long does it take for half the energy of the
system to be lost when C has the value found in part (c)?

3. In the problem of the pendulum string hitting a peg (Exercise 13) the
pendulum is released from rest at $\theta = 90°$. (*a*) Find the speed of the bob in
terms of g, L, and x when it is directly above the peg and moving in a circle of
radius $L - x$. (*b*) What is the least value of its acceleration at that point if the
string is not to go slack? (*c*) Show that the bob will not reach this point with
the string still taut unless x is greater than or equal to $3L/5$.

4. A particle moves under the influence of a conservative force along the x axis
with total energy E. The potential energy associated with the force is $U(x)$. (*a*)
Show that the speed of the particle is given by $v = \sqrt{2/m} \; \sqrt{E - U}$. (*b*) Show
that the distance dx traveled during the time interval dt is given by

$$\frac{dx}{\sqrt{E - U(x)}} = \sqrt{\frac{2}{m}} \; dt$$

(*c*) For a mass on a spring the potential energy is $U = \frac{1}{2}kx^2$, where k is the force
constant, and the total energy is $E = \frac{1}{2}kA^2$, where A is the maximum displacement. By integrating your result in part (*b*) find the total time taken for a mass
on a spring to move from $x = -A$ to $x = +A$.

5. A 2-kg mass is released on a smooth incline 4 m from a spring of constant
$k = 100$ N/m. The spring is fixed along the plane inclined at $\theta = 30°$ (Figure
9-21). (*a*) Find the maximum compression of the spring, assumed to be
massless. (*b*) If the incline is not smooth but the coefficient of friction between
it and the mass is 0.2, find the maximum compression. (*c*) For the rough
incline, how far up the incline will the mass travel after leaving the spring? (*d*)
Describe the subsequent motion of the mass for the rough incline.

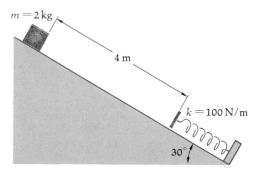

Figure 9-21
Problem 5.

6. A Superball can bounce to 90 percent of its original height. (*a*) How much
energy is lost after a 30-gm ball is bounced once from an initial height of 3 m?
(*b*) A ball dropped from an original height of H makes N bounces. Find a general expression for the maximum height of the ball after N bounces for a ball of
mass m dropped from height H. (*c*) About how many bounces are required if
the maximum height after the Nth bounce is 1 percent of the original height?

7. For the arrangement in Figure 9-16 with general masses m_1 and m_2, (*a*) find
an expression for the speed of the masses after m_2 has fallen a distance y. (*b*)
Differentiate this expression to obtain an expression for the acceleration of the
masses.

8. A space probe launched from the surface of the earth accelerates for a short
distance Δr and then moves under the influence of gravity only. Since it is

already a short distance Δr above the surface of the earth, its escape velocity is slightly less than that at the earth's surface by amount Δv. (a) Find Δv in terms of Δr using the differential approximation $\Delta v \approx dv$ and $\Delta r \approx dr$. (b) Find the escape velocity for $\Delta r = 200$ mi.

9. (a) Show that the total energy of a satellite in a circular orbit of radius r is given by $E = -GM_E m/2r$, where M_E is the mass of the earth and m the mass of the satellite, and that the speed v is given by $v = \sqrt{GM_E/r}$. (b) If you wish to decrease the radius of the circular orbit of a satellite, must you increase or decrease the kinetic energy of the satellite? How do the potential energy and total energy change if the radius is decreased? (c) Using the differential approximation $\Delta v \approx dv$ and $\Delta r \approx dr$, show that a small change in radius must be accompanied by a small change in speed given by

$$\frac{\Delta v}{v} = -\frac{1}{2}\frac{\Delta r}{r}$$

10. In the Bohr model of the hydrogen atom the electron moves in a circular orbit around the proton. Only certain orbits are allowed, given by $r = N^2 r_0$, where N is a positive integer and $r_0 = 0.529 \times 10^{-10}$ m is the radius of the smallest allowed orbit. (a) Show that the allowed total energies of the hydrogen atom are given by $E_N = -|E_1|/N^2$, where $E_1 = -\frac{1}{2}ke^2/r_0 = -13.6$ eV is the lowest energy (see Exercise 22). Make a diagram to scale of the allowed energies by labeling them on a vertical energy line for $N = 1, 2, 3, 4, 10$, and ∞. (b) Normally the hydrogen atom is in its lowest energy state $N = 1$, the ground state. If it absorbs just the right amount of energy, it can be excited to a higher energy state. Calculate the amount of energy needed to raise the hydrogen atom from its ground state $N = 1$ to its first excited state $N = 2$. Do the same for $N = 1$ to the second excited state $N - 3$. (c) When the hydrogen atom changes from an excited state to the ground state or an excited state of lower total energy, the excess energy is given off, usually as electromagnetic radiation. Calculate the energy given off when the atom makes the following changes: $N = 3$ to $N = 1$, $N = 2$ to $N = 1$, $N = 3$ to $N = 2$.

11. (a) With the following information, graph the potential-energy curve of the NaCl molecule versus separation distance r of the ions Na^+ and Cl^-:

1. The equilibrium separation is 2.4×10^{-10} m.
2. The dissociation energy is 4.2 eV.
3. When the potential energy is -1.0 eV, the separation is either 1.5×10^{-10} m or 7.0×10^{-10} m.
4. When the potential energy is $+5.0$ eV, the separation is 1.0×10^{-10} m.
5. When the potential energy is -3.0 eV, the separation is either 1.8×10^{-10} m or 3.5×10^{-10} m.
6. At infinite separation the energy of the Na^+ and Cl^- ions is $+1.3$ eV. (This is not zero because it takes 1.3 eV to make ions from the original atoms. The dissociation energy is measured from the minimum potential energy to zero.)

(b) From your graph find the maximum and minimum separation of the ions when the total energy is -2 eV.

12. A fountain designed to spray a column of water 36 ft into the air has a $\frac{1}{2}$-in-diameter nozzle at ground level. The water pump is 10 ft below the ground. The pipe to the fountain has a diameter of 1 in. Find the necessary pump pressure.

13. In an aspirator pump water is pumped through a pipe at pressure 30 lb/in² and a speed of 10 ft/sec. The pipe narrows at the same level, so that the pressure drops. A tube connects the narrow part of the pipe to a vacuum chamber, in which the pressure is to be 1 lb/in². What is the ratio of the areas of the pipe?

14. A large beer keg of height H and area A_1 is filled with beer. The top is open to atmospheric pressure. At the bottom is a spigot opening of area A_2, much smaller than A_1. (a) Show that the velocity of beer leaving the spigot is

approximately $\sqrt{2gh}$ when the height of the beer is h. (b) Show that in the approximation A_2 much less than A_1, the rate of change of height h of the beer is given by $dh/dt = -(A_2/A_1)\sqrt{2gh}$. (c) Show that the result of part (b) can be written $-dh/\sqrt{h} = (A_2/A_1)\sqrt{2g}\ dt$ and integrate to obtain an expression for h as a function of time. Take $h = H$ at $t = 0$ to determine the constant of integration. (d) Find the total time needed to drain the keg if $H = 2$ m, $A_1 = 0.8$ m², and $A_2 = 10^{-4} A_1$.

15. In calculating the escape velocity in Example 9-7 the rotation of the earth was neglected. The velocity that must be given to a body relative to the ground will be less if the body is launched in the direction of the earth's rotation or more if it is launched away from it. Take the earth's rotation into account for a body launched on the equator in the direction of motion of the earth's surface (horizontally eastward). (a) What is the speed relative to the center of the earth of an object at rest on the equator? (b) Relative to the earth's surface, what speed must be given to a body for it to escape the earth? (c) By what percentage is the work required to accelerate the body reduced because of the earth's rotation?

16. (a) For Figure 9-19 find the distance x at which the water strikes the ground. Express your result in terms of h and H. (b) Show that there are two values of h equidistant from the point $h = \frac{1}{2}H$ which give the same distance x. (c) Show that x is a maximum when $h = \frac{1}{2}H$. What is the value of the maximum distance x?

CHAPTER 10   Many-Particle Systems

So far we have been concentrating on the motion of a single particle. We now turn to more complicated problems dealing with two or more particles. Although the detailed description of the complete motion is quite complicated, some general methods, which follow directly from Newton's laws, are useful in dealing with a wide variety of problems.

We begin by defining the center of mass for a system of particles. We then show that the motion of this point is quite simple, no matter how complicated the motions of all the individual particles in the system. Figure 10-1 is a multiflash photograph of a baton thrown into the air. Although the motion of the baton is complicated, the motion of one point, the center of mass, is simple. While the baton is in the air, the center of mass follows a parabolic path. We shall show in general that the acceleration of the center of mass of a system of particles equals the resultant external force divided by the total mass of the system. For the baton thrown into the air, the acceleration of the center of mass is g downward.

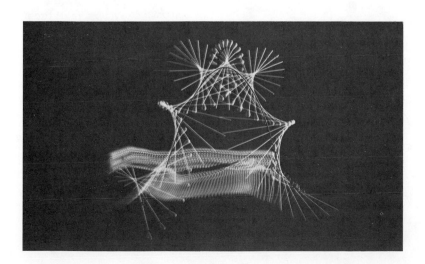

Figure 10-1
Multiflash photograph of a baton. One point on the baton, the center of mass, follows the same simple parabolic path as if it were a single point particle. (*Courtesy of Harold E. Edgerton.*)

10-1 Center of Mass

Let us first consider a simple system of two particles in one dimension. Let x_1 and x_2 be the coordinates of the particles relative to some arbitrary origin. The center-of-mass coordinate x_{CM} is then defined by

$$Mx_{CM} = m_1x_1 + m_2x_2 \qquad\qquad 10\text{-}1$$

Center of mass

where $M = m_1 + m_2$ is the total mass of the system. For this case of just two particles, the center of mass lies on the line joining the particles at some point between them (Figure 10-2a). This is easily seen if we choose our origin on one of the particles, say m_1 (Figure 10-2b). Then x_2 is the distance d between the two particles. The center-of-mass coordinate for this choice of origin is, from Equation 10-1,

$$Mx_{CM} = m_1x_1 + m_2x_2 = m_1(0) + m_2d$$

$$x_{CM} = \frac{m_2}{M}\,d = \frac{m_2}{m_1 + m_2}\,d \qquad\qquad 10\text{-}2$$

For equal masses, the center of mass is halfway between the particles. Otherwise, the center of mass is closer to the particle of greater mass (Figure 10-2c). For example if $m_1 = 4$ kg and $m_2 = 1$ kg, the center of mass is a distance $\frac{1}{5}d$ from m_1.

We obtain the velocity of the center of mass v_{CM} by differentiating Equation 10-1:

$$M\frac{dx_{CM}}{dt} = m_1\frac{dx_1}{dt} + m_2\frac{dx_2}{dt}$$

or

$$Mv_{CM} = m_1v_1 + m_2v_2 \qquad\qquad 10\text{-}3$$

Velocity of the center of mass

The right side of Equation 10-3 is just the total momentum of our two-particle system. Thus

The velocity of the center of mass times the total mass equals the total momentum of the system.

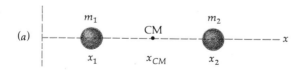

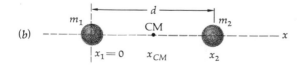

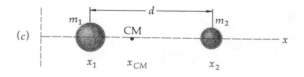

Figure 10-2
The center of mass of a two-particle system. (a) If the particles have equal mass, the center of mass is midway between them. (b) If the origin is chosen at m_1, the position of the center of mass is given by $x_{CM} = m_2d/(m_1 + m_2)$. (c) If the particles have unequal mass, the center of mass is closer to the more massive particle.

We obtain the acceleration of the center of mass by differentiating Equation 10-3:

$$M \frac{dv_{CM}}{dt} = m_1 \frac{dv_1}{dt} + m_2 \frac{dv_2}{dt}$$

or

$$Ma_{CM} = m_1 a_1 + m_2 a_2 = \Sigma F_{x1} + \Sigma F_{x2} = \Sigma F_x \qquad \text{10-4}$$

where $\Sigma F_{x1} = m_1 a_1$ is the resultant force acting on particle 1, $\Sigma F_{x2} = m_2 a_2$ is the resultant force acting on m_2, and $\Sigma F_x = \Sigma F_{x1} + \Sigma F_{x2}$ is the resultant force acting on the two-particle system. The forces acting on the system can be put in two categories, internal forces and external forces. The *internal forces* are the forces of interaction between the two particles: their mutual gravitational attraction, electrostatic attraction or repulsion, contact forces, and forces exerted by strings or springs which connect the particles. *External forces* are any forces exerted by agents outside the system. Figure 10-3 shows two particles connected by a spring and resting on a table. Our system is defined by the shading in this figure. The internal forces are those exerted by the spring, the gravitational attraction of the two bodies (which is so small that we can usually neglect it), and electrostatic attraction or repulsion if the bodies are charged. These forces are all exerted by agents within the system so defined. The external forces in this example are the weights of the bodies, exerted by the earth; the normal forces, exerted upward by the table; the frictional forces exerted by the table; and the tension $\mathbf{T}$. According to Newton's third law, for each force exerted by m_1 on m_2 there is an equal but opposite force exerted by m_2 on m_1. Thus the internal forces occur in pairs. When we sum all the forces acting on the two-particle system, the internal forces cancel out. Thus the total force ΣF_x acting on the system is just the total external force $\Sigma F_{x,\text{ext}}$. Equation 10-4 can thus be written

$$\Sigma F_{x,\text{ext}} = Ma_{CM} = M \frac{dv_{CM}}{dt} = \frac{dP}{dt} \qquad \text{10-5}$$

where $P = Mv_{CM} = m_1 v_1 + m_2 v_2$ is the total momentum. Equation 10-5 states that the resultant (external) force on the system equals the total mass times the acceleration of the center of the mass of the system. This has the same form as Newton's second law for a *single particle* of mass M located at the center of mass and under the influence of resultant force $\Sigma F_{x,\text{ext}}$.

The center of mass moves like a particle of mass $M = m_1 + m_2$ under the influence of the resultant external force on the system.

Motion of the center of mass

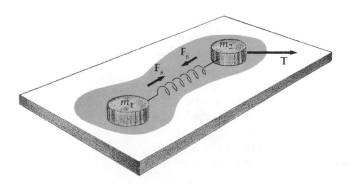

Figure 10-3
A system consisting of two masses connected by a spring on a table. The shaded region indicates the system. The forces external to this system are $\mathbf{T}$, the normal and frictional forces exerted by the table, and the weights of the masses.

We can easily generalize from our special case of two particles in one dimension to many particles in three dimensions. If we have N particles, the x coordinate of the center of mass is defined by

$$Mx_{CM} = m_1x_1 + m_2x_2 + m_3x_3 + \cdots = \Sigma m_i x_i \qquad \text{10-6}$$

where again, $M = \Sigma m_i$ is the total mass of the system. Similar equations define the y and z coordinates:

$$My_{CM} = \Sigma m_i y_i \qquad \text{10-7}$$

and

$$Mz_{CM} = \Sigma m_i z_i \qquad \text{10-8}$$

It is convenient to locate the center of mass by its *position vector* $\mathbf{r}_{CM}$, whose components are the rectangular coordinates x_{CM}, y_{CM}, and z_{CM}:

$$\mathbf{r}_{CM} = x_{CM}\mathbf{i} + y_{CM}\mathbf{j} + z_{CM}\mathbf{k}$$

If we use the definitions given in Equations 10-6 to 10-8, we have

$$\mathbf{r}_{CM} = \frac{\Sigma m_i x_i}{M}\mathbf{i} + \frac{\Sigma m_i y_i}{M}\mathbf{j} + \frac{\Sigma m_i z_i}{M}\mathbf{k}$$

$$M\mathbf{r}_{CM} = \Sigma m_i(x_i\mathbf{i} + y_i\mathbf{j} + z_i\mathbf{k})$$

But since the quantity in parentheses is just the position vector $\mathbf{r}_i$ of particle m_i,

$$M\mathbf{r}_{CM} = \Sigma m_i \mathbf{r}_i \qquad \text{10-9}$$

Equation 10-9 is the three-dimensional many-particle extension of Equation 10-1. We obtain the extension of Equation 10-3 for the total momentum of the system by differentiating Equation 10-9:

$$M\mathbf{v}_{CM} = \Sigma m_i \mathbf{v}_i = \mathbf{P} \qquad \text{10-10}$$

where, as before, the total momentum $\mathbf{P}$ is just the sum of the momenta of the particles. Differentiating again, we obtain

$$M\mathbf{a}_{CM} = \Sigma m_i \mathbf{a}_i = \frac{d\mathbf{P}}{dt} \qquad \text{10-11}$$

This is written in the form of Equation 10-5 by noting that $m_i\mathbf{a}_i = \Sigma\mathbf{F}_i$, where $\Sigma\mathbf{F}_i$ is the net force on the ith particle, and that in the sum $\Sigma\mathbf{F}_i$ the internal forces add to zero so that the sum is just the resultant external force on the system. Thus

$$\Sigma\mathbf{F}_{i,\text{ext}} = \frac{d\mathbf{P}}{dt} = M\mathbf{a}_{CM} \qquad \text{10-12}$$

Again

The center of mass of a system of particles moves like a single particle of mass $M = \Sigma m_i$ under the influence of the resultant external force acting on the system.

This theorem is important because it shows us how to describe the motion of one point, the center of mass, for any system of particles, no matter how extensive the system may be. The center of mass of the system behaves just like a single particle subject to the external forces only. The individual motions of the members of the system generally are much more complex. For example, even if air resistance is negligible, the motion of something like a pair of masses connected by a spring and thrown into the air is quite involved. The masses tumble and turn as they move and oscillate along the line joining them. But the center of mass of the system, which lies between the two masses, moves just as if it were a single particle: it follows a simple parabolic trajectory.

The justification for our earlier treatment of large objects as point particles is actually this theorem on center-of-mass motion. All large objects should be considered to be made up of many small masses whose motions are governed by Newton's three laws. The motion of such a system generally includes rotations and oscillations relative to the center of mass. To work out all details of the motion would be very difficult, but often we are satisfied with solving the motion of the center of mass. Our theorem states that this can be treated as a problem of the motion of a single point particle.

For the baton thrown into the air (Figure 10-1) the only external force acting on the object is its weight. Thus the center of mass follows the parabolic path of a particle in projectile motion predicted by Equation 10-12. This equation does not give a complete description of the complicated motion of the baton. We have described, so far, only the motion of one point, the center of mass of the system.

Some of the various methods of locating the center of mass of a system of particles are discussed in Section 10-5. In particular we shall see that the center of mass of a regular uniform object is at the geometric center of the object; the center of mass of a uniform disk is at the center of the disk, that of a uniform cylinder is on the axis of the cylinder, halfway between the faces, etc.

Example 10-1 A projectile explodes into two equal pieces, each of mass m, at the top of its flight. One of the pieces drops straight down while the other moves off horizontally so that they land simultaneously. Where does the second piece land?

Considering the projectile to be the system (whether it is in one piece or two), the only external force before the masses hit the ground is that due to gravity. The forces exerted in the explosion are internal forces, which do not affect the motion of the center of mass. After the explosion the center of mass traces out the rest of the parabola just as if there had been no explosion. Since the center of mass is halfway between equal masses and we know that one mass drops straight down, the other must land at an equal distance from the center of mass, as shown in Figure 10-4.

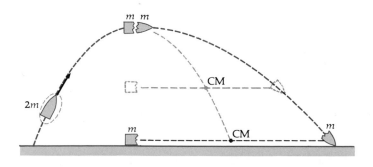

Figure 10-4
A projectile explodes into two equal fragments at the top of its flight so that the fragments land simultaneously. The internal forces of the explosion do not affect the motion of the center of mass. It follows the same parabolic path it would have followed if there had been no explosion. The center of mass is midway between the two equal fragments.

Example 10-2 Two equal masses connected by a spring rest on a smooth table with an external force **T** applied to one of the masses. Describe the motion. This situation is the same as that in Figure 10-3 except that there is no friction.

We cannot describe the detailed motion of each mass unless we know the spring constant. Even then, the problem is complicated. However, the motion of the center of mass is simple. Its acceleration is given by

$$\mathbf{a}_{CM} = \frac{\mathbf{T}}{m + m}$$

assuming the mass of the spring to be negligible. The forces exerted by the spring are internal to the system and do not affect the motion of the center of mass. (We can think of the spring as allowing each mass to exert a force on the other.) Although the motion of each mass may be quite complicated, the motion of the center of mass, which again is halfway between the equal masses, is that of uniform acceleration in the direction of the external tension $\mathbf{T}$.

Example 10-3 Consider the same example with no spring; i.e., just two equal masses on a smooth table with a force $\mathbf{T}$ exerted on one of them. It is of course easy to describe the motion of each separately. One remains at rest, and the other has acceleration $\mathbf{a}_2 = \mathbf{T}/m$. However, the center-of-mass motion is the same as in the previous example, $\mathbf{a}_{CM} = \mathbf{T}/(m + m) = \mathbf{T}/2m$. We include this simple example to show that the particles of a system need not be interacting and that external forces may be acting on only some of the particles in the system. Remember that the motion of the center of mass is only part of the description of the motion of a system of particles. Equation 10-12 says nothing about the motion of the particles relative to the center of mass. In this example the motion is quite different from that in the previous one, though the motion of the center of mass is the same.

Example 10-4 Consider a cylinder of mass M resting on a rough piece of paper on a table. A force $\mathbf{f}$ is applied to the cylinder by pulling the paper to the right, as in Figure 10-5. Which way does the cylinder move?

We shall show (Section 10-5) that the center of mass of a uniform cylinder is at its geometric center, i.e., on the axis, halfway between the faces. Taking the cylinder as our system, we have

$$\mathbf{f} = M\mathbf{a}_{CM}$$

The acceleration of the center of mass of the cylinder is in the direction of the force $\mathbf{f}$ and has the magnitude $\mathbf{f}/M$. Since $\mathbf{f}$ is in the direction the paper is pulled, $\mathbf{a}_{CM}$ is to the right for the case illustrated in Figure 10-5. This result may be somewhat surprising at first because the paper also accelerates and the cylinder appears to move back in the

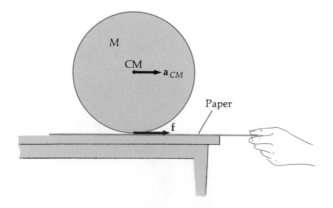

Figure 10-5
Cylinder rolling on a paper pulled to the right. Since the net external force on the cylinder is the frictional force to the right, the cylinder accelerates to the right. The cylinder rolls backward relative to the paper because the acceleration of the paper is greater than that of the cylinder.

direction opposite **f**. The forward acceleration of the cylinder is easily demonstrated, however, with a piece of chalk if the initial position of the chalk is marked on the *table*. The chalk accelerates back *relative to the paper,* because the acceleration of the paper is greater than that of the chalk. Relative to the table, the chalk accelerates forward in the direction of **f**. Note again that Equation 10-12 does not describe the rotation of the cylinder but only the motion of its center of mass.

Questions

1. Two masses are connected by a light string which passes over a pulley to form an Atwood's machine. Consider the two masses and the string as a system. What are the external forces? What are the internal forces?

2. A self-made man is said to have pulled himself up by his bootstraps. Discuss this from the point of view of internal and external forces and the motion of the center of mass of a system.

3. If only external forces can cause the center of mass of a system to accelerate, how can an automobile be accelerated by its motor?

4. A man is at rest at the center of a large, frictionless ice rink. Can he get off it? How?

5. A rocket ship initially is at rest (relative to the distant stars) in empty space when its engines are turned on. What happens to the center of mass of the rocket and its original contents? How can the rocket move?

10-2 Conservation of Linear Momentum

In an important special case of Equation 10-12 the resultant external force is zero. Then

$$\frac{d\mathbf{P}}{dt} = M\mathbf{a}_{CM} = 0$$

and

$$\mathbf{P} = M\mathbf{v}_{CM} = \Sigma m_i \mathbf{v}_i = \text{constant} \qquad \text{10-13}$$

The total momentum of the system and the velocity of the center of mass are constant. If a system is isolated from its surroundings so that no external forces are acting on it, the total momentum of the system does not change in time. This important result is known as the law of *conservation of linear momentum.* (The quantity $\mathbf{p} = m\mathbf{v}$, which we have been calling momentum, is often also called linear momentum to distinguish it from angular momentum, defined in Chapter 13.) This result is the generalization to many-particle systems of the conservation-of-momentum law for two particles discussed in Chapter 4. (The conservation of momentum for two particles with no external forces was, of course, the original experimental observation from which Newton postulated his third law.) This law is one of the most important in physics. It is more generally applicable than the law of conservation of mechanical energy because internal forces exerted by one particle in the system on another are often not conservative. Thus they can change the total mechanical energy of the system, but since

Conservation of momentum

these internal forces always occur in pairs, they cannot change the total linear momentum of the system. The conservation of momentum is particularly useful in studying collisions (Chapter 11).

Example 10-5 A man weighing 150 lb and a boy weighing 75 lb are standing together on a smooth ice surface for which friction is negligible. If after they push each other apart, the man moves away with speed 1 ft/sec, relative to the ice, how far apart are they after 5 sec (Figure 10-6)?

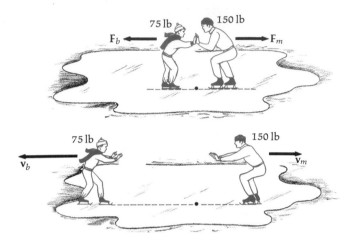

Figure 10-6
Man and boy on smooth ice push each other (Example 10-5). Since there are no external horizontal forces on the man-boy system, the total horizontal momentum which is zero initially does not change. The center of mass remains at its original position.

We take the man and boy together as the system. The weight of each is balanced by a corresponding normal force of the ice. Since there is no friction, the resultant force on the system is zero. The force exerted by the man on the boy is equal and opposite to that exerted by the boy on the man. The total momentum of the system is zero since they are standing at rest initially. Therefore, after they push each other, they must have equal and opposite momentum. Since the man has twice the mass of the boy, the boy must have twice the speed of the man. Since the man moves in one direction with speed 1 ft/sec, the boy moves in the opposite direction with speed 2 ft/sec. After 5 sec the man has moved 5 ft and the boy 10 ft and they are 15 ft apart. The center of mass of this system remains at rest in its original position, the point where they originally stood. Note that the mechanical energy of this system is not conserved. The force exerted by each on the other is not conservative. In this case the mechanical energy is increased, since the kinetic energy was initially zero and the potential energy did not change.

Example 10-6 A bullet of mass 10 gm moves horizontally with speed 400 m/sec and embeds itself in a block of mass 390 gm initially at rest on a smooth table. What is the final velocity of the bullet and block (Figure 10-7)?

Figure 10-7
Example 10-6. Collision of bullet and block resting on smooth surface. There are no external horizontal forces on the bullet-block system so the horizontal momentum is conserved. After the collision the bullet and block move together with velocity v_x given by conservation of momentum $(m_1 + m_2)v_x = m_1 v_{1x}$.

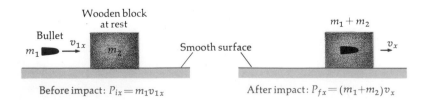

Before impact: $P_{ix} = m_1 v_{1x}$　　　After impact: $P_{fx} = (m_1 + m_2)v_x$

Since there are no horizontal forces on the bullet-block system, the horizontal component of the total momentum of the system is conserved. (There is a small vertical resultant force on the system, the weight of the bullet before it strikes the block. The bullet accelerates toward the earth before it strikes the block. We shall ignore this slight vertical motion.)

The total initial horizontal momentum P_{ix} before the bullet strikes the block is just that of the bullet:

$$P_{ix} = m_1 v_{1x} = (10 \text{ gm})(400 \text{ m/sec}) = 4000 \text{ gm-m/sec} = 4 \text{ kg-m/sec}$$

Afterward the bullet and block move together with a common velocity v_x. The total final momentum P_{fx} is

$$P_{fx} = (m_1 + m_2)v_x = (10 \text{ gm} + 390 \text{ gm})v_x = (0.4 \text{ kg})v_x$$

Since the total momentum is conserved, the final momentum equals the initial momentum.

$$(0.4 \text{ kg})v_x = 4 \text{ kg-m/sec}$$

$$v_x = 10 \text{ m/sec}$$

Since the bullet and block both move together with this velocity, the center of mass must move with this velocity. We could have found the velocity of the center of mass (which is constant) before the collision from Equation 10-10:

$$Mv_{CM,x} = \Sigma m_i v_{ix} = m_1 v_{1x} + m_2(0)$$

$$v_{CM,x} = \frac{m_1 v_{1x}}{m_1 + m_2} = \frac{4 \text{ kg-m/sec}}{0.4 \text{ kg}} = 10 \text{ m/sec}$$

This is the same calculation used to find v_x. A simple computation of the initial and final energies shows that again mechanical energy is not conserved. The original kinetic energy is $\frac{1}{2}m_1 v_{1x}^2 = \frac{1}{2}(0.01 \text{ kg})(400 \text{ m/sec})^2 = 800$ J, and the final energy is $\frac{1}{2}(0.4 \text{ kg})(10 \text{ m/sec})^2 = 20$ J. In this case most of the original kinetic energy (780 J out of 800 J) is lost because large nonconservative forces between the bullet and the block deform the bodies.

10-3   The Center-of-Mass Reference Frame

As we have seen, when the resultant external force on a system is zero, the velocity of the center of mass is constant. It is often convenient to choose a coordinate system with the origin at the center of mass. Then relative to the original coordinate system, this coordinate system moves with constant velocity $\mathbf{v}_{CM}$. If the original coordinate system is in an inertial reference frame, the one with origin at the center of mass is also. The reference frame with the origin at the center of mass is called the *center-of-mass reference frame*. Relative to this frame the velocity of the center of mass is of course zero. Since the total momentum of a system of particles equals the total mass times the velocity of the center of mass, the total momentum is zero in the center-of-mass reference frame. This frame is also called the *zero-momentum reference frame*. In the analysis of many problems it is convenient to transform from a given frame to the center-of-mass frame and back.

The laboratory reference frame.   The CM reference frame.

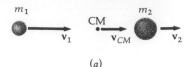   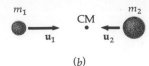

$$v_1 \qquad v_{CM} \qquad v_2$$

(a)   (b)

Figure 10-8
(a) Two particles moving in a general reference frame. In this frame the center of mass has velocity $\mathbf{v}_{CM} = (m_1\mathbf{v}_1 + m_2\mathbf{v}_2)/(m_1 + m_2)$. (b) In the center-of-mass, or zero-momentum, reference frame the center of mass is at rest, and the total momentum is zero. The velocities in the two frames are related by $\mathbf{u}_1 = \mathbf{v}_1 - \mathbf{v}_{CM}$ and $\mathbf{u}_2 = \mathbf{v}_2 - \mathbf{v}_{CM}$.

For example, the analysis of most collisions is easiest in the center-of-mass reference frame, as we shall see in the next chapter. However, many collisions are observed in other frames; e.g., one particle at rest initially is struck by another. The transformation between the center-of-mass frame and another frame is not difficult. Consider a simple two-particle system with one particle of mass m_1 moving with velocity $\mathbf{v}_1$ and the second of mass m_2 moving with velocity $\mathbf{v}_2$ (Figure 10-8a). The total momentum and velocity of the center of mass are given by Equation 10-10:

$$\mathbf{P} = m\mathbf{v}_{CM} = m_1\mathbf{v}_1 + m_2\mathbf{v}_2$$

We transform to the center-of-mass reference frame by subtracting the velocity $\mathbf{v}_{CM}$ from each particle, as illustrated in Figure 10-8. Let $\mathbf{u}_1$ and $\mathbf{u}_2$ be the velocities of the particles relative to the center of mass. Then

$$\mathbf{u}_1 = \mathbf{v}_1 - \mathbf{v}_{CM} \qquad \text{and} \qquad \mathbf{u}_2 = \mathbf{v}_2 - \mathbf{v}_{CM} \qquad\qquad 10\text{-}14$$

Velocity relative to center of mass

We can check that the total momentum is zero in the center-of-mass frame by computing the total momentum using the transformation Equation 10-14. We have

$$m_1\mathbf{u}_1 + m_2\mathbf{u}_2 = m_1(\mathbf{v}_1 - \mathbf{v}_{CM}) + m_2(\mathbf{v}_2 - \mathbf{v}_{CM})$$
$$= m_1\mathbf{v}_1 + m_2\mathbf{v}_2 - (m_1 + m_2)\mathbf{v}_{CM} = 0$$

since, by Equation 10-10, $(m_1 + m_2)\mathbf{v}_{CM}$ equals $m_1\mathbf{v}_1 + m_2\mathbf{v}_2$.

If the particles interact, as in a collision, their velocities will change; but if there is no resultant external force, the center of mass will remain at rest and the total momentum will not change. We can transform from the center-of-mass frame to any other frame in which the center of mass moves by adding the velocity $\mathbf{v}_{CM}$ to each velocity $\mathbf{u}_1$ and $\mathbf{u}_2$.

Example 10-7 Find the velocities and the total kinetic energy of the bullet and block of Example 10-6 in the center-of-mass reference frame (Figure 10-9).

In that example the initial velocity of the 10-gm bullet was 400 m/sec and the 390-gm block was initially at rest. We found the velocity of the center of mass to be 10 m/sec. If u_{1x} is the velocity of the bullet and u_{2x} the velocity of the block in the center-of-mass frame, we have

$$u_{1x} = 400 - 10 = 390 \text{ m/sec}$$

and

$$u_{2x} = 0 - 10 = -10 \text{ m/sec}$$

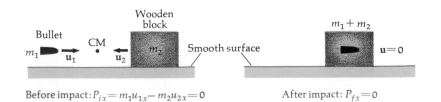

Before impact: $P_{ix} = m_1 u_{1x} - m_2 u_{2x} = 0$ After impact: $P_{fx} = 0$

Figure 10-9
Example 10-7. Collision of bullet and block of Example 10-6 viewed in the center-of-mass frame. After the collision the bullet and block are at rest in this frame.

The velocities after the collision are zero in this reference frame. The kinetic energies of the bullet and block before the collision are

$$\tfrac{1}{2}(0.01 \text{ kg})(390 \text{ m/sec})^2 + \tfrac{1}{2}(0.39 \text{ kg})(10 \text{ m/sec})^2$$
$$= 760.5 \text{ J} + 19.5 \text{ J} = 780 \text{ J}$$

After the collision the energy is zero in the center-of-mass frame because the particles are at rest. The energy lost, 780 J, is the same as that calculated in the original frame. We shall show in the next section that this is a general result.

___

Questions

6. Describe how the center-of-mass frame can be at rest even though both bodies in a collision are moving.

7. Can the center of mass have a greater speed than both bodies in a collision?

10-4   Kinetic Energy of a System of Particles

Although the total momentum of a system of particles must be constant if the resultant external force on the system is zero, the total mechanical energy of the system may change. Internal forces, which cannot change the total momentum, may be nonconservative and thus change the total mechanical energy. In Example 10-6 the forces exerted by the bullet and block on each other were evidently nonconservative since the energy of the system decreased. In such a case, some of the mechanical energy goes into thermal energy of the system, some of it goes into producing permanent deformation of the block and bullet, and a small amount goes into sound waves produced when the bullet strikes the block. Sometimes the total mechanical energy of a system will *increase* even if there is no external force on the system. This happened in Example 10-5 with the man and boy pushing each other. Another example is the explosion of a bomb, in which internal chemical energy is converted into kinetic energy of the fragments.

When the 10-gm bullet collided with the 390-gm block, the kinetic energy lost was 780 J. In the center-of-mass reference frame this was the total kinetic energy. In the original frame with the block initially at rest the 780-J energy loss was only part of the original energy. In that frame all the kinetic energy could not be lost because then the objects would be at rest and momentum would not be conserved. The kinetic energy remaining in that frame was just $\tfrac{1}{2}Mv_{CM}^2 = \tfrac{1}{2}(0.4)(10)^2 = 20$ J, where M was the total mass of the system. This example is a special case of a general theorem which we shall now derive.

The kinetic energy of a system of particles can be written as the sum of two terms: (1) the energy associated with center of mass motion $\tfrac{1}{2}Mv_{CM}^2$, where M is the total mass of the system, and (2) the energy of motion relative to the center of mass $\Sigma \tfrac{1}{2}m_i u_i^2$, where $\mathbf{u}_i$ is the velocity of the ith particle relative to the center of mass.

Consider a two-particle system in the center-of-mass reference frame and let $\mathbf{u}_1$ and $\mathbf{u}_2$ be the velocities of the two particles relative to the center of mass, which is at rest in this frame. Since the total momentum is zero in this frame, we have

$$m_1\mathbf{u}_1 + m_2\mathbf{u}_2 = 0 \qquad\qquad 10\text{-}15$$

The kinetic energy in this frame is

$$E_{kr} = \tfrac{1}{2}m_1u_1{}^2 + \tfrac{1}{2}m_2u_2{}^2 \qquad\qquad 10\text{-}16$$

(We use the subscript r to indicate that this energy is relative to the center of mass.) We now consider the system in another reference frame in which the center of mass is moving with velocity $\mathbf{v}_{CM}$. We transform to this frame by adding the velocity $\mathbf{v}_{CM}$ to the velocity of each particle according to Equation 10-14. In this frame the particles move with velocity $\mathbf{v}_1$ and $\mathbf{v}_2$, given by

$$\mathbf{v}_1 = \mathbf{u}_1 + \mathbf{v}_{CM} \qquad \mathbf{v}_2 = \mathbf{u}_2 + \mathbf{v}_{CM}$$

The kinetic energy in this frame is

$$\begin{aligned}
E_k &= \tfrac{1}{2}m_1v_1{}^2 + \tfrac{1}{2}m_2v_2{}^2 \\
&= \tfrac{1}{2}m_1(\mathbf{u}_1 + \mathbf{v}_{CM})^2 + \tfrac{1}{2}m_2(\mathbf{u}_2 + \mathbf{v}_{CM})^2 \\
&= \tfrac{1}{2}m_1(u_1{}^2 + v_{CM}{}^2 + 2\mathbf{v}_{CM}\cdot\mathbf{u}_1) + \tfrac{1}{2}m_2(u_2{}^2 + v_{CM}{}^2 + 2\mathbf{v}_{CM}\cdot\mathbf{u}_2) \\
&= \tfrac{1}{2}m_1u_1{}^2 + \tfrac{1}{2}m_2u_2{}^2 + \tfrac{1}{2}(m_1 + m_2)v_{CM}{}^2 + \mathbf{v}_{CM}\cdot(m_1\mathbf{u}_1 + m_2\mathbf{u}_2)
\end{aligned}$$

The first two terms are just the kinetic energy relative to the center of mass. The third term is $\tfrac{1}{2}Mv_{CM}{}^2$, where M is the total mass of the system. The last term is zero by Equation 10-15. Thus the kinetic energy in this frame is

$$E_k = \tfrac{1}{2}Mv_{CM}{}^2 + E_{kr}$$

where E_{kr} is the kinetic energy relative to the center of mass given by Equation 10-16. This result is easily generalized to a system of many particles:

$$E_k = \tfrac{1}{2}Mv_{CM}{}^2 + E_{kr} \qquad\qquad 10\text{-}17$$

where $M = \Sigma m_i$ is the total mass and $E_{kr} = \Sigma\tfrac{1}{2}m_iu_i{}^2$ is the kinetic energy of motion relative to the center of mass. We shall find many occasions to use this result in the following chapters.

If there are no external forces acting on the system, $\mathbf{v}_{CM}$ is constant. Then the first term in Equation 10-17 does not change; only the relative energy can increase or decrease. Since the relative energy is the same in all reference frames, the change in energy produced by internal forces is the same in all frames. This result is particularly useful in the analysis of collisions (Chapter 11).

10-5 Finding the Center of Mass

In this section we discuss methods of locating the center of mass for various systems of particles. Often the center of mass can be found by intuitive methods or by a simple experiment, avoiding involved calculations.

For a discrete system of particles the x coordinate of the center of mass is given by Equation 10-6:

$$Mx_{CM} = \Sigma m_ix_i$$

with similar equations for the y and z coordinates. If the system is a continuous body, we replace the sum by an integral:

$$Mx_{CM} = \int x \, dm$$

where dm is an element of mass. Let us consider again the simple system of two point masses. As we have seen, the center of mass lies between the masses. If we choose our origin at the center of mass, so that $x_{CM} = 0$, the coordinates of the particles are related by

$$m_1 x_1 + m_2 x_2 = 0$$

or

$$\frac{|x_1|}{|x_2|} = \frac{m_2}{m_1} \qquad\qquad 10\text{-}18$$

(One of the coordinates x_1 or x_2 must be negative since the origin is between the two particles.)

If the two point masses are connected by a light rod of negligible mass, the system balances at any orientation if pivoted at the center of mass (Figure 10-10a). This follows from the fact that the potential energy of a system of particles is the same as if the total mass were at the center of mass. Let y_i be the height above the earth of the ith particle in a general system. The potential energy of the system is

$$U = \Sigma m_i g y_i = g \Sigma m_i y_i$$

But $\Sigma m_i y_i = M y_{CM}$, and so

$$U = M g y_{CM} \qquad\qquad 10\text{-}19$$

If we try to balance our two particles on a light rod at some point other than the center of mass, the system will rotate until the potential energy is a minimum. This occurs when the center of mass is at its lowest point consistent with the restraint of the pivot. This point is directly below the pivot. Thus the system will rotate until the center of mass lies below the pivot (Figure 10-10b). However, if the pivot is at the center of mass, all orientations have the same potential energy and the system will balance.

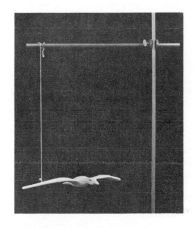

This bird looks symmetrical, but evidently its mass distribution is not. Where is its center of mass? (*Courtesy of Larry Langrill, Oakland University.*)

Example 10-8 Center of Mass of an Irregularly Shaped Plane Figure A simple experimental method based on the above analysis can be used to find the center of mass of any plane figure, e.g., a board or metal sheet. If we suspend the object from a frictionless pivot at any point on the object, it will hang in equilibrium with its center of mass directly below the pivot. If we draw a vertical line from the pivot (this can be done with the aid of a plumb bob), the center of mass must lie

Figure 10-10
(*a*) Two masses connected by a light rod pivoted at the center of mass. The gravitational potential energy depends only on the height of the center of mass, which is at the pivot for any orientation. The system balances at any orientation. (*b*) If the pivot is not at the center of mass, the potential energy is least when the center of mass is directly below the pivot.

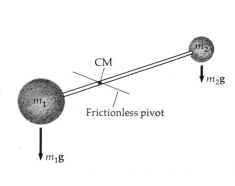

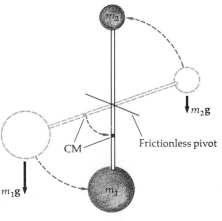

(*a*)

(*b*)

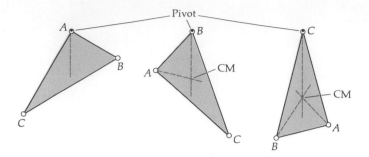

Figure 10-11
The center of mass of an irregularly shaped object lies
directly below any point
about which it is pivoted.
This property can be used to
locate the center of mass.

on this line. We can then suspend the object from another point and draw a second line, as shown in Figure 10-11. The center of mass will lie on the intersection of the two lines. The third suspension shown in Figure 10-11c confirms this result.

Example 10-9 Three Point Masses The center-of-mass position vector for three point masses is given by

$$M\mathbf{r}_{CM} = \Sigma m_i \mathbf{r}_i = m_1 \mathbf{r}_1 + m_2 \mathbf{r}_2 + m_3 \mathbf{r}_3 \qquad 10\text{-}20$$

where $M = m_1 + m_2 + m_3$ is the sum of the masses. A method useful for three point masses and many other systems is to break this sum into two or more parts. For example, the first two terms in the sum are related to the center of mass of the first two masses by

$$m_1 \mathbf{r}_1 + m_2 \mathbf{r}_2 = (m_1 + m_2)\mathbf{r}'_{CM}$$

where $\mathbf{r}'_{CM}$ is the center-of-mass position vector for masses m_1 and m_2 alone. Thus we can write Equation 10-20 as

$$M\mathbf{r}_{CM} = (m_1 + m_2)\mathbf{r}'_{CM} + m_3 \mathbf{r}_3$$

We have reduced the problem to finding first the center of mass for the two particles m_1 and m_2 and then finding the center of mass for two particles, m_3 and the system $m_1 + m_2$. This solution is illustrated in Figure 10-12. The center of mass for the masses m_1 and m_2 is at point P_{12}. It lies on the line joining m_1 and m_2 with distances given by Equation 10-18. The center of mass for the three-particle system is then on the line joining mass m_3 and point P_{12}. The distance d_3 from m_3 is related to the distance d_{12} from P_{12} by Equation 10-18

$$\frac{d_3}{d_{12}} = \frac{m_1 + m_2}{m_3}$$

We can see from this analysis that the center of mass for three equal masses lies at the intersection of the medians of the triangle formed by the three masses.

The technique used in Example 10-9 can be used for many systems consisting of several bodies. Consider, for example, two uniform sticks. The center-of-mass position vector for this system $\mathbf{r}_{CM}$ is defined by

$$M\mathbf{r}_{CM} = \sum_{\substack{\text{both}\\\text{sticks}}} m_i \mathbf{r}_i$$

where the sum is over all the mass elements of the two sticks. We can write this sum in two parts:

$$\sum_{\substack{\text{both}\\\text{sticks}}} m_i \mathbf{r}_i = \sum_{\text{stick 1}} m_i \mathbf{r}_i + \sum_{\text{stick 2}} m_i \mathbf{r}_i$$

The first sum on the right is just $M_1 \mathbf{r}_{CM1}$, where M_1 is the mass of stick

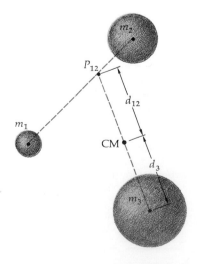

Figure 10-12
Center of mass of three point
particles. Point P_{12} is the
center of mass of the two-particle system $m_1 + m_2$. These
two particles can be replaced
by a single particle of mass
$m_1 + m_2$ at P_{12} for the calculation of the center of mass of
the three particles. The distances d_3 and d_{12} are related
by $m_3 d_3 = (m_1 + m_2)d_{12}$.

1 and $\mathbf{r}_{CM1}$ is the position vector to its center of mass, which is at the center of the stick since it is uniform. Similarly the second term is $M_2\mathbf{r}_{CM2}$, and

$$M\mathbf{r}_{CM} = M_1\mathbf{r}_{CM1} + M_2\mathbf{r}_{CM2}$$

The center of mass of the two-stick system lies on the line joining the centers of mass of the separate sticks, as shown in Figure 10-13. We see from this type of reasoning that we can treat any part of a complicated system as a point mass located at the center of mass of the part.

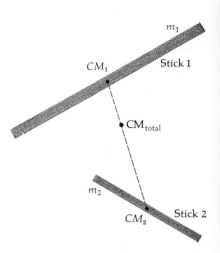

Figure 10-13
The center of mass of a system of two sticks can be found by treating each stick as a point particle at its individual center of mass.

Example 10-10 Symmetrical Bodies It is clear from our arguments about balance that the center of mass of a regular object must lie on any line of symmetry the object may have. For example, the center of mass of a uniform stick lies on the axis of the stick halfway between the ends. The center of mass of a uniform cylinder lies on its axis midway between the faces. We can prove this as follows. Consider first a thin uniform disk. The center of mass of the disk is clearly at the center, as can be seen by balancing or by the method of Example 10-8. If we have two parallel disks of equal mass, we can use the result discussed above. The center of mass of the two disks is on the line joining their individual centers of mass midway between the disks. A cylinder can be thought of as a set of uniform disks; thus its center of mass lies on the axis in the middle of the cylinder. This type of argument can be used to find the center of mass of other solid objects such as blocks.

Since any line through the diameter of a uniform sphere is a line of symmetry, the center of mass of the sphere is at its center.

10-6   Using Integration to Find the Center of Mass

If the center-of-mass coordinates of a continuous body are to be calculated, the sum $\Sigma m_i x_i$ must be replaced by the integral $\int x\, dm$, where dm is an element of mass. We then have

$$Mx_{CM} = \int x\, dm \qquad\qquad 10\text{-}21$$

In most cases, one or more of the coordinates can be found by symmetry without calculation. We give two examples here of finding the center of mass of a continuous body.

Example 10-11 Center of Mass of a Uniform Stick of Mass M and Length L (Figure 10-14) This very simple example, in which the result is known from symmetry beforehand, illustrates the technique of setting up the integration indicated in Equation 10-21. Let us call the linear density, or mass per unit length, λ. Since the density is uniform, $\lambda = M/L$. In

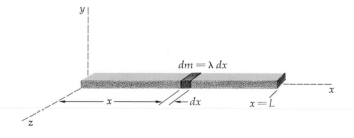

Figure 10-14
Calculation of the center of mass of a uniform stick. The mass element $dm = \lambda\, dx$ at distance x from the end is treated as a point particle.

Figure 10-14 we have chosen the origin at one end of the stick. We need only compute x_{CM}. A mass element of length dx is indicated. Its mass is $dm = \lambda \, dx$, and it is a distance x from the origin. Thus

$$Mx_{CM} = \int x \, dm = \int_0^L x\lambda \, dx = \frac{\lambda x^2}{2} \Big]_0^L = \frac{\lambda L^2}{2}$$

Using $\lambda = M/L$, we obtain the expected result:

$$x_{CM} = \frac{\lambda L^2}{2M} = \frac{M}{L}\frac{L^2}{2M} = \tfrac{1}{2}L$$

Example 10-12 Center of Mass of a Semicircular Hoop Figure 10-15 shows our choice of coordinate system for this calculation. We have chosen the origin to be on a line of symmetry of the figure. Clearly, $x_{CM} = 0$ because for every mass element at $+x$ there is an equal one at $-x$. However, y_{CM} is not zero. The center of mass is not at the origin, which is at the center of curvature of the hoop. All the mass is at positive y; thus $\int y \, dm$ cannot be zero. In the figure we have indicated a mass element of length $R \, d\theta$ at height $R \sin \theta$. The mass of this element is $dm = \lambda R \, d\theta$, where $\lambda = M/\pi R$ is the mass per unit length. We thus have

$$My_{CM} = \int y \, dm = \int_0^\pi R \sin \theta \, \lambda R \, d\theta = R^2 \lambda \int_0^\pi \sin \theta \, d\theta = 2R^2\lambda$$

since

$$\int_0^\pi \sin \theta \, d\theta = -\cos \theta \Big]_0^\pi = +2$$

Using $\lambda = M/\pi R$, we have

$$My_{CM} = 2R^2 \frac{M}{\pi R} \quad \text{or} \quad y_{CM} = \frac{2R}{\pi}$$

The center of mass is not within the body in this case. Figure 10-16 shows how to locate the center of mass by suspending a semicircular hoop from one end and then from another point.

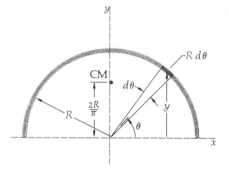

Figure 10-15
Calculation of the center of mass of a semicircular hoop. The center of mass lies on the y axis.

Questions

8. Must there be mass at the center of mass of a system? Explain.

9. Must the center of mass of a solid body be in the interior of the body? If not, give examples.

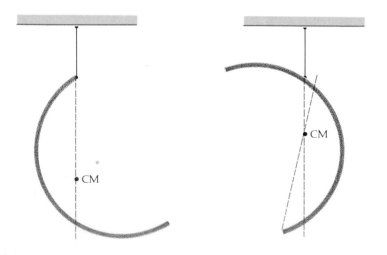

Figure 10-16
The center of mass of a semicircular hoop found experimentally by suspending it from two points.

Review

A. Define, explain, or otherwise identify:

Center of mass, 256

Internal forces, 257

External forces, 257

Conservation of momentum, 261

Center-of-mass reference frame, 263

Zero-momentum reference frame, 263

Relative kinetic energy, 265

Center-of-mass kinetic energy, 265

B. True or false:

1. The center of mass of a two-particle system is always closer to the more massive particle.

2. The total mass of a system times the velocity of the center of mass equals the total momentum of the system.

3. The momentum of a system can be conserved even if the mechanical energy is not.

4. Internal forces do not affect the motion of the center of mass of a system.

5. In the center-of-mass reference frame, the total momentum is zero.

6. The kinetic energy of a system of particles is different in different reference frames.

Exercises

Section 10-1, Center of Mass

1. A 2-kg mass is on the x axis at $x = 3$ m, and a 4-kg mass is on the x axis at $x = 6$ m. Find the center of mass of this system.

2. In Exercise 1 the 2-kg mass is moving to the right at 4 m/sec, and the 4-kg mass is moving to the left at 3 m/sec. (a) What is the velocity of the center of mass? (b) What is the total momentum of the system?

3. Three point masses are located in the xy plane as follows: a 1-kg mass is at the origin, a second 1-kg mass is at $x = 4$ m on the x axis, and a 2-kg mass is at the point $x = 2$ m, $y = 2$ m. Find the center of mass.

4. Three point masses have the following velocities along the x axis: a 1-kg mass is moving to the right with $v_x = 5$ m/sec, a 2-kg mass is moving to the left with $v_x = -3$ m/sec, and a 4-kg mass is moving to the right with $v_x = 4$ m/sec. Find (a) the total momentum of the system and (b) the velocity of the center of mass.

5. Two 3-kg masses have velocities $\mathbf{v}_1 = 2\mathbf{i} + 3\mathbf{j}$ m/sec and $\mathbf{v}_2 = 4\mathbf{i} - 6\mathbf{j}$ m/sec. Find (a) the velocity of the center of mass and (b) the total momentum of the system.

6. A 5-kg mass is at rest at the origin, and a second 5-kg mass is initially at rest at $x = 0$ and $y = 4$ m. A constant force $\mathbf{F} = (10 \text{ N})\mathbf{i}$ is applied to this second mass while no force acts on the first mass. (a) Find the acceleration of the second mass and its position at the times $t = 0, 1, 2$, and 3 sec. (b) On a diagram, indicate these positions and the position of the center of mass at these times. (c) Find the acceleration of the center of mass and from it the position of the center of mass at these times.

7. A frictional force of 500 lb acts on the surface of each of the rear tires of a 3000-lb car. (a) Draw a diagram indicating these forces and any others acting on the car on a level road at low speed. (b) Find the acceleration of the car neglecting wind resistance.

8. A wheel of mass 3 kg rolls down a plane inclined at $\theta = 30°$. The plane exerts a frictional force on the wheel up the plane of magnitude 4.9 N to keep the wheel from sliding. Find the acceleration of the center of mass of the wheel.

9. A 150-lb man stands at one end of a 60-ft-long log weighing 1500 lb. The log is floating in a millpond at right angles to the shore, the end opposite the man just touching the shore. The man walks along the log toward the shore. Assume that the water exerts no tangential force on the log. (a) On a drawing, indicate the initial position of the center of mass of the man-log system. (b) On a second drawing, directly below the first, indicate the position of the man, the log, and the center of mass when the man is at the middle of the log. (c) On a third drawing below the second, indicate the position of the man, the log, and the center of mass when the man is at the other end of the log. How far is he from shore at this time?

10. Two bodies have masses $m_1 = 1$ gm and $m_2 = 5$ gm. Initially they are at rest 10 cm apart on a frictionless surface. They are electrically charged so that they attract each other. They accelerate toward each other and collide. How far are they from the original position of m_1 when they meet?

11. A 10-gm mass and a 25-gm mass are compressed against opposite ends of a spring which rests on a horizontal table. The masses are not attached to the spring. The spring is released with both masses initially together at $x = 0$ at rest. At some time t after the masses leave the spring the 10-gm mass is observed at $x_1 = 20$ cm traveling at 5 cm/sec. (a) Where is the center of mass initially? (b) Where is the 25-gm mass, and how fast is it traveling at time t?

12. Some physics students find themselves in the middle of a box car teetering on the edge of a cliff (Figure 10-17). The car is on level ground and free to roll, but near the edge the ground is crumbling away. Which way should the students run to save themselves? (The ground is hard except just at the edge of the cliff.)

Figure 10-17
Exercise 12.

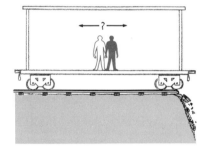

Section 10-2, Conservation of Linear Momentum

13. A cart of mass 10 kg is rolling along a level floor at a speed of 5 m/sec. A 4-kg mass dropped from rest lands in the cart. (a) What was the momentum of the cart before the mass was dropped? (b) What is the momentum of the cart and mass after the mass was dropped into the cart? (c) What is the speed of the cart and mass?

14. Two railroad cars have masses of 6×10^4 and 4×10^4 kg. They are initially rolling along the track in the same direction with the lighter car in front moving at 0.5 m/sec and the heavier car trailing but moving at 1.0 m/sec. They collide and couple. (a) What is the total momentum of the two-car system before the collision? (b) What is the total momentum of the two-car system after the collision? (c) What is the speed of the two cars after the collision? (d) Find the total kinetic energy before and after the collision.

15. An open-topped freight car weighing 20 tons is rolling along a track at 5 ft/sec. Rain is falling vertically downward into the car. After the car has collected 2 tons of water, what is its speed? *Hint:* You need not convert these weights to masses if you write mass as w/g.

16. A small 2000-lb automobile traveling north at 50 mi/h collides at an intersection with a 6000-lb truck traveling east at 40 mi/h. The car and truck stick together. (a) What is the total momentum of the car-truck system before the collision? (b) Find the magnitude and direction of the velocity of the combined wreckage just after the collision.

17. A 20-gm bullet is fired horizontally with velocity 250 m/sec from a 1.5-kg gun. If the gun were held loosely in the hand, what would be its recoil velocity?

18. Two masses are held at rest initially on a horizontal frictionless surface. They are pushed together, compressing a small spring between them which is not attached to either mass. When the masses are released the spring accelerates each, giving mass m_1 a velocity 5 m/sec to the left and m_2 a velocity 15 m/sec to the right. (a) What is the total momentum of the system before the masses are released? After they are released? (b) What is the ratio m_1/m_2?

Section 10-3, The Center-of-Mass Reference Frame, and Section 10-4, Kinetic Energy of a System of Particles

19. (a) In Exercise 14 find the velocity of the center of mass of the two-railroad-car system. (b) Find the velocity of each car relative to the center of mass before the collision. What are their velocities relative to the center of mass after the collision? (c) Find the kinetic energies of the cars relative to the center of mass before and after the collision.

20. In Exercise 18 let m_1 be 12 kg and m_2 be 4 kg. (a) Find the kinetic energy of the masses after they are released from the spring. (b) Find the velocity of each mass in a reference frame in which the center of mass is moving to the right with velocity 3 m/sec. (c) Find the total kinetic energy before and after the masses are released in the reference frame of part (b). How much energy do the masses gain from the spring in this frame?

21. A 3-kg mass moves at 5 m/sec to the right. It is chasing a second 3-kg mass moving at 1 m/sec also to the right. (a) Find the total kinetic energy of the two masses in this frame. (b) Find the velocity of the center of mass. (c) Find the velocities of the masses in the center-of-mass frame. (d) Find the kinetic energy of the masses in the center-of-mass frame. (e) Show that your answer for part (a) is greater than that for part (d) by the amount $\frac{1}{2}Mv_{CM}^2$, where v_{CM} is the velocity of the center of mass in the original frame and M is the total mass.

22. A 2-kg mass is moving to the right with a speed of 5 m/sec toward a stationary mass of 3 kg. (a) What is the kinetic energy of this system in this frame? (b) Find the velocity of the center of mass in this frame and the center-of-mass energy $\frac{1}{2}Mv_{CM}^2$. (c) Find the velocity of each mass in the center-of-mass reference frame. (d) Find the total kinetic energy of the two masses in the center-of-mass reference frame.

23. Two equal masses are connected by a light rigid rod. (a) How can they move so that their total kinetic energy is just the energy of center-of-mass motion? (b) How can they move so that their total kinetic energy is entirely energy of motion relative to the center of mass?

24. Describe how a solid ball can move so that (a) its total kinetic energy is just the energy of center-of-mass motion and (b) its total kinetic energy is energy of motion relative to the center of mass.

25. Two masses are connected by a spring. Describe how the masses can move so that (a) the total kinetic energy is just energy of center-of-mass motion and (b) the kinetic energy is entirely energy of motion relative to the center of mass.

Section 10-5, Finding the Center of Mass

26. Find the center of mass of the system shown in Figure 10-18.

27. Find the potential energy for the system in Exercise 26 directly by summing mgy_i over all the particles and show that the result equals Mgy_{CM}.

28. Each of the objects shown in Figure 10-19 is suspended from a ceiling by a thread attached to the point marked x on the object. Describe with a diagram the orientation of each suspended object.

Figure 10-18
Exercise 26.

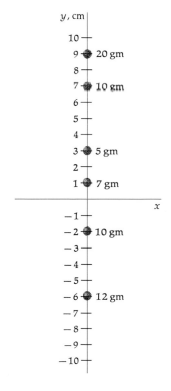

Figure 10-19
Exercise 28.

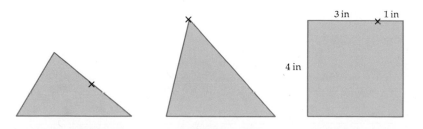

29. An equilateral triangle is made of three uniform sticks of equal length L (Figure 10-20). Find the center of mass of this system (a) by using symmetry arguments and (b) by first finding the center of mass of two of the sticks and then combining this system with the third stick.

Figure 10-20
Exercise 29.

30. Find the center-of-mass coordinates x_{CM} and y_{CM} for the figure shown in Figure 10-21. *Hint:* Consider the figure to be three squares and replace each square by a point mass at the center of mass of the square.

Figure 10-21
Exercise 30.

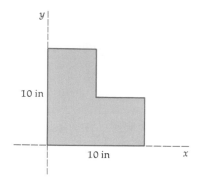

10-6, Using Integration to Find the Center of Mass

31. A nonuniform thin rod of length L lies along the x axis with one end at the origin. It has a linear mass density λ kg/m, given by $\lambda = \lambda_0(1 + x/L)$. The density is thus twice as great at one end as at the other. (a) Use $M = \int dm$ to find the total mass. (b) Find the center of mass of the rod.

Problems

1. The ratio of the mass of the earth to the mass of the moon is $M_E/m_m = 81.3$. The radius of the earth is about 4000 mi, and the distance to the moon is about 240,000 mi. (a) Locate the center of mass of the earth-moon system relative to the surface of the earth. (b) What external forces act on the earth-moon system? In what direction is the acceleration of the center of mass of this system? (c) Assume that the center of mass of this system moves in a circular orbit around the sun. How far must the center of the earth move in the radial direction (toward or away from the sun) during the 14 d between the time the moon is farthest from the sun (full moon) and when it is closest to the sun (new moon)?

2. A projectile is launched at 20 m/sec at an angle of 30° with the horizontal. In the course of its flight it explodes, breaking into two parts, one of which has twice the mass of the other. The two fragments land simultaneously. The lighter fragment lands 20 m from the launch point in the direction the projectile was fired. Where does the other fragment land?

3. A man of mass 80 kg is riding on a small cart of mass 40 kg which is rolling along a level floor at speed of 2 m/sec. He jumps off the back of the cart so that his speed relative to the ground is 1 m/sec in the direction opposite to the motion of the cart. (a) What is the speed of the center of mass of the man-cart system before and after he jumps? (b) What is the speed of the cart after the man jumps? (c) What is the speed of the center of mass after the man hits the ground and comes to rest? (d) What force is responsible for the change in the velocity of the center of mass? (e) Transform all velocities to the center-of-mass reference frame and indicate on a diagram the initial and final velocities of the man and cart in this frame. How much energy was expended by the man in jumping?

4. Use integration to find the center of mass of a uniform plate of material cut in the shape of a half circle.

5. A circular plate of radius r has a circular hole cut out with radius $\frac{1}{2}r$ (Figure 10-22). Find the center of mass of the plate. *Hint:* The hole can be represented by two disks superimposed; one of mass m and the other of mass $-m$.

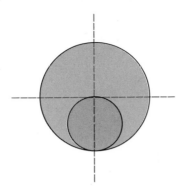

Figure 10-22
Problem 5.

6. Using the hint from Problem 5, find the center of mass of a solid sphere of radius r which has a spherical hole of radius $\frac{1}{2}r$, as shown in Figure 10-23.

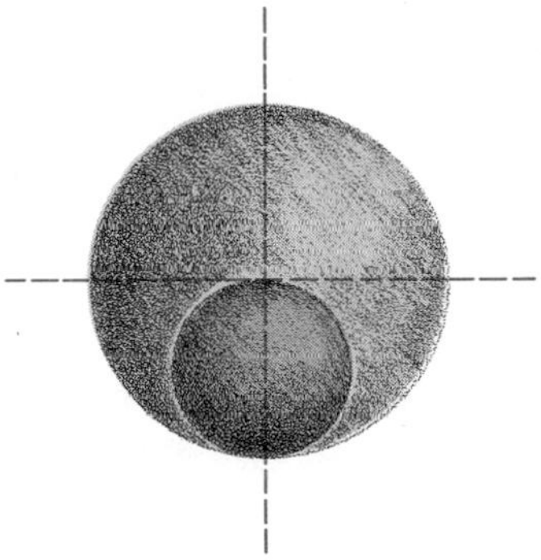

Figure 10-23
Problem 6.

7. A massless spring of force constant 1000 N/m is compressed a distance of 20 cm between masses of 8 and 2 kg. The spring is released on a smooth table. Since the masses are not attached to the spring, they move away with speeds v_1 and v_2. (a) Show that the kinetic energy of the masses when they leave the spring can be written $E_k = \frac{1}{2}p^2(1/m_1 + 1/m_2)$, where p is the momentum of either mass. Use conservation of energy to find p. (b) Find the velocity of each mass as it leaves the spring. (c) Find the velocity of each mass if the system is given an initial velocity of 4 m/sec perpendicular to the spring, as shown in Figure 10-24a. (d) What is the energy of center-of-mass motion in this case? What is the energy of motion relative to the center of mass? (e) Find the velocity of each mass if the system is given an initial velocity of 4 m/sec in the direction along the spring, as shown in Figure 10-24b. What is the energy of center-of-mass motion in this case? What is the energy of motion relative to the center of mass?

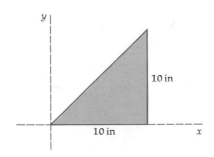

(a)   (b)

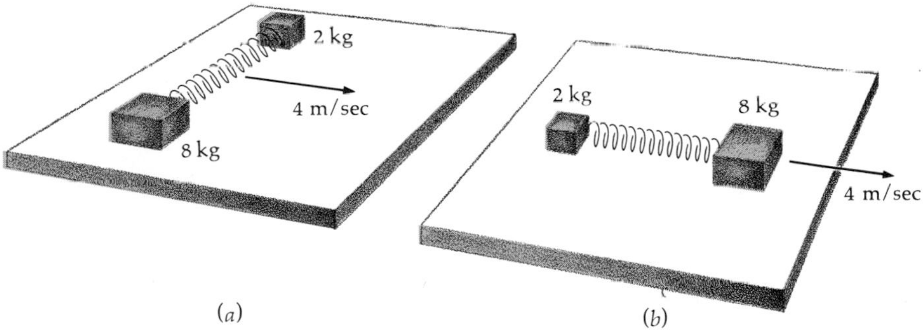

Figure 10-24
Problem 7.

8. An 80-lb boy gets on his 20-lb wagon on level ground along with two 10-lb bricks. He throws the bricks horizontally off the back of the wagon one at a time at a speed of 20 ft/sec relative to himself. How fast does he finally go?

9. Use integration to find the center of mass of the right isosceles triangle shown in Figure 10-25.

Figure 10-25
Problem 9.

CHAPTER 11   Collisions and Reactions

When two bodies come together and interact strongly for a short time, the event is called a *collision*. Often the forces exerted by one body on the other are much stronger than any other external forces present, and the time of the collision is so short that the bodies do not move appreciably during the interaction.

When the physical or chemical nature of the bodies is changed by the collision, the event is called a *reaction*, e.g., if a proton collides with a hydrogen atom so that the atom is ionized, i.e., the electron is removed, leaving two protons and a free electron after the collision. We consider some of the general properties of collisions and reactions in this chapter and apply our results to some useful examples.

11-1 Impulse

When a golf club hits a ball, the force $\mathbf{F}$ on the ball is zero just as contact is made. It then increases to a very large value and finally decreases to zero as the ball leaves the club (Figure 11-1). The total time of contact Δt is very small, perhaps only about 0.001 sec. The quantity of interest is not the instantaneous acceleration of the ball, which also varies with time (see Figure 11-1), but the velocity of the ball $\mathbf{v}_f$ or the momentum of the ball just as it leaves the club $\mathbf{p}_f = m\mathbf{v}_f$. These are related to the impulse $\mathbf{I}$ of the force $\mathbf{F}$ for the time interval Δt. The impulse of a force for the time interval from t_i to t_f is a vector defined by

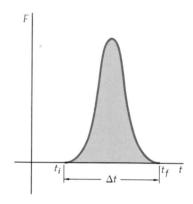

Figure 11-1
The force exerted by the club on a golf ball plotted as a function of time. The impulse exerted on the ball for the interval from t_i to t_f is the area below the F-versus-t curve for that interval. It equals the change in momentum of the ball.

$$\mathbf{I} = \int_{t_i}^{t_f} \mathbf{F}\, dt \qquad\qquad 11\text{-}1 \qquad \textit{Impulse defined}$$

The magnitude of the impulse is just the area under the F-versus-t curve. Assuming that $\mathbf{F}$ is the resultant force and using Newton's law $\mathbf{F} = d\mathbf{p}/dt$, we see that the impulse equals the total change in momentum during the time interval $t_f - t_i$:

$$\mathbf{I} = \int_{t_i}^{t_f} \mathbf{F} \, dt = \int_{t_i}^{t_f} \frac{d\mathbf{p}}{dt} \, dt = \mathbf{p}_f - \mathbf{p}_i = \Delta\mathbf{p} \qquad \text{11-2}$$

Impulse equals change in momentum

From Equation 11-2 the units of impulse are newton-seconds or kilogram-meters per second.

For a general force $\mathbf{F}$, the impulse given by Equations 11-1 and 11-2 depends on the choice of times t_i and t_f. But the forces which occur in collisions are zero except during a small time interval, as shown in Figure 11-1. For these forces, the impulse does not depend on the time interval as long as t_i is any time before the force occurs and t_f is any time after the force has decreased to zero. It is for this type of force that the concept of impulse is most useful.

For the golf club hitting a ball initially at rest, the change in momentum of the ball is just its final momentum,

$$\Delta\mathbf{p} = \mathbf{I} = \mathbf{p}_f$$

The impulse exerted by the ball on the club is equal in magnitude and opposite in direction to that exerted by the club on the ball since the forces are equal and opposite according to Newton's third law.

Golf club hitting a golf ball. Note the deformation of the golf ball due to the large force exerted by the club during the brief time of contact. As the ball leaves the club, it springs back to its original shape, converting the elastic potential energy of deformation into kinetic energy. (*Courtesy of Harold E. Edgerton.*)

Example 11-1 Find the impulse exerted by the block on the bullet for Example 10-6, in which a 10-gm bullet traveling at 400 m/sec embeds itself in a 390-gm block initially at rest.

In that example we found the final speed of the bullet and block to be $v_{CM} = 10$ m/sec. The change in momentum of the bullet is thus

$$\Delta p = p_f - p_i = (10^{-2} \text{ kg})(10 \text{ m/sec}) - (10^{-2} \text{ kg})(400 \text{ m/sec})$$
$$= -3.9 \text{ kg-m/sec} = -3.9 \text{ N-sec}$$

The impulse exerted on the bullet is thus 3.9 N-sec to the left. The bullet exerts an equal and opposite impulse on the block, as can be seen from its change in momentum:

$$\Delta p_{\text{block}} = (0.39 \text{ kg})(10 \text{ m/sec}) - 0 = 3.9 \text{ kg-m/sec} = 3.9 \text{ N-sec}$$

11-2   Time Average of a Force

The time-average force over the interval $\Delta t = t_f - t_i$ is defined by

$$\mathbf{F}_{av} = \frac{1}{\Delta t} \int_{t_i}^{t_f} \mathbf{F} \, dt = \frac{\mathbf{I}}{\Delta t} \qquad \text{11-3}$$

$\mathbf{F}_{av}$ is the constant force which gives the same impulse in the time interval Δt. F_{av} is indicated in Figure 11-2. It is often useful to compute the average force in a collision in order to compare it to such other forces as frictional forces or the force of gravity.

Example 11-2 What are reasonable magnitudes for the impulse I, average force F_{av}, and collision time Δt for a golf club hitting a golf ball?

Let us take the mass of a typical golf ball as $m = 45$ gm. The velocity of the ball when it leaves the club is related to the range by

$$R = \frac{v_0{}^2}{g} \sin 2\theta_0$$

where θ_0 is the angle of the projectile and air resistance is neglected. Taking $\theta_0 = 45°$ for simplicity (corresponding to the maximum range)

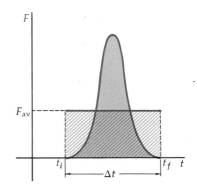

Figure 11-2
The average force F_{av} for some time interval is the constant force which gives the same impulse during that interval. The rectangular area $F_{av} \, \Delta t$ is the same as that under the F-versus-t curve.

and $R = 160$ m as a reasonable range (this is about 173 yd), we find

$$v_0{}^2 = Rg \approx 160 \times 10 = 1600$$

$$v_0 = 40 \text{ m/sec}$$

using $g \approx 10$ m/sec² to simplify the calculation. The magnitude of the impulse is thus

$$I = \int F\,dt = \Delta p = mv_0 = (45 \times 10^{-3} \text{ kg})(40 \text{ m/sec})$$
$$= 1.8 \text{ kg-m/sec} = 1.8 \text{ N-sec}$$

We can estimate the time of the collision as follows. The magnitude of the average acceleration is

$$a_{av} = \frac{F_{av}}{m} = \frac{I}{m\,\Delta t}$$

If the acceleration were constant and equal to the average acceleration, the velocity v_0 would be related to the distance traveled by the ball *during the collision* by the constant-acceleration expression

$$v_0{}^2 = 2a_{av}x$$

A reasonable estimate for the distance traveled by the ball while it is in contact with the golf club is the radius of the golf ball, about $x = 2$ cm. Then the average acceleration is

$$a_{av} = \frac{v_0{}^2}{2x} = \frac{1600}{2(0.02)} = 40{,}000 \text{ m/sec}^2$$

The average force is thus estimated to be

$$F_{av} = ma_{av} = (0.045 \text{ kg})(40{,}000 \text{ m/sec}^2) = 1800 \text{ N}$$

The estimated collision time is then

$$\Delta t = \frac{I}{F_{av}} = \frac{1.8 \text{ N-sec}}{1800 \text{ N}} = 0.001 \text{ sec}$$

We could also have estimated Δt from $v_0 = a_{av}\,\Delta t$. Then

$$\Delta t = \frac{v_0}{a_{av}} = \frac{40 \text{ m/sec}}{40{,}000 \text{ m/sec}^2} = 0.001 \text{ sec}$$

We see that the average force exerted by the club, 1800 N, is much larger than any other forces acting on the ball. For example, the weight of the ball is only about 0.45 N, and the frictional forces exerted by the grass or tee on the ball are even less, assuming the coefficient of friction to be less than 1.

Example 11-3 A machine gun fires R bullets per second. Each bullet has mass m and speed v. The bullets strike a fixed target, where they stop. What is the average force exerted on the target over a time long compared to the time between bullets?

Figure 11-3 shows a graph of the instantaneous force on the target. The average force is also indicated. The area under each impulse curve equals the change in momentum of each bullet. This change is just the initial momentum mv since the final momentum is zero. In time Δt, the number of bullets that hit the target is $R\,\Delta t$. Each bullet exerts an impulse mv on the target. The total impulse exerted in time Δt by the $R\,\Delta t$ bullets is consequently $R\,\Delta t\,mv$. The average force is thus

$$F_{av} = \frac{R\,\Delta t\,mv}{\Delta t} = Rmv \qquad\qquad 11\text{-}4$$

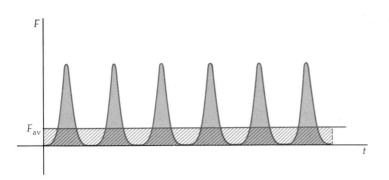

Equation 11-4 also gives the average recoil force exerted by the bullets on the gun. Each bullet receives an impulse of magnitude mv from the gun, and, by Newton's third law, each bullet exerts an impulse back on the gun of magnitude mv. Since in time Δt, $R \Delta t$ bullets leave the gun, the total impulse exerted on the gun in time Δt is $mvR \Delta t$ and the average recoil force is Rmv.

Questions

1. A ball is dropped from a second-story window, and an identical ball is simultaneously thrown up from the ground so that it just reaches the window. Compare the impulses of gravity on the two balls as they travel between the ground and the window.

2. Could two forces deliver the same impulse in a given time even though one reverses its direction during that time? Explain.

3. Can a force of large magnitude give a body a smaller impulse than a force of smaller magnitude? How?

4. How would the average of a force which fluctuates above and below a value F_0 compare with the average of a force that stays nearly steady at value F_0?

5. If a particle is under the influence of a single force which has an average value of zero during some time, must the acceleration of the particle be zero?

11-3 Pressure of a Gas

The force exerted by the bullets on the target in Example 11-3 is similar to the force exerted by gas molecules on the walls of a container. In the gas there are usually so many molecules that their individual impulses overlap in time and only the average force is observable. Another important difference is that the molecules do not stop when they hit the wall of the container but bounce back with their component of momentum perpendicular to the wall reversed. Let m be the mass of each molecule. If we take the x axis to be perpendicular to the wall, the x component of momentum of a molecule is $+mv_x$ before it hits the wall and $-mv_x$ afterward. The change in momentum during the collision with the wall is the final momentum minus the initial momentum:

$$\Delta p_x = -mv_x - (+mv_x) = -2mv_x$$

the minus sign indicating that the change in momentum is to the left, away from the wall. The magnitude of the momentum change is thus $2mv_x$. The total change in the momentum of all the molecules in some time interval Δt is $2mv_x$ times the number that hit the wall in this time interval. Let us consider a rectangular container of volume V with the right wall having area A (Figure 11-4). Let N_i be the number of gas molecules with x component of velocity v_{xi}. The number of these molecules hitting the right wall in time Δt is the number within a distance $v_{xi} \Delta t$ and traveling to the right. Since there are N_i such molecules in volume V, the number in volume $v_{xi} \Delta t\, A$ is just $(N_i/V)(v_{xi} \Delta t\, A)$. If we assume for the moment that v_{xi} is positive, the number that hit the right wall in time Δt is

$$\frac{N_i}{V} v_{xi} \Delta t\, A$$

The impulse exerted by the wall on these molecules equals the total change in momentum of these molecules, which is $2mv_{xi}$ times the number that hit:

$$I_i = \left(\frac{N_i v_{xi} \Delta t\, A}{V}\right) 2mv_{xi} = \frac{2N_i m v_{xi}^2 A\, \Delta t}{V}$$

This also equals the magnitude of the impulse exerted by these molecules *on* the wall. We obtain the average force exerted by these molecules by dividing the impulse by the time interval Δt. The pressure is this average force divided by the area A. The pressure exerted by these molecules is thus

$$P_i = \frac{I_i}{\Delta t\, A} = \frac{2N_i m v_{xi}^2}{V}$$

The total pressure exerted by all the molecules is obtained by summing over all the x components of velocity v_{xi} that are positive. Since on the average, at any time half the molecules will be moving to the right (positive v_{xi}) and half to the left (negative v_{xi}), we can sum over all the molecules and multiply by $\frac{1}{2}$.

$$P = \tfrac{1}{2}\Sigma P_i = \tfrac{1}{2}\Sigma \frac{2N_i m v_{xi}^2}{V} = \frac{m}{V}\, \Sigma N_i v_{xi}^2$$

We can write this in terms of the average value of v_x^2, defined to be

$$(v_x^2)_{av} = \frac{1}{N}\, \Sigma N_i v_{xi}^2$$

where $N = \Sigma N_i$ is the total number of molecules. Thus we can write for the pressure,

$$P = \frac{Nm}{V}\, (v_x^2)_{av} \qquad\qquad 11\text{-}5$$

If there is no preferred direction for the molecules to move, $(v_x^2)_{av}$ must be the same as $(v_y^2)_{av}$ and $(v_z^2)_{av}$. The square of the speed is

$$v^2 = v_x^2 + v_y^2 + v_z^2$$

Hence

$$(v^2)_{av} = (v_x^2)_{av} + (v_y^2)_{av} + (v_z^2)_{av} = 3(v_x^2)_{av}$$

Thus we can write the pressure in terms of the average square speed:

$$P = \frac{1}{3}\frac{N}{V}\, m(v^2)_{av} = \frac{2}{3}\frac{N}{V}\, (\tfrac{1}{2}mv^2)_{av} \qquad\qquad 11\text{-}6 \qquad\qquad \textit{Pressure of a gas}$$

showing that the pressure is proportional to the number of molecules per unit volume and to the average kinetic energy of the molecules.

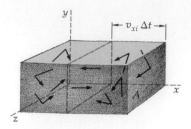

Figure 11-4
Gas molecules in a rectangular container. If a molecule has velocity component v_{xi}, it will hit the right face in the time interval Δt if it is within the distance $v_{xi} \Delta t$ of the face and if v_{xi} is positive. The number of molecules within this distance of the face is proportional to v_{xi} and to the number of molecules per unit volume.

Example 11-4 If 1 mole of a gas occupies 22.4 liters $(\ell) = 22.4 \times 10^{-3}$ m³ at standard temperature and pressure of 1 atmosphere (atm) $= 1.01 \times 10^5$ N/m², find the average square speed of oxygen molecules under standard conditions.

The number N of molecules in a mole times the mass m of each molecule is the *molecular weight*, which for oxygen gas is 32 gm/mole. According to Equation 11-6, the average square speed is

$$(v^2)_{av} = \frac{3PV}{Nm} = \frac{3(1.01 \times 10^5)(22.4 \times 10^{-3})}{32 \times 10^{-3}} = 2.12 \times 10^5 \text{ (m/sec)}^2$$

The square root of $(v^2)_{av}$ is called the root-mean-square (rms) speed v_{rms},

$$v_{rms} = \sqrt{(v^2)_{av}} = 460 \text{ m/sec}$$

This is slightly greater than the speed of sound in air (about 331 m/sec) but of the same order of magnitude.

11-4 The Impulse Approximation

According to Equation 11-3, the impulse for some time interval is related to the average force by

$$\mathbf{I} = \mathbf{F}_{av} \Delta t$$

In many collisions the average force exerted by one body on the other is very large, and the time of the collision is very small. We can then neglect other forces which act on the bodies and to a good approximation assume that the momentum of the two-body system is conserved during the collision. We can also neglect the motion of the bodies during the collision if Δt is very small. For example, the average force exerted by the golf club on the ball in Example 11-2 was 1800 N, much larger than the weight of the ball or frictional forces exerted by the grass or the tee on the ball. During the collision time of 0.001 sec in that example, the ball moved only 2 cm, which is a small distance compared with the total horizontal range of 160 m. Neglecting this motion, we assume that the ball is at rest during the collision and then projected with an initial speed $v = I/m$ (40 m/sec in that example). The approximation that F_{av} is very much larger than any other force and Δt is so small that there is essentially no motion during the collision is called the *impulse approximation*.

Example 11-5 For the 10-gm bullet moving with speed 400 m/sec and embedding itself in a 390-gm block initially at rest, estimate the collision time and check the validity of the impulse approximation, assuming that the bullet travels 5 cm into the block.

We first neglect the motion of the block. We estimate the collision time by assuming the acceleration to be constant. The easiest way to do this is to realize that the average velocity for constant acceleration is just the mean between the initial and final velocity. Since the bullet has initial velocity 400 m/sec and final velocity $v_{CM} = 10$ m/sec, as calculated in previous examples, its average velocity during the collision is 205 m/sec. To travel 5 cm $= 0.05$ m at 205 m/sec requires a time $\Delta t = (0.05 \text{ m})/(205 \text{ m/sec}) = 2.44 \times 10^{-4}$ sec. This is our estimate of the collision time. Since the block accelerates from rest to a final velocity of 10 m/sec, its average velocity is 5 m/sec during this time, assuming constant acceleration. It thus moves a distance of only (5 m/sec)(2.44 $\times$ 10^{-4} sec) = 1.22×10^{-3} m = 0.122 cm. In the impulse approximation, we neglect this small distance compared with the 5 cm moved by the bullet in the time of the collision. We could also estimate the distance moved by the block by viewing the collision in the center-of-mass reference frame. Since the block has 39 times the mass of the bullet, in this frame it travels with one-thirty-ninth the velocity. In a given time it thus travels $\frac{1}{39}$ times the distance traveled by the bullet. Thus if the bullet travels 5 cm, the block travels $\frac{1}{39}$(5 cm) = 0.128 cm. These two estimates differ slightly because in the first we assumed that the bullet traveled a total distance of 5 cm into the block while in the second we assumed that it traveled slightly farther, 5 + 0.128 cm.

Question

6. Name several situations in which the impulse approximation applies.

11-5 Energy Changes in Collisions; Reference Frames

In Chapter 10 we saw that the kinetic energy of a system of particles can be written as the sum of the kinetic energy associated with the motion of the center of mass and the kinetic energy of motion relative to the center of mass

$$E_k = \tfrac{1}{2}Mv_{CM}^2 + E_{kr}$$

where M is the total mass of the system, v_{CM} is the velocity of the center of mass, and $E_{kr} = \Sigma \tfrac{1}{2}m_i u_i^2$ is the energy of motion relative to the center of mass. As we have seen in this chapter, external forces are usually negligible compared with the internal forces the bodies exert on each other in a collision. Thus the total momentum of the two-body system is conserved, the velocity of the center of mass is constant, and the part of the kinetic energy associated with center-of-mass motion $\tfrac{1}{2}Mv_{CM}^2$ does not change. However, since the internal forces may be nonconservative, E_{kr} may change. The kinetic energy of center of mass motion $\tfrac{1}{2}Mv_{CM}^2$ is constant in any collision. If the kinetic energy of relative motion E_{kr} is also constant, the total kinetic energy remains unchanged in the collision and the collision is said to be *perfectly elastic*. On the other hand, if all the energy of relative motion is lost in the collision, the collision is said to be *perfectly inelastic*. Since there is no motion relative to the center of mass after a perfectly inelastic collision, the bodies move together after the collision with velocity v_{CM}. The bullet embedding itself in a block is a typical example of a perfectly inelastic collision. In many collisions only part of the kinetic-energy motion relative to the center of mass is lost; these collisions are neither perfectly elastic nor perfectly inelastic.

Many collisions are more easily analyzed in the center-of-mass, or zero-momentum, reference frame. For example, in an inelastic collision in this frame, the two bodies are at rest after the collision. As we shall see in the next section, in an elastic collision in one dimension, the velocities of the bodies are merely turned around in the center-of-mass frame. Although this frame may be more convenient for analysis, experiments are more often done in a reference frame in which one of the particles is initially at rest, called the *laboratory frame*. For example, in the famous alpha-particle scattering experiment from which Rutherford deduced the existence of the atomic nucleus, gold nuclei at rest in a thin gold foil were bombarded with alpha particles from a radioactive source. In a typical modern nuclear-physics experiment, nuclei at rest in a target in the laboratory are bombarded by particles accelerated in some type of nuclear accelerator. It is therefore important to know how to transform results obtained in the center-of-mass frame to the laboratory frame and vice versa. We accomplish these transformations using Equations 10-14. Let m_1 be the mass of the incident particle, $\mathbf{v}_1$ its velocity in the laboratory frame, and m_2 the mass of the stationary particle. The velocity of the center of mass in this frame is thus $\mathbf{v}_{CM} = m_1\mathbf{v}_1/(m_1 + m_2)$. The velocities of the particles in the center-of-mass frame are

Laboratory frame

$$\mathbf{u}_1 = \mathbf{v}_1 - \mathbf{v}_{CM} = \mathbf{v}_1 - \frac{m_1\mathbf{v}_1}{m_1 + m_2} = \frac{m_2\mathbf{v}_1}{m_1 + m_2}$$

11-7

$$\mathbf{u}_2 = -\mathbf{v}_{CM} = -\frac{m_1\mathbf{v}_1}{m_1 + m_2}$$

On the other hand, if we know the velocities $\mathbf{u}_1$ and $\mathbf{u}_2$ in the center-of-mass frame, we transform to the laboratory frame by adding $\mathbf{v}_{CM} = -\mathbf{u}_2$ to each velocity so that particle 2 is at rest.

11-6   Perfectly Elastic Collisions in One Dimension

Let us consider the collision in the center-of-mass reference frame first. We assume that the bodies are moving along the x axis. Let $\mathbf{u}_{1i}$ and $\mathbf{u}_{2i}$ be the initial velocities of the bodies. Their initial momenta are then $\mathbf{p}_{1i} = m_1\mathbf{u}_{1i}$ and $\mathbf{p}_{2i} = m_2\mathbf{u}_{2i}$. (We are using vector notation even though we are considering only one-dimensional collisions here in order to keep track of the signs. Of course if $\mathbf{u}_{1i}$ is to the right, $\mathbf{u}_{2i}$ must be to the left because the total momentum is zero in this frame.) Since the total momentum is zero in this frame, $\mathbf{p}_{1i} = -\mathbf{p}_{2i}$, and the magnitudes p_{1i} and p_{2i} are equal. The total kinetic energy E_{ri} is

$$E_{ri} = \tfrac{1}{2}m_1u_{1i}{}^2 + \tfrac{1}{2}m_2u_{2i}{}^2 = \frac{p_{1i}{}^2}{2m_1} + \frac{p_{2i}{}^2}{2m_2} = p_{1i}{}^2\left(\frac{1}{2m_1} + \frac{1}{2m_2}\right)$$

11-8

After the collision the energy is

$$E_{rf} = \frac{p_{1f}{}^2}{2m_1} + \frac{p_{2f}{}^2}{2m_2} = p_{1f}{}^2\left(\frac{1}{2m_1} + \frac{1}{2m_2}\right)$$

11-9

where p_{1f} and p_{2f} are the magnitudes of the final momenta of the bodies. We have used the fact that the total momentum after the collision is still zero, and hence the two masses still have momenta of equal magnitude though opposite direction. This is true to the extent that we can neglect any external force on the system during the collision. Setting the final energy equal to the initial energy for a perfectly elastic collision, we obtain

$$p_{1f}{}^2 = p_{1i}{}^2$$

11-10

or

$$\mathbf{p}_{1f} = \pm\mathbf{p}_{1i} \quad \text{and} \quad \mathbf{p}_{2f} = \pm\mathbf{p}_{2i}$$

11-11

The plus sign in Equation 11-11 corresponds to no collision at all, as would be the case if the masses were initially heading away from each other. Since only conservation of momentum and conservation of energy were used to derive Equation 11-11, it is not surprising that the result includes the case of no collision. If the bodies are originally heading toward each other and collide, Equation 11-11 gives for each body

$$\mathbf{p}_{1f} = -\mathbf{p}_{1i} \qquad \mathbf{p}_{2f} = -\mathbf{p}_{2i} \qquad\qquad 11\text{-}12$$

Since the velocity of each body is just its momentum divided by its mass, Equation 11-12 also implies that the final velocity of each body is related to its initial velocity by

$$\mathbf{u}_{1f} = -\mathbf{u}_{1i} \quad \text{and} \quad \mathbf{u}_{2f} = -\mathbf{u}_{2i} \qquad\qquad 11\text{-}13$$

The one-dimensional perfectly elastic collision is very simple in the zero-momentum reference frame. Each body is merely turned around and leaves with the speed and energy it had before the collision.

In other reference frames, the result does not look as simple. In general, even though the total kinetic energy is conserved, some or all of the kinetic energy of one body is transferred to the other body. In the familiar example of one body colliding with another of equal mass originally at rest, the first body comes to rest, transferring all its kinetic energy to the second. This result can easily be shown by first considering the collision of equal masses in the zero-momentum frame and then transforming to the laboratory frame, as illustrated in Figure 11-5.

We often want to calculate the energy transferred in an elastic collision from one body to the other in a reference frame different from the zero-momentum frame, e.g., the laboratory frame. It is always possible to do so by writing down the equations for conservation of energy and momentum and solving them for the two unknown final velocities in terms of the known masses and initial velocities. However, this procedure usually results in cumbersome algebra because the equation for conservation of kinetic energy is quadratic in the speeds. A much simpler procedure is to use the conservation-of-momentum equation and a result which we now state and prove:

The velocity of one body relative to the other is merely reversed in a one-dimensional elastic collision in all reference frames.

Elastic collisions

Another way of stating this theorem is that the relative speed of recession after the collision equals the relative speed of approach before the

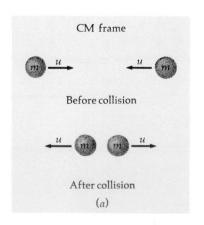

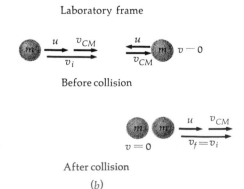

Figure 11-5
Perfectly elastic collision of equal masses in (a) the center-of-mass reference frame and (b) the laboratory reference frame. In the laboratory frame all the kinetic energy of the moving body is transferred to the initially stationary body. This occurs only when the masses are equal.

collision. This is easily proved by first considering the velocities in the zero-momentum frame. This result follows directly from Equation 11-13 in the center-of-mass frame. The velocity of one body *relative to the other* is $\mathbf{u}_2 - \mathbf{u}_1$ (or $\mathbf{u}_1 - \mathbf{u}_2$). According to Equation 11-13,

$$\mathbf{u}_{2f} - \mathbf{u}_{1f} = -\mathbf{u}_{2i} - (-\mathbf{u}_{1i}) = -(\mathbf{u}_{2i} - \mathbf{u}_{1i}) \qquad 11\text{-}14a$$

This is the theorem stated in the center-of-mass frame. We now consider another reference frame in which the center of mass moves with velocity $\mathbf{v}_{CM}$. According to the transformation Equations 10-14, a velocity $\mathbf{v}_{1i}$ in this frame is obtained by adding $\mathbf{v}_{CM}$ to the velocity $\mathbf{u}_{1i}$ in the center-of-mass frame. Thus the initial and final velocities in this frame are

$$\mathbf{v}_{1i} = \mathbf{u}_{1i} + \mathbf{v}_{CM} \qquad \mathbf{v}_{2i} = \mathbf{u}_{2i} + \mathbf{v}_{CM}$$

$$\mathbf{v}_{1f} = \mathbf{u}_{1f} + \mathbf{v}_{CM} \qquad \mathbf{v}_{2f} = \mathbf{u}_{2f} + \mathbf{v}_{CM}$$

When we add $\mathbf{v}_{CM}$ to each term in Equation 11-14a, the equation becomes

$$\mathbf{v}_{2f} - \mathbf{v}_{1f} = -(\mathbf{v}_{2i} - \mathbf{v}_{1i}) \qquad 11\text{-}14b$$

Our stated result holds for *any* reference frame moving with constant velocity $\mathbf{v}_{CM}$ relative to the center-of-mass frame.

Example 11-6 A 2-kg block slides with a speed of 10 m/sec along a smooth table and collides with an 8-kg block initially at rest. What are the final velocities of the blocks if the collision is elastic?

We shall illustrate two methods of solution. In the first, we use Equation 11-14 and work in the given frame, which is the laboratory frame. Let the moving block be body 1 and the stationary one block 2. Since we are working in one dimension, we can omit the vector notation and indicate the direction of the velocities with a positive or negative sign. The relative velocity before the collision is $v_{2i} - v_{1i} = 0 - 10 = -10$ m/sec. After the collision we have

$$v_{2f} - v_{1f} = -(-10) = +10 \text{ m/sec}$$

Conservation of momentum gives us a second relation between the final velocities. The total momentum before the collision is (2 kg)(10 m/sec) = 20 kg-m/sec. Thus

$$m_1 v_{1f} + m_2 v_{2f} = 20 \text{ kg m/sec}$$

$$2v_{1f} + 8v_{2f} = 20 \text{ m/sec}$$

Solving these two equations for the two unknown final velocities gives

$$v_{2f} = 4 \text{ m/sec} \qquad \text{and} \qquad v_{1f} = -6 \text{ m/sec}$$

It is left as an exercise to check that this solution agrees with energy conservation.

We now work this problem by first transforming into the center-of-mass reference frame (Figure 11-6). The velocity of the center of mass in the original frame is

$$v_{CM} = \frac{m_1 v_1}{m_1 + m_2} = \frac{(2 \text{ kg})(10 \text{ m/sec})}{10 \text{ kg}} = 2 \text{ m/sec}$$

We transform to the center-of-mass frame by subtracting this velocity from that of each body, as illustrated in Figure 11-6b.

$$u_{1i} = 10 - 2 = 8 \text{ m/sec}$$

$$u_{2i} = 0 - 2 = -2 \text{ m/sec}$$

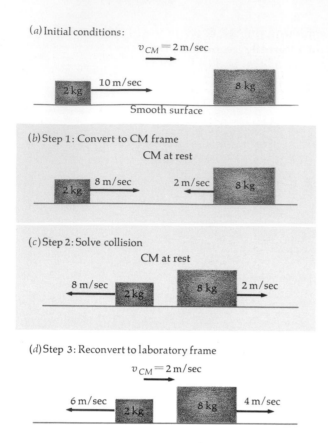

(a) Initial conditions:

$v_{CM} = 2$ m/sec

2 kg   10 m/sec   8 kg

Smooth surface

(b) Step 1: Convert to CM frame

CM at rest

2 kg   8 m/sec   2 m/sec   8 kg

(c) Step 2: Solve collision

CM at rest

8 m/sec   2 kg   8 kg   2 m/sec

(d) Step 3: Reconvert to laboratory frame

$v_{CM} = 2$ m/sec

6 m/sec   2 kg   8 kg   4 m/sec

Figure 11-6
Calculation of final velocities after an elastic collision (Example 11-6). (a) In the laboratory frame one body is initially at rest, and the center of mass moves to the right at 2 m/sec. (b) Transformation to the center-of-mass frame is accomplished by subtracting v_{CM} from each body. (c) In the center-of-mass frame the bodies are reversed by the elastic collision. (d) Transformation back to the laboratory frame is accomplished by adding v_{CM} to each body.

It is easily checked that the total momentum is zero in this frame. Since in this frame the velocity of each body is just turned around, we have

$$u_{1f} = -8 \text{ m/sec} \quad \text{and} \quad u_{2f} = 2 \text{ m/sec}$$

We now transform back into the original laboratory frame by adding the velocity $v_{CM} = 2$ m/sec to that of each body:

$$v_{1f} = -8 \text{ m/sec} + 2 \text{ m/sec} = -6 \text{ m/sec}$$

$$v_{2f} = 2 \text{ m/sec} + 2 \text{ m/sec} = 4 \text{ m/sec}$$

Example 11-7 A very large mass m_1 moving with speed v_{1i} collides elastically with a very small mass m_2 initially at rest (Figure 11-7). What is the speed of the small mass after the collision assuming that m_1 is much greater than m_2?

It should be intuitively clear that the large mass will not be affected very much by a collision with a very small mass. A cannonball will hardly be slowed down if it collides with a stationary piece of shot. Before the collision the relative velocity of approach is v_{1i}. Then after the collision the velocity of separation must be v_{1i}. For a first approximation we neglect any change in the velocity of m_1. Since it continues

Figure 11-7
A very large mass m_1 moving with velocity v_{1i} collides elastically with a stationary small mass m_2. Since the velocity of recession after the collision equals the velocity of approach before the collision and the velocity of the large mass is approximately unchanged, the final velocity of the small mass must be approximately $2v_{1i}$.

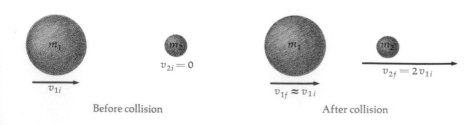

m_1 m_2 m_1 m_2

$v_{2i} = 0$

v_{1i} $v_{1f} \approx v_{1i}$ $v_{2f} = 2v_{1i}$

Before collision   After collision

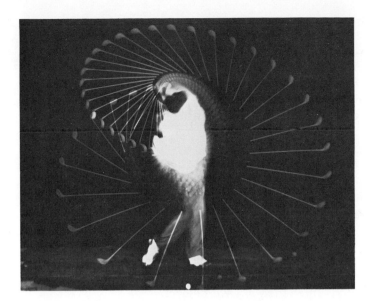

Figure 11-8
Multiflash photograph of a
golf club hitting a ball.
Comparison of the distances
traveled by the club and ball
between successive flashes
shows that the speed of the
ball after the collision is
approximately twice that of
the club. (*Courtesy of Harold
E. Edgerton.*)

to move with velocity v_{1i}, the velocity of the small mass m_2 must be $2v_{1i}$.

An example of such a collision is that of a golf club with the ball. The effective mass of the club is large because it is held by the player, who is in firm contact with the ground. Careful measurement of the distance between successive flashes in Figure 11-8 shows that the initial velocity of the ball is indeed twice that of the club.

The energy transferred is just the energy of the small mass after the collision, $\frac{1}{2}m_2(2v_{1i})^2 = 2m_2v_{1i}^2$. We can use the conservation of momentum to get a correction to our first approximation for the final velocity of the large mass. Since the small mass gains momentum $2m_2v_{1i}$ to the right, the large mass must lose this much momentum. If v_{1f} is the final velocity of the large mass, we have $m_1v_{1f} = m_1v_{1i} - 2m_2v_{1i}$, or $v_{1f} \approx v_{1i} - (2m_2/m_1)v_{1i}$.

We now calculate the energy transferred from the moving body to the stationary one for a general elastic collision in the laboratory frame. Let mass m_1 have initial velocity v_{1i} toward mass m_2, which is originally stationary. After the collision the bodies move with velocities v_{1f} and v_{2f}. The final energy of m_2 is the energy transferred since it is originally at rest. We thus wish to find v_{2f} in terms of v_{1i}. We can use either of the methods illustrated in Example 11-6. The second method, in which we transfer into and out of the center-of-mass frame, is probably the easier. The velocity of the center of mass in the laboratory frame is

$$v_{CM} = \frac{m_1v_{1i}}{m_1 + m_2}$$

Again, we transform to the center-of-mass frame by subtracting v_{CM} from each velocity in the laboratory frame. In the center-of-mass frame the mass m_2 moves to the left with this velocity before the collision. After the collision in this frame m_2 moves to the right with velocity v_{CM}. Thus in the *laboratory frame*, m_2 moves with velocity $2v_{CM}$ after the collision,

$$v_{2f} = 2v_{CM} = \frac{2m_1v_{1i}}{m_1 + m_2}$$

The kinetic energy of m_2 after the collision in this frame is thus

$$\tfrac{1}{2}m_2 v_{2f}^2 = \tfrac{1}{2}m_2 \frac{4m_1^2 v_{1i}^2}{(m_1 + m_2)^2} = \frac{4m_1 m_2}{(m_1 + m_2)^2}\tfrac{1}{2}m_1 v_{1i}^2 \qquad 11\text{-}15$$

The energy transferred is proportional to the original energy of the incident body. The energy gained by particle 2 equals that lost by particle 1. If we call the energy transferred ΔE, the fractional energy loss of the incident particle is

$$\frac{\Delta E}{E} = \frac{4m_1 m_2}{(m_1 + m_2)^2} \qquad 11\text{-}16$$

An elastic collision of two equal cannonballs from Johann Marcus Marci, *De Proportione Motus*, 1639. All of the kinetic energy of cannonball *a* is transferred to cannonball *b*. (*Courtesy of the Deutsches Museum, Munich.*)

The fractional energy transferred has a maximum value of 1 if the masses are equal, as previously discussed. Then the incident particle is at rest after the collision. If the masses of the particles are very different, the fractional energy transfer is small. In the limit of one mass much larger than the other the fractional energy transfer approaches 4 times the ratio of the small mass to the large one, as can be seen from Equation 11-16. For example if m_1 is much larger than m_2, we neglect m_2 in the denominator and obtain $\Delta E/E \approx 4m_1 m_2/m_1^2 = 4m_2/m_1$.

The result that the energy transfer in an elastic collision is large only if the masses of the two bodies are about the same has an important application in slowing down neutrons in a nuclear reactor. The probability of inducing fission in ^{235}U is large only if the neutron is moving very slowly, whereas the neutrons emitted in the fission are fast-moving. In order to sustain a chain reaction, these fast neutrons must be slowed down before they have a chance to leak out of the reactor. The main process for slowing neutrons down is elastic collisions. From our calculations we have seen that efficient transfer of neutron energy by elastic collision can occur only in collisions with particles having masses near that of the neutron. Neither electrons in the reactor, with a mass about $\frac{1}{2000}$ times that of a neutron, nor the nuclei of nonfissionable ^{238}U (which makes up about 99 percent of natural uranium), with a mass about 238 times the mass of a neutron, can efficiently absorb energy from neutrons in collisions. Thus some material containing light nuclei is placed in a reactor to slow the neutrons down. This material, called a *moderator,* is usually water (containing hydrogen and oxygen nuclei) or carbon.

11-7   Elastic Collisions in Three Dimensions

In the zero-momentum frame, Equations 11-8 to 11-10 also hold for three-dimensional collisions, but Equation 11-10 in terms of rectangular coordinates is now

$$p_{fx}^2 + p_{fy}^2 + p_{fz}^2 = p_{ix}^2 + p_{iy}^2 + p_{iz}^2$$

for each body. The *magnitude* of the momentum and the speed of each body are not changed by the collision, but the directions are not merely reversed, as in the one-dimensional case. Of course the final momenta of the bodies must be antiparallel to each other after the collision because the total momentum before and after the collision is zero in this frame. Figure 11-9 shows an elastic collision of spheres in the zero-momentum frame. The angle θ made by the final and initial velocity of each body depends on the perpendicular distance between

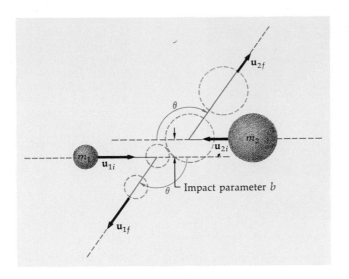

Figure 11-9
Off-center elastic collision of two spheres in the zero-momentum reference frame. The perpendicular distance between the original lines of motion of the centers is called the impact parameter b. Energy and momentum conservation require that the speed of each body be unchanged by the collision.

the lines of the initial velocities (Figure 11-9). This distance is called the *impact parameter b.* If b is zero, the collision is head on, the motion of both bodies is always along a line, and the collision is one-dimensional. If b is greater than the sum of the two radii of the spheres, there is no collision. If b is just smaller than this sum, a glancing collision results and the deflection of each body is small.

Impact parameter

In order to transform into the laboratory frame (Figure 11-10), we must add the velocity vector $\mathbf{v}_{CM} = -\mathbf{u}_{2i}$ to each body's velocity before and after the collision. In this frame body 1 moves with velocity $\mathbf{v}_{1i}$ a distance b from the parallel line through the center of body 2 while body 2 is stationary. The angle of deflection of body 1 is not the same in the laboratory frame as in the zero-momentum frame. (Since body 2 is stationary before the collision, we do not speak of its deflection in the laboratory frame.)

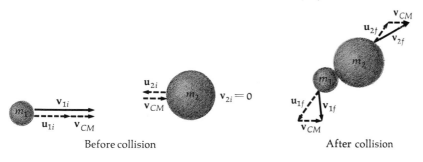

Before collision   After collision

Figure 11-10
Collision of Figure 11-9 viewed in the laboratory reference frame. In this frame one body is initially at rest, and the center of mass moves with velocity $\mathbf{v}_{CM} = -\mathbf{u}_{2i}$. The velocity vectors in this frame are obtained by adding $\mathbf{v}_{CM}$ to the velocity vectors in the zero-momentum frame.

We omit further complications of collisions in three dimensions except for the special case of equal masses, which we consider in the laboratory frame. Figure 11-11 shows the geometry of the collision. If $\mathbf{p}_{1i}$ and $\mathbf{p}_{1f}$ are the initial and final momentum vectors of body 1, the final

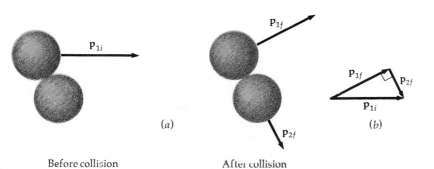

(a)   (b)

Before collision   After collision

Figure 11-11
Off-center elastic collision of two equal spheres in the laboratory reference frame. (a) After the collision, the spheres move at right angles to each other. (b) The momentum vectors for this collision form a right triangle.

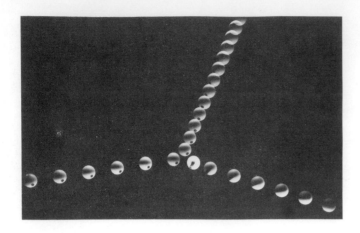

Multiflash photograph of an elastic off-center collision of two balls of equal mass. The dotted ball is incident from the left and strikes the striped ball initially at rest. The final velocities of the two balls are perpendicular to each other. (*From* PSSC Physics, *2d ed., p. 384, D. C. Heath and Company, Lexington, Mass., 1965.*)

momentum of body 2 is given by the conservation of momentum,

$$\mathbf{p}_{2f} = \mathbf{p}_{1i} - \mathbf{p}_{1f} \qquad \text{or} \qquad \mathbf{p}_{1f} + \mathbf{p}_{2f} = \mathbf{p}_{1i} \qquad \text{11-17}$$

Thus final momentum vectors add to form the triangle shown in the figure. Conservation of energy for this collision gives

$$\frac{p_{2f}^2}{2m} + \frac{p_{1f}^2}{2m} = \frac{p_{1i}^2}{2m} \qquad \text{or} \qquad p_{2f}^2 + p_{1f}^2 = p_{1i}^2 \qquad \text{11-18}$$

where we have canceled out the masses because they are equal. Equation 11-18 is just the pythagorean theorem for a right triangle formed by the vectors $\mathbf{p}_{1f}$, $\mathbf{p}_{2f}$, and $\mathbf{p}_{1i}$ with $\mathbf{p}_{1i}$ as the hypotenuse of the triangle. Thus $\mathbf{p}_{1f}$ and $\mathbf{p}_{2f}$ are perpendicular, and so are the final velocities $\mathbf{v}_{1f}$ and $\mathbf{v}_{2f}$. We see that for the special case of an elastic collision of equal masses in the laboratory frame, the final velocities of the bodies are perpendicular to each other.

11-8   Perfectly Inelastic Collisions

In a perfectly inelastic collision, all the energy of motion relative to the center of mass is lost. Thus in the zero-momentum frame, the bodies are at rest together with no kinetic energy. In the laboratory frame or any other reference frame, the bodies move together with the velocity of the center of mass in this frame. We have already considered a typical inelastic collision, a bullet embedding itself in a block. As we have seen, the *amount* of energy lost is the same in all reference frames. However, we are often interested in the *fraction* of the energy in the laboratory frame that is lost. We shall calculate this for the general case of a particle of mass m_1 with initial speed v_{1i} colliding inelastically with a particle of mass m_2 initially at rest.

The initial energy is

$$E_i = \tfrac{1}{2}m_1 v_{1i}^2 = \frac{p_{1i}^2}{2m_1} \qquad \text{11-19}$$

where $p_{1i} = m_1 v_{1i}$ is the initial momentum of m_1. The latter form, $p_{1i}^2/2m_1$, is more convenient in this case because the final momentum is the same as the initial momentum. After the collision, both bodies move together with momentum $p_f = p_{1i}$. Thus the final energy is

$$E_f = \frac{p_f^2}{2(m_1 + m_2)} = \frac{p_{1i}^2}{2(m_1 + m_2)} \qquad \text{11-20}$$

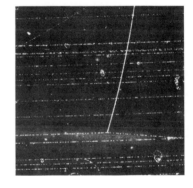

Proton-proton collision in a liquid-hydrogen bubble chamber. The incident proton enters from the left and interacts with a stationary proton in the chamber. The two particles move off at right angles after the collision. The slight curvature of the tracks is due to a magnetic field. (*Courtesy of Brookhaven National Laboratory.*)

The ratio of the final energy to the initial energy is just $m_1/(m_1 + m_2)$. Thus

$$E_f = \frac{m_1}{m_1 + m_2} E_i \tag{11-21}$$

and

$$E_i - E_f = \left(1 - \frac{m_1}{m_1 + m_2}\right) E_i = \frac{m_2}{m_1 + m_2} E_i$$

If we call the energy loss $-\Delta E$, we have

$$-\frac{\Delta E}{E_i} = \frac{m_2}{m_1 + m_2} \tag{11-22}$$

Perfectly inelastic collisions are often used to measure the initial velocity v_{1i} when it is very large and difficult to measure directly, e.g., that of a rifle bullet. The bullet is fired into a large block which is suspended in an arrangement called a *ballistic pendulum* (Figure 11-12). In the impulse approximation, the block remains at rest during the collision and then moves with velocity v_f, which is related to v_{1i} by the conservation of momentum,

$$v_f = \frac{m_1}{m_1 + m_2} v_{1i}$$

After the collision energy is conserved, and the block plus bullet swing to a height h given by

$$(m_1 + m_2)gh = \tfrac{1}{2}(m_1 + m_2)v_f^2 = \frac{m_1^2 v_{1i}^2}{2(m_1 + m_2)}$$

or

$$v_{1i} = \frac{m_1 + m_2}{m_1} \sqrt{2gh} \tag{11-23}$$

By measuring h and the masses we can obtain v_{1i}.

This collision, which will probably be inelastic, is observed by Laurel in the laboratory frame.

11-9   Reaction Threshold

When two atoms or molecules collide and the objects emerging from the collision are different from the original ones, the collision is called a *reaction*. Atomic and molecular reactions are characterized by an exchange between kinetic energy and internal potential energy, and energy is sometimes radiated away from the system.

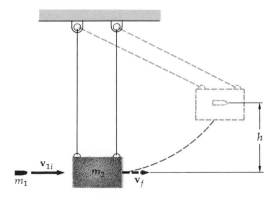

Figure 11-12
Ballistic pendulum. The bullet m_1 is stopped in the large block m_2, which swings up through a height h. The initial speed of the bullet is determined from measurement of h and the masses m_1 and m_2.

In this section we consider certain reactions as inelastic collisions. Consider the reaction in which a hydrogen atom is ionized by being struck by a proton, written

$$p + {}^1\text{H} \rightarrow p + p + e + Q \qquad\qquad 11\text{-}24$$

where Q is the excess kinetic energy in the zero-momentum reference frame of the final particles (those on the right) over the initial particles (those on the left). If Q is positive, the reaction is called *exothermic;* energy is given off, and the reaction may occur even if there is no initial kinetic energy (in the zero-momentum frame). If Q is negative, initial kinetic energy of the particles on the left side of the reaction is needed to make the reaction occur and the reaction is *endothermic.* For the reaction in Equation 11-24 $Q = -13.6$ eV: it takes 13.6 eV to remove the electron from the hydrogen atom.

In the zero-momentum frame the proton and ^{1}H atom approach each other with equal and opposite momentum. If the total kinetic energy is less than 13.6 eV, the ^{1}H atom cannot be ionized. If the total energy is greater than 13.6 eV, the atom can be ionized and the excess energy appears as kinetic energy of the final particles (two protons and one electron). If the energy is just 13.6 eV, the atom can be ionized with the particles relatively at rest after the reaction. Thus the minimum energy needed to produce this reaction in the zero-momentum reference frame is 13.6 eV. In the laboratory frame, where the hydrogen atom is initially at rest, the minimum kinetic energy of the incoming proton must be greater than 13.6 eV because not all of the original kinetic energy of the proton can be lost in ionizing the hydrogen atom. As we have seen, only the kinetic energy of motion relative to the center of mass can be lost. The minimum kinetic energy needed to produce the reaction in the laboratory frame is called the *reaction-threshold energy.* The reaction cannot take place in the laboratory frame at energies below the reaction threshold. We can use our results for perfectly inelastic collisions to calculate the reaction threshold, since at threshold the reaction products have no kinetic energy of relative motion but move together like the particles in a perfectly inelastic collision. Let the incoming object have mass m_1 and the stationary one have mass m_2. The energy loss in a perfectly inelastic collision is given by Equation 11-22,

$$-\Delta E = \frac{m_2}{m_1 + m_2} E_i$$

where E_i is the initial energy. In our case the initial energy E_i is the threshold energy E_{th}. The energy loss equals the magnitude of Q. Thus

$$\frac{m_2}{m_1 + m_2} E_{th} = |Q|$$

$$E_{th} = \frac{m_1 + m_2}{m_2} |Q| \qquad \text{for } Q \text{ negative} \qquad 11\text{-}25$$

Of course if Q is positive, there is no threshold energy.

Example 11-8 Reaction of Equation 11-24 Since the mass of the electron is $\frac{1}{1836}$ times that of the proton, the masses of a hydrogen atom and a proton are approximately equal: $m_1 \approx m_2$. Using $Q = -13.6$ eV gives

$$E_{th} = \frac{m + m}{m} 13.6 \text{ eV} = 2 \times 13.6 \text{ eV} = 27.2 \text{ eV}$$

In the laboratory frame, only half the initial kinetic energy can be used to ionize the hydrogen atom because half the energy is associated with center-of-mass motion, which cannot change. If the initial velocity of the proton is v_{1i} the velocity of the center of mass is $v_{CM} = \frac{1}{2} v_{1i}$ and the center-of-mass energy is

$$\tfrac{1}{2}(2m)v_{CM}^2 = \tfrac{1}{2}(2m)(\tfrac{1}{2}v_{1i})^2 = \tfrac{1}{2}(\tfrac{1}{2}mv_{1i}^2)$$

Thus in order for the proton to lose 13.6 eV to ionize the hydrogen atom, it must have an initial energy of 27.2 eV in the laboratory frame. The threshold energy is thus 27.2 eV.

Example 11-9 A hydrogen atom at rest in the laboratory is ionized by bombarding it with electrons. What is the threshold energy?

For this case, m_1 is very small compared with m_2. We have

$$E_{th} = \frac{m_1 + m_2}{m_2} |Q| = \left(1 + \frac{m_1}{m_2}\right)|Q| = (1 + \tfrac{1}{1836})|Q|$$

We can show in another way that the threshold energy is just slightly greater than 13.6 eV. Because the mass of the hydrogen atom is so much greater than that of the electron, the center of mass is practically at the hydrogen atom and moves with the very low velocity of

$$v_{CM} \approx \frac{m_e}{m_p} v_{1i}$$

where m_e is the electron mass, m_p the proton mass, and v_{1i} the initial velocity of the electron. Since v_{CM} is so small, the laboratory frame and the center-of-mass frame are almost the same; thus nearly all the initial kinetic energy is available for the reaction.

Example 11-10 What is the threshold in the laboratory for producing the nuclear reaction

$$n + {}^3\text{He} \rightarrow {}^2\text{H} + {}^2\text{H} + Q$$

by bombarding stationary ^{3}He atoms with neutrons?

The Q value of this reaction is $Q = -3.27$ MeV. (The inverse reaction, the combination of two deuterons to produce ^{3}He plus a neutron, ${}^2\text{H} + {}^2\text{H} \rightarrow {}^3\text{He} + n + 3.27$ MeV, is an important fusion reaction which produces energy.) Since the mass of ^{3}He is about 3 times that of the neutron, the quantity $(m_1 + m_2)/m_2$ in Equation 11-25 is $\frac{4}{3}$ and the threshold energy is

$$E_{th} = \frac{m_1 + m_2}{m_2}|Q| = \tfrac{4}{3}(3.27 \text{ MeV}) = 4.36 \text{ MeV}$$

In this case only one-fourth the energy of the neutron is wasted in center-of-mass motion. Again, if the particle initially at rest in the laboratory is much more massive than the bombarding particle, the laboratory frame is nearly the same as the center-of-mass frame and nearly all the initial energy is available for the reaction.

Question

7. What would be the threshold for the reaction in Example 11-10 if stationary neutrons could be bombarded by ^{3}He atoms?

11-10 Coefficient of Restitution

In general, collisions of macroscopic bodies are neither perfectly elastic nor perfectly inelastic. Some fraction of the initial energy of motion relative to the center of mass is lost. Consider a general collision in the zero-momentum frame. Figure 11-13 shows the force exerted by either body on the other. The total area under the curve is the impulse, which equals the total change in momentum of either particle. The time t_0 is the time at which the bodies are at rest. Before t_0, the bodies are moving toward each other and deforming each other. The maximum deformation of the bodies occurs at t_0. After t_0, the bodies move away from each other. It is convenient to break up the impulse into two parts, the impulse of deformation I_D, defined as the area under the curve before time t_0, and the impulse of restoration I_R, defined as the area after time t_0. Let p_i and p_f be the magnitudes of the initial and final momenta of either particle. (Since we are in the center-of-mass frame, the particles have equal and opposite momenta at all times.) The magnitude of the impulse of deformation equals p_i since each particle is brought to rest. Similarly the magnitude of the impulse of restoration equals p_f. Thus if t_i is any time before the force occurs and t_f is any time after the force has decreased to zero, the impulse of deformation is

$$I_D = \int_{t_i}^{t_0} F\, dt = p_i \qquad\qquad 11\text{-}26$$

and the impulse of restoration is

$$I_R = \int_{t_0}^{t_f} F\, dt = p_f \qquad\qquad 11\text{-}27$$

The ratio of the magnitudes of these impulses is called the *coefficient of restitution*:

$$\epsilon = \frac{I_R}{I_D} = \frac{p_f}{p_i} \qquad\qquad 11\text{-}28$$

Coefficient of restitution

This coefficient is a measure of the degree to which the bodies return to their original shape after the collision. If the collision is perfectly elastic, $I_R = I_D$ and $\epsilon = 1$. The work done in deforming the bodies from $t = t_i$ to $t = t_0$ is momentarily stored as potential energy and returned to the kinetic energy of the bodies, almost as if a spring between the bodies were expanding to its original length after being compressed. If

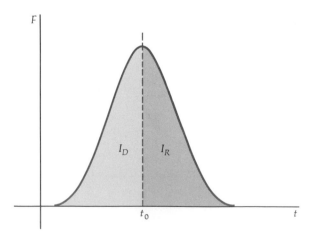

Figure 11-13
Force versus time for collision of two bodies showing impulse of deformation I_D and impulse of restoration I_R. The time t_0 is the instant at which there is no relative motion of the bodies. Before t_0 the bodies move toward each other, and deformation occurs. After t_0 they move away from each other. In a perfectly elastic collision $I_R = I_D$. In a perfectly inelastic collision $I_R = 0$.

the collision is perfectly inelastic, $I_R = 0$, $\epsilon = 0$, and the maximum deformation is permanent. The bodies remain at rest after the collision (in the zero-momentum frame). The original kinetic energy has been lost to deformation and radiated energy. The coefficient of restitution and the magnitudes of the initial and final moments of either particle are related by

$$p_f = \epsilon p_i \qquad\qquad 11\text{-}29$$

The final energy of the system is related to the initial energy by

$$E_{rf} = \frac{p_f^2}{2m_1} + \frac{p_f^2}{2m_2} = \epsilon^2 \left(\frac{p_i^2}{2m_1} + \frac{p_i^2}{2m_2} \right)$$

or

$$E_{rf} = \epsilon^2 E_{ri}$$

The fractional energy loss in the zero momentum frame is thus

$$\frac{-(E_{rf} - E_{ri})}{E_{ri}} = \frac{-\Delta E_{kr}}{E_{ri}} = 1 - \epsilon^2 \qquad\qquad 11\text{-}30$$

Equation 11-30 gives the fractional loss of the energy of motion relative to the center of mass E_{kr}, not the fractional loss in total kinetic energy in any other system. As with perfectly elastic collisions, it is convenient to work with the velocity of one body relative to the other in treating problems in other reference frames. From Equation 11-29 the speed of either particle after the collision is ϵ times its initial speed. Since the direction of each particle is reversed by the collision, the final velocities of the particles in the zero-momentum frame are related to their initial velocities by

$$u_{1f} = -\epsilon u_{1i} \qquad \text{and} \qquad u_{2f} = -\epsilon u_{2i}$$

Thus

$$u_{2f} - u_{1f} = -\epsilon(u_{2i} - u_{1i})$$

In another reference frame moving with velocity $-v_{CM}$ relative to the zero-momentum frame, we obtain the velocities of the particles by adding v_{CM}; thus

$$v_{2f} - v_{1f} = -\epsilon(v_{2i} - v_{1i}) \qquad\qquad 11\text{-}31$$

holds in all reference frames.

Example 11-11 The coefficient of restitution for steel on steel is measured by dropping a steel ball on a steel plate rigidly attached to the earth. If the ball is dropped from height h_i and rebounds to height h_f, what is the coefficient of restitution?

The speed of the ball just before it strikes the plate is

$$v_i = \sqrt{2gh_i}$$

For the ball to reach height h_f, the speed just after the collision must be $v_f = \sqrt{2gh_f}$. Since the plate is attached to the earth, we can neglect its rebound velocity. The relative speeds in this case are thus v_i and v_f

$$v_f = \epsilon v_i$$

$$\sqrt{2gh_f} = \epsilon \sqrt{2gh_i}$$

$$\epsilon = \sqrt{\frac{h_f}{h_i}}$$

Review

A. Define, explain, or otherwise identify:

B. True or false. If the statement is true only under certain conditions, write false and state the conditions under which it is true.

1. Only forces of very short duration exert impulses.

2. In a perfectly inelastic collision, all the kinetic energy of the particles is lost.

3. In a perfectly elastic collision, the energy of each particle is the same before and after the collision.

4. In a perfectly elastic collision, the relative velocity of recession after the collision equals the relative velocity of approach before the collision.

Exercises

Section 11-1, Impulse

1. Figure 11-14 shows the time variation of a force F_x acting on a 2-kg particle initially at rest. Find (*a*) the impulse of this force and (*b*) the final velocity of the mass.

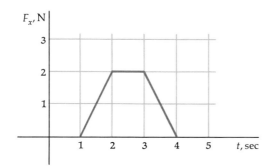

Figure 11-14
Exercises 1 and 6.

2. Figure 11-15 shows the time variation of a force which acts on a 3-kg particle. (*a*) Find the impulse of the force. (*b*) If the particle is initially moving in the *x* direction with velocity $v_x = 2$ m/sec, what is its velocity after the impulse?

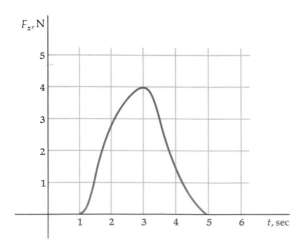

Figure 11-15
Exercises 2 and 7.

3. A 0.4-kg particle initially moving at 20 m/sec is stopped by a force which has the constant magnitude of 50 N for a short time Δt. (a) What is the impulse of the force? (b) Find the time interval Δt.

4. Find the impulse necessary to reduce the velocity of a 3000-lb car from 60 mi/h to zero.

5. A 10-lb object moving in a straight line is accelerated by a constant force of 50 lb. Its speed increases from 5 to 55 ft/sec. (a) What is the impulse delivered by the force in this time? (b) How long did it take for this increase in speed?

Section 11-2, Time Average of a Force

6. Find the average value of the force shown in Figure 11-14 for the time interval $t_i = 1$ sec to $t_f = 4$ sec.

7. Find the average value of the force shown in Figure 11-15 for the time interval $t_i = 1$ sec to $t_f = 5$ sec.

8. A particle of mass 2 kg is moving along the x axis at 4 m/sec, and 2 sec later it is moving along the y axis with the same speed. Find (a) the impulse that acted on the particle and (b) the average force acting on the particle for the 2-sec interval.

9. A 1-lb ball is dropped from a height of 10 ft above the floor. It rebounds to a height of 6 ft. (a) Find the impulse exerted by the floor on the ball. (b) If the ball was in contact with the floor for 10^{-3} sec, find the average force exerted by the floor on the ball during that time.

10. A man can exert an average force of 200 N on his machine gun. His gun fires 20-gm bullets at 1000 m/sec. How many bullets can he fire per minute?

Section 11-3, Pressure of a Gas

11. An object of mass m and energy E_k bounces perpendicularly off a wall without energy loss. Find an expression for the impulse I provided by the wall.

12. Within 60 sec, 50 golf balls are shot at a speed of 40 m/sec toward a net. The mass of each ball is 45 gm. They strike the net, loose all their energy, and fall to the ground. Find the average force exerted by the net.

13. During a 1-min interval one thousand 0.01-gm raindrops fall vertically on a car roof with an area of 3.2 m². The drops are falling at 5 m/sec when they strike the roof. Find the average force on the roof and the pressure.

14. In Exercise 13, replace the raindrops by hailstones of mass 2 gm, which bounce vertically off the roof without energy loss. Find the average force and the pressure.

15. Using the fact that 1 mole of gas occupies 22.4 ℓ at standard temperature and pressure of 1 atm $= 1.01 \times 10^5$ N/m² and contains Avogadro's number of molecules, find the average kinetic energy of a gas molecule under standard conditions.

16. Find the rms speed of an H_2 molecule if 1 mole of hydrogen gas is confined to a 10-ℓ container at a pressure of 3 atm.

Section 11-4, The Impulse Approximation

17. In Example 11-2, the weight of the golf ball was neglected. As a further check on the validity of the impulse approximation for this example, (a) find the distance the ball falls under the influence of gravity while it is in contact with the club during the collision, which lasts 0.001 sec, and (b) compare this distance with that traveled during the impulse due to the force of the club.

18. A 10-gm bullet traveling 1000 m/sec embeds itself in a 3-kg block. (*a*) Estimate the collision time assuming the bullet moves 10 cm before being brought to rest relative to the block. (*b*) How far does the block travel during this time? (*c*) Find the average force exerted by the bullet on the block during this time and compare it with the weight of the block. Is the impulse approximation a good approximation for this problem?

19. A 0.32-lb baseball travels 120 ft/sec toward home plate. It is hit by the bat, and its velocity is reversed, so that it travels at 120 ft/sec in the opposite direction. (*a*) Find the impulse exerted by the bat on the ball. (*b*) If the collision time is 0.001 sec, find the average force on the ball and compare it with the weight of the ball. Is the impulse approximation good for this problem?

Section 11-5, Energy Changes in Collisions; Reference Frames

20. A 10-gm particle collides with a 2-kg particle. The velocity of the 10-gm particle in the center-of-mass reference frame is 500 m/sec. (*a*) What is the velocity of the 2-kg particle in this frame? (*b*) Find the velocity and kinetic energy of the 10-gm particle in the laboratory frame, in which the 2-kg particle is at rest.

21. Draw a diagram indicating the initial and final velocities of each of the particles in the center-of-mass frame for each of the collisions shown in Figure 11-16.

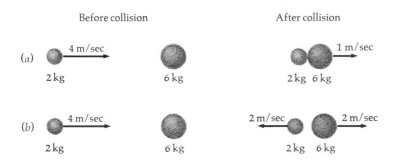

Figure 11-16
Collisions in the laboratory frame (Exercises 21 and 22).

22. (*a*) For each of the collisions shown in Figure 11-16, find the total kinetic energy of the particles before and after the collision in the laboratory frame shown. (*b*) Find the total kinetic energy of the particles before and after the collision in the center-of-mass frame. (*c*) Show that your answer in part (*a*) is greater than in part (*b*) by the amount $\frac{1}{2}(m_1 + m_2)v_{CM}^2$, where v_{CM} is the velocity of the center of mass in the laboratory frame.

23. Draw a diagram indicating the initial and final velocities of each of the particles in the laboratory frame (in which the more massive particle is initially at rest) for the collisions in Figure 11-17.

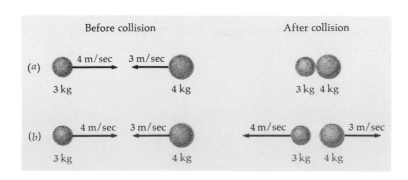

Figure 11-17
Collisions in the zero-momentum frame (Exercises 23 and 24).

24. For the collisions in Figure 11-17 find the initial and final total kinetic energies of the particles (a) in the center-of-mass frame and (b) in the laboratory frame. (c) Show that the energy in the laboratory frame is greater than that in the center-of-mass frame by $\frac{1}{2}(m_1 + m_2)v_{CM}^2$, where v_{CM} is the center-of-mass velocity in the laboratory frame.

Section 11-6, Perfectly Elastic Collisions in One Dimension

25. Which of the collisions shown in Figure 11-18 are perfectly elastic? (All masses are in kg.)

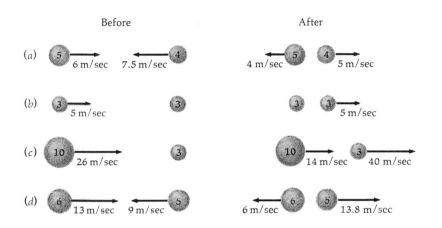

Before   After

(a)   5 → 6 m/sec   7.5 m/sec ← 4   ← 5 4 →
                                          4 m/sec   5 m/sec

(b)   3 → 5 m/sec   3   3   3 →
                                                  5 m/sec

(c)   10 → 26 m/sec   3   10 → 14 m/sec   3 → 40 m/sec

(d)   6 → 13 m/sec   9 m/sec ← 6   ← 6 5 →
                                          6 m/sec   13.8 m/sec

Figure 11-18
Exercise 25.

26. A 4-lb object moving at 5 ft/sec makes a perfectly elastic head-on collision with a 1-lb object initially at rest. Find the final velocities of each of the objects.

27. A 15-gm mass is moving to the right with speed of 3 m/sec toward a 6-gm mass moving to the left with speed of 7.5 m/sec. Find the final velocities of each if they make a perfectly elastic head-on collision.

28. A 10-kg object traveling with a speed of 5 m/sec collides elastically with a 100-gm object initially at rest. (a) Find the velocities of each object after the collision. (b) Find the fraction of the initial kinetic energy lost by the 10-kg object.

29. Two blocks are sliding toward each other on a smooth surface. One block, of mass 10 kg, is coming from the left at 5 m/sec, and the other, of mass 6 kg, is coming from the right at 3 m/sec. The two blocks make a head-on elastic collision. Find (a) the velocity of the center of mass before the collision, (b) the initial velocities of the blocks in the center-of-mass frame, (c) the velocities of the blocks in the center-of-mass frame after the collision, and (d) the velocities of the blocks after the collision in the original frame.

30. Two particles $m_1 = 4$ kg and $m_2 = 6$ kg are moving to the right along the x axis, with m_1 chasing m_2. The velocity of m_1 is 10 m/sec, and that of m_2 is 5 m/sec. The particles make an elastic collision. Find the velocity of (a) the center of mass of the system, (b) each particle in the center-of-mass system before and after the collision, (c) each particle after the collision in the original frame. (d) Calculate the initial and final kinetic energies in the original frame and confirm that the total kinetic energy is conserved. (e) Calculate the total kinetic energy in the center-of-mass frame and verify that it is less than your answer to part (d) by $\frac{1}{2}(m_1 + m_2)v_{CM}^2$.

Section 11-7, Elastic Collisions in Three Dimensions

31. Figure 11-19 shows the results of an elastic collision between two equal masses as seen in the center-of-mass reference frame. (a) Find the velocity vectors before and after the collision as seen in the laboratory frame in which

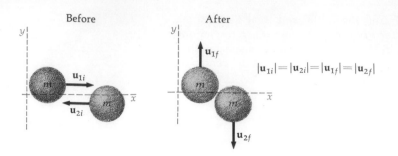

Figure 11-19
Elastic collision of equal masses (Exercise 31).

one of the masses is initially at rest and show them in a drawing. (*b*) Express the final velocities $\mathbf{v}_1$ and $\mathbf{v}_2$ of the masses in the laboratory frame in terms of the unit vectors $\mathbf{i}$ and $\mathbf{j}$ and show explicitly that $\mathbf{v}_1 \cdot \mathbf{v}_2 = 0$.

32. Figure 11-20 shows the results of a collision of two unequal masses in the laboratory frame. (*a*) Show that momentum is conserved in this collision. (*b*) Show that this collision is elastic by computing the initial and final kinetic energies. (*c*) Find the velocity of the center of mass in this frame and make a drawing indicating the initial and final velocities of the masses in the center-of-mass reference frame.

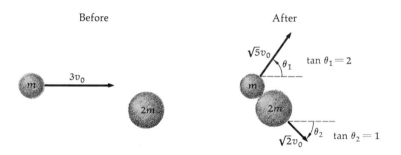

Figure 11-20
Exercise 32.

Section 11-8, Perfectly Inelastic Collisions

33. An 8-lb fish is swimming at 2 ft/sec to the right. It swallows a $\frac{1}{4}$-lb fish swimming toward him at 6 ft/sec to the left (Figure 11-21). What is the velocity of the larger fish immediately after his lunch?

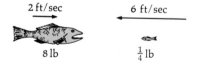

Figure 11-21
Inelastic collision of two fish (Exercise 33).

34. A 150-lb baseball player jumps vertically into the air to catch a 0.32-lb baseball traveling horizontally at 140 ft/sec. If the velocity of the baseball player is 1 ft/sec upward just before he catches the ball, what is his velocity just after he catches the ball?

35. When a 10-gm bullet hits a small stationary target, 60 percent of the mechanical energy is lost. Find the mass of the target.

36. A 10-gm bullet is fired with velocity of 300 m/sec into a pendulum bob which has a mass of 990 gm. How high does the pendulum bob (plus bullet) swing after the collision?

37. A 15-gm bullet is fired into the bob of a ballistic pendulum which has a mass of 1.5 kg. With the bob at maximum height the strings make an angle of 60° with the vertical. The length of the pendulum is 2 m. Find the velocity of the bullet.

38. A 20-ton railway car stands on a hill with its brakes set. The brakes are released, and the car rolls down to the bottom of the hill 16 ft below its original position (Figure 11-22). It collides with a 10-ton car resting at the bottom of the track (with brakes off). The two cars couple together and roll up the track to a height H. Find H.

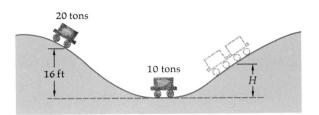

20 tons

16 ft

10 tons

H

Figure 11-22
Inelastic collision of railway cars (Exercise 38).

39. A 1-kg block of wood is attached to a spring of force constant 200 N/m and rests on a smooth surface, as shown in Figure 11-23. A 20-gm bullet is fired into the block, and the spring compresses 13.3 cm. (a) Find the original velocity of the bullet before the collision. (b) What fraction of the original mechanical energy is lost in this collision?

$k = 200\,\text{N/m}$

20 gm

1 kg

Figure 11-23
Exercise 39.

Section 11-9, Reaction Threshold, and Section 11-10, Coefficient of Restitution

40. The energy that must be supplied in the center-of-mass frame to separate a deuteron into its constituents, a neutron and a proton, is 2.2 MeV. What is the reaction threshold for the reaction $n + d \rightarrow n + n + p - 2.2$ MeV, where n stands for a neutron and d for the deuteron, at rest in the laboratory frame?

41. The Q value for the ionization of a lithium atom is $Q = -5.39$ eV. Find the threshold energy for the ionization of a lithium atom by bombardment with protons. (The mass of lithium is 7 times that of a proton.)

42. The energy needed to separate the two protons in an H_2 molecule is 4.48 eV. This process of separation is called *dissociation*. Suppose it is to be accomplished by a collision of H_2 with a proton. (a) Which takes more energy, bombarding stationary protons with H_2 molecules or bombarding stationary H_2 molecules with protons? (b) Calculate the threshold energy for each of these reactions.

43. A ball bounces to 80 percent of its original height. (a) What fraction of its mechanical energy is lost in each bounce? (b) Where does this energy go? (c) What is the coefficient of restitution of the ball-floor system?

44. A 2-kg object moving at 3 m/sec to the right collides with a 3-kg object moving at 2 m/sec to the left. The coefficient of restitution is 0.6. (a) Find the velocity of each object after the collision. (b) Transform this problem to the laboratory frame, in which the 3-kg object is originally at rest. Diagram the initial and final velocities in this frame.

Problems

1. A force varies with time according to $F = F_0 + Ct$, where F_0 and C are constants. Show both graphically and analytically that the average force during the time interval $t = 0$ to $t = t_1$ is $\frac{1}{2}(F_0 + F_1)$, where F_1 is the force at time t_1.

2. A 2-kg mass is accelerated from rest along a straight line. After 5 sec its speed is 20 m/sec. (a) What was the impulse of the force during the time $t = 0$ to $t = 5$ sec? (b) What was the average force for this interval? (c) If the force decreased linearly with time from some initial value to zero in 5 sec, what was the initial value?

3. Figure 11-24 shows the force acting on a 5-kg mass which moves along the x axis. At time $t = 0$, the mass is at rest at $x = 0$. (a) Make a graph of the impulse delivered by the force from $t = 0$ to time t as a function of t. (b) Make a graph of the speed of the mass as a function of time. (c) What is F_{av} for the period from $t = 0$ to $t = 12$ sec? (d) What is F_{av} for the period from $t = 0$ to $t = 6$ sec? From $t = 6$ sec to $t = 12$ sec?

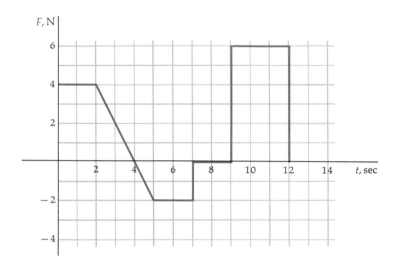

Figure 11-24
Force acting on 5-kg mass
(Problem 3).

4. A 3000-lb car traveling 60 mi/h crashes into a concrete wall. Estimate the time of the collision by assuming that the center of the car travels halfway to the wall with uniform acceleration and therefore average speed of 30 mi/h. (Use any reasonable length for the car.) What is the impulse exerted by the wall on the car? Using your estimate of the time for the collision, find the average force exerted by the wall on the car.

5. You throw a 0.32-lb ball upward so that it reaches a height of 100 ft. Use a reasonable value for the distance the ball moves while it is in your hand to calculate the average acceleration of the ball in the hand, the impulse given the ball, the average force exerted while the ball is in your hand, and the time the ball is accelerating. Can you neglect the weight of the ball while it is being thrown?

6. A 6-ft 180-lb man steps off a ledge and falls 16 ft to the ground, making an inelastic collision with it. (a) How fast was he traveling just before he hit? (b) What was the impulse exerted on man by the ground? (c) Estimate the time of collision assuming that the force on the man is constant and that he travels 3 ft during the collision. (d) Using your result for part (c), find the average force exerted by the ground on the man. (e) Discuss the validity of the impulse approximation for this problem.

7. A handball of mass 300 gm is thrown straight against a wall with a speed of 8 m/sec. It rebounds with the same speed. (a) What impulse was delivered to the wall? (b) If the ball was in contact with the wall for 0.003 sec, what average force was exerted on the wall? (c) The ball is caught by a player who brings it to rest. In the process his hand moves back 0.5 m. What is the impulse received by the player? (d) What was the average force exerted on the player?

8. A stream of glass beads comes out of a horizontal tube at 100 per second and strikes a balance pan, as shown in Figure 11-25. They fall a distance of 0.5 m to the balance and bounce back to the same height. Each bead has mass

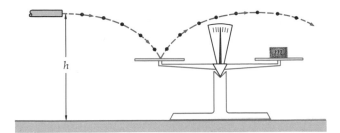

Figure 1
Glass beads
tically with ba
(Problem 8).

0.5 gm. How much mass m must be placed in the other pan of the balance to keep the pointer reading zero?

9. A block of wood weighing 5 lb is at rest on a level floor. A lump of putty weighing 1 lb is thrown at the block so that traveling horizontally it hits and sticks to the block. The block and putty slide 2.5 ft along the floor. If the coefficient of friction is 0.4, what was the original speed of the putty?

10. A particle has speed v_0 originally. It collides with a second particle at rest and is deflected through angle ϕ. Its speed after the collision is v. The second particle recoils, its velocity making an angle θ with the initial direction of the first particle. Show that

$$\tan \theta = \frac{v \sin \phi}{v_0 - v \cos \phi}$$

Do you have to assume that this collision was elastic or inelastic to get this result?

11. Two particles, $m_1 = 4$ kg and $m_2 = 6$ kg, slide without friction along the x axis. Initially m_1 has velocity 10 m/sec, and m_2 has velocity 5 m/sec. Particle m_1 overtakes m_2. The coefficient of restitution is 0.5. (a) What is the velocity of the center of mass? (b) What is the initial momentum of each particle in the center-of-mass reference frame? (c) What is the final momentum of each particle in the center-of-mass frame? (d) What is the final speed of each particle in the original frame? (e) How much kinetic energy was lost in this collision?

12. A 60-gm tennis ball is served at a speed of 50 m/sec. What is the impulse delivered by the racket? Use reasonable estimates for the distance traveled by the ball while in contact with the racket to estimate the collision time and the average force exerted by the racket.

13. A bullet of mass m_1 is fired with speed v into the bob of a ballistic pendulum, which has mass m_2. The bob is attached to a very light rod of length L, which is pivoted at the other end. The bullet is stopped in the bob. Find v such that the bob will swing through a complete circle.

14. A bullet of mass m_1 is fired with speed v into the bob of a ballistic pendulum of mass m_2. Find the maximum height attained by the bob if the bullet passes through the bob and emerges with speed $\frac{1}{4}v$.

15. A neutron of mass m makes an elastic collision with a carbon atom of mass $12m$. The carbon atom is initially at rest. (a) What is the energy of the neutron after the collision if its original energy is E_1? (b) How many collisions will a neutron need to make to reduce its energy from E_1 to $10^{-6} E_1$? (Assume head-on collisions.)

CHAPTER 12   Rotation of a Rigid Body about
a Fixed Axis

We now turn to the study of a special kind of motion of a system of
particles, rotational motion. Consider for example, the rotation of a
wheel about its axis. We cannot analyze this motion by considering
the wheel as a single particle because different parts of the wheel have
different velocities and accelerations. We must consider the wheel as a
system of particles; i.e., we consider the wheel to consist of a large
number of parts, each so small that we can neglect the variation in
velocity or acceleration over it. Thus each part can be treated as a particle.

No matter how the wheel moves, the distance between any two particles within it remains constant. Such a system is called a *rigid body*.
Although all objects in nature are deformable to some degree, the
rigid-body approximation is often very good, and it greatly simplifies
the analysis of the motion of the system. In general, the change in
position of a rigid body can be considered to consist of a translation of
the center of mass plus a rotation of the body about an axis through
the center of mass, as illustrated in Figure 12-1. We have already seen
how to treat the motion of the center of mass of any system; its acceleration equals the resultant force on the system divided by the total
mass. The rotational motion of a body is usually very complicated
because, in general, the axis of rotation changes direction as the body
moves.

Rigid body defined

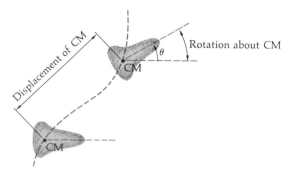

Figure 12-1
The general motion of a rigid
body consists of translation of
the center of mass plus rotation about the center of mass.

In this chapter we concentrate on the simplest kind of rotation, rotation of a rigid body about a fixed axis. In the next chapter we shall extend our discussion to systems in which the distance between any two particles can vary and consider the situation when the axis of rotation is not fixed in space.

12-1 Angular Velocity and Angular Acceleration

Consider one particle of a wheel rotating about its axis of symmetry, which is fixed in space. We can specify the position of the particle P_i by the distance r_i from the center of the wheel and the angle θ_i between a line from the center to the particle and a reference line fixed in space, as shown in Figure 12-2. In a small time dt, the particle moves along the arc of a circle a distance ds_i given by

$$ds_i = v_i\, dt \qquad\qquad 12\text{-}1$$

where v_i is the speed of the particle. The angle swept out by the line from the center to the particle, in radians, is this distance divided by the radius r_i:

$$d\theta = \frac{ds_i}{r_i} = \frac{v_i\, dt}{r_i} \qquad\qquad 12\text{-}2$$

Although the distance ds_i varies from particle to particle, the angle $d\theta$ swept out in a given time is the same for all the particles. For example, when one particle moves through a complete circle, so do all the other particles and $\Delta\theta = 2\pi$ rad for all the particles. The rate of change of angle with respect to time $d\theta/dt$ is the same for all the particles of the wheel. It is called the *angular velocity* ω of the wheel,

$$\omega = \frac{d\theta}{dt} \qquad\qquad 12\text{-}3 \qquad \textit{Angular velocity}$$

From Equation 12-2 the speed of the ith particle is related to its radius r_i and the angular velocity of the wheel by

$$v_i = r_i \frac{d\theta}{dt} = r_i \omega \qquad\qquad 12\text{-}4$$

Whether ω is positive or negative depends on whether θ is increasing or decreasing. The units of angular velocity are radians per second.

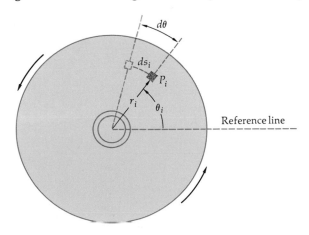

Figure 12-2
A wheel rotating about a fixed axis through its center. The distance ds_i moved by the ith particle in some time depends on r_i, but the angular displacement $d\theta$ is the same for all particles of the wheel.

Star tracks in a time exposure of the night sky. How can you decide whether these circular tracks are due to rotation of the earth or motion of the stars? (*Courtesy of Yerkes Observatory.*)

Since a radian is a dimensionless unit of angle, the dimensions of angular velocity are those of reciprocal seconds (sec^{-1}). Although angular velocity is frequently expressed in other units than radians per second, it is important to remember that expressions like Equations 12-3 and 12-4 and the other results we obtain for rotational motion are valid only when angles are expressed in radians.

Since angles are often expressed in degrees, angular velocity is sometimes expressed in degrees per second or degrees per minute. In such cases, the angular velocity can be reexpressed in radians by using

$$2\pi \text{ rad} = 360° \qquad \text{or} \qquad \frac{\pi \text{ rad}}{180°} = 1$$

Angular velocity is also commonly expressed in revolutions per second and revolutions per minute. Conversion to radians is accomplished by using

$$2\pi \text{ rad} = 1 \text{ rev} \qquad \text{or} \qquad \frac{2\pi \text{ rad}}{1 \text{ rev}} = 1$$

The rate of change of angular velocity with respect to time is called the *angular acceleration* α. For rotation about a fixed axis,

$$\alpha = \frac{d\omega}{dt} = \frac{d^2\theta}{dt^2} \qquad \qquad 12\text{-}5 \qquad \textit{Angular acceleration}$$

The units of angular acceleration are radians per second per second. The relation between the *tangential* linear acceleration of the ith particle of the wheel and the angular acceleration is obtained by taking the derivative of the speed v_i in Equation 12-4 with respect to t:

$$a_{it} = \frac{dv_i}{dt} = r_i \frac{d\omega}{dt}$$

Thus

$$a_{it} = r_i\alpha \qquad \qquad 12\text{-}6$$

Each particle of the wheel also has a *radial* linear acceleration, the centripetal acceleration:

$$a_{ir} = -\frac{v_i^2}{r_i} = -r_i \omega^2 \qquad\qquad 12\text{-}7$$

The three quantities angular displacement θ, angular velocity ω, and angular acceleration α are analogous to the linear displacement x, linear velocity v_x, and linear acceleration a_x, which we met in studying one-dimensional motion in Chapter 2. Because of the similarity of the definitions of the rotational and linear quantities, much of what we learned in Chapter 2 will be useful in treating problems of rotation about a fixed axis. For example, if the angular acceleration is constant, we can develop constant-angular-acceleration expressions which are exactly of the form of the constant-linear-acceleration expressions (Equations 2-11 to 2-13). If, for example, we have

$$\alpha = \frac{d\omega}{dt} = \alpha_0 = \text{constant}$$

then

$$\omega = \omega_0 + \alpha_0 t \qquad\qquad 12\text{-}8$$

Expressions for constant angular acceleration

where ω_0 is the angular velocity at $t = 0$. Integrating again, we obtain

$$\theta = \theta_0 + \omega_0 t + \tfrac{1}{2}\alpha_0 t^2 \qquad\qquad 12\text{-}9$$

As with the constant-linear-acceleration formulas, we can eliminate time from these equations to obtain an equation relating the angular displacement, angular velocity, and angular acceleration:

$$\omega^2 = \omega_0^2 + 2\alpha_0(\theta - \theta_0) \qquad\qquad 12\text{-}10$$

Example 12-1 A wheel rotates with constant angular acceleration of $\alpha_0 = 2$ rad/sec^2. If the wheel starts from rest, how many revolutions does it make in 10 sec?

The number of revolutions is related to the angular displacement by the fact that each revolution is an angular displacement of 2π rad. Thus we need to find the angular displacement $\theta - \theta_0$ in radians for a time 10 sec and multiply by the conversion factor 1 rev/2π rad.

Equation 12-9 relates the angular displacement to the time. We are given $\omega_0 = 0$ (the wheel starts from rest). Let us choose $\theta_0 = 0$. We have

$$\theta = \tfrac{1}{2}\alpha_0 t^2 = \tfrac{1}{2}(2 \text{ rad/sec}^2)(10 \text{ sec})^2 = 100 \text{ rad}$$

The number of revolutions is thus

$$100 \text{ rad} \times \frac{1 \text{ rev}}{2\pi \text{ rad}} = \frac{50}{\pi} \text{ rev} = 15.9 \text{ rev}$$

It is convenient to assign directions to the rotational quantities θ, ω, and α. We cannot describe the rotation of a wheel by any direction in the plane of the wheel because, by symmetry, all directions in this plane are equivalent. The direction in space uniquely associated with the rotation is the direction of the axis of rotation. We therefore choose the direction for the angle of rotation along the axis of rotation. Since ω equals the rate of change of θ, it will also be along the axis of rotation. Similarly, since we are considering only rotation about axes which are fixed in space, the rate of change of ω, which is the angular acceleration α, must also be in this direction.

Direction of θ, ω, and α

The rotation of this nonrigid body, a galaxy in Ursa Major, is indicated by its spiral shape. (*Courtesy of Hale Observatories.*)

Consider a wheel rotating in the clockwise sense, as in Figure 12-3*a*. We have chosen the direction for ω along the axis of rotation, but we must still decide the sense of ω, that is, for this rotation, whether ω points into or out of the diagram. The choice is made by convention. If the rotation is clockwise, as in the figure, ω is inward. If it is counterclockwise, ω is outward. This arbitrary decision is called the *right-hand rule*, as illustrated in Figure 12-3*a*. When the axis of rotation is grasped by the right hand with the fingers following the rotation, the extended thumb points in the direction of ω. The direction of ω is also that of the advance of a rotating right-handed screw, as shown in Figure 12-3*b*. The sense of the angular acceleration (for rotation about a fixed axis) depends on whether ω is increasing or decreasing. If ω is increasing, α is parallel to ω. If ω is decreasing, α is antiparallel to ω.

Question

1. Two points are on a wheel turning at constant angular velocity, one point on the rim and the other halfway between the rim and the axis. Which moves the greatest distance in time Δt? Which turns through the greater angle? Which has the greatest speed? Angular velocity? Acceleration? Angular acceleration?

12-2 The Vector Nature of Rotation

Having assigned directions to the rotational quantities θ, ω, and α, we now ask whether these quantities are vectors. A quantity having magnitude and direction is a vector only if it also obeys the law of vector addition. The significant property of the law of vector addition for this discussion is that the sum of two vectors is independent of the order in which they are added:

$$\mathbf{A} + \mathbf{B} = \mathbf{B} + \mathbf{A}$$

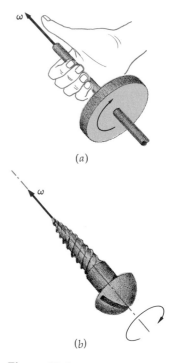

(a)

(b)

Figure 12-3
(*a*) The right-hand rule for determining the direction of the angular velocity ω. (*b*) The direction of ω is also the direction of advance of a rotating right-handed screw.

Consider, for example, the two successive 90° rotations illustrated in Figure 12-4*a*. The axes of the two rotations are different. Figure 12-4*b* shows these same rotations with the order reversed, demonstrating that the final position of the object depends on the order in which the two rotations occur. Angular displacement is not a vector quantity because it does not obey the addition law for vectors. Figure 12-5 shows the same object undergoing much smaller successive rotations of about 20°. The difference in the final positions for the two sequences is much less than for 90°. As the magnitude of the successive changes in angle approaches zero, the difference in final positions also approaches zero. Thus in the limit as $\Delta\theta \to 0$ the angular displacement obeys the vector-addition law. For general rotations we define the angular velocity by

$$\boldsymbol{\omega} = \frac{d\boldsymbol{\theta}}{dt} = \lim_{\Delta t \to 0} \frac{\Delta\boldsymbol{\theta}}{\Delta t}$$

(Here we have written $\Delta\boldsymbol{\theta}$ in boldface to indicate that it has a direction even though it is not a vector.) Since $\Delta\boldsymbol{\theta} \to 0$ as $\Delta t \to 0$, angular velocity does obey the vector-addition law and is a vector quantity. Similarly, we define the angular acceleration for general rotations by

$$\boldsymbol{\alpha} = \frac{d\boldsymbol{\omega}}{dt}$$

Since the derivative of a vector is also a vector, $\boldsymbol{\alpha}$ is a vector.

Figure 12-4
(*a*) The book is rotated 90° about the *x* axis and then 90° about the *y* axis. (*b*) The body is rotated first about the *y* axis and then about the *x* axis. The result of the two rotations depends on the order.

Figure 12-5
The same as Figure 12-4 except for a small angle of rotation. (*a*) A rotation about the *x* axis followed by a rotation about the *y* axis. (*b*) A rotation about the *y* axis followed by a rotation about the *x* axis. The result is nearly the same for the two cases.

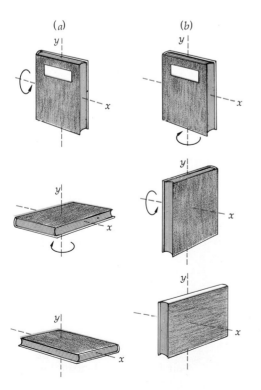

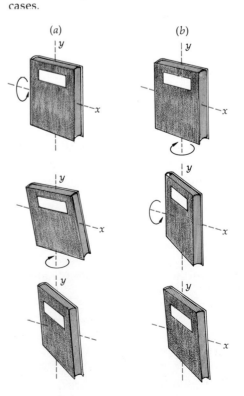

For rotations about a fixed axis, which we are considering in this chapter, the vector nature of the rotational quantities is not important. The angular velocity ω and the angular acceleration α have only one component, such as ω_z and α_z, for a wheel rotating about the z axis. Thus they can change only in magnitude. We are considering a simplification of general rotational motion, which resembles the simplification of linear motion in Chapter 2, where we restricted our discussion to one-dimensional motion. We consider some more complicated rotations in three dimensions in the next chapter.

Questions

2. A wheel is mounted on a vertical axis. A point on the east side is moving north, and a point on the west side is moving south. What is the direction of ω? What is the direction of α if the wheel is speeding up? If it is slowing down?

3. The axle of a wheel is mounted horizontally on a north-south line. The angular velocity points north. What is the direction of motion of a point on the rim at the top of the wheel? The angular acceleration α points south. Is the angular speed of the wheel increasing or decreasing?

12-3 Rotational Kinetic Energy and Moment of Inertia

The kinetic energy of a system of particles is just the sum of the kinetic energies of the individual particles. For a rotating wheel, we have

$$E_k = \Sigma \tfrac{1}{2} m_i v_i^2 = \Sigma \tfrac{1}{2} m_i (r_i \omega)^2 = \tfrac{1}{2} \omega^2 \Sigma m_i r_i^2 \qquad \text{12-11}$$

We have taken the angular velocity out of the sum because it is the same for each particle. The sum $\Sigma m_i r_i^2$ is a property of the wheel called the *moment of inertia I*. The moment of inertia depends on the mass distribution relative to the axis of rotation of the wheel. In terms of I, the kinetic energy can be written

Moment of inertia

$$E_k = \tfrac{1}{2} I \omega^2 \qquad \text{12-12}$$

If we compare this expression with that for the kinetic energy of translation of a single particle $\tfrac{1}{2} mv^2$, we see that in rotational motion, moment of inertia is analogous to mass in linear motion, just as angular velocity is analogous to linear velocity. Note that the distance r_i in Equation 12-11 is the perpendicular distance from the ith particle to the *axis* of rotation. In general, it is not the distance from the ith particle to the center of mass even though in a two-dimensional figure rotating about an axis perpendicular to the figure through the center of mass they are the same. An example should clarify this point.

Example 12-2 Four equal masses are connected by light (massless) rods to form a rectangle of sides $2a$ and $2b$, as shown in Figure 12-6. The system rotates about an axis in the plane of the figure through the center. Find the moment of inertia about this axis and the kinetic energy if the angular velocity is ω.

Since each mass is moving in a circle of radius a, the linear velocity

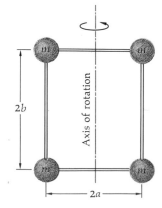

Figure 12-6
Four equal masses connected by light rods rotating about an axis in the plane of the masses through the center of mass.

of each particle is $v = a\omega$. Thus its kinetic energy is

$$\tfrac{1}{2}mv^2 = \tfrac{1}{2}m(a\omega)^2 = \tfrac{1}{2}ma^2\omega^2$$

The total kinetic energy is the sum over the four particles:

$$E_k = \Sigma\tfrac{1}{2}m_iv_i^2 = 4(\tfrac{1}{2}ma^2\omega^2) = \tfrac{1}{2}(4ma^2)\omega^2$$

Therefore the moment of inertia is $4ma^2$.

To avoid confusing r_i with its usual meaning (the distance from the ith particle to the origin) we shall use ρ_i for the perpendicular distance from the ith particle to the axis of rotation. Then the moment of inertia is

$$I = \Sigma m_i\rho_i^2 \qquad\qquad 12\text{-}13$$

Example 12-3 Find the moment of inertia of the system of Example 12-2 for rotation about an axis parallel to the first axis but passing through two of the masses, as shown in Figure 12-7.

For this rotation, two of the masses are at a distance $2a$ from the axis of rotation and two are at zero distance. The moment of inertia is thus

$$I = \Sigma m_i\rho_i^2 = m(0)^2 + m(0)^2 + m(2a)^2 + m(2a)^2 = 8ma^2$$

The moment of inertia is larger about this axis than the one parallel to it through the center of mass.

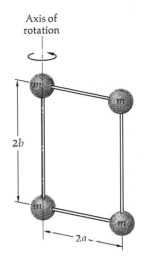

Figure 12-7
Four equal masses connected by light rods rotating about an axis through two of the masses.

Questions

4. Can the moment of inertia of an object by specified without describing where the axis of rotation is? Can a body have more than one moment of inertia?

5. How does the kinetic energy of a rotating body change if its angular velocity is doubled?

6. A disk is made by surrounding a wooden core with an iron rim. Another disk, of the same diameter, thickness, and total mass, is made by surrounding an iron core with a wooden rim. Which has the greater kinetic energy when rotated with angular velocity ω about its axis of symmetry?

12-4 Calculating the Moment of Inertia

We can often simplify the calculation of the moments of inertia for various bodies by using general theorems relating the moment of inertia about one axis of the body to that about another axis. *Steiner's theorem*, or the *parallel-axis theorem*, relates the moment of inertia about an axis through the center of mass to that about a second, parallel axis. Let I_{CM} be the moment of inertia about an axis through the center of mass of a body and I be that about a parallel axis a distance h away. The parallel-axis theorem states that

$$I = I_{CM} + Mh^2 \qquad\qquad 12\text{-}14 \qquad\qquad \textit{Parallel-axis theorem}$$

where M is the total mass of the body. Examples 12-2 and 12-3 illustrate a special case of this theorem with $h = a$, $M = 4m$, and $I_{CM} = 4ma^2$.

We can prove the theorem using the result (developed in Chapter 10) that the kinetic energy of a system of particles is the sum of the kinetic energy of center-of-mass motion plus the energy of motion relative to the center of mass,

$$E_k = \tfrac{1}{2}Mv_{CM}^2 + E_{kr} \qquad\qquad 12\text{-}15$$

Consider a rigid body rotating with angular velocity about an axis a distance h from a parallel axis through the center of mass, as in Figure 12-8. When the system rotates through an angle $\Delta\theta$ measured about the axis of rotation, it rotates through the same angle $\Delta\theta$ measured about any other parallel axis. The motion of the body relative to the center of mass is thus a rotation about the center-of-mass axis with the same angular velocity. The kinetic energy of motion relative to the center of mass is thus

$$E_{kr} = \tfrac{1}{2}I_{CM}\omega^2$$

The velocity of the center of mass relative to any fixed point on the axis of rotation is

$$v_{CM} = h\omega$$

The kinetic energy of motion of the center of mass is thus

$$\tfrac{1}{2}Mv_{CM}^2 = \tfrac{1}{2}M(h\omega)^2 = \tfrac{1}{2}M\omega^2 h^2$$

When the total kinetic energy of rotation is written $\tfrac{1}{2}I\omega^2$, Equation 12-15 becomes

$$E_k = \tfrac{1}{2}I\omega^2 = \tfrac{1}{2}M\omega^2 h^2 + \tfrac{1}{2}I_{CM}\omega^2 = \tfrac{1}{2}(Mh^2 + I_{CM})\omega^2$$

Thus

$$I = Mh^2 + I_{CM}$$

The *plane-figure theorem* relates the moments of inertia about two perpendicular axes in a plane figure to the moment of inertia about a third axis perpendicular to the figure. If x, y, and z are perpendicular axes for a figure which lies in the xy plane, the moment of inertia about the z axis equals the sum of the moments of inertia about the x and y axes. This is not difficult to prove. Figure 12-9 shows a figure in the xy plane. The moment of inertia about the x axis is

$$I_x = \Sigma m_i y_i^2$$

where the sum is taken over all the elements of mass m_i. (In practice, as we shall see, the sum is often done by integration.) The moment of inertia about the y axis is

$$I_y = \Sigma m_i x_i^2$$

The moment of inertia about the z axis is

$$I_z = \Sigma m_i \rho_i^2$$

But for each element m_i, $\rho_i^2 = x_i^2 + y_i^2$. Thus

$$I_z = \Sigma m_i \rho_i^2 = \Sigma m_i (x_i^2 + y_i^2) = \Sigma m_i x_i^2 + \Sigma m_i y_i^2$$

or

$$I_z = I_y + I_x \qquad\qquad 12\text{-}16$$

Example 12-4 What is the moment of inertia about an axis through one of the masses and perpendicular to the plane of the masses shown in Figure 12-10?

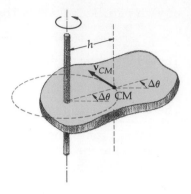

Figure 12-8
Parallel-axis theorem. The moment of inertia about any axis equals the moment of inertia about a parallel axis through the center of mass plus Mh^2, where M is the total mass and h is the distance between the axes.

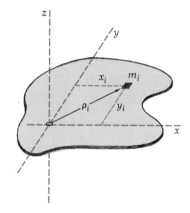

Figure 12-9
For a plane figure $I_z = \Sigma m_i \rho_i^2 = \Sigma m_i (x_i^2 + y_i^2) = \Sigma m_i x_i^2 + \Sigma m_i y_i^2 = I_y + I_x$.

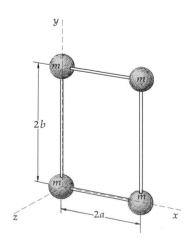

Figure 12-10
Example 12-4.

The moment of inertia calculated about the y axis in this figure was found in Example 12-3 to be $8ma^2$. By similar reasoning, the moment of inertia about the x axis in this figure is $8mb^2$. Thus $I_z = I_x + I_y = 8ma^2 + 8mb^2$. This result is easily checked by direct calculation.

For continuous bodies the sum in Equation 12-13 defining the moment of inertia is replaced by an integral. Let dm be a mass element a distance x from the axis of rotation. The moment of inertia for this axis is then

$$I = \int x^2 \, dm$$

12-17   *Moment of inertia*

Some examples will demonstrate how to calculate moments of inertia using Equation 12-17 and the two theorems discussed above (Equations 12-14 and 12-16). Moments of inertia of uniform bodies of varied shapes are given in Table 12-1.

Example 12-5 Find the moment of inertia of a uniform stick about an axis perpendicular to the stick through one end.

The mass element dm is shown in Figure 12-11. Since the total mass M is uniformly distributed along the length L, $dm = (M/L)\, dx$. The moment of inertia about the y axis is

$$I_y = \int_0^L x^2 \frac{M}{L} \, dx = \frac{M}{L} \int_0^L x^2 \, dx = \frac{M}{L} \left. \tfrac{1}{3} x^3 \right]_0^L$$

$$= \frac{M}{L} \frac{L^3}{3} = \tfrac{1}{3} M L^2$$

12-18

The moment of inertia about the z axis is also $\tfrac{1}{3} M L^2$, and that about the x axis is zero, assuming that all the mass is right on the x axis.

Example 12-6 Find the moment of inertia of a uniform stick about the y' axis through the center of mass (Figure 12-12).

Since the axis is through the center of mass of the stick, the integration extends from $-L/2$ to $+L/2$. Thus

$$I_{CM} = \int_{-L/2}^{+L/2} x'^2 \frac{M}{L} \, dx' = 2 \int_0^{L/2} x'^2 \frac{M}{L} \, dx'$$

$$= \frac{2M}{L} \frac{(L/2)^3}{3} = \tfrac{1}{12} M L^2$$

12-19

Figure 12-11
Calculation of I for a uniform stick with axis of rotation through one end.

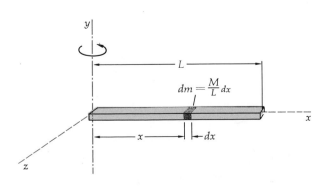

Figure 12-12
Calculation of I for a uniform stick rotating about an axis through the center of mass.

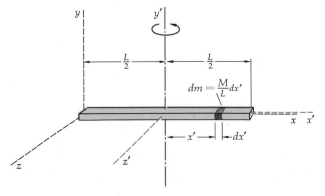

Table 12-1
Moments of inertia of uniform bodies of various shapes

Cylindrical shell about axis		$I = MR^2$
Solid cylinder about axis		$I = \frac{1}{2}MR^2$
Hollow cylinder about axis		$I = \frac{1}{2}M(R_1{}^2 + R_2{}^2)$
Cylindrical shell about diameter through center		$I = \frac{1}{2}MR^2 + \frac{1}{12}ML^2$
Solid cylinder about diameter through center		$I = \frac{1}{4}MR^2 + \frac{1}{12}ML^2$
Thin rod about perpendicular line through center		$I = \frac{1}{12}ML^2$
Thin rod about perpendicular line through one end		$I = \frac{1}{3}ML^2$
Thin spherical shell about diameter		$I = \frac{2}{3}MR^2$
Solid sphere about diameter		$I = \frac{2}{5}MR^2$
Solid rectangular parallelepiped about axis through center perpendicular to face		$I = \frac{1}{12}M(a^2 + b^2)$

This is smaller than that about the axis through the end, as would be expected because, on the average, the mass is closer to this axis than to the end. This result can also be obtained from the result of Example 12-5 and the parallel-axis theorem. In this case $h = \frac{1}{2}L$, and

$$I = I_{CM} + M(\tfrac{1}{2}L)^2 = \tfrac{1}{3}ML^2$$

or

$$I_{CM} = \tfrac{1}{3}ML^2 - \tfrac{1}{4}ML^2 = \tfrac{1}{12}ML^2$$

Example 12-7 Find the moment of inertia of a hoop about an axis through the center and perpendicular to the plane of the hoop (Fig. 12-13a).

Since all the mass is at a distance R from this axis, the moment of inertia is simply $I = MR^2$.

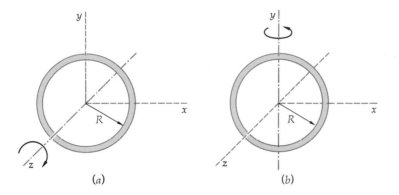

(a)   (b)

Figure 12-13
(a) Hoop rotating about an axis perpendicular to the plane of the hoop through its center. Since all the mass of the hoop is at distance R from this axis, the moment of inertia is just MR^2. (b) Hoop rotating about a diameter (Example 12-8). Since $I_y = I_x$ by symmetry and $I_y + I_x = I_z = MR^2$, it follows that $I_y = \frac{1}{2}MR^2$.

Example 12-8 Find the moment of inertia of a hoop about a diameter of the hoop (Fig. 12-13b).

Instead of solving this problem directly we use the plane-figure theorem. If we take the hoop to be in the xy plane with the center at the origin, by symmetry we have $I_x = I_y$. Since we have already found I_z to be MR^2, we have

$$I_z = I_x + I_y = 2I_x = MR^2$$

Therefore

$$I_x = I_y = \tfrac{1}{2}MR^2$$

Example 12-9 Find the moment of inertia of a uniform disk about the axis through its center and perpendicular to the plane of the disk.

We expect that I will be smaller than MR^2 since all the mass is less than the distance R from the axis. We can calculate I by taking mass elements as shown in Figure 12-14. Each mass element is a hoop with moment of inertia $r^2\, dm$. Since the area of each element is $2\pi r\, dr$, the mass of the element is $dm = (M/A)2\pi r\, dr$, where $A = \pi R^2$ is the area of the disk. Thus we have

$$I = \int r^2\, dm = \int_0^R r^2 \frac{M}{A} 2\pi r\, dr = \frac{2\pi M}{\pi R^2} \int_0^R r^3\, dr$$

$$= \frac{2M}{R^2} \frac{R^4}{4} = \tfrac{1}{2}MR^2 \qquad\qquad 12\text{-}20$$

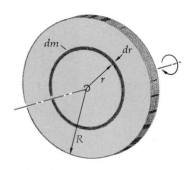

Figure 12-14
Calculation of I for a uniform disk about an axis through its center and perpendicular to the plane of the disk.

Example 12-10 Find the moment of inertia of a cylinder about its axis (Figure 12-15).

We can consider the cylinder to be a set of disks each with mass m_i and moment of inertia $\frac{1}{2}m_i R^2$. Then the moment of inertia of the complete cylinder is

$$I = \Sigma \tfrac{1}{2} m_i R^2 = \tfrac{1}{2} R^2 \Sigma m_i = \tfrac{1}{2} MR^2$$

where $M = \Sigma m_i$ is the total mass of the cylinder.

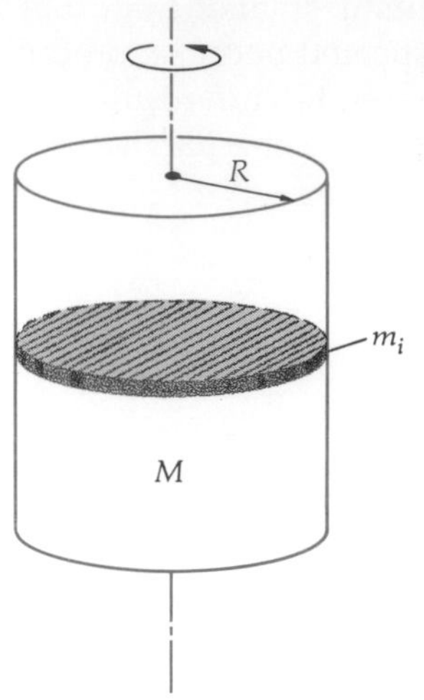

Example 12-11 Find the moment of inertia of a sphere about an axis through its center.

We can calculate this moment of inertia by treating the sphere as a set of disks. Consider the disk element in Figure 12-16 at height x above the center. The radius of the disk is

$$r = \sqrt{R^2 - x^2}$$

The mass of the disk is

$$dm = \frac{M}{V}\, \pi r^2\, dx = \frac{M}{V}\, \pi (R^2 - x^2)\, dx$$

where M is the total mass of the sphere and

$$V = \tfrac{4}{3}\pi R^3$$

is the volume of the sphere. The moment of inertia of the disk is

$$dI = \tfrac{1}{2} dm\ r^2 = \frac{1}{2}\frac{M}{V}\, \pi (R^2 - x^2)^2\, dx$$

The total moment of inertia of the sphere is

$$I = \int_{-R}^{+R} \frac{1}{2}\frac{M}{V}\, \pi (R^2 - x^2)^2\, dx = \frac{1}{2}\frac{M}{V}\, \pi 2 \int_{0}^{R} (R^2 - x^2)^2\, dx$$

$$= \frac{8\pi MR^5}{15V} = \tfrac{2}{5} MR^2 \qquad\qquad\qquad\text{12-21}$$

using $V = \tfrac{4}{3}\pi R^3$.

The moment of inertia is often given in handbooks in terms of the *radius of gyration k,* defined by

$$k^2 = \frac{I}{M} \qquad\qquad\qquad\text{12-22}$$

where M is the total mass of the body. The radius of gyration is thus the distance from the axis at which a point mass M would have the same moment of inertia about the axis. For example, the moment of inertia about an axis through one end of a stick is

$$Mk^2 = \tfrac{1}{3}ML^2 \quad \text{or} \quad k = \frac{L}{\sqrt{3}}$$

The radius of gyration is a measure of the distribution of mass of a body relative to a given axis of rotation. A large radius of gyration means that, on the average, the mass is relatively far from the axis.

Figure 12-15
Cylinder rotating about its axis. The cylinder can be thought of as a stack of disks of mass m_i. The moment of inertia about this axis is $I = \Sigma \frac{1}{2} m_i R^2 = \frac{1}{2}MR^2$, where $M = \Sigma m_i$ is the total mass.

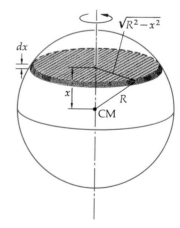

Figure 12-16
Calculation of the moment of inertia of a uniform sphere rotating about a diameter (Example 12-11). The sphere can be thought of as a set of disks of varying radius.

Questions

7. Through what point in a body must the axis of rotation pass if the moment of inertia is to be a minimum?

8. About what axis does a cube of a uniform substance have its smallest moment of inertia?

9. A disk, hoop, and sphere have the same radius and are uniform in density. Which has the greatest and which the least radius of gyration for an axis through the center of each and perpendicular to the plane of the disk and hoop? (Answer without looking up the moments of inertia.)

12-5 Torque

We now consider the dynamics of the rotation of a rigid body. In order to change the rotational kinetic energy of a rigid body, a force must do work on the body. Consider the forces F_1 and F_2 acting at the rim of a wheel, as in Figure 12-17. The force F_1 points along a radial line of the wheel; it is perpendicular to the motion of the point on the rim of the wheel at which the force acts. Thus it does no work on the wheel and cannot affect the wheel's kinetic energy. Since the kinetic energy is $\frac{1}{2}I\omega^2$, this force cannot affect the angular velocity of the wheel. For example, if the wheel is not rotating at all, a radial force F_1 will not produce rotation in either direction. However, the force F_2 has a component parallel or antiparallel to the motion of the rim of the wheel (depending upon how the wheel is rotating when the force is applied). This force does do work on the wheel which changes the wheel's kinetic energy. For example, if the wheel is rotating counterclockwise (ω out), the tangential component of F_2 will be in the direction of motion of the rim and the force will do positive work, increasing the kinetic energy and thus increasing the magnitude of the angular velocity. If the wheel is rotating in the opposite sense, F_2 will do negative work, decreasing ω and the kinetic energy.

Of course forces F_1 and F_2 shown in Figure 12-17 are not the only forces acting on the wheel. If the wheel is rotating about a fixed axis through the center of mass, the center of mass must be at rest. Then the resultant force acting on the wheel must be zero. The other forces not shown in this figure are exerted by the pivot at the axis of rotation and balance F_1 and F_2 so that the center of mass does not accelerate. If the pivot is frictionless, these forces are radial and do not affect the rotation. We shall ignore these forces for this discussion.

Let us now consider a single force F acting on a wheel which is already rotating in the counterclockwise sense (ω out), as in Figure 12-18. The force makes an angle ϕ with the radial line from the axis of rotation to the point of application of the force. The tangential component of the force is $F_t = F \sin \phi$. (We use ϕ here to avoid confusing this angle with the angle of rotation θ.) In a small time interval dt, the wheel rotates through an angle $d\theta = \omega \, dt$. The point of application of the force moves a distance $ds = r \, d\theta = r\omega \, dt$. The work done by the force is thus

$$dW = \mathbf{F} \cdot d\mathbf{s} = F_t \, ds = F \sin \phi \, r \, d\theta = Fr \sin \phi \, \omega \, dt \qquad 12\text{-}23$$

The rate at which the force does work is the power input, which equals the rate of change of kinetic energy:

$$P = \frac{dW}{dt} = F \sin \phi \, r\omega = \frac{d}{dt}(\tfrac{1}{2}I\omega^2) = I\omega \frac{d\omega}{dt} \qquad 12\text{-}24$$

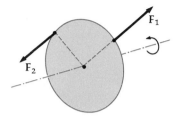

Figure 12-17
Two forces acting on a rotating wheel. F_1 is perpendicular to the motion and does no work on the wheel.

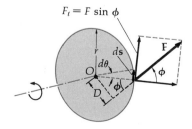

Figure 12-18
Only the tangential component F_t does work on the wheel.

Canceling ω in Eq. 12-24, we have

$$F \sin \phi \, r = I \frac{d\omega}{dt} = I\alpha \qquad\qquad 12\text{-}25$$

The quantity $Fr \sin \phi$ is called the *torque* τ or moment of the force **F** about the point O. Note that the distance $D = r \sin \phi$ is the perpendicular distance between the line of action of the force **F** and the point O on the axis of rotation. If the line of action of the force goes through the axis of rotation, the torque is zero:

$$\tau = Fr \sin \phi = FD = F_t r \qquad\qquad 12\text{-}26$$

Torque defined

The distance $D = r \sin \phi$ is called the *lever arm* of the force **F**. In terms of the torque, Equations 12-23 and 12-25 are written

$$dW = \tau \, d\theta = \tau \omega \, dt \qquad\qquad 12\text{-}27$$

and

$$\tau = I\alpha \qquad\qquad 12\text{-}28$$

Again it is convenient to assign a direction to the torque. (In the next chapter we shall see that in general torque is a vector.) We have chosen a case where the torque increases the magnitude of ω, and we want the direction of torque for this case to be in the same direction as **ω**. We define the direction of torque to be the direction of rotation **ω** which that torque acting alone will produce if the body starts from rest.

Equation 12-28 can then be written as a vector equation. If several forces exert torques on the wheel, it is the work done by the resultant torque that equals the change in kinetic energy. Thus we substitute $\Sigma\tau$ for τ in the above equations and obtain

$$\Sigma\boldsymbol{\tau} = I\boldsymbol{\alpha} \qquad\qquad 12\text{-}29$$

Equation 12-29 is the rotational analog of Newton's second law $\Sigma\mathbf{F} = m\mathbf{a}$ for linear motion. The resultant torque is analogous to the resultant force, the moment of inertia is analogous to the mass, and the angular acceleration is analogous to the linear acceleration.

Culver Pictures

Torque exerted by a wrench on a nut is proportional to the force and the lever arm. Charlie could exert a greater torque with the same force if he held the wrenches nearer their ends. What is the direction of the torque exerted by each wrench when he pulls on it?

Figure 12-19
String wrapped around a disk
(Example 12-12).

Example 12-12 A string is wound around the rim of a uniform disk pivoted to rotate about a fixed axis without friction. The mass of the disk is 3 kg, and its radius R is 25 cm. The string is pulled with a force of 10 N (Figure 12-19). If the disk is initially at rest, what is its angular velocity after 5 sec?

The moment of inertia about the axis of the disk is

$$I = \tfrac{1}{2}mR^2 = \tfrac{1}{2}(3 \text{ kg})(0.25 \text{ m})^2 = 0.0938 \text{ kg-m}^2$$

Since the direction of the string as it leaves the rim of the disk is always tangent to the disk, the lever arm of the force it exerts is just R. The applied torque is thus

$$\tau = FR = (10 \text{ N})(0.25 \text{ m}) = 2.5 \text{ N-m}$$

From Equation 12-29

$$\alpha = \frac{\Sigma\tau}{I} = \frac{2.5 \text{ N-m}}{0.0938 \text{ kg-m}^2} = 26.7 \text{ rad/sec}^2$$

Since α is constant, we find ω after 5 sec from Equation 12-8, setting $\omega_0 = 0$:

$$\omega = \omega_0 + \alpha t = 0 + (26.7 \text{ rad/sec}^2)(5 \text{ sec}) = 133 \text{ rad/sec}$$

This angular speed can be expressed in more familiar terms as

$$\omega = \frac{133 \text{ rad}}{1 \text{ sec}} \frac{60 \text{ sec}}{1 \text{ min}} \frac{1 \text{ rev}}{2\pi \text{ rad}} = 1270 \text{ rev/min}$$

Example 12-13 A mass m is tied to a light string wound around a wheel of moment of inertia I and radius R (Figure 12-20). The wheel bearing is frictionless. Find the tension in the string, the acceleration of m, and its speed after it has fallen a distance h from rest.

The only force which exerts a torque on the wheel is the tension T in the string. It has lever arm R. Hence

$$TR = I\alpha \tag{12-30}$$

Two forces act on the suspended mass, the upward tension **T** and the downward weight $m\mathbf{g}$. Hence $\Sigma\mathbf{F} = m\mathbf{a}$ for this body gives

$$mg - T = ma \tag{12-31}$$

(where we have taken down as the positive direction).

There are three unknowns, T, a, and α, in these two equations. However, the string provides a constraint by which a and α can be related. As the wheel turns through angle θ, an amount of string of length $s = R\theta$ unwraps and the mass m falls an equal distance $y = R\theta$. Thus we have

$$y = R\theta \tag{12-32}$$

$$\frac{dy}{dt} = R\frac{d\theta}{dt} \quad \text{or} \quad v = R\omega \tag{12-33}$$

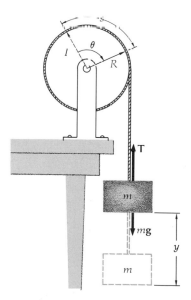

Figure 12-20
Mass tied to string which is wrapped around a rotating wheel (Example 12-13). When the wheel turns through an angle θ, a length of string $R\theta$ unwraps and the mass drops a distance $y = R\theta$.

Differentiating again gives

$$a = R\alpha \tag{12-34}$$

Thus, Equation 12-30 can be written

$$TR = I\frac{a}{R} \quad \text{or} \quad Ia = TR^2 \tag{12-35}$$

Solving Equations 12-31 and 12-35 for a and T, we obtain

$$a = \frac{m}{m + I/R^2}g \qquad T = \frac{I/R^2}{m + I/R^2}mg \tag{12-36}$$

Since the acceleration is constant, the speed after m has moved down a distance h from rest can be computed from the constant-acceleration relationship,

$$v^2 - v_0^2 = 2as$$

or

$$v^2 - 0 = 2\frac{m}{m + I/R^2}gh$$

$$v^2 = \frac{2m}{m + I/R^2}gh \tag{12-37}$$

Energy is conserved for this system, since there is no friction. We can therefore obtain the same result using the conservation-of-energy principle. As m moves down the distance h, the potential energy of the system decreases by amount mgh. This must be balanced by the simultaneous increase in kinetic energy of the system. Since the system starts at rest, the increase in kinetic energy is the final kinetic energy:

$$E_k = \tfrac{1}{2}mv^2 + \tfrac{1}{2}I\omega^2 = \tfrac{1}{2}mv^2 + \tfrac{1}{2}I\frac{v^2}{R^2} = \tfrac{1}{2}\left(m + \frac{I}{R^2}\right)v^2$$

Equating this to mgh gives

$$\tfrac{1}{2}\left(m + \frac{I}{R^2}\right)v^2 = mgh$$

or

$$v^2 = \frac{2m}{m + I/R^2}gh$$

Questions

10. If force **F** produces no torque, does it have any effect on the motion of a body? Could it be removed without making any change in the motion?

11. What are two ways **F** can be applied to a body so that it produces no torque about the axis of rotation?

12. If the torque of **F** is zero about one point, will it necessarily be zero about another point? *Can* it be zero about any other point?

13. How can a small force produce a greater angular acceleration of a wheel than a large force?

14. If a body has no angular velocity at a given instant, can it have a torque acting on it? If a body has no angular acceleration at a given instant, can it have a torque acting on it?

12-6 Static Equilibrium of a Rigid Body

If an object remains stationary, it is said to be in *static equilibrium*. An important problem is to determine the forces acting on a body when it is in static equilibrium. In Chapter 5 we found that a necessary condition for a particle to be stationary is that the resultant force be zero. Then the particle does not accelerate, and if its velocity is initially zero, it remains zero. Since the acceleration of the center of mass of a body equals the resultant force divided by the total mass, this condition is also necessary for a rigid body to be in equilibrium. However, even if the center of mass of a body is at rest, the body may rotate. Thus it is also necessary that the resultant torque about the center of mass be zero. If the center of mass of a body is at rest and there is no rotation about it, there is no rotation about any point. The necessary conditions for static equilibrium of a rigid body are therefore:

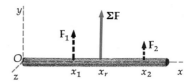

Figure 12-21
The two parallel forces F_1 and F_2 can be replaced by a single resultant force ΣF which has the same effect. The point of application of ΣF is x_r, given by $|\Sigma F|x_r = F_1 x_1 + F_2 x_2$.

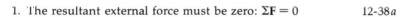

1. The resultant external force must be zero: $\Sigma F = 0$ 12-38a

2. The resultant external torque about any point must be zero: $\Sigma \tau = 0$ 12-38b

Conditions for static equilibrium

In giving examples of the application of these conditions we shall often wish to replace two or more forces acting on a body by a single equivalent force. When two parallel forces F_1 and F_2 act on a rod as in Figure 12-21, the single force having the same effect on the translation of the center of mass of the body is the sum $\Sigma F = F_1 + F_2$. If this single force is to have the same effect on rotation, we must choose the point of application so that the torque it exerts about any point equals the sum of the torques exerted by the original forces. Let x_1 and x_2 be the lever arms for the two original forces about the point O. The resultant force ΣF will produce the same torque about O if it is applied at distance x_r given by

$$x_r \Sigma F = F_1 x_1 + F_2 x_2 \qquad 12\text{-}39$$

We can use this result to show that the force of gravity exerted on the various parts of a body can be replaced by a single force, the total weight, acting at the center of mass. Figure 12-22 shows an object made of several small particles, each of weight w_i. The total weight is $W = \Sigma w_i$. Generalizing Equation 12-39 to the case of several parallel forces and using $\Sigma F = W$, we have for the point of application of the resultant force x_g,

$$x_g W = \Sigma w_i x_i \qquad 12\text{-}40$$

Equation 12-40 defines the coordinate of the *center of gravity*. If the acceleration of gravity does not vary over the body (nearly always the case), we write $w_i = m_i g$ and $W = Mg$ and cancel the common factor g. Then

$$x_g Mg = \Sigma m_i g x_i$$

or

$$Mx_g = \Sigma m_i x_i$$

Usually the center of gravity is the same as the center of mass.

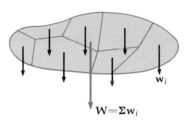

Figure 12-22
The weights of all the particles of a body can be replaced by the total weight acting at the center of gravity (which coincides with the center of mass if g does not vary over the body).

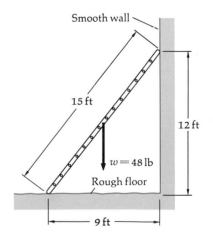

Figure 12-23
Ladder on rough floor leaning against a smooth wall (Example 12-14).

Example 12-14 A uniform 15-ft ladder weighs 48 lb and leans against a smooth vertical wall (Figure 12-23). The foot of the ladder is 9 ft from the wall. What is the minimum coefficient of friction necessary between the ladder and the floor if the ladder is not to slip?

The three forces acting on the ladder are (1) the weight $\mathbf{w}$ at the center of mass acting down, (2) the force $\mathbf{F}_1$ exerted horizontally by the wall (since the wall is smooth it exerts only a normal force), and (3) the force exerted by the floor, which is the vector sum of the normal force $\mathbf{N}$ and the frictional force $\mathbf{f}$. From our first condition we have

$$N = w = 48 \text{ lb} \qquad \text{and} \qquad F_1 = f$$

Since we know neither f nor F_1, we must use the second condition of equilibrium: the resultant torque about any point is zero. The most convenient point to choose for considering rotation is the point of contact between the ladder and the floor. Then since neither $\mathbf{N}$ nor $\mathbf{f}$ exerts any torque about this point, they will not appear in the equation. The torque exerted by the weight about this point is 48 lb times the lever arm, 4.5 ft. This torque tends to rotate the ladder clockwise and is thus directed inward. The torque exerted by $\mathbf{F}_1$ is in the opposite direction and has the magnitude F_1 times its lever arm, 12 ft. Thus the second condition gives

$$w(4.5) - F_1(12) = 0$$

$$F_1 = \frac{4.5w}{12} = \frac{4.5(48)}{12} = 18.0 \text{ lb}$$

This equals the magnitude of the frictional force, but since f is related to the normal force by

$$f \leq \mu_s N$$

we have

$$\mu_s \geq \frac{f}{N} = \frac{18.0}{48} = 0.375$$

where μ_s is the coefficient of static friction.

Another way of solving this problem is perhaps easier. Let $\mathbf{F}_2 = \mathbf{f} + \mathbf{N}$ be the force exerted by the floor. The ratio f/N is just the cotangent of the angle between the force $\mathbf{F}_2$ and the horizontal. We can find this angle as follows. If we extend the lines of action of the forces $\mathbf{w}$ and $\mathbf{F}_1$, we note that they meet at point P_1 in Figure 12-24. The torques exerted by these two forces about this point are zero. Since $\mathbf{F}_2$ is the only other force acting, it must also exert zero torque about point P_1. Thus its line of action must also pass through this point. From the figure one sees that if this is so, the cotangent of the angle made by $\mathbf{F}_2$ with the horizontal is $(4.5 \text{ ft})/(12 \text{ ft}) = 0.375 = f/N \leq \mu_s$. We note that if a body is in static equilibrium under the influence of three nonparallel coplanar forces, the lines of action of these forces must intersect at one point.

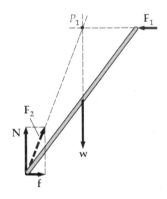

Figure 12-24
Forces acting on the ladder in Example 12-14. Point P_1 is the intersection of the lines of action of $\mathbf{F}_1$ and $\mathbf{w}$. The line of action of the force exerted by the floor $\mathbf{F}_2 = \mathbf{N} + \mathbf{f}$ must also pass through P_1 so that there is no resultant torque about that point.

If two forces are antiparallel, we cannot replace them with a single force which has the same effect. Two equal but opposite forces acting as shown in Figure 12-25 are called a *couple*. The torque produced by these forces about point O is

$$\tau = Fx_2 - Fx_1 = F(x_2 - x_1) = FD$$

where F is the magnitude of either force and $D = x_2 - x_1$ is the distance between them. This result does not depend on the choice of the point O, and the torque produced by a couple is the same about all points in space.

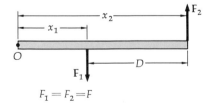

$$F_1 = F_2 = F$$

Figure 12-25
Two forces equal in magnitude but opposite in direction constitute a couple. The couple exerts the same torque FD about any point in space. Since the resultant force of a couple is zero, it cannot be replaced by a single force.

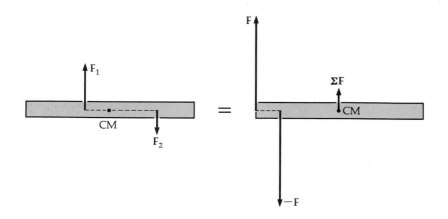

Figure 12-26
Two antiparallel forces of unequal magnitude can be replaced by force $\Sigma F = F_1 + F_2$ acting through the center of mass plus a couple whose torque equals that of the original forces about the center of mass.

The two unequal antiparallel forces in Figure 12-26 can be replaced by a single force equal to the resultant force plus a couple, as shown in the figure. In general, any number of parallel and antiparallel forces can be replaced by a single resultant force plus a couple. Since the resultant force exerted by a couple is zero, the only way it can be balanced is by a second couple which exerts equal but opposite torque. For example, the forces **N** and **w** in Figure 12-24 form a couple of moment (48 lb) (4.5 ft) = 216 lb-ft. It is balanced by the forces **f** and F_1, which also form a couple. Since the separation of these forces is $D = 12$ ft, their magnitude must be (216 lb-ft)/(12 ft) = 18 lb as found in Example 12-14.

Review

A. Define, explain, or otherwise identify:

Rigid body, 304
Angular velocity, 305
Angular acceleration, 306
Moment of inertia, 310
Parallel-axis theorem, 311

Radius of gyration, 316
Torque, 318
Center of gravity, 321
Couple, 322

B. True or false:

1. Angular velocity and linear velocity have the same dimensions.

2. All parts of a rotating wheel have the same angular velocity.

3. All parts of a rotating wheel have the same angular acceleration.

4. Any quantity that has magnitude and direction is a vector.

5. The moment of inertia of a body depends on the location of the axis of rotation.

6. The moment of inertia of a body depends on the angular velocity of the body.

7. Two forces can be equal in magnitude and direction but produce different torques about some point.

8. Whenever the resultant force acting on a body is zero, the body is in static equilibrium.

9. If a body is in static equilibrium, the resultant torque must be zero about any point.

10. A couple can be replaced by a single force which has the same effect.

Exercises

Section 12-1, Angular Velocity and Angular Acceleration

1. Find a formula to convert radians per second to revolutions per minute.

2. A 12-in-diameter record rotates at $33\frac{1}{3}$ rev/min. (a) What is its angular velocity in radians per second? (b) Find the speed and (linear) acceleration of a point on the rim of the record.

3. A particle moves in a circle of radius 100 m with constant speed of 20 m/sec. (a) What is its angular velocity in radians per second about the center of the circle? (b) How many revolutions does it make in 30 sec?

4. A wheel starts from rest and has a constant angular acceleration of 2 rad/sec². (a) What is its angular velocity after 5 sec? (b) Through what angle has the wheel turned after 5 sec? (c) How many revolutions has it made in 5 sec? (d) After 5 sec, what is the speed and acceleration of a point 1.5 ft from the axis of rotation?

5. A turntable rotating at $33\frac{1}{3}$ rev/min is shut off. It brakes with constant angular acceleration and comes to rest in 2 min. (a) Find the angular acceleration. (b) What is the average angular velocity of the turntable? (c) How many revolutions does it make before stopping?

6. A disk of radius 10 cm rotates about its axis from rest with constant angular acceleration of 10 rad/sec². At $t = 5$ sec what is (a) the angular velocity of the disk and (b) the tangential and radial acceleration a_t and a_r of a point on the edge of the disk?

7. A spot on a record rests on the radial line $\theta = 0°$ (fixed in space). At $t = \frac{1}{4}$ sec after the turntable is turned on, the spot has advanced to $\theta = 10°$. Assuming constant angular acceleration, how long will it be before the record is rotating at 33.3 rev/min?

Section 12-2, The Vector Nature of Rotation, and Section 12-3, Rotational Kinetic Energy and Moment of Inertia

8. A coordinate system is fixed to a car. The rear axle coincides with the z axis, which is positive to the right when viewed from behind the car; the x axis is parallel to the ground pointing toward the front of the car; the y axis is vertically upward. (a) The car starts from rest and accelerates uniformly. What are the directions of $\boldsymbol{\omega}$ and $\boldsymbol{\alpha}$? (b) What are the directions of $\boldsymbol{\omega}$ and $\boldsymbol{\alpha}$ while the car slows down?

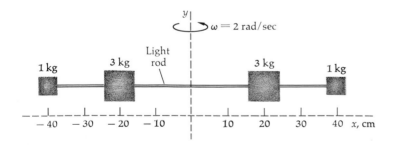

Figure 12-27
Exercise 9.

9. The masses in Figure 12-27 are connected by a very light rod whose moment of inertia may be neglected. They rotate about the y axis with angular velocity of 2 rad/sec. (a) Find the speed of each mass and use it to calculate the kinetic energy of this system directly from $\Sigma\frac{1}{2}m_iv_i^2$. (b) Find the moment of inertia about the y axis and calculate the kinetic energy from $E_k = \frac{1}{2}I\omega^2$.

10. Four point masses are at the corners of a square connected by massless rods, $m_1 = m_3 = 3$ kg and $m_2 = m_4 = 4$ kg. The length of the side of the square is $L = 2$ m (Figure 12-28). (a) Find the moment of inertia about an axis perpendicular to the plane of the masses and passing through m_4. (b) How much work is required to produce a rotation of 2 rad/sec about this axis?

11. Four 2-kg point masses are located at the corners of a rectangle of sides 3 and 2 m (Figure 12-29). (a) Find the moment of inertia of this system about an axis perpendicular to the plane of the masses and through one of the masses. (b) The system is set rotating about this axis with kinetic energy of 184 J. Find the number of revolutions the system makes per minute.

12. An irregularly shaped object rotates about an axis at 50 rev/min. The object does 300 J of work as it stops. What is the moment of inertia of the object?

Section 12-4, Calculating the Moment of Inertia

13. Use the parallel-axis theorem to find the moment of inertia of the four-mass system in Figure 12-28 about an axis perpendicular to the plane of the masses and through the center of mass. Check your result by direct computation.

14. (a) Find the moment of inertia I_x for the four-mass system of Figure 12-28 about the x axis which passes through m_3 and m_4. (b) Find I_y for that system about the y axis which passes through m_1 and m_4. (c) Use Equation 12-16 to calculate the moment of inertia I_z about the z axis through m_4 and perpendicular to the plane of the figure.

15. (a) Use the parallel-axis theorem to find the moment of inertia about an axis parallel to the z axis and through the center of mass of the system in Figure 12-29. (b) Let x' and y' be axes in the plane of the figure through the center of mass and parallel to the sides of the rectangle. Compute $I_{x'}$ and $I_{y'}$ and use your results and Equation 12-16 to check your result for part (a).

16. Find the moment of inertia of a disk of mass M and radius R about an axis through the edge of the disk parallel to the axis of the disk (see Figure 12-30).

17. Use Equation 12-16 to find the moment of inertia of a disk of radius R and mass M about an axis in the plane of the disk through its center (Figure 12-31).

18. Use the parallel-axis theorem and Equation 12-21 to find the moment of inertia of a solid sphere about an axis tangent to the sphere (Figure 12-32).

19. Find the moment of inertia about an axis tangent to a hoop of mass M and radius R.

20. Find the moment of inertia about an axis intersecting a hoop and parallel to the axis of the hoop.

Figure 12-28
Exercises 10, 13, and 14.

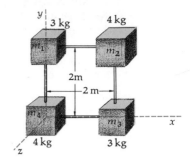
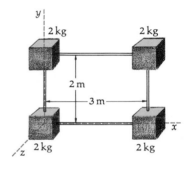

Figure 12-29
Exercises 11 and 15.

Figure 12-30
Exercise 16.

Axis of rotation

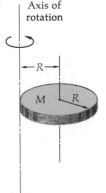

Figure 12-31
Exercise 17.

Figure 12-32
Exercise 18.

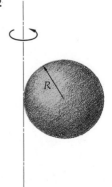

Section 12-5, Torque

21. A uniform disk of radius 0.12 m and mass 5 kg is pivoted so that it rotates freely about its axis. A string wrapped around the disk is pulled with a force of 20 N (Figure 12-33). (a) What is the torque exerted on the disk? (b) What is the angular acceleration of the disk? (c) If the disk starts from rest, what is its angular velocity after 3 sec? (d) What is its kinetic energy after 3 sec? (e) Find the total angle θ the disk turns through in 3 sec and (f) show that the work done by the torque $\tau\theta$ equals the kinetic energy.

22. A disk-shaped grindstone of mass 2 kg and radius 7 cm is spinning at 700 rev/min. When the power is shut off, a man continues to sharpen his ax by holding it against the grindstone for 10 sec until the grindstone stops rotating. (a) Find the angular acceleration of the grindstone assuming it to be constant. (b) What is the torque exerted by the ax on the grindstone? (Assume no other frictional torques.)

23. In order to start a playground merry-go-round rotating, a rope is wrapped around it and pulled. A force of 200 N is exerted on the rope for 10 sec. During this time the merry-go-round, which has a radius of 2 m, makes one complete rotation. (a) Find the angular acceleration of the merry-go-round assuming it to be constant. (b) What torque is exerted by the rope on the merry-go-round? (c) What is the moment of inertia of the merry-go-round?

24. A force F is exerted tangential to a wheel of radius R. When the wheel turns through an angle $d\theta$, the force acts through a distance $ds = R\,d\theta$. Show that the power input of the force is $\tau\omega$, where τ is the torque exerted by the force and ω is the angular velocity.

25. A rope is wrapped around a 3-kg cylinder of radius 10 cm which is free to turn about its axis. The rope is pulled with a force of 15 N. The cylinder is initially at rest at $t = 0$. (a) Find the torque exerted by the rope, the moment of inertia of the cylinder, and the angular acceleration of the cylinder. (b) Find the angular velocity of the cylinder at time $t = 4$ sec. What is the power input of the force at this time (see Exercise 24)?

26. A 1000-lb block is lifted by a steel cable which passes over a pulley to a motor-driven winch (see Figure 12-34). The diameter of the winch drum is 1 ft, and the moment of inertia of the pulley is negligible. (a) What force must be exerted by the cable to lift the block at a constant velocity of 3 in/sec? (b) What torque does the cable exert on the winch drum? (c) What is the angular velocity of the winch drum? (d) What power must be developed by the motor to drive the winch drum?

27. An engine develops 400 lb-ft of torque at 3500 rev/min. Find the horsepower developed by the engine.

28. Figure 12-35 shows a number of forces applied to an object pivoted on an axis through point O and perpendicular to the figure. Compute the torque of each force and the resultant torque. If the object is released from rest in the position shown, which way will it turn?

Figure 12-33
Exercise 21.

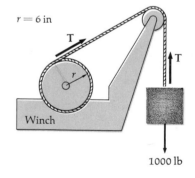

Figure 12-34
Exercise 26.

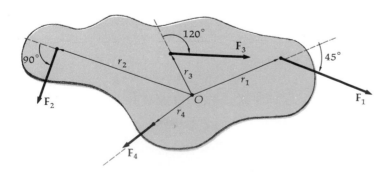

Figure 12-35
Exercise 28. The magnitudes of the vectors shown are

$F_1 = 10$ lb	$r_1 = 1.0$ ft
$F_2 = 6$ lb	$r_2 = 1.5$ ft
$F_3 = 8$ lb	$r_3 = 0.5$ ft
$F_4 = 4$ lb	$r_4 = 0.5$ ft

29. The system in Figure 12-36 is released from rest. The 30-kg mass is 1 m above the floor. The pulley is a uniform disk with a radius of 10 cm and mass 5 kg. Find the velocity of the 30-kg mass just before it hits the floor. *Hint:* Use conservation of energy; be sure to include the energy of rotation of the pulley.

Section 12-6, Static Equilibrium of a Rigid Body

30. An 80-lb child and a 60-lb child sit balanced on a seesaw. The total length of the board is 10 ft. Use the fact that the resultant torque about the fulcrum must be zero to find the position of the fulcrum.

31. A 20-lb board that is 12 ft long rests on two supports each 1 ft from the end of the board. An 80-lb block is placed on the board 3 ft from one end (and thus 2 ft from one support) (Figure 12-37). Find the force exerted by each support on the board.

Figure 12-36
Exercise 29.

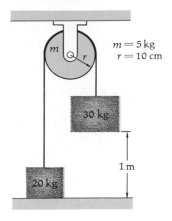

$m = 5\,\mathrm{kg}$
$r = 10\,\mathrm{cm}$

Figure 12-37
Exercise 31.

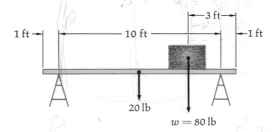

32. A 500-lb weight is supported by a cable attached to a strut hinged at point A (Figure 12-38). The strut is supported by a second cable under tension T_2, as shown. The mass of the strut is negligible. (*a*) What are the three forces acting on the strut? (*b*) Show that the vertical component of the tension T_2 must equal 500 lb. (*c*) Find the force exerted on the strut by the hinge.

33. Find the force exerted by the hinge A on the strut for the arrangement in Figure 12-39. (Neglect the mass of the strut.)

Figure 12-38
Exercise 32.

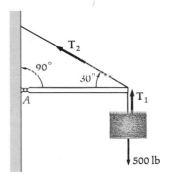

Problems

1. A wheel mounted on a horizontal axle has a radial line painted on it which is used to determine its angular position relative to the vertical. The wheel has a constant angular acceleration. At time $t = 0$ the line points straight up. At $t = 1$ sec, the line is again vertical, having turned through one full revolution. At $t = 2$ sec, the line points straight down, the wheel having turned through $1\frac{1}{2}$ rev more since $t = 1$ sec. (*a*) What is the angular acceleration of the wheel? (*b*) What was its angular velocity at $t = 0$?

2. Calculate the kinetic energy of rotation of the earth and compare it with the kinetic energy of the earth's center-of-mass motion. Assume the earth to be a homogeneous sphere of mass 6.0×10^{24} kg and radius 6.4×10^6 m. The radius of the earth's orbit is 1.5×10^{11} m.

3. A flywheel has a mass of 100 kg and a radius of gyration of 0.5 m and rotates with angular velocity of 1200 rev/min. (*a*) A constant tangential force is applied at a radial distance of 0.5 m. What work must this force do to stop the wheel? (*b*) If the wheel is brought to rest in 2 min, what torque does the force produce? What is the magnitude of the force? (*c*) How many revolutions does the wheel make in these 2 min?

4. A wheel is mounted on an axis that is not frictionless. It is at rest initially. A constant external torque of 50 N-m is applied to the wheel for 20 sec. At the end of the 20 sec the wheel has an angular velocity of 600 rev/min. The external torque is then removed and the wheel comes to rest after 120 sec more. (*a*) What is the moment of inertia of the wheel? (*b*) What is the frictional torque (assumed to be constant)?

Figure 12-39
Exercise 33.

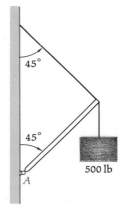

5. A uniform cylinder of mass m_1 and radius R is pivoted on frictionless bearings. A string wrapped around the cylinder connects to a mass m_2, which is on a frictionless incline of angle θ, as shown in Figure 12-40. This system is released from rest with m_2 a height h above the bottom of the incline. (a) What is the acceleration of m_2? (b) What is the tension in the string? (c) What is the total energy of the system when m_2 is at height h? (d) What is the total energy when m_2 is at the bottom of the incline and has speed v? (e) What is the speed v? (f) Evaluate your answers for the extreme cases of $\theta = 0$, $\theta = 90°$, and $m_1 = 0$.

Figure 12-40
Problem 5.

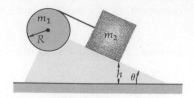

6. An Atwood's machine has two masses, $m_1 = 500$ gm and $m_2 = 510$ gm, connected by a string of negligible mass which passes over a frictionless pulley (Figure 12-41). The pulley has mass 50 gm and radius of gyration $k = 3$ cm. The radius at which the string passes over the pulley is 4 cm. The string does not slip on the pulley. (a) Find the acceleration of the masses. (b) What is the tension in the string supporting m_1? In the string supporting m_2? By how much do they differ? (c) What would your answers have been if you had neglected the motion of the pulley?

Figure 12-41
Problem 6.

7. A body of mass m is pivoted on a horizontal frictionless axle with its center of mass a distance h from the pivot, as shown in Figure 12-42. Its moment of inertia is $I = mk^2$ about this axis, where k is the radius of gyration. The line from the center of mass to the pivot makes an angle θ_0 with the vertical ($\theta_0 < 90°$). The body is released from rest. At some later time the angle is θ. (a) Show that the magnitude of the angular acceleration is

$$\alpha = \frac{gh \sin \theta}{k^2}$$

and that the tangential component of the acceleration of the center of mass is

$$a_t = \frac{gh^2 \sin \theta}{k^2}$$

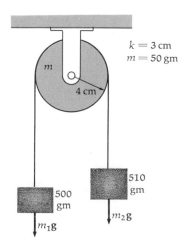

$k = 3$ cm
$m = 50$ gm

4 cm

510 gm

500 gm

$m_2 g$

$m_1 g$

(b) Use conservation of energy to find the angular velocity and show that the magnitude of the radial acceleration of the center of mass is

$$a_r = 2 \frac{gh^2}{k^2} (\cos \theta - \cos \theta_0)$$

(c) Show that the force exerted by the pivot has radial and tangential components given by

$$F_r = mg \cos \theta + 2 \frac{mgh^2}{k^2} (\cos \theta - \cos \theta_0)$$

and

$$F_t = mg \frac{h^2}{k^2} \sin \theta - mg \sin \theta$$

(d) For the case of a simple pendulum show that part (c) gives the expected results for F_t at any θ and for $\dot{F}_r$ at $\theta = 0$. (e) Find F_r and F_t for a uniform stick pivoted at one end.

Figure 12-42
Problem 7.

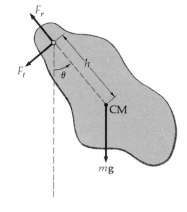

F_r

F_t

θ

h

CM

mg

8. A uniform rectangular plate has mass m and sides a and b. (a) Show by integration that its moment of inertia about an axis perpendicular to the plate and through one corner is $\frac{1}{3}m(a^2 + b^2)$. (b) What is the moment of inertia about an axis at the center of mass and perpendicular to the plate?

9. A cube of uniform material has mass m and side L. By integration, find its moment of inertia about an axis parallel to an edge and through the center of mass.

10. Find by integration the moment of inertia of a spherical shell of radius R and mass m about an axis through the center of the shell.

11. A hollow cylinder has mass m, outside radius R_2, and inside radius R_1.

Show that its moment of inertia about its symmetry axis is given by $I = \frac{1}{2}m(R_2^2 + R_1^2)$.

12. Figure 12-43 shows a pair of uniform spheres each of mass 500 gm and radius 5 cm. They are mounted on a uniform rod which has length $L = 30$ cm and mass 60 gm. (*a*) Calculate the moment of inertia of this system about an axis perpendicular to the rod through the center of the rod using the approximation that the two spheres can be treated as point masses a distance 20 cm from the axis of rotation and the mass of the rod is negligible. (*b*) Calculate the moment of inertia exactly and compare your result with your approximate value.

13. A 6 by 3 by 3 ft box of uniform mass is placed on end on a rough, hinged plank, as shown in Figure 12-44. The plank is inclined at an angle θ, which is slowly increased. The coefficient of friction is great enough to prevent the box from sliding. Find the greatest angle that can be applied without tipping the box over. *Hint:* First show that if the center of gravity of the box does not lie above any portion of the base, the conditions for static equilibrium cannot be satisfied.

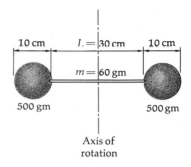

Figure 12-43
Problem 12.

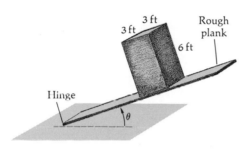

Figure 12-44
Problem 13.

14. A 200-lb boy sits on top of a ladder of negligible weight which rests on a frictionless floor. There is a cross brace halfway up the ladder (Figure 12-45). The angle at the apex is $\theta = 30°$. (*a*) What is the force exerted by the floor on each leg of the ladder? (*b*) Find the tension in the cross brace. (*c*) If the cross brace is moved down toward the bottom of the ladder (with the same angle θ), will its tension be greater or less?

15. A wheel of mass M and radius R rests on a horizontal surface against a step of height h ($h < R$). The wheel is to be raised over the step by a horizontal force $\mathbf{F}$ applied to the axle of the wheel. Find the force $\mathbf{F}$ necessary to raise the wheel over the step.

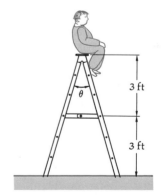

Figure 12-45
Problem 14.

CHAPTER 13 Rotation in Space and Angular Momentum

In a general motion, the axis of rotation of a body is not fixed in space. The motion is complex because the directions as well as the magnitudes of the angular velocity and acceleration change. Another complication arises for systems which are not rigid. The moment of inertia may change in time. In describing these complex motions of a system of particles we shall find it useful to introduce the concept of angular momentum. As you might expect, angular momentum plays a role in rotational motion analogous to that of (linear) momentum in linear motion. In particular, we shall see that the time rate of change of the angular momentum of a system equals the resultant torque acting on the system.

13-1 Torque as a Vector Product

We begin by considering a single particle of mass m moving with velocity $\mathbf{v} = d\mathbf{r}/dt$ and momentum $\mathbf{p} = m\mathbf{v}$, where $\mathbf{r}$ is the radius vector of the particle from the origin O. We assume that the point O is in an inertial reference frame so that Newton's law $\Sigma\mathbf{F} = d\mathbf{p}/dt$ holds, where $\Sigma\mathbf{F}$ is the resultant force. For the moment let us consider a single force $\mathbf{F}$ acting on the particle (Figure 13-1). The torque exerted by $\mathbf{F}$ relative to the point O is defined to be a vector of magnitude $Fr \sin \phi$, where ϕ is the angle between $\mathbf{F}$ and $\mathbf{r}$, and direction perpendicular to the plane formed by $\mathbf{F}$ and $\mathbf{r}$ shown in Figure 13-1. This is the same definition we gave in the last chapter for torque. Here we want to emphasize that torque is defined *relative to a point in space* (and not relative to an axis, like the moment of inertia of a body). The torque is conveniently written as the *vector product*, or *cross product*, of $\mathbf{r}$ and $\mathbf{F}$:

$$\tau = \mathbf{r} \times \mathbf{F} \qquad \text{13-1}$$

In general, the cross product of two vectors $\mathbf{A}$ and $\mathbf{B}$ is defined to be:

Figure 13-1
Force $\mathbf{F}$ acting on a particle at position $\mathbf{r}$. The torque τ is perpendicular to both $\mathbf{F}$ and $\mathbf{r}$ and has the magnitude $Fr \sin \phi$.

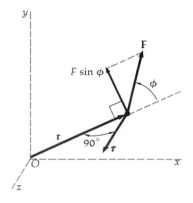

*a vector whose magnitude equals the area of the parallelogram formed by the two vectors (Figure 13-2a) and whose direction is perpendicular to the plane containing **A** and **B** and in the sense given by the right-hand rule as **A** is rotated into **B** through the smallest angle between the vectors.*

Cross product defined

If ϕ is the angle between the two vectors and $\hat{\mathbf{n}}$ is a unit vector perpendicular to each and in the sense described, the cross product of **A** and **B** is

$$\mathbf{A} \times \mathbf{B} = AB \sin \phi \, \hat{\mathbf{n}} \qquad \text{13-2}$$

The relation between the vectors **A**, **B**, and **A** × **B** is shown in Figure 13-2b.

If **A** and **B** are parallel, **A** × **B** is zero. From the definition, Equation 13-2, it follows that

$$\mathbf{A} \times \mathbf{A} = 0 \qquad \text{13-3}$$

and

$$\mathbf{A} \times \mathbf{B} = -\mathbf{B} \times \mathbf{A} \qquad \text{13-4}$$

It should be especially noted that the order in which the two vectors are multiplied is significant. Unlike the multiplication of ordinary numbers, changing the order of two vectors in a cross product changes the result. In fact, as indicated in Equation 13-3, changing the order of the factors in a cross product changes the sign of the result. The cross products of rectangular unit vectors are easily found. For example, $\mathbf{i} \times \mathbf{j} = |\mathbf{i}||\mathbf{j}| \sin 90° \, \mathbf{k} = \mathbf{k}; \, \mathbf{k} \times \mathbf{k} = 0$. Similarly,

$$\mathbf{i} \times \mathbf{i} = 0 \qquad \mathbf{i} \times \mathbf{j} = \mathbf{k} \qquad \mathbf{i} \times \mathbf{k} = -\mathbf{j}$$

$$\mathbf{j} \times \mathbf{i} = -\mathbf{k} \qquad \mathbf{j} \times \mathbf{j} = 0 \qquad \mathbf{j} \times \mathbf{k} = \mathbf{i}$$

$$\mathbf{k} \times \mathbf{i} = \mathbf{j} \qquad \mathbf{k} \times \mathbf{j} = -\mathbf{i} \qquad \mathbf{k} \times \mathbf{k} = 0$$

Some properties of the cross product of two vectors follow:

1. Associative law

$$\mathbf{A} \times (\mathbf{B} + \mathbf{C}) = \mathbf{A} \times \mathbf{B} + \mathbf{A} \times \mathbf{C} \qquad \text{13-5}$$

2. Derivative: if **A** and **B** are functions of some variable such as t, the derivative of **A** × **B** follows the usual product rule for derivatives:

$$\frac{d}{dt} (\mathbf{A} \times \mathbf{B}) = \mathbf{A} \times \frac{d\mathbf{B}}{dt} + \frac{d\mathbf{A}}{dt} \times \mathbf{B} \qquad \text{13-6}$$

We must keep the order straight here since, for example, $\mathbf{B} \times d\mathbf{A}/dt = -(d\mathbf{A}/dt) \times \mathbf{B}$.

From Equation 13-1 we see that the magnitude of the torque $Fr \sin \phi$ can be thought of as the product of F with the distance $r \sin \phi$, which is the perpendicular distance from the line of the force to the origin, called the lever arm. Alternatively, we can think of it as the product of r and $F \sin \phi$, which is the component of the force perpendicular to **r**.

Figure 13-2
(a) The vector product **A** × **B** is a vector **C** perpendicular to both **A** and **B** and of magnitude $AB \sin \phi$, which is the area of the parallelogram shown. (b) The direction of **A** × **B** is the direction of rotation when **A** is rotated into **B** through the angle ϕ.

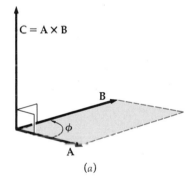

(a)

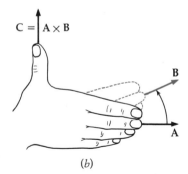

(b)

Questions

1. Will a force acting on a body produce a torque if the body is not mounted on an axle?

2. A body is mounted on pivots so that it is free to turn about the z axis. A force **F** is applied at point (x,y,z). Does the z component of the torque depend on what value z has? On what value F_z has?

13-2   Angular Momentum of a Particle

We define the angular momentum **L** of a particle relative to the origin O to be the cross product of the position vector **r** and the linear momentum **p** (see Figure 13-3):

$$\mathbf{L} = \mathbf{r} \times \mathbf{p} \qquad\qquad 13\text{-}7$$

Angular momentum defined

Like torque, angular momentum is defined relative to a point in space. The magnitude is the magnitude of the momentum p times $r \sin \phi$, which is the perpendicular distance from the line of motion to the origin. Alternatively, L is the distance r times $p \sin \phi$, which is the component of the momentum perpendicular to **r**. We shall now show that Newton's second law implies that the rate of change of angular momentum equals the resultant torque acting on the particle. If we have a number of forces acting on a particle, the resultant torque relative to the point O is the sum of the torques due to each force:

$$\Sigma \boldsymbol{\tau} = \mathbf{r} \times \mathbf{F}_1 + \mathbf{r} \times \mathbf{F}_2 + \cdots = \mathbf{r} \times \Sigma \mathbf{F}$$

According to Newton's law, the resultant force equals the rate of change of linear momentum $d\mathbf{p}/dt$. Thus

$$\Sigma \boldsymbol{\tau} = \mathbf{r} \times \Sigma \mathbf{F} = \mathbf{r} \times \frac{d\mathbf{p}}{dt} \qquad\qquad 13\text{-}8$$

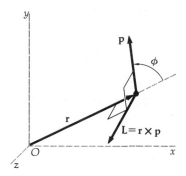

Figure 13-3
A particle with momentum **p** at position **r** has angular momentum relative to the origin $\mathbf{L} = \mathbf{r} \times \mathbf{p}$, which is perpendicular to both **r** and **p**.

We can compute the rate of change of the angular momentum using the product rule for derivatives:

$$\frac{d\mathbf{L}}{dt} = \frac{d}{dt} \, (\mathbf{r} \times \mathbf{p}) = \frac{d\mathbf{r}}{dt} \times \mathbf{p} + \mathbf{r} \times \frac{d\mathbf{p}}{dt}$$

The first term on the right of this equation is zero because

$$\frac{d\mathbf{r}}{dt} \times \mathbf{p} = \mathbf{v} \times m\mathbf{v} = 0$$

Thus

$$\frac{d\mathbf{L}}{dt} = \mathbf{r} \times \frac{d\mathbf{p}}{dt} \qquad\qquad 13\text{-}9$$

Comparing Equations 13-8 and 13-9, we have

$$\Sigma \boldsymbol{\tau} = \frac{d\mathbf{L}}{dt} \qquad\qquad 13\text{-}10$$

Resultant torque on a particle

Equation 13-10 is the rotational analog of Newton's second law $\Sigma \mathbf{F} = d\mathbf{p}/dt$. It follows directly from that law and from the definitions of torque and angular momentum. We shall give two examples of the calculation of the angular momentum for a single particle.

Example 13-1 Find the angular momentum for a point mass moving in a circle of radius r. The speed of the particle v and the magnitude of its angular velocity are related by $v = r\omega$ (see Equation 12-4). Choosing the xy plane to be the plane of the circle (Figure 13-4), we have for the

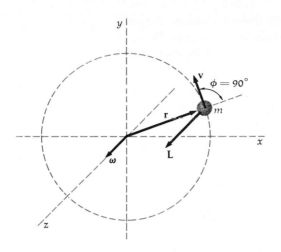

Figure 13-4
Particle moving in a circle
(Example 13-1). The angular
momentum is along the z axis
parallel to the angular velocity
$\boldsymbol{\omega}$ and has magnitude $mvr =$
$mr^2(v/r) = I\omega$.

angular momentum relative to the center of the circle,

$$\mathbf{L} = \mathbf{r} \times \mathbf{p} = \mathbf{r} \times m\mathbf{v} = rmv \sin 90° \, \mathbf{k} = rmv\mathbf{k} = mr^2\boldsymbol{\omega}\mathbf{k}$$

Note that in this case the angular momentum is in the same direction as the angular velocity. Since the quantity mr^2 is the moment of inertia I for a single particle about the z axis, we have

$$\mathbf{L} = mr^2\boldsymbol{\omega} = I\boldsymbol{\omega}$$

If the only force acting on the particle is the centripetal force directed toward the origin, there will be no torque about the origin. Then, according to Equation 13-10, the angular momentum will be constant. This implies that ω and hence the speed will be constant.

Example 13-2 Find the angular momentum relative to the origin of a particle moving in a straight line a perpendicular distance b from the origin with constant speed.

This situation is shown in Figure 13-5. The angular momentum is

$$\mathbf{L} = \mathbf{r} \times \mathbf{p} = -rmv \sin \phi \, \mathbf{k}$$

The distance $r \sin \phi$ is just the perpendicular distance b from the line of motion to the origin, as can be seen from the figure. Thus

$$\mathbf{L} = -mvb\mathbf{k} \qquad\qquad 13\text{-}11$$

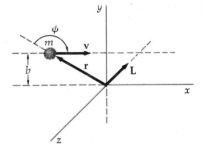

Figure 13-5
Particle moving in a straight
line parallel to the x axis and
a distance b from it with constant speed v. The angular
momentum relative to the ori-
gin is $\mathbf{L} = \mathbf{r} \times m\mathbf{v} =$
$-rmv \sin \phi \, \mathbf{k} = -mvb\mathbf{k}$.

Although we usually associate angular momentum with rotational motion, Example 13-2 illustrates that even a particle moving in a straight line has angular momentum about a point not on the line. We shall use this result in Examples 13-6 and 13-8.

Questions

3. What is the angle between a particle's linear momentum $\mathbf{p}$ and its angular momentum $\mathbf{L}$?

4. Two particles have the same momentum $\mathbf{p}$. Are their angular momenta also equal? Is it possible for two particles with the same momentum $\mathbf{p}$ also to have the same angular momentum if they are not at the same place?

5. A particle moves along a straight line at constant speed. How does its angular momentum about any point vary in time?

6. Two particles, A and B, move along parallel lines a distance D apart. They have the same speed v but travel in opposite directions. What is magnitude of A's angular momentum about a point on its own path? About a point on B's path? About particle B?

7. A particle moving at constant velocity has zero angular momentum about a particular point. Show that the particle either has passed through that point or will pass through it.

13-3   Torque and Angular Momentum for a System of Particles

The total angular momentum of a system of particles is the sum of the angular momenta of the individual particles.

$$\mathbf{L} = \Sigma \mathbf{L}_i$$

Let $\boldsymbol{\tau}_i$ be the resultant torque acting on the ith particle. The resultant torque acting on the system is then

$$\Sigma \boldsymbol{\tau}_i = \Sigma \frac{d\mathbf{L}_i}{dt} = \frac{d}{dt}(\Sigma \mathbf{L}_i) = \frac{d\mathbf{L}}{dt} \qquad 13\text{-}12$$

In Equation 13-12 the sum of the torques may include internal torques as well as those due to forces external to the system. We shall now show that if the internal forces between any two particles act parallel to the line joining the particles, the sum of the internal torques is zero. Then

$$\Sigma \boldsymbol{\tau}_{\text{ext}} = \frac{d\mathbf{L}}{dt} \qquad 13\text{-}13$$

Consider two particles as shown in Figure 13-6. Although $\mathbf{F}_1 = -\mathbf{F}_2$ by Newton's third law, it is not immediately obvious that the internal torques cancel. The sum of the two torques in this figure is

$$\boldsymbol{\tau}_1 + \boldsymbol{\tau}_2 = \mathbf{r}_1 \times \mathbf{F}_1 + \mathbf{r}_2 \times \mathbf{F}_2 = \mathbf{r}_1 \times \mathbf{F}_1 + \mathbf{r}_2 \times (-\mathbf{F}_1)$$
$$= (\mathbf{r}_1 - \mathbf{r}_2) \times \mathbf{F}_1$$

The vector $\mathbf{r}_1 - \mathbf{r}_2$ is along the line joining the two particles. Thus, if the force $\mathbf{F}_1$ acts parallel to the line joining m_1 and m_2, $\mathbf{F}_1$ and $\mathbf{r}_1 - \mathbf{r}_2$ are either parallel or antiparallel and

$$(\mathbf{r}_1 - \mathbf{r}_2) \times \mathbf{F}_1 = 0$$

If this is true for all the internal forces, the internal torques cancel in pairs.

Equation 13-13 is the rotational analog of the result for linear motion (obtained in Chapter 10) that the resultant external force equals the rate of change of the total linear momentum of a system. Before we give examples of its use we discuss the problem of finding the total angular momentum for some typical systems. In many cases of a body rotating with angular velocity $\boldsymbol{\omega}$ the angular momentum is parallel to $\boldsymbol{\omega}$ and has the magnitude $I\omega$, where I is the moment of inertia about the axis of rotation. This is true whenever the body is rotating about a symmetry axis through the center of mass and the angular momentum is taken relative to the center of mass. Thus in these cases,

$$\mathbf{L} = I\boldsymbol{\omega} \qquad 13\text{-}14$$

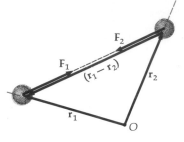

Figure 13-6
The internal forces $\mathbf{F}_1$ and $\mathbf{F}_2$ produce no net torque about O if they act along the line joining the particles.

Multiflash photograph of a diver. The center of mass moves in a parabolic path after the diver leaves the board. The angular momentum is provided by the initial torque due to the force of the board. This force does not pass through the center of mass if the diver leans forward as he jumps. If the diver wanted to undergo $1\frac{1}{2}$ rev in the air, he would draw in his arms and legs, decreasing his moment of inertia to increase his angular velocity. (*Courtesy of Harold E. Edgerton.*)

Equation 13-13 is then

$$\Sigma \tau_{\text{ext}} = \frac{d}{dt}(I\boldsymbol{\omega}) \qquad \text{13-15}$$

In the case of a rigid body the moment of inertia is constant. Then Equation 13-15 reduces to Equation 12-29:

$$\Sigma \tau_{\text{ext}} = \frac{d}{dt}(I\boldsymbol{\omega}) = I\frac{d\boldsymbol{\omega}}{dt} = I\boldsymbol{\alpha} \qquad \text{13-16}$$

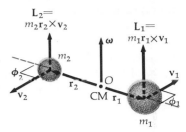

Figure 13-7
Particles m_1 and m_2 rotating about their center of mass with angular velocity $\boldsymbol{\omega}$. Their total angular momentum about O is $\mathbf{L} = I\boldsymbol{\omega}$.

Example 13-3 Find the angular momentum of a dumbbell rotating about an axis through the center of mass and perpendicular to the line joining the masses.

Let the angular velocity be $\boldsymbol{\omega}$ along the axis of rotation, which we call the z axis, as shown in Figure 13-7. The total angular momentum about the center of mass at the origin O is

$$\mathbf{L} = \mathbf{L}_1 + \mathbf{L}_2 = \mathbf{r}_1 \times m_1\mathbf{v}_1 + \mathbf{r}_2 \times m_2\mathbf{v}_2 = (r_1 m_1 v_1 + r_2 m_2 v_2)\mathbf{k}$$

The speeds of the particles are related to the angular velocity by $v = r\omega$. Thus

$$\mathbf{L} = (m_1 r_1{}^2\omega + m_2 r_2{}^2\omega)\mathbf{k} = (m_1 r_1{}^2 + m_2 r_2{}^2)\omega\mathbf{k} = I\boldsymbol{\omega}$$

where $I = m_1 r_1{}^2 + m_2 r_2{}^2$ is the moment of inertia.

Example 13-4 Find the angular momentum of a disk rotating about an axis through the center of mass and perpendicular to the plane of the disk.

The calculation of the angular momentum of the disk about the center of mass is the same as that for the dumbbell in Example 13-3 except that there are more particles (Figure 13-8). The angular momentum of each particle is along the axis of rotation parallel to the angular velocity $\boldsymbol{\omega}$. Also, the speed of each particle is related to the angular velocity by $v_i = r_i\omega$. Thus we again obtain for the total angular momentum

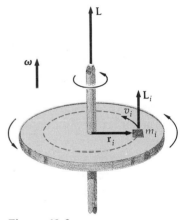

Figure 13-8
Disk rotating about its axis (Example 13-4). The angular momentum is $\mathbf{L} = I\boldsymbol{\omega}$.

$$\mathbf{L} = \Sigma m_i r_i v_i \mathbf{k} = \Sigma m_i r_i (r_i\omega)\mathbf{k} = (\Sigma m_i r_i{}^2)\omega\mathbf{k} = I\boldsymbol{\omega}$$

(where $\mathbf{k}$ is the unit vector in the direction of $\boldsymbol{\omega}$).

Example 13-5 A disk rotates about an axis perpendicular to the plane of the disk but a distance h from the symmetry axis, as shown in Figure 13-9. Calculate the angular momentum about the origin O' on the disk at the axis of rotation.

Again the angular momentum of each particle is parallel to ω. The calculation is the same as in the previous example except that the momentum of inertia I' is now about the new axis of rotation.

$$\mathbf{L} = I'\boldsymbol{\omega}$$

We can use the parallel-axis theorem to relate the angular momentum relative to O' in Example 13-5 to that about the center of mass found in Example 13-4. According to this theorem, the moment of inertia I' about axis z' is related to that about the z axis through the center of mass a distance h away by

$$I' = I_{CM} + Mh^2$$

If we multiply each term by ω, we obtain

$$I'\omega = I_{CM}\omega + Mh^2\omega \qquad \text{13-17}$$

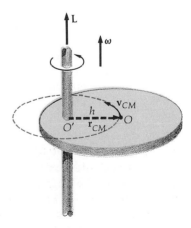

Figure 13-9
Disk rotating about an axis perpendicular to the disk but at a distance h from the symmetry axis. Again, the angular momentum is $\mathbf{L} = I'\boldsymbol{\omega}$, where I' is the moment of inertia of the disk about the axis of rotation.

The quantity $I'\omega$ is the magnitude of the angular momentum about the point O'. The quantity $I_{CM}\omega$ is the magnitude of the angular momentum about the center of mass O. The speed of the center of mass relative to O' is $v_{CM} = h\omega$. Let $\mathbf{r}_{CM}$ be the vector from O' to the center of mass O. This vector is perpendicular to $\mathbf{v}_{CM}$ and has the magnitude h. Then the term $Mh^2\omega$ is the magnitude of the vector $\mathbf{r}_{CM} \times M\mathbf{v}_{CM}$. This vector is the angular momentum of a particle of mass M (the total mass of the disk) moving with velocity $\mathbf{v}_{CM}$. Equation 13-17 is thus

$$\mathbf{L} = \mathbf{L}_{CM} + \mathbf{r}_{CM} \times M\mathbf{v}_{CM} \qquad\qquad 13\text{-}18$$

This result, obtained by considering rotation of a disk about two parallel axes, holds in general.

The angular momentum about any point in space O' is the sum of the angular momentum about the center of mass of the system plus the angular momentum associated with center-of-mass motion about the point O'.

This theorem follows directly from the definition of angular momentum and the properties of the center of mass. Its general derivation will be outlined in Section 13-8.

One result of Equation 13-18 is that if the only motion of a body is rotation about an axis through its center of mass, the angular momemtum of the body about any point in space is $\mathbf{L}_{CM}$. This follows from the fact that in this case $\mathbf{v}_{CM} = 0$. The angular momentum of a body about its center of mass is often called the *spin* of the body. For example, in the motion of a gyroscope (Section 13-6) the angular momentum consists of two parts, $\mathbf{L}_{CM}$, due to the spinning of the wheel, and $\mathbf{r}_{CM} \times m\mathbf{v}_{CM}$, due to the precession of the gyroscope. Equation 13-18 is also used to find the angular momentum of an electron in an atom. The electron has an intrinsic angular momentum, called *spin*, just as if it were a tiny spinning ball. The total angular momentum of the electron is the sum of this spin angular momentum and the angular momentum associated with the orbital motion of the electron around the nucleus.

Spin defined

We conclude this section with an example of the use of Equation 13-13.

Example 13-6 Two masses m_1 and m_2 in an Atwood's machine are connected by a string passing over a pulley, as shown in Figure 13-10. The pulley has radius R and moment of inertia I about its axis of rotation. Find the acceleration of the masses and the angular acceleration of the pulley.

Let us compute the angular momentum about the center of the pulley. We note that although the masses are not rotating, they do have angular momentum. In Example 13-2 we saw that relative to some point O the angular momentum of a particle moving with velocity v along a line is mvb, where b is the distance from the line to the point. In this case $b = R$, the radius of the pulley. The angular momentum of each mass is in the same direction, the direction of the angular velocity of the pulley, which is into the page. The magnitude of the total angular momentum of the two masses plus pulley is

$$L = m_1vR + m_2vR + I\omega$$

where v is the speed of the masses. If we assume that the string does not slip on the pulley, the speed of the string v must equal the speed

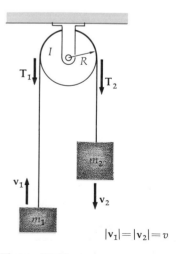

Figure 13-10
Atwood's machine. The angular momentum of each mass and that of the pulley about the center of the pulley is into the page. The rate of change of the total angular momentum equals the resultant torque about this point, which is just $m_2gR - m_1gR$.

of the rim of the pulley $R\omega$. We eliminate ω from the equation:

$$L = (m_1 + m_2)vR + I\frac{v}{R}$$

The only external forces acting on this system are the weights of the masses and pulley and the supporting force at the axis of the pulley. This force and the weight of the pulley do not exert any torque about the center of the pulley. The resulting external torque on this system is thus $m_2gR - m_1gR$. The direction of the torque is into the page, parallel to the angular momentum. Therefore Equation 13-13 is

$$\Sigma\tau_{\text{ext}} = (m_2 - m_1)gR = \frac{dL}{dt} = \left[(m_2 + m_2)R + \frac{I}{R}\right]\frac{dv}{dt}$$

and the acceleration is

$$a = \frac{dv}{dt} = \frac{m_2 - m_1}{m_1 + m_2 + I/R^2}g$$

The effect of taking into account the rotation of the pulley is to reduce the acceleration of the masses. The angular acceleration of the pulley is

$$\alpha = \frac{d\omega}{dt} = \frac{1}{R}\frac{dv}{dt} = \frac{a}{R}$$

Questions

8. If the total angular momentum of a system of particles is zero, are all the particles at rest?

9. If the total angular momentum of a system is constant, can we conclude that no net force acts on the system?

10. If no net force acts on a system, will there be no torque on the system?

11. Two external forces act on an extended body and, in general, will produce a net torque on the body. Are there any points about which the total torque of these forces is zero?

13-4 Conservation of Angular Momentum

If the resultant external torque acting on a system is zero, the total angular momentum is constant. We thus have a third conservation law for isolated systems. Energy, linear momentum, and angular momentum are conserved. The law of conservation of angular momentum is a fundamental law of nature. Even on a microscopic scale in atomic and nuclear physics, in which newtonian mechanics does not hold, the angular momentum of an isolated system is constant in time.

This section gives several examples of how the law of conservation of angular momentum is used.

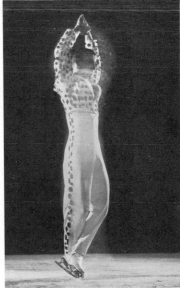

Bradley Smith/Photo Researchers, Inc.

Because the torque exerted by the ice is small, the angular momentum of the skater is approximately constant. When he reduces his moment of inertia by drawing in his arms, his angular velocity increases.

Example 13-7 A disk with moment of inertia I_1 is rotating with angular velocity ω_i about a frictionless shaft. It drops onto another disk with moment of inertia I_2 initially at rest (see Figure 13-11). Because of surface friction, the two disks eventually attain a common angular velocity. What is it?

Each disk exerts a torque on the other, but there is no torque external to the two-disk system. The angular momentum of the first disk about its center of mass is $I_1\boldsymbol{\omega}_i$. Since the center of mass is at rest, this is also the angular momentum about any point. When both disks are rotating together, the total angular momentum is

$$I_1\boldsymbol{\omega}_f + I_2\boldsymbol{\omega}_f = (I_1 + I_2)\boldsymbol{\omega}_f = I_1\boldsymbol{\omega}_i$$

Thus the final angular velocity is

$$\boldsymbol{\omega}_f = \frac{I_1}{I_1 + I_2}\,\boldsymbol{\omega}_i$$

This interaction of the disks is analogous to an inelastic collision of two masses in one dimension. Energy is not conserved in this collision. We can see this best by writing the energy in terms of the angular momentum. The initial kinetic energy is

$$E_{ki} = \tfrac{1}{2}I_1\omega_i^2 = \frac{(I_1\omega_i)^2}{2I_1} = \frac{L_i^2}{2I_1}$$

The final energy is

$$\tfrac{1}{2}(I_1 + I_2)\omega_f^2 = \frac{[(I_1 + I_2)\omega_f]^2}{2(I_1 + I_2)} = \frac{L_f^2}{2(I_1 + I_2)}$$

Since $L_f = L_i$, the final kinetic energy is less than the initial energy by the factor $I_1/(I_1 + I_2)$.

Example 13-8 A playground merry-go-round is at rest, pivoted about a frictionless axis. A child runs along a path tangential to the rim with initial speed v_i and jumps onto the merry-go-round (Figure 13-12). What is the angular velocity of the merry-go-round and child?

We cannot expect that energy will be conserved because the child makes an inelastic collision with the rim of the merry-go-round. Linear momentum is not conserved either. The pivot of the merry-go-round exerts an impulse during the collision, but because it is frictionless, it cannot exert any torque. Thus, angular momentum about the pivot is conserved. This problem again demonstrates that we need not have circular motion to have angular momentum. Figure 13-12 shows that the initial angular momentum of the child about the pivot point is

$$\mathbf{L}_i = \mathbf{r} \times m\mathbf{v}_i = mv_iR\mathbf{k}$$

where R is the perpendicular distance between the pivot and the line of motion of the child, which is also the radius of the merry-go-round, and m is the mass of the child. The direction of the angular momentum is shown in Figure 13-12. After the child is on the merry-go-round, the angular momentum is

$$mv_fR\mathbf{k} + I\omega\mathbf{k} = (mR\omega R + I\omega)\mathbf{k} = (mR^2 + I)\omega\mathbf{k}$$

where we have used $v_f = R\omega$ because the child is at rest relative to the rim of the merry-go-round after having jumped on. Since $\mathbf{L}_f = \mathbf{L}_i$, we have

$$(mR^2 + I)\omega\mathbf{k} = mv_iR\mathbf{k} \qquad \text{or} \qquad \omega = \frac{mv_iR}{mR^2 + I}$$

Example 13-9 Show that a planet moving around the sun sweeps out equal areas in equal times.

Figure 13-13 shows a planet moving in an orbit about the sun (here

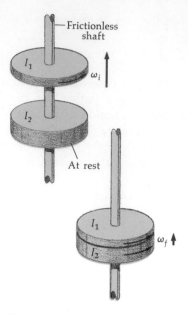

Figure 13-11
Inelastic rotational collision (Example 13-7). The disk I_1, initially rotating with angular velocity $\boldsymbol{\omega}_i$, drops onto disk I_2, which is initially at rest. Because of the internal frictional forces between the disks, the two disks eventually turn together with a common angular velocity $\boldsymbol{\omega}_f$. If the shaft is frictionless, angular momentum is conserved, and $I_1\boldsymbol{\omega}_i = (I_1 + I_2)\boldsymbol{\omega}_f$.

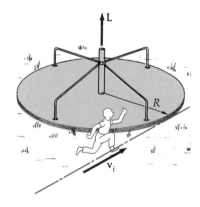

Figure 13-12
Example 13-8. A child runs tangentially to the rim of a merry-go-round and jumps on. If the pivot is frictionless, angular momentum is conserved.

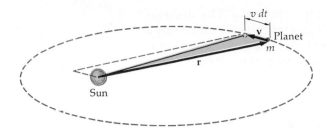

Figure 13-13
Planet orbiting sun (Example
13-9). The area swept out in
time dt is half the area of the
parallelogram indicated,
which is the magnitude of the
cross product of $\mathbf{r}$ and $\mathbf{v}\ dt$.
Thus $dA = \frac{1}{2}|\mathbf{r} \times \mathbf{v}\ dt| =$
$(1/2m)|\mathbf{r} \times m\mathbf{v}|\ dt = (1/2m)L\ dt$.

we are not using the approximation of a circular orbit). Since the force
on the planet is directed toward the sun, there is no torque about the
sun and the angular momentum about the sun is conserved. (We shall
assume the sun to be at rest.) In time dt the planet moves a distance
$v\ dt$ and sweeps out the area indicated in Figure 13-13. This is half the
area of the parallelogram formed by the vectors $\mathbf{r}$ and $\mathbf{v}\ dt$, which is
$\mathbf{r} \times \mathbf{v}\ dt$. Thus the area dA swept out in dt by the radius vector $\mathbf{r}$ is

$$dA = \tfrac{1}{2}|\mathbf{r} \times \mathbf{v}\ dt| = \frac{1}{2m}|\mathbf{r} \times m\mathbf{v}|\ dt = \frac{1}{2m}L\ dt$$

or

$$\frac{dA}{dt} = \frac{L}{2m}$$

where L is the angular momentum of the planet. Since L is a constant,
the rate at which the radius vector of the planet sweeps out area is also
constant. Expressed differently, the radius vector sweeps out equal
areas in equal times. This fact, deduced from astronomical observation
early in the seventeenth century by Johannes Kepler, is one of three
empirical laws known as *Kepler's laws of planetary motion*. It is a
consequence of the fact that the force of gravity is a central force. We
see also that since L is constant in direction as well as magnitude, the
motion of a planet must remain in the plane formed by $\mathbf{r}$ and $\mathbf{v}$.

Questions

12. A man sits on a spinning piano stool with his arms folded. If he
extends his arms, what happens to his angular velocity?

13. Does a particle moving at constant speed along a straight line
sweep out equal areas in equal times with respect to any arbitrary
point?

14. A particle moves under the influence of a central force $\mathbf{F} = f(r)\hat{\mathbf{r}}$.
Will its angular momentum about the center of force be constant?
Would Kepler's law of equal areas apply to this motion?

15. It is said that a cat always lands on its feet. If a cat starts falling feet
up, how can it land on its feet without violating the law of conservation of angular momentum?

13-5 Translation and Rotation

In general, the motion of a system of particles is very complicated
when the axis of rotation is not fixed. As we have mentioned, we can
consider any such motion as a combination of a translation of the

center of mass plus a rotation about the center of mass. The description of the translation is not difficult. The center of mass moves as a point particle under the influence of the net external force. The analysis of the rotational motion is simplified by an important theorem concerning the angular momentum relative to the center of mass:

The resultant torque about the center of mass equals the rate of change of angular momentum relative to the center of mass no matter how the center of mass is moving:

$$\Sigma\tau_{CM} = \frac{d\mathbf{L}_{CM}}{dt} \qquad\qquad 13\text{-}19$$

Torque relative to the center of mass

Equation 13-19, stating that the resultant torque equals the rate of change of the angular momentum, is the same as Equation 13-13 except that now the torques and angular momentum are taken relative to the center of mass. Since we used Newton's second law to derive Equation 13-13, our derivation is valid only if the torques and angular momentum are taken relative to a point fixed in an inertial reference frame. This theorem states that the result is also valid if the torques and angular momentum are taken relative to the center of mass even if the center of mass is accelerating. The proof is given in Section 13-8.

In this section we apply this theorem to a special class of translational and rotational motion which can be easily treated: motion of a rigid body rotating about an axis of symmetry which moves so that it remains parallel to a line fixed in space, e.g., a rolling ball or cylinder.

Consider a cylinder rolling down an inclined plane, as in Figure 13-14. The forces on the cylinder are its weight, the normal force of the incline, and friction exerted by the incline. (If there were no friction, a cylinder released from rest at the top with no rotation would slide down the plane without rotating.) As for the general motion of a system of particles, the resultant force equals the total mass times the acceleration of the center of mass of the cylinder. If the cylinder rolls without slipping, the linear distance s moved by the center of mass is related to the angle of rotation of the cylinder ϕ by

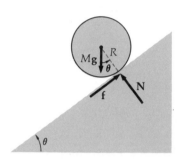

Figure 13-14
Forces on a cylinder rolling down an incline.

$$s = R\phi \qquad\qquad 13\text{-}20$$

Condition for rolling without slipping

(see Figure 13-15). Thus the linear velocity of the center of mass is related to the angular velocity of rotation by

$$v_{CM} = \frac{ds}{dt} = R\,\frac{d\phi}{dt} = R\omega \qquad\qquad 13\text{-}21$$

and the acceleration of the center of mass is

$$a_{CM} = \frac{dv_{CM}}{dt} = R\,\frac{d\omega}{dt} = R\alpha \qquad\qquad 13\text{-}22$$

where α is the angular acceleration. As the cylinder accelerates down the incline, the angular velocity of rotation must increase if it is to roll without slipping. Thus the angular momentum of the cylinder about the center of mass increases. This increase is due to the torque exerted by the frictional force on the cylinder. (The weight $M\mathbf{g}$ and the normal force $\mathbf{N}$ act through the center of mass and therefore exert no torque about this point.) The magnitude of the torque exerted by friction is fR. According to Equation 13-19, we can set this torque equal to the rate of change of angular momentum about the center of mass even though the center of mass is accelerating. The magnitude of the angu-

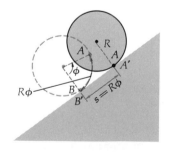

Figure 13-15
When the cylinder rolls without slipping, the point of contact moves the same distance $s = R\phi$ whether measured from A to B along the cylinder or from A' to B' along the plane. This distance is also the distance moved by the center of mass.

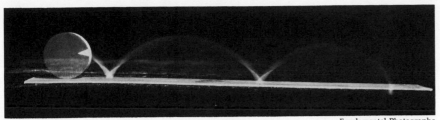

Fundamental Photographs

Path of a point on the rim of a rolling checker. The motion is simpler when viewed in the center-of-mass reference frame, which moves with the center of the checker. In this frame the point on the rim moves in a circle.

lar momentum about the center of mass is

$$L_{CM} = I_{CM}\omega \qquad\qquad 13\text{-}23$$

where I_{CM} is the moment of inertia about the axis of rotation, which is constant. Since the direction of the axis of rotation does not change (the axis remains parallel to its original direction), only the magnitude of the angular momentum changes. Thus

$$\frac{dL_{CM}}{dt} = I_{CM}\frac{d\omega}{dt} = I_{CM}\alpha \qquad\qquad 13\text{-}24$$

Setting the torque equal to the rate of change of the angular momentum, we obtain

$$fR = I_{CM}\alpha \qquad\qquad 13\text{-}25$$

The linear acceleration of the center of mass of the cylinder is parallel to the incline. The resultant force in this direction is $Mg \sin\theta - f$. Thus $\Sigma F = Ma_{CM}$ gives

$$Mg \sin\theta - f = Ma_{CM} \qquad\qquad 13\text{-}26$$

We can solve Equations 13-25 and 13-26 for any of the unknowns α, a_{CM}, and f using the condition for rolling without slipping (Equation 13-22). We first eliminate $\alpha = a_{CM}/R$.

$$fR = I_{CM}\frac{a_{CM}}{R}$$

$$f = \frac{I_{CM}}{R^2}\,a_{CM} \qquad\qquad 13\text{-}27$$

Putting this into Equation 13-26 gives

$$Mg \sin\theta - \frac{I_{CM}}{R^2}a_{CM} = Ma_{CM}$$

$$a_{CM} = \frac{Mg \sin\theta}{M + I_{CM}/R^2} \qquad\qquad 13\text{-}28$$

For a cylinder, $I_{CM} = \tfrac{1}{2}MR^2$. Then

$$a_{CM} = \frac{Mg \sin\theta}{M + \tfrac{1}{2}M} = \tfrac{2}{3}g \sin\theta \qquad\qquad 13\text{-}29$$

We can use this to find the force of friction and the angular acceleration

$$f = \frac{\tfrac{1}{2}MR^2}{R^2}\,a_{CM} = \tfrac{1}{3}Mg \sin\theta \qquad\qquad 13\text{-}30$$

$$\alpha = \frac{a_{CM}}{R} = \frac{2}{3}\frac{g}{R}\sin\theta \qquad\qquad 13\text{-}31$$

The frictional force in Equation 13-30 was found from the total mass, the moment of inertia, and the angle of inclination without any mention of the coefficient of friction. Since the cylinder is rolling without slipping, the surfaces of the cylinder and plane in contact are always instantaneously at rest relative to each other. Thus the friction is static friction, which in general is *not* equal to its maximum, limiting value $\mu_s N$. If μ_s is the coefficient of static friction, we have

$$f \leqslant \mu_s N = \mu_s Mg \cos \theta$$

Using $f = \frac{1}{3} Mg \sin \theta$ from Equation 13-30 for rolling without slipping, we have

$$\tfrac{1}{3} Mg \sin \theta \leqslant \mu_s Mg \cos \theta \qquad \text{or} \qquad \tan \theta \leqslant 3\mu_s$$

If the incline is too steep, or if the coefficient of static friction between it and the cylinder is too small, the cylinder will slip as it moves down the incline.

The linear acceleration is less than $g \sin \theta$, of course, because of the frictional force directed up the incline. Since the acceleration is constant, we can find the velocity of the cylinder at the bottom using the constant-acceleration formulas. If the original height of the center of mass of the cylinder is h, the total distance traveled is $s = h/(\sin \theta)$. The velocity at the bottom is then given by

$$v_{CM}{}^2 = 2a_{CM}s = 2(\tfrac{2}{3}g \sin \theta) \frac{h}{\sin \theta} = \tfrac{4}{3}gh$$

$$v_{CM} = \sqrt{\tfrac{4}{3}gh} \qquad\qquad\qquad 13\text{-}32$$

We can also obtain this result from energy considerations. We note that since the friction is static, there is no mechanical-energy dissipation. Thus the total mechanical energy, the kinetic energy plus the potential energy, is conserved. At the top of the incline the total energy is the potential energy Mgh. At the bottom the total energy is kinetic energy, which in general can be written $\frac{1}{2}Mv_{CM}{}^2 + E_{kr}$. Here, the energy relative to the center of mass is just the rotational energy $\frac{1}{2}I_{CM}\omega^2$. Conservation of energy gives

$$\tfrac{1}{2}Mv_{CM}{}^2 + \tfrac{1}{2}I_{CM}\omega^2 = Mgh \qquad\qquad 13\text{-}33$$

Using the rolling condition $v_{CM} = R\omega$, we can eliminate either v_{CM} or ω. Eliminating ω gives

$$\tfrac{1}{2}Mv_{CM}{}^2 + \tfrac{1}{2}I_{CM} \left(\frac{v_{CM}}{R}\right)^2 = Mgh$$

or

$$v_{CM}{}^2 = \frac{2Mgh}{M + I_{CM}/R^2}$$

For $I_{CM} = \frac{1}{2}MR^2$ for a cylinder, we obtain

$$v_{CM}{}^2 = \frac{2Mgh}{M + \frac{1}{2}M} = \tfrac{4}{3}gh$$

in agreement with the result found from the acceleration.

Question

16. A wheel rolls along a level surface at speed v_0 without slipping. Relative to the surface, what is the speed of the bottom of the wheel? Of the center of the wheel? Of the top of the wheel? Is there any frictional force acting on the wheel?

13-6   Motion of a Gyroscope

We now consider an example of a gyroscope or symmetric top, in which the axis of rotation changes direction. In general such motions are very complicated. Figure 13-16 shows such a system consisting of a bicycle wheel free to turn on an axle which is pivoted at a point a distance D from the center of the wheel but free to turn in any direction.

When the axle is held horizontal and released, if the wheel is not spinning it simply falls. The torque about the point O is MgD in the direction into the diagram in Figure 13-16. As the wheel falls, its angular momentum due to center-of-mass motion is also into the page. Since the center of mass accelerates downward, the upward force $\mathbf{F}$ exerted by the support at O is evidently less than Mg.

Now assume that the wheel is spinning. The angular momentum relative to point O is the angular momentum relative to the center of mass, in this case the spin, plus the angular momentum due to motion of the center of mass. It is quite easy in practice to make the spin angular momentum very large so that to a first approximation we can neglect the contribution due to motion of the center of mass. From Figure 13-16 we see that the torque is perpendicular to the angular momentum. In order for the angular momentum to change in the direction of the torque, the axle must move in the horizontal plane, as shown. In time dt the angular-momentum change has the magnitude

$$dL = \tau \, dt = MgD \, dt$$

The angle $d\phi$ through which the axle moves is

$$d\phi = \frac{dL}{L} = \frac{MgD \, dt}{L}$$

The movement of the axle is called precession. The angular velocity of precession is

$$\omega_p = \frac{d\phi}{dt} = \frac{MgD}{L} \qquad\qquad 13\text{-}34$$

Since the center of mass does not drop, the support point evidently exerts an upward force equal to Mg.

The observation that the wheel moves in a horizontal plane instead of falling is at first surprising. We are more familiar with situations like a falling stick, where there is no initial angular momentum, so that the direction of the change in angular momentum is also the direction of the angular momentum. There is an analogous situation in circular motion, e.g., that of the moon about the earth. The earth exerts a force on the moon toward the earth. Why doesn't the moon move toward the earth and hit it? If the moon were held with no initial momentum and released, a change in momentum (from zero) toward the earth

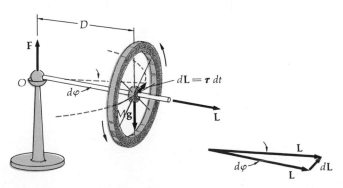

Figure 13-16
Gyroscope. The weight $M\mathbf{g}$ produces a torque about the pivot into the paper which causes a *change* in the angular momentum in that direction. If the wheel is initially spinning so there is initial angular momentum $\mathbf{L}$ along the axle, the change is perpendicular to $\mathbf{L}$ and the axle moves in the direction of the torque. This motion is called precession.

would result in the moon's moving toward the earth. However, since the moon has an initial momentum perpendicular to the line joining the earth, a change in momentum toward the earth merely results in the moon's being deviated from straight-line motion into a circular arc. Thus though $d\mathbf{p}$ is always toward the earth, $\mathbf{p}$ is tangential to the circle. For the gyroscope, if there is no initial angular momentum, the torque (into the page) causes an angular momentum (into the page) associated with center-of-mass motion as it falls; but if there is a large initial angular momentum along the axis of the wheel, this same torque merely deflects the angular momentum into the page. In this case the pivot exerts an upward force sufficient to prevent the center of mass from falling.

If the axle is not horizontal but makes an angle θ with the vertical, as in Figure 13-17, the torque about point O is $MgD \sin \theta$. The angle of precession in time dt is now

$$d\phi = \frac{dL}{L \sin \theta} = \frac{MgD \sin \theta \, dt}{L \sin \theta} = MgD \, dt \qquad \text{13-35}$$

The angular velocity of precession ω_p is thus independent of angle θ.

A careful observation of the motion of a gyroscope reveals that if the axle is held horizontally and released from rest, the motion of the axle is not confined to the horizontal plane. It dips down at first, and as the gyroscope precesses, there is a small vertical oscillation called *nutation*. This effect was neglected in our previous discussion because we neglected the contribution of the center-of-mass motion to the total angular momentum about the pivot. If the spin angular momentum is much greater than that due to the motion of the center of mass during precession, the nutation is very small. The precessional motion of the center of mass results in a small component of angular momentum $MD^2\omega_p$ in the vertical direction. However, there is no torque in this direction. In order for the axle to precess in the horizontal plane without nutation, it must be given an angular impulse $MD^2\omega_p$ as it is released. If this is not done, the vertical component of the total angular momentum must remain zero. Then as the axle begins to precess, it must dip down so that there is a downward component of *spin* angular momentum to cancel the vertical angular momentum due to center-of-mass motion. We can analyze the motion qualitatively from the moment of release.

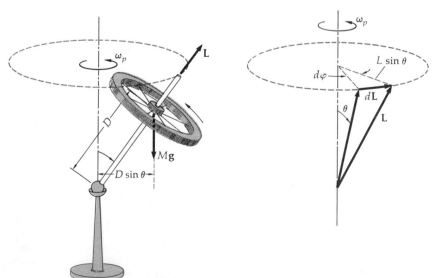

Figure 13-17
The rate of precession of a gyroscope is independent of the angle its axis makes with the vertical.

Just before the axle is released, the force of support at O is $Mg/2$ and that at the hand is $Mg/2$. At the time of release, therefore, the center of mass must accelerate downward. At this time the angular velocity of precession increases from zero. As the center of mass falls and the axle begins to precess, the force at O increases. When the force equals Mg, the acceleration of the center of mass is zero but it is moving downward. It thus overshoots its equilibrium position, the force at O becomes greater than Mg, and the center of mass eventually stops its downward motion and moves up again until it is horizontal.

Optional

13-7 Static and Dynamic Imbalance

In all the examples of rotation we have considered so far, the angular-momentum vector **L** has been parallel to the angular-velocity vector **ω**. This result is not general but holds only for rotation about certain axes called *principal axes*. In this section we consider some examples in which the angular-momentum and angular-velocity vectors are not parallel.

Figure 13-18 shows a simple dumbbell consisting of equal masses connected by a light rod rotating about an axis through the center of mass but not perpendicular to the line of centers. (In Example 13-3 we considered a similar system rotating about an axis perpendicular to the line of centers, in which case the angular-momentum vector is parallel to the angular-velocity vector.) Each mass is moving in a circle of radius $r \sin \theta$ with speed $r \sin \theta \, \omega$, where θ is the angle between the rod and the axis of rotation and ω is the angular velocity. The angular momentum of the mass m_1 is $\mathbf{L}_1 = \mathbf{r}_1 \times m_1 \mathbf{v}_1$. Its direction is in the plane of the paper perpendicular to the rod, and its magnitude is $L_1 = mr^2 \sin \theta \, \omega$. The angular momentum of mass m_2 is $\mathbf{L}_2 = \mathbf{r}_2 \times m_2 \mathbf{v}_2$. Since $\mathbf{r}_2 = -\mathbf{r}_1$ and $\mathbf{v}_2 = -\mathbf{v}_1$, $\mathbf{L}_2 = \mathbf{L}_1$. The total angular momentum of the system is thus $2mr^2 \sin \theta \, \omega$ in the direction perpendicular to the rod, as shown. It is not parallel to **ω**, which is along the axis of rotation. As the dumbbell rotates, the angular-momentum vector rotates with its tip moving in a circle, as shown in Figure 13-19. If the angular velocity is constant, the angular momentum has a constant magnitude but its direction changes in time. Thus even if the angular velocity is

Figure 13-18
Equal masses rotating about an axis through their center of mass, which is not a symmetry axis. The angular momentum **L** is perpendicular to the symmetry axis and makes an angle ϕ with the angular velocity **ω**. Since **L** changes direction as the masses rotate, a torque must be exerted by the bearings on the system.

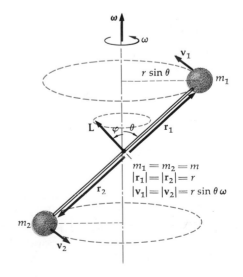

$$m_1 = m_2 = m$$
$$|\mathbf{r}_1| = |\mathbf{r}_2| = r$$
$$|\mathbf{v}_1| = |\mathbf{v}_2| = r \sin \theta \, \omega$$

Figure 13-19
The angular-momentum vector for the rotating system of Figure 13-18. The tip of the angular-momentum vector moves in a circle. Only the component $L_\rho\hat{\boldsymbol{\rho}}$ perpendicular to the axis changes. When the system rotates through an angle $d\beta = \omega\, dt$, the angular-momentum change is given by $|dL_\rho\hat{\boldsymbol{\rho}}|/L_\rho = d\beta$.

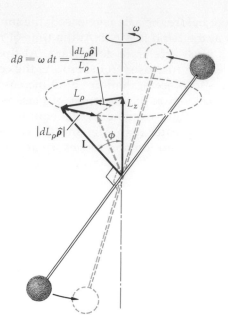

constant, the angular-momentum vector changes in time and there must be a resultant torque on the system. This torque is exerted by the bearings which support the system. At the instant shown in Figure 13-18, the torque must be out of the page since that is the direction of the change in **L**. We can calculate the magnitude of this torque by calculating the rate of change of the angular momentum.

Let us write the angular momentum vector **L** as the sum of a vector $L_z\mathbf{k}$ parallel to the axis of rotation and a vector $L_\rho\hat{\boldsymbol{\rho}}$ perpendicular to the axis of rotation, where **k** and $\hat{\boldsymbol{\rho}}$ are unit vectors. From Figure 13-19 we have for the magnitudes of these vectors,

$$L_z = L \cos\phi = L \sin\theta = 2mr^2 \sin^2\theta\,\omega$$

and

$$L_\rho = L \sin\phi = L \cos\theta = 2mr^2 \sin\theta \cos\theta\,\omega$$

The vector $L_z\mathbf{k}$ is constant in magnitude and direction if $\boldsymbol{\omega}$ is constant. The vector $L_\rho\hat{\boldsymbol{\rho}}$ is constant in magnitude, but its direction changes because the direction of the unit vector $\hat{\boldsymbol{\rho}}$ changes as the system rotates. From Figure 13-19 we have for the change in $L_\rho\hat{\boldsymbol{\rho}}$ in time dt,

$$\frac{|dL_\rho\hat{\boldsymbol{\rho}}|}{L_\rho} = d\beta$$

where

$$d\beta = \omega\, dt$$

Then

$$\frac{|dL_\rho\hat{\boldsymbol{\rho}}|}{dt} = \omega L_\rho = 2mr^2 \sin\theta \cos\theta\,\omega^2 \qquad\text{13-36}$$

Since $L_z\mathbf{k}$ is constant, the rate of change of **L** is just that of $L_\rho\hat{\boldsymbol{\rho}}$. Equation 13-36 thus gives the magnitude of $d\mathbf{L}/dt$, which in turn equals the magnitude of the torque. Note that the rate of change of angular momentum is proportional to ω^2 and is zero when $\theta = 0$, corresponding to rotation about the line joining the masses, and when $\theta = 90°$, corresponding to rotation about an axis perpendicular to the line joining the masses. The torque which produces this change in

Figure 13-20
This arrangement of two rotating masses is dynamically balanced. The angular momentum vector is parallel to the angular-velocity vector. Since **L** does not change during the rotation, no torque need be exerted on the system.

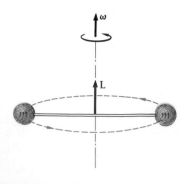

Figure 13-21
This arrangement of four rotating masses is dynamically balanced. The total angular-momentum vector L is now parallel to the angular-velocity vector **ω**, and so L does not change during the rotation.

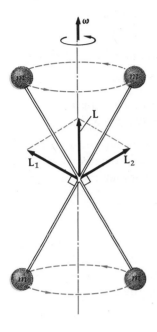

angular momentum is exerted by the bearings. Its direction is perpendicular to the axis of rotation, and its effect is to maintain a constant direction for the angular velocity **ω** as the direction of the angular-momentum vector changes. The rotating system exerts an equal and opposite torque on the support bearings, causing wear. Such a system is said to be *dynamically imbalanced.* The greater the angular velocity the greater the wear on the bearings. Note that there is no static imbalance. If we pivot the dumbbell at the center of mass, it will be in static equilibrium at any orientation. If we have a system of unknown mass distribution, e.g., an automobile wheel, we cannot detect or correct dynamic imbalance by static balance methods.

We can correct the dynamic imbalance of our two-mass system by making the angle θ equal to 90°, as in Figure 13-20, or by adding masses, as in Figure 13-21. Then L and **ω** will be parallel, and no torque will be required to maintain a constant angular velocity.

Figure 13-22 shows a wheel rotating about an axis which is off center but parallel to the symmetry axis through the center of mass and a distance h from it. The angular momentum about a point on the axis of rotation in the plane of the wheel is the sum of the angular momentum about the center of mass I_{CM} and the angular momentum associated with motion of the center of mass, both parallel to the axis of rotation. The angular momentum is just $I\omega$, where I is the moment of inertia about the axis of rotation. If the wheel rotates with constant angular velocity, the angular momentum is constant and no torque is required. However, since the center of mass moves in a circle of radius h, there must be a net *force* on the wheel of magnitude $Mv_{CM}^2/h = M(h\omega)^2/h = Mh\omega^2$ and directed toward the axis of rotation from the center of mass. This force is exerted by the bearings. Again, the equal but opposite force exerted on the bearings causes wear on the bearings, particularly if ω is large. This *static imbalance* can be detected and corrected by pivoting the wheel on a horizontal axle and balancing so that the axis of rotation passes through the center of mass.

Figure 13-22
Static imbalance. When the wheel rotates about this axis, the angular momentum L is parallel to the angular velocity **ω**. The bearings do not need to exert a torque to maintain rotation, but they must exert a force on the wheel because the center of mass of the wheel moves in a circle and therefore accelerates.

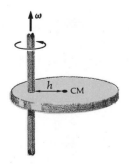

13-8 Derivation of Relations for Torque and Angular Momentum about the Center of Mass

In this section we give general derivations of the important theorems about the angular momentum relative to the center of mass expressed in Equations 13-18 and 13-19. Figure 13-23 shows a general system of particles. We write the position vector of the ith particle relative to origin O as the sum of the position vector of the center of mass $\mathbf{r}_{CM}$ and the vector from the center of mass to the particle $\mathbf{r}_{ir}$,

$$\mathbf{r}_i = \mathbf{r}_{CM} + \mathbf{r}_{ir} \qquad 13\text{-}37$$

The vector $\mathbf{r}_{CM}$ is defined by

$$M\mathbf{r}_{CM} = \Sigma m_i \mathbf{r}_i$$

Substituting $\mathbf{r}_{CM} + \mathbf{r}_{ir}$ for $\mathbf{r}_i$ in this equation, we have

$$M\mathbf{r}_{CM} = \Sigma m_i(\mathbf{r}_{CM} + \mathbf{r}_{ir}) = \mathbf{r}_{CM}\Sigma m_i + \Sigma m_i \mathbf{r}_{ir} = M\mathbf{r}_{CM} + \Sigma m_i \mathbf{r}_{ir}$$

Thus

$$\Sigma m_i \mathbf{r}_{ir} = 0 \qquad 13\text{-}38$$

Figure 13-23
Relationship among $\mathbf{r}_i$, $\mathbf{r}_{CM}$, and $\mathbf{r}_{ir}$.

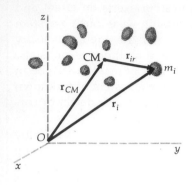

The velocity of the ith particle is

$$\mathbf{v}_i = \frac{d\mathbf{r}_i}{dt} = \frac{d\mathbf{r}_{CM}}{dt} + \frac{d\mathbf{r}_{ir}}{dt} = \mathbf{v}_{CM} + \mathbf{u}_i$$

where $\mathbf{u}_i$ is the velocity relative to the center of mass. We note from differentiating Equation 13-38 that

$$\Sigma m_i \mathbf{u}_i = 0 \qquad \text{13-39}$$

The angular momentum of the ith particle about O is

$$\mathbf{L}_i = \mathbf{r}_i \times m_i\mathbf{v}_i = (\mathbf{r}_{CM} + \mathbf{r}_{ir}) \times m_i(\mathbf{v}_{CM} + \mathbf{u}_i)$$
$$= \mathbf{r}_{CM} \times m_i\mathbf{v}_{CM} + \mathbf{r}_{ir} \times m_i\mathbf{u}_i + \mathbf{r}_{CM} \times m_i\mathbf{u}_i + \mathbf{r}_{ir} \times m_i\mathbf{v}_{CM}$$

The total angular momentum is the sum of $\mathbf{L}_i$ over all the particles

$$\mathbf{L} = \Sigma\mathbf{L}_i = (\mathbf{r}_{CM} \times M\mathbf{v}_{CM}) + (\Sigma\mathbf{r}_{ir} \times m_i\mathbf{u}_i) + (\mathbf{r}_{CM} \times \Sigma m_i\mathbf{u}_i)$$
$$+ (\Sigma m_i\mathbf{r}_{ir}) \times \mathbf{v}_{CM}$$

The last two terms in this expression are zero by Equations 13-38 and 13-39. The first term is the angular momentum associated with center-of-mass motion, and the second term is the angular momentum relative to the center of mass $\mathbf{L}_{CM}$. We thus have

$$\mathbf{L} = \mathbf{L}_{CM} + \mathbf{r}_{CM} \times M\mathbf{v}_{CM} \qquad \text{13-18}$$

We derive Equation 13-19 relating the torque about the center of mass to the rate of change of $\mathbf{L}_{CM}$ by first considering the torque about point O, which we assume to be in an inertial reference frame. If $\mathbf{F}_i$ is the external force acting on the ith particle, the net external torque about O is

$$\Sigma\,\tau_{\text{ext}} = \Sigma\,\mathbf{r}_i \times \mathbf{F}_i = \Sigma\,(\mathbf{r}_{CM} + \mathbf{r}_{ir}) \times \mathbf{F}_i$$
$$= \mathbf{r}_{CM} \times \Sigma\mathbf{F}_i + \Sigma\mathbf{r}_{ir} \times \mathbf{F}_i$$

The second term in this equation is the resultant torque about the center of mass, which we shall call $\Sigma\tau_{CM}$. Thus

$$\Sigma\tau_{\text{ext}} = \Sigma\tau_{CM} + \mathbf{r}_{CM} \times \Sigma\mathbf{F}_i \qquad \text{13-40}$$

Computing the rate of change of the angular momentum about the origin O using Equation 13-18 gives

$$\frac{d\mathbf{L}}{dt} = \frac{d\mathbf{L}_{CM}}{dt} + \frac{d\mathbf{r}_{CM}}{dt} \times M\mathbf{v}_{CM} + \mathbf{r}_{CM} \times M\frac{d\mathbf{v}_{CM}}{dt}$$

The second term in this equation is zero because $d\mathbf{r}_{CM}/dt = \mathbf{v}_{CM}$ and $\mathbf{v}_{CM} \times M\mathbf{v}_{CM} = 0$. Writing $\mathbf{a}_{CM}$ for $d\mathbf{v}_{CM}/dt$, we have

$$\frac{d\mathbf{L}}{dt} = \frac{d\mathbf{L}_{CM}}{dt} + \mathbf{r}_{CM} \times M\mathbf{a}_{CM} \qquad \text{13-41}$$

When we set the resultant torque about O (Equation 13-40) equal to the rate of change of angular momentum about O (Equation 13-41), we get

$$\Sigma\tau_{\text{ext}} = \Sigma\tau_{CM} + \mathbf{r}_{CM} \times \Sigma\mathbf{F}_i = \frac{d\mathbf{L}}{dt} = \frac{d\mathbf{L}_{CM}}{dt} + \mathbf{r}_{CM} \times M\mathbf{a}_{CM}$$

But since $\Sigma\mathbf{F}_i = M\mathbf{a}_{CM}$,

$$\Sigma\tau_{CM} = \frac{d\mathbf{L}_{CM}}{dt} \qquad \text{13-19}$$

which is the theorem to be proved.

Review

A. Define, explain, or otherwise identify:

Cross product, 330

Vector product, 330

Angular momentum, 332

Spin, 336

Precession, 343

Nutation, 344

Dynamic imbalance, 347

Static imbalance, 347

B. True or false:

1. $\mathbf{A} \times \mathbf{B}$ is the same as $\mathbf{B} \times \mathbf{A}$.

2. Angular momentum and torque are defined relative to a point in space.

3. A particle must move in a circle to have angular momentum.

4. If the resultant torque on a body is zero, its angular momentum must be zero.

5. The angular momentum of a particle is always perpendicular to its velocity.

6. If the angular momentum of a body does not change, the resultant torque on the body must be zero.

7. In general, when a body rolls down an incline without slipping, the frictional force has the magnitude $\mu_s N$.

Exercises

Section 13-1, Torque as a Vector Product

1. A force of magnitude F is applied horizontally in the negative x direction to the rim of a disk of radius R, as shown in Figure 13-24. Write $\mathbf{F}$ and $\mathbf{r}$ in terms of the unit vectors $\mathbf{i}$, $\mathbf{j}$, and $\mathbf{k}$ and compute the torque produced by the force about the origin at the center of the disk.

2. A 0.5-kg particle falls under the influence of gravity. (a) It is at $y = 10$ m and $x = 2$ m at time t_1. What is the torque about the origin exerted by gravity on the particle at this time? (b) At some later time the particle is at $y = 0$, $x = 2$ m. What is the torque about the origin at this time?

3. Compute the torque about the origin for the force $\mathbf{F} = -mg\mathbf{j}$ and $\mathbf{r} = x\mathbf{i} + y\mathbf{j}$ and show that this torque is independent of the coordinate y.

4. If $\mathbf{A} \times \mathbf{B} = 0$, what can be said about the vectors $\mathbf{A}$ and $\mathbf{B}$?

5. Find $\mathbf{A} \times \mathbf{B}$ for (a) $\mathbf{A} = 5\mathbf{i}$ and $\mathbf{B} = 5\mathbf{i} + 5\mathbf{j}$, (b) $\mathbf{A} = 5\mathbf{i}$ and $\mathbf{B} = 5\mathbf{i} + 5\mathbf{k}$, (c) $\mathbf{A} = 2\mathbf{i} + 2\mathbf{j}$ and $\mathbf{B} = -2\mathbf{i} + 2\mathbf{j}$.

6. Under what conditions is the magnitude of $\mathbf{A} \times \mathbf{B}$ equal to $\mathbf{A} \cdot \mathbf{B}$?

7. Show that for general vectors $\mathbf{A} = A_x\mathbf{i} + A_y\mathbf{j} + A_z\mathbf{k}$ and $\mathbf{B} = B_x\mathbf{i} + B_y\mathbf{j} + B_z\mathbf{k}$,

$$\mathbf{A} \times \mathbf{B} = \begin{vmatrix} \mathbf{i} & \mathbf{j} & \mathbf{k} \\ A_x & A_y & A_z \\ B_x & B_y & B_z \end{vmatrix}$$

$$= (A_yB_z - A_zB_y)\mathbf{i} + (A_zB_x - A_xB_z)\mathbf{j} + (A_xB_y - A_yB_x)\mathbf{k}$$

Section 13-2, Angular Momentum of a Particle

8. A 2-kg particle moves with constant speed of 3 m/sec in the xy plane in the y direction along the line $x = 5$ m. (a) Find the angular momentum $\mathbf{L}$ relative to the origin. (b) What torque about the origin is needed to maintain this motion?

9. A body of mass 3 kg moves at constant speed of 4 m/sec around a circle of radius 5 m. (a) What is its angular momentum about the center of the circle?

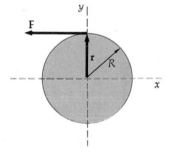

Figure 13-24
Exercise 1.

(b) What is its moment of inertia about an axis through the center of the circle and perpendicular to the plane of the motion? (c) What is the angular velocity of the particle?

10. A 3-kg body moves at constant speed of 4 m/sec along a straight line. (a) What is its angular momentum about a point 5 m from the line? (b) Describe qualitatively how its angular velocity about that point varies with time.

11. A particle travels in a circular path. (a) If its linear momentum p is doubled, how is its angular momentum affected? (b) If the radius of the circle is doubled but the speed is unchanged, how is the angular momentum of the particle affected?

12. A particle of mass m moves with speed v in a circle of radius r. Show that the kinetic energy of the particle can be written $E_k = L^2/2mr^2 = L^2/2I$, where L is the angular momentum of the particle and $I = mr^2$ is its moment of inertia.

13. A 2-kg mass moves in a circle of radius 3 m. Its angular momentum relative to the center of the circle depends on time according to $L = 4t$, where t is in seconds and L in kg-m²/sec. (a) Find the torque acting on the particle. (b) Find the angular velocity as a function of time.

14. A planet moves in an ellipitical orbit about the sun with the sun at one focus of the ellipse. (a) What is the torque produced by the gravitational force of attraction of the sun for the planet? (b) At position A in Figure 13-25 the planet is a distance r_1 from the sun and is moving with speed v_1 perpendicular to the line from the sun to the planet. At position B it is at distance r_2 moving with speed v_2, again perpendicular to the line from the sun to the planet. What is the ratio of v_1 and v_2 in terms of r_1 and r_2?

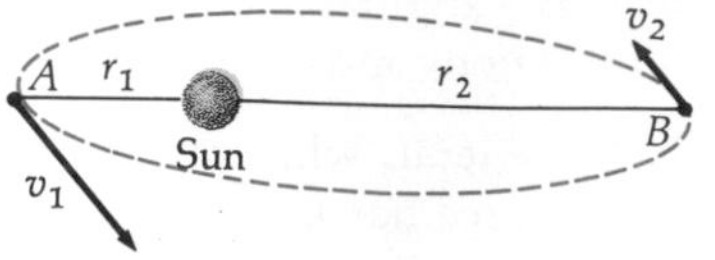

Figure 13-25
Planet moving in elliptical path about the sun (Exercise 14).

Section 13-3, Torque and Angular Momentum for a System of Particles

15. Calculate the angular momentum of the earth spinning about its axis and compare it with the angular momentum the earth has about the sun. (Assume the earth to be a homogeneous sphere of mass 6.0×10^{24} kg and radius 6.4×10^6 m.) The radius of the earth's orbit is 1.5×10^{11} m.

16. A homogeneous cylinder of mass 100 kg and radius 0.3 m is mounted so that it turns without friction on its fixed axis of symmetry. It is rotated by a drive belt that wraps around its perimeter and exerts a constant torque. At time $t = 0$, its angular velocity is zero. At time $t = 30$ sec, the angular velocity is 600 rev/min. (a) What is its angular momentum at that time? (b) At what rate is the angular momentum increasing? What is the torque acting on the cylinder? (c) What is the magnitude of the force acting on the rim of the cylinder?

17. A 15-gm coin of diameter 1.5 cm is spinning about a vertical diameter at a fixed point on a tabletop at 10 rev/sec. (a) What is the angular momentum of the coin about its center of mass? (b) What is its angular momentum about a point on the table 10 cm from the coin?

18. The coin in Exercise 17 spins about a vertical diameter at 10 rev/sec but also travels in a straight line across the tabletop at 5 cm/sec. (a) What is the angular momentum of the coin about a point on the line of motion? (b) What is the angular momentum of the coin about a point 10 cm from the line of motion? (There are two answers to this question. Explain why and give both.)

19. A 4-kg mass rests on a frictionless horizontal table. It is connected to a second mass of 2 kg by a string which passes over a frictionless pulley, as shown in Figure 13-26. The pulley is a uniform cylinder of radius 4 cm and mass 1 kg. (a) What is the net external torque on the system (two masses plus pulley) about the center of the pulley? (b) If the masses are moving with speed v with the pulley turning at angular velocity $\omega = v/r$, what is the total angular momentum of the system about the center of the pulley? (Assume that the center of mass of the 4-kg-mass is at the height of the string.) (c) Find the acceleration

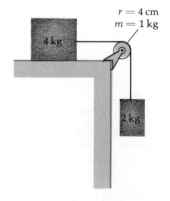

$r = 4$ cm
$m = 1$ kg
4 kg
2 kg

Figure 13-26
Exercise 19.

of the masses by differentiating your result for L in part (b) and setting dL/dt equal to the resultant torque.

20. (a) Assuming that the incline is smooth in Figure 13-27 and that the string passes through the center of mass of m_2, find the resultant torque acting on the system about the center of the pulley. (b) Write an expression for the total angular momentum of the system about the center of the pulley when the masses are moving with speed v. Assume the pulley has a moment of inertia I and radius r. (c) Find the acceleration of the masses from your results of parts (a) and (b) by setting the resultant torque equal to the rate of change of the angular momentum of the system.

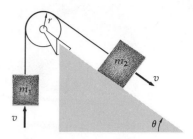

Figure 13-27
Exercise 20.

21. A uniform stick of length l and mass M lies on a smooth table. It rotates with angular velocity about an axis perpendicular to the table and through one end of the stick. (a) What is the angular momentum of the stick about the end? (b) What is the speed of the center of mass of the stick? (c) What is the angular momentum of the stick about the center of mass? (d) Show that Equation 13-18 is satisfied for this situation, where L is the angular momentum about the end of the stick.

Section 13-4, Conservation of Angular Momentum

22. A merry-go-round of radius 2 m and moment of inertia 500 kg-m² is rotating at 0.25 rev/sec. A child of mass 25 kg sitting at the center crawls out to the rim. Find (a) the new angular velocity of the merry-go-round and (b) the initial and final kinetic energy.

23. A disk is rotating freely at 1800 rev/min about a vertical axis through its center. A second disk mounted on the same shaft above the first is initially at rest. The moment of inertia of the second disk is twice that of the first. The second disk is dropped onto the first one, and the two eventually rotate together with a common angular velocity. (a) Find the new angular velocity. (b) Show that kinetic energy is lost during the "collision" of the two disks.

24. A circular platform is mounted on a vertical frictionless axle. Its radius is $r = 5$ ft, and its moment of inertia is $I = 200$ slug-ft². It is at rest initially. A 150-lb man stands on the edge of the platform and begins to walk along the edge at a speed $v_0 = 3$ ft/sec *relative to the ground*. (a) What is the angular velocity of the platform? (b) When the man has walked once around the platform so that he is at his original position on it, what is his angular displacement relative to the ground?

25. A man stands at the center of a circular platform holding his arms extended horizontally with an 8-lb weight in each hand. He is set rotating about a vertical axis with angular velocity of 0.5 rev/sec. The moment of inertia of the man plus platform is 4 slug-ft², assumed constant. The weights are 3 ft from the axis of rotation. He now pulls the weights in toward his body until they are 6 in from the axis of rotation. Find (a) his new angular velocity and (b) the initial and final kinetic energy of the man and platform. (c) How much work must the man do to pull in the weights?

26. A particle is traveling with a constant velocity **v** along a line which is a distance b from the origin O. Let dA be the area swept out by the position vector from O to the particle in time dt (Figure 13-28). Show that dA/dt is constant in time and equal to $\frac{1}{2}L/m$, where L is the angular momentum of the particle about the origin.

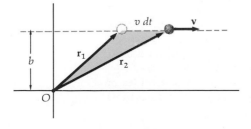

Figure 13-28
Area swept out in time dt by a particle moving in a straight line (Exercise 26).

Helicopter with two rotors
(Exercise 27).

Russ Kinne/Photo Researchers

27. Explain why a helicopter with just one main rotor has a second smaller rotor mounted on a horizontal axis at the rear. Describe the resultant motion of the helicopter if this rotor fails during a flight.

Section 13-5, Translation and Rotation

28. A sphere, disk, and hoop made of homogeneous materials have the same radius (10 cm) and mass (3 kg). They are released from rest at the top of a 30° incline and roll down without slipping through a vertical distance of 2 m. (*a*) What are their speeds at the bottom? (*b*) Find the frictional force f in each case. (*c*) If they start together at $t = 0$, at what time does each reach the bottom?

29. A ball rolls without slipping along a horizontal plane. (*a*) Show that the frictional force on the ball must be zero. *Hint:* Consider a possible direction for a frictional force and what effect such a force would have on the velocity of the center of mass and on the angular velocity (*b*) If the plane is not horizontal, is the frictional force zero?

30. A 100-kg uniform disk of radius 0.60 m is placed flat on some smooth ice. Two skaters wind ropes around the disk in the same sense. Each skater thus pulls on his rope and skates away, exerting constant forces for 5 sec (Figure 13-29). Describe the motion of the disk; i.e., what is the acceleration, velocity, and the position of the center of mass as a function of time, and what are the angular acceleration and angular velocity as a function of time?

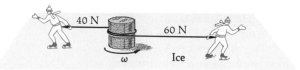

Figure 13-29
Exercise 30.

31. A homogeneous cylinder of radius 6 in weighs 100 lb. It is rolling along the floor without slipping at a speed of 15 ft/sec. How much work is needed to stop the disk?

Section 13-6, Motion of a Gyroscope

32. A bicycle wheel is mounted in the middle of a 2-ft-long axle. The tire and rim weigh 8 lb and have a radius of 12 in. The wheel is spun at 10 rev/sec, and the axle is then placed in a horizontal position with one end resting on a pivot. (*a*) What is the angular momentum due to the spinning of the wheel? (Treat the wheel as a hoop.) (*b*) What is the angular velocity of precession? How long does it take for the axle to swing through 360° around the pivot? (*c*) What is the angular momentum associated with the motion of the center of mass, i.e., due to the precession? In what direction is this angular momentum?

33. A uniform disk of mass 2 kg and radius 6 cm is mounted in the center of a 10-cm axle and spun at 900 rev/min. The axle is then placed in a horizontal position with one end resting on a pivot. The other end is given an initial horizontal velocity so that the precession is smooth with no nutation. (a) Find the angular velocity of precession. (b) What is the speed of the center of mass during the precession? (c) What are the magnitude and direction of the acceleration of the center of mass? (d) What are the vertical and horizontal components of the force exerted by the pivot?

Section 13-7, Static and Dynamic Imbalance

34. A uniform disk of radius 30 cm, thickness 3 cm, and mass 5 kg rotates at $\omega = 10$ rad/sec about an axis parallel to the symmetry axis but 0.5 cm from that axis. (a) Find the net force on the bearings due to this imbalance. (b) Where should a 100-gm mass be placed on the disk to correct this problem?

35. A 2-kg mass attached to a string of length 1 m moves in a horizontal circle as a conical pendulum (Figure 13-30). The string makes an angle $\theta = 30°$ with the vertical. (a) Show that the angular momentum of the mass about the point of support P has a horizontal component toward the center of the circle as well as a vertical component and find these components. (b) Find the magnitude of $d\mathbf{L}/dt$ and show that it equals the magnitude of the torque exerted by gravity about the point of support.

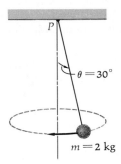

Figure 13-30
Conical pendulum (Exercise 35).

Problems

1. A particle moves in a circle of radius r with angular velocity $\boldsymbol{\omega}$. (a) Show that its velocity is $\mathbf{v} = \boldsymbol{\omega} \times \mathbf{r}$. (b) Show that its centripetal acceleration is $\mathbf{a}_r = \boldsymbol{\omega} \times \mathbf{v} = \boldsymbol{\omega} \times (\boldsymbol{\omega} \times \mathbf{r})$.

2. A particle located at $\mathbf{r} = x\mathbf{i} + y\mathbf{j} + z\mathbf{k}$ has momentum $\mathbf{p} = p_x\mathbf{i} + p_y\mathbf{j} + p_z\mathbf{k}$. A force $\mathbf{F} = F_x\mathbf{i} + F_y\mathbf{j} + F_z\mathbf{k}$ acts on it. (a) Calculate the components of the angular momentum of the particle and of the torque about the origin in terms of the components of these three vectors. (b) Compute the time derivative of the z component of angular momentum and show that $dL_z/dt = v_x p_y - v_y p_x + xF_y - yF_x$, where v_x and v_y are the x and y components of the velocity of the particle. Is this expression the same as that for the z component of the torque? Explain.

3. A wheel of radius R rolls without slipping at speed V. (a) Show that the x and y coordinates of point P in Figure 13-31 are $r_0 \cos\theta$ and $R + r_0 \sin\theta$. (b) Show that the total velocity $\mathbf{v}$ of point P has components $v_x = V + (r_0 V \sin\theta)/R$ and $v_y = -(r_0 V \cos\theta)/R$. (c) Show that $\mathbf{v}$ and $\mathbf{r}$ are perpendicular to each other by calculating their scalar product. (d) Show that $v = r\omega$, where $\omega = V/R$ is the angular velocity of the wheel. (e) These results demonstrate that in the case of rolling without slipping, the motion is the same as if the rolling object were instantaneously rotating about the point of contact with angular speed $\omega = V/R$. Calculate the kinetic energy of the wheel assuming that it is in pure rotation about point O and show that the result is the same as calculated from the sum of the translational kinetic energy of the center of mass and the rotational kinetic energy of rotation about the center of mass.

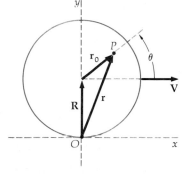

Figure 13-31
Problem 3.

4. A mass m is attached to a light string which passes through a small hole in a frictionless tabletop. Initially the mass is sliding with speed v_0 in a circle of radius r_0 about the hole. A man under the table now begins to pull the string in slowly. (a) Show that when the mass is moving in a circle of radius r, the tension in the string is

$$T = L_0^2/mr^3$$

where $L_0 = mv_0 r_0$ is the initial angular momentum. (b) The string is pulled in until the radius of the circular orbit is r_f. Using the result of part (a), calculate the work done by integrating $T\,dr$ and show that the work done is

$(L_0^2/2m)(r_f^{-2} - r_0^{-2})$. (c) What is the velocity v_f when the radius of the circle is r_f? Show that the work done as calculated in part (b) equals the change in kinetic energy $\frac{1}{2}mv_f^2 - \frac{1}{2}mv_0^2$. (d) If $m = 0.5$ kg, $r_0 = 1$ m, and $v_0 = 3$ m/sec and the maximum tension the string can withstand without breaking is 200 N, find the minimum value of r_f before the string breaks.

5. A uniform ball of radius r rolls without slipping along the loop-the-loop track in Figure 13-32. It starts at rest at height h above the bottom of the loop. If the ball is not to leave the track at the top of the loop, what is the least value h can have (in terms of the radius R of the loop)? What would h have been if the ball were to slide along a frictionless track instead of rolling?

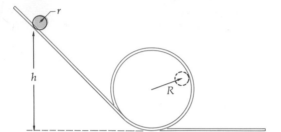

Figure 13-32
Problem 5.

6. A heavy homogeneous cylinder has mass m and radius R. It is accelerated by a force T, which is applied through a rope wound around a light drum of radius r attached to the cylinder (Figure 13-33). The coefficient of static friction is sufficient for the cylinder to roll without slipping. (a) Find the friction force. (b) Find the acceleration a of the center of the cylinder. (c) Is it possible to choose r so that a is greater than T/m? How? (d) What is the direction of the friction force in the circumstances of part (c)?

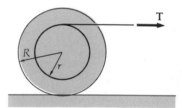

Figure 13-33
Problem 6.

7. Assuming the earth to be a homogeneous sphere of radius r and mass m, show that the period T of rotation about its axis is related to its radius by $T = (4\pi m/5L)r^2$, where L is the angular momentum of the earth due to its rotation. Suppose that the radius r changes by a very small amount Δr due to some internal effect, e.g., thermal expansion. (a) Show that the fractional change in the period T is given approximately by $\Delta T/T = 2\Delta r/r$. *Hint:* Use differentials dr and dT to approximate the changes in these quantities. (b) By how many miles would the earth need to expand for the period to change by $\frac{1}{4}$ d/y, so that leap year would not be needed?

8. A particle moves with constant speed v in the xy plane along a line parallel to the x axis a distance b from this axis. Let **r** be the radius vector from the origin to the particle at some instant, and let θ be the angle between **r** and the x axis. (a) Let v_r be the component of the velocity along the radius vector and v_t be the component perpendicular to the radius vector. Show that $v_r = v \cos \theta$ and $v_t = v \sin \theta$. (b) Show that the angular velocity of the particle about the origin is $\omega = v_t/r = (v/b) \sin^2 \theta$. Using this result and the moment of inertia of the particle about the z axis $I = mr^2$, show that mbv and $I\omega$ are equivalent expressions for the angular momentum of the particle about the origin.

9. Prove that $\mathbf{A} \times (\mathbf{B} + \mathbf{C}) = \mathbf{A} \times \mathbf{B} + \mathbf{A} \times \mathbf{C}$ by writing out both sides in terms of their rectangular components and the unit vectors **i**, **j**, and **k**.

10. Prove that $d(\mathbf{A} \times \mathbf{B})/dt = \mathbf{A} \times d\mathbf{B}/dt + (d\mathbf{A}/dt) \times \mathbf{B}$ by writing out $\mathbf{A} \times \mathbf{B}$ in terms of the components of the vectors and the unit vectors **i**, **j**, and **k** and differentiating directly.

CHAPTER 14   Oscillations

Motion which repeats itself is called *periodic*, the time required for each repetition being the *period*. Systems which display periodic motion occur frequently, e.g., the motion of a mass on a spring, the motion of the moon about the earth, the oscillation of a pendulum, the oscillations of atoms in a molecule, the vibration of a string of a violin. The study of periodic motion has many applications. In addition to the examples listed above, periodic motion occurs in many types of wave motion. For example, in a sound wave, air molecules oscillate along the line of propagation of the wave. Light and other electromagnetic waves are characterized by the oscillation of electric and magnetic field vectors perpendicular to the line of propagation of the wave.

14-1 Simple Harmonic Motion

We shall concentrate on the simplest form of oscillation, called *simple harmonic motion*. In simple harmonic motion in one dimension, the displacement x of the particle from equilibrium as a function of time is given by an equation such as

$$x = A \cos (\omega t + \delta) \qquad \text{14-1}$$

Position function for simple harmonic motion

where A, ω, and δ are constants. This function is plotted in Figure 14-1. The maximum displacement A is called the *amplitude*. The quantity $\omega t + \delta$ is called the *phase* of the motion, and the constant δ is called the *phase constant*. When the phase increases by 2π from its value at time t, the particle again has the same position and velocity as at time t, since $\cos (\omega t + \delta + 2\pi) = \cos (\omega t + \delta)$. During the time that the phase increases by 2π, the particle performs a full cycle of the motion. We can determine the *period T* of the motion by noting that the phase at time $t + T$ is just 2π plus the phase at time t:

Amplitude, phase, and phase constant defined

Period defined

$$\omega(t + T) + \delta = 2\pi + \omega t + \delta \qquad \text{or} \qquad \omega T = 2\pi$$

or

$$T = \frac{2\pi}{\omega} \qquad \text{14-2}$$

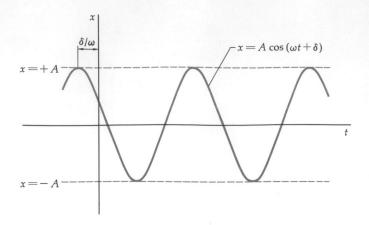

Figure 14-1
Plot of x versus t for simple harmonic motion.

The number of vibrations per unit time $1/T$ is called the *frequency f:*

Frequency and angular frequency defined

$$f = \frac{1}{T} = \frac{\omega}{2\pi} \qquad\qquad 14\text{-}3$$

The constant $\omega = 2\pi f$ is the *angular frequency.* It has the units radians per second, the same units as angular velocity, also designated ω. In Section 14-2 we shall see that angular frequency is very closely related to the angular velocity of a circular motion which is associated with simple harmonic motion.

In terms of the frequency or period, Equation 14-1 is

$$x = A \cos\left(2\pi f\, t + \delta\right) = A \cos\left(\frac{2\pi t}{T} + \delta\right) \qquad\qquad 14\text{-}4$$

The phase constant δ depends on our choice of zero time. If we choose $t = 0$ when $x = A$, the phase constant is zero. On the other hand, if we choose $t = 0$ when $x = 0$, δ is either $\pi/2$ or $3\pi/2$, depending on whether x is increasing or decreasing at $t = 0$. For example, if $\delta = 3\pi/2$,

$$x = A \cos\left(\omega t + \frac{3\pi}{2}\right) = A \sin \omega t$$

Then x is increasing at $t = 0$, as can be seen from a graph of $x = A \sin \omega t$ versus t.

We can find the general relation between the initial position x_0 and the constants A and δ by setting $t = 0$ in Equation 14-1:

$$x_0 = A \cos \delta \qquad\qquad 14\text{-}5$$

The phase constant is not very important for the description of one-dimensional simple harmonic motion of a single particle since we can always choose our zero time to make $\delta = 0$. If we have two particles undergoing simple harmonic motion, or if we consider the combination of simple harmonic motion of a particle in two dimensions, we have two phase constants. If they are not equal, we can make only one of them equal to zero by our choice of zero time. Then the other one, the relative phase constant of the two motions, is important in the description of the motion.

The velocity of a particle undergoing simple harmonic motion is

$$v_x = \frac{dx}{dt} = -A\omega \sin\left(\omega t + \delta\right)$$

$$= A\omega \cos\left(\omega t + \delta + \frac{\pi}{2}\right) \qquad\qquad 14\text{-}6$$

The phase of the velocity differs from that of the position by $\pi/2$ rad, or $90°$. When $\cos(\omega t + \delta) = 1$, $\sin(\omega t + \delta) = 0$, and so when x is its maximum or minimum value, the velocity is zero. Similarly when $\sin(\omega t + \delta) = 1$, $\cos(\omega t + \delta) = 0$. The speed is maximum when the particle is at its equilibrium position $x = 0$. According to Equation 14-6, the initial velocity is related to A and ω by

$$v_0 = -A\omega \sin \delta \qquad\qquad 14\text{-}7$$

Since Equations 14-5 and 14-7 can be solved for A and δ in terms of the initial position x_0 and initial velocity v_0, the amplitude A and the phase constant δ are completely determined by the initial conditions.

The acceleration of the particle is

$$a_x = \frac{d^2x}{dt^2} = \frac{dv_x}{dt} = -\omega^2 A \cos(\omega t + \delta)$$

or

$$a_x = -\omega^2 x \qquad\qquad 14\text{-}8$$

In simple harmonic motion, *the acceleration is proportional to the displacement and in the opposite direction.*

Acceleration is proportional to displacement in simple harmonic motion

According to our discussion of one-dimensional motion in Chapter 2, if the acceleration of a particle is known, the position $x(t)$ is completely determined by the initial conditions, e.g., the initial position and velocity. As we have seen, these are related to the constants A and δ in Equation 14-1. Thus we could begin with Equation 14-8, relating the acceleration to the position, and work backward to obtain Equation 14-1 (see Problem 13). That is, the general solution of the differential equation 14-8 is of the form of Equation 14-1; the two equations are equivalent. We can therefore take either Equation 14-1 and 14-8 as our definition of simple harmonic motion.

Equation 14-8 is important because the acceleration of a particle is just the resultant force divided by the mass. *Whenever a force is proportional to the displacement (and in the opposite direction), we can conclude that the motion is simple harmonic.* For example, if a mass m is acted upon by a spring of force constant k, the force is $-kx$ and the acceleration is

Conditions for simple harmonic motion

$$a_x = \frac{F_x}{m} = -\frac{k}{m} x \qquad\qquad 14\text{-}9$$

This equation is the same as Equation 14-8 with k/m replacing ω^2. The solution of this equation is Equation 14-1 with $\omega = \sqrt{k/m}$ and the constants A and δ determined by the initial position and velocity of the mass. Note that the angular frequency ω and thus the period $T = 2\pi/\omega$ are determined only by the force constant and the mass. They do not depend on the amplitude A. This is an important property of simple harmonic motion.

In simple harmonic motion, the frequency and period are independent of the amplitude.

Frequency and period are independent of the amplitude

We can write the general position function $x(t)$ for simple harmonic motion in terms of the initial position and velocity, rather than the amplitude and phase constant as in Equation 14-1. We use the trigonometric identity for the cosine of the sum of two angles,

$$\cos(\theta + \phi) = \cos \theta \cos \phi - \sin \theta \sin \phi$$

Then Equation 14-1 can be written

$$A \cos (\omega t + \delta) = A \cos \delta \cos \omega t - A \sin \delta \sin \omega t$$

But according to Equations 14-4 and 14-7, $A \cos \delta = x_0$ and $-A \sin \delta = v_0/\omega$. Then

$$x(t) = A \cos (\omega t + \delta) = x_0 \cos \omega t + \frac{v_0}{\omega} \sin \omega t \qquad \text{14-10}$$

Position function in terms of the initial conditions

Questions

1. What is the phase constant δ in Equation 14-1 if the position of the oscillating particle at time $t = 0$ is (a) 0, (b) $-A$, (c) A, (d) $A/2$?

2. How far does a particle oscillating with amplitude A move in one full period?

3. How far apart are the extreme positions of a particle oscillating with amplitude A?

4. If you know that the speed of an oscillator of amplitude A is zero at certain times, can you say exactly what its displacement is at those times? Can you say what direction it will be moving in a moment? If not, how close can you come?

5. If you are told that the speed of an oscillator of amplitude A has its maximum value at a certain time, can you say exactly what its displacement is? Can you say in what direction it is moving?

6. Can the acceleration and displacement of a simple harmonic oscillator ever be in the same directions? The acceleration and the velocity? The velocity and the displacement? Explain.

7. What is the magnitude of the acceleration of an oscillator of amplitude A and angular frequency ω when its speed is a maximum? When its displacement is a maximum?

14-2 Circular Motion and Simple Harmonic Motion

There is a simple relation between circular motion with constant angular velocity and simple harmonic motion. Consider a particle moving in a circle of radius A with constant angular velocity ω, as shown in Figure 14-2. The angular displacement of the particle relative to the x axis is given by

$$\theta = \omega t + \delta \qquad \text{14-11}$$

where δ is the angular displacement at $t = 0$. From Figure 14-2 we see that the x component of the position of the particle is given by

$$x = A \cos \theta = A \cos (\omega t + \delta) \qquad \text{14-12}$$

Thus the projection on a straight line of uniform circular motion at angular velocity ω is simple harmonic motion with angular frequency ω. The frequency and period of the circular motion are the same as the corresponding frequency and period of the projected simple harmonic motion. The relation between circular motion and simple harmonic motion can be demonstrated with a turntable and a mass on a spring (Figure 14-3).

Projection of circular motion on a straight line is simple harmonic motion

The projection of the circular motion on the y axis is $y = A \cdot \sin \theta = A \sin (\omega t + \delta) = A \cos (\omega t + \delta - \pi/2)$ (Figure 14-2). We can therefore consider the circular motion of a particle to be the combination of perpendicular simple harmonic motions having the same amplitude and frequency but having a relative phase difference of $\pi/2$. We shall consider combinations of perpendicular simple harmonic motions with other phase differences or with different amplitudes in a later section.

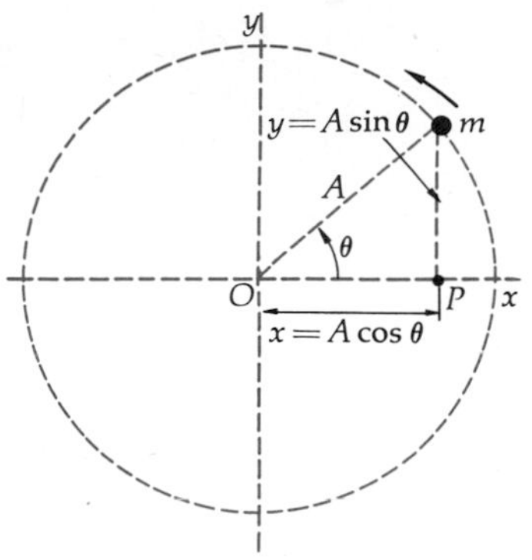

Question

8. At what points does the mass traveling on the circle in Figure 14-2 have the same displacement as the oscillating particle? The same speed? The same acceleration?

Figure 14-2
Particle moving with constant angular velocity ω in a circle of radius A. Its angular displacement is $\theta = \omega t + \delta$. Point P on the x axis moves with simple harmonic motion $x = A \cos (\omega t + \delta)$. The projection on the y axis, $y = A \sin (\omega t + \delta)$, is also simple harmonic motion but differs in phase by 90° from x.

14-3 A Mass on a Spring

If a mass is attached to a spring stretched a distance x from its equilibrium length, the force on the mass is $F_x = -kx$, where k is called the *force constant* of the spring. Newton's second law for the motion of the mass is

$$F_x = m \frac{d^2 x}{dt^2} = -kx$$

or

$$\frac{d^2 x}{dt^2} = a_x = -\frac{k}{m} x \qquad \text{14-13}$$

This is the same as Equation 14-8 with

$$\omega^2 = \frac{k}{m} \qquad \text{14-14}$$

The mass oscillates with simple harmonic motion. The position of the mass is given by either Equation 14-1 or 14-10:

$$x = A \cos (\omega t + \delta) = x_0 \cos \omega t + \frac{v_0}{\omega} \sin \omega t$$

The velocity of the mass is

$$v_x = \frac{dx}{dt} = -\omega A \sin (\omega t + \delta) = -\omega x_0 \sin \omega t + v_0 \cos \omega t$$

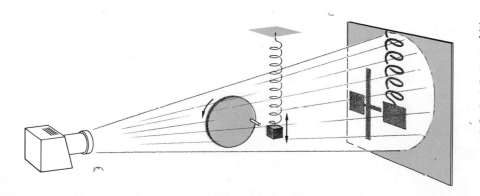

Figure 14-3
The shadow of a peg on a turntable is projected on the screen along with the shadow of a mass on a spring. When the period of rotation of the turntable equals the period of oscillation of the mass, the shadows move together.

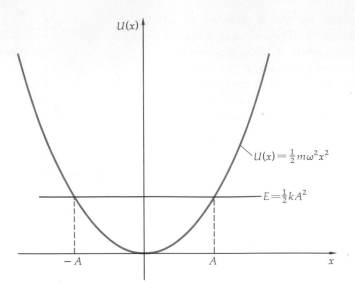

Figure 14-4
The potential-energy function
$U(x) = \frac{1}{2}kx^2 = \frac{1}{2}m\omega^2x^2$ for a
mass on a spring. The hori-
zontal line represents the total
energy $\frac{1}{2}kA^2$ for amplitude A.

The potential energy of the mass on the spring is

$$U(x) = \tfrac{1}{2}kx^2 = \tfrac{1}{2}m\omega^2x^2 \qquad\qquad 14\text{-}15$$

where we use $\omega^2 = k/m$. This is the parabolic potential-energy function discussed in Example 9-5. It is sketched again in Figure 14-4.

The total energy is

$$E = U(x) + \tfrac{1}{2}mv_x^2 = \tfrac{1}{2}m\omega^2x^2 + \tfrac{1}{2}m\left(\frac{dx}{dt}\right)^2$$

Using Equations 14-1 and 14-6 for x and dx/dt, we obtain

$$\begin{aligned}
E &= \tfrac{1}{2}m\omega^2A^2\cos^2(\omega t + \delta) + \tfrac{1}{2}m\omega^2A^2\sin^2(\omega t + \delta)\\
&= \tfrac{1}{2}m\omega^2A^2\left[\cos^2(\omega t + \delta) + \sin^2(\omega t + \delta)\right]\\
&= \tfrac{1}{2}m\omega^2A^2 = \tfrac{1}{2}kA^2 \qquad\qquad 14\text{-}16
\end{aligned}$$

The total energy is thus constant and equal to the maximum potential energy or the maximum kinetic energy. The total energy is shown by the horizontal line in Figure 14-4. As discussed in Chapter 9, the kinetic energy is represented on this graph by the vertical distance between the total energy and the potential energy. It is easily seen from the figure that the two turning points are at equal distances from the equilibrium point $x = 0$ and that the maximum velocity occurs at $x = 0$.

Example 14-1 A 2-kg mass stretches a spring 10 cm when it hangs vertically in equilibrium. The mass is then attached to the spring resting on a smooth table and fixed at one end, as shown in Figure 14-5. The mass is held a distance 5 cm from the equilibrium position and released. Find the frequency f, angular frequency ω, period T, amplitude A, and phase constant δ for the resulting simple harmonic motion. What is the maximum speed of the mass, and when does it occur?

The force constant of the spring is determined from the first measurement. We have for the vertical-equilibrium case $ky = mg$, where y is the amount the spring is stretched. Using $g \approx 10$ m/sec², $m = 2$ kg, and $y = 10$ cm $= 0.1$ m, we have

$$k = \frac{mg}{y} = \frac{(2\text{ kg})(10\text{ m/sec}^2)}{0.1\text{ m}} = 200\text{ N/m}$$

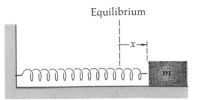

Figure 14-5
Mass m is attached to a spring
and slides without friction on
a horizontal surface.

We are given that the initial position of the mass is $x_0 = 5$ cm $= 0.05$ m and that the initial velocity is zero. Then from Equations 14-5 and 14-7 we have $\delta = 0$ and $A = 0.05$ m. The equation of motion of the mass is

$$x = (0.05 \text{ m}) \cos \omega t$$

where $\omega = \sqrt{k/m} = \sqrt{200/2} = 10$ rad/sec. The period, found from $\omega T = 2\pi$, is

$$T = \frac{2\pi}{10} = 0.63 \text{ sec}$$

The reciprocal of the period is the frequency. The units of frequency are reciprocal seconds, often written cycles per second or vibrations per second for easier reading. The unit cycles per second is now called the hertz (Hz). Thus

$$f = \frac{1}{T} = 1.59 \text{ cycles/sec} = 1.59 \text{ Hz}$$

The velocity of the mass is

$$v_x = \frac{dx}{dt} = -\omega A \sin \omega t = -(10 \text{ rad/sec})(0.05 \text{ m}) \sin \omega t$$

$$= -(0.5 \text{ m/sec}) \sin \omega t$$

The maximum speed is 0.5 m/sec. This occurs first when $\omega t_1 = \pi/2$ or $t_1 = \frac{1}{4}T = 0.16$ sec. At this time, $x = 0$ and v_x is negative, indicating that the motion is to the left. The speed is maximum again one-half cycle later, when $x = 0$ but the mass is moving to the right. The functions $x(t)$ and $v_x(t)$ are sketched in Figure 14-6.

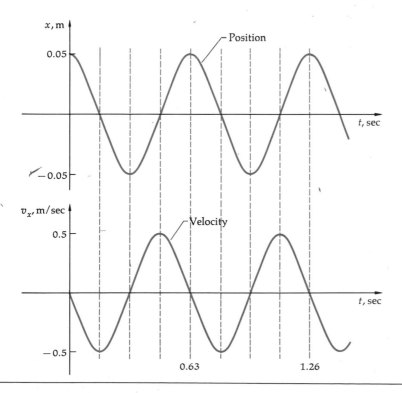

Figure 14-6
Position and velocity of mass versus time for Example 14-1.

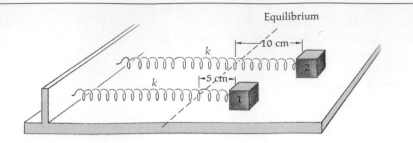

Figure 14-7
Example 14-2. Two identical mass-spring systems starting at rest with different displacements. Since the period does not depend on the amplitude, both masses reach their equilibrium positions at the same time.

Example 14-2 A second spring identical to that in Example 14-1 is attached to a second particle, also of mass 2 kg. This spring is stretched a distance of 10 cm from equilibrium and released at the same time as the first, which is stretched to 5 cm. Which mass reaches the equilibrium position first?

Figure 14-7 shows the initial positions of the masses, and Figure 14-8 is a sketch of the two position functions $x_1(t)$ and $x_2(t)$. Since both the springs have the same force constant and the masses are equal, the angular frequencies and periods of the two motions are the same. Only the amplitudes differ. *Thus they both reach the equilibrium position at the same time.* Particle 2 has twice as far to go to reach equilibrium as particle 1, but it begins with twice the initial acceleration. The equations of motion of the two particles are

$$x_1 = (0.05 \text{ m}) \cos 10t \qquad x_2 = (0.10 \text{ m}) \cos 10t$$

where t must be in seconds because we put in the value $\omega = 10$ rad/sec.

This example illustrates that the angular frequency and period in simple harmonic motion are independent of the amplitude. Thus the time $\frac{1}{4}T$ to reach the equilibrium position is the same for each particle. Of course the energies of the two particles are different: more work is required to stretch the second spring 10 cm than to stretch the first one 5 cm. The energies are

$$E_1 = \tfrac{1}{2}kA_1{}^2 = \tfrac{1}{2}(200)(0.05^2) = 0.25 \text{ J}$$

$$E_2 = \tfrac{1}{2}kA_2{}^2 = \tfrac{1}{2}(200)(0.1^2) = 1.0 \text{ J}$$

You should show that at the equilibrium position the velocity of the second mass is twice that of the first.

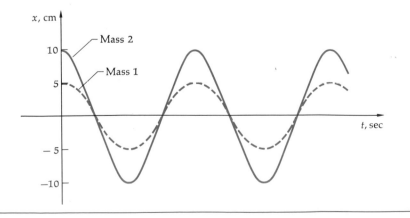

Figure 14-8
Plot of position versus time for the masses in Figure 14-7.

Example 14-3 A mass on a spring at rest at $x = 0$ is struck a blow with a hammer, which delivers an impulse I. What is the subsequent motion?

We choose $t = 0$ just after the impulse. The mass is thus at $x = 0$ with an initial velocity $v_0 = p_0/m = I/m$, where p_0, the initial momentum, is equal to the impulse. The form of Equation 14-10 is useful here since x_0 is zero. Thus

$$x = \frac{v_0}{\omega} \sin \omega t = \frac{I}{m\omega} \sin \omega t$$

Questions

9. A mass m oscillates at frequency f when it is attached to a horizontal spring as in Figure 14-5. It is released from rest when its displacement is A. If a second mass m is attached to the first and the two are released at the same displacement, how will the frequency be changed? The angular frequency? The period? The amplitude? The maximum speed? The maximum acceleration? The maximum force exerted by the spring?

10. Explain how each of the following quantities will change if the impulse I in Example 14-4 is doubled: the amplitude, the frequency, the period, the maximum speed, the total energy, the phase constant.

11. The effect of the mass of a spring on the motion of an object attached to it is usually neglected; describe qualitatively its effect when it is not neglected.

14-4 The Simple Pendulum

Another important example of periodic motion is that of a simple pendulum. If the angle made by the string with the vertical is not too great, the motion of the bob of the pendulum is simple harmonic.

Consider a mass on a string of length L, as shown in Figure 14-9. The forces on the mass are the weight mg and the tension T in the string. The tangential force has magnitude $mg \sin \theta$ and is in the direction of decreasing θ. Let s be the arc length measured from the bottom

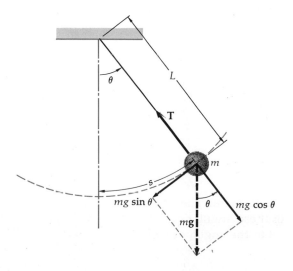

Figure 14-9
Forces on a simple pendulum.

of the circle. The arc length is related to the angle measured from the vertical by

$$s = L\theta \tag{14-17}$$

The tangential acceleration is d^2s/dt^2. The tangential component of $\Sigma\mathbf{F} = m\mathbf{a}$ is

$$\Sigma F_t = -mg \sin\theta = m\frac{d^2s}{dt^2}$$

or

$$\frac{d^2s}{dt^2} = -g \sin\theta = -g \sin\frac{s}{L} \tag{14-18}$$

If s is much less than L, the angle $\theta = s/L$ is small and we can approximate $\sin\theta$ by the angle θ. Using $\sin(s/L) \approx s/L$ in Equation 14-18, we have

$$\frac{d^2s}{dt^2} = -\frac{g}{L}s \tag{14-19}$$

Motion of a pendulum is simple harmonic for small amplitudes

We see that for small angles for which the approximation $\sin\theta \approx \theta$ is valid, the acceleration is proportional to the displacement. The motion of the pendulum is simple harmonic for small displacement. If we write ω^2 for g/L, Equation 14-19 becomes

$$\frac{d^2s}{dt^2} = -\omega^2 s$$

$$\omega^2 = \frac{g}{L} \tag{14-20}$$

The solution of this equation is

$$s = s_0 \cos(\omega t + \delta) \tag{14-21}$$

where s_0 is the maximum displacement measured along the arc of the circle. The period of the motion is

$$T = \frac{2\pi}{\omega} = 2\pi\sqrt{\frac{L}{g}} \tag{14-22}$$

It is often more convenient to measure the displacement of the pendulum in terms of the angle made with the vertical $\theta = s/L$. Equations 14-18, 14-19, and 14-21 can be written in terms of the angular displacement and the angular acceleration $d^2\theta/dt^2$ by dividing each side by L, giving

$$\frac{d^2\theta}{dt^2} = -\frac{g}{L}\sin\theta \tag{14-18a}$$

which for small θ becomes

$$\frac{d^2\theta}{dt^2} = -\frac{g}{L}\theta \tag{14-19a}$$

The solution of this equation is

$$\theta = \theta_0 \cos(\omega t + \delta) \tag{14-21a}$$

where $\theta_0 = s_0/L$ is the maximum angular displacement. The criterion for simple harmonic motion stated in terms of these angular quantities is that the angular acceleration be proportional to the angular displacement, as in Equation 14-19a.

Example 14-4 What is the period of a pendulum 1 m long when $g = 9.8$ m/sec²?

The angular frequency is given by Equation 14-20,

$$\omega = \sqrt{\frac{g}{L}} = \sqrt{\frac{9.8 \text{ m/sec}^2}{1 \text{ m}}} = 3.13 \text{ rad/sec}$$

The period is

$$T = \frac{2\pi}{\omega} = \frac{2\pi}{3.13} = 2.01 \text{ sec}$$

A time of this magnitude is easily measured accurately. For example, with a watch which can measure time intervals to ±1 sec, we can measure such a period to ±1 part in 200 by measuring the time for 100 vibrations.

The potential energy of the simple pendulum is the gravitational potential energy mgy of the mass, where y is the height measured from some reference level. It is convenient to choose $y = 0$ at the lowest point of the mass at $\theta = 0$. Then y is related to θ by

$$y = L - L \cos \theta = L(1 - \cos \theta)$$

as can be seen from Figure 14-10. The potential energy is then

$$U = mgL(1 - \cos \theta) \qquad\qquad 14\text{-}23$$

and the total energy is

$$E = \tfrac{1}{2}mv^2 + U = \tfrac{1}{2}mv^2 + mgL(1 - \cos \theta) \qquad\qquad 14\text{-}24$$

For small angles, we can approximate $\cos \theta$ by

$$\cos \theta = 1 - \frac{\theta^2}{2} + \cdots \approx 1 - \frac{\theta^2}{2} \qquad\qquad 14\text{-}25$$

Then $1 - \cos \theta \approx \tfrac{1}{2}\theta^2$, and the potential energy is

$$U(\theta) \approx \tfrac{1}{2}mgL\theta^2 \qquad\qquad 14\text{-}26a$$

According to Equation 14-15, the potential energy for a mass on a spring can be written $U = \tfrac{1}{2}m\omega^2 x^2$. If we use $\theta = s/L$ and $\omega^2 = g/L$, we have for the potential energy of a simple pendulum,

$$U = \tfrac{1}{2}mgL \left(\frac{s}{L}\right)^2 = \tfrac{1}{2}m\,\frac{g}{L}\,s^2 = \tfrac{1}{2}m\omega^2 s^2 \qquad\qquad 14\text{-}26b$$

The simple pendulum was used for early determinations of the acceleration of gravity because both the period and the length are easily measured. Direct measurement by observation of free fall, in con-

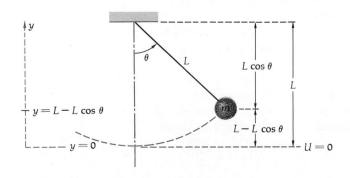

Figure 14-10
The height of the mass above its equilibrium position is $y = L - L \cos \theta$. Its potential energy is $U = mgy = mgL(1 - \cos \theta)$ relative to $U = 0$ at $\theta = 0$.

trast, is difficult because the time of fall over reasonable distances is too short for easy measurement. From Equation 14-22 we have, in terms of L and T,

$$g = \frac{4\pi^2 L}{T^2}$$ 14-27

The pendulum mass is not involved in the determination of g because, according to Newton's law of gravitation, the force of gravity on an object is proportional to its mass or inertia m. This characteristic is unique to gravity among all the forces we know and is a matter of considerable interest. One consequence is that all objects near the earth fall with the same acceleration if air resistance is negligible, a fact that has seemed surprising to men since ancient times. (The well-known story of how Galileo demonstrated this by dropping weights from the Leaning Tower of Pisa is one example of the excitement this discovery aroused in the sixteenth century.)

Another consequence is that the period of a simple pendulum depends only on its length and not on the mass of the pendulum bob. Newton noted this and used the simple pendulum as an experimental test of his conclusion that the force of gravity is proportional to the mass of the object upon which it acts. He reasoned as follows: suppose the weight of an object were proportional not to m but to some other property of the body. That is, suppose

$$w = g m_G$$

where g is a constant and m_G stands for a measure of this property. (It has been customary in recent years to call m_G the gravitational mass to distinguish it from the inertial mass m.) The acceleration of this body in free fall would then be

$$a = \frac{w}{m} = \frac{m_G}{m} g$$

Thus one would expect different objects to have different free-fall accelerations depending on the ratio m_G/m. It would be reasonable to suppose that m_G/m would depend on such things as the chemical composition of the body, or its temperature, or other physical characteristics.

The experimental fact, however, is that a is the same for all bodies. This means that for every body m_G/m has the same value. That is, m_G must be proportional to m in all circumstances. Since this is the case, we need not maintain the distinction between m_G and m, and indeed this is our custom. We must keep in mind, however, that this is an experimental conclusion limited by the accuracy of experiment. As an experimental law, the statement that m_G is proportional to m is known as the *principle of equivalence*. Experiments testing it were carried out   *Principle of equivalence* by Simon Stevin in the 1580s (he also discovered the law of vector addition for forces). Galileo publicized this law widely, and his contemporaries made considerable improvements on the experimental accuracy with which the law was established.

The most precise early experiments made to back up the principle of equivalence were those of Newton, using simple pendulums. He noted that if m_G were different from the inertial mass m, derivation of the formula for the period of the pendulum would lead not to Equation 14-18 but to

$$T = 2\pi \sqrt{\frac{mL}{m_G g}}$$

Thus, one would expect the periods of pendulums with the same length but different masses and made of different materials to differ. This would be the case unless m_G is exactly proportional to m for every body. Newton conducted careful measurements with different types of pendulum bobs and found to high accuracy that the pendulum period is independent of the mass or composition of the bob. His measurements with the pendulum were much more precise than earlier experiments with falling bodies. Newton was able to establish the proportionality of m_G and m to an accuracy of about 1 part in 1000. Experiments comparing gravitational mass with inertial mass have been improved steadily over the years, until recently the proportionality of m_G and m has been demonstrated to an accuracy of about 1 part in 10^{12}. This accuracy means that the principle of equivalence is one of the best established of all physical laws.

Since the proportionality of m_G and m is so well established, we never distinguish between them, and put $m_G = m$. (This amounts to choosing the constant of proportionality equal to 1, which in turn determines the units of G in the law of gravity.)

The motion of a simple pendulum is simple harmonic only if the angular displacement is small so that $\sin \theta \approx \theta$ is valid. Figure 14-11 shows a plot of θ and $\sin \theta$. Even for the rather large angle $\theta = \pi/6 = 0.523$ (which is 30°), the difference between θ and $\sin \theta = \sin 30° = 0.500$ is only 4.6 percent. If the angular displacement is large, the angular acceleration given by Equation 14-18a is not proportional to the angular displacement. The motion of the pendulum for large angles is not simple harmonic. However, the motion is periodic, though the period is not independent of the amplitude, as in simple harmonic motion. From Figure 14-11 we see that at large angles $\sin \theta$ is less than θ. The force which accelerates the mass back toward equilibrium has the magnitude $mg \sin \theta$; this is less than $mg\theta$, which would produce simple harmonic motion. Thus the acceleration at large angles is less than it must be for simple harmonic motion, and the period is slightly longer. We can see this also from Figure 14-12; it compares the actual potential energy, which is proportional to $1 - \cos \theta$, with the approximation, proportional to $\frac{1}{2}\theta^2$. For a given total energy E, the pendulum travels farther between a turning point and equilibrium in the actual potential energy than it would in the parabolic potential-energy function for simple harmonic motion. The period is therefore

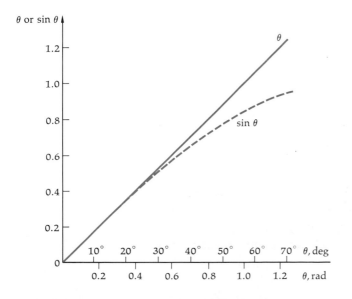

Figure 14-11
Comparison of θ and $\sin \theta$. For sufficiently small angles the curves are approximately the same.

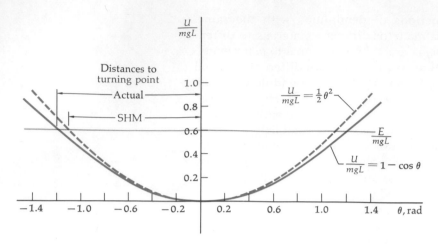

Figure 14-12
Comparison of the potential-
energy function $U = mgL(1 - \cos\theta)$ with the approximation
$U = \frac{1}{2}mgL\theta^2$. SHM stands for
simple harmonic motion.

slightly longer. We omit the details of the difficult solution of Equation
14-18a for large angles. The period can be expressed in the power
series

$$T = T_0\left(1 + \frac{1}{2^2}\sin^2\tfrac{1}{2}\theta_0 + \frac{1}{2^2}\frac{3^2}{4^2}\sin^4\tfrac{1}{2}\theta_0 + \cdots\right)$$ 14-28

*Large-amplitude motion of a
pendulum*

where θ_0 is the maximum angular displacement and $T_0 = 2\pi\sqrt{L/g}$ is
the period in the limit of small angles.

Because the actual period of a simple pendulum does vary with
amplitude θ_0, the simple pendulum makes a poor clock unless it is
modified to compensate for energy loss due to friction and air resistance.

Example 14-5 A simple pendulum clock is calibrated to keep accurate
time at an amplitude of $\theta_0 = 10°$. When the amplitude has decreased so
that it is very small, $\theta_0 \ll 10°$, how much time does the clock gain
in 1 d?

The original period is approximately

$$T \approx T_0(1 + \tfrac{1}{4}\sin^2 5°)$$

since $\sin^4\tfrac{1}{2}\theta_0 \ll \sin^2\tfrac{1}{2}\theta_0$ in Equation 14-28. After the amplitude has
decreased, the period is just T_0. Since this is less, the frequency is
greater and the clock gains time. The fractional change in the period is

$$-\frac{\Delta T}{T} = \frac{T - T_0}{T_0} = \tfrac{1}{4}\sin^2 5° = \tfrac{1}{4}(0.0872)^2 \approx 2 \times 10^{-3}$$

The number of minutes in 1 d is

$$\frac{24\text{ h}}{1\text{ d}}\frac{60\text{ min}}{1\text{ h}} = 1440\text{ min/d}$$

Thus a fractional change of 2×10^{-3} corresponds to an accumulated
error in 1 d of $2 \times 10^{-3} \times 1440$ min $= 2.88$ min ≈ 3 min. Though the
fractional change in period is only 2×10^{-3}, corresponding to a percentage change of 0.2 percent, this results in an intolerable error in
timekeeping by today's standards for clocks. For this reason pendulum
clocks are designed to maintain constant amplitude.

Question

12. The string or wire supporting a pendulum increases slightly in
length when its temperature is raised. How would this affect a clock
operated by the pendulum?

14-5 The Physical Pendulum

Any rigid body suspended from some point other than the center of mass will oscillate when displaced from its equilibrium position. Consider a plane figure suspended from a point a distance D from the center of mass, as shown in Figure 14-13. The torque about the suspension point is $MgD \sin \theta$ in a direction tending to decrease θ. The angular acceleration of the body is related to the torque by

$$\Sigma\tau = I\alpha = I \frac{d^2\theta}{dt^2}$$

where I is the amount of inertia about the point of suspension. Therefore

$$-MgD \sin \theta = I \frac{d^2\theta}{dt^2} \qquad 14\text{-}29$$

For a simple pendulum this is the same as Equation 14-18a since $I = ML^2$ and $D = L$. Once again, the motion is simple harmonic only if the angular displacements are small, so that the approximation $\sin \theta \approx \theta$ holds. For this case, we have

$$\frac{d^2\theta}{dt^2} = -\frac{MgD}{I} \theta = -\omega^2\theta \qquad 14\text{-}30$$

where $\omega^2 = MgD/I$. The period is

$$T = \frac{2\pi}{\omega} = 2\pi \sqrt{\frac{I}{MgD}} \qquad 14\text{-}31$$

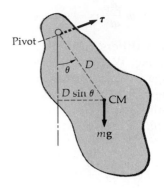

Figure 14-13
Physical pendulum. The weight produces a torque of magnitude $MgD \sin \theta$ about the pivot, tending to decrease the angle θ.

For large angles, the period is given by Equation 14-28, where T_0 is now given by Equation 14-31. The result we have just obtained can be used to measure the moment of inertia of a plane figure. The center of mass can be located by suspending the body from two different points, as discussed previously. To find the moment of inertia about some point P we suspend the body from point P and measure the period of oscillation. The moment of inertia is then obtained from

$$I = \frac{MgDT^2}{4\pi^2} \qquad 14\text{-}32$$

Example 14-6 What is the period of oscillation of a stick pivoted at one end if the amplitude of oscillation is small?

The moment of inertia of a uniform stick of length L about an axis at one end was shown in Chapter 12 to be $I = \frac{1}{3}ML^2$. The distance from the pivot to the center of mass is $\frac{1}{2}L$. Substituting these values in Equation 14-31 gives

$$T = 2\pi \sqrt{\frac{\frac{1}{3}ML^2}{Mg(\frac{1}{2}L)}} = 2\pi \sqrt{\frac{2L}{3g}}$$

If $L = 1$ m, the value of T obtained from this result is 1.64 sec.

14-6 A Mass on a Vertical Spring

When a mass hangs from a vertical spring, as in Figure 14-14, there is a force mg downward in addition to the force of the spring $-ky$ (y is measured downward from the unstretched position of the spring). The motion of the mass is given by

$$m \frac{d^2y}{dt^2} = -ky + mg \qquad \text{14-33}$$

which differs from Equation 14-13 in containing the constant-force term mg. We handle this extra term in a simple way. When the mass hangs in equilibrium from the spring, the spring is stretched by y_0, where $ky_0 = mg$. Let us measure the extension of the spring from this equilibrium point; i.e., let $y' = y - y_0$. Then $dy'/dt = dy/dt$ and $d^2y'/dt^2 = d^2y/dt^2$. Multiplying y' by k, we obtain

$$ky' = k(y - y_0) = ky - ky_0 = ky - mg$$

Thus $ky = ky' + mg$, and Equation 14-33 becomes

$$m \frac{d^2y'}{dt^2} = -ky' \qquad \text{14-34}$$

which has the familiar solution $y' = A \cos(\omega t + \delta)$ with $\omega^2 = k/m$. The mass oscillates about the equilibrium point where the total force on the mass is zero.

The potential energy of the spring is not $\frac{1}{2}ky'^2$ but $\frac{1}{2}ky^2 = \frac{1}{2}k(y' + y_0)^2$. When the spring is stretched from $y = y_0$ to some value y, the potential energy of the *spring* is increased by

$$\Delta U_{sp} = \frac{1}{2}ky^2 - \frac{1}{2}ky_0^2 = \frac{1}{2}ky'^2 + ky_0y' + \frac{1}{2}ky_0^2 - \frac{1}{2}ky_0^2$$
$$= \frac{1}{2}ky'^2 + mgy' \qquad \text{14-35}$$

but the potential energy due to the earth's gravity is decreased by mgy'. When the spring is stretched from y_0 to y, the total potential-energy change is

$$\Delta U = \frac{1}{2}ky^2 - \frac{1}{2}ky_0^2 - mgy' = \frac{1}{2}ky'^2 \qquad \text{14-36}$$

That is, when the spring is stretched by amount y' from equilibrium, the total change in potential energy, *including gravitational potential energy*, is $\frac{1}{2}ky'^2$. Thus if we measure from the equilibrium position, we can forget about the effect of the earth. The potential-energy functions $\frac{1}{2}ky^2$ and $-mgy$ are shown as dashed curves in Figure 14-15. Their sum is the solid curve.

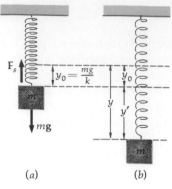

Figure 14-14
Mass on a vertical spring. In equilibrium the spring is stretched by the amount $y_0 = mg/k$. The effect of gravity can be neglected if the position of the mass is measured from its equilibrium position.

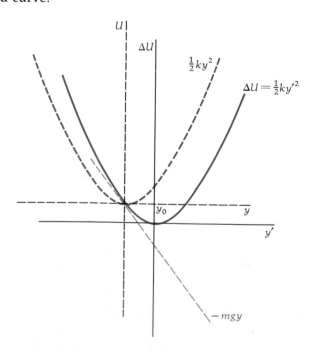

Figure 14-15
Shown as dashed lines are the spring potential energy $\frac{1}{2}ky^2$ and the gravitational potential energy $-mgy$ for a mass on a vertical spring with y taken to be positive downward. The total potential energy, the sum of these, is $\frac{1}{2}ky'^2 - \frac{1}{2}mgy_0$, where y' is measured from the equilibrium position of the mass. If ΔU is measured from the constant value $-\frac{1}{2}mgy_0$, then $\Delta U = \frac{1}{2}ky'^2$.

14-7 General Motion near Equilibrium

Whenever a particle is displaced from a position of stable equilibrium, the motion of the particle is simple harmonic if the displacements are small enough. Figure 14-16 shows some arbitrary resultant force as a function of position. At positions x_1 and x_2 the force is zero; these are positions of equilibrium. However, the equilibrium at x_2 is unstable, for if the particle is displaced slightly in the positive x direction, the force is positive, while if it is displaced slightly in the negative x direction, the force is negative. In both cases, the force accelerates the particle away from equilibrium. The position x_1 is one of stable equilibrium. If the particle is displaced slightly in either direction from equilibrium, the force accelerates it back toward equilibrium. The particle thus oscillates about this equilibrium point. As we can see from Figure 14-16b, the plot of force versus displacement is approximately

Figure 14-16
(a) An arbitrary force F_x versus x. Points x_1 and x_2 are equilibrium points where the force is zero. At x_1 the equilibrium is stable because for small displacements from equilibrium the force is toward x_1. For small displacements from x_2 the force is away from x_2 so the equilibrium is unstable at that point. (b) Near x_1 the force F_x can be approximated by a straight line, $F_x = -k(x - x_1)$. For small displacements $\epsilon = x - x_1$ the motion is approximately simple harmonic.

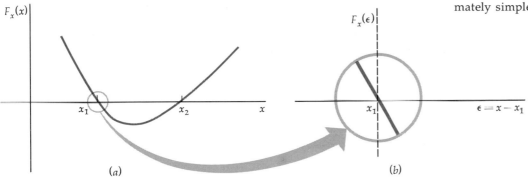

(a)   (b)

a straight line if we consider only small displacements from the equilibrium point x_1. Let ϵ measure the displacement from equilibrium:

$\epsilon = x - x_1$

The equation for the force for small ϵ is

$F_x = -k\epsilon$

where k is the magnitude of the slope of F_x versus x near x_1. Since the force is proportional to the displacement, the motion will be simple harmonic.

We can also examine the motion from the point of view of the potential-energy function $U(x)$ associated with the force. Figure 14-17 shows $U(x)$ versus x. As discussed in Chapter 8, the maximum at x_2 corresponds to unstable equilibrium, whereas the minimum at x_1 cor-

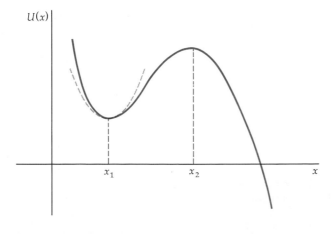

Figure 14-17
Potential-energy function $U(x)$ corresponding to the force in Figure 14-16. The minimum at x_1 indicates stable equilibrium whereas the maximum at x_2 indicates unstable equilibrium. Near x_1 the curve is approximately parabolic in shape, as in simple harmonic motion.

responds to stable equilibrium. The dashed curve in this figure is a parabolic curve which approximately fits $U(x)$ near the stable equilibrium point x_1. As long as the displacements from point x_1 are not too great, $U(x)$ is approximately parabolic and the motion is simple harmonic.

Questions

13. Give several examples of familiar motions which are either exactly or approximately simple harmonic motions.

14. The force on a particle is represented by the potential-energy function $U(x)$. The particle may execute approximate simple harmonic motion about a point x_1 only if $dU/dx = 0$ at that point. Explain why. The motion will be approximately simple harmonic for small displacements from x_1 *only* if d^2U/dx^2 is positive at x_1. Explain.

15. A ball bounces up and down on a table, reaching a maximum height h above the table and making perfectly elastic collisions with the table. Is this motion periodic? If so, what are the period and frequency? How does the motion differ from simple harmonic motion?

14-8 Combinations of Two Simple Harmonic Motions

The phase constant δ is important only if there are two or more simple harmonic motions. The simple example of two particles oscillating along the same direction with the same frequency and amplitude illustrates the role played by the phase constant in such cases. The equations of motion of the two particles are

$$x_1 = A \cos (\omega t + \delta_1) \qquad x_2 = A \cos (\omega t + \delta_2)$$

The difference in phase of the two motions at any time is

$$(\omega t + \delta_2) - (\omega t + \delta_1) = \delta_2 - \delta_1 = \delta_r$$

where δ_r is called the *relative phase* of the motion. Let us choose our zero time so that δ_1 is zero. Since the relative phase is independent of time, the phase constant δ_2 must then be δ_r. The equations of motion are then

$$x_1 = A \cos \omega t \qquad \text{and} \qquad x_2 = A \cos (\omega t + \delta_r)$$

It is interesting to look at the motion of one particle relative to the other. Let us imagine that we are sitting on particle 1 and observing particle 2. The position of particle 2 relative to particle 1 is

$$x_r = x_2 - x_1 = A \cos (\omega t + \delta_r) - A \cos \omega t$$

We can simplify this by using the trigonometric identity

$$\cos A - \cos B = -2 \sin \tfrac{1}{2}(A - B) \sin \tfrac{1}{2}(A + B)$$

Then

$$x_r = -A \, (2 \sin \tfrac{1}{2}\delta_r) \sin (\omega t + \tfrac{1}{2}\delta_r)$$
$$= 2A \sin \tfrac{1}{2}\delta_r \cos (\omega t + \tfrac{1}{2}\delta_r + \tfrac{1}{2}\pi) \qquad\qquad 14\text{-}37$$

where we have also used the identity $\cos (\theta + \tfrac{1}{2}\pi) = -\sin \theta$. Equation 14-37 gives simple harmonic motion with amplitude $2A \sin \tfrac{1}{2}\delta_r$. For example, if δ_r is zero, $x_r = 0$. There is no motion of one particle relative to the other; they move back and forth in phase. On the other hand, if the relative phase is 180°, the relative position is given by

$$x_r = 2A \sin 90° \cos (\omega t + \tfrac{1}{2}\pi + \tfrac{1}{2}\pi) = -2A \cos \omega t$$

Since we have chosen our zero time so that particle 1 is at its maximum displacement, particle 2 will be at its maximum negative displacement at zero time for $\delta_r = 180°$. Thus the relative motion is simple harmonic motion with amplitude $2A$.

   Another interesting combination of two simple harmonic motions is the motion of a particle in a plane such that both its x and y coordinates vary as in simple harmonic motion. If we consider only cases in which the frequencies and amplitudes are equal,

$$x = A \cos (\omega t + \delta_1) \qquad y = A \cos (\omega t + \delta_2)$$

Again we can choose one of the phase constants to be zero by our choice of zero time. If we choose δ_1 to be zero and call $\delta_2 - \delta_1 = \delta_r$ the relative phase, as before, we have

$$x = A \cos \omega t \tag{14-38}$$

$$y = A \cos (\omega t + \delta_r) \tag{14-39}$$

We have already noted (Section 14-2) that when δ_r is $\pm 90°$, these equations are those for motion of a particle in a circle of radius A. Then

$$y = A \cos (\omega t \pm 90°) = \mp A \sin \omega t$$

and

$$x^2 + y^2 = A^2 \cos^2 \omega t + A^2 \sin^2 \omega t = A^2$$

which is the equation for a circle in the xy plane.
   If the relative phase is zero, we have

$$y = A \cos \omega t = x$$

The motion is thus along the straight line $y = x$. Figure 14-18 shows the path of the particle described by Equations 14-38 and 14-39 for $\delta_r = 30°$. For relative phase angles between 0 and 90° the motion is an ellipse, of which the major axis is the line $y = x$. (The case $\delta_r = 0$ is a special case of an ellipse with zero minor axis; similarly, $\delta_r = 90°$ is the special case of equal major and minor axes, the circle.) This is also the case for relative phases between 270 and 360°. For δ_r between 90 and 270° the motion is an ellipse with major axis along the line $x = -y$. The special case of $\delta_r = 180°$ describes linear motion along this line, as can be seen from

$$y = A \cos (\omega t + 180°) = -A \cos \omega t = -x$$

   If amplitudes are not equal, the combination of two simple harmonic motions of equal frequency describes the x and y coordinates for elliptical motion, the orientation and eccentricity depending on the difference in amplitudes and on δ_r. For example, if $\delta_r = 90°$ and the amplitudes are different, we have

$$x = A \cos \omega t \qquad y = B \cos \omega t \qquad y = B \cos (\omega t + 90°) = -B \sin \omega t$$

$$\frac{x^2}{A^2} + \frac{y^2}{B^2} = \cos^2 \omega t + \sin^2 \omega t = 1$$

which is the equation for an ellipse with major axis along the x or y axis, depending on whether A is greater or less than B.

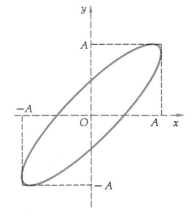

Figure 14-18
Plot of $x = A \cos \omega t$, $y = A \cos (\omega t + \delta_r)$ for $\delta_r = 30°$.

Review

A. Define, explain, or otherwise identify:

Periodic motion, 355
Simple harmonic motion, 355
Phase, 355
Amplitude, 355

Phase constant, 355
Period, 355
Angular frequency, 356
Principle of equivalence, 366

B. True or false:

1. All periodic motion is simple harmonic motion.

2. All simple harmonic motion is periodic motion.

3. In simple harmonic motion, the period is proportional to the amplitude.

4. In simple harmonic motion, the total energy is proportional to the square of the amplitude.

5. In simple harmonic motion, the phase constant depends on the initial conditions.

6. The motion of a simple pendulum is simple harmonic for any initial angular displacement.

7. The motion of a simple pendulum is periodic for any initial angular displacement.

8. In simple harmonic motion the acceleration is proportional to the displacement and in the opposite direction.

9. Any motion in which the acceleration is proportional to the displacement and in the opposite direction is simple harmonic motion.

10. Motion near a position of stable equilibrium is generally simple harmonic if the displacements are small.

Exercises

Section 14-1, Simple Harmonic Motion

1. A particle has displacement x given by $x = 3 \cos (5\pi t + \pi)$, where x is in meters and t in seconds. (a) What are the frequency f and period T of the motion? (b) What is the greatest distance the particle travels from equilibrium? (c) Where is the particle at time $t = 0$? At time $t = \frac{1}{2}$ sec?

2. (a) Find an expression for the velocity of the particle whose position is given in Exercise 1. (b) What is the maximum velocity? (c) When does this maximum velocity occur?

3. A particle in simple harmonic motion is at rest a distance of 6 in from its equilibrium position at time $t = 0$. Its period is 2 sec. Write expressions for its position x, its velocity v, and its acceleration a as functions of time.

4. The position of a particle traveling in one dimension is given by $x = 5 \cos 4\pi t$, where x is in meters and t in seconds. (a) Find the amplitude, angular frequency, frequency, and period of the motion. (b) Find an expression for the velocity of the particle at any time t. (c) What is the velocity at time $t = 0$?

5. The position of a particle is given by $x = 4 \sin 2t$, where x is in feet and t is in seconds. (a) What is the maximum value of x? What is the first time after $t = 0$ at which this maximum value occurs? (b) Find an expression for the velocity of the particle as a function of time. What is the velocity at time $t = 0$? (c) Find an expression for the acceleration of the particle as a function of time. What is the acceleration at time $t = 0$? What is the maximum value of the acceleration?

6. A particle oscillates with simple harmonic motion with period $T = 2$ sec. It

is initially at its equilibrium position and moving with speed 4 m/sec in the direction of increasing x. Write expressions for its position x, its velocity v, and its acceleration a as functions of time.

Section 14-2, Circular Motion and Simple Harmonic Motion, and Section 14-3, A Mass on a Spring

7. A particle travels in a circle in the xy plane with center at the origin. The radius of the circle is 40 cm, and the speed of the particle is 80 cm/sec. (a) What is the angular velocity of the particle? (b) What are the frequency f and period T of the circular motion? (c) Write the x and y components of the position vector **r** as functions of time.

8. A 2-kg mass is attached to a spring of force constant $k = 5 \times 10^3$ N/m. The spring is stretched 10 cm from equilibrium and released. (a) Find the angular frequency, the frequency f, and the amplitude of oscillation. (b) Find $x(t)$, $v(t)$, and $a(t)$. (c) Write expressions for the kinetic and potential energies as functions of time. What is the total energy?

9. A 3-kg mass is attached to a spring and oscillates with amplitude of 10 cm and frequency $f = 2$ Hz. (a) What is the force constant of the spring? (b) What is the total energy of the motion? (c) Write an equation $x(t)$ describing the position of the mass relative to its equilibrium position. Can the phase constant be determined from the information given?

10. A particle is attached to a spring. At $t = 0$ it is in its equilibrium position and has speed 5 cm/sec toward the negative side of the x axis. Its frequency is $f = 3$ Hz. (a) At what value of t does it next come to rest? (b) Where is it at that time? (c) What is its acceleration at that time? (d) Write an expression for its position $x(t)$ at any time t.

11. A 100-gm mass executes simple harmonic motion with a frequency of 20 Hz and amplitude 0.5 cm. (a) What is the constant k for the force acting on it? (b) What is the maximum acceleration? (c) What is the total energy of the motion?

12. A mass $m = 500$ gm executes simple harmonic motion with a period of 0.5 sec. Its total energy is 5 J. (a) What is the amplitude of the oscillation? (b) What is the maximum speed? (c) What is the maximum acceleration?

13. When the displacement of a mass oscillating on a spring is half its amplitude, what fraction of its total energy is its kinetic energy? At what displacement are its kinetic and potential energies equal?

14. A 100-gm particle is attached to a spring and oscillates with a frequency of 10 Hz and total energy of 1.25 J. (a) What is the force constant of the spring? (b) What is the amplitude of the oscillation?

Section 14-4, The Simple Pendulum

15. Find the length of a simple pendulum if the period of the pendulum is 1 sec at a point where $g = 9.8$ m/sec^2.

16. What would be the period of the pendulum in Exercise 15 on the moon, where the acceleration of gravity is one-sixth that on earth?

17. If the period of a pendulum 70 cm long is 1.68 sec, what is the value of g at the location of the pendulum?

18. A pendulum set up in the stairwell of a 10-story building consists of a heavy weight suspended on a 110-ft wire. What is its period of oscillation ($g = 32$ ft/sec^2)?

19. Derive Equation 14-18a by setting the torque on the mass about the support point equal to $I\alpha$.

Section 14-5, The Physical Pendulum

20. A thin disk of mass 5 kg and radius 20 cm is suspended by an axis perpendicular to the disk through its rim. It is displaced slightly from equilibrium and released. Find the period of the subsequent simple harmonic motion.

21. A circular hoop of radius 50 cm is hung on a narrow horizontal rod and allowed to swing in the plane of the hoop. What is the period of its oscillation assuming that the amplitude is small?

22. A 3-kg plane figure is suspended at a point 10 cm from its center of mass. When it is set oscillating with small amplitude, the period of oscillation is 2.6 sec. Find the moment of inertia I about an axis perpendicular to the plane of the figure and through the pivot point.

Section 14-6, A Mass on a Vertical Spring

23. A 4-lb weight is suspended from a vertically hanging spring of force constant $k = 24$ lb/ft. (a) Find the distance y_0 the spring is stretched when the weight hangs in equilibrium and the potential energy of the spring relative to the unstretched length. (b) The weight is then pulled downward a distance $y' = 3$ in below its equilibrium point. Find the change in the potential energy of the spring, the change in the gravitational potential energy, and the total change in potential energy. Show that the total change in potential energy equals $\frac{1}{2}ky'^2$. (c) The weight is then released. Find the period, frequency, and amplitude of the subsequent oscillation.

24. A 3-kg mass oscillates on a vertical spring of force constant $k = 500$ N/m. Its maximum kinetic energy is 5 J. (a) If the total potential energy is chosen to be zero at equilibrium, what is the maximum potential energy of oscillation? (b) What are the amplitude and period of oscillation? (c) What is the maximum stretching of the spring from its natural length?

Section 14-7, General Motion near Equilibrium

There are no exercises for this section.

Section 14-8, Combinations of Two Simple Harmonic Motions

25. Two identical mass-spring systems are situated as in Figure 14-19. Each mass is 4 kg, and the spring constants are $k_1 = k_2 = 100$ N/m. Each spring is initially stretched 4 cm, but the masses are released separately. (a) What must the relative phase be for the relative motion to be simple harmonic with maximum amplitude? What is the maximum amplitude? (b) If the first mass is released at time $t = 0$, at what time t should the second be released if the relative amplitude is to be a maximum?

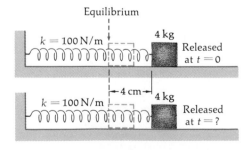

Equilibrium

$k = 100$ N/m   4 kg   Released at $t = 0$

$\leftarrow$ 4 cm $\rightarrow$ 4 kg

$k = 100$ N/m   Released at $t = ?$

Figure 14-19
Exercise 25.

26. (a) For the situation of Exercise 25, what should the time of release of the second mass be for the amplitude of the relative motion to be 4 cm? (b) If the second mass is released just as the first passes through equilibrium for the first time, what is the amplitude of the relative motion?

27. A particle moves so that its x and y coordinates are given by $x = 10 \cos \pi t$ and $y = 5 \cos (\pi t + \pi/2)$. Sketch the path of the particle in the xy plane.

Problems

1. The period of an oscillating particle is 8 sec. At $t = 0$ it is at its equilibrium position. (a) How does the distance it travels in the first 4 sec compare with that it travels in the second 4 sec? (b) The distance traveled in the first 2 sec and the second 2 sec? (c) The first second and the second second?

2. A block of wood slides on a smooth horizontal surface. It is attached to a spring and oscillates with a period of 0.3 sec. A second block rests on top of it. The coefficient of static friction between the blocks is $\mu_s = 0.25$. (a) If the amplitude of oscillation is 1 cm, will the block on top slip? (b) What is the greatest amplitude of oscillation for which the blocks will not slip?

3. A mass suspended on a vertical spring oscillates with a period of 0.5 sec. When the mass is attached to the spring and allowed to hang at rest, by how much is the spring stretched?

4. The equation of motion of a certain oscillating particle is $x = 6.0 \cos (9.5t + 1.0)$, where x is in centimeters and t is in seconds. (a) Find the speed v as a function of time. What is the maximum speed of the particle? When does it occur? (b) Find the acceleration a as a function of time. What is a_{max}? When does it occur? (c) Sketch the acceleration a versus position x.

Figure 14-20
Problem 5.

5. (a) A small particle of mass m slides without friction in a spherical bowl of radius r. Show that the motion of the mass is the same as if it were attached to a string of length r. (b) A mass m_1 is displaced a small distance s_1 from the bottom of the bowl, where s_1 is much smaller than r. A second mass m_2 is displaced in the opposite direction a distance $s_2 = 3s_1$ (s_2 is also much smaller than r) (Figure 14-20). If the masses are released at the same time, where do they meet? Explain.

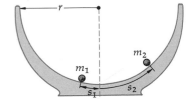

6. Show that for the situations shown in Figures 14-21a and b the mass oscillates with angular frequency $\omega = \sqrt{k_{eff}/m}$, where k_{eff} is given by (a) $k_{eff} = k_1 + k_2$ and (b) $1/k_{eff} = 1/k_1 + 1/k_2$.

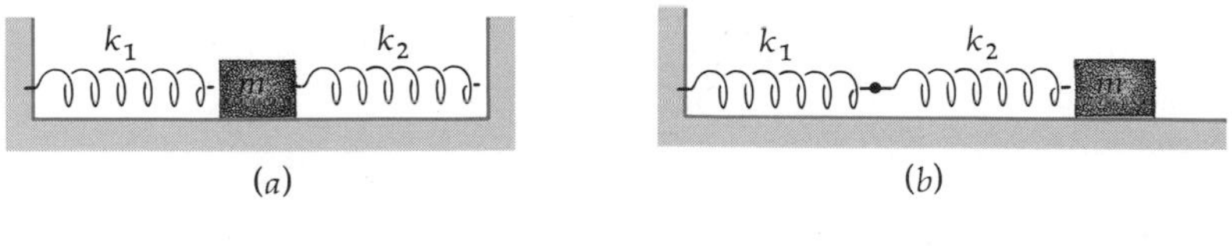

(a)

(b)

Figure 14-21
Problem 6.

7. A mass m_1 slides on a smooth horizontal surface and is attached to a spring of force constant k. It oscillates with amplitude A. When the spring is at its greatest extension and the mass is instantaneously at rest, a second mass m_2 is placed on top of m_1. (a) What is the smallest value the coefficient of static friction μ_s can have without permitting m_2 to slip on m_1? (b) Explain how the total energy E, the amplitude A, the angular frequency ω, and the period T are changed by placing m_2 on m_1 in this way, assuming no slipping. Check your results with the expression for the total energy $E = \frac{1}{2}m\omega^2 A^2$.

8. A weight is suspended from a spring and set into a vertical oscillation with a frequency of 4 Hz and an amplitude of 7 cm. A very small bit of rock is placed on top of the oscillating weight just as it reaches its lowest point. (a) At what distance above the equilibrium position does the rock lose contact with the weight? (b) What is the velocity of the rock when it leaves the weight? (c) What is the greatest distance above the equilibrium position reached by the rock? (Assume that the rock has no effect on the oscillation and use the fact that the maximum downward acceleration of the rock is g.)

9. A spring has force constant $k = 150$ N/m. It is suspended from a rigid support. A 1-kg mass is attached at the bottom end and released from rest when the spring is unstretched. (a) How far down does the mass move before it starts upward again? (b) How far below the starting point is the equilibrium position for the 1-kg mass? (c) What is the period of the oscillation? (d) Relative to this equilibrium position, what is the total energy of the mass-spring system? (e) What is the speed of the mass when it first reaches the equilibrium position? How long after it started does the mass have this speed? (f) If instead of being attached to the spring the mass had been dropped, when would it have reached the (previous) equilibrium level, and what would be its speed?

10. The acceleration of gravity g varies with location on the earth because of the earth's rotation and because the earth is not exactly spherical. This was first discovered in the seventeenth century, when it was noted that a pendulum clock carefully adjusted to keep correct time in Paris lost about 90 sec/d near the equator. (a) Show that a small change in g produces a small change in the period of a pendulum T given by

$$\frac{\Delta T}{T} \approx -\frac{1}{2}\frac{\Delta g}{g}$$

(Use differentials to approximate ΔT and Δg.) (b) How great a change in g is needed to account for a change in period of 90 sec/d?

11. A physical pendulum consists of a spherical bob of radius r and mass m suspended from a string of length $L - r$ (Figure 14-22). (The distance from the center of the sphere to the point of support is L.) Often when r is much less than L, this situation is treated as a simple pendulum of length L. (a) Using the parallel-axis theorem, find the moment of inertia of this pendulum about the point of support. Show that the period can be written

$$T = 2\pi \sqrt{\frac{L}{g}} \sqrt{1 + \frac{2r^2}{5L^2}} = T_0 \sqrt{1 + \frac{2r^2}{5L^2}}$$

where $T_0 = \sqrt{L/g}$ is the period of a simple pendulum of length L. (b) Using the binomial approximation $(1 + x)^n \approx 1 + nx$ for $x \ll 1$, show that for r much smaller than L, the period is $T \approx T_0(1 + r^2/5L^2)$. (c) If L is 1 m and $r = 2$ cm, find the error made in using the approximation $T = T_0$ for this pendulum. How large must the radius of the bob be for the error to be 1 percent?

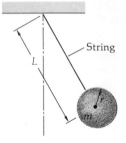

Figure 14-22
Problem 11.

String

L

m

12. A 2-kg mass is attached to a spring of force constant 10 N/m and rests on a smooth horizontal surface. A second mass of 1 kg slides along the surface toward the first at 6 m/sec. (a) Find the amplitude of oscillation if the masses make a perfectly inelastic collision and remain together on the spring. What is the period of oscillation? (b) Find the amplitude and period of oscillation if the collision is perfectly elastic. (c) For each case, write down the position x as a function of time t for the mass attached to the spring assuming that the collision occurs at time $t = 0$. What is the impulse given to the 2-kg mass in each case?

13. In this problem you are asked to derive Equation 14-1 from the application of Newton's laws of motion and the conservation of energy to the problem of a mass m which is influenced by a single force $-kx$, where k is a constant. (a) From Newton's second law write down the acceleration d^2x/dt^2 as a function of displacement from equilibrium x. (b) The solution of the differential equation found in part (a) requires two integrations. The first results in an equation which is equivalent to the conservation of energy. From the conservation of energy, write an equation relating the speed v to the displacement x and the total energy E. Show that the total energy can be written $E = \frac{1}{2}m\omega^2 A^2$, where $\omega = \sqrt{k/m}$ and A is the maximum value of x. (c) In terms of the quantities defined in part (b) show that the speed v can be written $v = \omega\sqrt{A^2 - x^2}$. Using $v = dx/dt$, obtain the equation

$$\frac{dx}{\sqrt{A^2 - x^2}} = \omega\, dt$$

(d) Using the substitution $x = A \sin \theta$, show that the equation obtained in part (c) becomes $d\theta = \omega \, dt$, which has the solution $\theta = \omega t + \theta_0$. Compare your solution for $x(t)$ with Equation 14-1.

14. A pendulum clock that has run down to a very small amplitude gains 5 min/d. What is the proper (angular) amplitude for the pendulum to keep correct time?

15. A simple pendulum of length L is released from rest from an angle of θ_0. (a) Assuming the motion to be simple harmonic, find the velocity of the pendulum as it passes through $\theta = 0$. (b) Using conservation of energy, find this velocity exactly. (c) Show that your results for parts (a) and (b) are the same when θ_0 is small. (d) Find the difference in your result for $\theta_0 = 10°$ and $L = 1$ m.

16. Derive the equation $T = 2\pi \sqrt{mL/m_G g}$ for a simple pendulum where m_G is the gravitational mass, defined by $w = m_G g$, and w is the weight of the bob. Discuss the possibility of detecting m/m_G if the gravitational and inertial masses are proportional.

17. An object has moment of inertia I about its center of mass. When pivoted at point P_1 (Figure 14-23), it oscillates about the pivot with period T. There is a second point P_2 on the opposite side of the center of mass about which the object can be pivoted and for which the period of oscillation is also T. Show that $h_1 + h_2 = gT^2/4\pi^2$.

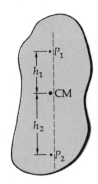

Figure 14-23
Problem 17.

18. Figure 14-24 shows a tube of constant cross section A open to the atmosphere. It is filled to the level shown by a dashed line with an incompressible liquid which flows through the tube with negligible friction. The total length of the liquid column is L. Show that if the liquid surface is depressed on one side and then released, the liquid level will oscillate about the equilibrium position in simple harmonic motion with a period given by $T = 2\pi \sqrt{L/2g}$.

19. The following expression is sometimes used to represent the intermolecular potential energy:

$$U(x) = \tfrac{1}{2} U_0 x_0^6 \left(\frac{x_0^6}{x^{12}} - \frac{2}{x^6} \right)$$

(a) Show that there is an equilibrium point at $x = x_0$. (b) What is the value of $U(x_0)$? (c) What is the frequency of small oscillations about the equilibrium point if the oscillating mass is m?

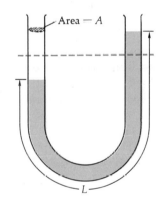

Figure 14-24
Problem 18.

20. Figure 14-25 shows two point masses m_1 and m_2 at their equilibrium positions. The period of oscillation of each mass is 4 sec. The two masses are pulled to the point $x = 25$ cm and released. If m_1 is released at time $t = 0$ and m_2 is released at time $t = t_0$, describe the motion of m_1 relative to m_2 for (a) $t_0 = 0$, (b) $t_0 = 1$ sec, (c) $t_0 = 2$ sec. Find the distance of closest approach and the first time it occurs (after $t = 0$) in each case.

21. A simple pendulum 50 cm long is suspended from the roof of a cart accelerating in the horizontal direction with $a = 7$ m/sec² (Figure 14-26). Find the period of small oscillations of the pendulum about its equilibrium angle.

Figure 14-25
Problem 20.

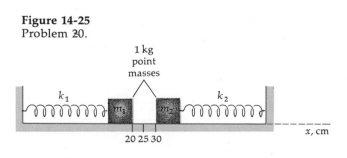

Figure 14-26
Problem 21.

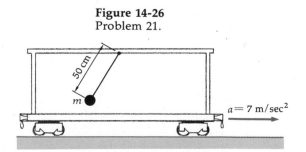

CHAPTER 15   Damped and Forced Oscillations

In all real oscillatory motion, mechanical energy is dissipated because of some kind of frictional force. Left to itself, a mass on a spring or a pendulum will eventually stop oscillating. When the mechanical energy of oscillatory motion decreases with time, the motion is said to be *damped*. If the frictional or damping forces are small, the motion is nearly periodic except that the amplitude decreases slowly with time, as in Figure 15-1.

A forced oscillator is one that is driven by an external force. A simple example is a swing given a push once each period. Here the external force is a periodic impulsive force. Its effect is to increase the amplitude of oscillation until a steady state is reached, at which time the work done by the external force during each cycle equals the energy lost due to damping during the cycle. In Sections 15-2 and 15-3 we shall study the effects of an external simple harmonic driving force of angular frequency ω on an oscillator of natural frequency ω_0. We shall see that in the steady state, the amplitude and therefore the energy of the oscillator depend on the frequency of the driving force. The amplitude and energy are large when the frequency of the driving force is close to the natural frequency. This phenomenon is called *resonance*.

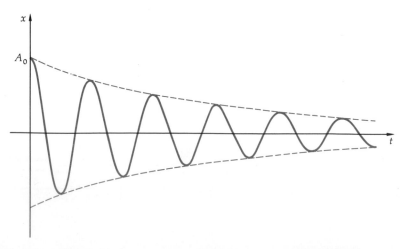

Figure 15-1
Displacement versus time for a slightly damped oscillator. The amplitude decreases slowly with time.

15-1 Damped Oscillations

In Figure 15-2 the oscillation of the mass is damped because of the motion of the plunger in the liquid. Mechanical energy of the mass-spring system is dissipated into heat in the liquid. The rate of energy loss can be varied by changing the size of the plunger or the viscosity of the liquid. Although the detailed analysis of such a damping force is usually complicated, often we can represent such a force by an empirical expression for which the mathematics is comparatively simple and which is in reasonable accord with experiment. The simplest and most common representation of such a force is one proportional to the velocity of the mass but in the opposite direction,

$$\mathbf{F}_d = -b\mathbf{v} \qquad\qquad 15\text{-}1$$

where b is a constant which describes the degree of damping. Many rather complicated phenomena can be approximately described with such a force. This force is clearly nonconservative because it depends on the velocity. Since it is always directed opposite the direction of motion, the work done by the force is always negative; thus it always decreases the mechanical energy of the system. We can also see this by computing the power input of this force:

$$P = \mathbf{F}_d \cdot \mathbf{v} = -bv^2 \qquad\qquad 15\text{-}2$$

Newton's second law, $\Sigma\mathbf{F} = m\mathbf{a}$, applied to the motion of the mass on a spring of force constant k and with damping force $-bv$ is

$$\Sigma F_x = -kx - bv = m\,\frac{dv}{dt} \qquad\qquad 15\text{-}3$$

The total mechanical energy of the system is

$$E = \tfrac{1}{2}mv^2 + \tfrac{1}{2}kx^2$$

The rate of change of mechanical energy is

$$\frac{dE}{dt} - \tfrac{1}{2}m\left(2v\,\frac{dv}{dt}\right) + \tfrac{1}{2}k\left(2x\,\frac{dx}{dt}\right) = v\left(m\,\frac{dv}{dt} + kx\right) \qquad\qquad 15\text{-}4$$

using $dx/dt = v$. But from Equation 15-3 we see that $m\,dv/dt + kx = -bv$. Thus

$$\frac{dE}{dt} = -bv^2 \qquad\qquad 15\text{-}5$$

The rate of change of mechanical energy is just the power input by the damping force, which is always negative.

We can get some understanding of the behavior of a damped oscillator without solving Equation 15-3 in detail. If the damping is small, we expect the mass to oscillate with a frequency ω' which is approximately equal to the *natural frequency* of undamped motion ω_0:

$$\omega' \approx \omega_0 = \sqrt{\frac{k}{m}} \qquad\qquad 15\text{-}6$$

We expect the amplitude of the oscillation to decrease slowly. The energy change in one period T is

$$\Delta E = \int_0^T \frac{dE}{dt}\,dt = \int_0^T -bv^2\,dt = -b\int_0^T v^2\,dt = -bT(v^2)_{\text{av}} \qquad\qquad 15\text{-}7$$

where the average value of v^2 over one period is defined by

$$(v^2)_{\text{av}} = \frac{1}{T}\int_0^T v^2\,dt \qquad\qquad 15\text{-}8$$

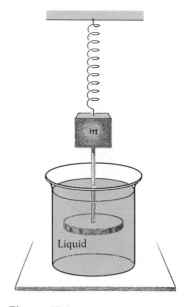

Figure 15-2
Physical arrangement of a damped oscillator. The motion of the mass is damped by the plunger immersed in the liquid. The damping can be varied by varying the size of the plunger or the viscosity of the liquid.

In simple harmonic motion, the energy oscillates between potential and kinetic energy. The average value of either the kinetic or the potential energy is one-half the total energy. Thus

$$\tfrac{1}{2}m(v^2)_{\text{av}} = (E_k)_{\text{av}} = \tfrac{1}{2}E \qquad\qquad 15\text{-}9$$

or

$$(v^2)_{\text{av}} = \frac{E}{m} \qquad\qquad 15\text{-}10$$

Combining Equations 15-7 and 15-10, we obtain for the energy loss per period,

$$-\frac{\Delta E}{T} = \frac{b}{m}\,E \qquad\qquad 15\text{-}11$$

A reasonable criterion for small damping is that the energy loss in one period be a small fraction of the total energy:

$$-\Delta E \ll E$$

or

$$-\frac{\Delta E}{E} = \frac{bT}{m} \ll 1 \qquad\qquad 15\text{-}12$$

In terms of the angular frequency $\omega' = 2\pi/T$ this criterion for small damping can be written

$$\frac{b}{m} \ll \frac{\omega'}{2\pi} \qquad\qquad 15\text{-}13$$

We can use Equation 15-11 to guess the time dependence of the energy. Let us write Δt for the time interval T in Equation 15-11. Then

$$\frac{\Delta E}{\Delta t} = -\frac{b}{m}\,E$$

Since the energy loss in one period is a small fraction of the total energy, the loss of energy per period cannot differ much from the instantaneous rate of energy loss. We assume therefore that the differences in the last equation can be replaced by differentials:

$$\frac{dE}{dt} = -\frac{b}{m}\,E \qquad\qquad 15\text{-}14$$

The function whose derivative is proportional to the function itself is the exponential. The solution of Equation 15-14 is

$$E(t) = E_0 e^{-(b/m)t} \qquad\qquad 15\text{-}15$$

where E_0 is the energy at time $t = 0$. We can use this equation to find the time dependence of the amplitude since the energy is proportional to the square of the amplitude. If $A(t)$ is the amplitude at time t and A_0 is that at $t = 0$, we have

$$\frac{E(t)}{E_0} = \frac{A^2(t)}{A_0{}^2}$$

Thus from Equation 15-15

$$\frac{A^2(t)}{A_0{}^2} = e^{-(b/m)t}$$

or

$$A(t) = A_0 e^{-(b/2m)t} \qquad\qquad 15\text{-}16$$

The exact solution of Equation 15-3 can be found by standard methods for solving differential equations. The solution for the case of small damping, $b < 2m\omega_0$, is

$$x = A_0 e^{-(b/2m)t} \cos{(\omega' t + \delta)} \qquad\qquad 15\text{-}17$$

where A_0 is the original amplitude and the frequency ω' is the angular frequency with which the oscillator passes through its equilibrium position ($x = 0$); ω' is given by

$$\omega' = \omega_0 \sqrt{1 - \left(\frac{b}{2m\omega_0}\right)^2} \qquad\qquad 15\text{-}18$$

in which $\omega_0 = \sqrt{k/m}$ is the angular frequency without damping, the natural frequency. We see that our qualitative observations were correct. For small damping, the frequency is very nearly equal to the natural frequency, and the time dependence of the amplitude is given by Equation 15-16, as expected.

Example 15-1 If the energy loss per cycle is one-tenth the total energy, by how much does the frequency differ from the natural frequency?
  According to Equation 15-11, we have

$$-\frac{\Delta E}{E} = \frac{b}{m} T = \frac{b2\pi}{m\omega'} = \frac{1}{10}$$

or

$$\left(\frac{b}{2m\omega'}\right)^2 = \left(\frac{1}{40\pi}\right)^2 = \frac{1}{16{,}000}$$

using $\pi^2 \approx 10$. Since ω' is so close to ω_0, we can use this in Equation 15-18, to give

$$\omega' = \omega_0 \sqrt{1 - \frac{1}{16{,}000}}$$

For small ϵ, $\sqrt{1 - \epsilon} \approx 1 - \frac{1}{2}\epsilon$. Thus

$$\omega' \approx \omega_0 \left(1 - \frac{1}{32{,}000}\right) \qquad \text{and} \qquad \omega' - \omega_0 \approx \frac{-\omega_0}{32{,}000}$$

or

$$\frac{\omega' - \omega_0}{\omega_0} \approx -\frac{1}{32{,}000} = -0.003\%$$

We see that for this rather large damping, the frequency decreases by only about 0.003 percent. Such a small change in frequency is difficult to observe. Even when the damping is large enough to reduce the amplitude by half in each oscillation, the frequency is shifted by only 0.6 percent from the natural frequency ω_0. Hence the distinction between ω' and ω_0 can usually be ignored.

Figure 15-1 shows a plot of x given by Equation 15-17 versus t. This solution is valid if b is less than $2m\omega_0$. The value

$$b_c = 2m\omega_0 \qquad\qquad 15\text{-}19$$

is called the *condition for critical damping.* If b is greater or equal to this critical value, the system does not oscillate at all but merely returns to its equilibrium position. The greater the damping the longer it takes for the system to return to equilibrium. When $b = b_c$, the system is said to be *critically damped.* The mass returns to its equilibrium posi-

Critical damping

tion in the shortest possible time with no oscillation. When b is greater than b_c, the system is *overdamped*. Figure 15-3 shows the displacement versus time for a critically damped and overdamped oscillator.

In many practical applications critical damping or nearly critical damping is used to avoid oscillations and yet have the system return to equilibrium quickly. One example is the use of shock absorbers to damp the oscillations of an automobile on its wheels. Such a system is usually slightly underdamped (b slightly less than the critical value); you can see this by pushing down on the front or back of the car and observing that one or two oscillations occur before the system comes to rest.

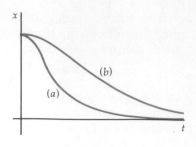

Figure 15-3
Displacement versus time for (*a*) critically damped and (*b*) overdamped oscillator.

Questions

1. Two oscillators have the same mass and the same original amplitude and are subject to the same damping force. One, however, has twice the frequency of the other. Which will be damped to half the original amplitude first?

2. Explain qualitatively why you would expect the frequency of an oscillator to decrease because of the presence of a damping force.

15-2   The Forced Oscillator

In many cases an oscillator is subject to an applied force or external force in addition to the damping force and the restoring force. The external force increases the mechanical energy of the oscillator if it acts in the direction of motion of the mass, and it absorbs energy from the oscillator if it acts opposite to the direction of motion. We saw a simple example of this in Chapter 14, a mass on a vertical spring. Here the external force is the weight, which is constant in magnitude and direction. The work done by this force is positive when the mass is moving down and negative when the mass is moving up. The net work done during one cycle is zero, and on the average this constant force does not change the energy of the mass-spring system. A constant force merely changes the position of equilibrium of the system.

A particularly important kind of external force is one which varies harmonically with time such as

$$F_{ext} = F_0 \sin \omega t \qquad\qquad 15\text{-}20$$

where ω is the angular frequency of the force, which is generally not related to the natural angular frequency of the system $\omega_0 = \sqrt{k/m}$. A mass m attached to a spring of force constant $k = m\omega_0^2$ subject to a damping force $-bv$ and to an external force $F_0 \sin \omega t$ then obeys the equation of motion given by

$$\Sigma F = -m\omega_0^2 x - bv + F_0 \sin \omega t = m\,\frac{dv}{dt}$$

or

$$m\,\frac{d^2x}{dt^2} + b\,\frac{dx}{dt} + m\omega_0^2 x = F_0 \sin \omega t \qquad\qquad 15\text{-}21$$

We first consider the special case of no damping because it is easier to solve. Equation 15-21 becomes, for $b = 0$,

$$m\,\frac{d^2x}{dt^2} + m\omega_0^2 x = F_0 \sin \omega t \qquad\qquad 15\text{-}22$$

We can show by substitution that we can satisfy Equation 15-22 with the function

$$x = A \sin \omega t$$

Then

$$\frac{d^2x}{dt} = -\omega^2 A \sin \omega t$$

Substituting this into Equation 15-22, we obtain

$$m(-\omega^2 A \sin \omega t) + m\omega_0^2 A \sin \omega t = F_0 \sin \omega t$$

which is satisfied for any time if A is given by

$$A = \frac{F_0}{m(\omega_0^2 - \omega^2)}$$

It is customary to write the position function $x(t)$ in terms of an amplitude A which is intrinsically positive. We can do this by introducing a phase angle δ. We write

$$x = A \sin(\omega t - \delta) \tag{15-23}$$

where

$$A = \frac{F_0}{m|\omega_0^2 - \omega^2|} \tag{15-24}$$

and the phase angle δ is 0 if $\omega < \omega_0$ and $\pm\pi$ if $\omega > \omega_0$.

The mass oscillates in phase with the driving force if ω is less than ω_0 and 180° out of phase with the driving force if ω is greater than ω_0. The amplitude of the oscillation is large if the frequency of the external force is near the natural frequency of the oscillator. For the case of no damping, Equation 15-24 implies that A approaches infinity as ω approaches ω_0. In all real situations there is some damping, and the amplitude is finite when $\omega = \omega_0$. Since the energy of the oscillator is proportional to A^2, the energy is large when ω is near ω_0. (For no damping the energy is infinite at $\omega = \omega_0$.) This phenomenon, called *resonance*, will be discussed more fully in the next section.

15-3   The Forced Oscillator with Damping

The solution of Equation 15-21 with damping is more complicated. We shall discuss the result first and then show that it is a solution of this equation. Equation 15-21 can be satisfied by the function $x(t)$ given by

$$x(t) = A \sin(\omega t - \delta) \tag{15-25}$$

where the amplitude A is given by

$$A = \frac{F_0}{\sqrt{m^2(\omega_0^2 - \omega^2)^2 + b^2\omega^2}} \tag{15-26}$$

and the phase constant δ by

$$\tan \delta = \frac{b\omega}{m(\omega_0^2 - \omega^2)} \tag{15-27}$$

The phase angle δ is shown in Figure 15-4, from which we see that

$$\sin \delta = \frac{b\omega}{\sqrt{m^2(\omega_0^2 - \omega^2)^2 + b^2\omega^2}} = \frac{b\omega A}{F_0} \tag{15-28}$$

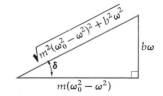

Figure 15-4
The phase angle δ in Equation 15-25.

When the driving frequency equals the natural frequency, $\omega = \omega_0$, the amplitude A is large, but it is not infinite because of the presence of the damping constant b in the denominator of Equation 15-26.

The function $x(t)$ given by Equation 15-25 is not the complete solution of Equation 15-21 for a forced oscillator with damping. There are no constants of integration in this solution, which can be adjusted to satisfy the initial conditions, i.e., the initial position and velocity of the mass on the spring. The complete solution of Equation 15-21 is

$$x(t) = A'e^{-(b/2m)t} \cos(\omega't + \delta') + A\sin(\omega t - \delta) \qquad 15\text{-}29$$

where A' and δ' are arbitrary constants which can be adjusted to meet the initial conditions and ω' is the frequency of the damped but unforced oscillator given by Equation 15-18.

The first term is just the solution of the unforced damped oscillator given by Equation 15-17. This part of the solution is called the *transient solution*. After a long time, so that $bt/2m \gg 1$, this part of the entire solution becomes negligible because the amplitude has decreased exponentially with time. The second term, given by Equation 15-25, is called the *steady-state solution*. The amplitude A, given by Equation 15-26, does not decrease with time. After a long time ($bt/2m \gg 1$) we can neglect the transient solution. The position is then given by Equation 15-25 no matter what the initial conditions were.

The velocity of the particle in the steady state is obtained from the position function (Equation 15-25) by differentiation:

$$v = \frac{dx}{dt} = A\omega \cos(\omega t - \delta) \qquad 15\text{-}30$$

The power input of the driving force in the steady state is

$$\begin{aligned} P = Fv &= (F_0 \sin \omega t) \, A\omega \cos(\omega t - \delta) \\ &= A\omega F_0 \sin \omega t \cos(\omega t - \delta) \end{aligned}$$

Using the identity $\cos(\omega t - \delta) = \cos \omega t \cos \delta + \sin \omega t \sin \delta$, we can write this as

$$P = \omega A F_0 \sin \delta \sin^2 \omega t + \omega A F_0 \cos \delta \sin \omega t \cos \omega t \qquad 15\text{-}31$$

The power input varies with time over a cycle. During one cycle, the second term in Equation 15-31 is negative as often as it is positive and has an average value of zero. The average value of $\sin^2 \omega t$ over one cycle is $\frac{1}{2}$ (see Figure 15-5). The average power is therefore

$$P_{av} = \tfrac{1}{2}\omega A F_0 \sin \delta$$

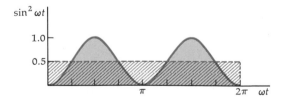

Figure 15-5
Plot of $\sin^2 \omega t$ versus ωt. The average value over a cycle is $\frac{1}{2}$. This result can be obtained from the fact that $\cos^2 \omega t$ has the same average value over a cycle and from $\sin^2 \omega t + \cos^2 \omega t = 1$.

Using $\sin \delta = b\omega A/F_0$ from Equation 15-28, we can eliminate either $\sin \delta$ or A. We have

$$P_{av} = \tfrac{1}{2}b\omega^2 A^2 = \frac{1}{2}\frac{F_0^2}{b}\sin^2 \delta \qquad 15\text{-}32$$

Alternatively, we can write either A or $\sin \delta$ in terms of the driving frequency ω:

$$P_{\text{av}} = \frac{1}{2} \frac{b\omega^2 F_0^2}{m^2(\omega_0^2 - \omega^2)^2 + b^2\omega^2}$$ 15-33

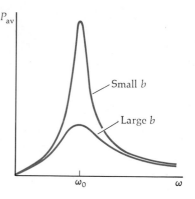

Figure 15-6 shows plots of the average power input versus driving frequency for two different values of the damping constant b. These curves are called *resonance curves*. The maximum power input occurs at the resonance frequency $\omega = \omega_0$. We can see this from Equations 15-32 and 15-27 since $\sin \delta$ has its maximum value of 1 at $\delta = 90°$. But according to Equation 15-27, when $\delta = 90°$, $\omega = \omega_0$ because $\tan 90°$ is infinite. The value of the maximum average power input is

$$P_{\text{av}} = \frac{1}{2} \frac{F_0^2}{b} \sin^2 90° = \frac{1}{2} \frac{F_0^2}{b}$$ 15-34

Figure 15-6
Plot of the average power input of the forcing term versus driving frequency ω for different values of the damping constant b. P_{av} is maximum in each case when $\omega = \omega_0$. The smaller the damping the greater the value of P_{av} at resonance and the narrower the resonance curve, i.e., the sharper the resonance.

At resonance the displacement is 90° out of phase with the driving force, but the velocity is in phase with the driving force. Thus at resonance the particle is always moving in the direction of the driving force, as would be expected for maximum power input. When the damping is small, the power input at resonance is large and the resonance is sharp; i.e., the resonance curve is narrow, indicating that the power input is large only near the resonance frequency. When the damping is large, the power input at resonance is small and the resonance curve is broad.

For any driving frequency the total energy is the sum of the potential and kinetic energies,

$$E = \tfrac{1}{2}kx^2 + \tfrac{1}{2}mv^2$$

Using $k = m\omega_0^2$ and Equations 15-25 and 15-30 for x and v, we have for the total energy in the steady state,

$$E = \tfrac{1}{2}m\omega_0^2 A^2 \sin^2 (\omega t - \delta) + \tfrac{1}{2}m\omega^2 A^2 \cos^2 (\omega t - \delta)$$ 15-35

Except at resonance, the total energy is not constant in time but varies during each cycle. The average total energy is large near resonance when the amplitude A is large. At the resonance frequency $\omega = \omega_0$ the total energy is constant in the steady state and has the value

$$E = \tfrac{1}{2}m\omega_0^2 A^2 = \tfrac{1}{2}m\omega_0^2 \frac{F_0^2}{\omega_0^2 b^2} = \frac{1}{2} \frac{mF_0^2}{b^2}$$ 15-36

At resonance the total energy varies inversely with the square of the damping constant.

Optional

Let us now show that $x = A \sin (\omega t - \delta)$, with A given by Equation 15-26 and δ by Equation 15-27, is indeed a solution of Equation 15-21. We have

$$x = A \sin (\omega t - \delta)$$

$$\frac{dx}{dt} = \omega A \cos (\omega t - \delta) \qquad \frac{d^2x}{dt} = -\omega^2 A \sin (\omega t - \delta)$$

Putting these into Equation 15-21 gives

$$m(\omega_0^2 - \omega^2) A \sin (\omega t - \delta) + b\omega A \cos (\omega t - \delta) = F_0 \sin \omega t$$ 15-37

This equation must be satisfied for all values of the time t. This is possible only if the constants A and δ have certain values. We can find them by choosing the convenient values $t = 0$ and $t = \delta/\omega$ for the time.

For example, when $t = 0$, Equation 15-37 becomes

$$m(\omega_0{}^2 - \omega^2) A \sin(-\delta) + b\omega A \cos(-\delta) = 0$$

or since $\sin(-\delta) = -\sin\delta$ and $\cos(-\delta) = \cos\delta$, we have

$$\tan\delta = \frac{b\omega}{m(\omega_0{}^2 - \omega^2)}$$

in agreement with Equation 15-27. Similarly, when $\omega t = \delta$ in Equation 15-37, we have

$$\sin(\omega t - \delta) = 0 \qquad \cos(\omega t - \delta) = 1 \qquad \sin\omega t = \sin\delta$$

Thus

$$b\omega A = F_0 \sin\delta \qquad \text{or} \qquad A = \frac{F_0 \sin\delta}{b\omega}$$

Using Equation 15-28 for $\sin\delta$ obtained from Figure 15-4, we have

$$A = \frac{F_0}{\sqrt{m^2(\omega_0{}^2 - \omega^2)^2 + b^2\omega^2}}$$

Questions

3. Give several examples of common situations which might be described as an oscillator driven by a sinusoidal driving force.

4. Under what conditions will the resonance curve of an oscillator have a high, narrow peak at $\omega = \omega_0$? Under what conditions will it have a broad, low peak?

5. When you push a swing, resonance occurs when the driving frequency equals the natural frequency and also when it is $\frac{1}{2}$, $\frac{1}{3}$, ... times the natural frequency, i.e., if you push every other swing, every third swing, etc. Explain how this differs from a harmonic driving force, for which resonance occurs only at the natural frequency.

Review

A. Define, explain, or otherwise identify:

Damped oscillations, 380   Transient solution, 386
Resonance, 380   Steady-state solution, 386
Critical damping, 383

B. True or false:

1. The frequency of a slightly damped (unforced) oscillator is nearly equal to the undamped frequency.

2. The frequency of a damped (unforced) oscillator decreases as the amplitude decreases.

3. The amplitude of a damped (unforced) oscillator decreases exponentially with time.

4. In the steady state, the frequency of a forced oscillator equals the frequency of the forcing term.

5. In the steady state the motion of a forced oscillator is independent of the initial conditions.

6. At resonance, the displacement of a forced oscillator is in phase with the forcing term.

7. At resonance, the velocity of a forced oscillator is in phase with the forcing term.

8. The resonance frequency of a forced oscillator depends on the damping.

9. The sharpness of the resonance of a forced oscillator depends on the damping.

10. In a forced oscillator, the power input of the forcing term is maximum when $\omega = \omega_0$.

Exercises

Section 15-1, Damped Oscillations

1. Show that the SI units of the damping constant b are kilograms per second.

2. From the equation $x = A \cos \omega t$ for an undamped oscillator and the fact that the average value of either $\sin^2 \omega t$ or $\cos^2 \omega t$ over a cycle is $\frac{1}{2}$, show that $v_{av}^2 = \frac{1}{2} A^2 \omega^2 = E/m$.

3. A 2-kg mass oscillates on a spring of force constant $k = 400$ N/m with initial amplitude of 3 cm. (a) Find the period and the total initial energy. (b) If the energy decreases by 1 percent per period, find the damping constant b.

4. An oscillator has a mass of 50 gm and period of 2 sec. Its amplitude decreases by 5 percent each cycle. (a) What is the damping constant b? (b) What fraction of the oscillator's energy is dissipated in each cycle?

5. The amplitude of a damped oscillator decreases by 50 percent each cycle. (a) By what fraction does the energy decrease during each cycle? (b) Use Equation 15-16 to show that $bT/2m = \ln 2$ and $b/2m = (\ln 2)\omega/2\pi$, where T is the period of the oscillator and ω is the angular frequency. (c) Using this result in Equation 15-18, show that the fractional decrease in the frequency is about 0.006.

6. Show that the ratio of amplitudes for two successive oscillations is constant in a damped oscillator.

Section 15-2, The Forced Oscillator

7. A 1-kg mass is attached to a spring of force constant 500 N/m. The oscillator is driven by an external force $F = 10 \sin 2\pi t$ N, where t is in seconds. Assuming negligible damping, find the period and the amplitude of the oscillations of the system.

8. A 5-lb weight is attached to a spring of force constant $k = 100$ lb/ft. It is driven by a force $F = F_0 \sin 2\pi f t$. It oscillates at frequency of 5 Hz and with amplitude of 2 ft. Find F_0 and f_0. (Assume negligible damping.)

9. A mass m rests on a frictionless horizontal table. It is acted on by a single force $F = F_0 \sin \omega t$. (a) Find the acceleration $a(t)$ and the velocity $v(t)$ assuming the mass to be at rest at time $t = 0$. (b) Find the position $x(t)$ assuming that $x = 0$ at time $t = 0$. Show that the amplitude of this simple harmonic motion is $A = F_0/m\omega^2$. (c) This mass is now attached to a spring of force constant k and is again driven by the above force. Compare its motion with that when it was not attached to the spring.

10. Find the resonant frequencies for the three systems in Figure 15-7.

11. The resonant (angular) frequency of a certain system is 60π rad/sec. This system is forced to oscillate at $\omega_1 = 40\pi$ rad/sec and at $\omega_2 = 80\pi$ rad/sec. Find the ratio A_1/A_2. (Assume the same maximum force in each case.)

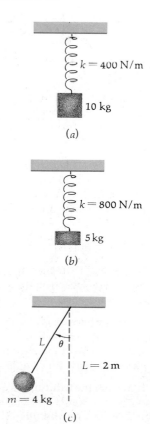

Figure 15-7
Exercise 10.

$k = 400$ N/m

10 kg

(a)

$k = 800$ N/m

5 kg

(b)

L θ

$L = 2$ m

$m = 4$ kg

(c)

12. If an oscillator is driven by a force $F = F_0 \sin \omega t$, show that the power input of the external force is

$$P = F_0 A \omega \sin \omega t \cos \omega t = \frac{F_0^2 \omega}{2m(\omega_0^2 - \omega^2)} \sin 2\omega t$$

assuming no damping and $\omega < \omega_0$.

13. Show that the maximum value of the kinetic energy of a forced oscillator with no damping is not the same as the maximum value of the potential energy but is less or greater depending on whether ω is less or greater than ω_0. Explain this qualitatively by considering the relative directions of the external force and the velocity when the particle is moving toward its equilibrium position for (a) $\omega < \omega_0$ and (b) $\omega > \omega_0$. Hint: Note that x and F are in phase if $\omega < \omega_0$ and 180° out of phase if $\omega > \omega_0$.

Section 15-3, The Forced Oscillator with Damping

14. Show that at resonance ($\omega = \omega_0$, $\delta = \pi/2$) the driving force is equal in magnitude to the damping force at all times and 180° out of phase.

15. A 2-kg mass oscillates on a spring of force constant $k = 400$ N/m. The damping constant is $b = 2.00$ kg/sec. It is driven by a sinusoidal force of maximum value 10 N and angular frequency 10 rad/sec. (a) What is the amplitude of the oscillations? (b) If the frequency of the driving force is varied, at what frequency will resonance occur? (c) Find the amplitude of the vibrations at resonance.

16. Show that the average power dissipated in the damping material is equal to the average power delivered by the external force at all frequencies of the external force.

Problems

1. A damped oscillator loses one-twentieth of its energy during each cycle. How many cycles elapse before half the original energy is dissipated? During this time, by how much is the amplitude reduced? (Use the approximation of Equation 15-15).

2. Show that for a weakly damped oscillator, the maximum value of the damping force in any cycle is approximately $b/m\omega_0$ times the maximum value of the restoring force during that cycle.

3. Suppose that the amplitude of an oscillator decreases by a factor R each cycle. That is, if A_1 is the amplitude after the first cycle, $A_2 = RA_1$ is that after the second cycle, etc. A driving force $F = kB \sin \omega t$ is applied to the oscillator, where k is the force constant of the oscillator. Show that if ω is the resonant frequency, the amplitude is given by

$$A = \frac{\pi B}{\ln (1/R)}$$

4. A 1-kg mass attached to a certain spring is critically damped by an external viscous force. Describe the resultant motion if the same viscous force damps a 2-kg mass attached to the same spring.

5. A 3-kg sphere dropped in air has a terminal velocity of 25 m/sec (see Equation 5-38). It is attached to a spring of force constant $k = 400$ N/m and oscillates in air with an initial amplitude of 20 cm. (a) When will the amplitude be 10 cm? (b) How much energy will have been lost when the amplitude is 10 cm?

6. Show that the value of the driving frequency for which the amplitude A is maximum is $\omega = \sqrt{\omega_0^2 - b^2/2m^2}$. Hint: Find ω for which the denominator of Equation 15-26 is a minimum.

7. A 0.5-kg mass oscillates on a spring of force constant $k = 300$ N/m. During the first 10 sec it loses 0.5 J to friction. If the initial amplitude was 15 cm, (*a*) how long before the energy is reduced to 0.1 J? (*b*) What is the frequency of oscillation?

8. Show by direct substitution that Equation 15-17 is an exact solution of Equation 15-3.

9. A damped oscillator oscillates with frequency ω', which is 10 percent less than its undamped frequency. By what factor is its amplitude decreased in each cycle? By what factor is its energy reduced in each cycle?

10. Let ω_1 and ω_2 be the angular frequencies for which A^2 is half its value at resonance. Show that for small damping $|\omega_1 - \omega_2| \approx b/m$.

CHAPTER 16   Gravity

Gravity is one of the four basic interactions, discussed briefly in Chapter 6. Although of negligible importance in the interactions of elementary particles, gravity is of primary importance in the interactions of large objects. It is gravity that binds us to the earth and that binds the earth and the other planets to the solar system. The gravitational force plays an important role in the evolution of stars and in the behavior of galaxies. In a sense, it is gravity that holds the universe together.

In this chapter we shall study the force of gravity in some detail. In particular we shall show how Newton's law describing the gravitational force exerted by one particle on another can be applied to the description of the force exerted by one extended body on another. In doing this we shall introduce the concepts of a field and of lines of force, concepts used again later in the discussion of electric and magnetic forces.

16-1 Kepler's Laws

The gravitational force exerted by a particle of mass m_1 on another particle of mass m_2 is given by Newton's law of gravitation, discussed briefly in Chapter 6:

$$\mathbf{F}_{12} = -\frac{Gm_1m_2}{r_{12}^2}\,\hat{\mathbf{r}}_{12} \qquad 16\text{-}1$$

where $\mathbf{r}_{12}$ is the vector from mass m_1 to m_2 and $\hat{\mathbf{r}}_{12} = \mathbf{r}_{12}/r_{12}$ is a unit vector.

Newton justified this law of gravitation primarily by demonstrating that with it he could derive Kepler's laws of planetary motion, which summarize the kinematics of the motion of planets about the sun. These laws were deduced between 1600 and 1620 by Johannes Kepler, who used the precise astronomical observations of the planets made in the late 1500s by Tycho Brahe. Kepler's laws are:

Castle in the Pyrenees, by René Magritte. (*Collection of Harry Torczyner, New York;* photograph by G. D. Hackett.)

1. All planets move in elliptical orbits with the sun at one focus.

2. A line joining any planet to the sun sweeps out equal areas in equal times.

3. The square of the period of any planet is proportional to the cube of the planet's mean distance from the sun.

We have already discussed the relation between law 2 and angular momentum in Chapter 13. The rate of area swept out by a planet is proportional to the angular momentum of the planet about the sun. Thus Kepler's second law is equivalent to the statement that the angular momentum of a planet is constant. This implies that there is no torque exerted on the planet about the sun, i.e., that any force exerted on the planet is a central force. That is, the force is along the radial line from the planet to the sun, as with Newton's law of gravitation.

Newton showed that the general path of a planet moving under the influence of an inverse-square-law force is an ellipse with the center of force at one focus if the path is closed. (For a comet which does not return, the path is a parabola or hyperbola.) We shall consider the simpler special case of a circular orbit to show that Kepler's third law implies that the force varies inversely with the square of the distance.

Consider a planet moving with speed v about the sun in a circle of radius r. In the time of one period T the planet travels a distance of $2\pi r$. Thus

$$T = \frac{2\pi r}{v} \qquad\qquad 16\text{-}2$$

For a circular orbit, the mean distance from the sun is just the radius r. (For a general elliptical path with the sun at one focus, the mean distance equals the semimajor axis of the ellipse.) Kepler's third law for this case is then

$$T^2 = Cr^3 \qquad\qquad 16\text{-}3$$

where C is a proportionality constant which may depend on the properties of the sun but is the same for all planets. We have then

$$T^2 = \left(\frac{2\pi r}{v}\right)^2 = Cr^3 \qquad\qquad 16\text{-}4$$

or

$$v^2 = \frac{4\pi^2}{Cr} \qquad\qquad 16\text{-}5$$

Since the planet is in a circular orbit, it has centripetal acceleration v^2/r. If we call the mass of the planet m_p, Newton's second law applied to the planet is

$$F = m_p a = \frac{m_p v^2}{r} \qquad\qquad 16\text{-}6$$

where F is the central force exerted on the planet by the sun. Substituting v^2 from Equation 16-5 into Equation 16-6, we have

$$F = \frac{m_p(4\pi^2/Cr)}{r} = \frac{4\pi^2}{C}\frac{m_p}{r^2} \qquad\qquad 16\text{-}7$$

The force on the planet is proportional to the mass of the planet and inversely proportional to the square of the distance from the sun. By the law of action and reaction, this equals in magnitude the force ex-

Johannes Kepler (1571–1630).

Earth rise as seen from the moon.

NASA

erted on the sun by the planet. Since the force exerted by the planet on the sun is proportional to the mass of the planet, symmetry suggests that the force exerted by the sun on the planet is proportional to the mass M_s of the sun.[1] Writing $4\pi^2/C = GM_s$, where G is a constant independent of the mass of the sun or the planets we have

$$F = \frac{Gm_pM_s}{r^2}$$ 16-8

As discussed in Section 6-1, Newton compared the acceleration of the moon in its orbit with the acceleration of objects near the surface of the earth for further confirmation of the inverse-square law of gravitation.

Questions

1. The sun, as seen from the earth, moves more rapidly against the background of stars in the winter than in the summer. On the basis of this fact and Kepler's laws, what can you say about the relative distances of the earth from the sun during these two seasons?

2. The distance from the center of the earth to the moon is 3.84×10^5 km and the period of the moon 27.3 d. How could you use these data to determine the period of a satellite just above the earth's surface (orbit radius 6400 km)?

3. Two of Jupiter's moons have orbit radii which differ by a factor close to 2. How do their periods compare?

[1] If the law of gravity derived from these laws of planetary motion were a special law applying only to the force between the sun and each planet, there would be no reason why the force should be proportional to M_s. Newton, however, argued that the same law should apply to any two masses in the universe. He supported this idea by noting that Kepler's third law also applies to such motions as the moons of Jupiter except that the constant of proportionality is different. This of course would be the case if the gravitational force is proportional to the masses of both bodies and the constant G is a universal constant of nature.

Comets

Stephen P. Maran
NASA-Goddard Space Flight Center

Our present knowledge of comets indicates that if we set out to make an artificial comet and launch it into space, we might well begin with a huge lump of ice embedded with extremely fine sand. Indeed, a small version of such an experiment was contemplated for the final manned Skylab flight in 1973–1974 but not executed. It appears that the solid part of a comet, called the *nucleus,* is nothing more than a large ball of ice and other frozen gases, interspersed with rock particles, most of which are microscopic. Physical and chemical characteristics of the various parts of a comet are summarized in Table 1.

As a comet approaches the sun, solar heating sublimates the frozen matter at the surface of the nucleus, releasing gas that flows outward. This gas forms a large cloud, the *coma,* known popularly as the *head* of the comet. The gases also blow off some of the rock particles, or dust. Close to the nucleus, the gases retain the same molecular form they exhibited in the frozen state, but as these *parent molecules* diffuse outward, they are gradually broken up into *daughter products,* i.e., smaller molecules and individual atoms, by ultraviolet radiation from the sun. This radiation also ionizes some of the molecules and atoms, producing the *cometary plasma,* electrified material that is influenced by the solar wind pervading interplanetary space. The wind consists of charged particles that are constantly streaming away from the outermost region of the sun, the *corona.* The wind sweeps the cometary plasma out of the coma in a direction away from the sun, forming the *ionic,* or *plasma, tail,* which stretches for tens of millions of miles.

Table 1
Physical and chemical characteristics of comets

Part of comet	Physical characteristics	Chemical composition
Nucleus	Typical diameter 0.3–30 km Mass 10^{13}–10^{19} gm	No observations but many theories; the most probable major constituents are frozen water and carbon monoxide, but other substances must be present
Coma	Diameter 10^4–10^6 km (and $>10^7$ km when observed in the ultraviolet light emitted by hydrogen atoms of the coma)	Parent molecules include water vapor† (H_2O), methyl cyanide† (CH_3CN), and hydrogen cyanide† (HCN), all as gases Silicate particles‡ Daughter products: C, C_2, C_3, H, O, OH, CH, CN, NH, NH_2
Tail	Widths up to 10^6 km	
Dust	Length up to 10^7 km	Silicate particles‡
Plasma	Length up to 10^8 km	OH^+, H_2O^+, CO^+, CO_2^+, CH^+, N_2^+

† Found in only one comet as of late 1974.

‡ The presence of silicate particles is shown by infrared spectra of dust in the coma and tail.

Comet Kohoutek (1973), photographed with a wide-angle Schmidt telescope at the Joint Observatory for Cometary Research. Streaming back from the bright coma are the smooth-looking dust tail (below) and a striking plasma tail (above). Note the large disturbance in the plasma tail near the right edge of the picture; it was observed to move rapidly down the tail in the direction opposite the sun. On the original photographic plate a short "antitail," believed to consist of relatively large dust particles, extends to the left from the coma and appears (due to a perspective effect) to point toward the sun. (*Courtesy of the Joint Observatory for Cometary Research.*)

The visible light of the sun plays a similar role; its radiation pressure exerts a significant force on the dust particles of the coma, also in a direction away from the sun. The acceleration of a dust particle away from the sun depends on its surface area and mass, whereas acceleration toward the sun, due to solar gravity, is the same for all particles. Thus the less dense dust has a large component of motion away from the sun, while the largest particles have only a small component and tend to stream out roughly back along the orbit of the nucleus. The total effect is usually to form a smooth, curved yellow dust tail, which contrasts strikingly with the kinks, wave patterns, and rayed formations of the highly structured plasma tail. The color of the dust tail comes from the reflection of sunlight. The plasma tail looks blue, due to molecular band emission by the carbon monoxide ion (CO^+). The plasma tail acts as a sort of cosmic wind sock, allowing earth-bound astronomers to infer the speed and direction of the solar wind at various places in the solar system by observing the plasma tails of comets.

Since the material in the coma, dust tail, and plasma tail of a comet is not gravitationally bound to the small mass of the nucleus, it escapes into space. Thus, on each new approach to the sun, a comet evolves a new coma and tails.

Nearly all the comets that have been observed follow elliptical orbits that take them well within the orbit of Mars, and some go inside that of Mercury. Indeed, we rarely detect comets beyond Jupiter because they are so dim. When one tries to calculate how much material a comet loses each time it nears the sun and forms a coma, a problem arises. In a typical case, a bright comet like Halley's will vanish as a result of the sublimation process in a time (say, less than 1 million years) that is very short compared with the age of the solar system (5 billion years). Why then do we see any comets at all? Why weren't they all eliminated before man evolved on earth?

Two obvious possibilities are that new comets are continuously being formed, presumably on the outskirts of the solar system, or that huge numbers of comets were left over from the origin of the solar system and are stored somewhere on the outskirts. (This second hypothesis is actually the leading idea at present.) A third possibility, namely, that comets come to us from distant regions of the galaxy, seems to be excluded by the fact (old books to the contrary) that no confirmed hyperbolic orbits have been observed among comets approaching the sun.

The *Oort cloud* is the name given to the region, extending more than 2 light-years from the sun, in which perhaps 100 billion comets are thought to exist. In orbit around the sun, but usually never coming closer to it than the orbit of Jupiter, they are not significantly heated and can remain indefinitely in celestial cold storage. However, the cross-sectional area of the cloud is so great that occasionally stars must move through it on their way around the galaxy.

Indeed, roughly 3000 stars must have passed through the inner cloud (radius 4.7×10^{12} mi) since the solar system formed. The gravitation of a passing star perturbs cometary orbits. Some comets are accelerated and may escape from the cloud into interstellar space. Others are sent into the inner solar system, where, thanks to additional gravitational perturbations (especially by Jupiter), they are sometimes trapped in much smaller orbits that return them repeatedly to the vicinity of the sun. This is the source of the periodic comets well known to astronomers, such as Halley's (76 years) and Encke's (3.3 years), and it presumably explains why we still see comets near the sun billions of years after solar heat ought to have dissipated them.

16-2 The Cavendish Experiment

The first measurement of the gravitation constant G was made by Henry Cavendish in 1798. Figure 16-1 shows a schematic sketch of the apparatus he used to measure the gravitational force between two known masses.

The two small masses m_2 are at the ends of a light rod suspended by a fine fiber. A torque is required to turn the two masses through the angle θ from their equilibrium position because the fiber must be twisted. Careful measurement shows that the torque required to turn the fiber through a given angle is proportional to the angle. The constant of proportionality can be determined, and the fiber and the suspended masses can be used to measure very small torques. This arrangement, called a *torsion balance,* was invented in the eighteenth century, by John Michell. The French physicist Charles Augustin de Coulomb used a torsion balance in 1785 to determine the law of electric force known by his name. Cavendish used a refined and especially sensitive torsion balance in his determination of G.

In Cavendish's experiment two large masses m_1 are placed near the small masses m_2, as shown in Figure 16-1. The apparatus is allowed to come to equilibrium in this position; because the apparatus is so sensitive and because the gravitational force is so small, this takes hours. Instead of measuring the deflection angle directly, Cavendish reversed the positions of the large masses, as shown by the dashed lines in the figure. If the balance is allowed to come to equilibrium again, it will turn through angle 2θ in response to the reversal of the torque. From the measurement of the angle and of the torsion constant, the forces between the masses m_1 and m_2 can be determined. When their masses and their separations are known, G can be calculated. Cavendish obtained a value for G within about 1 percent of the present value:

$$G = 6.67 \times 10^{-11} \text{ N-m}^2/\text{kg}^2$$

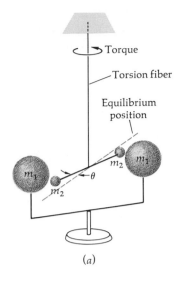

(a)

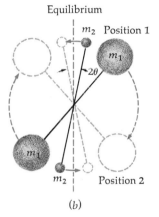

(b)

Figure 16-1
Schematic drawing of the Cavendish apparatus for determining the gravitational constant G. (a) Because of the gravitational attraction of the large masses m_1 for the nearby small masses m_2, the fiber is turned through a very small angle θ from its equilibrium position. (b) The large masses are reversed so that they are at the same distance from the equilibrium position of the balance but on the other side. The fiber then turns through the angle 2θ. Measurement of this angle and of the torsion constant of the fiber makes it possible to determine the force exerted by m_1 on m_2, which in turn allows the constant G to be determined.

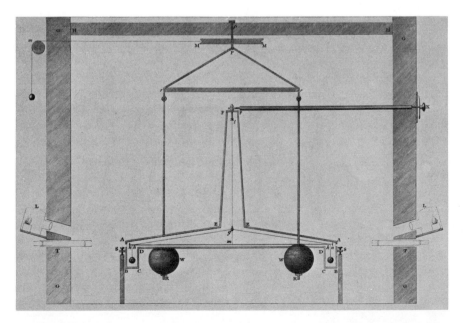

Original drawing of Cavendish's apparatus. (*Courtesy of the Deutsches Museum, Munich.*)

Questions

4. Cavendish presented his result in a paper entitled Weighing the Earth. Explain how the earth's mass can be calculated from the values of G, g, and R_E.

16-3   The Gravitational Field

The gravitational force exerted by two masses m_1 and m_2 on a third mass m_0 at some point is just the (vector) sum of the forces exerted by m_1 and m_2 individually; i.e., the presence of a second mass m_2 has no effect on the force exerted by m_1 on m_0. (We could imagine, for example, that the second mass m_2 might shield m_0 from the force exerted by m_1 or influence it in some other way, but no such effects are observed.) If there are several masses, as in Figure 16-2, the gravitational force they exert on a mass m_0 at some point P is given by

$$\mathbf{F} = -\frac{Gm_1m_0}{r_{10}{}^2}\,\hat{\mathbf{r}}_{10} - \frac{Gm_2m_0}{r_{20}{}^2}\,\hat{\mathbf{r}}_{20} - \cdots \frac{Gm_im_0}{r_{i0}{}^2}\,\hat{\mathbf{r}}_{i0} + \cdots \qquad \text{16-9}$$

where $\hat{\mathbf{r}}_{i0}$ is the unit vector directed from the ith mass to point P. The force per unit mass acting on a mass at point P is called the *gravitational field* at that point:

$$\mathbf{g} = \frac{\mathbf{F}}{m_0} = \Sigma\left(-\frac{Gm_i}{r_{i0}{}^2}\,\hat{\mathbf{r}}_{i0}\right) \qquad \text{16-10}$$

The gravitational force on mass m_0 at point P can be written

$$\mathbf{F} = m_0\mathbf{g} \qquad \text{16-11}$$

The gravitational field is a vector function of position (and of time if the positions of the other masses vary with time). By moving the test mass m_0 from point to point and measuring the force on it we can map out this function.

As we discussed in Chapter 4, the gravitational field is a particularly useful concept in dealing with the action-at-a-distance problem. We consider the interaction of the masses m_i with the mass m_0 in Figure 16-2 as a two-step process. The masses m_i set up a condition in space, the gravitational field, and this field in turn exerts the force on m_0. Similarly, m_0 contributes to the gravitational field at the location of

Radio Times/Hulton Picture Library

Henry Cavendish (1731–1810).

Gravitational field defined

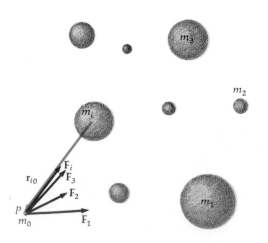

Figure 16-2
The force on the test mass m_0 is the vector sum of the forces exerted by each mass in the system. The resultant force on m_0 divided by m_0 is the gravitational field $\mathbf{g}$ at point P. The force on any other mass placed at point P is then the product of the mass and the gravitational field $\mathbf{g}$.

each mass m_1. If we suddenly move one of the masses such as m_3, the field at point P does not change immediately. There must be time for the propagation of the disturbance from m_3 to m_0. The change in the field due to a change in position of m_3 propagates as a gravitational wave with the speed 3×10^8 m/sec, the same as the speed of light. These statements, of course, do not follow from Newton's theory of gravitation but are features of the modern theory developed by Einstein in our own century. Newton's theory can be thought of as an approximation valid when the propagation time is negligible. In fact, it is sufficient for almost all known purposes.

Example 16-1 Two equal point masses are located on the y axis at points $y = +a$ and $y = -a$, as shown in Figure 16-3. Find the gravitational field on the x axis.

At a general point (x,y,z), not necessarily on the x axis, the gravitational field has three components each of which is a function of x, y, and z: $g_x(x,y,z)$, $g_y(x,y,z)$, and $g_z(x,y,z)$. Figure 16-3 illustrates the calculation of the field at a point in the xy plane. Here the field has only two components, $g_x(x,y)$ and $g_y(x,y)$; the z component is zero. In this example, we restrict our attention to the calculation of the field on the x axis, as in Figure 16-4. By symmetry, the y component of the field is zero. Since the point P is equidistant from each mass, the magnitudes of the contributions from each mass are equal. If we call this magnitude g_1, the x component of the field is

$$g_x = -2g_1 \cos \theta = -2 \frac{Gm}{r^2} \frac{x}{r} = -2Gm \frac{x}{(x^2 + a^2)^{3/2}} \qquad \text{16-12}$$

using $r^2 = x^2 + a^2$ and $\cos \theta = x/r$. The minus sign is included because for positive values of x the field is toward the left.

Figure 16-5 is a sketch of the field $g_x(x)$ versus x. It is left for you to show (Problem 1) that as $x \to \infty$, $g_x \to -2Gm/x^2$ and that the field is maximum at $x = a/\sqrt{2}$.

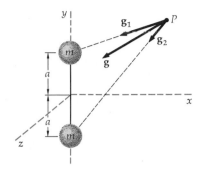

Figure 16-3
The gravitational field at point P in the xy plane due to the two masses on the y axis generally has both x and y components.

Figure 16-4
The gravitational field at a point P on the x axis due to the two masses on the y axis has only an x component because the y components of $\mathbf{g}_1$ and $\mathbf{g}_2$ cancel.

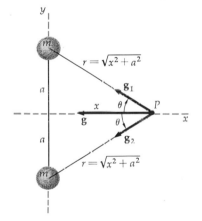

Figure 16-5
Sketch of g_x versus x for Example 16-1.

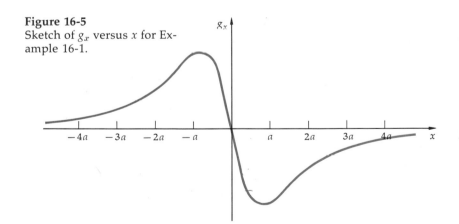

The gravitational field of an extended body (Figure 16-6) can be calculated from Equation 16-10 by treating the body as consisting of a great number of parts, each small enough to be considered a particle. In Figure 16-6 the body has been divided into N parts each of volume ΔV_i and mass $\Delta m_i = \rho \Delta V_i$, where ρ is the mass per unit volume. The gravitational field at point P is then approximately

$$\mathbf{g} \approx \sum \left(-\frac{G \, \Delta m_i}{r_{i0}{}^2} \, \hat{\mathbf{r}}_{i0} \right)$$ 16-13

As N goes to infinity and Δm_i goes to zero, the sum becomes an integral:

$$\mathbf{g} = \int \left(-\frac{G \, dm}{r'^2} \, \hat{\mathbf{r}}' \right)$$ 16-14

where r' is the distance from the mass element dm to the point where the field is to be calculated. The integral in Equation 16-14 ranges over all mass in the system.

Questions

5. What are the dimensions of the gravitational field?

6. If a mass m_0 is placed on the x axis at the point $x = 0$ in Example 16-1, it will be in equilibrium because $g_x = 0$ at that point. Is this equilibrium stable or unstable?

16-4 Gravitational Potential

The potential energy of a mass m_0 in the presence of a distribution of masses is usually easier to calculate than the force on m_0 because the potential energy is a scalar quantity whereas the force is a vector. Consider again the distribution of masses shown in Figure 16-2. In Chapter 8 we found the gravitational potential energy of a mass m_0 a distance r_{i0} from a mass m_i to be $-Gm_im_0/r_{i0}$ (see Equation 8-14). The total gravitational potential energy of the mass m_0 in Figure 16-2 is then

$$U = -\frac{Gm_1m_0}{r_{10}} - \frac{Gm_2m_0}{r_{20}} - \cdots \frac{Gm_im_0}{r_{i0}} - \cdots$$ 16-15

The potential energy of the mass m_0 divided by the mass m_0, called the *gravitational potential,* is a scalar function of position

$$V = \frac{U}{m_0} = \sum \left(-\frac{Gm_i}{r_{i0}} \right)$$ 16-16

Gravitational potential defined

Since, in general, the force is the negative derivative of the potential-energy function, the gravitational field is the negative derivative of the potential function. When the distribution of masses has perfect spherical symmetry, the gravitational potential depends only on the radial coordinate r. Then the gravitational field has only a radial component given by[1]

$$g_r = -\frac{dV(r)}{dr}$$ 16-17

For a continuous distribution of mass, the gravitational potential is

$$V = \int \left(-\frac{G \, dm}{r'} \right)$$ 16-18

---

[1] For the general case in which V depends on all three coordinates x, y, and z, the rectangular components of the gravitational field are related to the partial derivatives of the potential function (Section 8-10).

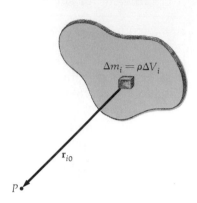

Figure 16-6
Calculation of the gravitational field at point P due to an extended body. The body is considered to consist of small mass elements m_i each of which is treated as a point mass. The field at P is then the sum of the fields due to the mass elements. This sum is found by integration.

$\Delta m_i = \rho \Delta V_i$

$\mathbf{r}_{i0}$

P

16-5   Lines of Force

It is convenient to picture a field by drawing *lines of force* to indicate the direction of the field at each point. The field vector is tangent to the line at each point. Of course, there is an infinite number of points in space, and so only a few representative lines are drawn. Since we could choose any point to indicate the direction of the field by a line, the lines need not be continuous, but it is both customary and useful to draw continuous lines ending at a point mass. (The lines end rather than begin at the mass because the field near a point mass points toward the mass.)

Figure 16-7 shows the lines of force for a single point mass. They are equally spaced because of spherical symmetry. Consider a spherical surface a distance r from the mass. For a fixed number of lines, the number of lines per unit area of the sphere is inversely proportional to the area of the sphere $4\pi r^2$. Thus the *density of lines* (lines per unit area) decreases with distance as $1/r^2$, just as the magnitude of the field does, and the strength of the gravitational field is indicated by the density of lines.

If we have more than one point mass, we can indicate the field strength by choosing the number of lines to each mass to be proportional to the mass. Figure 16-8 shows the lines of force for two equal point masses separated by a distance $2a$. Without calculating the field at each point, we can construct this pattern as follows. Since the masses are equal, we choose an equal number of lines to each mass. At points very near one of the masses, the field is approximately due to the nearby mass only, since the field varies inversely as the square of the distance. Thus the lines through the surface of a sphere of very small radius ϵ ($\epsilon \ll a$) about one mass are equally spaced. At a very large distance from the masses $D \gg a$, the field should look like that from a single point mass of strength $2m$; the lines from the two masses are equally spaced through the surface of a sphere of very large radius. We can see by merely looking at Figure 16-18 that the gravitational field in the space between the masses is weak because there are few lines in this region compared with the region just to the right or left of the masses, where the lines are more closely spaced. This information can of course also be obtained by direct calculation of the field at points in these regions.

We can apply the above reasoning to drawing the lines of force for any system of point masses. Let d be the greatest separation between any two masses in the system. If we are very far from the system, the

Figure 16-7
Lines of force indicating the gravitational field near a point mass. The direction of the lines indicates the direction of the field at any point; the density of the lines indicates the magnitude of the field.

How to sketch lines of force

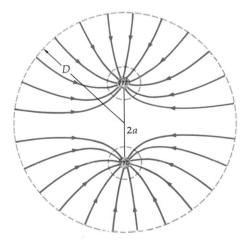

Figure 16-8
Lines of force indicating the gravitational field produced by two point masses. Very near either mass the lines are the same as if there were a single point mass m. Very far from the two masses, the lines are the same as if the two masses were together as a single mass $2m$.

detailed structure of the system cannot be important. The gravitation field is thus the same as that due to a single point mass equal to the total mass of the system. On a sphere of radius r, where r is much greater than d, the lines of force are equally spaced and point in toward the system. The number of lines is chosen to be proportional to the total mass of the system. At each point mass in the system, a number of lines proportional to the mass converge symmetrically. The lines cannot intersect at a point in space where there is no mass because there can be only one direction for the gravitational field at such a point. There are no points like that in Figure 16-9 from which the lines of force diverge. If such a point existed, a test mass placed anywhere near that point would experience a force away from the point, implying a negative mass at that point; however, negative mass which repels other mass has never been observed.[1] It is important to realize that *the convention indicating the field strength by lines of force works only because the gravitational field varies inversely as the square of the distance from a point mass.* Since the electric field of a point charge also varies inversely as the square of the distance, the concept of lines of force is also very useful in picturing the electrostatic field.

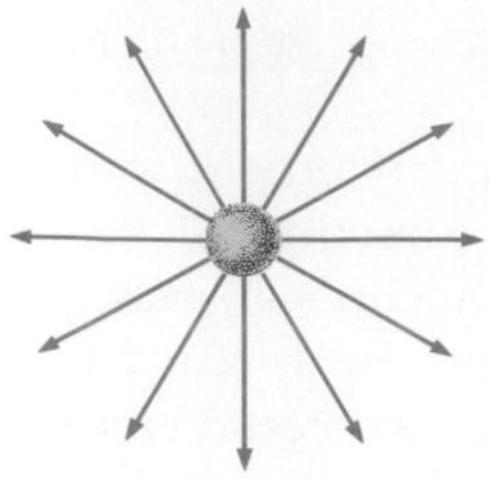

Figure 16-9
Lines of force diverging from a point in space. Such a situation is not observed in nature for the gravitational force. According to this pattern, a test mass placed near this point would be repelled from the point as if negative mass were at the point.

16-6   The Gravitational Field of a Spherically Symmetric Mass Distribution

A particularly important special mass distribution is one which is spherically symmetric. To a good approximation, the earth is such a distribution. We shall consider first a spherically symmetric shell of radius R and mass M (we calculate the field produced by such a system using integral calculus in Section 16-7). In this section, we shall investigate the properties of such a field by considering the lines of force. The only unique direction in a system with spherical symmetry is radial. The only possible direction for lines of force is along radii, either toward the center of symmetry or away from it. We choose the origin at the center of the shell. At very great distances compared with the shell radius R, the field must look like that of a point mass M. Thus the lines are radial and equally spaced far from the shell. Because of the symmetry of this system, the lines remain radial and equally spaced at all distances from the shell. The lines of force are shown for this system in Figure 16-10. Outside the shell the lines are exactly the same as those for a point mass M at the origin. Thus the field produced outside the shell is just

$$g_r = -\frac{GM}{r^2} \qquad\qquad 16\text{-}19$$

The lines end on the shell. There can be no lines inside the shell. If, for example, there were lines radially inward inside the shell, they would converge at the origin, but since there is no mass at the origin, they cannot converge there. If they were radially outward, they would have

[1] It has been proposed that antiparticles such as positrons or antiprotons may have negative gravitational mass and exhibit gravitational repulsion of ordinary mass. Because of the extreme weakness of the gravitational interaction compared with other interactions between particles, the validity of this proposal is difficult to test directly by experiment. There are theoretical reasons for expecting the gravitational mass to be positive for antiparticles as well as for ordinary particles.

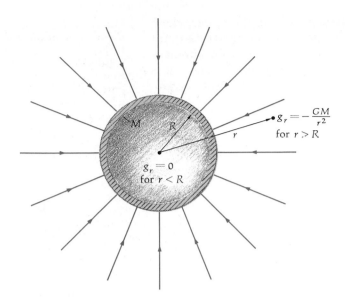

Figure 16-10
Lines of force for a spherical-shell mass distribution. Outside the shell the lines are the same as if there were a point mass at the center of the shell. Inside the shell there are no lines, indicating that the gravitational field is zero inside the shell.

to diverge from the origin, an impossibility, as we have already discussed. We thus have the remarkable result that a spherically symmetric shell of mass produces no gravitational field inside it:

$$g_r = 0 \qquad r < R \qquad\qquad\qquad 16\text{-}20$$

If a test mass m_0 is placed anywhere inside the shell, it will experience no gravitational attraction by the shell. This is easily understood for m at the center of the sphere, since then for any mass element of the sphere there is an equal mass element an equal distance on the opposite side. Their attractions for m_0 would cancel. That this is also true at every other point inside the sphere is not at first obvious. It is a consequence of the inverse-square character of the force.[1] As we shall see later, Franklin's discovery that the same thing occurs for the electric force led Priestley to conclude that the electric force is also an inverse-square force.

   We can now use the results of Equations 16-19 and 16-20 to find the gravitational field inside and outside a solid sphere whose density depends only on the distance r from the center. We treat the solid sphere as a set of spherical shells. The gravitational field outside the sphere is just

$$g_r = -\frac{GM}{r^2} \qquad\qquad\qquad 16\text{-}21$$

where M is the total mass of the sphere since each shell acts as if it were a point mass at the center of the sphere. In other words, the field which any spherically symmetric mass produces outside its own surface is just the same as if all the mass were concentrated at the center of the sphere. It is this result which in Chapter 6 justified our calculating the force of gravity of the earth at its surface by treating the earth as a point mass concentrated at the center of the earth.

   Derivation of the gravitational field inside the solid sphere is more complicated. The field inside the sphere depends upon how the density varies with r. Consider the case of constant density:

$$\rho = \rho_0 = \frac{M}{V} = \frac{M}{\frac{4}{3}\pi R^3}$$

[1] It is only because the force varies as $1/r^2$ that the field strength can be indicated by our system of lines of force.

We wish to find the field at a point a distance r from the center where $r < R$. As we have seen, a spherical shell produces no field at a point within the shell. Thus the only contribution to the field at this point is due to the mass M' within the radius r (see Figure 16-11). This mass produces a field which is the same as if this mass were at the origin. The mass within the distance r is

$$M' = \rho_0 \left(\tfrac{4}{3}\pi r^3\right) = \frac{M}{\tfrac{4}{3}\pi R^3}\, \tfrac{4}{3}\pi r^3 = M\,\frac{r^3}{R^3}$$

and the gravitational field at a point inside the sphere is

$$g_r = -\frac{GM'}{r^2} = -\frac{GMr^3/R^3}{r^2} = -\frac{GM}{R^3}\,r \qquad\qquad 16\text{-}22$$

The magnitude of the field increases with distance for $r < R$. Figure 16-12 is a plot of g_r versus r for a solid sphere of constant density.

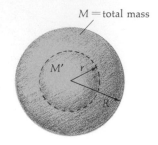

Figure 16-11
Calculation of the gravitational field inside a uniform-solid-sphere mass distribution. At distance r, only the part of the mass M' contributes to the field. The field at r is proportional to M'/r^2, which is proportional to the distance r because M' is proportional to the volume $\tfrac{4}{3}\pi r^3$.

$M =$ total mass

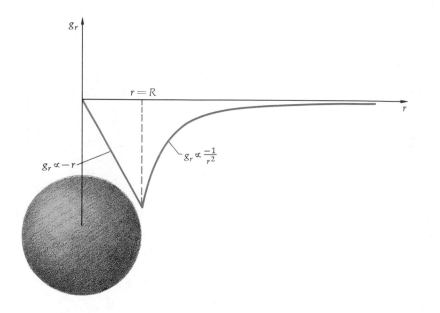

Figure 16-12
Sketch of g_r versus r for a uniform-solid-sphere mass distribution. The magnitude of the field increases with r inside the sphere and decreases as $1/r^2$ outside the sphere, as indicated.

Example 16-2 A particle slides along a smooth straight tunnel dug through the earth (Figure 16-13). Show that the motion is simple harmonic and find the period.

Let the x axis be along the tunnel and choose the y axis through the center of the earth, as shown in Figure 16-13. When the mass is at position x, the gravitational force acting on it is, according to Equation 16-22,

$$\mathbf{F} = m\mathbf{g} = -m\,\frac{GM_E}{R_E^{\,3}}\,r\hat{\mathbf{r}}$$

where M_E is the mass of the earth and R_E is the radius of the earth. (We have assumed that the density of the earth is uniform.) The y component of this force is balanced by the normal force of the tunnel. The x component of this force is

$$F_x = -\frac{GM_E m}{R_E^{\,3}}\,r\sin\theta$$

But $\sin\theta = x/r$. Thus

$$F_x = -\frac{GM_E m r}{R_E^{\,3}}\,\frac{x}{r} = -\frac{GM_E m}{R_E^{\,3}}\,x$$

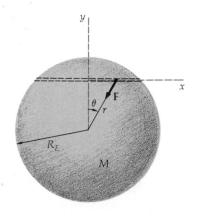

Figure 16-13
Tunnel through the earth for Example 16-2. The y axis is chosen through the center of the earth and perpendicular to the tunnel, which is taken along the x axis. The x component of the gravitational force on a mass in the tunnel is proportional to x and toward $x = 0$.

The acceleration in the x direction is thus

$$a_x = \frac{F_x}{m} = -\frac{GM_E}{R_E{}^3} x = -\omega^2 x \qquad 16\text{-}23$$

where

$$\omega^2 = \frac{GM_E}{R_E{}^3} \qquad 16\text{-}24$$

Equation 16-23 is the equation for simple harmonic motion with angular frequency ω. We can write ω in terms of the acceleration of gravity near the earth's surface:

$$g = \frac{GM_E}{R_E{}^2}$$

Thus

$$\omega^2 = \frac{g}{R_E}$$

The period is

$$T = \frac{2\pi}{\omega} = 2\pi \sqrt{\frac{R_E}{g}} = 2\pi \sqrt{\frac{3960 \times 5280 \text{ ft}}{32.2 \text{ ft/sec}^2}} = 5.06 \times 10^3 \text{ sec} = 84.4 \text{ min}$$

The period of free fall through a straight tunnel is the same as that of a satellite orbiting near the surface of the earth. It is independent of the length of the tunnel, which means that a transit system working on this principle could provide trips from one city to any other city on earth, with every trip taking about 42 min regardless of the distance involved. (Although it still seems too expensive, serious proposals for a tunnel between New York and Washington, D.C. have been made.)

Optional

16-7 Derivation by Integral Calculus of the Gravitational Field of a Spherical Shell

We can calculate the gravitational field produced at any point by a spherically symmetric shell either directly, from Equation 16-14, or indirectly, by first calculating the gravitational potential from Equation 16-18 and then finding the field from $g_r = -dV/dr$. The second method is easier because the potential is a scalar function, whereas the gravitational field is a vector function.

Consider a spherical shell of radius R, thickness t much smaller than R, and mass M. We are interested in the gravitational potential produced by this shell at some point P which may be either inside or outside the shell. By symmetry, the potential can depend only on the radial distance r from the center of the shell to the point P. (The point P at which the field or potential is to be calculated is called the *field point.*) Nearly all the difficulty in this calculation is due to unfamiliar spherical geometry and the use of spherical coordinates r, θ, and ϕ to set up the integral in Equation 16-18. The actual evaluation of the integral is quite easy.

Let us first consider a field point outside the shell, as in Figure

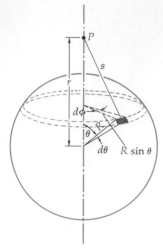

Figure 16-14
Calculation of the gravitational field of a uniform spherical shell by integration. The area of the element indicated is $R^2 \sin\theta \, d\theta \, d\phi$. The distance from this element to the field point is related to r, R, and θ by the law of cosines: $s^2 = r^2 + R^2 - 2rR \cdot \cos\theta$.

16-14. A small area element indicated on the spherical shell has sides $R \, d\theta$ and $R \sin\theta \, d\phi$. Its area is

$$dA = R \, d\theta \, R \sin\theta \, d\phi = R^2 \sin\theta \, d\theta \, d\phi$$

The mass of this area element (which has thickness t) is

$$dm = \rho t \, dA = \rho t R^2 \sin\theta \, d\theta \, d\phi$$

where ρ is the mass per unit volume. Let s be the distance from this area element to the field point. The gravitational potential due to this element is

$$dV = -\frac{G \, dm}{s} = -\frac{G\rho t R^2 \sin\theta \, d\theta \, d\phi}{s} \qquad 16\text{-}25$$

Note how much simpler this is than the gravitational *field* due to this element, which is a vector pointing along the line from the mass element to the field point (Figure 16-15). In order to calculate the gravitational *field* directly, we would have to resolve the field due to the element dm into components and calculate each component by integration over all mass elements. To calculate the scalar gravitational potential we need only sum the scalar contribution given by Equation 16-25 over all mass elements. We can immediately sum over angles ϕ since the distance s does not depend on ϕ. Figure 16-14 shows a circular strip of circumference $2\pi R \sin\theta$, width $R \, d\theta$, and thickness t. The area of this strip is thus $2\pi R^2 \sin\theta \, d\theta$ (which can also be obtained from the area $R^2 \sin\theta \, d\theta \, d\phi$ by integrating $d\phi$ from $\phi = 0$ to $\phi = 2\pi$). The gravitational potential at point P due to this strip is

$$dV = -\frac{G \, dm}{s} = -\frac{G t \rho 2\pi R^2 \sin\theta \, d\theta}{s} \qquad 16\text{-}26$$

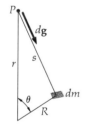

Figure 16-15
The gravitational field $d\mathbf{g}$ at point P due to the mass element dm on the spherical shell. A direct calculation of the gravitational field due to the shell would require summing the vectors $d\mathbf{g}$ due to each element dm over the shell. It is easier to find the gravitational potential first because it is a scalar function.

The total potential produced by the complete shell is obtained by summing over all the circular strips on the shell, i.e., by integrating from $\theta = 0$ to $\theta = 180°$. The quantities s and θ are related by the law of cosines:

$$s^2 = r^2 + R^2 - 2rR \cos\theta \qquad 16\text{-}27$$

Computing the derivative with respect to θ of each term in Equation 16-27 gives

$$2\frac{s \, ds}{d\theta} = -2rR(-\sin\theta)$$

or

$$\sin\theta \, d\theta = \frac{s}{rR} \, ds \qquad 16\text{-}28$$

Instead of substituting s from Equation 16-27 into Equation 16-26 and integrating over θ it is easier to substitute $\sin\theta \, d\theta$ from Equation 16-28 into Equation 16-26 and integrate from $s = r - R$ at $\theta = 0$ to $s = r + R$ at $\theta = 180°$. We then have

$$dV = -\frac{G t \rho 2\pi R^2 (s/rR) \, ds}{s} = -\frac{G t \rho 2\pi R}{r} \, ds$$

and

$$V = -\frac{G t \rho 2\pi R}{r} \int_{r-R}^{r+R} ds = -\frac{G t \rho 2\pi R}{r} \left[r + R - (r - R) \right]$$

$$= -\frac{G t \rho 2\pi R(2R)}{r} = -\frac{G(\rho t 4\pi R^2)}{r}$$

The area $4\pi R^2$ times the thickness t is the total volume of the shell. Then $\rho t 4\pi R^2$ is just the total mass M of the spherical shell. Thus we have

$$V = -\frac{GM}{r} \qquad \text{for } r > R \qquad \qquad 16\text{-}29$$

The gravitational potential due to a spherical shell is the same at a point *outside* the shell as it would be if there were only a point mass M at the center of the shell and the shell were absent.

The calculation of the potential at a field point *inside* the shell is the same except that now s varies from $s = R - r$ at $\theta = 0$ to $s = R + r$ at $\theta = 180°$. Thus for a point inside the shell,

$$V = -\frac{Gt\rho 2\pi R}{r} \int_{R-r}^{R+r} ds = -\frac{Gt\rho 2\pi R}{r}[R + r - (R - r)]$$

$$= -\frac{Gt\rho 2\pi R}{r} 2r = -G(\rho t 4\pi R) = -\frac{G(\rho t 4\pi R^2)}{R}$$

Again, $\rho t 4\pi R^2 = M$, and thus

$$V = -\frac{GM}{R} \qquad \text{for } r < R \qquad \qquad 16\text{-}30$$

The gravitational potential *inside* the shell is constant, independent of r. Equations 16-29 and 16-30 both give the same result for the point $r = R$.

We can now easily calculate the gravitational field:

$$g_r = \begin{cases} -\dfrac{dV}{dr} = -\dfrac{d}{dr}\left(-\dfrac{GM}{r}\right) = -\dfrac{GM}{r^2} & r > R \\[2ex] -\dfrac{dV}{dr} = -\dfrac{d}{dr}\left(-\dfrac{GM}{R}\right) = 0 & r < R \end{cases} \qquad 16\text{-}31$$

Review

A. Define, explain, or otherwise identify:

Kepler's laws, 393

Gravitational constant, 394

Torsion balance, 397

Gravitational field, 398

Gravitational potential, 400

Lines of force, 401

B. True or false:

1. Kepler's law of equal areas implies that gravity varies inversely with the square of the distance.

2. The planet closest to the sun on the average has the shortest period of revolution about the sun.

3. Kepler's laws can be derived from Newton's law of gravitation combined with Newton's laws of motion.

4. The force on a test mass m_0 in a gravitational field is the product of m_0 and the gravitational field at that point.

5. The gravitational potential energy of a test mass m_0 at some point is the product of m_0 and the gravitational potential at that point.

6. A spherically symmetric shell produces no gravitational field anywhere.

7. Both the magnitude and direction of the gravitational field are indicated by lines of force.

8. There are no lines of force inside a spherically symmetric shell.

9. Lines of force cannot begin at some point in space and radiate outward in all directions from it.

10. The gravitational field propagates instantaneously through space.

Exercises

Section 16-1, Kepler's Laws

1. Suppose a small planet were discovered with a period of 5 y. What would be its mean distance from the sun?

2. Halley's comet has a period of about 76 y. What is its mean distance from the sun?

3. The comet Kohoutek has a period estimated to be at least 10^6 y. What is its mean distance from the sun?

4. The mean distance of Jupiter from the sun is 5.22 times that of the earth. What is the period of Jupiter?

5. The radius of the earth's orbit is 1.49×10^{11} m and that of Uranus is 2.87×10^{12} m. What is the period of Uranus?

6. Use Kepler's law $T^2 = Cr^3$ for satellites orbiting the earth to find the period for a satellite just above the surface of the earth from the knowledge that the period of the moon is 27.3 d and the distance to the moon is 60 times the radius of the earth. What should the radius be for a satellite which orbits the earth with a period of 1 d?

Section 16-2, The Cavendish Experiment

7. If the value of G is known, it is possible to find the mass of the earth from the radius of the moon's orbit and its period T. Calculate the mass of the earth using $r_m = 3.84 \times 10^5$ km and $T = 27.3$ d.

8. The masses in a Cavendish apparatus are $m_1 = 10$ kg and $m_2 = 10$ gm, the separation of their centers is 5 cm, and the rod separating the two small masses is 20 cm long. What is the force of attraction between the large and small spheres? What torque must be exerted by the suspension to balance these forces?

Section 16-3, The Gravitational Field

9. A point mass m is on the x axis at $x = +a$ and another equal mass is on the x axis at $x = -a$. (a) What is the gravitational field halfway between the masses at $x = 0$? (b) Find the gravitational field on the x axis at points $x > a$.

10. At what point on the line from the earth to the moon is the gravitational field due to these two bodies zero? The distance from the earth to the moon is 3.84×10^5 km, and the mass of the moon is 0.0123 times the mass of the earth. Are there other points for which the gravitational field due to the earth and the moon is zero?

11. The gravitational field at some point in space is $\mathbf{g} = 2\mathbf{i} + 3\mathbf{j}$ N/kg. (a) What is the magnitude of the force exerted on a 3-kg mass placed at that point? (b) What is the direction of the force? (c) Write an expression for the gravitational force on a 5-kg mass placed at that point.

12. What is the magnitude and direction of the gravitational field due to the earth at the surface of the earth?

13. A 1-gm point mass is moving in a circle of radius 100 m. Its speed is 0.5 m/sec. The only force on the mass is due to gravity. What is the gravitational field at the position of the mass due to other masses?

Section 16-4, Gravitational Potential

14. If the potential energy of a particle depends on x only, the force is given by $F_x = -dU/dx$. From this fact, show that in this situation the gravitational field is related to the gravitational potential by $g_x = -dV/dx$.

15. Two point masses m are situated on the y axis at $y = +a$ and $y = -a$. (a) Find the gravitational potential $V(x)$ for points on the x axis. (b) Sketch $V(x)$ versus x. (c) Find the gravitational field $g_x = -dV/dx$ from your result in part (a). At what point is $g_x = 0$?

16. For the two-point mass distribution in Exercise 15, find the work required to bring a test mass m_0 from $x = \infty$ to the point $x = 0$.

17. A uniform thin ring of mass M and radius r lies in the yz plane with its center at the origin. (a) Find the gravitational potential $V(x)$ for points on the x axis (which is the axis of the ring). (b) Sketch $V(x)$ versus x. (c) At what point does $V(x)$ have maximum magnitude? What is the gravitational field at that point?

18. Two point masses m are situated on the x axis at $x = +a$ and $x - -a$. Sketch the gravitational potential $V(x)$ versus x for points on the x axis. At the point $x = 0$ the gravitational field is zero. Does the potential function $V(x)$ have a maximum or minimum point at $x = 0$?

Section 16-5, Lines of Force

19. Two point masses $m_1 = m$ and $m_2 = 2m$ are a distance D apart. Sketch the lines of force for this configuration.

20. Figure 16-16 shows lines of force in the xy plane due to a disk mass. What information about the gravitational field due to the disk can be obtained from the fact that (a) the lines are much more dense near the disk than far from the disk and (b) in the region near the center of the face of the disk the lines are uniformly spaced?

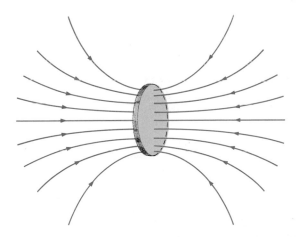

Figure 16-16
Lines of force for a uniform disk mass (Exercise 20). Only the lines in the plane of the axis are shown.

21. Explain why the lines of force shown in Figure 16-17a to c are not possible representations of any gravitational field.

Figure 16-17
Exercise 21.

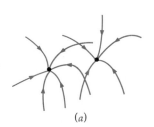

(a)

(b)

(c)

Section 16-6, The Gravitational Field of a Spherically Symmetric Mass Distribution

22. (a) Use Equation 16-17 to show that the gravitational potential $V(r)$ is constant inside a spherical-shell mass distribution. (b) How does $V(r)$ vary with r outside a spherical-shell mass distribution? (c) Sketch $V(r)$ versus r for such a distribution, using the fact that the potential must be continuous at $r = R$, where R is the radius of the shell.

23. A spherical shell has a radius $R = 2$ m and mass 12 kg. What is the gravitational field (a) just outside and (b) just inside the shell?

24. Find the gravitational potential V just inside and just outside the spherical-shell mass distribution of Exercise 23.

25. (a) Show that the gravitational potential $V(r)$ outside a solid spherical mass of constant density is given by $V(r) = -GM/r$, where M is the total mass of the sphere. (b) Show that the gravitational potential just outside the earth's surface $V(R_E)$ can be written $V(R_E) = -gR_E$, where g is the acceleration of gravity and R_E is the radius of the earth. (c) Evaluate the gravitational potential energy just outside the earth's surface in joules per kilogram. (d) The kinetic energy needed for a mass m to escape from the earth is just $m|V(R_E)|$, where $V(R_E)$ is the gravitational potential near the earth's surface. Find the energy needed for an 80-kg astronaut to escape from the earth's gravitational field. At 3 cents per kilowatt-hour, how much would this energy cost?

Problems

1. For the mass distribution in Example 16-1 show that g_x is approximately $-2GMm/x^2$ when x is much greater than a. Show also that the maximum value of $|g_x|$ occurs at the point $x = a/\sqrt{2}$.

2. The gravitational potential $V(x)$ was found for points along the x axis for a ring mass in Exercise 17. (a) Calculate the gravitational field from $g_x = -dV/dx$. (b) At what points is the magnitude of the gravitational field maximum? (c) Calculate the gravitational field on the axis of a ring directly from Equation 16-14 by first finding g_x due to some element of the ring Δm_i and then summing over all the elements.

3. Two planets of equal mass orbit a much more massive star. Planet m_1 moves in a circular orbit of radius 1×10^8 km and period 2 y. Planet m_2 moves in an elliptical orbit with closest distance $r_1 = 1 \times 10^8$ km and farthest distance $r_2 = 1.8 \times 10^8$ km, as shown in Figure 16-18. (a) Using the fact that the mean radius of an elliptic orbit is the length of the semimajor axis, find the period of m_2's orbit. (b) What is the mass of the star? (c) Which planet has the greater speed at point P? Which has the greater total energy? (d) How does the speed of m_2 at point P compare with that at A? Hint: Use conservation of angular momentum.

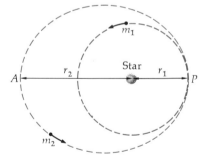

Figure 16-18
Problem 3.

4. The restoring torque for a torsional balance is proportional to the angular displacement $\tau = -k\theta$, where k is the torque constant. (a) From the equation $\tau = I\alpha$, where I is the moment of inertial for rotation about the axis of suspension, show that twisting motion of the balance is simple harmonic motion with period $T = 2\pi \sqrt{I/k}$. (b) A Cavendish apparatus uses two 25-gm metal spheres attached to the ends of a 40-cm rod with negligible mass. The rod is suspended by a fine fiber attached to its center. When the fiber is twisted, the rod oscillates with simple harmonic motion with a period of 1000 sec. Find the torsion constant of the fiber. (c) When two large masses m are placed near the smaller masses, the fiber twists through an angle $\theta = 2$ mrad. What is the force of attraction between each large mass m and the nearby 25-gm mass?

5. (a) Show that the gravitational field of a uniform ring of mass is zero at the center of the ring. (b) Figure 16-19 shows a point P in the plane of the ring but not at the center. Consider two elements of the ring of length s_1 and s_2 (in-

dicated in the figure) at a distance r_1 and r_2 from point P, respectively. What is the ratio of the masses of these elements? Which produces the greater field at point P? What is the direction of the field at point P due to these elements? (c) What is the direction of the gravitational field at point P? (d) Suppose that the gravitational field due to a point mass varied as $1/r$ rather than $1/r^2$. What then would be the resultant gravitational field at point P due to the elements shown? (e) How would your answers to parts (b) and (c) differ if point P were inside a spherical shell of mass rather than a plane circular ring?

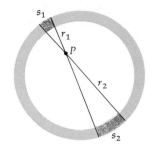

Figure 16-19
Problem 5.

6. A uniform thin disk has mass M and radius R. (a) What is the density of the disk in terms of M and R? (b) What is the gravitational potential on the axis of the disk a distance x from its center produced by a ring element of the disk of radius r and width dr? (c) Integrate your result for part (b) to obtain the potential for the whole disk. (d) From the potential, find the gravitational field strength at any point on the axis of the disk.

7. A mass m slides through the tunnel discussed in Example 16-2, starting at rest at the surface of the earth. Show that the maximum speed m reaches is given by $x\sqrt{g/R_E}$, where x is the initial displacement from equilibrium shown in Figure 16-13. Evaluate this speed for a tunnel joining two cities 200 mi apart. Use the approximation $2x = 200$ mi.

8. A thick spherical shell has an inner radius R_1 and an outer radius R_2. It has mass M and uniform density. Find the gravitational field g_r as a function of distance r from the center of the sphere for the ranges $0 \le r \le R_1$, $R_1 \le r \le R_2$, and $R_2 \le r < \infty$.

9. A uniform rod of mass M and length L lies along the x axis with center at the origin. Consider an element of length dx a distance x from the origin. Show that this element produces a gravitational field at a point on the axis x_0 (x_0 greater than $\frac{1}{2}L$) given by

$$\frac{-GM\ dx/L}{(x_0 - x)^2}$$

Integrate this result over the rod to find the total gravitational field at the point x_0 due to the rod.

10. Calculate the gravitational potential at point x_0 due to the uniform rod of Problem 9. Use your result to find the gravitational field from $g_x = -dV/dx$.

11. A straight, smooth tunnel is dug through a spherical planet whose mass density ρ_0 is constant. The tunnel passes through the center of the planet and is perpendicular to the planet's axis of rotation, which is fixed in space. The planet rotates with an angular velocity so that objects in the tunnel have no acceleration relative to the tunnel. Find the relation between ρ_0 and ω for this to be true.

CHAPTER 17   Temperature

Thermodynamics is the study of energy transfers involving temperature between macroscopic bodies. In the following three chapters we shall define the concepts of temperature, heat, and internal energy just as we carefully defined such concepts as mass, force, potential energy, etc., in our study of mechanics. Until we do however, we shall use the words temperature and heat in their ordinary meanings: temperature is what we measure with an ordinary thermometer; heat is energy which is transferred because of a temperature difference.

17-1 Macroscopic State Variables

Consider a body of gas enclosed in some volume V and at some pressure P. There are two different approaches to the description of such a system. One way, the *microscopic* method, involves a description in terms of the many particles, the molecules, composing the gas. It requires many assumptions about the particles which are very difficult to test directly. For example, we assume that the gas is composed of N molecules each moving about in a random way making elastic collisions with other molecules and with the walls of the container. Because of the huge size of N (for example, 1 mole, or 32 gm, of oxygen contains Avogadro's number $N_A \approx 6 \times 10^{23}$ molecules), it is impossible to apply Newton's laws of motion separately to each molecule or even to list the coordinates of each molecule. Entering just one coordinate for each molecule of 1 mole of material into a computer at the rate of 1 μs per molecule would take 6×10^{17} sec $\approx 2 \times 10^{10}$ y, which is the same order of magnitude as the age of the universe. However, just because N is so large, statistical methods of treatment are very accurate. In Chapter 11 we showed by a simple analysis that the pressure exerted by a gas on the wall of its container is proportional to the average kinetic energy of the molecules in the gas. This type of calculation is an example of a microscopic calculation of a macroscopic variable, the pressure P.

Microscopic description

Macroscopic description

In the *macroscopic* approach characteristic of thermodynamics, the system is described by only a few variables, e.g., pressure, temperature, volume, and internal energy. These variables (except internal energy) are closely related to our senses and are easily measured. (Compare these few variables with the $6N$ coordinates and velocity components of the molecules and additional variables describing rotation or oscillation of molecules like O_2 which contain more than one atom.) Few assumptions are made in the macroscopic description of matter. The laws of thermodynamics, which, like Newton's laws of motion, are elegant and compact generalizations of the results of experience, are therefore quite general and independent of any particular molecular assumptions made in the microscopic approach. In fact, much of thermodynamics was developed before the molecular model of matter was completely accepted.

The minimum number of macroscopic variables needed to describe a system depends on the kind of system, but it is always a small number. Usually we must specify the composition (if the system is not homogeneous, e.g., a mixture of two or more gases), the mass of each part, and only two more variables such as the pressure P and the volume V of a gas. If we restrict our discussion to homogeneous systems of constant mass, we usually need only two variables.

We cannot say that a gas is at pressure P and temperature t unless the system is in *thermal equilibrium* with itself. Consider, for example, a gas in an enclosure of volume V, isolated from its surroundings. If we stir the gas rapidly in one corner of the container, we cannot assign a pressure and temperature to the whole gas until the gas settles down. If we measure the temperature at various parts of the gas just after stirring, we get different results, which change with time. *When the macroscopic properties of an isolated system become constant in time, the system is in thermal equilibrium with itself.* We can then describe some property such as pressure with a single variable P for the whole system.

Suppose we have a fixed mass of gas originally at some pressure P_0 and volume V_0. Let the temperature of the gas be t_0 (measured in any familiar way). We now do various things to the gas, e.g., compress it, heat it, let it expand against a piston, and cool it, but finally return it to the original pressure P_0 and volume V_0. When the gas is again at equilibrium with itself at the original pressure and volume, we find that its temperature is again t_0. Every other macroscopic property which we can measure is also the same as it was originally. The variables P and V thus specify a *macroscopic state* of the system. We could also specify the state of this system using the variables P and t. For a constant-mass gas system, an equilibrium state is specified by any two macroscopic variables.

There are many other kinds of macroscopic systems whose states can be described by two variables (assuming constant mass). For example, a long rod can be described by its length L and the pressure P (which is usually atmospheric pressure unless the rod is enclosed in some way). Similarly, the state of a conducting wire can be described by the pressure and its electric resistance. Even such a complicated system as an electric cell can often be described quite well in terms of only two variables, such as the emf[1] of the cell and the charge. In the following chapters we shall define temperature carefully and other mac-

[1] We will discuss emf in Chapter 34.

roscopic variables, e.g., internal energy and entropy, which are also *state variables*. We repeat the meaning of the concept of a macroscopic equilibrium state: if a system is in a given equilibrium state at some time and after various processes it is brought back to that equilibrium state, all the macroscopic properties of the system will be the same as they were originally. Note that a *macroscopic* equilibrium state is quite different from a *microscopic* state, which is specified by the positions, velocities, and internal coordinates of all the molecules in the system. In a simple gas in thermal equilibrium with itself the microscopic state is continually changing because the molecules are changing position and making collisions which change their velocities; but in equilibrium, the macroscopic state remains the same. There are many different microscopic states which correspond to a single macroscopic state.

State variables

17-2   Adiabatic and Diathermic Walls

Before we give a rigorous definition of temperature, let us consider some common experiences with properties of materials using temperature as measured by a common thermometer. Consider, for example, a given mass of gas with volume V and at pressure P. If we heat the gas (say with a bunsen burner), the pressure or volume of the gas will change. If the volume is kept constant, the pressure increases as the temperature is raised. Alternatively, if the pressure is kept constant by allowing the gas to expand against a piston, as in Figure 17-1, the volume increases as the temperature is raised. An increase in the temperature of the gas is indicated by an increase in either the pressure P or the volume V when the other variable is kept constant.

Let us consider some simple experiments with constant-volume gas systems. We place two systems, originally at different temperatures, in close contact by separating them only by a thin metal wall. The temperature of each system changes until the two systems reach a common temperature between the original temperatures. The pressure of one system increases as its temperature increases, and the pressure of the other decreases as its temperature decreases, until eventually the pressures reach steady values when the systems reach their final equilibrium temperature. When the pressures stop changing, the systems are in *thermal equilibrium* with each other. The time it takes to reach equilibrium depends on the masses and types of the gases, on the original temperatures, and on the kind of contact made between the two systems. If we replace the thin metal wall with a thick asbestos wall, the time is very long. Over rather long periods of time with this separation, the two systems remain at nearly their original pressures and temperatures. (We are assuming that both systems are otherwise isolated from other systems.) If we heat one system, stir it, or do anything to it, very little if anything happens to the other system.

We can do similar experiments with other types of systems. For example, instead of constant-volume gas systems, we might use long metal rods at constant pressure. In that case, the increase or decrease in the length of a rod indicates an increase or decrease in its temperature. From these experiments, we can define two kinds of ideal separations *without any prior reference to temperature*. Two systems are said to be separated by an *adiabatic wall* if we can arbitrarily change the variables of one system without having any effect on the other. For example, if two constant-volume gas systems are separated by an adiabatic wall, we can increase the pressure of one by heating it, or by

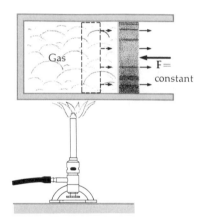

Figure 17-1
By confining a gas in a cylinder with a movable piston it is possible to heat the gas at constant pressure. As the gas is heated, the piston moves to the right against a constant force, which equals the pressure times the area.

Adiabatic and diathermic walls

some other means, and the other system will not be affected. (The common name for an adiabatic wall is a perfectly insulating wall. An adiabatic wall does not allow heat flow. However, we do not need to define heat or temperature in order to define an adiabatic wall.) The opposite extreme is a *diathermic wall*. If two systems are connected by a diathermic wall, a change in one of the thermodynamic variables of one system will influence the variables of the other system. For example, if two constant-volume gas systems are connected by a diathermic wall, heating one gas will cause an increase in pressure in the other gas (as well as in the one heated directly). The common name for a diathermic wall is, of course, a perfect heat conductor. Systems connected by a diathermic wall are said to be in *thermal contact*. In practice thin metal walls make excellent approximations to a perfect diathermic wall. Similarly, an adiabatic wall can be approximated with thick asbestos.

Thermal contact

17-3 The Zeroth Law of Thermodynamics

We now use our definitions of adiabatic and diathermic walls to define temperature. Suppose that two systems, originally separated, are put in contact by a diathermic wall and isolated from their surroundings by adiabatic walls (Figure 17-2). In general, when the systems are put in thermal contact, the thermodynamic variables will change from their original values. For example, the pressures P_1 and P_2 of two gas systems held at constant volume will change. One will increase, and the other will decrease. When the pressures reach their final equilibrium values, the two systems are said to be in *thermal equilibrium with each other*.

Suppose that when the two systems (originally separated) are put in contact by a diathermic wall, no change is observed in any of the variables of either system. The connected systems are in equilibrium, and they are said to be in thermal equilibrium even if they are separated. We thus generalize the concept of thermal equilibrium to include systems which are not in contact by a diathermic wall but which would be in equilibrium if they were so connected.

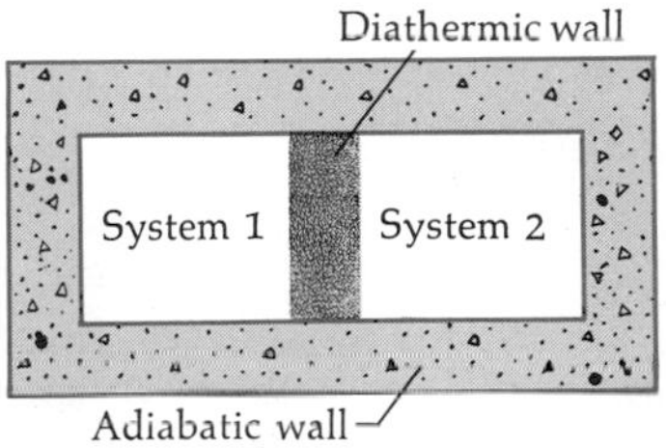

Figure 17-2
Systems 1 and 2 are connected to each other by a diathermic wall but separated from their surroundings by adiabatic walls. In general the thermodynamic variables of the systems will change when they are in thermal contact (connected by a diathermic wall). When the variables cease to change, the systems are in thermal equilibrium with each other.

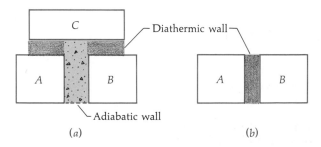

(a)   (b)

Figure 17-3
(a) Systems A and B are in thermal contact with system C. When system A and system B are in thermal equilibrium with system C, they are in thermal equilibrium with each other. This can be checked by putting them in contact with each other, as in (b). This experimental result is known as the zeroth law of thermodynamics.

Suppose system *A* is in thermal equilibrium with system *C* and system *B* is also in thermal equilibrium with system *C*, as in Figure 17-3a. If we now place system *A* and *B* in thermal contact, i.e., connect them by a diathermic wall (Figure 17-3b), we find that they are in thermal equilibrium with each other. This result is known as the zeroth law of thermodynamics:

If two systems are in thermal equilibrium with a third system, the two systems are in thermal equilibrium with each other.

Zeroth law

The zeroth law of thermodynamics, which is an experimental law, shows that the relation between two systems implied by the statement that they are in thermal equilibrium is a *transitive relation*. That is, if *A* is in thermal equilibrium with *C* and *B* with *C*, then *A* is in thermal equilibrium with *B*. This transitive property is needed in order to define temperature.

The first step in defining temperature is to determine a rule for saying when two systems in equilibrium have the same temperature. This resembles our procedure in mechanics, where the first law of motion defined the condition of motion without force. We adopted the rule that there is no force when velocity does not change. In the case of temperature, we shall say that two systems have the same temperature if none of their macroscopic variables change when they are connected with a diathermic wall. This definition of temperature equality is more succinctly expressed as follows:

Two systems in thermal equilibrium with each other are at the same temperature.

Two systems which are not in equilibrium with each other are at different temperatures. Suppose the two systems are constant-volume gas systems and that P_1 decreases and P_2 increases when they are put in thermal contact. We could arbitrarily define either system to be at a higher temperature and still have a self-consistent concept of temperature. However, since it is important to make scientific definitions of quantities as close to common usage as possible, we want a system which feels hot to the touch to be at a higher temperature than one which feels cold. (It is not always possible to distinguish the relative hotness or coldness of two objects by sense of touch, but there is a quite general agreement for a wide choice of systems such as ice water and water at 80°F.) In our example of two gas systems, system 1 is originally "warmer" than system 2. We associate *decreasing pressure* of a constant-volume gas system with *decreasing temperature* and *increasing pressure* with *increasing temperature*. If we place a constant-volume gas system in thermal contact with a metal rod when the systems are not in thermal equilibrium, we find that if the pressure of the gas increases, the length of the rod decreases and vice versa. Thus increasing rod length at constant pressure indicates that the temperature of the rod is increasing.

Questions

1. Consider the relation expressed by "A loves B." Is this relation transitive?

2. Suppose we define two countries to be in equilibrium with each other when they are at peace with each other. Is this kind of equilibrium situation transitive?

3. Give two other examples of relations between two objects which are transitive and two which are not.

4. As we have defined it, does temperature have any meaning applied to a system not in equilibrium with itself?

5. How could you determine whether two bodies have the same temperature if it is impossible to put them into contact with each other?

17-4   Temperature Scales and Thermometers

We have just defined a method for comparing any two systems and deciding whether they are at the same temperature or, if not, which system is at the greater temperature. We note that if A is at a greater temperature than B and B is at a greater temperature than C, it follows that A is at a greater temperature than C. If we have N systems, we can order them along a line such that all those to the right of a given system have a greater temperature than that system and all those to the left have a smaller temperature. We are now in position to define a *temperature scale*, which we can use to assign a number to each system so that greater numbers indicate greater temperatures.

We first choose a system to be our thermometer. We then define the temperature to be any monotonic function of one of the state variables of the system with the other held constant. For example, let us choose a given mass of mercury contained in a narrow glass tube. When we put this system in thermal contact with another, e.g., a constant-volume gas system, we note that the length of the mercury column increases when the temperature increases. Thus the mercury system behaves essentially the same as a rod at constant pressure. (The pressure on the mercury is usually not kept constant, but the compressibility of mercury is so small that even a rather large change in pressure has little effect on the volume or length of the mercury column.) The simplest functional relation between temperature and length of the mercury column that we can choose is a linear one. We define the temperature t to be related to the length L by

$$t = aL + b \qquad\qquad\qquad 17\text{-}1$$

The slope a and the intercept b can be determined by arbitrarily defining the temperature of two reproducible states of some standard system. Alternatively, we can choose the intercept b to be zero and determine a by arbitrarily defining the temperature of a single standard state.

The first method, which was universally used before 1954, is still of some practical use today. We choose for one of our states the ice point of water, i.e., water and ice in equilibrium with air at atmospheric pressure (also called the *normal melting point*), and define its temperature to be zero degrees Celsius,[1] written 0°C. Similarly we define the temperature of the steam point, i.e., water and water vapor in equilibrium at atmospheric pressure (also called the *normal boiling point*), to be 100°C. Let L_i be the length of the mercury column when it is in thermal equilibrium with a system at the ice point and L_s be that when it is in thermal equilibrium with a system at the steam point. We then have from Equation 17-1,

$$t_i = aL_i + b = 0°C \qquad \text{and} \qquad t_s = aL_s + b = 100°C \qquad 17\text{-}2$$

These equations can be solved for a and b in terms of L_i and L_s, giving

$$a = \frac{100°C}{L_s - L_i} \qquad \text{and} \qquad b = -\frac{(100°C)L_i}{L_s - L_i} \qquad 17\text{-}3$$

Having determined these constants, we can measure the temperature of any other system by noting the length L of the mercury when the

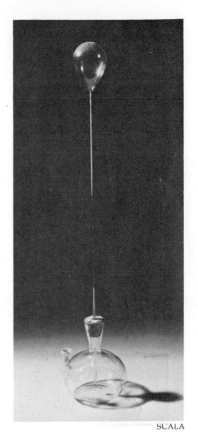

Galileo's thermoscope, one of the earliest thermometers, used by Galileo to detect temperature changes. (*Courtesy of the Science Museum, Florence.*)

---

[1] The Celsius scale was formerly called the centigrade scale.

thermometer is in equilibrium with that system and using Equation 17-1. Figure 17-4 illustrates this temperature scale.

If we choose a different system to be our thermometer, we define the temperature measured by this thermometer in a similar way. For example, if we choose a constant-volume gas thermometer (Figure 17-5), we define the temperature to vary linearly with the pressure

$$t = aP + b \qquad\qquad 17\text{-}4$$

Again we can find a and b in terms of the pressure P_i at the ice point and P_s at the steam point. We obtain equations similar to Equation 17-3:

$$a = \frac{100°C}{P_s - P_i} \quad\text{and}\quad b = -\frac{(100°C)P_i}{P_s - P_i} \qquad 17\text{-}5$$

Other thermometers can be similarly calibrated. For example, we can use a copper rod at constant pressure and define the temperature to be a linear function of the length or a resistance thermometer at constant pressure, for which the temperature is defined to vary linearly with electric resistance. The constants a and b are different for different thermometers.

We now ask whether the various thermometers defined in this way will agree when measuring the temperature of some system other than at the ice point or steam point of water. For example, if we measure the normal boiling point of sulfur (at about 444°C), we get different numbers using different thermometers. The discrepancies are particularly bad at temperatures far from the two fixed points. Similarly, if we measure thermal properties of some material, the results depend on the particular thermometer used. For example, if we use a copper-rod thermometer to measure the length of another piece of copper versus temperature, we find of course that the length of the copper varies linearly with temperature because the temperature scale was defined by this property. But if we measure the length of the same rod versus temperature using a mercury thermometer, we find that the expansion of the copper is not quite linear with temperature.

There is one group of thermometers for which the measured temperatures are in very close agreement at all temperatures. These are the gas thermometers, either constant-volume thermometers (discussed above) or constant-pressure thermometers, for which the temperature is taken to vary linearly with volume. If we measure the temperature of a given state of some system with different gas thermometers, we get close agreement no matter what kind of gas is used as long as the pressure in the thermometer is not too high and the temperature is not too low. If we reduce the pressure in all the thermometers by using less gas, the agreement improves.

Let us consider the following procedure. We measure some temperature, say the normal boiling point of sulfur, with a constant-volume gas thermometer, choosing the amount of gas so that the pressure at the steam point of water is $P_s = 1000$ millimeters of mercury (mm Hg; common units of pressure are described in Section 17-5). We then remove some of the gas in the thermometer so that the steam-point pressure P_s is now only 500 mm Hg and again measure the temperature of the normal boiling point of sulfur. We repeat this process with smaller and smaller amounts of gas and plot the measured temperature versus P_s. Figure 17-6 shows such a plot for various gases used in a constant-volume thermometer. As the pressure approaches zero, all the thermometers approach the same value, $t = 444.60°C$. We obtain similar results with a constant-pressure thermometer using the volume of the gas to measure the temperature. As the mass of the gas is

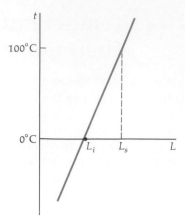

Figure 17-4
Temperature scale based on two fixed points. The temperature is assumed to vary linearly with the length of a mercury column. L_i and L_s are the lengths when the column is at the ice point and steam point of water, respectively. The scale is then determined by defining these temperatures to be 0°C and 100°C.

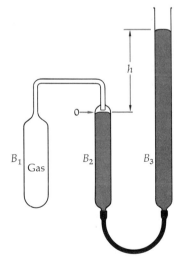

Figure 17-5
A constant-volume gas thermometer. The temperature is chosen to be proportional to the pressure of the gas in the bulb B_1. The tube B_3 is open to the atmosphere at the top and connected to the tube B_2 by a flexible hose. The volume is maintained constant by raising or lowering B_3 until the mercury in B_2 is at the zero mark. The pressure of the gas is then determined from the height h.

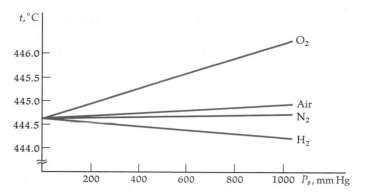

Figure 17-6
Temperature of the boiling point of sulfur measured with constant-volume gas thermometers filled with various gases. P_s, the pressure at the steam point of water, is varied by varying the amount of gas in the thermometer. As the pressure is reduced, the measured temperatures approach the value 444.60°C for all thermometers.

reduced so that the pressure becomes smaller, the temperatures measured by different gas thermometers tend to a single limit.

This agreement of all gas thermometers at low pressures means that the intercept b in Equation 17-4 and given by Equation 17-5 is the same for all gas thermometers. Measurements of b at low pressures yield values of about $-273.15°C$. There is some disagreement between various laboratories where this number is measured because of the experimental difficulty in measuring the ice-point and steam-point temperatures. Both states are very difficult to duplicate experimentally. Because of these difficulties, a temperature scale based on a single fixed point with b defined to be zero was adopted in 1954 by the International Committee on Weights and Measures. A reference state which is much more easily reproduced than either the ice point or steam point is the *triple point of water*, the single temperature and pressure at which water, water vapor, and ice coexist in equilibrium. The triple-point pressure is 4.58 mm Hg, and the temperature is about 0.01°C. The temperature of this state is conventionally chosen to be 273.16° (note that this is not degrees Celsius). Setting b equal to zero in Equation 17-4, we have for a constant-volume thermometer,

$$t = aP \qquad\qquad\qquad 17\text{-}6$$

Let P_3 be the pressure when the thermometer is in equilibrium with a system containing water, water vapor, and ice; then

$$273.16° = aP_3 \quad \text{or} \quad a = \frac{273.16°}{P_3}$$

Thus

$$t = \frac{273.16°}{P_3}\, P \qquad\qquad\qquad 17\text{-}7$$

In Section 17-7 and in Chapter 18 we shall discuss the behavior of real gases at very low pressures in terms of a simple model called an *ideal gas*. The temperature scale defined by Equation 17-7 in the limit of very low pressures is called the *ideal-gas temperature θ*. For a constant-volume gas thermometer, the ideal-gas temperature is defined to be

$$\theta = 273.16° \lim_{P_3 \to 0} \frac{P}{P_3} \qquad \text{constant } V \qquad\qquad 17\text{-}8a$$

Similarly, for a constant-pressure thermometer the ideal-gas temperature is

$$\theta = 273.16° \lim_{P_3 \to 0} \frac{V}{V_3} \qquad \text{constant } P \qquad\qquad 17\text{-}8b \qquad \textit{Ideal-gas temperature}$$

This temperature scale has the following advantages: (1) the measured temperature of any state is the same no matter what kind of gas is used, and (2) the reference state (the triple point of water) is easily reproduced throughout the world, so that temperatures measured in various laboratories are in good agreement with each other. This temperature scale depends on the properties of gases but not on the properties of any particular gas. Any substance can be used as long as it remains a gas. The lowest temperature that can be measured with a gas thermometer is about 1° using helium for the gas. Below this temperature helium liquefies. All other gases liquefy at higher temperatures.

In Chapter 19 we show that we can define a temperature scale which is independent of the properties of any material, called the *Kelvin scale,* or *absolute temperature scale.* We shall show that for temperatures above 1° for which the ideal-gas scale can be defined, the ideal-gas and absolute temperature scales are identical. In anticipation of this, we write[1] K after temperatures defined by the ideal-gas scale, and we shall use the conventional symbol T for the absolute temperature scale even though we cannot define this scale until we study the second law of thermodynamics.

Figure 17-7 shows the temperature of the steam point of water measured with constant-volume gas thermometers using various gases at various pressures P_3. As P_3 is reduced to zero by reducing the amount of gas in the various thermometers, the measurements approach the common value of 373.15 K, the ideal-gas temperature, which is the same as the absolute temperature. The ice-point temperature measured in a similar way is 273.15 K.

The Bettmann Archive

Lord Kelvin (1824–1907) with his compass for use on board iron ships.

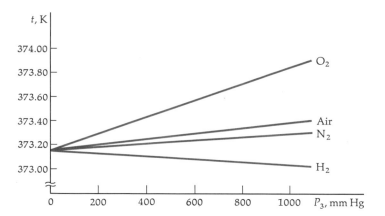

Figure 17-7
Temperature of the steam point of water measured with constant-volume gas thermometers filled with various gases. The temperature is plotted as a function of P_3, the pressure at the triple point of water, which is varied by varying the amount of gas in the thermometer. As the pressure approaches zero, the measured temperatures approach the value 373.15 K for all thermometers.

17-5   Units of Pressure

Pressure is defined as force per unit area and has SI units of newtons per square meter. As we showed in Chapter 5, the pressure difference between the top and bottom of a column of incompressible liquid is $P = \rho g h$. In other words, the pressure difference is proportional to the height of the liquid column. The height h thus can be used as a measure of pressure. For example, a pressure can be specified by saying it is the same as the pressure difference between the top and the bottom of a column of mercury 500 mm high. A pressure so specified could be expressed in newtons per square meter by using the relationship $P = \rho g h$ and the actual values of ρ and g.

In practice, pressure is often measured in millimeters of mercury (commonly called Torr, after the Italian physicist Torricelli) and in

---

[1] The name of the unit of temperature was changed by the 13th General Conference on Weights and Measures in 1967 from degree Kelvin (°K) to kelvin (K), defined to be 1/273.16 of the absolute temperature of the triple point of water.

inches and feet of water (written in H₂O and ft H₂O). Another common unit is the atmosphere (atm), which is approximately air pressure at sea level. The atmosphere, millimeters of mercury, and newton per square meter are related as follows:

$$1 \text{ atm} = 760 \text{ mm Hg} = 760 \text{ Torr} = 1.013 \times 10^5 \text{ N/m}^2$$

$$1 \text{ mm Hg} = 1 \text{ Torr} = 1.316 \times 10^{-3} \text{ atm} = 133.3 \text{ N/m}^2$$

Other units commonly used on weather maps are the bar and the millibar, defined by

$$1 \text{ bar} = 10^3 \text{ millibar} = 10^5 \text{ N/m}^2$$

A pressure of 1 bar is just slightly less than 1 atm.

Example 17-1 The pressure in a pipe in a water system is 50 ft H₂O. Express this pressure in millimeters of mercury, atmospheres, and newtons per square meter.

To say that the pressure is 50 ft H₂O means that the pressure is $P = \rho_w g h_w$, where ρ_w is the density of water and $h_w = 50$ ft. The same pressure is related to the height h_m of mercury column by $P = \rho_m g h_m$, where ρ_m is the density of mercury. Thus

$$\rho_m g h_m = \rho_w g h_w \qquad \text{or} \qquad h_m = \frac{\rho_w}{\rho_m} h_w$$

When we use $\rho_w = 1$ gm/cm³ and $\rho_m = 13.6$ gm/cm³, this gives for $h_w = 50$ ft,

$$h_m = 3.68 \text{ ft} = 1120 \text{ mm}$$

Thus $P = 1120$ mm Hg. We can express P in other units using the conversion factors given above:

$$P = 1120 \text{ mm Hg} \frac{1.316 \times 10^{-3} \text{ atm}}{1 \text{ mm Hg}} = 1.47 \text{ atm}$$

and

$$P = 1120 \text{ mm Hg} \frac{133.3 \text{ N/m}^2}{1 \text{ mm Hg}} = 1.49 \times 10^5 \text{ N/m}^2$$

Otto von Guericke's demonstration of the large forces due to atmospheric pressure. The force exerted by the horses was not great enough to separate the two hollow hemispheres, which had been placed together and evacuated. (*Courtesy of the Deutsches Museum, Munich.*)

Most pressure gauges read the difference between the absolute pressure and atmospheric pressure. This difference is called *gauge pressure*. Absolute pressure is obtained from gauge pressure by adding the atmospheric pressure P_{atm}:

$$P = P_{gauge} + P_{atm}$$

17-6   Celsius, Rankine, and Fahrenheit Scales

The Celsius temperature scale (formerly called the centigrade scale) is defined as follows. The temperature t_C on the Celsius scale is related to a temperature T on the Kelvin, or absolute, scale by

$$t_C = T - 273.15 \qquad\qquad 17\text{-}9$$

The size of the degree on the Celsius scale is the same as that on the Kelvin scale. That is, a temperature difference of 10 kelvins is the same as a temperature difference of 10 Celsius degrees. The difference between the Kelvin and Celsius scales lies only in the choice of zero temperature.

The Celsius temperatures of the ice point and steam point are found by subtracting 273.15 from the Kelvin temperatures (273.15 K for the ice point and 373.15 K for the steam point). The results are 0.00°C for the ice point and 100.00°C for the steam point. The triple-point temperature of water (273.16 K) is 0.01°C. Table 17-1 lists the normal-boiling-point (NBP) and normal-melting-point (NMP) temperatures on the Celsius scale for several substances. These temperatures can be used to calibrate practical thermometers.

Two other scales in use in this country and Great Britain are the Rankine scale, defined by

$$T_R = \tfrac{9}{5}T \qquad\qquad 17\text{-}10$$

and the Fahrenheit scale, which has the same size degree as the Rankine scale but its zero is shifted:

$$t_F = T_R - 459°R \qquad\qquad 17\text{-}11$$

The Fahrenheit and Celsius scales can be related by eliminating T_R from Equations 17-10 and 17-11 and then eliminating T from the result and Equation 17-9. The result is

$$t_F = \tfrac{9}{5}t_C + 32°F \qquad \text{or} \qquad t_C = \tfrac{5}{9}(t_F - 32°F) \qquad 17\text{-}12$$

The temperatures of the ice point and steam point on the Fahrenheit scale are 32 and 212°F, respectively. The temperature 0°F is roughly the coldest temperature that can be obtained with a salt-water solution before freezing, and the temperature 100°F is roughly that of the human body. The Fahrenheit degree is smaller than the Celsius degree. The difference in temperature between the steam point and ice point is 180 Fahrenheit degrees, which equals 100 Celsius degrees.

In specifying a temperature it is important, of course, to say what scale is used. Temperatures specified on the absolute scales (Kelvin or Rankine) are customarily indicated with the letter T. Their numerical values are followed by K (kelvins) or °R (degrees Rankine). For example, the triple point of water is 273.16 K = 491.69°R. Temperatures

Table 17-1

Normal melting-point (NMP) and normal boiling-point (NBP) temperatures for various substances

	NMP, °C	NBP, °C
Alcohol, ethyl	−114	78
Bismuth	271	920
Bromine	−7	60
Gold	1063.00	2600
Lead	330	1170
Lithium	186	1336
Mercury	−39	358
Nitrogen	−210	−196
Oxygen	−219	−182.97
Silver	960.80	1950
Sulfur	119.0	444.60
Sulfuric acid	8.6	326
Water	0.00	100.00
Zinc	420	918

expressed on the Celsius or Fahrenheit scales generally are indicated with the letter t. Their numerical values are followed by °C (degrees Celsius) or °F (degrees Fahrenheit). For example, the steam point of water is 100.00°C = 212.0°F.

The Celsius and Fahrenheit scales are the ones in everyday use. The absolute scales, as we shall see, are much more convenient for scientific purposes, partly because many formulas are more simply expressed when absolute scales are used and partly because the absolute temperature can be given a more fundamental interpretation.

Example 17-2 What is the Kelvin temperature corresponding to 70°F? From Equation 17-12 the Celsius temperature equal to 70°F is

$$t_C = \tfrac{5}{9}(70°F - 32°F) = \tfrac{5}{9}(38) = 21°C$$

From Equation 17-9,

$$T = t_C + 273.15 \text{ K} = 21 + 273 = 294 \text{ K}$$

Example 17-3 At what temperature do t_C and t_F have the same numerical value?

Setting $t_F = t_C$ in Equation 17-12 gives

$$t_C = \tfrac{9}{5}t_C + 32$$

Solving for t_C gives

$$t_C = -40°C$$

In other words, −40°C and −40°F are the same temperature.

Questions

6. Do the temperature scales we have defined allow negative temperatures?

7. Can you suggest why the Celsius and Fahrenheit scales are considered more convenient for ordinary nonscientific purposes?

8. Are there any circumstances in which the Kelvin and Celsius temperatures are not significantly different? The Kelvin and Rankine temperatures? The Rankine and Fahrenheit temperatures?

17-7 Equations of State; Ideal Gases

According to Equation 17-8, the absolute temperature (or ideal-gas temperature) of a gas at low pressures is proportional to the pressure at constant volume and to the volume at constant pressure. Thus at low pressures, the product PV is approximately proportional to the temperature T,

$$PV \approx CT$$

where C is a constant of proportionality appropriate to a particular body of gas. Since we expect the pressure (see Chapter 11) to be proportional to the amount of gas, it is convenient to write C as

$$C = Nk$$

where N is the number of molecules of gas and k is a constant which is found experimentally to have the same value for any kind or amount of gas. We then have

$$PV = NkT \quad \text{or} \quad \frac{PV}{N} = kT \qquad\qquad 17\text{-}13$$

It is often convenient to write the amount of gas in terms of the number of moles. A mole of any substance is the amount of that substance that contains Avogadro's number of molecules, where Avogadro's number N_A (approximately 6.022×10^{23}) is defined to be the number of carbon atoms in 12 gm of ^{12}C. The mass of 1 mole is called the *molecular weight M*. The molecular weight of ^{12}C is, by definition, 12 gm $= 12 \times 10^{-3}$ kg. If we have n moles of a substance, the number of molecules is

Mole defined

$$N = nN_A \qquad\qquad 17\text{-}14$$

Equation 17-13 can also be written

$$\frac{PV}{n} = N_A kT = RT \qquad\qquad 17\text{-}15$$

where

$$R = kN_A \qquad\qquad 17\text{-}16$$

Figure 17-8 shows plots of PV/nT versus pressure P for several gases. We see that as the pressure approaches zero, the quantity PV/nT approaches the same value R, called the *universal gas constant*, for all gases. The numerical value of R depends of course on the units chosen. In addition to the international system, in which the pressure is expressed in newtons per square meter, the volume in cubic meters, and thus PV in newton-meters, i.e., joules, we often express P in atmospheres and V in liters (1 $\ell = 10^3$ cm^3 $= 10^{-3}$ m^3). The values of R in these units are

Universal gas constant

$$R = 8.31 \text{ J/mole-K} = 0.0821 \ \ell\text{-atm/mole-K} \qquad\qquad 17\text{-}17$$

In terms of the heat-energy unit, the calorie (cal), which we define in the next chapter, the gas constant is

$$R = 1.99 \text{ cal/mole-K} \qquad\qquad 17\text{-}18$$

The constant k, called *Boltzmann's constant*, can be thought of as the gas constant per molecule,

Boltzmann's constant

$$k = \frac{R}{N_A} = 1.38 \times 10^{-23} \text{ J/K} \qquad\qquad 17\text{-}19$$

From Figure 17-8 we see that for real gases, PV/nT is very nearly constant over a rather large range of pressure. We define an *ideal gas* to

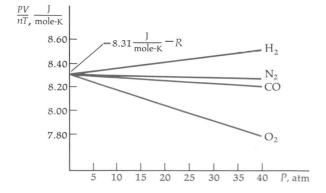

Figure 17-8
Plot of PV/nT versus P for real gases. As the pressure approaches zero, PV/nT approaches the value $R = 8.31$ J/mole-K for all gases. The ideal-gas equation $PV = nRT$ is a good approximation for all real gases for low pressures up to a few atmospheres.

be one for which PV/nT is constant for all pressures. Thus for an ideal gas, the pressure, volume, and temperature are related by

$$PV = nRT \qquad\qquad 17\text{-}20$$

Ideal-gas law

The concept of an ideal gas is an extrapolation of the behavior of real gases at low pressures to an ideal behavior. However, as can be seen from Figure 17-8, many gases differ little from an ideal gas at reasonably low pressures. Thus Equation 17-20, called an *equation of state,* is quite useful in describing properties of real gases. By our definition of temperature, to every state P_0V_0 there corresponds a temperature T_0. Thus there must be some function $T(P,V)$ for every gas, even at high pressures. Instead of thinking of T as a function of the two variables P and V, we could think of P as a function of T and V or V as a function of P and T. In any case, there is always a functional relationship between these three variables; given any two, the third is determined. However, it is not usually easy to find the relationship, and it is not possible to write down any single equation like Equation 17-20 which describes real gases over all ranges of the variables. For other thermodynamic systems, such as a metal rod, the state can be specified by the pressure P and length L. Thus the temperature T is a function of P and L. The equation which relates these variables over some range is also called an equation of state. Equations of state of systems are often found by fitting experimental data empirically. For example, there have been hundreds of equations of state proposed for various ranges of pressure for various gases. In thermodynamics, knowing that an equation of state exists is often sufficient to relate various measurable quantities such as the coefficient of thermal expansion and the compressibility even though the form of the equation is not known.

Equation of state

Example 17-4 What volume is occupied by 1 mole of gas at standard temperature of 0°C and normal atmospheric pressure?

From Equation 17-9, the absolute temperature corresponding to 0°C is 273 K. Using this and $P = 1$ atm, we find for the volume

$$V = \frac{nRT}{P} = \frac{(1 \text{ mole})(0.0821 \text{ } \ell\text{-atm/mole-K})(273 \text{ K})}{1 \text{ atm}}$$

$$= 22.4 \text{ } \ell$$

Example 17-5 How many molecules are there in 1 cm³ of gas at 0°C and atmospheric pressure?

Here $P = 1$ atm, $V = 1$ cm$^3 = 10^{-3}$ ℓ, and $T = 273$ K. From Equation 17-15,

$$n = \frac{PV}{RT} = \frac{(1 \text{ atm})(10^{-3} \text{ } \ell)}{(0.0821 \text{ } \ell\text{-atm/mole K})(273 \text{ K})}$$

$$= 4.46 \times 10^{-5} \text{ mole}$$

From Equation 17-14,

$$N = nN_A = (4.46 \times 10^{-5} \text{ mole})(6.02 \times 10^{23} \text{ molecules/mole})$$

$$= 2.68 \times 10^{19} \text{ molecules}$$

Questions

9. How does the pressure of an ideal gas change when any one of the following quantities is doubled while the others are kept fixed: (*a*) the absolute temperature, (*b*) the volume, (*c*) the number of molecules?

10. A certain rod has length L_0 at temperature T_0. Its length is found to increase linearly with temperature and to be practically independent of pressure. Write an approximate equation of state for the rod.

17-8   Molecular Interpretation of Temperature

In Chapter 11 we calculated the pressure exerted by gas molecules making elastic collisions with the walls of their container. Using the fact that the number of molecules that hit a wall in the yz plane in a given time is proportional to the x component of velocity v_x and to the number of molecules per unit volume N/V and that the change in momentum of each molecule is $2mv_x$, we found that the average pressure on the wall is proportional to N/V and mv_x^2. The relation we derived is

$$P = \frac{N}{V} m(v_x^2)_{av}$$

or

$$PV = 2N(\tfrac{1}{2}mv_x^2)_{av} \qquad\qquad 17\text{-}21$$

Comparing this with the equation of state for an ideal gas,

$$PV = nRT = NkT$$

we have

$$(\tfrac{1}{2}mv_x^2)_{av} = \tfrac{1}{2}kT \qquad\qquad 17\text{-}22$$

The average kinetic energy associated with translational motion in the x direction is $\tfrac{1}{2}kT$. Since x was an arbitrary direction, we could write similar equations for $(\tfrac{1}{2}mv_y^2)_{av}$ and $(\tfrac{1}{2}mv_z^2)_{av}$:

$$(\tfrac{1}{2}mv_y^2)_{av} = \tfrac{1}{2}kT \qquad (\tfrac{1}{2}mv_z^2)_{av} = \tfrac{1}{2}kT$$

Adding these and writing $v^2 = v_x^2 + v_y^2 + v_z^2$, we have

$$(\tfrac{1}{2}mv^2)_{av} = \tfrac{3}{2}kT \qquad\qquad 17\text{-}23$$

The absolute temperature is thus a measure of the average translational kinetic energy of the molecules. (We include the word translational because a molecule may have other kinds of kinetic energy, e.g., rotational or vibrational. Only the translational kinetic energy comes into the calculation of the pressure exerted on the walls of the container.) The total translational kinetic energy of n moles of a gas containing N molecules is

$$E_k = N(\tfrac{1}{2}mv^2)_{av} = \tfrac{3}{2}NkT = \tfrac{3}{2}nRT \qquad\qquad 17\text{-}24$$

The translational kinetic energy is $\tfrac{3}{2}kT$ per molecule, or $\tfrac{3}{2}RT$ per mole.

We can use these results to estimate the order of magnitude of the kinetic energies and speeds of the molecules in a gas. The average kinetic energy of translation of a molecule is $\tfrac{3}{2}kT$. At a temperature $T = 300$ K $= 27°C$ ($= 81°F$), $\tfrac{3}{2}kT$ has the value

$$\tfrac{3}{2}kT = 6.21 \times 10^{-21} \text{ J} = 0.0388 \text{ eV}$$

Thus the average kinetic energy of a molecule in a gas at this temperature is about 0.04 eV. The square root of $(v^2)_{av}$ is the root-mean-square (rms) speed. The average value of v^2 is, by Equation 17-23,

$$(v^2)_{av} = \frac{3kT}{m} = \frac{3N_A kT}{N_A m} = \frac{3RT}{M}$$

where $M = N_A m$ is the mass of 1 mole, or the molecular weight. Thus the rms speed is

$$v_{rms} = \sqrt{(v^2)_{av}} = \sqrt{\frac{3kT}{m}} = \sqrt{\frac{3RT}{M}} \qquad \text{17-25}$$

For oxygen gas at $T = 300$ K we have, using $M = 32 \times 10^{-3}$ kg/mole,

$$v_{rms} = \sqrt{\frac{3(8.31 \text{ J/mole-K})(300 \text{ K})}{32 \times 10^{-3} \text{ kg/mole}}} = 483 \text{ m/sec}$$

Questions

11. How does the average translational kinetic energy of a molecule in a gas change if the pressure is doubled while the volume is kept constant? If the volume is doubled while the pressure is kept constant?

12. By what factor must the absolute temperature of a gas be increased to double the rms speed of its molecules?

13. Would you expect all the molecules in a body of gas to have the same speed?

14. Two bodies of gas have the same temperature. Do their molecules necessarily have the same rms speeds? Explain.

17-9 Thermal Expansion

When the temperature of a body increases, it usually expands (one exception to this is water between 0 and 4°C, which contracts when the temperature increases). Let us first consider a long rod. Let its length be L_0 when the temperature is T_0 and $L = L_0 + \Delta L$ when the temperature is $T = T_0 + \Delta T$. We define the average coefficient of linear expansion for the interval ΔT to be the ratio of the fractional change in length divided by the temperature change when the pressure is constant:

$$\bar{\alpha} = \frac{\Delta L/L_0}{\Delta T} \qquad \text{17-26}$$

Coefficient of linear expansion

Thus $\Delta L = \bar{\alpha} L_0 \Delta T$. The coefficient of linear expansion at temperature T_0 is the limit of Equation 17-26 as ΔT approaches zero:

$$\alpha = \lim_{\Delta T \to 0} \frac{\Delta L/L_0}{\Delta T} \qquad \text{17-27}$$

The coefficient of expansion of a solid or liquid usually does not vary much with pressure, but it does vary appreciably with temperature. Thus the expression

$$\Delta L = \alpha L_0 \Delta T \qquad \text{17-28}$$

where α at the initial temperature is used in place of $\bar{\alpha}$, holds approximately if the temperature interval is so small that α and its average value do not differ appreciably.

The coefficient of volume expansion β, sometimes called the expansivity, is defined similarly as the limit of the ratio of the fractional change in volume to the temperature interval as the temperature interval approaches zero at constant pressure:

$$\beta = \lim_{\Delta T \to 0} \frac{\Delta V / V_0}{\Delta T} \qquad \text{17-29}$$

Coefficient of volume expansion

For a given material, the coefficient of volume expansion is just 3 times the coefficient of linear expansion. Consider a box of dimensions L_1, L_2, and L_3. The volume at temperature T is

$$V_0 = L_1 L_2 L_3$$

At temperature $T + \Delta T$ the volume is $V_0 + \Delta V$, where

$$\begin{aligned}
V_0 + \Delta V &= (L_1 + \Delta L_1)(L_2 + \Delta L_2)(L_3 + \Delta L_3) \\
&= (L_1 + \alpha L_1 \, \Delta T)(L_2 + \alpha L_2 \, \Delta T)(L_3 + L_3 \, \Delta T) \\
&= L_1 L_2 L_3 (1 + \alpha \, \Delta T)^3 \\
&= V_0 [1 + 3\alpha \, \Delta T + 3(\alpha \, \Delta T)^2 + (\alpha \, \Delta T)^3]
\end{aligned}$$

Thus

$$\Delta V = V_0 [3\alpha \, \Delta T + 3(\alpha \, \Delta T)^2 + (\alpha \, \Delta T)^3]$$

and

$$\frac{\Delta V / V_0}{\Delta T} = 3\alpha + 3\alpha^2 \, \Delta T + \alpha^3 (\Delta T)^2$$

Hence

$$\beta = \lim_{\Delta T \to 0} \frac{\Delta V / V}{\Delta T} = 3\alpha \qquad \text{17-30}$$

Figure 17-9, which shows the volume occupied by 1 gm of water versus temperature, illustrates that β is in general a function of temperature. The increase in size of any part of a body for a given temperature change is proportional to the original size of that part of the body. If we increase the temperature of a steel ruler, for example, the effect will be similar to that of a (very slight) photographic enlargement. Previously equally spaced lines will still be equally spaced but with correspondingly larger spaces. Values of α and β for various substances are given in Table 17-2.

Table 17-2

Approximate values of thermal expansion coefficients

Material	α, per °C	Material	β, per °C
Aluminum	24×10^{-6}	Acetone	1.5×10^{-3}
Brass	19×10^{-6}	Alcohol	1×10^{-3}
Carbon, diamond	1.2×10^{-6}	Mercury	0.18×10^{-3}
graphite	7.9×10^{-6}	Air	3.67×10^{-3}
Glass, ordinary	9×10^{-6}		
Pyrex	3.2×10^{-6}		
Ice	51×10^{-6}		
Invar	1×10^{-6}		
Steel	11×10^{-6}		

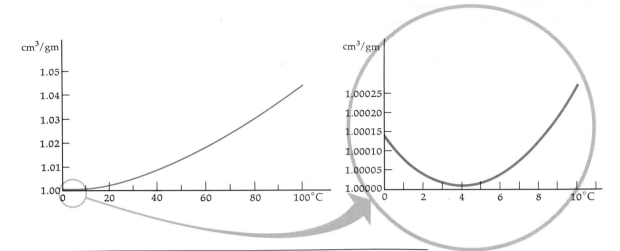

Figure 17-9
Volume of 1 gm of water at atmospheric pressure versus temperature. The minimum volume corresponding to maximum density occurs at 4°C. Since the variation in volume is only about 4 percent from $t = 0°C$ to $t = 100°C$, the thermal expansion or contraction of water can often be neglected.

Example 17-6 The temperature of a metal sphere is changed by a small amount ΔT. Using differentials to approximate small changes, show that $\beta = 3\alpha$.

The volume of a sphere of radius r is

$$V = \tfrac{4}{3}\pi r^3$$

Let Δr and ΔV be the changes in the radius and volume associated with the temperature change ΔT. Approximating these changes by differentials, we have

$$dV = \frac{dV}{dr}\, dr = 4\pi r^2\, dr$$

$$\frac{dV}{V} = \frac{4\pi r^2\, dr}{\tfrac{4}{3}\pi r^3} = 3\,\frac{dr}{r}$$

The coefficient of volume expansion is then

$$\beta = \frac{dV/V}{dT} = 3\,\frac{dr/r}{dT} = 3\alpha$$

Example 17-7 Find the coefficient of expansion of an ideal gas.

For a given temperature change, a gas at constant pressure expands by much more than a liquid or solid. We can compute β for any system if we know its equation of state. For an ideal gas we have

$$PV = nRT \quad \text{or} \quad V = \frac{nR}{P}\,T$$

Since P is constant, we can easily find dV/dT,

$$\frac{dV}{dT} = \frac{nR}{P} \quad \text{and} \quad \beta = \frac{1}{V}\frac{dV}{dT} = \frac{nR}{PV} = \frac{nR}{nRT} = \frac{1}{T}$$

The expansivity is thus just the reciprocal of the absolute temperature.

Questions

15. If a metal sheet with a hole in it expands, does the hole get larger or smaller?

16. If mercury and glass had the same coefficient of expansion, could a mercury thermometer be built?

17. What are the dimensions of α? Of β?

18. Near the bottom of a very deep lake, the water is compressed to its maximum density by the weight of the water above it. What value do you expect for the water temperature near the bottom of a deep lake?

19. If the maximum density of water occurred at 0°C, would you expect ice to form first at the top or bottom of a lake?

Review

A. Define, explain, or otherwise identify:

Microscopic description, 412
Macroscopic description, 413
System in thermal equilibrium with itself, 413
Macroscopic state of a gas, 413
State variables, 414
Adiabatic wall, 414
Diathermic wall, 415
Thermal contact, 415
Thermal equilibrium, 415
Zeroth law of thermodynamics, 415
Two systems at the same temperature, 416
Normal boiling point, 417
Normal melting point, 417

Triple point of water, 419
Ideal-gas temperature, 419
Pressure, 420
Kelvin temperature scale, 420
Gauge pressure, 422
Celsius temperature scale, 422
Rankine temperature scale, 422
Fahrenheit temperature scale, 422
Mole, 424
Universal gas constant, 424
Boltzmann's constant, 424
Ideal-gas law, 425
Equation of state, 425
Coefficient of linear expansion, 427

B. True or false:

1. An adiabatic wall is an insulating wall.

2. A diathermic wall is a heat-conducting wall.

3. The zeroth law of thermodynamics results from logical deduction rather than from experiment.

4. Two objects in thermal equilibrium with each other must be in thermal equilibrium with any third object.

5. All thermometers give the same result when measuring the temperature of any particular system.

6. The absolute temperature of a gas is a measure of the average translational kinetic energy of the gas molecules.

Exercises

Section 17-1, Macroscopic State Variables; Section 17-2, Adiabatic and Diathermic Walls; and Section 17-3, The Zeroth Law of Thermodynamics

1. A thermometer and pressure gauge are placed in the corner of a 25-ℓ flexible container of gas. At time $t = 0$, the system begins to expand, so that at $t = 5$ sec, when the volume is 30 ℓ, the pressure reading is 1.8 atm and the thermometer reads 20.0°C. At $t = 10$ sec, the system begins to contract, so that at $t = 14$ sec the volume is again 30 ℓ; the reading is then 1.8 atm, and the thermometer reads $t = 19$°C. If P and V define a state, explain the discrepancies in the temperature readings.

2. A 75-gm gaseous system is in thermal equilibrium with itself in the state $T = 30$°C, $V = 10 \ \ell$, and $P = 3$ atm. The volume is increased to 15 ℓ while the pressure is held at 3 atm. Can the temperature be maintained at 30°C? If so, what must be done?

3. In each of the cases below, systems A and B are initially in thermal equilibrium with each other and are connected by a wall. Explain in each case whether the wall must be diathermic or adiabatic or if it may be either. (a)

When P_A is increased at constant volume, P_B and V_B do not change. (b) When P_A is increased at constant volume, V_B increases at constant pressure. (c) When P_A is increased while V_A is decreased, P_B and V_B do not change.

4. A class of 30 students meets in a classroom with 30 numbered seats. The teacher gives a macroscopic description of the room one day by saying that 28 of the seats are occupied. Discuss possible microscopic descriptions of the classroom on that day, and show that there are many microscopic "states" corresponding to the macroscopic state of 28 seats occupied.

5. If system A is not in thermal equilibrium with B, and B is not in thermal equilibrium with C, what conclusions, if any, can be drawn about the temperatures T_A, T_B, and T_C? In particular, can you conclude that T_A and T_C are not equal?

6. If system A is not in thermal equilibrium with system B and system B *is* in thermal equilibrium with system C, what conclusions can be drawn about the temperatures T_A, T_B, and T_C?

Section 17-4, Temperature Scales and Thermometers

7. The length of the column of a mercury thermometer is 4.0 cm when the thermometer is immersed in ice water and 24.0 cm when the thermometer is placed in boiling water. (a) What should the length be at room temperature of 22.0°C? (b) The mercury column is 25.4 cm long when the thermometer is placed in a boiling chemical solution. What is the temperature of the solution?

8. The pressure of a constant-volume gas thermometer is 0.400 atm at the ice point and 0.546 atm at the steam point. (a) When the pressure is 0.100 atm, what is the temperature? (b) What is the pressure at the boiling point of sulfur (444.6°C)?

9. A constant-volume gas thermometer reads 50 mm Hg pressure at the triple point of water. (a) What will the pressure be when the thermometer measures a temperature of 300 K? (b) What ideal-gas temperature corresponds to a pressure of 68 mm Hg?

10. A constant-volume gas thermometer has a pressure of 30 mm Hg when it reads a temperature of 373 K. (a) What is its triple-point pressure P_3? (b) What temperature corresponds to a pressure of 0.175 mm Hg?

Section 17-5, Units of Pressure

11. (a) Find the height in feet of a water column corresponding to a pressure of 1 atm. (b) Find the pressure in atmospheres and in millimeters of mercury which supports a water column of height 30.0 ft.

12. Compute a conversion factor for pressure in atmospheres to pounds per square inch.

13. What is normal atmospheric pressure in inches of mercury?

14. The lowest atmospheric pressure recorded at sea level was 25.9 in Hg in a hurricane in 1958. (a) What was the pressure in pounds per square inch? (b) A bottle of wine was stoppered on the day before the hurricane, when the barometer read 30.0 in Hg. If the diameter of the cork was 1 in, what frictional force was necessary to hold the cork in during the hurricane?

15. (a) What is the pressure on a swimmer 20 ft below the surface of water? (b) What is the gauge pressure at that depth?

16. Water flows through a pipe at sea level under a gauge pressure of 40 lb/in². What is the absolute pressure (a) in pounds per square inch and (b) in inches of mercury?

17. A tabletop is 3 by 6 ft. What is the force on the table due to the air pressure down on it? Give your answer in pounds.

Section 17-6, Celsius, Rankine, and Fahrenheit Scales

18. The boiling point of O_2 is $-182.86°C$. Express this temperature in (a) kelvins, (b) in degrees Fahrenheit, (c) in degrees Rankine.

19. The boiling point of tungsten is $5900°C$. Find this temperature in (a) kelvins and (b) degrees Fahrenheit.

20. The highest and lowest temperature ever recorded in the United States are $134°F$ (in California in 1913) and $-80°F$ (in Alaska in 1971). Express these temperatures on the Celsius scale.

21. What is the Celsius temperature corresponding to the normal temperature of the human body of $98.6°F$?

22. In Portugal, in September 1933, the temperature rose to $70°C$ for 2 min. What is that temperature in degrees Fahrenheit?

23. The temperature of the interior of the sun is about 10^7 K. What is this temperature (a) on the Celsius scale, (b) on the Rankine scale, and (c) on the Fahrenheit scale?

Section 17-7, Equations of State; Ideal Gases

24. One mole of a gas occupies a volume of 10 ℓ at a pressure of 1 atm. (a) What is the temperature of the gas? (b) The container is fitted with a piston so that the volume can change. The gas is heated at constant pressure and expands to a volume of 20 ℓ. What is its temperature in kelvins? In degrees Celsius? (c) The volume is now fixed at 20 ℓ and the gas is heated at constant volume until its temperature is 350 K. What is its pressure?

25. One mole of gas is in a container fitted with a piston and the initial pressure and temperature are 2 atm and 300 K. (a) What is the initial volume of the gas? (b) The gas is allowed to expand at constant temperature until the pressure is 1 atm. What is the new volume? (c) The gas is now compressed and heated at the same time until it is back to its original volume, at which time the pressure is 2.5 atm. What is its temperature then?

26. A cylindrical container has a cross-sectional area of 3 in² and height of 8 in and is open at the top. It contains air at 1 atm pressure. A tightly fitting piston weighing 2.0 lb is inserted and gradually lowered until the increased pressure in the container supports the weight of the piston. (a) What is the force exerted on the top of the piston due to atmospheric pressure? (b) What is the force that must be exerted by the gas in the container on the bottom of the piston to keep it in equilibrium? What is the pressure in the container? (c) Assuming that the temperature of the gas in the container does not change, what is the height of the equilibrium position of the piston?

27. A gas is kept at constant pressure. If its temperature is changed from 50 to $100°C$, by what factor does its volume change?

28. A pressure as low as 1×10^{-8} mm Hg can be achieved by an oil-diffusion pump. How many molecules are there in 1 cm³ at this pressure if the temperature is 300 K?

29. A cubic metal box, of side 20 cm, contains air at pressure of 1 atm and temperature 300 K. It is sealed so that the volume is constant and heated to a temperature of 400 K. Find the net force on each wall of the box.

30. The mass of Avogadro's number of ^{12}C atoms is 12 gm. Each ^{12}C atoms contains six protons and six neutrons. Since the proton and neutron have approximately equal mass, the mass of Avogadro's number of protons or neutrons is 1 gm. The mass of the proton or neutron (in grams) is thus approximately the reciprocal of Avogadro's number. Calculate this mass and compare your answer with the mass given in Appendix Table B-1.

31. A 10-ℓ vessel contains gas at 0°C and a pressure of 4 atm. How many moles of gas are there in the vessel? How many molecules?

Section 17-8, Molecular Interpretation of Temperature

32. Show that $\sqrt{3RT/M}$ has the correct units meters per second if R is in joules per mole-kelvin, T in kelvins, and M in kilograms per mole.

33. Find v_{rms} of an argon atom if 1 mole of the gas is confined to a 1-ℓ container at 10 atm. ($M = 40 \times 10^{-3}$ kg/mole.) Compare this with that of helium gas under the same conditions (the molecular weight of helium is 4×10^{-3} kg/mole).

34. Find the total translational kinetic energy of 1 ℓ of oxygen gas held at a temperature of 0°C and at atmospheric pressure.

35. At what temperature will the rms speed of an H_2 molecule equal 332 m/sec?

Section 17-9, Thermal Expansion

36. A steel ruler has a length of 1 ft at 20°C. What is its length at 100°C?

37. Use Table 17-2 to find the coefficient of volume expansion for steel.

38. A 100-m-long bridge is built of steel. If it is built as a single continuous structure, how much will its length change from the coldest winter days (−30°C) to the hottest summer days (40°C)?

39. The experimental value of β for N_2 gas at 0°C and 1 atm is 0.003673/°C. Compare this with the theoretical value $\beta = 1/T$, assuming that N_2 is an ideal gas.

40. Define a coefficient of area expansion. Calculate it for a square and a circle, and show that it is 2 times the coefficient of linear expansion.

Problems

1. Gas is confined in a steel cylinder at 20°C and at a pressure of 5 atm. (a) If the cylinder is surrounded by boiling water and allowed to come to thermal equilibrium, how great will the gas pressure be? (b) If the gas is then allowed to escape until the pressure is again 5 atm, what fraction of the original gas (by weight) will escape? (c) If the temperature of the remaining gas in the cylinder is returned to 20°C, what will be its final pressure?

2. A constant-volume gas thermometer with a triple-point pressure of $P_3 = 500$ mm Hg is used to measure the boiling point of some substance. When it is placed in thermal contact with the boiling substance, its pressure is 734 mm Hg. Some of the gas in the thermometer is then allowed to escape so that its triple-point pressure is 200 mm Hg. When it is placed in thermal contact with the boiling substance, its pressure is 293.4 mm Hg. Again, some of the gas is removed from the thermometer so that its triple-point pressure is 100 mm Hg. Its pressure when in contact with the boiling substance is then 146.65 mm Hg. Find the ideal-gas temperature of the boiling substance to one place beyond the decimal point.

3. An automobile tire is filled to a gauge pressure of 28 lb/in² when it is at an air temperature of 70°F. After the car has been driven at high speed, the tire temperature has increased to 120°F. (a) Assuming that the volume of the tire has not changed, find the gauge pressure of the air in the tire (assume air to be an ideal gas). (b) Calculate the gauge pressure if the tire expands so that the volume increases by 10 percent.

4. An open steel 10-ℓ container is filled with acetone. The temperature rises from 0 to 40°C. How much acetone overflows from the container?

5. (a) If 2 moles of hydrogen gas (1 mole = 2 gm) are at atmospheric pressure and room temperature (20°C), what is the average translational kinetic energy of a hydrogen molecule in this gas? (b) What is v_{rms} for the hydrogen molecules? (c) What is the total kinetic energy of translation of all the molecules in this amount of hydrogen gas? (d) If a solid object of the same mass had this kinetic energy, what would its speed be?

6. At the surface of the sun the temperature is about 6000 K, and all materials are gaseous. From data given by the spectrum of light it is known that most elements are present on the sun. (a) What is the average kinetic energy of translation of an atom at the surface of the sun? Express your answer in both joules and electron volts. (b) What is the range of v_{rms} at the surface of the sun if the atoms present range from hydrogen ($M = 1$ gm/mole) to uranium ($M = 238$ gm/mole)?

7. In Chapter 9 it was shown that the escape velocity for a planet of radius R and acceleration of gravity g at the surface is given by $v_e{}^2 = 2Rg$. (a) At what temperature does v_{rms} for O_2 equal the escape velocity for the earth? (b) At what temperature is v_{rms} for H_2 the escape velocity for the earth? (c) Temperatures in the upper atmosphere reach about 1000 K. Can this account for the low abundance of hydrogen in the earth's atmosphere? (d) Compute the corresponding temperatures for the moon, where g has about one-sixth its value on the earth and $R = 1738$ km. Can you account in this way for the absence of an atmosphere on the moon?

8. One way to construct a device with two points whose separation remains the same in spite of temperature changes is to use rods with different coefficients of expansion, in the arrangement shown in Figure 17-10. The two rods are bolted together at one end. Show that the distance L will not change with temperature if the lengths L_A and L_B are chosen so that $L_A/L_B = \alpha_B/\alpha_A$. If material B is steel, material A is brass, and $L_A = 250$ cm at 0°C, what is the value of L? If both rods had been made of brass, how much would L increase when the rods are heated to 100°C?

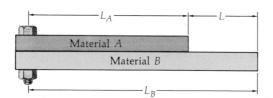

Figure 17-10
Problem 8.

9. A steel tube has an outside diameter of 3.000 cm at room temperature (20°C). A brass tube has an inside diameter of 2.997 cm at the same temperature. To what temperature must the ends of the tubes be heated if the steel tube is to be inserted into the brass tube?

10. A clock pendulum is made of brass and has a period of 1.00 sec at a winter temperature 16°C. How much will the same clock gain or lose each day in summer if the average temperature then is 26°C?

11. Ten objects have the speeds tabulated:

Speed, m/sec	2	5	6	8
Number of objects	3	3	3	1

Calculate the rms speed for these objects.

12. At 127°C, 2.4 moles of a certain gas occupy a 4-ℓ container at a density of 1.68×10^{-2} gm/cm³. (a) Find the total translational kinetic energy. (b) Find the average kinetic energy per molecule. (c) What is the pressure of the gas? (d) What is the molecular weight of the gas? (e) What diatomic gas could this be?

CHAPTER 18

Heat, Work, and the First Law of Thermodynamics

When two systems at different temperatures are placed in thermal contact, heat flows from the hotter system to the colder one until the two systems reach equilibrium at a common temperature. In the seventeenth century, Galileo, Newton, and other scientists generally supported the theory of the ancient Greek atomists, who considered heat to be a manifestation of molecular motion. In the next century, methods were developed for making quantitative measurements of the amount of heat that leaves or enters the body, and it was found that often when two bodies are in thermal contact, the amount of heat that leaves one body equals the amount of heat that enters the other body. An apparently successful theory of heat as a conserved material substance developed, in which heat, believed to be an invisible, weightless fluid, called *caloric,* was neither created nor destroyed but merely transferred from one body to another. The caloric theory of heat served quite well in the description of heat transfer but eventually was discarded when it was observed that caloric apparently could be created endlessly by friction with no corresponding disappearance of caloric somewhere else. In other words, the principle of conservation of caloric, which had been the experimental foundation of this theory of heat, proved to be false.

The first clear experimental observations showing that caloric cannot be conserved were made at the end of the eighteenth century by Benjamin Thompson, an American loyalist who emigrated to Europe, became the director of the Bavarian Arsenal, and was given the title Count Rumford. Thompson supervised the boring of cannon for Bavaria. Because of the heat generated by the boring tool, water was used for cooling. It had to be replaced continuously because it boiled away during the boring. According to the caloric theory, as the metal from the bore was cut into small chips, its ability to retain caloric was decreased. Therefore it released caloric to the water, heating it and causing it to boil. Thompson noticed, however, that even when the drill was too dull to cut the metal, the water still boiled away as long as the drill was turned. Apparently caloric was produced merely by friction and could be produced endlessly. He suggested that

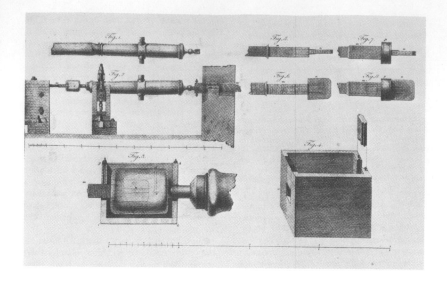

Rumford's apparatus. Figure 3 shows a close-up of the borer (dotted blunt shape at left) within the metal being bored. The vertical dotted lines indicate a hole for the insertion of a thermometer. The box in Figures 3 and 4 contained water and was placed around the metal and borer. It was used in later experiments to counteract the criticism that air somehow had caused the temperature increase. (*From* Philosophical Transactions of the Royal Society, *vol. 88, 1798.*)

heat is not a substance that is conserved but some form of motion communicated from the bore to the water. He showed, in fact, that the heat produced is approximately proportional to the work done by the boring tool.

The caloric theory of heat continued to be the leading theory for some 40 years after Thompson's work, but it was gradually weakened as more and more examples of nonconservation of heat were observed. Not until the 1840s did the modern mechanical theory of heat emerge. In this view heat is another form of energy, exchangeable at a fixed rate with the various forms of mechanical energy already discussed. The most varied and precise experiments demonstrating this were performed, starting in the late 1830s, by James Joule (1818–1889), after whom the metric unit of energy is named. Joule showed that the appearance or disappearance of a given quantity of heat is always accompanied by the disappearance or appearance of an equivalent quantity of mechanical energy. The experiments of Joule and others showed that neither heat nor mechanical energy is conserved independently but that the mechanical energy lost always equals the heat produced (measured in the same units). What is conserved is the total of mechanical energy and heat energy.

18-1   Heat Capacity and Specific Heat

Although we do not need a special unit for heat once we recognize that heat is a form of energy and can be measured in joules, foot-pounds, electron volts, or any other energy unit, the historical unit of heat, the calorie, is still used. The calorie was originally defined to be the amount of heat energy needed to raise the temperature of one gram of water one Celsius degree. After careful measurement showed that this energy depends slightly on the temperature of the water, the calorie was defined[1] more precisely to be the amount of energy needed to raise the temperature of one gram of water from 14.5 to 15.5°C. The

Historical definition of calorie

---

[1] The calorie is now defined in terms of mechanical energy units, as discussed in Section 18-3. It differs slightly from the historical calorie discussed here, but the difference can usually be neglected in practice.

amount of heat energy needed to raise the temperature of one kilogram of water from 14.5 to 15.5°C is the *kilocalorie* (1 kcal = 10^3 cal).
(The "calorie" used in measuring the energy equivalent of foods is
actually the kilocalorie.)

Using this definition of the calorie, we can measure the amount of
heat energy needed to raise the temperature of any object by any
amount. The amount of heat energy needed to raise the temperature of
an object one Celsius degree (or one kelvin since the kelvin and the
Celsius degree are the same size) is called the *heat capacity C* of the object. The heat capacity of one gram of water at 15°C is, by definition, 1
cal/°C = 1 cal/K. The heat capacity of an object is proportional to its
mass. The ratio of the heat capacity to the mass is the *specific heat c*,

Heat capacity

$$c = \frac{C}{m} \qquad\qquad 18\text{-}1$$

Specific heat

The heat capacity per mole is called the *molar heat capacity C'*,

$$C' = \frac{C}{n} \qquad\qquad 18\text{-}2$$

Molar heat capacity

where n is the number of moles. Since the mass of 1 mole is the molecular weight M, the molar heat capacity and specific heat are related by

$$C' = \frac{C}{n} = \frac{mc}{n} = Mc \qquad\qquad 18\text{-}3$$

Specific heats and molar heat capacities have been determined for
many substances and are listed in tables. The heat energy Q involved
in a temperature change ΔT of a system of mass m is

$$Q = C\,\Delta T = mc\,\Delta T \qquad\qquad 18\text{-}4$$

The temperature change is positive for heat put into the system and
negative for heat taken out of the system.

Suppose we wish to measure the specific heat of water at 80°C. We
need only place a given mass of water m_1 at 80°C in thermal contact
with a second mass of water m_2 at 14.5°C and isolate the two systems
from their surroundings by an adiabatic wall. When the temperature
of the second mass of water has risen to 15.5°C, the amount of heat
transfered is m_2 cal/gm. The specific heat of the water originally at

The complete arrangement of
Joule's original apparatus for
demonstrating the equiva-
lence of heat and mechanical
energy (see Figure 18-2, page
441, for a schematic diagram).
(*Crown copyright; courtesy of
the Science Museum, London.*)

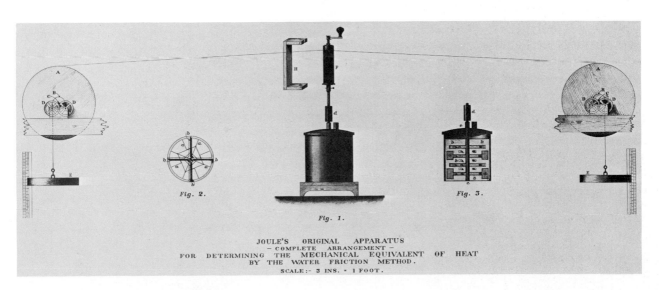

Fig. 2.

Fig. 3.

Fig. 1.

JOULE'S   ORIGINAL   APPARATUS
– COMPLETE   ARRANGEMENT –
FOR   DETERMINING   THE   MECHANICAL   EQUIVALENT   OF   HEAT
BY   THE   WATER   FRICTION   METHOD.
SCALE :- 3 INS. = 1 FOOT.

80°C can then be calculated from Equation 18-4 using the measured temperature change ΔT. This method measures the *average* specific heat over the temperature interval ΔT. If we suspect that the specific heat varies appreciably over the interval ΔT, we can reduce the temperature change by adjusting the masses in the experiment (either increasing m_1 or decreasing m_2). The specific heat of water varies by only about 1 percent from 0 to 100°C at a pressure of 1 atm. We can usually neglect this variation and take the specific heat of water to be 1.00 cal/gm-°C over the whole range 0 to 100°C.

Because the specific heat of water varies so slightly with temperature, the heat capacity or specific heat of an object can be conveniently measured by heating it to some easily measurable temperature, placing it in a water bath of known mass and temperature, and measuring the final equilibrium temperature. If the whole system is isolated from its surroundings, the heat which leaves the body equals the heat which enters the water. Let m be the mass of the body, c be its specific heat, and T_1 its initial temperature. Let m_w be the mass of the water bath, $c_w = 1$ cal/gm-K be its specific heat, T_0 its initial temperature, and T_f the final equilibrium temperature of the system. The heat absorbed by the water is $m_w c_w (T_f - T_0)$, and that lost by the body is $mc(T_1 - T_f)$. Since these amounts of heat are equal, c can be calculated from

$$mc(T_1 - T_f) = m_w c_w (T_f - T_0) \qquad 18\text{-}5$$

Because only differences in temperature are important in Equation 18-5, and because the kelvin and the Celsius degree are the same size, the temperatures can be measured on either the Celsius or Kelvin scale without affecting the result.

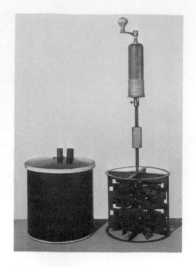

Closeup view of the water-containing section of Joule's apparatus showing the paddle wheel (right) and its adiabatic container (left). (*Loaned to the Science Museum, London, by Dr. Joule, Manchester.*)

Example 18-1 When 100 gm of aluminum shot is heated to 100°C and placed in 500 gm of water initially at 18.3°C, the final equilibrium temperature of the mixture is 21.7°C. What is the specific heat of aluminum?

The temperature change of the water is $21.7°C - 18.3°C = 3.4°C$. Thus the heat absorbed by the water is

$$Q_w = m_w c_w \, \Delta T_w = (500 \text{ gm})(1 \text{ cal/gm-°C})(3.4°C) = 1700 \text{ cal}$$

The temperature change of the aluminum is $100°C - 21.7°C = 78.3°C$. Its average specific heat over this interval is

$$c_{Al} = \frac{Q_w/\Delta T_{Al}}{m_{Al}} = \frac{1700 \text{ cal}}{(100 \text{ gm})(78.3°C)} = 0.217 \text{ cal/gm-°C}$$

The specific heat of aluminum (and of most materials) is considerably less than that of water. In practice, it is useful to choose the initial temperature of the water bath (using a rough knowledge of the specific heat to be measured) so that the final and initial temperatures of the water differ from room temperature by about the same amount. Then, even if the water is not perfectly isolated from its surroundings, it will gain about as much heat from the surroundings when it is below room temperature as it loses when it is above room temperature. For this example, the initial temperature was chosen for a room temperature of $20°C = 68°F$.

In general, the amount of heat needed to raise the temperature of a body at constant pressure is not the same as the amount needed at constant volume. We must therefore distinguish between the specific heat at constant pressure c_p and the specific heat at constant volume c_v. For solids and liquids this difference amounts to a few percent or less and can be ignored except in the most precise experiments. Since it is quite easy to keep the pressure on a solid or liquid constant whereas it is extremely difficult to adjust the pressure to keep the volume constant, it is c_p that is measured for a solid or liquid. (This is somewhat unfortunate since it is c_v that is most easily related to the theory of solids and liquids. However, if c_p is known, c_v can be calculated from it by methods of thermodynamics which are too complicated to discuss here.) Table 18-1 lists the specific heat and molar heat capacity for several metals. Heat capacity per mole is about 6 cal/mole-K for most of these metals, a result known as the *Dulong-Petit law*. At sufficiently high temperatures, the molar heat capacity for all metals is about $3R \approx 6$ cal/mole-K. Later we shall discuss the significance of the Dulong-Petit law and the fact that at low temperatures the heat capacities of metals are all less than this value. Figure 18-1 shows the temperature dependence of C_v' for iron. The shape of this curve is similar for all solids. The temperature at which C_v' approaches the Dulong-Petit limit differs for different materials.

Table 18-1

Specific heat and molar heat capacity for various metals at room temperature

	Specific heat, cal/ gm-K	Molar heat capacity, cal/ mole-K
Aluminum	0.215	5.82
Bismuth	0.0294	6.15
Copper	0.0923	5.85
Gold	0.0301	6.11
Lead	0.0305	6.32
Silver	0.0558	6.02
Tungsten	0.0321	5.92
Zinc	0.0925	6.03

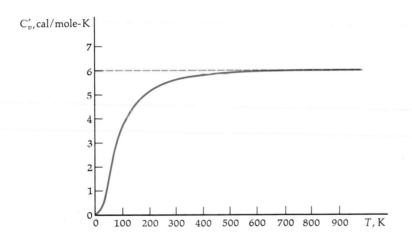

Figure 18-1
Molar heat capacity versus temperature for iron. All solids show this characteristic behavior, the values of the temperature scale varying from one material to another. At sufficiently high temperatures, the heat capacity is $3R \approx 6$ cal/mole-K, the Dulong-Petit law. Iron obeys the Dulong-Petit law for temperatures above about 350 K.

Questions

1. A large body of water, such as a lake or an ocean, tends to moderate the variations of temperature near it. Why? Would this be true if the specific heat of water had a value closer to that of other common substances?

2. Explain the circumstances under which the following situations could arise. (*a*) A massive body at high temperature is brought into thermal contact with a light object at low temperature. The final equilibrium temperature is closer to the original temperature of the massive body. (*b*) A massive body at high temperature is brought into thermal contact with a light body at a low temperature. The final temperature is closer to the original temperature of the light body.

3. Body *A* has twice the mass and twice the specific heat of body *B*. If they are supplied with equal amounts of heat, how do their temperature changes compare?

18-2   Latent Heat

In most circumstances, when heat is supplied to a body at constant pressure, the result is an increase in the temperature of the body. Sometimes, however, a body may absorb large amounts of heat without any change in temperature. This happens when the physical condition, or *phase*, of the material is changing from one form to another, e.g., from liquid to gas, from solid to liquid, or from one crystalline form to another. Thus the effect of heat is not only to change the temperature of a body but sometimes also to change the phase without any temperature change.

Phase

These phenomena are easily understood in terms of the mechanical or molecular theory of heat. When the temperature of a body increases because of heat added, it reflects an increase of the kinetic energy of motion of the molecules. When a material changes from liquid to gaseous form, the molecules, which originally were held together by their natural attractions, are moved far apart from each other. This requires that work be done against the attractive forces; i.e., it requires that energy be supplied to the molecules to separate them even though their kinetic energies (and the temperature) do not increase in the process. From this model we would expect the change in phase from liquid to gas to require heat even though there is no rise in temperature. Similarly, other changes, e.g., from solid to liquid or from one solid form to another, are rearrangements of molecules and involve work done against the forces they exert on each other. It is not surprising that heat energy may be required to accomplish these rearrangements at constant temperature.

For pure substances, changes of phase occur, for any given pressure, only at particular temperatures. For example, pure water at atmospheric pressure changes from solid to liquid at 0°C and from liquid to gas at 100°C. The first temperature is called the *melting point*, the second the *boiling point*.

For pure substances, a specific quantity of heat is required to change the phase of given amount of the substance. The required heat is proportional to the mass of the material,

$$Q = mL \qquad\qquad 18\text{-}6$$

where L is a constant characteristic of the substance and the kind of phase change involved. If the phase change is from solid to liquid, L_f is called the *latent heat of fusion* of the substance. For water at atmospheric pressure, the latent heat of fusion is 79.7 cal/gm. If the phase change is from liquid to gas, L_v is called the *latent heat of vaporization*. For water at atmospheric pressure, the latent heat of vaporization is 540 cal/gm.

Latent heat defined

Example 18-2 If 500 gm of ice at 0°C is heated at atmospheric pressure until all the ice is turned to gas, how much heat is required?

The amount of heat required to melt the ice is

$$Q_1 = mL_f = (500 \text{ gm})(79.7 \text{ cal/gm}) = 3.98 \times 10^4 \text{ cal}$$

The amount of heat required to raise the temperature of the liquid from 0 to 100°C is

$$Q_2 = mc\,\Delta T = (500 \text{ gm})(1 \text{ cal/gm-°C})(100°\text{C}) = 5.00 \times 10^4 \text{ cal}$$

Finally, the amount of heat required to vaporize the liquid is

$$Q_3 = mL_v = (500 \text{ gm})(540 \text{ cal/gm}) = 27.0 \times 10^4 \text{ cal}$$

The total heat required is

$$Q = Q_1 + Q_2 + Q_3 = 36.0 \times 10^4 \text{ cal}$$

Most of the heat supplied in this process was needed to change the phase of the water, not to increase its temperature.

18-3 The First Law of Thermodynamics

If we add heat to water, we raise its temperature, but we can also raise the temperature of water by stirring it or doing work on it in some other way without adding heat. Figure 18-2 is a schematic illustration of Joule's most famous experiment to determine the amount of work required to produce a given amount of heat. (Once the experimental equivalence of heat and energy has been established, one can describe Joule's work as determining the size of the calorie in energy units.) The water is enclosed by adiabatic walls so that its surroundings cannot affect its temperature by heat conduction. The weights, falling at constant speed, turn a paddle wheel, which does work against the water. If no energy is lost through friction in the bearings, etc., the work done by the paddle wheel against the water equals the loss in mechanical energy. The latter is just the potential energy $2mgh$ lost as the weights fall through height h.

The work needed to raise the temperature of 1 gm of water 1 Celsius degree when no heat is conducted was first measured accurately by Joule. Figure 18-3 shows the variation in this work with temperature. This figure is essentially a plot of the specific heat of water in joules per gram-kelvin versus temperature. As mentioned in Section 18-1, the specific heat of water varies by only about 1 percent from 0 to 100°C. At 15°C it has the value 4.186 J/gm-K. Thus the calorie defined as the heat energy needed to raise 1 gm of water from 14.5 to 15.5°C is equivalent to 4.186 J. Once it is recognized that the calorie is just another unit of energy, there is no need to define it experimentally. It is much more convenient to define the calorie in terms of other energy units already defined. The calorie (called the *thermochemical calorie*) is now defined to be exactly 4.184 J:

$$1 \text{ cal} = 4.184 \text{ J} \qquad 18\text{-}7$$

The calorie defined by Equation 18-7 differs slightly from the historical calorie (sometimes called the 15-degree calorie). It is approximately the energy needed to raise the temperature of water from 16.5 to 17.5°C. This difference is negligible in most applications. The size of the calorie in mechanical-energy units is called the *mechanical equivalent of heat*.

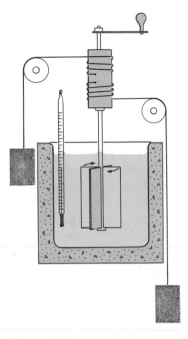

Figure 18-2
Schematic diagram of one of Joule's experiments to determine the amount of work required to produce a given increase in the temperature of water. The work done on the system is determined by the loss in gravitational potential energy of the masses.

Mechanical equivalent of heat

Figure 18-3
Specific heat of water versus temperature.

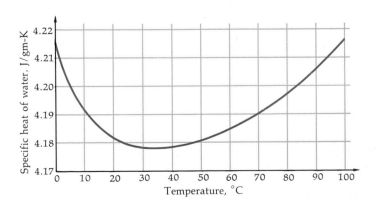

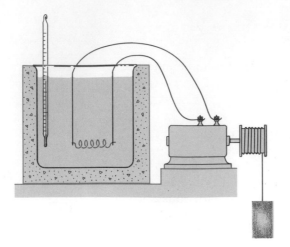

Figure 18-4
Another method of doing
work on an adiabatically
shielded system consisting of
water plus resistor. Electrical
work is done on the system
by the generator, which is
driven by the falling weight.
The amount of work needed
to raise the temperature a
given amount is the same no
matter how the work is done.

There are other ways of doing work on a system. In Figure 18-4 an electrical resistor is placed in a water bath and connected to a generator driven by a falling weight. (This is another of Joule's early experiments.) The water as a system is not thermally insulated because it is in thermal contact with the resistor and can absorb heat from it. If we consider the resistor plus water as the thermodynamic system, this system is shielded from its surroundings by an adiabatic wall. No heat can be conducted into this system. The electrical work needed to raise the temperature of the water by some amount is the same as the mechanical work needed to raise the temperature by the same amount. This type of experiment can be performed on various thermodynamic systems such as gases, solids, etc. We find that if the system is adiabatically shielded from its surroundings, the work required to change the system from state 1 to state 2 is *independent of how the work is performed* (stirring, compressing and expanding, electrical work, etc.) and depends only on the initial and final states. This experimental result expresses one form of the first law of thermodynamics.

James Prescott Joule
(1818–1889).

First law (first version)

The adiabatic work done on a system in taking the system from state 1 to state 2 is independent of the manner of doing the work and depends only on the initial and final states of the system.

Adiabatic work means work done on a system which neither gains nor loses heat to its surroundings while the work is done; i.e., it is work done on a system surrounded by adiabatic walls. Since the adiabatic work depends on the initial and final states of the system, we can define a function U of the state of the system such that the change in U equals the adiabatic work done. This new state function is called the *internal energy* of the system:

$$U_2 - U_1 = \text{adiabatic work done } on \text{ the system bringing it}$$
$$\text{from state 1 to state 2} \qquad \qquad 18\text{-}8$$

Internal energy defined

If the system is not adiabatically shielded from its surroundings, the work done on the system as it goes from state 1 to state 2 does depend on the process involved. For example, a given temperature increase of the system might be accomplished with no work done by placing the system in thermal contact with another system at a higher temperature. However, if we also measure the heat put into the system, the sum of the net heat put in plus the work done on the system going from state 1 to state 2 is the same for all processes and equal to the adiabatic work necessary to bring the system from state 1 to state 2 (assuming, of course, that the heat is expressed in energy units). Thus the sum of the heat put into a system plus the work done on the system equals the increase in internal energy of the system for any process. It is customary to write W for the work done *by* the system on its surroundings. Then $-W$ is the work done *on* the system by the surroundings. For example, if a gas expands against a piston, W is positive; if the gas is compressed, W is negative. Also, the heat put *into* a system is usually written Q. Then we have

$$Q + (-W) = \Delta U \qquad\qquad 18\text{-}9$$

usually written

$$Q = \Delta U + W \qquad\qquad 18\text{-}10 \qquad \textit{First law}$$

This equation is the generalization of our first statement of the first law to include nonadiabatic work. It is another way of stating the first law of thermodynamics.

Equation 18-10 is sometimes taken to be the definition of heat. This point deserves elaboration, because, like Newton's laws of motion, the first law of thermodynamics is often misunderstood. From experiments with adiabatic work alone, we can define a state function, the internal energy of the system. We can thus prove the existence of the function U with no reference to heat at all. (Recall that we have already defined what is meant by an adiabatic wall without having to define temperature.) The first law of thermodynamics states that the state function U exists. This statement of the first law is analogous to our statement of the zeroth law, namely, that a state function called the temperature exists. Having shown that U exists, and given a method of measuring ΔU for any two states (one must shield the system adiabatically from its surroundings and measure the work needed to bring the system from state 1 to state 2), we find that if the system is not adiabatically shielded, the work done on the system does not equal the change in U. We can then define the heat put into the system as the difference between the change in U and the work done *on* the system. We find that heat defined in this way has all the characteristics we have already discussed. For example, 4.186 J of heat will raise the temperature of 1 gm of water from 14.5 to 15.5°C if no work is done.

We shall usually refer to Equation 18-10 as the statement of the first law of thermodynamics. However, we should always keep in mind that the real "content" of this law is that a state function U exists and can be found from Equation 18-10 or by the method as discussed above.

Questions

4. The experiments of Joule discussed in this section involved the conversion of mechanical energy into heat. Can you give examples in which heat is converted into mechanical energy?

5. Can a system absorb heat with no change in its internal energy?

6. The first law of thermodynamics is sometimes stated as follows: the total energy, including heat energy, of an isolated system is conserved. How is this statement related to $Q = \Delta U + W$?

18-4   Work and the *PV* Diagram for a Gas

Consider a gas which can be expanded or compressed by the motion of a tightly fitting frictionless piston, as shown in Figure 18-5. When the piston moves, the volume of the gas changes and either the pressure or temperature or both must change since these three variables are related by an equation of state (such as $PV = nRT$ for an ideal gas). If the piston moves suddenly, the gas will not be in an equilibrium state just afterward. For example, if we suddenly push the piston in to compress the gas, the pressure near the piston will be greater than that far from the piston initially. Until the gas settles down, we cannot define equilibrium macroscopic state variables for the gas such as T, P, or U. However, if we move the piston slowly in small steps, waiting after each step for equilibrium to be reestablished, we can compress or expand the gas in such a way that the gas is never far from an equilibrium state. This kind of a process is called a *quasi-static process*. In a quasi-static expansion, the piston moves infinitely slowly with no acceleration, so that the gas moves through a series of equilibrium states. In practice, it is possible to approximate quasi-static processes fairly well.

Let us begin with a gas at a fairly high pressure and let it expand quasi-statically. The force exerted by the gas on the piston is PA, where A is the area of the piston and P is the gas pressure, which in general changes as the gas expands (the pressure could be kept constant by heating the gas as it expands). Since in a quasi-static expansion the piston does not accelerate, there must be an external force pushing against the piston also equal to PA. The piston does work on the agent providing this external force. If the piston moves a distance dx, the work done *by the gas* is

$$dW = F\,dx = PA\,dx = P\,dV \qquad\qquad 18\text{-}11$$

where $dV = A\,dx$ is the increase in volume of the gas. Equation 18-11 holds also for a quasi-static compression. Then dV is negative, indicating that the gas does negative work. In order to calculate the work done by the gas for an expansion from volume V_1 to volume V_2 we need to know how the pressure varies during the expansion; i.e., we cannot integrate $P\,dV$ to find the total work unless we know the function $P(V,T)$, which in general is not constant. The states of the gas can be represented on a P-versus-V diagram as in Figure 18-6. Since any point (P_0,V_0) determines the thermodynamic state of the gas, each point on this diagram corresponds to a definite state. For example, the set of states corresponding to a constant pressure of 1 atm is represented by a horizontal line parallel to the V axis. *The work done by the gas in expanding a small amount dV is $P\,dV$ represented by the area under the P-versus-V curve as shown.*

Consider the gas initially in the state P_1V_1 and expanding to the final state P_2V_2. Let us suppose that these two states are at the same temperature. (If the gas is ideal, $P_1V_1 = P_2V_2$.) Several possible kinds of expansion are illustrated in Figure 18-7 by different curves connecting these states. Along path A (Figure 18-7a), the pressure is kept constant until the gas reaches the volume V_2, and then the volume is kept constant while the pressure is reduced to its final value P_2. The work done along this path is $P_1(V_2 - V_1)$ for the horizontal (constant-pressure) part of the path and zero along the constant volume. This work is indicated by the shaded area under the curve. (In order to reduce the

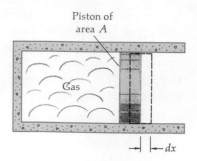

Figure 18-5
Gas confined in adiabatically shielded cylinder with movable piston. When the piston moves a distance dx, the volume changes by $dV = A\,dx$. The work done by the gas is $PA\,dx = P\,dV$, where P is the pressure.

Quasi-static process

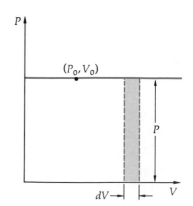

Figure 18-6
A particular state P_0V_0 is represented by a point on a PV diagram. The area $P\,dV$ on this diagram represents the work done by the gas as it expands.

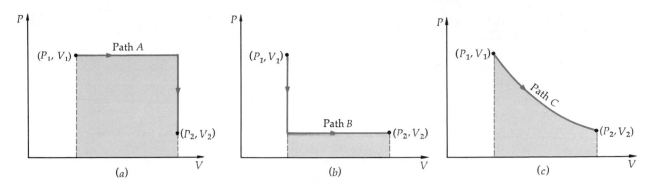

(a)   *(b)*   *(c)*

pressure at constant volume from P_2 to P_1, the gas must be allowed to cool; i.e., heat is transferred *from* the gas.) Along path B (Figure 18-7b), the gas is first cooled at constant volume to pressure P_2 and then allowed to expand at constant pressure until it reaches volume V_2. Clearly, the work done along this path is less than that along path A, as indicated by the area under the curve for this path. The work along path B is $P_2(V_2 - V_1)$. Along path C (Figure 18-7c) both the pressure and volume change for each part of the path. The work done is again the area under this curve connecting the initial and final states. We cannot compute this work unless we know the shape of this curve, which is the function $P(V)$.

It should be clear that the work done going from state 1 to state 2 depends on how the gas goes from one state to the other. Thus, even though we have written dW for a small amount of work, there is no function W which is determined by the state. That is, dW does not stand for the differential of any function. (Note the similarity of this discussion with that for a nonconservative force.) In this case the work done by a system depends on the path on the PV diagram connecting the initial and final states. Thus the work cannot be a function of the state. Consider, for example, letting the system go from state 1 to state 2 along path A in Figure 18-8 and returning to state 1 along path B. Although the system has returned to its original state, the net work done is not zero. It is, in fact, the difference between the area under curve A and the area under curve B. This difference is just the area enclosed by the total path of the cycle in the PV diagram. This work is positive if the system moves clockwise around the path.

According to the first law of thermodynamics, the internal energy of the gas *is* a state function. Thus for the cyclic path indicated in Figure 18-8, the net change in internal energy must be zero since the system is in its original state at the end. Thus an amount of heat Q equal to the work done must have been added to the gas during the cycle. If the gas traverses a cyclic path in the opposite direction (counterclockwise), the system does negative work and must give off an equal amount of heat. Since the work done *does* depend on the path taken while the internal energy change does not, the heat added must also depend on the path. In our example of a cyclic path, the net heat added or subtracted was not zero but equal in magnitude to the work done. Thus heat is *not* a function of state either. When we write dQ for a small amount of heat added to the system, we must remember that this is not the differential of a function Q. No such function indicating the heat in a system in each state exists. To remind ourselves of the fact that there are no state functions W and Q, we shall write dW and dQ to indicate small amounts of work done or heat added. Of course, if we know the path, we can calculate the work done and the heat added. For systems other than gases, the work done may take a different form because variables other than P and V define the state of the system.

Figure 18-7
Three paths in the *PV* diagram connecting an initial state P_1V_1 and a final state P_2V_2. (*a*) Along path A, the gas first expands at constant pressure from volume V_1 to volume V_2, and then the pressure is reduced (by cooling) at constant volume. The work done is the area indicated. (*b*) Along path B, the gas is first cooled at constant volume to reduce the pressure from P_1 to P_2 and then the gas expands at constant pressure. The work done indicated by the shaded area is less than that done along path A. (*c*) Along path C, both the pressure and volume vary. The work done is the area indicated.

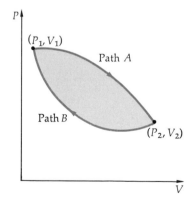

Figure 18-8
A gas goes from state 1 to state 2 along path A and returns to its original state along path B. The net work done for the complete trip is indicated by the shaded area. Since this work is not zero, work is not a state function. Since the net change in internal energy for the complete trip must be zero, an amount of heat equal to the work done must be added. Heat is not a state function either.

For example, an electric cell is described by its charge q and its emf ε. The work done by an electric cell is $dW = \varepsilon\, dq$. In the expansion of a solid or liquid, the work done is also $P\, dV$, but this is usually very small because dV is usually small. In any case, the discussion of work and heat for gases applies to all thermodynamic systems. Neither work nor heat is a function of the state of a system. Both the work done and the heat added depend on how a system goes from one state to another. The first law says that the heat added to the system minus the work done *by* the system depends not on how the system goes from one state to another but only on the initial and final states. In terms of our new notation for small amounts of work and heat, the first law is written

$$dQ = dU + dW \qquad\qquad 18\text{-}12$$

Note that we *do* write dU rather than dU because U *is* a function of the state of the system. Thus it is incorrect to say that a system contains a lot of heat when it is hot. We can increase the temperature of a system by adding heat or by doing work on the system or by some combination of both. It is correct to say that a hot system contains a lot of internal energy because internal energy is a property of the state of a system.

Example 18-3 An ideal gas expands at constant temperature. What work is done by the gas in the expansion?

An *isothermal* process is one which takes place at constant temperature. An isothermal expansion of a gas can be carried out by placing the gas in thermal contact with a second system of very large heat capacity and at the same temperature as the gas. This second system is known as a *heat reservoir*. Since the heat capacity of a heat reservoir is very large, it can supply heat to the gas or receive heat from the gas without appreciable change in temperature. We shall assume that the heat reservoir has an infinite heat capacity so that its temperature does not change at all. In practice there are many heat reservoirs available, e.g., lakes or large conducting masses. To be effective, a heat reservoir must be able to conduct heat to and from a system quickly so that a constant temperature is maintained in the system.

We can calculate the work done for this expansion because we know the path in the PV diagram. We are given that the temperature is constant and that the gas is ideal so that it obeys the equation of state $PV = nRT$. Then

$$P\, dV = \frac{nRT}{V}\, dV \qquad\qquad 18\text{-}13$$

The work done by the gas as it expands isothermally from volume V_1 to V_2 is thus

$$W = \int_{V_1}^{V_2} P\, dV = nRT \int_{V_1}^{V_2} \frac{dV}{V} = nRT \ln \frac{V_2}{V_1} \qquad\qquad 18\text{-}14$$

T has been removed from the integration because it is constant. This work is indicated by the shaded area under the PV curve in Figure 18-9. We cannot compute the heat absorbed from the heat reservoir unless we know how the internal energy changes during an isothermal expansion. When we discuss this point in the next section, we shall see that the internal energy of an *ideal* gas depends only on the temperature and so it does not change in an isothermal expansion. For this case, the heat removed from the reservoir is equal to the work done by the gas.

Isothermal process defined

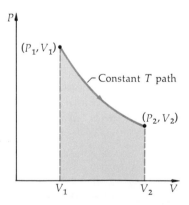

Figure 18-9
Isothermal expansion of an ideal gas. The curve in the PV diagram for which T is constant is called an isotherm. For an ideal gas $PV = $ constant along an isotherm. The work done by the gas undergoing an isothermal expansion is indicated by the shaded area and calculated in Example 18-3.

Example 18-4 If the volume of 1 mole of nitrogen is expanded at room temperature (20°C) from 10 to 20 ℓ, how much heat must be supplied to keep the temperature of the gas from dropping?

As mentioned in Example 18-3, the internal energy of an ideal gas depends only on its temperature. Hence, during an expansion at constant temperature, its internal energy does not change, or $\Delta U = 0$. Then, according to the first law (Equation 18-10), the heat Q absorbed by the gas equals the work W done by the system. The work W is given by Equation 18-14. Substituting the given values, we have

$$W = nRT \ln \frac{V_2}{V_1} = (1 \text{ mole})(8.31 \text{ J/mole-K})(293 \text{ K}) \ln\left(\tfrac{20}{10}\right)$$

$$= 1.69 \times 10^3 \text{ J} = \frac{1.69 \times 10^3}{4.18} \text{ cal} = 404 \text{ cal}$$

This work equals the heat that must be supplied to keep the temperature constant.

Question

7. When systems like confined bodies of gas or solids in contact with the air expand or contract, they do work on their surroundings. Can you suggest examples of a system that does work without changing its volume?

18-5 Internal Energy of a Gas

In Chapter 17 we saw that according to a simple molecular model of a gas, the temperature of a gas is associated with the average kinetic energy of translation of the molecules of the gas by the relation

$$PV = \tfrac{2}{3}E_k = nRT \qquad \text{or} \qquad E_k = \tfrac{3}{2}nRT$$

where n is the number of moles and $E_k = N(\tfrac{1}{2}mv^2)_{av}$ is the total translational kinetic energy of the N molecules. If this translational energy is taken to be the total internal energy of the gas, the internal energy will depend only on the temperature of the gas and not on the volume or pressure. Writing U for E_k, we have

$$U = \tfrac{3}{2}nRT \qquad\qquad 18\text{-}15$$

If the internal energy of the gas includes other kinds of energy in addition to kinetic energy of translation, the internal energy will be different from that given by Equation 18-15. It may or may not depend on the pressure and volume. Consider, for example, the molecular model of a diatomic gas, such as O_2, known as the rigid-dumbbell model. In this model the molecule of O_2 is pictured as two spheres connected by a rigid (massless) rod. In addition to translational energy of the center of mass of the molecules, there can be rotational energy of the atoms about the center of mass. If we picture a gas of this kind of molecule, we can expect a kind of equilibrium between the average energy of translation and the average energy of rotation. For example, imagine the gas to have initially only translational energy with no molecule rotating. This state cannot persist for long because when two molecules collide with each other, the molecules will surely be set in rotation. The greater the speed of translation of the molecules the greater will be the energy transferred to rotational energy if there is no initial rotation. However, if a molecule with a very large rotational energy collides with another molecule, it stands a very good chance of losing some of its rotational energy to translational energy. We thus expect some type of equilibrium between translational kinetic energy and rotational kinetic energy to result from molecular collisions. If we raise the temperature of the gas, the translational kinetic energy also rises. According to this discussion, we would expect the rotational kinetic energy to rise proportionally to maintain the equilibrium between these two forms of kinetic energy. It thus seems likely that rotational kinetic energy will also be just a function of the temperature, since changing the volume or pressure will change the number of collisions in a given time but will not change the nature of the energy-transfer process occurring in each collision.

If there is an attraction or repulsion between molecules in a gas even when they are far apart, we would not expect the internal energy to be a function of just the temperature. Consider the case of attraction. Since work is required to increase the separation of two molecules, the potential energy associated with the attractive force between them increases with separation. If the volume of the gas increases at constant temperature, the kinetic energy remains constant (since it depends only on the temperature) but the potential energy increases. Therefore, the total internal energy depends on the volume as well as on the temperature.

Even if there is a force of attraction between molecules of the gas and a contribution to the internal energy due to the associated potential energy, this contribution will be small if the pressure of the gas is small because then the average separation of any two molecules is very great. In the limit of very low pressures, this potential energy will be negligible. Since the equation of state for an ideal gas $PV = nRT$ holds for real gases only at very low pressures, we place the additional restriction on the definition of an ideal gas that the internal energy be a function of temperature only[1]:

$$PV = nRT \qquad U = U(T) \text{ only} \qquad\qquad 18\text{-}16$$

Ideal gas defined

Whether or not the internal energy of a *real* gas depends on the volume or pressure was first investigated by Joule, using a process called *free adiabatic expansion of a gas*. The system shown in Figure 18-10 consists of a gas in a compartment of volume V_1 and an adjacent compartment which is evacuated. The compartments are connected by a stopcock initially closed. The whole system is surrounded by a rigid adiabatic wall so that no heat can be transferred into or out of the system and no work can be done by the system or on the system. The temperature of the gas in the left compartment is measured, and then the stopcock is opened. The gas rushes into the evacuated chamber.

[1] This restriction is not as arbitrary as it appears. It can be shown that if the equation of state is $PV = nRT$, the first and second laws of thermodynamics (Chapter 19) imply that U is a function only of temperature.

Figure 18-10
Free adiabatic expansion of a gas. When the stopcock is opened, the gas expands rapidly into the evacuated chamber. Since no work is done and the whole system is adiabatically insulated, the final internal energy of the gas equals its initial internal energy. If the final temperature equals the initial temperature, the internal energy does not depend on the volume but only on the temperature.

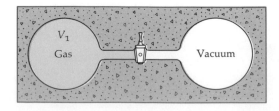

This expansion is not a quasi-static process. It is adiabatic because no heat can flow in or out during the expansion. Eventually the gas reaches an equilibrium state at the new volume V_2, which is the total volume of both compartments. The internal energy of the final state is the same as that of the initial state because no work was done by the gas and no heat was added. Joule's experiment consisted of trying to determine whether the final temperature is the same as the initial temperature. If the temperature does not change, the internal energy must depend only on the temperature and not on the volume or pressure.[1]

The Joule experiment is very difficult. Joule himself was unable to measure any temperature difference between the initial and final states. Later experiments of this type and others show that if the initial pressure is very high, the temperature after a free expansion is slightly lower than before the expansion. This indicates that for a real gas at high pressure, there is some attraction between the molecules. When the average separation of the molecules is increased (as in a free expansion), the potential energy increases (becomes less negative), and since the total energy does not change, the kinetic energy of translation, indicated by the temperature, must decrease. We can think of this effect by a simple model. Consider two balls which attract each other initially moving away from each other, so that their separation is increasing. When they are farther apart, their velocity has decreased. Their potential energy has increased, but because of energy conservation their kinetic energy has decreased.

Questions

8. When a bottle of compressed gas is allowed to expand into the atmosphere, the temperature of the expanding gas drops sharply. Why isn't this regarded as evidence that the internal energy of a gas depends on more than T?

9. In Joule's adiabatic free-expansion experiment if the gas temperature increased on expansion, what would you conclude about the intermolecular forces in the gas? Explain.

10. At a given temperature, would you expect a mole of the diatomic gas H_2 to have the same, more, or less internal energy than a mole of the monatomic gas He? Explain.

18-6 Heat Capacities of an Ideal Gas

When a gas is heated at constant volume, no work is done, so that all the heat goes into increasing the internal energy (and therefore the temperature). On the other hand, when the gas is heated at constant pressure, it expands and does work, so that only part of the heat goes into an increase in the internal energy and the temperature. It therefore takes more heat at constant pressure than at constant volume for a given increase in internal energy or for a given increase in tempera-

---

[1] We can see this as follows. A state of the system is determined by any two variables P, V, or T. We can thus think of the internal energy U as a function of any two of these variables. Consider U to be a function of the temperature T and the volume V. Suppose that regardless of the size of the second compartment, the temperature does not change in the free expansion. Since U does not change either, this means that we can change V in any way and not change U or T. Then U cannot be a function of V. By similar reasoning, we can conclude that U does not depend on P either.

ture. Thus C_p is greater than C_v. In this section we shall use the equation of state for an ideal gas to develop a simple and useful expression relating these heat capacities.

Let $(dQ)_v$ be a small amount of heat put into a gas at constant volume. Since no work is done, the first law gives $dU = (dQ)_v$. The heat capacity at constant volume is then

$$C_v = \frac{(dQ)_v}{dT} = \frac{dU}{dT} \qquad\qquad 18\text{-}17$$

If $(dQ)_p$ is a small amount of heat put into a gas at constant pressure, we have for the heat capacity at constant pressure,

$$C_p = \frac{(dQ)_p}{dT} \qquad\qquad 18\text{-}18$$

According to the first law, $(dQ)_p$ is related to the increase in internal energy by $(dQ)_p = dU + (dW)_p$, where $(dW)_p = P\, dV$ is the work done at constant pressure. Then

$$C_p = \frac{dU}{dT} + \frac{(dW)_p}{dT} = C_v + \frac{(dW)_p}{dT} \qquad\qquad 18\text{-}19$$

Equation 18-19 holds for any system—ideal gas, real gas, liquid, or solid. The work done at constant pressure $(dW)_p = P\, dV$ can be related to the change in temperature using the equation of state. For an ideal gas we have

$$PV = nRT$$

Then taking the differential of both sides gives

$$P\, dV + V\, dP = nR\, dT \qquad\qquad 18\text{-}20$$

At constant pressure $dP = 0$. Then the work is

$$(dW)_p = P\, dV = nR\, dT$$

Substituting this into Equation 18-19 gives

$$C_p = C_v + \frac{nR\, dT}{dT} = C_v + nR \qquad\qquad 18\text{-}21$$

The heat capacity of an ideal gas at constant pressure is greater than that at constant volume by the amount nR.

In general, for all materials that expand when heated, the heat capacity at constant pressure is greater than that at constant volume. This is because at constant volume no work is done, so that all the added heat goes into increasing the internal energy (and therefore the temperature) whereas at constant pressure some of the heat energy is used to perform work. For solids, there is very little expansion, and so little work is done when the solid is heated at constant pressure. Then, according to Equation 18-19, C_p and C_v are approximately equal.

18-7 The Equipartition of Energy

If the internal energy of a gas consists of translational kinetic energy only, the internal energy of n moles of the gas (from Equation 18-15) is

$$U = \tfrac{3}{2}nRT$$

The heat capacities are then

$$C_v = \frac{dU}{dT} = \tfrac{3}{2}nR \qquad\qquad 18\text{-}22$$

and

$$C_p = C_v + nR = \tfrac{5}{2}nR \qquad\qquad 18\text{-}23$$

Table 18-2 lists the measured heat capacities per mole for several gases. The ideal-gas prediction $C_p - C_v = nR$ holds quite well for all gases ($R = 1.99$ cal/mole-K), and the prediction $C_v = \tfrac{3}{2}nR$ holds quite well for most monatomic gases but not for other types.

In Section 18-5 we mentioned that we expect other kinds of kinetic energy, e.g., rotational energy, also to depend on the temperature. We discussed qualitatively how we might expect some kind of equilibrium between various kinds of energy, e.g., rotational and translational kinetic energy of a dumbbell molecule, to be established as a result of molecular collisions. Let us explore this equilibrium further.

Consider a smooth hard-sphere model of a monatomic gas. Suppose that at some time all the molecules are moving in the x direction only; that is, v_y and v_z are zero, and the kinetic energy of each is $\tfrac{1}{2}mv_x^2$. Such an unusual state is not an equilibrium state. Because of off-center collisions the molecules will acquire components v_y and v_z. On the average, such a collision at first will tend to increase the average value of v_y^2 and v_z^2 until these quantities have the same average values as v_x^2. When equilibrium is established, the average values of the energies associated with each kind of translation, $\tfrac{1}{2}mv_x^2$, $\tfrac{1}{2}mv_y^2$, and $\tfrac{1}{2}mv_z^2$, will be equal to each other. Since we have seen that the average translational energy per molecule is $\tfrac{3}{2}kT$, where $k = R/N_A$ is Boltzmann's constant, each of these terms must have an average value of $\tfrac{1}{2}kT$ in equilibrium. (The assumption of smooth-sphere molecules rules out any exchange of rotational energy since the forces in a collision will be along the line of centers of the molecules and will not change the angular momentum of the molecules.)

The example we have just considered is somewhat artificial because it is very unlikely that all the molecules would be moving only in the x

---

Table 18-2

Molar heat capacities of various gases, cal/mole-K

	C_v'	C_p'	$C_p' - C_v'$	$\gamma = C_p'/C_v'$
Monatomic gases:				
Helium	3.00	4.98	1.98	1.66
Neon				1.64
Argon	3.00	5.00	2.00	1.67
Krypton				1.68
Xenon				1.66
Diatomic gases:				
N_2	4.96	6.95	1.99	1.40
O_2	4.96	6.95	1.99	1.40
CO	4.93	6.95	2.02	1.41
HCl	5.02	7.08	2.06	1.41
Polyatomic gases:				
CO_2	6.74	8.75	2.01	1.30
H_2O	6.46	8.46	2.00	1.31
CH_4	6.48	8.49	2.01	1.31

direction at one time. However, there are similar situations in which the average value of the kinetic energy associated with motion in the x direction is momentarily greater than that associated with motion in the y or z direction. Consider a gas which is being compressed by a piston moving with a very slow speed in the negative x direction, so that the compression is approximately quasi-static. Suppose further that the gas is insulated so that the compression is adiabatic. From the molecular point of view, the energy of the gas increases because of the elastic collisions of the molecules with the moving piston. If v_x is the x component of a molecule moving toward the piston and u is the speed of the piston, the x component of velocity of the molecule after col-liding with the piston will be $v'_x = -|v_x + 2u|$. (The molecule moves with speed $v_x + u$ relative to the piston; the motion is toward the piston before the collision and away from the piston after the colli-sion.) The energy associated with motion in the x direction after the collision is $\frac{1}{2}mv'^2_x = \frac{1}{2}m(v_x + 2u)^2 \approx \frac{1}{2}mv_x^2 + 2muv_x$ (neglecting u^2 compared with v_x^2). The moving piston directly increases the energy associated with motion in the x direction but has no effect on that as-sociated with motion in the y or z directions (assuming the piston to be smooth). But immediately after colliding with the piston, molecules collide with other nearby molecules and a new equilibrium is es-tablished, $\frac{1}{2}mv_x^2$, $\frac{1}{2}mv_y^2$, and $\frac{1}{2}mv_z^2$ each having the same average value of $\frac{1}{2}kT$, which is increased by the work done on the gas by the moving piston. (See Problem 13 for the completion of this analysis in showing that the increase in molecular energy is indeed equal to the work done by the moving piston.)

Sharing the energy equally between the three terms in the transla-tional energy is a special case of the *equipartition theorem,* a result which can be derived from more advanced statistical mechanics. Each coordinate, velocity component, angular-velocity component, etc., which appears in the expression for the energy of a molecule is called a *degree of freedom.* The equipartition theorem states that

In equilibrium, there is associated with each degree of freedom an average energy of $\frac{1}{2}kT$ per molecule.   *Equipartition theorem*

For example, take a rigid-dumbbell model of a molecule (Figure 18-11) which can translate in the x, y, and z directions and can rotate about axes x' and y' through the center of mass perpendicular to the z' axis along the line joining the two masses.[1] The kinetic energy for this dumbbell-model molecule is then

$$E_k = \tfrac{1}{2}mv_x^2 + \tfrac{1}{2}mv_y^2 + \tfrac{1}{2}mv_z^2 + \tfrac{1}{2}I_{x'}\omega_{x'}^2 + \tfrac{1}{2}I_{y'}\omega_{y'}^2 \qquad 18\text{-}24$$

The equipartition theorem states that the average energy associated with each of the five degrees of freedom is $\frac{1}{2}kT$, so that the total energy for N molecules is

$$U = N(5)(\tfrac{1}{2}kT) = \tfrac{5}{2}NkT \qquad 18\text{-}25$$

Writing $N = nN_A$, where n is the number of moles, N_A is Avogadro's

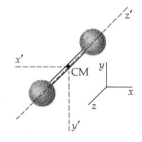

Figure 18-11
Rigid-dumbbell model of a diatomic gas. The motion can be described as translation of the center of mass along the x, y, or z axis plus rotation about the x', y', or z' axis. If the spheres are smooth or points, rotation about the z' axis can be neglected because this energy cannot be changed by collisions.

---

[1] We rule out rotation about the z' axis of the dumbbell by assuming either that the atoms are points and the moment of inertia about this axis is therefore zero or that the atoms are hard smooth spheres, in which case rotation about this axis cannot be changed by collisions and therefore does not participate in the absorption of energy. Either of these assumptions also rules out the possibility of rotation of a monatomic molecule.

number, and $N_A k = R$, we have

$$U = \tfrac{5}{2} nRT$$

The heat capacity at constant volume is then

$$C_v = \frac{dU}{dT} = \tfrac{5}{2} nR \qquad\qquad 18\text{-}26$$

and

$$C_p = C_v + nR = \tfrac{7}{2} nR \qquad\qquad 18\text{-}27$$

From the observation that C_v/n for nitrogen and oxygen is approximately $\tfrac{5}{2}R$, Clausius speculated (about 1880) that these gases must be diatomic gases which can rotate about two axes as well as translate.

If a diatomic molecule can vibrate as well as rotate and translate, there are two more degrees of freedom, corresponding to the kinetic and potential energy of vibration along the line joining the atoms. The equipartition theorem thus predicts that the internal energy per mole will be $\tfrac{7}{2}RT$. According to Table 18-2, diatomic gases apparently do not vibrate. The calculation of the number of degrees of freedom for a molecule with more than two atoms is more complicated if all possible vibrations are allowed.

Although the equipartition theorem is quite successful in relating the molar heat capacities of a gas to its atomic structure, it fails in several ways. There is no hint from the equipartition theorem why diatomic molecules do not vibrate or why the heat capacities vary with temperature. According to the equipartition theorem, there should be no temperature variation. The equipartition theorem fails because classical mechanics itself breaks down when applied to atomic and molecular systems and must be replaced by quantum mechanics. In the quantum-mechanical theory of motion, the possible energies of an atom or molecule are quantized. In particular there is a gap between the lowest possible energy and the next lowest. For the vibrational energies of a diatomic molecule, this energy gap is of the order of several tenths of an electron volt, which is much higher than $kT \approx 0.026$ eV (the value at ordinary temperatures). A collision between molecules having energies of the order of kT cannot increase their energy from the lowest possible energy to the next allowed energy, and consequently vibrational energy does not play a role in the internal energy at low temperatures.

The rotational energy of a molecule is also quantized, spacing between possible energies depending inversely on the moment of inertia. For rotation of monatomic molecules and rotation of diatomic molecules about the line joining the atoms the moment of inertia is so small that the energy spacing is large compared with kT at ordinary temperatures. For these cases, rotational energy plays no role in the internal energy at ordinary temperatures. On the other hand, the moment of inertia for rotation of diatomic molecules about axes perpendicular to the line joining the atoms is so large that the spacing of possible energies is small compared with kT at ordinary temperatures. In this case, energy quantization is not important, and the predictions of the classical equipartition theory are correct.

Using the equipartition theorem and a simple model of a solid, we can understand the Dulong-Petit law for the heat capacities of a solid. (We need not distinguish between C_v and C_p here because for a solid they are almost equal.) Let us assume that a solid consists of a regular

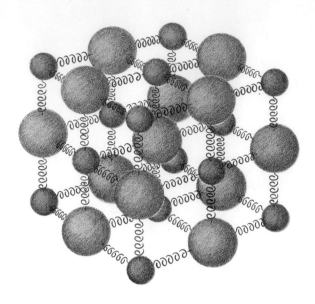

Figure 18-12
Model of a solid consisting of atoms connected to each other by springs. The internal energy of the solid then consists of kinetic and potential vibrational energy.

array of molecules each having a fixed equilibrium position and being connected by springs to each of its neighbors, as in Figure 18-12. Each molecule can oscillate in the x, y, and z directions. Its total energy is thus

$$E_k = \tfrac{1}{2}mv_x{}^2 + \tfrac{1}{2}mv_y{}^2 + \tfrac{1}{2}mv_z{}^2 + \tfrac{1}{2}Kx^2 + \tfrac{1}{2}Ky^2 + \tfrac{1}{2}Kz^2$$

where K is the effective force constant of the springs. The molecule thus has six degrees of freedom and an average energy of $6 \times \tfrac{1}{2}kT$ according to the equipartition theorem. Thus the total energy of 1 mole of a solid is

$$U = 3RT$$

and the molar heat capacity is $C_v' = dU/dT = 3R$, which is the Dulong-Petit law. This result holds for many solids but not for all at room temperature. Again, the reason for disagreement between this prediction and experiment is the breakdown of classical physics in the atomic realm.

Example 18-5 One mole of oxygen gas is heated from room temperature of 20°C and pressure of 1 atm to a temperature of 100°C. How much heat must be supplied if the volume is kept constant during the heating? How much heat must be supplied if the pressure is kept constant? How much work is done if the pressure is constant? (Assume that oxygen is an ideal gas.)

The heat capacity at constant volume of oxygen is

$$C_v = \tfrac{5}{2}nR = 5 \text{ cal/K}$$

using $n = 1$ mole and $R = 2$ cal/mole-K. The heat added to raise the temperature from 293 to 373 K is then

$$Q_1 = C_v\,\Delta T = (5 \text{ cal/K})(80 \text{ K}) = 400 \text{ cal}$$

Since no work is done when the volume is constant, this is also the increase in the internal energy of the gas $\Delta U = 400$ cal. Since the internal energy depends only on the temperature, this is the increase in internal energy when the gas temperature changes from 293 to 373 K for any process.

If we keep the pressure constant, the heat that must be added is

$$Q_2 = C_p\,\Delta T = \tfrac{7}{2}nR\,\Delta T = (7 \text{ cal/K})(80 \text{ K}) = 560 \text{ cal}$$

again using $R = 2$ cal/mole-K. This is greater because although the internal energy change is again 400 cal, work is done when the gas expands. The work done at constant pressure is

$$W = P \, dV = P(V_2 - V_1)$$

We can find the initial and final volumes from the result of Example 17-4 that the volume at 1 atm and 273 K is 22.4 ℓ. Since $PV = nRT$, the volume at constant pressure is just proportional to the temperature. The volumes V_1 and V_2 are thus

$$V_1 = (22.4 \, \ell)(\tfrac{293}{273}) = 24.0 \, \ell \qquad V_2 = (22.4 \, \ell)(\tfrac{373}{273}) = 30.6 \, \ell$$

The work done is thus

$$W = P(V_2 - V_1) = (1 \text{ atm})(30.6 - 24.0 \, \ell) = 6.6 \, \ell\text{-atm}$$

We can convert this energy unit to calories or joules using the fact that 1 atm $= 1.013 \times 10^5$ N/m² and 1 $\ell = 10^3$ cm³ $= 10^{-3}$ m³. Then

$$1 \, \ell\text{-atm} \times \frac{1.013 \times 10^5 \text{ N/m}^2}{1 \text{ atm}} \frac{10^{-3} \text{ m}^3}{1 \, \ell} = 101.3 \text{ N-m} = 101.3 \text{ J}$$

or

$$1 \, \ell\text{-atm} = 101.3 \text{ J} = 24.2 \text{ cal}$$

Thus

$$W = 6.6 \, \ell\text{-atm} \times \frac{24.2 \text{ cal}}{1 \, \ell\text{-atm}} = 160 \text{ cal}$$

The heat added at constant pressure, 560 cal, equals the change in internal energy, 400 cal, plus the work done, 160 cal. These two processes are illustrated on a PV diagram in Figure 18-13.

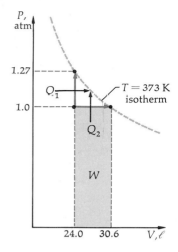

Figure 18-13
Example 18-5. Two methods of raising the temperature of a gas from 293 to 373 K. The dashed gray line is the isotherm at 373 K. The internal energy of the two different final states is the same because they have the same temperature. The heat added at constant pressure Q_2 is therefore greater than that added at constant volume Q_1 by the amount of work done W.

Question

11. Between 0 and 4°C water contracts as its temperature increases. Which is larger for water in this temperature range, C_p or C_v?

18-8 Quasi-static Adiabatic Expansion of an Ideal Gas

In a quasi-static adiabatic expansion of an ideal gas, the gas expands slowly against a piston, doing work on it. The gas is insulated from its surroundings so that no heat enters or leaves the gas. The work done by the gas equals the decrease in internal energy of the gas, and the temperature of the gas decreases. The curve representing this process on a PV diagram is shown in Figure 18-14. This curve is steeper than for an isothermal expansion; i.e., when the volume increases by dV, the pressure in an adiabatic expansion decreases by more than it does in an isothermal expansion because the temperature decreases.

The work done by a gas undergoing a quasi-static adiabatic expansion equals the area under the curve shown in Figure 18-14. The calculation of this work is important in determining efficiencies of heat engines (Chapter 19), and for this purpose we need to know the relation between P and V for this curve. This relation can be found by applying the first law to the adiabatic process and using the equation

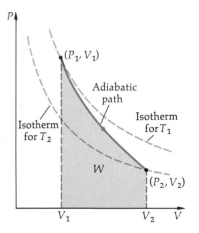

Figure 18-14
Quasi-static adiabatic expansion of an ideal gas. The curve on the PV diagram connecting the initial and final states (called an adiabatic path) is steeper than the isotherms because the temperature drops during the adiabatic expansion. This curve is given by $PV^\gamma = $ constant.

of state for an ideal gas. From the first law we have

$$dQ = dU + dW = dU + P\ dV = 0 \qquad\qquad \text{18-28}$$

According to Equation 18-17, the change in internal energy of the gas is related to the change in temperature by

$$dU = C_v\ dT \qquad\qquad \text{18-29}$$

(Although this result was obtained considering a constant-volume process, it relates the state variables U and T and is therefore valid independent of the process.) Then

$$C_v\ dT + P\ dV = 0 \qquad\qquad \text{18-30}$$

From the equation of state $PV = nRT$ we have (Equation 18-20)

$$P\ dV + V\ dP = nR\ dT$$

Using $dT = - (P/C_v)\ dV$ from Equation 18-30, we have

$$P\ dV + V\ dP = nR \left(-\frac{P\ dV}{C_v} \right)$$

or

$$C_v P\ dV + C_v V\ dP = -nRP\ dV$$

Rearranging gives

$$(C_v + nR)P\ dV + C_v V\ dP = 0$$

But $C_v + nR$ is just C_p. Using this fact and dividing each term by $C_v PV$, we obtain

$$\frac{C_p}{C_v} \frac{dV}{V} + \frac{dP}{P} = 0$$

or

$$\gamma \frac{dV}{V} + \frac{dP}{P} = 0 \qquad\qquad \text{18-31}$$

where

$$\gamma = \frac{C_p}{C_v} \qquad\qquad \text{18-32}$$

is the ratio of heat capacities. Integrating each term gives

$$\gamma \ln V + \ln P = \text{constant} \qquad \ln PV^\gamma = \text{constant}$$

or

$$PV^\gamma = \text{constant} \qquad\qquad \text{18-33}$$

This is the equation relating P and V in a quasi-static adiabatic expansion. This equation also applies to a quasi-static adiabatic compression in which the piston does work on the gas.

Example 18-6 A quantity of air expands adiabatically and quasi-statically from an initial pressure of 2 atm and volume 2 ℓ at room temperature (20°C) to twice its original volume. What are the final pressure and temperature? For air, $\gamma = 1.4$.

According to Equation 18-33, the quantity PV^γ remains unchanged during a quasi-static adiabatic expansion. Thus, if P_1 and V_1 are the initial pressure and volume and P_2 and V_2 are the final pressure and volume, we have

$$P_2 V_2{}^{\gamma} = P_1 V_1{}^{\gamma} \qquad \text{or} \qquad P_2 = P_1 \left(\frac{V_1}{V_2}\right)^{\gamma}$$

For the given values we have

$$P_2 = (2 \text{ atm}) \left(\frac{2\ \ell}{4\ \ell}\right)^{1.4} = 0.758 \text{ atm}$$

Since we always have $PV = nRT$ or $PV/T = nR$, we must have for this fixed number of moles of gas

$$\frac{P_2 V_2}{T_2} = \frac{P_1 V_1}{T_1}$$

or

$$T_2 = T_1 \frac{P_2 V_2}{P_1 V_1} = (293 \text{ K}) \frac{(0.758 \text{ atm})(4\ \ell)}{(2 \text{ atm})(2\ \ell)} = 222 \text{ K}$$

The corresponding Celsius temperature is

$$t_C = 222 \text{ K} - 273 \text{ K} = -51°\text{C}$$

Review

A. Define, explain, or otherwise identify:

Calorie, 437
Heat capacity, 437
Specific heat, 437
Molar heat capacity, 437
Phase, 440
Latent heat of fusion, 440
Latent heat of vaporization, 440
Adiabatic work, 442
Internal energy, 442
Heat, 443

Quasi-static process, 444
PV diagram, 444
Isothermal process, 446
Ideal gas, 448
Free adiabatic expansion
  of a gas, 448
Equipartition theorem, 452
Quasi-static adiabatic expansion
  of a gas, 455

B. True or false:

1. The heat capacity of a body is the amount of heat it can store at a given temperature.

2. A quasi-static process is one in which no work is done.

3. When a system goes from state 1 to state 2, the heat added is the same for all processes.

4. When a system goes from state 1 to state 2, the work done by the system is the same for all processes.

5. When a system goes from state 1 to state 2, the change in internal energy of the system is the same for all processes.

6. The internal energy of an ideal gas depends only on the temperature.

7. C_p is greater than C_v for any material which expands when heated.

8. In a free adiabatic expansion of an ideal gas the final state is the same as the initial state.

9. In an isothermal expansion of an ideal gas the work done by the gas equals the heat absorbed.

10. Equipartition of energy says that in equilibrium all molecules of a gas have the same energy.

Exercises

Section 18-1, Heat Capacity and Specific Heat

1. How many calories of heat energy are required to raise the temperature of 20 kg of water from 10 to 20°C?

2. A man typically consumes food with a total energy value of 3000 kcal each day. (a) How many joules of energy is this? (b) Calculate his power output in watts assuming this energy is dissipated at a steady rate during 24 h.

3. A common unit of heat energy is the British thermal unit (Btu), defined to be the amount of energy needed to raise the temperature of one pound of water one Fahrenheit degree. (a) Find the number of calories in 1 Btu. (b) Find the number of joules in 1 Btu. (c) Find the number of Btu needed to raise the temperature of 1 gal of water from 32 to 212°F (1 gal contains 8 pints; 1 pint of water weighs 1 lb).

4. A 200-gm piece of lead is heated to 90°C and dropped into 500 gm of water initially at 20°C. Neglecting the heat capacity of the container, find the final temperature of the lead and water.

5. The specific heat of a certain metal is determined by measuring the temperature change which occurs when a heated piece of the metal is placed in an insulated container made of the same material and containing water. The piece of metal has mass 100 gm and an initial temperature of 100°C. The container has a mass of 200 gm and contains 500 gm of water at an initial temperature of 17.3°C. The final temperature is 22.7°C. What is the specific heat of the metal?

Section 18-2, Latent Heat

6. How much heat must be removed when 100 gm of steam at 150°C is cooled and frozen into 100 gm of ice at 0°C? (Take the specific heat of steam to be 0.48 cal/gm-°C.)

7. A 200-gm piece of ice at 0°C is put into 500 gm of water at 20°C. The system is in a container of negligible heat capacity and insulated from its surroundings. (a) What is the final equilibrium temperature of the system? (b) How much of the ice melts?

8. A 50-gm piece of ice at 0°C is placed in 500 gm of water at 20°C, as in Exercise 7. What is the final temperature of the system assuming no heat loss to the surroundings?

9. Liquid nitrogen boils at −196°C and has a heat of vaporization of 48 cal/gm. A 50-gm piece of aluminum at 20°C is cooled to −196°C by placing it in a large container of liquid nitrogen. How much nitrogen is vaporized? (Assume that the specific heat of aluminum is constant and equal to 0.22 cal/gm-°C.)

Section 18-3, The First Law of Thermodynamics

10. (a) How much work must be done on 1 kg of water to raise its temperature from 20 to 25°C assuming the water is adiabatically shielded from its surroundings? (b) In an actual experiment, the work done is 2.25×10^4 J. How much heat escapes to the surroundings? Express your answer both in joules and in calories.

11. (a) Express the heat capacity of water in joules per kilogram-kelvin. (b) Express the latent heat of fusion of water in joules per kilogram.

12. A 1-kg piece of ice is dropped from a height of 40 m into a box of wood shavings. (a) Assuming that all the mechanical energy lost goes into internal energy of the ice, how much ice is melted? (b) Does the fraction of the mass of ice that is melted in this experiment depend on the original mass of ice?

13. A box of lead shot is thrown vertically into the air to a height of 4 m and allowed to fall to the floor in a lecture demonstration. The original temperature

of the lead is 20°C. Five such throws are made, and the temperature is then measured. What result do you expect?

14. A lead bullet with speed 200 m/sec is stopped in a block of wood. Assuming that all the energy goes into heating the bullet, find the final temperature of the bullet if the initial temperature is 20°C.

15. At Niagara Falls, water drops a distance of 162 ft. (*a*) If all the change in potential energy goes into internal energy of the water, compute the increase in temperature. (*b*) Do the same for Yosemite Falls, where the water drops 2.42×10^3 ft.

16. A body of water is heated at constant pressure from 20 to 40°C. Explain carefully why it is incorrect to say that the water at 40°C contains more heat than it did at 20°C. Is it correct to say that the water has more internal energy at 40°C than at 20°C? Why, or why not?

Section 18-4, Work and the *PV* Diagram for a Gas

17. Use the values of the gas constant R to find conversion factors for energy in liter-atmospheres to joules and to calories.

In Exercises 18 to 21, 1 mole of a gas is originally in state $P_1 = 3$ atm, $V_1 = 1 \ell$, and $U_1 = 456$ J. Its final state is $P_2 = 2$ atm, $V_2 = 3 \ell$, and $U_2 = 912$ J. All processes are quasi-static.

18. The gas is allowed to expand at constant pressure to a volume of 3ℓ. It is then cooled at constant volume until its pressure is 2 atm. (*a*) Indicate this process on a *PV* diagram and calculate the work done by the gas. (*b*) Find the heat added during the entire process.

19. The gas is first cooled at constant volume until its pressure is 2 atm. It is then allowed to expand at constant pressure until its volume is 3ℓ. (*a*) Indicate this process on a *PV* diagram and calculate the work done by the gas. (*b*) Find the heat added during this process.

20. The gas is allowed to expand isothermally until its volume is 3ℓ and its pressure is 1 atm. It is then heated at constant volume until its pressure is 2 atm. (*a*) Indicate this process on a *PV* diagram and calculate the work done by the gas. (*b*) Find the heat added during this process.

21. The gas expands and heat is added such that the gas follows a straight-line path on the *PV* diagram from its initial state to its final state. (*a*) Indicate this process on a *PV* diagram and calculate the work done by the gas. (*b*) Find the heat added for this process.

Section 18-5, Internal Energy of a Gas

22. An ideal gas is originally at pressure $P_1 = 2$ atm, volume $V_1 = 1 \ell$, and temperature $T_1 = 300$ K. It is allowed to expand at constant pressure until its volume is 4ℓ. (*a*) How much work is done during this expansion? (*b*) What is the temperature of the gas after this expansion? (*c*) The gas is now cooled at constant volume until its pressure is 0.5 atm. What is its temperature now? (*d*) What is the net amount of heat added to the gas during the complete process of expansion and cooling?

23. A certain ionized gas is composed of ions which repel each other. The gas undergoes a free adiabatic expansion. How does the temperature of the gas change? Why?

Section 18-6, Heat Capacities of an Ideal Gas, and Section 18-7, The Equipartition of Energy

24. The heat capacity at constant volume of a certain monatomic gas is $C_v = 11.9$ cal/K. (*a*) Find the number of moles of the gas. (*b*) What is the in-

ternal energy of this gas at $T = 300$ K? (c) What is the heat capacity at constant pressure?

25. For a certain gas, the heat capacity at constant pressure is greater than that at constant volume by 5.97 cal. (a) How many moles of the gas are there? (b) If the gas is monatomic, what are C_v and C_p? (c) If the gas molecules are diatomic molecules which rotate but do not vibrate, what are C_v and C_p?

26. For air, $\gamma = C_p/C_v = 1.40$. Which is the best model for an air molecule: (a) an elastic sphere, (b) a rigid dumbbell, or (c) a nonrigid dumbbell which vibrates and rotates? Explain.

27. Find the molar heat capacities C'_v and C'_p and the ratio $\gamma = C_p/C_v$ for ideal gases which are: (a) elastic spherical atoms, (b) rigid dumbbell-shaped molecules, (c) nonrigid dumbbell-shaped molecules which vibrate and rotate.

28. For each of the gases with the following $\gamma = C_p/C_v$ give a possible physical model of the molecule: (a) $\gamma = 1.29$, (b) $\gamma = 1.40$, (c) $\gamma = 1.67$.

29. One mole of an ideal monatomic gas is heated at constant volume from $T = 300$ K to $T = 600$ K: (a) find the increase in internal energy, the work done W, and the heat added Q. (b) Find these same quantities if this same gas is heated from 300 to 600 K at constant pressure. (c) For the constant-pressure heating in part (b), calculate the work done by the gas directly from $W = \int P\, dV = P\, \Delta V$ for $P = 1$ atm. Hint: Use the equation of state to find the initial and final volume of the gas.

30. Repeat each part of Exercise 29 for a gas containing diatomic molecules which rotate but do not vibrate.

31. The specific heat of helium at constant pressure is measured to be 1.248 cal/gm-K. Use the fact that helium is a monatomic gas to calculate the molecular weight of helium from this result.

32. (a) Show that when an ideal gas undergoes a temperature change, its internal-energy change is $U = C_v\, \Delta T$ no matter what the process is. (b) Show explicitly that this expression holds for the expansion of an ideal gas at constant pressure. Do this by calculating the work done and showing that it can be written $W = nR\, \Delta T$, where ΔT is the temperature change, and subtracting this from the heat added, $Q = C_p\, \Delta T$.

Section 18-8, Quasi-static Adiabatic Expansion of an Ideal Gas

33. Show that during a quasi-static adiabatic expansion of an ideal gas, $TV^{\gamma-1} = $ constant, where T is the temperature.

34. Show that during a quasi-static adiabatic expansion of an ideal gas, $T^\gamma/P^{\gamma-1} = $ constant.

35. One mole of an ideal gas ($\gamma = \frac{5}{3}$) expands adiabatically from a pressure of 10 atm and temperature 0°C to a final state of pressure 2 atm. Find (a) the initial and final volume and (b) the final temperature.

36. An ideal gas at room temperature (20°C) is compressed quasi-statically and adiabatically to half its original volume. (a) If the gas is monatomic, find its final temperature. (b) If the gas is diatomic (but the molecules do not vibrate), find its final temperature.

Problems

1. A piece of ice is dropped from a height H. Find the minimum value of H such that the ice melts when it makes an inelastic collision with the ground. Is it reasonable to neglect the variation in the acceleration of gravity in doing this problem? Comment on the reasonableness of neglecting air resistance. What effect would air resistance have on your answer?

2. An insulated cylinder with a movable piston (to maintain constant pressure) initially contains 100 gm of ice at $-10°C$. Heat is supplied to the cylinder at a constant rate by a 100-W heater. Make a graph showing the temperature of the cylinder contents as a function of time starting at time $t = 0$, when the temperature is $-10°C$, and ending when the temperature is $110°C$. (Use $c = 0.50$ cal/gm-°C for the average specific heat of ice from -10 to $0°C$ and for steam from 100 to 110°C.)

3. Compare the specific heat of ice with that calculated from the Dulong-Petit law. How might you account for the difference in these values on the basis of a mechanical model of ice? Explain on the basis of the equipartition theorem how the specific heat of steam (at constant pressure) can be about 0.50 cal/gm-°C.

4. One mole of water at 100°C is vaporized at a constant pressure of 1 atm. (a) Find the amount of heat added. (b) Find the change in volume and the work done by the system in expanding against atmospheric pressure (treat the steam as an ideal gas). (c) Find the change in internal energy of the water. (d) If 1 mole of ice at 0°C and 1 atm is completely liquefied, is the internal-energy change greater or less than $18L_f$? Explain.

5. A 200-gm aluminum calorimeter can contains 500 gm of water at 20°C; 300 gm of aluminum shot is heated to 100°C and placed in the calorimeter. Using the value of the specific heat given in Table 18-1, find the final temperature of the system (assuming no heat losses to the surroundings). If this calorimeter is to be used to make an accurate measurement of the specific heat of aluminum, what should the initial temperature of the water and container be to minimize the error due to heat transfer to and from the surroundings if the room temperature is 20°C?

6. (a) A 200-gm aluminum calorimeter can contains 500 gm of water at 20°C. A 100-gm piece of ice cooled to $-20°C$ is placed in the can. Find the final temperature of the system assuming no heat losses (assume the specific heat of ice is 0.50 cal/gm-°C). (b) A second 200-gm piece of ice at $-20°C$ is added. How much ice remains in the system after it reaches equilibrium? (c) Would your answer to part (b) be different if both pieces were added at the same time?

7. An ideal gas at initial pressure P_1 and volume V_1 expands quasi-statically and adiabatically to volume V_2 and pressure P_2. Show by direct integration of $P\,dV$, using $PV^\gamma = $ constant, that the work done is

$$W = \frac{P_1V_1 - P_2V_2}{\gamma - 1}$$

Show that this is consistent with the general result that the internal-energy change is $\Delta U = C_v\,\Delta T$.

8. Using the ideal-gas law and $PV^\gamma = $ constant for an adiabatic expansion, show that the slope of the adiabatic-expansion curve at any point on the PV diagram is steeper (greater magnitude) than the slope of the isothermal-expansion curve.

9. A solid material has a density ρ, coefficient of linear expansion α, and mass m. (a) Show that at pressure P the heat capacities C_p and C_v are related by $C_p - C_v = 3\alpha mP/\rho$. (b) Find the difference (in calories per mole-kelvin) in the molar heat capacities of aluminum at atmospheric pressure using $\alpha = 23 \times 10^{-6}$ per kelvin, $M = 27$ gm/mole, and $\rho = 2.7$ gm/cm³.

10. One mole of an ideal gas is in an initial state of $P = 2$ atm and $V = 10\ \ell$ indicated by point a on the PV diagram of Figure 18-15. It expands at constant pressure to point b, where its volume is 30 ℓ, and is then cooled at constant volume until its pressure is 1 atm at point c. It is then compressed at constant pressure to its original volume at point d and finally heated at constant volume until it is back at its original state. (a) Find the temperature of each state a, b, c,

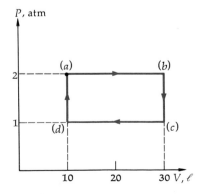

Figure 18-15
PV diagram for Problem 10.

and d. (b) Assuming that the gas is monatomic, find the heat added along each path of the cycle. (c) Calculate the work done along each path. (d) Find the internal energy of each state a, b, c, and d. (e) What is the net work done by the gas for the complete cycle? How much heat is added during the complete cycle?

11. One mole of N_2 gas is maintained at room temperature (20°C) and at a pressure of 5 atm. It is allowed to expand adiabatically and quasi-statically until its pressure equals the room pressure of 1 atm. It is then heated at constant pressure until its temperature is again 20°C. During this heating, the gas expands. After it reaches room temperature, it is heated at constant volume until its pressure is 5 atm. It is then compressed at constant pressure until it is back in its original state. (a) Construct an accurate PV diagram showing each process in the cycle. (b) *From your graph* determine the work done by the gas during the complete cycle. (c) How much heat was added or subtracted from the gas during the complete cycle? (d) Check your graphical determination of the work in part (b) by calculating the work done during each process of the cycle.

12. An ideal gas of n moles is at pressure P_1, volume V_1, and temperature T_h. It undergoes an isothermal expansion until its pressure and volume are P_2 and V_2. It then expands adiabatically until its temperature is T_c and its pressure and volume are P_3 and V_3. It is then compressed isothermally until it is at volume V_4, which is related to its initial volume V_1 by $T_c V_4^{\gamma-1} = T_h V_1^{\gamma-1}$. It is then compressed adiabatically until it is in its original state, P_1V_1 and T_h. (a) Assuming that each process is quasi-static, plot this cycle on a PV diagram (this cycle is known as a Carnot cycle for an ideal gas, to be discussed in the next chapter). (b) Show that the heat added during the isothermal expansion at T_h is $Q_h = nRT_h \ln (V_2/V_1)$. (c) Show that the heat rejected by the gas during the isothermal compression at T_c is $Q_c = nRT_c \ln (V_3/V_4)$. (d) Using the result of Exercise 33 that for an adiabatic expansion or compression, $TV^{\gamma-1}$ is constant, show that $V_2/V_1 = V_3/V_4$. (e) The efficiency of such a cycle is defined to be the net work done divided by the heat absorbed Q_h. Show from the first law of thermodynamics that the efficiency is $1 - Q_c/Q_h$. Using your results of the previous parts of this problem, show that $Q_c/Q_h = T_c/T_h$.

13. In Section 18-7 the collisions of gas molecules with a piston moving at a small speed u along the x axis were discussed. It was shown that in colliding with the moving piston, each molecule gains energy equal to $2m|v_x|u$, where m is the mass of the molecule and v_x is its x component of velocity. Suppose that the gas has molecular weight M, density ρ, and temperature T. The area of the piston is A. (a) Show that the number of molecules which collide with the piston in a short time dt is given by $(N_A \rho A |v_x| \, dt)/2M$. (Assume that all molecules have the same magnitude of v_x, half moving to the left and half to the right.) (b) What is the total kinetic energy gained by all the molecules that hit the piston during the time dt? (c) From your result for part (b) show that the rate at which the internal energy of the gas increases because of collisions with the moving piston is

$$\frac{dU}{dt} = \tfrac{1}{2}mv_x^2 \, \frac{2N_A\rho A}{M} \, u$$

(d) Show that the last result also can be written

$$\frac{dU}{dt} = \frac{\rho RT}{M} \, Au$$

(e) If $|dx|$ is the distance the piston moves in time dt, show that the increase in internal energy when the piston moves this distance is

$$dU = \frac{\rho RT}{M} \, A|dx|$$

(f) Use the last result, the gas law, and the fact that $dV = -A|dx|$ is the change in volume of the gas to show that $dU = -P \, dV$. This shows that the work done *on* the gas by the piston equals the increase in internal energy.

CHAPTER 19   The Availability of Energy

Our study of the first law of thermodynamics showed that it is possible to increase the internal energy of a body either by adding heat to the body or by doing work on it. In this sense, work and heat are equivalent: 10 J of either work or heat added to a system increases the internal energy of the system by 10 J. On the other hand, if we decrease the internal energy of a system and try to use this energy to do work on another system or to add heat to another system, we find an important difference between heat and work. There is no difficulty in taking internal energy out of a system in the form of heat, but it is impossible to convert this internal energy completely into work without any other change taking place in the surroundings. There is thus an asymmetry in the roles played by heat and work which is not evident from the first law.

It is quite easy to convert work completely into heat without any change in the surroundings (except in the body which absorbs the heat). A common example is the work done against friction, which can go completely into heat. This process can go on indefinitely as long as there is a supply of work but the reverse is often impossible. We cannot easily convert heat completely into work. One situation in which heat is converted into work is in the isothermal expansion of an ideal gas (or in an expansion of a real gas along a path of constant internal energy). If the expansion is allowed to proceed freely, the heat absorbed by the gas is completely converted into work until the pressure of the gas equals atmospheric pressure and the expansion stops. If the gas system is to be used again, however, work must be done on it to compress it.

Besides the conversion of heat into work, there are other conceivable processes which are consistent with the first law of thermodynamics but do not occur in nature. For example, heat does not go from a cold body to a hot body by itself. Processes which conserve energy but which never occur are all related to processes which are irreversible. For example, a block with initial kinetic energy slides along a rough table. The work against friction goes completely into heat, which increases the internal energy of the block and the table. The reverse process never occurs. The internal energy of the block and

Irreversible processes

table is never spontaneously converted into kinetic energy of the block, sending it sliding along the table while the table and block cool off. Similarly, heat can flow from a hot body to a cold body until the two bodies are at the same temperature. The reverse never happens. Two bodies in thermal contact at the same temperature remain at the same temperature. Heat is never spontaneously conducted from a cold body to a hot body. A third type of irreversible process not involving heat is the free adiabatic expansion of a gas. Whether or not the gas is ideal, the final energy equals the initial energy. Left to itself, a gas will never contract so that it fills only one of two connected compartments.

The second law of thermodynamics summarizes the fact that processes of this type do not occur. There are many different ways of stating the second law; we shall study several and show them to be equivalent. Before giving precise statements of the second law we shall consider heat engines and refrigerators briefly because the study of heat-engine efficiency gave rise to the first clear statements of the second law of thermodynamics.

19-1 Heat Engines and the Second Law of Thermodynamics

The first practical heat engine was the steam engine, invented in the seventeenth century for pumping water out of coal mines. Today the primary use of steam engines is in generating electric energy. We need not consider details of the operation of the steam engine or any other type because we can represent all types of engines—steam engines, internal-combustion engines, diesel engines—schematically. A substance or system, called the *working substance* (water in the case of the steam engine), absorbs a quantity of heat Q_h from a heat reservoir at temperature T_h. It does work W and rejects heat Q_c to a heat reservoir at a lower temperature T_c. The working substance then returns to its original state. (We use the subscripts h and c to indicate quantities associated with the hot and cold reservoirs.) The engine, therefore, is a cyclic device. At various parts of its cycle it absorbs or rejects heat, and it does work, the purpose being to gain some work at the expense of heat absorbed in each cycle.

Working substance

Early steam engine designed by James Watt in the mid-1700s.

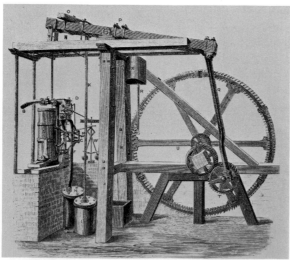

The Granger Collection

Since the initial and final states of the engine are the same, the final internal energy must equal the initial internal energy. The first law of thermodynamics thus relates the heat absorbed Q_h, the heat rejected Q_c, and the work done W by

$$Q_h - Q_c = W \qquad\qquad 19\text{-}1$$

(In this discussion we temporarily abandon our convention that the heat is positive if absorbed by the system and negative if rejected. We shall use Q_h and Q_c to represent positive amounts of heat energy and use a minus sign, as in Equation 19-1. Later in this chapter we revert to the original convention and write $Q_h + Q_c$ for the net heat absorbed, allowing Q_c to be a negative number.)

Figure 19-1 shows a schematic representation of a heat engine. Heat Q_h enters from reservoir at temperature T_h (hot reservoir), and heat Q_c is rejected to the heat reservoir at temperature T_c (cold reservoir). Work W is done by the engine. When the cycle is completed, the process repeats. The efficiency ϵ of the engine is defined to be the ratio of the work done to the heat absorbed:

$$\epsilon = \frac{W}{Q_h} = \frac{Q_h - Q_c}{Q_h} = 1 - \frac{Q_c}{Q_h} \qquad\qquad 19\text{-}2$$

Since the heat Q_h is usually produced by burning coal, oil, or some other kind of fuel which must be paid for, one tries to design a heat engine with the greatest possible efficiency. We can see from Equation 19-2 that we want to reject as small a fraction of the heat absorbed as possible. For perfect efficiency ($\epsilon = 1 = 100$ percent), $Q_c = 0$ and no heat is rejected in a cycle. In that case, all the heat absorbed from the first reservoir would be converted into work done. None would be rejected or lost to the second reservoir.

Although the efficiency of heat engines has been greatly increased since the early steam engines, it is impossible to make a heat engine which is 100 percent efficient, i.e., which would reject no heat to a reservoir at a lower temperature. This experimental result is known as the *Kelvin-Planck statement of the second law of thermodynamics:*

It is impossible for an engine working in a cycle to produce no other effect than that of extracting heat from a reservoir and performing an equivalent amount of work.

Note the words "in a cycle"; it is possible to extract heat from a single reservoir and transform it completely into work. An ideal gas undergoing an isothermal expansion does just this, but after the expansion the gas is not in its original state. In order to bring the gas back to its original state, heat must be rejected to another reservoir at a lower temperature, or the original work obtained must be put back into the gas as in an isothermal compression, in which case the net heat absorbed and net work done are both zero. If the Kelvin-Planck statement were not true, it would be possible to design a heat engine for a ship which would extract heat from the ocean (a convenient heat reservoir) and use it to power the ship without using a second heat reservoir at a lower temperature to receive any of the heat.

Figure 19-2 is a schematic representation of a refrigerator, which is similar to a heat engine but absorbs heat from a cold reservoir and rejects heat to a hot one. In order to do this, work W must be done *on* the refrigerator. If Q_c is the heat absorbed and W the work done on the

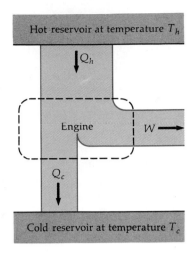

Figure 19-1
Schematic representation of a heat engine which removes heat energy Q_h from a hot reservoir at temperature T_h, does work W, and rejects heat energy Q_c to a cold reservoir at temperature T_c.

Second law (Kelvin-Planck statement)

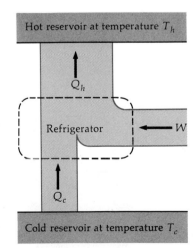

Figure 19-2
Schematic representation of a refrigerator which removes heat energy Q_c from a cold reservoir, rejects heat energy Q_h to the hot reservoir, and absorbs work W, which must be done on the refrigerator.

refrigerator, the heat rejected is

$$Q_h = W + Q_c$$

The purpose of a refrigerator is to transfer heat from a cold body to a hot body. It is desirable to do this with the least possible work. It is found from experience that some work W must always be done. This result is the Clausius statement of the second law of thermodynamics:

Refrigerators

It is impossible for a refrigerator working in a cycle to produce no other effect than the transfer of heat from a colder body to a hotter body.

Second law (Clausius statement)

If the Clausius statement were not true, it would be possible in principle to cool our homes in the summer with a refrigerator which pumped heat to the outside without using any electricity or other energy.

A measure of the performance of a refrigerator is the ratio Q_c/W, called the *coefficient of performance* η:

$$\eta = \frac{Q_c}{W} \qquad\qquad 19\text{-}3$$

The greater the coefficient of performance the better the refrigerator. Typical refrigerators have coefficients of performance of about 5 or 6. In terms of this ratio, the Clausius statement of the second law is that the coefficient of performance of a refrigerator cannot be infinite.

Questions

1. What plays the role of the high-temperature reservoir in the steam engine? The low-temperature reservoir? What plays these roles in an internal-combustion engine?

2. How does friction in an engine affect its efficiency?

Brown Brothers

Rudolf Julius Emanuel Clausius (1822–1888).

19-2 Equivalence of the Kelvin-Planck and Clausius Statements

Although the two statements of the second law we have given may seem to be quite different, they are in fact equivalent; i.e., if either statement is true, the other must be true also. We can prove this by showing that if either statement is false, the other must be false also. This is sufficient to prove equivalence.

We first assume that the Clausius statement is false, i.e., that it is possible to transfer heat from a cold reservoir to a hot reservoir without any other effects. Suppose an ordinary engine removes 100 J of energy from a hot reservoir, does 40 J of work, and exhausts 60 J of energy to the cold reservoir. This engine has an efficiency of 40 percent. If the Clausius statement were not true, we could use a perfect refrigerator to remove 60 J of energy from the cold reservoir and transfer it to the hot reservoir, doing no work in the process. The net effect of our perfect refrigerator working along with the ordinary engine would be to remove 40 J from the hot reservoir and do 40 J of work, with no energy rejected (Figure 19-3). This violates the Kelvin-Planck statement of the second law. Thus if the Clausius statement were false, the Kelvin-Planck statement would be too, as illustrated in Figure 19-4 for general values of Q_h and Q_c used by the ordinary engine.

We now assume that the Kelvin-Planck statement is false and show that this implies the Clausius statement is false. Consider an ordinary refrigerator which removes 100 J of energy from a cold reservoir, uses 50 J of work, and rejects 150 J to the hot reservoir (Figure 19-5). If the Kelvin-Planck statement were false, we could remove energy from a single reservoir and convert it completely into work with 100 percent efficiency. We could thus use such a perfect engine to remove 50 J of energy from the hot reservoir and do 50 J of work (Figure 19-5b). The net result of the ordinary refrigerator combined with the perfect engine would be to transfer 100 J from the cold reservoir to the hot reservoir without any work being done, contradicting the Clausius statement. (Figure 19-6 illustrates this combination of an ordinary refrigerator and perfect engine for general values of Q_h, W, and Q_c used by the refrigerator.) Hence, if the Kelvin-Planck statement were false, the Clausius statement would be too. We have thus shown that these two statements of the second law of thermodynamics are equivalent.

Figure 19-3
(a) An ordinary heat engine which removes 100 J from a hot reservoir, performs 40 J of work, and rejects 60 J to the cold reservoir. If the Clausius statement of the second law were false, a perfect refrigerator (b) could remove 60 J from the cold reservoir and reject it to the hot reservoir without requiring any work. (c) The net effect of the ordinary engine and perfect refrigerator in (a) and (b) is a perfect engine which violates the Kelvin-Planck statement of the second law by removing 40 J from the hot reservoir and converting it completely into work with no other effects.

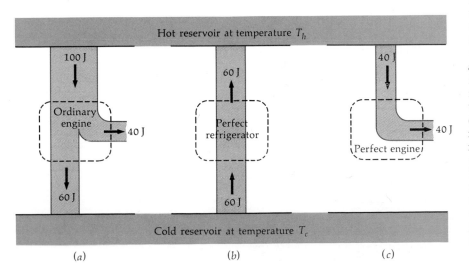

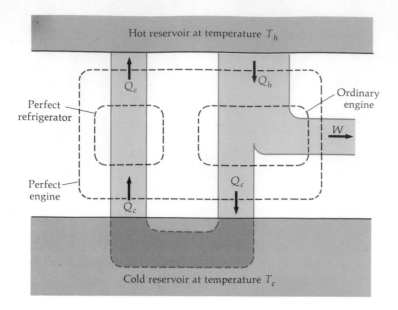

Figure 19-4
An ordinary heat engine combined with a perfect refrigerator, resulting in a perfect heat engine. The combination removes heat $Q_h - Q_c$ from the hot reservoir and converts it completely into work.

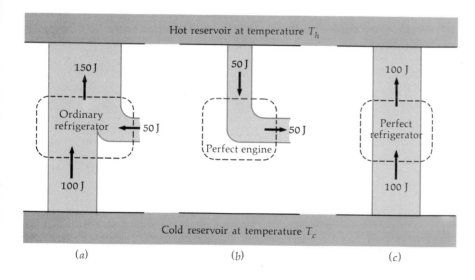

Figure 19-5
(a) An ordinary refrigerator which removes 100 J from a cold reservoir and rejects 150 J to a hot reservoir, requiring 50 J of work. If the Kelvin-Planck statement of the second law were false, a perfect engine (b) could remove 50 J from the hot reservoir and convert it completely into work with no other effects. (c) The net effect of the ordinary refrigerator and perfect engine in (a) and (b) is a perfect refrigerator which violates the Clausius statement of the second law.

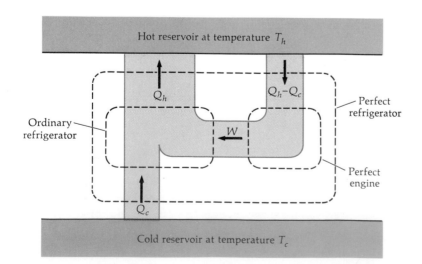

Figure 19-6
Ordinary refrigerator combined with a perfect heat engine, resulting in a perfect refrigerator which removes heat Q_c from the cold reservoir and transfers it to the hot reservoir without requiring any work.

19-3   Reversibility

Consider the simple process of conduction of a certain amount of heat Q from a hot reservoir to a cold reservoir. According to the Clausius statement of the second law of thermodynamics, it is impossible to transfer this heat back to the hotter reservoir without other changes in the surroundings. We could use a refrigerator to remove the heat Q from the lower reservoir and transfer heat $Q + W$ to the upper reservoir, where W is the work done on the refrigerator, but then energy W has been lost by the surroundings and an extra amount of heat W is in the upper reservoir. According to the Kelvin-Planck statement, we cannot remove this heat from the hot reservoir and convert it completely into work. In any cycle some heat must be rejected to the surroundings. The process of heat conduction from a hot body to a cold body is *irreversible:* we cannot bring the system back to its original state without making a permanent change in the surroundings.

We define any process to be irreversible if the system and surroundings cannot be brought back to their initial states. Irreversibility is intimately connected with the second law of thermodynamics. We can find many irreversible processes in nature. Any process that converts mechanical energy into internal energy is irreversible; e.g., the process of a block sliding along a rough table until it stops due to friction is irreversible because we cannot extract the increased internal energy of the block and table and convert it completely back into mechanical energy. Similarly, if we drop a block of wood into a lake, mechanical energy mgh is converted into internal energy of the lake (and block). We cannot reverse this process without doing work. Another irreversible process we have discussed is the free adiabatic expansion of a gas from one compartment into an evacuated compartment. Although no work is done, no heat is transferred, and the internal energy does not change, it is impossible to reverse the process.

This irreversibility is also related to the second law of thermodynamics. Let us suppose that the gas at temperature T freely expands from volume V_1 of the original compartment to the final volume V_2. As we have mentioned, the final temperature is also T since the internal energy of the ideal gas did not change. We can bring this gas back to its original volume by isothermal compression, in which we do work W and an equal amount of heat $Q = W$ is rejected into a heat reservoir at temperature T. Such an isothermal compression is itself reversible, but the combination of the free expansion plus the isothermal compression puts the gas back in its original state, with an amount of work W converted into energy in the heat reservoir. Since it is impossible to convert this energy back into work with no other changes, the complete process is irreversible. Thus the free adiabatic expansion is irreversible (Figure 19-7). The gas, once expanded, will never spontaneously contract to its original volume.

From these considerations and our statements of the second law of thermodynamics, we can list some conditions necessary for a process to be reversible:

1. There must be no work done by friction, viscous forces, or other dissipative forces which produce heat.

2. There can be no heat conduction due to a temperature difference.

3. The process must be quasi-static so that the system is always in an equilibrium state (or infinitesimally near an equilibrium state).

Irreversibility defined

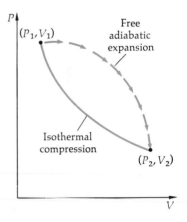

Figure 19-7
A free adiabatic expansion of an ideal gas is indicated by the arrows. It cannot be represented by a curve on the PV diagram because the process is not quasi-static and the gas does not pass through equilibrium states. The gas can be returned to its original state by a reversible isothermal compression during which work is converted completely into heat. The combined cycle cannot be reversed because heat cannot be converted completely into work with no other change. The free adiabatic expansion is therefore irreversible.

Conditions for reversibility

Any process which violates any of the above conditions is irreversible. Most processes in nature are irreversible. In order to have a reversible process, great care must be taken to eliminate frictional and other dissipative forces and to make the process quasi-static. Since this can never be done completely, a reversible process seems impossible in practice. Nevertheless, one can come very close to a reversible process, and the concept is very important in theory.

Consider, for example, a quasi-static expansion of a gas from volume V_1 to volume V_2 at constant pressure, shown in Figure 19-8. If this process is to be reversible, the heat must be absorbed isothermally so that there is no heat conduction across a finite temperature difference. We can approximate the constant-pressure path by an alternate reversible path consisting of the series of isotherms and quasi-static, adiabatic paths indicated. Heat is then absorbed along each isotherm from a reservoir at the temperature of the isotherm. With a very large number of heat reservoirs, we can make this alternate path approximate the constant-pressure path as closely as we wish. Similarly, any quasi-static process represented by a curve in the PV diagram can be approximated by a reversible path consisting of a series of isotherms and quasi-static, adiabatic paths.

In the next sections, we shall see how the investigation of reversible processes leads to a theoretical limit on the efficiency of a heat engine operating between two reservoirs of fixed temperatures and to the definition of an absolute temperature scale which does not depend on the properties of any material.

Figure 19-8
Quasi-static expansion at constant pressure. This process can be considered reversible because it can be approximated as closely as desired by a series of reversible isothermal and adiabatic curves as shown. Any quasi-static process can be performed approximately reversibly using a large number of heat reservoirs so that the heat is absorbed or rejected approximately isothermally.

Questions

3. Give several examples of irreversible processes in everyday life.

4. If a system changes from state A to state B by an irreversible process, does that mean it can never be returned to state A again?

5. A bottle of ink is slowly, carefully, and thoroughly mixed with a tub of water. Is this a reversible or an irreversible process?

19-4   The Carnot Engine

In 1824, before the first law of thermodynamics was established, a young French engineer, Carnot, described an ideal reversible engine working in a simple cycle between two heat reservoirs and found the theoretical limit for the efficiency of an engine in terms of the temperatures of the heat reservoirs. Such an ideal engine is now called a *Carnot engine* and its cycle a *Carnot cycle.* Carnot's results are given in a simple but important theorem, called the *Carnot theorem:*

The Granger Collection

Drawing by L. L. Boilly of Nicolas Léonard Sadi Carnot (1796–1832), aged seventeen.

No engine working between two given heat reservoirs can be more efficient than a reversible engine working between those reservoirs.

Carnot's theorem

This theorem is easily proved by using the second law of thermodynamics. Consider a *reversible* engine working between two reservoirs. Let it remove 100 J of energy from the hot reservoir, do 40 J of work, and reject 60 J of energy into the cold reservoir (Figure 19-9*a*). Its efficiency is thus 40 percent. Since the engine is reversible, it can be operated in reverse, absorbing 60 J of energy from the cold reservoir

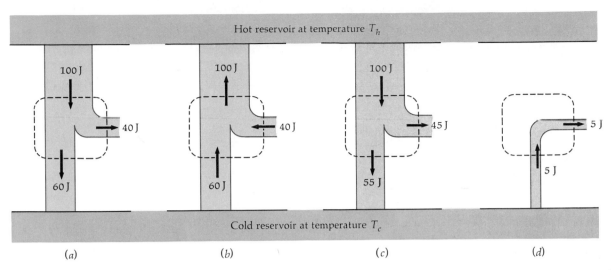

and rejecting 100 J into the hot reservoir while work of 40 J is done on it (Figure 19-9*b*). We now consider a second engine, which may or may not be reversible, and wish to show that if its efficiency is greater than that of the first engine, the second law can be violated. Suppose we use this engine to remove 100 J of energy from the hot reservoir. Since we are assuming that its efficiency is greater than 40 percent, it does more than 40 J of work and rejects less than 60 J to the cold reservoir. Suppose it has an efficiency of 45 percent. Then 45 J of work is done, and 55 J of energy is rejected (Figure 19-9*c*). The net effect of these two engines working together (with the reversible engine operating as a refrigerator) is that 5 J of energy has been removed from the cold reservoir and changed completely into work (Figure 19-9*d*). This violates the Kelvin-Planck statement of the second law. Figure 19-10 illustrates this proof for general quantities of heat removed and rejected and work done. In this figure we choose the heat removed by the second engine Q'_h to be equal to that rejected by the first engine operating as a refrigerator. If the efficiency of the second engine is greater than that of the first, the work done W' is greater than the work W needed to operate the reversible engine as a refrigerator and the heat rejected Q'_c is less than that absorbed from the cold reservoir by the reversible engine. Thus the net effect of the two engines is that heat $Q_c - Q'_c$ is removed from the cold reservoir and changed completely into work. (On the other hand, if the efficiency of the second

Figure 19-9
Illustration of the Carnot theorem. (*a*) A reversible heat engine with 40 percent efficiency. (*b*) The same heat engine run backward as a refrigerator. (*c*) An assumed heat engine working between the same two reservoirs with efficiency of 45 percent. (*d*) The combination of the two engines with the reversible engine operating as a refrigerator results in a perfect heat engine, violating the second law.

Figure 19-10
Illustration of the Carnot theorem for arbitrary amounts of heat and work. The reversible engine on the right is run backward as a refrigerator, removing heat Q_c from the cold reservoir and rejecting heat Q_h to the hot reservoir while requiring work W. If the engine on the left has a greater efficiency than the reversible one, it can remove heat $Q'_h = Q_h$ and perform work W' which is greater than W. If part of this work is used to run the refrigerator, the net effect of both engines is the removal of heat $Q_c - Q'_c$ from the cold reservoir and complete conversion of this heat into work.

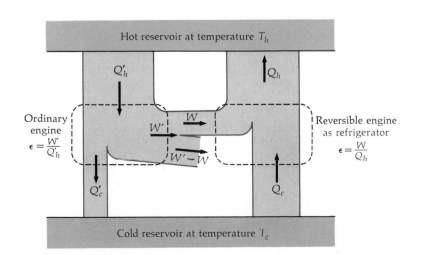

engine is less than that of the reversible engine, there is no contradiction of the second law of thermodynamics. The net result is that net work has been converted into internal energy in the cold reservoir.) We have shown that the efficiency of the reversible engine must be greater than or equal to the efficiency of any other engine working between the same reservoirs, which is the Carnot theorem.

A corollary to the Carnot theorem reads

All reversible engines working between the same two reservoirs have the same efficiency.

This follows from Carnot's theorem because for any two reversible engines, the theorem requires that each have an efficiency greater than or equal to the efficiency of the other. Since they cannot both have greater efficiency than the other, their efficiencies must be equal.

Carnot's theorem and corollary can also be expressed in terms of refrigerators:

No refrigerator working between two given heat reservoirs can have a greater coefficient of performance than a reversible refrigerator working between those reservoirs.

Similarly,

All reversible refrigerators working between the same two reservoirs have the same coefficient of performance.

The proof that these statements follow from the second law of thermodynamics is left as an exercise.

It is not hard to see what the cycle of a Carnot engine must be. Since the cycle is reversible, heat cannot be transferred at a temperature difference, because this is always irreversible. The engine must therefore absorb and reject heat isothermally. Figure 19-11 shows a Carnot cycle for a gas used as the working substance of a Carnot engine. We assume that the gas is originally at the temperature of the hot reservoir T_h. (If not, we can always perform an adiabatic expansion or compression until the gas is at this temperature.) The gas first expands isothermally to position 2 on the PV diagram. During this expansion the gas does some work and absorbs heat Q_h from the hot reservoir. Since this heat is absorbed isothermally and quasi-statically, the expansion is reversible (assuming no friction). If the gas were

Carnot cycle

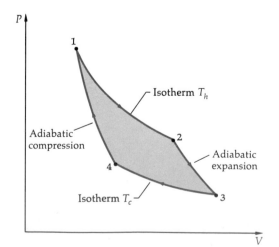

Figure 19-11
A Carnot cycle for a gas. The gas first expands isothermally from state 1 to state 2, absorbing heat from the reservoir at T_h and performing work. It then expands adiabatically until its temperature is T_c in state 3. Work is performed but no heat is absorbed or rejected. The gas is then compressed isothermally to state 4 and adiabatically back to its original state. All parts of the cycle are quasi-static. The net work performed during the cycle is indicated by the enclosed area.

ideal, the work done would equal the heat absorbed, but this is not generally true of a real gas. After the gas is removed from contact with the hot reservoir, it undergoes a quasi-static adiabatic expansion until the temperature of the gas is that of the cold reservoir T_c at position 3. Some work is done, but no heat is transferred. The gas then is compressed isothermally while it is in contact with the lower-temperature reservoir to state 4. During this compression, work is done on the gas, and heat Q_c is rejected to the reservoir. Finally, the gas is compressed adiabatically and quasi-statically until it reaches its original temperature T_h at state 1, and the cycle is completed. The net work done is indicated by the enclosed area in Figure 19-11, i.e., the area defined by the two isothermal curves and two adiabatic curves, and is equal to $Q_h - Q_c$ by the first law of thermodynamics. The efficiency of the engine can be computed if the equation of state is known so that the work can be computed under the curves in the PV diagram. We shall compute the efficiency for an ideal gas in the next section.

19-5 The Absolute Temperature Scale

The efficiency is the same for all reversible engines working between the same two heat reservoirs, whatever the working substance—ideal gas, real gas, or water. As long as the engine performs in a reversible cycle, which means absorbing and rejecting the heat isothermally (and quasi-statically) and changing its temperature adiabatically (and quasi-statically), the efficiency is the same no matter what the system is. Since the only characteristic of a heat reservoir is its temperature, the efficiency of a Carnot engine must depend only on the temperatures of the two heat reservoirs. The heats absorbed and rejected by a Carnot engine are related to the efficiency by Equation 19-2, which can be written

$$\frac{Q_c}{Q_h} = 1 - \epsilon_R$$

where we have added the subscript R to indicate that ϵ_R is the efficiency of a reversible engine. Thus Q_c/Q_h must also be a function only of the temperatures of the reservoirs. We use this result to define an absolute temperature scale, defining the ratio of absolute temperatures of the heat reservoirs by

$$\frac{T_c}{T_h} = \frac{Q_c}{Q_h}$$ 19-4 *Absolute temperature defined*

To use this definition to measure the ratio of two temperatures we would have to set up a reversible engine to operate between two reservoirs with the two temperatures and carefully measure the heats absorbed from or rejected to those reservoirs in each cycle of the engine.

Equation 19-4 defines only the ratio of two absolute temperatures. The temperature scale is completely determined by choosing one fixed point, as was done for the other temperature scales. We define the absolute temperature of the triple point of water to be 273.16 K. In terms of the absolute temperatures of the heat reservoirs the efficiency of a Carnot engine is

$$\epsilon_R = 1 - \frac{T_c}{T_h}$$ 19-5 *Efficiency of a Carnot engine*

This definition of an absolute temperature scale is completely independent of the properties of any materials. Although such a scale might seem impossible to use in practice, that is not the case. At very low temperatures, where the ideal-gas scale cannot be defined, temperatures are actually measured by performing a Carnot cycle as carefully as possible between the two heat reservoirs and measuring the heats Q_h and Q_c. In the temperature range in which the ideal-gas temperature scale can be defined, the absolute and ideal-gas temperature scales are identical, as will be shown in Example 19-1.

Example 19-1 The Carnot cycle shown in Figure 19-11 is performed using an ideal gas. Show that the ratio of the heat rejected to the heat absorbed equals the ratio of the ideal-gas temperatures of the heat reservoirs.

The equation of state of an ideal gas is $PV = nR\theta$, where we now use θ for the ideal-gas temperature to avoid assuming the result we wish to prove. We want to show that $Q_c/Q_h = \theta_c/\theta_h$. We first consider the isothermal expansion from state 1 to state 2. Since the internal energy of an ideal gas does not change if the temperature is constant, the work done by the gas in this isothermal expansion equals the heat absorbed Q_h. The work done is

$$W_{1-2} = \int_1^2 P\, dV = \int_1^2 \frac{nR\theta_h}{V}\, dV = nR\theta_h \ln \frac{V_2}{V_1}$$

Then

$$Q_h = nR\theta_h \ln \frac{V_2}{V_1}$$

Similarly, the heat rejected to the cold reservoir Q_c equals the work done *on* the gas in the isothermal compression at temperature θ_c from state 3 to state 4. This work is of the same magnitude as that done *by* the gas if it expands from state 4 to state 3. The heat rejected is thus

$$Q_c = nR\theta_c \ln \frac{V_3}{V_4}$$

The ratio of these heats is then

$$\frac{Q_c}{Q_h} = \frac{\theta_c \ln (V_3/V_4)}{\theta_h \ln (V_2/V_1)} \qquad\qquad \text{19-6}$$

We can relate the volumes V_1, V_2, V_3, and V_4 using the equation for the adiabatic expansion and compression derived in Section 18-8. Along these curves we have

$$PV^\gamma = \text{constant}$$

If we eliminate the pressure using the equation of state, we have

$$\frac{nR\theta}{V} V^\gamma = \text{constant}$$

or

$$\theta V^{\gamma-1} = \text{constant} \qquad\qquad \text{19-7}$$

Applying Equation 19-7 to states 2 and 3 connected by the adiabatic expansion, we have

$$\theta_h V_2^{\gamma-1} = \theta_c V_3^{\gamma-1} \qquad\qquad \text{19-8}$$

Similarly for the states 4 and 1 connected by the adiabatic compression,

$$\theta_h V_1^{\gamma-1} = \theta_c V_4^{\gamma-1} \qquad\qquad 19\text{-}9$$

Dividing Equation 19-8 by Equation 19-9 gives

$$\left(\frac{V_2}{V_1}\right)^{\gamma-1} = \left(\frac{V_3}{V_4}\right)^{\gamma-1} \qquad\qquad 19\text{-}10$$

and so $V_2/V_1 = V_3/V_4$ and $\ln (V_2/V_1) = \ln (V_3/V_4)$. After canceling the logarithmic terms in Equation 19-6, our result is proved:

$$\frac{Q_c}{Q_h} = \frac{\theta_c}{\theta_h} \qquad\qquad 19\text{-}11$$

Comparing this result with Equation 19-4, we see that the ratio of the ideal-gas temperatures of the two reservoirs is the same as the ratio of the absolute temperatures. Since both temperature scales have the same fixed value, 273.16 K for the temperature of the triple point of water, the ideal-gas and absolute temperatures are equal everywhere the ideal-gas temperature scale is defined.

Example 19-2 An engine works between reservoirs at 400 and 300 K, extracting 100 cal from the hot reservoir during each cycle. What is the greatest efficiency possible for this engine, and how much work can it perform during each cycle?

A Carnot engine working between these reservoirs has efficiency

$$\epsilon_R = 1 - \frac{T_c}{T_h} = 1 - \frac{300 \text{ K}}{400 \text{ K}} = 0.25 = 25\%$$

The work done by a Carnot engine in one cycle is found from $\epsilon_R = W/Q_h$. Then $W = \epsilon_R Q_h = (0.25)(100 \text{ cal}) = 25 \text{ cal}$. Any other engine working between these two reservoirs will have efficiency less than 25 percent because of irreversibility due to friction, non-quasi-static processes, etc.

Example 19-3 What is the greatest possible coefficient of performance of a refrigerator working between reservoirs at 400 and 300 K?

From Example 19-2 we saw that a Carnot engine performed 25 cal of work for each 100 cal of heat removed from the upper reservoir. It must therefore reject 75 cal to the lower-temperature reservoir. When this engine is run as a refrigerator, it removes 75 cal from the reservoir at 300 K and rejects 100 cal to the reservoir at 400 K. The coefficient of performance of a Carnot refrigerator between these reservoirs is thus

$$\eta = \frac{Q_c}{W} = \frac{75 \text{ cal}}{25 \text{ cal}} = 3$$

Any other refrigerator will have a smaller coefficient of performance because of irreversibility due to friction, non-quasi-static processes, etc.

Questions

6. Why do power-plant designers try to increase the temperature of the steam fed to engines as much as possible?

7. Will the efficiency of the engines in a power plant be improved if more cooling water is used so that the temperature rise in the cooling water is smaller?

Power Plants and Thermal Pollution

Laurent Hodges
Iowa State University

The scientist's interest in a heat engine is focused primarily on the work obtained from a given heat input. Society, however, is also interested in the waste heat output. When it leads to undesirable effects, it is called *thermal pollution.*

The largest heat engines in modern society—those from which thermal pollution can be the most damaging—are the steam engines used in electric power plants to drive an ac generator and produce electricity. About 85 percent of the electricity in the United States is produced by such steam engines, the other 15 percent being generated by water power in hydroelectric plants.

The operation of a steam electric power plant is explained in Figure 1. The *heat source* is usually either the combustion of a fossil fuel (coal, oil, or natural gas) or the fission of uranium 235; occasionally geothermal power or the combustion of solid wastes are used. The *boiler* heats water, converting it into high-temperature, high-pressure steam. A nuclear power plant has a reactor core in which water is heated into steam (boiling-water reactor, or BWR) or into pressurized water (pressurized-water reactor, or PWR); in the PWR, heat is then transferred to a steam cycle in a heat exchanger. Steam expands and cools as it passes through the *turbine,* converting heat energy into the mechanical energy of a rotating shaft. A *generator* converts the mechanical energy into electric energy (alternating current). The *condenser* cools the steam exhausted from the turbine and reduces its pressure, thereby increasing the efficiency of the power plant.

Typical temperatures and efficiencies for steam electric plants are shown in Table 19-1. The average efficiency for all power plants in the United States is about 34 percent, but the best fossil-fuel plants have about 40 percent efficiency and many smaller or older plants have efficiencies of 25 percent or

Figure 1
Major components of a steam electric power plant (schematic).

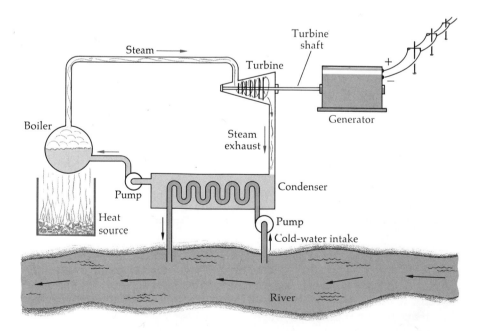

Table 19-1

Typical higher and lower temperatures and efficiencies for steam electric power plants

	Temperature, °C		Efficiency, %	
	High	Low	Carnot	Actual
Large fossil-fuel plant	380	40	52	40
Boiling-water reactor	285	40	44	34
Pressurized-water reactor	315	40	47	34

less. The efficiencies of BWR and PWR nuclear plants are about 34 percent, but those of some experimental nuclear plants (cooled by gas or liquid sodium) are closer to 40 percent.

Existing power plants must dispose of waste heat amounting to twice the electrical energy generated. A popular method, *once-through cooling*, involves passing water from a river or lake through the condenser once and then returning it at a warmer temperature. This works well only if adequate water is present. Many plants of 1000 MW_e (e = electric power) are being built today, sometimes several at one location. Even on a large river the resulting temperature rise may be significant: a 3000 MW_e–9000 MW_t (t = thermal power) installation using a flow of 120 m^3/sec (the average yearly minimum flow of the Missouri River at its mouth) would cause a 12°C rise by its 6000 MW of waste heat.

Thermal pollution is the assortment of undesirable effects resulting from waste heat. Although the temperature changes produced in a river by once-through cooling may not be as large as natural daily or seasonal fluctuations, possible adverse ecological effects include greater growth of bacteria, pathogens, or undesirable blue-green algae or physiological and behavioral effects on aquatic life (such as fish). As an example, large-mouth bass acclimated to 30°C show a 50 percent mortality rate within 72 h if the water is warmed to 34°C.

Warmer water also contains less dissolved oxygen necessary for aquatic life or for the decomposition of organic wastes; e.g., the dissolved oxygen at saturation in water is 11.3 mg/ℓ at 10°C and 6.6 mg/ℓ at 40°C. Warmer water is also less viscous, permitting faster deposition of the sediment load, which can affect aquatic food supplies.

Most of the cooling water used in the United States is used by electric power plants. The rapid growth of electric energy consumption, averaging 7 percent annually for many decades, has led to greater use of alternatives to once-through cooling. One of these is the cooling pond, really a private lake, but the land requirements for such a pond are often too expensive to consider.

The most common alternative, the cooling tower, transfers the waste heat to the atmosphere (Figure 2). Usually these are evaporative towers, which use the heat to evaporate water. Evaporative cooling towers are either large hyperbolic types, with a natural draft, or mechanical-draft (fan) types. Nonevaporative cooling towers, which transfer heat to the air by heat exchangers, are expensive but may be needed in moist climates. At ordinary temperatures the evaporation of water requires 580 cal/gm, and so the water requirements are much less than in once-through cooling, although the water is consumed in the process. The evaporated water can cause fog, increased precipitation, and (in certain climates) icing of nearby roads, and the cooling tower and its plume may be blots on the landscape, so that evaporative towers are not a completely satisfactory answer to the problem of thermal pollution.

Figure 2
Hyperbolic natural-draft evaporative cooling towers at the Fort Martin Power Station, Fairmont, West Virginia. (*The Marley Company, Mission, Kansas.*)

19-6 Entropy

We used the zeroth law of thermodynamics to define a new thermody-namic state function, the temperature, which is a measure of the hotness or coldness of a system. In the molecular picture, the tempera-ture measures the average kinetic energy of translation of the mole-cules. Similarly the first law allowed us to define the internal-energy function U. Now we consider a new thermodynamic state function, the entropy S, related to the second law of thermodynamics. The exis-tence of this state function is less obvious than that of T and U, and it is less easily related to the second law than they were to the zeroth and first laws. The reason is that the second law is more complicated than the first and zeroth laws in that it describes processes that do *not* occur. We shall first look at the special case of an ideal gas. With this simple system it is easy to show that there is a new state function which is related to the heat absorbed by the system and to the temper-ature at which the heat is absorbed.

Consider an arbitrary quasi-static reversible process in which an ideal gas absorbs an amount of heat dQ. According to the first law, dQ is related to the internal-energy change dU and the work done $dW = P\, dV$ by

$$dQ = dU + P\, dV$$

For our ideal gas we can substitute nRT/V for P from the equation of state and write the change in internal energy in terms of the heat capacity $dU = C_v\, dT$:

$$dQ = C_v\, dT + nRT\, \frac{dV}{V} \qquad\qquad \text{19-12}$$

When the system goes from state 1 to state 2, we must know the path followed in order to calculate the total heat added Q. This is clear mathematically from the fact that we cannot integrate Equation 19-12. There is no problem with the first term because $C_v = dU/dT$ is a func-tion of temperature only. The integration of this term gives the change in internal energy. The problem is with the second term, the work done by the gas. This point was discussed in Chapter 18. Both the heat added and the work done depend on the path. Thus neither is a state function. However, if we divide each term in Equation 19-12 by T, we can integrate both terms on the right-hand side:

$$\frac{dQ}{T} = C_v\, \frac{dT}{T} + nR\, \frac{dV}{V} \qquad\qquad \text{19-13}$$

Even if C_v is not constant, it can depend only on T, so that whatever function of T it is, the first term can be integrated. Thus if the system is carried from state 1 to state 2 along any quasi-static path, we can find the sum (integral) of dQ/T. It is in fact given by

$$\int_1^2 \frac{dQ}{T} = C_v\, \ln\frac{T_2}{T_1} + nR\, \ln\frac{V_2}{V_1} \qquad\qquad \text{19-14}$$

assuming C_v to be constant. The right side of Equation 19-14 depends only on the states 1 and 2 and not on the path. If in addition the path is reversible, we get the same magnitude for the sum of dQ/T by reversing the path and going from state 2 to 1, except that the sum is the negative of that obtained from state 1 to state 2. Thus if we travel along one reversible path from 1 to 2 and back along another reversible path from state 2 to 1, the sum of dQ/T is zero. This implies that there is a state function whose change is given by dQ/T. We call this function the *entropy S:*

Entropy defined

$$\Delta S = S_2 - S_1 = \int_1^2 \frac{dQ_R}{T} \qquad \text{19-15}$$

or

$$dS = \frac{dQ_R}{T} \qquad \text{19-16}$$

We have used the subscript R on dQ to remind us that the heat must be absorbed reversibly.

A Carnot cycle is an example of a reversible closed path. During the isothermal expansion the entropy change of the gas is

$$\int_1^2 \frac{dQ_R}{T_h} = \frac{1}{T_h} \int dQ = \frac{Q_h}{T_h} \qquad \text{19-17}$$

During the reversible adiabatic expansion and compression the entropy change is zero because no heat is absorbed or rejected by the gas. During the isothermal compression the entropy change is

$$\int \frac{dQ_R}{T_c} = \frac{1}{T_c} \int dQ_R = \frac{Q_c}{T_c} \qquad \text{19-18}$$

This is negative because, according to our sign convention, dQ is negative when heat leaves the system. (We now return to our original sign convention, in which the heat out of a system is negative, and use Q_c for the heat *absorbed* from the cold reservoir. Since heat is actually rejected, Q_c is negative.) Since the total entropy change must be zero for a complete cycle, we have

$$\frac{Q_h}{T_h} + \frac{Q_c}{T_c} = 0 \qquad \text{19-19}$$

or

$$\frac{-Q_c}{Q_h} = \frac{T_c}{T_h}$$

This is the same as the defining equation (19-4) for the ratio of absolute temperatures except for the sign of Q_c because now the symbols Q_h and Q_c stand for heat absorbed and are negative if heat is rejected.

Although we have been considering a special system, the ideal gas, it can be shown that the sum of dQ_R/T for any system undergoing a *reversible* change is independent of path and depends only on the initial and final states. (See the general proof in Section 19-10.) Thus Equations 19-15 and 19-16, which define the entropy function, are general. It is important to understand, however, that only for a reversible process is $\int dQ/T$ independent of path. For an irreversible process, change in the entropy of the system will not be given by $\int dQ/T$. However, since entropy is a state function, we can find the entropy change in the system undergoing an irreversible process from state 1 to state 2 by considering any *reversible* process connecting the same two states and computing $\int dQ_R/T$.

Example 19-4 An ideal gas undergoes a free adiabatic expansion from volume V_1 to volume V_2. What is the change in entropy of the gas?

As we have discussed previously, the final energy of the gas is the same as the initial energy, and the final temperature is the same as the initial temperature. *During the free expansion,* we cannot describe the gas as having a temperature or internal energy since the gas does not pass through equilibrium states. In this process no heat is absorbed or rejected, so that $\Delta Q = 0$ for the whole process. We might expect, therefore, that there is no entropy change, but this is not correct. The change in entropy is *not* given by the sum of dQ/T because this process is not reversible. We cannot even define T during the process. However, changes in a state function are independent of the process. The entropy change, $S_2 - S_1$, is independent of the process connecting states 1 and 2. We can compute $S_2 - S_1$ by considering a *reversible* process connecting these two states. Any reversible process will do since the entropy change does not depend on the process. The convenient process for this example is a quasi-static isothermal expansion from volume V_1 to V_2. Since the temperature is constant for this reversible process, the entropy change is just Q/T, where Q is the total heat absorbed. This heat absorbed equals the work done by the gas (since the internal energy is constant). Thus

$$Q = W = \int_1^2 P\, dV = nRT \int_1^2 \frac{dV}{V} = nRT \ln \frac{V_2}{V_1} \qquad \text{19-20}$$

The entropy change is

$$S_2 - S_1 = \frac{Q}{T} = \frac{nRT \ln (V_2/V_1)}{T} = nR \ln \frac{V_2}{V_1} \qquad \text{19-21}$$

This same result is obtained by integrating Equation 19-13 with $dT = 0$. The entropy of the system increases even though no heat is transferred. In order for the entropy of a gas to decrease by a free adiabatic "expansion" the final volume would have to be less than the initial volume. An adiabatic free contraction never occurs in nature.

Example 19-5 The melting point of a substance is T_0, and its latent heat is L. What entropy change takes place when mass m of the substance melts at this temperature?

To compute the entropy change we must consider a reversible process for the melting and calculate the integral of dQ_R/T for that process. Such a process would be the melting produced by putting the solid material into contact with a reservoir whose temperature is only very slightly higher so that melting would proceed very slowly. The process could be reversed by lowering the reservoir temperature very slightly, causing any molten material to freeze again. As we have seen, for heat conduction to be reversible the temperature difference must be vanishingly small.

From Equation 19-15,

$$\Delta S = \int \frac{dQ_R}{T} = \frac{1}{T_0} \int dQ_R = \frac{mL}{T_0}$$

The temperature T_0 is removed from the integral since the melting takes place at the constant temperature T_0. The remaining integral $\int dQ_R$ is just the heat required to melt mass m, or mL.

To illustrate our result numerically, we calculate the entropy increase when 100 gm of mercury melts at $-39°C$ ($= 234$ K) and atmospheric pressure. The latent heat of fusion is 2.8 cal/gm:

$$S = \frac{mL}{T_0} = \frac{(100 \text{ gm})(2.8 \text{ cal/gm})}{234 \text{ K}} = 1.2 \text{ cal/K}$$

Questions

8. Does the entropy change in a physical system passing from state 1 to state 2 depend on the path taken between those states?

9. Is it possible to choose the value of the entropy of a system at some particular state and thereafter assign a value to the entropy of the system for every other state?

10. Give an example of a system whose entropy decreases with time.

11. A gas undergoes a quasi-static adiabatic expansion to twice its original volume. How does its entropy change?

19-7 Entropy Change of the Universe

We have shown that the entropy state function exists for an ideal gas,[1] and we have seen how to calculate changes in entropy for a system, but it is not immediately evident what use this function is. In this section we calculate the entropy change of the universe for several processes, where by universe we mean the system and surroundings, e.g., heat reservoirs and other systems, which interact with the system. In these examples we shall often be calculating the entropy change of a heat reservoir, which is easy to do.

The state of a heat reservoir is determined by its internal energy and its temperature, which, by definition, is constant. If a heat reservoir absorbs heat Q by some reversible process, the entropy of the reservoir increases by Q/T, where T is the temperature of the reservoir. If it absorbs the heat in an irreversible process, we need only replace the irreversible process by a reversible process to calculate the change in entropy. Again the entropy increases by Q/T. If the reservoir gives up heat Q to the system or another reservoir, the entropy of the reservoir *decreases* by Q/T since Q is now leaving the reservoir.

Let us first consider the entropy change of the universe for a reversible process. A system absorbs heat Q from a reservoir at temperature T; since the process is reversible, the absorption must occur isothermally. The system has an increase in entropy of Q/T, and the reservoir has a decrease in entropy by the same amount. Thus the total change in the entropy of the system plus reservoir is zero. No matter how many systems and reservoirs take part in a reversible process, for each heat absorption or rejection the entropy change of the reservoir is exactly the negative of that for the system. Thus *the entropy change of the universe is zero for a reversible process.*

Example 19-6 Show that the entropy change of the universe is zero during a Carnot cycle.

The entropy change of the engine after a complete cycle must be zero because the engine is back in its original state. We thus have only to calculate the entropy change of the two heat reservoirs. The hot reservoir gives up heat Q_h at temperature T_h; its entropy therefore decreases. The entropy change of the hot reservoir is

$$\Delta S_h = -\frac{|Q_h|}{T_h} \qquad\qquad 19\text{-}22$$

---

[1] The proof for a general system is given in Section 19-10.

The cold reservoir absorbs heat Q_c at temperature T_c. Its entropy change is positive, given by

$$\Delta S_c = + \frac{|Q_c|}{T_c} \qquad\qquad 19\text{-}23$$

Since this is a Carnot cycle, the heats $|Q_h|$ and $|Q_c|$ are related by Equation 19-4:

$$\frac{|Q_c|}{|Q_h|} = \frac{T_c}{T_h}$$

If we substitute $|Q_h|T_c/T_h$ for $|Q_c|$ in Equation 19-23, we have

$$\Delta S_c = + \frac{|Q_h|}{T_h}$$

Thus the sum of ΔS_c and ΔS_h is zero.

We now consider the entropy change of the universe for processes that are irreversible. In Section 19-10 we show that for a general irreversible process the entropy change of the universe must be positive. Here we shall merely investigate some typical irreversible processes and show in each case that the entropy change of the universe must be greater than zero.

Let us consider heat conduction from a hot reservoir to a cold reservoir; i.e., an amount of heat Q is conducted from a reservoir at temperature T_h to a reservoir at temperature T_c. The entropy of the hot reservoir decreases by Q/T_h, and that of the cold reservoir increases by Q/T_c. Since T_c is less than T_h, the increase in the entropy of the cold reservoir is greater in magnitude than the decrease in the entropy of the hot reservoir. Thus the total entropy change of the universe is greater than zero and is given by

$$\Delta S_u = \frac{Q}{T_c} - \frac{Q}{T_h} > 0 \qquad\qquad 19\text{-}24$$

Next let us consider a process that is irreversible because mechanical energy is lost to internal energy due to friction or other dissipative forces — say a block with kinetic energy E_k sliding along a rough table until it is stopped by friction. We assume for simplicity that the table is large so that it can be considered a reservoir. (This is not important to the conclusion but it simplifies the calculation.) Because of the work done by friction, the internal energy of the table (and block if it is not much smaller than the table) and air surroundings increases by the amount E_k. The entropy change of the table and air surroundings is the same as if heat $Q = E_k$ were transferred reversibly. Thus the entropy change of the table and air is $+ E_k/T$. This is the entropy change of the universe:

$$\Delta S_u = + \frac{E_k}{T} > 0 \qquad\qquad 19\text{-}25$$

Finally we consider the third kind of irreversible process discussed in Section 19-3, a non-quasi-static process, again taking the free adiabatic expansion of an ideal gas. Since no heat is transferred and no work done, the surroundings suffer no change whatsoever. We have already calculated the entropy change of the gas in Example 19-4. This is also the entropy change of the universe. According to Equation 19-21, the entropy change of the gas and universe is $nR \ln (V_2/V_1)$, which is positive because V_2 is greater than V_1.

In each of these typical irreversible processes the change in entropy of the universe is greater than zero. The reason is that a negative change in entropy of the universe would violate the second law. For example, in heat conduction, the entropy change of the universe could be negative only if the heat left the cold reservoir and entered the hot reservoir, violating the Clausius statement of the second law of thermodynamics. Similarly, in the conversion of internal energy into mechanical energy, the entropy change would be negative if heat were removed from the surroundings and converted completely into mechanical energy, violating the Kelvin-Planck statement of the second law. We have already discussed the fact that the irreversibility of the free adiabatic expansion is connected with the second law. If the entropy of the universe were to decrease in a free adiabatic "expansion," the gas would have to contract to a smaller volume by itself.

As a result we have a new statement of the second law of thermodynamics, equivalent to the other statements:

For any process, the entropy change of the universe must be greater than or equal to zero. The entropy change is equal to zero only if the process is reversible.
Second law (entropy form)

The entropy function gives us a numerical measure of the irreversibility of a given process. Suppose, for example, we consider a real attempt to produce an isothermal quasi-static reversible expansion of a gas. We can compute the entropy change of the heat reservoir by measuring the heat which leaves the reservoir. The entropy of the gas can be found in tables. If there is friction in the piston and some turbulence in the gas (because the piston has a slight acceleration or because the expansion happens too fast for the gas to remain near equilibrium states), the total entropy change will not be zero but some positive number. If we reduce the friction or make the process more nearly quasi-static, the total entropy change will be less. We thus have a numerical measure of the degree of irreversibility of any given process.

19-8 Entropy and the Availability of Work

Let us consider a single heat reservoir at temperature T. If there are no other heat reservoirs around and no systems at a lower temperature which can be used as heat reservoirs, we cannot use any of the internal energy of the reservoir for the performance of work. The internal energy of the reservoir is unavailable. To extract some of it and do work requires a lower-temperature reservoir or system to accept the heat rejected. The second law states that we cannot remove some of the internal energy and convert it completely into work; in other words, the conversion of some of the internal energy completely into work would decrease the entropy of the universe, which is not allowed.

Whenever an irreversible process occurs, some energy is made unavailable for use as work which was available before the process took place. For the block sliding along the rough table, if it were not for the mechanical energy lost because of friction, the original kinetic energy of the block could be used to do some useful work, e.g., lifting a weight or turning a generator. After the block's energy has been converted into internal energy of the surroundings, all this potential for doing work is lost. The entropy increase of the universe in this case is $E_k/T = \Delta S_u$. The energy which is made unavailable in this case is $T \Delta S_u$, where T is the temperature of the available reservoir and ΔS_u is the increase in entropy of the universe. This is a general result.

In an irreversible process in which the entropy change of the universe is ΔS_u some energy becomes unavailable for use to perform work. The amount of energy made unavailable equals $T \Delta S_u$, where T is the temperature of the lowest-temperature reservoir available.

In another example, the free adiabatic expansion of an ideal gas, we assume that we have one reservoir at the temperature of the gas, which is the same after the expansion as before. The entropy change when the gas expands freely from volume V_1 to volume V_2 is $nR \ln (V_2/V_1)$, as calculated in Example 19-4. Since there are no effects on the surroundings, this is the entropy change of the universe. According to the result stated above, the energy made unavailable for doing work is T times this entropy change, or $nRT \ln (V_2/V_1)$. This is just the work that could have been done if we had allowed the gas to expand isothermally and quasi-statically. During this reversible expansion, the gas does work $W = nRT \ln (V_2/V_1)$, as previously calculated, while absorbing an equal amount of heat from the reservoir. Thus the result of the irreversible adiabatic free expansion is the loss of the opportunity of converting this internal energy of the reservoir into work.

In our final example the loss of ability to do work occurs in the irreversible process of conducting heat Q from a hot reservoir to a cold reservoir. The entropy change of the universe for this process was computed in the previous section:

$$\Delta S_u = \frac{Q}{T_c} - \frac{Q}{T_h}$$

Q is *not* the amount of energy made unavailable for work because, according to the second law, we could not have removed this heat from the hot reservoir and changed it completely into work. However, some of this energy could have been converted into work, the amount depending on the efficiency of the heat engine used. The greatest amount of work that can be obtained from the heat energy Q is the amount done using a Carnot engine. The efficiency of a Carnot engine is

$$\epsilon_R = 1 - \frac{T_c}{T_h}$$

The work that can be done with heat Q entering a Carnot engine is thus

$$W = \epsilon_R Q = Q - \frac{QT_c}{T_h} = T_c \left(\frac{Q}{T_c} - \frac{Q}{T_h} \right)$$

Again the work that could have been done is just the temperature of the lowest-temperature heat reservoir available T_c times the entropy change of the universe.

19-9   Molecular Interpretation of Entropy

There is a fundamental difference between the internal energy of a system and the macroscopic mechanical energy of a system. Figure 19-12 shows a schematic illustration of the velocities of the molecules in a system whose center of mass is at rest. If the system is a solid, the molecules are vibrating about their equilibrium positions; if it is a gas, the molecules move throughout the volume. In either case, the velocity vectors $\mathbf{u}_i$ point in all directions. In a gas the velocity of any one molecule points in a direction unrelated to that of any other molecule unless the two molecules have just collided. The total linear momentum and the total angular momentum of the system are zero. The average kinetic energy associated with each degree of freedom is $\frac{1}{2}kT$ per molecule. Suppose we now add a velocity v in the x direction to the velocity of each molecule. The center of mass of the system then moves with speed v in the x direction, and the total kinetic energy is the sum of the center-of-mass energy $\frac{1}{2}Mv^2$ and the energy relative to the center of mass,

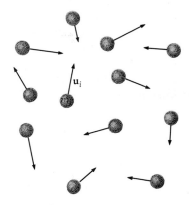

Figure 19-12
In a body at rest, the velocities of the molecules $\mathbf{u}_i$ are randomly directed.

$$E_k = \tfrac{1}{2}Mv^2 + \Sigma \tfrac{1}{2}m_i u_i{}^2 \qquad\qquad 19\text{-}26$$

where M is the total mass. The two terms in the expression for kinetic energy are quite different. The motion giving rise to the term $\frac{1}{2}Mv^2$ is an ordered motion. Each molecule has the same component v in the same direction. Thus all the molecules are moving together. The center of mass moves with velocity v. This *mechanical kinetic energy* has nothing to do with the internal energy or the temperature. On the other hand, the second term (the energy relative to the center of mass) is the original internal kinetic energy. It is associated with the random, or *disordered*, motion of the molecules. The ordered part of the total energy of the system, the mechanical kinetic energy, can easily be changed completely into other forms, e.g., potential energy or rotational kinetic energy (we could attach a string to the moving system and use it to lift a weight or turn a wheel, which could even generate electric energy).

On the other hand, the disordered part of the total energy, the internal energy, cannot easily be converted into mechanical energy. As we have seen, only part of this energy can be extracted and converted into work, which requires that some of the random motion be converted into ordered motion. We can do this with a heat engine, say with gas as the working substance. We place the gas system in contact with the system whose internal energy we want to convert into mechanical energy, calling this system the reservoir, in keeping with previous usage. If the gas is at a slightly lower temperature, some internal energy is transferred to the gas by conduction. In the molecular picture of conduction the molecules in the walls of the gas enclosure are vibrating with less average energy than those of the walls of the reservoir because the gas temperature is lower. When the two systems are in contact, some of the kinetic energy of vibration of molecules in the reservoir walls is transferred to kinetic energy of the molecules in the gas-container walls. (This energy is quickly brought to equilibrium with the potential energy of vibration of the wall molecules.) Now the gas container is not in equilibrium with the gas molecules, and when the gas molecules collide with the container walls, energy is transferred on the average to the gas molecules. These molecules in turn collide with others and transfer kinetic energy, until the temperature of the gas and wall has increased to that of the reservoir. The pressure of the gas is thus increased. This random internal energy can now be transferred into ordered mechanical energy by letting the gas at higher pressure expand against a piston. Since the piston can move in only one direction, the transfer of energy of the gas molecules to energy of the piston amounts to transfer of random kinetic energy of the gas molecules to ordered motion of the piston.

We have seen in this chapter that ordered mechanical energy is generally more useful than unordered internal energy. The second law of thermodynamics can be stated: The entropy of the universe always increases (or in reversible processes remains the same). This can be related to the fact that molecular motion always tends toward disordered rather than ordered motion. Imagine a block with mechanical kinetic energy $\frac{1}{2}Mv^2$ moving toward a wall and making an inelastic collision, thereby changing the kinetic energy into internal kinetic energy of disordered motion of the molecules in the block, wall, and air. Our experience and intuition tell us that the reverse will not happen. Disordered motion of the molecules will never spontaneously change into ordered motion. Thus entropy increase is an increase in the disordered motion of the molecules in the universe. When the entropy of the universe reaches its maximum value, no more processes can occur (except absolutely reversible processes). Heat will not be conducted because the whole universe will be at the same temperature. No more energy will be available to perform work.

Questions

12. When a gas is compressed isothermally, it gives up heat and its entropy decreases. In what sense is the motion of the gas molecules more orderly after compression?

13. When you pick up a spilled deck of cards and put them in a neat pile, you have increased the order of the universe. What compensating decrease in order accompanied your action?

14. One liter each of oxygen gas and nitrogen gas are initially in separate containers at room temperature and atmospheric pressure. Then the containers are connected, allowing the gases to mix. What happens to the entropy of the system consisting of the two gases?

Optional

19-10 Proof that Entropy Exists and that the Entropy of the Universe Can Never Decrease

The general proof of the existence of the entropy function is not difficult but rather abstract. It is interesting, however, because it shows that entropy and its increase are directly related to the second law of thermodynamics. We first state and prove the *Clausius inequality*.

Let a system undergo any cycle in which it absorbs and rejects heat from any number of different heat reservoirs, performs positive or negative work, and returns to its original state. Let ΔQ_i be the heat absorbed from the ith reservoir at temperature T_i. If heat is rejected to the reservoir, ΔQ_i will be negative. The Clausius inequality states that the sum of $\Delta Q_i/T_i$ is less than or equal to zero:

$$\sum \frac{\Delta Q_i}{T_i} \leq 0 \tag{19-27}$$

To prove this inequality we use an auxiliary reservoir at temperature T_0 and as many Carnot cycles as we need. We run a Carnot engine between each reservoir and our auxiliary reservoir so that the original reservoir is returned to its original state. That is, if ΔQ_i was removed from the ith reservoir, we run a Carnot engine between it and the auxiliary reservoir to replace ΔQ_i in the ith reservoir. The heat extracted from the auxiliary reservoir when the engine runs between it and the ith reservoir is given by Equation 19-4,

$$\Delta Q_{0i} = \frac{T_0}{T_i} \Delta Q_i \tag{19-28}$$

We note that if ΔQ_i is positive, ΔQ_{0i} is positive, and if ΔQ_i is negative, ΔQ_{0i} is negative, indicating that heat was put into the auxiliary reservoir. The total heat *extracted* from the auxiliary reservoir is

$$\Delta Q_0 = \sum \Delta Q_{0i} = T_0 \sum \frac{\Delta Q_i}{T_i}$$

After this procedure all the original reservoirs are in their original state. The net result is that heat ΔQ_0 has been extracted from the auxiliary reservoir and, according to the first law of thermodynamics, an equal amount of work has been done. The second law of thermodynamics tells us that ΔQ_0 must be less than or equal to zero, so that either no heat was extracted or negative work was done and heat was put into the auxiliary reservoir. Thus $\Delta Q_0 \leq 0$ and therefore

$$\sum \frac{\Delta Q_i}{T_i} \leq 0$$

The Clausius inequality has thus been proved. If the system undergoes a reversible cycle, absorbing heat ΔQ_i from the ith reservoir, the equality in Equation 19-27 holds. We can see this by merely reversing the cycle, putting ΔQ_i into each reservoir, which merely changes the sign of ΔQ_i, giving

$$-\sum \frac{\Delta Q_i}{T_i} \leq 0$$

The only possibility for a reversible cycle is thus

$$\sum \frac{\Delta Q_i}{T_i} = 0 \qquad \text{reversible cycle} \tag{19-29}$$

We are now in position to define entropy for a general system. We let the system move reversibly from state A to state B along one path and move back to state A along a different *reversible* path. Any number of reservoirs can be used. According to Equation 19-29, we have

$$\sum_{\substack{A \\ \text{path 1}}}^{B} \frac{\Delta Q_i}{T_i} + \sum_{\substack{B \\ \text{path 2}}}^{A} \frac{\Delta Q_i}{T_i} = 0$$

Since each path is reversible, we have

$$\sum_{\substack{B \\ \text{path 2}}}^{A} \frac{\Delta Q_i}{T_i} = -\sum_{\substack{A \\ \text{path 2}}}^{B} \frac{\Delta Q_i}{T_i}$$

so that

$$\sum_{\substack{A \\ \text{path 1}}}^{B} \frac{\Delta Q_i}{T_i} - \sum_{\substack{A \\ \text{path 2}}}^{B} \frac{\Delta Q_i}{T_i} = 0 \quad \text{or} \quad \sum_{\substack{A \\ \text{path 1}}}^{B} \frac{\Delta Q_i}{T_i} = \sum_{\substack{A \\ \text{path 2}}}^{B} \frac{\Delta Q_i}{T_i}$$

Thus the sum $\Sigma(\Delta Q_i/T_i)$ is the same for all reversible paths connecting state A to state B. Figure 19-13 shows an arbitrary quasi-static process for a gas. In order to make this process reversible, we must absorb heat isothermally. We do so by using a very large number of heat reservoirs at different temperatures and following the alternative path, consisting of a series of isotherms and adiabatic paths. In the limit of an infinite number of reservoirs, we can make this alternate path approach the original path as closely as desired. Thus for a general reversible path we need a large number of heat reservoirs, and the heat absorbed from each is very small. We thus write $đQ_R$ for ΔQ_i and indicate the sum by an integral sign, leaving off the subscripts i

$$\int_{\text{path 1}} \frac{đQ_R}{T} = \int_{\text{path 2}} \frac{đQ_R}{T}$$

Since the integral of $đQ_R/T$ is independent of path between two states, it can depend only on the states. We thus can define a new function of state, the entropy S, by

$$S_B - S_A = \int_A^B \frac{đQ_R}{T}$$

where again we put a subscript R on $đQ$ to remind us that the process must be reversible. For a very small change we can write

$$dS = \frac{đQ_R}{T}$$

These are the same as Equations 19-15 and 19-16.

We now consider a system which undergoes an *irreversible* change from state A to state B and then is brought back to state A by a reversible process. This cycle is indicated in Figure 19-14, where the wiggly line indicates the irreversible process, for which we cannot plot a curve because the system is not in equilibrium. According to the Clausius inequality, we have for the complete cycle

$$\underbrace{\int_A^B \frac{đQ}{T}}_{\substack{\text{Irreversible} \\ \text{path}}} + \int_B^A \frac{đQ_R}{T} < 0 \qquad\qquad 19\text{-}30$$

or

$$\underbrace{\int_A^B \frac{đQ}{T}}_{\substack{\text{Irreversible} \\ \text{path}}} < -\int_B^A \frac{đQ_R}{T} = \int_A^B \frac{đQ_R}{T}$$

where we have used

$$\int_A^B \frac{đQ_R}{T} = -\int_B^A \frac{đQ_R}{T}$$

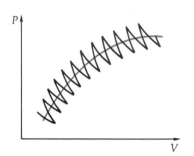

Figure 19-13
An arbitrary path can be approximated as accurately as desired by a series of alternating quasi-static adiabatic and isothermal paths.

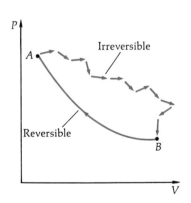

Figure 19-14
An irreversible transition from A to B followed by a reversible transition from B to A.

for the *reversible* path. By our definition of entropy we have

$$S_B - S_A = \int_A^B \frac{dQ_R}{T}$$

Thus

$$\int_A^B \frac{dQ}{T} < S_B - S_A$$
Irreversible
path

In particular, if a system is isolated from its surroundings, no heat can enter or leave. When an isolated system changes from state A to state B, $dQ = 0$. Then if the process is irreversible,

$$0 < S_B - S_A \qquad S_B > S_A$$

We can consider the universe to be an isolated system. We have thus shown that the entropy of the universe always increases for any irreversible process. The defining equation (19-15) for entropy when applied to the universe with $dQ_R = 0$ shows that for reversible processes the entropy change of the universe is zero.

Review

A. Define, explain, or otherwise identify:

Heat engine, 464
Efficiency of a heat engine, 465
Kelvin-Planck statement of the
  second law, 465
Clausius statement of the
  second law, 466
Coefficient of performance, 466
Equivalent statements, 467
Reversible and irreversible
  processes, 469

Carnot engine, 470
Carnot cycle, 472
Absolute temperature scale, 473
Carnot efficiency, 473
Entropy, 478
Available energy, 483
Ordered and disordered motion, 485

B. True or false:

1. Work can never be converted completely into heat.

2. Heat can never be converted completely into work.

3. All heat engines have the same efficiency.

4. It is impossible to transfer a given quantity of heat from a cold reservoir to a hot reservoir.

5. The coefficient of performance of a refrigerator cannot be greater than 1.

6. All reversible processes are quasi-static.

7. All quasi-static processes are reversible.

8. All Carnot cycles are reversible.

9. The entropy of a system can never decrease.

10. The entropy of the universe can never decrease.

Exercises

Section 19-1, Heat Engines and the Second Law of Thermodynamics

1. An engine with 20 percent efficiency does 100 J of work in each cycle. (*a*) How much heat is absorbed in each cycle? (*b*) How much heat is rejected?

2. An engine absorbs 100 cal of heat and does 120 J of work in each cycle. (a) What is its efficiency? (b) How much heat is rejected in each cycle?

3. An engine absorbs 100 cal and rejects 60 cal in each cycle. (a) What is its efficiency? (b) If each cycle takes 0.5 sec, find the power output of this engine in watts.

4. A refrigerator absorbs 5000 cal from a cold reservoir and rejects 8000 cal. (a) Find the coefficient of performance Q_c/W of the refrigerator. (b) The refrigerator is reversible and is run as a heat engine ($Q_h = 8000$ cal; $Q_c = 5000$ cal). What is its efficiency?

5. An engine with an output of 200 W has an efficiency of 30 percent. It works at 10 cycles/sec. How much heat is absorbed and how much rejected in each cycle?

Section 19-2, Equivalence of the Kelvin-Planck and Clausius Statements

6. Suppose that there are two statements A and B, each of which may be either true or false. (a) List the four possible combinations of truth or falsity for the two statements. (b) What possibilities are left if it is shown that if A is true, B must also be true? Does this prove that the statements are equivalent? (c) Suppose that it is also shown that if A is false, B is false. Do these two results prove that the two statements are equivalent?

7. A certain engine running at 30 percent efficiency draws 200 cal of heat from a hot reservoir. Assume the Clausius statement to be false and show how this engine combined with a perfect refrigerator can violate the Kelvin-Planck statement of the second law.

8. A certain refrigerator takes in 500 J of heat from a cold reservoir and rejects 800 J to a hot reservoir. Assume that the Kelvin-Planck statement is false and show how a perfect engine working with this refrigerator can violate the Clausius statement.

Section 19-3, Reversibility

9. Show that the reversibility of the following processes would violate the second law of thermodynamics: (a) an object is dropped from a height H to the ground and comes to rest; (b) a car traveling at high speed brakes to a stop; (c) a hot object is placed in a cold-water bath, and the system comes to an equilibrium temperature.

Section 19-4, The Carnot Engine

10. A reversible engine working between reservoirs at temperatures T_h and T_c has efficiency of 30 percent. When working as a heat engine it rejects 140 cal of heat to the cold reservoir. A second engine working between the same two reservoirs also rejects 140 cal to the cold reservoir. Show that if the second engine has an efficiency of greater than 30 percent, the two engines can work together to violate the Kelvin-Planck statement of the second law.

11. A reversible engine working between reservoirs at temperatures T_h and T_c has an efficiency of 20 percent. When working as a heat engine, it does 100 J of work in each cycle. A second engine working between the same two reservoirs also does 100 J of work in each cycle. Show that if the efficiency of the second engine is greater than 20 percent, the two engines can work together to violate the Clausius statement of the second law.

12. A Carnot engine works between two heat reservoirs as a refrigerator. It removes 100 cal from the cold reservoir and rejects 150 cal to the hot reservoir during each cycle. Its coefficient of performance is $\eta = Q_c/W = 100/50 = 2$. (a) What is the efficiency of the Carnot engine when working as a heat engine

between the same two reservoirs? (*b*) Show that no other engine working as a refrigerator between the same two reservoirs can have a coefficient of performance greater than 2.

Section 19-5, The Absolute Temperature Scale

13. A Carnot engine works between two heat reservoirs at temperatures $T_h = 300$ K and $T_c = 200$ K. (*a*) What is its efficiency? (*b*) If it absorbs 100 cal from the hot reservoir during each cycle, how much work does it do? How much heat does it reject during each cycle? (*c*) What is the coefficient of performance of this engine when working as a refrigerator between these two reservoirs?

14. An engine works between the steam point 100°C and the ice point 0°C. What is its greatest possible efficiency?

15. Which has the greater effect on increasing the efficiency of a Carnot engine, a 5-K increase in the temperature of the hot reservoir or a 5-K decrease in the temperature of the cold reservoir (in each case keeping the other reservoir fixed)?

16. A refrigerator works between 0°C and room temperature of 20°C. (*a*) What is the greatest possible coefficient of performance? (*b*) If the refrigerator is to be cooled to −10°C, what is the greatest coefficient of performance assuming the same room temperature of 20°C?

17. An engine draws heat from a reservoir at a temperature of 100°C. If the engine is to be 40 percent efficient, what is the warmest possible temperature of the cold reservoir?

18. An engine is designed to exhaust heat to the atmosphere at $t = 20$°C. What is the least possible temperature of the upper reservoir if the engine is to have an efficiency of 20 percent?

Section 19-6, Entropy, and Section 19-7, Entropy Change of the Universe

19. Two moles of an ideal gas at $T = 400$ K expand quasi-statically and isothermally from an initial volume of 40 ℓ to a final volume of 80 ℓ. (*a*) Find the entropy change of the gas. *Hint:* Use Equation 19-21. (*b*) What is the entropy change of the universe for this process?

20. The gas in Exercise 19 is taken from the same initial state ($T = 400$ K, $V_1 = 40$ ℓ) to the same final state ($T = 400$ K, $V_2 = 80$ ℓ) by a non-quasi-static process. (*a*) Is the entropy change of the gas the same as, greater than, or less than that found in Exercise 19? (*b*) Is the entropy change of the universe greater than, less than, or the same as that found in Exercise 19?

21. A system absorbs 200 cal of heat reversibly from a reservoir at 300 K and rejects 100 cal reversibly to a reservoir at 200 K as it moves from state A to state B. During this process, which is quasi-static, 50 cal of work is done by the system. (*a*) What is the change in internal energy of the system? (*b*) What is the change in entropy of the system? (*c*) What is the change in entropy of the universe? (*d*) If the system went from state A to state B by a non-quasi-static process, how would your answers to parts (*a*), (*b*), and (*c*) differ?

22. A system absorbs 300 cal from a reservoir at 300 K and 200 cal from a reservoir at 400 K. It returns to its original state, doing 100 cal of work and rejecting 400 cal of heat to a reservoir at temperature T. (*a*) What is the entropy change of the system for the complete cycle? (*b*) If the cycle is reversible, what is the temperature T?

23. Two moles of an ideal gas originally at $T = 400$ K and $V = 40$ ℓ undergoes a free adiabatic expansion to twice its volume. What is (*a*) the entropy change of the gas and (*b*) the entropy change of the universe?

24. A 5-kg block is dropped from rest at a height of 6 m above the ground. It hits the ground and comes to rest. The block, ground, and atmosphere are all at 300 K initially. What is the entropy change of the universe for this process?

25. If 500 cal of heat is conducted from a reservoir at 400 K to one at 300 K, what is the entropy change of the universe?

26. A 200-kg block of ice at 0°C is placed in a large lake. The temperature of the lake is just slightly higher than 0°C, and the ice melts. (a) What is the entropy change of the ice? (b) What is the entropy change of the lake? (c) What is the entropy change of the universe (ice plus lake)?

27. Calculate the entropy change of 1 gm of water at 100°C when it changes to steam under standard conditions.

Section 19-8, Entropy and the Availability of Work

28. A heat engine works in a cycle between reservoirs at 400 and 200 K. The engine absorbs 1000 cal of heat from the hot reservoir and does 200 cal of work in each cycle. (a) What is the efficiency of this engine? (b) Find the entropy change of the engine, each reservoir, and of the universe for each cycle. (c) What is the efficiency of a Carnot engine working between the same two reservoirs? How much work could be done by a Carnot engine in each cycle if it absorbed 1000 cal from the hot reservoir? (d) Show that the difference in the work done by the Carnot engine and the original engine is $T_c \Delta S_u$, where ΔS_u is that calculated in part (b).

29. A refrigerator works between two reservoirs at 200 and 400 K. In each cycle it absorbs 200 cal from the cold reservoir and rejects 600 cal to the hot reservoir. (a) How much work is put into the refrigerator in each cycle? (b) What is the entropy change of the universe in each cycle? (c) If a Carnot engine is used as a refrigerator between these two reservoirs, how much heat would be rejected if it absorbed 200 cal from the cold reservoir? How much work is needed? Hint: The easiest way to find these is to use the fact that the entropy change of the universe is zero for a reversible process. (d) Show that the additional work required by the first refrigerator is $T_h \Delta S_u$, where ΔS_u is that found in part (b). (e) The additional work required by the first engine is not completely wasted because the energy is stored in the hot reservoir. Find how much of this additional work could be recovered using a Carnot engine between the reservoirs and show that the amount of work completely wasted by the first engine is equal to $T_c \Delta S_u$.

30. How much of the 500 cal of heat taken from a reservoir at 400 K in Exercise 25 could have been converted into work using a cold reservoir at 300 K?

31. One mole of an ideal gas first undergoes an adiabatic free expansion from $V_1 = 12.3 \ \ell$, $T_1 = 300$ K to $V_2 = 24.6 \ \ell$, $T_2 = 300$ K. It is then compressed isother- mally and quasi-statically back to its original state. (a) What is the entropy change of the universe for the complete cycle? (b) How much work is wasted in this cycle? (c) Show that the work wasted is $T \Delta S_u$.

32. Which process is more wasteful: (a) a block moving with 500 J of kinetic energy being slowed to rest by friction (temperature of the atmosphere = 300 K) or (b) 1000 J of heat conducted from a reservoir at 400 K to one at 300 K? Hint: How much of the 1000 J of heat could be converted into work in an ideal situation? (c) Compute the entropy change of the universe in each case.

Section 19-9, Molecular Interpretation of Entropy

33. Air consists mainly of diatomic molecules (N_2 and O_2) which can rotate about two axes. The total kinetic energy of 1 mole of air can be written $E_k = \frac{1}{2} M v_{CM}^2 + \frac{5}{2} RT$, where $M = 29.0$ gm/mole is the molecular weight and v_{CM} is

the velocity of the center of mass. Which has more total energy, 1 mole of air at rest at 25°C or 1 mole of air at 20°C with a wind speed of 10 m/sec (about 22.4 mi/h)? Which could be used to run a windmill?

Problems

1. An engine operates with a working substance which is 1 mole of an ideal gas with $C_v = \frac{3}{2}R$ and $C_p = \frac{5}{2}R$. The cycle begins at $P_1 = 1$ atm and $V_1 = 24.6$ ℓ. The gas is heated at constant volume to $P_2 = 2$ atm. It then expands at constant pressure until $V_2 = 49.2$ ℓ. During these two steps, heat is absorbed. The gas is then cooled at constant volume until its pressure is again 1 atm. It is then compressed at constant pressure back to its original state. During the last two steps heat is rejected. All steps are quasi-static and reversible. (a) Indicate this cycle on a PV diagram. Find the work done, the heat added, and the internal-energy change for each step of the cycle. (b) Find the efficiency of this cycle.

2. An engine using 1 mole of an ideal gas with $C_v = \frac{5}{2}R$ and $C_p = \frac{7}{2}R$ performs a cycle consisting of three steps: (1) an adiabatic expansion from initial pressure of 2.64 atm and volume 10 ℓ to a final pressure of 1 atm and volume 20 ℓ, (2) a compression at constant pressure to its original volume of 10 ℓ, (3) heating at constant volume to its original pressure of 2.64 atm. Find the efficiency of this cycle.

3. The behavior of a gasoline engine can be approximated by an ideal cycle called an *Otto cycle* (Figure 19-15). The steps are tabulated. (a) Compute the heat input and the heat output and show that the efficiency can be written

$$\epsilon = 1 - \frac{T_e - T_b}{T_d - T_c}$$

where T_e is the temperature of state e, etc. (b) Using the relation for adiabatic expansion or compression $TV^{\gamma-1} = \text{constant}$, show that $\epsilon = 1 - (V_1/V_2)^{\gamma-1}$, where $V_1 = V_c = V_d$ and $V_2 = V_b = V_e$. (c) The ratio V_2/V_1 is called the *compression ratio*. Find the efficiency of this cycle for a compression ratio of 8 (use $\gamma = 1.5$ for ease in calculation). (Compression ratios much greater than this cannot be achieved in real engines because of preignition problems.) (d) Explain why the efficiency of a real gasoline engine might be much less than that calculated in part (c).

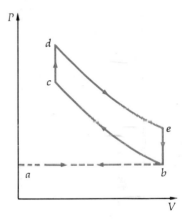

Figure 19-15
Otto cycle for Problem 3.

No.	Step	Action	Comment
1	$a \to b$	Air intake at constant pressure	Number of moles varies from 0 to n; volume varies from 0 to V_b, given by $P_0 V_b = nRT_b$, where T_b = temperature of outside air
2	$b \to c$	Adiabatic compression to pressure P_c, temperature T_c, and volume V_c	
3	$c \to d$	Heating at constant volume	To approximate effect of explosion in gasoline engine
4	$d \to e$	Adiabatic expansion to volume $V_e = V_b$	The power stroke in the cycle
5	$e \to b$	Cooling at constant volume until pressure is again atmospheric pressure P_0	
6	$b \to a$	Exhaust of gas at constant pressure	

4. An engine using 1 mole of an ideal gas initially at $V_1 = 24.6 \ \ell$ and $T = 400$ K performs in a cycle which consists of four steps: (1) isothermal expansion at $T = 400$ K to twice its volume, (2) cooling at constant volume to $T = 300$ K, (3) isothermal compression to its original volume, and (4) heating at constant volume to its original temperature of 400 K. Assume that $C_v = 5$ cal/K. Sketch the cycle on a PV diagram and calculate its efficiency.

5. A steam engine takes in superheated steam at 270°C and discharges condensed steam from its cylinder at 50°C. Its efficiency is 30 percent. (a) How does this efficiency compare with the best efficiency possible for these temperatures? (b) If the useful power output of the engine is 200 kW, how much heat does the engine discharge to the surroundings in 1 h?

6. One mole of an ideal gas for which $\gamma = 1.4$ is carried through a Carnot cycle. The high temperature and pressure are 400 K and 4 atm; the low temperature and pressure are 300 K and 1 atm. (a) What is the volume when $T = 400$ K and $P = 4$ atm? (b) What are the values of PV and PV^γ for the isothermal and adiabatic lines through this point? (c) What is the volume when $T = 300$ K and $P = 1$ atm? (d) What are the values of PV and PV^γ for the isothermal and adiabatic lines through this point? (e) Make an accurate plot of the Carnot cycle.

7. If two adiabatic curves on a PV diagram intersected, a cycle could be completed using an isothermal path between the two adiabatics, as shown in Figure 19-16. Show that such a cycle could violate the second law of thermodynamics.

Figure 19-16
Problem 7.

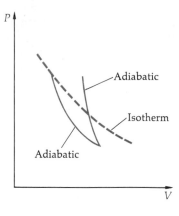

8. (a) Show that the coefficient of performance of a Carnot refrigerator working between two reservoirs at temperatures T_h and T_c is related to the efficiency of a Carnot engine by $\eta = T_c/\epsilon T_h$. (b) Show that the coefficient of performance of a Carnot refrigerator can be written

$$\eta = \frac{T_c}{T_h - T_c}$$

(c) Which gives the greater increase in coefficient of performance of a refrigerator, increasing the temperature of the cold reservoir by 5 K or decreasing the temperature of the hot reservoir by 5 K?

9. The cooling compartment of a refrigerator and its contents are at 5°C. Their heat capacity is 20 kcal/°C. The refrigerator exhausts heat to the room at 25°C. What is the smallest power motor which could be used to operate the refrigerator if it is to reduce the temperature of the interior and the contents by 1°C in 1 min?

10. (a) Show that if the Clausius statement of the second law were not true, the entropy of the universe could decrease. (b) Show that if the Kelvin-Planck statement of the second law were not true, the entropy of the universe could decrease. (c) An alternative statement of the second law is that the entropy of the universe cannot decrease. Have you just proved that this statement is equivalent to the Clausius and Kelvin-Planck statements?

11. Equation 19-14 gives the entropy change of an ideal gas when the volume and temperature increase reversibly. Use this equation and the relation $TV^{\gamma-1}$ = constant to show explicitly that the entropy change is zero for an adiabatic expansion from state $V_1 T_1$ to state $V_2 T_2$.

12. A substance is heated reversibly at constant pressure from temperature T_1 to temperature T_2. Show that the entropy of the substance increases by the amount $\Delta S = C_p \ln (T_2/T_1)$, where C_p is the heat capacity at constant pressure. Is this also the entropy change of the substance if the heating is done irreversibly?

13. A 100-gm piece of ice at 0°C is placed in an insulated container with 100 gm of water at 100°C. (a) When equilibrium is established, what is the

final temperature of the water? (*b*) Find the entropy change of the universe for this process (see Problem 12). (Ignore the heat capacity of the container.)

14. A block of metal has a heat capacity $C_p = 100$ cal/K. It is to be heated from $T_1 = 200$ K to $T_2 = 400$ K. In Problem 12 it was shown that the entropy increase of the metal is $\Delta S = C_p \ln (T_2/T_1) = 100 \ln 2 = 69.3$ cal/K. In this problem you are to investigate the entropy change in the universe when the block is heated in several different ways. (*a*) The block originally at 200 K is placed in contact with a reservoir at temperature 400 K. How much heat is needed to raise the temperature of the block to 400 K? What is the entropy change in the reservoir when it loses this much heat? What is the entropy change of the universe for this process? (*b*) The block originally at 200 K is first placed in contact with a reservoir at 300 K until it comes to thermal equilibrium. It is then placed in contact with a reservoir at 400 K until its temperature is 400 K. Compute the entropy change of each reservoir and of the universe and compare your result with the process in part (*a*). (*c*) The block originally at 200 K is heated by first placing it in contact with a reservoir at 250 K until it comes to equilibrium. It is heated to 300 K, then to 350 K, and finally to 400 K, using successively reservoirs at 300, 350, and 400 K. Compute the entropy change of each reservoir and the entropy change of the universe for this process. Explain how reversible heating can be approximately accomplished in practice.

CHAPTER 20   Wave Pulses

Wave motion can be thought of as the transport of energy and momentum from one point in space to another without the transport of matter. In mechanical waves, e.g., water waves, waves on a string, or sound waves, the energy and momentum are transported by means of a disturbance in the medium which is propagated because the medium has elastic properties. On the other hand, in electromagnetic waves, the energy and momentum are carried by electric and magnetic fields which can propagate through vacuum. (We shall define an electric field and magnetic field in later chapters. For our study of waves, we need only say that these fields are represented by vectors, are produced by electric charges, and exert forces on other charges. These fields are similar to the gravitational field, which is produced by masses and exerts forces on other masses.)

The variety of wave phenomena observed in nature is immense. Yet in this immense variety are many features common to all kinds of waves and others shared by a wide range of wave phenomena. In the next few chapters we shall study these general properties of waves, illustrating them with different wave phenomena observed in nature.

We first consider wave motion in one dimension. As we found in our study of mechanics, much can be learned before introducing the complications of two and three dimensions. In this chapter we shall study the motion of wave pulses with examples in waves on strings and sound waves. We shall calculate the speed of these waves in terms of properties of the transmitting media by applying the laws of mechanics. We shall study the properties of harmonic waves in Chapter 21 and of standing waves in Chapter 22. In Chapter 23 we consider the superposition of waves of different frequency. The general properties of waves in two and three dimensions will be considered in Chapters 24 and 25. Chapters 26 and 27 deal with the especially important topics of light and other electromagnetic waves. Chapter 28, a brief introduction to the special theory of relativity, is a revision of newtonian mechanics of special importance for motion at great speeds. It is discussed at that point because its discovery was intimately related to research on light waves.

20-1 Wave Pulses

When a string stretched under tension is given a flip, as in Figure 20-1, the shape of the string changes in time in a regular way. The bump which is produced at the end by the flip and which travels down the string is called a *wave pulse*. It is a disturbance in the string, i.e., a distortion of the shape of the string from its normal, or equilibrium, shape. The pulse travels down the string at a definite speed which depends on the nature of the string and the tension. As it moves, the pulse usually changes shape, gradually spreading out. This effect, called *dispersion*, occurs to some extent in all waves (except electromagnetic waves in vacuum), but in many important examples the dispersion is negligible and the wave pulse travels with approximately the same shape. The fate of the pulse at the other end of the string depends on how the string is fastened at that end. If it is tied to a rigid support, the pulse will be reflected and return inverted, as in Figure 20-2. If the support is not rigid, some or all of the pulse will be absorbed or it may be reflected without inversion.

Dispersion defined

A pulse traveling down a stretched string is only one example of a wave pulse: the report of a gunshot is a wave pulse of sound; a lightning flash is a wave pulse of light; a tidal wave is a wave pulse in water.

A chief characteristic of a wave pulse is that it has a beginning and an end. It is a disturbance of limited extent. At any instant only a limited region of space is disturbed. At any point, the wave pulse passes by in a limited time. We can produce a light pulse, for example, by quickly opening and closing a shutter in front of a continuous light source. Figure 20-3*a* shows such a pulse with an upward and a downward bump, easily produced on a string. A wave which has many oscillations up and down, as in Figure 20-3*b*, is usually not called a wave

Wave pulse

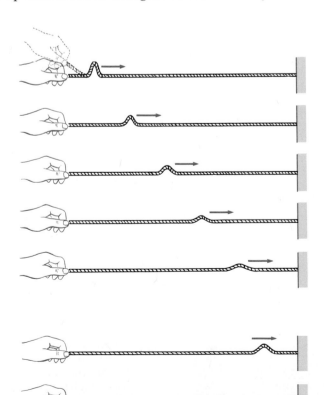

Figure 20-1
Wave pulse moving to the right on a string. The change in shape of the pulse as it moves is called dispersion.

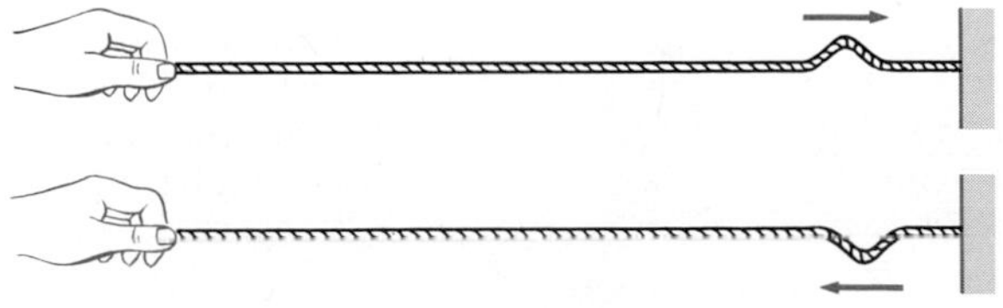

Figure 20-2
A pulse arriving at the rigid support of a stretched string is reflected and inverted.

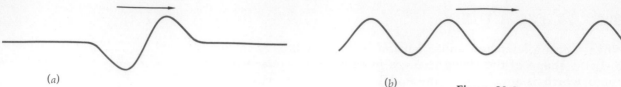

Figure 20-3
(*a*) A wave pulse may include both positive and negative displacement. (*b*) A wave with many alternating positive and negative displacements ordinarily is not called a pulse.

pulse. To a good approximation such a wave can be treated as an endless series of alternate up and down disturbances. Unlike a wave pulse of limited extent, such a wave does not spread out as it travels. This kind of wave is also common in nature. We shall study it in Chapter 21.

We can demonstrate that energy and momentum are transported by a wave pulse by hanging a weight on a string under tension, as shown in Figure 20-4, and giving the string a flip at one end. When the pulse arrives at the weight, the weight will be lifted momentarily. The action of the hand in flipping one end of the string is transmitted along the string until the weight is lifted at the other end. This is usually described by saying that the momentum introduced by the hand flipping the end of the string is transmitted along the string and is received by the weight. Similarly, the work done by the hand in lifting the string goes into kinetic and potential energy of the string at the pulse. (We shall calculate this energy in the string later.) This energy is transmitted along the string and is converted into work lifting the weight. Thus *both energy and momentum are transmitted by such a wave pulse.*

It is not the mass elements of the string that are transported but the *disturbance in the shape* caused by flipping one end. The mass elements of the string, in fact, move in a direction perpendicular to the string and thus perpendicular to the direction of motion of the pulse. A wave in which the disturbance is perpendicular to the direction of the wave is called a *transverse wave.* Other examples of transverse waves are any electromagnetic wave (light, radar, radio, television, etc.) produced by a vibrating charge system which produces an alternating electric and magnetic field. The electric and magnetic field vectors are perpendicular to the direction of propagation of the electromagnetic wave (and, as we shall see later, perpendicular to each other). Sound waves in air, on the other hand, are not transverse waves. In sound waves, a disturbance in the pressure and density of air is set up by the vibration of a body (say a tuning fork or a violin string), and the disturbance is propagated through the air by the collisions of air molecules. The motion of the air molecules is in the same direction as the propagation. Any wave in which the disturbance is parallel to the direction of propagation is called a *longitudinal wave.* Sound consists of longitudinal waves. A longitudinal pulse in a spring analogous to a sound pulse can be produced by suddenly compressing the spring (Figure 20-5).

Water waves are neither completely transverse nor completely longitudinal but a combination of the two. Figure 20-6 shows the motion of a water particle in a water wave. The water particle moves in a

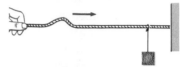

Figure 20-4
Wave pulses transmit both energy and momentum, indicated here by the upward motion of the weight when the pulse arrives.

Waves transmit momentum and energy

Transverse and longitudinal waves defined

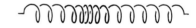

Figure 20-5
Longitudinal wave pulse in a spring. In a longitudinal wave, the disturbance is in the direction of motion of the wave.

Figure 20-6
Surface waves on water. The water particles on the surface move in nearly circular paths having both longitudinal and transverse components.

nearly circular path, its motion having both transverse and longitudinal components.

Photograph of a longitudinal wave pulse in a Slinky.

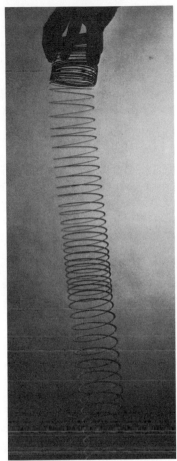

Questions

1. Give examples of wave pulses in nature in addition to those mentioned in the text. In each case, explain what kind of disturbance from equilibrium occurs.

2. Consider a long line of cars equally spaced by one car length and moving with the same speed. One car suddenly slows to avoid a dog and then speeds up until it is again one car length behind the car ahead. Discuss how the space between cars propagates back along the line. How is this like a wave pulse? Is there a transport of energy and momentum? Is the wave transverse or longitudinal? What does the speed of propagation depend on? (It may be useful to consider this situation in a reference frame in which all the cars are initially at rest.)

20-2   The Wave Function; Interference

Consider a pulse on a string at $t = 0$, as in Figure 20-7. The shape of the string at time $t = 0$ can be represented by some function $y = f(x)$. At some later time, the pulse is farther down the string, so that the shape at that time is some other function of x. Let us assume that the pulse does not vary in shape, i.e., is not dispersed, and introduce a new coordinate system with origin O' which moves with the speed v of the pulse. In this reference frame the pulse is stationary. The shape of the string is $y' = f(x')$ for all time. The coordinates of the two reference frames are related by

$$y = y' \quad \text{and} \quad x = x' + vt$$

Thus the displacement of the string in frame O can be written

$$y = f(x - vt) \qquad \text{wave moving right} \tag{20-1}$$

This same line of reasoning for a pulse moving to the left leads to

$$y = f(x + vt) \qquad \text{wave moving left} \tag{20-2}$$

In each of these expressions v is the speed of propagation of the wave. The function $y = f(x - vt)$ is called the *wave function*. For waves on a string, the wave function is the vertical displacement of the string at point x and time t. This wave function is thus a function of two variables. Analogous wave functions for sound waves are the pressure $P(x - vt)$ or a related function, the displacement from the equilibrium position of gas molecules. The wave function for electromagnetic waves is the electric field vector $\mathbf{E}(x - vt)$ or the magnetic field vector $\mathbf{B}(x - vt)$, which is related to the electric field.

Wave function defined

Figure 20-7
A wave pulse moving in the positive x direction with speed v relative to the origin O. In the primed coordinate system moving with the same speed as the pulse, the wave function is $y' = f(x')$ for all time, assuming no change in the shape of the pulse. In the unprimed system, the wave function is $y = f(x - vt)$.

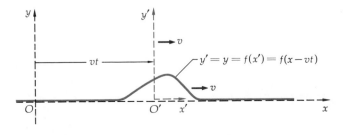

Consider two pulses on a string moving in opposite directions, as in Figure 20-8. The shape of the string when they meet can be found by adding the displacements produced by each pulse separately, as in the figure. For the special case of two pulses identical except that one is inverted relative to the other (Figure 20-9) there will be one time when the pulses exactly overlap and add to zero. At this time the string is horizontal, but it is not at rest. A short time later the pulses emerge, and each continues in its original direction. Combining separate waves to produce the resultant wave is called *interference*. Interference is the characteristic property of wave motion. It occurs whenever two waves meet in the same region of space. There is no analogous situation in particle motion. Two particles never overlap or add together in this way. Interference is unique to wave motion.

If we have two pulses on a string, as in Figures 20-8 and 20-9, we cannot write the shape of the string as a simple function of either $x - vt$ or $x + vt$ alone. However, the shape is still a function of the two variables x and t. Let us write $y(x,t)$ for the wave function of the string on which there are two pulses. That is, $y(x,t)$ is the vertical displacement of the point x at time t. We can treat the phenomenon of interference mathematically by noting that if we call the wave function for the single pulse moving to the right $y_1(x - vt)$ and that for the pulse moving to the left $y_2(x + vt)$, the total wave function is just the algebraic sum of the individual wave functions,

$$y(x,t) = y_1(x - vt) + y_2(x + vt) \qquad \text{20-3}$$

For example, if we have two pulses of the same shape but one inverted, there will be some time $t = t_1$ when they overlap. At this time $y_1(x - vt_1) = -y_2(x + vt_1)$, so that $y(x,t_1) = 0$.

The mathematical addition of two wave functions to form the resulting wave function, as in Equation 20-3, is called *superposition*. The statement that the resultant wave function is the algebraic sum of the individual wave functions is called the *principle of superposition*. In most wave phenomena small wave pulses obey this principle. Electromagnetic waves in vacuum always obey it. For other wave phenomena, the principle of superposition does not hold for very large pulses; i.e., the wave function produced by the interference of two very large pulses is not merely the algebraic sum of the individual wave pulses. Such waves are called *nonlinear* and are not studied in this book.

If one of the two pulses is inverted relative to the other, as in Figure 20-9, the algebraic addition when the pulses meet amounts to an arithmetic subtraction. The resultant pulse is smaller than the larger pulse and perhaps smaller than either. The pulses tend to cancel each other when they overlap. Such interference is called *destructive interference*. (Complete cancellation occurs only if the shapes are identical.)

On the other hand, if the displacement of the two pulses is in the same direction, the resultant pulse when they overlap is greater than either pulse by itself. This is called *constructive interference*.

Questions

3. When two waves moving in opposite directions interfere, does either impede the progress of the other?

4. Can you give an example where the principle of superposition is not obeyed in the interference of waves?

Figure 20-8
Two wave pulses moving in opposite directions on a string. The shape of the string when the pulses meet can be found by adding the displacements of each separate pulse. This kind of wave superposition is called constructive interference.

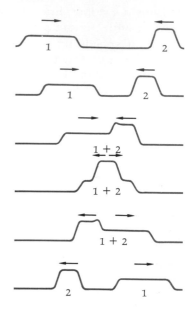

Figure 20-9
Two pulses having opposite displacements traveling in opposite directions on a string. Here the algebraic addition of the displacements of the separate pulses amounts to a subtraction of the magnitudes. This kind of wave superposition is called destructive interference.

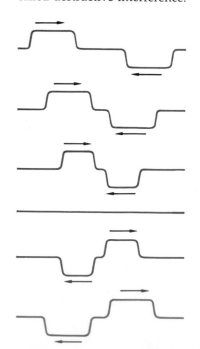

20-3   Velocity of Waves on a String

A general property of waves is that their velocity depends on the properties of the medium and is independent of the motion of the source relative to the medium. (There need not be a medium for light and other electromagnetic waves, which can propagate through a vacuum. The velocity of the electromagnetic waves in vacuum is fixed, about 3×10^8 m/sec.) For example, the velocity of a sound wave produced by a train whistle depends only on the properties of the air and not on the motion of the train. For wave pulses on a string which do not change their shape we can derive from laws of mechanics an expression for the velocity in terms of the properties of the string.

Consider a pulse moving to the right along a string with speed v. The speed v is related to the tension T in the string and the mass μ per unit length of the string by

$$v = \sqrt{\frac{T}{\mu}} \qquad\qquad 20\text{-}4$$

Velocity of transverse waves on a string

We shall give a derivation of this result which is somewhat artificial but illustrates that Equation 20-4 follows from the application of Newton's laws to the parts of the string. A more general derivation will be given in Section 21-8.

Let us consider the pulse in a reference frame O', in which it is at rest. In this frame, the string moves to the left with speed v, and the pulse is stationary. It is useful to imagine a glass tube of the same shape as the pulse and at rest in this frame, as in Figure 20-10a. We then have the remarkable phenomenon of the string passing through the tube without hitting the sides. A small segment of the string (Figure 20-10b) is moving in an approximately circular arc of radius r. The tension is just great enough to give the segment the centripetal acceleration v^2/r. If we increased the velocity of the string without changing the tension, the segment would hit the top of the tube; if we decreased the velocity, the segment would hit the bottom of the tube. Equation 20-4 is derived by setting the resultant force on the segment due to the tension equal to the mass of the segment times its centripetal acceleration.

Let θ be the angle subtended by the segment. Figure 20-10b shows that the tension T at each end of the segment makes an angle $\frac{1}{2}\theta$ with the tangent line at the center of the segment, which is horizontal for this segment. The radial component of the tension (vertical in this figure) at each end is $T \sin \frac{1}{2}\theta$. Since this tension acts at both ends, the resultant radial force which points toward the center of the circle has the magnitude

$$F_r = 2T \sin \tfrac{1}{2}\theta \approx 2T(\tfrac{1}{2}\theta) = T\theta$$

Figure 20-10
(a) A string under tension is pulled through a tube with speed v so that the string does not touch the tube. In the frame in which the string has no horizontal velocity, the tube and pulse move to the right along the string without changing shape. (b) Forces on a string segment moving in a circular arc. The centripetal acceleration of the segment is provided by the radial components of the tension.

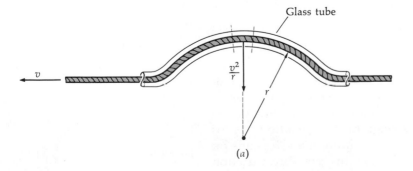

(a)

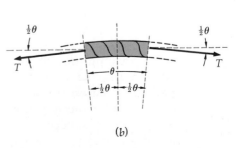

(b)

assuming that the angle θ is small enough so that the approximation $\sin \frac{1}{2}\theta \approx \frac{1}{2}\theta$ can be used. The length of the segment is $r\theta$, where r is the radius of the circle. Hence the mass of the segment is

$$m = \mu L = \mu r\theta$$

where μ is the mass per unit length of the string (the linear mass density). Setting the resultant force $T\theta$ equal to the mass $\mu r\theta$ times the acceleration v^2/r, we obtain

$$T\theta = \mu r\theta \frac{v^2}{r}$$

or

$$T = \mu v^2 \qquad v = \sqrt{\frac{T}{\mu}}$$

Since this velocity is independent of r and θ, this result holds for all segments of the string. However, the derivation depends on the assumption that the angle θ is small. This will be true if the height of the pulse is small compared with its length or, alternatively, if the slope of the string is small.

In our original reference frame, the string is of course fixed, and the pulse moves with velocity $v = \sqrt{T/\mu}$ along the string without changing shape.

Example 20-1 The tension in a string is provided by hanging a mass of 3 kg at one end, as in Figure 20-11. The mass density of the string is 0.02 kg/m. What is the velocity of waves on the string?

The tension in the string is

$$T = mg = 3 \times 9.81 = 29.4 \text{ N}$$

The velocity is therefore

$$v = \sqrt{\frac{T}{\mu}} = \sqrt{\frac{29.4}{0.02}} = 38.3 \text{ m/sec}$$

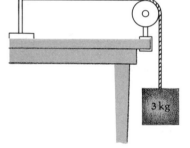

Figure 20-11
Example 20-1. The speed of a wave on the string can be found from $v = \sqrt{T/\mu}$.

Questions

5. Two strings are stretched between the same two posts. One weighs twice as much as the other. How should they be adjusted so that waves travel through both at the same speed?

6. The strings of a piano for low notes generally have copper wire wound around them like an outside covering. What effect does this have on the speed of wave pulses on the string?

20-4   Velocity of Sound Waves

Sound waves are longitudinal waves of compression and rarefaction of a material medium such as air. The wave function $y(x \pm vt)$ can represent either the variation in the pressure from its normal value or the displacement of air molecules from their equilibrium position. As for waves on a string, the speed of a sound wave through air depends on properties of the medium. For waves in which the pressure variation is not too great, the velocity of sound v is given by

$$v = \sqrt{\frac{B}{\rho_0}} \qquad\qquad 20\text{-}5 \qquad \textit{Velocity of sound}$$

where ρ_0 is the equilibrium density of the air and B is the *bulk modulus*, defined to be the ratio of the change in pressure to the fractional decrease in volume,

$$B = \frac{\Delta P}{-\Delta V/V} \qquad\qquad 20\text{-}6 \qquad \textit{Bulk modulus defined}$$

(The minus sign is included in the definition to make B positive since an increase in pressure produces a decrease in volume.)

Comparing Equation 20-5 with Equation 20-4 for the speed of waves on a string, we see that in both the wave speed depends on an elastic property of the medium (the tension T for string waves and the bulk modulus B for compression waves) and on an inertial property of the medium (the linear mass density or the volume mass density). Equation 20-5 follows from the application of Newton's laws to fluids. We shall consider a simple but somewhat artificial derivation of this relation for a compression wave in a fluid confined to a long tube so that the wave pulse moves in one dimension. First, however, let us apply the result to an ideal gas. The pressure and volume of an ideal gas are related by the equation of state (17-20), repeated here for convenience:

$$PV = nRT \qquad\qquad 20\text{-}7$$

In order to compute the bulk modulus for such a gas we need to relate the change in the volume to the change in pressure. Differentiating Equation 20-7, we obtain

$$P \, dV + V \, dP = nR \, dT$$

If the compression and rarefaction of the gas occur at constant temperature ($dT = 0$), the changes in pressure and volume are related by

$$P \, dV + V \, dP = 0$$

or

$$dP = \frac{-P \, dV}{V} \qquad \text{constant temperature} \qquad\qquad 20\text{-}8$$

The isothermal bulk modulus is thus

$$B_{\text{iso}} = \frac{-dP}{dV/V} = P \qquad\qquad 20\text{-}9$$

and the speed of a wave is

$$v = \sqrt{\frac{B}{\rho_0}} = \sqrt{\frac{P}{\rho_0}} \qquad\qquad 20\text{-}10$$

We can write this in terms of the temperature using Equation 20-7:

$$P = \frac{nRT}{V} = \frac{\rho_0 RT}{M}$$

where M is the molecular weight and $\rho_0 = nM/V$ is the mass density. Then

$$v = \sqrt{\frac{RT}{M}} \qquad\qquad 20\text{-}11$$

This expression, first obtained by Newton, gives the correct temperature dependence for the speed of sound, but it yields values for v

about 20 percent too small compared with experimental measurements.[1]

The reason for the inaccuracy of Equation 20-11 lies in the assumption of a constant temperature for the compressions and rarefactions of the gas. Since compression of a gas tends to increase and rarefaction to decrease the temperature, these processes can be isothermal only if sufficient heat is conducted away during the compression and back during rarefaction. However, air and other gases are poor heat conductors. To a good approximation, no heat is exchanged between parts of the gas during the compressions or rarefactions occurring in a sound wave. The compressions and rarefactions are essentially adiabatic. For an adiabatic process in an ideal gas the pressure and volume are related by Equation 18-33, repeated here for convenience:

$$PV^\gamma = \text{constant} \qquad\qquad 20\text{-}12$$

where γ is the ratio of the specific heat at constant pressure to that at constant volume. For air $\gamma = 1.4$. Differentiating Equation 20-12 gives

$$P\gamma V^{\gamma-1}\, dV + V^\gamma\, dP = 0 \quad\text{or}\quad \frac{\gamma\, dV}{V} = -\frac{dP}{P}$$

The adiabatic bulk modulus is thus

$$B_{\text{adiab}} = -\frac{dP}{dV/V} = \gamma P$$

or, substituting $\rho_0 RT/M$ for P,

$$B_{\text{adiab}} = \frac{\gamma \rho_0 RT}{M}$$

and the velocity of sound is

$$v = \sqrt{\frac{\gamma RT}{M}} \qquad\qquad 20\text{-}13 \qquad \textit{Velocity of sound in a gas}$$

This differs from Equation 20-11 by the factor γ inside the square root. For air $M = 29.0$ gm/mole. Using $R = 8.31$ J/mole-K and $\gamma = 1.4$, we find for $T = 300$ K,

$$v = 347 \text{ m/sec} = 1138 \text{ ft/sec}$$

Equation 20-13 should be compared with Equation 17-25 for the rms speed of the molecules in the gas $v_{\text{rms}} = \sqrt{3RT/M}$. These two expressions have the same dependence on the temperature and the molecular weight of the gas and differ only in the constants 3 and γ.

The fact that these speeds are nearly equal is to be expected. A compression wave is propagated in a gas by the collisions of gas molecules with each other. If the speed of the gas molecules increases, the frequency of collisions increases and so does the speed of propagation of the wave.

Questions

7. How would you expect the bulk modulus of a solid to compare with that of a gas?

---

[1] Newton was not content to be within 20 percent of the experimental value for the speed of sound. He fudged his original result of 979 ft/sec by introducing various correction factors until he obtained 1142 ft/sec, which agreed exactly with experiment, and claimed an accuracy of 0.1 percent. See R. S. Westfall, Newton and the Fudge Factor, *Science*, vol. 179, p. 751, February 1973.

8. Although the density of most solids is more than 1000 times that of air, the speed of sound in a solid is usually greater than that in air. Why?

9. Equation 20-13 is obtained using the equation for quasi-static adiabatic processes in a gas. It works well even when the wave frequencies are thousands of compressions and expansions per second. How can such rapid processes be treated as quasi-static?

Optional

Derivation of Equation 20-5

Consider a fluid of density ρ_0 and pressure P in a long tube, as in Figure 20-12. We suddenly compress the fluid by increasing the pressure by an amount ΔP at the left, causing the piston to move to the right for a short time Δt. The motion of the piston is transmitted to the molecules in the fluid by collisions of the molecules with the piston and each other. We shall make the simplifying assumption that the piston moves with a constant speed u for the time Δt and that the speed u is much less than the wave speed v. (It is important not to confuse these two speeds.) In time Δt the piston moves a distance $u\,\Delta t$, and the wave pulse moves a distance $v\,\Delta t$. We shall also assume that the action of the piston gives a speed u to all the fluid from the piston to the edge of the pulse a distance $v\,\Delta t$ from the original position of the piston. The justification for this assumption is that the wave speed v is the speed at which disturbances are propagated through the fluid. Thus in time Δt the greatest distance the disturbance can penetrate into the fluid ahead of the piston is $v\,\Delta t$. The assumption that all the fluid in this region moves with constant speed u amounts to the assumption of a rectangular shape for the pulse. We shall calculate the speed of the pulse by setting the change in momentum of the fluid equal to the impulse acting on the fluid due to the increased pressure acting for the time Δt.

If A is the area of the piston, the impulse is

$$\text{Impulse} = (A\,\Delta P)\,\Delta t$$

The mass of fluid set into motion is the density ρ_0 times the volume $Av\,\Delta t$. The change in momentum is this mass times the velocity u:

$$\text{Momentum change} = \rho_0(Av\,\Delta t)u$$

Equating this momentum change to the impulse, we obtain

$$A(\Delta P)\,\Delta t = \rho_0(Av\,\Delta t)u$$

or

$$\Delta P = \rho_0 vu \qquad\qquad 20\text{-}14$$

Figure 20-12
Analysis of a longitudinal wave pulse produced by suddenly moving the piston to the right with speed u. After a short time Δt, the piston has moved a distance $u\,\Delta t$, and the wave pulse has moved a distance $v\,\Delta t$. If a rectangular pulse is assumed, the molecules in the shaded region of length $v\,\Delta t$ are all moving with speed u. Application of Newton's laws to these molecules gives $v = \sqrt{B/\rho_0}$ for the speed of the wave pulse in the gas.

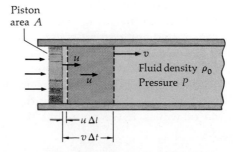

Piston
area A

u

v

u

Fluid density ρ_0
Pressure P

$u\,\Delta t$

$v\,\Delta t$

The change in pressure is related to the decrease in volume of the fluid by the bulk modulus defined in Equation 20-6:

$$\Delta P = B\,\frac{-\,\Delta V}{V}$$

The original volume of fluid under consideration is $V = Av\,\Delta t$, and the change in volume is the volume swept out by the piston $\Delta V = -Au\,\Delta t$. Thus

$$-\frac{\Delta V}{V} = \frac{Au\,\Delta t}{Av\,\Delta t} = \frac{u}{v} \tag{20-15}$$

and

$$\Delta P = \frac{Bu}{v} \tag{20-16}$$

Using this result for ΔP in Equation 20-14, we obtain

$$\frac{Bu}{v} = \rho_0 vu \qquad \text{or} \qquad v = \sqrt{\frac{B}{\rho_0}}$$

20-5 Reflection and Transmission of Pulses

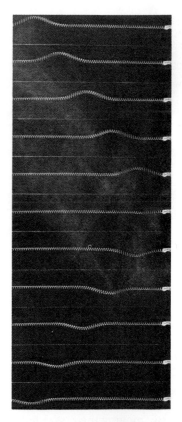

Consider a pulse moving down a string which is fixed at the end. When the pulse reaches the end, it is reflected and inverted; i.e., an identical but inverted pulse moves back along the string from the fixed end. When the pulse arrives at the fixed end, it exerts a force on the support and the support pulls back on the string, keeping the end point fixed and thus setting up the reflected, inverted pulse. The shape of the string while the pulse is being reflected can be constructed by adding to the original pulse an inverted image pulse, as in Figure 20-13.

We can also easily determine the behavior of a pulse which arrives at a free end of a string as in Figure 20-14. In this case, the free end has the string pulling on only one side. It overshoots and sends back a reflected pulse which is not inverted. Again, we can find the shape of the string with use of an image pulse, as shown in the figure.

We often have a boundary which is neither perfectly rigid nor perfectly free. Consider a light string fastened to a heavier string, as in Figure 20-15a. In the limit of an infinitely heavy string, a pulse arriving at the boundary is inverted when it is reflected because the boundary is essentially fixed due to the large inertia of the heavy string. However, if the string to the right is not infinitely heavy, part of the pulse is transmitted and part reflected and inverted (Figure 20-15c). The fraction of energy transmitted depends on the relative densities of the two strings. If the strings are the same, there is of course no boundary and all the energy of the pulse is transmitted. If a pulse hits a boundary at which the string becomes lighter (less dense), again part will be transmitted and part reflected, but the reflected pulse will not be inverted because the heavier string tends to overshoot as if it were a free end. Again, the fraction of energy transmitted and reflected depends on how similar the strings are in density.

This information obtained experimentally with strings can be applied to other kinds of waves if we note that the speed of a string

Reflection of a pulse from a fixed end. The reflected pulse is inverted. (*From* PSSC Physics, 2d ed., p. 260, D. C. Heath and Co., Lexington, Mass., 1965.)

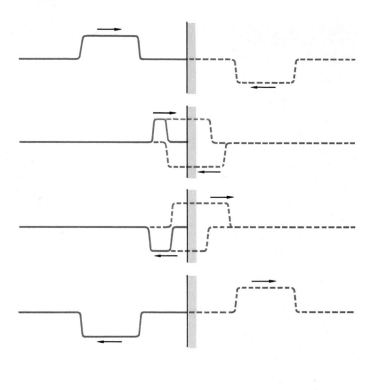

Figure 20-13
Reflection of a wave pulse at the fixed end of a string. The shape of the string at any time can be found by adding an inverted image pulse to the incident pulse. This results in the end of the string being at rest at all times and in a reflected pulse which is inverted.

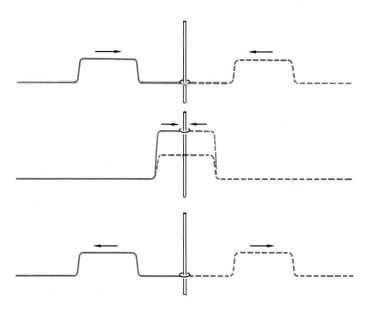

Figure 20-14
Reflection of a wave pulse at the free end of a string. Here tension is maintained at the free end by tying it to a ring of negligible mass which is free to slide along a post. (Another method is to tie the free end to a very light thread that is free to move.) The shape of the string at any time can be found by adding an erect image pulse to the incident pulse. At the free end, the string overshoots the original pulse amplitude, resulting in a reflected pulse which is not inverted.

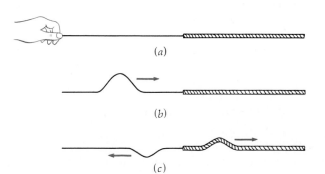

Figure 20-15
(a) A light string attached to a heavy string. (b) A wave pulse on the light string incident on the boundary. (c) After the pulse strikes the boundary, part of it is transmitted and part is reflected. The reflected pulse is inverted.

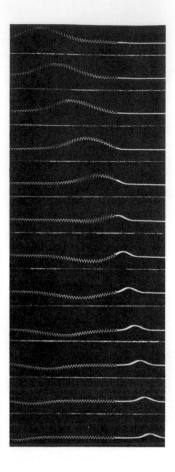 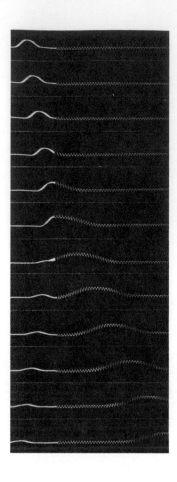

(Left) Pulse passing from a light spring to a heavy spring. The wave is partially transmitted and partially reflected. The reflected pulse is inverted. (Right) Pulse passing from a heavy spring to a light spring. The wave is partially transmitted and partially reflected. The reflected pulse is not inverted in this case. (*From PSSC Physics, 2d ed., pp. 261 and 262, D. C. Heath and Co., Lexington, Mass., 1965.*)

pulse is smaller in the heavier string. When a pulse travels from a light string to a heavy string, it enters a medium in which the speed is slower and, as we have noted, the reflected pulse is inverted. We can state a general rule:

If a wave pulse enters a region where the velocity is smaller, the reflected pulse is inverted. If it enters a region where the velocity is greater, the reflected pulse is not inverted.

Pulse inversion on reflection

Review

A. Define, explain, or otherwise identify:

Dispersion, 497
Transverse wave, 498
Longitudinal wave, 498
Wave function, 499

Superposition, 500
Interference, 500
Bulk modulus, 503

B. True or false:

1. Wave pulses on strings are transverse waves.

2. Sound waves in air are longitudinal waves.

3. When two wave pulses interfere destructively so that there is complete cancellation, the energy momentarily disappears.

4. If the absolute temperature is doubled, the speed of sound waves in air is doubled.

5. When a wave pulse is reflected, it is always inverted.

Exercises

Section 20-1, Wave Pulses

1. A pulse moving through a medium in which there is energy absorption but no dispersion gets smaller, but its relative shape stays the same, i.e., does not spread out. Sketch such a wave pulse at several equally spaced times.

2. Figure 20-16 shows a pulse wave at time $t = 0$. The pulse moves to the right with speed 1 cm/sec, and it spreads out (disperses) in such a way that the area under the pulse stays approximately constant. Sketch the shape of the string at times $t = 1$, 2, and 3 sec.

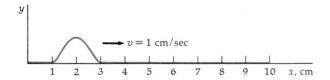

Figure 20-16
Pulse on string for Exercises 2, 3, 4, 22, and 24.

3. Assume that the wave pulse on a string shown in Figure 20-16 is moving to the right without changing shape. At this particular time, which segments of the string are moving up? Which are moving down? Is there any segment of the string at the pulse which is instantaneously at rest? Answer these questions by sketching the pulse at a slightly later time and a slightly earlier time to see how the string segments are moving.

4. Make a sketch of the velocity of each string segment versus position for the pulse shown in Fig. 20-16.

Section 20-2, The Wave Function; Interference

5. The wave function for a pulse on a string is

$$y(x,t) = \frac{0.4a^3}{a^2 + (x - vt)^2}$$

where a and v are constants. Sketch the shape of the string at times $t = 0$, 1, and 2 sec for $a = 4$ cm and $v = 3$ cm/sec. (Take $x = 0$ in the middle of the string.)

6. Repeat Exercise 5 for the wave function

$$y(x,t) = \frac{0.4a^3}{a^2 + (x + vt)^2}$$

7. A possible wave function for a pulse is $y = Ae^{-(x+vt)^2/a^2}$, where A and a are constants. Sketch this pulse for times $t = 0$ and $t = 1$ sec for the values $A = 1$ cm, $a = 2$ cm, and $v = 3$ cm/sec.

8. Figure 20-17 shows two wave pulses in a stretched string, each moving with speed 1 m/sec but in opposite directions. Sketch the shape of the string at 4, 5, 6, 7, 8, and 9 msec.

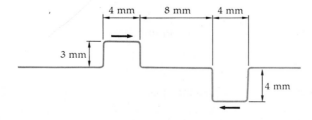

Figure 20-17
Pulses for Exercise 8.

9. Figure 20-18 shows a wave pulse on a string moving to the right. Sketch a second pulse moving to the left which can completely cancel this pulse at some

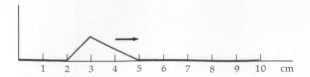

Figure 20-18
Pulse for Exercises 9, 26, and
27.

time. At the time of complete cancellation, indicate with arrows the direction of the instantaneous velocity of the string segments. (The direction of the instantaneous velocity of the string segments can be found by sketching the shape of the string at a time just after there is complete cancellation.)

10. Figure 20-19 shows two wave pulses in a stretched string, starting at $t = 0$. The pulses are moving in opposite directions each with speed 1 cm/sec. Sketch the shape of the string at $t = 1, 1.5, 2, 2.5, 3$, and 4 sec.

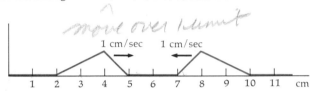

Figure 20-19
Pulses for Exercise 10.

Section 20-3, Velocity of Waves on a String

11. (a) Show that the units of $\sqrt{T/\mu}$ are meters per second if T is in newtons and μ in kilograms per meter. (b) What are the units of T and μ if the units of $\sqrt{T/\mu}$ are feet per second?

12. A steel piano wire is 0.7 m long and has a mass of 5 gm. It is stretched with a tension of 500 N. (a) What is the speed of transverse waves in the wire? (b) To reduce the wave speed by a factor of 2 without changing the tension, what mass of copper wire would have to be wrapped around the steel wire?

13. A common lecture demonstration of pulse waves uses a piece of rubber tubing tied at one end to a fixed post and passed over a pulley to a suspended weight at the other end. Suppose that the distance from the fixed support to the pulley is 30 ft, the weight of this length of tubing is 1.5 lb, and the suspended weight is 25 lb. If the tubing is given a transverse blow at one end, how long will it take the resulting pulse to reach the other end?

14. A long glass tube has a semicircular bend of radius 8 cm. A string of mass density 0.04 kg/m under tension of 20 N is pulled through the tube. (a) At what speed should the string be pulled if it is to pass through the tube without touching the sides? (b) What is the acceleration of the string segments passing through the semicircular bend in this case? (c) Draw a diagram indicating the forces acting on a string segment in the bend which provide this acceleration.

15. A steel wire 20 ft long weighs 0.15 lb. It is under tension of 200 lb. What is the speed of a transverse wave pulse on this wire?

Section 20-4, Velocity of Sound Waves

16. (a) What are the SI units of bulk modulus? (b) Show that $\sqrt{B/\rho}$ has the units of meters per second if B has the correct SI units and ρ is in kilograms per cubic meter.

17. Find the speed of sound in air at $T = 300$ K in miles per hour.

18. The bulk modulus for water is 2.1×10^9 N/m^2. Use this to find the speed of sound in water.

19. The bulk modulus for aluminum is 6.9×10^{10} N/m^2. The density of aluminum is 2.7×10^3 kg/m^3. Find the speed of sound in aluminum.

20. Calculate the speed of sound waves in helium gas ($M = 4$ gm/mole) and compare it with the rms speed of helium molecules. (Take $T = 300$ K and $\gamma = 1.67$ for helium.)

21. The speed of sound in mercury is 1450 m/sec. What is the bulk modulus for mercury ($\rho = 13.6 \times 10^3$ kg/m³)?

Section 20-5, Reflection and Transmission of Pulses

22. A pulse is moving 1 cm/sec to the right on a 10-cm string (Figure 20-16). Assume that the string is fixed at both ends and that there is no loss upon reflection. (*a*) When is the next time the string will have the shape shown? (*b*) Which way will the pulse be moving the next time the string has the shape shown?

23. The motion described in Exercise 22 is periodic, i.e., repeats itself after some time T. Write a general expression for the period in terms of the length of the string L and the pulse speed v. Does the period depend on the initial position of the pulse?

24. A pulse is moving at 1 cm/sec to the right on a 10-cm string with its maximum originally 2 cm from the left end (Figure 20-16). Assume that the left end of the string is fixed but the right end is attached to a very light thread so that it is free to move and there is no loss upon reflection. (*a*) When is the next time the string will have the shape shown? (*b*) Which way will the pulse be moving at that time? (*c*) When is the next time the string will have the shape shown *and* the pulse moving to the right?

25. The motion in Exercise 24 is periodic. Write a general expression for the period of this motion in terms of the length of the string L and the pulse speed v.

26. A pulse of the shape shown in Figure 20-18 is moving to the right at 1 cm/sec along a string which is fixed at the right end. Sketch the shape of the string at 5, 6, 7, 8, and 9 sec. *Hint:* At 5 sec the leading edge of the pulse has just reached the end of the string. Sketch an inverted image pulse to the right of the end of the string. After each additional second, the real pulse moves to the right 1 cm, and the image pulse moves 1 cm to the left. The shape of the string is found by algebraic addition of these pulses.

27. Repeat Exercise 26 for the right end of the string attached to a very light thread so that it is free to move. (In this case the image pulse is not inverted.)

Problems

1. A common rule of thumb for deciding the distance to a lightning flash is to begin counting when the flash is observed and stop when the thunder clap is heard. The number of seconds counted is then divided by 5 to get the distance in miles. Why is this justified? What is the velocity of sound in miles per second? How accurate is this procedure? Is the correction for the time taken for the light to reach you important? (The speed of light is about 3×10^8 m/sec.)

2. A method for measuring the speed of sound using an ordinary watch (with a second hand) is to stand some distance from a large flat wall and clap your hands rhythmically in such a way that the echo from the wall is heard halfway between each two claps. Show that the speed of sound is given by $v = 4LN$, where L is the distance to the wall and N is the number of claps per unit time. What is a reasonable value for L for this experiment to be feasible? (If you have access to a flat wall outdoors somewhere, try this method and compare your result with the standard value.)

3. A man drops a stone from a high bridge and hears it strike the water below exactly 4 sec later. (*a*) Estimate the distance to the water on the assumption that the travel time for the sound to reach the man is negligible. (*b*) Improve

your estimate by using your result from part (*a*) for this distance to estimate the time for sound to travel this distance. Then calculate the distance the rock falls in 4 sec minus this time. (*c*) Calculate the exact distance and compare with your previous estimates.

4. (*a*) Compute the derivative of the speed of a wave on a string with respect to the tension dv/dT and show that the differentials dv and dT obey $dv/v = \frac{1}{2}dT/T$. (*b*) A wave moves with speed 300 m/sec on a wire which is under tension of 500 N. Using dT to approximate the change in tension, find how much the tension must be changed to increase the speed to 312 m/sec.

5. Compute the derivative of the velocity of sound with respect to the absolute temperature and show that the differentials dv and dT obey $dv/v = \frac{1}{2}dT/T$. Use this to compute the percentage change in the velocity of sound when the temperature changes from 0 to 27°C. If the speed of sound is 331 m/sec at 0°C, what is it (approximately) at 27°C? How does this approximation compare with an exact calculation?

6. A coiled spring such as a Slinky is stretched to a length L. It has a force constant k and mass m. Show that the velocity of longitudinal compression waves along the spring is given by $v = L\sqrt{k/m}$. Show that this is also the velocity of transverse waves along the spring if the natural length of the spring is much less than L.

7. A heavy rope 3 m long is attached to the ceiling and allowed to hang freely. (*a*) Show that the speed of transverse waves in the rope is independent of its mass and length but does depend on the distance y from the bottom according to the formula $v = \sqrt{gy}$. (*b*) If the bottom end of the rope is given a sudden sideways displacement, how long does it take the resulting wave pulse to go to the ceiling, reflect, and return to the bottom of the rope?

8. If a loop of chain is spun at high speed, it will roll like a hoop without collapsing. Consider a chain of linear mass density μ which is rolling without slipping at a high speed v_0. (*a*) Show that the tension in the chain is $T = \mu v_0^2$. (*b*) If the chain rolls over a small bump, a transverse wave pulse will be generated in the chain. At what speed will it travel along the chain? (*c*) How far around the loop (in degrees) will a transverse wave pulse travel in the time the hoop rolls through one complete revolution?

CHAPTER 21 Harmonic Waves in One Dimension

An important special kind of wave function is a sine or cosine function, such as

$$y(x,t) = y_0 \sin k(x - vt) \qquad \text{21-1}a$$

where k is called the *wave number* and y_0 is the *amplitude*. This function can also be written

$$y(x,t) = y_0 \sin (kx - kvt) = y_0 \sin (kx - \omega t) \qquad \text{21-1}b$$

where

$$\omega = kv \qquad \text{21-2}$$

is called the *angular frequency*. This type of wave, called a *harmonic wave*, can be produced, for example, by attaching one end of a stretched string to a tuning fork which vibrates with simple harmonic motion at angular frequency ω in a direction perpendicular to the string. In this case, $y(x,t)$ is the displacement of the string at position x and time t.

We can produce harmonic sound waves by vibrating a piston or loudspeaker cone with simple harmonic motion, causing the air molecules next to the piston or speaker to oscillate with simple harmonic motion about their equilibrium position. These molecules collide with neighboring molecules, causing them to oscillate. If we replace $y(x,t)$ and y_0 in Equation 21-1 with the displacement $s(x,t)$ and the maximum displacement s_0, the equation describes the displacement of the molecules from their equilibrium positions as a function of x and t. These displacements, however, are along the x axis, the direction of propagation of the wave.

Harmonic waves have one unusual property that makes them particularly important. As we mentioned in the last chapter, wave pulses change their shape as they propagate; i.e., they are dispersed, or spread out, as their energy and momentum are absorbed. Harmonic waves, however, do not change their shape even when they are propagated in a highly dispersive medium. The only thing that happens to a harmonic wave is that the amplitude y_0 may decrease as the wave is propagated.

The proof that harmonic waves do not change shape is beyond the scope of this text, but we can see why this is an important feature. In general, the shape of a wave changes as it progresses, and the definition of its speed becomes problematical. What point in the changing wave shape can be chosen as the typical one whose speed is the wave speed? There is no unambiguous answer. For harmonic waves, however, the shape is constant. Any point on the wave can be chosen as the one whose speed is the wave speed v. Indeed, it is only for harmonic waves that the wave speed v can be precisely defined.

For this reason any detailed treatment of waves with other shapes and propagating in dispersive media is carried out by resolving those waves into a superposition of harmonic waves. The mathematical technique used, called *Fourier analysis,* is applied in virtually every branch of physical science and engineering. Although we shall not discuss Fourier analysis in detail, we mention it because its frequent practical applications are a chief reason for studying harmonic waves. (The infinite train of alternating crests and valleys of a harmonic wave may otherwise seem rather artificial.)

Importance of harmonic waves

Harmonic waves are also important because many cases occur in nature where a harmonic wave shape is a good approximation to the true wave shape. For example, the waves corresponding to a single note played on a musical instrument may contain from around 30 up to several thousand successive crests and valleys. Over much of its length such a pulse wave is very much like a segment of a harmonic wave.

From our study of simple harmonic motion it should be clear that that we could just as well choose a cosine function in Equation 21-1 since the sine and cosine functions differ only in phase, $\sin \theta = \cos (\theta - \pi/2)$. We also could have reversed kx and ωt in Equation 21-1 since $\sin (kx - \omega t) = \sin (\omega t - kx + \pi)$. We have omitted a constant phase in Equation 21-1 for simplicity.

21-1   Wavelength, Frequency, and Velocity of Harmonic Waves

Figure 21-1 shows the shape of a harmonic wave at some instant t_0. This figure could be obtained by taking a snapshot of a harmonic wave on a string. The wavelength λ is the distance between two successive crests, i.e., the distance in space within which the wave function repeats itself at a given time. The shape of this curve is obtained from Equation 21-1 by setting the time equal to a constant t_0. Then

Wavelength defined

$$y = y_0 \sin (kx - \omega t_0) \qquad\qquad 21\text{-}3$$

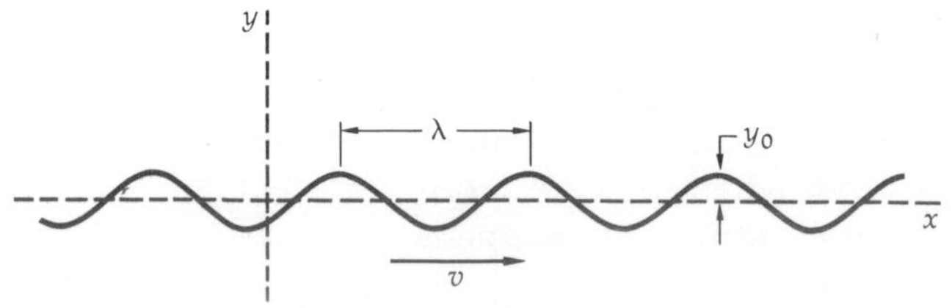

Figure 21-1
Harmonic wave at some instant in time. λ is the wavelength and y_0 is the amplitude.

We can simplify the appearance of this equation by noting that the quantity ωt_0 is merely a phase constant; calling it δ, we have

$$y = y_0 \sin (kx - \delta) \qquad \text{21-4}$$

We can use Equation 21-4 to relate the wavelength λ to the wave number k. Let the position of one wave crest be x_1 and the position of the next crest be x_2. In the distance $x_2 - x_1 = \lambda$ the wave function repeats itself. Thus the argument of the sine function $kx - \delta$ changes by 2π:

$$(kx_2 - \delta) - (kx_1 - \delta) = 2\pi$$

$$k(x_2 - x_1) = 2\pi$$

or

$$k\lambda = 2\pi$$

The wavelength λ and the wave number k are related by

$$k = \frac{2\pi}{\lambda} \qquad \lambda = \frac{2\pi}{k} \qquad \text{21-5}$$

The wave number k is 2π times the number of waves per unit length.

If we look instead at the motion of one point on the string in time, we see that it is simple harmonic motion. Setting $x = x_0$ in Equation 21-1, we have

$$y(x_0, t) = y_0 \sin (kx_0 - \omega t) = y_0 \sin (\omega t - \delta_1)$$

where $\delta_1 = kx_0 + \pi$ is again just a phase constant. The amplitude of this simple harmonic motion is y_0, and the angular frequency is ω. The period T and the frequency f are related to the angular frequency in the usual way:

$$T = \frac{2\pi}{\omega} \qquad \text{21-6}$$

and

Period and frequency defined

$$f = \frac{1}{T} = \frac{\omega}{2\pi} \qquad \text{21-7}$$

There is a simple relation between the frequency f, the wavelength λ, and the velocity v of a harmonic wave. Consider, for example, vibrating the end of a string with frequency f for a time t. In this time the number of waves generated is $N = ft$. The first wave generated travels a distance vt. The ratio of this distance to the number of waves in this distance is the wavelength λ. Thus

$$\lambda = \frac{vt}{N} = \frac{vt}{ft} = \frac{v}{f}$$

or

$$v = f\lambda \qquad \text{21-8}$$

This relation can also be obtained by combining Equation 21-2 with Equations 21-5 and 21-7:

$$\omega = kv \qquad 2\pi f = \frac{2\pi}{\lambda} v$$

or

$$v = f\lambda$$

Example 21-1 A stretched string is vibrated by a tuning fork which oscillates with a frequency of 60 Hz. Waves are produced with wavelength 10 cm. How far does a wave travel in 3 sec?

In 3 sec, $3 \times 60 = 180$ waves are produced with wavelength 10 cm. The first wave has traveled $180(10 \text{ cm}) = 18$ m. The velocity of the waves is

$$v = f\lambda = (60 \text{ sec}^{-1})(10 \text{ cm}) = 600 \text{ cm/sec} = 6 \text{ m/sec}$$

Questions

1. How do two harmonic waves of the same amplitude, frequency, and wavelength differ if their phase difference is π? 2π? 3π?

2. How is the wave number k of a harmonic wave related to the number of wave crests that lie in a length L along the direction of wave propagation?

3. In time Δt how many wave crests pass a given point if the wave frequency is f?

4. When a harmonic wave is transmitted across the boundary between two media with different wave speeds, the frequency of the wave does not change. Explain why.

5. A harmonic wave strikes a boundary and is both reflected and transmitted. The reflected wave is inverted. Is the wavelength of the transmitted wave greater or less than that of the original wave?

21-2 Polarization

In any transverse wave, the vibration is perpendicular to the direction of propagation of the wave and can therefore be resolved into rectangular components in the plane perpendicular to the direction of propagation. For example, in a wave propagating in the x direction along a string, the vibration of each element of the string is in the yz plane. If the vibration of each element of the string is along a line, the wave is said to be *linearly polarized*. The harmonic wave given by Equation 21-1 is linearly polarized in the y direction. In this wave, the string vibrates in the xy plane. Linear polarization is also called *plane polarization*. A harmonic wave propagating in the x direction and polarized in the z direction would be represented by an equation similar to 21-1 with y replaced by z. *Linear, or plane, polarization*

There are other kinds of polarization of transverse waves. If one end of a string is moved with constant speed in a circle, a wave will be propagated along the string in which each element of the string moves in a circle. Such a wave is called a *circularly polarized wave*. A general harmonic wave propagating in the x direction has both y and z vibrational components, which can be written *Circular polarization*

$$y = y_0 \sin (kx - \omega t)$$

$$z = z_0 \sin (kx - \omega t + \delta)$$

21-9

where δ is the phase difference between the two components. Such a wave is said to be polarized if the phase difference is constant. If the phase difference is zero, the wave is linearly polarized along the line making an angle θ with the z axis given by $\tan \theta = y/z = y_0/z_0$. If the phase is $\pi/2$, we can write $z = z_0 \sin (kx - t + \tfrac{1}{2}\pi) = z_0 \cos (kx - \omega t)$. *General definition of polarization*

Then the y and z components of the vibration are related by

$$\frac{y^2}{y_0^2} + \frac{z^2}{z_0^2} = \sin^2 (kx - \omega t) + \cos^2 (kx - \omega t) = 1 \qquad 21\text{-}10$$

This is the equation of an ellipse. Such a wave is said to be *elliptically polarized*. Circular polarization is a special case of elliptical polarization with $y_0 = z_0$.

Elliptical polarization

Most waves produced by a single source, e.g., those produced by vibrating one end of a string, are polarized. Waves produced by many sources are usually unpolarized. A light wave, for example, is usually produced by millions of atoms acting independently. The electric field for a light wave propagating in the x direction can be resolved into y and z components at any instant, but the phase difference δ varies in a random way because there is no correlation between the electric field vectors produced by the different atoms. Such a light wave is unpolarized.

Since there is only one possible direction of vibration for a longitudinal wave, namely, along the direction of propagation, polarization has no meaning for a longitudinal wave.

Questions

6. How could you produce unpolarized waves on a string?

7. Discuss the polarization of water waves on the surface of a lake.

21-3 Harmonic Sound Waves

We can produce harmonic sound waves in one dimension by vibrating a piston with simple harmonic motion against an air column enclosed in a long narrow tube. The motion of the piston produces in the nearby air mass a displacement which is transmitted to other air masses in the tube by molecular collisions. Let us write $s(x,t)$ for the displacement of a tiny volume of the gas from its equilibrium position. If the piston vibrates with simple harmonic motion of angular frequency ω, the displacement $s(x,t)$ will be a harmonic wave with the same angular frequency and with wave number k related to the angular frequency and the velocity of the wave by Equation 21-2. We can write the displacement function

$$s(x,t) = s_0 \sin (kx - \omega t) \qquad 21\text{-}11$$

where s_0 is the maximum displacement of the gas from its equilibrium position. This displacement is along the direction of motion of the wave; sound is a longitudinal wave.

The displacement of the gas molecules causes the density of the gas at a point to increase or decrease, depending on whether the gas is displaced toward the point or away from it. If the density increases (or decreases) at some point, the pressure also increases (or decreases) at that point, giving a pressure wave which is related to the displacement wave. We can use Figure 21-2a, showing a sketch of the function $s(x,t)$ at some time t_0, to see that the pressure wave is 90° out of phase with the displacement wave. Points P_1 and P_3 are points of zero displacement at this time. Just to the left of point P_1 the displacement is negative, indicating that the gas is displaced to the left away from the point. Just to the right of P_1 the displacement is positive, indicating that the gas is displaced to the right, again away from this point. Thus

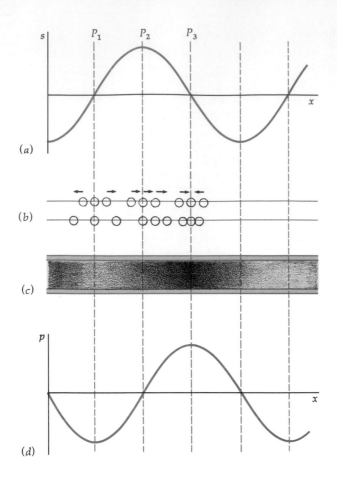

(a)

(b)

(c)

(d)

Figure 21-2
(a) The displacement from equilibrium s of air molecules in a harmonic sound wave at some instant. At P_1 and P_3 the air molecules are at their equilibrium positions. At P_2 they have maximum displacement. (b) Equilibrium position (top) and displacement (bottom) of three molecules near each of the points P_1, P_2, and P_3. (c) Density of molecules at this time. (d) Plot of pressure change versus position at this time. At the position of maximum displacement, the pressure change is zero. At points of zero displacement, the pressure change is either maximum or minimum. The pressure change is 90° out of phase with the displacement.

the density is a minimum at point P_1 because the gas is displaced away from this point on both sides. At point P_3 the density is a maximum because to the right of this point the displacement is negative, indicating displacement to the left toward the point, and to the left of the point the displacement is positive, indicating displacement to the right, toward the point. The displacements near these points are indicated in Figure 21-2b. At point P_2, the displacement function is a maximum. The displacements to the left and to the right of this point are both positive and of equal magnitude. Thus the density does not change near this point. Figure 21-2c indicates the change in density for this displacement wave. Since the pressure is maximum when the density is maximum, the pressure varies with distance at this time, as indicated in Figure 21-2d.

We shall show below that a displacement wave given by Equation 21-11 implies a pressure wave given by

Sound as a pressure wave

$$p = p_0 \sin (kx - \omega t - 90°) \qquad 21\text{-}12$$

where p stands for the *change* in pressure from the equilibrium pressure with no wave and the amplitude p_0 is the maximum value of this change. We shall also show that the pressure amplitude p_0 is related to the displacement amplitude s_0 by

$$p_0 = \rho \omega v s_0 \qquad 21\text{-}13$$

where v is the speed of propagation and ρ is the equilibrium density of the gas. We can think of a sound wave as either a displacement wave or a pressure wave, the displacement and pressure being 90° out of phase with each other.

Optional

Derivation of Equations 21-12 and 21-13

We can relate the changes in pressure to the displacement by noting that a displacement changes the volume of a given mass of gas and that changes in pressure and volume are related by the bulk modulus defined by Equation 20-6. Since we are using the notation that p stands for a *change* in pressure here, we have

$$p = -B \frac{\Delta V}{V} \qquad\qquad 21\text{-}14$$

We can express the bulk modulus in terms of the wave speed v and the gas density using Equation 20-5:

$$B = \rho v^2 \qquad\qquad 21\text{-}15$$

Thus the change in pressure p is related to the change in volume V by

$$p = -\rho v^2 \frac{\Delta V}{V} \qquad\qquad 21\text{-}16$$

Consider a given mass of gas initially between points x_1 and x_2 occupying a volume $V = A(x_2 - x_1) = A\,\Delta x$, where A is the area of the tube of gas and $\Delta x = x_2 - x_1$. This volume is shown in Figure 21-3. In general, a displacement of the gas causes a change in the volume of a given mass of gas. If the displacement is the same at points x_2 and x_1, there will be no change in the volume. For example, a positive displacement at point x_2 increases the volume, but a positive displacement at point x_1 decreases the volume. There will be a change in the volume of the gas only if there is a difference in displacement at these two points. Let us call the difference in displacement Δs:

$$\Delta s = s(x_2, t_0) - s(x_1, t_0) \qquad\qquad 21\text{-}17$$

The change in volume of the gas equals this difference in displacement times the area A:

$$\Delta V = A\,\Delta s$$

This change in volume is related to the change in pressure p by Equation 21-16:

$$p = -\rho v^2 \frac{A\,\Delta s}{A\,\Delta x} = -\rho v^2 \frac{\Delta s}{\Delta x} \qquad\qquad 21\text{-}18$$

In the limit of small Δx the ratio $\Delta s/\Delta x$ becomes the derivative of s with respect to x. This is a partial derivative because we are considering a single time t_0. Replacing $\Delta s/\Delta x$ by the notation for partial derivative $\partial s/\partial x$, we have

$$p = -\rho v^2 \frac{\partial s}{\partial x} \qquad\qquad 21\text{-}19$$

Figure 21-3
Change in volume of a given mass of gas because of variation in displacement with position. The original equilibrium volume is $V = A\,\Delta x$, where A is the area of the tube and $\Delta x = x_2 - x_1$. The change in volume is $\Delta V = A(s_2 - s_1) = A\,\Delta s$, where s_2 and s_1 are the displacements at x_2 and x_1. The fractional change in volume is then $\Delta V/V = \Delta s/\Delta x$.

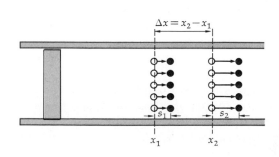

We compute $\partial s / \partial x$ from Equation 21-11:

$$s = s_0 \sin (kx - \omega t_0)$$

$$\frac{\partial s}{\partial x} = k s_0 \cos (kx - \omega t_0)$$

Thus

$$p = -\rho v^2 s_0 k \cos (kx - \omega t_0) = +k\rho v^2 s_0 \sin (kx - \omega t_0 - 90°)$$
$$= p_0 \sin (kx - \omega t_0 - 90°) \qquad\qquad 21\text{-}20$$

where

$$p_0 = k\rho v^2 s_0$$

But according to Equation 21-2, $kv = \omega$. Thus

$$p_0 = \rho \omega v s_0 \qquad\qquad 21\text{-}21$$

and Equations 21-20 and 21-21 are identical to Equations 21-12 and 21-13.

21-4   Superposition and Interference of Harmonic Waves

According to the principle of superposition, when there are two or more waves, the resultant wave is just the algebraic sum of the waves. We shall now show that the superposition of two harmonic waves of the same amplitude, frequency, and wavelength traveling in the same direction but differing in phase gives a harmonic wave of the same frequency and wavelength. The amplitude and phase of the resultant wave depend on the phase difference between the original two waves.

Let y_1 and y_2 be the wave functions for the two original waves:

$$y_1 = y_0 \sin (kx - \omega t) \qquad y_2 = y_0 \sin (kx - \omega t + \delta)$$

The resultant wave is the sum

$$y_3 = y_1 + y_2 = y_0 \sin (kx - \omega t) + y_0 \sin (kx - \omega t + \delta)$$

We can simplify this by using the relation

$$\sin A + \sin B = 2 \sin \tfrac{1}{2}(A + B) \cos \tfrac{1}{2}(A - B) \qquad\qquad 21\text{-}22$$

For this case $A = kx - \omega t$ and $B = kx - \omega t + \delta$, so that

$$\tfrac{1}{2}(A + B) = kx - \omega t + \tfrac{1}{2}\delta \qquad \text{and} \qquad \tfrac{1}{2}(A - B) = -\tfrac{1}{2}\delta$$

Thus

$$y_3 = 2y_0 \cos \tfrac{1}{2}\delta \sin (kx - \omega t + \tfrac{1}{2}\delta) \qquad\qquad 21\text{-}23$$

where we have used $\cos(-\tfrac{1}{2}\delta) = \cos \tfrac{1}{2}\delta$. We see that the resultant of these two waves is another harmonic wave with the same frequency and wavelength. It differs in phase from both the original waves, and its amplitude is $2y_0 \cos \tfrac{1}{2}\delta$.

If $\delta = 0$, the two waves are in phase, $\cos \tfrac{1}{2}\delta = 1$, and the resultant amplitude is just twice that of either wave. If $\delta = \pi$, $\cos \tfrac{1}{2}\delta = 0$ and the two waves cancel each other totally. These two extremes are *perfectly constructive interference* and *perfectly destructive interference*.

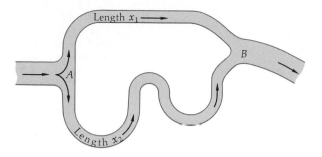

Figure 21-4
A wave enters from the left
and splits equally at point A.
At point B, where the waves
recombine, there is a phase
difference because of the path
difference. If λ is the wavelength and Δx the path difference, the phase difference
is $\delta = 2\pi \, \Delta x/\lambda$.

A common cause of phase difference between two waves is a difference in the path length traveled by two waves from the same
source. In Figure 21-4 we assume that the wave splits equally at point
A; that is, the amplitudes of the waves in the two tubes are equal.
When they recombine at point B, there will be a phase difference due
to the longer path traveled by the wave in the lower tube.

If the path difference is exactly 1 wavelength, the phase difference
between the two waves will be 2π, which is the same as no phase difference, and the two waves will interfere constructively. Perfectly constructive interference will also occur if the path difference is any integer times the wavelength. On the other hand, if the path difference is
exactly $\frac{1}{2}$ wavelength, $\frac{3}{2}$ wavelength, or any odd number of half
wavelengths, the phase difference will be π and the interference will
be destructive. For any other path difference Δx the phase difference
will be

$$\delta = 2\pi \, \frac{\Delta x}{\lambda} \qquad\qquad 21\text{-}24$$

21-5   Vector Method of Addition of Harmonic Waves

A simple geometric interpretation of harmonic wave functions leads to
a method of addition of harmonic waves of the same frequency by
geometric construction. This method lets us find the sum of two or
more harmonic waves geometrically without having to remember the
trigonometric identity of Equation 21-22. It is useful even if the amplitudes of the waves are different or if there are more than two waves.
The method is based on the fact that the y (or x) component of the
resultant of two vectors equals the sum of the y (or x) components of
the vectors.

Let $y_1 = A_1 \sin (kx - \omega t)$ and $y_2 = A_2 \sin (kx - \omega t + \delta)$ be the wave
functions of the two waves. We wish to add these waves at some point
x and some time t. We can simplify our notation by writing θ for the
quantity $kx - \omega t$. Our problem is then to find the sum

$$y_1 + y_2 = A_1 \sin \theta + A_2 \sin (\theta + \delta)$$

Consider a vector of magnitude A_1 making an angle θ with the x axis
(Figure 21-5). The y component of this vector is $A_1 \sin \theta$, which is the
wave function y_1. Similarly, the wave function $y_2 = A_2 \sin (\theta + \delta)$ is
the y component of a vector of magnitude A_2 making an angle $\theta + \delta$
with the x axis. By the laws of vector addition, the sum of these components equals the y component of the resultant vector, as shown in
Figure 21-5. The y component of the resultant vector, $A' \sin (\theta + \delta')$, is

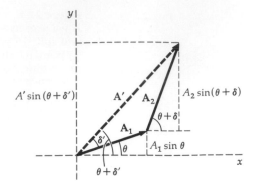

Figure 21-5
The wave function $A_1 \sin \theta$ is the y component of the vector $\mathbf{A}_1$ making angle θ with the x axis. The wave function $A_2 \cdot \sin (\theta + \delta)$ is the y component of the vector $\mathbf{A}_2$ making angle $\theta + \delta$ with the x axis. The sum of these wave functions is $A' \cdot \sin (\theta + \delta')$, which is the y component of the resultant vector $\mathbf{A}' = \mathbf{A}_1 + \mathbf{A}_2$.

a harmonic wave function which is the sum of the two original wave functions,

$$A_1 \sin \theta + A_2 \sin (\theta + \delta) = A' \sin (\theta + \delta') \qquad \text{21-25}$$

where $\theta = kx - \omega t$ and A' (the amplitude of the resultant wave) and δ' (the phase of the resultant wave relative to the first wave) are found by adding the vectors representing the waves, as in Figure 21-5. As time varies, θ varies. The vectors representing the two wave functions and the resultant vector representing the resultant wave function rotate in space, but their relative positions do not change because all the vectors rotate with the same angular velocity ω.

Example 21-2 Use the vector method of addition to derive Equation 21-23 for the superposition of two waves of the same amplitude.

Figure 21-6 shows the vectors representing two waves of equal amplitude A and the resultant wave of amplitude A'. These three vectors form an isosceles triangle in which the two equal angles are δ'. Since the sum of these angles equals the exterior angle δ, we have

$$\delta' = \tfrac{1}{2}\delta$$

The amplitude A' can be found from the right triangle shown in Figure 21-6b formed by bisecting the resultant vector. From this triangle we have

$$\cos \tfrac{1}{2}\delta = \frac{\tfrac{1}{2}A'}{A}$$

Therefore the amplitude is given by $A' = 2A \cos \tfrac{1}{2}\delta$, and the resultant wave is

$$A' \sin (\theta + \delta') = 2A \cos \tfrac{1}{2}\delta \sin (kx - \omega t + \tfrac{1}{2}\delta)$$

in agreement with Equation 21-23.

Example 21-3 Find the resultant of the two waves

$$y_1 = 4 \sin (kx - \omega t) \qquad \text{and} \qquad y_2 = 3 \sin (kx - \omega t + 90°)$$

Figure 21-7 shows the vector diagram for this addition. The vectors make an angle of 90° with each other. The resultant of these two vectors has the magnitude 5 and makes an angle of 37° with the first vector, as shown. The sum of these two waves is

$$y_1 + y_2 = 5 \sin (kx - \omega t + 37°)$$

This vector method of addition will prove very useful in our study of waves in the following chapters.

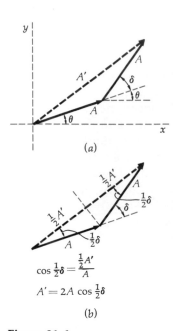

$$\cos \tfrac{1}{2}\delta = \frac{\tfrac{1}{2}A'}{A}$$

$$A' = 2A \cos \tfrac{1}{2}\delta$$

(b)

Figure 21-6
The vector addition for two waves of equal amplitude A having phase difference δ. (a) The vectors at a particular time at which $\theta = \omega t$. (b) Construction for finding the amplitude of the resultant wave. The amplitude A' is found from $\cos \tfrac{1}{2}\delta = \tfrac{1}{2}A'/A$, giving $A' = 2A \cos \tfrac{1}{2}\delta$.

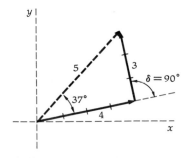

Figure 21-7
Vector diagram for the addition of the two waves in Example 21-3.

21-6 Energy and Intensity of a Harmonic Wave on a String

An important property of waves is that they transport energy and momentum. The transport of energy in a wave ordinarily is described in terms of the *wave intensity*, defined to be the average rate at which the wave transmits energy per unit area normal to the direction of propagation. That is, the intensity at any point in a wave is the average incident energy per unit time per unit area. Since the energy per unit time is the power, the intensity is the average power incident per unit area[1]:

$$I = \frac{(\Delta E / \Delta t)_{\mathrm{av}}}{A} = \frac{P_{\mathrm{av}}}{A} \qquad\qquad 21\text{-}26$$

Wave intensity defined

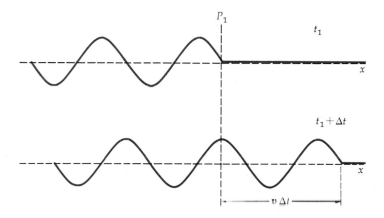

Figure 21-8
Energy transmitted by a wave on a string. At time t_1 the wave has just reached point P_1. In an additional time Δt the wave travels a distance $v\,\Delta t$. The energy transmitted past point P_1 is $\eta Av\,\Delta t$, where η is the average energy per unit volume and A is the cross-sectional area of the string.

There is a simple relation between the intensity of a wave and the energy per unit volume in the medium carrying the wave. Suppose that a wave traveling to the right on a string has just reached point P_1 at time t_1, as shown in Figure 21-8. The part of the string to the left of P_1 contains energy because each segment is oscillating with simple harmonic motion. At this time t_1 there is no energy to the right of P_1 because the wave has not yet reached that part of the string. After a time Δt the wave moves past point P_1 a distance $v\,\Delta t$, where v is the wave velocity. The total energy of the string is thus increased by the amount of energy in the string past point P_1. This increase in energy can be written

$$\Delta E = \eta Av\,\Delta t$$

where η is the average energy per unit volume of the string, A the cross-sectional area, and $Av\,\Delta t$ the volume of the string past point P_1 which now contains energy.[2] The rate of increase of energy is the power passing point P_1. (The source of this energy is of course at the left end of the string, where there must be an external force doing

---

[1] For waves on a string, the area is just the cross-sectional area of the string. We use the term power per unit area even though it is somewhat artificial for a string because our discussion can then be carried over to any kind of waves.

[2] We assume that the amplitude of the wave is small so that $v\,\Delta t$ is approximately the length of string containing the wave beyond point P_1.

work on the end of the string.) Thus the average incident power is

$$P_{av} = \frac{\Delta E}{\Delta t} = \eta A v \qquad\qquad 21\text{-}27$$

The intensity of the wave at a point is found by dividing the incident power by the area of the string:

$$I = \frac{P_{av}}{A} = \eta v \qquad\qquad 21\text{-}28$$

Thus the *intensity equals the product of the wave velocity v and the average energy per unit volume, or the average energy density η*. This result is applicable to all waves.

To calculate the average energy density η for waves on a string, consider a harmonic wave of angular frequency ω and amplitude y_0 traveling along a stretched string. Each element of the string is undergoing simple harmonic motion of angular frequency ω and amplitude y_0. The energy of a segment of the string of mass Δm is of the same form as Equation 14-16 for the energy of a mass oscillating on a spring,

$$\Delta E = \tfrac{1}{2}(\Delta m)\omega^2 y_0^2 \qquad\qquad 21\text{-}29$$

The mass of the segment is just the mass density ρ times the volume of the segment ΔV. Substituting $\Delta m = \rho\, \Delta V$ into Equation 21-29, we have

$$\Delta E = \tfrac{1}{2}\rho\omega^2 y_0^2 \,\Delta V$$

The average energy density

$$\eta = \frac{\Delta E}{\Delta V} = \tfrac{1}{2}\rho\omega^2 y_0^2 \qquad\qquad 21\text{-}30$$

Average energy density for waves on a string

is proportional to the square of the frequency and to the square of the amplitude. Substituting this result into Equation 21-28, we have for the intensity of a wave on a string,

$$I = \eta v = \tfrac{1}{2}\rho\omega^2 y_0^2 v \qquad\qquad 21\text{-}31$$

This result, that the *intensity of a wave is proportional to the square of the frequency and to the square of the amplitude,* is a general property of all harmonic waves. For example, the amplitude of a light wave is the maximum value of the electric field E_0. The intensity of a light wave is proportional to E_0^2 and to the square of the frequency. Similarly, as we shall see in Section 21-7, the intensity of sound waves is proportional to the square of either the displacement amplitude or the pressure amplitude.

Questions

8. If the amplitude of harmonic waves is doubled, how does the rate at which energy is used to generate the waves change?

9. Transverse waves of the same frequency and amplitude are sent along two parallel stretched wires. The wires are of the same kind except that one has a greater tension than the other. Over a long time, which wire transmits the greater energy? For which wire is the wave intensity greater?

21-7 Energy and Intensity of Harmonic Sound Waves

Let Δm be the mass of a small portion of gas whose displacement is given by Equation 21-11. Since this motion is simple harmonic at any given time, the average energy of vibration of this gas is given by Equation 21-29,

$$\Delta E = \tfrac{1}{2}\Delta m\ \omega^2 s_0^2$$

where ω is the angular frequency and s_0 is the maximum displacement amplitude. If the element of gas has equilibrium volume ΔV, the average energy per unit volume, or average energy density, is

$$\eta = \frac{1}{2}\frac{\Delta m}{\Delta V}\ \omega^2 s_0^2 = \tfrac{1}{2}\rho\omega^2 s_0^2$$

Energy density for sound waves

where $\rho = \Delta m/\Delta V$ is the mass density. We can relate this to the pressure amplitude p_0 using Equation 21-13 relating p_0 to s_0. We have

$$\eta = \tfrac{1}{2}\rho\omega^2\left(\frac{p_0}{\rho\omega v}\right)^2 = \frac{1}{2}\frac{p_0^2}{\rho v^2} \qquad\qquad \text{21-32}$$

Thus, using Equation 21-31, we have for the intensity of a sound wave in a gas,

$$I = \eta v = \tfrac{1}{2}\rho v\omega^2 s_0^2 = \frac{1}{2}\frac{p_0^2}{\rho v} \qquad\qquad \text{21-33}$$

Intensity of sound waves

The human ear can accommodate a rather large range of sound-wave intensities from about 10^{-12} W/m², which is usually taken to be the threshold of hearing, to about 1 W/m², which produces a sensation of pain in most people. The pressure amplitudes for these extreme sound intensities can be easily calculated from Equation 21-33. Taking $\rho = 1.29$ kg/m³ for the density of air and $v = 331$ m/sec for the speed of sound, we find for the pressure amplitude corresponding to the threshold of hearing intensity of 10^{-12} W/m²:

$$p_0^2 = 2\rho vI = 2(1.29)(331)(10^{-12}) = 8.54 \times 10^{-10}$$

and

$$p_0 = 2.92 \times 10^{-5} \text{ N/m}^2$$

Similarly for an intensity of 1 W/m² at the pain threshold, the pressure amplitude is

$$p_0 = \sqrt{2(1.29)(331)(1)} = 29.2 \text{ N/m}^2$$

These very small changes in pressure are superimposed on the normal constant atmospheric pressure level of about 1.01×10^5 N/m².

Because of the enormous range of intensities to which the ear is sensitive, and because the psychological sensation of loudness varies with intensity not directly but apparently more nearly logarithmically, a logarithmic scale is used to describe the intensity level of a sound wave. The intensity level β measured in *decibels* (dB) is defined by

$$\beta = 10 \log \frac{I}{I_0} \qquad\qquad \text{21-34}$$

Decibel intensity level for sound waves

where I is the intensity corresponding to the level β and I_0 is a reference level which we shall take to be the threshold of hearing:

$$I_0 = 10^{-12} \text{ W/m}^2 \qquad\qquad 21\text{-}35$$

On this scale, the threshold of hearing is

$$\beta = 10 \log \frac{I_0}{I_0} = 10 \log 1 = 0 \text{ dB} \qquad\qquad \textit{Threshold of hearing}$$

Similarly, the pain threshold is

$$\beta = 10 \log \frac{1}{10^{-12}} = 10 \log 10^{12} = 10(12) = 120 \text{ dB} \qquad \textit{Pain threshold}$$

Table 21-1 lists the intensity level of some common sounds.

Table 21-1
Decibel intensity level of some common sounds

Source of sound	dB	Description
Large rocket engine (nearby)	180	
Jet takeoff (nearby)	150	
Pneumatic riveter; machine gun	130	
Rock concert with amplifiers (2 m); jet takeoff (60 m)	120	Pain threshold
Construction noise (3 m)	110	
Subway train	100	
Heavy truck (15 m); Niagara Falls	90	Constant exposure endangers hearing
Noisy office with machines; average factory	80	
Busy traffic	70	
Normal conversation (1 m)	60	
Quiet office	50	Quiet
Library	40	
Soft whisper (5 m)	30	Very quiet
Rustling leaves	20	
Normal breathing	10	Barely audible
	0	Hearing threshold

Questions

10. A source of sound waves oscillates with a fixed amplitude and frequency. Assuming the atmospheric pressure to remain constant, how does the rate at which sound waves carry energy away from the source change if the air temperature goes up?

11. Two tuning forks produce sound waves in air of the same amplitude. One has frequency 256 Hz and the other frequency 512 Hz. Which makes the louder sound, i.e., produces sound with the greater intensity?

Sonic Booms

Laurent Hodges
Iowa State University

One of the most common examples of a shock wave resulting from a source traveling faster than the wave-propagation speed is the sonic boom generated by an airplane whose speed exceeds the speed of sound. A common misconception is that a sonic boom is produced when the airplane first accelerates past the speed of sound ("crashes through the sound barrier"). In fact the shock wave exists during the whole time that the airplane is traveling supersonically, and a boom is heard every time the shock wave sweeps over a person with good hearing.

The airplane actually has two shock waves, associated with its front and back (Figure 1). These bow and tail shock waves have nearly the shape of cones (*Mach cones*) with apex at the aircraft's bow and tail, respectively. (Some deviation from perfect cone shape results from variations in sound speeds in different parts of the air.) The cones have a half angle θ given by $\sin \theta = v_s/v_a$, where v_s is the speed of sound and v_a is the speed of the airplane.

The air pressure in the region of overpressure between the two shock waves differs from normal atmospheric pressure, the bow and tail waves being associated with pressures higher and lower than normal, respectively. This pressure deviation produces the sensation of sound in an observer. As the bow wave sweeps over the observer, the pressure increases by the *overpressure* ΔP in a rise time τ. It then decreases to approximately ΔP below normal atmospheric pressure before suddenly returning to normal at the tail wave. A plot of the time dependence of the pressure at a point has the shape of a slanted N and is therefore called the *N signature* (Figure 2).

The major pressure changes experienced at the ear occur as the bow and tail shock waves reach the observer, each producing an explosive sound. This

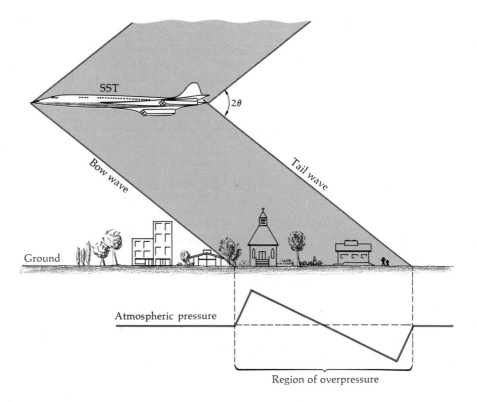

Figure 1
Cross-sectional representation of a supersonic transport in flight, showing bow and tail shock waves, Mach cone angle 2θ, and a plot of the pressure at ground level. The deviations from normal atmospheric pressure occur in the region of overpressure between the two shock waves.

SST

2θ

Bow wave

Tail wave

Ground

Atmospheric pressure

Region of overpressure

Figure 2
A plot of the N signature, or time dependence, of the pressure at a given point which experiences a sonic boom. Shown are the overpressure ΔP, the rise time τ, and the time duration T.

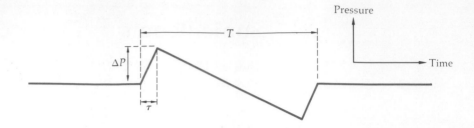

explains the double crack of a sonic boom. A supersonic bullet also has bow and tail shock waves, but its two booms are so close together in time that only one crack is heard.

The startling nature of a sonic boom and the resulting annoyance are a function of the rise time, the total duration of the N signature, and the overpressure. The rise time is of the order of 3 ms, and the time duration varies from about 0.1 sec for a supersonic fighter to about 0.4 sec for the British-French Concorde supersonic transport (SST). A typical overpressure for an SST 20 km overhead would be 100 N/m² (about 2 lb/ft²). The overpressure is less at places on either side of the plane's path, but booms can be heard over a zone typically 80 km wide.

It is not possible to assign a decibel reading to a sonic boom because of its short duration and its highly nonsinusoidal form. A sinusoidal sound wave with a rms pressure of 100 N/m² would correspond to 134 dB, but the overpressure is not the only relevant parameter. Shorter rise times, for example, are associated with "louder" booms. U.S. Air Force studies have shown that an overpressure of 100 N/m² corresponds to 110 to 120 on the perceived-noise decibel scale. (This modification of the decibel scale incorporates the psychological effects of the frequency spectrum of aircraft noises.)

There have been several systematic studies of the effects of sonic-boom exposures, including some carried out by the U.S. Air Force, which has tried to make the sonic boom psychologically acceptable by referring to it as the "sound of freedom." Besides startling people and animals, sonic booms can rattle dishes, shatter glass, and even lead to structural damage to some buildings. Actual experience indicates that approximately $600 in damage claims would result from every million person-booms, and Shurcliff[1] has estimated that each SST might experience 1,000 claims totaling $500,000 annually.

Some people are not bothered much by sonic booms. Many others will presumably learn to tolerate them if supersonic flight becomes commonplace, but even they will suffer the adverse physiological effects that result from any loud noise. Perhaps 25 to 50 percent of the United States population will not be able to adapt to sonic booms, and frequent supersonic flights will doubtless encounter stiff public opposition.

Soviet Russia and a British-French consortium have SSTs for sale as of 1975, but only the state-owned airlines of the three producing countries have purchased the aircraft. The SST program ended in the United States in 1971 after Congress cut off federal funding, although there are occasional attempts to revive the project. Many countries have instituted restrictions, e.g., no supersonic flights over land, that may well make the SST highly uneconomical, so that its future is highly uncertain.

No aircraft design exists at present which permits supersonic flight without generating the shock waves responsible for sonic booms. Even if such a design were feasible, SSTs present other formidable problems:

---

[1] William A. Shurcliff, *S/S/T and Sonic Boom Handbook*, Ballantine Books, New York, 1970. For further information see Harvey H. Hubbard, Sonic Booms, *Physics Today*, vol. 21, p. 31, February 1968; Herbert A. Wilson, Jr., Sonic Boom, *Scientific American*, vol. 206, p. 36, January 1962; and Karl D. Kryter, Sonic Booms from Supersonic Transport, *Science*, vol. 163, p. 359, 1969.

1. They are extremely noisy apart from the boom (United States engineers tried for a noise limit of 120 perceived-noise decibels 3 mi away from the SST at the start of its takeoff).

2. Their exhaust pollutants (especially nitrogen oxides) may reduce the ozone concentration in the stratosphere, permitting an increase in harmful ultraviolet radiation at the earth's surface.

3. SSTs are very inefficient from the point of view of energy consumption.

Optional

21-8 The Wave Equation

A general wave function $y(x,t)$ is a solution of a differential equation called the *wave equation*. The wave equation relates the second derivative of the wave function with respect to x to the second derivative with respect to t. Because there are two variables, these are partial derivatives. We can obtain the wave equation by recalling that the function

$$y(x,t) = y_0 \sin (kx - \omega t)$$

is a particular solution for harmonic waves. It is useful to write the angular frequency ω in terms of the velocity v and wave number k, that is, $\omega = kv$. Then the harmonic wave function is

$$y(x,t) = y_0 \sin (kx - kvt) \tag{21-36}$$

The derivative with respect to x, holding t constant, is

$$\frac{\partial y}{\partial x} = ky_0 \cos (kx - kvt)$$

Similarly, the second derivative with respect to x is

$$\frac{\partial^2 y}{\partial x^2} = -k^2 y_0 \sin (kx - kvt) = -k^2 y(x,t) \tag{21-37}$$

The derivative of $y(x,t)$ with respect to t, holding x constant, is

$$\frac{\partial y}{\partial t} = -kvy_0 \cos (kx - kvt)$$

and

$$\frac{\partial^2 y}{\partial t^2} = -k^2 v^2 y_0 \sin (kx - kvt) = -k^2 v^2 y(x,t) \tag{21-38}$$

Combining Equations 21-37 and 21-38, we obtain

Wave equation

$$\frac{\partial^2 y}{\partial x^2} = \frac{1}{v^2} \frac{\partial^2 y}{\partial t^2} \tag{21-39}$$

the wave equation. If y is the displacement of a vibrating string, this equation describes string waves. If we interpret y as the increase or decrease in the pressure or density of a gas, this equation describes sound waves. The same equation also describes electromagnetic waves, in which y is the electric or magnetic field.

Equation 21-39 is satisfied by any wave in one dimension which is propagated without dispersion or change of shape. We showed earlier that in general such a wave has a wave function which can be expressed as a function of either $x + vt$ or $x - vt$. We can easily show that any function of $x - vt$ or of $x + vt$ satisfies Equation 21-39. Let $\theta = x - vt$ and consider any wave function

$$y = y(x - vt) = y(\theta)$$

The derivative of y with respect to θ we shall call y'. Then by the chain rule for derivatives

$$\frac{\partial y}{\partial x} = \frac{\partial y}{\partial \theta}\frac{\partial \theta}{\partial x} = y'\frac{\partial \theta}{\partial x} \qquad \text{and} \qquad \frac{\partial y}{\partial t} = \frac{\partial y}{\partial \theta}\frac{\partial \theta}{\partial t} = y'\frac{\partial \theta}{\partial t}$$

Since $\partial \theta / \partial x = 1$ and $\partial \theta / \partial t = -v$, we have

$$\frac{\partial y}{\partial x} = y' \qquad \text{and} \qquad \frac{\partial y}{\partial t} = -vy'$$

Taking the second derivatives, we obtain

$$\frac{\partial^2 y}{\partial x^2} = y''$$

$$\frac{\partial^2 y}{\partial t^2} = -v\frac{\partial y'}{\partial t} = -v\frac{\partial y'}{\partial \theta}\frac{\partial \theta}{\partial t} = +v^2 y''$$

Thus again

$$\frac{\partial^2 y}{\partial x^2} = \frac{1}{v^2}\frac{\partial^2 y}{\partial t^2}$$

Of course other equations relating the derivatives of these wave functions can be obtained by taking derivatives of the functions.

Equation 21-39 is important because it is a direct consequence of Newton's second law, $\Sigma \mathbf{F} = m\mathbf{a}$, applied to a segment of a string. For sound waves, the identical equation (with the reinterpretation of y mentioned) is derived from Newton's laws applied to fluids. Similarly, an equation of the same form can be derived for electromagnetic waves from Maxwell's equations for electric and magnetic fields. We discuss the case of string waves only, to illustrate that the wave equation is a consequence of newtonian mechanics.

In Figure 21-9 we have isolated one segment of a string. Our derivation will apply only if the wave is of small enough amplitude for the angle between the string and the horizontal (the original direction of the string with no wave) to be small, in which case the length of the segment is approximately Δx and the mass is $\mu \Delta x$. The string moves vertically with acceleration $\partial^2 y / \partial t^2$. The net vertical force is

$$\Sigma F = T \sin \theta_2 - T \sin \theta_1 \qquad\qquad 21\text{-}40$$

where θ_2 and θ_1 are the angles indicated in Figure 21-9 and T is the tension in the string. Again, since the angles are assumed to be small, we can approximate $\sin \theta$ by $\tan \theta$. The tangent of the angle made by

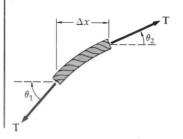

Figure 21-9
The segment of a stretched string used for the derivation of the wave equation. The net vertical force on the segment is $T \sin \theta_2 - T \sin \theta_1$, where T is the tension. The wave equation is derived by setting this force equal to the mass of the segment times its acceleration.

the string with the horizontal is just the slope of the curve formed by the string,

$$\tan \theta = S = \frac{\partial y}{\partial x}$$

Thus the net force on the string segment can be written

$$\Sigma F = T(\sin \theta_2 - \sin \theta_1) \approx T(\tan \theta_2 - \tan \theta_1) = T(S_2 - S_1) = T \Delta S$$

where $S = \tan \theta = \partial y / \partial x$ is the slope and ΔS is the change in slope. Setting this resultant force equal to the mass times the acceleration gives

$$T \Delta S = \mu \, \Delta x \, \frac{\partial^2 y}{\partial t^2} \qquad T \frac{\Delta S}{\Delta x} = \mu \, \frac{\partial^2 y}{\partial t^2}$$

In the limit $\Delta x \to 0$, we have

$$\lim_{\Delta x \to 0} \frac{\Delta S}{\Delta x} = \frac{\partial S}{\partial x} = \frac{\partial}{\partial x} \left(\frac{\partial y}{\partial x} \right) = \frac{\partial^2 y}{\partial x^2}$$

Thus

$$\frac{\partial^2 y}{\partial x^2} = \frac{\mu}{T} \frac{\partial^2 y}{\partial t^2} \qquad\qquad\qquad 21\text{-}41$$

It is important to realize that this wave equation for a stretched string holds only for small angles and thus for small displacements $y(x,t)$. Comparing Equations 21-39 and 21-41, we see again that the velocity of the wave is

$$v = \sqrt{\frac{T}{\mu}}$$

An important property of the wave equation 21-39 is that it is linear; i.e., the function $y(x,t)$ and its derivatives occur only to the first power. There are no terms like y^2, $(\partial y / \partial x)^2$, $y \, \partial^2 y / \partial t^2$, or $(\partial^2 y / \partial t^2)^2$. An important property of linear equations is that if $y_1(x,t)$ and $y_2(x,t)$ are two solutions of the equation, the linear combination

$$y_3(x,t) = C_1 y_1(x,t) + C_2 y_2(x,t) \qquad\qquad 21\text{-}42$$

where C_1 and C_2 are any constants, is also a solution. This is easily shown by direct substitution of y_3 into the equation, using

$$\frac{\partial^2 (y_1 + y_2)}{\partial x^2} = \frac{\partial^2 y_1}{\partial x^2} + \frac{\partial^2 y_2}{\partial x^2} \qquad \text{and} \qquad \frac{\partial^2 (y_1 + y_2)}{\partial t^2} = \frac{\partial^2 y_1}{\partial t^2} + \frac{\partial^2 y_2}{\partial t^2}$$

Equation 21-42 is the mathematical statement of the superposition principle. If any two waves satisfy the wave equation 21-39, their sum also satisfies the same wave equation.

The superposition principle for waves on a string holds only if the amplitudes of the waves are small, so that the approximation used in deriving the wave equation $\sin \theta \approx \tan \theta$ holds. If the amplitudes are large, so that this approximation does not hold, the resulting equation relating the time derivatives and spatial derivatives of $y(x,t)$ is not linear. In that case, the sum of two solutions is not a solution and the principle of superposition does not hold.

A similar small-amplitude approximation must be made to derive the wave equation for sound waves from Newton's laws for a fluid. Thus the principle of superposition for sound waves is also limited to waves in which the amplitude of the pressure or displacement variations is small.

> The wave equation for electromagnetic waves follows directly from the laws of electricity and magnetism without any small-amplitude approximation. Thus the principle of superposition holds for electromagnetic waves under all conditions.

Review

A. Define, explain, or otherwise identify:

Harmonic wave, 513
Wave number, 513
Amplitude, 513
Angular frequency, 513
Wavelength, 514
Polarization, 516
Linearly polarized wave, 516

Circularly polarized wave, 516
Elliptically polarized wave, 517
Intensity, 523
Energy density, 524
Intensity level, 525
Decibel, 525
Wave equation, 529

B. True or false:

1. The speed of a harmonic wave equals the frequency times the wavelength.

2. Harmonic waves are dispersionless.

3. Plane polarization means the same as linear polarization.

4. Sound waves can be polarized.

5. The displacement and pressure in a sound wave are 180° out of phase.

6. The sum of two harmonic waves of the same wavelength and frequency is itself a harmonic wave.

7. The energy in a harmonic wave is proportional to the square of the amplitude of the wave.

8. The units of intensity are watts per square meter.

9. The intensity of a wave equals the product of the energy density and the speed.

10. A 60-dB sound has twice the intensity of a 30-dB sound.

Exercises

Use the following data in these exercises and problems unless otherwise specified: speed of sound in air = 340 m/sec = 1110 ft/sec; speed of sound in water = 1500 m/sec; speed of light $c = 3 \times 10^8$ m/sec.

Section 21-1, Wavelength, Frequency, and Velocity of Harmonic Waves

1. Equation 21-1b expresses the displacement of a harmonic wave as a function of x and t in terms of the wave parameters k and ω. Write equivalent expressions which instead of k and ω contain the following pairs of parameters: (a) k and v, (b) λ and f, (c) λ and T, (d) λ and v, (e) f and v.

2. The ear is sensitive to sound frequencies in the range of about 20 to 20,000 Hz. (a) What are the wavelengths in air corresponding to these frequencies? (b) What are the wavelengths in water?

3. (a) Middle C in the musical scale has a frequency of 262 Hz. What is the wavelength of this note in air? (b) The frequency of the C an octave above middle C is twice that of middle C. What is the wavelength of this note?

4. The eye is sensitive to electromagnetic waves whose wavelengths are in the range of about 4×10^{-7} to 7×10^{-7} m. What are the corresponding frequencies of these light waves?

5. Typical frequencies of electromagnetic radio waves are 100 kHz for AM and 100 MHz for FM. Calculate the wavelengths corresponding to these frequencies. (All electromagnetic waves travel with the speed of light.)

6. The wave function for a harmonic wave in a string is $y(x,t) = 0.001 \cdot \sin(62.8x + 314t)$, where y and x are in meters and t is in seconds. (a) What direction does this wave travel, and what is its speed? (b) Find the wavelength, frequency, and period of this wave. (c) What is the maximum displacement of any string segment?

7. A wave travels to the right along a string with speed 10 m/sec. Its frequency is 60 Hz, and its amplitude is 0.02 m. (a) Write a suitable wave function for this wave. (b) Is this wave function the only function that could describe this wave? Explain.

Section 21-2, Polarization

8. One end of a long string moves with frequency f and amplitude A in random directions in a plane perpendicular to the string, thus setting up unpolarized waves in the string. The string passes through a long narrow vertical slot in a board. (a) Describe the wave that passes through the slot in the board. The string now passes through a second board with a long narrow slot. Describe the wave on the other side of the second board (b) if the slot is horizontal and (c) if the slot makes an angle of 45° with the vertical.

9. What kind of polarization results when the phase δ in Equation 21-9 is (a) 180° and (b) 270°?

10. Explain why sound waves cannot be polarized.

Section 21-3, Harmonic Sound Waves

11. A typical loud sound wave with a frequency of 1000 Hz would have a pressure amplitude of about 10^{-4} atm. (a) At $t = 0$, the pressure is maximum at some point x_1. What is the displacement at that point at $t = 0$? (b) What is the maximum value of the displacement at any time and place? (Take the density of air to be 1.29 kg/m³.)

12. What is the displacement amplitude for a sound wave of frequency 100 Hz and pressure amplitude 10^{-4} atm?

13. The displacement amplitude for a sound wave of frequency 300 Hz is 10^{-7} m. What is the pressure amplitude for this wave?

14. (a) Find the displacement amplitude for a sound wave of frequency 1000 Hz at the pain-threshold pressure amplitude of 29 N/m². (b) Find the displacement amplitude for a sound wave with the same pressure amplitude but of frequency 500 Hz.

15. (a) Find the displacement amplitude for a sound wave of frequency 500 Hz at the threshold-of-hearing pressure amplitude of 2.9×10^{-5} N/m². (b) Find the displacement amplitude for a wave of the same pressure amplitude but of frequency 1000 Hz.

Section 21-4, Superposition and Interference of Harmonic Waves, and Section 21-5, Vector Method of Addition of Harmonic Waves

16. Two waves with the same frequency, wavelength, and amplitude are traveling in the same direction. If they differ in phase by 90° and each has amplitude 0.05 m, find the amplitude of the resultant wave.

17. Two waves traveling on a string in the same direction have frequency 100 Hz, wavelength 2 cm, and amplitude 0.02 m. They differ in phase by 60°. What is the amplitude of the resultant wave?

18. Two sound sources oscillate in phase with the same amplitude. They are separated in space by $\frac{1}{3}\lambda$. What is the amplitude of the resultant wave from the two sources at a point on the line joining the sources if the amplitude due to each source separately is A? (Assume that the point is not between the sources.)

19. Two sound sources oscillate in phase with frequency 100 Hz. At a point 5.00 m from one source and 5.85 m from the other the amplitude of the sound from each source separately is A. (a) What is the difference in phase of the sound waves from the two sources at that point? (b) What is the amplitude of the resultant wave at that point?

20. Use the vector method of wave addition to find the wave function for the sum of the two waves $y_1 = 3 \sin (kx - \omega t)$ and $y_2 = 4 \sin (kx - \omega t + \delta)$ for (a) $\delta = 90°$ and (b) $\delta = 60°$. (Find the amplitude and phase of the resultant wave using a ruler and protractor or use trigonometry.)

21. Use the vector method of addition to find graphically the sum of the following three waves: $y_1 = 5 \sin (kx - \omega t)$, $y_2 = 5 \sin (kx - \omega t + 120°)$, and $y_3 = 5 \sin (kx - \omega t + 240°)$.

Section 21-6, Energy and Intensity of a Harmonic Wave on a String, and Section 21-7, Energy and Intensity of Harmonic Sound Waves

22. Show that the power passing a point on a string equals the product of the energy per unit length and the wave velocity.

23. A 20-m string has a mass of 0.06 kg and is under a tension of 50 N. Waves of frequency 200 Hz and amplitude 1 cm move along the string from left to right. (a) What is the total energy of the waves in the string? (b) What is the power transmitted past a given point in the string (see Exercise 22)?

24. Two parallel tubes with the same diameter are filled with gas at the same pressure and temperature. One tube contains H_2, and the other contains O_2. (a) If sound waves traveling in the tubes have the same displacement amplitude and frequency, how do the intensities compare? (b) If the waves have the same frequency and pressure amplitude, how do the intensities compare? (c) If the waves have the same frequency and intensity, how do the pressure amplitudes and displacement amplitudes compare?

25. A piston at one end of a long tube filled with air at room temperature and normal pressure oscillates with frequency 500 Hz and amplitude 0.1 mm. The area of the piston is 100 cm². (a) What is the pressure amplitude of the sound waves generated in the tube? (b) What is the intensity of the waves? (c) What average power is required to keep the piston oscillating (neglecting friction)?

26. What is the intensity level in decibels for a sound wave (a) of intensity of 10^{-10} W/m², (b) of intensity of 10^{-2} W/m²?

27. Find the intensity if (a) $\beta = 10$ dB and (b) $\beta = 3$ dB. (c) Find the pressure amplitude for sound waves in air at standard temperature and pressure for each of these intensities.

28. Show that if the intensity is doubled, the intensity level increases by $10 \log 2 = 3.0$ dB.

29. What fraction of the acoustic power of a noise would have to be eliminated to lower its sound intensity level from 90 to 70 dB?

Section 21-8, The Wave Equation

30. Show that the function $y_3(x,t)$ given by Equation 21-42 satisfies the wave equation if $y_1(x,t)$ and $y_2(x,t)$ do.

31. Show explicitly that the following functions satisfy the wave equation: (a) $y(x,t) = (x + vt)^3$; (b) $y(x,t) = Ae^{ik(x-vt)}$, where A and k are constant and $i = \sqrt{-1}$; (c) $y(x,t) = \ln k(x - vt)$.

32. (a) Show that the function $y = A \sin kx \cos \omega t$ satisfies the wave equation. (b) Use the trigonometric identity $\sin A + \sin B = 2 \sin \frac{1}{2}(A + B) \cos \frac{1}{2}(A - B)$ to show that the wave function in part (a) is the sum of a wave traveling to the right and a wave traveling to the left.

Problems

1. Derive the general relation given by Equation 21-24 between δ and Δx, where δ is the phase difference due to a path difference Δx. Do this by letting x_1 be the path length for one wave and $x_2 = x_1 + \Delta x$ be that for the other wave. Then show that $y_2 = y_0 \sin (kx_2 - \omega t)$ can be written $y_2 = y_0 \sin (kx_1 - \omega t + \delta)$, where δ is given by Equation 21-24.

2. For the wave function given in Exercise 6 find the maximum speed and acceleration experienced by any point on the string.

3. A tuning fork attached to a stretched wire generates transverse waves. The vibration of the fork is perpendicular to the string. Its frequency is 440 Hz, and its amplitude of oscillation is 0.50 mm. The wire has linear mass density of 0.01 kg/m and is under a tension of 1000 N. (a) Find the period and frequency of waves in the wire. (b) What is the speed of the waves? (c) What are the wavelength and wave number? (d) Write a suitable wave function for the waves on the wire. (e) Calculate the maximum speed and acceleration of a point in the wire. (f) At what average rate must energy be supplied to the fork to keep it oscillating at a steady amplitude?

4. Two loudspeakers oscillate in phase but are separated by a distance d. The speakers and a listener are in a straight line. The listener is a distance x from the nearer speaker (and $x + d$ from the farther speaker). The speakers put out a sound wave of frequency 500 Hz. The intensity at the listener for each speaker acting separately is I_0. (a) For what values of d will the listener hear no sound from the two speakers? (b) For what values of d will the listener hear the greatest sound intensity from the speakers? For these values, what will the intensity be? (Give your answer in terms of I_0.) (c) What intensity will be heard by the listener if $d = 17$ cm?

5. Two wires of different densities are soldered together end to end and then stretched under tension T (the same in both wires). The wave speed in the first wire is twice that in the second wire. When a harmonic wave traveling in the first wire is reflected at the junction of the wires, the reflected wave has half the amplitude of the transmitted wave. (a) Assuming no loss in the wire, what fraction of the incident power is reflected at the junction and what fraction is transmitted? (b) If the incident-wave amplitude is A, what are the reflected- and transmitted-wave amplitudes?

6. Power is to be transmitted along a stretched wire by means of transverse harmonic waves. The wave speed is 100 m/sec, and the linear density is 0.01 kg/m. The power source oscillates with amplitude of 0.50 mm. (a) What average power is transmitted along the wire if the frequency is 400 Hz? (b) The power transmitted can be increased by increasing the tension in the wire, the frequency of the source, or the amplitude of the waves. If just one of these quantities were changed, how would each quantity have to be changed to effect an increase in power by a factor of 100? (c) Which of the changes would probably be accomplished most easily?

7. (a) Use the vector model of addition to find the amplitude of the sum of three harmonic waves $y_1 = y_0 \sin \theta$; $y_2 = y_0 \sin (\theta + \delta)$, and $y_3 = y_0 \cdot \sin (\theta + 2\delta)$, where $\theta = kx - \omega t$ for $\delta = 30°$ and $y_0 = 2$ cm. (b) What is the smallest value of δ for which the resultant amplitude is 0? (c) If the inten-

sity of each wave separately is I_0, what is the intensity of the resultant wave when $\theta = 0$? For what values of δ other than 0 is the intensity again this value? (d) Find the other value of δ less than 360° for which the resultant intensity is 0. Make a sketch of the resultant intensity versus δ from $\delta = -4\pi$ to $\delta = +4\pi$.

8. This problem is essentially Problem 7 with four waves $y_1 = y_0 \sin \theta$, $y_2 = y_0 \cdot \sin (\theta + \delta)$, $y_3 = y_0 \sin (\theta + 2\delta)$, and $y_4 = y_0 \sin (\theta + 3\delta)$. (a) Find the values of δ for which the resultant amplitude is greatest, and find the intensity (in terms of I_0 the intensity of one wave separately) at these maxima. (b) Find the three values of δ between 0 and 360° for which the resultant amplitude is 0. (c) Make a sketch of the resultant intensity versus δ.

9. (a) Make a sketch of y versus θ for each of the three waves in Problem 7 for $\delta = 120°$. Add the wave functions graphically to show that the sum of the three waves is zero for all θ. (b) Make a similar sketch for $\delta = 180°$ and find the sum graphically.

10. Three noise sources produce intensity levels of 70, 73, and 80 dB when acting separately. When acting together, the intensities of the sources add. (There is no interference between the amplitudes from the different sources because the relative phase changes randomly.) Find the sound intensity level in decibels when the three sources act at the same time. Discuss the usefulness of eliminating the two least intense sources in order to reduce the intensity level of the noise.

11. An article on noise pollution claims that the sound intensity level in large cities has been increasing by about 1 dB annually. What percentage increase in intensity does this correspond to? Does this increase seem reasonable?

CHAPTER 22   Standing Waves

When waves are confined in space, like waves on a piano string or in an organ pipe, there are reflections at both ends and therefore waves traveling in both directions. These waves combine according to the general law of wave interference. For a given string or pipe, there are certain frequencies for which this interference results in a stationary vibration pattern called a *standing wave*. The study of standing waves has many applications in the field of music and in nearly all areas of science and technology.

22-1 Standing Waves on a String Fixed at Both Ends

Consider a string fixed at one end and tied at the other end to a tuning fork, which vibrates with frequency f and small amplitude. The waves produced by the tuning fork travel down the string, are inverted by reflection at the fixed end, and travel back to the fork. Since the fork is vibrating with a small amplitude, it acts essentially as a fixed end as far as reflection is concerned. The waves are thus inverted once again upon reflection at the fork and start back down the string.

Let us look at a particular wave crest generated at the left end of the string by the fork. It travels to the right end of the string, is reflected back, and is reflected again at the fork. Since it has been reflected twice, it has been inverted twice and now differs from the next wave crest coming from the fork only in that the first has already traveled a distance $2L$, where L is the length of the string. If this distance is exactly equal to the wavelength λ, the twice-reflected wave will be in phase with the second wave and the two will interfere constructively. The resultant wave has an amplitude twice that of either wave (assuming no loss by reflection). When this resulting wave travels the distance $2L$ to the fixed end and back and has been reflected twice, it will exactly overlap the third wave generated by the tuning fork. Thus each

Doug Fulton/Photo Researchers

Standing waves are produced on a violin string by bowing the string and sometimes by plucking the string.

new wave is in phase with the waves reflected at the tuning fork, and the amplitude continues to increase as the string absorbs energy from the tuning fork. This continues until a maximum amplitude is reached due to various damping effects, e.g., loss of energy due to reflection or imperfect flexibility of the string. This maximum amplitude is much larger than that of the tuning fork. Thus the tuning fork is in *resonance* with the string when the tuning-fork frequency is such that the wavelength in the string equals twice the length of the string.

When the wavelength just equals the length of the string, the distance $2L$ equals 2 wavelengths. When the first wave has traveled the distance $2L$ and is reflected at the fork, it will be exactly in phase with the third wave generated at the fork. Again a large amplitude builds up in the string. We can see, in general, that resonance will occur if the distance $2L$ is any integer times the wavelength. Thus the condition for resonance is

$$2L = n\lambda \qquad \qquad 22\text{-}1$$

Resonance, or standing-wave, condition for a string fixed at both ends

where n is any integer. In terms of the frequency of the waves $f = v/\lambda$, the condition for resonance is

$$2L = n\frac{v}{f}$$

or

$$f_n = n\frac{v}{2L} = nf_1 \qquad \qquad 22\text{-}2$$

where

$$f_1 = \frac{v}{2L} = \frac{1}{2L}\sqrt{\frac{T}{\mu}} \qquad \qquad 22\text{-}3$$

is the lowest resonance frequency, called the *fundamental frequency*, and we have used Equation 20-4 for the speed of waves on the string.

Fundamental frequency defined

It is sometimes convenient to think of the resonance condition in terms of the time necessary for the first wave to travel to the end and back. Since this distance is $2L$, the time will be $2L/v$, where v is the wave speed. If this time equals the period of vibration of the fork, the first wave will add constructively to the second wave. Resonance will also result if this time equals any integral number of periods. Thus we can write the resonance condition

$$\frac{2L}{v} = nT = \frac{n}{f} \qquad \text{or} \qquad f = n\frac{v}{2L}$$

which is the same as that found by fitting an integral number of wavelengths into the distance $2L$. The frequencies given by Equation 22-2 are called the *natural frequencies* of the string.

Example 22-1 A string is stretched between two fixed supports 1 m apart, and the tension is adjusted until the fundamental frequency of the string is 440 Hz. What is the speed of transverse waves on the string?

From the condition for resonance (Equation 22-1) the wavelength for the fundamental, or $n = 1$, resonance frequency is $\lambda = 2L = 2$ m. Hence, the wave speed is

$$v = \lambda f = (2 \text{ m})(440 \text{ sec}^{-1}) = 880 \text{ m/sec}$$

What happens when the frequency of the tuning fork is *not* equal to one of the natural frequencies of the string? When the first wave has traveled the distance $2L$ and is reflected from the fork, it differs in phase from the wave being generated at the fork (Figure 22-1). Suppose the phase difference is δ. As we saw in Section 21-4, if each wave has amplitude y_0, the resultant of the new wave and the twice-reflected wave is a wave of amplitude $2y_0 \cos \frac{1}{2}\delta$ differing in phase by $\frac{1}{2}\delta$ from either of the two waves. When this resultant wave travels a distance of $2L$ and is again reflected at the fork, it will differ in phase from the next wave from the fork by $\delta + \frac{1}{2}\delta = \frac{3}{2}\delta$. The incoming waves and reflected waves will have a phase difference which depends on the number of reflections. The interference is thus sometimes constructive, in which case the amplitude of the wave on the string will increase, and sometimes destructive, in which case the amplitude of the wave will decrease. On the average, the amplitude will not increase but remain that of the first wave generated; i.e., the amplitude of the resultant wave will equal that of the tuning fork, which is small. The string does not absorb energy on the average. Only when the frequency of the tuning fork equals one of the natural frequencies of the string given by Equation 22-2 will the waves add in phase and the amplitude build up.

This resonance phenomenon is analogous to the resonance of a simple harmonic oscillator with a harmonic driving force. If the frequency of the driving force equals the natural frequency of a simple harmonic oscillator, the oscillator absorbs the maximum amount of energy from the driving force. Note, however, that a string fixed at both ends has not just one natural frequency but a sequence of natural frequencies, which are integral multiples of the fundamental frequency. This series is called a *harmonic series*. The second frequency $f_2 = 2f_1$ is called the second harmonic; the nth is called the nth harmonic. In another terminology often used in music, the second harmonic is called the first overtone; the third harmonic is the second overtone; etc. Thus the nth harmonic is also the $(n-1)$st overtone.

When the string is vibrated at any one of its natural frequencies, it absorbs energy from the external force and the amplitude increases until it reaches a maximum value; then the rate at which energy is lost to damping effects equals the rate at which energy is absorbed. The wave produced is called a *standing wave*. The change in the shape of the

Figure 22-1
Waves on a string produced by a tuning fork whose frequency is not in resonance with the natural frequencies of the string. The wave leaving the tuning fork for the first time (*dashed line*) and the wave leaving for the second time after being reflected twice (*gray line*) are not in phase and do not interfere constructively. The resultant wave is indicated by the black line.

Harmonic series

Winds set up standing waves in the Tacoma Narrows Suspension Bridge, leading to its collapse on Nov. 7, 1940, only 4 months after it had been opened for traffic.

Wide World

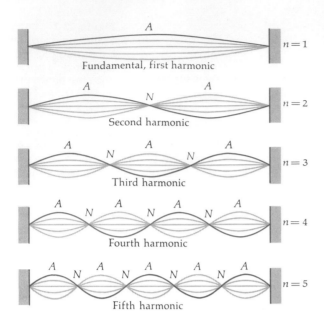

Figure 22-2
Standing waves on a string
fixed at both ends. The points
labeled A are antinodes, those
labeled N are nodes. In general the nth harmonic has n
antinodes and $n - 1$ nodes.

string is indicated in Figure 22-2 for the fundamental and the first
few harmonics. The energy of the string is constant, but it oscillates
from potential energy, when the string has its maximum displacement
and is at rest, to kinetic energy, when the string is in its original shape
but each segment has its maximum speed.

Questions

1. Give some examples of standing waves in one dimension other
than on stretched strings.

2. The wires for low notes on a piano are wound with copper to
increase their mass. Why?

3. When the tension in a string is increased without changing its
length, how do the resonance frequencies change? How do the wavelengths corresponding to these frequencies change?

22-2 Standing-Wave Functions

We can find the mathematical form of the wave functions illustrated in
Figure 22-2 and also rederive the resonance condition, or standing-
wave condition (Equation 22-1), by considering the resultant wave
formed by the addition of a wave traveling to the right and a wave
traveling to the left. Let us call the displacement of the wave moving to
the right y_R and that of the wave moving to the left y_L and assume that
the amplitudes are equal. Then

$$y_R = y_0 \sin (kx - \omega t) \quad \text{and} \quad y_L = y_0 \sin (kx + \omega t)$$

where $k = 2\pi/\lambda$ is the wave number and $\omega = 2\pi f$ is the angular
frequency. The sum of these two waves is

$$y(x,t) = y_R + y_L = y_0 \sin (kx - \omega t) + y_0 \sin (kx + \omega t)$$

or

$$y(x,t) = 2y_0 \cos \omega t \sin kx \qquad \qquad 22\text{-}4 \qquad \textit{Standing-wave function}$$

where again we have used Equation 21-22 for the sum of two sine functions.[1]

If the string is fixed at $x = 0$ and $x = L$, we have the following *boundary conditions* on the wave functions:

$$y(x = 0, t) = 0 \qquad \text{22-5}$$

and

Boundary conditions

$$y(x = L, t) = 0 \qquad \text{22-6}$$

for all times t. The first boundary condition is automatically met because $\sin kx = 0$ at $x = 0$. The second boundary condition is met only for those particular values of the wave number which satisfy

$$\sin kL = 0 \qquad \text{22-7}$$

The values k_n which satisfy this equation are given by

$$k_n L = n\pi \qquad \text{22-8}$$

where n is any integer. In terms of the wavelength $\lambda = 2\pi/k$, Equation 22-8 is

$$\frac{2\pi}{\lambda_n} L = n\pi$$

or

$$n\lambda_n = 2L \qquad \text{22-9}$$

This is the same condition we derived earlier. Twice the length of the string must equal an integral number of wavelengths.

The standing-wave functions are sine functions whose wavelengths are given by Equation 22-9. The easiest way to remember the condition for standing waves is to draw a line of length L and fit sine waves so that the displacement is zero at each end. We note that in addition to $x = 0$ and $x = L$, there are other points which are at rest if n is greater than 1. For example, for the wave for the second harmonic, the midpoint of the string is at rest. Such a point is called a *node*. In the third harmonic there are two points at rest (Figure 22-2). In general there are $n - 1$ nodes (not counting $x = 0$ and $x = L$) for the nth harmonic.

Nodes and antinodes defined

The points where the amplitude of vibration is maximum are called *antinodes*. There is an antinode halfway between each pair of nodes, including the ends. There are n antinodes when the string is vibrating in its nth harmonic.

22-3 Standing Waves on a String Fixed at One End

We can also produce standing waves in a string with one end free instead of fixed. A nearly free end results if the string is tied to a very long light thread, as in Figure 22-3. The analysis differs only slightly.

---

[1] We might think that we should subtract the two waves since the wave is inverted by reflection. It makes no difference whether we add or subtract the waves. Using $\sin A - \sin B = 2 \sin \frac{1}{2}(A - B) \cos \frac{1}{2}(A + B)$, we have $y_L - y_R = 2y_0 \sin \omega t \cos kx$. This differs from Equation 22-4 only in the phases of the functions. Equation 22-4 is more convenient because we do not have to add a phase constant in order to meet the boundary conditions.

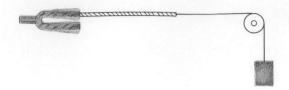

Figure 22-3
An approximation to a string
fixed at one end and free at
the other end can be pro-
duced by connecting the
"free" end of the string to a
very long, light thread. Since
the amplitude of the tuning
fork is small, the end attached
to the fork is approximately
fixed.

When the first wave is reflected at the free end, it is not inverted; however, it is inverted when it is reflected at the tuning fork. Thus if it arrives back at the fork in a time equal to *half* the period of the fork, it
will add constructively to the next wave (actually the last half of the
first wave). Constructive interference will also occur if the time taken
for the first wave to travel twice the length of the string equals $\frac{3}{2}$ times
the period, $\frac{5}{2}$ times the period, or any odd number of half periods.
Thus the resonance condition for a standing wave on a string of length
L fixed at one end (the tuning fork) and free at the other end is

$$\frac{2L}{v} = \frac{nT}{2} \qquad \text{where } n = 1, 3, 5, 7, \ldots \qquad\qquad 22\text{-}10$$

or, in terms of the frequency $f = 1/T$,

$$f = n\frac{v}{4L} \qquad n = 1, 3, 5, 7, \ldots \qquad\qquad 22\text{-}11$$

The natural frequencies of this system occur in the ratios $1:3:5:7$; that
is, the frequencies are given by

$$f_n = nf_1 \qquad \text{where } n = 1, 3, 5, 7, \ldots \qquad\qquad 22\text{-}12$$

where

$$f_1 = \frac{v}{4L} \qquad\qquad 22\text{-}13$$

is the fundamental frequency. The even harmonics are missing.

We can obtain the natural frequencies for this system and the wave
functions as we did for the string fixed at both ends from Equation
22-4:

$$y = 2y_0 \cos \omega t \sin kx$$

The boundary conditions are now that $y = 0$ at $x = 0$, which occurs
automatically, and that y is a maximum or minimum at $x = L$. That is,
the point $x = L$ is an antinode. This occurs if

$$\sin k_n L = \pm 1$$

or

$$k_n L = \frac{n\pi}{2} \qquad \text{where } n = 1, 3, 5, 7, \ldots \qquad\qquad 22\text{-}14$$

In terms of the wavelength $\lambda_n = 2\pi/k_n$, this condition is

$$\frac{2\pi}{\lambda_n} L = \frac{n\pi}{2}$$

or

$$n\lambda_n = 4L \qquad\qquad 22\text{-}15$$

This is the same as Equation 22-11 since the frequency is $f = v/\lambda$.
Again, the easiest way to remember the standing-wave condition is to
draw a string of length L and fit sine waves so that the end at $x = L$ is
an antinode. The first few standing-wave modes for a string fixed at
one end are shown in Figure 22-4.

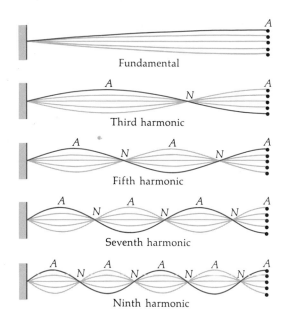

Fundamental

Third harmonic

Fifth harmonic

Seventh harmonic

Ninth harmonic

Figure 22-4
Standing waves on a string fixed at only one end. The points labeled A are antinodes; those labeled N are nodes.

22-4 Standing Sound Waves

Much of what we have learned about standing waves on strings can be applied to standing sound waves. Figure 22-5 shows a tube of air closed at the right end and fitted with a movable piston at the left. Because the air cannot vibrate past the closed end at $x = L$, this point must be a displacement node. If the vibration of the piston at the left end is of small amplitude, that end is also approximately a displacement node. (This approximation is similar to that for the vibrating string.) Let us call the speed of sound v. The standing-wave condition for this system is the same as for the string fixed at both ends, and all the same equations apply. The distance $2L$ must contain an integral number of wavelengths:

$$n\lambda = 2L$$

The allowed wavelengths are those which can be fitted into the length of the tube with displacement nodes at each end. Since the pressure variation is 90° out of phase with the displacement, these displacement nodes are pressure antinodes.

Margot Granitas/Photo Researchers

Standing waves are set up in a flute by blowing across the mouthpiece.

Example 22-2 If the speed of sound is 340 m/sec, what are the allowed frequencies and wavelengths for standing waves in a closed tube 1 m long?

The fundamental frequency is the lowest allowed frequency and corresponds to the longest wavelength $\lambda = 2L = 2$ m. Thus the fundamental frequency is $f_1 = v/\lambda = (340 \text{ m/sec})/(2 \text{ m}) = 170$ Hz. The frequency and wavelength of the nth harmonic are

$$f_n = nf_1 = 170n \text{ Hz} \quad \text{and} \quad \lambda_n = \frac{2L}{n} = \frac{2}{n} \text{ m}$$

If the end of the tube at the right in Figure 22-5 is not closed but open to the atmosphere, the open end must remain at atmospheric pressure. The open end is therefore a pressure node and a displacement antinode. The condition for the resonant frequencies and wavelengths for this system are the same as those for a string with one

Air

Movable piston

Figure 22-5
Air confined to a tube closed at both ends. There is a displacement node at both ends if the amplitude of the piston is small. The standing-sound-wave condition is thus the same as for a string fixed at both ends.

end fixed and one end free. The wavelength of the fundamental node is 4 times the length of the tube, and only odd harmonics are present. For the tube of length 1 m (Example 22-2) the wavelength of the fundamental is 4 m, and its frequency is $f_1 = \frac{340}{4} = 85$ Hz. The other allowed frequencies are

$$f_3 = 3f_1 = 255 \text{ Hz} \qquad f_5 = 5f_1 = 425 \text{ Hz}$$

and so on.

The result that the open end of the tube is a pressure node and displacement antinode is based on the assumption that the sound wave in the tube is a one-dimensional wave, which is approximately true if the diameter of the tube is very small compared with the wavelength of the sound wave. In practice, the displacement antinode and pressure node are slightly beyond the open end of the tube. The effective length of the tube L_{eff} is thus somewhat longer than the true length L. We can write the effective length of the tube as the sum of the true length L and an end correction ΔL:

$$L_{\text{eff}} = L + \Delta L$$

The end correction ΔL is of the order of the radius R of the tube. Note that the distance between two successive nodes or antinodes is still $\frac{1}{2}\lambda$ even though the distance from the open end of the tube to the first displacement node is slightly less than $\frac{1}{4}\lambda$ because of the end correction.

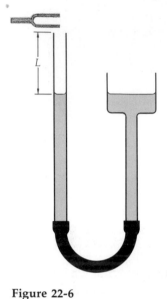

Figure 22-6
Apparatus for determining the speed of sound in air (Example 22-3). Sound waves of the frequency of the tuning fork are excited in the tube on the left. The length of the tube is varied by adjusting the water level. Resonance occurs when the effective length of the tube (length plus end correction) equals $\frac{1}{4}\lambda, \frac{3}{4}\lambda, \frac{5}{4}\lambda \ldots$, where λ is the wavelength of the sound. Measuring the difference in water level for two successive resonances gives $\frac{1}{2}\lambda$, from which the speed of sound can be found using $v = f\lambda$.

Example 22-3 Figure 22-6 shows an apparatus for measuring the speed of sound in air. A narrow vertical tube is partially filled with water; its level can be adjusted, and therefore the length of the air column can be adjusted. A tuning fork of frequency f_0 vibrates at the top open end of the tube and sends sound waves down the tube, which are reflected at the adjustable bottom, the water level. When the water level is such that the standing-wave condition is met, resonance occurs and the increased energy in the sound wave can be detected by the ear. In this experiment, the wavelength is $\lambda_0 = v/f_0$, where v is the velocity of sound in air. It is fixed since v is determined by the air temperature and f_0 by the tuning fork. If ΔL is the correction term for the tube, the tube will resonate if its length is L_1, given by

$$L_1 + \Delta L = \tfrac{1}{4}\lambda_0$$

It will also resonate at lengths $L_3, L_5, \ldots, L_n$, given by

$$L_3 + \Delta L = \tfrac{3}{4}\lambda_0$$

$$L_5 + \Delta L = \tfrac{5}{4}\lambda_0$$

$$L_n + \Delta L = \frac{n}{4}\lambda_0 \qquad n = 1, 3, 5, 7, \ldots$$

The wavelength, and thus the speed of sound, can be measured by measuring the distance between two consecutive resonances, e.g.,

$$L_3 - L_1 = \tfrac{1}{2}\lambda_0 = \frac{v}{2f_0}$$

An organ pipe is a familiar example of the use of standing waves in air columns. In the flue-type organ pipe shown in Figure 22-7 a stream of air is directed against the sharp edge of an opening (point A

in the figure). The complicated swirling motion of the air near the edge sets up vibrations in the air column. The resonance frequencies of the pipe depend on the length of the pipe and on whether the other end is closed or open.

In a closed organ pipe there is a pressure antinode at the closed end and a pressure node near the opening (point *A* in Figure 22-7). The resonance frequencies for such a pipe are therefore those for a tube open at one end and closed at the other. The wavelength of the fundamental is approximately 4 times the length of the pipe, and only odd harmonics are present.

In an open organ pipe, there is a pressure node near both ends of the pipe. The resonance frequencies for a pipe open at both ends are the same as for one closed at both ends except that there is an end correction at each end. The wavelength of the fundamental is 2 times the effective length of the pipe, and all harmonics are present.

Figure 22-7
Flue-type organ pipe. A stream of air is blown against the edge, causing a swirling motion of the air near point *A*, which excites standing waves in the pipe. There is a pressure node near point *A*, which is open to the atmosphere. The resonance frequencies of the pipe depend on the length of the pipe and on whether the other end is open or closed.

Question

4. How do the resonance frequencies of an organ pipe change when the air temperature increases?

Review

A. Define, explain, or otherwise identify:

Fundamental frequency, 538	Overtone, 539
Standing wave, 539	Node, 541
Harmonic, 539	Antinode, 541
Harmonic series, 539	Boundary condition, 541

B. Sketch the wave functions for the fundamental and first three overtones for a string fixed at both ends. From your sketches, write down the wavelengths for these harmonics. Do the same for a string fixed at one end and free at the other end.

C. Sketch both the displacement and pressure standing-wave functions for the fundamental and first three overtones in an organ pipe open at both ends. Do the same for a pipe open at only one end.

D. True or false:

1. A standing wave can be considered to be the superposition of a wave traveling to the right plus one traveling to the left.

2. The frequency of the fifth harmonic is 5 times the frequency of the fundamental.

3. In a pipe open at one end and closed at the other end, the even harmonics cannot be excited.

4. Because of the end correction, the pressure antinodes are not separated by $\frac{1}{2}$ wavelength in an open pipe.

5. The fundamental frequency of an organ pipe open at one end and closed at the other is lower than that of a pipe of the same length open at both ends.

Exercises

Section 22-1, Standing Waves on a String Fixed at Both Ends

1. A 5-gm steel wire 1 m long is under tension of 968 N. (*a*) Find the speed of transverse waves in the wire. (*b*) Find the wavelength and frequency of the fundamental. (*c*) Find the frequency of the second and third harmonics.

2. Middle C on the equal-temperament scale used by modern instrument makers has a frequency of 261.63 Hz. If this is the fundamental frequency of a 7-gm piano wire 80 cm long, what should the tension in the wire be?

3. A piano string without windings has a fundamental frequency of 200 Hz. When it is wound with wire, its linear mass density is doubled. What is its fundamental frequency then?

4. A string fixed at both ends is 3 m long. It resonates in its second harmonic at a frequency of 60 Hz. What is the speed of transverse waves in the string?

5. The length of the B string on a certain guitar is 60 cm. It vibrates at 247 Hz. (a) What is the speed of transverse waves on the string? (b) If the linear mass density is 0.01 gm/cm, what should the tension be when it is in tune?

Section 22-2, Standing-Wave Functions

6. Show that the wave function for the nth harmonic on a string of length L fixed at both ends can be written $y(x,t) = y_0 \cos (n\pi vt/L) \sin (n\pi x/L)$, where v is the wave speed.

7. A string 3 m long fixed at both ends is vibrating in its third harmonic. The maximum displacement of any point on the string is 4 mm. The speed of transverse waves on this string is 50 m/sec. (a) What are the wavelength and frequency for this wave? (b) Write the wave function for this wave.

8. The wave function for a certain standing wave on a string fixed at both ends is $y(x,t) = 0.30 \sin 0.20x \cos 300t$, where y and x are in centimeters and t in seconds. (a) What are the wavelength and frequency of these waves? (b) What is the speed of transverse waves on this string? (c) If the string is vibrating in its fourth harmonic, how long is it?

9. The wave function for a certain standing wave on a string fixed at both ends is $y(x,t) = 0.5 \sin 0.025x \cos 500t$, where y and x are in centimeters and t in seconds. (a) Find the speed and amplitude of the two traveling waves which result in this standing wave. (b) What is the distance between successive nodes in the string? (c) What is the shortest possible length of the string?

10. A string 2.51 m long has the wave function given in Exercise 9. (a) Sketch the position of the string for the times $t = 0$; $t = \frac{1}{4}T$; $t = \frac{1}{2}T$; $t = \frac{3}{4}T$, where $T = 1/f$ is the period of the vibration. (b) Find T in seconds. (c) When the string is horizontal, what has become of the energy in the wave?

Section 22-3, Standing Waves on a String Fixed at One End

11. A 160-gm rope 4 m long is fixed at one end and tied to a light string at the other end. Its tension is 400 N. (a) What are the wavelengths of the fundamental and the first two overtones? (b) What are the frequencies of these standing waves?

12. A string fixed at one end only is vibrating in its fundamental mode. The wave function is $y(x,t) = 0.02 \sin 2.36x \cos 377t$, where y and x are in meters and t in seconds. (a) What is the wavelength of the wave? (b) What is the length of the string? (c) What is the speed of transverse waves on the string?

13. A string 5 m long fixed at one end only is vibrating in its fifth harmonic with frequency 400 Hz. The maximum displacement of any segment of the string is 3 cm. (a) What is the wavelength of this wave? What is the wave number k? (b) What is the angular frequency? (c) Write the wave function for this standing wave.

14. Three successive resonant frequencies for a certain string are 75, 125, and 175 Hz. (a) Find the ratios of each pair of successive resonant frequencies. (b) How can you tell that these frequencies are for a string fixed at one end only rather than for a string fixed at both ends? (c) What is the fundamental

frequency? (*d*) Which harmonics are these resonance frequencies? (*e*) If the speed of transverse waves on this string is 400 m/sec, find the length of the string.

Section 22-4, Standing Sound Waves

Unless otherwise specified, take the speed of sound to be 340 m/sec = 1115 ft/sec.

15. Calculate the fundamental frequency for a 32-ft organ pipe which is open at both ends and for one that is closed at one end.

16. The normal range of hearing is from about 20 Hz to about 20,000 Hz. What is the greatest length of an organ pipe that would have its fundamental note in this range: (*a*) if it is closed at one end; (*b*) if it is open at both ends?

17. The shortest pipes used in organs are about 3 in long. (*a*) What is the fundamental frequency of a pipe this long which is open at both ends? (*b*) What is the highest harmonic for such a pipe which is within the audible range (see Exercise 16)?

18. The space above the water in the tube shown in Figure 22-6 is 120 cm long. Near the open end is a loudspeaker driven by an audio oscillator whose frequency can be varied from 10 to 5000 Hz. (*a*) What is the lowest frequency of the oscillator which will resonate with the tube? (*b*) What is the greatest frequency which will resonate? (*c*) How many different frequencies of the oscillator will produce resonance?

Problems

1. The G string on a violin is 30 cm long. When played without fingering, it vibrates at frequency 196 Hz. The next higher notes on the scale are A (220 Hz), B (247 Hz), C (262 Hz), and D (294 Hz). How far from the end of the string must a finger be placed to play these notes?

2. A steel piano wire is 40 cm long, has a mass of 2 gm, and is under tension of 600 N. (*a*) What is the fundamental frequency? (*b*) What is the wavelength in air of sound produced when the wire vibrates at its fundamental frequency? (*c*) If the highest frequency a certain listener can hear is 14,000 Hz, what is the highest harmonic produced by the wire that he can hear?

3. An early method for determining the speed of sound in different gases was as follows. A cylindrical glass tube is placed horizontally, and a quantity of light powder or fine wood shavings is spread along the bottom of the tube. One end is closed by a piston which can be attached to an oscillator of known frequency *f* (such as a tuning fork). The other end is closed by a piston whose position can be varied. While the opposite piston is made to oscillate at frequency *f*, the movable piston's position is adjusted until resonance occurs. When this happens, the powder collects in piles equally spaced along the bottom of the tube. (This is known as *Kundt's method.*) (*a*) Explain why the powder collects in this way. (*b*) Derive a formula which gives the speed of sound in the gas in terms of *f* and the distance between piles of powder. (*c*) Give suitable values for the frequency *f* and the length *L* of the tube for which the speed of sound could be measured using either air or helium.

4. A string with mass density of 4×10^{-3} kg/m is under tension of 360 N and is fixed at both ends. One of its resonance frequencies is 375 Hz. The next higher resonance frequency is 450 Hz. (*a*) What is the fundamental resonance frequency? (*b*) Which harmonics are the ones given? (*c*) What is the length of the string?

5. A string fastened at both ends has successive resonances with wavelengths of 0.54 m for the *n*th harmonic and 0.48 m for the (*n* + 1)st harmonic. (*a*) Which harmonics are these? (*b*) What is the length of the string? (*c*) What is the wavelength of the fundamental?

6. Three successive resonance frequencies in an organ pipe are 1310, 1834, and 2358 Hz. (*a*) Is the pipe closed at one end or open at both ends? (*b*) What is the fundamental frequency? (*c*) What is the length of the pipe?

7. A wire of mass 1 gm and length 50 cm is stretched with a tension of 440 N. It is placed near the open end of the tube in Figure 22-6 and stroked with a violin bow so that it oscillates with its fundamental frequency. The water level in the tube is lowered until a resonance is first obtained, at 18 cm below the top of the tube. Use these data to determine the speed of sound in air. Why is this method not very accurate?

8. A 40-cm-long wire of mass 0.01 kg vibrates in its second harmonic. When it is placed near the open end of the tube in Figure 22-6, resonance with the fundamental of the tube occurs if the water level is 1 m below the top of the tube. Assuming the speed of sound in air to be 340 m/sec, find: (*a*) the frequency of oscillation of the air column in the tube, (*b*) the speed of waves along the wire, and (*c*) the tension in the wire.

9. With a tuning fork of frequency 500 Hz held above the tube in Figure 22-6, resonances are found when the water level is at distances 16, 50.5, 85, and 119.5 cm. Use these data to calculate the speed of sound in air. How far outside the open end of the tube is the pressure node?

10. In a lecture demonstration of standing waves, a string is attached to a tuning fork which vibrates at 60 Hz and sets up transverse waves of that frequency on the string. The other end of the string passes over a pulley, and the tension is varied by attaching weights to that end. The string has approximate nodes at the tuning fork and at the pulley. If the string has mass density 8 gm/m and is 2.5 m long (from tuning fork to pulley), what must the tension be for the string to oscillate in its fundamental mode? Find the tensions for vibration in the first three overtones.

11. (*a*) For the wave function given in Exercise 12, find the velocity of a string segment at some point x as a function of time. (*b*) Which point has the greatest speed at any time? What is the maximum speed of this point? (*c*) Find the acceleration of a string segment at some point x as a function of time. Which point has the greatest acceleration? What is the maximum acceleration of this point?

12. A 2-m string is fixed at one end and vibrating in its third harmonic. The greatest displacement of any segment of the string is 3 cm. The frequency of vibration is 100 Hz. (*a*) Write the wave function for this vibration. (*b*) Write an expression for the kinetic energy of a segment of the string of length dx at point x at some time t. At what time is this kinetic energy maximum? What is the shape of the string at this time? (*c*) Find the maximum kinetic energy of the string by integrating your expression over the total length of the string. (*d*) Find the potential energy of a segment of the string and compute the maximum potential energy of the string by integration. *Hint:* Remember that the potential energy of a mass m in simple harmonic motion of angular frequency ω is $\frac{1}{2}m\omega^2 y^2$, where y is the displacement.

13. Show that if the tension in a string fixed at both ends is changed by a small amount ΔT, the frequency of the fundamental is changed approximately by Δf, where $\Delta f/f = \frac{1}{2}\Delta T/T$. (Use the differential approximation.) Does this result apply to all harmonics? Use this to find the percentage change in tension needed to raise the frequency of the fundamental of a piano wire from 260 to 262 Hz.

14. Show that if the temperature changes by a small amount ΔT, the fundamental frequency of an organ pipe changes by approximately Δf, where $\Delta f/f = \frac{1}{2}\Delta T/T$. Suppose an organ pipe closed at one end has a fundamental frequency of 200 Hz when the temperature is 68°F. What will its fundamental frequency be when the temperature is 90°F? (Ignore any change in the length of the pipe due to thermal expansion.)

CHAPTER 23	The Superposition of Waves of Different Frequency

In Chapters 21 and 22 we considered superposition of harmonic waves of the same frequency and wavelength. We now turn to superposition and interference of waves with different frequencies and wavelengths, considering first two waves with nearly equal frequencies and wavelengths. A familiar example of the interference of two waves is the beats produced by two sound waves of nearly equal frequency. We then discuss the harmonic analysis of complex waves and the inverse problem of synthesis of harmonic waves to form a complex wave. The related phenomena of dispersion and the superposition of many waves to form a wave packet are discussed qualitatively in Sections 23-4 and 23-5. Finally, we consider the general motion of a vibrating string in terms of the superposition of standing harmonic waves. Since a complete detailed discussion of these subjects requires methods of Fourier analysis, which most students do not study until their junior or senior years, we must be content with a qualitative discussion.

23-1 Superposition of Two Waves of Nearly Equal Frequency and Wavelength

Consider two waves of slightly different frequencies and wavelengths but having equal amplitudes and moving in the same direction. Let us choose the waves to be in phase at $t = 0$ and $x = 0$:

$$y_1 = y_0 \cos (k_1 x - \omega_1 t) \tag{23-1}$$

$$y_2 = y_0 \cos (k_2 x - \omega_2 t) \tag{23-2}$$

where $k = 2\pi/\lambda$ and $\omega = 2\pi f$. The resultant wave can be found by graphical addition or by algebraic addition of the functions using a trigonometric identity for the sum of two cosine functions. Figure 23-1*a*

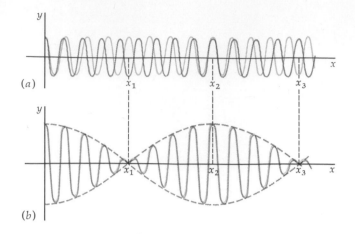

(a)

(b)

Figure 23-1
(a) Two waves of different wavelengths at a time when they are in phase at $x = 0$. At x_1 they are 180° out of phase and cancel completely. At x_2 they are in phase again, and at x_3 they are 180° out of phase. (b) The resultant of the two waves shown in part (a). The wavelength of the rapid oscillation is about the same as that of the original waves, but the amplitude is modulated, as indicated by the dashed envelope. The amplitude is maximum at $x = 0$ and $x = x_2$ and zero at $x = x_1$ and $x = x_3$.

shows the waves at time $t = 0$, when the waves are in phase and add constructively at $x = 0$. Because of the difference in wavelength, they are out of phase at other places. At some distance x_1, the waves are 180° out of phase and add to zero. At an equal distance beyond this point, at x_2, the waves are in phase again. The greater the difference in wavelength, the shorter the distance x_1 in which they become out of phase. Figure 23-1 also could represent the time dependence of the waves at the fixed point $x = 0$. Initially they are in phase, but because of the difference in frequency, they are 180° out of phase and cancel at some time later t_1. Later they are again in phase and interfere constructively. The greater the difference in frequency, the sooner they will be out of phase and cancel.

We can add the waves algebraically by using $\cos A + \cos B = 2 \cos \frac{1}{2}(A - B) \cos \frac{1}{2}(A + B)$. Let $y(x,t)$ be the resultant wave function. Then

$$y(x,t) = y_2 + y_1 = y_0 \cos (k_2 x - \omega_2 t) + y_0 \cos (k_1 x - \omega_1 t)$$
$$= 2y_0 \cos \left[\tfrac{1}{2}(k_2 - k_1)x - \tfrac{1}{2}(\omega_2 - \omega_1)t\right] \cos \left[\tfrac{1}{2}(k_2 + k_1)x - \tfrac{1}{2}(\omega_2 + \omega_1)t\right] \quad \text{23-3}$$

This result is simplified if we use the notation Δk and $\Delta \omega$ for the differences in wave number and frequency and $\bar{k}$ and $\bar{\omega}$ for their averages:

$$\Delta k = k_2 - k_1 \qquad \Delta \omega = \omega_2 - \omega_1$$
$$\bar{k} = \tfrac{1}{2}(k_2 + k_1) \qquad \bar{\omega} = \tfrac{1}{2}(\omega_2 + \omega_1)$$

23-4

Then Equation 23-3 becomes

$$y = 2y_0 \cos (\tfrac{1}{2}\Delta k \, x - \tfrac{1}{2}\Delta \omega \, t) \cos (\bar{k}x - \bar{\omega}t) \quad \text{23-5}$$

The resultant wave is sketched in Figure 23-1b for frequencies and wavelengths nearly equal so that $\Delta \omega$ and Δk are small and $\bar{\omega}$ and $\bar{k}$ nearly equal to the frequency and wave number of either wave. The result is a wave of about the same frequency and wavelength as the original waves but with the amplitude *modulated* by the factor $\cos (\tfrac{1}{2}\Delta k \, x - \tfrac{1}{2}\Delta \omega \, t)$. The velocity of the resultant wave $v = \bar{\omega}/\bar{k}$ is nearly the same as that of the individual waves. This velocity is called the *phase velocity*. The envelope (dashed curve in Fig. 23-2b) travels as a wave of wave number $\tfrac{1}{2}\Delta k$ and angular frequency $\tfrac{1}{2}\Delta \omega$. We can find the velocity of the envelope by considering the modulating factor in

Phase velocity

Equation 23-5,

$$\cos\left(\tfrac{1}{2}\Delta k\, x - \tfrac{1}{2}\Delta\omega\, t\right) = \cos\tfrac{1}{2}\Delta k\left(x - \frac{\Delta\omega}{\Delta k}\, t\right) = \cos\tfrac{1}{2}\Delta k\, (x - v_g t)$$

where

$$v_g = \frac{\Delta\omega}{\Delta k} \tag{23-6}$$

The velocity of the envelope $v_g = \Delta\omega/\Delta k$ is called the *group velocity*. The relation between the group velocity and phase velocity depends on the medium through which the wave is transmitted. We shall show in Section 23-5 that in a medium in which the phase velocity does not depend on the frequency of the wave, the group and phase velocities are equal. Such a medium is called a *dispersionless medium*. Examples are waves on a perfectly flexible string, sound waves in air, and light waves in vacuum. On the other hand, if the phase velocity does depend on the frequency or wavelength, the group velocity and phase velocity are not equal. Examples are water waves, light waves in glass or water, and waves on a string which is not perfectly flexible. A medium for which the phase velocity depends on the frequency is called a *dispersive medium*.

Group velocity

23-2 Beats

When two sound sources of comparable intensity have nearly equal frequencies, the interference of the sound waves at the ear alternates between constructive and destructive interference. The sound heard alternates between loud and soft. Each pulsation is called a *beat*. The frequency of oscillation between loud and soft, or *beat frequency*, is just equal to the difference between the frequencies of the two sound sources. We can see this from Equation 23-5, which gives the resultant interference pattern of the two waves. If we have two tuning forks of frequency f_1 and f_2 and let the ear be at some fixed point x_1, the quantities $\tfrac{1}{2}\Delta k\, x_1$ and $\bar{k}x_1$ in Equation 23-5 are just phase constants. If we ignore these phase constants, we can write for the wave at the ear,

$$y(t) = 2y_0 \cos\left(\tfrac{1}{2}\Delta\omega\, t\right)\cos\bar\omega t = 2y_0 \cos\left(2\pi\tfrac{1}{2}\Delta f\, t\right)\cos 2\pi\bar f t \tag{23-7}$$

The ear thus hears the average frequency $\bar f = \tfrac{1}{2}(f_1 + f_2)$ with the amplitude $2y_0 \cos\left(2\pi\tfrac{1}{2}\Delta f\, t\right)$. Since the energy of the wave is proportional to the square of the amplitude, the sound is loud whenever the amplitude is either maximum or minimum, i.e., whenever $\cos 2\pi\tfrac{1}{2}\Delta f\, t$ is $+1$ or -1. Since the maximum amplitude occurs with frequency $\tfrac{1}{2}\Delta f$, the frequency of maximum and minimum amplitudes is just twice this, or Δf. For example, if the tuning forks have frequencies 241 and 243 Hz, the ear will hear the frequency 242 Hz and the sound will be loud twice each second.

The phenomenon of beats is often used to compare an unknown frequency with a known frequency, as in tuning a piano with a tuning fork. The ear can detect beats up to about 10 per second. Above this the fluctuations in loudness are too rapid to be heard. Beats are also used to detect small frequency changes like those produced in a radar beam reflected from a moving car. The shift in frequency of the reflected beam is due to the doppler effect (Chapter 24). The velocity of

Figure 23-2
Moiré pattern showing beats produced by two sets of parallel lines when the spacing of one set differs slightly from that of the other.

the car is found by measuring the change in frequency, determined by measuring the beats produced by the reflected beam and the original radar source. There are many interesting related phenomena, such as the moiré pattern shown in Figure 23-2, which can be thought of as the beats produced by two sets of parallel lines when the spacing of one set differs slightly from that of the other.

Questions

1. Two violinists are practicing the same piece of music together. Can a listener hear the beats as they play? If beats are heard, what can be done to prevent them?

2. When musical notes are sounded together to make chords, beats are produced. Why are they not noticed as such?

3. Discuss how the phenomenon of beats is used to tune musical instruments.

4. Pianos generally have more than one wire for each note. The wires are struck simultaneously by the same hammer when the note is sounded. What is the effect if these wires are slightly out of tune?

23-3 Harmonic Analysis and Synthesis

Many complex periodic waves are mixtures of harmonic waves of several frequencies. Figure 23-3 shows the waveforms $y(t)$ produced by a tuning fork, clarinet, and cornet playing the same musical note. All the waves have the same period because they correspond to the same note, or pitch, but the waveforms are quite different. The difference in tone quality is related to the difference in the waveforms. A useful mathematical result, called *Fourier's theorem*, allows us to analyze any periodic function in terms of sines and cosines. According to Fourier's theorem, any periodic function can be represented with arbitrary accuracy by a sum of sine and cosine functions. Consider some periodic function $y(t)$, which might describe the displacement of air particles near a sound source. Fourier's theorem states that this function can be written

$$y(t) = \sum_n A_n \sin \omega_n t + B_n \cos \omega_n t \qquad \text{23-8}$$

The lowest angular frequency ω_1 corresponds to the period of the func-

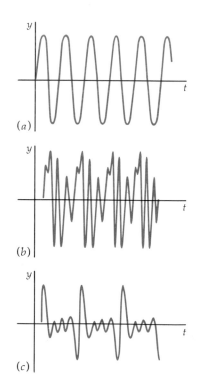

Figure 23-3
Waveforms of (*a*) a tuning fork, (*b*) clarinet, and (*c*) cornet each at frequency of 440 Hz and at approximately the same intensity. (*Redrawn from Charles A. Culver, Musical Acoustics, 4th ed., p. 103. Copyright © 1956 by McGraw-Hill Book Company, Inc.; by permission of McGraw-Hill Book Company.*)

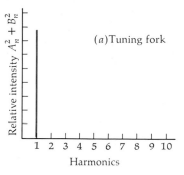

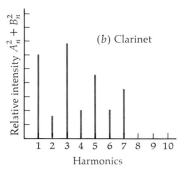

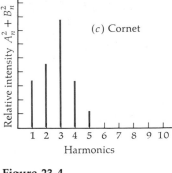

tion; i.e., the lowest frequency is $\omega_1 = 2\pi/T$, where T is the period. The other frequencies are integral multiples of the lowest or fundamental frequency

$$\omega_n = n\omega_1 = n\frac{2\pi}{T} \qquad\qquad 23\text{-}9$$

The relative values of the constants A_n and B_n are related to the waveform, i.e., the shape of the function, and can be determined by a method called *Fourier analysis* or *harmonic analysis* if the function $y(t)$ is given. The value of $A_n{}^2 + B_n{}^2$ is proportional to the intensity of the nth harmonic component of the function. Figure 23-4 illustrates the harmonic analysis of the waveforms shown in Figure 23-3. Both figures show that the clarinet, for example, is much richer in harmonics than the tuning fork.

The construction of an arbitrary periodic waveform from its component harmonics is called *synthesis*. Figure 23-5 shows a square wave

Figure 23-4
Relative intensities of the harmonics in the waveform shown in Figure 23-3 for the tuning fork, clarinet, and cornet. (*Redrawn from Charles A. Culver,* Musical Acoustics, *4th ed., p. 104. Copyright © 1956 by McGraw-Hill Book Company, Inc.; by permission of McGraw-Hill Book Company.*)

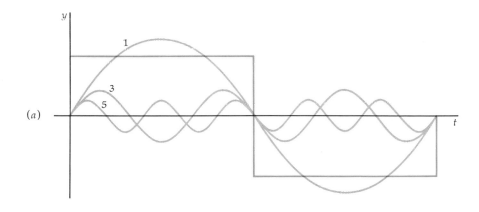

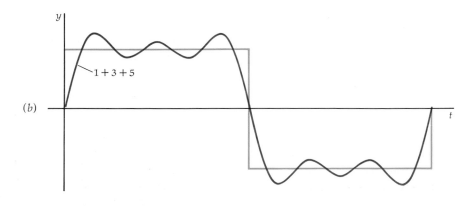

Figure 23-5
(*a*) Square wave and first three harmonics used to synthesize the wave. (*b*) Synthesis of square wave using the first three odd harmonics.

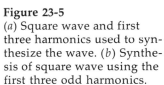

Synthesizer designed by Heinrich von Helmholtz and constructed by Rudolph Koenig. (*From Adolphe Ganot,* Elementary Treatise on Physics, *9th ed., William Wood and Publishers, Ltd., Great Britain, 1910.*) Each electrically driven tuning fork has a resonator which can be mechanically opened or closed to vary the intensity of that harmonic. In the more modern Moog synthesizer, shown here with Robert Moog, the harmonics are produced and mixed electronically.

Figure 23-6
Relative amplitudes of the harmonics needed to synthesize the square wave shown in Figure 23-5. For even n, $A_n = 0$, whereas for odd n, A_n is proportional to $1/n$. The relative energy in the harmonics is proportional to A_n^2.

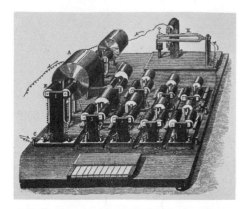

Norlin Music, Inc.

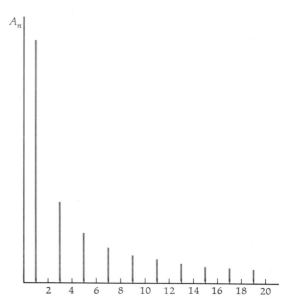

which can be produced by an electronic square-wave generator and its approximate synthesis using only three harmonics. This waveform can be synthesized with sine functions ($B_n = 0$ for all n) containing only odd harmonics. The relative values of the constants A_n are shown in Figure 23-6. The more harmonics used the better the approximation to the actual waveform.

23-4 Wave Packets

The complicated waveforms discussed in the previous section are periodic in time. Functions which are not periodic, e.g., pulses, can also be represented by harmonic functions (sines and cosines), but a continuous distribution of frequencies is required. The sum in Equation 23-8 then becomes an integral over frequency, and the constants A_n and B_n

become functions of frequency $A(\omega)$ and $B(\omega)$. Often $A(\omega)$ and $B(\omega)$ are nearly zero except in a certain frequency range $\Delta\omega$.

The characteristic feature of a wave pulse which distinguishes it from a periodic wave or harmonic wave of a single frequency is that the pulse has a beginning and an end whereas a harmonic wave of a single frequency repeats over and over. To send a signal with a wave, we need some kind of pulse rather than a harmonic wave of a single frequency. There is an important relation between the distribution of frequencies of the harmonic functions which make up a pulse and the time duration of the pulse. If the time duration Δt is very short, the range of frequencies $\Delta\omega$ is very large. The general relation between Δt and $\Delta\omega$ is

$$\Delta\omega\,\Delta t \approx 1 \qquad\qquad\qquad 23\text{-}10$$

The exact value of this product depends on just how the quantities Δt and $\Delta\omega$ are defined. For any reasonable definitions, $\Delta\omega$ is of the order of $1/\Delta t$.

The wave pulse produced by a source of short duration Δt has a narrow width in space $\Delta x = v\,\Delta t$, where v is the wave speed. Each harmonic wave of frequency ω has a wave number $k = \omega/v$. A range of frequencies $\Delta\omega$ implies a range of wave numbers $\Delta k = \Delta\omega/v$ in the resulting wave. (For simplicity here we assume that v is the same for each harmonic wave component; i.e., we neglect dispersion.) Substituting $\Delta k\,v$ for $\Delta\omega$ in Equation 23-10 gives

$$\Delta k\,v\,\Delta t \approx 1$$

or

$$\Delta k\,\Delta x \approx 1 \qquad\qquad\qquad 23\text{-}11$$

where again the exact value for the product depends on the precise definition of Δk and Δx.

The relations expressed by Equations 23-10 and 23-11 are important characteristics of wave pulses which apply to all types of waves. They are of particular importance in communications theory and the theory of quantum mechanics. Since information cannot be transported by a harmonic wave which has no beginning or end in time, the transmission of short pulses depends on the capability of transmitting a wide range of frequencies. In quantum mechanics, the position of a particle is described by a *wave packet* whose width reflects the uncertainty in the location of the particle. The possible values for the measurement of the momentum of the particle are proportional to the wave numbers k of the harmonic waves making up the wave packet. Thus a particle which is well localized in space, as evidenced by a narrow wave packet, must contain a wide range of possible momentum values, as evidenced by the large range of wave numbers. This property of wave packets is at the heart of the famous Heisenberg uncertainty principle.

Example 23-1 Show that the separation in space between successive zeros of the envelope function in Figure 23-1b varies inversely with the difference in the wave numbers of the two waves. Show also that the time between successive zeros in the envelope at some point in space varies inversely with the difference in frequency of the two waves.

The envelope in this figure resembles a wave pulse except that the envelope repeats again and again in space. Consider a point x_1 at which the envelope is zero at some time t_0. Let x_3 be the next point at which the envelope is zero at this time. This next point occurs when the argument of the modulated cosine function in Equation 23-5 changes by π:

$$\tfrac{1}{2}\Delta k\, x_3 - \tfrac{1}{2}\Delta k\, x_1 = \pi$$

or

$$(x_3 - x_1)\, \Delta k = \Delta x\, \Delta k = 2\pi \qquad\qquad 23\text{-}12$$

If we examine the envelope at a single point x_0 at successive times t_1 and t_2 at which the envelope is zero, we have

$$\tfrac{1}{2}\Delta\omega\, t_2 - \tfrac{1}{2}\Delta\omega\, t_1 = \pi$$

or

$$\Delta\omega\, (t_2 - t_1) = \Delta\omega\, \Delta t = 2\pi \qquad\qquad 23\text{-}13$$

These relations are similar to Equations 23-10 and 23-11 except that the products equal 2π instead of 1, Δx is not really the width of the wave packet, nor is Δt its duration since the envelope repeats endlessly.

23-5 Dispersion

We found in Section 23-1 that the velocity of the envelope of two waves, the group velocity, is given by

$$v_g = \frac{\Delta\omega}{\Delta k} \qquad\qquad 23\text{-}14$$

In the more general case of a wave packet, the group velocity v_g is the speed with which the packet as a whole moves. As we have seen, such a packet can be thought of as a superposition of many harmonic waves with wave numbers k and angular frequencies ω distributed over limited ranges. Within these ranges ω and k do not vary independently. If any k is known, the corresponding value of ω can be determined. In other words, ω is a function of k. The group velocity of the packet is then defined by

$$v_g = \frac{\Delta\omega}{\Delta k} \approx \frac{d\omega}{dk} \qquad\qquad 23\text{-}15$$

where $d\omega/dk$ is evaluated at the midpoint of the range of ω or k. Equation 23-15 is a generalization of Equation 23-14, which was found for just two waves.

Each harmonic wave that contributes to a wave packet has its own wave number k and angular frequency ω. It travels at a speed v_p, called its *phase velocity*, given by

$$v_p = \frac{\omega}{k} = f\lambda \qquad\qquad 23\text{-}16$$

By substituting $\omega = kv_p$ into Equation 23-15 we can obtain a relationship between the group velocity of a wave packet and the phase velocities of its constituent harmonic waves

$$v_g = \frac{d\omega}{dk} = \frac{d}{dk}(kv_p) = v_p + k\frac{dv_p}{dk} \qquad\qquad 23\text{-}17$$

Relationship between group and phase velocity

The group velocity differs from the phase velocity if the phase velocity depends on the wave number k or on the wavelength or frequency since these quantities are all related. If the medium is such that the phase velocity does not depend on the frequency (or wavelength or wave number), the group and phase velocities are equal and the medium is said to be *nondispersive*. An example is a wave on a perfectly flexible string for which the phase velocity is given by $v_p = \sqrt{T/\mu}$. Since the tension and mass density are independent of the frequency or wave number, $dv_p/dk = 0$. Other examples of waves in nondispersion media are sound waves in air and light waves (or any electromagnetic waves) in vacuum.

An important characteristic of a nondispersive medium is that a pulse will maintain the same shape as it travels. We have seen from our discussion of Fourier analysis that we can consider a pulse to consist of a distribution of harmonic waves of different frequencies. If the pulse is to maintain the same shape as it travels, all the component harmonic waves must travel with the same phase velocity. Conversely, if the phase velocity is different for different frequencies, the shape of a pulse will change as it travels.

If the phase velocity does depend on the wavelength (and thus on the frequency), the group velocity and phase velocity differ and the medium is called *dispersive*. Examples are waves on a real string or wire which is not perfectly flexible and light waves in any material such as glass or water.

For example, when a beam of sunlight enters a piece of glass from air, the light is bent, or refracted (we shall study this phenomenon in Chapter 26). The angle of refraction depends on the relative phase velocity of the light in glass and in air. The phase velocity of light in glass depends slightly on the wavelength of light, being somewhat greater for the longer wavelengths (red) than for short wavelengths (blue). The long wavelengths are bent less than the short wavelengths, and the light beam is spread out, or dispersed, into its component colors, or wavelengths.

Example 23-2 Dispersion of Deep Water Waves of Long Wavelength The angular frequency ω for water waves whose wavelength is greater than about 2 cm but much smaller than the depth of the water is related to the wave number k by

$$\omega = \sqrt{gk} \qquad\qquad 23\text{-}18$$

where g is the acceleration of gravity. Such a relation between the angular frequency and the wave number is called a *dispersion relation*. Given the dispersion relation for waves in a medium, we can calculate the phase and group velocities of the waves. The phase velocity is simply ω/k (Equation 23-16):

Dispersion relations

$$v_p = \frac{\omega}{k} = \sqrt{\frac{g}{k}}$$

We can calculate the group velocity from either Equation 23-15 or 23-17. Using Equation 23-15, we obtain

$$v_g = \frac{d\omega}{dk} = \frac{1}{2}\sqrt{\frac{g}{k}} = \tfrac{1}{2}v_p$$

We thus find that the group velocity is just half the phase velocity. This phenomenon is not too difficult to observe. If a short train of

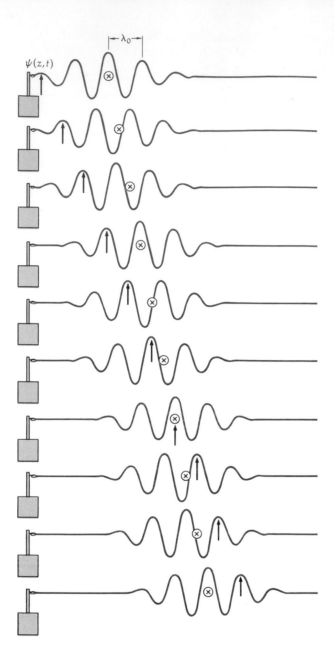

Figure 23-7
Wave packet for which the group velocity is half the phase velocity. The arrow travels at the phase velocity following a point of constant phase for the dominant wavelength λ_0. The cross at the center of the group travels at the group velocity. (*Adapted from Frank S. Crawford, Jr.,* Waves and Oscillations, *p. 294, Berkeley Physics Course, vol. 3, McGraw-Hill Book Company, New York, 1965. Courtesy of Education Development Center, Inc., Newton, Mass.*)

crests and valleys is produced on water, the crests can be seen to move ahead faster than the group as a whole. When a crest reaches the front of the group, it disappears. New crests continuously form at the rear, move through the group, and disappear at the front. Such a group is illustrated in Figure 23-7.

For deep water waves of wavelength of the order of a few centimeters, the phase velocity is influenced by the surface tension as well as by gravity. With very short wavelengths, the phase velocity is determined mainly by the surface tension, and the phase velocity is less than the group velocity (see Exercise 12 and Problem 6).

Question

5. Under what conditions is group velocity greater than phase velocity? Less than phase velocity? The same?

Optional

Example 23-3 Dispersion in a Real Piano Wire The result that the phase velocity of a wave on a string is equal to $\sqrt{T/\mu}$ is based on the assumption that the string is perfectly flexible. When such a string is deformed by a wave, there is a restoring force proportional to the tension. The greater the tension the greater the restoring force and the greater the phase velocity. A real piano wire is not perfectly flexible. Even if the wire is not under tension, there is a tendency for the wire to return to its original shape because of its stiffness. The phase velocity of a wave on a wire which is not perfectly flexible is somewhat greater than $\sqrt{T/\mu}$ because of the stiffness of the wire. The dispersion relation for a real piano wire can be written

$$\frac{\omega^2}{k^2} = \frac{T}{\mu} + \alpha k^2 \qquad 23\text{-}19$$

where α is a small positive constant which depends on the stiffness of the string. For a perfectly flexible string α is zero. For a piano wire, α is usually so small that the quantity αk^2 is much less than T/μ. Using Equation 23-16, we have for the phase velocity

$$v_p = \frac{\omega}{k} = \sqrt{\frac{T}{\mu} + \alpha k^2} = \sqrt{\frac{T}{\mu}} \sqrt{1 + \frac{\alpha k^2 \mu}{T}}$$

$$\approx \sqrt{\frac{T}{\mu}} \left(1 + \frac{\alpha k^2 \mu}{2T} + \cdots \right) \qquad \text{for small } \alpha \qquad 23\text{-}20$$

We see that the phase velocity does depend on k. The derivative of the phase velocity with respect to wave number is

$$\frac{dv_p}{dk} = \sqrt{\frac{T}{\mu}} \frac{\alpha k \mu}{T} = \alpha k \sqrt{\frac{\mu}{T}}$$

Thus the group velocity is

$$v_g = v_p + k \frac{dv_p}{dk} = v_p + \alpha k^2 \sqrt{\frac{\mu}{T}} \qquad 23\text{-}21$$

The fact that the phase velocity of waves on a piano string depends on the wavelength is apparently largely responsible for the distinctive tone of a piano.[1] Although the wavelengths for the standing waves on a real piano wire are still given by $\lambda_1 = 2L$, $\lambda_2 = 2L/2$, $\lambda_3 = 2L/3$, ... , the frequencies of the overtones are not just integral multiples of the fundamental because $f_n = v_p/\lambda_n$ and v_p depends on the wavelength. Since the phase velocity increases as the wavelength decreases, as indicated by Equation 23-20 ($\lambda = 2\pi/k$), the overtones are slightly sharp compared with those of a perfectly flexible string. The overtones are not true harmonics.

Optional

23-6 Superposition of Standing Waves

In our consideration of a perfectly flexible string fixed at both ends, we found that the standing-wave functions have the form

$$y_n(x,t) = A_n \cos(\omega_n t + \delta_n) \sin k_n x \qquad 23\text{-}22$$

where A_n is the amplitude and δ_n is a phase constant which depends on the choice of zero time. The boundary conditions $y = 0$ at $x = 0$ and

[1] See E. Donnell Blackham, The Physics of the Piano, *Scientific American*, December 1965.

Figure 23-8
If a string of length L fixed at both ends is to vibrate with only its fundamental frequency, it must have the initial shape shown.

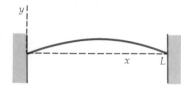

$x = L$ restrict the allowed frequencies and wave numbers to those obeying the standing-wave condition

$$k_n L = n\pi \qquad \text{and} \qquad \omega_n = 2\pi f_n = n2\pi \frac{v}{2L} \qquad \text{23-23}$$

where v is the phase velocity (which is the same as the group velocity because we are considering a perfectly flexible string). If Equation 23-22 is to describe the standing-wave function for a vibrating string, the string must have the shape of a sine function at any instant. At time $t = 0$, the wave function, according to this equation, is

$$y_n(x,0) = A_n \cos \delta_n \sin k_n x$$

For example, if the string is at rest at $t = 0$, it will vibrate in its fundamental mode with frequency f_1 corresponding to the wavelength $\lambda_1 = 2L$ only if the string has the initial shape indicated in Figure 23-8.

In general, a string fixed at both ends does *not* vibrate in a *single* harmonic or mode. The general vibration of a string contains a mixture of the allowed harmonic functions which satisfy the standing-wave conditions. The general wave function for waves on a perfectly flexible string fixed at both ends is a sum of harmonic wave functions:

$$y(x,t) = \sum_n A_n \cos(\omega_n t + \delta_n) \sin k_n x \qquad \text{23-24}$$

Since each term in the sum is zero at both $x = 0$ and $x = L$, the function $y(x,t)$ also obeys the boundary conditions for any values of the amplitudes A_n and the phase constants δ_n. The particular values of these constants depend on the initial shape and initial velocity of the string. The velocity of any point on the string is obtained by differentiating Equation 23-24 with respect to time:

$$u_y = \frac{\partial y}{\partial t} = \sum_n - \omega_n A_n \sin(\omega_n t + \delta_n) \sin k_n x \qquad \text{23-25}$$

At time $t = 0$, the shape of the string and the velocity of each element are given by

$$y(x,0) = \sum_n A_n \cos \delta_n \sin k_n x = \sum_n A_n \cos \delta_n \sin \frac{n\pi x}{L} \qquad \text{23-26}$$

and

$$u_y(x,0) = \sum_n - \omega_n A_n \sin \delta_n \sin k_n x = -\sum_n \omega_n A_n \sin \delta_n \sin \frac{n\pi x}{L} \qquad \text{23-27}$$

where we have substituted $n\pi/L$ for k_n from the standing-wave condition. The constants A_n and δ_n are determined from the initial conditions by Fourier analysis. We shall discuss the result of such analysis for a particular example as an illustration of the general vibration of a string.

Suppose the string is initially at rest with the shape illustrated in Figure 23-9, which results from plucking the string in the center to a height b. The initial shape can be described by the function

$$f(x) = \begin{cases} \dfrac{2b}{L} x & \text{for } 0 < x < \dfrac{L}{2} \\[2mm] 2b - \dfrac{2b}{L} x & \text{for } \dfrac{L}{2} < x < L \end{cases}$$

Since the string is at rest at $t = 0$, the constants δ_n are zero from Equa-

Figure 23-9
String plucked at the center to
a height b.

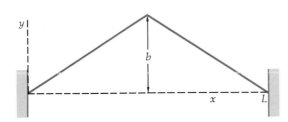

Figure 23-10
Synthesis of the string plucked as in Figure 23-9
using only the first three odd harmonics. The
height b is exaggerated in this drawing to show the
relative amplitudes of the harmonics. The heavy
colored line is the approximation to the original
shape of the string using just these three harmonics. (*By permission from Robert M. Eisberg,* Fundamentals of Modern Physics, *p. 200. Copyright* ©
1961 by John Wiley & Sons, Inc.)

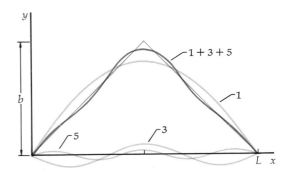

tion 23-27. The constants A_n are then determined by

$$f(x) = \sum_n A_n \sin \frac{n\pi x}{L}$$

Figure 23-10 illustrates the approximation using only the first three
nonzero terms. Detailed analysis yields for these constants,

$$A_1 = \frac{8b}{\pi^2} \qquad A_2 = 0 \qquad A_3 = -\frac{8b}{9\pi^2} = -0.111 A_1$$

$$A_4 = 0 \qquad A_5 = \frac{8b}{25\pi^2} = 0.040 A_1$$

[All the constants for even n are zero because the initial shape is symmetric about the line $x = \frac{1}{2}L$, whereas $\sin(n\pi x/L)$ is antisymmetric
about this line for n even.]

Review

A. Define, explain, or otherwise identify:

Phase velocity, 550
Group velocity, 551
Beats, 551
Harmonic analysis, 553
Synthesis, 553

Wave packets, 555
Dispersion relation, 557
Nondispersive medium, 557
Dispersive medium, 557

B. True or false:

1. The beat frequency between two sound waves of nearly equal frequency
equals the difference in the frequency of the individual sound waves.

2. Information cannot be transported by a single harmonic wave.

3. Phase velocity and group velocity are never equal.

4. If the phase velocity is the same for all wavelengths, a wave pulse will
maintain its shape as it propagates.

5. When a violin string is bowed, it vibrates with a single frequency equal to
its fundamental frequency.

Exercises

Section 23-1, Superposition of Two Waves of Nearly Equal Frequency and Wavelength

1. Two harmonic waves travel simultaneously along a long wire. Their wave functions are

$$y_1 = 0.002 \cos (6.0x - 600t) \quad \text{and} \quad y_2 = 0.002 \cos (5.8x - 580t)$$

where y and x are in meters and t is in seconds. (a) What is the greatest displacement that occurs in the wire? (b) Write the wave function for the resultant wave in the form of Equation 23-5. What is the phase velocity of the resultant wave? (c) What is the group velocity? (d) What is the separation in space of successive crests of the group? (e) Are these waves dispersionless or dispersive?

2. Repeat Exercise 1 for the two wave functions

$$y_1 = 0.003 \cos (8.0x - 400t) \quad \text{and} \quad y_2 = 0.003 \cos (7.8x - 380t)$$

3. Two sound waves of frequency 500 and 505 Hz travel at 340 m/sec along the x axis in air. The displacement amplitude of each wave is s_0. (a) Write wave functions $s_1(x,t)$ and $s_2(x,t)$ for each wave assuming that they are in phase at $x = 0$ and $t = 0$, and write the wave function $s(x,t)$ for the resultant wave. (b) Sketch the resultant wave function as a function of t for some fixed value of x.

Section 23-2, Beats

4. Two tuning forks have frequencies 256 and 260 Hz. What is the beat frequency if the forks both vibrate at the same time?

5. When a violin string is played (without fingering) simultaneously with a tuning fork of frequency 440 Hz, beats are heard at the rate of three beats per second. When the tension in the string is increased slightly, the beat frequency decreases. What was the initial frequency of the violin string?

6. Two tuning forks are struck simultaneously, and four beats per second are heard. The frequency of one fork is 500 Hz. (a) What are the possible values for the frequency of the other fork? (b) A piece of wax is placed on one of the forks to lower its frequency slightly. Explain how the measurement of the new beat frequency can be used to determine which of your answers to part (a) is the correct frequency of the second fork.

Section 23-3, Harmonic Analysis and Synthesis

There are no exercises for this section.

Section 23-4, Wave Packets

7. Information for use by computers is transmitted along a cable in the form of short electric pulses at the rate of 100,000 pulses per second. (a) What is the maximum duration in time of each pulse such that two pulses do not overlap? (b) What is the range of frequencies to which the receiving equipment must respond?

8. A tuning fork of frequency f_0 begins vibrating at time $t = 0$ and is stopped after a time interval Δt. The sound waveform at some later time is shown as a function of x in Figure 23-11. Let N be the (approximate) number of cycles in this waveform. (a) How are N, f_0, and Δt related? (b) If Δx is the length in space of this wave group, what is the wavelength in terms of Δx and N? (c) What is the wave number k in terms of N and Δx? (d) The number of cycles N is uncertain by approximately ± 1 cycle. Explain why (see Figure 23-11). (e)

Figure 23-11
Exercise 8.

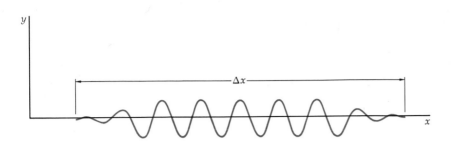

What is the uncertainty in the frequency Δf due to the uncertainty in N? (f) Show that the uncertainty in the wave number due to the uncertainty in N is $2\pi/\Delta x$.

9. Show that a small spread in wave numbers Δk implies a spread in wavelengths $\Delta \lambda$ related to Δk by

$$|\Delta\lambda| \approx \frac{2\pi}{k^2} |\Delta k|$$

Hint: Differentiate $\lambda = 2\pi/k$.

10. What is Δk for the two waves in Exercise 1? Find the distance Δx between two successive zeros of the resultant wave function for the group of two waves in that exercise.

Section 23-5, Dispersion

11. The dispersion relation for sound waves in air is

$$\omega = \sqrt{\frac{\gamma RT}{M}}\, k$$

Find the phase velocity and the group velocity. Are these waves dispersive?

12. The dispersion relation for water waves of very short wavelength in deep water is

$$\omega^2 = \frac{T}{\rho} k^3$$

where T is the surface tension and ρ is the density. (a) What is the phase velocity for these waves in terms of T, ρ, and k? (b) What is the group velocity? (c) Is the group velocity greater or less than the phase velocity?

Section 23-6, Superposition of Standing Waves

There are no exercises for this section.

Problems

1. Show that Equation 23-14, which gives the group velocity of the envelope of two waves exactly, leads to $v_g = v_p$ when the phase velocities of the two waves are equal.

2. A tuning fork of frequency f_0 can be "stopped" by looking at it in a dark room with a strobe light of the same frequency. If the strobe frequency is slightly different from f_0, the fork appears to vibrate in slow motion. Discuss how this phenomenon is related to beats.

3. The use of a vernier scale is related to beats. It is easiest to learn how such a scale works by constructing one. Along the edge of a card (such as a 3 by 5 index card) make a set of 9 marks equally spaced with spacing 0.9 cm and number the marks (call the edge 0). Your scale can now be used to interpolate

between centimeter marks on a scale which is not divided into millimeters. Place your card along a scale ruled in centimeters so that the edge is somewhere between the 2- and 3-cm marks. Then note which mark on your card is aligned with a centimeter mark on the scale. If, for example, the fourth mark on your scale is aligned with a centimeter mark, the edge of the card is at 2.4 cm. Explain how this works and how it is related to beats.

4. When two piano wires are in tune, they have the same tension T_0 and fundamental frequency f_0. What fractional change $\Delta T/T_0$ in one of the wires will produce audible beats with frequency f_B when the wires vibrate simultaneously? With what accuracy must the tension in the wires be adjusted so that when the wires are tuned to middle C ($f_0 = 261$ Hz), the beat frequency will be less than one beat in 5 sec?

5. Consider the waves on a stretched string with wave functions given by Equations 23-1 and 23-2. Show that the intensity of the resultant wave (wave function given by Equation 23-5) equals the sum of the individual intensities of the component waves. (Remember that the intensity is the power per unit area *averaged over time*.)

6. The general dispersion relation for water waves can be written

$$\omega^2 = (gk + \frac{T}{\rho} k^3) \tanh kH$$

where g is the acceleration of gravity, ρ is the density of water, T is the surface tension, and H is the depth of the water, and the hyperbolic tangent is defined by

$$\tanh x = \frac{e^x - e^{-x}}{e^x + e^{-x}} = \frac{e^{2x} - 1}{e^{2x} + 1}$$

(a) Show that the hyperbolic tangent has the properties that for $x \gg 1$ $\tanh x \approx 1$ and for $x \ll 1$ $\tanh x \approx x$. Use these results to write separate dispersion relations for deep-water waves ($kH \gg 1$) and shallow-water waves ($kH \ll 1$). (b) Show that in shallow water, the group velocity and the phase velocity are both equal to $\sqrt{gH}$ if the wavelength is long enough to ensure that $Tk^2/\rho = 4\pi^2 T/\lambda^2\rho \ll g$. (c) Show that for deep water the phase velocity is given by $v_p = \sqrt{g/k + Tk/\rho}$, and find the group velocity. (d) For water, $\rho = 10^3$ kg/m^3 and $T = 0.075$ N/m. Evaluate v_p and v_g in deep water for small ripples with $\lambda = 1$ cm and for large waves with $\lambda = 1$ m. For what wavelength are the phase and group velocities equal in deep water?

CHAPTER 24 Spherical and Circular Waves

Many of the interesting properties of sound and light, such as diffraction or refraction, depend on wave properties in three dimensions and are not evident in one-dimensional waves, but they can be observed with two-dimensional waves on the surface of water and are often produced by the demonstration device known as a *ripple tank*.

It is possible to treat such phenomena mathematically by first deriving a wave equation in two or three dimensions analogous to Equation 21-39 and finding solutions appropriate to the boundary conditions of the particular problem. Such a mathematical treatment is too difficult to consider here and is not necessary for understanding many of the phenomena we consider. We shall concentrate instead on a more phenomenological description.

We shall introduce some of the terminology of three-dimensional waves and then study the changes in observed frequency which occur when the wave source and observer are in relative motion. This phenomenon, known as the *doppler effect*, is most familiar for sound waves. We then consider some important special cases of interference of waves from two or more sources separated in space. We shall have occasion to apply the results of Section 21-4 for the interference due to phase differences caused by path differences. The final section is devoted to a discussion of coherence.

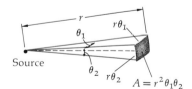

Figure 24-1
Narrow cone defined by the angles θ_1 and θ_2. If these angles are very small, any variation with angle of the power radiated into the cone can be neglected. Since the area subtended by the cone at a distance r is proportional to r^2, the intensity varies as $1/r^2$.

24-1 Wavefronts

Consider a point source of sound or light emitting harmonic waves in three dimensions. Let us assume for the moment that the energy from the source spreads out equally in all directions. At a distance r from the source, the energy is uniformly distributed on a sphere of area $4\pi r^2$. The intensity (average energy per unit area per unit time) thus decreases as $1/r^2$. Such a wave is called a *uniform spherical wave*.

If the energy is not radiated equally in all directions, we consider a narrow cone, as shown in Figure 24-1. If the angles θ_1 and θ_2 are very small, we can neglect any variation in the energy with angle in the cone. The area shown is $A = (r\theta_1)(r\theta_2) = r^2\theta_1\theta_2$. Again, the area increases as r^2, and so the intensity decreases as $1/r^2$.

In studying waves in one dimension we saw that the intensity of a wave is proportional to the square of the amplitude. This is a general property of waves and holds for spherical waves also. Since the intensity is proportional to the square of the amplitude and varies as $1/r^2$, the amplitude must vary as $1/r$. We can write the wave function ψ for an outgoing spherical harmonic wave of frequency $f = \omega/2\pi$ and wavelength $\lambda = 2\pi/k$ as

$$\psi(r,t) = \frac{A_0}{r} \sin{(kr - \omega t + \delta)} \qquad \text{24-1}$$

Wave function for spherical wave

where A_0 is independent of r and δ is a phase constant. (We use ψ for wave function rather than y because y is employed as a coordinate.) At any time t_0, the phase of the wave, $kr - \omega t_0 + \delta$, is constant on a spherical surface. Such a surface of constant phase is called a *wavefront*. It is customary to represent a wave in three dimensions by drawing the wavefronts corresponding to maximum displacement, as in Figure 24-2.

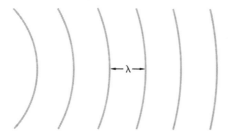

Figure 24-2
Representation of wavefronts from a point source. The circular arcs represent spherical surfaces of maximum displacement at a given time. The distance between these surfaces is the wavelength.

The radial distance between successive maxima is of course the wavelength. The phase velocity v is related to the wavelength and frequency in the usual way:

$$v = f\lambda \qquad \text{24-2}$$

The motion of the wavefront can be indicated by directed lines perpendicular to the front, called *rays* (Figure 24-3). For a spherical wave, the rays are radial lines. At a great distance from a point source a small part of the wavefront can be approximated by a plane, and the rays are approximately parallel lines. Such a wave is called a *plane wave*.

Rays defined

Figure 24-4 shows circular waves from a point source in a ripple tank. These are the two-dimensional analogs of spherical waves. The wavefronts are circles, and the rays are radial lines. At great distances from the source, a small part of the wavefront is approximately a line analogous to a plane wave (Figure 24-5). Such plane waves, or line waves, can also be produced by a line source, as in Figure 24-6.

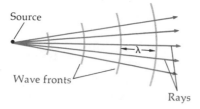

Source

Wave fronts

Rays

Figure 24-3
Representation of the motion of wavefronts by rays perpendicular to the wavefronts. For a point source, the rays are radial lines diverging from the source.

Figure 24-4
Circular wavefronts diverging
from a point source in a
ripple tank. The circles are
two-dimensional wavefronts.
(*From* PSSC Physics, *2d ed.,
p. 272, D. C. Heath and Company, Lexington, Mass., 1965.*)

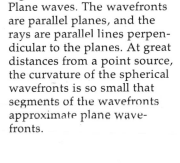

Figure 24-5
Plane waves. The wavefronts
are parallel planes, and the
rays are parallel lines perpendicular to the planes. At great
distances from a point source,
the curvature of the spherical
wavefronts is so small that
segments of the wavefronts
approximate plane wave-
fronts.

Consider a plane wave moving through space. Let us take the x
direction for the direction of the rays. The wavefronts are then planes
parallel to the yz plane. Since, by definition, the phase is constant on
the wavefront, the phase must depend only on x and t and not on y
and z. Thus the wave function is a function of x and t. The wavelength
is the distance along the ray parallel to the x axis between successive
maxima. The wave functions for plane waves are the same as for one-
dimensional waves,

$$\psi(x,t) = A_0 \sin (kx - \omega t + \delta) \qquad 24\text{-}3$$

If the medium through which the plane wave is traveling contains no
objects or apertures which disturb the yz-plane symmetry, the plane
wave behaves like a one-dimensional wave and we can apply what we
have learned about such waves to plane waves. For example, if the
wave meets an infinite plane boundary in the yz plane, it is reflected
just as a wave on a string is reflected. However, if there are obstacles
or apertures, we must treat the plane wave as a three-dimensional
wave. For example, if a plane wave is incident on a spherical obstacle,
the plane symmetry is destroyed and the resulting reflection from the
sphere, refraction (if the sphere is transparent to the wave), or diffraction leads to waves which are not plane waves. We shall see how to
treat some of these more complicated cases in the next chapter, when
we study how a wave propagates in three dimensions.

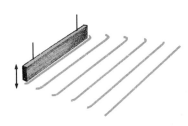

Figure 24-6
Two-dimensional analog of
plane waves can be generated
in a ripple tank with a flat
board oscillating up and
down in the water, producing
wavefronts which are straight
lines.

24-2 The Doppler Effect

If a wave source and a receiver are moving relative to each other, the
frequency observed by the receiver is not the same as the source
frequency. When they are moving toward each other, the observed
frequency is greater than the source frequency, and when they are
moving away from each other, the observed frequency is less than the
source frequency. This is called the *doppler effect*. A familiar example is

Doppler effect

the change in pitch of a train whistle or car horn when the train or car is approaching or receding.

Before we calculate the doppler shift in frequency, we consider the reception of waves by a receiver at rest relative to the source. Figure 24-7 shows a source sending out waves of frequency f_0. This figure can represent either two-dimensional circular waves, e.g., surface water waves, or a two-dimensional drawing of three-dimensional spherical waves, e.g., waves from a point source. Let N be the number of waves emitted in a time Δt. Then $N = f_0 \Delta t$. If v is the speed of the wave relative to the medium, which we assume to be at rest relative to both the source and receiver, these waves will be contained in a distance $v \Delta t$. The wavelength of the waves is the ratio of this distance and the number of waves,

$$\lambda = \frac{v\,\Delta t}{N} = \frac{v\,\Delta t}{f_0\,\Delta t} = \frac{v}{f_0} \qquad 24\text{-}4$$

The frequency observed by the receiver is the number of waves that pass him per unit time. The number that pass the receiver in time Δt is the number in the distance $v \Delta t$, which is $v \Delta t/\lambda$. The frequency observed by the receiver f_R is then

$$f_R = \frac{v\,\Delta t/\lambda}{\Delta t} = \frac{v}{\lambda} = f_0 \qquad 24\text{-}5$$

This is, of course, just the frequency of the wave source.

If the medium is not at rest relative to the source and receiver, e.g., there is a wind blowing in the case of sound waves, the wavelength changes but the frequency received is still equal to the source frequency. Let u_w be the speed of the wind, which we assume to be blowing from source to receiver. The speed of the waves relative to the source or receiver is then $v' = v + u_w$. The wavelength λ' between source and receiver is now greater than that with no wind because the same number of waves $N = f_0 \Delta t$ is now contained in the distance $v' \Delta t = (v + u_w) \Delta t$. The wavelength is

$$\lambda' = \frac{v'\,\Delta t}{N} = \frac{(v + u_w)\,\Delta t}{f_0\,\Delta t} = \frac{v + u_w}{f_0} = \frac{v + u_w}{v}\,\lambda_0 \qquad 24\text{-}6$$

where $\lambda_0 = v/f_0$ is the original wavelength with no wind. The number of waves received in time Δt is now the number in the distance $v' \Delta t$, which is just $f_0 \Delta t$. Thus the frequency received is again f_0, in agreement with our common experience that the wind does not affect the observed frequency of horns, whistles, or musical instruments. Note that the wavelength and observed frequency with a wind blowing are given by Equations 24-4 and 24-5 with v replaced by v'.

Figure 24-8 shows a point source moving with speed u_S relative to a medium in which the wave speed is v. In front of the source the wavefronts are closer together, and behind the source they are farther apart than for a stationary source. The calculation of the wavelengths is similar to Equations 24-4 and 24-6. In time Δt, the source emits $N = f_0 \Delta t$ waves. The first wavefront travels a distance $v \Delta t$ while the source travels a distance $u_S \Delta t$. In front of the source the N wavefronts occupy the distance $v \Delta t - u_S \Delta t$; behind the source they occupy the distance $v \Delta t + u_S \Delta t$. The wavelength in front of the source is therefore

$$\lambda'_f = \frac{v\,\Delta t - u_S \Delta t}{f_0\,\Delta t} = \frac{v - u_S}{f_0} = \frac{v}{f_0}\left(1 - \frac{u_S}{v}\right) \qquad 24\text{-}7$$

Figure 24-7
Representation of wavefronts from a point source at rest relative to the medium. The diagram can represent either spherical surfaces in three dimensions or circular wavefronts in two dimensions. The wavelength λ_0 is related to the frequency f_0 and the velocity of the wave v by $\lambda_0 = v/f_0$.

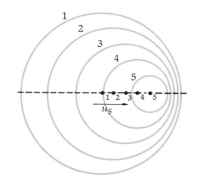

Figure 24-8
Successive wavefronts emitted by a point source moving with speed u_S to the right. Each numbered wavefront was emitted when the source was at the correspondingly numbered position. The wavelength is shorter than λ_0 in front of the source and greater than λ_0 behind the source.

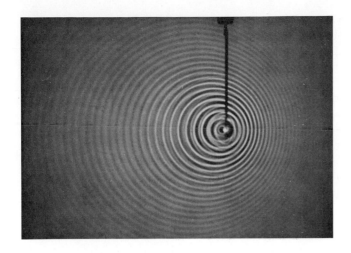

Waves in a ripple tank produced by a source moving to the right with speed less than the wave speed. (*Courtesy of Film Studio, Education Development Center, Newton, Mass.*)

Behind the source the wavelength is

$$\lambda'_b = \frac{v\,\Delta t + u_S\,\Delta t}{f_0\,\Delta t} = \frac{v + u_S}{f_0} = \frac{v}{f_0}\left(1 + \frac{u_S}{v}\right) \qquad 24\text{-}8$$

When we use $\lambda_0 = v/f_0$ for the wavelength of the source at rest, Equations 24-7 and 24-8 can be written

$$\lambda' = \lambda_0\left(1 \pm \frac{u_S}{v}\right) \qquad 24\text{-}9$$

Wavelength in front of or behind moving source

The speed v of the waves depends only on the properties of the medium and not on the motion of the source. The frequency at which the waves pass a point at rest relative to the medium is thus

$$f' = \frac{v}{\lambda'_f} = \frac{f_0}{1 - u_S/v} \qquad 24\text{-}10$$

Frequency in front of moving source

for the source approaching the receiver and

$$f' = \frac{v}{\lambda'_b} = \frac{f_0}{1 + u_S/v} \qquad 24\text{-}11$$

Frequency behind moving source

for the source receding from the receiver.

If the source is at rest and the receiver is moving relative to the medium, there is no change in the wavelength but the frequency of the waves passing the receiver is increased when the receiver moves toward the source and decreased when he moves away from the source. The number of waves that pass a stationary receiver in time Δt is the number in the distance $v\,\Delta t$, which is $v\,\Delta t/\lambda_0$. When the receiver moves toward the source with speed u_R, he passes an additional number $u_R\,\Delta t/\lambda_0$ (Figure 24-9). The total number of waves passing the receiver in time Δt is then

$$N = \frac{v\,\Delta t + u_R\,\Delta t}{\lambda_0} = \frac{v + u_R}{\lambda_0}\,\Delta t$$

The frequency observed is this number divided by the time interval,

$$f' = \frac{N}{\Delta t} = \frac{v + u_R}{\lambda_0} = \frac{v + u_R}{v}\,\frac{v}{\lambda_0}$$

or

$$f' = f_0\left(1 + \frac{u_R}{v}\right) \qquad 24\text{-}12$$

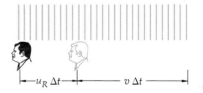

Figure 24-9
The number of waves which pass a stationary receiver in time Δt is the number in the distance $v\,\Delta t$, where v is the wave speed. If the receiver moves toward the source with speed u_R, he passes the additional number which are in the distance $u_R\,\Delta t$. If λ is the wavelength, the total number of waves passed by the receiver in time Δt is $(v\,\Delta t + u_R\,\Delta t)/\lambda$ and the observed frequency is $f' = (v + u_R)/\lambda$.

If the receiver moves away from the source with speed u_R, similar reasoning leads to the observed frequency,

$$f' = f_0 \left(1 - \frac{u_R}{v}\right) \qquad \text{24-13}$$

Frequency observed by receding receiver

The results expressed in Equations 24-10 to 24-13 can be combined when both source and receiver are moving relative to the medium. If the medium is moving, the wave speed v is replaced by $v' = v \pm v_w$, where v_w is the speed of the medium.

Example 24-1 The frequency of a car horn is 400 Hz. What frequency is observed if the car moves toward a stationary receiver with speed $u_S = 100$ ft/sec (about 68 mi/h)?

The velocity of sound relative to air is about 1100 ft/sec. According to Equation 24-7, λ is

$$\lambda_f' = \frac{v - u_S}{f_0} = \frac{1100 - 100 \text{ ft/sec}}{400 \text{ sec}^{-1}} = \frac{1000}{400} \text{ ft} = 2.5 \text{ ft}$$

The frequency observed is

$$f' = \frac{v}{\lambda_f'} = \frac{1100 \text{ ft/sec}}{2.5 \text{ ft}} = 440 \text{ sec}^{-1} = 440 \text{ Hz}$$

Of course we could use Equation 24-10 directly:

$$f' = \frac{f_0}{1 - u_S/v} = \frac{400 \text{ Hz}}{1 - 100/1100} = \frac{1100}{1000} \, 400 \text{ Hz} = 440 \text{ Hz}$$

We performed the intermediate step of calculating the wavelength first because this effect is easy to remember.

Example 24-2 The horn of a stationary car has a frequency of 400 Hz. What frequency is observed by a receiver moving toward the car at 100 ft/sec?

For a moving receiver, the wavelength does not change; the receiver merely passes more waves in a given time. The observed frequency, according to Equation 24-12, is

$$f' = f_0 \left(1 + \frac{u_R}{v}\right) = (400 \text{ Hz}) \left(1 + \frac{100}{1100}\right) = \frac{1200}{1100} (400 \text{ Hz}) = 436 \text{ Hz}$$

Example 24-3 Work Example 24-2 in the reference frame of the receiver.

In the reference frame of the receiver the car is moving toward the receiver with speed of 100 ft/sec. This situation is not identical to that in Example 24-1 because here the receiver observes an apparent wind of 100 ft/sec in the direction from the car to receiver. The speed of sound waves relative to the receiver is $v' = v + u_w = 1100$ ft/sec + 100 ft/sec = 1200 ft/sec. The wavelength of the waves in front of the car (which is moving relative to the receiver) is given by Equation 24-7 with v' replacing v. Thus

$$\lambda_f' = \frac{v' - u_S}{f_0} = \frac{v + u_w - u_S}{f_0}$$

But the wind speed u_w and the source speed u_S are the same. Both are 100 ft/sec in the frame of the receiver. The wavelength is thus

$$\lambda_f' = \frac{v}{f_0} = \lambda_0 = \frac{1100}{400} \text{ ft} = 2.75 \text{ ft}$$

The observed frequency is

$$f' = \frac{v'}{\lambda_f} = \frac{1200}{2.75} \text{ Hz} = 436 \text{ Hz}$$

The fact that the wavelength and frequency are the same as in Example 24-2 should not be surprising. In these two examples, identical problems were worked in two different reference frames moving with constant velocity relative to each other.

From the above examples we note that the doppler shift in frequency depends on whether the source or receiver moves relative to the medium. In Example 24-1 the source moved with speed 100 ft/sec relative to the still air, and the frequency shifted from 400 to 440 Hz. In Examples 24-2 and 24-3 the receiver moved with speed 100 ft/sec relative to the still air, and the frequency shifted from 400 to 436 Hz. These shifts are approximately but not exactly equal.

It is interesting to compare these doppler shifts when the speed of the source or receiver u is much less than the wave speed v. According to Equation 24-10, the frequency observed for a source moving toward a receiver with speed u is

$$f' = \frac{f_0}{1 - u/v} = f_0 \left(1 - \frac{u}{v}\right)^{-1} \qquad \text{24-14}$$

We can compare this with Equation 24-12 for a moving receiver using the binomial expansion

$$(1 + n)^n = 1 + nx + \frac{n(n-1)}{2} x^2 + \cdots$$

For x much less than 1 we may approximate, using only the first few terms. With $x = -u/v$ and $n = -1$, we have

$$\left(1 - \frac{u}{v}\right)^{-1} = 1 + (-1)\left(-\frac{u}{v}\right) + \frac{(-1)(-2)}{2}\left(-\frac{u}{v}\right)^2 + \cdots$$

$$\approx 1 + \frac{u}{v} + \frac{u^2}{v^2}$$

Then Equation 24-14 is approximately

$$f' \approx f_0 \left(1 + \frac{u}{v} + \frac{u^2}{v^2}\right) \qquad \text{24-15}$$

On the other hand, if a receiver moves toward a stationary source with speed u, the observed frequency is (Equation 24-12)

$$f' = f_0 \left(1 + \frac{u}{v}\right)$$

These results differ only in terms of order u^2/v^2 or higher.

In many practical situations this difference can be neglected, but the difference is real and of theoretical importance because it shows that these two situations are really different. Not only is the *relative* motion of the source and receiver important, but also the "absolute" motion of both relative to the medium. Thus if we can measure the doppler shift in frequency to order u^2/v^2, we can tell whether it is the source that is moving relative to the medium or the receiver. For sound waves in air, for example, we can also tell which is moving by noting whether it is the source or receiver that feels the wind in his own reference frame. However, a problem arises for light or other electromagnetic waves which propagate through a vacuum. Our equations for the doppler effect seem to imply that we could detect absolute motion relative to the vacuum if we could measure the doppler shift accurately enough, but this contradicts the principle of relativity. Thus either these equations are not correct, or the principle of relativity does not hold. We shall see in our study of special relativity in Chapter 28 that a small but important correction must be made for these equations describing the doppler effect. The correct expression for the doppler shift of light and other electromagnetic waves is

$$f' = \frac{\sqrt{1 - u^2/c^2}}{1 \pm u/c} f_0 \qquad\qquad 24\text{-}16$$

Relativistic doppler effect for light

where u is the relative velocity of source and observer and c is the velocity of light. There is no way of distinguishing which is moving; only the relative velocity is important. Comparing this expression with Equations 24-10 and 24-11, we see that the expressions differ by the factor $\sqrt{1 - u^2/c^2}$. This factor arises in the relativistic derivation because the time interval between the emission of the first wave and the Nth wave is not the same when measured by the source and by the receiver, though we have, of course, assumed this in our classical derivation.

In our derivation of the doppler-shift expressions we have assumed that the velocity u of the source or receiver is less than the wave velocity v. If the source moves with velocity greater than the wave velocity, there will be no waves in front of the source. The waves behind the source will be increased in length according to Equation 24-8, but they will be confined to a cone which narrows as u increases. We can easily calculate the angle of this cone. Consider the source at position P_1 at time t_1, as shown in Figure 24-10. After a time Δt, the wave emitted from this point will have traveled a distance $v \, \Delta t$. The source will have traveled a greater distance $u \, \Delta t$ and will be at point P_2. The line from this new position of the source to the wavefront emitted when the source was at P_1 makes an angle θ with the path of the source given by

$$\sin \theta = \frac{v \, \Delta t}{u \, \Delta t} = \frac{v}{u} \qquad\qquad 24\text{-}17$$

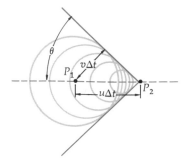

Figure 24-10
A source moving from P_1 to P_2 with speed u greater than the wave speed v in the medium. There are no wavefronts ahead of the source, and so the envelope of the wavefronts forms a cone with the source at the apex. The angle this cone makes with the direction of motion of the source is given by $\sin \theta = v/u$.

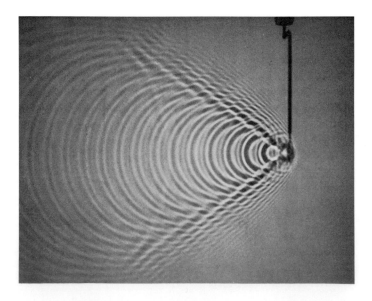

Waves in a ripple tank produced by a source moving to the right with speed greater than the wave speed. (*Courtesy of Film Studio, Education Development Center, Newton, Mass.*)

Equation 24-17 applies to electromagnetic radiation given off when a charged particle moves in a medium with speed u which is greater than the speed of light v in that medium.[1] Called *Čerenkov radiation*, it is confined to the cone with angle given by Equation 24-17.

If the receiver moves toward the source faster than the wave speed, there is no problem. Equation 24-12 holds for the observed frequency. If the receiver moves away from the source faster than the wave speed, the waves never reach the receiver. For light waves, the relativistic expression for the doppler effect given by Equation 24-16 becomes imaginary for u greater than c because of the square-root term in the numerator. According to the special theory of relativity, the relative speed between the source and receiver cannot exceed the speed of light c in vacuum.

Questions

1. Suppose the velocity vector of a receiver makes an angle with the line between him and the source. How will the doppler shift differ from the case discussed in this section? What if the velocity vector makes a right angle with the line joining source and receiver?

2. If the source and receiver are at rest relative to each other but the wave medium is moving relative to them, will the receiver detect any wavelength or frequency shift?

24-3 Interference of Two Point Sources

Figure 24-11 shows the wave pattern produced by two point sources S_1 and S_2 a distance d apart oscillating in phase, each producing circular waves of frequency f, wavelength λ, and speed $v = f\lambda$.

[1] It is impossible for a particle to move faster than c, the speed of light in a vacuum. In a medium such as glass, however, electrons and other particles can move faster than the speed of light in that medium.

Figure 24-11
Interference of two point sources in a ripple tank. (*From PCCS Physics, 2d ed., p. 285, D. C. Heath and Company, Lexington, Mass., 1965*)

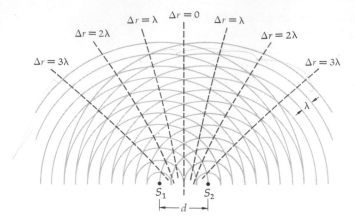

Figure 24-12
Geometric construction of the
interference pattern similar to
Figure 24-11. The wavefronts
add constructively at the
points of intersection. Such
points occur whenever the
path lengths from the two
sources differ by an integral
number of wavelengths.

We can construct a similar pattern (Figure 24-12) with a compass by
drawing circular arcs representing wave crests from each source at
some particular time. At the points where the crests from each source
overlap, the waves add constructively. At these points the paths for
the waves from the two sources are either equal in length or differ by
an integral number of wavelengths. These path differences are indicated in Figure 24-12. The line through a set of interference maxima
for which the path difference is some integral number of wavelengths
is a hyperbola. (A hyperbola is in fact defined as the locus of points
whose distances from two fixed points differ by a constant amount.)
Between each set of interference maxima are interference minima,
where the path difference is an odd number of half wavelengths.
These lines along which the waves completely cancel are called *nodes*
or *nodal lines.* They are also hyperbolas.

We can understand this pattern by applying the results of Section
21-4 for the combination of two harmonic waves having equal
frequency, wavelength, and amplitude but a phase difference δ. We
found that the resultant of two such harmonic waves is itself a harmonic wave with amplitude

$$A = 2A_0 \cos \tfrac{1}{2}\delta \qquad\qquad\qquad 24\text{-}18$$

where A_0 is the amplitude of each wave separately. The phase difference is related to the path difference Δr by

$$\delta = 2\pi \frac{\Delta r}{\lambda} \qquad\qquad\qquad 24\text{-}19$$

The interference of two sound sources can be demonstrated by driving
two separated speakers with the same amplifier which is fed by an
audio signal generator. Moving about the room enables one to detect
by ear the positions of constructive or destructive interference. The
sound intensity will not actually be zero at the points of destructive
interference of the direct sound waves because of sound reflections
from the walls and other objects in the room.

In optics, we often observe the intensity pattern on a screen placed
parallel to the line of the sources and far away from the sources compared with their separation d. This is called *Young's experiment,* after
Thomas Young, who in 1801 used the interference pattern produced
by two light sources to demonstrate the wave nature of light. We can
calculate the intensity pattern observed when the screen is far from the
sources.

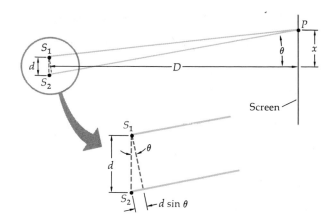

Figure 24-13
Geometry for calculation of
interference pattern observed
on a screen a distance D from
two sources in phase separated by a much smaller distance d. When the screen is
far away, the rays from the
sources to some point P are
approximately parallel and
the path difference is approximately $d \sin \theta$. Constructive
interference occurs when
$d \sin \theta$ is an integral number
of wavelengths. For small θ the
approximation $\sin \theta \approx \tan \theta$
$= x/D$ can be used.

At very large distances from the sources, the lines from the two
sources to some point P on the screen are approximately parallel, and
the path difference is approximately $d \sin \theta$, as shown in Figure 24-13.
We thus have interference maxima at angles given by

$$d \sin \theta = m\lambda \qquad\qquad 24\text{-}20$$

and minima at

$$d \sin \theta = (m + \tfrac{1}{2})\lambda \qquad\qquad 24\text{-}21$$

*Two-source interference
maxima and minima*

where m is any integer (0, 1, 2, . . .). The phase difference at a point P
is $2\pi/\lambda$ times the path difference $d \sin \theta$:

$$\delta = \frac{2\pi}{\lambda} d \sin \theta \qquad\qquad 24\text{-}22$$

Phase difference at angle θ

The distance x measured along the screen from the central point to
point P is related to θ by $x = D \tan \theta$. We are usually interested in
points for which θ is very small, so that $\sin \theta \approx \tan \theta \approx \theta$. Then
$\sin \theta \approx x/D$. Substituting this into Equation 24-22, we have for the
phase difference at distance x,

$$\delta = \frac{2\pi x d}{\lambda D} \qquad\qquad 24\text{-}23$$

The amplitude at this distance is

$$2A_0 \cos \tfrac{1}{2}\delta = 2A_0 \cos \frac{\pi x d}{\lambda D}$$

The intensity is proportional to the square of the amplitude and thus
is proportional to

$$4A_0{}^2 \cos^2 \frac{\pi x d}{\lambda D}$$

Figure 24-14 is a plot of the intensity versus $\sin \theta$, which is equivalent
to a plot of intensity versus x for small θ since $\sin \theta \approx x/D$.

Since the average value of a cosine-squared function over a cycle is
$\tfrac{1}{2}$, the average intensity is proportional to $2A_0{}^2$, indicated by the dashed
line.[1] This is just the intensity which would arise from the two sources

---

[1] For any angle α the identity $\sin^2 \alpha + \cos^2 \alpha = 1$ holds. It is easily seen from a plot of
$\sin^2 \alpha$ and $\cos^2 \alpha$ that the average values of these functions over a complete cycle are
equal. Thus the average value of either $\sin^2 \alpha$ or $\cos^2 \alpha$ over a cycle is $\tfrac{1}{2}$.

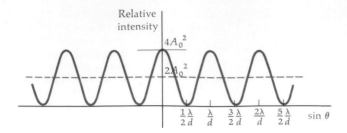

Figure 24-14
Plot of relative intensity versus sin θ observed on a screen far away from two sources in phase as in Figure 24-13. The maximum intensity is proportional to $4A_0^2$ at sin $\theta = m\lambda/d$, where m is an integer and A_0 is the amplitude from each source separately. This is 4 times the intensity of either source acting separately. The average relative intensity $2A_0^2$, indicated by the dashed line, equals twice that of each source acting separately. Since sin $\theta \approx x/D$ for small θ, this is also a plot of relative intensity versus x.

acting separately without interference. We see thus that as a result of the interference of waves from the two sources, the energy is redistributed in space. At a maximum point, the energy is 4 times that from a single source, whereas at a minimum point there is no energy at all; but if we average over many interference maxima and minima, we get the same energy as if the sources acted separately without interference. This is what we should expect from the conservation of energy.

Questions

3. If two nearby sources have the same amplitude but different frequencies, is it possible for them to produce an interference pattern with fixed points of complete destructive interference?

4. The two sources whose resultant amplitude is given by Equation 24-18 were assumed to be vibrating in phase. Discuss what differences might be expected if they differed in phase.

24-4   Interference Pattern of Three or Four Equally Spaced Sources

If we have three or more sources which are equally spaced and in phase with each other, the intensity pattern on a screen far away is similar to that due to two sources, but there are important differences. The position on the screen of the intensity maxima is the same no matter how many sources we have (assuming that they are equally spaced and in phase), but these maxima have much greater intensity and are much sharper if there are many sources. In this section we shall study the pattern produced by three or four sources. The extension of this study to the problem of finding the intensity pattern of a large number of equally spaced sources is not difficult and has important applications for diffraction and the diffraction grating, discussed in the next two chapters.

We first consider the case of three sources, as shown in Figure 24-15. The geometry is the same as for two sources. At a great distance

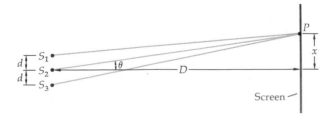

Figure 24-15
Geometry for calculating the intensity pattern far from three equally spaced sources in phase.

from the sources, the rays from the sources to a point P on the screen are approximately parallel. The path difference between the first and second source is then $d \sin \theta$, as before, and between the first and third source the path difference is $2d \sin \theta$. The wave at point P is the sum of three waves. Let $\alpha = kr_1 - \omega t_0$ be the phase of the first wave at point P, where t_0 is the time of observation and r_1 is the distance from the first source to point P. We thus have the problem of addition of the three waves of the form

$$\psi_1 = A_0 \sin \alpha \qquad \psi_2 = A_0 \sin (\alpha + \delta) \qquad \psi_3 = A_0 \sin (\alpha + 2\delta) \qquad 24\text{-}24$$

where

$$\delta = \frac{2\pi}{\lambda} d \sin \theta \qquad 24\text{-}25$$

It is easiest to analyze the resulting pattern in terms of the phase angle δ between the first and second sources or between the second and third sources rather than directly in terms of the space angle θ. If we know the resultant amplitude due to the three waves at some point P corresponding to a particular phase angle δ, we can relate this phase angle to θ by Equation 24-25.

The addition of these waves is easiest with the vector model of addition (Section 21-5). We shall be most interested in the points of perfectly constructive interference and those of perfectly destructive interference, i.e., the interference maxima and minima.

At the central maximum point $\theta = 0$, the phase angle δ is zero, and the amplitude of the resultant wave is 3 times that of each individual wave. Since the intensity is proportional to the square of the amplitude, the intensity at this central maximum is 9 times that from each source acting separately. As we move away from this central maximum, the angle θ increases and the phase angle δ also increases. Figure 24-16 shows the vector addition of three waves for a phase angle δ of about $30° = \pi/6$ rad. (This corresponds to a point P on the screen for which θ is given by $\sin \theta = \lambda \delta/2\pi d = \lambda/12d$.) The resultant amplitude is considerably less than 3 times that of each source. As the phase angle δ increases, the resultant amplitude decreases until the amplitude is zero at $\delta = 120°$. For this phase difference, the three vectors form an equilateral triangle (Figure 24-17). This first interference minimum occurs at a smaller phase angle (and therefore at a smaller angle θ) than the $180°$ for only two sources. As δ increases from $120°$, the resultant amplitude increases, reaching a secondary maximum when δ is $180°$. At the phase angle $\delta = 180°$ the amplitude is the same as that from a single source since the waves from the first two sources cancel each other, leaving only the third. The intensity of the secondary maximum is one-ninth that of the central maximum. As δ increases beyond $180°$, the amplitude again decreases and is zero at $\delta = 180° + 60° = 240°$. For δ greater than $240°$ the amplitude increases and is again 3 times that of each source when $\delta = 360°$. This phase angle corresponds to a path difference of 1 wavelength for the waves from the first two sources and 2 wavelengths for the waves from the first and third sources. The three waves are thus in phase at this point. Figure 24-18 shows the intensity pattern on a screen far from three equally spaced sources. The maxima are at the same positions as for just two sources, i.e., at points corresponding to the angles θ given by

$$d \sin \theta = m\lambda$$

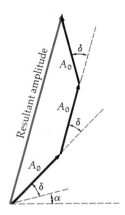

Figure 24-16
Vector addition to determine the resultant amplitude due to three waves, each of amplitude A_0, which have phase differences of δ and 2δ due to path differences of $d \sin \theta$ and $2d \sin \theta$. The angle $\alpha = kx - \omega t$ varies with time but does not affect the calculation of the resultant amplitude.

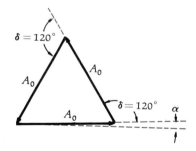

Figure 24-17
The resultant amplitude for the waves from three sources is zero when δ is $120°$. This interference minimum occurs at a smaller angle θ than the first minimum for two sources, which occurs when δ is $180°$.

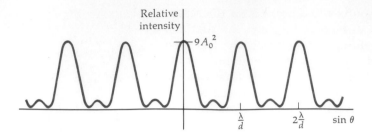

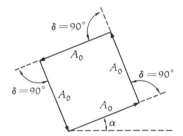

Figure 24-18
A plot of the relative-intensity pattern on a screen far from three equally spaced sources in phase. The major maxima occur when $d \sin \theta = m\lambda$ at the same position as for two sources, but for three sources the maxima are narrower and of greater intensity. Between successive major maxima there is a secondary maximum with one-ninth the intensity of the major maxima.

where m, called the *order number*, is 0, 1, 2, ... These maxima are stronger and narrower than those for two sources.

These results can be generalized to more sources. For example, if we have four equally spaced sources in phase, the interference maxima are again given by Equation 24-20 but the maxima are still narrower and there are two small secondary maxima between each pair of principal maxima. At $\theta = 0$, the intensity is 16 times that from a single source. The first interference minimum occurs when δ is 90°, as can be seen by the vector diagram of Figure 24-19. The first secondary maximum is near $\delta = 120°$, where the waves from three of the sources cancel, leaving only the wave from the fourth source. The intensity of the secondary maximum is approximately one-sixteenth that of the central maximum. There is another minimum at $\delta = 180°$, a secondary maximum near $\delta = 240°$, and another minimum at $\delta = 270°$ before the next principal maximum at $\delta = 360°$. This discussion shows that as we increase the number of sources, the intensity becomes more and more concentrated in the maxima given by Equation 24-20 and that the maxima become narrower. For N sources, the intensity at the maxima is N^2 times that of a single source, and the first minimum occurs at a phase angle of $\delta = 360°/N$. In that case the N vectors form a closed polygon of N sides. There is a secondary maximum between each two minimum points. These secondary maxima are very weak compared with the principal maxima.

Figure 24-19
Vector diagram for the first minimum from four equally spaced sources in phase. The amplitude is zero when the phase difference of the waves from adjacent sources is 90°.

Example 24-4 Four equally spaced sources produce light of wavelength 5×10^{-7} m = 500 nm and are separated by a distance $d = 0.1$ mm. The sources are in phase, and the interference pattern is viewed on a screen 1 m away. The intensity of each source separately on the screen is I_0. Find the positions of the interference maxima and compare the central maximum with that for just two sources with the same spacing.

According to Equation 24-20, the maxima are at angles given by

$$\sin \theta = m \frac{\lambda}{d} = m \frac{5 \times 10^{-7} \text{ m}}{1 \times 10^{-4} \text{ m}} = m(5 \times 10^{-3} \text{ rad})$$

where $m = 0$, 1, 2, 3, ... Since θ is small, we can approximate $\sin \theta \approx \tan \theta \approx \theta$. The distance x measured along the screen from the central maximum is related to θ by

$$x = D \tan \theta \approx D\theta$$

The position of the mth maximum is thus

$$x_m = D\theta_m = m(1 \text{ m})(5 \times 10^{-3}) = m(5 \text{ mm})$$

The maxima are thus separated by 5 mm on the screen. The first minimum occurs when the phase difference between two adjacent sources is $\delta = 90° = \pi/4$. This corresponds to a path difference of $\lambda/4$.

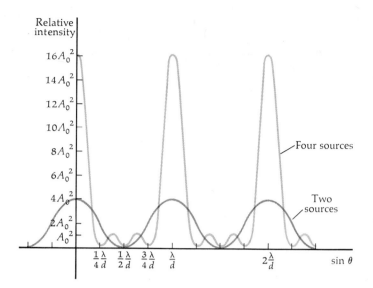

Figure 24-20
Comparison of relative-intensity pattern for four sources with that for two sources when the spacing d is the same. The maxima for four sources are stronger and narrower than those for two sources.

The angle θ of this minimum is given by $d \sin \theta = \lambda/4$ or

$$\sin \theta = \frac{\lambda}{4d} = \frac{5 \times 10^{-7} \text{ m}}{4 \times 10^{-4} \text{ m}} = 1.25 \times 10^{-3}$$

The position x of this minimum is

$$x = R\theta = (1 \text{ m})(1.25 \times 10^{-3}) = 1.25 \text{ mm}$$

We shall take for the width of the maximum the distance between the first minimum above the maximum to the first minimum below the maximum, or $2x = 2.5$ mm. If we had only two sources of this same spacing, the maxima would be at the same points but the first minimum would be at an angle θ corresponding to a path difference of $\lambda/2$. The width of this maximum would be twice as great as with four sources (Figure 24-20).

24-5   Coherence

Two or more sources need not be in phase to produce an interference pattern. Figure 24-21 shows the pattern produced by two sources 180° out of phase. The pattern is the same as Figure 24-14 except that the maxima and minima are interchanged. Points equidistant from the sources or those for which the distance differs by an integral number of wavelengths are nodes because the waves are 180° out of phase. At points where the path distance differs by one-half wavelength (or $\frac{3}{2}\lambda$, $\frac{5}{2}\lambda$, . . .), the waves are in phase because the 180° phase difference of the source is offset by the 180° phase difference due to the path dif-

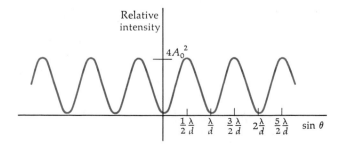

Figure 24-21
Relative-intensity pattern on a screen far from two equal point sources which are 180° out of phase. When the path difference $d \sin \theta$ equals an integral number of wavelengths, the waves are 180° out of phase and interfere destructively. The pattern is the same as in Figure 24-14 except that it is shifted by $\frac{1}{2}\lambda/d$. The effect on the intensity pattern of a constant phase difference between two sources is merely to shift the pattern.

ference. It should be evident that similar interference patterns will be produced no matter what the phase difference between the sources is as long as the phase difference is constant in time. Two sources which are in phase or have a constant phase difference are called *coherent sources*. Coherent sources of water waves in a ripple tank are easy to produce by driving both sources by the same motor. There are many examples of sound and light sources whose phase difference is not constant but varies randomly from 0 to 2π. For two sources of equal strength, the intensity at any instant is proportional to $4A_0{}^2\cos^2\frac{1}{2}\delta$, where now the phase difference δ is a function of time. If δ varies rapidly, only the time average $4A^2(\cos^2\frac{1}{2}\delta)_{av} = 2A_0{}^2$ will be observed. There is then no observable interference pattern. Such sources are said to be *incoherent*.

Coherent and incoherent sources

As an example of incoherent sources, consider two candles. The light from each candle is the result of millions of independent atomic transitions. Each source, which we can consider to be a point on the macroscopic scale, is in reality the sum of millions of microscopic sources, the individual atoms. Since it is impossible to predict the exact time a particular atom will make a transition, the phase fluctuates randomly. Many fluctuations occur during a time interval as short as 10^{-8} sec. The eye or other detection device cannot respond to such rapid intensity changes as would be indicated by the time variation of $4A_0{}^2\cos^2\frac{1}{2}\delta$. Thus only the average value of $2A_0{}^2$ is seen, and there is no possibility of observing an interference pattern with two independent light sources. It is, in fact, rather difficult to obtain coherent light sources except by splitting a single source into duplicates in ways to be discussed later.

Review

A. Define, explain, or otherwise identify:

B. True or false:

1. If both source and receiver move with the same speed in the same direction, there is no doppler shift.

2. The doppler shift of sound waves depends only on the relative velocity of source and receiver.

3. The doppler shift of light waves in vacuum depends only on the relative velocity of source and receiver.

4. The waves from two sources in phase interfere constructively everywhere in space.

5. Two sources which are out of phase by 180° are incoherent.

6. Interference patterns are observed only for coherent sources.

Exercises

Section 24-1, Wavefronts

1. A certain high-fidelity speaker system is designed to radiate power equally in all directions. It has a power output of 0.01 W. (*a*) What is the intensity at a

distance of 3 m? (*b*) Find the approximate intensity level in decibels at this point.

2. The energy from the sun falling on a unit area at normal incidence at the earth, called the solar constant, is equal to 1370 W/m². Use the fact that the earth is 1.49×10^{11} m from the sun to find the total power radiated from the sun.

3. A point source radiates circular waves equally in all directions in the plane of the water surface. (*a*) How does the intensity depend on the distance from the source? (*b*) How does the amplitude of the waves depend on distance from the source? (For two-dimensional waves, the intensity is the power per unit length.)

Section 24-2, The Doppler Effect

4. This exercise is a doppler-effect analogy. A conveyor belt moves to the right with speed $v = 1000$ ft/min. A very fast pieman puts pies on the belt at a rate of 20 per minute, and they are received at the other end by a pie eater. (*a*) If the pieman is stationary, find the spacing λ between the pies and the frequency f with which they are received by a stationary pie eater. (*b*) The pieman now walks with speed 100 ft/min toward the receiver while he continues to put pies on the belt at 20 per minute. Find the spacing of the pies and the frequency with which they are received by the stationary pie eater. (*c*) Repeat your calculations for a stationary sender and a pie eater who moves toward him at 100 ft/min.

5. For the situation described in Exercise 4 derive general expressions for the spacing of the pies and the frequency with which they are received in terms of the speed of the belt v, the speed of the sender u_S, and the speed of the receiver u_R.

In Exercises 6 through 11 the source emits sound of frequency 200 Hz *which moves through still air with speed* 340 m/sec.

6. The source moves with speed 80 m/sec relative to still air toward a stationary listener. (*a*) Find the wavelength of the sound between source and listener. (*b*) Find the frequency heard by the listener.

7. Consider the situation in Exercise 6 from the reference frame in which the source is at rest. In this frame the listener moves toward the source with speed 80 m/sec, and there is a wind of speed 80 m/sec blowing from the listener to the source. (*a*) What is the speed of sound from source to listener in this frame? (*b*) Find the wavelength of the sound between source and listener. (*c*) Find the frequency heard by the listener.

8. The source moves with speed 80 m/sec away from the stationary listener. (*a*) Find the wavelength of the sound waves between the source and listener. (*b*) Find the frequency heard by the listener.

9. The listener moves with speed 80 m/sec relative to still air toward a stationary source. (*a*) What is the wavelength of the sound between the source and listener? (*b*) What is the frequency heard by the listener?

10. Consider the situation in Exercise 9 in the reference frame in which the listener is at rest. (*a*) What is the wind velocity in this frame? (*b*) What is the speed of sound from source to listener in this frame, i.e., relative to the listener? (*c*) Find the wavelength of the sound between the source and listener in this frame. (*d*) Find the frequency heard by the listener.

11. The listener moves with speed 80 m/sec relative to the still air away from a stationary source. Find the frequency heard by the listener.

12. A whistle of frequency 500 Hz moves in a circle of radius 1 m making 3 rev/sec. What are the maximum and minimum frequencies heard by a stationary listener?

Section 24-3, Interference of Two Point Sources

13. With a compass, draw circular arcs representing wave crests for each of two point sources a distance d apart for $d = 6$ cm and $\lambda = 1$ cm (see Figure 24-12). Connect the intersections corresponding to points of constant path difference and label the path difference for each line. (If you do not have a ruler marked in centimeters, you may use $d = 3.0$ in and $\lambda = 0.5$ in.)

14. Two sound speakers are separated by a distance of 6 ft. A listener sits directly in front of one speaker a distance of 8 ft from it so that the two speakers and listener form a right triangle. Find the two lowest frequencies for which the path difference is an odd number of half-wavelengths. Why might these frequencies be heard even if the speakers are driven in phase by the same amplifier? (Use $v = 1100$ ft/sec for the speed of sound.)

15. Two sound speakers are driven in phase by an audio amplifier at frequency 600 Hz. The speed of sound is 340 m/sec. The speakers are on the y axis, one at $y = +1.00$ m and the other at $y = -1.00$ m. A listener begins at $y = 0$ and walks along a line parallel to the y axis at a very large distance x away. (a) At what angle θ (between the line from the origin to the listener and the x axis) will he first hear a minimum in the sound intensity? (b) At what angle will he first hear a maximum (after $\theta = 0$)? (c) How many maxima can he possibly hear if he keeps walking in the same direction?

16. Two sound sources, driven in phase by the same amplifier, are 2 m apart on the y axis. At a point a very large distance from the y axis, constructive interference is heard at the angle $\theta_1 = 8°$ and next at $\theta_2 = 16°10'$ with the x axis. If the speed of sound is 340 m/sec, (a) what is the wavelength of the sound waves from the sources, and (b) what is the frequency of the sources? (c) At what other angles is constructive interference heard? (d) What is the smallest angle for which the sound waves completely cancel?

17. Two point sources are in phase and separated by a distance d. The interference pattern of the sources is detected along a line parallel to that through the sources and a large distance D from the sources, as in Figure 24-13. Show that the mth interference maximum is at a distance x from the central maximum point where x is given approximately by

$$x = m\frac{D\lambda}{d}$$

Section 24-4, Interference Pattern of Three or Four Equally Spaced Sources

18. With a compass, draw circular arcs representing wave crests for each of three point sources a distance d apart for $d = 6$ cm and $\lambda = 1$ cm. Connect the intersections corresponding to points of constant path difference and label the path difference for each line. (If you do not have a ruler marked in centimeters, you may use $d = 3.0$ in and $\lambda = 0.5$ in.) How do the lines of constructive interference differ from those for just two sources of the same spacing?

19. Three sound speakers are driven in phase by an audio amplifier of frequency 1700 Hz. They are situated along the y axis with separation $d = 60$ cm. The speed of sound is 340 m/sec. (a) At what angles is constructive interference heard at a great distance from the speakers? (b) At what angles is there complete cancellation? (You may give your answers in terms of $\sin \theta$.)

20. A fourth speaker of the same frequency and in phase with the other three speakers is added along the line of the speakers in Exercise 19 a distance of 60 cm from one of the end speakers. How does this change your answers to parts (a) and (b) in Exercise 19?

21. Five coherent sources are equally spaced along a line and in phase. At some point far away, the phase difference due to the path difference between any two adjacent sources is δ. What is the least value of δ such that there is complete cancellation at that point?

Section 24-5, Coherence

22. Two violinists are standing a few feet apart and playing the same notes. Are there places in the room at which certain notes are not heard because of destructive interference? Explain.

23. Two speakers separated by some distance emit sound of the same frequency. At some point P the intensity due to each speaker separately is I_0. The path distance from P to one of the speakers is $\frac{1}{2}\lambda$ greater than that from P to the other speaker. What is the intensity at P if (a) the speakers are coherent and in phase; (b) the speakers are incoherent; and (c) the speakers are coherent but have a phase difference of 180°?

24. Answer the questions of Exercise 23 for the point P' for which the distance to the far speaker is 1λ greater than the distance to the near speaker. Again assume that the intensity at point P' is I_0 due to each speaker separately.

25. Two speakers separated by some distance emit sound waves of the same frequency, but speaker 1 leads speaker 2 in phase by 90°. Let r_1 be the distance from some point to speaker 1 and r_2 be the distance to speaker 2. Find $r_2 - r_1$ such that the sound at the point will be (a) maximum and (b) minimum. (Express your answers in terms of the wavelength.)

Problems

1. A car moves with speed 40 ft/sec toward a stationary wall. Its horn emits 200-Hz sound waves which move at 1100 ft/sec. (a) Find the wavelength of the sound in front of the car and the frequency with which the waves strike the wall. (b) Since the waves reflect off the wall, the wall acts as a source of sound waves at the frequency found in part (a). What frequency does the man in the car hear reflected from the wall? (c) What is the beat frequency heard by the man in the car between the direct sound and reflected sound?

2. Work Problem 1 in the reference frame in which the car is stationary and the wall moves toward the car at 40 ft/sec.

3. An automobile has an acoustic noise output of 0.10 W. If the sound is radiated isotropically, i.e., equally in all directions, (a) What is the intensity at a distance of 30 m? (b) What is the sound intensity level in decibels at this distance? (c) At what distance is the sound intensity level 40 dB?

4. (a) Show that if the relative velocity of source and receiver u is much less than v, the doppler shift in frequency and wavelength can be written

$$\frac{\Delta f}{f} = \pm\frac{u}{v} \quad \text{and} \quad \frac{\Delta\lambda}{\lambda} = \mp\frac{u}{v}$$

(b) Show that these expressions also hold for the relativistic doppler shift given by Equation 24-16. (c) In the light received from a certain galaxy, the hydrogen spectrum is present, but all the wavelengths are greater than those emitted by excited hydrogen atoms in the laboratory by 1 percent. This is known as the red shift. Calculate the speed with which the galaxy is moving away from the earth.

5. The ratio of the frequencies of one note to the semitone above it on the diatonic scale is 15:16. Find the speed of a car such that the tone of its horn drops a semitone as it passes you. (Take 1100 ft/sec for the speed of sound.)

6. Two point sources of coherent light are in phase and separated by a distance d. Their interference pattern is to be observed on a screen a very large distance D from the sources. Calculate the spacing x of the maxima on the screen for light of wavelength $\lambda = 5 \times 10^{-7}$ m, $D = 1$ m, and $d = 1$ cm (see Exercise 17). Would you expect to observe the interference of light on the screen for this situation? How close together should you place these sources so that the maxima are separated by 1 mm for this wavelength and screen distance?

7. (a) Show that the positions of the interference minima on a screen a large distance D away from three equally spaced sources (spacing d with $d \gg \lambda$) are given approximately by

$$x = n \frac{\lambda D}{3d} \qquad \text{where } n = 1, 2, 4, 5, 7, 8, 10, \ldots$$

that is, n is not a multiple of 3. (b) For $D = 1$ m, $\lambda = 5 \times 10^{-7}$ m, and $d = 0.1$ mm, calculate the width of the principal interference maxima (distance between successive minima) for three sources.

8. Show that the positions of the interference minima on a screen a large distance D from four equally spaced sources are given by

$$x = n \frac{\lambda D}{4d} \qquad \text{where } n = \text{integer not a multiple of 4}$$
$$d \gg \lambda$$

Compare the width of the principal interference maxima for four sources with that for two sources for $d = 0.1$ mm, $D = 2$ m, and $\lambda = 6 \times 10^{-7}$ m.

9. Five coherent sources are equally spaced on a line. Analyze and sketch the expected interference pattern on a screen far away. Include the intensity at the maximum points in terms of the intensity I_0 of each source separately and the points of zero intensity.

10. Show that the angles θ_1 and θ_2 in Figure 24-22 are approximately equal if the distance between the line of sources and the screen D is much greater than the spacing of the sources d. For definiteness in your calculations take $D = 10^4 d$. (a) Show that if both angles θ_1 and θ_2 are small, so that θ, $\sin \theta$, and $\tan \theta$ are all approximately equal, the difference between the two angles is about d/D. What is this difference in degrees for $D = 10^4 d$? (b) In the general case, show $\tan \theta_1 - \tan \theta_2 = d/D$. Use the trigonometric identity

Figure 24-22
Problem 10.

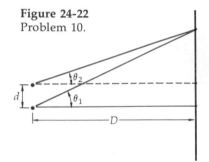

$$\tan A - \tan B = \frac{\sin (A - B)}{\cos A \cos B}$$

to show that $\sin (\theta_1 - \theta_2)$ is much less than 1 so that it can be approximated by $\theta_1 - \theta_2$. Using this, show that even if θ_1 and θ_2 are not small, the difference is of the order of d/D.

11. Two identical speakers emit sound waves of frequency 680 Hz uniformly in all directions with a total audio output of 1.00×10^{-3} W each. The speed of sound in air is 340 m/sec. A point P is a distance 2.00 m from one speaker and 3.00 m from the other. (a) Find the intensities I_1 and I_2 from each speaker at point P separately. (b) If the speakers are driven coherently and in phase, what will be the intensity at point P? (c) If they are driven coherently but out of phase by 180°, what will be the intensity at point P? (d) If the speakers are incoherent, what will be the intensity at point P?

12. Two sources have a phase difference δ_0 which is proportional to time $\delta_0 = Ct$, where C is a constant. The amplitude of the wave from each source at some point P is A_0. (a) Write the wave functions for each of the two waves at point P assuming that this point is a distance x_1 from one source and $x_1 + \Delta x$ from the other. Find the resultant wave function and show that its amplitude is $2A_0 \cos \frac{1}{2}(\delta + \delta_0)$, where δ is the phase difference at P due to the path difference. (b) Sketch the intensity at point P versus time for a zero path difference. (Let I_0 be the intensity due to each wave separately.) What is the time average of the intensity? (c) Make the same sketch for the intensity at a point for which the path difference is half a wavelength.

13. Instructions for connecting stereo speakers to an amplifier correctly so that they are in phase are as follows: "After both speakers are connected, play a monophonic record or program with the bass control turned up and the treble turned down. While listening to the speakers, turn the balance control so that first one speaker is heard separately, than the two together, and then the other

separately. If the bass is stronger when both speakers play together, they are connected properly. If the bass is weaker when both play together compared with each separately, interchange the connections on one speaker." Explain why this method works. In particular, explain why a stereo source is not used and why only the bass is compared.

14. A radio telescope consists of two antennas separated by a distance of 200 m. Each antenna is tuned to a particular frequency such as 20×10^6 Hz. The signals from each antenna are fed into a common amplifier, but one signal first passes through a phase adjuster, which delays the phase by an amount chosen so that the telescope can "look" in different directions. With zero phase delay, plane radio waves incident vertically produce signals which add constructively at the amplifier. What should the phase delay be so that signals coming from an angle $\theta = 10°$ with the vertical (in the plane formed by the vertical and the line joining the antennas) add constructively at the amplifier?

CHAPTER 25   Wave Propagation

The description of the propagation of a wave is quite simple if the wave encounters no obstructions in space. A spherical wave can be described by the wave function

$$\psi(r,t) = \frac{A}{r} \sin (kr - \omega t) \qquad \text{25-1}$$

for all r and t, provided it travels through a homogeneous space with no obstacles. In this case the wavefronts move out radially. Each new wavefront emerging from the source is a spherical surface concentric with the previous wavefronts. For plane waves moving in the x direction, a one-dimensional wave function such as

$$\psi(x,t) = A \sin (kx - \omega t) \qquad \text{25-2}$$

can be used, the same wave function as for one-dimensional waves on a string. Each new wavefront that passes any point is a plane parallel to the previous wavefront. We can construct the new wavefront by finding the surface perpendicular to the rays drawn as straight lines perpendicular to the previous wavefront.

25-1 Diffraction

When a portion of the wave is cut off by an obstruction, the propagation of the wave is more complicated. The portion of the wavefront which is not obstructed does not simply propagate in the direction of the straight rays, as might be expected. The situation is illustrated in Figure 25-1, which shows plane waves in a ripple tank meeting a barrier with a small opening. The waves to the right of the barrier are not confined to the narrow angle of the rays from the source which can pass through the opening but are circular waves, as if there were a source at the opening. We can understand this by noting that the motion of the water at the opening due to the incoming waves is no dif-

Figure 25-1
Plane waves in a ripple tank meeting a barrier with a small opening. The waves to the right of the barrier are circular waves concentric about the opening, just as if there were a point source at the opening. (*Courtesy of Film Studio, Education Development Center, Newton, Mass.*)

Figure 25-2
Comparison of the transmission through a narrow opening in a barrier of (*a*) a beam of particles and (*b*) a wave. In (*a*) the transmitted particles are confined to a narrow angle subtended by the opening. In (*b*) the opening acts as a point source of circular waves, which are radiated to the right through a much wider angle than that subtended by the opening. The bending of the waves resulting from the limiting of the wavefront at the barrier is called diffraction.

Figure 25-3
Plane waves in a ripple tank meeting a barrier with an opening large compared with the wavelength. The effect of the barrier is noticeable only near the edges. (*Courtesy of Film Studio, Education Development Center, Newton, Mass.*)

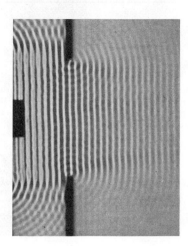

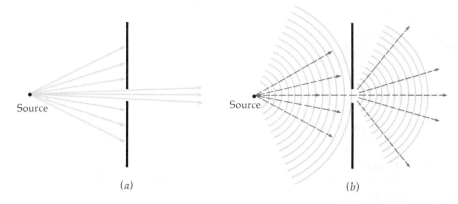

(*a*) (*b*)

ferent from that produced by a point source at the opening (assuming the opening to be very small).

The propagation of a wave is thus quite different from the propagation of a stream of particles. If the rays shown in Figure 25-2a indicate the direction of particles streaming out of a source, the barrier would stop any particles hitting it. Those getting through the opening would be confined to the narrow angle indicated. On the other hand, if the rays indicate the direction of propagation of a wave (Figure 25-2b), the rays appear to bend around the edges of the barrier, as in Figure 25-1. This bending of the rays, which always occurs to some extent when part of the wavefront is limited, is called *diffraction.*

In Figure 25-1 the opening in the barrier was chosen to be much smaller than the wavelength and could be considered to be a point. Figure 25-3 shows plane waves hitting a barrier with an opening much larger than the wavelength. The wave beyond the barrier is similar to a plane wave in the region far from the edges of the opening. Near the edges the wavefront is distorted, and the wave appears to bend slightly. In the limit of an infinitely large opening, the situation is the same as if there were no barrier. If a wave is limited by a barrier with an opening of the order of the size of the wavelength, the situation is more complicated: the transmitted wave is neither approximately a plane wave nor a circular wave from a point source (Figure 25-4).

25-2 Huygens' Construction

Although calculation of the wave pattern of an obstructed wave from a wave equation such as Equation 21-39 is too complicated to consider here, we can describe the propagation of the wave using a geometric method discovered by Christian Huygens about 1678. Huygens' method considers each point on a given wavefront to be a point source of waves. The new wavefront at some time Δt later is then the surface

Diffraction: The bending of waves around obstacles

Figure 25-4
Plane waves in a ripple tank meeting a barrier with an opening with a width 5 times the wavelength. When the opening is neither much smaller than the wavelength, as in Figure 25-1, nor much larger than the wavelength, as in Figure 25-3, the wave pattern to the right of the barrier is more difficult to analyze. (*Courtesy of Film Studio, Education Development Center, Newton, Mass.*)

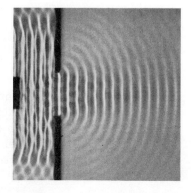

Portrait of Christian Huygens, the Dutch physicist and astronomer (1629–1695).

Brown Brothers

Figure 25-5
Huygens' construction for the propagation of (*a*) plane waves to the right and (*b*) outgoing spherical, or circular, waves.

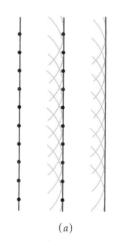

(*a*)

(*b*)

which is the envelope of all the little spherical wavelets emitted by these point sources and which have expanded during the time Δt. Figure 25-5 illustrates the Huygens construction for the propagation of unobstructed plane waves and spherical waves. Of course if each point on a wavefront were really a point source, there would be waves in the backward direction. These waves were ignored by Huygens. (In the refinement of Huygens' method, derived by G. R. Kirchhoff in the nineteenth century, the intensity of the wavelets depends on angle and is zero in the backward direction.)

Huygens' construction, or *principle*, seems quite reasonable for mechanical waves such as water waves. Each point on a wavefront does in fact oscillate much like a point source. We indicated in the discussion of Figure 25-1 that if all the points but one on a plane wavefront are obstructed, a circular wave is indeed propagated beyond the barrier. Although it is not so easy to see why Huygens' principle should be valid for electromagnetic waves (such as light) propagating in free space, it does indeed apply to all types of waves and can be related to the general wave equation governing them. We shall use Huygens' principle to calculate the diffraction pattern of a single slit and to derive the laws of reflection and refraction of waves.

25-3 Diffraction Pattern of a Single Slit

Let us consider the problem of a plane wave incident on a small opening which is of the order of the size of the wavelength. We shall consider the problem in two dimensions, e.g., water waves in a ripple tank. The analogous problem for light is a three-dimensional plane wave incident on a long narrow slit. If the length of the slit is much greater than the width, the geometry is approximately cylindrically symmetric and a two-dimensional analysis is valid. We thus refer to the opening as a single *slit* although in the case of the two-dimensional ripple tank it is merely a gap in a barrier.

According to Huygens' principle, we can find the wave to the right of the opening by assuming that each point along the line of the opening is a source of circular waves. The wave pattern to the right will then be the interference pattern produced by this linear array of sources. If we confine ourselves to finding the intensity pattern on a line parallel to the barrier and a great distance away from it, the calculation is merely an extension of those done in Chapter 24 for the interference pattern produced by two, three, or four equally spaced sources.

Let us divide the opening or slit into N equal intervals and assume that there is a point source of waves at the midpoint of each interval (Figure 25-6). If d is the distance between two adjacent sources and a the width of the opening, we have

$$d = \frac{a}{N}$$

For convenience we shall take the number of sources N to be an even number. Since the line on which we are calculating the intensity is very far from the sources, the rays from the sources to a point P on the line are approximately parallel. The path difference between any two adjacent sources is then $d \sin \theta$, and the phase difference is $\delta = (2\pi/\lambda)d \sin \theta$. If A_0 is the amplitude due to a single source, the amplitude at the central maximum point $\theta = 0$, where all the waves are in phase, is $A_{\max} = NA_0$ (Figure 25-7). We find the amplitude at some

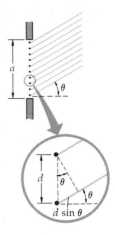

Figure 25-6
Diagram for the calculation of the interference pattern far from a narrow slit. The slit of width a is assumed to contain a large number of in-phase point sources separated by a distance d. The rays from these sources to a point very far away are approximately parallel. The path difference for the waves from adjacent sources is then $d \sin \theta$.

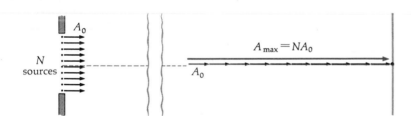

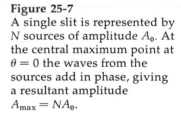

Figure 25-7
A single slit is represented by N sources of amplitude A_0. At the central maximum point at $\theta = 0$ the waves from the sources add in phase, giving a resultant amplitude $A_{\max} = NA_0$.

other point P at an angle θ by using the vector method of addition of waves. As in Chapter 24 for the addition of two, three, or four waves, the intensity is zero at a point such that the vectors representing the waves form a closed polygon. In this case the polygon has N sides (Figure 25-8). At point P the wave from the first source near the top of the opening and that from the source just below the middle of the opening are 180° out of phase. Consider, for example, 100 sources. These sources give zero intensity at a point P when the waves from the first source and the fifty-first source are out of phase by 180° and cancel (Figure 25-9). Then the waves from the second and fifty-second are also out of phase by 180° and cancel, as do those from the third and fifty-third sources, etc. In this case the waves from the sources near the top and bottom of the opening differ in phase by nearly 360°. (The phase difference is, in fact, 360° − 360°/N.) Thus if the number of sources is very large, *we get complete cancellation when the waves from the first and last sources are out of phase by 360°, corresponding to a path difference of 1 wavelength.* Since this path difference is $a \sin \theta$, where a

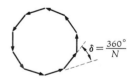

Figure 25-8
Vector diagram for the first minimum in the single-slit diffraction pattern. When the waves from the N sources completely cancel, the N vectors form a closed polygon. The phase difference between waves from adjacent sources is then $\delta = 360°/N$. When N is very large, the waves from the first and last sources are approximately in phase.

is the width of the opening, the condition for the first minimum is

$$a \sin \theta = \lambda$$

Location of first minimum

We get a second minimum when the path difference for the waves from the top and bottom of the opening is 2 wavelengths. We can see this by dividing the group of N sources into two parts. The waves from the top half of the sources add to zero, as we have just seen. Similarly the waves from the bottom half of the sources also add to zero. Thus the condition for a minimum in the diffraction pattern is

$$a \sin \theta = m\lambda \qquad m = 1, 2, 3, \ldots \qquad \text{25-3}$$

We now calculate the amplitude at a general point for which the waves from two adjacent sources differ in phase by δ. Figure 25-10 shows the vector diagram for the addition of N waves which differ in phase from the first wave by $\delta, 2\delta, \ldots, (N-1)\delta$. When N is very large and δ is very small, the vector diagram is approximately an arc of a circle. The resultant amplitude A is the length of the chord of this arc. We calculate this resultant amplitude in terms of the phase difference ϕ between the first and last waves. From Figure 25-10 we have

$$\sin \tfrac{1}{2}\phi = \frac{\tfrac{1}{2}A}{r} \qquad \text{or} \qquad A = 2r \sin \tfrac{1}{2}\phi \qquad \text{25-4}$$

where r is the radius of the arc. Since the length of the arc is $A_{max} = NA_0$ and the angle subtended is ϕ, we have

$$\phi = \frac{A_{max}}{r} \qquad \text{25-5}$$

or

$$r = \frac{A_{max}}{\phi}$$

Substituting this into Equation 25-4 gives

$$A = \frac{2A_{max}}{\phi} \sin \tfrac{1}{2}\phi = A_{max} \frac{\sin \tfrac{1}{2}\phi}{\tfrac{1}{2}\phi} \qquad \text{25-6}$$

Since the amplitude at the central maximum point ($\theta = 0$) is A_{max}, the ratio of the intensity at any other point to that at the central maximum point is

$$\frac{I}{I_0} = \frac{A^2}{A^2_{max}} = \left(\frac{\sin \tfrac{1}{2}\phi}{\tfrac{1}{2}\phi} \right)^2$$

or

$$I = I_0 \left(\frac{\sin \tfrac{1}{2}\phi}{\tfrac{1}{2}\phi} \right)^2 \qquad \text{25-7}$$

Single-slit diffraction pattern

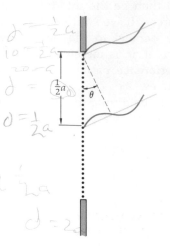

Figure 25-9
At the first diffraction minimum of a single slit, the waves from the source near the top and those from the source just below the middle of the slit are 180° out of phase and cancel. The waves from all N sources then cancel in pairs, giving totally destructive interference.

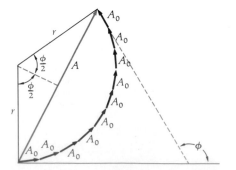

Figure 25-10
Vector model for the calculation of the amplitude of the waves from N sources in terms of the phase difference ϕ between the waves from the first source near the top of the slit and the last source near the bottom of the slit. For large N, the resultant amplitude A is the chord of a circular arc of length NA_0.

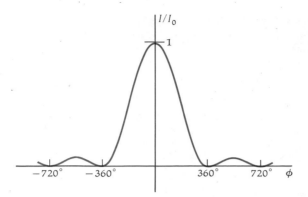

Figure 25-11
Diffraction pattern of a single slit. The relative intensity I/I_0 is plotted versus the phase difference ϕ between the waves from the top and bottom of the slit. Since $\phi = (2\pi a \sin\theta)/\lambda$, the positions of the minima on the screen depend on the ratio of the slit width a to the wavelength λ. The pattern shown here assumes that a is greater than λ. If $a = \lambda$, $\phi = 360°$ at $\theta = 90°$ and only one minimum is observed.

The phase difference ϕ between the first and last waves is $2\pi/\lambda$ times the path difference $a \sin\theta$ between the top and bottom of the opening:

$$\phi = \frac{2\pi}{\lambda} a \sin\theta \qquad 25\text{-}8$$

The function I/I_0 is plotted in Figure 25-11. The first minimum occurs when $\phi = 360°$, at the point where the waves from the top and bottom of the opening have a path difference of λ and are in phase. The second minimum occurs at $\phi = 720°$, where the waves from the top and bottom of the opening have a path difference of 2λ.

Approximately midway between these minima is a secondary maximum. Figure 25-12 shows the vector diagram for determining the relative intensity of this secondary maximum. The phase difference between the first and last waves is approximately $360° + 180°$. The vectors thus complete $1\frac{1}{2}$ circles. The resultant amplitude is the diameter A of a circle with a circumference C which is two-thirds the total length $NA_0 = A_{\max}$. Thus

$$C = \tfrac{2}{3} A_{\max} = \pi A \qquad A = \frac{2}{3\pi} A_{\max} \qquad A^2 = \frac{4}{9\pi^2} A_{\max}^2 \qquad 25\text{-}9$$

and the intensity at this point is

$$I = \frac{4}{9\pi^2} I_0 = \frac{1}{22.2} I_0 \qquad 25\text{-}10$$

The intensity pattern consists of a broad central maximum which has intensity I_0 at $\theta = 0$ and drops to zero intensity at the first diffraction minimum at an angle θ given by

$$a \sin\theta = \lambda \qquad 25\text{-}11$$

There is then a secondary maximum of intensity approximately $I_0/22.2$ at an angle θ given approximately by

$$a \sin\theta = \tfrac{3}{2}\lambda \qquad 25\text{-}12$$

The second minimum occurs at an angle θ given by $a \sin\theta = 2\lambda$.

We can check on the validity of Huygens' principle by experimentally comparing the wave pattern produced by a plane wave incident on a small opening of width a with the wave pattern produced by N point sources along a line with spacing $d = a/N$. Such a comparison is shown in Figure 25-13.

The intensity pattern of Figure 25-11 is called a *Fraunhofer diffraction pattern* of a single slit. The assumptions made in deriving Equation 25-7 describing the pattern were (1) that plane waves are incident normally on the slit (we assumed that the amplitudes and phases of the

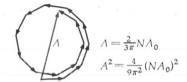

Circumference $\times \tfrac{3}{2} = NA_0 = \pi A \times \tfrac{3}{2}$

$$A = \frac{2}{3\pi} NA_0$$
$$A^2 = \frac{4}{9\pi^2}(NA_0)^2$$

Figure 25-12
Vector diagram for calculating the amplitude of the first secondary maximum of the single-slit diffraction pattern. This secondary maximum occurs when the N vectors complete approximately $1\frac{1}{2}$ circles.

Fraunhofer diffraction

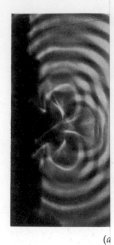

(a

many Huygens'
(on a screen) at a
the openings. (V
on the screen a
Another conditio
width of the slit
slit, is much wi
imum will be ve
will occur at an a
is extremely far a
angle made by
central maximun
ivation.

When the dif
called a *Fresnel da*
is much more d
ference between

Questions

1. If the frequenc
tern is slowly an
tion pattern char

2. If the width o
slowly and stead

25-4   Valid
        Appr

We can indicate t
by drawing strai
wave the energy
straight line, mu

¹ See Richard E. Hask
Physics, vol. 38, p. 10:

The description of the propagation of electrons and other particles can also be considered as an application of the ray approximation. According to the modern theory of quantum mechanics, the motion of electrons and other particles is governed by the laws of wave propagation. The wavelength associated with the motion of a particle of mass m is related to its momentum p and its kinetic energy E_k by an equation first suggested by Louis de Broglie in 1924:

$$\lambda = \frac{h}{p} = \frac{h}{\sqrt{2mE_k}} \qquad\qquad 25\text{-}13$$

where h, called Planck's constant, has the value

$$h = 6.63 \times 10^{-34} \text{ J-sec}$$

Because h is so small, the de Broglie wavelength given by Equation 25-13 is often very small. Then the ray approximation is valid, and the wave nature of the particle is not observed. (For example, the de Broglie wavelength of an electron with kinetic energy of 1 keV is only about 4×10^{-9} cm.) Diffraction and interference of electrons were first observed by Clinton J. Davisson and Lester H. Germer at the Bell Laboratories in 1927, confirming de Broglie's hypotheses. Although Equation 25-13 is valid for any particle, the wavelength associated with a macroscopic particle or even with most molecules is so small that the ray approximation is always valid and diffraction and interference are not observed.

25-5   Huygens' Principle Applied to Reflection

Figure 25-15 shows the reflection of plane waves by a plane barrier. When the incident waves strike the barrier, new waves are generated which move away from the barrier. Experimentally it is found that the rays of the incident waves and those of the reflected waves make equal angles with the normal to the barrier. This *law of reflection* can be stated by saying that

The angle of incidence equals the angle of reflection.

The law of reflection can be derived from Huygens' principle. Figure 25-16 shows a plane wavefront AA' striking a barrier at point A. The angle θ_i between the ray corresponding to this wavefront and the normal to the barrier is called the *angle of incidence*. As can be seen from the figure, the angle of incidence θ_i equals the angle ϕ_i between the incident wavefront and the barrier. The position of the wavefront after a time Δt is found by constructing wavelets of radius $v \Delta t$ with centers on the wavefront AA'. Wavelets which do not strike the barrier form the portion of the new wavefront BB'. Wavelets which do strike the barrier are reflected and form the portion of the new wavefront BB''. By a similar construction, the wavefront $C''CC'$ is obtained from the Huygens' wavelets originating on the wavefront $B''BB'$. Figure 25-17 is an enlargement of a portion of Figure 25-16 showing a part of the original wavefront AP which strikes the barrier in time Δt. In this time the wavelet from point P reaches the barrier at point B, and the wavelet from point A reaches point B''. The reflected wavefront BB''

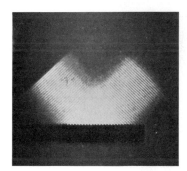

Figure 25-15
Ultrasonic plane waves in water reflecting from a steel plate. (*Battelle-Northwest Photography.*)

rays would travel. If the wave strikes a barrier with an opening much larger than the wavelength, the wave (to a good approximation) still propagates along straight lines through the opening like a stream of particles (except for some diffraction near the edges of the opening). This description of a wave moving in a straight line in the direction of the rays is called the *ray approximation*. For light waves it is called the *geometrical-optics approximation*. The approximation is valid as long as the portion of wavefront considered is large compared with the wavelength, in which case diffraction can be ignored. If, on the other hand, the opening is not much larger than the wavelength, or if we are interested in the region of space within a few wavelengths of an obstacle, diffraction is very important and the wave does not travel in the straight lines indicated by the ray approximation. From another viewpoint, if we have a given set of openings, obstacles, apertures, etc., the ray approximation will be valid in the limit of very short wavelengths and will not be valid if the wavelength is large.

The characteristic phenomena distinguishing wave propagation from the propagation of a beam of particles are interference and diffraction. If the wavelength is very small, the bending of the wave around corners, i.e., diffraction, is not easy to observe. Similarly, the interference of two point sources cannot be observed easily if the wavelength is very small compared with the separation of the sources, because the interference maxima are very close together. Thus if we have a wave phenomenon with very small wavelength, the ray approximation is a good approximation and the propagation of the wave cannot be easily distinguished from the propagation of a beam of particles. The wavelength of visible light ranges from about 400 to 700 nm. Because this is usually much smaller than objects, apertures, and separation of sources (and because coherent-light sources are difficult to obtain), interference and diffraction of light are not easy to see. It was not until the early 1800s, when the interference pattern produced by two slits close together was observed by Thomas Young and others, that light was generally accepted as a wave phenomenon. For many optical applications we can neglect interference and diffraction and use the ray approximation. On the other hand, the wavelengths of audible sound corresponding to frequencies from 20 to 20,000 Hz are about 17 m to 1.7 cm. Except for very high frequencies, these wavelengths are usually large compared with obstacles and apertures available. Thus the ray approximation is seldom valid for sound waves except at very high frequencies. The bending of sound waves around corners is a very common experience.

The ray approximation is often valid for high-energy electromagnetic radiation such as x-rays, discovered in 1895 by Wilhelm Röntgen. He named the mysterious new kind of radiation he discovered x-rays because he was unable to determine whether the radiation consisted of a beam of particles such as electrons or protons or a wave such as light but with very short wavelengths. The rays passed through strong electric and magnetic fields without the deflection which would be expected if the rays were charged particles. The rays also passed through small apertures without exhibiting the interference or diffraction which would be expected if they were waves. X-rays were later found to be a form of electromagnetic radiation similar to light but with much smaller wavelength. The wavelength of a typical x-ray is of the order of 10^{-8} cm, which accounts for the difficulty in observing interference or diffraction.

Battelle-Northwest Photography

When waves are incident on a barrier having two slits with size and separation of the order of the wavelength, both diffraction and interference can be observed. Here plane ultrasonic waves in water are incident on slits of width 2λ and separation 8λ. The photograph was made using a laser and a Schlieren technique, which is sensitive to the small changes in the optical index of refraction of the water introduced by the compressions and rarefactions of the ultrasonic waves.

The description of the propagation of electrons and other particles can also be considered as an application of the ray approximation. According to the modern theory of quantum mechanics, the motion of electrons and other particles is governed by the laws of wave propagation. The wavelength associated with the motion of a particle of mass m is related to its momentum p and its kinetic energy E_k by an equation first suggested by Louis de Broglie in 1924:

$$\lambda = \frac{h}{p} = \frac{h}{\sqrt{2mE_k}} \qquad\qquad 25\text{-}13$$

where h, called Planck's constant, has the value

$$h = 6.63 \times 10^{-34} \text{ J-sec}$$

Because h is so small, the de Broglie wavelength given by Equation 25-13 is often very small. Then the ray approximation is valid, and the wave nature of the particle is not observed. (For example, the de Broglie wavelength of an electron with kinetic energy of 1 keV is only about 4×10^{-9} cm.) Diffraction and interference of electrons were first observed by Clinton J. Davisson and Lester H. Germer at the Bell Laboratories in 1927, confirming de Broglie's hypotheses. Although Equation 25-13 is valid for any particle, the wavelength associated with a macroscopic particle or even with most molecules is so small that the ray approximation is always valid and diffraction and interference are not observed.

25-5 Huygens' Principle Applied to Reflection

Figure 25-15 shows the reflection of plane waves by a plane barrier. When the incident waves strike the barrier, new waves are generated which move away from the barrier. Experimentally it is found that the rays of the incident waves and those of the reflected waves make equal angles with the normal to the barrier. This *law of reflection* can be stated by saying that

The angle of incidence equals the angle of reflection.

The law of reflection can be derived from Huygens' principle. Figure 25-16 shows a plane wavefront AA' striking a barrier at point A. The angle θ_i between the ray corresponding to this wavefront and the normal to the barrier is called the *angle of incidence*. As can be seen from the figure, the angle of incidence θ_i equals the angle ϕ_i between the incident wavefront and the barrier. The position of the wavefront after a time Δt is found by constructing wavelets of radius $v\,\Delta t$ with centers on the wavefront AA'. Wavelets which do not strike the barrier form the portion of the new wavefront BB'. Wavelets which do strike the barrier are reflected and form the portion of the new wavefront BB''. By a similar construction, the wavefront $C''CC'$ is obtained from the Huygens' wavelets originating on the wavefront $B''BB'$. Figure 25-17 is an enlargement of a portion of Figure 25-16 showing a part of the original wavefront AP which strikes the barrier in time Δt. In this time the wavelet from point P reaches the barrier at point B, and the wavelet from point A reaches point B''. The reflected wavefront BB''

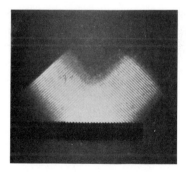

Figure 25-15
Ultrasonic plane waves in water reflecting from a steel plate. (*Battelle-Northwest Photography.*)

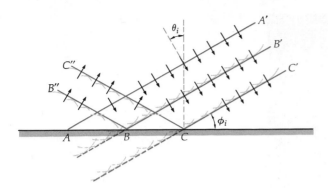

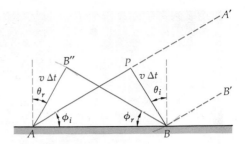

Figure 25-16
Plane wave reflected at a
plane barrier. The angle θ_i
between the incident ray and
the normal to the barrier is
the angle of incidence. It is
the same as the angle ϕ_i
between the incident wavefront and the barrier.

Figure 25-17
Geometry of Huygens' con-
struction for the calculation of
the law of reflection. The wavefront AP originally intersects
the barrier at point A. After a
time Δt, the Huygens' wavelet from P strikes the barrier at point B, and the one
from A reaches point B''.
Since the triangles ABP and
BAB'' are congruent, the
angles θ_i and θ_r are equal.

makes an angle ϕ_r with the barrier which is equal to the angle of
reflection θ_r between the reflected ray and the normal to the barrier.
The triangles ABP and BAB'' are both right triangles with a common
side AB and equal sides $AB'' = BP = v\,\Delta t$. Hence these triangles are
congruent, and the angles ϕ_i and ϕ_r are equal, implying that the angle
of reflection θ_r equals the angle of incidence θ_i:

$$\theta_i = \theta_r \hspace{6cm} \text{25-14}$$ *Law of reflection*

This is just the law of reflection. Having derived this rule from the
consideration of the propagation of wavefronts and Huygens' principle, it is now much easier to consider only the rays, assuming that we
need not worry about diffraction effects.

25-6 Huygens' Principle Applied to Refraction

Figure 25-18 shows a plane wave incident on a line where the velocity
of propagation changes abruptly from v_1 to v_2. In a ripple tank this is
accomplished by changing the depth of the water. As with waves on a
string arriving at a point where the velocity changes suddenly because
of a change in the density of the string, there is a transmitted wave
and a reflected wave. The angle of reflection equals the angle of incidence, as when the wave reflects from an impenetrable barrier. In
Figure 25-19 we apply Huygens' construction to find the wavefront of
the transmitted wave, in much the same way as we did for reflection.
AP is a portion of a wavefront in medium 1 at angle of incidence θ_1
which intersects the line dividing the media at point A.

In time Δt the wavelet from P travels the distance $v_1\,\Delta t$ and reaches
the point B on the line AB separating the two media, while the wavelet
from point A travels the distance $v_2\,\Delta t$ into the second medium. The

Figure 25-18
Refraction of plane waves in a ripple tank at a boundary at which the wave speed changes because the depth of water changes. Note that reflection also occurs at the boundary. (*Courtesy of Film Studio, Education Development Center, Newton, Mass.*)

new wavefront BB' is not parallel to the original wavefront AP because the speeds v_1 and v_2 are different. From the triangle APB we have

$$\sin \phi_1 = \frac{v_1 \, \Delta t}{AB} \qquad \text{or} \qquad AB = \frac{v_1 \, \Delta t}{\sin \phi_1} = \frac{v_1 \, \Delta t}{\sin \theta_1}$$

using the fact that the angle ϕ_1 equals the angle of incidence θ_1. Similarly, from triangle $AB'B$ we have

$$\sin \phi_2 = \frac{v_2 \, \Delta t}{AB} \qquad \text{or} \qquad AB = \frac{v_2 \, \Delta t}{\sin \phi_2} = \frac{v_2 \, \Delta t}{\sin \theta_2}$$

where $\theta_2 = \phi_2$ is the angle between the refracted ray and the normal to the line separating the media. θ_2 is called the *angle of refraction*. Equating the two values of AB gives

$$\frac{\sin \theta_1}{v_1} = \frac{\sin \theta_2}{v_2} \qquad\qquad 25\text{-}15$$

This relation is called the *law of refraction*. For a light wave the ratio of the velocity of light in vacuum to that in a medium is called the *index of refraction n* of the medium:

$$n = \frac{c}{v} \qquad\qquad 25\text{-}16$$

The index of refraction is a characteristic of the medium. The law of refraction is usually written

$$n_1 \sin \theta_1 = n_2 \sin \theta_2 \qquad\qquad 25\text{-}17$$

for light waves; this is known as *Snell's law*. Note that if v_2 is less than v_1, the angle of refraction θ_2 is less than the angle of incidence θ_1. The ray is then bent toward the normal, the line perpendicular to the boundary surface. For example, when light passes from air to glass, the ray is bent toward the normal; when it passes from glass to air, the ray is bent away from the normal.

The frequency of the transmitted (and reflected) wave is the same as that of the incident wave. This can be understood in terms of Huygens' construction, where we think of the incident wave at the barrier as the source of the transmitted wave. Since the velocity of the transmitted wave differs from that of the incident wave, their wavelengths also differ, as can be seen from Figure 25-18.

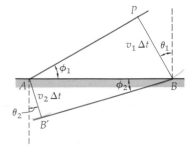

Figure 25-19
Application of Huygens' principle to the refraction of plane waves at the surface separating a medium in which the wave speed is v_1 from a medium in which the wave speed is v_2. For the case shown, in which v_2 is less than v_1, the rays are bent toward the normal to the surface.

Index of refraction

Snell's law

25-7 Total Internal Reflection

Consider a plane wave passing from a medium in which the speed is v_1 into a medium in which the speed v_2 is greater than v_1. If θ_1 is the angle of incidence, the angle of refraction θ_2 is given by

$$\sin \theta_2 = \frac{v_2}{v_1} \sin \theta_1 \qquad\qquad 25\text{-}18$$

This equation cannot be satisfied if the right-hand side is greater than 1. We define the *critical angle* of incidence $\theta_1 = \theta_{\text{crit}}$ such that the angle of refraction θ_2 is 90°. Then

$$\sin \theta_2 = 1 \qquad\qquad 25\text{-}19$$

and

$$\sin \theta_{\text{crit}} = \frac{v_1}{v_2} \qquad\qquad 25\text{-}20$$

Critical angle

For angles of incidence greater than the critical angle, the law of refraction cannot be satisfied. This situation is illustrated in Figure 25-20, where the speeds have been chosen such that $v_1/v_2 = \frac{2}{3}$. The critical angle is then given by $\sin \theta_{\text{crit}} = \frac{2}{3}$, $\theta_{\text{crit}} = 41.8°$. When the angle of incidence is greater than this angle, there is no refracted wave in the second medium. All the energy is reflected. (This phenomenon is called *total internal reflection* because in optics the incident light is usually inside glass and reflected from the glass-air surface.)

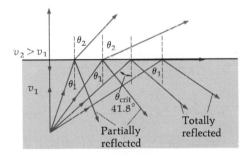

Figure 25-20
Total internal reflection. At the critical angle θ_{crit}, given by $\sin \theta_{\text{crit}} = v_1/v_2$, the angle of refraction is 90°, and the refracted ray emerges parallel to the boundary surface. At angles of incidence greater than the critical angle, there is no refracted ray. If the first medium is glass and the second air, v_1/v_2 is approximately $\frac{2}{3}$ and the critical angle is 41.8°.

Although all the energy of a wave is reflected if the angle of incidence is greater than the critical angle, it is not correct to conclude that there is no disturbance in the second medium. That is, the wave function does not drop to zero immediately at the boundary but decreases exponentially with distance and becomes negligible within a few wavelengths of the boundary. Instantaneously, some energy may flow into or out of the second medium, but the time average of the energy flow into the second medium is zero.

The penetration of the wave function into the second medium can be observed if we place a third medium similar to the first just beyond the reflecting surface. We then have a narrow barrier formed by the second medium. In a ripple tank it is a narrow section of deeper water in which the wave velocity is greater than in the shallower water on either side. In Figure 25-21 we see that if the barrier is very narrow, some of the waves get through even though the angle of incidence is greater than the critical angle. The simple interpretation of Equation 25-18 that all the rays are reflected when the angle of incidence exceeds the critical angle is not quite correct because of the penetration of the

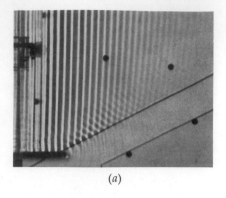 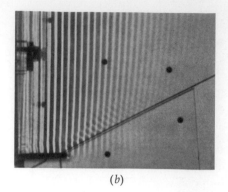

(a)   (b)

Figure 25-21
Barrier penetration by waves
in a ripple tank. In (a) the
waves are totally reflected
from a gap of deeper water.
When the gap is very narrow,
as in (b), a transmitted wave
is observed. (*Courtesy of Film
Studio, Education Development
Center, Newton, Mass.*)

wave into the second medium. *Barrier penetration* is of great importance in the theory of quantum mechanics.

25-8   Bragg Scattering

When a plane wave is incident on an obstacle that is small compared
with the wavelength the obstacle acts as a point source of waves
because of reflection. This situation is of interest because of its similarity to that of an electromagnetic wave incident on an atom which
absorbs some of the energy and reradiates (this is called *scattering*). Of
particular interest is the scattering of a wave by a large number of
regularly spaced scattering centers. We can analyze this scattering by
finding under what conditions the waves from the various scattering
centers will be in phase.

We first consider the scattering of a wave by an array of scattering
centers whose separation is much less than the wavelength, e.g., the
scattering of light (of typical wavelength 500 nm) by a regular array of
atoms in a transparent crystal (separation of about 0.1 nm). Here the
phase difference between waves scattered from nearby centers can be
neglected, and the calculation of the wave scattered back is the same
as Huygens' construction for reflection. The angle of incidence equals
the angle of reflection. This is, in fact, a microscopic interpretation of
reflection. The refracted wave results from the combination and interference of the incident wave with the waves scattered in the general
direction of the incident wave. The detailed calculation of this interference is quite complicated, but the result is that the combination of the
scattered waves and incident wave produces a wave which lags in
phase behind the incident wave; i.e., the wavefront does not travel as
far into the medium in a given time as the incident wave would. This
is interpreted as a smaller velocity for the wave in the medium.

If the separation of the scattering centers is of the same order as the
wavelength, as when x-rays are incident on a crystal or other matter,
for example, the waves scattered from nearby centers will differ appreciably in phase. Figure 25-22 shows a single row of scattering centers
and incident and scattered rays from two successive centers. The scattered waves will be in phase if the path differences Δr_1 and Δr_2 are
equal. From the symmetry of the figure we see that this occurs when
the scattering angle equals the angle of incidence. The angle of reflection equals the angle of incidence whether the reflection is from a continuous barrier, a set of point reflectors, or a set of point reradiators (as
with reflection of light).

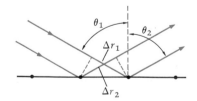

Figure 25-22
Scattering from a line of scattering centers. The waves
scattered by two successive
centers will be in phase if the
path differences Δr_1 and Δr_2
are equal; this occurs if the
scattered angle θ_2 equals
the angle of incidence θ_1.

We now consider another layer of scatterers, as in Figure 25-23. The scattering angle for the wave scattered from this second layer will also equal the incident angle. What are the conditions for the waves scattered from the two layers to be in phase? Figure 25-23 shows rays from atoms in these two layers. The path difference for these rays is $2d$ sin ϕ, where $\phi = 90° - \theta$ is the complement of the angle of incidence. If this path difference is an integral number of wavelengths, the scattering from these two layers will be in phase. In fact, if this condition holds, the scattering from all layers will be in phase when the layers are equally spaced. Thus there will be a strong scattered wave at angles which satisfy

$$2d \sin \phi = m\lambda \qquad \text{25-21}$$

where m is an integer. This is called the *Bragg scattering condition,* after William Lawrence Bragg, who applied this type of analysis in 1912 to the scattering of x-rays by a crystal. At other angles, the scattered waves from various layers tend to cancel out because of destructive interference.

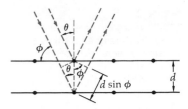

Figure 25-23
Scattering from two successive planes of scattering centers. The waves scattered from the two planes will be in phase if the path difference $2d$ sin ϕ is an integral number of wavelengths.

Radar Astronomy

G. H. Pettengill
Massachusetts Institute of Technology

Figure 1
Steerable parabolic reflector, 84 ft in diameter, used for radar astronomy at 23-cm wavelength by M.I.T. Lincoln Laboratory, Westford, Massachusetts. [*From M. I. Skolnik (ed.),* Radar Handbook, *p. 33-18, McGraw-Hill Book Company, New York, 1970, by permission of the publishers.*]

Essentially all we know about celestial objects has come from measuring the electromagnetic radiation they emit or reflect. In passing through the earth's atmosphere, however, much of this radiation is absorbed. In fact, we can observe celestial objects from the ground only at wavelengths lying within two major spectral windows, where our atmosphere is relatively transparent. The shorter-wavelength, or optical, window shows us the splendor of the night sky and has made the development of optical astronomy possible. The longer-wavelength, or radio, window has been increasingly exploited in recent years and provides the basis for radio astronomy.

Radio waves can be collected and focused by antennas using large parabolic reflecting surfaces, as shown in Figure 1. This particular antenna uses an additional reflector to redirect the incident energy collected by the paraboloid onto a feed horn, visible near the center of the primary reflector. The same antenna can be used for radiating energy by reversing the receiving process. In this case, a transmitter is attached to the feed horn, and its radiation is modified by reflection into a plane wave extending across the aperture of the large paraboloid. At great distances from the antenna, diffraction transforms this circular section of a plane wave into a spherical wavefront with a cone angle (beam width) proportional to the ratio of the radio wavelength to the reflector's diameter.

If the radio transmitter is sufficiently powerful (typically several hundred kilowatts) and the antenna sufficiently large (100 ft or so in diameter), it is possible to detect radar echoes from the moon and the nearer planets. Since the time, frequency, and polarization characteristics of the coherent transmitted waveform are precisely known, the alteration of these characteristics as seen in the echo can be related to properties of the distant scattering surface and the propagating medium en route. The study of celestial objects in this way is called *radar astronomy.*

How does the distant surface scatter the incident radar energy? If the surface contains regions, or facets, which are many wavelengths across and flat within plus or minus the wavelength, the reflection is coherent, or specular,

Figure 2
Schematic representation of radar scattering and the dispersion of echo power in delay and frequency introduced by finite target size and rotation. [*After M. I. Skolnik (ed.),* Radar Handbook, *p. 33-3, McGraw-Hill Book Company, New York, 1970, by permission of the publishers.*]

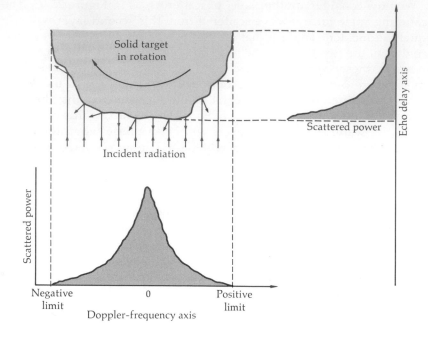

Figure 3
Radar map of the moon at 70-cm wavelength, made by T. W. Thompson using the Arecibo radar in Puerto Rico. Note the unusually strong diffuse scattering by the crater Tycho in the bottom center of the map, indicating that it has an exceptionally rough surface. (*From T. W. Thompson, Atlas of Lunar Radar Maps at 70-cm Wavelength,* The Moon, *vol. 10, no. 1, p. 54, May 1974.*)

i.e., it emerges at an angle with respect to the facet normal which is equal to the angle of incidence. Since the radar observes only backscattered radiation, only surface facets oriented at right angles to the line of sight are seen, as diagrammed schematically in Figure 2. If the surface is rough, so that its height varies rapidly by a significant fraction of the wavelength as one moves across it, the scattering is diffuse and depends only slightly on the angle of incidence to the surface.

Rotation of the target introduces a doppler frequency shift which varies linearly across the target as shown in Figure 2. Therefore, if one analyzes the echo power in frequency, one can establish a direct correspondence between the echo intensity observed in a given frequency interval and the scattering efficiency of a calculable location on the surface which imparts those particular values of frequency shift. Similarly, the distribution of echo power in time delay can be related to scattering from identifiable regions of the surface lying at different distances from the radar.

By analyzing the echo power simultaneously as a function of delay and frequency, radar scattering maps of the surface can be obtained, as shown in Figure 3. These radar maps clearly show the existence of large-scale surface features, such as craters and mountain ranges, and are of particular interest in the study of the cloud-shrouded surface of Venus, which is not otherwise accessible to study from afar. This technique can also be turned around and used to determine the planet's rotation. In this way it was discovered by radar that Mercury does not always keep one face turned to the sun, as was formerly thought, but rotates 50 percent faster than its average orbital rate.

By timing the total round-trip echo delay, the distance between the radar and the surface of the distant celestial object is obtained, since the velocity of radio waves is known. In many current experiments, the timing accuracy is better than one-millionth of a second, while the total round-trip echo delay exceeds 1000 sec. Thus, we measure distance with a fractional accuracy of 1×10^{-9}, or a thousand times more accurately than is possible using optical observations alone. Not many measurements in physics achieve this level of accuracy, which is sufficient to make observable results of many extremely small perturbing effects on the orbits of the inner planets. Several of these effects are identifiable with relativistic "corrections" and have been used successfully to confirm the accuracy of Einstein's formulation of the theory of general relativity.

Review

A. Define, explain, or otherwise identify:

Diffraction, 587   Index of refraction, 596
Huygens' principle, 588   Snell's law, 596
Fraunhofer diffraction, 591   Critical angle, 597
Fresnel diffraction, 592   Barrier penetration, 597
Ray approximation, 593   Bragg scattering, 599
Law of reflection, 595

B. True or false:

1. Diffraction occurs whenever the wavefront is limited.

2. Diffraction occurs only for transverse waves.

3. Diffraction by a single slit is caused by the scattering of the waves by the edges of the slit.

4. Fresnel diffraction differs from Fraunhofer diffraction only in the region of observation of the pattern.

5. According to Huygens' principle, each point on a wavefront can be considered to be a source of waves.

6. The ray approximation is never valid for sound waves.

7. The ray approximation is always valid for light waves.

8. The first diffraction minimum for a single slit occurs at a point where the waves from the top and bottom of the slit are out of phase by 180°.

9. The law of reflection is the same for sound, water, and light waves.

10. The propagation of waves of extremely short wavelength is indistinguishable from the propagation of a beam of particles.

Exercises

Section 25-1, Diffraction

1. Discuss the relationship between diffraction and interference. Can diffraction occur without interference? Can interference occur without diffraction?

2. If the wavelength is much larger than a sound speaker, the speaker radiates in all directions much like a point source. On the other hand, if the wavelength is much smaller than the speaker, the sound travels approximately in a straight line in front of the speaker. Find the frequency of sound for which the wavelength is (a) 10 times and (b) one-tenth the diameter of a 12-in speaker. Do the same for a 3-in-diameter speaker. (Take 1100 ft/sec for the speed of sound.)

Section 25-2, Huygens' Construction

3. Use Huygens' construction to explain qualitatively the wave pattern produced by plane waves incident on a semi-infinite straight barrier (Figure 25-24).

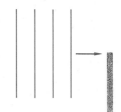

Figure 25-24
Exercise 3.

Section 25-3, Diffraction Pattern of a Single Slit

4. Equation 25-3, $a \sin \theta = m\lambda$, and Equation 24-20, $d \sin \theta = m\lambda$, are sometimes confused. For each equation, define the symbols and explain its application.

5. Plane waves of wavelength 2 cm in a ripple tank hit a barrier with a 6-cm gap. At what angle is (a) the first diffraction minimum and (b) the second minimum?

6. Light of wavelength 600 nm is incident on a long narrow slit. Find the angle of the first diffraction minimum if the width of the slit is (a) 0.1 mm, (b) 1 mm, (c) 0.01 mm.

7. The single-slit diffraction pattern of light is observed on a screen a large distance D from the slit. (a) Show that if the angle θ is small, the distance x from the central maximum on the screen to the first minimum is given by $x = D\lambda/a$. (b) Find x for $D = 2$ m, $\lambda = 600$ nm, and $a = 0.1$ mm.

8. Plane microwaves are incident on a long slit 5 cm wide. The first diffraction minimum is observed at $\theta = 37°$. What is the wavelength of the microwaves?

Section 25-4, Validity of the Ray Approximation

9. When is the propagation of a wave like the propagation of a beam of particles?

10. Why is the ray approximation more useful for light than for sound? Is the ray approximation ever valid for sound? Explain.

11. Discuss the validity of the ray approximation for AM radio waves, FM radio waves, and TV waves.

12. Use Equation 25-13 to calculate the de Broglie wavelength for an electron whose kinetic energy is 13.6 eV, and compare your result with the circumference of the first Bohr orbit in hydrogen, which is 3.33×10^{-10} m. (The mass of the electron is about 9.11×10^{-31} kg.)

13. Use Equation 25-13 to calculate the de Broglie wavelength for a particle of mass 1 gm moving with speed 1 cm/sec. Could you ever observe diffraction of waves of this wavelength?

Section 25-5, Huygens' Principle Applied to Reflection

There are no Exercises for Section 25-5.

Section 25-6, Huygens' Principle Applied to Refraction

14. A beam of light is incident at $\theta = 45°$ on a plane water surface. The index of refraction of water is 1.33. At what angle is the refracted beam?

15. A plane sound wave is incident on a water surface at a small angle with the vertical. Is the direction of motion of the sound wave bent toward the normal or away from it?

16. Plane water waves in a ripple tank traveling at 60 cm/sec hit a boundary line, where the speed changes to 40 cm/sec. The direction of propagation of the incident waves makes an angle of 30° with the normal to the boundary line. What is the direction of propagation of the refracted waves?

17. A beam of microwaves of wavelength 3 cm in air is incident on a flat piece of paraffin at an angle of 30° to the normal. The angle of refraction is 20°. (a) What is the index of refraction of paraffin for these microwaves? (b) What is the wavelength of the microwaves in paraffin? ($n = 1$ in air for microwaves as well as for visible light.)

Section 25-7, Total Internal Reflection

18. What is the critical angle for total internal reflection for light traveling from water ($n = 1.33$) to air ($n = 1.00$)?

19. Sound waves traveling at 340 m/sec enter water, where their speed is 1480 m/sec. What is the critical angle for total reflection?

Section 25-8, Bragg Scattering

20. What should be the spacing of scatterers in a ripple tank so that waves of wavelength 4 cm will be strongly scattered at an angle of incidence of $\theta = 45°$?

21. In a demonstration of Bragg scattering of microwaves, tiny steel balls are set in a piece of styrofoam with spacing of 4 cm. (The styrofoam has no effect on the microwaves.) Microwaves of wavelength 3 cm are then scattered from the steel balls. At what angles relative to the planes of the steel balls do you expect to observe scattering?

Problems

1. A band is marching down a football field with a constant speed v_1. About midfield, the band comes to a section of muddy ground, which has a sharp boundary making an angle of 30° with the 50-yd line, as shown in Figure 25-25. In the mud, the marchers move with speed $v_2 = \frac{1}{2}v_1$. Diagram how each line of marchers is bent as it encounters the muddy section of the field so that eventually the band is marching in a different direction. Indicate the original direction by a ray and the final direction by a second ray, and find the angles between these rays and the line perpendicular to the boundary line. Is their direction of motion bent toward the perpendicular to the boundary line or away from it?

2. The depth of water in a ripple tank changes suddenly along a V-shaped boundary line, as shown in Figure 25-26. Plane waves move in the deeper water toward the boundary line, at which point their speed decreases. Indicate graphically the shape of the wavefronts of the waves in the shallow-water region.

3. Investigate the effect on critical angle of a thin film of water on a glass surface for rays which originate in the glass. Take $n = 1.5$ for glass and $n = 1.33$ for water. (a) What is the critical angle for total internal reflection at the glass-water surface? (b) Is there any range of incident angles greater than θ_{crit} for glass to air refraction for which light rays will leave the glass and the water and pass into the air medium?

4. Show that a light ray transmitted through a glass slab emerges parallel to the incident ray but displaced from it. For an incident angle of 60°, index of refraction of glass $n = 1.5$, and slab thickness of 10 cm, find the displacement measured along the glass surface from the point at which the incident ray would emerge if there were no slab.

5. At the second secondary maximum of the diffraction pattern of a single slit, the phase difference between the waves from the top and bottom of the slit is approximately 5π. The vectors used to calculate the amplitude at this point complete $2\frac{1}{2}$ circles. Show that the intensity of the second secondary maximum is given by $I = (4/25\pi^2)I_0$, where I_0 is the intensity of the central maximum.

6. Equation 25-7 for the Fraunhofer single-slit intensity pattern can be written $I = (I_0 \sin^2 \alpha)/\alpha^2$, where $\alpha = \frac{1}{2}\phi$ is half the phase difference between waves from the top and bottom of the slit. The central maximum occurs when $\alpha = 0$, and the minima occur when $\sin \alpha = 0$ and $\alpha \neq 0$. The positions of the secondary maxima can be found by setting $dI/d\alpha = 0$. (a) Show that $dI/d\alpha = 2I_0\alpha^{-3} \sin \alpha \, (\alpha \cos \alpha - \sin \alpha)$. (b) Show that the secondary maxima occur when $\tan \alpha = \alpha$. (c) Find α and ϕ for the first secondary maximum and compare your result with the approximation that ϕ is midway between the first minimum $\phi = 2\pi$ and the second minimum $\phi = 4\pi$. (d) Calculate I/I_0 at the first secondary maximum and compare your result with the approximate value given by Equation 25-10.

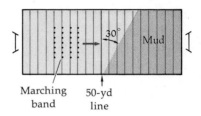

Figure 25-25
Problem 1.

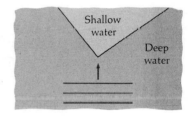

Figure 25-26
Problem 2.

CHAPTER 26   Light

We now turn to a very important special case of wave motion, light. To gain historical perspective, we look briefly at some early theories and experiments which help us understand light as a wave motion and then discuss some of the general properties of electromagnetic waves. These are important in explaining the phenomena of reflection, refraction, diffraction, interference, and polarization of light. We shall find that most features of these phenomena can be understood in terms of the properties of waves already discussed. The application of the laws of reflection and refraction to the formation of images by mirrors and lenses, a subject essential for understanding optical instruments and for laboratory work, will be postponed until Chapter 27.

26-1 Waves or Particles?

The controversy over the nature of light is one of the most interesting in the history of science. Early theories considered light to be a stream of particles which emanated from a source and caused the sensation of vision upon entering the eye. The most influential proponent of this particle theory of light was Newton. Using it, he was able to explain the laws of reflection and refraction.

   The law of reflection of light from a plane boundary is easily explained by the particle theory. Figure 26-1 shows a particle bouncing off a hard, plane surface. If there is no friction, the component of par-

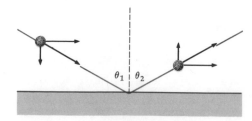

Figure 26-1
Reflection of a particle. If the surface is smooth, the angle of reflection θ_2 equals the angle of incidence θ_1.

ticle momentum parallel to the surface is not changed by the collision
but the component perpendicular to the wall is reversed (assuming
that the mass of the wall is much greater than that of the particle and
that the collision is elastic). Thus the angle of reflection equals the
angle of incidence, as is observed for light rays reflecting from a plane
mirror.

Figure 26-2 illustrates Newton's explanation of refraction, e.g., at an
air-glass or air-water surface. Newton assumed that the light particles
are strongly attracted to the glass or water so that when they approach
the surface, they receive a momentary impulse which increases the
component of momentum perpendicular to the surface. Thus the direction of the momentum of the light particle changes, the light beam
being bent toward the normal to the surface, in agreement with observation. An important feature of this theory is that the velocity of light
must be greater in water or glass than in air to account for the bending
of the beam toward the normal when it enters the water or glass and
away from the normal when it enters air from either of those media.
Experimental determination of the velocity of light in water did not
come until about 1850, nearly 200 years later.

The chief proponents of the wave theory of light propagation were
Huygens and Hooke. We have already seen how Huygens' theory is
applied to derive the law of reflection and Snell's law of refraction by
assuming that light travels more slowly in a glass or water medium
than in air. Newton saw the virtues of the wave theory of light, particularly as it explained the colors formed by thin films, which Newton
studied extensively. However, he rejected the wave theory because of
the observed straight-line propagation of light.

> To me the fundamental supposition itself seems impossible, namely,
> that the waves or vibrations of any fluid can, like the rays of light, be
> propagated in straight lines, without a continual and very extravagant
> spreading and bending every way into the quiescent medium, where
> they are terminated by it. I mistake if there be not both experiment
> and demonstration to the contrary.[1]

Because of Newton's great reputation and authority, this reluctant
rejection of the wave theory of light, based on lack of evidence of
diffraction, was strictly adhered to by Newton's followers. Even after
evidence of diffraction was available, they sought to explain it as scattering of light particles from the edges of slits. Newton's particle
theory of light was accepted for more than a century.

In 1801, Thomas Young revived the wave theory of light. He was
one of the first to introduce the idea of interference as a wave phenomenon in both light and sound. Figure 26-3 shows his drawing of
the combination of waves from two sources. His observation of interference with light was a clear demonstration of the wave nature of
light. Young's work, however, went unnoticed by the scientific community for more than a decade. Perhaps the greatest advance in the
general acceptance of the wave theory of light was due to the French
physicist Augustin Fresnel (1788–1827), who performed extensive
experiments on interference and diffraction and put the wave theory
on a mathematical basis. He showed, for example, that the observed
rectilinear propagation of light is a result of the very short wavelength
of visible light. An interesting triumph of the wave theory was an

[1] Newton, *Opticks*, 1704.

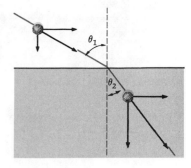

Figure 26-2
Refraction of a particle from
air into water or glass. According to Newton's particle
theory of light, the light particle is attracted by the water
or glass, and its component of
momentum normal to the surface is increased, bending the
path toward the normal to the
surface.

The Science Museum, London

Portrait of Thomas Young,
British physicist, physician,
and egyptologist (1773–1829).

Interference of light waves

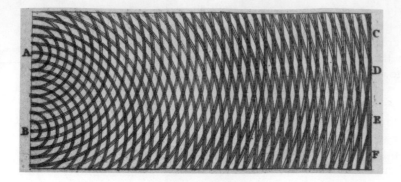

Figure 26-3
Thomas Young's drawing of a
two-source interference pattern. (*Courtesy of the Niels
Bohr Library, American Institute of Physics.*)

experiment suggested to Fresnel by Poisson, who sought to discredit
the wave theory. Poisson noted that if an opaque disk is illuminated
by light from a source on its axis, the Fresnel wave theory predicts that
light waves bending around the edge of the disk should meet and interfere constructively on the axis, producing a bright spot in the center
of the shadow of the disk. Poisson considered this to be a ridiculous
contradiction of fact, but Fresnel's immediate demonstration that such
a spot does in fact exist convinced many doubters that the wave theory
of light is valid (see Figure 26-4).

In 1850, Foucault measured the speed of light in water and showed
that it is less than that in air, thus ruling out Newton's particle theory.
In 1860, James Clerk Maxwell published his mathematical theory of
electromagnetism, which accounted for all observed electric and magnetic phenomena and also led to a wave equation for the propagation
of electromagnetic waves. The velocity of these waves, predicted from
electric and magnetic constants measurable in static laboratory experiments, is the same as the speed of light within the accuracy of the
various experiments. Maxwell correctly suggested that this agreement
is not accidental but indicated that light is an electromagnetic wave.
Maxwell's theory was confirmed in 1887 by Hertz, who produced and
detected waves in the laboratory by strictly electrical means. Hertz
used a spark gap in a tuned circuit to generate the waves and another
similar circuit to detect them. In the latter half of the nineteenth century, Kirchhoff and others applied Maxwell's equations to explain the
interference and diffraction of light and other electromagnetic waves
and put Huygens' empirical methods of construction on a firm mathematical basis.

This discussion of the wave-particle controversy of light would not
be complete if we failed to mention the discovery in the twentieth century that although the wave theory is generally correct in describing
light and other electromagnetic waves, it fails to account for all their
observed properties. One of the ironies of the history of science is that
in his famous experiment of 1887, which confirmed Maxwell's wave
theory, Hertz also discovered the *photoelectric effect*, which Einstein,
only a few years later, showed can be explained only by a particle
model of light. Hertz noticed that the passage of sparks in the receiving gap of his apparatus was facilitated by light from the generating gap. Investigating this effect in 1900, Lenard found that light
falling on a metal surface ejects electrons and that the energies of these
electrons do *not* depend on the intensity of the light. This result was
quite surprising since the intensity is the energy per second per unit
area falling on the metal surface. In 1905 Einstein demonstrated that

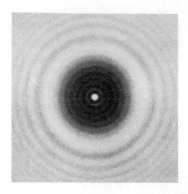

Figure 26-4
Diffraction of light by an
opaque disk, showing the
bright spot in the center of
the shadow. (*From M. Cagnet,
M. Françon, and J. C. Thrierr,
Atlas of Optical Phenomena,
plate 33, Springer-Verlag,
Berlin, 1962.*)

this result can be explained by the assumption that the energy of a light wave is *quantized* into small bundles, called *photons.* The energy of a photon is proportional to the frequency of the wave. According to Einstein, the energy of a photon is

$$E = hf \qquad\qquad 26\text{-}1$$

where $h = 6.63 \times 10^{-34}$ J-sec is Planck's constant and f is the frequency of the light wave.

An electron ejected from a metal surface exposed to light receives its energy from a single photon. When the intensity of light of a given frequency is increased, more photons fall on the surface, ejecting more electrons, but the energy of each electron is not increased. Using this model, Einstein predicted that the maximum energy of an electron ejected in the photoelectric effect and escaping from the metal would increase linearly with the frequency of incident light and that if the frequency of the incident light were below a certain threshold frequency f_t (which depends on the kind of metal), the photoelectric effect would not occur, no matter what the intensity, because no single photon would have enough energy to eject an electron. These predictions were accurately confirmed about 10 years later in difficult experiments performed by the American physicist R. A. Millikan. Thus a particle model of light was reintroduced.

Complete understanding of this dual nature of light did not come until the 1920s, when the diffraction of electrons was discovered by Davisson and Germer and the theory of quantum mechanics was worked out by Schrodinger, Heisenberg, Dirac, and others. The behavior of fundamental quantities such as light, electrons, and other subatomic particles is correctly described by the modern theory of quantum mechanics, which differs from both a classical wave theory and a classical particle theory. However, in some circumstances, the quantum theory is similar to a classical wave theory, and in others it is similar to a classical particle theory. For example, the propagation of these fundamental quantities can always be described as a wave propagation exhibiting the usual wave effects of interference and diffraction. On the other hand, the exchange of energy between these fundamental quantities, as in the photoelectric effect, is usually best described in terms of particle mechanics. As discussed in Chapter 25, when the wavelength of any wave is very small compared with the various obstacles and apertures it encounters, the ray approximation is valid and the propagation is in straight lines. The wave nature of electrons and other so-called "particles" is not easily observed because of their extremely short wavelengths—the same difficulty which prevented Newton from observing the wave character of light. On the other hand, the particle nature of the energy exchanges involving light is often not noticed because such an enormous number of photons is involved and the energy of each photon is so small. This difficulty is not unlike that of observing the particle nature of gas molecules which exert pressure on the walls of a container.

Nineteenth-century engraving of Augustin Jean Fresnel (1788–1827).

Questions

1. The spreading of a beam of light after passing through a very small opening was observed before Newton's time, but it was argued that this was to be expected if the beam consisted of particles moving in not quite perfectly parallel paths. Explain.

2. Supporters of the particle theory of light were able to explain the slight diffraction of light observed, but they could not hope to explain Young's observation of interference on the basis of the particle theory. Why?

3. According to Newton's theory of light, reflection and refraction are the result of attractive forces between the particles of light and the atoms of the refracting material. Explain why this attraction affects the path of a light beam only as it enters or leaves the refracting material and not in the center of the material.

4. Newton concluded that the force between particles of light and atoms of matter must be an attractive force. Can you see why? How would absorption of light by matter be explained in Newton's theory?

26-2   Electromagnetic Waves

Electromagnetic waves include light, radio, radar, x-rays, gamma rays, microwaves, and others, all of which involve the propagation of waves of electric and magnetic fields through space with velocity $c = 3 \times 10^8$ m/sec (in vacuum). All electromagnetic waves are generated when electric charge is accelerated. The differences between the various types of electromagnetic waves are in their frequency and wavelength. Table 26-1 shows the electromagnetic spectrum with the names usually associated with the various frequency and wavelength ranges, which often are not well defined and sometimes overlap. For example, electromagnetic waves with wavelength about 10^{-10} m are called x-rays if their origin is atomic and gamma rays if their origin is nuclear. The human eye is sensitive to electromagnetic radiation of wavelengths from about 4×10^{-7} m to about 7×10^{-7} m, the range called *visible light*.[1] The word light is also used for wavelengths just beyond the visible range. Ultraviolet light is electromagnetic radiation just to the shorter-wavelength side of the visible range, and infrared light is that just to the longer-wavelength side. There are no limits on the wavelengths of electromagnetic radiation: all frequencies are theoretically possible.

The electromagnetic spectrum

A complete description of electromagnetic waves is based on the laws of electricity and magnetism as expressed in Maxwell's theory, but even after we have studied electricity and magnetism in Chapters 29 to 41, a detailed description of that kind is too difficult for an introductory course. For our purposes, we shall merely state that the wave function for electromagnetic waves is the electric field $\mathbf{E}(x,y,z,t)$ (Figure 26-5). (We could also take the associated magnetic field to be the wave

---

[1] A convenient unit for the wavelength of light is the nanometer. Since 1 nm = 10^{-9} m, the visible range is 400 to 700 nm.

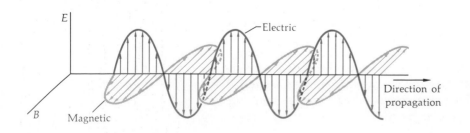

Figure 26-5
The electric and magnetic field vectors are in phase, perpendicular to each other, and perpendicular to the direction of propagation of the wave.

Table 26-1

Electromagnetic spectrum

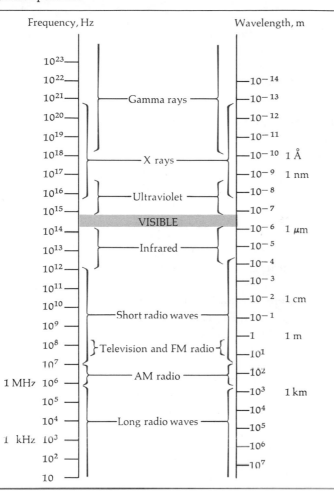

Frequency, Hz		Wavelength, m
10^{23}		
10^{22}		10^{-14}
10^{21}	Gamma rays	10^{-13}
10^{20}		10^{-12}
10^{19}		10^{-11}
10^{18}	X rays	10^{-10} 1 Å
10^{17}		10^{-9} 1 nm
10^{16}	Ultraviolet	10^{-8}
10^{15}		10^{-7}
10^{14}	VISIBLE	10^{-6} 1 μm
10^{13}	Infrared	10^{-5}
10^{12}		10^{-4}
10^{11}		10^{-3}
10^{10}		10^{-2} 1 cm
10^{9}	Short radio waves	10^{-1}
10^{8}	Television and FM radio	1 1 m
10^{7}		10^{1}
1 MHz 10^{6}	AM radio	10^{2}
10^{5}		10^{3} 1 km
10^{4}	Long radio waves	10^{4}
1 kHz 10^{3}		10^{5}
10^{2}		10^{6}
10		10^{7}

function, but it is less convenient because most detectors, including the human eye, are sensitive mainly to the electric field rather than the magnetic field.) The vector **E** is perpendicular to the direction of propagation. Thus electromagnetic waves are transverse waves and can be polarized. We shall study some of the properties of these fields in later chapters. For the present we need know only some rudimentary facts. Electric fields are produced by electric charges, and they exert forces on other charges. As with all wave phenomena, the energy density and intensity of the waves are proportional to the square of the amplitude of the wave. The resultant wave at any point is the vector sum of the amplitudes of individual waves.

Although charges at rest or moving with constant velocity produce electric fields, the alternating electric and magnetic fields associated with electromagnetic radiation are produced only by charges which accelerate. According to classical electromagnetic theory, for example, a charged particle which oscillates with simple harmonic motion of frequency f radiates energy in the form of electromagnetic waves of the same frequency. Figure 26-6 shows a simple and common type of radiating charge system, two equal but opposite charges whose separation varies harmonically with time. Such an arrangement is called an

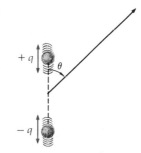

Figure 26-6
Electric dipole. The intensity of radiation is proportional to $\sin^2 \theta$. It is zero along the direction of the dipole and maximum perpendicular to this direction.

electric dipole, and the radiation given off is called *electric-dipole radia-tion.* Some characteristics of such radiation are important for the un-derstanding of light:

The dipole radiator

1. The magnitude of the electric field at large distances varies in-versely as the distance and is proportional to sin θ, where θ is the angle between the dipole axis and the point of observation, as shown in Figure 26-6. Since the intensity is proportional to E^2, the intensity of electric-dipole radiation varies as $1/r^2$ and is proportional to $\sin^2 \theta$.

2. The radiation from a single dipole is polarized with the electric field in the plane of the paper of Figure 26-6.

Many electromagnetic waves exhibit the characteristics of electric-dipole radiation. For example, radio waves are produced by currents oscillating in a straight antenna, an arrangement equivalent to an elec-tric dipole. Light waves come from charge oscillations in atoms. Although the classical theory of electromagnetic radiation has been modified by the theory of quantum mechanics, many of the features of light and other electromagnetic waves can be understood in terms of the classical theory.

All electromagnetic waves are similar except for their wavelength and frequency, but this difference is very important. For example, it is not possible to generate light waves using ordinary circuits to produce the acceleration of the charges because the eye is sensitive only to elec-tromagnetic waves with frequency of the order of 10^{14} Hz, much greater than can be attained with ordinary circuits. As we know, the behavior of waves closely depends on the relative sizes of the wave-length and of the physical objects and apertures the waves en-counter. Since the wavelength of light is in the rather narrow range of about 400 to 700 nm, it is much smaller than most obstacles and aper-tures and the ray approximation often is valid.

Questions

5. What sort of radiation is an electromagnetic wave whose frequency is 10^5 Hz? 10^{10} Hz? 10^{15} Hz? 10^{20} Hz?

6. A dipole antenna on top of a high broadcasting tower has a vertical axis. Describe qualitatively the variation in intensity you would ob-serve if you walked on the ground along a straight path which passes directly under the antenna, starting a great distance from the tower and continuing to a great distance on the other side.

7. If you want to broadcast radio waves from a dipole antenna, would you place the axis of the antenna horizontally or vertically? Why?

26-3   The Speed of Light

The first effort to measure the speed of propagation of light was made by Galileo. He and a partner situated themselves on hilltops about a mile apart, each with a lantern and a shutter to cover it. Galileo proposed to measure the time for light to traverse twice the distance between the experimenters. *A* would uncover his lantern, and when *B* saw the light, *B* would uncover his. The time between *A*'s un-covering his lantern and seeing the light from *B*'s would be the time

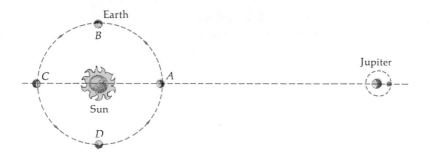

for light to travel back and forth between the experimenters. Though this method is sound in principle, the speed of light is so great that the time interval to be measured is much smaller than fluctuations in the human response time and Galileo was unable to obtain any value for the speed of light.

The first indication of the true magnitude of the speed of light came from astronomical observations of the period of one of the moons of Jupiter. This period is determined by measuring the time between eclipses (when the moon disappears behind Jupiter). The eclipse period is about 42.5 h, but measurements made when the earth is moving away from Jupiter, as from point A to C in Figure 26-7, give a greater time for this period than measurements made when the earth is moving toward Jupiter, as from point C to A in the figure. Since these measurements differ by only about 15 sec from the average value, the discrepancies were difficult to measure accurately. In 1675 the astronomer Roemer attributed these discrepancies to the fact that the velocity of light is not infinite. During the 42.5 h between eclipses of Jupiter's moon, the distance between the earth and Jupiter changed, making the path for the light longer or shorter. Roemer devised the following method for measuring the cumulative effect of these discrepancies. We neglect the motion of Jupiter (it is much slower than that of the earth). When the earth is at the nearest point A, the period of the moon is measured. The time when an eclipse should occur $\frac{1}{2}$ y later, when the earth is at point C, is computed. The observed eclipse time is about 16 min later than predicted. This is the time it takes light to travel a distance equal to the diameter of the earth's orbit.

The first nonastronomical measurement of the velocity of light was made by the French physicist Fizeau in 1849. His method is illustrated in Figure 26-8. This method was greatly improved by Foucault, who replaced the toothed wheel by a rotating mirror. In about 1850, Foucault measured the speed of light in air and in water, showing that it is less in water. Precise measurements of the speed of light using essentially the same method were performed by the American physicist A. A. Michelson from 1880 to 1930.

Figure 26-7
Roemer's method of measuring the speed of light. The time between eclipses of Jupiter's moon appears greater when the earth is moving from A to C than when it is moving from C to A. The difference is due to the time it takes light to travel the distance traveled by the earth along the line of sight during one period of the moon.

Figure 26-8
Fizeau's method of measuring the speed of light. Light from the source is reflected by mirror B and transmitted through a gap in the toothed wheel to mirror A. The speed of light is determined by measuring what angular speed of the wheel will permit the reflected light to be transmitted by the next gap in the toothed wheel so that an image of the source is observed. (*Redrawn from Bernard Jaffe,* Michelson and the Speed of Light. *Copyright © 1960 by Doubleday & Company, Inc. Used by permission of the publisher.*)

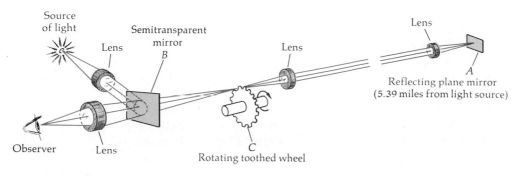

Another method does not involve light directly but is based on Maxwell's theory that light is an electromagnetic wave. In this theory, the velocity of an electromagnetic wave in vacuum is related to an electrical constant, which can be determined by an accurate measurement of the capacitance of a parallel-plate capacitor, and to a magnetic constant, which is related to the SI unit of electric current. Thus a measurement of capacitance and a calibration experiment involving the measurement of magnetic force yields two electromagnetic constants which are related to the speed of light. Accurate measurements using this method were made by Rosa and Dorsey of the U.S. Bureau of Standards in 1906.

There is general agreement between the results of these various methods for determining the speed of light and many others which we shall not discuss. The accepted value for the speed of light is now

$$c = 2.997925 \pm 0.000003 \times 10^8 \text{ m/sec}$$

The value 3×10^8 m/sec is sufficiently accurate for nearly all calculations.

Question

8. Estimate the time required for light to make the round trip in Galileo's experiment to determine c.

26-4   Reflection

Since light propagates as a wave, the law of reflection derived for general wave motion from Huygens' principle in Chapter 25 applies. The angle of reflection equals the angle of incidence. This result can also be derived from an interesting principle enunciated in the seventeenth century by the French mathematician Fermat, stated for our purposes as follows:

The path taken by light in traveling from one point to another is such that the time of travel is a minimum when compared with nearby paths.

Fermat's principle

In the next section we shall apply this principle to derive Snell's law of refraction. The application to the reflection of light is not difficult. Consider Figure 26-9. We wish to find the path taken by light in traveling from point A to the mirror and then to point B. The problem of reflection stated for the application of Fermat's principle is: At what point P in Figure 26-9 must light strike the mirror so that it will travel from point A to point B in the least time? Since the light always travels in the same medium for this problem, the time will be minimum when the distance is minimum. In Figure 26-9 the distance APB is the same as the distance $A'PB$, where A' is the image point of the source A. Point A' lies along the perpendicular from A to the mirror and is equidistant behind the mirror. Obviously, as we vary point P, the distance $A'PB$ is least when the points A', P, and B lie on a straight line. We can see from the figure that when this occurs, the angle of incidence equals the angle of reflection.

Although we have derived the law of reflection in several different ways, we have not discussed the detailed mechanism by which light is reflected by atoms, nor have we developed expressions for the relative

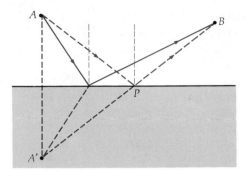

Figure 26-9
Geometry for deriving the
law of reflection by Fermat's
principle.

intensities of the reflected and transmitted light. A complete discussion of the reflection of light taking into account the fact that light is an electromagnetic wave of very great frequency and applying the laws of electricity and magnetism to such a wave cannot be given in our elementary treatment. The results of such a treatment are in full accord with experiment, and we shall mention a few.

The fraction of energy reflected at a boundary between two media depends on the angle of incidence, the indexes of refraction of the two media, and the polarization of the incident light wave. The general expression for the fraction of energy reflected is quite complicated. When light is incident normally, the reflected intensity is

$$I = \left(\frac{n_2 - n_1}{n_2 + n_1}\right)^2 I_0 \qquad\qquad 26\text{-}2$$

where I_0 is the incident intensity and n_1 and n_2 are the indexes of refraction of the two media, which are related to the speeds of the waves in the media by

$$n = \frac{c}{v}$$

For a typical case of reflection from an air-glass surface for which $n_1 = 1$ and $n_2 = 1.5$, Equation 26-2 gives $I = \frac{1}{25}I_0$; only about 4 percent of the energy is reflected, the rest being transmitted.

For light reflected from an air-glass surface, the reflected wave is 180° out of phase with the incident wave when the incident wave is in air, but the two waves are in phase when the incident wave is in glass. We can understand this from our discussion of the reflection of pulse waves on a string. In Chapter 20 we found that when a pulse is incident on a boundary beyond which the velocity of the pulse is slower, the pulse is inverted but it is not inverted if the velocity of the pulse is greater beyond the boundary. If we consider a harmonic wave to be a series of pulses, an inversion of the pulse is equivalent to a 180° phase difference between the incident and reflected wave. This phase difference can be observed experimentally (see Section 26-7).

Intensity and phase of reflected light

The mechanism of reflection by glass can be understood in terms of a model in which the glass atoms absorb the incident light and reradiate it. In Chapter 25 we found that the waves from a row of equally spaced scatterers interfere constructively at points for which the angle of reflection equals the angle of incidence. In our discussion of Bragg scattering we also found that if the spacing of the scatterers is of the order of the wavelength, the waves from scatterers beneath the surface gave constructive interference only if the angle of incidence satisfies the Bragg condition (Equation 25-21). However, for light reflected from glass or any other material, the spacing of the atoms is of the order of 0.1 nm, whereas the wavelength of light is about 5000 times greater. Thus the path difference for light from neighboring rows of atoms is negligible, and reflection occurs for any angle of incidence.

26-5   Refraction

As with reflection, we can apply Huygens' principle to the refraction of light. The result is *Snell's law*,

$$n_1 \sin \theta_1 = n_2 \sin \theta_2 \qquad\qquad 26\text{-}3$$

where θ_1 is the angle of incidence, θ_2 is the angle of refraction, and n_1 and n_2 are the indexes of refraction in the two media. This result can also be obtained from *Fermat's principle*. Figure 26-10 shows possible paths for light traveling from point A in air to point B in glass. Point P_1 is on the straight line between A and B, but this path is not the one for the shortest travel time because light travels with a smaller velocity in the glass. If we move slightly to the right of P_1, the total path length is greater but the distance traveled in the slower medium is less than for the path through P_1. It is not apparent from the figure which path is that of least time, but it is not surprising that a path slightly to the right of the straight-line path takes less time because the time gained by traveling a shorter distance in the glass more than makes up for the time lost traveling a longer distance in air. As we move the point of intersection of the possible path to the right of point P_1, the total time for the travel from A to B decreases until we reach a minimum point $P_{\min}$. Beyond this point, the time saved by traveling a shorter distance in the glass does not compensate for the greater time required for the greater distance traveled in air.

Figure 26-11 shows the geometry for finding the path of least time. If L_1 is the distance traveled in medium 1 with index of refraction n_1 and L_2 the distance in medium 2 with index of refraction n_2, the time for light to go along this path is

$$t = \frac{L_1}{v_1} + \frac{L_2}{v_2} = \frac{L_1}{c/n_1} + \frac{L_2}{c/n_2} = \frac{n_1 L_1}{c} + \frac{n_2 L_2}{c} \qquad 26\text{-}4$$

The quantity nL is called the *optical-path length*. According to Fermat's principle, light will travel along the path for which the time is minimum. According to Equation 26-4, this is the path for which the optical-path length is a minimum. We can derive Snell's law of refraction by considering the time as a function of a single parameter indicating the position of point $P_{\min}$. In terms of the distance x in Figure 26-11, we have

$$L_1{}^2 = a^2 + x^2 \qquad \text{and} \qquad L_2{}^2 = b^2 + (d - x)^2 \qquad 26\text{-}5$$

Figure 26-12 shows the time t as a function of x. At the value of x for which the time is a minimum, the slope of this graph is zero, or

$$\frac{dt}{dx} = 0$$

Differentiating Equation 26-4 gives

$$\frac{dt}{dx} = \frac{1}{c}\left(n_1 \frac{dL_1}{dx} + n_2 \frac{dL_2}{dx} \right)$$

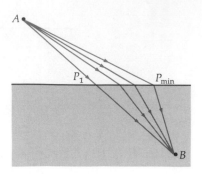

Figure 26-10
Geometry for deriving Snell's law of refraction from Fermat's principle. The point $P_{\min}$ is the point at which light must strike the glass for travel time from A to B to be a minimum.

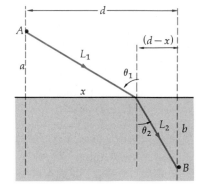

Figure 26-11
Geometry for deriving Snell's law from Fermat's principle.

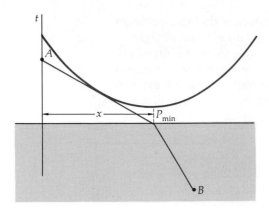

Figure 26-12
Sketch of time for light to travel from point A to B in Figure 26-10 versus distance x measured along the refracting surface. The time is minimum at the point for which the angles of incidence and refraction obey Snell's law.

For the path of minimum time we have

$$n_1 \frac{dL_1}{dx} + n_2 \frac{dL_2}{dx} = 0 \qquad\qquad 26\text{-}6$$

We can compute these derivatives from Equations 26-5. We have

$$2L_1 \frac{dL_1}{dx} = 2x \qquad \text{or} \qquad \frac{dL_1}{dx} = \frac{x}{L_1}$$

But x/L_1 is just $\sin\theta_1$, where θ_1 is the angle of incidence. Thus

$$\frac{dL_1}{dx} = \sin\theta_1$$

Similarly,

$$2L_2 \frac{dL_2}{dx} = 2(d-x)(-1) \qquad \text{or} \qquad \frac{dL_2}{dx} = -\frac{d-x}{L_2} = -\sin\theta_2$$

where θ_2 is the angle of refraction. Thus Equation 26-6 is

$$n_1 \sin\theta_1 + n_2(-\sin\theta_2) = 0 \qquad \text{or} \qquad n_1 \sin\theta_1 = n_2 \sin\theta_2$$

We recognize this result as Snell's law.

As with reflection, a complete discussion of refraction of light based on the laws of electricity and magnetism is beyond the scope of this text. The intensity of the transmitted light can always be found by conservation of energy if the intensity of the reflected light is known. The sum of the energy transmitted and reflected must equal the energy incident.

Understanding the mechanism of refraction by atoms is also more difficult than for reflection. We can consider the transmitted wave as the resultant of the interference of the incident wave and a wave radiated from the atoms which absorb and reradiate some of the light. For light entering glass from air, there is a phase lag between the reradiated wave and the incident wave. Thus the resultant wave also lags in phase behind the incident wave. This phase lag means that the position of a wave crest of the transmitted wave is retarded relative to the position of a wave crest of the incident wave in the medium. Thus in a given time, the transmitted wave does not travel as far in the medium as the original incident wave; i.e., the phase velocity of the transmitted wave is less than that of the incident waves. Thus the index of refraction, defined to be the ratio of the velocity of light in vacuum to that in glass, is greater than 1.

Figure 26-13 shows the index of refraction of a glass as a function of

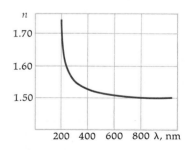

Figure 26-13
Typical dispersion curve for light in glass. The index of refraction decreases slightly as the wavelength increases, indicating that the speed of light in the glass depends on wavelength and frequency. A beam of white light incident at some angle will be slightly spread out into its component colors.

wavelength. The slight decrease in n as λ increases is the phenomenon of *dispersion,* discussed in previous chapters. When a beam of light is incident at some angle on a glass surface, the angle of refraction of the shorter wavelengths (toward the blue end of the visible spectrum) is slightly smaller than that of the longer wavelengths (toward the red end of the spectrum). The formation of a rainbow is a familiar example of dispersion of sunlight by refraction in water droplets.

26-6 Diffraction and Resolution

Diffraction of light, like that of any wave, occurs when a portion of the wavefront is limited by an obstacle or aperture of some kind. The intensity of light at any point in space can be computed using Huygens' principle by taking each point on the wavefront to be a point source and computing the resulting interference pattern, known as a *diffraction pattern.* If the pattern is observed at a great distance from the obstacle or aperture, so that the rays reaching that point are approximately parallel, it is called a *Fraunhofer diffraction pattern.* Such a pattern is often achieved by using a lens to focus parallel rays on the viewing screen. When the diffraction pattern is viewed from a point close to the source, it is called a *Fresnel diffraction pattern.* The diffrac-

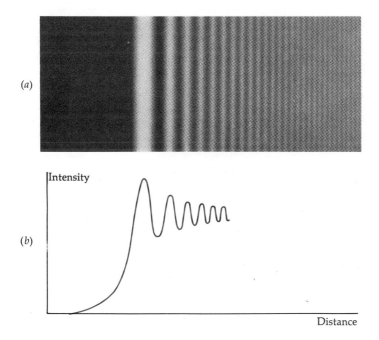

(a)

(b)

Figure 26-14
(a) Fresnel diffraction of a straight edge. (*From M. Cagnel, M. Françon, and J. C. Thrierr,* Atlas of Optical Phenomena, *plate 32, Springer-Verlag, Berlin, 1962.*) (b) Intensity versus distance along a line perpendicular to the edge.

Figure 26-15
Fresnel diffraction of a circular aperture. Compare this pattern with Figure 26-4, the Fresnel diffraction pattern of an opaque circular disk. The patterns are complements of each other. (*From M. Cagnet, M. Françon, and J. C. Thrierr,* Atlas of Optical Phenomena, *plate 33, Springer-Verlag, Berlin, 1962.*)

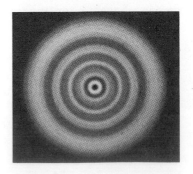

tion of light is often difficult to observe because the wavelength of light is so small or because the light intensity is not great enough.

Figure 26-14a shows the Fresnel diffraction pattern of a straight edge illuminated by light from a point source. A graph of the intensity versus distance measured along a line perpendicular to the edge is shown in Figure 26-14b. The light intensity does not fall abruptly to zero in the geometric shadow but decreases rapidly and is negligible within a few wavelengths of the edge. The Fresnel diffraction pattern of a circular aperture is shown in Figure 26-15. Note the similarity of this pattern and the diffraction pattern of an opaque disk shown in

Figure 26-4. The Fresnel diffraction pattern of a rectangular opening is shown in Figure 26-16. These patterns cannot be seen with broad light sources like an ordinary light bulb because the dark fringes of the pattern produced by light from one point on the source overlap the bright fringes of the pattern produced by light from another point. As mentioned previously, the calculation of the intensity patterns for Fresnel diffraction is very difficult and beyond the scope of this book.

Figure 26-17 shows the Fraunhofer diffraction pattern produced by a single slit and the relative intensity distribution calculated in Chapter 25. Such a pattern can be produced in the classroom using a very small slit and a laser which provides a narrow light beam of great intensity. The width of the central maximum of this pattern depends on the ratio of the wavelength of the light to the width of the slit. The narrower the slit, the wider the central maximum. We found in Chapter 25 that the slit width a, the wavelength λ, and the angle θ between the ray to the center of the pattern and the ray to the first minimum are related by

$$\sin \theta = \frac{\lambda}{a} \qquad \text{26-7}$$

The Fraunhofer pattern of a circular aperture, shown in Figure 26-18, has important applications to the resolution of optical in-

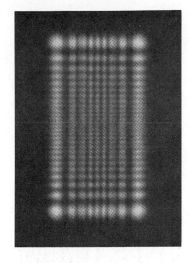

Figure 26-16
Fresnel diffraction of a rectangular opening. (*From M. Cagnet, M. Françon, and J. C. Thrierr,* Atlas of Optical Phenomena, *plate 34, Springer-Verlag, Berlin, 1962.*)

(a)

(b)

Intensity

π 2π ϕ

Figure 26-17
(a) Fraunhofer diffraction pattern of a single slit. (*From M. Cagnet, M. Françon, and J. C. Thrierr,* Atlas of Optical Phenomena, *plate 18, Springer-Verlag, Berlin, 1962.*) (b) Plot of intensity versus phase difference for the pattern in (a).

struments because many such instruments contain circular apertures. The angle θ subtended by the first diffraction minimum is related to the wavelength λ and the diameter of the opening D by

$$\sin \theta = 1.22 \frac{\lambda}{D} \qquad \text{26-8}$$

Equation 26-8 is similar to Equation 26-7 except for the factor 1.22; it arises from the mathematical analysis, which is similar to that for a single slit but more complicated because of the cylindrical geometry.

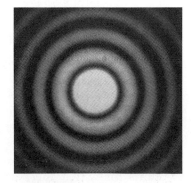

Figure 26-18
Fraunhofer diffraction pattern of a circular aperture. (*From M. Cagnet, M. Françon, and J. C. Thrierr,* Atlas of Optical Phenomena, *plate 16, Springer-Verlag, Berlin, 1962.*)

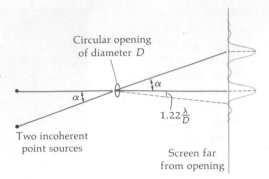

In many applications, the angle θ is small, so that $\sin \theta$ can be replaced by θ. Then

$$\theta \approx 1.22 \frac{\lambda}{D} \qquad\qquad 26\text{-}9$$

Figure 26-19 shows two incoherent point sources which subtend an angle α at a circular aperture which is far from the sources. The Fraunhofer diffraction intensity patterns are also sketched in this figure. If α is much greater than $1.22\,\lambda/D$, there will be little overlap in the diffraction patterns of the two sources and they will be seen as two sources. However, as α is decreased, the overlap of the diffraction patterns increases and it becomes difficult to distinguish the two sources from one source. At the critical angular separation

$$\alpha_c - 1.22 \frac{\lambda}{D} \qquad\qquad 26\text{-}10$$

the first minimum of the diffraction pattern of one source falls at the central maximum of the other source. These objects are said to be just resolved by Rayleigh's criterion for resolution. Figure 26-20 shows the diffraction patterns for two sources for $\alpha > \alpha_c$ and $\alpha = \alpha_c$.

26-7 Interference

Interference patterns of light from two or more sources can be observed only if the sources are coherent. We have mentioned that the randomness of the radiation process of atoms means that two different light sources are generally incoherent. There are several interesting ways of obtaining coherent sources. One arrangement, illustrated in Figure 26-21, is called *Lloyd's mirror*. The original source S has an image S' behind the mirror. The two sources are the original source S and its image behind the mirror S'. The interference pattern on a screen far from these sources is that for two coherent point sources which differ in phase by 180°. These sources differ in phase by 180° because of the phase shift produced by reflection. Thus the central band at the edge of the mirror is dark. Constructive interference occurs at points for which the path difference is $\frac{1}{2}$ wavelength or any odd number of half wavelengths. At these points, the 180° phase difference due to the path difference combines with the 180° phase difference of the sources to produce constructive interference.

Another method of producing two coherent light sources is to illuminate two parallel slits with a single source. Let us first consider the

Figure 26-19
Two distant sources which subtend an angle α at a circular aperture are easily resolved if α is greater than $1.22\lambda/D$, where λ is the wavelength of light and D the diameter of the aperture. Then the diffraction patterns have little overlap. If α is not greater than $1.22\lambda/D$, the overlap of the diffraction patterns makes it difficult to distinguish two sources from one.

Rayleigh's criterion for resolution

Figure 26-20
Diffraction patterns of a circular aperture and two incoherent point sources for (a) α greater than $1.22\lambda/D$ and (b) the limit of resolution $\alpha = 1.22\lambda/D$. (From M. Cagnet, M. Françon, and J. C. Thrierr, Atlas of Optical Phenomena, plate 16, Springer-Verlag, Berlin, 1962.)

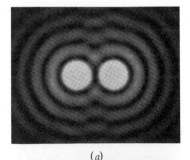

(a)

(b)

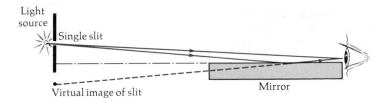

Light source

Single slit

Mirror

Virtual image of slit

case where the width of each slit is small compared with the wavelength. Each slit then acts like a line source (equivalent to a point source in two dimensions). The interference pattern is the same as that obtained with Lloyd's mirror except that the sources are in phase, so that points equidistant or differing in path by a whole number of wavelengths are bright. On a screen far away, so that the rays from each slit are nearly parallel, the interference maxima are at an angle θ given by Equation 24-20:

$$d \sin \theta = m\lambda \qquad 26\text{-}11$$

This equation follows from the fact that the path difference is $d \sin \theta$, where d is the separation of the slits. If the width a of each slit is not small compared with the wavelength, we must also take into account the diffraction pattern of each slit separately. The amplitude A of the light from each slit at some point on the screen at angle θ is less than the amplitude A_{max} at the central maximum point $\theta = 0$. According to Equation 25-6, the amplitude of light from a single slit is

$$A = A_{max} \frac{\sin \frac{1}{2}\phi}{\frac{1}{2}\phi} \qquad 26\text{-}12$$

where ϕ is the phase difference between the light from the top and bottom of the slit. This phase difference is related to the angle θ by

$$\phi = \frac{2\pi a}{\lambda} \sin \theta \qquad 26\text{-}13$$

Thus Equations 26-12 and 26-13 give the amplitude of the light from each slit at angle θ. At points for which light from the two slits has a path difference $d \sin \theta$ equal to a whole number of wavelengths, the amplitude of the resultant wave from the two slits is $2A$. At other points, the light from the two slits has a phase difference of $\delta = (2\pi d \sin \theta)/\lambda$ and the resultant amplitude is $2A \cos \frac{1}{2}\delta$. Figure 26-22 shows the combined interference and diffraction produced by two slits whose separation d is 10 times the width a of each slit. This pattern is merely the two-source interference pattern modulated by the diffraction pattern of each slit. The central diffraction maximum contains 19 interference maxima, the central interference maximum and 9 maxima on either side. The tenth interference maximum on either side of the central one is at the angle θ given by $\sin \theta = 10\lambda/d = \lambda/a$, since $d = 10a$. This coincides with the position of the first diffraction minimum, and so this interference maximum is not seen. At this point the light from the two slits would be in phase and interfere constructively, but there is no light from either slit because the point is a diffraction minimum.

Figure 26-23 shows the central portion of the interference patterns produced by two, three, and four equally spaced slits. These patterns are the same as those calculated in Chapter 24 for two, three, and four equally spaced point sources except that the intensity is modulated by

Figure 26-21
Lloyd's mirror for observation of double source interference with light. The two sources (light source and its image) are coherent and out of phase by 180° because of the phase change on reflection. The central interference band at points equidistant from the sources is dark.

Figure 26-22
Interference-diffraction pattern of two slits with separation d equal to 10 times their width a. The tenth interference maximum on either side of the central interference maximum is missing because it falls at the first diffraction minimum. (*From M. Cagnet, M. Françon, and J. C. Thrierr, Atlas of Optical Phenomena, plate 18, Springer-Verlag, Berlin, 1962.*)

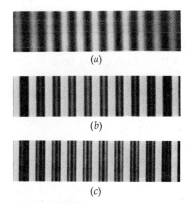

(a)

(b)

(c)

Figure 26-23
Interference patterns for (a) two, (b) three, and (c) four coherent sources. In (b) there is a secondary maximum between each pair of principal maxima, and in (c) there are two such secondary maxima, as calculated in Chapter 24. (*From M. Cagnet, M. Françon, and J. C. Thrierr, Atlas of Optical Phenomena, plate 19, Springer-Verlag, Berlin, 1962.*)

diffraction. Note the presence of a secondary maximum between two successive principal maxima in the three-slit pattern and two secondary maxima in the four-slit pattern, as calculated in Chapter 24.

Perhaps the most commonly observed case of interference of light is that due to reflection of light from the two surfaces of a thin film of water, air, or oil. Consider viewing, at small angles with the normal, a thin film of water (such as a soap bubble), as in Figure 26-24. Part of the light is reflected from the upper surface. Since light travels more slowly in water than in air, there is a 180° phase change in this reflection. Part of the light enters the film and is refracted and partially reflected by the bottom water-air surface. There is no phase change in this reflection. If the light is nearly perpendicular to the surfaces, both the ray reflected from the top surface and the one reflected from the bottom surface can enter the eye at point P in the figure. The path difference between these two rays is $2t$, where t is the thickness of the film. This path difference produces a phase difference of $2\pi(2t/\lambda')$, where λ' is the wavelength of the light in the film, which is related to the wavelength λ_0 in air by

Interference with thin films

$$\lambda' = \frac{v}{f} = \frac{c/n}{f} = \frac{\lambda_0}{n} \qquad\qquad 26\text{-}14$$

The phase difference between these two rays is thus 180° plus that due to the path difference. Destructive interference occurs when the path difference $2t$ is zero or a whole number of wavelengths (in the film). Constructive interference occurs if the path difference is an odd number of half wavelengths.

When a thin water film lies on a glass surface (Figure 26-25), the ray which reflects from the lower water-glass surface also suffers a 180° phase change because the index of refraction of glass (about 1.5) is greater than that of water (about 1.33). Thus both the rays shown in the figure suffer a 180° phase change upon reflection. The phase difference between these rays is thus due solely to the path difference and is given by $\delta = 2\pi(2t/\lambda')$.

A common example of this type of interference occurs when there is a thin film of oil on a water surface or street. Because the thickness of the film varies from point to point, constructive interference occurs for different wavelengths at different points, thus giving rise to colored bands.

Figure 26-24
(a) Light rays reflected from the top and bottom surfaces of a thin film are coherent and produce interference. (b) If the film thickness is much smaller than the wavelength of light, the interference will be destructive because one of the rays suffers a change in phase of 180° upon reflection.

Figure 26-25
Interference of light reflected from a thin film of water resting on a glass surface. In this case, both rays suffer a change in phase of 180° upon reflection.

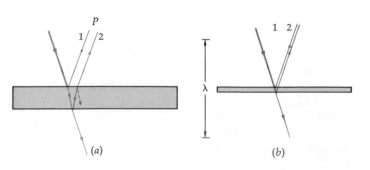

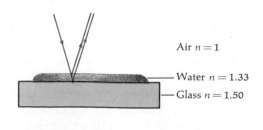

Figure 26-26 illustrates the interference pattern observed when light is reflected from an air film between a spherical glass surface and a plane glass surface in contact. These circular interference fringes are known as *Newton's rings*. Near the point of contact, where the path difference between the ray reflected from the upper glass-air surface and the lower air-glass surface is essentially zero or at least small compared with the wavelength of light, the interference is perfectly destructive because of the 180° phase shift of the ray reflected from the lower air-glass surface. This region is therefore dark. The first bright fringe occurs at a radius such that the path difference contributes a phase difference of 180°, which adds to that due to the phase shift upon reflection and produces a total phase difference of 360°, or zero. The computation of the fringe spacing in terms of the radius of curvature of the spherical piece of glass is left as an exercise.

Figure 26-26
Newton's rings observed with light reflected from a thin film of air between a plane glass and convex glass surface. At the center the thickness of the air film is negligible and the interference is destructive because of the phase change of one of the rays. (*Bausch & Lomb.*)

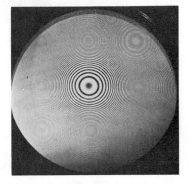

Questions

9. Why must the film used to observe thin-film interference colors be thin?

10. Nonreflecting coatings for lenses are made by coating the lens with an extremely thin layer of material with an index of refraction smaller than that of the glass. Explain how this works. Why does the coating have to be thin compared with the wavelength of visible light?

11. The spacing of Newton's rings decreases rapidly as the diameter of the rings increases. Explain qualitatively.

12. The flatness of finely polished sheets of glass can be tested by placing the glass on another sheet known to be very flat and looking at the light reflected from the thin air film between the sheets. Explain how this procedure works.

26-8 Diffraction Gratings

A useful tool for the analysis of light is the diffraction grating, which consists of a large number of equally spaced slits. Such a grating can be made by cutting parallel, equally spaced grooves on a glass or metal plate with a precision ruling machine. A grating with 20,000 slits per inch is not uncommon. The spacing of the slits for such a grating is $d = 1 \text{ in}/20,000 = 5 \times 10^{-5} \text{ in} = 1.27 \times 10^{-4} \text{ cm}$. Consider a plane light wave incident normally on such a grating and assume that the width of each slit is so small that we can neglect any decrease in intensity due to the diffraction of each slit. The interference pattern produced on a screen a large distance from the grating is then just that due to a large number of equally spaced line sources. We studied some of the properties of such a pattern in Section 24-4. The interference maxima are at angles θ given by

$$d \sin \theta = m \lambda \qquad \text{26-15}$$

The position of an interference maximum does not depend on the number of sources, but the more sources, the sharper the maximum.

Figure 26-27 shows a typical spectroscope, which uses a diffraction grating to analyze light from a source, usually a tube containing atoms of a gas, e.g., helium or sodium vapor. The atoms are excited because

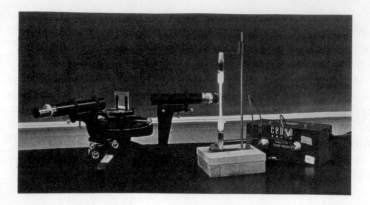

Figure 26-27
Typical student spectroscope.
Light from a slit near a source
is made parallel by a lens and
falls on a grating. The diffracted light is viewed with a
telescope at an angle which
can be accurately measured.
The wavelength of the light is
found from Equation 26-15.
(*Courtesy of Larry Langrill,
Oakland University.*)

of bombardment by electrons accelerated by the high voltage across
the tube. The light emitted by such a source does not consist of a continuous spectrum but contains only certain wavelengths characteristic
of the atoms in the source. Light from the source passes through a
narrow slit and is made parallel by a converging lens. Parallel light
from the lens is incident on the grating. Rather than falling on a screen
a large distance away, the parallel light from the grating is focused by
a telescope and viewed by the eye. The telescope is mounted on a rotating platform which is calibrated so that the angle θ can be measured. At angle $\theta = 0$ the central maximum for all wavelengths is seen.
If light of a particular wavelength λ is emitted by the source, the first
interference maximum is seen at angle θ, given by Equation 26-15 with
$m = 1$. Each wavelength emitted by the source produces a separate
image of the slit in the spectroscope and is called a *line*. Thus by
measuring the angle θ and knowing the spacing of the grating d, we
can measure the wavelengths emitted by the source.

An important feature of such a spectroscope is its ability to measure
light of two nearly equal wavelengths λ_1 and λ_2. For example, the two
prominent yellow lines in the spectrum of sodium have wavelengths
589.00 and 589.59 nm, which can be seen as two separate wavelengths
if their interference maxima do not overlap. According to the Rayleigh
criterion of resolution, these wavelengths are resolved if the angular
separation of their interference maxima is greater than the angular
separation between one interference maximum and the first interference minimum on either side of it. The resolving power of a diffraction grating is defined to be

$$R = \frac{\lambda}{\Delta\lambda}$$ 26-16   *Resolving power defined*

where $\Delta\lambda$ is the difference between two nearby wavelengths each
approximately equal to λ. For example, in order to resolve the two
yellow lines in the sodium spectrum the resolving power must be

$$R = \frac{589.00}{589.59 - 589.00} = \frac{589}{0.59} = 998$$

The resolving power of a grating is just the product of the number of
slits N and the order of the interference maximum m:

$$R = mN$$ 26-17

Thus in order to resolve the two yellow sodium lines in the first order
($m = 1$) we need a grating containing about 1000 slits in the area illuminated by the light.

Optional

Derivation of Equation 26-17

The mth-order interference maximum corresponding to some wavelength λ is at an angle θ given by Equation 26-15:

$$\sin \theta = \frac{m\lambda}{d}$$

If we change the wavelength by a small amount $d\lambda$, the angular change $d\theta$ is found by differentiating this expression,

$$\cos \theta \; d\theta = \frac{m}{d} \; d\lambda$$

or

$$\Delta \theta \approx \frac{m \, \Delta \lambda}{d \cos \theta} \qquad\qquad 26\text{-}18$$

This is the angular separation of the mth-order interference maximum for two nearly equal wavelengths. The phase difference ϕ between the light from two adjacent slits is given by

$$\phi = \frac{2\pi d}{\lambda} \sin \theta \qquad\qquad 26\text{-}19$$

This phase difference is $m2\pi$ for the mth-order interference maximum. The first interference *minimum* occurs when the vectors representing the N sources form a closed polygon, as in Chapter 25. The phase difference between two adjacent sources is then $2\pi/N$. If we change the angle θ by a small amount $d\theta$, the change in phase is

$$d\phi = 2\pi d \cos \theta \; \frac{d\theta}{\lambda}$$

Setting this change in phase equal to $2\pi/N$, we obtain for the angular separation $d\theta$ between the interference maximum and the first minimum,

$$2\pi d \cos \theta \; \frac{d\theta}{\lambda} = d\phi = \frac{2\pi}{N}$$

or

$$d\theta = \frac{\lambda}{Nd \cos \theta} \approx \Delta \theta \qquad\qquad 26\text{-}20$$

According to the Rayleigh criterion of resolution, these wavelengths are resolved if the angle $\Delta \theta$ between the interference maximum and minimum given by Equation 26-20 equals the angle $\Delta \theta$ between the interference maxima of the two wavelengths given by Equation 26-18. Thus the wavelengths are resolved if

$$\frac{\lambda}{Nd \cos \theta} = \frac{m \, \Delta \lambda}{d \cos \theta} \qquad \text{or} \qquad \frac{\lambda}{\Delta \lambda} = mN$$

26-9   Polarization

Since light, like any electromagnetic wave, is a transverse wave, it can be polarized. The wave function $\mathbf{E}(x,y,z,t)$ is perpendicular to the direction of propagation. If the direction of $\mathbf{E}$ is always parallel to a fixed line in space, the light is said to be *linearly polarized*. Consider, for example, light propagating in the z direction. The electric field vector then has components E_x and E_y. If one of these components is zero, or if the ratio E_x/E_y does not depend on time, the light is linearly polarized. Elliptical polarization and circular polarization occur when the components E_x and E_y differ in phase by 90°. For example, if the components E_x and E_y are given by

$$E_x = E_0 \sin (kz - \omega t)$$

$$E_y = E_0 \cos (kz - \omega t) = E_0 \sin (kz - \omega t + 90°)$$

the wave is circularly polarized. At any point in space, the tip of the **E** vector rotates in a circle. If the magnitudes of E_x and E_y are not equal and the components differ in phase by 90°, the tip of the **E** vector moves in an ellipse. If the phase difference between E_x and E_y varies randomly in time, the light is unpolarized.

As we have mentioned, the radiation from a single atomic electric dipole is linearly polarized, but in general light sources consist of immense numbers of atoms which act independently. The resultant light is unpolarized. Four phenomena produce polarized light from unpolarized light:

1. Absorption   *Means of polarization*

2. Reflection

3. Scattering

4. Birefringence

We shall look briefly at each in this section.

Absorption

A common method of polarization is absorption in a sheet of commercial material called Polaroid, invented by E. H. Land in 1938. This material contains long-chain hydrocarbon molecules which are aligned when the sheet is stretched in one direction during the manufacturing process. These chains become conducting (at optical frequencies) when the sheet is dipped in a solution containing iodine. When light is incident with its electric field vector **E** parallel to the chains, electric currents are set up along the chains and the light energy is absorbed. If the electric field **E** is perpendicular to the chains, the light is transmitted. The direction perpendicular to the chains is called the *transmission axis*. We shall make the simplifying assumption that all the light is transmitted when **E** is parallel to the transmission axis and all the light is absorbed when **E** is perpendicular to the transmission axis.

Consider a light beam in the z direction incident on a Polaroid   *Polarization by absorption*
which has its transmission axis in the y direction. On the average, half of the incident light has its **E** vector in the y direction and half in the x direction. Thus half the intensity is transmitted, and the transmitted light is linearly polarized with its **E** vector in the y direction.

Suppose we have a second piece of Polaroid whose transmission axis makes an angle θ with that of the first, as in Figure 26-28. The **E** vector of the light between the Polaroids can be resolved into two components, one parallel and one perpendicular to the transmission axis of the second Polaroid. If we call the direction of the transmission axis of the second Polaroid y',

$$E_{y'} = E \cos \theta \quad \text{and} \quad E_{x'} = E \sin \theta$$

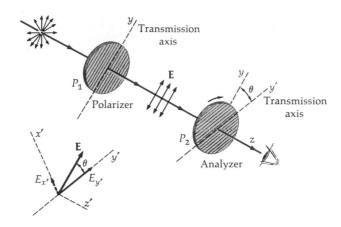

Figure 26-28
Two Polaroids with their
transmission directions
making an angle θ with each
other. Only the component
$E \cos \theta$ is transmitted through
the second Polaroid. If the
original intensity is I_0, the intensity between the Polaroids
is $\frac{1}{2}I_0$ and that transmitted by
both Polaroids is $\frac{1}{2}I_0 \cos^2 \theta$.

Only the component $E_{y'}$ is transmitted by the second Polaroid. The transmitted intensity is proportional to the square of the transmitted amplitude. Thus, if I_1 is the intensity between the two Polaroids, the intensity transmitted by both Polaroids is

$$I_{\text{total}} = I_1 \cos^2 \theta \qquad\qquad 26\text{-}21$$

Intensity transmitted by two polaroids

(The intensity incident on the second Polaroid is of course half that incident on the first.) When two polarizing elements are placed in succession in a beam of light as described here, the first is called the *polarizer* and the second is called the *analyzer*. If the polarizer and analyzer are crossed, i.e., have their axes perpendicular to each other, no light gets through when the absorption is complete. In practice, usually some weak red or blue light can be observed through crossed Polaroids, indicating that the absorption is not complete over the entire visible spectrum. Equation 26-21 is known as *Malus' law* after its discoverer, E. L. Malus (1775–1812). It applies to any two polarizing elements whose transmission directions make an angle θ with each other.

Polarization of electromagnetic waves by absorption can be demonstrated with microwaves (electromagnetic waves with wavelengths of the order of centimeters). In a typical microwave generator, polarized waves are radiated by a dipole antenna. An absorber can be made of a screen of parallel straight wires (Figure 26-29). An electric field parallel to the wires sets up currents in the wires, which absorb energy. If the waves are perpendicular to the wires, no currents are set up and the wave is transmitted. The transmission axis of this polarizer is perpendicular to the direction of the wires. When the axis is parallel to the dipole antenna (the wires are perpendicular to the dipole), the microwaves are transmitted. If the absorber is rotated 90° from this position, the waves are absorbed.

Reflection

When unpolarized light is reflected from a plane surface, e.g., that separating air and glass or air and water, the reflected light is partially polarized. The degree of polarization depends on the angle of incidence and the indexes of refraction of the two media. When the angle of incidence is such that the reflected and refracted rays are perpendicular to each other, the reflected light is completely polarized. This result was discovered experimentally by Sir David Brewster in 1812.

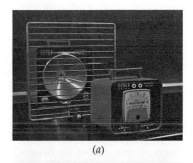

(a)

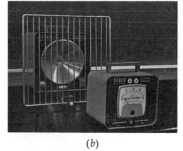

(b)

Figure 26-29
Polarization of microwaves. The electric field of the microwaves is vertical, parallel to the vertical dipole radiator. (a) When the metal wires are horizontal (transmission axis vertical), the waves are transmitted, as indicated by the high reading of the detector. (b) When the wires are vertical (transmission axis horizontal), the waves are absorbed, as indicated by the low reading on the detector. (*Courtesy of Larry Langrill, Oakland University.*)

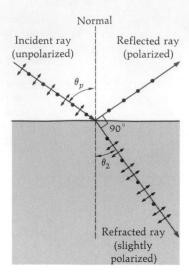

Normal

Incident ray
(unpolarized)

Reflected ray
(polarized)

θ_p

90°

θ_2

Refracted ray
(slightly
polarized)

Figure 26-30
Polarization by reflection. The
incident wave is unpolarized
and has components of **E** parallel to the plane of incidence
indicated by the arrows and
components perpendicular to
the plane of incidence indicated by the dots. For incidence at the polarizing
angle θ_p shown, the reflected
light is completely polarized
with **E** perpendicular to the
plane of incidence as in-
dicated by the dots. Then the
transmitted beam is partially
polarized because only a
small fraction of the light is
reflected. At other angles of
incidence, the reflected light
is partially polarized.

Figure 26-30 shows light incident at the polarizing angle θ_p for
which the reflected light is completely polarized. The electric field
vector **E** of the incident light can be resolved into components parallel
and perpendicular to the plane containing the incident ray, the normal
to the surface, and the reflected ray. This plane is called the *plane of incidence*. The reflected light is completely polarized with its electric field
vector perpendicular to the plane of incidence. We can relate the
polarizing angle θ_p to the indexes of refraction of the media using
Snell's law. If n_1 is the index of refraction of the first medium and n_2
that of the second medium, we have

$$n_1 \sin \theta_p = n_2 \sin \theta_2$$

where θ_2 is the angle of refraction. From Figure 26-30 we see that the
sum of the angle of reflection and the angle of refraction is 90°. Since
the angle of reflection equals the angle of incidence, we have

$$\theta_2 = 90° - \theta_p$$

Then

$$n_1 \sin \theta_p = n_2 \sin (90° - \theta_p) = n_2 \cos \theta_p$$

or

$$\tan \theta_p = \frac{n_2}{n_1} \qquad\qquad 26\text{-}22$$

Equation 26-22 is known as *Brewster's law*. Although the reflected light
is completely polarized when the incident angle is θ_p, the transmitted
light is only partially polarized because only a small fraction of the incident light is reflected. If the incident light itself is polarized with its
electric field vector in the plane of incidence, there is no reflected light
when the angle of incidence is θ_p (Figure 26-31). There is no simple
method of deriving Brewster's law; it can be derived from the electromagnetic theory of light, but such a derivation is beyond the scope
of this book. We can understand the result qualitatively from Figure
26-31. If we consider the molecules of the second medium to be oscillating in the direction of the electric field of the refracted ray, they
cannot radiate energy along the direction of oscillation, which would
be the direction of the reflected ray.

Because of the polarization of reflected light, sun glasses made of

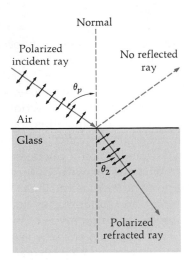

Normal

Polarized
incident ray

No reflected
ray

θ_p

Air

Glass

θ_2

Polarized
refracted ray

Figure 26-31
Polarized light incident at the
polarizing angle. If the incident light has no component of **E** perpendicular to the
plane of incidence, there is no
reflected light.

polarizing material can be very effective in cutting out glare. If the light is reflected from a horizontal surface such as a lake or snow on the ground, the plane of incidence will be vertical and the electric field vector of the reflected light will be predominately horizontal. Sun glasses with their transmission axis vertical will then reduce the glare by absorbing much of the reflected light.

Scattering

The phenomenon of absorption and reradiation is called *scattering*. Reflection of light is actually a scattering of light by a large number of scattering centers closely spaced compared with the wavelength. Refraction is a similar phenomenon in which the scattered light interferes with the incident light. The term scattering, however, usually refers to the situation in which the scattering centers are separated by distances not small compared with the wavelength of light. A familiar example of light scattering is that from clusters of air molecules (due to random fluctuations in the density of air) which tend to scatter short wavelengths more than long wavelengths, thus giving the sky its blue color.

Scattering can be demonstrated by adding a small amount of powdered milk to a container of water. The milk particles absorb light and reradiate it as dipole radiators. Consider a beam of light in the z direction (Figure 26-32). The **E** vector is therefore in the x and y directions. The scattering particles oscillate parallel to the **E** vector of the incident wave, i.e., in the x and y directions but not in the z direction. If we look perpendicular to the beam, say in the x direction, we see light radiated by charges in the scattering centers oscillating in the y direction but we do not see radiation due to the oscillations in the x direction because no light is radiated in the direction along the line of the dipoles. We also see no radiation with **E** in the z direction because the scattering particles do not oscillate in that direction. The light radiated in the x direction is thus polarized with its **E** vector in the y direction. This is easily demonstrated by polarizing the incident beam with a Polaroid. If the incident light contains only **E** vectors in the x direction, no scattered light will be observed in the x direction whereas light will be observed in the y direction. By rotating the Polaroid 90° the situation will be reversed: no scattered light in the y direction and scattered light in the x direction.

Polarization by scattering

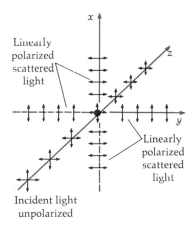

Linearly
polarized
scattered
light

x

z

y

Linearly
polarized
scattered
light

Incident light
unpolarized

Figure 26-32
Polarization by scattering.
Unpolarized light prop-
agating in the z direction is
incident on a scattering center
at the origin. The light scattered in the x direction is polarized with its **E** vector in the
y direction, while the light
scattered in the y direction is
polarized in the x direction.

Optional

Polarization by birefringence

Figure 26-33
Huygens' wavelets for the *O* and *E* rays when light is incident on a birefringent crystal whose optic axis is in the plane of incidence. The *E* ray is polarized with its electric field in the plane of incidence whereas the *O* ray has its electric field perpendicular to the plane of incidence. The *O* ray obeys Snell's law of refraction, but the *E* ray does not.

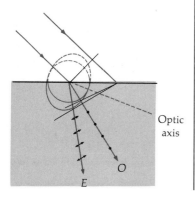

Birefringence

Birefringence is a complicated phenomenon occurring in materials, e.g., calcite, which are anisotropic, i.e., have a preferred direction. Although most materials are isotropic (all directions are equivalent), some crystals, because of their atomic structure, have different optical properties for light traveling in different directions through the crystal. The velocity of light in calcite and some other materials is not the same in all directions. Such materials are called *double-refracting* or *birefringent*. When a light ray is incident on a calcite crystal, it may be separated into two rays, which travel with different velocities and are polarized in mutually perpendicular directions. The Huygens wavelets corresponding to one ray are spherical, as in propagation through ordinary isotropic materials, and this ray is called the *ordinary (O) ray*. In the *extraordinary (E) ray* the Huygens wavelets are ellipsoids of revolution, indicating that the velocity of propagation depends on direction. Along the axis of revolution, the two rays propagate with the same speed. The Huygens wavelets are tangent to each other in this direction, called the *optic axis*. Perpendicular to the optic axis, the difference in speed of the two rays is maximum. In calcite, the extraordinary ray travels faster than the ordinary ray perpendicular to the optic axis. The index of refraction for the ordinary ray in calcite is about $n_O = 1.66$. For the extraordinary ray n_E varies from 1.66 along the optic axis to 1.49 perpendicular to it.

The propagation of both the ordinary ray and extraordinary ray through a birefringent crystal can be traced by construction of the Huygens' wavelets for the rays. For a general orientation of the crystal this is extremely complex. We shall restrict our discussion to the cases in which the optic axis of the crystal is either in the plane of incidence or perpendicular to it. Then both the *O* and *E* rays are in the plane of incidence, and the polarization of the rays is relatively simple. The ordinary ray is polarized with its electric field perpendicular to the plane formed by the optic axis and the direction of propagation, whereas the extraordinary ray has its electric field in the plane containing the optic axis and the direction of propagation. The Huygens' wavelets and polarizations are shown in Figures 26-33 and 26-34 for

Figure 26-34
Huygens' wavelets for the *O* and *E* rays when unpolarized light is incident on a birefringent crystal whose optic axis is perpendicular to the plane of incidence. In this case both rays obey Snell's law of refraction.

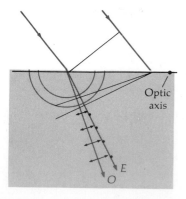

Figure 26-35
Huygens' wavelets for light incident normally on a birefringent crystal whose optic axis is perpendicular to the surface. Both rays propagate with the same speed and in the same direction, and so there is no separation of the rays.

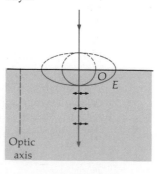

Figure 26-36
Huygens' wavelets for light incident normally on a birefringent crystal whose optic axis is not perpendicular to the surface. The *E* ray is deflected even at normal incidence.

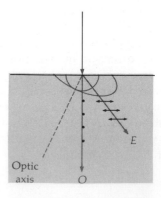

the optic axis in the plane of incidence and perpendicular to the plane of incidence, respectively. Note the spatial separation of the rays.

In Figure 26-35 a light beam is incident normally on the face of a crystal whose optic axis is perpendicular to the surface. The two rays propagate along the optic axis with the same speed, and no separation of the rays is observed. In Figure 26-36 the light is incident normally on a crystal whose optic axis is not perpendicular to the surface. In this case, even for normal incidence the extraordinary ray is deflected. The two rays are separated in space and emerge from the crystal as parallel rays separated by an amount that depends on the thickness of the crystal (Figure 26-37). This effect can be demonstrated by placing a calcite crystal over a small source and observing the two images of the source through the crystal. A Polaroid analyzer can be used to show that the two rays are polarized in mutually perpendicular directions. If the crystal is rotated about the line of the incident ray, the extraordinary ray is rotated through space.

There are several interesting applications of double refraction. One is the Nicol prism, which is used as a polarizer. Figure 26-38 shows the geometry of such a prism. A crystal is cut so that the two rays are separated in space. The crystal is cut along the diagonal and cemented together with a transparent cement which has an index of refraction ensuring that one ray suffers total internal reflection whereas the other is transmitted. The advantage of a Nicol prism over Polaroid is that the emerging light is completely polarized over the entire visible spectrum.

Double refraction can also be used to produce a rotation of the plane of polarization of linearly polarized light and to produce circularly polarized light from linearly polarized light. Figure 26-39 shows a light beam incident normally on the surface of a crystal whose optic axis is parallel to the surface. The Huygens' wavelets for the O and E rays have circular cross sections but of different radii. The two rays travel in the same direction as the incident ray but with different speeds and different polarizations. In calcite, for example, the extraordinary ray travels faster than the ordinary ray so that its wavefront gets ahead of that for the O ray.

Figure 26-37
A narrow beam of unpolarized light incident normally on a birefringent crystal oriented as in Figure 26-36 is split into two beams (exaggerated in this figure). When the crystal is rotated in space, the E ray rotates in space.

Figure 26-38
A Nicol prism made by cutting a birefringent crystal and gluing it back together. The O ray is totally reflected at the surface of the crystal and glue, but the E ray is not. The emerging beam is polarized.

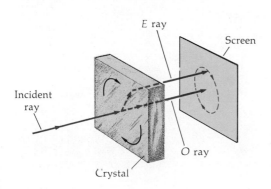

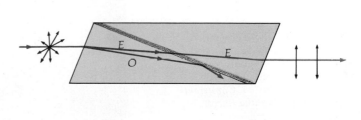

Figure 26-39
Huygens' wavelets for unpolarized light incident normally on a birefringent crystal
whose optic axis is parallel to
the surface. The *O* and *E* rays
propagate in the same direction but with different
speeds. The two beams
emerge from the crystal with
a phase difference which depends on the indexes of refraction for the two rays and
on the thickness of the
crystal.

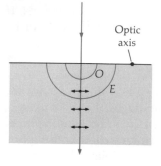

The beams emerge with a phase difference which depends on the thickness of the plate. In a *half-wave plate*, the thickness is such that the number of wavelengths of the ordinary ray is greater than the number of wavelengths of the extraordinary ray by an odd number of half wavelengths. The phase difference is thus 180°. (Since the ordinary ray travels slower, it has a smaller wavelength, $\lambda = v/f$, than the extraordinary ray. A given thickness thus has more ordinary waves in it than extraordinary waves.) If the incident light is polarized in the direction at 45° to the polarizations of the two rays, the emerging light will have its polarization rotated as shown in Figure 26-40. In a *quarter-wave plate* the thickness is such that there is a 90° phase difference between the rays when they emerge. The resulting electric field is of the form $E_x = E_0 \sin \omega t$, $E_y = E_0 \cos \omega t$. The **E** vector thus rotates in a circle, and the wave is circularly polarized.

Many common substances, e.g., cellophane and transparent tape, are birefringent. Interesting and beautiful patterns can be observed by placing such materials between crossed Polaroids. With no substance between the Polaroids, no light is transmitted. However, a sheet of cellophane acts as a half-wave plate for light of a certain color depending on the thickness of the cellophane. Thus it rotates the plane of polarization of the light between the Polaroids, and some light of that color is transmitted.

Figure 26-40
Rotation of the direction of
polarization by a half-wave
plate birefringent crystal.
Light from the Polaroid is linearly polarized with its **E**
vector making an angle of 45°
with the optic axis, which is
parallel to the surface of the
crystal, as in Figure 26-39.
The thickness of the crystal is
such that the *E* and *O* rays
differ in phase by 180° upon
emergence from the crystal.
The effect is to rotate the
direction of polarization by
90°.

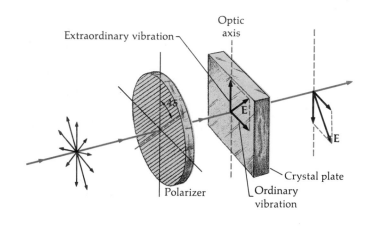

Double image produced by
birefringent calcite crystal.
(*From Richard T. Weidner and
Robert L. Sells,* Elementary
Classical Physics, *vol. II, 2d
ed., fig. 36.17, p. 743. Copyright
© 1973 by Allyn and Bacon,
Inc. Used by permission.*)

The Expanding Universe

Martin Rees
Institute of Astronomy, Cambridge University

Over 400 years has passed since Copernicus argued that the earth must be dethroned from the privileged central position accorded to it by Ptolemy's cosmology, and described the general layout of the solar system in the form accepted today. Only within the twentieth century, however, have we fully recognized the sun's status as just one of the 100 billion (10^{11}) stars in the Milky Way. Harlow Shapley, Jan Oort, and others showed that the Milky Way is a flat disk-shaped system of stars spinning around its axis about once every 2×10^8 years. The sun lies near the edge of the disk, roughly 30,000 light-years from its center. Furthermore, the Milky Way, our own galaxy, is merely one of the 100 billion galaxies on the cosmic scene. If our galaxy were viewed from a distance of several million light-years, it would look rather like the spiral galaxies illustrated (Figures 1 and 2).

Photographs of the sky taken with a large telescope may show many thousands of galaxies, some so far away that the light we receive has taken billions of years on its journey toward us. Although individual stars cannot be resolved, it is possible to obtain spectra of the integrated light from all the stars in a remote galaxy. Such spectra imply that the galaxies contain more or less the same mixture of chemical elements as the sun and that the basic physical constants (e.g., the masses and charges of electrons and protons and Planck's constant) whose values determine the properties of the radiation from stellar surfaces do indeed have uniform values throughout the observed part of the universe.

Such spectra do, however, reveal a significant systematic trend: the spectral lines are all shifted toward the red, i.e., to longer wavelengths. This effect was studied by Edwin Hubble, who interpreted the red shift as a doppler effect and inferred that distant galaxies are receding from us. Moreover, Hubble showed that the recession velocity of a given galaxy is proportional to its distance. The velocity-distance relationship is also found to be the same in all parts of the sky; in other words, the apparent "expansion of the universe" is *isotropic* (the same in all directions). At first sight, one might wonder whether this implies that the earth (or at least our galaxy) is in a privileged central position; but it is easy to convince oneself that this is not so and that a hypothetical astronomer on another galaxy could also have discovered Hubble's law: he would find all the other galaxies apparently receding isotropically from *him*.

If the galaxies had always had the same relative velocities, Hubble's law implies that at some definite time in the past (about 15 billion years ago) they would all have been "squashed together" at one point. Georges Lemáitre and others proposed that the universe actually evolved from a primordial state of high density. The primary aim of observational cosmology since the 1930s has therefore been to test whether there was indeed a "big bang" or whether (as postulated by a rival theory) the universe has existed forever in a steady state, new material and new galaxies being continually created so that its average properties remain unaltered despite the Hubble expansion.

By observing distant objects, one can in principle probe the early history of the universe and perhaps learn whether galaxies were more crowded together or looked systematically different in the past. But ordinary galaxies become invisibly faint, even with the largest optical telescopes, before one reaches the distance where such effects might really be expected to show up. Fortunately, however, galaxies occasionally undergo mysterious violent explosions, in the course of which a small central region flares up until it outshines all the rest of the galaxy by a factor of 100. These rare bright sources of light, called *quasars*, permit optical astronomers to probe much deeper into space—and thus much farther into the past—than is possible using normal galaxies. These "galactic

Figure 1
The *Andromeda* galaxy, Messier 31. (*Courtesy of Mount Wilson and Palomar Observatories.*)

explosions'' are sometimes strong sources of radio-frequency emission and are one of the important cosmic phenomena that can be investigated by the techniques of radio astronomy. Very little is understood about the physical nature of quasars. The problems they pose are among the most important and challenging that confront modern astrophysicists. However, by comparing the number of quasars observed at different distances one can infer that quasars (whatever they may actually be) were very much commoner in early epochs than they are now. This is quite incompatible with the steady-state model: in an evolutionary cosmology, however, there are reasons for expecting that galaxies might indeed have been more prone to violent explosions when they were young.

In 1965, Arno Penzias and Robert Wilson made a discovery (more or less by accident) which is generally regarded as the most compelling evidence for a big bang. They found that the earth is apparently bathed in microwave radiation which comes from all directions but has no apparent source. Subsequent measurements have established that this microwave background seems to have the spectrum of blackbody radiation at a temperature of 2.7 K (see Section 42-1). Most cosmologists interpret this background as a relic of a hot, dense early phase when the universe was opaque and any radiation would have established the blackbody spectrum characteristic of thermal equilibrium. This possibility had in fact been predicted by George Gamow in 1948.

By synthesizing the disparate strands of evidence bearing on the problem most cosmologists have reached a tentative consensus on how the universe might have evolved to its present state. About 15 billion years ago, according to this picture, all the material in the universe — all the stuff of which galaxies are now composed — constituted an exceedingly compressed and hot gas (hotter, in fact, than the center of the sun). The intense radiation in this fireball, though cooled and diluted by the expansion, would still be around, pervading the whole universe: this is the interpretation of the microwave background radiation discovered by Penzias and Wilson — an ''echo,'' as it were, of the ''explosion'' which initiated the universal expansion. The early universe would not have been *completely* smooth and homogeneous (it may even have been turbulent), and the primordial irregularities eventually developed into galaxies. It may, however, have taken a billion years for these fluctuations to condense out into gravitationally bound systems. The most distant quasars are so far away that the radiation now reaching us has been traveling toward us for 80 or even 90 percent of the time elapsed since the initial big bang. That is, it set out toward us when the universe was only 10 or 20 percent of its present age. The universe would have appeared much more violent and active at these early epochs: the cosmic radio background would have been 100 times more intense; and an astronomer then would have found his nearest quasar perhaps 50 times closer to him than in these relatively quiescent times.

The microwave background is a "fossil" from even earlier stages in the expansion. There seems to be another important vestige of the primordial fireball in the present universe: the element helium, which constitutes about a quarter of the mass of most stars, including the sun. The other chemical elements could have been synthesized via the nuclear reactions which provide the power source in the cores of ordinary stars. It is believed that all the carbon, nitrogen, oxygen, and iron on the earth—and indeed in our own bodies—was manufactured in stars which exhausted their energy supply and exploded as *supernovae* before the sun formed. The solar system then condensed from gas contaminated by debris ejected from early generations of stars. One of the major achievements of theoretical astrophysics has been to understand some quantitative details of these processes of *cosmic nucleosynthesis,* e.g., the relative abundances of different elements. But it proved hard to account in this fashion for all the observed helium; it was therefore gratifying both to cosmologists and to experts on nucleosynthesis when the expected composition of material emerging from the big bang was calculated and found to be about 75 percent hydrogen and 25 percent helium. The synthesis of helium would have occurred within only a *few minutes* of the big bang. This epitomizes how the discovery of the microwave background has extended the scope of cosmology by bringing remote eras that were previously entirely speculative within the scope of quantitative scientific discussion. Such discussions do, however, entail extrapolating the locally determined physical laws into domains where one cannot be overwhelmingly sure of their continued validity.

The consistency of most of the available data (limited though it is) with this general picture encourages most cosmologists to adopt the hot big-bang theory as a working hypothesis, which should form the basis for interpreting new observations until some better theory emerges or until some glaring contradiction reveals itself.

Having drawn some tentative conclusions about how the universe has evolved from the primordial fireball to its present state, one is tempted to speculate about its future and its eventual fate. The conventional theories offer two alternative scenarios: either the universe will continue expanding forever, or else the expansion is slowing down to such an extent that it will eventually stop and be followed by a recontraction. This question can in principle be tackled observationally by extending Hubble's work out to very large distances: the velocity-distance relationship obviously refers to the velocity of the galaxies at the time when the light we now receive was emitted, so that any deceleration should be measurable if one observes sufficiently remote objects, In practice, however, this technique has yet to provide reliable results, because the observational difficulties are severe and also because of various uncertain corrections which must be incorporated in the analysis.

But there is another, more indirect, way of trying to determine how much the universal expansion is slowing down. Imagine that a big sphere is shattered by an explosion, the debris flying off in all directions. Each fragment feels the gravitational pull of all the others, and this causes the expansion to decelerate. If the explosion were sufficiently violent, the debris would fly apart forever; but if the fragments were not moving quite so fast, gravity might bind them together strongly enough to bring the expansion to a halt. The material would then collapse again. More or less the same argument probably holds for the universe. One might feel somewhat uneasy about applying a

Figure 2
NGC 4594, spiral galaxy in
Virgo, seen edge on. (*Courtesy
of Hale Observatories.*)

result based on Newton's theory of gravity to the whole universe. But even though one cannot describe the global properties of the universe properly (or the propagation of light) without using a more sophisticated theory such as Einstein's general relativity, the dynamics of the expansion are thought to be the same as in Newton's theory. One can therefore rephrase our earlier question: Does the universe have the escape velocity or not?

In the case of the galaxies (which, for the purposes of this argument, are regarded as fragments of the expanding universe) we know the expansion velocity from Hubble's law. What we do *not* know is the amount of gravitating matter that is causing the deceleration. It is straightforward procedure to calculate how much material would be needed in order to halt the expansion: it works out at about one atom per cubic meter. If the average concentration of material were *below* this so-called critical density, we would expect the universe to continue expanding forever; but if the mean density *exceeded* the critical density, the universe would seem destined eventually to recontract. There are various ways of estimating the masses of individual galaxies. Also, of course, one knows roughly how many galaxies there are in a typical volume of space. These estimates are bedeviled by many uncertainties, but it looks as though the material in galaxies, if spread uniformly through space, would fall short of the critical density by a factor of at least 30. At first sight, one might accept this as evidence that the universe will go on expanding forever; but this inference would really be unjustified, because there may be a lot *more* material embodied in some form *other than* ordinary galaxies. Galaxies are, admittedly, the most prominent features in the sky when we look with an optical telescope, but there is no reason to believe that everything in the universe shines. There may be many objects so cool that they radiate predominantly in the infrared and *absorb* light instead of emitting it—"dead" galaxies, for example, whose stars have all exhausted their nuclear energy, or objects shrouded by opaque clouds of dust. Alternatively, some objects may be so hot that their emission is concentrated in the ultraviolet and x-ray bands of the electromagnetic spectrum. This kind of astronomical work was not feasible until quite recently because since the air is very opaque in these wavebands, one must send equipment above the earth's atmosphere to make observations.

Space astronomy is still in its pioneering stage, and one would not be surprised if it were to disclose many unsuspected sorts of objects. So our present inventory of the contents of the universe may well prove exceedingly biased and incomplete. For instance, there may be a large amount of diffuse gas *between* the galaxies. (There is, after all, no reason to expect all, or even most, of the primordial hydrogen and helium to have condensed into galaxies.) This intergalactic gas may in fact be responsible for emitting most of the cosmic x-rays reaching us from beyond our own galaxy.

At present, therefore, there is no definite reason for believing that there is enough gravitating material in the universe to bring the expansion to a halt. But it remains quite conceivable that the universe contains a great deal of stuff even more elusive than intergalactic gas. For instance, the critical density could be provided by neutrinos, gravitational waves, or black holes without there being the slightest chance of detection by present techniques.

We therefore cannot rule out the possibility that eventually the expansion will stop and turn into a contraction. Distant galaxies, displaying *blue* shifts instead of red shifts, would eventually collide and merge with one another. As the contraction proceeded further, the sky would become brighter and brighter; and eventually all the stars would explode (because the sky would be hotter than the fuel in their interiors!), everything in the universe being finally engulfed in a fireball like that from which, according to most cosmologists, it emerged: the ultimate, universal, gravitational collapse. But there is no immediate cause for concern; our breathing space before this cataclysm should be billions of years, at the very least. If the universe did *not* have the critical density, it would continue to expand forever. Each galaxy would fade to a dull glow as its constituent stars exhausted their available energy and the supply of gas from which new bright stars can condense was inexorably depleted.

An important aim of current cosmological research is to decide between these two contrasting fates for the universe. Also, one hopes eventually to fill in more details of the big-bang picture and to understand how the primordial material aggregates into galaxies; why galaxies have the sizes and shapes that are observed; how they evolve; and why they sometimes explode. But one should remain aware that present data in cosmology are still limited, ambiguous, and fragmentary; and they all depend on complex instruments stretched right to the limits of their sensitivity and performance. Therefore, even if this general picture seems consistent with what is known at the moment, it would be rash to bet *too* heavily on its being correct. Moreover, self-consistency is, of course, no guarantee of truth in itself. Many more observations are essential before we can be sure whether we really know, even in outline, the basic overall structure of the physical universe or whether our current ideas will eventually be discarded and superseded as surely as Ptolemy's epicycles were.

Review

A. Identify the contributions the following people made to the understanding of light:

Young   Fermat
Fresnel   Brewster
Foucault   Roemer
Maxwell   Fizeau
Hertz

B. Define, explain, or otherwise identify:

Photoelectric effect, 606   Rayleigh criterion for resolution, 618
Photon, 607   Lloyd's mirror, 618
Electric-dipole radiation, 610   Newton's rings, 621
Fermat's principle, 612   Brewster's law, 626
Dispersion of light, 616

C. Questions

1. Discuss two different methods of measuring the speed of light.

2. When is the propagation of wave motion like that of particle motion?

3. What are the characteristics of electric-dipole radiation?

4. What four phenomena produce polarized light?

5. Why are diffraction and interference effects difficult to observe for light?

6. What proof is there that an 180° phase change occurs when light is reflected from glass?

7. Why do we see colored bands from an oil film?

8. Why can you see objects beneath a water surface better with Polaroid sunglasses?

Exercises

Section 26-1, Waves or Particles?

1. Show that if the wavelength of an electromagnetic wave is given in nanometers (1 nm = 10^{-9} m), the photon energy in electron volts is given by $E = (1240 \text{ eV-nm})/\lambda$.

2. Use the result of Exercise 1 to calculate the range of photon energies in the visible spectrum of wavelengths from about 400 to 700 nm.

3. What is the energy of an x-ray photon whose wavelength is 0.1 nm?

Section 26-2, Electromagnetic Waves

4. What is the frequency of a 3-cm microwave?

5. What is the frequency of an x-ray of wavelength 0.1 nm?

6. Find the wavelength for a typical AM radio frequency and for a typical FM radio frequency.

7. What is the wavelength of a photon whose energy is 1 eV? In what part of the electromagnetic spectrum is it?

8. A radiating electric dipole lies along the z axis. Let I_1 be the intensity of the radiation at a distance $r = 10$ m and at angle $\theta = 90°$. Find the intensity (in terms of I_1) at (a) $r = 30$ m, $\theta = 90°$; (b) $r = 10$ m, $\theta = 45°$; (c) $r = 20$ m, $\theta = 30°$.

9. (*a*) For the situation described in Exercise 8, at what angle is the intensity at $r = 5$ m also equal to I_1? (*b*) At what distance is the intensity equal to I_1 at $\theta = 45°$?

Section 26-3, The Speed of Light

10. The distance light travels in 1 year is called a light-year. (*a*) Calculate the number of miles in a light-year. (The number of seconds in a year is about 3.16×10^7. (*b*) Find a conversion factor from light-years to kilometers.

11. The spiral galaxy in the Andromeda constellation is about 2×10^{19} km away from us. How long does it take light from that galaxy to reach us?

12. How long does it take light to travel from the sun to the earth, a distance of about 93×10^6 mi? What is the distance to the sun in light-minutes?

13. On a rocket sent to Mars to take pictures, the camera is triggered by radio waves which (like all electromagnetic waves) travel with the speed of light. What is the time delay between sending and receiving the signal from the earth to Mars? (Take the distance to Mars to be 6×10^7 mi.)

Section 26-4, Reflection

14. Calculate the fraction of light energy reflected from water at normal incidence. ($n = 1.33$ for water.)

15. Light is incident normally on a slab of glass of index of refraction $n = 1.5$. Reflection occurs at both surfaces of the slab. About what percentage of the incident light energy is transmitted by the slab?

16. A physics student playing pocket billiards wishes to strike his cue ball so that it hits a cushion and then hits the eight ball squarely. He chooses several points on the cushion and for each point he measures the distance from it to the cue ball and to the eight ball. He aims at the point for which the sum of these distances is least. Will his cue ball hit the eight ball? How is this method related to Fermat's principle?

Section 26-5, Refraction

17. A beam of light strikes a plane glass surface at an angle of incidence of 45°. The index of refraction of the glass varies with wavelength according to the graph in Figure 26-13. How much smaller is the angle of refraction for blue light of wavelength 400 nm than for red light of wavelength 700 nm?

18. Find the speed of light in water ($n = 1.33$) and in glass ($n = 1.5$).

19. A beam of monochromatic red light of wavelength 700 nm in air travels in water. What is the wavelength in water? Does a swimmer under water observe the same color or a different color for this light?

20. The critical angle for total internal reflection in diamond is about 24°. What is the index of refraction of diamond?

Section 26-6, Diffraction and Resolution

21. In a lecture demonstration of diffraction, a laser beam of wavelength 700 nm passes through a vertical slit 0.5 mm wide and hits a screen 6 m away. Find the horizontal length of the principal diffraction maximum on the screen; i.e., find the distance between the first minimum on the left and the first minimum on the right of the central maximum.

22. Two sources of wavelength 700 nm are separated by a horizontal distance x. They are 5 m from a vertical slit of width 0.5 mm. What is the least value of x

permitting the diffraction pattern of the sources to be resolved by the Rayleigh criterion?

23. Light of wavelength 700 nm is incident on a pinhole of diameter 0.1 mm. (a) What is the angle between the central maximum and the first diffraction minimum for Fraunhofer diffraction? (b) What is the distance between the central maximum and the first diffraction minimum on a screen 8 m away?

24. Two light sources ($\lambda = 700$ nm) are 10 m away from the pinhole in Exercise 23. How far apart must the sources be for their diffraction patterns to be resolved by the Rayleigh criterion?

25. (a) How far apart must two objects be on the moon to be resolved by the eye? Take the diameter of the pupil of the eye to be 5.0 mm, $\lambda = 600$ nm, and $d = 385,000$ km for the distance to the moon. (b) How far apart must the objects on the moon be to be resolved by a telescope which has a 5.00-m-diameter mirror?

26. Estimate the maximum distance (in miles) that you can be from a car and still resolve the headlights, assuming the diameter of your pupils to be 5.0 mm.

Section 26-7, Interference

27. A long, narrow, horizontal slit lies 1 mm above a plane mirror. The interference pattern produced by the slit and its image is viewed on a screen a distance 1 m from the slit. The wavelength of the light is 600 nm. (a) Find the distance above the mirror of the first maximum. (b) How many dark bands per centimeter are seen on the screen?

28. A two-slit Fraunhofer diffraction-interference pattern is observed with light of wavelength 500 nm. The slits have a separation $d = 0.1$ mm and width a. Find the width a if the fifth-interference maximum is at the same angle as the first diffraction minimum. For this case, how many bright fringes will be seen in the central diffraction maximum?

29. A two-slit Fraunhofer diffraction-interference pattern is observed with light of wavelength 700 nm. The slits have a width $a = 0.01$ mm and are separated by $d = 0.2$ mm. How many bright fringes will be seen in the central diffraction maximum?

30. Light of wavelength 500 nm is incident normally on a film of water 10^{-4} cm thick. The index of refraction of water is 1.33. (a) What is the wavelength of the light in the water? (b) How many wavelengths are contained in the distance $2t$, where t is the film thickness? (c) What is the phase difference between the wave reflected from the top of the film and the one reflected from the bottom after it has traveled this distance?

31. A wedge-shaped film of air is made by placing a small slip of paper between the edges of two flat pieces of glass. Light of wavelength 700 nm is incident normally on the glass plates and interference bands are observed by reflection. (a) Is the first band near the point of contact of the plates dark or bright? Why? (b) There are five dark bands per centimeter. What is the angle of the wedge?

32. A loop of wire is dipped in soapy water and held so that the soap film is vertical. When viewed by reflection with white light, the top of the film appears black. Explain why. Below the black region are colored bands. Is the first band red or blue?

33. A thin layer of a transparent material of index of refraction 1.30 is used as a nonreflective coating on the surface of glass of index of refraction 1.50. What should the thickness be for the film to be nonreflecting for light of wavelength 600 nm (in vacuum)?

Section 26-8, Diffraction Gratings

34. A diffraction grating with 2000 slits per centimeter is used to measure the wavelengths emitted by hydrogen gas. At what angles θ would you expect to find the two blue lines of wavelength 434 and 410 nm?

35. With the grating used in Exercise 34, two other lines in the hydrogen spectrum are found in first order at angles $\theta_1 = 9.72 \times 10^{-2}$ rad and $\theta_2 = 1.32 \times 10^{-1}$ rad. Find the wavelengths of these lines.

36. Repeat Exercise 34 for a grating with 15,000 lines per centimeter.

37. A grating of 2000 slits per centimeter is used to analyze the spectrum of mercury. (a) Find the angular deviation in first order of the two lines of wavelength 579.0 and 577.0 nm. (b) How wide must the beam be on the grating for these lines to be resolved?

38. What is the longest wavelength that can be observed in fifth order using a grating with 4000 slits per centimeter?

Section 26-9, Polarization

39. Two Polaroid sheets have their transmission directions crossed so that no light gets through. A third sheet is inserted between the two so that its transmission direction makes an angle with the first sheet. Unpolarized light of intensity I_0 is incident on the first sheet. Find the intensity transmitted through all three sheets if (a) $\theta = 45°$; (b) $\theta = 30°$.

40. Two Polaroid sheets are inserted between two other Polaroid sheets which have their transmission directions crossed, so that the angle between each successive pair of sheets is 30°. Find the transmitted intensity if the original light is unpolarized with intensity I_0.

41. The polarizing angle for a certain substance is 60°. (a) What is the angle of refraction of light incident at this angle? (b) What is the index of refraction of this substance?

42. The critical angle for total internal reflection for a substance is 45°. What is the polarizing angle for this substance?

43. What is the polarizing angle for glass with $n = 1.5$?

44. A beam of linearly polarized light strikes a calcite crystal such that its electric vector makes an angle of 30° with the optic axis (Figure 26-41). What is the ratio of the intensities of the ordinary and extraordinary rays?

45. Light of wavelength λ in air is incident on a slab of calcite so that the ordinary and extraordinary rays travel in the same direction as in Figure 26-39. Show that the phase difference between these rays after traversing a thickness t is

$$\delta = \frac{2\pi}{\lambda}(n_O - n_E)t$$

Figure 26-41
Exercise 44.

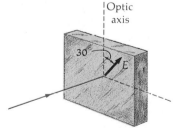

Problems

1. Show that the transmitted intensity through a glass slab of index of refraction n for normally incident light is approximately given by

$$I_T = I_0 \left[\frac{4n}{(n+1)^2}\right]^2$$

Actually a bit more is transmitted because this ignores multiple reflections inside the glass.

2. Light is incident normally upon one face of a prism of glass whose index of

refraction is n (Figure 26-42). The light is totally reflected at the right side. (a) What is the minimum value n can have? (b) When this prism is immersed in a liquid whose index of refraction is 1.15, there is still total reflection, but in water, whose index is 1.33, there no longer is total reflection. Use this information to put bounds upon n.

Figure 26-42
Problem 2.

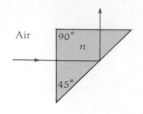

3. Light of wavelength λ is diffracted through a single slit of width a, and the resulting pattern is viewed on a screen a long distance D away from the slit. (a) Show that the width of the principal maximum on the screen is approximately given by $2D\lambda/a$. (b) If a slit of width $2D\lambda/a$ is cut in the screen and illuminated, show that the width of its principal maximum at the same distance D, that is, back on the slit plane, is a to the same approximation. This reciprocity is no accident and is in fact exact. It is studied in detail in advanced courses in optics.

4. Two slits of width a are separated by a distance d. Show that the intensity for a Fraunhofer pattern is given by

$$I = I_0 \left(\cos \frac{\delta}{2} \right)^2 \left(\frac{\sin \phi/2}{\phi/2} \right)^2$$

where $\delta = (2\pi d/\lambda) \sin \theta$ and $\phi = (2\pi a/\lambda) \sin \theta$.

5. For a diffraction grating one is interested not only in resolving power R, which is the ability of the grating to separate two close wavelengths, but also the dispersion D of the grating. This is defined by $D = \Delta\theta_m/\Delta\lambda$ in mth order. (a) Show that D can be written

$$D = \frac{m}{\sqrt{d^2 - m^2\lambda^2}}$$

where d is the slit spacing. (b) If a diffraction grating with 2000 slits per centimeter is to resolve the two sodium yellow lines (wavelengths 589.0 and 589.6 nm) in second order, how many slits must be illuminated by the beam? (c) What would the separation be between these resolved yellow lines if the pattern were viewed on a screen 4 m from the grating?

6. The ceiling of your lecture hall is probably covered with acoustic tile which has small holes separated by about 6.0 mm. (a) Using $\lambda = 500$ nm, how far could you be from this tile and still resolve these holes? The diameter of the pupil of your eye is about 5.0 mm. (b) Could you "see" these holes better with red or blue light?

7. The telescope on Mount Palomar has a diameter of about 5.0 m. Assuming "ideal" sky conditions, the resolution would be diffraction-limited. Suppose a double star were 4 light-years away. What would the stellar separation have to be for their images to be resolved?

8. Suppose that the central diffraction maximum for two slits contained 17 interference fringes for some wavelength. How many interference fringes would you expect in the first secondary diffraction maximum?

9. A Newton's-ring apparatus consists of a glass lens of radius of curvature R which rests upon a flat glass plate, as shown in Figure 26-43. Usually both pieces of glass have the same index of refraction n, and so the thin film is air of variable thickness. The pattern is viewed by reflected light. (a) Show that for a thickness t the condition for a bright (constructive) interference fringe is $t = \frac{1}{2}(m + \frac{1}{2})\lambda$, $m = 0, 1, 2, \ldots$ (b) Show that so long as $t/R \ll 1$, the radius r of a bright circular fringe is given by $r = \sqrt{(m + \frac{1}{2})\lambda R}$, $m = 0, 1, 2, \ldots$ (c) How would the transmitted pattern look in comparison with the reflected one? (d) Use $R = 10$ m and a diameter of 4.0 cm for the lens. How many bright fringes would you see if the apparatus was illuminated by yellow sodium light ($\lambda \approx 590$ nm) and viewed by reflection? (e) What would be the diameter of the sixth bright fringe? (f) If the glass used in the apparatus has index of refrac-

Figure 26-43
Problem 9.

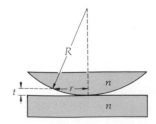

tion $n = 1.5$ and water is placed between the two pieces of glass, what change will take place for the bright fringes?

10. The diameter of fine wires can be very accurately measured by interference patterns. Two accurately flat pieces of glass of length D are arranged with the wire as shown in Figure 26-44. The setup is illuminated by monochromatic light, and the resulting interference fringes are detected. Suppose $D = 20$ cm and one uses yellow sodium light for illumination ($\lambda \approx 590$ nm). If there are 19 bright fringes seen along this 20-cm distance, what are the limits on the diameter of the wire? *Hint:* The nineteenth might not be right at the end, but you do not see 20 fringes.

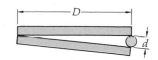

Figure 26-44
Problem 10.

11. A thin film of index of refraction of 1.5 is surrounded by air. It is illuminated normally by white light and viewed by reflection. Analysis of the resulting reflected light shows that the wavelengths 360, 450, and 600 nm are the only missing wavelengths near the visible portion of the spectrum. That is, for these wavelengths there is destructive interference. (a) What is the thickness of the film? (b) What visible wavelengths would be extra bright in the reflected interference pattern? (c) If this film is supported on glass whose index of refraction is 1.6, what wavelengths in the visible spectrum will be missing from the reflected light?

12. A camera lens is made of glass whose index of refraction is 1.6. This lens is coated with a magnesium fluoride film ($n = 1.38$) to enhance its light transmission. This film is to produce a zero reflection for light of wavelength 540 nm. Treat the lens surface as a flat plane and the film as a uniformly thick flat film. (a) How thick must the film be to accomplish its objective in first order? (b) Would there be destructive interference for any other visible wavelengths? (c) By what factor would the reflection be reduced by this film for 400 and 700 nm? Neglect the variation in the reflected light amplitudes from the two surfaces.

13. A very important device which utilizes interference phenomena is Michelson's interferometer, shown schematically in Figure 26-45. Light from an extended source strikes a diagonal mirror which reflects 50 percent of the incident light and transmits 50 percent. The two beams strike the mirrors 1 and 2, which are made precisely perpendicular to one another. The light finally is viewed after the beams combine. The interference is that of a thick slab of air of thickness $d_1 - d_2$. Essentially circular interference fringes are seen because of slight differences in the angle of incidence from the extended source. One of the mirrors is movable to change the effective thickness of this air film. (a) If mirror 1 moves so that the first bright fringe replaces the bright center spot, show that the path length has changed by 1 wavelength and that mirror 1 has moved $\frac{1}{2}$ wavelength. (b) Suppose that a hollow cell of length l with glass windows at both ends is introduced into each arm of the interferometer. The mirrors are adjusted to give a bright fringe at the center. Now the air is pumped out of one of these cells, and there is a shift of N fringes. The illuminating light has wavelength λ. Show that the index of refraction of air is given by $n_{air} = 1 + N\lambda/2l$. (c) If, as in (b), $l = 10.00$ cm, $\lambda = 600$ nm, and $N = 100$, what is n_{air}? (d) Why are identical cells placed in both arms, or is this necessary?

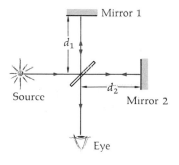

Figure 26-45
Problem 13.

14. A thin film of index of refraction $n = 1.5$ for light of wavelength 600 nm is inserted in one arm of a Michelson interferometer (see Problem 13). (a) If a fringe shift of 12 fringes occurs, what is the thickness of this film? (b) If the illuminating light is changed to 400 nm, the fringe shift as this film is inserted becomes 24 fringes. What is the index of refraction of this film to light of wavelength 400 nm?

15. Show that if the ordinary and extraordinary waves which were separated by a crystal were combined, they could not produce interference effects. Do this by considering the general problem of two waves of equal amplitude with perpendicular polarizations and show that they cannot interfere by showing

that their combined intensity averaged over one or many cycles is independent of their phase difference.

16. A prism is made of calcite cut so that its upper face contains the optic axis. The incident ray is unpolarized and normally incident upon the surface (Figure 26-46). (*a*) What must the angle of the prism be for the ordinary ray to be totally reflected internally? (*b*) Show that the extraordinary ray will not then be internally reflected, so that the emerging light will be plane-polarized.

17. The indexes for ordinary and extraordinary waves given for calcite are for green light ($\lambda = 540$ nm). (*a*) What would be the minimum thickness for a quarter-wave plate? (*b*) What would be the minimum thickness for a half-wave plate? (*c*) Would you expect these to be quarter- and half-wave plates for, say, red light? Why or why not?

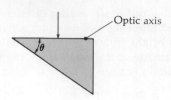

Figure 26-46
Problem 16.

CHAPTER 27 Geometric Optics

The wavelength of light is often very small compared with the size of any obstacles or apertures it encounters, and diffraction effects can be neglected. The study of such situations, for which the ray approximation is valid and light propagates in straight lines, is known as *geometric optics*. In this chapter, we shall apply the laws of reflection and refraction discussed in Chapter 26 to study the formation of images by mirrors and lenses.

27-1 Plane Mirrors

Figure 27-1 shows a narrow bundle of light rays from a point source P reflected from a plane mirror. After reflection, the rays diverge exactly as if they came from a point P' behind the plane of the mirror. The point P' is called the *image* of the object P. When these rays enter the eye, they cannot be distinguished from rays diverging from a source at P' with no mirror. The image is called a *virtual image* because the light

Virtual image

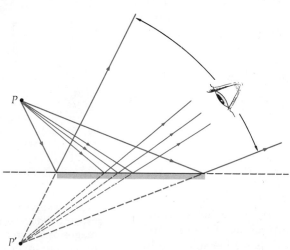

Figure 27-1
Image formed by a plane mirror. The rays from point P, which strike the mirror and enter the eye, appear to come from the image point P' behind the mirror. The image can be seen by the eye anywhere in the region indicated.

does not actually emanate from the image but only appears to. Geometric construction using the law of reflection shows that the image
point lies on the line through the object perpendicular to the plane of
the mirror and at a distance behind the plane equal to the distance
from the plane to the object. The image can be seen by an eye anywhere in the region indicated, in which a line from the image to the
eye passes through the mirror. From the figure we note that the object
need not be directly in front of the mirror. An image can be seen as
long as the object is above the plane of the mirror.

Figure 27-2 shows an image of an extended object formed by a
plane mirror. The size of the image is the same as that of the object.

Figure 27-3 illustrates the formation of multiple images by two

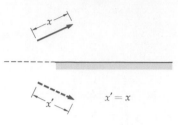

Figure 27-2
Image of an extended object
in a plane mirror. The image
is the same size as the object.

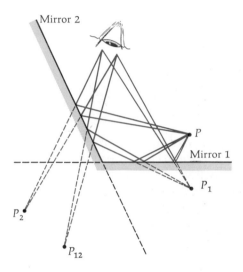

Figure 27-3
Images formed by two plane
mirrors. P_1 is the image of the
object P in mirror 1, and P_2 is
the image of the object in
mirror 2. Point P_{12} is the
image of P_1 in mirror 2 seen
when light rays from the object reflect first from mirror 1
and then from mirror 2, as
shown. The image P_2 does
not have an image in mirror 1
because it is behind that
mirror.

plane mirrors making an angle with each other. Light reflected from
mirror 1 strikes mirror 2 just as if it came from the image point P_1. The
image P_1 is called the *object point* for mirror 2. Its image is at point P_{12}.
This image will be formed whenever the image point P_1 is in front of
the plane of mirror 2. The image at point P_2 is due to rays from the object which reflect directly from mirror 2. Since P_2 is behind the plane
of mirror 1, it cannot serve as an object point for a further image in
mirror 1. The number of multiple images formed by two mirrors
depends on the angle between the mirrors and the position of the object.

Question

1. How tall must a mirror be for a standing person to see his entire
reflection in it?

27-2   Spherical Mirrors

Figure 27-4 shows a bundle of rays from a point on the axis of a concave spherical mirror reflecting from the mirror and converging at
point P'. The rays then diverge from this point just as if there were an
object at that point. This image is called a *real image* because the light
actually does emanate from the image point. It can be seen by an eye

The lake serves as a plane
mirror producing virtual
images of the trees. The
images are inverted and the
same size as the trees.

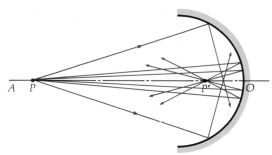

Figure 27-4
Rays from a point object P on
the axis AO of a concave
spherical mirror form an
image at P' if the rays strike
the mirror near the axis. Nonparaxial rays striking the
mirror at points far from O
are not reflected through the
image point P'.

placed to the left of the image looking into the mirror. It could also be
observed on a ground-glass viewing screen or photographic film
placed at the image point. A virtual image cannot be observed on a
screen at the image point because there is no light there. Despite this
distinction between real and virtual images, the light rays diverging
from a real image and those appearing to diverge from a virtual image
are identical, so that no distinction is made by the eye when viewing
either a real or a virtual image.

From Figure 27-4 we see that only rays which strike the mirror at
points near the axis AO are reflected through the image point. Such
rays are called *paraxial rays*. Because other rays converge to different
points near the image point, the image appears blurred, an effect
called *spherical aberration*. The image can be sharpened by reducing
the size of the mirror so that nonparaxial rays do not strike the mirror.
Although the image is then sharper, its brightness is reduced because
less light intensity is reflected.

Spherical aberration

Using the law of reflection and elementary geometry, we can relate
the image distance s' to the object distance s and the radius of curvature r. The geometry is shown in Figure 27-5. The result is

$$\frac{1}{s} + \frac{1}{s'} = \frac{2}{r} \qquad\qquad 27\text{-}1$$

The derivation of this equation assumes that angles made by the incident and reflected rays with the axis are small, an assumption equivalent to that of paraxial rays.

When the object distance is much greater than the radius of curvature of the mirror, the term $1/s$ in Equation 27-1 can be neglected,
resulting in $s' = \frac{1}{2}r$ for the image distance. This distance is called
the *focal length f* of the mirror, and the image point is called the *focal
point F:*

Focal length

$$f = \tfrac{1}{2}r \qquad\qquad 27\text{-}2$$

In terms of the focal length f the mirror equation is

$$\frac{1}{s} + \frac{1}{s'} = \frac{1}{f} \qquad\qquad 27\text{-}3$$

Figure 27-5
Geometry for calculating the
image distance s' from the
object distance s and the
radius of curvature r. The
angle β is an exterior angle to
the triangle PAC and is therefore equal to the sum of α and
θ; $\beta = \alpha + \theta$. Similarly, from
triangle PAP', we have
$\gamma = \alpha + 2\theta$. Eliminating θ
from these equations gives
$2\beta = \gamma + \alpha$. When these
angles are small, they are related to the image distance,
object distance, and radius of
curvature by $\alpha \approx l/s$, $\beta \approx l/r$,
and $\gamma \approx l/s'$. Thus

$$\frac{2l}{r} - \frac{l}{s'} + \frac{l}{s} \quad \text{or} \quad \frac{1}{s} + \frac{1}{s'} - \frac{2}{r}$$

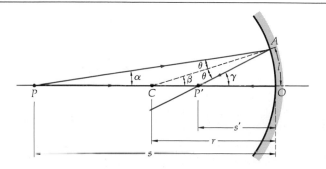

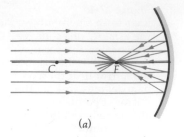

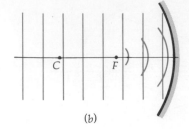

Figure 27-6
(*a*) Parallel rays strike a concave mirror and are reflected through the focal point at a distance $\frac{1}{2}r$. (*b*) The incoming wavefronts are plane waves; upon reflection they become spherical waves, which converge at the focal point.

(*a*)   (*b*)

The focal point is the point at which parallel rays corresponding to plane waves from infinity are focused, as illustrated in Figure 27-6. (Again, only paraxial rays are focused at a single point.) Figure 27-7 shows rays from a point source at the focal point which strike the mirror and are reflected parallel to the axis. This illustrates a property of waves called *reversibility*. If we reverse the direction of a reflected ray, the law of reflection assures that the reflected ray will be along the original incoming ray but in the opposite direction. Reversibility holds also for refracted rays. If we have a real image of a source formed by a reflecting or refracting surface, we can place a source at the image point and the new image will be formed at the position of the original source.

A useful method of locating images is by geometric construction of a ray diagram, as illustrated in Figure 27-8, where the object is a human figure perpendicular to the axis a distance *s* from the mirror. By a judicious choice of rays from the head of the figure we can quickly locate the image. A ray from the head parallel to the axis is reflected through the focal point a distance $\frac{1}{2}r$ from the mirror, as shown. Another ray, through the center of curvature of the mirror, strikes the mirror perpendicular to the surface and is reflected back along its original path. The intersection of these two rays locates the image point of the head. This can be checked by a third ray through the focal point, which is reflected back parallel to the axis.

We see from the figure that the image is inverted and is not the same size as the object. The magnification of the optical system (the spherical mirror in this case) is defined to be the ratio of the image size to the object size. Comparison of the triangles in Figure 27-9 shows that the magnification is just the ratio of the distances *s′* and *s*.

Figure 27-7
Illustration of reversibility. Rays diverging from a point source at the focal point of a concave mirror are reflected from the mirror as parallel rays. The rays are the same as in Figure 27-6*a* but in the reverse direction.

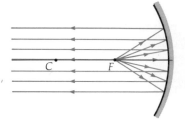

Ray diagram for spherical mirrors

Figure 27-9
Geometry for finding the magnification of a spherical mirror. From the upper triangle, $\tan \theta = y/s$, and from the lower triangle, $\tan \theta = -y'/s'$, where the negative sign is introduced because y' is negative. Then the magnification is $m = y'/y = -s'/s$.

Figure 27-8
Ray diagram for location of image by geometric construction.

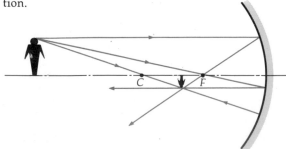

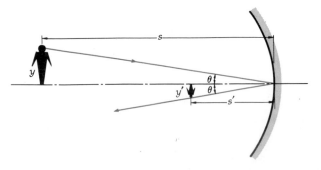

Figure 27-10
When the object is between the focal point and the mirror, the image is virtual and behind the mirror. Here it is located by a ray along the line from the focal point through the head of the figure which is reflected parallel to the axis and a ray from the center of curvature through the head which is reflected back on itself. These rays diverge from a point behind the mirror. A third ray (not shown) could be drawn from the head parallel to the axis. It is reflected through the focal point, and its extension behind the mirror intersects the other rays at the image point.

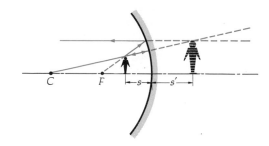

Figure 27-11
Ray diagram for a convex mirror. The ray parallel to the axis is reflected as if it came from the focal point to the right of the mirror, and the ray toward the center of curvature is reflected back on itself. A third ray (not shown) could be drawn toward the focal point. It would be reflected parallel to the axis, and its extension behind the mirror would intersect the other two rays at the image.

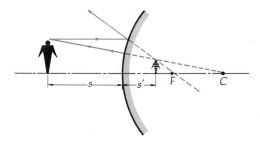

When the object is between the mirror and its focal point, the rays reflected from the mirror do not converge but appear to diverge from a point behind the mirror, as illustrated in Figure 27-10. In this case the image is virtual and erect. If s is less than $\frac{1}{2}r$ in Equation 27-1, the image distance s' turns out to be negative. We can apply Equations 27-1 and 27-3 to this case and to convex mirrors if we adapt a convenient sign convention. Whether the mirror is convex or concave, real images can be formed only on the same side of the mirror as the object and virtual images are formed on the opposite side, where there are no actual light rays. Distances to points on the real side are taken to be positive; distances to points on the virtual side are taken to be negative. Thus for the concave mirror, s and r are positive, and s' is positive or negative depending on whether the image is real or virtual. For a convex mirror (Figure 27-11) the center of curvature is on the virtual side, and so r is taken to be negative. The focal length is also negative. For either case, Equation 27-1 gives the image distance s' in terms of the object distance and radius of curvature. The lateral magnification of the image is given by

Sign convention

$$m = \frac{y'}{y} = -\frac{s'}{s} \qquad 27\text{-}4$$

A negative magnification, which occurs when both s and s' are positive, indicates that the image is inverted. Although these equations with this sign convention are relatively easy to use, practical work in optics often requires only knowledge of whether the image is real or virtual and an approximate knowledge of its location. This can most easily be obtained by constructing a ray diagram. A similar method for locating images produced by a lens and various lens combinations (Section 27-4) is often of more practical use than the equations relating the image position to the object position.

Fundamental Photographs
The Granger Collection

Convex mirror resting on paper with equally spaced parallel stripes. The image of each point in front of the mirror is virtual and behind the mirror. Note the reduction in size and distortion in the shape of the image and the large number of stripes that are seen. Convex mirrors are useful for wide-angle viewing when the distortion in shape is not important.

Example 27-1 An object 2 cm high is 10 cm from a convex mirror with a radius of curvature of 10 cm. Locate the image and find its height.

Since the center of curvature of a convex mirror is on the virtual-image side of the mirror, the focal length is negative:

$$f = \tfrac{1}{2}r = \tfrac{1}{2}(-10 \text{ cm}) = -5 \text{ cm}$$

Using Equation 27-3 to find the image distance gives

$$\frac{1}{10 \text{ cm}} + \frac{1}{s'} = \frac{1}{f} = -\frac{1}{5 \text{ cm}}$$

$$\frac{1}{s'} = -\frac{2}{10 \text{ cm}} - \frac{1}{10 \text{ cm}} = -\frac{3}{10 \text{ cm}}$$

$$s' = -\tfrac{10}{3} \text{ cm} = -3\tfrac{1}{3} \text{ cm}$$

The image distance is negative, indicating a virtual image behind the mirror. The magnification is

$$m = -\frac{s'}{s} = -\frac{-3\tfrac{1}{3}}{10} = +\frac{1}{3}$$

The image is erect and one-third the size of the object. Its size is

$$y' = my = \tfrac{1}{3}(2 \text{ cm}) = \tfrac{2}{3} \text{ cm}$$

The ray diagram for this example is similar to Figure 27-11.

Questions

2. Under what circumstances will a concave mirror produce an erect image? A virtual image? An image smaller than the object? An image larger than the object?

3. Answer Question 2 for a convex mirror.

27-3 Images Formed by Refraction

Figure 27-12 illustrates the formation of an image by refraction from a spherical surface separating two media with indexes of refraction n_1 and n_2. Again, only paraxial rays converge to one point. We can derive an equation relating the image distance to the object distance, the radius of curvature, and the indexes of refraction by applying Snell's law of refraction to these rays and using the small-angle approximation. This derivation is given in Figure 27-13. The resulting equation is

$$\frac{n_1}{s} + \frac{n_2}{s'} = \frac{n_2 - n_1}{r} \qquad \text{27-5}$$

Image distance related to object distance for refraction

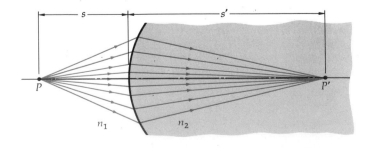

Figure 27-12
Image formed by refraction at a spherical surface.

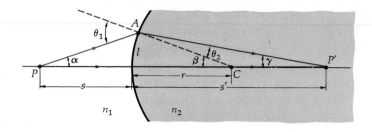

Figure 27-13
Geometry for the derivation
of Equation 27-5. The angles
θ_1 and θ_2 are related by Snell's
law, which for small angles
can be written $n_1\theta_1 = n_2\theta_2$.
From the triangle ACP' we
obtain

$$\beta = \theta_2 + \gamma = \frac{n_1}{n_2}\,\theta_1 + \gamma \qquad \text{or}$$

$$n_1\theta_1 = n_2\beta - n_2\gamma$$

From triangle PAC we obtain
$\theta_1 = \alpha + \beta$. Eliminating θ_1
gives $n_1\alpha + n_2\gamma = (n_2 - n_1)\beta$.
Equation 27-5 follows from
substituting the small-angle
approximations $\alpha \approx l/s$,
$\gamma \approx l/s'$, and $\beta \approx l/r$.

We can use the same sign convention for this equation, but we must note that for refraction real images are formed to the right of the surface (if the object is to the left) and virtual images to the left. Thus s' and r are taken to be positive if the image and center of curvature lie to the right of the surface.

We can apply Equation 27-5 to find the apparent depth of an object under water when viewed from directly overhead. For this case the surface is a plane surface, the radius of curvature is infinite, and the image and object distances are related by

$$\frac{n_1}{s} + \frac{n_2}{s'} = 0$$

$$s' = -\frac{n_2}{n_1}\,s \qquad\qquad 27\text{-}6 \qquad \textit{Apparent depth}$$

The negative sign indicates that the image is virtual and on the same side of the refracting surface as the object, as shown in the ray diagram in Figure 27-14.

Figure 27-14
The image P' appears at the
depth s', which is less than
the actual depth s of the object P. When viewed in air
from directly overhead, the
apparent depth equals the
real depth divided by the
index of refraction of water.

Example 27-2 Find the apparent depth of a fish resting 1 m below the surface of water which has an index of refraction $n = \frac{4}{3}$.

Using $n_1 = \frac{4}{3}$ and $n_2 = 1$ in Equation 27-6, we obtain

$$s' = -\tfrac{3}{4}\,(1 \text{ m}) = -75 \text{ cm}$$

The apparent depth is three-fourths the actual depth. Note that this result holds only when the object is viewed from directly overhead so that the rays are paraxial.

27-4   Lenses

The most important application of Equation 27-5 is in the finding of the position of the image formed by a lens. We do this by considering the refraction at each surface separately. We shall consider a glass lens of index of refraction n with air on both sides. Let the radii of curvature of the surfaces of the lens be r_1 and r_2. If an object is at a distance s from the first surface, application of Equation 27-1 gives for the distance of the image due to the refraction at the first surface,

$$\frac{1}{s} + \frac{n}{s'_1} = \frac{n-1}{r_1} \qquad\qquad 27\text{-}7$$

This image is usually not formed (unless the lens is extremely thick) because the light is again refracted at the second surface. Consider, for example, that the image distance s'_1 is negative, indicating a virtual image to the left of the first surface. The light leaving this surface

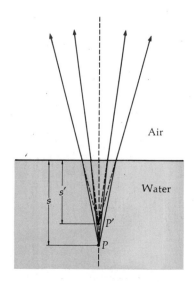

strikes the second surface *as if* it came from an object at the image position. If the thickness of the lens is t, the distance of this point from the second surface is $-s'_1 + t$. (For this case $-s'_1$ is positive because s'_1 is negative.) We can find the final image position due to both refractions by using this distance for the object distance for the second surface. If s'_1 turns out to be positive, indicating a real image to the right of the first surface, the image point is to the right of the second surface if s'_1 is greater than the thickness t and to the left of the surface if t is greater than s'_1. In either case the distance from the surface is $-s'_1 + t$. Even if this image is to the right of the second surface, we can use this distance in Equation 27-5 to find the image distance of the final image. In this case the object distance for the second refraction is negative, indicating a virtual object on the right side of the surface. Thus for all possible values of the first image distance s'_1, the image formed by refraction at the second surface is at a distance s' from this surface given by

$$\frac{n}{-s'_1 + t} + \frac{1}{s'} = \frac{1-n}{r_2} = -\frac{n-1}{r_2} \qquad \text{27-8}$$

For this refraction the light is traveling from the glass medium of index n into the air medium of index 1.

For a general lens of thickness t, it is usually easier to find the distance s'_1 numerically from Equation 27-7 and use this result in Equation 27-8 to find s' than to eliminate s'_1 from these two equations. However, in many cases, the thickness t is much smaller than any of the other distances involved. For such a *thin lens* we can neglect t in Equation 27-7 and easily eliminate s'_1 from these equations. Solving for n/s'_1 in each equation, we obtain

$$\frac{n}{s'_1} = \frac{n-1}{r_1} - \frac{1}{s} = \frac{1}{s'} + \frac{n-1}{r_2}$$

or

$$\frac{1}{s} + \frac{1}{s'} = (n-1)\left(\frac{1}{r_1} - \frac{1}{r_2}\right) \qquad \text{27-9}$$

Equation 27-9 gives the image distance s' in terms of the object distance s and the properties of the thin lens r_1, r_2 and the index of refraction n. As with mirrors, the focal length of a thin lens is defined to be the image distance when the object distance is very large. Setting s equal to infinity and writing f for the image distance s', we obtain

$$\frac{1}{f} = (n-1)\left(\frac{1}{r_1} - \frac{1}{r_2}\right) \qquad \text{27-10} \qquad \textit{Focal length for thin lens}$$

and

$$\frac{1}{s} + \frac{1}{s'} = \frac{1}{f} \qquad \text{27-11} \qquad \textit{Thin-lens equation}$$

Equation 27-10 is sometimes called the *lens-maker's equation* because it gives the focal length of a thin lens in terms of the properties of the lens. Equation 27-11 is the same as that for a spherical mirror. Note that one or both of the radii of curvature can be negative according to our sign convention. For example, in the double convex lens shown in Figure 27-15, the first radius r_1 is positive because the center of curvature lies on the right side of the surface, which is the real side, but r_2 is negative because the center of curvature of the second surface lies on the left, or virtual, side of the surface. In this case, both surfaces tend to bend a light ray toward the axis of the lens. This lens is a *converging lens*. Since both $1/r_1$ and $-1/r_2$ are positive and $n-1$ is positive, the focal length f given in Equation 27-10 must be positive, indicating that parallel light from the left is focused on the right, or real, side of the lens. A converging lens has a positive focal length and is therefore called a *positive lens*.

Figure 27-16 shows a double concave lens. Both surfaces tend to diverge light rays away from the lens axis. Since both $1/r_1$ and $-1/r_2$ are negative, the focal length is negative, indicating that parallel light from the left is diverged as if it came from a point on the left of the lens. A *diverging lens* has a negative focal length and is called a *negative lens*. For any other lens in which both r_1 and r_2 are positive or both negative, the lens is converging or diverging depending on which radius of curvature has the greatest magnitude. If we turn any lens around, we interchange r_1 and r_2 and change their signs; i.e., the new radii of curvature for a lens which is turned around are related to the old by $r_1' = -r_2$ and $r_2' = -r_1$. We see from Equation 27-10 that the new focal length f' has the same magnitude and sign as the old. Thus, for example, if parallel light strikes a double convex lens from the *right*, it is focused at a point on the left a distance f from the lens. The two points on the left and right of any thin lens a distance f from the lens

Parallel light rays incident on a double convex lens. The refracted rays converge at the second focal point of the lens. Reflected rays from each surface of the lens can also be seen. The first surface acts as a convex mirror producing diverging reflected rays, whereas the second surface acts as a concave mirror producing converging reflected rays.

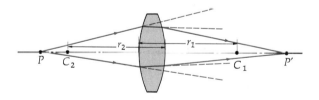

Parallel light rays incident on a double concave lens. The refracted rays diverge as if they came from the first focal point to the left of the lens. As in the previous photograph, reflected rays from each surface of the lens can also be seen. Here, the first surface acts as a concave mirror producing converging reflected rays, whereas the second surface acts as a convex mirror producing diverging reflected rays.

Figure 27-16
Double concave lens. Both surfaces bend the light rays away from the axis. Here r_1 is negative and r_2 is positive in Equation 27-10, resulting in a negative focal length.

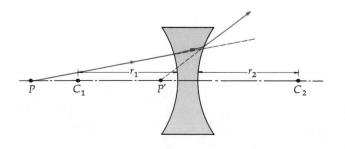

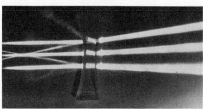

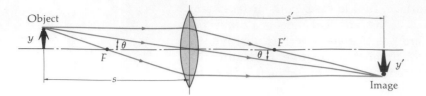

Figure 27-17
Ray diagram for a thin converging lens. For simplicity
we assume all the bending to
take place at a line. The ray
through the center of the lens
is undeflected because the
lens surfaces there are nearly
parallel and close together.

are the first and second focal points of the lens, designated by F and F'. Using the reversibility property of light rays, we can see that light diverging from either of the focal points and striking the lens will leave the lens as a parallel beam. If we set the object distance s in Equation 27-10 equal to the focal distance f, we obtain $s' = \infty$, indicating that the light is not focused but emerges as a parallel beam.

If we have two or more thin lenses, we can find the final image produced by the system by finding the first image distance and using it along with the distance between lenses to find the object distance for the second lens. That is, we consider each image, whether it is formed or not, as the object point for the next lens.

As with images formed by mirrors, it is convenient to locate the image by graphical methods. Figure 27-17 illustrates the graphical method for a converging lens. Three convenient rays from the head of the figure are (1) a ray parallel to the axis of the lens which is bent through the focal point on the right of the lens, (2) a ray through the center of the lens (vertex) which is undeflected, and (3) a ray through the first focal point of the lens which emerges parallel to the axis. These three rays converge to the image point, as indicated. In this case the image is real and inverted. From Figure 27-17 we have $\tan \theta = y/s = -y'/s'$. The lateral magnification is then

Ray diagram for lenses

$$m = \frac{y'}{y} = -\frac{s'}{s} \qquad\qquad 27\text{-}12$$

This expression is the same as that for mirrors. Again, a negative magnification indicates that the image is inverted.

Example 27-3 A double convex thin lens made of glass of index of refraction $n = 1.5$ has both radii of curvature of magnitude 20 cm. An object 2 cm high is placed 10 cm from the lens. Find the focal length of the lens, locate the image, and find its size.

We first calculate the focal length from Equation 27-10, noting that $r_1 = 20$ cm and $r_2 = -20$ cm by our sign convention. The focal length is thus

$$\frac{1}{f} = (1.5 - 1)\left(\frac{1}{20 \text{ cm}} - \frac{1}{-20 \text{ cm}}\right) = \frac{1}{20 \text{ cm}} \qquad \text{or} \qquad f = 20 \text{ cm}$$

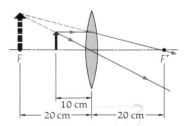

Figure 27-18
Ray diagram for Example 27-3. When the object is between the first focal point and a converging lens, the image is virtual and erect.

Figure 27-18 shows a ray diagram for a small object 10 cm from the lens. The ray parallel to the axis is bent through the focal point, as shown. The ray through the vertex is undeflected. These rays are diverging on the right of the lens. The image is thus located by extending the rays back until they meet. These two rays are sufficient to locate the image. (As a check, we could draw a third ray from the object along a line through the focal point in front of the lens. This ray would then leave the lens parallel to the axis.) We can see immediately from this drawing that the image is virtual, erect, and enlarged. It is on the same side of the lens as the object and is about twice as far away from the lens. Since it is quite easy to make an error in the calculation

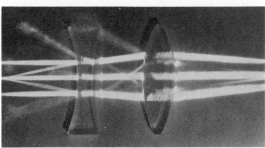

Bausch & Lomb

Parallel light rays incident on a diverging lens followed by a converging lens. The net effect of such a lens system can be converging rays, diverging rays, or parallel rays, depending on the lens separation and the relative size of the focal lengths of the two lenses. Here the net effect is to produce converging rays from parallel rays.

of the image distance from Equation 27-11, it is always a good idea to check your result with a ray diagram.

The image distance is found algebraically from Equation 27-11:

$$\frac{1}{10 \text{ cm}} + \frac{1}{s'} = \frac{1}{20 \text{ cm}}$$

$$\frac{1}{s'} = \frac{1}{20 \text{ cm}} - \frac{1}{10 \text{ cm}} = -\frac{1}{20 \text{ cm}} \quad \text{or} \quad s' = -20 \text{ cm}$$

The image distance is negative, indicating that the image is virtual and to the left of the lens. The magnification is $m = -s'/s = -(-20 \text{ cm})/(10 \text{ cm}) = +2$. The image is thus twice as large as the object, and it is erect. Since the size of the object is given as 2 cm, the size of the image is 4 cm.

Example 27-4 A second lens of focal length $+10$ cm is placed 20 cm to the right of the lens in Example 27-3. Locate the final image.

Figure 27-19 shows a ray diagram for this example. The rays used to locate the image of the first lens are not necessarily convenient rays for the second lens. In this case one of the rays is, the one through the focal point of the first lens which passes through the center of the second lens and is undeflected. We add a second ray from the virtual image of the first lens parallel to the axis. This ray is bent through the focal point of the second lens. We see that the final image is real, inverted, and just outside the focal point of the second lens. We locate its position algebraically by noting that the virtual image of the first lens is 20 cm to the left of that lens and 40 cm to the left of the second lens. Using $s = 40$ cm and $f = 10$ cm, we have

$$\frac{1}{40} + \frac{1}{s'} = \frac{1}{10}$$

giving

$$s' = \tfrac{40}{3} \text{ cm} = 13.3 \text{ cm}$$

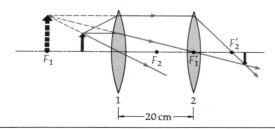

Figure 27-19
Ray diagram for Example 27-4. The image due to the first lens serves as the object for the second lens. The final image is located by drawing two rays from the first image. In this case one of the original rays used to locate the first image is undeflected by the second lens. A second ray from the first image parallel to the axis locates the final image.

Example 27-5 Two lenses of focal length 10 cm each are 15 cm apart. Find the final image of an object 15 cm from one of the lenses.

In the ray diagram of Figure 27-20 the image of the first lens would be 30 cm from the lens if the second lens were not there. We calculate this using $s = 15$ cm and $f = 10$ cm in the thin-lens equation:

$$\frac{1}{15} + \frac{1}{s'} = \frac{1}{10}$$

$$\frac{1}{s'} = \frac{1}{10} - \frac{1}{15} = \frac{1}{30} \qquad \text{or} \qquad s' = 30 \text{ cm}$$

This image is not formed because the light rays strike the second lens before they reach the image position. We can locate the final image graphically by choosing rays which are heading toward the unformed image when they strike the lens. A ray parallel to the axis and one through the center of the second lens are sufficient. We see that the final image is between the second lens and its focal point. The final image can be located algebraically by using the first image as the object for the second lens. Since this image is not formed because it is to the right of the second lens, it is a *virtual object.* Since it is 15 cm to the right of the second lens, the object distance is $s = -15$ cm. Then

$$\frac{1}{-15} + \frac{1}{s'} = \frac{1}{f} = \frac{1}{10}$$

$$\frac{1}{s'} = \frac{1}{10} + \frac{1}{15} = \frac{5}{30} = \frac{1}{6} \qquad \text{or} \qquad s' = 6 \text{ cm}$$

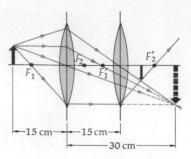

Figure 27-20
Ray diagram for Example 27-5. The image due to the first lens is to the right of the second lens. The rays do not actually converge to that image because they are refracted by the second lens. That image therefore serves as a virtual object for the second lens. The final image is found by drawing rays toward the first image, as shown. Here, a ray parallel to the axis and one through the center of the second lens are used.

Questions

4. A fish underwater is viewed from a point not directly overhead. Is its apparent depth greater or less than three-fourths its actual depth?

5. Under what conditions will the focal length of a thin lens be positive? Negative?

6. Under what conditions will the image formed by a positive thin lens be erect? Inverted? Real? Virtual? Larger than the object? Smaller than the object?

7. The focal length of a lens is different for different colors of light. Why?

27-5   Aberrations

When all the rays from a point object are not focused at a single image point, the resulting blurring of the images is called an *aberration.* The blurring of the image of an object point on the axis due to rays which strike a mirror or lens at points far from the axis is called *spherical aberration.* The similar blurring of the image of a point off the axis due to the different magnification of different parts of the lens or mirror is called *coma,* after the comet-shaped image observed. The imaging of a point source off axis into two mutually perpendicular lines at different locations is called *astigmatism.* The distortion in shape of the image of an extended object due to the fact that the magnification depends on the distance of the object point from the axis is called *distortion.* These aberrations, common to both mirrors and lenses, are the result of the laws of reflection and refraction applied to spherical surfaces. They are not evident in our simple equations for mirrors and lenses because we have used small-angle approximations in the derivation of these equations. Some of the aberrations can be eliminated or at least partially corrected by using nonspherical surfaces for the mirror or lens, but this is usually difficult and costly compared with the production of spherical surfaces. One example of a nonspherical reflecting surface is the parabolic mirror illustrated in Figure 27-21. Parallel rays incident on the parabolic surface are reflected and focused at a common point no matter how far the rays are from the axis. Parabolic reflecting surfaces are important in large astronomical telescopes, in which a large reflecting surface is needed to make the intensity of the image as great as possible. A parabolic surface can also be used in a searchlight to produce a parallel beam of light from a small source placed at the focus of the surface.

An important aberration of lenses not found in mirrors is *chromatic aberration* due to the variation in the index of refraction with wavelength. From Equation 27-10 we see that even for paraxial rays, the focal length of a lens depends on the index of refraction and is therefore slightly different for different wavelengths. Chromatic and other aberrations can be partially corrected by using combinations of lenses instead of a single lens. For example, a positive lens and a negative lens of greater focal length can be used together to produce a converging lens system which has much less chromatic aberration than a single lens of the same focal length as the system.

27-6 The Eye

The optical system of prime importance is the eye, shown in Figure 27-22. Light enters the eye through a variable aperture, the pupil, and is focused by the cornea-lens system on the retina, a film of nerve fibers which covers the back surface. The retina contains tiny sensing structures called *rods* and *cones*, which receive the image and transmit the information along the optic nerve to the brain. The shape of the crystalline lens can be slightly altered by the action of the ciliary muscle. When the eye is focused on an object far away, the muscle is relaxed and the cornea-lens system has its maximum focal length, about 2.5 cm, the distance from the cornea to the retina. When the object is brought closer to the eye, the ciliary muscle tenses, increasing the curvature of the lens slightly and decreasing its focal length, so that the image is again focused on the retina. This process is called *accommodation*. If the object is too close to the eye, the lens cannot focus the light on the retina and the image is blurred. The closest point for which the lens can focus the image on the retina is called the *near point*. The distance from the eye to the near point varies greatly from one person to another and with age. At the age of 10 years, the near point may be as close as 7 cm, whereas at 60 years it may recede to 200 cm because of the loss of flexibility of the lens. The standard value taken for the near point is 25 cm.

If the relaxed eye images distant objects behind the retina, the person is said to be farsighted. Farsightedness is corrected with a converging lens (Figure 27-23). On the other hand, if light from a distant

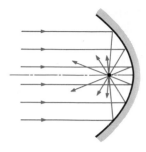

Figure 27-21
Spherical aberration can be eliminated by using the surface of a paraboloid of revolution, which reflects all parallel rays through a common focal point.

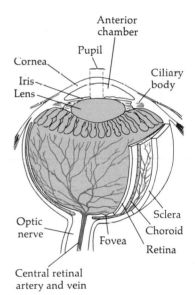

Figure 27-22
Cutaway view of the human eye. The amount of light entering the eye is controlled by the iris, which regulates the size of the pupil. Because the lens is surrounded by material of nearly equal index of refraction, most of the refraction occurs at the cornea. The lens thickness is controlled by the ciliary muscle. The cornea and lens together focus the image on the retina, which contains about 125 million receptors called rods and cones and about 1 million optic-nerve fibers. (*From H. Curtis,* Biology, *2d ed., p. 696. Worth Publishers, Inc., New York, 1975.*)

object is focused in front of the retina by the relaxed eye, the person is nearsighted. Nearsightedness is corrected with a diverging lens (Figure 27-24).

Example 27-6 By how much must the focal length of the cornea-lens system of the eye change when the object is moved from infinity to the near point at 25 cm? Assume that the distance from the cornea to the retina is 2.5 cm.

When the object is at infinity, the focal length of the cornea-lens system is 2.5 cm. When the object is at 25 cm, the focal length f must be such that the image distance is 2.5 cm. From Equation 27-11 we have

$$\frac{1}{25 \text{ cm}} + \frac{1}{2.5 \text{ cm}} = \frac{1}{f}$$

$$\frac{1}{f} = \frac{1}{25 \text{ cm}} + \frac{10}{25 \text{ cm}} = \frac{11}{25 \text{ cm}}$$

$$f = \frac{25 \text{ cm}}{11} = 2.27 \text{ cm}$$

The focal length must therefore decrease by 0.23 cm.

The apparent size of an object is determined by the size of the image on the retina. The larger the image on the retina, the greater the number of rods and cones activated. From Figure 27-25 we see that the size of the image on the retina is greater when the object is close than it is when the object is far away. Even though the actual size of the object does not change, its apparent size is greater when it is brought closer to the eye. Since the near point is the closest point to the eye for which a sharp image can be formed on the retina, the distance to the near point is called the *distance of most distinct vision.* A convenient measure of the size of the image on the retina is the angle subtended by the object at the eye (Figure 27-25). When an object of height y is at the near point, the angle subtended is

$$\theta_0 = \frac{y}{25 \text{ cm}} \qquad\qquad 27\text{-}13$$

where we have taken 25 cm to be the distance to the near point and have used the small-angle approximation $\theta_0 \approx \tan \theta_0$.

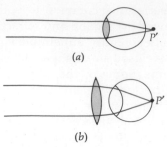

(a)

(b)

Figure 27-23
(a) The lens of a farsighted eye focuses parallel rays to form an image behind the retina. (b) A converging lens corrects this defect by bringing the image onto the retina.

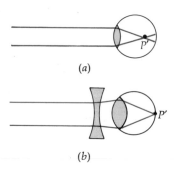

(a)

(b)

Figure 27-24
(a) The lens of a nearsighted eye focuses parallel rays to form an image in front of the retina. (b) A diverging lens corrects this defect.

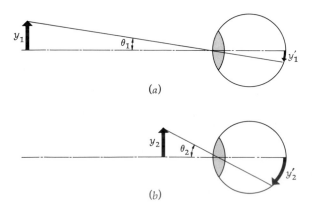

(a)

(b)

Figure 27-25
(a) A distant object of height y_1 looks small because the size of the image on the retina is small. (b) When the same object is closer, it looks larger because the retinal image is larger. The size of the image on the retina is approximately proportional to the angle θ subtended by the object.

Nerve fibers

Receptor layer
Rods   Cones

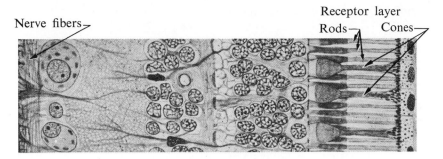

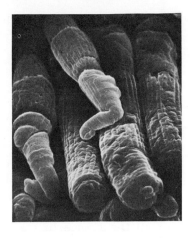

(*Above*) Section of the retina of the human eye. (*From F. W. Sears and M. W. Zemansky*, University Physics, *4th ed., p. 582. Addison-Wesley Publishing Company, Inc., Reading, Mass., 1970.*) (*Right*) Rods (cylindrical structures extending into lower foreground) and cones (upper middle and left) as seen through a scanning electron microscope, magnified 1600 times. (*Courtesy of E. R. Lewis.*)

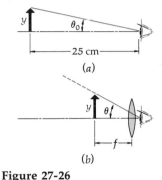

(*a*)

(*b*)

Figure 27-26
(*a*) An object at the near point subtends an angle θ_0 at the eye. (*b*) When the object is at the focal point of the converging lens, the image (not shown) is at infinity and can be viewed by the relaxed eye. When f is less than 25 cm, the converging lens acts as a simple magnifier by allowing the object to be brought closer to the eye, increasing the angle subtended by the object to θ and thereby increasing the image size on the retina.

Magnification of magnifier, image at infinity

27-7 The Simple Magnifier

The apparent size of an object can be increased by using a converging lens called a *simple magnifier*. In Figure 27-26, a small object of height y is at the first focal point of a thin converging lens of focal length f, which is less than the distance of most distinct vision (taken to be 25 cm). The image formed by the converging lens is virtual, erect, and at infinity. It subtends an angle θ given approximately by

$$\theta = \frac{y}{f}$$

Viewed through the lens, the object looks larger because the image on the retina is larger by the factor θ/θ_0, where θ_0, given by Equation 27-13, is the angle subtended by the object at the near point. The ratio θ/θ_0 is called the *angular magnification* or *magnifying power M* of the lens:

$$M = \frac{\theta}{\theta_0} = \frac{25 \text{ cm}}{f} \qquad\qquad 27\text{-}14$$

The angular magnification of a simple magnifier can be increased if the object is moved closer to the magnifier, since the image moves in from infinity and the angle subtended increases. The largest usable magnification occurs when the image is at the near point of the eye, 25 cm. The angle subtended is then y/s, where s is the object distance for the magnifier. We can calculate s from the thin-lens equation 27-11. Using $s' = -25$ cm, we have

$$\frac{1}{s} + \frac{1}{-25} = \frac{1}{f}$$

$$\frac{1}{s} = \frac{1}{f} + \frac{1}{25} = \frac{25 + f}{25f} \qquad s = \frac{25f}{25 + f}$$

Then

$$\theta = \frac{y}{s} = \frac{25 + f}{25f} y \qquad \text{and} \qquad \frac{\theta}{\theta_0} = \frac{25 + f}{f} = 1 + \frac{25}{f}$$

When the image of the magnifier is viewed at the near point, the magnification is

$$M = 1 + \frac{25 \text{ cm}}{f} \qquad \qquad 27\text{-}15$$

Magnification of magnifier, image at near point

It is left as an exercise to show that in this case the angular magnification is equal to the linear magnification $m = y/y'$.

Simple magnifiers are used as eyepieces or oculars in compound microscopes and telescopes to view the image formed by another lens or lens system. To correct aberrations, combinations of lenses with a resulting short positive focal length may be used as a simple magnifier in place of a single lens, but the principle is the same. For convenience we shall use $25 \text{ cm}/f$ for the magnification of a simple magnifier, though the choice between Equations 27-14 and 27-15 depends on whether the image is viewed at infinity or the near point. Of course, if the distance to the near point is much greater than 25 cm, the effective magnification of a simple magnifier is greater. For example, if the near point for a person is 50 cm from his eye, a simple magnifier of focal length f gives an angular magnification of $50 \text{ cm}/f$ when the image is viewed at infinity.

27-8   The Compound Microscope and the Telescope

The compound microscope and the telescope are interesting applications of the principles of geometric optics. In their simplest forms each consists of two converging lenses. The lens nearest the object, called the *objective*, forms a real image of the object. The lens nearest the eye, called the *ocular* or *eyepiece,* is used as a simple magnifier to view the image formed by the objective.

The compound microscope (Figure 27-27) is used to look at very small objects which are close by. The distance between the second focal point of the objective and the first focal point of the ocular is called the *tube length l* and is usually fixed. The object is placed just outside the focal point of the objective so that an enlarged image is formed at the first focal point of the ocular a distance $l + f_o$ from the objective where f_o is the focal length of the objective. From Figure 27-27, $\tan \beta = y/f_o = -y'/l$. The lateral magnification of the objective is therefore

$$m_o = \frac{y'}{y} = -\frac{l}{f_o} \qquad \qquad 27\text{-}16$$

The overall magnification of the compound microscope is the product

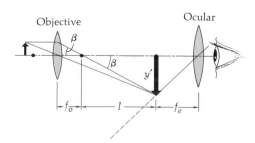

Objective

Ocular

Figure 27-27
Schematic diagram of a compound microscope, consisting of two positive lenses, the objective of focal length f_o, and the ocular, or eyepiece, of focal length f_e, separated by a distance $l + f_o + f_e$. The real image of the object formed by the objective is viewed by the ocular, which acts as a simple magnifier.

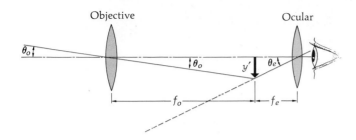

Figure 27-28
Schematic diagram of an
astronomical telescope. The
objective forms a real image
of a distant object near its
second focal point, which
coincides with the first focal
point of the ocular. The ocular
serves as a simple magnifier
which views this image.

of the lateral magnification of the objective and the angular magnification of the ocular $M_e = (25 \text{ cm})/f_e$, where f_e is the focal length of the ocular, or eyepiece:

$$m = m_o M_e = -\frac{l}{f_o} \frac{25 \text{ cm}}{f_e} \qquad \text{27-17}$$

Magnification of compound microscope

The astronomical telescope, illustrated schematically in Figure 27-28, is used to view objects which are far away and often large. The purpose of the objective is to form a real image at the first focal point of the ocular. Since the object is usually a great distance from the telescope, the image is at the second focal point of the objective. The objective and ocular are therefore separated by a distance $f_o + f_e$, where f_o and f_e are the focal lengths of the objective and ocular, respectively.

Let θ_o be the angle subtended by the object when it is viewed directly by the unaided eye. This is the same as the angle subtended by the object at the objective shown in the figure. The angle θ_e in the figure is that subtended by the final image. The angular magnification of the telescope is θ_e/θ_o. Using the small-angle approximation in Figure 27-28, we have

$$\tan \theta_o = -\frac{y'}{f_o} \approx \theta_o$$

where we have introduced the negative sign to make θ_o positive when y' is negative. Similarly,

$$\tan \theta_e = \frac{y'}{f_e} \approx \theta_e$$

Since y' is negative, θ_e is negative, indicating that the image is inverted. The angular magnification of the telescope is then

$$M = \frac{\theta_e}{\theta_o} = -\frac{f_o}{f_e} \qquad \text{27-18}$$

Magnification of telescope

Review

A. Define, explain, or otherwise identify:

B. True or false:

1. Virtual and real images look the same to the eye.

2. A virtual image cannot be displayed on a screen.

3. Aberrations occur only for real images.

4. A negative image distance implies that the image is virtual.

5. All rays parallel to the axis of a spherical mirror are reflected through a single point.

6. A diverging lens cannot form a real image from a real object.

7. The image distance for a positive lens is always positive.

8. Chromatic aberration does not occur with mirrors.

9. The image formed by the objective of a telescope is inverted and larger than the object.

10. The final image of a telescope is virtual.

Exercises

Section 27-1, Plane Mirrors

1. The image of the point object shown in Figure 27-29 is viewed by an eye as shown. Draw a bundle of rays from the object which reflect from the mirror and enter the eye. For this object position and mirror, indicate the region of space in which the eye can be placed to see the image.

Figure 27-29
Exercise 1.

2. A point object is placed between parallel mirrors separated by 30 cm. The object is 10 cm from the lower mirror and 20 cm from the upper mirror. (*a*) Find the distances from the upper mirror to the first four images which occur above the mirror. (*b*) Find the distances from the lower mirror to the first four images which occur below the lower mirror.

3. Two plane mirrors make an angle of 90°. Show that there are three images for any position of the object. Draw appropriate bundles of rays from the object to the eye for viewing each image.

4. Two plane mirrors make an angle of 60° with each other. Consider a point object on the bisector of the angle between the mirrors. Show on a sketch the location of all the images formed.

5. Repeat Exercise 4 for two mirrors at an angle of 120°.

Section 27-2, Spherical Mirrors

6. A concave spherical mirror has a radius of curvature of 50 cm. Draw ray diagrams to locate the image (if one is formed) for an object at a distance of (*a*) 100 cm, (*b*) 50 cm, (*c*) 25 cm, and (*d*) 10 cm from the mirror. For each case state whether the image is real or virtual; erect or inverted; and enlarged, reduced, or the same size as the object.

7. Use the mirror Equations 27-2 to 27-4 to locate and describe the images for the object distances and mirror of Exercise 6.

8. Repeat Exercise 6 with a convex mirror of the same radius of curvature.

9. Repeat Exercise 7 with a convex mirror of the same radius of curvature.

10. Show that a convex mirror cannot form a real image of a real object no matter where the object is placed by showing that s' is always negative for positive s.

11. Show that a concave mirror forms a virtual image of a real object when $s < f$ and a real image when $s > f$ (see Exercise 10).

12. A certain telescope uses a concave spherical mirror of radius 8 m as its objective. Find the location and diameter of the image of the moon, which has a diameter of 3.5×10^6 m and is 3.8×10^8 m from the mirror.

Section 27-3, Images Formed by Refraction

13. A sheet of paper with writing on it is protected by a thick glass plate having an index of refraction of 1.5. If the plate is 2 cm thick, at what distance beneath the top of the plate does the writing appear when viewed from directly overhead?

14. A very long glass rod has one end ground to a convex hemispherical surface of radius 5 cm. Its index of refraction is 1.5. (a) A point object is on the axis in air a distance 20 cm from the surface. Find the image and state whether it is real or virtual. (b) Repeat for an object at 5 cm from the surface. (c) Repeat for an object very far from the surface. Draw a ray diagram for this case.

15. At what distance from the rod of Exercise 14 should an object be placed so that the light rays in the rod are parallel to the axis? Draw a ray diagram for this situation.

16. Repeat Exercise 14 for a glass rod with a concave hemispherical surface of radius $(-)$ 5 cm.

17. Repeat Exercise 14 for the glass rod immersed in water of index of refraction 1.33; this means that the objects are in water.

18. Repeat Exercise 14 for a glass rod with a concave surface ($r = -5$ cm) immersed in water (index of refraction 1.33).

Section 27-4, Lenses

19. The following thin lenses are made of glass with an index of refraction of 1.5. Make a sketch and find the focal length in air for each: (a) double convex. $r_1 = 10$ cm, $r_2 = -21$ cm; (b) plano-convex: $r_1 = \infty$, $r_2 = -10$ cm; (c) double concave: $r_1 = -10$ cm, $r_2 = +10$ cm; (d) plano-concave: $r_1 = \infty$, $r_2 = +20$ cm.

20. Glass with an index of refraction of 1.6 is used to make a thin lens which has radii of equal magnitude. Find the radii of curvature and make a sketch if the focal length in air is to be (a) $+5$ cm, (b) -5 cm.

21. The following thin lenses are made of glass of index of refraction 1.5. Make a sketch and find the focal length in air for each: (a) $r_1 = 20$ cm, $r_2 = 10$ cm; (b) $r_1 = 10$ cm, $r_2 = 20$ cm; (c) $r_1 = -10$ cm, $r_2 = -20$ cm.

22. For the following object distances and focal lengths of thin lenses in air find the image distance and the magnification and state whether the image is real or virtual, erect or inverted. (a) $s = 40$ cm, $f = 20$ cm; (b) $s = 10$ cm, $f = 20$ cm; (c) $s = 40$ cm, $f = -30$ cm; (d) $s = 10$ cm, $f = -30$ cm.

23. What is meant by a negative object distance? How can it occur? Find the image distance and magnification and state whether the image is virtual or real, erect or inverted, for a thin lens in air when (a) $s = -20$ cm, $f = +20$ cm; (b) $s = -10$ cm, $f = -30$ cm.

24. An object 3.0 cm high is placed 20 cm from a thin lens of focal length 5.0 cm. Draw a careful ray diagram to find the position and size of the image and check your result using the thin-lens equation.

25. Repeat Exercise 24 for an object 1 cm high placed 10 cm from a thin lens of focal length 5.0 cm.

26. Repeat Exercise 24 for an object 1.0 cm high placed 10 cm from a thin lens whose focal length is −5.0 cm.

27. A thin converging lens of focal length 10 cm is used to obtain an image which is twice as large as a small object. (a) Where should the object be placed? (b) Where is the image? (c) Is the image real or virtual? Erect or inverted?

28. Show that a diverging lens can never form a real image from a real object. *Hint:* Show that s′ is always negative.

29. Show that a converging lens forms a real image from a real object when s is greater than f. *Hint:* Show that s′ is then positive.

30. Show that if two thin lenses of focal lengths f_1 and f_2 are placed in contact, they are equivalent to a single thin lens of focal length $f = f_1 f_2/(f_1 + f_2)$.

31. Two converging lenses each of focal length 10 cm are separated by 35 cm. An object is 20 cm to the left of the first lens. Find the final image using both ray diagrams and the lens equation. What is the overall lateral magnification?

Section 27-5, Aberrations

32. A concave spherical mirror has a radius of curvature 6.0 cm. A point object is on the axis at 9.0 cm from the mirror. Construct a careful ray diagram showing rays from the object which make angles of 5, 10, 30, and 60° with the axis, strike the mirror, and are reflected back across the axis. (Use a compass to draw the mirror, and use a protractor to measure the angles needed to find the reflected rays.) What is the spread along the axis of the image for these rays?

33. For the mirror of Exercise 32, draw rays parallel to the axis at distances 0.5, 1, 2, and 4 cm and find the points at which the reflected rays cross the axis.

34. A double convex lens of radii $r_1 = +10$ cm and $r_2 = −10$ cm is made from a glass with an index of refraction of 1.53 for blue light and 1.47 for red light. Find the focal lengths of this lens for red and blue light.

Section 27-6, The Eye

In the following exercises take the distance from the cornea-lens system of the eye to the retina to be 2.5 cm.

35. Suppose the eye were designed like a camera with a lens of fixed focal length f = 2.5 cm which could move toward or away from the retina. Approximately how far would the lens have to move to focus the image on the retina of an object 25 cm from the eye? *Hint:* Find the distance behind the retina of the image for an object at 25 cm.

36. Since the index of refraction of the lens of the eye is not very different from that of the surrounding material, most of the refraction takes place at the cornea, where n changes abruptly from 1.0 in air to about 1.4. Assuming the cornea to be homogeneous with an index of refraction of 1.4, calculate its radius if it focuses parallel light on the retina a distance 2.5 cm away.

37. If two point objects close together are to be seen as two distinct objects, the images must fall on two different cones on the retina which are not adjacent; i.e., there must be an unactivated cone between them. The separation of the cones is about 1 μm = 10^{-6} m. What is the smallest angle the two points can subtend (see Figure 27-30)? How close can two points be if they are 20 m from the eye?

Figure 27-30
Exercise 37.

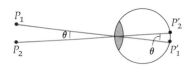

38. The resolution of two point objects is also limited by diffraction. Use the Rayleigh criterion ($\theta = 1.22\lambda/D$) to calculate this limit on the angle subtended by two point objects if the diameter of the pupil is 5 mm and λ = 500 nm.

Compare this with the limitation due to the separation of the cones (see Exercise 37).

Section 27-7, The Simple Magnifier

39. Show that when the image is viewed at the near point, the lateral and angular magnifications of a simple magnifier are equal.

40. What is the magnifying power of a lens of focal length 5 cm when (a) the image is viewed at infinity and (b) the image is viewed at the near point 25 cm from the eye?

41. A lens of focal length 10 cm is used as a simple magnifier with the image at infinity by one person whose near point is at 25 cm and by another whose near point is at 50 cm from the eye. What is the effective magnifying power of the lens for each person? Compare the size of the image on the retina when each looks at the same object with the magnifier.

Section 27-8, The Compound Microscope and the Telescope

42. A crude symmetric hand-held microscope consists of two converging lenses each of focal length 5.0 cm fastened in the ends of a tube 30 cm long. (a) What is the "tube length"? (b) What is the lateral magnification of the objective? (c) What is the overall magnification of the microscope? (d) How far from the objective should the object be placed?

43. Repeat Exercise 42 for the same two lenses separated by 40 cm.

44. A simple telescope has an objective with a focal length of 100 cm and an ocular with a focal length of 5 cm. It is used to look at the moon, which subtends an angle of about 0.009 rad. (a) What is the size of the image formed by the objective? (b) What angle is subtended by the final image at infinity?

45. The objective lens of the refracting telescope at the Yerkes Observatory has a focal length of about 19.5 m. What is the diameter of the image of the moon formed by the objective? (The angle subtended by the moon is about 0.009 rad.)

Problems

1. (a) Show that the focal length f' for a thin lens of index of refraction n_2 in a medium of index n_1 is given by

$$\frac{1}{f'} = \frac{n_2 - n_1}{n_1}\left(\frac{1}{r_1} - \frac{1}{r_2}\right)$$

(b) Show that if f is the focal length of a thin lens in air, its focal length in water is f', given by

$$f' = \frac{n_1(n_2 - 1)}{n_2 - n_1} f$$

where n_1 is the index of refraction of water and n_2 is that of the lens material.

2. A glass rod 96 cm long with an index of refraction of 1.6 has its ends ground to convex spherical shapes of radii 8 cm and 16 cm. A point object is in air on the axis 20 cm from the 8-cm-radius end. (a) Find the image distance due to refraction at the first surface. (b) Find the final image due to refraction at both surfaces. (c) Is the final image real or virtual?

3. Repeat Problem 2 for a point object in air 20 cm from the 16-cm-radius end.

4. Find the focal length of a thick double convex lens with an index of refraction of 1.5, thickness 4 cm, and radii +20 and −20 cm.

5. Show that as an object moves away from a spherical mirror, its image always moves in the opposite direction. *Hint:* Compute ds' in terms of ds from the mirror equation.

6. A small object is 20 cm from a thin positive lens of focal length 10 cm. To the right of the lens is a plane mirror which crosses the axis at the focal point of the lens and is tilted so that the reflected rays do not go back through the lens (see Figure 27-31). Sketch a ray diagram showing the final image. Is this image real or virtual?

Figure 27-31
Problem 6.

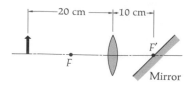

7. (*a*) Show that to obtain magnification of magnitude m with a thin positive lens of focal length f, the object distance must be given by

$$s = \frac{m+1}{m} f$$

(*b*) A 35-mm camera has a lens with a 50-mm focal length. It is used to take a picture of a person 175 cm tall so that the image size on the film is 35 mm. How far from the camera should the person stand?

8. A convenient form of the thin-lens equation due to Newton measures the object and image distances from the focal points. Show that if $x = s - f$ and $x' = s' - f$, the thin-lens equation can be written $xx' = f^2$ and the lateral magnification is given by $m = -x'/f = -f/x$. Indicate x and x' on a sketch.

9. Show that a small change dn in the index of refraction of a lens material produces a small change in the focal length df given approximately by

$$\frac{df}{f} = -\frac{dn}{n-1}$$

Use this to find the focal length of a thin lens for blue light, for which $n = 1.53$, if the focal length for red light, for which $n = 1.47$, is 20 cm.

10. The lateral magnification of a spherical mirror or a thin lens is given by $m = -s'/s$. Show that for objects of small horizontal extent, the longitudinal magnification is approximately $-m^2$. Do this by showing that $ds'/ds = -s'^2/s^2$.

11. A disadvantage of the astronomical telescope for terrestial use, e.g., at a football game, is that the image is inverted. A galilean telescope uses a converging lens as its objective but a diverging lens as its eyepiece. The image formed by the objective is beyond the diverging lens at its second focal point so that the final image is virtual and at infinity. Show with a ray diagram that the final image is erect. Show that the angular magnification is $M = -f_o/f_e$, where f_o is the focal length of the objective and f_e is the focal length of the ocular; f_e is negative, making M positive.

12. A galilean telescope is designed so that the final image is at the near point, 25 cm from the eye. The focal length of the objective is 100 cm and that of the ocular is -5 cm. (*a*) If the object distance is 30 m, where is the image of the objective? (*b*) What is the object distance for the ocular so that the final image is at the near point? (*c*) How far apart are the lenses? (*d*) If the object height is 1 m, what is the height of the final image? What is the angular magnification? (See Problem 11.)

CHAPTER 28   Special Relativity

The special theory of relativity is one of the great intellectual triumphs of the twentieth century. Since we have time for only a brief introduction, we discuss relativistic kinematics in some detail and merely outline relativistic dynamics, omitting derivations.[1] The theory of special relativity can be derived from two postulates proposed by Einstein in a paper on the electrodynamics of moving bodies when he was only 26.[2] These two postulates, simply stated, are:

---

1. Absolute, uniform motion cannot be detected.

2. The speed of light is independent of the motion of the source.

Einstein postulates

Although each postulate seems quite reasonable, many of the implications of the two together are quite surprising and contradict what is often called common sense. For example, a direct consequence is that all observers measure the same number for the speed of light in vacuum independent of their relative motion. We shall derive this and other results later. First we shall look at a historically important experiment related to the theory of special relativity, the Michelson-Morley experiment.

---

[1] There are several good books on relativity at the elementary level which the interested student may want to consult to pursue this subject in more detail, e.g., R. Resnick, *Introduction to Special Relativity*, Wiley, New York, 1968; A. P. French, *Special Relativity*, Norton, New York, 1968; and C. Kacser, *Introduction to the Special Theory of Relativity*, Prentice-Hall, Englewood Cliffs, N.J., 1967.

[2] *Annalen der Physik*, vol. 17, p. 841, 1905. For a translation from the original German, see W. Perrett and G. B. Jeffery (trans.), *The Principle of Relativity: A Collection of Original Memoirs on the Special and General Theory of Relativity* by H. A. Lorentz, A. Einstein, H. Minkowski, and H. Weyl, Dover, New York, 1923.

28-1   The Michelson-Morley Experiment

From our study of wave motion we know that all waves except electromagnetic waves require a medium for their propagation. The speed of waves depends only on properties of the medium. With sound waves, for example, absolute motion, i.e., motion relative to the still air, can be detected. The doppler effect for sound depends not only on the relative motion of the source and listener but on the absolute motion of each relative to the air. It was natural to expect that some kind of medium supports the propagation of light and other electromagnetic waves. Such a medium, called the *ether*, was proposed in the nineteenth century. The ether as proposed must have unusual properties. Although it would require great rigidity to support waves of such high velocity (recall that the velocity of waves on a string depends on the tension of the string), it must introduce no drag force on the planets, as their motion is fully accounted for by the law of gravitation.

It was of considerable interest to determine the velocity of the earth relative to the ether. Maxwell pointed out that in measurements of the speed of light (such as Fizeau's toothed-wheel method), the earth's speed v relative to the ether appears only in the second order v^2/c^2, an effect then considered too small to measure. In these measurements, the time for a light pulse to travel to and from a mirror is determined. Figure 28-1 shows a light source and mirror a distance L apart. If we assume both are moving with speed v through the ether, classical theory predicts that the light will travel toward the mirror with speed $c - v$ and back with speed $c + v$ (both speeds relative to the mirror and light source). The time for the total trip will be:

$$t_1 = \frac{L}{c - v} + \frac{L}{c + v} = \frac{2cL}{c^2 - v^2} = \frac{2L}{c}\left(1 - \frac{v^2}{c^2}\right)^{-1} \qquad \text{28-1}$$

For v much less than c, we can expand this result using the binomial expansion

$$(1 + x)^n \approx 1 + nx + \cdots \qquad \text{for } x \ll 1$$

Then

$$t_1 \approx \frac{2L}{c}\left(1 + \frac{v^2}{c^2} + \cdots\right) \qquad \text{28-2}$$

If we take the orbital speed of the earth about the sun for an estimate of v, we have $v \approx 3 \times 10^4$ m/sec $= 10^{-4}c$ and $v^2/c^2 = 10^{-8}$. Thus the correction for the earth's motion is small indeed. Michelson realized that though this effect is too small to be measured directly, it should be possible to determine v^2/c^2 by a difference measurement. Figure 28-2 is a diagram of his apparatus, called a *Michelson interferometer*.

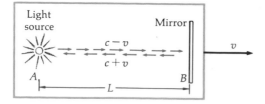

Figure 28-1
Light source and mirror moving with speed v relative to the "ether." According to classical theory, the speed of light relative to the source and mirror would be $c - v$ toward the mirror and $c + v$ away from the mirror.

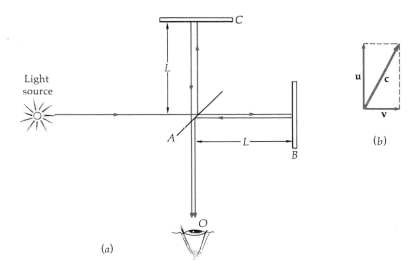

(a)

Figure 28-2
(a) Schematic drawing of the Michelson interferometer. (b) According to classical theory, if the interferometer moves to the right with velocity **v** relative to the ether, the light must move with velocity **c** in the direction shown (relative to the ether) to strike the upper mirror. Its velocity relative to the interferometer is then **u** = **c** − **v**, and its speed is $u = \sqrt{c^2 - v^2}$.

Light from the source is partially reflected and partially transmitted by mirror A. The transmitted beam travels to mirror B and is reflected back to A. The reflected beam travels to mirror C and is reflected back to A. The two beams recombine and form an interference pattern, which is viewed by an observer at O. Equation 28-2 gives the classical result for the round-trip time t_1 for the transmitted beam. Since the reflected beam travels (relative to the earth) perpendicular to the earth's velocity, the velocity of this beam relative to earth (according to classical theory) is the vector difference **u** = **c** − **v**. The magnitude of u is $\sqrt{c^2 - v^2}$; so the round-trip time for this beam is

$$t_2 = \frac{2L}{\sqrt{c^2 - v^2}} = \frac{2L}{c}\left(1 - \frac{v^2}{c^2}\right)^{-1/2} \approx \frac{2L}{c}\left(1 + \frac{1}{2}\frac{v^2}{c^2} + \cdots\right) \qquad 28\text{-}3$$

where again the binomial expansion has been used. There is thus a time difference:

$$\Delta t = t_1 - t_2 \approx \frac{2L}{c}\left(1 + \frac{v^2}{c^2}\right) - \frac{2L}{c}\left(1 + \frac{1}{2}\frac{v^2}{c^2}\right) = \frac{Lv^2}{c^3} \qquad 28\text{-}4$$

The time difference is to be detected by observing the interference of the two beams of light. Because of the difficulty of making the two paths of equal length to the precision required, the interference pattern of the two beams is observed and then the whole apparatus rotated 90°. The rotation produces a time difference given by Equation 28-4 for each beam. The total time difference of $2\Delta t$ is equivalent to a path difference of $2c\,\Delta t$. The interference fringes observed in the first orientation should thus shift by a number of fringes ΔN given by

$$\Delta N = \frac{2c\,\Delta t}{\lambda} = \frac{2L}{\lambda}\frac{v^2}{c^2} \qquad 28\text{-}5$$

A student-type Michelson interferometer. The fringes are produced on a ground-glass screen by light from a laser. (*Courtesy of Dr. L. Velinsky.*)

where λ is the wavelength of the light. In Michelson's first attempt, in 1881, L was about 1.2 m and λ was 590 nm. For $v^2/c^2 = 10^{-8}$, ΔN was expected to be 0.04 fringe. Even though the experimental uncertainties were estimated to be about this same magnitude, when no shift was observed, Michelson reported this as evidence that the earth did not move relative to the ether. In 1887, when he repeated the experiment with Edward W. Morley, he used an improved system for rotating the apparatus without introducing a fringe shift because of mechanical strains, and he increased the effective path length L to about 11 m by a

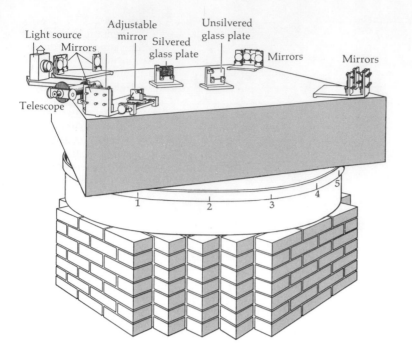

Light source
Adjustable mirror
Silvered glass plate
Unsilvered glass plate
Mirrors
Mirrors
Mirrors
Telescope

1   2   3   4   5

Figure 28-3
Drawing of Michelson-Morley apparatus used in their 1887 experiment. The optical parts were mounted on a sandstone slab 5 ft square, which was floated in mercury, thereby reducing the strains and vibrations that had affected the earlier experiments. Observations could be made in all directions by rotating the apparatus in the horizontal plane. (*From R. S. Shankland, The Michelson-Morley Experiment, Copyright © November 1964 by Scientific American, Inc. All rights reserved.*)

series of multiple reflections. Figure 28-3 shows the configuration of the Michelson-Morley apparatus. For this attempt, ΔN was expected to be about 0.4 fringe, about 20 to 40 times the minimum possible to observe. Once again, no shift was observed. The experiment has since been repeated under various conditions by a number of people, and no shift has ever been found.

The null result of the Michelson-Morley experiment is easily understood in terms of the Einstein postulates. According to postulate 1, absolute uniform motion cannot be detected. We can consider the whole apparatus and the earth to be at rest. No fringe shift is expected when the apparatus is rotated 90° since all directions are equivalent. It should be pointed out that Einstein did not set out to explain this experiment. His theory arose from his considerations of the theory of electricity and magnetism and the unusual property of electromagnetic waves, namely, they propagate in a vacuum. In his first paper, which contains the complete theory of special relativity, he made only a passing reference to the Michelson-Morley experiment, and in later years he could not recall whether he was aware of the details of this experiment before he published his theory.

Albert A. Michelson in his laboratory. (*Courtesy of the Niels Bohr Library, American Institute of Physics.*)

28-2 Consequences of Einstein's Postulates

An immediate consequence of the two Einstein postulates is that:

Every observer measures the same value for the speed of light independent of the relative motion of the sources and observers.

Consider a light source S and two observers, R_1 at rest relative to S and R_2 moving toward S with speed v, as shown in Figure 28-4a. The

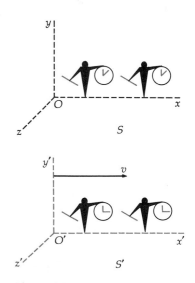

(a)   (b)

Figure 28-4
If absolute motion cannot be
detected, the reference frame
shown in (a), in which the
receiver R_1 and source S are
stationary and receiver R_2 is
moving toward the source
with speed v, is equivalent to
the reference frame shown in
(b), in which R_2 is stationary
and the source S and receiver
R_1 are moving with speed v.
Since the speed of light does
not depend on the source
speed, receiver R_2 measures
the same value for that speed
as receiver R_1.

speed of light measured by R_1 is $c = 3 \times 10^8$ m/sec. What is the speed
measured by R_2? The answer is *not* $c + v$. By postulate 1, Figure 28-4a
is equivalent to Figure 28-4b, in which R_2 is pictured at rest and the
source S and R_1 are moving with speed v. That is, since absolute motion has no meaning, it is not possible to say which is really moving
and which is at rest. By postulate 2, the speed of light from a moving
source is independent of the motion of the source. Thus, looking at
Figure 28-4, we see that R_2 measures the speed of light to be c, the
same as measured by R_1.

This result—that all observers measure the same value for the speed
of light—contradicts our intuitive ideas about relative velocities. If a
car moves at 30 mi/h away from an observer and another car moves at
70 mi/h in the same direction, the velocity of the second car relative to
the first car is 40 mi/h. This result is easily measured and conforms to
our intuition. However, according to Einstein's postulates, if a light
beam is moving in the direction of the cars, observers in both cars will
measure the same speed for the light beam. We shall see later that our
intuitive ideas about the combination of velocities are approximations
which hold only when the speeds are very small compared with the
speed of light. In practice, even in a plane moving with the speed of
sound, it is not possible to measure the speed of light accurately
enough to distinguish the difference between the result c and $c \pm v$,
where v is the speed of the plane. In order to make such a distinction,
we must either move with a very great velocity (much greater than
that of sound) or make extremely accurate measurements, as in the
Michelson-Morley experiment.

Einstein recognized that his postulates have important consequences for measuring time intervals and space intervals as well as
relative velocities. He showed that the size of time and space intervals
between two events depends on the reference frame in which the
events are observed. The changes in such measurements from one reference frame to another are called *time dilation* and *length contraction*.
Instead of developing the general formalism of relativity, known as the
Lorentz transformation, we shall derive these famous relativistic effects
directly from the Einstein postulates by considering some simple
special cases of measurement of time and space intervals. We shall also
study the problems of clock synchronization and simultaneity, which
are of central importance in understanding special relativity. In these
discussions we shall be comparing measurements made by observers
who are moving relative to each other. We shall use a rectangular coordinate system xyz with origin O, called the S *reference frame*, and
another system $x'y'z'$ with origin O', called the S' *frame*, which is
moving with constant velocity relative to the S frame. For simplicity,
we shall consider the S' frame to be moving with speed v along the x
(or x') axis relative to S. In each frame we assume that there are as
many observers as needed equipped with clocks, metersticks, etc.,
which are identical when compared at rest (see Figure 28-5).

Figure 28-5
Coordinate reference frames S
and S' moving with relative
speed v. In each frame are observers with metersticks and
clocks.

28-3   Time Dilation and Length Contraction

Since the results of measurement of the time and space intervals between events do not depend on the kind of apparatus used for the measurements or on the events, we are free to choose any events and measuring apparatus which will help us understand the application of the Einstein postulates to the results of the measurement. Convenient events in relativity are those which produce light flashes. A convenient clock is a light clock, pictured schematically in Figure 28-6. A photocell detects the light pulse and sends a voltage pulse to an oscilloscope giving a vertical deflection of the trace on the scope. The phosphorescent material on the face of the oscilloscope tube gives a persistent light that can be observed visually or photographed. The time between two light flashes is determined by measuring the distance between pulses on the scope and knowing the sweep speed of the scope. Such a clock, which can easily be calibrated and compared with other types of clocks, is often used in nuclear-physics experiments.

We first consider an observer A' at rest in frame S' a distance D from a mirror, as shown in Figure 28-7a. He explodes a flash gun and measures the time interval $\Delta t'$ between the original flash and the return flash from the mirror. Since light travels with speed c, this time is

$$\Delta t' = \frac{2D}{c} \qquad\qquad 28\text{-}6$$

We now consider these same two events, the original flash of light and the returning flash, as observed in reference frame S, where observer A' and the mirror are moving to the right with speed v. The events happen at two different places x_1 and x_2 in frame S because between the original flash and the return flash observer A' has moved a horizontal distance $v \, \Delta t$, where Δt is the time interval between the events measured in S. In Figure 28-7b we see that the path traveled by the light is longer in S than in S'. However, by Einstein's postulates, light travels with the same speed c in frame S as it does in frame S'. Since it travels farther in S at the same speed, it takes longer in S to reach the mirror and return. The time interval in S is thus longer than

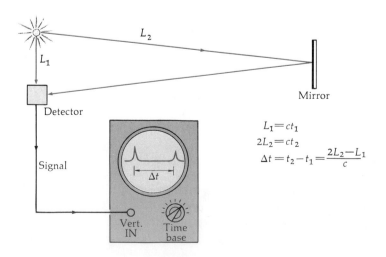

$$L_1 = ct_1$$
$$2L_2 = ct_2$$
$$\Delta t = t_2 - t_1 = \frac{2L_2 - L_1}{c}$$

Figure 28-6
Light clock for measuring time intervals. The time is measured by reading the distance between pulses on the oscilloscope after calibrating the sweep speed.

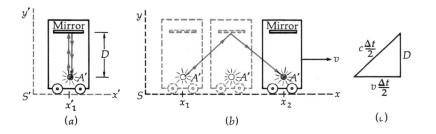

Figure 28-7
(a) A' and mirror are at rest in
S'. The time for a light pulse
to reach the mirror and return
is measured to be $2D/c$ by A'.
(b) In frame S, A' and the
mirror are moving with speed
v. If the speed of light is the
same in both reference
frames, the time for the light
to reach the mirror and return
is longer than $2D/c$ in S
because the distance traveled
is greater than $2D$. (c) Right
triangle for computing the
time Δt in frame S.

it is in S'. We can easily calculate Δt in terms of $\Delta t'$. From the triangle in Figure 28-7c we have

$$\left(\frac{c\,\Delta t}{2}\right)^2 = D^2 + \left(\frac{v\,\Delta t}{2}\right)^2 \qquad\qquad 28\text{-}7$$

or

$$\Delta t = \frac{2D}{\sqrt{c^2 - v^2}} = \frac{2D}{c}\frac{1}{\sqrt{1 - v^2/c^2}}$$

Using $\Delta t' = 2D/c$, we have

$$\Delta t = \frac{\Delta t'}{\sqrt{1 - v^2/c^2}} = \gamma\,\Delta t' \qquad \text{where } \gamma = \frac{1}{\sqrt{1\ v^2/c^2}} \geqslant 1 \qquad 28\text{-}8 \qquad \textit{Time dilation}$$

Observers in S would say that the clock held by A' runs slow since he claims a shorter time interval for these events.

A' measures the times of the light flash and return at the same point in S', while in S these events happen at two different places. A single clock can be used in S' to measure the time interval, but in S, two synchronized clocks are needed, one at x_1 and one at x_y. The time between events that happen at the *same place* in a reference frame (as with A' in S' in this case) is called *proper time*. The time interval measured in any other reference frame is always longer than the proper time. This expansion is called *time dilation*.

Time dilation is closely related to another phenomenon, *length contraction*. The length of an object measured in the reference frame in which the object is at rest is called its *proper length*. In a reference frame in which the object is moving, the measured length is shorter than its proper length. We can see this from our previous example using the light clocks. Suppose that x_1 and x_2 in that example are at the ends of a measuring rod of length $L_0 = x_2 - x_1$ measured in frame S in which the rod is at rest. Since A' is moving relative to this frame with speed v, the distance moved in time Δt is $v\,\Delta t$. Since he moves from point x_1 to point x_2 in this time, this distance is $L_0 = x_2 - x_1 = v\,\Delta t$. According to observer A' in frame S' (Figure 28-8), the measuring rod moves with speed v and takes a time $\Delta t'$ to move past him. The length of the rod in his frame is $L' = v\,\Delta t'$. Since the time interval $\Delta t'$ is less than Δt, the length L' is less than L_0. These lengths are related by

$$L' = v\,\Delta t' = \frac{v\,\Delta t}{\gamma} = \frac{L_0}{\gamma} = \sqrt{1 - \frac{v^2}{c^2}}\,L_0 \qquad 28\text{-}9 \qquad \textit{Length contraction}$$

(Both A and A' measure the same relative velocity. Otherwise there would be a lack of symmetry, and postulate 1 would be violated; i.e., we could choose the frame with the smaller or greater relative velocity to be a preferred frame.) Thus the length of the rod is smaller when

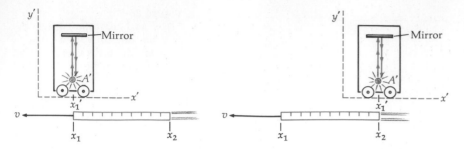

Figure 28-8
Measuring the length of a moving object. The rod has length $L_0 = x_2 - x_1 = v\,\Delta t$ as measured in frame S, in which it is at rest. In frame S' it moves past a fixed point in time $\Delta t' = \sqrt{1 - v^2/c^2}\,\Delta t$, so its length $L' = v\,\Delta t'$ is shorter in this frame.

measured in a frame in which it is moving. Since this contraction is just the amount proposed by Lorentz and FitzGerald to explain the Michelson-Morley experiment, it is often called the *Lorentz-FitzGerald contraction.*

An interesting example of the observation of these phenomena is afforded by the appearance of muons as secondary radiation from cosmic rays. Muons decay according to the statistical law of radioactivity,

$$N(t) = N_0 e^{-t/T} \qquad\qquad 28\text{-}10$$

where N_0 is the number at time $t = 0$, $N(t)$ is the number at time t, and T is the mean lifetime, which is about 2 μsec for muons at rest. Since they are created (from the decay of π mesons) high in the atmosphere, usually several thousand meters above sea level, few muons should reach sea level. A typical muon moving with speed of $0.998c$ would travel only about 600 m in 2 μsec. However, the lifetime of the muon measured in the earth's reference frame is increased by the factor $1/\sqrt{1 - v^2/c^2}$, which is 15 for this particular speed. The mean-lifetime measured in the earth's reference frame is therefore 30 μsec, and a muon of this speed travels about 9000 m in this time. From the muon's point of view, it lives only 2 μsec, but the atmosphere is rushing past it with a speed of $0.998c$. The distance of 9000 m in the earth's frame is thus contracted to only 600 m, as indicated in Figure 28-9.

It is easy to distinguish experimentally between the classical and relativistic predictions of the observation of muons at sea level. Suppose that we observe 10^8 muons at an altitude of 9000 m in some time interval with a muon detector. How many would we expect to observe at sea level in the same time interval? According to the nonrelativistic prediction, the time taken for these muons to travel 9000 m is (9000 m)/$0.998c \approx 30$ μsec, which is 15 lifetimes. Putting $N_0 = 10^8$ and $t = 15T$ into Equation 28-10, we obtain

$$N = 10^8 e^{-15} = 30.6$$

We would thus expect all but about 31 of the original 100 million muons to decay before reaching sea level.

According to the relativistic prediction, the earth must travel only the contracted distance of 600 m in the rest frame of the muon. This takes only 2 μsec $= T$. Thus the number expected at sea level is

$$N = 10^8 e^{-1} = 3.68 \times 10^7$$

Thus relativity predicts that we would observe 36.8 million muons in the same time interval. Experiments of this type have confirmed the relativistic predictions.

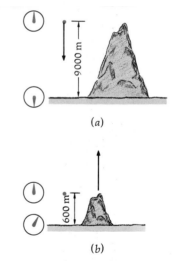

(a)

(b)

Figure 28-9
Although muons are created high above the earth and their mean lifetime is only about 2 μsec when at rest, many appear at the earth's surface. (*a*) In the earth's reference frame a typical muon moving at $0.998c$ has a mean lifetime of 30 μsec and travels 9000 m in this time. (*b*) In the reference frame of the muon, the distance traveled by the earth is only 600 m in the muon's lifetime of 2 μsec.

Question

1. You are standing on a corner and a friend is driving past in an automobile. Both note the times when the car passes two different intersections. You each determine from your watch readings the time that elapses between these two events. Which of you has determined the proper time interval?

28-4 Clock Synchronization and Simultaneity

At first glance, time dilation and length contraction seem contradictory not only to our intuition but to our ideas of self-consistency. If each reference frame can be considered at rest with the other moving, the clocks in the "other" frame should run slow. How can there be any self-consistency if each observer sees the clocks of the other run slow? The answer to this puzzle lies in the problems of clock synchronization and in the concept of simultaneity. We note that the time intervals Δt and $\Delta t'$ considered in Section 28-3 were measured in quite different ways. The events in frame S' happened at the same place, and the times could be measured on a single clock; but in S, the two events happened at different places. The time of each event was measured on a different clock and the interval found by subtraction. This procedure requires that the clocks be synchronized. We shall show in this section that

Two clocks synchronized in one reference frame are not synchronized in any other frame moving relative to the first frame.

A corollary to this result is that

Two events which are simultaneous in one reference frame are not simultaneous in another frame moving relative to the first

(unless the events and clocks are in the same plane perpendicular to the relative motion).

Comprehension of these facts usually resolves all relativity paradoxes. Unfortunately, the intuitive (and incorrect) belief that simultaneity is an absolute relation is difficult to get rid of.

How can we synchronize two clocks separated in space? The problem is not difficult but requires some thought. One obvious method is to bring the clocks together and set them to read the same time, then move them to their original positions. This method has the drawback that, as we have seen, each clock runs slowly during the time it is moved, according to the time-dilation result. Suppose the clocks are at rest at points A and B a distance L apart in frame S. If an observer at A looks at the clock at B and sets his clock to read the same time, the clocks will not be synchronized because of the time L/c it takes light to travel from one clock to another. To synchronize the clocks, the observer at A must set his clock ahead by the time L/c. Then he will see that the clock at B reads a time which is L/c behind the time on his clock, but he will calculate that the clocks are synchronized when he allows for the time L/c for the light to reach him. All observers except those midway between the clocks will see the clocks reading different times, but they will also compute that the clocks are synchronized when they correct for the time it takes the light to reach them. An equivalent method for the synchronization of two clocks would be for a third observer C at a point midway between the clocks to send a light signal and for observers at A and B to set their clocks to some prearranged time when they receive the signal.

We now examine the question of simultaneity. Suppose A and B agree to explode bombs at t_0 (having previously synchronized their clocks). Observer C will see the light from the two explosions at the same time, and since he is equidistant from A and B, he will conclude that the explosions were simultaneous. Other observers in S will see the light from A or B first, depending on their location, but after correcting for the time the light takes to reach them, they also will conclude that the explosions were simultaneous. *We shall thus define two events in a reference frame to be simultaneous if the light signals from the events reach an observer halfway between the events at the same time.* To show that two events which are simultaneous in frame S are not simultaneous in another frame S' moving relative to S we use an example introduced by Einstein. A train is moving with speed v past the station platform. We have observers A', B', and C' at the front, back, and middle of the train. (We shall consider the train to be at rest in S' and the platform in S.) We now suppose that the train and platform are struck by lightning at the front and back of the train and that the lightning bolts are simultaneous in the frame of the platform (S) (Figure 28-10). That is, an observer C halfway between the positions A and B, where the lightning strikes, observes the two flashes at the same time. It is convenient to suppose that the lightning scorches the train and platform so that the events can be easily located in each reference frame. Since C' is in the middle of the train, halfway between the places on the train which are scorched, the events can be simultaneous in S' only if C' sees the flashes at the same time. However, C' sees the flash from the front of the train before he sees the flash from the back. In frame S, when the light from the front flash reaches him, he has moved some distance toward it, so that the flash from the back has not yet reached him, as indicated in Figure 28-11. He must therefore conclude that the events are not simultaneous. The front of the train was struck before the back. As we have discussed above, all observers in S' on the train will agree with C' when they have corrected for the time it takes light to reach them.

Let L_0 be the distance between the scorch marks on the platform, which is also the length of the train measured in frame S. This distance is smaller than the proper length of the train L_T' because of length contraction. Figure 28-12 shows the situation in frame S', in which the train is at rest and the platform is moving. In this frame the

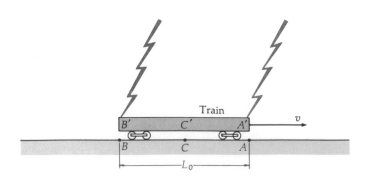

Figure 28-10
Simultaneous lightning bolts strike the ends of a train traveling with speed v in frame S attached to the platform. The light from these simultaneous events reaches observer C midway between the events at the same time. The distance between the bolts is L_0, which is also the length of the train measured in frame S.

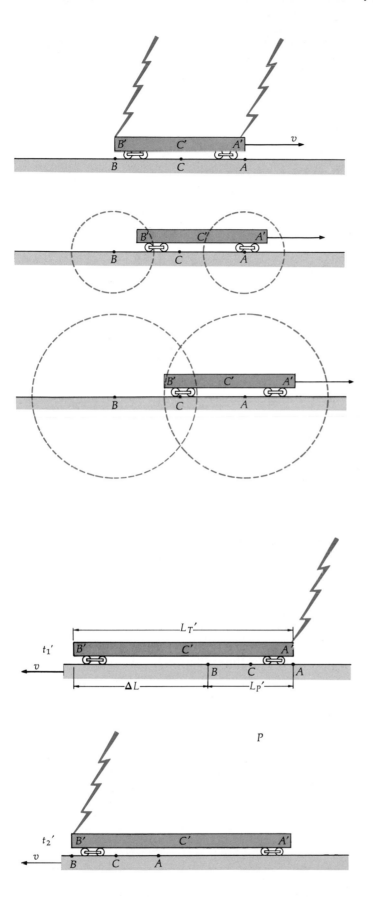

Figure 28-11
In frame S the light from the bolt at the front of the train reaches observer C' in the middle of the train before light from the bolt at the rear of the train because the train is moving. Since C' is midway between the events (which occur at the front and rear of the train), these events are not simultaneous for him.

Figure 28-12
Lightning bolts of Figures 28-10 and 28-11 as seen in reference frame S' fixed to the train. In this frame the distance $L_{P}' = BA$ along the moving platform is contracted, and the train length $L_{T}' = B'A'$ is its proper length. At time t_1' lightning strikes the front of the train when A' and A are coincident. At a later time t_2' the lightning strikes the rear of the train when B' and B are coincident. During the time between flashes the platform moves the distance $L_{T}' - L_{P}'$.

distance between the burns on the platform is contracted and is related to L_0 by

$$L_p' = \frac{L_0}{\gamma} \qquad\qquad 28\text{-}11$$

Similarly the train length is the proper length related to L_0 by

$$L_T' = \gamma L_0 \qquad\qquad 28\text{-}12$$

(The drawing of this figure has been made for $\gamma = 1.5$.) The time interval in S' between these two events is the time it takes the platform
to move the distance ΔL, where

$$\Delta L = L_T' - L_p' = \gamma L_0 - \frac{L_0}{\gamma} = \left(1 - \frac{1}{\gamma^2}\right) \gamma L_0 = \frac{v^2}{c^2} \gamma L_0$$

since

$$1 - \frac{1}{\gamma^2} = 1 - \left(1 - \frac{v^2}{c^2}\right) = \frac{v^2}{c^2}$$

Thus

$$t_2' - t_1' = \frac{\Delta L}{v} = \frac{\gamma L_0 v}{c^2}$$

During this time (according to observers in S'), the clocks at A and B
ticked off a time interval which is smaller because of time dilation:

$$\Delta t_s = \frac{1}{\gamma} (t_2' - t_1') = \frac{L_0 v}{c^2} \qquad\qquad 28\text{-}13$$

This is the time interval by which the clocks in S are unsynchronized
according to observers in S'. That is, the clock at A is ahead of that at B
by the amount $L_0 v / c^2$. This result is worth remembering.

---

*If two clocks are synchronized in the frame in which they are at rest, they
will be out of synchronization in another frame. In the frame in which they
are moving, the "chasing clock" leads by an amount $\Delta t_s = L_0 v / c^2$, where L_0
is the proper distance between the clocks.*

*Lack of synchronization of
moving clocks*

Example 28-1 A numerical example should help clarify time dilation,
clock synchronization, and the internal consistency of these results.
Let the light clock used in Section 28-3 be moving with speed $v = 0.8c$.
Then

$$1 - \frac{v^2}{c^2} = 1 - 0.64 = 0.36 \quad \text{and} \quad \gamma = \frac{1}{\sqrt{1 - v^2/c^2}} = \frac{1}{\sqrt{0.36}} = \frac{1}{0.6} = \frac{5}{3}$$

Also, let the distance $x_2 - x_1$ be 40 light-minutes, that is, the distance
light travels in 40 min (a convenient notation for this unit is c-min):

$$x_2 - x_1 = 40 \ c\text{-min}$$

The time for the light clock to travel this distance is Δt given by

$$\Delta t = \frac{x_2 - x_1}{v} = \frac{40 \ c\text{-min}}{0.8c} = 50 \ \text{min}$$

(Note that the c in the unit c-min cancels, giving the time in
minutes.)

The proper time interval in S' is shorter by the factor γ,

$$\Delta t' = \frac{\Delta t}{\gamma} = \frac{50 \text{ min}}{\frac{5}{3}} = 30 \text{ min}$$

(Thus we have taken $D = 15$ c-min for the distance from the flash gun to the mirror in Figure 28-7.) The distance traveled is

$$\Delta x' = v\,\Delta t' = (0.8c)\,(30 \text{ min}) = 24 \text{ } c\text{-min}$$

This is just the distance $x_2 - x_1$ which is contracted in S':

$$L' = \frac{L_0}{\lambda} = \frac{40 \text{ } c\text{-min}}{\frac{5}{3}} = 24 \text{ } c\text{-min}$$

Figure 28-13 shows the situation viewed in S'. We assume that the clock at x_1 reads noon at the time of the light flash. The clocks at x_1 and x_2 are synchronized in S but not in S'. In S', the clock at x_2, which is chasing the one at x_1, leads by

$$\frac{L_0 v}{c^2} = \frac{(40 \text{ } c\text{-min}) \,(0.8c)}{c^2} = 32 \text{ min}$$

When the light clock coincides with x_2, the clock there reads 50 min past noon. Thus the time between events is 50 min in S. Note that according to observers in S', this clock ticks off $50 - 32 = 18$ min for a trip which takes 30 min in S'. Thus this clock runs slow by the factor $30/18 = \frac{5}{3}$.

Thus each observer sees clocks in the other frame run slow. According to observers in S who measure 50 min for the time interval, the time interval in S' is too small (30 min) because the single clock in S' runs too slow by the factor $\frac{5}{3}$. According to the observers in S', the observers in S measure a time which is too *long* despite the fact that their clocks run too slow, because they are out of synchronization. The clocks tick off only 18 min, but the second one leads the first by 32 min, so the time interval found by subtraction is 50 min.

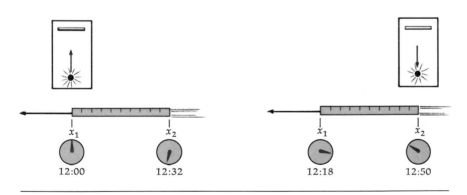

x_1 12:00 x_2 12:32 x_1 12:18 x_2 12:50

Figure 28-13
Example 28-1. In frame S' the rod passes the light clock in a time $\Delta t' = 30$ min. During this time the clocks at each end of the rod tick off $30/\gamma$ min $= 18$ min. But the clocks are unsynchronized, with the chasing clock leading by $L_0 v/c^2 = 32$ min, where L_0 is the proper separation of the clocks. The time interval in S is therefore 32 min + 18 min $= 50$ min, which is longer than 30 min.

Questions

2. Two observers are in relative motion. In what circumstances can they agree on the simultaneity of two different events?

3. If event A occurs before event B in some frame, might it be possible for there to be a reference frame in which event B occurs before event A?

4. Two events are simultaneous in a frame in which they also occur at the same point in space. Are they simultaneous in other reference frames?

28-5   The Doppler Effect

In deriving the doppler effect (Chapter 24) we found that the change in frequency for a given velocity v depends on whether it is the source or receiver that is moving with that speed. Such a distinction is possible for sound because there is a medium (the air) relative to which the motion takes place, and so it is not surprising that the motion of the source or the receiver relative to the still air can be distinguished. Such a distinction between motion of the source or receiver cannot be made for light or other electromagnetic waves in vacuum. The expressions we have derived for the doppler effect cannot be correct for light. We now derive the relativistic doppler effect (Equation 24-16).

We shall consider a source moving toward a receiver with velocity v and work in the frame of the receiver. Let the source emit N waves. If the source is moving toward the receiver, the first wave will travel a distance $c\,\Delta t_R$ and the source will travel $v\,\Delta t_R$ in the time Δt_R measured in the frame of the receiver. The wavelength will be $\lambda' = [c\,\Delta t_R - (v\,\Delta t_R)]/N$. The frequency f' observed by the receiver will therefore be

$$f' = \frac{c}{\lambda'} = \frac{c}{c-v}\frac{N}{\Delta t_R} = \frac{1}{1 - v/c}\frac{N}{\Delta t_R}$$

If the frequency of the source is f_0, it will emit $N = f_0\,\Delta t_S$ waves in time Δt_S, measured by the source. Here Δt_S is the proper time interval (the first wave and the Nth wave are emitted at the same place in the source's reference frame). Times Δt_S and Δt_R are related by the usual time-dilation equation $\Delta t_S = \Delta t_R/\gamma$. Thus we obtain for the doppler effect for a moving source,

$$f' = \frac{1}{1 - v/c}\frac{f_0\,\Delta t_S}{\Delta t_R} = \frac{f_0}{1 - v/c}\frac{1}{\gamma} = \frac{\sqrt{1 - v^2/c^2}}{1 - v/c}f_0 \qquad 28\text{-}14a$$

which differs from our classical equation only in the time-dilation factor.

We now do the calculation in the reference frame of the source. That is, we assume that the source is at rest and the receiver moves with velocity v toward the source. In time Δt_S in the frame of the source, the receiver encounters all the waves in the distance $v\,\Delta t_S$ in addition to the waves in the distance $c\,\Delta t_S$, just as in the classical calculation. The number of waves encountered is thus

$$N = \frac{c\,\Delta t_S + v\,\Delta t_S}{\lambda} = \frac{(c+v)\,\Delta t_S}{c/f_0} = \left(1 + \frac{v}{c}\right)f_0\,\Delta t_S$$

where we have used $\lambda = c/f_0$ for the wavelength. In the classical calculation, we need only divide N by the time interval Δt to obtain the frequency observed by the moving receiver, but here we must be careful to divide by the time interval in the receiver's frame to find the observed frequency. In this case the receiver's time interval Δt_R is proper time, i.e., the time interval between encountering the first wave and the Nth wave. Both events occur at the receiver. Thus $\Delta t_R = \Delta t_S/\gamma$, and the frequency observed is

$$f' = \frac{N}{\Delta t_R} = \left(1 + \frac{v}{c}\right)f_0\frac{\Delta t_S}{\Delta t_R} = \gamma\left(1 + \frac{v}{c}\right)f_0 = \frac{1 + v/c}{\sqrt{1 - v^2/c^2}}f_0 \qquad 28\text{-}14b$$

It is left to the student to show that this result is identical to Equation 28-14a.

When the source and receiver move away from each other with relative speed v, the frequency received is given by

$$f' = \frac{1 - v/c}{\sqrt{1 - v^2/c^2}}\, f_0 = \frac{\sqrt{1 - v^2/c^2}}{1 + v/c}\, f_0 \qquad\qquad \text{28-15}$$

Relativistic doppler effect

Question

5. How do the relativistic equations for the doppler effect for sound waves differ from those for light waves?

28-6 The Lorentz Transformation

We now consider the general relation between the coordinates x, y, z, and t of an event as seen in reference frame S and the coordinates x', y', z', and t' of the same event as seen in reference frame S', which is moving with uniform velocity relative to S. We shall consider only the simple special case in which the origins are coincident at time $t = t' = 0$ and S' is moving with speed v along the x (or x') axis. The classical relation, called the *galilean transformation*, is

$$x = x' + vt' \qquad y = y' \qquad z = z' \qquad \text{and} \qquad t = t' \qquad\qquad \text{28-16}$$

with the inverse transformation

$$x' = x - vt \qquad y' = y \qquad z' = z \qquad \text{and} \qquad t' - t$$

(For the rest of this discussion we shall ignore the equations for y and z, which do not change for this special case of motion along the x and x' axes.) These equations are consistent with experiment as long as v is much less than c. They lead to the familiar classical addition law for velocities. If a particle has velocity $u_x - dx/dt$ in frame S, its velocity in frame S' is

$$u_x' = \frac{dx'}{dt'} = \frac{dx}{dt} - v = u_x - v \qquad\qquad \text{28-17}$$

If we differentiate this equation again, we find that the acceleration of the particle is the same in both frames: $a_x = du_x/dt = du_{x}'/dt = a_{x}'$.

It should be clear that this transformation is not consistent with the Einstein's postulates of special relativity. If light moves along the x axis with speed c in S, these equations imply that the speed in S' is $u_x' = c - v$ rather than $u_x' = c$, which is consistent with Einstein's postulates and with experiment. The classical transformation equations must therefore be modified to be consistent with Einstein's postulates and to reduce to the classical equations when v is much less than c. We shall give a brief outline of one method of obtaining the relativistic transformation which is called the *Lorentz transformation*. We assume that the equation for x is of the form

$$x = K(x' + vt')$$

where K is a constant which can depend on v and c but not on the coordinates. For this equation to reduce to the classical one, K must approach 1 as v/c approaches zero. The inverse transformation must look the same except for the sign of the velocity:

$$x' = K(x - vt)$$

Let us assume that a light pulse starts at the origin at $t = 0$. Since we have assumed that the origins are coincident at $t = t' = 0$, the pulse also starts out at the origin of S' at $t' = 0$. The equation for the wavefront of the light pulse is $x = ct$ in frame S and $x' = ct'$ in S'. Substituting these in our transformation equations gives

$$ct = K(ct' + vt') = K(c + v)t'$$

and

$$ct' = K(ct - vt) = K(c - v)t$$

We can eliminate either t' or t from these two equations and determine K. The result is $K^2 = (1 - v^2/c^2)^{-1}$, and K is the same as γ:

$$K = \gamma = \frac{1}{\sqrt{1 - v^2/c^2}}$$

The transformation is therefore $x = \gamma(x' + vt')$. We can obtain equations for t and t' by combining this transformation with its inverse

$$x' = \gamma(x - vt)$$

Using this value for x' in the above equation, we get

$$x = \gamma \left[\gamma(x - vt) + vt' \right]$$

which can be solved for t' in terms of x and t. The complete Lorentz transformation is:

$$x = \gamma(x' + vt') \qquad y = y'$$

$$t = \gamma \left(t' + \frac{vx'}{c^2} \right) \qquad z = z'$$

28-18 *Lorentz transformation*

with the inverse

$$x' = \gamma(x - vt) \qquad y' = y$$

$$t' = \gamma \left(t - \frac{vx}{c^2} \right) \qquad z' = z$$

28-19

Example 28-2 Two events occur at the same place x_0 at times t_1 and t_2 in S. What is the time interval between them in S'?

$$t_1' = \gamma \left(t_1 - \frac{vx_0}{c^2} \right) \qquad t_2' = \gamma \left(t_2 - \frac{vx_0}{c^2} \right)$$

$$t_2' - t_1' = \gamma(t_2 - t_1)$$

This is our familiar result for time dilation since $t_2 - t_1$ is the proper time interval.

We can obtain the velocity transformation by differentiating the Lorentz transformation equations or merely by taking differences. We shall do the latter. Suppose a particle moves a distance Δx in time Δt in frame S. Its velocity is $u_x = \Delta x/\Delta t$. Using the transformation equations, we obtain

$$\Delta x' = \gamma(\Delta x - v\,\Delta t) \qquad \text{and} \qquad \Delta t' = \gamma \left(\Delta t - \frac{v\,\Delta x}{c^2} \right)$$

The velocity in S' is

$$u_x' = \frac{\Delta x'}{\Delta t'} = \frac{\gamma(\Delta x - v\,\Delta t)}{\gamma \left(\Delta t - \frac{v\,\Delta x}{c^2} \right)} = \frac{\frac{\Delta x}{\Delta t} - v}{1 - \frac{v}{c^2} \frac{\Delta x}{\Delta t}}$$

or

$$u_x' = \frac{u_x - v}{1 - vu_x/c^2}$$ 28-20   *Velocity addition*

Similarly the inverse velocity transformation is

$$u_x = \frac{u_x' + v}{1 + vu_x'/c^2}$$ 28-21

If a particle has components of velocity along y and z, it is not difficult to find the components in S'. In these cases, $\Delta y' = \Delta y$ and $\Delta z' = \Delta z$, with the same relation between $\Delta t'$ and Δt. Thus we obtain

$$u_y' = \frac{\Delta y'}{\Delta t'} = \frac{\Delta y}{\gamma \left(\Delta t - \dfrac{v\, \Delta x}{c^2} \right)} = \frac{\Delta y/\Delta t}{\gamma \left(1 - \dfrac{v}{c^2} \dfrac{\Delta x}{\Delta t} \right)}$$

or

$$u_y' = \frac{u_y}{\gamma(1 - vu_x/c^2)}$$ 28-22

with a similar result for u_z'. The inverse is

$$u_y = \frac{u_y'}{\gamma(1 + vu_x'/c^2)}$$ 28-23

Example 28-3 Light moves along the x axis with speed $u_x - c$. What is its speed in S'?

$$u_x' = \frac{c - v}{1 - vc/c^2} = \frac{c(1 - v/c)}{1 - v/c} = c$$

as required by the postulates.

Question

6. The Lorentz transformation for y and z is the same as the classical result: $y = y'$ and $z = z'$. Yet the relativistic velocity transformation does not give the classical result $u_y = u_y'$ and $u_z = u_z'$. Explain.

28-7 The Twin Paradox

Homer and Ulysses are identical twins. Ulysses travels at high speed to a planet beyond the solar system and returns while Homer remains at home. When they are together again, which twin is older, or are they the same age? The correct answer is that Homer, the twin who stays at home, is older. This problem, with variations, has been the subject of spirited debate for decades, though there are very few who disagree with the answer.[1]

The problem is a paradox because of the seemingly symmetric roles played by the twins with the asymmetric result in their aging. The paradox is resolved when the asymmetry of the twins' roles is noted. The relativistic result conflicts with common sense based on our strong but incorrect belief in absolute simultaneity. We shall consider

[1] A collection of some important papers concerning this paradox can be found in *Special Relativity Theory, Selected Reprints,* American Association of Physics Teachers, New York, 1963.

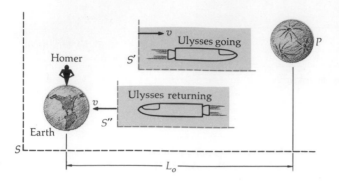

a particular case with some numerical magnitudes which, though impractical, make the calculations easy.

Let planet P and Homer on earth be fixed in reference frame S a distance L_0 apart, as in Figure 28-14. We neglect the motion of the earth. Reference frames S' and S'' are moving with speed v toward and away from the planet, respectively. Ulysses quickly accelerates to speed v, then coasts in S' until he reaches the planet, when he stops and is momentarily at rest in S. To return he quickly accelerates to speed v toward earth and coasts in S'' until he reaches earth, where he stops. We can assume that the acceleration times are negligible compared with the coasting times. (Given the time needed to reach the speed v, we can formulate the problem with L_0 large enough to meet this condition.) We use the following values for illustration: $L_0 = 8$ light-years and $v = 0.8c$; then $\sqrt{1 - v^2/c^2} = \frac{3}{5}$.

It is easy to analyze the problem from Homer's point of view. According to his clock, Ulysses coasts in S' for a time $L_0/v = 10$ years and in S'' for an equal time. Thus Homer is 20 years older when Ulysses returns. The time interval in S' between leaving earth and arriving at the planet is shorter because it is proper time; the time to reach the planet by Ulysses' clock is

$$\Delta t' = \sqrt{1 - \frac{v^2}{c^2}} \frac{L_0}{v} = \frac{3}{5} \times 10 = 6 \text{ years}$$

Since the same time is required for the return trip, Ulysses will have recorded 12 years for the round trip and will be 8 years younger than Homer.

From Ulysses' point of view, the calculation of his trip time is not difficult. The distance from earth to planet is contracted and is only

$$L_0 \sqrt{1 - \frac{v^2}{c^2}} = \frac{3}{5} \times 8 = 4.8 \text{ light-years}$$

At $v = 0.8c$, it takes only 6 years each way. The real difficulty in this problem is for Ulysses to understand why his twin ages 20 years during his absence. If we consider Ulysses at rest and Homer moving away, doesn't Homer's clock measure proper time? When Ulysses measures 6 years, Homer should measure only $\frac{3}{5}(6) = 3.6$ years. Then why shouldn't Homer age only 7.2 years during the round trip? This, of course, is the paradox. The difficulty with the analysis from the point of view of Ulysses is that he does not remain in an inertial frame. What happens while Ulysses is stopping and starting? To investigate this problem in detail, we would need to treat accelerated reference frames, a subject dealt with in the study of general relativity but beyond the scope of this book. However, we can get some insight into the problem by considering the lack of synchronization of moving clocks.

Suppose that there is a clock at the planet P synchronized in S with Homer's clock at earth. In reference frame S', these clocks are un-synchronized by the amount $L_0 v/c^2$. For our example this is 6.4 years. Thus when Ulysses is coasting in S' near the planet, the clock at the planet leads that at the earth by 6.4 years. After he stops, he is in the frame S in which these two clocks are synchronized. Thus in the negligible time (according to Ulysses) it takes him to stop, the clock at earth must gain 6.4 years. Accordingly, his twin on earth ages 6.4 years. This 6.4 years plus the 3.6 years that Homer aged during the coasting makes him 10 years older by the time Ulysses is stopped in frame S. When Ulysses is in frame S'' coasting home, the clock at earth leads that at the planet by 6.4 years, and it will run another 3.6 years before he arrives home. We do not need to know the detailed behavior of the clocks during the acceleration in order to know the cumulative effect; the special relativity theory is enough to show us that if the clocks on earth and at the planet are synchronized in S, the clock on earth lags that at P by $L_0/c^2 = 6.4$ years when viewed in S' and the clock on earth leads that at P by this amount when viewed in S''.

The difficulty in understanding the analysis of Ulysses lies in the difficulty of giving up the idea of absolute simultaneity. Suppose Ulysses sends a signal calculated to arrive at earth just as he arrives at P. If the arrival of this signal and the arrival of Ulysses at the planet are simultaneous in S', they are not simultaneous in S; in fact, the signal arrives at earth 6.4 years before Ulysses arrives at the planet, ac-cording to observers in S. The roles of the twins are not symmetric because Ulysses does not remain in an inertial reference frame but must accelerate.

It is instructive to have the twins send regular signals to each other so that they can record the other's age continuously. If they arrange to send a signal once a year, the age of the other can be determined merely by counting the signals received. The arrival frequency of the signals will not be 1 per year because of the doppler shift. The frequency observed will be given by Equation 28-14 or 28-15, which for our example is 3 per year or $\frac{1}{3}$ per year, depending on whether the source is approaching or receding.

Consider the situation first from the point of view of Ulysses. During the 6 years it takes him to reach the planet (remember that the distance is contracted in his frame), he receives signals at the rate of $\frac{1}{3}$ per year, and so he receives 2 signals. As soon as he turns around and starts back to earth, he receives 3 signals per year; in the 6 years it takes him to return he receives 18 signals, giving a total of 20 for the trip. He accordingly expects his twin to have aged 20 years.

We now consider the situation from Homer's point of view. He receives signals at the rate of $\frac{1}{3}$ per year not only for the 10 years it takes Ulysses to reach the planet but also for the 8 years it takes for the last signal sent by Ulysses from S' to get back to earth. (He cannot know that Ulysses has turned around until the signals reach him.) During the first 18 years, Homer receives 6 signals. In the final 2 years before Ulysses arrives, Homer receives 6 signals, or 3 per year. (The first signal sent after Ulysses turns around takes 8 years to reach earth, whereas Ulysses, traveling at $0.8c$, takes 10 years to return and there-fore arrives just 2 years after Homer begins to receive signals at the faster rate.) Thus Homer expects Ulysses to have aged 12 years. In this analysis, the asymmetry of the twins' roles is apparent. Both twins agree that when they are together again, the one who has been accelerated will be younger than the one who stayed home.

The predictions of the special theory of relativity concerning the twin paradox have been tested many times using small particles which can be accelerated to such large speeds that γ is appreciably greater than 1. Unstable particles can be accelerated and trapped in circular orbits in a magnetic field, for example, and their lifetimes compared with those of identical particles at rest. In all such experiments the accelerated particles live longer on the average than those at rest, as predicted. These predictions are also confirmed by the results of an experiment using high-precision atomic clocks flown around the world in commercial airlines, but the analysis of this experiment is complicated by the necessity of including gravitational effects treated in the general theory of relativity.[1]

28-8 Relativistic Momentum

Our discussion of Newton's second law in Chapter 4 showed that if $\Sigma \mathbf{F} = m\mathbf{a}$ holds in one reference frame, it holds in any other reference frame moving with constant velocity relative to the first. According to the galilean transformation, the accelerations in the two frames are equal, $a_x' = a_x$, and forces, e.g., those due to stretching of springs, are also the same in the two frames. However, according to the Lorentz transformation, accelerations are not the same in two such reference frames. If a particle has acceleration a_x' and velocity u_x' in frame S', its acceleration in S obtained by computing du_x/dt from Equation 28-21 is

$$a_x = \frac{a_x'}{\gamma^3 (1 + u_x' v/c^2)^3}$$

Thus either the force transforms in a similar way, or $\Sigma \mathbf{F} = m\mathbf{a}$ does not hold. It is reasonable to expect that $\Sigma \mathbf{F} = m\mathbf{a}$ does not hold at high speeds, for this equation implies that a constant force will accelerate a particle to unlimited velocity if it acts for a long time. However, if a particle's velocity were greater than c in some reference frame S', we could not transform from S' to the rest frame of the particle because γ becomes imaginary when $v > c$. We can see from the velocity transformation that if a particle's velocity is less than c in some frame S, it is less than c in all frames moving relative to S with $v < c$. Thus it seems reasonable to conclude that particles never have speeds greater than c.

We can avoid the question of how to transform forces by considering a problem in which the total force is zero, namely, a collision of two masses. In classical mechanics, the total momentum $\Sigma m_i \mathbf{u}_i$ is conserved. We can see by a simple example that this quantity, the classical total momentum, is not conserved relativistically. That is, the conservation of the quantity $\Sigma m_i \mathbf{u}_i$ is an approximation which holds only at low speeds.

Consider an observer in frame S with a ball A and one in S' with a ball B. The balls each have mass m and are identical when compared at rest. Each observer throws his ball along his y or y' axis with speed

[1] The details of these tests can be found in J. C. Hafele and Richard E. Keating, Around-the-World Atomic Clocks: Predicted Relativistic Time Gains and Around-the-World Atomic Clocks: Observed Relativistic Time Gains, *Science*, July 14, 1972, p. 166.

u_0 (measured in his own frame) so that the balls collide. Assuming the balls to be perfectly elastic, each observer will see his ball rebound with its original speed u_0. If the total momentum is to be conserved, the y component must be zero because the momentum of each ball is merely reversed by the collision. However, if we consider the relativistic velocity transformations, we can see that the quantity mu_y does not have the same magnitude for each ball as seen by either observer.

Let us consider the collision as seen in frame S (Figure 28-15). In this frame ball A moves along the y axis with velocity $u_{yA} = u_0$. Ball B has an x component of velocity $u_{xB} = v$ and a y component $u_{yB} = u_{yB}'/\gamma = -u_0\sqrt{1 - v^2/c^2}$. Here we have used the velocity-transformation Equations 28-21 and 28-23 and the facts that u_{yB}' is just $-u_0$ and $u_{xB}' = 0$. We see that the y component of velocity of ball B is smaller in magnitude than that of ball A. The factor $\sqrt{1 - v^2/c^2}$ comes from the time-dilation factor. The time taken for ball B to travel a given distance along the y axis in S is greater than the time measured in S' for the ball to travel this same distance. Thus in S, the total y component of classical momentum is not zero. Since the velocities are reversed in an elastic collision, momentum as defined by $\mathbf{P} = \Sigma m\mathbf{u}$ is not conserved in S. Analysis of this problem in S' leads to the same conclusion (Figure 28-15b). In the classical limit, $v \ll c$, momentum is conserved, of course, because $\gamma \approx 1$.

The reason for defining momentum to be $\Sigma m\mathbf{u}$ in classical mechanics was that this quantity is conserved when there are no external forces, as in collisions. We now see that this quantity is conserved only in the approximation $v \ll c$. We shall define *relativistic momentum* $\mathbf{p}$ of a particle to have the following properties:

1. $\mathbf{p}$ is conserved in collisions.

2. $\mathbf{p}$ approaches $m\mathbf{u}$ as u/c approaches zero.

We shall state without proof that the quantity meeting these conditions is

$$\mathbf{p} = \frac{m\mathbf{u}}{\sqrt{1 - u^2/c^2}} \qquad \text{28-24}$$

where u is the speed of the particle. We thus take this equation for the definition of relativistic momentum of a particle. It is clear that this definition meets our second criterion because the denominator approaches 1 when u is much less than c. Proof that it also meets the first criterion involves much tedious algebra. From this definition, the momenta of the two balls A and B as seen in S is

$$p_{yA} = \frac{mu_0}{\sqrt{1 - u_0^2/c^2}} \qquad p_{yB} = \frac{mu_{yB}}{\sqrt{1 - (u_{xB}^2 + u_{yB}^2)/c^2}}$$

where $u_{yB} = -u_0\sqrt{1 - v^2/c^2}$ and $u_{xB} = v$. We shall leave the details of showing that $p_{yB} = -p_{yA}$ as an exercise. Because of the similarity of the factor $1/\sqrt{1 - u^2/c^2}$ and γ in the Lorentz transformation, Equation 28-24 is often written

$$\mathbf{p} = \gamma m\mathbf{u} \qquad \text{with } \gamma = 1/\sqrt{1 - u^2/c^2} \qquad \text{28-25}$$

This use of the symbol γ for two different quantities can cause some confusion. The notation is standard, however, and simplifies many of

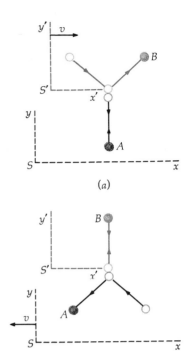

(a)

(b)

Figure 28-15
(a) Elastic collision of two identical balls as seen in frame S. The vertical component of the velocity of ball B is u_0/γ in S if it is u_0 in S'. (b) The same collision as seen in S'. In this frame ball A has vertical component of velocity u_0/γ.

Relativistic momentum defined

the equations. We shall use the notation except when considering transformations between reference frames. Then, to avoid confusion, we shall write out the factor $\sqrt{1 - u^2/c^2}$ and reserve γ for $1/\sqrt{1 - v^2/c^2}$, where v is the relative speed of the frames.

One interpretation of Equation 28-24 is that the mass of an object increases with speed. The quantity

$$\frac{m}{\sqrt{1 - u^2/c^2}} = \gamma m$$

is sometimes called the *relativistic mass*, written $m(u)$. The mass of the body in its rest frame, written m_0, is called the *rest mass*. Although this makes the expression $m(u)\mathbf{u}$ for relativistic momentum similar to the nonrelativistic expression, the use of relativistic mass often leads to mistakes. For example, the expression $\frac{1}{2}m(u)u^2$ is not the correct relativistic expression for kinetic energy. The measurement of relativistic mass involves measuring the force needed to produce a given change in momentum. The experimental evidence often cited as verification that mass depends on velocity can also be interpreted merely as evidence of the validity of the assumption that the force equals the time rate of change of relativistic momentum. We shall avoid using a symbol for relativistic mass. The symbol m in this book always refers to the rest mass.

28-9   Relativistic Energy

We have seen that the quantity $m\mathbf{u}$ is not conserved in collisions but that $\gamma m\mathbf{u}$ is, with $\gamma = 1/\sqrt{1 - u^2/c^2}$. Evidently Newton's law in the form $\Sigma\mathbf{F} = m\mathbf{a}$ cannot be correct relativistically, since it leads to the conservation of $m\mathbf{u}$. We can get a hint of the correct form of Newton's second law by writing it $\Sigma\mathbf{F} = d\mathbf{p}/dt$. Let us assume that this equation is correct if relativistic momentum $\mathbf{p}$ is used. The validity of this assumption can be determined only by examining its consequences, since an unbalanced force on a high-speed particle is measured by its effect on momentum and energy. We are essentially defining force by the equation $\Sigma\mathbf{F} = d\mathbf{p}/dt$. As in classical mechanics, we shall define kinetic energy as the work done by an unbalanced force in accelerating a particle from rest to some velocity. Considering one dimension only, we have

$$E_k = \int_{u=0}^u \Sigma F \, ds = \int_0^u \frac{d(\gamma mu)}{dt} \, ds = \int_0^u u \, d(\gamma mu) \qquad 28\text{-}26$$

using $u = ds/dt$. The computation of the integral in Equation 28-26 is not difficult but requires some messy algebra. It is left as a problem to show that

$$d(\gamma mu) = m \left(1 - \frac{u^2}{c^2}\right)^{-3/2} du$$

Substituting this into the integrand, in Equation 28-26, we obtain

$$E_k = \int_0^u u \, d(\gamma mu) = \int_0^u m \left(1 - \frac{u^2}{c^2}\right)^{-3/2} u \, du$$

$$= mc^2 \left(\frac{1}{\sqrt{1 - u^2/c^2}} - 1\right)$$

or

$$E_k = \gamma mc^2 - mc^2 \qquad\qquad\qquad 28\text{-}27$$

We can check this expression for low speeds by noting that for $u/c \ll 1$,

$$\gamma = \left(1 - \frac{u^2}{c^2}\right)^{-1/2} \approx 1 + \frac{1}{2}\frac{u^2}{c^2} + \cdots$$

thus

$$E_k \approx mc^2\left(1 + \frac{1}{2}\frac{u^2}{c^2} + \cdots - 1\right) = \tfrac{1}{2}mu^2$$

The expression for kinetic energy consists of two terms. One term, γmc^2, depends on the speed of the particle (through the factor γ), and the other term, mc^2, is independent of the speed. The quantity mc^2 is called the *rest energy* of the particle. The total energy E is then defined to be the sum of the kinetic energy and the rest energy,

$$E = E_k + mc^2 = \gamma mc^2 = \frac{mc^2}{\sqrt{1 - u^2/c^2}} \qquad\qquad 28\text{-}28$$

Thus the work done by an unbalanced force increases the energy from the rest energy mc^2 to γmc^2 (or increases the mass from m to γm).

The identification of the term mc^2 as rest energy is not merely a convenience. The rest energy of a system of particles is related to the potential energy of the system. For example, if two particles are bound together by attracting forces so that the potential energy U is negative (relative to zero potential energy at infinite separation), the rest mass of the system is less by the amount $|U|/c^2$ than the rest mass of the two particles when separated. The following example should help to clarify the relation between mass and energy.

Consider two particles moving toward each other each with speed u and colliding with a spring which is compressed and locks shut (Figure 28-16). (The spring is just an artificial device for visualizing the storage of energy.) In the newtonian-mechanics description, the original kinetic energy is converted into potential energy of the spring. When the spring is unlocked, the potential energy reappears as kinetic energy. The total energy of the system is conserved. According to Equation 28-28, the total energy from the point of view of relativity consists of kinetic energy and rest energy. If the total energy is conserved and the kinetic energy disappears, the rest energy must increase. It is possible to show in detail that this does happen. According to relativity theory, the mass of the two particles connected by a compressed spring should be greater than the mass of the two free particles. This result is borne out by experiment. The difference in mass is U/c^2, where U is the total potential energy. For ordinary macroscopic masses and springs, this change in mass is too small to be observed, but in nuclear-energy transformations such a change in mass is often observed. In radioactive decay, e.g., a radium nucleus

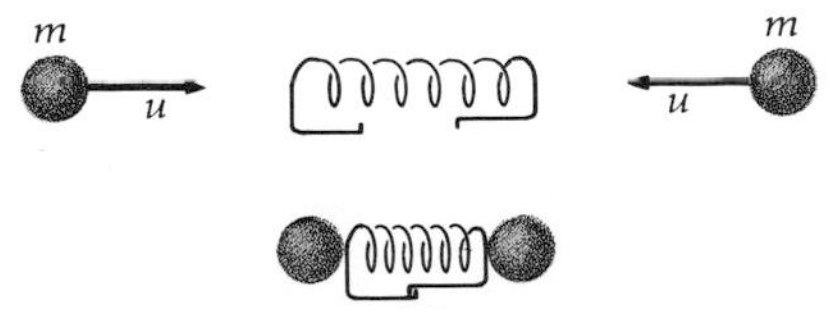

Figure 28-16
Two objects colliding with massless spring, which locks shut. The total rest mass M after the collision is greater than that before the collision. $M = 2m + E_k/c^2$, where E_k is the original kinetic energy. The total energy, rest energy plus kinetic energy, does not change.

decaying by emission of an alpha particle, the energy given off is an appreciable fraction of the rest energy of the original nucleus. The sum of the masses of the alpha particle and the final nucleus is less than the mass of the original radium nucleus by E_k/c^2, where E_k is the kinetic energy of the decay particles.

In practical applications the momentum or energy of a particle is often known rather than the speed. Equation 28-24 for the relativistic momentum and Equation 28-28 for the relativistic energy can be combined to eliminate the speed u. The result (see Exercise 24) is

$$E^2 = p^2c^2 + (mc^2)^2 \qquad\qquad 28\text{-}29$$

If the energy of a particle is much greater than its rest energy mc^2, the second term on the right of Equation 28-29 can be neglected, giving the useful approximation

$$E \approx pc \qquad \text{for } E \gg mc^2 \qquad\qquad 28\text{-}30$$

Equation 28-30 is an exact relation between energy and momentum for particles with no rest mass, e.g., photons and neutrinos. If the kinetic energy of a particle is small compared with its rest energy, the term pc is small compared with mc^2. From Equation 28-29 we then have

$$E = [p^2c^2 + (mc^2)^2]^{1/2} = mc^2 \left(1 + \frac{p^2}{m^2c^2}\right)^{1/2}$$

$$\approx mc^2 \left(1 + \frac{1}{2}\frac{p^2}{m^2c^2} + \cdots\right)$$

$$= mc^2 + \frac{p^2}{2m} + \cdots \qquad\qquad 28\text{-}31$$

where we have used the approximation for small x, $(1 + x)^{1/2} \approx 1 + \frac{1}{2}x$. This result agrees with the classical result for the kinetic energy, $E_k = E - mc^2 = p^2/2m$. The velocity of a particle is often most easily calculated from its total energy and momentum. If we multiply Equation 28-28 by $\mathbf{u}$ and compare with Equation 28-24, we obtain the useful result

$$\mathbf{u}E = \mathbf{p}c^2$$

or

$$\mathbf{u} = \frac{\mathbf{p}c^2}{E} \qquad\qquad 28\text{-}32$$

The most convenient unit for expressing the energy of an electron or other subatomic particle is the electron volt (eV) and its multiples, defined in Section 8-6. The electron volt is related to the joule by

$$1 \text{ eV} = 1.602 \times 10^{-19} \text{ J}$$

Example 28-4 An electron with rest energy 0.511 MeV moves with speed $u = 0.8c$. Find its total energy, kinetic energy, and momentum.
From Equation 28-28,

$$\gamma = \frac{1}{\sqrt{1 - 0.64}} = \frac{5}{3} = 1.67$$

The total energy is then

$$E = \gamma mc^2 = 1.67(0.511 \text{ MeV}) = 0.853 \text{ MeV}$$

The kinetic energy is the total energy minus the rest energy:

$$E_k = E - mc^2 = 0.853 \text{ MeV} - 0.511 \text{ MeV} = 0.342 \text{ MeV}$$

The momentum can be calculated from either Equation 28-24 or 28-32. From Equation 28-24 we have for its magnitude

$$p = \gamma mu = \gamma m(0.8c) = \frac{0.8\gamma \, mc^2}{c} = \frac{(0.8)(1.67)(0.511 \text{ MeV})}{c}$$

$$= 0.683 \, \frac{\text{MeV}}{c}$$

The unit MeV/c is a convenient momentum unit.

Example 28-5 An electron has a total energy 5 times its rest energy. What is its momentum and speed?

From Equation 28-29,

$$p^2c^2 = E^2 - (mc^2)^2 = (5mc^2)^2 - (mc^2)^2 = 24(mc^2)^2$$

$$pc = \sqrt{24} \, mc^2 = (4.90)(0.511 \text{ MeV}) = 2.50 \text{ MeV}$$

$$p = 2.50 \, \frac{\text{MeV}}{c}$$

We find its speed from Equation 28-32:

$$u = \frac{pc^2}{E} = \frac{\sqrt{24} \, mc^2}{5mc^2} c - 0.980c$$

28-10 Mass and Binding Energy

Some numerical examples from atomic and nuclear physics will illustrate changes in rest mass and rest energy. A unit of mass convenient for discussing atomic and nuclear masses is the *unified mass unit* u, defined as one-twelfth the mass of the neutral carbon atom consisting of the ^{12}C nucleus and six electrons. (This unit replaces the older atomic mass unit based on the oxygen atom.) Since 1 mole of carbon contains Avogadro's number of atoms and has a mass of 12 gm, the relation between the unified mass unit and the gram is

$$1 \text{ u} = \frac{1 \text{ gm}}{6.0220 \times 10^{23}} = 1.6606 \times 10^{-24} \text{ gm}$$

$$= 1.6606 \times 10^{-27} \text{ kg} \qquad\qquad 28\text{-}33$$

For purposes of rough calculations, we can write

$$1 \text{ u} = 1.66 \times 10^{-24} \text{ gm} = 1.66 \times 10^{-27} \text{ kg}$$

The rest energy of 1 gm is

$$(1 \text{ gm})c^2 = (10^{-3} \text{ kg})(3 \times 10^8 \text{ m/sec})^2$$
$$= 9 \times 10^{13} \text{ J} = 5.61 \times 10^{32} \text{ eV} \qquad\qquad 28\text{-}34$$

The rest energy of a unified mass unit is

$$(1 \text{ u})c^2 = 931.5 \text{ MeV} \qquad\qquad 28\text{-}35$$

The rest masses and rest energies of some elementary particles and light nuclei are given in Table 28-1, from which we can see that the mass of a nucleus is not the same as the sum of the masses of its parts.

Table 28-1

Rest energies of some elementary particles and light nuclei

Particle	Symbol	Rest energy, MeV
Photon	γ	0
Neutrino (antineutrino)	ν $(\bar{\nu})$	0
Electron (positron)	e or e^- (e^+)	0.5510
Muon	μ	105.7
Pi meson	π^0	135
	$\pi^\pm$	139.6
Proton	p	938.280
Neutron	n	939.573
Deuteron	^{2}H or d	1875.628
Triton	^{3}H	2808.944
Alpha	^{4}He or α	3727.409

Example 28-6 The simplest example is that of the deuteron ^{2}H, consisting of a neutron and a proton bound together. Its rest energy is 1875.63 MeV. The sum of the rest energies of the proton and neutron is $938.28 + 939.57 = 1877.85$ MeV. Since this is greater than the rest energy of the deuteron, the deuteron cannot spontaneously break up into a neutron and a proton. The binding energy of the deuteron is $1877.85 - 1875.63 = 2.22$ MeV. In order to break up the deuteron into a proton and a neutron, at least 2.22 MeV must be added. This can be done by bombarding deuterons with energetic particles or electromagnetic radiation.

If a deuteron is formed by combination of a neutron and proton, energy must be released. When neutrons from a reactor are incident on protons, some neutrons are captured. The nuclear reaction is $n + p \rightarrow d + \gamma$. Most of these reactions occur for the low-energy neutrons (kinetic energy less than 1 eV). The energy of the photon plus the kinetic energy of the deuteron is 2.22 MeV.

Example 28-7 A free neutron decays into a proton plus an electron plus an antineutrino

$$n \rightarrow p + e + \bar{\nu}$$

What is the kinetic energy of the decay products?

Here rest energy is converted into kinetic energy. Before decay, $(mc^2)_n = 939.57$ MeV. After decay, $(mc^2)_p + (mc^2)_e + (mc^2)_{\bar{\nu}} = 938.28 + 0.511 + 0 = 938.79$ MeV. Thus rest energy of $939.57 - 938.79 = 0.78$ MeV has been converted into kinetic energy of the decay products.

Example 28-8 The binding energy of the hydrogen atom (energy to remove the electron from the atom) is 13.6 eV. How much mass is lost when an electron and a proton form a hydrogen atom?

The mass of a proton plus that of an electron must be greater than that of the hydrogen atom by

$$\frac{13.6 \text{ eV}}{931.5 \text{ MeV/u}} = 1.46 \times 10^{-8} \text{ u}$$

This mass difference is so small that it is usually neglected.

Example 28-9 How much energy is needed to remove one proton from a ^{4}He nucleus?

Removal of one proton from ^{4}He leaves ^{3}H. From Table 28-1, the rest energy of ^{3}H plus that of a proton is $2808.94 + 938.28 = 3747.22$ MeV. This is about 19.8 MeV greater than the rest energy of ^{4}He.

These examples show that because atomic binding energies are so small (of the order of 1 eV to 1 keV), the mass changes are negligible in atomic (or chemical) reactions, but the nuclear binding energies are quite large and involve appreciable changes in mass.

Albert Einstein (1879–1955)

Gerald Holton
Harvard University

A friend visiting Albert Einstein in his modest walkup apartment in Bern, Switzerland found him seated at a kitchen table, "pipe in his mouth, his left hand moving a children's carriage back and forth, in his right hand a shabby stub of a pencil." The year was 1905, and Einstein was writing his great paper on the special theory of relativity. Some of Einstein's personality traits can already be glimpsed here—the simplicity of his personal life and his ability to lift himself out of his surroundings by concentrating on the work before him.[1]

The paper of 1905 was the culmination of ideas that had started to preoccupy Einstein 10 years earlier, when he was about 16. At that age, he later wrote in his autobiography,[2] the following paradox occurred to him. As one knows from ample experience, galilean relativity holds in mechanics; if you throw a ball forward in a moving carriage, observing the motion of the ball cannot tell you how fast you and the carriage are moving. But when it comes to optics, it seems to be different. For example, if I move along "a beam of light with the velocity c [velocity of light in a vacuum], I should observe such a beam of light as a spatially oscillatory electromagnetic field." Looking back along the beam a distance of one whole wavelength, one should see that the local magnitudes of the electric and magnetic field vectors increase point by point from, say, zero to full strength, and then decrease again to zero, one wavelength away. Seeing such a curious field in free space would tell me that I am going at the speed of light with respect to absolute space, or the ether.

Einstein suspected that the imagined result of this *Gedanken experiment* (thought experiment) must somehow be in error. In any case, he said later, "from the very beginning it appeared to me intuitively clear that, judged from the standpoint of such an observer, everything would have to happen according to the same laws as for an observer who, relative to the earth, was at rest."

Einstein solved the apparent paradox in 1905 by showing that the expectation of what one would see in pursuing a light beam is false, being grounded in a wrong idea that "unrecognizedly was anchored in the unconscious" of all scientists of the time—namely the absolute character of time and of simultaneity. Instead, Einstein showed that a sound view of how physical nature operates can be gained by boldly postulating two principles.

[1] See Einstein's essay Motive of Research, pp. 224–227 in Sonja Bargmann (trans. and ed.), *Ideas and Opinions*, Crown, New York, 1954; look also at the other essays in this collection.

[2] Albert Einstein, *Autobiographical Notes,* in P. A. Schilpp (ed.), *Albert Einstein: Philosopher-Scientist,* Harper, New York, 1959.

The first principle of relativity for inertial systems generalizes galilean relativity to encompass not only mechanics but also optics and electromagnetism; the second principle says that light in a vacuum will always be found to move with the same speed *c*, regardless of the state of motion of the emitting body. The solutions to all paradoxes and experimentally puzzling observations at the time were derivable from these two principles.

At least as important to Einstein was the fact that this approach hugely simplified our view of nature, in two ways: it broke down barriers between hitherto entirely separate notions, and it cleansed physics of unnecessary conceptions that had produced pseudo problems. Now electromagnetism was on the same footing as mechanics, instead of allowing "privileged" systems. Time and space were found to have interpenetrating meaning—what really existed was a space-time continuum. Electric fields and magnetic fields were at bottom the same reality perceived in different experimental conditions. Mass and energy were equivalent. Even the boldly intuitive approach to scientific discovery and the rigorously rational method were joined into one powerful approach in his paper. Moreover, the notion of the ether was at last declared to be "superfluous." It had long been embedded in physics but had required the assumption of ever more puzzling properties. While he was the first to give up the idea of an ether, to the end of his life Einstein was dedicated to the theme of the *field* (or, in general, the continuum) as the basic conceptual tool for the fundamental explanation of phenomena.

The preoccupation with the continuum, with explaining mysterious orderliness, with finding pleasing simplicity—these were conceptions in physics which characterized Einstein's work from the beginning. But as with other highly creative persons, the power of his work was derived not merely from good physical ideas. Instead it came from a fusion of his scientific interests and his characteristics as a human being, a synthesis of his life-style and his perception of the laws of nature. Much of what was most daring or novel in his great work in physics was present in Einstein's ways of everyday thinking and behaving, even as a child or a young student. Take, for example, the ability to come back, again and again, for years, to a difficult puzzle. In early life, this trait showed up as what may have seemed mere obstinacy. He was commonly reported to have been withdrawn as a child, preferring to play by himself, erecting complicated constructions. Before he was 10, with infinite patience he was making fantastic card houses that had as many as 14 floors. He was unable or unwilling to talk until the age of 3. In school, he was not an exceptional student, preferring to follow his own thoughts. Later, he stuck to his ideas with the same persistence when the experiments of others seemed to disconfirm his theories (in time, those experiments usually turned out to be wrong). And during the last 30 years of his life, he persisted in his skepticism concerning the fundamental explanatory power of quantum mechanics and in his dedication to the problems of field theory, unlike most physicists of the time.

With this single-mindedness and concentration on his own revolutionary ideas went his deep suspicion of established authority, in science no less than in daily life. The same young student who quietly challenged the established ideas in physics also abhorred the current political, religious, and social conventions. From the age of 12 years on, he confessed later, he had "suspicion against every kind of authority." One result was that his teachers were quite delighted to see him drop out of high school at the age of 15½, when he no longer could stand the regimented and militaristic way of life in his native Germany and in his school. Moving to the freer, more democratic atmosphere of Switzerland, he found at last a school to his liking, and there had a glorious year before entering the Polytechnic Institute of Zurich. It was there, too, that he had his first ideas on relativity.

Related to this trait of uncompromisingly sticking to his own identity was his search for what is really *necessary*. In his early work in physics, he said, the thought of having to explain electric and magnetic field effects as "two fundamentally different cases was for me unbearable." The highest aim was

Einstein at 17. (*From Carl Seelig*, Albert Einstein: Eine Dokumentarische Biographie, *Europa Verlag, Zurich, 1954.*)

Einstein at 53. (*Courtesy of the Archives, California Institute of Technology.*)

nothing less than finding the most economical, simple principles, the barest bones of nature's frame, cleansed of everything that is ad hoc, redundant, unsymmetrical. "What really interests me," he once said, "is whether God has any choice in the creation of the world." Nature does not like anything that is unnecessary. Nor did Einstein, in his personal life—in his clothing, in his manner of speech and writing, in his behavior, from his preference for the classical music of Bach and Mozart down to his preference for using the same bar of soap for washing and shaving instead of complicating life unnecessarily by facing two kinds of soap every morning.

The preference for an egalitarian democracy characterized Einstein's political life just as it did his physics. The man who declared every inertial system to be created equal before all the laws of physics was also, from youth on, fiercely opposed to every antidemocratic and narrowly nationalistic political or social system.

Einstein kept an abiding interest in philosophy, which penetrated his work; his ideas, in turn, influenced the development of modern philosophy itself. Here, too, he was cutting across unnecessary barriers. This is not surprising. Genius discovers itself not in splendid solutions to little puzzles but in the struggle with deep and perhaps eternal problems at the point where science and philosophy join.[1]

[1] For further discussion on Einstein's relativity theory, its genesis and influence, see Gerald Holton, *Thematic Origins of Scientific Thought, Kepler to Einstein,* chaps. 5–10, Harvard University Press, Cambridge, Mass., 1973. Among recent biographies see Banesh Hoffmann, with the collaboration of Helen Dukas, *Albert Einstein: Creator and Rebel,* Viking, New York, 1972; Jeremy Bernstein, *Einstein,* Viking, New York, 1973; or the best of the older biographies: Philipp Frank, *Einstein: His Life and Times,* Knopf, New York, 1947.

Review

A. Define, explain, or otherwise identify:

Einstein postulates, 665
Ether, 666
Time dilation, 671
Length contraction, 671
Proper time, 671
Proper length, 671

Synchronized clocks, 673
Lorentz transformation, 680
Twin paradox, 681
Relativistic momentum, 685
Rest energy, 687
Unified mass unit, 689

B. True or false:

1. The speed of light is the same in all reference frames.

2. Proper time is the shortest time interval between two events.

3. Absolute motion can be determined by means of length contraction.

4. The light-year is a unit of distance.

5. A particle cannot be accelerated to a speed greater than the speed of light in vacuum.

6. Simultaneous events must occur at the same place.

7. If two events are not simultaneous in one frame, they cannot be simultaneous in any other frame.

8. $\Sigma \mathbf{F} = m\mathbf{a}$ holds at all speeds if the mass is the relativistic mass.

9. Rest mass can sometimes be converted into energy.

10. If two particles are tightly bound together by strong attractive forces, the mass of the system is less than the sum of the masses of the individual particles when separated.

Exercises

Section 28-1, The Michelson-Morley Experiment

1. In one series of measurements of the speed of light, Michelson used a path length L of 35.4 km (22 mi). (a) What is the time needed for light to make the round-trip distance of $2L$? (b) What is the classical correction term in seconds in Equation 28-2 assuming the earth's speed is $v = 10^{-4}c$? (c) From about 1600 measurements, Michelson quoted the result for the speed of light as 299,796 ± 4 km/sec. Is this experiment accurate enough to be sensitive to the correction term in Equation 28-2?

2. An airplane flies with speed c relative to still air from point A to point B and returns. Compare the time required for the round trip when the wind blows from A to B with speed v with that when the wind blows perpendicularly to the line AB with speed v.

Section 28-2, Consequences of Einstein's Postulates

There are no exercises for this section.

Section 28-3, Time Dilation and Length Contraction

3. Derive the following results for v much less than c and use when applicable in the exercises and problems.

(a) $\gamma \approx 1 + \dfrac{1}{2}\dfrac{v^2}{c^2}$

(b) $\dfrac{1}{\gamma} \approx 1 - \dfrac{1}{2}\dfrac{v^2}{c^2}$

(c) $\gamma - 1 \approx 1 - \dfrac{1}{\gamma} \approx \dfrac{1}{2}\dfrac{v^2}{c^2}$

4. How great must the relative speed of two observers be for their time-interval measurements to differ by 1 percent (see Exercise 3)?

5. The proper mean life of π mesons is 2.6×10^{-8} sec. If a beam of such particles has speed $0.9c$, (a) what would their mean life be as measured in the laboratory? (b) how far would they travel on the average before they decay? (c) what would your answer be to part (b) if you neglected time dilation?

6. (a) In the reference frame of the π meson in Exercise 5, how far does the laboratory travel in a typical lifetime of 2.6×10^{-8} sec? (b) What is this distance in the laboratory frame?

7. A meterstick moves with speed $v = 0.6c$ relative to you in the direction parallel to the stick. (a) Find the length of the stick measured by you. (b) How long does it take for the stick to pass you?

8. Supersonic jets achieve maximum speeds of about $3 \times 10^{-6}c$. (a) By what percentage would you see such a jet contracted in length? (b) During a time of $1\,y = 3.15 \times 10^7$ sec on your clock, how much time would elapse on the pilot's clock? How many minutes are lost by the pilot's clock in 1 y of your time?

Section 28-4, Clock Synchronization and Simultaneity

Exercises 9 to 13 refer to the following situation: an observer in S' lays out a distance $L' = 100$ c-min between points A' and B' and places a flashbulb at the midpoint C'. He arranges for the bulb to flash and for clocks at A' and B' to be started at 0 when the light from the flash reaches the clocks (see Figure 28-17). Frame S' is moving to the right with speed 0.6c relative to an observer C in S who is at the midpoint between A' and B' when the bulb flashes and sets his clock to zero at that time.

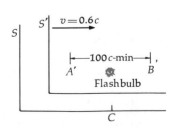

Figure 28-17
Exercises 9 to 13.

9. What is the separation distance between clocks A' and B' according to the observer in S?

10. As the light pulse from the flashbulb travels toward A' with speed c, A' travels toward C with speed $0.6c$. Show that the clock in S reads 25 min when the flash reaches A'.

11. Show that the clock in S reads 100 min when the light flash reaches B', which is traveling away from C with speed $0.6c$.

12. The time interval between reception of the flashes is 75 min according to the observer in S. How much time does he expect to have elapsed on the clock at A' during this 75 min?

13. The time interval calculated in Exercise 12 is the amount that the clock at A' leads that at B' according to observers in S. Compare this result with $L_0 v/c^2$.

Section 28-5, The Doppler Effect

14. (a) Show that Equations 28-14a and b are identical and can both be written

$$f' = f_0 \sqrt{\frac{1 + \beta}{1 - \beta}} \qquad \text{where } \beta = \frac{v}{c}$$

(b) Show that Equation 28-15 can be written

$$f' = f_0 \sqrt{\frac{1 - \beta}{1 + \beta}}$$

15. How fast must you move toward a red light ($\lambda = 650$ nm) for it to appear green ($\lambda = 525$ nm)?

16. A distant galaxy is moving away from us at speed 2×10^7 m/sec. Calculate the fractional red shift $(\lambda' - \lambda_0)/\lambda_0$ in the light from this galaxy.

17. A distant galaxy is moving away from the earth such that each wavelength is shifted by a factor of 2; that is, $\lambda' = 2\lambda_0$. What is the speed of the galaxy relative to us (see Exercise 14)?

Section 28-6, The Lorentz Transformation

18. Two events happen at the same point x_0' in frame S' at times t_1' and t_2'. (a) Use Equations 28-18 to show that in frame S the time interval between the events is greater than $t_2' - t_1'$ by the factor γ. (b) Why are Equations 28-19 less convenient than Equations 28-18 for this problem?

19. The length of an object moving in your frame is measured by recording the positions of each end x_2 and x_1 at the same time t_0. (a) Use Equations 28-19 to show that the result is smaller than the proper length $x_2' - x_1'$ measured in the frame in which the object is at rest. (b) Why are Equations 28-19 more convenient for this problem than Equations 28-18?

20. Two spaceships are approaching each other. (a) If the speed of each is $0.9c$ relative to the earth, what is the speed of one relative to the other? (b) If the speed of each relative to the earth is 30,000 m/sec (about 100 times the speed of sound), what is the speed of one relative to the other?

21. A light beam moves along the y' axis with speed c in frame S' which is moving to the right with speed v relative to frame S. (a) Find u_x and u_y, the x and y components of the velocity of the light beam in frame S. (b) Show that the magnitude of the velocity of the light beam in S is c.

22. A particle moves with speed $0.9c$ along the x'' axis of frame S'', which moves with speed $0.9c$ along the x' axis relative to frame S'. Frame S' moves with speed $0.9c$ along the x axis relative to frame S. (a) Find the speed of the particle relative to frame S'. (b) Find the speed of the particle relative to frame S.

Section 28-7, The Twin Paradox

There are no exercises for this section.

Section 28-8, Relativistic Momentum, and Section 28-9, Relativistic Energy

23. Show that $p_{yA} = -p_{yB}$, where p_{yA} and p_{yB} are the relativistic momenta of the balls in Figure 28-15, given by

$$p_{yA} = \frac{mu_0}{\sqrt{1 - u_0^2/c^2}} \qquad p_{yB} = \frac{mu_{yB}}{\sqrt{1 - (u_{xB}^2 + u_{yB}^2)/c^2}}$$

$$u_{yB} = -u_0 \sqrt{1 - \frac{v^2}{c^2}} \qquad \text{and} \qquad u_{xB} = v$$

24. Combine Equations 28-24 and 28-28 to derive the equation $E^2 = p^2c^2 + m^2c^4$.

25. Make a sketch of the total energy of an electron E as a function of its momentum p. (See Equations 28-30 and 28-31 for the behavior of E at the limits of large and small values of p.)

26. Make a sketch of the kinetic energy of an electron E_k versus its speed u (see Equation 28-27).

27. An electron of rest energy 0.511 MeV has a total energy of 5 MeV. (a) Find its momentum in units of MeV/c from Equation 28-29. (b) Find the ratio of its speed u to the speed of light.

28. How much energy would be required to accelerate a particle of mass m from rest to speeds of (a) $0.5c$, (b) $0.9c$, (c) $0.99c$? Express your answers as multiples of the rest energy.

29. The rest energy of a proton is about 938 MeV. If its kinetic energy is also 938 MeV, find (a) its momentum and (b) its speed.

30. If the kinetic energy of a particle equals its rest energy, what error is made by using $p = mu$ for its momentum?

Section 28-10, Mass and Binding Energy

31. How much energy is required to remove one of the neutrons from ^{3}H to yield ^{2}H plus the neutron?

32. The rest mass of ^{3}He (two protons and one neutron) is 3.01603 u. (a) What is the rest energy of ^{3}He in megaelectron volts? (b) How much energy is needed to remove a proton to make ^{2}H plus the proton?

33. The energy released when sodium and chlorine combine to form NaCl is 4.2 eV. (a) What is the increase in mass (in unified mass units) when a molecule of NaCl is dissociated into an atom of Na and an atom of Cl? (b) What percentage error is made in neglecting this mass difference? (The mass of Na is about 23 u, and that of Cl is about 35.5 u.)

34. In nuclear fusion, two ^{2}H atoms are combined to produce ^{4}He. (a) Calculate the decrease in rest mass in unified mass units. (b) How much energy is released in this reaction? (c) How many such reactions must take place per second to produce 1 W of power?

Problems

1. An airplane has speed c relative to still air. The wind blows toward the northeast, at speed v and angle θ east of north. (a) If the plane flies directly north relative to the earth's surface a distance L and returns directly south, show that the round-trip time is

$$T_0 \frac{\sqrt{1 - \beta^2 \sin^2 \theta}}{1 - \beta^2} \qquad \text{where } \beta = \frac{v}{c} \qquad \text{and} \qquad T_0 = \frac{2L}{c}$$

(b) If the plane flies directly east a distance L and back, show that the round-trip time is

$$T_0 \frac{\sqrt{1 - \beta^2 \cos^2 \theta}}{1 - \beta^2}$$

(c) Show that the time difference for the trips in parts (a) and (b) is $\Delta T \approx \frac{1}{2} T_0 \beta^2 \cos 2\theta$ for small β. (d) In the analogous Michelson-Morley experiment, ΔT is detected by observing a fringe shift in an interferometer. If θ happens to be 45° and the pattern is observed during rotation of 90°, as in the Michelson-Morley experiment, describe the classically expected result.

2. A plane flies at a speed of 1000 mi/h. How long must the plane fly before its clock loses 1 sec because of time dilation?

3. (a) Show that the speed u of a particle of mass m and total energy E is given by $u/c = [1 - (mc^2/E)^2]^{1/2}$ and that if E is much greater than mc^2, the approximation $u/c \approx 1 - \frac{1}{2}(mc^2/E)^2$ holds. (b) Find the speed of an electron of kinetic energy 0.51 MeV and that of an electron of kinetic energy 10 MeV.

4. What percent error is made in using $\frac{1}{2}mu^2$ for the kinetic energy of a particle if its speed is (a) $u = 0.1c$, (b) $u = 0.9c$?

5. Two spaceships each 100 m long when measured at rest travel toward each other with speeds $0.8c$ relative to earth. (a) How long is each ship as measured by someone on earth? (b) How fast is each ship traveling as measured by the other? (c) How long is one ship when measured by the other? (d) At some time $t = 0$ (on earth clocks) the fronts of the ships are together as they begin to pass each other. At what time (on earth clocks) are their backs together? (e) Sketch diagrams in the frame of one of the ships showing the passing of the other ship.

6. A stick has proper length L and makes an angle θ with the x axis in frame S. Show that the angle made with the x' axis in frame S' moving along the $+x$ axis with speed v is θ' given by $\tan \theta' = \gamma \tan \theta$ and that the length of the stick in S' is $L' = L[(\cos \theta/\gamma)^2 + \sin^2 \theta]^{1/2}$.

7. Show that if a particle moves at an angle θ with the x axis with speed u in frame S, it moves at an angle θ' with the x' axis in S' given by

$$\tan \theta' = \frac{\sin \theta}{\gamma (\cos \theta - v/u)}$$

8. For the special case of a particle moving with speed u along the y axis in S, show that the momentum and energy in frame S' are related to the momentum and energy in S by the transformation equations

$$p_x' = \gamma \left(p_x - \frac{vE}{c^2} \right) \qquad p_y' = p_y$$

$$p_z' = p_z \qquad \frac{E'}{c} = \gamma \left(\frac{E}{c} - \frac{vp_x}{c} \right)$$

Compare these equations with the Lorentz transformation for x', y', z', and t'. These show that the quantities p_x, p_y, p_z, and E/c transform in the same way as x, y, z, and ct.

9. The equation for a spherical wavefront of a light pulse which begins at the origin at time $t = 0$ is $x^2 + y^2 + z^2 - (ct)^2 = 0$. Using the Lorentz transformation equations, show that such a light pulse also has a spherical wavefront in frame S' by showing that $x'^2 + y'^2 + z'^2 - (ct')^2 = 0$ in S'.

10. In Problem 9 you showed that the quantity $x^2 + y^2 + z^2 - (ct)^2$ has the same value (0) in both S and S'. Such a quantity is called an *invariant*. From the results of Problem 8 the quantity $p_x^2 + p_y^2 + p_z^2 - (E/c)^2$ must also be invariant. Show that this quantity has the value $-m^2c^2$ in both S and S' reference frames.

11. Two observers agree to test time dilation. They use identical clocks, and one observer in frame S' moves with speed $v = 0.6c$ relative to the other observer in frame S. When their origins coincide, they start their clocks. They agree to send a signal when their clocks read 60 min and to send a confirmation signal when each receives the other's signal. (a) When does the observer in S receive the first signal from the observer in S'? (b) When does he receive the confirmation signal? (c) Make a table showing the times in S when the observer sent the first signal, received the first signal, and received the confirmation signal. How does this table compare with one constructed by the observer in S'?

12. A double star rotates about its center of mass such that one member is moving toward the earth and the other away from the earth. The center of mass is fixed relative to the earth. The hydrogen line of wavelength 656.3 nm in the laboratory has wavelength 660.0 nm from one of the stars and 650.0 nm from the other. What is the ratio of the masses of the stars? *Hint:* Show that the nonrelativistic doppler-effect equations have sufficient accuracy for this problem.

13. Two events in S are separated by a distance $D = x_2 - x_1$ and time $T = t_2 - t_1$. (a) Use the time-transformation equations to show that in frame S' moving with speed v relative to S the time separation is $t_2' - t_1' = \gamma(T - vD/c^2)$. (b) Show that the events can be simultaneous in frame S' only if D is greater than cT. (c) If one of the events is the *cause* of the other, the separation D must be less than cT since D/c is the smallest time a signal can take to travel from x_1 to x_2 in frame S. Show that if D is less than cT, t_2' is greater than t_1' in all reference frames. This shows that the cause must precede the effect in all reference frames (assuming that it does in one frame). (d) Suppose that a signal could be sent with speed c' greater than c so that in frame S the cause precedes the effect by the time $T = D/c'$, which is less than D/c. Show that there is then a reference frame moving at speed v less than c in which the effect precedes the cause.

14. Show that if v is much less than c, the doppler frequency shift is approximately given by $\Delta f/f = \pm v/c$, both classically and relativistically. A radar transmitter-receiver bounces a signal off an aircraft and observes a fractional increase in the frequency of $\Delta f/f = 8 \times 10^{-7}$. What is the speed of the aircraft? (Assume the aircraft to be moving directly toward the transmitter.)

15. Figure 28-16 shows an inelastic collision of two masses in the center-of-mass frame, which we shall take to be frame S. In this problem you are to show that if momentum is to be conserved in another reference frame, the rest mass of the system must increase by the amount E_k/c^2, where $E_k = 2mc^2(\gamma - 1)$ is the loss in kinetic energy in the collision. (a) Let frame S' move with speed u so that one of the masses is initially at rest and the other moves with speed u'. Show that $u' = 2u/(1 + u^2/c^2)$. Use this result to show that

$$\sqrt{1 - \frac{u'^2}{c^2}} = \frac{1 - u^2/c^2}{1 + u^2/c^2}$$

(b) Show that the initial momentum in frame S' can be written $P_i' = 2mu/(1 - u^2/c^2)$. (c) After the collision the two masses move with speed u in frame S' (since they are at rest in S). What is the total momentum after the collision in S' in terms of u and the new total mass of the system M? (d) Show that if the final momentum equals the initial momentum in S', the final mass is related to the initial mass by $M = 2m/\sqrt{1 - u^2/c^2}$ and that the increase in mass is $\Delta m = (\gamma - 1)2m$, where $\gamma = 1/\sqrt{1 - u^2/c^2}$. (e) Show that the increase in mass is just E_k/c^2, where E_k is the initial kinetic energy in frame S which is completely lost in the collision. (f) In each reference frame, find the total energy before and after the collision and show that the total energy is conserved.

CHAPTER 29   The Electric Field

The electrostatic force between two point charges was examined briefly in Chapter 6 in our discussion of forces in nature. We now begin a more detailed study of electricity. After a short history of the development of the concept of electric charge, we again state Coulomb's law for the force between two charges and define the electric field. We then discuss some general properties of electric fields, including Gauss' law, and look at the behavior of point charges and electric dipoles in electric fields. The calculation of electric fields from given charge distributions is left for Chapter 30.

29-1 Electric Charge

Observations of electrical attraction can be traced back to the ancient Greeks.[1] The Greek philosopher Thales of Miletus (640–546 B.C.) observed that when amber is rubbed, it attracts small objects such as straw or feathers. (This attraction was often confused with the magnetic attraction of lodestone for iron.) Little more of electric phenomena was understood until the sixteenth century, when the English physician William Gilbert (1540–1603) studied electric and magnetic phenomena systematically. Gilbert showed that many substances besides amber acquire an attractive property when rubbed. He was one of the first to clearly understand the distinction between this attraction and the magnetic attraction, and he introduced the terms electric force, electric attraction, and magnetic pole. (The word electric comes from the Greek *elektron*, meaning amber, and the word magnetic comes from *Magnesia*, the country where magnetic iron ore was found.) Gilbert is perhaps best known for his discovery that the behavior of a compass needle is due to the earth being itself a large magnet with

---

[1] Much of this discussion follows the excellent book, *A History of Theories of Aether and Electricity*, by Sir Edmund Whittaker, Nelson, London, 1953; Torchbook edition, Harper, New York, 1960.

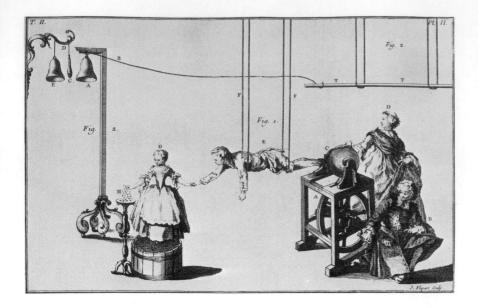

A 1745 woodcut showing electrical effects being transmitted through two persons before attracting feathers or bits of paper at table on left. (*Courtesy of the Deutsches Museum, Munich.*)

poles at the north and the south; the orientation of a compass needle near the earth is merely an example of the general phenomenon of the attraction or repulsion of the poles of two magnets. Gilbert apparently failed to observe electric repulsion.

Around 1729 Stephen Gray, an Englishman, discovered that electric attraction and repulsion can be transferred from one body to another if the bodies are connected by certain substances, particularly metals. This discovery was of great importance, since previously experimenters could electrify an object only by rubbing it. The discovery of electric conduction also implied that electricity has an existence of its own and is not merely a property somehow brought out in a body by rubbing. The existence of two kinds of electricity was suggested by Charles François Du Fay (1698–1739). He describes an experiment in which a gold leaf is attracted by a piece of glass rod previously rubbed. When the leaf is touched by the glass, it acquires the "electric virtue" and then repels the glass. "It is certain that bodies which have become electric by contact are repelled by those which have rendered them electric; but are they repelled likewise by other electrified bodies of all kinds?"[1] Du Fay answers this question by noting that the gold leaf which is repelled by the glass rod is attracted by an electrified piece of amber or resin. If the gold leaf is electrified by touching it to an electrified piece of amber, it is repelled by the amber but attracted by the electrified glass. He gave the two kinds of electricity the names vitreous and resinous and postulated the existence of two fluids which become separated by friction[2] and are neutralized when they combine.

In 1747 the great American statesman and scientist Benjamin Franklin proposed a one-fluid model of electricity, describing the following experiment. If person A standing on wax (to insulate him from the ground) rubs a glass tube with a piece of silk cloth and another person B, also standing on wax, touches the glass tube, both A

---

[1] *Mémoir de l'Académie* (1733), as quoted by Whittaker.

[2] We now know that the transfer of charge from one substance to another is not associated with friction but with the close contact of the substances achieved by rubbing them together.

and *B* become electrified. They can each give a spark to person *C* standing on the ground. However, if *A* and *B* touch each other before touching *C*, the electricity of *A* and *B* is neutralized. Franklin proposed that every body has a "normal" amount of electricity. When a body is rubbed against another, some of the electricity is transferred from one body to the other; thus one has an excess and the other an equal deficiency. The excess and deficiency can be described with plus and minus signs: one body is plus and the other minus. An important feature of Franklin's model is the implication of conservation of electricity, now known as the *law of conservation of charge*. The electric charge is not created by the rubbing; it is merely transferred. Since Franklin chose to call Du Fay's vitreous kind of electricity positive, Du Fay's resinous electricity was merely the lack of vitreous electricity, or negative. A glass rod when rubbed acquires an excess of electricity and is positive, whereas an amber rod loses electricity when rubbed and becomes negative.

Law of conservation of charge

Franklin's choice was unfortunate because we now know that it is electrons which are transferred in the rubbing process, and according to Franklin's convention the electrons have a negative charge. When glass is rubbed with silk, electrons are transferred from the glass to the silk, leaving the glass positive and the silk negative, and when amber is rubbed with fur, electrons are transferred from the fur to the amber.

In the course of his experiments, Franklin noticed that small cork balls inside a metal cup seemed to be completely unaffected by the electricity of the cup. He asked his friend Joseph Priestley (1733–1804) to check this fact, and Priestley began experiments which showed that there is no electricity on the inside surface of a hollow metal vessel (except near the opening). From this result, Priestley correctly deduced that the force between two charges varies as the inverse square of the distance between them, just like the gravitational force between two masses. (Recall that a spherical shell of mass exerts no gravitational force on a point mass inside the shell.)

The Granger Collection

Portrait of Charles Augustin de Coulomb.

The inverse-square force law for electricity was confirmed by experiments of Charles Coulomb (1736–1806) using a torsion balance of his own invention. [The torsion balance was independently invented by the English scientist John Michell (1724–1793) and was used by him to show that the force between two magnetic poles also varies as the inverse square of the distance between the poles.] Coulomb's experimental apparatus was essentially the same as that described (Chapter 16) for the Cavendish experiment, with the masses replaced by small charged balls. For the magnitudes of charges easily transferred by rubbing, the gravitational attraction of the balls is completely negligible compared with the electric attraction or repulsion. Coulomb also used his torsion balance to confirm that the force between two magnetic poles varies as the inverse square of the distance.

The nineteenth century saw rapid growth in the understanding of electricity and magnetism, culminating with the great experiments of Michael Faraday (1791–1867) and the mathematical theory of James Clerk Maxwell (1831–1879). The experiments at the beginning of the twentieth century should be mentioned in this brief historical discussion of electric charge. In 1897, the English physicist J. J. Thomson showed that all materials contain particles which have the same charge-to-mass ratio. We now know that these particles, called *electrons*, are a fundamental part of the makeup of all atoms. In 1909 the American physicist Robert Millikan discovered that electric charge

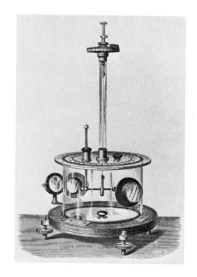

Coulomb's torsion balance. (*Courtesy of the Smithsonian Institution.*)

always occurs in integral amounts of a fundamental unit; i.e., the charge is *quantized*. Any charge can be written $q = Ne$, where N is an integer and e is the magnitude of the fundamental unit. The electron has charge $-e$, and the proton $+e$. The electron and the proton are very different particles. For example, the mass of the proton is about 2000 times that of the electron, and protons exert strong nuclear forces on neutrons, protons, pi mesons, etc., whereas electrons do not participate in the strong nuclear interaction. Yet the magnitude of the charge of the electron is exactly equal to that of the proton. All elementary particles have either no charge, e.g., the neutron, a charge of $+e$, or a charge of $-e$. Particles made up of combinations of elementary particles may have a charge of $2e$, $3e$, etc. For example, the alpha particle, which is the nucleus of the helium atom, consisting of two protons and two neutrons bound tightly together, has a charge of $+2e$. An atom with Z protons in its nucleus (Z is the *atomic number*) has Z electrons outside the nucleus and is electrically neutral.

In this brief historical survey we have mentioned three important properties of electric charge:

1. Charge is conserved.

2. Charge is quantized.

3. The force between two point charges varies as the inverse square of the distance between the charges.

Questions

1. Compare the properties of electric charge with those of gravitational mass. Discuss similarities and differences.

2. How might the world be different if the charge of the proton were slightly greater in magnitude than that of the electron?

29-2 Coulomb's Law

The experiments of Coulomb and others on the forces exerted by one point charge on another are summarized in *Coulomb's law*:

The force exerted by one point charge on another is along the line joining the charges. It is repulsive if the charges have the same sign and attractive if the charges have opposite signs. The force varies inversely as the square of the distance separating the charges and is proportional to the magnitude of each charge.

Coulomb's law can be written as a simple vector equation. Let q_1 and q_2 be two point charges separated by a distance r_{12}, which is the magnitude of the vector $\mathbf{r}_{12}$ pointing from charge q_1 to charge q_2 (Figure 29-1). The force exerted by charge q_1 on charge q_2 is then given by

$$\mathbf{F}_{12} = \frac{kq_1q_2}{r_{12}^2}\, \hat{\mathbf{r}}_{12}$$

29-1a   *Coulomb's law*

where k is a constant and $\hat{\mathbf{r}}_{12} = \mathbf{r}_{12}/r_{12}$ is the unit vector pointing from q_1 toward q_2. The force $\mathbf{F}_{21}$ exerted by q_2 on q_1 is the negative of $\mathbf{F}_{12}$ by Newton's third law. It can be written in a way symmetrical to Equation 29-1a. Let $\mathbf{r}_{21}$ be the vector from q_2 to q_1 whose magnitude is the separation distance. Then $\mathbf{r}_{21} = -\mathbf{r}_{12}$, and the force $\mathbf{F}_{21}$ is

$$\mathbf{F}_{21} = \frac{kq_2q_1}{r_{21}{}^2} \hat{\mathbf{r}}_{21} \qquad\qquad 29\text{-}1b$$

where $\hat{\mathbf{r}}_{21} = \mathbf{r}_{21}/r_{21} = -\hat{\mathbf{r}}_{12}$ is the unit vector pointing from charge q_2 toward q_1.

Equations 29-1 include the result that like charges repel and unlike charges attract. If both q_1 and q_2 are positive or both are negative, the force $\mathbf{F}_{21}$ on q_1 is in the direction $\hat{\mathbf{r}}_{21}$ away from charge q_2 and the force $\mathbf{F}_{12}$ on q_2 is in the direction $\hat{\mathbf{r}}_{12}$ away from q_1. That is, if both charges have the same sign, the force is repulsive. Similarly, if one charge is positive and the other negative, the product q_1q_2 is negative and the forces given by Equations 29-1 are attractive.

The constant k depends on the choice of units for charge. In SI units, the unit of charge is the *coulomb* (C), which is defined as the amount of charge which flows past a point in a wire in one second when the current in the wire is one ampere. The ampere (A) is defined in terms of a magnetic force measurement which we shall describe later. In this system of units, the quantities F_{12}, q_1, q_2, and r_{12} in Equation 29-1 are all defined independently, and the constant k is determined by experiment. The measured value of k is

$$k = 8.99 \times 10^9 \text{ N-m}^2/\text{C}^2 \qquad\qquad 29\text{-}2$$

In most calculations, it is convenient to use the approximation

$$k \approx 9 \times 10^9 \text{ N-m}^2/\text{C}^2 \qquad\qquad 29\text{-}3$$ *Coulomb constant*

With his torsion balance, Coulomb was able to show that the exponent of r_{12} in the force equation is 2 within a few percent uncertainty. Although this was the most accurate measurement then, experiments similar to Priestley's can now be done with much greater precision, and we now know that the exponent is 2 within about 2 parts in 10^9.

In a system of charges, each charge exerts a force on each other charge given by Equation 29-1. The resultant force on any charge is the vector sum of the individual forces exerted on that charge by all the other charges in the system.

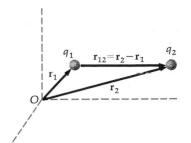

Figure 29-1
Charge q_1 at position $\mathbf{r}_1$ and charge q_2 at $\mathbf{r}_2$ relative to origin O. The force exerted by q_1 on q_2 is in the direction of the vector $\mathbf{r}_{12} = \mathbf{r}_2 - \mathbf{r}_1$ if both charges have the same sign, and in the opposite direction if the charges have opposite signs.

Example 29-1 Three positive point charges lie on the x axis; $q_1 = 25 \ \mu\text{C}$ is at the origin, $q_2 = 10 \ \mu\text{C}$ is at $x = 2$ m, and $q_3 = 20 \ \mu\text{C}$ is at $x = 3$ m (Figure 29-2). Find the resultant force on q_3.

The force on q_3 due to q_2, which is 1 m away, is in the positive x direction and has the magnitude

$$F_{23} = \frac{kq_2q_3}{r_{23}{}^2} = \frac{(9 \times 10^9 \text{ N-m}^2/\text{C}^2)(10 \times 10^{-6} \text{ C})(20 \times 10^{-6} \text{ C})}{(1 \text{ m})^2} = 1.8 \text{ N}$$

The force on q_3 due to q_1, which is 3 m away, is also in the positive x direction. Its magnitude is

$$F_{13} = \frac{kq_1q_3}{r_{13}{}^2} = \frac{(9 \times 10^9 \text{ N-m}^2/\text{C}^2)(25 \times 10^{-6} \text{ C})(20 \times 10^{-6} \text{ C})}{(3 \text{ m})^2} = 0.50 \text{ N}$$

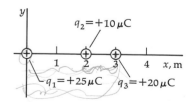

Figure 29-2
Three point charges on the x axis for Example 29-1.

Since both forces are in the positive x direction, the resultant is also in that direction and is just the sum of the magnitudes:

$$F = F_{23} + F_{13} = 2.3 \text{ N}$$

Note that the charge q_2, which is between q_1 and q_3, has no effect on the force $\mathbf{F}_{13}$ exerted by q_1 on q_3, just as the charge q_1 has no effect on the force exerted by q_2 on q_3.

Example 29-2 Charge $q_1 = +25 \ \mu\text{C}$ is at the origin, charge $q_2 = -10 \ \mu\text{C}$ is on the x axis at $x = 2$ m, and charge $q_3 = +20 \ \mu\text{C}$ is at the point $x = 2$ m, $y = 2$ m. Find the resultant force on q_3.

Since q_2 and q_3 have opposite signs, the force exerted by q_2 on q_3 is attractive in the negative y direction, as shown in Figure 29-3. Its magnitude is

$$F_{23} = \frac{(9 \times 10^9)(10 \times 10^{-6})(20 \times 10^{-6})}{2^2} = 0.45 \text{ N}$$

The distance between q_1 and q_3 is $2\sqrt{2}$ m. The force exerted by q_1 on q_3 is directed along the line from q_1 to q_3 and has the magnitude

$$F_{13} = \frac{(9 \times 10^9)(25 \times 10^{-6})(20 \times 10^{-6})}{(2\sqrt{2})^2} = 0.56 \text{ N}$$

The resultant force is the vector sum of these two forces. Since F_{13} makes an angle of $45°$ with the x and y axes, its x and y components are equal to each other and to $F_{13}/\sqrt{2} = 0.40$ N. The x and y components of the resultant force are therefore

$$F_x = F_{23x} + F_{13x} = 0 + 0.40 = 0.40 \text{ N}$$

$$F_y = F_{23y} + F_{13y} = -0.45 + 0.40 = -0.05 \text{ N}$$

The fundamental unit of electric charge e is related to the coulomb by

$$e = 1.60 \times 10^{-19} \text{ C} \qquad\qquad 29\text{-}4$$

Electron charge

Typical laboratory charges of 10 to 100 nC which can be produced by rubbing involve the transfer of many electrons. For example, the number of electrons in 10 nC is

$$10 \times 10^{-9} \text{ C} \times \frac{1 \text{ electron}}{1.60 \times 10^{-19} \text{ C}} = 6.25 \times 10^{10} \text{ electrons}$$

Such charges do not reveal that electric charge is quantized. A million electrons could be added to or subtracted from this charge without detection by ordinary instruments.

Example 29-3 In the hydrogen atom, the electron is separated from the proton by a distance of about 5.3×10^{-11} m on the average. What is the electrostatic force exerted by the proton on the electron?

Since the proton charge is $+e$ and the electron charge $-e$, the force is attractive and has the magnitude

$$\frac{ke^2}{r^2} = \frac{(9 \times 10^9)(1.6 \times 10^{-19})^2}{(5.3 \times 10^{-11})^2} = 8.2 \times 10^{-8} \text{ N}$$

This is much greater than the negligible gravitational force between the particles, which is

$$F_G = \frac{Gm_e m_p}{r^2} = \frac{(6.67 \times 10^{-11})(9.1 \times 10^{-31})(1.67 \times 10^{-27})}{(5.3 \times 10^{-11})^2} = 3.6 \times 10^{-47} \text{ N}$$

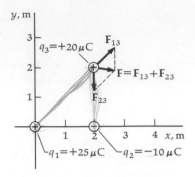

Figure 29-3
Force diagram for Example 29-2. The resultant force on charge q_3 is the vector sum of the forces $\mathbf{F}_{13}$ due to q_1 and $\mathbf{F}_{23}$ due to q_2.

Question

3. If the sign convention for charge were changed so that the electron charge was positive and the proton charge negative, would Coulomb's law be written the same or differently?

29-3 The Electric Field

In Chapter 16 we defined the gravitational field due to a set of masses to be the gravitational force exerted by these masses on a test mass m_0 divided by m_0. The electric field due to a set of charges is defined analogously. If we have a set of point charges q_i at various points in space and place a test charge q_0 at some point P, the force on it will be the vector sum of the forces exerted by the individual charges. Since each of these forces is proportional to the charge q_0, the resultant force is proportional to q_0. The electric field $\mathbf{E}$ at point P is defined to be this force divided by q_0[1]:

$$\mathbf{E} = \frac{\mathbf{F}}{q_0}$$ 29-5   *Electric field defined*

The SI unit of electric field is the newton per coulomb (N/C).

Example 29-4 When a 5-nC test charge is placed at a point, it experiences a force of 2×10^{-4} N in the x direction. What is the electric field $\mathbf{E}$ at that point?

From the definition, the electric field is

$$\mathbf{E} = \frac{2 \times 10^{-4} \mathbf{i} \text{ N}}{5 \times 10^{-9} \text{ C}} = 4 \times 10^4 \mathbf{i} \text{ N/C}$$

where $\mathbf{i}$ is the unit vector in the x direction.

We think of the electric field as a condition in space set up by the system of point charges. This condition is described by the vector $\mathbf{E}$. By moving the test charge q_0 from point to point we can find the electric field vector $\mathbf{E}$ at any point (except one occupied by a charge q_i). The electric field $\mathbf{E}$ is thus a vector function of position. The force exerted on a test charge q_0 at any point is related to the electric field at that point by

$$\mathbf{F} = q_0 \mathbf{E}$$ 29-6

Example 29-5 What is the force on an electron placed at the point in Example 29-4 where the electric field is $4 \times 10^4 \mathbf{i}$ N/C?

Since the charge of the electron is $-e = -1.6 \times 10^{-19}$ C, the force is

$$\mathbf{F} = (-1.6 \times 10^{-19} \text{ C})(4 \times 10^4 \mathbf{i} \text{ N/C}) = -6.4 \times 10^{-15} \mathbf{i} \text{ N}$$

Although it is customary to think of the test charge q_0 as positive, we see from Example 29-5 that the sign of q_0 has no effect on the definition of $\mathbf{E}$. If q_0 is positive, the force on q_0 is in the direction of the field $\mathbf{E}$; if q_0 is negative, the force is in the direction opposite to $\mathbf{E}$. The ratio $\mathbf{F}/q_0$ is the same in either case.

---

[1] In this discussion we assume that the presence of the test charge q_0 does not change the original distribution of the other charges. This is true in practice if there are no conductors present or if q_0 is so small that its influence on the original charge distribution is negligible.

Like the gravitational field, the electric field is more than a calculation device. This concept enables us to avoid the problem of action at a distance if we allow that the field is not propagated instantaneously. We thus think of the force exerted on charge q_0 at point P as being exerted *by the field at point P* rather than by the charges, which are some distance away. Of course the field at point P is produced by the other charges, but not instantaneously.

Consider a single point charge q_1 at the origin. If we place a test charge q_0 at some point given by the position vector $\mathbf{r}$, the force on this charge will be

$$\mathbf{F} = \frac{kq_1q_0}{r^2}\,\hat{\mathbf{r}}$$

The electric field at point $\mathbf{r}$ due to the charge q_1 is thus

$$\mathbf{E} = \frac{kq_1}{r^2}\,\hat{\mathbf{r}} \qquad\qquad 29\text{-}7$$

Electric field of a point charge

If charge q_1 is suddenly moved at time $t = 0$, the change in field at $\mathbf{r}$ is not instantaneous. The change in the field is propagated with the speed of light c, and so the test charge at $\mathbf{r}$ will not react to the change in position of the charge q_1 until a later time $t = r/c$, the time required for propagation of the change in the field. In this chapter we shall be concerned only with static electric fields.

The electric field due to a system of point charges can be found from Coulomb's law. Let $\mathbf{r}_{i0}$ be the vector from the ith charge q_i to point P. The force on a test charge q_0 at P is then

$$\mathbf{F} = \frac{kq_1q_0}{r_{10}{}^2}\,\hat{\mathbf{r}}_{10} + \frac{kq_2q_0}{r_{20}{}^2}\,\hat{\mathbf{r}}_{20} + \cdots + \frac{kq_iq_0}{r_{i0}{}^2}\,\hat{\mathbf{r}}_{i0} + \cdots$$

$$= q_0 \sum_i \frac{kq_i}{r_{i0}{}^2}\,\hat{\mathbf{r}}_{i0}$$

where $\hat{\mathbf{r}}_{i0} = \mathbf{r}_{i0}/r_{i0}$ is the unit vector pointing away from q_i toward point P. The electric field at point P due to the charges q_i (but excluding the test charge q_0) is then

$$\mathbf{E} = \frac{\mathbf{F}}{q_0} = \sum \frac{kq_i}{r_{i0}{}^2}\,\hat{\mathbf{r}}_{i0} \qquad\qquad 29\text{-}8$$

Electric field due to system of point charges

Example 29-6 A positive charge q_1 is at the origin, and a second positive charge q_2 is on the x axis at $x = a$. Find the electric field at points on the x axis (Figure 29-4).

In the region $x > a$, the electric field due to each charge is along the x axis in the positive x direction. The distance r_{10} to the charge q_1 is x, and the distance r_{20} to q_2 is $x - a$. The electric field is then

$$\mathbf{E} = \frac{kq_1}{x^2}\,\mathbf{i} + \frac{kq_2}{(x-a)^2}\,\mathbf{i} \qquad x > a$$

where $\mathbf{i}$ is the unit vector in the x direction. In the region $0 < x < a$, the field due to q_1 is in the positive direction, but the field due to q_2 is in the negative x direction. The distance to q_1 is again x, and that to q_2 is $a - x$. Since these distances are squared in calculating the field, we can write either $(a - x)^2$ or $(x - a)^2$. The electric field between the charges is thus

$$\mathbf{E} = \frac{kq_1}{x^2}\,\mathbf{i} - \frac{kq_2}{(x-a)^2}\,\mathbf{i} \qquad 0 < x < a$$

Figure 29-4
Point charges on x axis for Example 29-6.

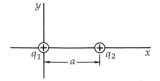

To the left of the origin, the field due to each charge is in the negative x direction. The distance to q_1 is $-x$, and that to q_2 is $-x + a$ (since x is a negative number). Then for negative x we have

$$\mathbf{E} = -\frac{kq_1}{x^2}\,\mathbf{i} - \frac{kq_2}{(x-a)^2}\,\mathbf{i} \qquad x < 0$$

Example 29-7 Find the electric field at points on the y axis for the charges in Example 29-6.

The fields due to each charge at a point on the positive y axis are shown in Figure 29-5. The field due to q_1 has magnitude $E_1 = kq_1/y^2$ and is in the $+y$ direction. The field due to q_2 has magnitude $E_2 = kq_2/(y^2 + a^2)$ and is in a direction making an angle θ with the y axis. The resultant electric field is the vector sum of these fields, $\mathbf{E} = \mathbf{E}_1 + \mathbf{E}_2$. The y component of the resultant electric field is

$$E_y = \frac{kq_1}{y^2} + \frac{kq_2}{y^2 + a^2}\cos\theta = \frac{kq_1}{y^2} + \frac{kq_2 y}{(y^2 + a^2)^{3/2}}$$

using $\cos\theta = y/\sqrt{y^2 + a^2}$. The x component of the resultant electric field is

$$E_x = -\frac{kq_2}{y^2 + a^2}\sin\theta = -\frac{kq_2 a}{(y^2 + a^2)^{3/2}}$$

using $\sin\theta = a/\sqrt{y^2 + a^2}$.

29-4   Lines of Force

The presence of an electric field can be indicated by drawing lines of electric force, just as the gravitational field was indicated by lines of gravitational force in Chapter 16. The electric field vector $\mathbf{E}$ is tangent to the line at each point. Since there is an infinite number of points in space, only a few representative lines are drawn and we indicate the field by drawing continuous lines which begin or end on charges.

Figure 29-6 shows the lines of force, or electric field lines, of a single positive point charge. Consider a spherical surface of radius r with its center at the charge. For a fixed number of lines emerging from the charge, the number of lines per unit area on the sphere is inversely proportional to the area of the sphere, $4\pi r^2$. Thus the density of lines decreases with distance as $1/r^2$, just as the magnitude of the electric field decreases. If we adopt the convention of drawing a fixed number of lines from a point charge, the number being proportional to the charge strength, and if we draw the lines symmetrically about the point charge, the field strength is indicated by the density of the lines. This result is the same as for the lines of force indicating the gravitational field, which also decreases in magnitude as the square of the distance from a point mass; however, there is one important difference. The lines of electric force diverge from a positive point charge, indicating that the electric field points away from a positive charge, and the lines converge on a negative point charge, indicating that the electric field points toward a negative charge. There is only one kind of gravitational mass. As discussed in Chapter 16, lines of gravitational force converge at a point mass, and there are no points in space where lines of gravitational force diverge. This difference is due of course to

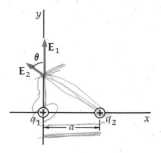

Figure 29-5
Example 29-7. On the y axis, the electric field $\mathbf{E}_1$ due to charge q_1 is along the y axis, and the field $\mathbf{E}_2$ due to charge q_2 makes an angle θ with the y axis. The resultant electric field is the vector sum $\mathbf{E} = \mathbf{E}_1 + \mathbf{E}_2$.

Figure 29-6
Electric field lines, or lines of force, of a single charge. The photograph shows bits of thread suspended in oil. The electric field of the charged object in the center induces opposite charges on the ends of each bit of thread, causing the thread to align itself parallel to the field. (*Courtesy of Harold M. Waage, Princeton University.*)

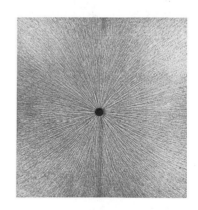

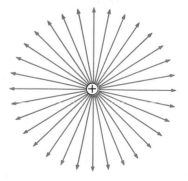

the fact that the electric force can be either attractive or repulsive whereas the gravitational force is always attractive.

For future reference we summarize our rules for drawing lines of electric force.

Rules for drawing lines of force to indicate electric field patterns

1. The number of lines leaving a positive point charge or entering a negative charge is proportional to the charge.

2. The lines are drawn symmetrically leaving or entering a point charge.

3. Lines begin or end only on charges.

4. The density of lines (number per unit area perpendicular to the lines) is proportional to the magnitude of the field.

5. No two field lines can cross.

Rule 5 follows from the fact that $\mathbf{E}$ has a unique direction at any point in space. If two lines crossed, two directions would be indicated for $\mathbf{E}$ at the point of intersection.

Example 29-8 Sketch the lines of force for two equal positive point charges q separated by a distance a.

We construct this pattern without calculating the field at each point. At points near one of the charges, the field is approximately due to that charge alone because of the inverse-square-distance dependence of the field.

Thus on a sphere of very small radius $(r \ll a)$ about one of the charges, the field lines are radial and equally spaced. Since the charges are equal, we draw an equal number of lines from each charge. At very large distances from the charges, the field is approximately the same as that due to a point charge of magnitude $2q$. So on a sphere of radius $r \gg a$, the lines are approximately equally spaced. The lines are shown in Figure 29-7. (In our two-dimensional drawings, these spheres are replaced by circles.) This figure is the same as Figure 16-8 for two equal point masses except for the direction of the lines. If the two point charges were negative, the lines would be identical to those for two point masses.

Electric lines of force of two equal charges of the same sign shown by bits of thread suspended in oil. (*Courtesy of Harold M. Waage, Princeton University.*)

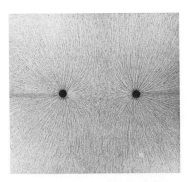

Figure 29-7
Example 29-8. Lines of force due to two positive point charges.

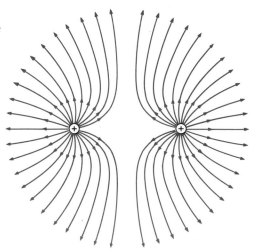

Figure 29-8
Example 29-9. Lines of force for an electric dipole.

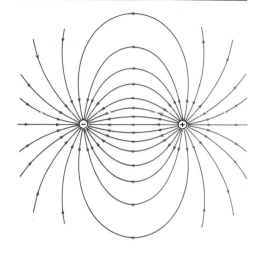

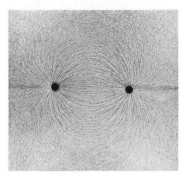

Electric lines of force of two equal but opposite charges shown by bits of thread suspended in oil. (*Courtesy of Harold M. Waage, Princeton University.*)

Example 29-9 Sketch the lines of force for two charges of equal magnitude and opposite signs separated by a distance a.

This system is called an *electric dipole*. Very near the positive charge, the lines are radial outward. Very near the negative charge, the lines are radial inward. Since the charges have equal magnitude, the number of lines that begin at the positive charge equals the number that end at the negative charge. At great distances from the charges, $r \gg a$, the separation of the charges becomes less important. The field of one charge tends to neutralize that of the other. The lines of force are shown in Figure 29-8.

Example 29-10 Sketch the lines of force for a negative charge $-q$ a distance a from a positive charge $+2q$.

Since the positive charge is twice the magnitude of the negative charge, twice as many lines leave the positive charge as enter the negative charge. At great distances from the charges the system looks like a single charge $+q$. Half the lines beginning on the positive charge leave the system. On a sphere of radius r, where r is much larger than the separation of the charges a, these lines are approximately symmetrically spaced and point radially outward, the same as the lines from a single positive point charge $+q$. The other half leaving the positive charge $+2q$ enter the negative charge. The field lines are shown in Figure 29-9.

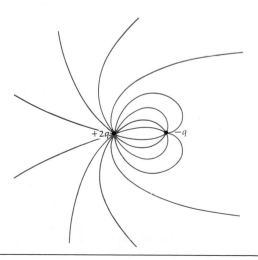

Figure 29-9
Example 29-10. Lines of force for a point charge $+2q$ and a second point charge $-q$. At great distances from the charges the lines are the same as for a single charge $+q$.

Question

4. If the electric force between two point charges varied as $1/r^3$ rather than $1/r^2$, could the same system of lines of force be used to indicate the magnitude of the electric field? Why or why not?

29-5   Electric Flux

The idea of lines of force can be put on a quantitative basis. Consider a mathematical surface enclosing the dipole charge distribution of Example 29-9. (The surface may be spherical or any other shape.) The number of lines of force coming from the positive charge and crossing the surface going out of the enclosure depends on where the surface is drawn, but this number is exactly equal to the number of lines entering the enclosure and ending on the negative charge. We count the number leaving as positive and the number entering as negative; the net number leaving (or entering) is zero. In the other examples, the net number of lines leaving any surface enclosing the charges is proportional to the net charge enclosed by the surface. The number is the same for all surfaces enclosing the charges. This result in its quantitative form, which we now develop, is known as *Gauss' law*.

Consider first, an electric field which is uniform in magnitude and direction over some region. The lines of force for such a field are illustrated in Figure 29-10. Consider the rectangular surface of area A, perpendicular to the field shown in the figure. Since the number of lines per unit area is proportional to the magnitude of the electric field, the number of lines through this surface is proportional to the product of the field E and the area A:

$$N \propto EA \qquad\qquad 29\text{-}9$$

The product of the electric field strength and the area of a surface perpendicular to the field is called the *flux ϕ* of the field through the surface:

$$\phi = EA \qquad\qquad 29\text{-}10$$

The units of flux are N-m²/C. The number of lines through the surface is proportional to the flux:

$$N \propto \phi \qquad\qquad 29\text{-}11$$

The proportionality constant depends upon the choice of the number of lines leaving or entering a unit charge.

In Figure 29-11 the surface of area A_2 is not perpendicular to the electric field $\mathbf{E}$. The number of lines that cross area A_2 is the same as the number that cross area A_1. The areas are related by

$$A_2 \cos \theta = A_1$$

where θ is the angle between $\mathbf{E}$ and the unit vector $\hat{\mathbf{n}}$ perpendicular to surface A_2, as shown. The flux through a surface not perpendicular to $\mathbf{E}$ is defined to be

$$\phi = \mathbf{E} \cdot \hat{\mathbf{n}}\, A = EA \cos \theta = E_n A \qquad\qquad 29\text{-}12$$

where $E_n = \mathbf{E} \cdot \hat{\mathbf{n}}$ is the component of the electric field vector perpendicular, or normal, to the surface.

We can generalize our definition of electric flux to curved surfaces over which the electric field may vary in magnitude or direction or

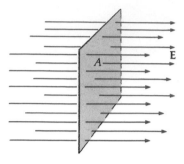

Figure 29-10
Lines of force for a uniform electric field crossing an area A perpendicular to the field. The product EA is called the electric flux.

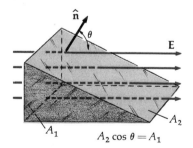

Figure 29-11
Lines of force for a uniform electric field which is perpendicular to the area A_1 but makes an angle θ with the unit vector $\hat{\mathbf{n}}$ normal to area A_2. When $\mathbf{E}$ is not perpendicular to the area, the flux through the area is $E_n A$, where $E_n = E \cos \theta$ is the component of $\mathbf{E}$ perpendicular to the area. The flux through A_2 is then the same as that through A_1.

both by dividing the surface up into a large number of very small elements. If each element is small enough, it can be considered to be a plane and the variation of the electric field $\mathbf{E}$ across the element can be neglected. Let $\hat{\mathbf{n}}_i$ be the unit vector perpendicular to such an element and ΔA_i be its area (Figure 29-12). (If the surface is curved, the unit vectors $\hat{\mathbf{n}}_i$ will have different directions for different elements.) The flux of the electric field through this element is

$$\Delta \phi_i = \mathbf{E} \cdot \hat{\mathbf{n}}_i \, \Delta A_i$$

The total flux through the surface is the sum of $\Delta \phi_i$ over all the elements. In the limit as the number of elements approaches infinity and the area of each element approaches zero, this sum becomes an integral. The general definition of electric flux is then

$$\phi = \lim_{\Delta A_i \to 0} \sum_i \mathbf{E} \cdot \hat{\mathbf{n}}_i \, \Delta A_i = \int \mathbf{E} \cdot \hat{\mathbf{n}} \, dA \qquad 29\text{-}13$$

As with a uniform electric field and a plane surface, the number of lines of force through any surface is proportional to the flux.

We are often interested in the flux of the electric field through a closed surface, i.e., a surface which separates space into two regions, one inside the surface and one outside. On a closed surface the unit normal vector $\hat{\mathbf{n}}$ is defined to be directed outward at each point. At a point where a line of force leaves the surface, $\mathbf{E}$ is directed outward, and $\mathbf{E} \cdot \hat{\mathbf{n}}$ is positive, but at a point where a line of force enters the surface, $\mathbf{E}$ is directed inward, and $\mathbf{E} \cdot \hat{\mathbf{n}}$ is negative. The total or net flux ϕ_{net} through the closed surface is positive or negative depending on whether $\mathbf{E}$ is predominantly outward or inward on the surface. Since the flux through any part of the surface is proportional to the number of lines through the surface, the net flux is proportional to the net number of lines of force leaving the surface, i.e., the number of lines going out of the surface minus the number going into the surface. The integral over a closed surface is indicated by the symbol $\oint$. The net flux through a closed surface is therefore written

$$\phi_{\text{net}} = \oint \mathbf{E} \cdot \hat{\mathbf{n}} \, dA = \oint E_n \, dA \qquad 29\text{-}14$$

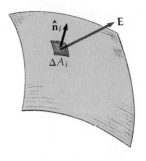

Figure 29-12
When $\mathbf{E}$ varies in either magnitude or direction, the area is divided into small area elements ΔA_i. The flux through the area is computed by summing $E_n \, \Delta A_i$ over all the area elements.

Example 29-11 A point charge q is at the center of a spherical surface of radius R. Calculate the net flux of the electric field through this surface.

The electric field at any point on the surface is radial, with magnitude

$$E = \frac{kq}{R^2}$$

Since $\hat{\mathbf{n}}$ is also radial on the surface, $\mathbf{E} \cdot \hat{\mathbf{n}} = kq/R^2$ is constant everywhere on the surface. The net flux through the spherical surface is therefore just the product of the radial component of $\mathbf{E}$ and the total area of the spherical surface $4\pi R^2$:

$$\phi_{\text{net}} = \oint \mathbf{E} \cdot \hat{\mathbf{n}} \, dA = \oint \frac{kq}{R^2} \, dA = \frac{kq}{R^2} \oint dA = \frac{kq}{R^2} \, 4\pi R^2$$

or

$$\phi_{\text{net}} = \oint \mathbf{E} \cdot \hat{\mathbf{n}} \, dA = 4\pi kq \qquad 29\text{-}15$$

29-6 Gauss' Law

In Example 29-11, we calculated the flux of a point charge through a spherical surface of radius R to be $4\pi kq$, independent of R. The number of lines of force going out through a spherical surface of radius R is also proportional to the charge q. This is consistent with our previous observations (Examples 29-8 to 29-10) that the net number of lines going out of a surface is proportional to the net charge inside the surface. This number of lines is the *same for all surfaces surrounding* the charge, independent of the shape of the surface. Since the number of lines and the flux are just proportional to each other, it follows that Equation 29-15 holds for the flux through any surface enclosing the point charge q. The net flux through any surface surrounding a point charge q equals $4\pi kq$.

We can extend this result to systems of more than one point charge. In Figure 29-13 the surface S encloses two point charges q_1 and q_2, and there is a third point charge q_3 outside the surface. Since the electric field at any point on the surface is the vector sum of the electric fields produced by each of the three charges, the net flux $\phi_{net} = \oint \mathbf{E} \cdot \hat{\mathbf{n}} \, dA$ through the surface is just the sum of the fluxes due to the individual charges. The flux through the surface S due to the charge q_3, which is outside the surface, is zero because every line of force from q_3 that enters the surface at one point leaves the surface at some other point. The net number of lines through the surface from a charge outside the surface is zero. The flux through the surface due to charge q_1 is $4\pi kq_1$, and that due to charge q_2 is $4\pi kq_2$. The net flux through the surface equals $4\pi k(q_1 + q_2)$, which may be positive, negative, or zero depending on the signs and magnitudes of the two charges. In general, for a system of charges $q_1, q_2, \ldots, q_i \ldots$ the net flux through any surface S equals $4\pi k$ times the net charge inside the surface:

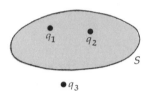

Figure 29-13
Three point charges q_1, q_2, and q_3 and a surface S which encloses q_1 and q_2.

$$\phi_{net} = \oint \mathbf{E} \cdot \hat{\mathbf{n}} \, dA = 4\pi kq_{inside} \qquad \text{29-16}$$

Gauss' law

This important result is Gauss' law. Its validity depends on the fact that the electric field due to a single point charge varies inversely with the square of the distance from the charge. It was this property of the electric field that made it possible to draw a fixed number of lines of force from a charge and have the density of lines be proportional to the field strength. It is customary to write the coulomb constant k in terms of another constant ϵ_0, called the *permittivity of free space*,

$$k = \frac{1}{4\pi\epsilon_0} \qquad \text{29-17}$$

With this notation, Coulomb's law and Gauss' law are written

$$\mathbf{F}_{12} = \frac{1}{4\pi\epsilon_0} \frac{q_1 q_2}{r_{12}^2} \hat{\mathbf{r}}_{12} \qquad \text{29-18}$$

and

$$\phi_{net} = \oint \mathbf{E} \cdot \hat{\mathbf{n}} \, dA = \frac{1}{\epsilon_0} q_{inside} \qquad \text{29-19}$$

The value of ϵ_0 in SI units is

$$\epsilon_0 = \frac{1}{4\pi k} = \frac{1}{4\pi(8.99 \times 10^9)} = 8.85 \times 10^{-12} \frac{\text{C}^2}{\text{N-m}^2} \qquad \text{29-20}$$

Karl Friedrich Gauss, German mathematician, astronomer, and physicist (1777–1855).

Radio Times/Hulton Picture Library

Questions

5. If the electric field **E** is zero everywhere on a closed surface, is the net flux through the surface necessarily zero? What then is the net charge inside the surface?

6. If the net flux through a closed surface is zero, does it follow that the electric field **E** is zero everywhere on the surface? Does it follow that the net charge inside the surface is zero?

29-7 Motion of Point Charges in Electric Fields

When a particle with charge q is placed in an electric field **E**, it experiences a force $q\mathbf{E}$. If this is the only force on it, the particle has an acceleration $q\mathbf{E}/m$, where m is the mass of the particle.[1] If the electric field is known, the charge-to-mass ratio of the particle can be determined from the measured acceleration. For example, for an electric field uniform in space and constant in time, the path of the particle is a parabola, similar to that of a projectile in a uniform gravitational field. The measurement of the deflection of electrons in a uniform electric field was used by J. J. Thomson in 1897 to demonstrate the existence of electrons and to measure their charge-to-mass ratio. (Since magnetic fields were also used in this measurement, we defer discussion of the Thomson experiment until Chapter 36.) We shall give some examples of motion of electrons in constant electric fields. Problems of this type can be worked using the constant-acceleration formulas from Chapter 2 or the equations for projectile motion from Chapter 3.

Example 29-12 An electron is projected into a uniform electric field $E = 1000$ N/C with initial speed $v_0 = 2 \times 10^6$ m/sec parallel to the field. How far does the electron travel before it is brought momentarily to rest?

Since the charge of the electron is negative, the force $-e\mathbf{E}$ is in the direction opposite the field. We thus have a constant-acceleration problem in which the acceleration is opposite to the initial velocity, and we are asked to find the distance traveled. We can use the constant-acceleration expression relating the distance to the velocity (Equation 2-13):

$$v^2 = v_0^2 + 2a(x - x_0)$$

Using $x_0 = 0$, $v = 0$, $v_0 = 2 \times 10^6$, and $a = -eE/m$, we find for the distance,

$$x = \frac{mv_0^2}{2eE} = \frac{(9.11 \times 10^{-31})(2 \times 10^6)^2}{2(1.6 \times 10^{-19})(1000)} = 1.14 \times 10^{-2} \text{ m}$$

Example 29-13 An electron is projected into a uniform electric field $E = 2000$ N/C with initial velocity $v_0 = 10^6$ m/sec perpendicular to the field. Compare the weight of the electron to the electric force on it. By how much is the electron deflected after it has traveled 1 cm?

The electric force on the electron is eE, and the gravitational force is

---

[1] We are assuming that the speed of the particle is small enough for us to use classical mechanics and neglect special relativity. This assumption is often *not* valid for the motion of electrons in electric fields.

mg. Their ratio is

$$\frac{F_{\text{electric}}}{F_{\text{weight}}} = \frac{eE}{mg} = \frac{(1.6 \times 10^{-19})(2000)}{(9.1 \times 10^{-31})(9.8)} = 3.6 \times 10^{13}$$

As in most common cases, the electric force is huge compared with the gravitational force, which is wholly negligible.

It takes the electron a time

$$t = \frac{x}{v_0} = \frac{10^{-2} \text{ m}}{10^6 \text{ m/sec}} = 10^{-8} \text{ sec}$$

to travel the distance of 1 cm perpendicular to the field. In this time it is deflected a distance antiparallel to the field given by

$$y = \tfrac{1}{2} at^2 = \tfrac{1}{2} \frac{eE}{m} t^2$$

Substituting the known values of e/m, E, and t gives

$$y = 1.76 \times 10^{-2} \text{ m} = 1.76 \text{ cm}$$

Questions

7. The direction of the force on a positive charge in an electric field is, by definition, in the direction of the field line passing through the position of the charge. Must the acceleration be in this direction? The velocity? Explain.

8. A positive charge is released from rest in an electric field. It starts out in the direction of a field line. Will it continue to move along the field line?

29-8 Electric Dipole in Electric Fields

Although atoms and molecules are electrically neutral, they are affected by electric fields because they contain positive and negative charges. In some molecules, the center of positive charge does not coincide with the center of negative charge. These *polar molecules* are said to have a permanent electric dipole moment. When such a molecule is placed in a uniform electric field, there is no net force on it, but there is a torque, which tends to rotate the molecule. In a nonuniform electric field the molecule experiences a net force because the field at the center of the positive charge is different from that at the center of the negative charge. An example of a polar molecule is NaCl, which is essentially a positive sodium ion of charge $+e$ combined with a negative chlorine ion of charge $-e$.

Atoms and molecules for which the centers of positive and negative charge coincide are also affected by an electric field. Because the electric force on the positive charge is in the direction opposite that on the negative charge, the electric field tends to separate, or polarize, these charges. These systems then have an induced dipole moment when they are in an electric field, and they also experience a net force in a nonuniform field. The force produced by a nonuniform electric field on an electrically neutral charge system is responsible for the familiar attraction of a charged comb for uncharged bits of paper.

Figure 29-14 shows a schematic picture of a polar molecule with the centers of positive and negative charge indicated. (The center of

Figure 29-14
Schematic diagram of a polar molecule. The centers of positive and negative charge are separated by a distance L. The molecule behaves like a electric dipole of moment $\mathbf{p} = q\mathbf{L}$, where q is the magnitude of the total positive or negative charge.

charge is defined analogously to the center of mass, with the mass replaced by the charge.) The behavior of such a molecule can be described by a vector **p**, called the *dipole moment*. Let the total positive charge of the molecule be q and the total negative charge $-q$. The dipole moment of the molecule is defined to be the product of the charge q and the displacement vector **L** pointing from the center of negative charge to the center of positive charge:

$$\mathbf{p} = q\mathbf{L} \qquad\qquad 29\text{-}21$$

Dipole moment defined

The diameter of an atom or molecule is of the order of 10^{-10} m = 1 Å. A convenient unit for electric dipole moments of atoms and molecules is the fundamental electronic charge e times the distance 1 Å. For example, the dipole moment of NaCl in these units has a magnitude of about 2 e-Å.

We can often simplify the description of the behavior of a polar molecule in an electric field by replacing the complicated charge distribution of the molecule with a simple electric dipole consisting of two charges q and $-q$ separated by a distance L and having the same dipole moment as the molecule.

Figure 29-15 shows a simple electric dipole whose dipole moment makes an angle θ with an external uniform electric field **E**. The forces acting on the dipole,

$$\mathbf{F}_1 = q\mathbf{E} \qquad \mathbf{F}_2 = -q\mathbf{E}$$

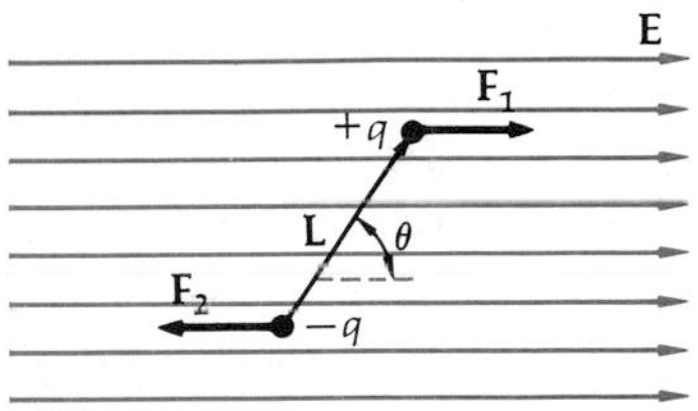

are opposite in direction and equal in magnitude since the field is uniform. The net force on the dipole is thus zero. However, the two forces produce a torque, which tends to rotate the dipole so that it points in the direction of the field. We found in Chapter 12 that the torque produced by two equal and opposite forces, called a couple, is the same about any point in space. From Figure 29-15 we see that the torque about the negative charge has the magnitude $F_1 L \sin\theta = qEL \sin\theta = pE \sin\theta$. The direction of the torque is into the paper and such that it rotates the dipole moment **p** into the direction of the electric field **E**. This torque can be conveniently written as the cross product of the dipole moment **p** and the electric field **E**:

$$\boldsymbol{\tau} = \mathbf{p} \times \mathbf{E} \qquad\qquad 29\text{-}22$$

Torque on an electric dipole

If the electric dipole is placed in a nonuniform electric field, as in Figure 29-16, there is a net force acting on the dipole in addition to the

Figure 29-15
Electric dipole in a uniform electric field. The net force on the dipole is zero, but there is a net torque $\mathbf{L} \times \mathbf{F}_1 = q\mathbf{L} \times \mathbf{E} = \mathbf{p} \times \mathbf{E}$ which tends to align the dipole in the direction of the field.

Figure 29-16
For a general orientation of an electric dipole in a nonuniform electric field there is both a net torque tending to align the dipole with the field and a net force on the dipole. Here, the net force has a component parallel to **E** and a small component downward. Its magnitude depends on the dipole moment **p**, its orientation, and on how rapidly the field varies in space.

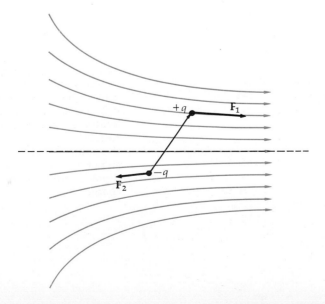

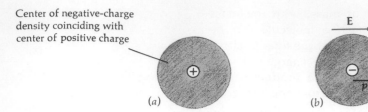

Center of negative-charge density coinciding with center of positive charge

(a)

E

(b)

torque tending to align the dipole with the field. The net force depends on the orientation of the dipole in the field, the dipole moment, and on how rapidly the field varies in space. For the situation pictured in Figure 29-16 the net force on the dipole has a component downward and one in the direction of increasing field.

Most molecules and all atoms are nonpolar; i.e., the centers of positive and negative charge coincide, and there is no permanent electric dipole moment. However, an external electric field produces a separation of the positive and negative charge distribution, thus inducing a dipole moment (Figure 29-17). Since the positive charge is displaced in the direction of the field and the negative charge is displaced in the opposite direction, the induced dipole moment is always in the direction of the external electric field. Thus no torque is exerted because the angle between the dipole moment **p** and the field **E** is zero. However, if the electric field is not uniform, there will be an external force acting on the dipole. Figure 29-18 shows a nonpolar molecule in an external electric field of a positive point charge Q. The induced dipole moment is parallel to **E** in the radial direction from the point charge. Since the field is stronger at the negative charge nearer the point charge, the force on the dipole is toward the point charge, and the dipole is attracted toward the point charge. If the point charge were negative, the induced dipole would be in the opposite direction and the dipole would again be attracted to the point charge.

Figure 29-17
Schematic diagram of a nonpolar molecule. (*a*) In the absence of an external electric field the center of the positive charge coincides with that of the negative charge, and there is no dipole moment. (*b*) In the presence of an external electric field the centers of positive and negative charge are displaced, producing a dipole moment parallel to the external field. The magnitude of the dipole moment depends on the magnitude of the electric field E.

Question

9. A small, light, conducting ball with no net electric charge is suspended from a thread. When a positive charge is brought near the ball, the ball is attracted toward the charge. How does this come about? If the ball is an insulator instead of a conductor, it still is attracted. How? Would the situation have been different if the charge brought near were negative instead of positive? How?

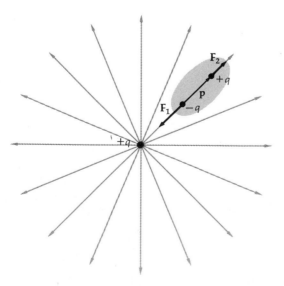

Figure 29-18
Nonpolar molecule in the electric field of a positive point charge. The induced dipole moment is parallel to the field **E**. Since the center of negative charge is closer to the point charge than the center of positive charge, the dipole is attracted to the point charge.

Optional

29-9 Mathematical Derivation of Gauss' Law

Gauss' law can be derived mathematically using the concept of the solid angle. Consider an area element ΔA on a spherical surface. The solid angle $\Delta \Omega$ subtended by ΔA at the center of the sphere is defined to be

$$\Delta \Omega = \frac{\Delta A}{r^2}$$

where r is the radius of the sphere. Since ΔA and r^2 both have dimensions of length squared, the solid angle is dimensionless. The unit of solid angle is the *steradian*. Since the total area of a sphere is $4\pi r^2$, the total solid angle subtended by a sphere is

$$\frac{4\pi r^2}{r^2} = 4\pi \text{ steradians}$$

There is a close analogy between the solid angle and the ordinary plane angle, which is defined to be the ratio of an element of arc length of a circle Δs divided by the radius of the circle. $\theta = \Delta s / r$ rad. The total (plane) angle subtended by a circle is 2π rad.

In Figure 29-19 the area element ΔA is not perpendicular to the radial lines from point O. The unit vector $\hat{\mathbf{n}}$ normal to the area element makes an angle θ with the unit radial vector $\hat{\mathbf{r}}$. In this case, the solid angle subtended by ΔA is

Solid angle

$$\Delta \Omega = \frac{\Delta A\,\hat{\mathbf{n}} \cdot \hat{\mathbf{r}}}{r^2} = \frac{\Delta A \cos \theta}{r^2} \qquad 29\text{-}23$$

Figure 29-20 shows a point charge q surrounded by a surface of arbitrary shape. To calculate the flux through this surface, we want to find $\mathbf{E} \cdot \hat{\mathbf{n}}\,\Delta A$ for each element of area on the surface and sum over the entire surface. The flux through the area element shown is

$$\Delta \phi = \mathbf{E} \cdot \hat{\mathbf{n}}\,\Delta A = \frac{kq}{r^2}\,\hat{\mathbf{r}} \cdot \hat{\mathbf{n}}\,\Delta A = kq\,\Delta \Omega$$

The solid angle $\Delta \Omega$ is the same as that subtended by the corresponding area element of a spherical surface of any radius. The sum of the flux through the entire surface is kq times the total solid angle subtended by the closed surface, which is 4π steradians:

$$\phi_{\text{net}} = \oint \mathbf{E} \cdot \hat{\mathbf{n}}\,dA = kq \oint d\Omega = 4\pi kq$$

Figure 29-19
Area element ΔA whose normal $\hat{\mathbf{n}}$ is not parallel to the radial line from O to the center of the element. The solid angle subtended by this element at O is $(\Delta A \cos \theta)/r^2$.

Figure 29-20
Point charge q enclosed by surface S. The flux through the area element ΔA is proportional to the solid angle subtended by the area element at the charge. The net flux through the surface found by summing over all area elements is proportional to the total solid angle 4π, which is independent of the shape of the surface.

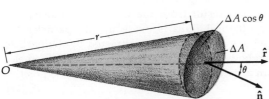

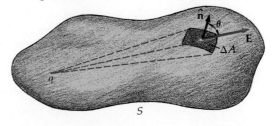

Benjamin Franklin (1706–1790)

I. Bernard Cohen
Harvard University

When Franklin's contemporaries wanted to express their admiration for his scientific achievements, they could think only of comparing him to Newton. Joseph Priestley wrote that Franklin's book on electricity "bid fare to be handed down to posterity as expressive of the true philosophy of electricity; just as the Newtonian philosophy is of the true system of nature in general."

In order to appreciate Franklin's contribution, we must remember that electricity is a young branch of physics. Whereas we can trace an early history of statics and dynamics, heat and light, and even atomic theory back to the Greeks, electricity emerged as a proper subject of scientific study only in the days of Newton. How meager the information about electricity was in the early eighteenth century can be seen from the fact that the fundamental distinction between conductors and nonconductors had not yet been made, nor had it been discovered that there are two kinds of electric charge. When Franklin took up this new subject, he was almost forty, past the age when we usually think of great scientific discoveries being made. Largely self-educated, Franklin had already obtained a solid grounding in experimental physics, having studied Newton's *Opticks* and many of the primary textbooks of newtonian experimental science. His printing and newspaper business was successful enough to allow him to retire from active participation, and he eagerly seized upon the new subject and began to work at it intensively, together with a small group of coworkers.

His interest had first been sparked when he attended some public lectures on science given by a Dr. Adam Spencer, first in Boston (where Franklin was visiting his family) and then in Philadelphia (where Franklin sponsored Spencer's lectures). He purchased Spencer's apparatus in order to perform experiments himself. Soon thereafter the Library Company of Philadelphia (of which Franklin was the principal founder) received a gift of electrical apparatus, with instructions for using it, from Peter Collinson, a London merchant.

Before long Franklin and his fellow experimenters realized that they had progressed in knowledge far beyond the literature that accompanied the apparatus. Franklin periodically sent reports of his work to Collinson and to other London correspondents, which eventually were assembled into a book entitled *Experiments and Observations on Electricity, Made at Philadelphia*, first published in England in 1751.

For his pioneering research in electricity, Franklin was elected a Fellow of the Royal Society of London, with the singular distinction of being forgiven the annual dues. Shortly thereafter he was awarded the Society's Copley Gold Medal, the highest scientific honor then being awarded in England. His book on electricity was a spectacular success, going through five editions in English and being translated into French (three editions in two different translations), Italian, and German. In 1773 Franklin was elected a Foreign Associate of the French Academy of Sciences, an extraordinary honor, since, according to the terms of the Academy's foundation, there could be only eight such foreign associates at any one time. No American was similarly honored again for another century.

Franklin's discoveries included not only important new experimental evidence but a new theory of electricity which made a science of the subject. This theory was based upon the fundamental postulate that all electrostatic phenomena (charging and discharging) result from the motion or transfer of a single electrical fluid. This hypothetical fluid is made up of "particles," or atoms, of electricity which repel one another but are attracted by the particles of "ordinary" matter. A charged body is one that has either lost or gained electrical fluid and accordingly is in a state which Franklin called "plus" (positive)

Title page of the first edition of Franklin's book on electricity, from the copy he presented to Harvard College. (*Courtesy of the Houghton Library, Harvard University.*)

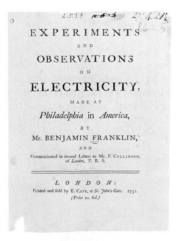

Portrait of Franklin. (*Courtesy of the Yale University Library*.)

or "minus" (negative). This postulate was closely associated with a major theoretical principle, known today as the law of conservation of charge: whatever charge is lost by one body must be gained by one or more other bodies, so that negative and positive charges always appear simultaneously or are simultaneously canceled out in equal amounts. This law implies that electrical effects are not the results of the "creation" of some mysterious entity (as had often been supposed before Franklin) but merely follow from an alteration or redistribution of the amount of electrical fluid in a body. A variety of experiments confirmed the universality of this principle of conservation of charge, which remains (along with conservation of momentum) one of the most fundamental principles of physical science.

One of Franklin's most startling experiments was the analysis of the Leyden jar, the first "condenser," or capacitor. The Leyden jar is a glass jar with a metal-foil outer coating. Inside, the jar has another metal-foil coating or contains water, or lead shot, which is in contact with a wire passing through the cork of the jar. The jar is usually charged by bringing the wire into contact with an electrostatic generator or a bit of rubbed amber, glass, or sulfur, while the outer coating is grounded. Franklin discovered that when a jar is charged, and the two conductors are then separated from the glass, neither the inner nor the outer conductor will show any sign of being charged; but when they are placed once again on the two sides of the jar, that jar will produce all the familiar phenomena of being charged. Franklin said that the whole charge "resides" in the glass, but today we refer to this phenomenon under the name of "polarization of the dielectric." Franklin also showed that such a device does not depend upon the shape of the bottle, and he invented the parallel plane capacitor.

In a variety of experiments, Franklin also showed the effects of grounding and insulation. He discovered that a grounded pointed conductor can actually

The sentry box experiment, from a manuscript of Franklin's with drawings. (*Courtesy of the American Academy of Arts and Sciences, Office of Charles and Ray Eames.*)

"draw off" the charge of a nearby charged object and that contrariwise a charged pointed conductor, however well-insulated, will "throw off" the charge through its point. This led him to study the electrical nature of the lightning discharge. Franklin's most important experiment with lightning was not made with the familiar kite but was the experiment of the sentry box. Franklin proposed that on a high building, a sentry box be erected, with a long pointed conductor rising up through the roof. To determine whether the pointed conductor were charged, an experimenter would stand on an insulated stool inside the sentry box and bring up to the pointed conductor a grounded conductor set in an insulating handle. His objective was to draw a spark. If, as Franklin supposed, thunderclouds are electrically charged, then the pointed conductor would always become charged by induction when such clouds passed overhead, and a spark could be obtained. In practice, this experiment not only proved that clouds are electrically charged, so that lightning is only an ordinary electrical discharge from clouds on a large scale, but such rods also attracted a stroke of lightning and showed it to be an electrical phenomenon. Described in his book on electricity, this experiment was first performed in France. Franklin had been waiting for the completion of the spire of Christ Church in Philadelphia, where he hoped to erect a sentry box in order to perform the experiment. The kite was thought of as an alternative, an afterthought.

The lightning experiments brought fame to Franklin far and wide. Today the historical importance of these experiments is often misunderstood. In proving that lightning is an electrical discharge, Franklin showed that the electrical experiments performed in the laboratory are directly related to events in the natural world on a large scale. Thereafter, any general science of nature that did not include electricity would obviously be incomplete. Furthermore, the lightning experiments led Franklin to the invention of the lightning rod. For the first time in history, research in pure science led to a practical invention of major consequence. Bacon had indeed been correct in predicting that pure scientific knowledge would lead men to practical applications which would enable them to control their environment.

When Franklin began his research in electricity, the great French scientist Buffon (1707–1788) pointed out that electricity was not yet a science, but only a collection of bizarre phenomena subject to no single law. After Franklin produced his theory of electricity, which explained and correlated the known phenomena and also predicted verifiable new ones, it was generally agreed that electricity had indeed become a science. In fact, electricity was the first new science (or branch of science) to arise since Newton.

It is often thought that Franklin was not really a pure scientist in the ordinary sense of this expression, that he was, rather, a gadgeteer and inventor whose claim to science was aggrandized by his success as a statesman. In point of fact the opposite is true. By 1776, when Franklin was sent to France as the American representative, he had already gained an international reputation for his scientific work, a factor of the greatest importance in his diplomatic success. This was no unknown local patriot, but one of the leading figures of the scientific world. We continually pay tribute to Franklin's scientific genius whenever we use the many words he introduced into the language of electricity: plus and minus, positive and negative, electric battery, and a host of others.

When Franklin retired from business, he hoped to devote the rest of his life to the peaceful pursuit of a career in science. The demands of his community and his country, however, all too soon drew him into a life of public service. Franklin's choice between his love of science and his duty to his fellow men was expressed by him as follows: "Had Newton been Pilot but of a single common Ship, the finest of his Discoveries would scarce have excused, or attoned for his abandoning the Helm one Hour in Time of Danger; how much less if she carried the Fate of the Commonwealth."

Review

A. Define, explain, or otherwise identify:

Charge conservation, 701

Charge quantization, 702

Coulomb's law, 702

Electric field, 705

Lines of force, 707

Electric dipole, 709

Flux, 710

Gauss' law, 712

Polar molecule, 714

Dipole moment, 715

B. True or false:

1. The electric field of a point charge always points away from the charge.

2. The charge of the electron is the smallest charge possible.

3. Electric lines of force never diverge from a point in space.

4. Electric lines of force cannot cross at a point in space.

5. If there is no charge in a region of space, the electric field must be zero everywhere on a surface surrounding the region.

Exercises

Section 29-1, Electric Charge, and Section 29-2, Coulomb's Law

1. Find the number of electrons in a charge of (a) 1 μC; (b) 10^{-12} C.

2. In electrolysis a quantity of electricity called the faraday $\mathcal{F}$, equal to Avogadro's number of electron charges, will deposit 1 gram ionic weight of monovalent ions, for example, 23 gm of Na and 35.5 gm of Cl in the decomposition of NaCl. Calculate the number of coulombs in a faraday. (Avogadro's number is 6.02×10^{23}.)

3. Two protons in the helium nucleus are about 10^{-15} m apart. Calculate the electrostatic force exerted by one proton on the other.

4. A charge $q_1 = 4.0$ μC is at the origin, and a charge $q_2 = 6.0$ μC is on the x axis at $x = 3.0$ m. (a) Find the force on charge q_2. (b) Find the force on q_1. (c) How would your answers differ if q_2 were -6.0 μC?

5. Three point charges are on the x axis; $q_1 = -6.0$ μC is at $x = -3.0$ m, $q_2 = 4.0$ μC is at the origin, and $q_3 = -6.0$ μC is at $x = 3.0$ m. Find the force on q_1.

6. Two equal charges of 3.0 μC are on the y axis, one at the origin and the other at $y = 6$ m. A third charge $q_3 = 2$ μC is on the x axis at $x = 8$ m. Find the force on q_3.

7. Three charges are at the corners of a square of side L. The two charges at the opposite corners are positive, and the other is negative. All have the same magnitude q. Find the force exerted by these charges on a fourth charge $+q$ placed at the remaining corner.

Section 29-3, The Electric Field

8. A charge of 4.0 μC is at the origin. What is the magnitude and direction of the electric field on the x axis at (a) $x = 5$ m and (b) $x = 10$ m? (c) Sketch the function E_x versus x for both positive and negative x. (Remember that E_x is negative when E points in the negative x direction.)

9. Two charges each $+4$ μC are on the x axis, one at the origin and the other at $x = 8$ m. (a) Find the electric field on the x axis at $x = 10$ m and at $x = 2$ m. (b) At what point on the x axis is the electric field zero? (c) What is the direction of E at points on the x axis just to the right of the origin? Just to the left

of the origin? (d) Sketch E_x versus x. (Remember that E_x is negative when **E** points in the negative x direction.)

10. Two equal positive charges of magnitude $q_1 = q_2 = 6.0$ nC are on the y axis at points $y_1 = +3$ cm and $y_2 = -3$ cm. (a) What is the magnitude and direction of the electric field at the point on the x axis at $x = 4$ cm? (b) What is the force exerted on a test charge $q_0 = 2$ nC placed on the axis at $x = 4$ cm?

11. Charge $q_1 = +6.0$ nC is on the y axis at $y = +3$ cm, and charge $q_2 = -6.0$ nC is on the y axis at $y = -3$ cm. (a) What is the magnitude and direction of the electric field on the x axis at $x = 4$ cm? (b) What is the force exerted on a test charge $q_0 = 2$ nC placed on the x axis at $x = 4$ cm?

12. When a test charge $q_0 = 2$ nC is placed at the origin, it experiences a force of 8.0×10^{-4} N in the positive y direction. (a) What is the electric field at the origin? (b) What would be the force on a charge -4 nC placed at the origin?

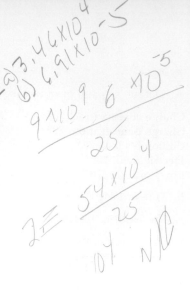

Figure 29-21
Lines of force for Exercise 13.

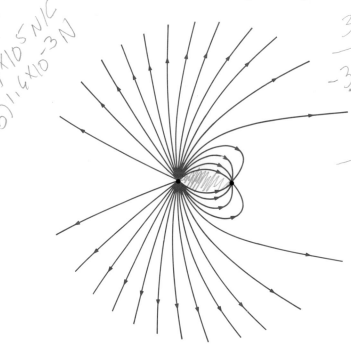

Section 29-4, Lines of Force

13. Figure 29-21 shows lines of force for a system of two point charges. (a) What are the relative magnitudes of the charges? (b) What are the signs of the charges? (c) In what regions of space is the electric field strong? In what regions is it weak?

14. Two charges $+q$ and $-3q$ are separated by a small distance. Draw lines of force for this system.

15. Three equal positive point charges are situated at the corners of an equilateral triangle. Sketch the lines of force in the plane of the triangle.

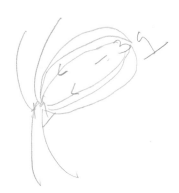

Section 29-5, Electric Flux, and Section 29-6, Gauss' Law

16. Consider a uniform electric field $\mathbf{E} = 2 \times 10^3\ \mathbf{i}$ N/C. (a) What is the flux of this field through a square 10 cm on a side whose plane is parallel to the yz plane? (b) What is the flux through the same square if the normal to its plane makes a 30° angle with the x axis?

17. What is the net flux of the uniform electric field of Exercise 16 through a cube of side 10 cm oriented so that its faces are parallel to the coordinate planes?

18. A point charge $+2 \ \mu C$ is at the center of a sphere of radius 0.5 m. (a) Find the surface area of the sphere. (b) Find the magnitude of the electric field at points on the surface of the sphere. (c) What is the flux of the electric field due to the point charge through the surface of the sphere? (d) Would your answer to part (c) change if the point charge were moved so that it was inside the sphere but not at the center? (e) What is the net flux through a cube of side 1 m which circumscribes the sphere?

19. A single point charge $+2 \ \mu C$ is at the origin. A spherical surface of radius 2.0 m has its center on the x axis at $x = 5$ m. (a) Sketch lines of force for the point charge. Do any lines enter the spherical surface? (b) What is the net number of lines that leave the spherical surface counting those that enter as negative? (c) What is the net flux of the electric field due to the point charge through the spherical surface?

20. A positive point charge q is at the center of a cube of side L. A large number N of lines of force are drawn from the point charge. (a) How many of the lines pass through the surface of the cube? (b) How many lines pass through each face (assuming none are on the edges or corners)? (c) What is the net outward flux of the electric field through the cubical surface? (d) Use symmetry arguments to find the flux of the electric field through one face of the cube. (e) Which if any of your answers would change if the charge were inside the cube but not at its center?

21. Careful measurement of the electric field at the surface of a black box indicates that the net outward flux through the surface of the box is 6.0×10^3 N-m²/C. (a) What is the net charge inside the box? (b) If the net outward flux through the surface of the box were zero, could you conclude that there were no charges inside the box? Why or why not?

22. An electric field is uniform in the positive x direction for positive x and uniform with the same magnitude but in the negative x direction for negative x. $E = 200 \ i$ N/C for $x > 0$ and $E = -200 \ i$ N/C for $x < 0$. A right circular cylinder of length 20 cm and radius 5 cm has its center at the origin and its axis along the x axis so that one face is at $x = +10$ cm and the other at $x = -10$ cm. (a) What is the outward flux through each face? (b) What is the flux through the side of the cylinder? (c) What is the net outward flux through the cylindrical surface? (d) What is the net charge inside the cylinder?

Section 29-7, Motion of Point Charges in Electric Fields

23. In finding the acceleration of an electron or other charged particle the ratio of the charge to mass of the particle is important. (a) Compute e/m for an electron. (b) What is the magnitude and direction of acceleration of an electron in a uniform electric field of magnitude 100 N/C? (c) Nonrelativistic mechanics can be used only if the speed of the electron is significantly less than the speed of light c. Compute the time it takes for an electron placed at rest in an electric field of magnitude 100 N/C to reach a speed 0.01c. (d) How far does the electron travel in that time?

24. (a) Compute e/m for a proton and find its acceleration in a uniform electric field of magnitude 100 N/C. (b) Find the time it takes for a proton initially at rest in such a field to reach the speed of 0.01c (see Exercise 23).

25. An electron has an initial velocity 2.0×10^6 m/sec in the x direction. It enters a uniform electric field which is in the y direction $E = 400 \ j$ N/C. (a) Find the acceleration of the electron. (b) How long does it take for the electron to travel 10 cm in the x direction? (c) By how much and in what direction is the electron deflected after traveling 10 cm in the x direction?

26. Calculate the magnitude and direction of the electric field which would be needed to balance the weight of (a) an electron, (b) a proton, (c) an oil drop

which has a mass of 2×10^{-10} gm and carries a charge of $+10e$, (d) a pingpong ball of mass 25 gm and charge $+0.01 \mu C$.

27. The earth has an electric field in its atmosphere which is about 150 N/C directed upward. Compare the upward electric force on a proton with the downward gravitational force.

28. An electron is projected with an initial speed of 3.0×10^5 m/sec at an upward angle of 30° with the horizontal. There is a vertically upward electric field of strength 10^5 N/C. (a) What is the highest point reached by the electron? (b) How long does it take to reach this height? (c) What is the horizontal range of the electron projectile?

Section 29-8, Electric Dipole in Electric Fields

29. Two point charges $q_1 = 2.0$ pC ($=2 \times 10^{-12}$ C) and $q_2 = -2.0$ pC are separated by 4 μm. What is the dipole moment of this pair of charges? Sketch the pair and indicate the direction of the dipole moment.

30. A dipole of moment 5.0 e-Å is placed in a uniform electric field of strength 4.0×10^4 N/C. What is the magnitude of the torque on the dipole when (a) the dipole is parallel to the electric field, (b) the dipole is perpendicular to the electric field, (c) the dipole makes an angle of 30° with the electric field?

Problems

1. Four charges of equal magnitude are arranged at the corners of a square of side L, as shown in Figure 29-22. (a) Find the magnitude and direction of the force exerted on the charge on the lower left corner by the other charges. (b) Show that the electric field due to the four charges at the midpoint of one of the sides of the square is directed along that side toward the negative charge and his magnitude

$$k \frac{8q}{L^2} \left(1 - \frac{\sqrt{5}}{25}\right)$$

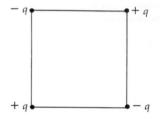

Figure 29-22
Charge distribution for Problem 1.

2. Two charges q_1 and q_2 are placed a distance L apart. (a) What must the relative values of q_1 and q_2 be (including sign) if the electric field is to be zero at some point on the line of the charges a distance D from q_2 and $D + L$ from q_1? (b) Is there a second point anywhere on the line of the charges at which the electric field is zero?

3. Two charges q_1 and q_2 when combined give a total charge of 6 μC. When they are separated by 3 m, the force exerted by one charge on the other has the magnitude 8×10^{-3} N. Find q_1 and q_2 if (a) both are positive so that they repel each other; (b) one is positive and the other negative so that they attract each other.

4. A positive charge Q is to be divided into two positive charges q_1 and q_2. Show that for a given separation D, the force exerted by one charge on the other is greatest if $q_1 = q_2 = \frac{1}{2}Q$.

5. Two equal positive charges q are on the y axis; one is at $y = a$ and the other at $y = -a$. (a) Show that the electric field on the x axis is along the x axis with E_x given by $E_x = 2kqx(x^2 + a^2)^{-3/2}$. (b) Show that near the origin, when x is much smaller than a, E_x is approximately $2kqx/a^3$. (c) Show that for x much larger than a, E_x is approximately $2kq/x^2$. Explain why you would expect this result even before calculating it.

6. (a) Show that the electric field for the charge distribution in Problem 5 has its greatest magnitude at the points $x = a/\sqrt{2}$ and $x = -a/\sqrt{2}$ by computing dE_x/dx and setting the derivative equal to zero. (b) Sketch the function E_x versus x using the results of (a) and parts (b) and (c) of Problem 5.

7. For the charge distribution in Problem 5 the electric field at the origin is zero. A test charge q_0 placed at the origin will therefore be in equilibrium. (a) Discuss the stability of the equilibrium for a positive test charge by considering small displacements from equilibrium along the x axis and small displacements along the y axis. (b) Repeat part (a) for a negative test charge. (c) Find the magnitude and sign of a charge q_0 which can be placed at the origin so that the net force on each of the three charges is zero. Consider what happens if any of the charges are displaced slightly from equilibrium.

8. An electric dipole of moment **p** makes an angle θ with a uniform electric field **E**. (a) How much work must be done by an external torque to twist the dipole by a small amount $d\theta$? (b) Show that the work required to rotate the dipole until it is perpendicular to the field is $W = pE \cos \theta$. (c) Use your result of part (b) to show that if the potential energy of the dipole is chosen zero when the dipole is perpendicular to the field, the potential energy at angle θ is $U(\theta) = -\mathbf{p} \cdot \mathbf{E}$.

9. An electric dipole consists of two charges $+q$ and $-q$ separated by a very small distance $2a$. Its center is on the x axis at $x = x_1$, and it points along the x axis toward positive x. It is in a nonuniform electric field which is also in the x direction given by $\mathbf{E} = Cx\mathbf{i}$, where C is a constant. (a) Find the force on the positive charge and that on the negative charge and show that the net force on the dipole is $Cp\mathbf{i}$. (b) Show that in general, if a dipole of moment **p** lies along the x axis in an electric field in the x direction, the net force on the dipole is given approximately by $(dE_x/dx)p\mathbf{i}$.

10. A positive point charge $+Q$ is at the origin and a dipole of moment **p** is a distance r away and in the radial direction, as in Figure 29-18. (a) Show that the force exerted by the electric field of the point charge on the dipole is attractive with approximate magnitude $2kQp/r^3$ (see Problem 9). (b) Consider now the dipole at the origin and a point charge Q a distance r away along the line of the dipole. From your result of part (a) and Newton's third law, show that the magnitude of the electric field of the dipole along the line of the dipole a distance r away is approximately $2kp/r^3$.

CHAPTER 30   Calculation of the Electric Field

The electric field produced by a given charge distribution can be calculated in a straightforward way from Coulomb's law. The field $\mathbf{E}_i$ at some point P due to a single point charge q_i is

$$\mathbf{E}_i = \frac{kq_i}{r_{i0}^2}\,\hat{\mathbf{r}}_{i0} \qquad\qquad 30\text{-}1$$

where r_{i0} is the distance from the charge to point P (the field point) and $\hat{\mathbf{r}}_{i0}$ is a unit vector pointing from the charge to P. This equation follows directly from Coulomb's law for the force exerted by charge q_i on a test charge q_0 at point P and the definition of electric field as the force on a test charge divided by the charge. We shall refer to Equation 30-1 as Coulomb's law for the electric field due to a single point charge. The field due to several point charges is found by finding the field due to each charge q_i and summing vectorially to obtain the total field (see Equation 29-8). Charge distributions often consist of many charges so close together that the charge can be considered to be continuously distributed over a surface or through a volume, even though, microscopically, electric charge is a discrete quantity. The use of a continuous charge density to describe a distribution of a large number of discrete charges is similar to the use of a continuous mass density to describe air, which in reality consists of a large number of discrete molecules. In either case, it is usually easy to find a volume element $\Delta\mathcal{V}$ large enough to contain many individual charges or molecules (billions) and yet small enough to ensure that replacing $\Delta\mathcal{V}$ by a differential $d\mathcal{V}$ and using calculus introduces negligible error. If charge ΔQ is distributed throughout a volume $\Delta\mathcal{V}$, the *charge density* ρ is defined by

$$\rho = \frac{\Delta Q}{\Delta\mathcal{V}} \qquad\qquad 30\text{-}2 \qquad\qquad \textit{Volume density of charge}$$

Often charge is distributed in a thin layer on the surface of an object. (We shall show later that a static charge on a conductor always resides on the surface of the conductor.) In these cases, it is convenient to define a *surface density* σ. Let t be the thickness of the layer of charge. Then in a volume element of area ΔA, the charge is

$$\Delta Q = \rho t\, \Delta A = \sigma\, \Delta A$$

where σ is the charge per unit area:

$$\sigma = \frac{\Delta Q}{\Delta A} = \rho t$$ 30-3 *Surface density of charge*

Similarly, if the charge is along a line, e.g., on a string of cross-sectional area A, we choose a volume element of length ΔL, $\Delta\mathscr{V} = A\,\Delta L$, and define the charge per unit length λ by

$$\Delta Q = \rho A\,\Delta L = \lambda\,\Delta L$$

or

$$\lambda = \frac{\Delta Q}{\Delta L} = \rho A$$ 30-4 *Linear density of charge*

It is sometimes possible to use Gauss' law to find the electric field **E** from a given charge distribution. This is not as straightforward as using Coulomb's law because in Gauss' law, **E** is in the integrand of a surface integral. When there is a high degree of symmetry in the charge distribution, however, such that the direction of **E** is known, and a surface can be found on which **E** is constant in magnitude, this magnitude can be found quite easily from Gauss' law. Of course in any such case, **E** can also be found directly from Coulomb's law.

30-1 Calculation of **E** from Coulomb's Law

Example 30-1 Calculate the electric field of an electric dipole at a point on the axis of the dipole and far away (Figure 30-1).

Let the dipole consist of a positive charge $+q$ at the origin and a negative charge $-q$ at $x = -a$. The field lines for this charge configuration were shown in Figure 29-8. At great distances ($r \gg a$) the field due to the positive charge is nearly canceled by that due to the negative charge. At points on the x axis, $E_y = E_z = 0$, and E_x is given by

$$E_x = \frac{kq}{x^2} + \frac{k(-q)}{(x+a)^2} = kq\left[\frac{1}{x^2} - \frac{1}{(x+a)^2}\right]$$

Putting the terms in brackets over a common denominator, we obtain

$$\frac{1}{x^2} - \frac{1}{(x+a)^2} = \frac{(x+a)^2 - x^2}{x^2(x+a)^2} = \frac{2ax + a^2}{x^2(x+a)^2} = \frac{2ax(1+a/2x)}{x^4(1+a/x)^2}$$

For $x \gg a$, we can neglect $a/2x$ compared with 1 in the numerator and a/x compared with 1 in the denominator. We then have

$$E_x \approx \frac{2kqa}{x^3}$$ 30-5

The field of a dipole thus decreases as $1/x^3$ at great distances. The

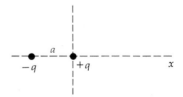

Figure 30-1
Electric dipole on the x axis.

product qa is the dipole moment. Writing $p = qa$ for this quantity, we have

$$E_x \approx \frac{2kp}{x^3}$$ 30-6

We can understand the x dependence of the dipole field in a slightly different way. This field is the sum of the fields due to nearby positive and negative charges. This is equivalent to the difference of the fields due to two nearby positive charges. Consider a positive charge at the origin. The field at a distance x is

$$E_x = \frac{kq}{x^2}$$

The field at a slightly greater distance $x + \Delta x$ is

$$E_x(x + \Delta x) = \frac{kq}{(x + \Delta x)^2}$$

This is equivalent to the field at x due to a second charge displaced a distance Δx to the left of the origin. If Δx is small, we can use the differential approximation:

$$E_x(x + \Delta x) \approx E_x + \frac{dE_x}{dx} \Delta x$$

Differentiating E_x, we obtain

$$\frac{dE_x}{dx} = -\frac{2kq}{x^3}$$

Thus

$$E_x(x + \Delta x) \approx E_x(x) - \frac{2kq\ \Delta x}{x^3}$$

If we now subtract this field from $E_x(x)$, we obtain

$$E_x(x) - E_x(x + \Delta x) \approx \frac{2kq\ \Delta x}{x^3}$$

which is the same as a dipole field of moment $p = q\ \Delta x$.

Example 30-2 Calculate the electric field on the axis of a uniform ring of charge.

Let Q be the total positive charge, which is distributed uniformly on a ring of radius a. We wish to find the field at a point on the axis of the ring a distance x from the center of the ring.

Figure 30-2 shows the part of the field $\Delta \mathbf{E}$ due to a portion of the charge ΔQ. This field has a component ΔE_x along the axis of the ring and a component $\Delta E_\perp$ perpendicular to the axis. From the symmetry of

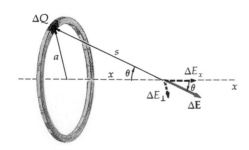

Figure 30-2
Charged ring of radius a. The electric field on the x axis is in the x direction.

the figure, we see that resultant field due to the entire ring must lie along the axis of the ring; i.e., the perpendicular components will sum to zero. In particular, the perpendicular component shown will be canceled by that due to another portion of the charge on the ring directly opposite the one shown.

The axial component due to the part of the charge shown is

$$\Delta E_x = \frac{k\,\Delta Q}{s^2}\cos\theta = \frac{k\,\Delta Q}{s^2}\frac{x}{s} = \frac{k\,\Delta Q x}{(x^2 + a^2)^{3/2}}$$

where

$$s^2 = x^2 + a^2 \quad \text{and} \quad \cos\theta = \frac{x}{s} = \frac{x}{\sqrt{x^2 + a^2}}$$

Since the distance from the field point, at which we are calculating the field, to all parts of the charge is the same and the angle θ is the same for all parts of the charge, the field due to the entire ring of charge is

$$E_x = \sum \frac{k\,\Delta Q x}{(x^2 + a^2)^{3/2}} = \frac{kx}{(x^2 + a^2)^{3/2}}\sum \Delta Q = \frac{kQx}{(x^2 + a^2)^{3/2}} \qquad 30\text{-}7$$

Example 30-3 Calculate the electric field on the perpendicular bisector of a uniform line charge.

Let λ be the linear density. At a point on a perpendicular bisector the field **E** will not have a component parallel to the line of charge.

Figure 30-3 shows the geometry of the problem. We have chosen a coordinate system such that the origin is at the center of the line charge, the charge is on the x axis, and our field point is on the y axis at P. The magnitude of the field produced by an element of charge $\Delta Q = \lambda\,\Delta x$ is

$$|\Delta\mathbf{E}| = \frac{k\lambda\,\Delta x}{s^2}$$

The perpendicular component (in this case, the y component) is

$$\Delta E_y = \frac{k\lambda\,\Delta x}{s^2}\cos\theta = \frac{k\lambda\,\Delta x}{s^2}\frac{y}{s} \qquad 30\text{-}8$$

The total field E_y is computed by summing over all Δx of the line charge. For our case, we can sum (integrate) from $x = 0$ to $x = L/2$ and multiply by 2 because, by symmetry, each half of the line charge gives an equal contribution. The integration is somewhat simplified if we change from the variable x to θ and integrate from $\theta = 0$ to θ_0, defined by

$$\tan\theta_0 = \frac{\frac{1}{2}L}{y}$$

Equation 30·8 in terms of the differentials dE_y and dx is

$$dE_y = \frac{ky\lambda}{s^3}\,dx$$

The distance x is related to the angle θ by

$$x = y\tan\theta$$

Then

$$dx = y\sec^2\theta\,d\theta = y\left(\frac{s}{y}\right)^2 d\theta$$

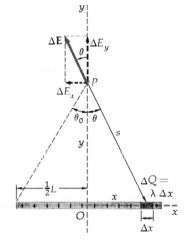

Figure 30-3
Line charge. The electric field on the y axis which bisects the charge is in the y direction.

and

$$dE_y = \frac{ky\lambda}{s^3}\frac{s^2}{y}\,d\theta = \frac{k\lambda}{y}\frac{y}{s}\,d\theta = \frac{k\lambda}{y}\cos\theta\,d\theta$$

The total y component of the field is twice the integral of this from $\theta = 0$ to $\theta = \theta_0$:

$$E_y = 2\int_{\theta=0}^{\theta=\theta_0} dE_y = \frac{2k\lambda}{y}\int_0^{\theta_0}\cos\theta\,d\theta = \frac{2k\lambda}{y}\sin\theta_0 \qquad\text{30-9}$$

where

$$\sin\theta_0 = \frac{\tfrac{1}{2}L}{\sqrt{(\tfrac{1}{2}L)^2 + y^2}}$$

The field produced by a line charge of infinite length is obtained from this result by setting $\theta_0 = 90°$. Then

$$E_y = \frac{2k\lambda}{y} = \frac{1}{2\pi\epsilon_0}\frac{\lambda}{y} \qquad\text{30-10}$$

where we have written $k = 1/4\pi\epsilon_0$.

Example 30-4 Calculate the electric field on the axis of a uniformly charged disk.

Let the disk have a radius R and contain a uniform charge per unit area σ. The electric field on the axis of the disk will be parallel to the axis. We can calculate this field by treating the disk as a set of concentric ring charges and using our result from Example 30-2. Consider a ring of radius r and width dr, as in Figure 30-4. The area of this ring is $2\pi r\,dr$, and its charge is $dq = 2\pi\sigma r\,dr$. The field produced by this ring is given by Equation 30-7, replacing Q by $2\pi\sigma r\,dr$ and a by r. Thus

$$dE_x = \frac{kx2\pi\sigma r\,dr}{(x^2 + r^2)^{3/2}}$$

The total field produced by the disk is found by integrating from $r = 0$ to $r = R$:

$$E_x = kx\pi\sigma\int_0^R (x^2 + r^2)^{-3/2}\,2r\,dr = kx\pi\sigma\left[\frac{(x^2 + r^2)^{-1/2}}{-\tfrac{1}{2}}\right]_0^R$$

$$= -2kx\pi\sigma\left(\frac{1}{\sqrt{x^2 + R^2}} - \frac{1}{x}\right) = 2\pi k\sigma\left(1 - \frac{x}{\sqrt{x^2 + R^2}}\right) \qquad\text{30-11}$$

The interesting and important result for the case of the field near an infinite plane of charge can be obtained from this result by letting either R go to infinity or x go to zero. Then

$$E_x = 2\pi k\sigma = \frac{\sigma}{2\epsilon_0} \qquad\text{30-12}$$

where again $k = 1/4\pi\epsilon_0$. Thus the field due to an infinite-plane charge distribution is uniform; i.e., the field does not depend on x. The lines of force for a disk are shown in Figure 30-5. Near the disk the field lines are equally spaced parallel lines, indicating that the field is uniform. At distances comparable with the radius of the disk the lines begin to spread. The density of lines near the axis is thus smaller, indicating that the field is weaker. At great distances from the disk, the field lines are approximately the same as those due to a point charge of magnitude $q = \sigma\pi R^2$.

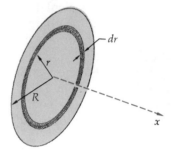

Figure 30-4
Disk carrying uniform surface charge density. The electric field at a point on the axis of the disk is along the axis. It can be calculated by finding the field due to a ring of radius r and area $2\pi r\,dr$ and summing over the rings from $r = 0$ to $r = R$, the radius of the disk.

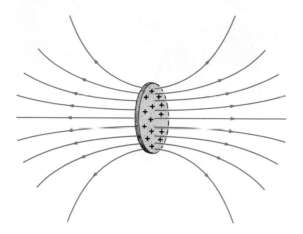

Figure 30-5
Lines of force for a uniformly charged disk. Only lines in a plane perpendicular to the disk containing a diameter are shown. Near the disk the field is uniform. Far from the disk the lines are the same as those due to a point charge.

Example 30-5 Find the electric field due to a spherical shell of charge.

The calculation of the electric field due to a spherical shell of charge is very difficult using Coulomb's law because of the difficult spherical geometry needed in setting up the problem. The result is analogous to that for the gravitational field due to a spherical shell of mass. Inside the shell of charge the electric field is zero. Outside the shell the field is the same as that due to a point charge at the center of the shell. In terms of the radius of the sphere R and the charge per unit area this result is

$$E_r = 0 \quad \text{for } r < R \qquad\qquad 30\text{-}13$$

and

$$E_r = \frac{kQ}{r^2} \quad \text{for} \quad r > R \qquad\qquad 30\text{-}14$$

where $Q = 4\pi R^2 \sigma$ is the total charge. This result is easily obtained using Gauss' law, as we shall show in the next section. The calculation from Coulomb's law is given below.

Optional

We first consider a field point P outside the shell. The geometry is shown in Figure 30-6. By symmetry, the field must be radial. We can consider the spherical shell to be a set of ring elements and use our result found in Example 30-2. We thus choose for our charge element the strip shown, which has circumference $2\pi R \sin \theta$ and width $R\, d\theta$. The area of this strip is $dA = 2\pi R^2 \sin \theta\, d\theta$, and the charge is

$$dQ = \sigma\, dA = \sigma 2\pi R^2 \sin \theta\, d\theta$$

The radial component of the field due to this ring is

$$dE_r = \frac{k\, dQ}{s^2} \cos \alpha = \frac{k\sigma 2\pi R^2 \sin \theta\, d\theta}{s^2} \cos \alpha \qquad\qquad 30\text{-}15$$

Before we integrate over the charge distribution, we must eliminate two of the three related variables s, θ, and α. It is convenient to write everything in terms of s, which varies from $s = r - R$ at $\theta = 0$ to $s = r + R$ at $\theta = 180°$. By the law of cosines, we have

$$s^2 = r^2 + R^2 - 2rR \cos \theta \qquad\qquad 30\text{-}16$$

Differentiating gives

$$2s\, ds = +2rR \sin \theta\, d\theta$$

Figure 30-6
Spherical shell of radius R carrying a uniform surface charge. The field at point P a distance r from the center of the sphere is found by first finding the field of a ring of width $R\,d\theta$ and radius $R\sin\theta$ and summing over the rings from $\theta = 0$ to $\theta = \pi$. The resulting field for $r > R$ is the same as if all the charge were at the origin. When point P is inside the shell, $r < R$, the field due to the shell is zero.

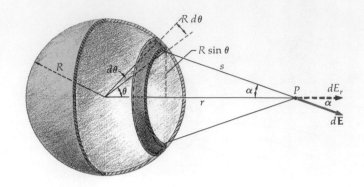

or

$$\sin\theta\,d\theta = \frac{s\,ds}{rR} \qquad\qquad 30\text{-}17$$

An expression for $\cos\alpha$ can also be obtained from the law of cosines applied to the same triangle. We have

$$R^2 = s^2 + r^2 - 2sr\cos\alpha$$

or

$$\cos\alpha = \frac{s^2 + r^2 - R^2}{2sr} \qquad\qquad 30\text{-}18$$

Substituting these results into Equation 30-15 then gives

$$dE_r = \frac{k\sigma 2\pi R^2}{s^2}\,\frac{s\,ds}{rR}\,\frac{s^2 + r^2 - R^2}{2sr}$$

$$= \frac{k\sigma\pi R}{r^2}\left(1 + \frac{r^2 - R^2}{s^2}\right)ds \qquad\qquad 30\text{-}19$$

The field due to the entire shell of charge is found by integrating from $s = r - R\ (\theta = 0)$ to $s = r + R\ (\theta = 180°)$:

$$E_r = \frac{k\sigma\pi R}{r^2}\int_{r-R}^{r+R}\left(1 + \frac{r^2 - R^2}{s^2}\right)ds$$

$$= \frac{k\sigma\pi R}{r^2}\left[s - \frac{r^2 - R^2}{s}\right]_{r-R}^{r+R} \qquad\qquad 30\text{-}20$$

Substitution of the upper and lower limits yields $4R$ for the quantity in brackets. Thus

$$E_r = \frac{k\sigma 4\pi R^2}{r^2} = \frac{kQ}{r^2} \qquad\qquad 30\text{-}21$$

where $Q = 4\pi R^2\sigma$ is the total charge. The field is therefore the same as that due to a point charge Q at the origin.

We now consider a field point inside the shell. The calculation for this case is identical except that s now varies from $R - r$ to $r + R$. Thus

$$E_r = \frac{k\sigma\pi R}{r^2}\left[s - \frac{r^2 - R^2}{s}\right]_{R-r}^{r+R} \qquad\qquad 30\text{-}22$$

Substitution of these upper and lower limits yields 0. Therefore

$$E_r = 0 \quad\text{ point inside shell} \qquad\qquad 30\text{-}23$$

Example 30-6 Calculate the electric field due to a solid sphere of constant charge density.

We treat this problem just as we did the corresponding gravitational problem. Let the charge density be $\rho = Q/\mathscr{V}$, where $\mathscr{V} = \frac{4}{3}\pi R^3$ is the volume of the sphere of charge. Considering the sphere to be a set of spherical shells, we can use the result of the previous example. For field points outside the sphere, each shell produces a field identical to that of a point charge at the center of the sphere. Then for any point outside the sphere of charge, $r > R$,

$$E_r = \frac{kQ}{r^2}$$

Consider now a point P inside the charge, $r < R$. In Figure 30-7 the dotted line encloses a sphere of radius r. Let q' be the portion of the charge inside this sphere. Since the charge density is uniform, q' is given by

$$q' = \rho\mathscr{V}' = \rho\frac{4}{3}\pi r^3 = \frac{Q}{\frac{4}{3}\pi R^3}\,\frac{4}{3}\pi r^3 = \frac{Qr^3}{R^3}$$

The field at P due to q' is the same as if q' were at the origin. The rest of the charge produces no field at point P. This part of a charge can be considered to be a set of spherical shells, and for each shell, point P is inside. The field at a distance $r < R$ is thus

$$E_r = \frac{kq'}{r^2} = \frac{kQr^3/R^3}{r^2} = \frac{kQ}{R^3}\,r \qquad\qquad 30\text{-}24$$

Figure 30-8 shows a sketch of E_r versus r. This function is sometimes used to describe the electric field of an atomic nucleus, which can be considered to be approximately a uniform sphere of charge.

Questions

1. Explain why the electric field increases with r rather than decreasing as $1/r^2$ as one moves out from the center of a spherical charge distribution of constant density.

2. Can a charge distribution which is infinite in extent produce an electric field that is finite everywhere?

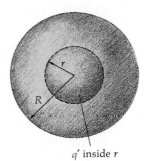

q' inside r

Figure 30-7
Solid ball of charge. The electric field at a point $r < R$ inside the ball is kq'/r^2, where $q' = (r^3/R^3)Q$ is that part of the total charge Q which is inside the sphere of radius r.

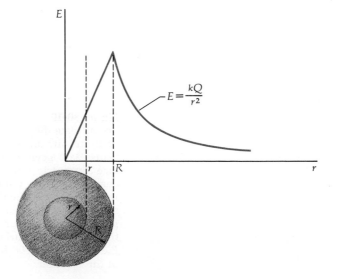

Figure 30-8
Plot of E_r versus r for a solid ball of charge of radius R. For r less than R the field increases linearly with r. Outside the charge, the electric field is kQ/r^2, the same as that due to a point charge Q at the origin.

30-2   Calculation of **E** from Gauss' Law

Gauss' law can be used to calculate the magnitude of **E** when there is a high degree of symmetry so that the direction of **E** is known everywhere and the magnitude of **E** is constant over some simple surface.

Example 30-7 Find the field of a point charge q.

We consider this simple example to illustrate the method and to show that Gauss' law is indeed equivalent to Coulomb's law. We choose our origin at the point charge and consider a spherical surface of radius r (Figure 30-9). This mathematical surface, used for calculation only, is called a *gaussian surface*. It is clear from symmetry that **E** must be radial and that its magnitude can depend only on the distance from the charge. The normal component $\mathbf{E} \cdot \hat{\mathbf{n}} = E_r$ has the same value everywhere on our spherical surface. The flux of **E** through this surface is thus

$$\phi_{net} = \oint \mathbf{E} \cdot \hat{\mathbf{n}}\, dA = \oint E_r\, dA = E_r \oint dA$$

But $\oint dA$ is just the total area of the spherical surface, $4\pi r^2$. Since the total charge inside the surface is just the point charge q, Gauss' law gives

$$E_x 4\pi r^2 = 4\pi k q \qquad \text{or} \qquad E_r = \frac{kq}{r^2} \qquad\qquad 30\text{-}25$$

We have thus derived Coulomb's law from Gauss' law. Since we originally derived Gauss' law from Coulomb's law, we have now shown the two laws to be equivalent.[1]

Example 30-8 Find the electric field due to a uniform line charge of infinite length along the x axis.

Because of the symmetry of this situation, **E** must be perpendicular to the line and directed away from the line charge. Also, the magnitude of **E** can depend only on the perpendicular distance of the field point from the line.

Figure 30-10 shows a cylindrical can coaxial with the line charge. Let r be the radius of the cylinder, which we use for our gaussian surface. Everywhere on the cylindrical part of the surface, **E** is perpendicular to the surface and constant in magnitude. On the ends of the can, **E** is parallel to the surface; thus the ends make no contribution to the flux.

---

[1] Gauss' law and Coulomb's law are strictly equivalent only for electrostatic fields. Nonelectrostatic electric fields, produced by changing magnetic fields, are not described by Coulomb's law. Because the lines of force for these fields are closed lines, which do not diverge from, or converge to, any point in space, the net flux of nonelectrostatic fields through any closed surface is zero. Hence, Gauss' law holds for all electric fields.

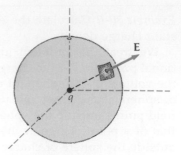

Figure 30-9
Calculation of the electric field of a point charge using Gauss' law. On a spherical surface surrounding the charge, the electric field **E** is perpendicular to the surface and constant in magnitude. The flux of the field through this surface is then $E_r 4\pi r^2$.

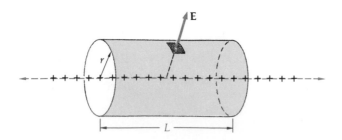

Figure 30-10
To calculate the electric field of an infinite line charge from Gauss' law a cylindrical surface concentric with the line charge is chosen. On the cylinder the electric field is constant in magnitude and perpendicular to the surface. The flux through the surface is $E_n 2\pi r L$, where L is the length of the cylinder.

The total charge inside the gaussian surface is λL, where λ is the charge per unit length of the line charge and L is the length of the can. Setting the flux through this surface equal to $4\pi k$ times the net charge inside, we obtain

$$\phi_{\text{net}} = \oint E_n \, dA = E_n \oint dA = 4\pi k\lambda L$$

Since the area of the cylindrical surface is $2\pi rL$,

$$E_n 2\pi rL = 4\pi k\lambda L \qquad \text{or} \qquad E_n = 2k\frac{\lambda}{r} \qquad\qquad 30\text{-}26$$

in agreement with our result for Example 30-3.

Consider now a finite line charge which extends beyond the cylindrical gaussian surface shown (Figure 30-11). Since the net charge inside the surface is still λL, it might seem that we would get the same result for **E** no matter how long the line charge is. However, if the line charge is not infinite, our symmetry arguments break down. One problem is that **E** is not perpendicular to the cylindrical surface except at points equidistant from the ends of the charge. Also, E_n is not constant everywhere on the surface; that is, E_n depends not only on r but also on the distance from the center of the line charge. For this case of a finite line charge and the gaussian surface shown, Gauss' law still holds. The integral of E_n over the gaussian surface (including the faces) equals $4\pi k$ times the net charge inside the surface λL. However, this law is *not useful* for calculating the field **E** because E_n is not constant on the gaussian surface.

This example is typical of many applications of Gauss' law. Whenever there is a high degree of symmetry (infinite line charge in this example), the field **E** is obtained from Gauss' law far more easily than by direct integration using Coulomb's law. In the absence of a high degree of symmetry, Gauss' law is of no use in obtaining **E**.

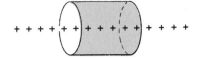

Figure 30-11
Cylindrical surface concentric with a line charge of finite length. Gauss' law is not useful for finding the electric field of a finite line charge because the electric field is not constant in magnitude on the surface and is not perpendicular to the surface.

Example 30-9 Find the electric field due to an infinite plane of uniform charge density σ.

Again, we can use symmetry to argue that the field **E** must be perpendicular to the plane, depend only on the distance from the plane to the field point, and have the same magnitude but opposite direction at points the same distance below as above the plane. We choose for our gaussian surface a pillbox-shaped cylinder with its axis perpendicular to the plane and with its center on the plane (Figure 30-12). Let each end of the cylinder be parallel to the plane and have area A. In this case, **E** is parallel to the cylindrical surface, and there is no flux through this curved surface. Since the flux out of each face is $\mathbf{E} \cdot \hat{n}A = E_nA$, the total flux is $2E_nA$. The net charge inside the surface is σA. From Gauss' law,

$$\phi_{\text{net}} = 2E_nA = 4\pi k\sigma A$$

$$E_n = 2\pi k\sigma = \frac{\sigma}{2\epsilon_0} \qquad\qquad 30\text{-}27$$

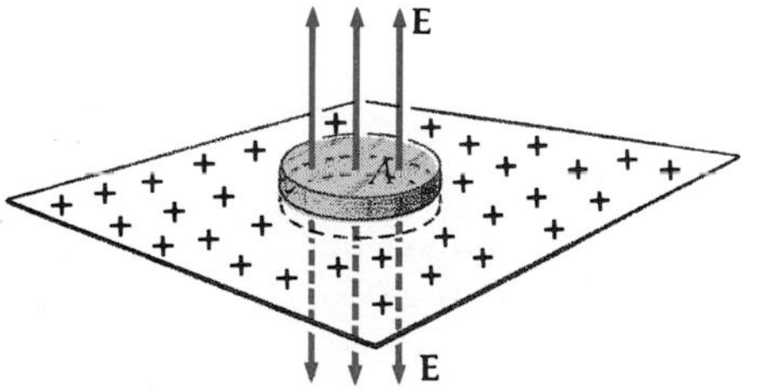

Figure 30-12
Gaussian surface for the calculation of the electric field due to an infinite plane of charge. On the upper and lower faces of this pillbox surface **E** is perpendicular to the surface and constant in magnitude. The flux through this surface is $2EA$, where A is the area of each face.

This result agrees with that obtained from Coulomb's law in Example 30-4.

As in Example 30-8, the symmetry arguments so necessary for the *use* of Gauss' law depend on the fact that the plane charge distribution is infinite in extent.

Example 30-10 Find the field due to a spherical shell of charge of radius R.

We choose a spherical gaussian surface of radius r, concentric with the charged shell. Whether r is greater than R or less than R, the symmetry of the problem implies that $\mathbf{E}$ is perpendicular to the gaussian surface and constant in magnitude on the surface. The flux through the gaussian surface is thus $4\pi r^2 E_r$. For a field point outside the shell, $r > R$,

$$\phi_{\text{net}} = 4\pi r^2 E_r = 4\pi kQ$$

where Q is the total charge on the shell. Then

$$E_r = \frac{kQ}{r^2} \quad \text{outside}$$

For a field point inside the shell, $r < R$, the total charge inside the gaussian surface is zero, so that the flux must be zero:

$$\phi_{\text{net}} = 4\pi r^2 E_r = 0$$

Therefore

$$E_r = 0 \quad \text{inside}$$

Since Newton's law and Coulomb's law have the same inverse-square-distance dependence, the results of all of these calculations of the electric field using Gauss' law or Coulomb's law can be applied directly to the analogous cases of calculation of the gravitational field. We need only replace the appropriate charge or charge density by the corresponding mass or mass density and change the sign, since the gravitational field of a point mass (which is always positive) points toward the mass, whereas the electric field of a (positive) point charge points away from the charge. Gauss' law for the flux of the gravitational field $\mathbf{g}$ is

$$\phi_{\text{net}} = \oint \mathbf{g} \cdot \hat{\mathbf{n}}\, dA = \oint g_n\, dA = -4\pi G m_{\text{inside}} \qquad \text{30-28}$$

where m_{inside} is the total mass enclosed by surface S.

Questions

3. Is the electric field $\mathbf{E}$ in Gauss' law the part of the electric field due to the charge inside the surface, or is it the net electric field due to all charges whether they are inside or outside the surface?

4. What information is needed in addition to the total charge inside a surface to use Gauss' law to find the electric field?

Review

A. Without looking back at the examples, draw a careful diagram and set up the integration to find (*a*) the electric field on the bisector of a line charge (Example 30-3) and (*b*) the electric field on the axis of a uniformly charged disk

(Example 30-4). Then check your diagram and integral against those in the examples.

B. True or false:

1. Gauss' law holds only for symmetric charge distributions.

2. The electric field on one side of a uniformly charged infinite sheet is uniform.

3. The electric field inside a uniformly charged spherical shell is zero.

4. At any point where ρ is not zero the electric field is infinite.

Exercises

Section 30-1, Calculation of E from Coulomb's Law

1. A dipole consists of a charge $+q$ on the x axis at $x = a$ and a charge $-q$ on the x axis at $x = -a$. (a) What direction is the electric field at a point on the y axis? (b) Find an expression for the electric field at a point on the y axis. (c) Show that for y much greater than a, the electric field on the y axis is given by $E_x \approx -kp/y^3$, where $p = q(2a)$ is the dipole moment.

2. Show that the electric field E_x on the axis of a ring charge of radius a approaches that of a point charge, as expected when the distance x from the plane of the ring is much greater than a (see Equation 30-7).

3. A 0.5-μC charge is uniformly distributed on a ring of radius 3 cm. Find the electric field on the axis of the ring at (a) 1 cm, (b) 2 cm, (c) 3 cm, and (d) 600 cm from the center of the ring.

4. Show that the electric field on the perpendicular bisector of a line charge approaches that of a point charge, as expected when the distance y from the charge is much greater than the length L. Hint: Use $\sin \theta_0 \approx \tan \theta_0$ in Equation 30-9.

5. A uniform line charge of density λ is on the x axis from $x = -a$ to $x = +b$. (a) Find an expression for the y component of the electric field at a point on the y axis. (b) Show that your expression is the same as Equation 30-9 when $a = b = \frac{1}{2}L$.

6. A line charge has density $\lambda = 2$ μC/m and a total length of 6 m. Use approximate expressions to find the electric field on the bisector of the line at distances 2 cm and 60 m.

7. (a) Using the fact that at very great distances a disk charge should look like a point charge, write down the expected form of the electric field on the axis of a uniformly charged disk of radius a and charge density σ. (b) Derive this result from Equation 30-11 by showing that $x/\sqrt{x^2 + a^2} = (1 + a^2/x^2)^{-1/2}$ and approximating this expression for a/x much less than 1 using the binomial expansion $(1 + \epsilon)^n \approx n\epsilon$ for small ϵ.

8. A disk of radius 5 cm carries a uniform charge density 3×10^{-5} C/m². Using reasonable approximations, find the electric field on the axis of the disk at distances (a) 0.01 cm, (b) 0.02 cm, (c) 0.03 cm, and (d) 500 cm.

9. Show that the electric field just outside a spherical shell of charge is σ/ϵ_0, where σ is the charge per unit area.

10. A spherical shell of radius 10 cm carries a charge of 2 μC uniformly distributed on its surface. Find the electric field at the following distances from the center of the shell: (a) 5 cm, (b) 9.99 cm, (c) 10.01 cm, (d) 20 cm, and (e) 40 cm.

11. A sphere of radius 10 cm has a charge 2 μC uniformly distributed throughout its volume. Find the electric field at the following distances from

the center of the sphere: (*a*) 5 cm, (*b*) 9.99 cm, (*c*) 10.01 cm, (*d*) 20 cm, and (*e*) 40 cm. Compare these results with those of Exercise 10.

Section 30-2, Calculation of E from Gauss' Law

12. Can you use Gauss' law to calculate the electric field on the axis of an electric dipole? If so, do it. If not, explain why not.

13. A sphere of radius R has a constant volume charge density ρ. Use Gauss' law to derive expressions for the electric field a distance r from the center of the sphere for (*a*) $r < R$ and (*b*) $r > R$.

14. A thin-walled cylindrical shell of radius R and infinite length carries a uniform surface charge density σ. (*a*) Show that the electric field is zero for $r < R$. (*b*) Show that for $r > R$ the electric field has the magnitude

$$E = \frac{\sigma R}{\epsilon_0} \frac{1}{r}$$

(*c*) Show that your result for part (*b*) is the same as for an infinite line charge of the same charge per unit length.

15. An infinitely long cylinder of radius R has a uniform volume charge density ρ. Show that the electric field has the magnitude

$$E = \begin{cases} \dfrac{R^2 \rho}{2\epsilon_0} \dfrac{1}{r} & r \geq R \\[2mm] \dfrac{\rho}{2\epsilon_0} r & r \leq R \end{cases}$$

Sketch E versus r.

Problems

1. Show that E_x on the axis of a ring charge (Equation 30-7) has its maximum and minimum values at $x = +a/\sqrt{2}$ and $x = -a/\sqrt{2}$. Sketch E_x versus x.

2. (*a*) Find the slope dE_x/dx at the origin for E_x due to a ring charge (Equation 30-7). (*b*) A bead of mass m and carrying a negative charge $-q$ slides along a stretched thread which is along the axis of a ring of radius a and total charge Q. Show that if the bead is displaced slightly from the origin and released, the motion is simple harmonic. Find the frequency of the motion. *Hint:* Show that the acceleration is approximately proportional to the displacement (in the opposite direction), and write the proportionality constant as $-\omega^2$.

3. A line charge of density λ has the shape of a square of side L which lies in the yz plane with its center at the origin. Find the electric field on the x axis at arbitrary distance x and compare your result to the field on the axis of a charged ring of approximately the same size carrying the same total charge.

4. A disk of radius 30 cm carries a uniform charge density σ. Compare the approximation $E = \sigma/2\epsilon_0$ with the exact expression for the electric field (Equation 30-11) by computing the neglected term as a percentage of the field for distances $x = 0.1$ cm, 0.2 cm, and 3 cm. At what distance is the neglected term 1 percent of $\sigma/2\epsilon_0$?

5. Two infinite planes are parallel to each other and separated by a distance d. Find the electric field to the left of the planes, to the right of the planes, and between the planes when (*a*) each plane contains a uniform charge density $\sigma = +\sigma_0$ and (*b*) the left plane has a uniform charge density $\sigma = +\sigma_0$ and the right plane $\sigma = -\sigma_0$. Draw the lines of force for each case.

6. A semi-infinite line charge of uniform density λ lies along the x axis from $x = 0$ to $x = \infty$. Find both E_x and E_y at a point on the y axis.

7. A sphere of radius R carries a volume charge density proportional to the distance from the center; $\rho = Ar$ for $r \leqslant R$, $\rho = 0$ for $r > R$, where A is a constant. (*a*) Find the total charge by summing the charges in shells of thickness dr and volume $4\pi r^2 \, dr$. (*b*) Find the electric field E_r both inside and outside the charge distribution.

8. Consider two infinitely long concentric cylindrical shells. The inner shell has radius R_1 and carries a uniform surface charge density σ_1 while the outer shell has radius R_2 and carries a uniform surface charge density σ_2. (*a*) Use Gauss' law to find the electric field in the regions $r < R_1$, $R_1 < r < R_2$, and $R_2 < r$. (*b*) What should the ratio σ_2/σ_1 and the relative sign be for the electric field to be zero at $r > R_2$? What then is the electric field between the shells? (*c*) Sketch lines of force for the situation in part (*b*).

9. A spherical shell of radius R_1 carries a total charge q_1 uniformly distributed on its surface. A second larger spherical shell of radius R_2 concentric with the first carries a charge q_2 uniformly distributed on its surface. (*a*) Use Gauss' law to find the electric field in the regions $r < R_1$, $R_1 < r < R_2$, and $R_2 < r$. (*b*) What should the ratio of the charges q_1/q_2 and their relative sign be for the electric field to be zero for $r > R_2$? (*c*) Sketch the lines of force for the situation in part (*b*).

10. A thin wire carries a uniform linear charge density λ and is bent into a circular arc which subtends an angle $2\theta_0$, as shown in Figure 30-13. Show that the electric field at the center of curvature of the arc has the magnitude $E = (2k\lambda \sin \theta_0)/R$. *Hint:* Consider an element of length $dl = R \, d\theta$ at some angle θ, find the component of the field along the line OC, and integrate from $\theta = -\theta_0$ to $\theta = +\theta_0$.

Figure 30-13
Problem 10.

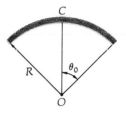

11. (*a*) Show that for both the infinite plane charge and the spherical shell, the electric field is discontinuous at the surface charge by the amount σ/ϵ_0. (*b*) Prove that in general, when there is a surface charge σ, the electric field component perpendicular to the surface is discontinuous by the amount σ/ϵ_0. Do this by constructing a gaussian pillbox with faces on each side of the surface. Use Gauss' law to find $E_2 - E_1$, where E_2 is the normal component of E on one side and E_1 is that on the other side of the surface.

CHAPTER 31 Conductors in Electrostatic Equilibrium

The great difference in the electrical behavior of conductors and insulators was noted even before the discovery of electric conduction. Gilbert had classified materials according to their ability to be electrified. Objects which could be electrified he called *electrics;* those which could not (metals and some other materials) he called *nonelectrics.* After Gray discovered conduction, Du Fay showed that all materials can be electrified but that care must be taken in insulating Gilbert's nonelectrics from the ground (or the experimenter) lest the charge be quickly conducted away. Using only physiological sensation for detection, Cavendish compared the conducting abilities of many substances.[1]

> It appears from some experiments, of which I propose shortly to lay an account before this Society, that iron wire conducts about 400 million times better than rain or distilled water — that is, the electricity meets with no more resistance in passing through a piece of iron wire 400,000,000 inches long than through a column of water only one inch long. Sea-water, or a solution of one part of salt in 30 of water, conducts 100 times, or a saturated solution of sea-salt about 720 times better than rain-water.

Because of the enormous variation in the ability to conduct electricity, it is possible and convenient to classify most materials as conductors or insulators (nonconductors). The ability of a material to conduct electricity is measured by its conductivity. The conductivity of a typical conductor is of the order of 10^{15} times that of a typical insulator, whereas within the group of conductors the conductivity varies only over several orders of magnitude. We shall give a precise definition of electric conductivity (and its reciprocal, resistivity) in Chapter 34, when we study electric currents, i.e., electric charges in motion. There we also discuss *semiconductors,* which are neither conductors nor insulators.

[1] Henry Cavendish, *Philosophical Transactions of the Royal Society of London,* vol. 66, p. 196 (1776).

In this chapter we shall be concerned only with the behavior of conductors in electrostatic equilibrium, i.e., when all electric charges are at rest. We shall find that (1) the electric field is zero inside a conductor in electrostatic equilibrium; (2) any excess charge on a conductor must reside entirely on the surface of the conductor; and (3) the electric field just outside the surface of the conductor is perpendicular to the surface and has the magnitude σ/ϵ_0, where σ is the surface charge density (which may vary from point to point on the surface).

31-1 Free Charge in Conductors

The property of a conductor important in studying electrostatic fields is the availability inside the conductor of charge that is free to move about. The source of this free charge is electrons in a conductor which are not bound to any atom. For example, in a single atom of copper, 29 electrons are bound to the nucleus by electrostatic attraction of the positively charged nucleus. The outermost electrons are more weakly bound than the innermost electrons because of the greater distance from the positive nucleus and because of the repulsion of the inner electrons. (This is called *screening*.) When a large number of copper atoms combine to form metallic copper, the electron binding of a single atom is changed by interaction with neighboring atoms. One or more of the outer electrons in an atom are no longer bound but are free to move throughout whole metal, much as a gas molecule is free to move about in a box. The number of free electrons depends on the particular metal but is of the order of one per atom.

In the presence of an external electric field, the free charge in a conductor moves about the conductor until it is so distributed that it creates an electric field which cancels the external field inside the conductor. Consider a charge q inside a conductor. If there is a field $\mathbf{E}$ inside the conductor, there will be a force $q\mathbf{E}$ on this charge; and if it is free to move, i.e., if it is not bound to an atom or molecule by a stronger force, it will accelerate. Thus electrostatic equilibrium is impossible in a conductor unless the electric field is zero everywhere inside the conductor.

Figure 31-1 shows a conducting slab placed in an external electric field $\mathbf{E}_0$. The free electrons are originally distributed uniformly throughout the slab. Since the slab is made up of neutral atoms, it is electrically neutral (assuming that no extra charge has been placed on it). If the external electric field is to the right, there will be a force on each electron $-e\mathbf{E}_0$ to the left because the electron has a negative charge, and the free electrons accordingly accelerate to the left. At the surface of the conductor, the conductor exerts forces on these electrons

Lines of force for an oppositely charged cylinder and plate shown by bits of fine thread suspended in oil. Note that the field lines are perpendicular to the conductors and that there are no lines inside the cylinder. (*Courtesy of Harold M. Waage, Princeton University.*)

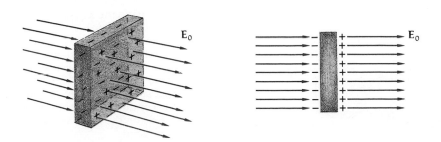

Figure 31-1
Conducting slab in an external electric field $\mathbf{E}_0$. A positive charge is induced on the right face and a negative charge on the left face, so that the resultant electric field inside the conductor is zero. The field lines then end on the left face and begin again on the right face.

which balance that due to the external field, so that the electrons are bound to the conductor. (If the external field is very strong, the electrons can be stripped off from the surface. In electronics this is called *field emission.* We assume here that the external field is not strong enough to overcome the forces binding the electrons to the surface.) The result is a negative surface charge density on the left side of the slab and a positive surface charge density on the right side because of the removal of some of the free electrons from that side. Both these charge densities produce an electric field inside the slab which is opposite the external field. These two fields cancel everywhere inside the conductor, so that there is no unbalanced force on the free electrons and electrostatic equilibrium results. The behavior of the free charge in a conductor placed in an external electric field is similar no matter what the conductor's shape. When an external field is applied, the free charge quickly distributes itself until an equilibrium distribution is reached such that the net electric field is zero everywhere inside the conductor. The time to reach equilibrium depends on the conductivity. We shall calculate this time in Section 34-6 and show that for copper and other good conductors it is less than about 10^{-16} sec. For all practical purposes, electrostatic equilibrium is reached instantaneously.

No electrostatic field in the interior of a conductor

Question

1. Distinguish between free charge in a conductor and net charge in a conductor.

31-2   Charge and Field at Conductor Surfaces

In this section we shall use Gauss' law to show that in electrostatic equilibrium (1) any net electric charge on a conductor resides on the surface of the conductor and (2) the electric field just outside the surface of a conductor is perpendicular to the surface and has the magnitude σ/ϵ_0, where σ is the local surface charge density at that point on the conductor.

To obtain the first result we consider a gaussian surface just inside the actual surface of a conductor as shown in Figure 31-2. In electrostatic equilibrium, since the electric field is zero everywhere inside the conductor it is zero everywhere on the gaussian surface, which is chosen to be completely within the conductor. Since $E_n = 0$ at all points on the gaussian surface, the net flux $\oint E_n\, dA$ through the surface must be zero. By Gauss' law this flux equals $4\pi k$ times the net charge inside the surface. Thus there can be no net charge inside any surface lying completely within the conductor. If there is any net charge on

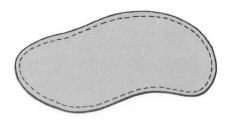

Figure 31-2
The dashed line indicates a gaussian surface chosen to be just inside the actual surface of the conductor. Since the electric field is zero everywhere on the gaussian surface, Gauss' law implies that there is zero net charge inside that surface. Any charge on the conductor must therefore reside on the surface of the conductor.

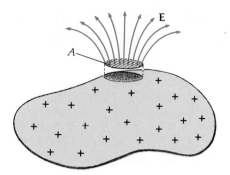

Figure 31-3
Pillbox gaussian surface with
one face just outside and the
other face just inside the surface of a charged conductor.
The flux through the gaussian
surface is E_nA, where E_n is the
electric field at the surface of
the conductor.

the conductor, it must be on the conductor surface.

To find the electric field just outside the surface of a conductor, we consider a portion of the conductor surface small enough to be considered flat with charge density σ, which has negligible variation over the portion. We construct a cylindrical pillbox gaussian surface (Figure 31-3) with one face just outside the conductor and parallel to its surface and the other face just inside the conductor. On the surface of the conductor, in equilibrium, the electric field must be perpendicular to the surface. If there were a tangential component of **E**, the free charge on the conductor would move until this component became zero. Since one face of the pillbox surface is just outside the conductor, we can take **E** to be perpendicular to this face. The other face of the pillbox is inside the conductor, where **E** is zero. There is no flux through the cylindrical surface of the pillbox because **E** is tangential to this surface. The flux through the pillbox is thus E_nA, where E_n is the field just outside the conductor surface and A is the area of the face of the pillbox. The net charge inside the gaussian surface is σA. Gauss' law gives

$$\phi_{net} = \oint E_n \, dA = E_n A = \frac{\sigma A}{\epsilon_0}$$

or

$$E_n = \frac{\sigma}{\epsilon_0} \qquad\qquad 31\text{-}1$$

Field at the surface of a conductor

This result is just twice the field produced by an infinite plane of charge. We can understand this result by comparing the flux lines for an infinite plane charge with those for a conducting slab carrying the same charge density. Figure 31-4 shows a very large charged plane sheet. The flux lines are perpendicular to the sheet and point away from it on both sides. If we construct a pillbox gaussian surface with one face to the right of the sheet and the other to the left of the sheet, the net flux through this surface is E_n2A, where A is the area of each face. Setting this net flux equal to $1/\epsilon_0$ times the net charge inside the pillbox σA, we obtain $E_n = \frac{1}{2}(\sigma/\epsilon_0)$ for the electric field of an infinite sheet.

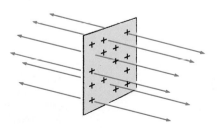

Figure 31-4
Electric field lines for a charge
density σ on an infinite
plane. The electric field has
the magnitude $\sigma/2\epsilon_0$.

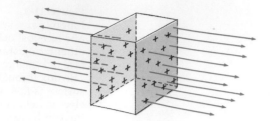

Figure 31-5
Electric field lines for a charge density σ on each face of a conducting slab. Inside the conductor, the electric field due to one face cancels that due to the other face. Outside the conductor, the field has the magnitude σ/ϵ_0 because there are two planes of charge.

Figure 31-5 shows a large conducting slab. If we put a net charge on this conductor, it will be distributed equally on both faces. The conducting slab is equivalent to two conducting sheets. The electric field due to each of these sheets is $\frac{1}{2}(\sigma/\epsilon_0)$. Between the sheets these fields are in opposite directions, so that their magnitudes subtract, giving zero net field inside the conductor. To the right or left of both sheets the two fields add, giving a magnitude of σ/ϵ_0 for the net electric field. As can be seen from the figures, the flux lines from the conducting slab are twice as dense as those for a single charged sheet if the charge *densities* are the same. (To produce equal charge densities on a large conducting slab and a large plane sheet, twice as much charge is needed on the slab.) It is important to realize that the electric field produced by the charge on the surface of a conductor is completely determined by the charge distribution. It has nothing to do with the fact that the charge distribution is on a conductor. This same charge distribution placed on an insulator or in space would produce the same electric field. Of course, the presence of the conductor affects the original distribution of the charge: it is distributed so that the electric field is zero everywhere inside the conductor.

Similar but slightly more complicated reasoning can be applied to a conductor of arbitrary shape. Consider point P just outside the surface of a conductor, as shown in Figure 31-6. We can consider the charge on the surface of a conductor to consist of two parts, (1) the charge in the immediate neighborhood of point P and (2) all of the rest of the charge. Since point P is just outside the surface, the charge in the immediate neighborhood looks like an infinite plane charge. It produces a field of magnitude $\frac{1}{2}(\sigma/\epsilon_0)$ at P and a field of equal magnitude just inside the conducting surface pointing away from the surface. The rest of the charge on the conductor (or elsewhere) must produce a field $\frac{1}{2}(\sigma/\epsilon_0)$ inside the conductor pointing toward the surface so that the net field inside the conductor is zero. Whereas the field due to this second part of the charge cancels the field *inside* the conductor, it adds to that produced by the neighboring charge just outside the conductor, giving a net field $\frac{1}{2}(\sigma/\epsilon_0) + \frac{1}{2}(\sigma/\epsilon_0) = \sigma/\epsilon_0$ just outside the conductor. (The part of the field due to the distant charges has the same magnitude and direction at points just inside and just outside the surface.) This argument was first given by Laplace in about 1800.

Figure 31-6
Arbitrarily shaped conductor carrying a charge on its surface. The charge in the vicinity of point P looks like an infinite plane sheet of charge if P is very close to the conductor. This charge produces an electric field of magnitude $\sigma/2\epsilon_0$ both inside and outside the conductor, as indicated by the solid arrows. Since the resultant field inside the conductor must be zero, the rest of the charge must produce a field of equal magnitude indicated by the dashed arrows. Inside the conductor these fields cancel, but outside at point P they add to give $E = \sigma/\epsilon_0$.

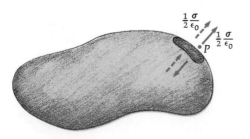

31-3 Charging by Induction

A simple and practical method of charging a conductor makes use of the free movement of charge in a conductor. In Figure 31-7 two uncharged metal spheres are in contact. When a charged rod is brought near the spheres, free electrons on one sphere flow to the other. Suppose, for example, the rod is positively charged. The rod attracts negatively charged electrons, and the sphere nearest the rod acquires electrons from the other, leaving the near sphere with a negative charge and the far sphere with an equal positive charge due to lack of electrons. If the spheres are separated before the rod is removed, they will have equal and opposite charges. A similar result is obtained, of course, with a negatively charged rod which drives electrons from the nearest sphere to the other. In each case, the spheres are charged without being touched by the rod, and the charge on the rod is undisturbed. This is called *electrostatic induction.*

A convenient large conductor is the earth itself. For most purposes we can consider the earth to be an infinitely large conductor. When a conductor is connected to the earth, it is said to be *grounded.* This is indicated by a connecting wire and parallel horizontal lines, as in Figure 31-8b. We can use the earth to charge a single conductor by induction. In Figure 31-8a a positively charged rod is brought near a neutral conductor, and the conductor becomes polarized as shown. Free electrons are attracted to the side near the positive rod, leaving the other side with a positive charge. If we ground the conductor while the charged rod is still present, the conductor becomes charged oppositely to the rod because electrons from the earth travel along the connecting wire and neutralize the positive charge on the far side of the conductor. The connection to ground is broken before the rod is removed, to complete the charging by induction.

Questions

2. An insulating rod is given a charge and then used to charge a set of conductors by induction. What practical limit is there on the number of times the rod can be used without recharging?

3. Can insulators as well as conductors be charged by induction?

Review

A. Define, explain, or otherwise identify:

Electrostatic equilibrium, 741
Free charge in a conductor, 741

Field emission, 742
Charging by induction, 745

B. True or false:

1. The electric field inside a conductor is always zero.

2. The result that **E** = 0 inside a conductor in equilibrium can be derived from Gauss' law.

3. A conductor has free electrons only if it has an excess negative charge.

4. If the net charge on a conductor is zero, the charge density σ must be zero at every point on the surface.

5. Half of the electric field at a point just outside the surface of a conductor is due to the charge on the surface in the immediate vicinity of that point.

Figure 31-7
Charging by induction. The two spherical conductors in contact become oppositely charged because the positively charged rod attracts electrons to the left side, leaving the right side positive. If the spheres are now separated with the rod in place, they retain equal and opposite charges.

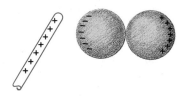

Figure 31-8
(a) The charge on a single conducting sphere is polarized because the electrons are attracted to the positively charged rod, leaving the opposite side positive. (b) When the conductor is grounded, i.e., connected to a very large conductor such as the earth, electrons from the earth neutralize the positive charge, leaving an excess negative charge. This charge remains if the ground connection is broken before the rod is removed.

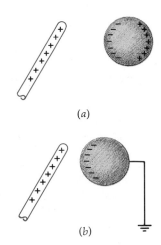

(a)

(b)

Exercises

Section 31-1, Free Charge in Conductors

1. A penny has a mass of 3 gm and is made of copper of molecular weight 63.5 gm/mole. Assume one free electron for each copper atom. (*a*) How many free electrons are there in a penny? (*b*) How much free charge (in coulombs) is there?

Section 31-2, Charge and Field at Conductor Surfaces

2. A penny is in an external electric field of magnitude 1000 N/C and direction perpendicular to its faces. (*a*) Find the charge densities on each face of the penny. (*b*) If the radius of a penny is taken to be 1 cm, what is the total charge on one face?

3. A metal slab has square faces of side 10 cm. It is placed in an external electric field which is perpendicular to its faces. What is the magnitude of the electric field if the total charge on one of the faces of the slab is 10^{-9} C?

4. A charge of 6×10^{-9} C is placed uniformly on a square sheet of nonconducting material of side 20 cm in the yz plane. (*a*) What is the charge density σ? (*b*) What is the magnitude of the electric field just to the right and just to the left of the sheet? (*c*) The same charge is placed on a conducting square slab of side 20 cm and thickness 1 mm. What is the charge density σ? (Assume the charge distributes itself uniformly on the large square surfaces.) (*d*) What is the magnitude of the electric field just to the right and just to the left of each face of the slab?

5. A nonconducting spherical shell of outer radius 15 cm carries a net charge of 2 μC uniformly distributed on its surface. (*a*) What is the charge density σ? What is **E** (*b*) just outside the surface of the sphere and (*c*) just inside the shell? (*d*) A portion of the shell is removed, leaving a small hole. What is **E** just outside and just inside the shell at the hole? Assume the rest of the charge on the shell is undisturbed.

6. An irregularly shaped conductor carries a surface charge. At some point P on the surface the charge density is $\sigma = 1 \, \mu\text{C/m}^2$. (*a*) What is the electric field just outside the surface at point P? (*b*) The conductor is now replaced by an insulator of exactly the same shape and carrying exactly the same charge density σ at each corresponding point. What now is the electric field just outside the surface at point P? (*c*) If your answers for parts (*a*) and (*b*) are different, explain why. If they are the same, in what way is the insulator different from the conductor?

Section 31-3, Charging by Induction

7. Explain, giving each step, how a positively charged insulating rod can be used to give a metal sphere (*a*) a negative charge and (*b*) a positive charge. (*c*) Can the same rod be used to give one sphere a positive charge and another sphere a negative charge without recharging the rod?

8. Figure 31-9 shows a device called an electroscope, which consists of two metal-foil leaves attached to a conducting rod with a conducting knob on top. When uncharged, the leaves hang together vertically. When charged, the leaves repel each other. The divergence of the leaves indicates the amount of charge. (*a*) A positively charged nonconducting rod is brought close to the knob of the uncharged electroscope, and the leaves diverge. Explain why, with a diagram showing the charge distribution. (*b*) If the rod is removed, the leaves come back together. If the knob is momentarily grounded with the rod close, the leaves also come together but then diverge when the rod is removed. Explain with a diagram. What sign of net charge is now on the electroscope?

Figure 31-9
Electroscope for Exercise 8.

(c) With the electroscope charged as in part (b), a negatively charged rod is brought close. Do the leaves come together or diverge more? Explain with a diagram.

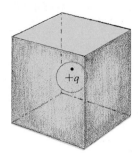

Problems

1. A solid conductor has a cavity as shown in Figure 31-10. A small charge $+q$ is placed in the cavity. (This can be done by placing a charge on a small piece of cork and suspending the cork with an insulating string.) (a) Prove that there is an induced charge density on the inner surface of the cavity such that the total induced charge is $q' = -q$ independent of the location of q. (b) Draw lines of force for this problem.

2. A spherical conducting shell with zero net charge has inner radius a and outer radius b. A point charge q is placed at the center in the cavity. (a) Use Gauss' law and the properties of conductors in equilibrium to find the electric field in each of the regions $r < a$, $a < r < b$, and $b < r$. (b) Draw lines of force for this situation. (c) Describe the charge density on the outer surface ($r = b$) of the sphere. How would this charge density be affected if the point charge in the cavity were moved away from the center? Sketch lines of force for the case in which the point charge is not at the center of the cavity.

3. The electrostatic force on a charge at some point is the product of the charge and the electric field due to all other charges. Consider a small charge on the surface of a conductor $\Delta q = \sigma \, \Delta A$. (a) Show that the electrostatic force on the charge is $\sigma^2 \, \Delta A / 2\epsilon_0$. (b) Explain why this is just half $\Delta q \, E$, where $E = \sigma/\epsilon_0$ is the electric field just outside the conductor at that point. (c) The force per unit area is called the electrostatic stress. Find the stress when a charge of 2 μC is placed on a conducting sphere of radius 10 cm.

CHAPTER 32   Electric Potential

In Chapter 8 we showed that a central force such as the electrostatic force given by Coulomb's law is conservative; i.e., the work done on a particle by this force as the particle moves from one point to another depends only on the initial and final positions and not on the path taken. There is thus a potential-energy function associated with this force. As with any conservative force, we define the change in the potential energy to be equal to the negative of the work done by the force. Then the work done by the force equals the decrease in the potential energy. If $d\mathbf{s}$ is a small displacement of a particle under the influence of a conservative force $\mathbf{F}$, the change in potential energy dU is defined by

$$dU = -\mathbf{F} \cdot d\mathbf{s} \qquad\qquad 32\text{-}1$$

We note again that only the *change* in the potential-energy function is defined by this equation. The absolute value of the potential-energy function is usually determined by choosing the potential energy to be zero at some convenient position.

32-1   Potential Difference

Consider a test charge q_0 in an electric field $\mathbf{E}$ produced by some system of charges. The force on q_0 is $q_0\mathbf{E}$. This force is the sum of the individual force exerted on q_0 by each charge in the system. Since each individual force is given by Coulomb's law and is therefore conservative, the resultant force $q_0\mathbf{E}$ is conservative. Therefore the work done by this force equals the decrease in the potential energy. If this is the only force that does work on the particle, the decrease in potential energy is accompanied by an increase in kinetic energy of equal magnitude. The change in electrostatic potential energy of a test charge q_0 when it is given a displacement $d\mathbf{s}$ is given by Equation 32-1 with the force $\mathbf{F}$ equal to $q_0\mathbf{E}$:

$$dU = -q_0\mathbf{E} \cdot d\mathbf{s} \qquad\qquad 32\text{-}2$$

The potential-energy change is proportional to the test charge q_0. The potential-energy change divided by the test charge q_0 is called the *potential difference dV:*

$$dV = \frac{dU}{q_0} = -\mathbf{E} \cdot d\mathbf{s} \qquad\qquad 32\text{-}3$$

The potential difference (Figure 32-1) between some point a and another point b is

$$V_b - V_a = \int_a^b dV = -\int_a^b \mathbf{E} \cdot d\mathbf{s} \qquad\qquad 32\text{-}4$$

The integral $\int_a^b \mathbf{E} \cdot d\mathbf{s}$ is the work done by the electric field in moving the test charge q_0 from a to b divided by the charge q_0. Since the force $q_0\mathbf{E}$ is conservative, this work does not depend on the path taken from a to b. If we wish to move a test charge q_0 from point a to point b without acceleration, we must exert an applied force $\mathbf{F}_{\mathrm{app}}$ which is equal and opposite to the force $q_0\mathbf{E}$ exerted by the field. The work done by such an *applied* force is the negative of that done by the electric field and is therefore equal to the *increase* in the potential energy of the charge. According to our definition of potential difference, the increase in potential energy is just q_0 times the potential difference ΔV:

$$\Delta U = q_0\, \Delta V \qquad\qquad 32\text{-}5$$

We thus have a simple physical interpretation of the potential difference $V_b - V_a$.

The potential difference $V_b - V_a$ is the work per unit charge necessary to move a test charge without acceleration from point a to point b.

The SI unit of potential difference is the joule per coulomb, called a volt (V):

$$1\ \mathrm{V} = 1\ \mathrm{J/C} \qquad\qquad 32\text{-}6$$

From Equation 32-3 we note that the dimensions of potential difference are also these of electric field times distance. Thus the unit of electric field $\mathbf{E}$, the newton per coulomb, is also equal to a volt per meter:

$$1\ \mathrm{N/C} = 1\ \mathrm{V/m}$$

For example, if we have a constant electric field of 10 V/m in the x direction, the potential decreases by 10 V in each meter in the x direction.

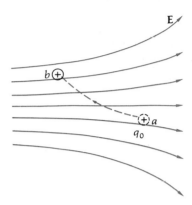

Figure 32-1
When a positive test charge q_0 moves from point a to point b, the electric field does negative work and the potential energy of the charge increases. The change in potential energy per unit charge is the potential difference $V_b - V_a = (U_b - U_a)/q_0$.

32-2 Electric Potential

Equation 32-4 and its differential form (Equation 32-3) define the change in the function V called the *electric potential*. Like the potential energy U, only the *change* in the electric-potential function V is significant. The value of the electric-potential function at any point is usually determined by arbitrarily choosing V to be zero at some convenient point. If V and U are chosen to be zero at the same point (the usual case), the electric potential at any point is just the potential energy of a charge q_0 at that point divided by the charge q_0. The potential V is more convenient than the potential-energy function U because V does not depend on the test charge q_0. The potential V is a scalar function of position which is determined by the charge distribution (and by the choice of the point at which $V = 0$). There is a close relation between the potential V and the electric field $\mathbf{E}$ which is also determined by the charge distribution. The electric field $\mathbf{E}$ and the potential V are related by Equation 32-3, which is the same as the general relation between a conservative force $\mathbf{F}$ and the potential energy U in Equation 32-1. The relation between a conservative force $\mathbf{F}$ and the potential energy U was discussed in some detail in Section 8-7. You should review that section. We shall repeat some of that discussion here.

In many cases of interest the electric field is just the negative derivative of the electric potential. Consider, for example, the case in which the electric field has only one component E_x. Then $\mathbf{E} \cdot d\mathbf{s} = E_x\, dx$, and Equation 32-3 gives

$$dV = -\mathbf{E} \cdot d\mathbf{s} = -E_x\, dx$$

or

$$E_x = -\frac{dV}{dx} \qquad\qquad 32\text{-}7$$

If we consider displacement vectors $d\mathbf{s}$ which are perpendicular to the x direction, the potential does not change because $\mathbf{E} \cdot d\mathbf{s}$ is zero. Thus if $\mathbf{E}$ has only the component E_x, the potential V is not a function of y or z.

We next consider a spherically symmetric charge distribution, e.g., a point charge at the origin, a uniformly charged spherical shell, or any spherical charge distribution for which the charge density depends only on the radial distance r. We have found that the electric field is radial; then

$$dV = -\mathbf{E} \cdot d\mathbf{s} = -E_r\, dr$$

where dr is the radial component of $d\mathbf{s}$. Then

$$E_r = -\frac{dV}{dr} \qquad\qquad 32\text{-}8$$

Again, if we choose $d\mathbf{s}$ perpendicular to the radial direction, the potential does not change. This implies that the potential function V is a function only of r.

Let us now consider various displacements of magnitude ds but with different directions in an arbitrary electric field $\mathbf{E}$. If $d\mathbf{s}$ is perpendicular to $\mathbf{E}$, the potential does not change. The greatest change in V occurs when $d\mathbf{s}$ is parallel or antiparallel to $\mathbf{E}$. When $d\mathbf{s}$ is parallel to $\mathbf{E}$, we have

$$dV = -E\, ds$$

$$E = -\frac{dV}{ds} \qquad d\mathbf{s}\|\mathbf{E} \qquad\qquad 32\text{-}9$$

As stated in Chapter 8, a vector which points in the direction of the greatest change in a scalar function and whose magnitude equals the derivative of the function with respect to distance in that direction is called the *gradient* of the function. Thus in general, the electric field is the negative gradient of the potential. The electric field lines point in the direction of greatest decrease in the potential function.

In general, the potential is a function of all three coordinates x, y, and z, and the electric field has components E_x, E_y, and E_z. Any one of these components, say E_x, is related to the change in the potential function $V(x,y,z)$ when the corresponding coordinate x varies and the other two coordinates y and z are fixed. Consider a displacement $d\mathbf{s} = dx\, \mathbf{i}$. Then

$$dV(x,y,z) = -\mathbf{E} \cdot d\mathbf{s} = -E_x\, dx \qquad y \text{ and } z \text{ held fixed}$$

Thus E_x is the negative derivative of V with respect to x with y and z held constant. Such a derivative is called a partial derivative and written $\partial V/\partial x$, as discussed in Section 8-10. In general the rectangular components of the electric field are related to the potential function by

$$E_x = -\frac{\partial V}{\partial x} \qquad E_y = -\frac{\partial V}{\partial y} \qquad E_z = -\frac{\partial V}{\partial z}$$

Questions

1. Explain in your own words the distinction between electric potential and electric potential energy.

2. If a charge is moved a small distance in the direction of an electric field, does its electric *potential energy* increase or decrease? Does your answer depend on the sign of the charge?

3. If a charge is moved a small distance in the direction of an electric field, does the electric potential increase or decrease? Does your answer depend on the sign of the charge?

4. What direction can you move relative to an electric field so that the electric potential does not change?

5. Under what circumstances can the displacement of an electric charge result in a simultaneous increase in its electric potential energy and decrease in the electric potential?

6. A positive charge is released from rest in an electric field. Will it move toward a region of greater or smaller electric potential?

32-3 Potential Due to a Point Charge and Electrostatic Potential Energy

The electric field at a distance r from a point charge q at the origin is

$$\mathbf{E} = \frac{kq}{r^2}\, \hat{\mathbf{r}}$$

where $k = 1/4\pi\epsilon_0$ is the coulomb constant and $\hat{\mathbf{r}}$ is a unit vector in the radial direction. We can calculate the potential function V from Equation 32-3:

$$dV = -\mathbf{E} \cdot d\mathbf{s} = -\frac{kq}{r^2}\, dr$$

Integrating gives

$$V = -\int \frac{kq}{r^2}\, dr = \frac{kq}{r} + C$$

where C is an arbitrary constant of integration which arises because only the potential difference is defined by Equation 32-3. The constant is determined by the choice of zero potential. The potential decreases as r increases and approaches the limit C as r approaches infinity. It is customary to choose the constant C to be zero so that the potential is zero at $r = \infty$; then

$$V = \frac{kq}{r}$$

32-10

Potential of a point charge at the origin

The potential is positive or negative depending on the sign of the charge q.

If we bring a test charge q_0 from a very large distance from the origin ($r = \infty$) to some distance r, the work we must do against the electric force equals the increase in potential energy. This increase in potential energy is just q_0V, where V is given by Equation 32-10 (Figure 32-2). The electric potential $V = kq/r$ is the work per unit charge that must be done to bring a positive test charge from infinity to the distance r without acceleration.

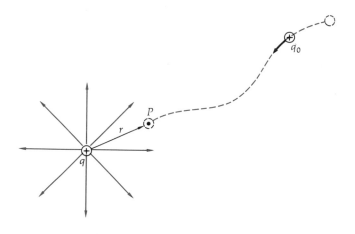

Figure 32-2
The work required to bring a test charge q_0 without acceleration from infinity to point P a distance r from a charge q at the origin is kqq_0/r. The work per unit charge is kq/r, the electric potential at point P relative to zero potential at infinity.

We note that the potential at some point due to a point charge at the origin depends only on the distance from the point to the origin. If we have a point charge q_1 at some point P_1 other than the origin, the potential at any other point P_2 is

$$V = \frac{kq_1}{r_{12}}$$

32-11

where r_{12} is the distance between points P_1 and P_2. The work needed to bring a second charge q_2 from an infinite distance away to point P_2 without acceleration is q_2V, or

$$W = \frac{kq_1q_2}{r_{12}}$$

This work is independent of the path taken by q_2 and is symmetric in the charges. We get the same result if we first place charge q_2 at point P_2 and then compute the work to bring charge q_1 from an infinite distance away to point P_1. This work is the *electrostatic potential energy* of the two-charge system relative to zero potential energy when the charges are infinitely separated. In general, the *electrostatic potential energy of a system of point charges is the work needed to bring the charges*

Electrostatic potential energy

from an infinite separation to their final positions without acceleration. The electrostatic potential energy of a two-charge system is

$$U = \frac{kq_1q_2}{r_{12}} \qquad\qquad 32\text{-}12$$

Example 32-1 What is the electric potential at the distance of the first Bohr orbit, $r = 0.529 \times 10^{-10}$ m from a proton? What is the potential energy of the electron and proton at this separation?

The charge of the proton is $q = 1.6 \times 10^{-19}$ C. Equation 32-10 gives

$$V = \frac{kq}{r} = \frac{(9 \times 10^9)(1.6 \times 10^{-19})}{0.529 \times 10^{-10}} = 27.2 \text{ V}$$

The charge of the electron is $-e = -1.6 \times 10^{-19}$ C. In SI units the potential energy of the electron a distance 0.529×10^{-10} m from a proton is

$$U = qV = (-1.6 \times 10^{-19} \text{ C})(27.2 \text{ V}) = -4.36 \times 10^{-18} \text{ J}$$

It is more convenient to find this potential energy in electron volts, the energy unit defined to be one volt times the magnitude of the charge of the electron (Section 8-6). The potential energy in electron volts is

$$U = qV = (-e)(27.2 \text{ V}) = -27.2 \text{ eV}$$

Example 32-2 Find the electrostatic potential energy of three point charges q_1, q_2, and q_3 at points P_1, P_2, and P_3, respectively.

Let r_{12} be the distance between q_1 and q_2, r_{13} the distance between q_1 and q_3, and r_{23} the distance between q_2 and q_3. The work needed to place q_1 at P_1 and q_2 at P_2 is kq_1q_2/r_{12}, as previously computed. To bring charge q_3 from an infinite distance away to point P_3 we must do work against the resultant electric field, which is just the vector sum of the individual fields due to q_1 and q_2. This work is therefore just the sum of the work that must be done against the field due to each charge separately. The work due to charge q_1 is kq_1q_3/r_{13}, and the work due to charge q_2 is kq_2q_3/r_{23}. The work needed to bring charge q_3 to point P_3 with charges q_1 and q_2 already at their final positions is therefore

$$W_3 = \frac{kq_1q_3}{r_{13}} + \frac{kq_2q_3}{r_{23}}$$

The total work to assemble the three charges is

$$W = U = \frac{kq_1q_2}{r_{12}} + \frac{kq_1q_3}{r_{13}} + \frac{kq_2q_3}{r_{23}}$$

This result is independent of the order in which the charges are brought to their final positions.

Questions

7. Justify the following statement: A pair of like charges has a positive potential energy, while a pair of unlike charges has a negative potential energy.

8. If q is negative, the potential at some point P is negative. How can negative potential be interpreted in terms of the work done by an applied force in bringing a positive test charge from infinity to the field point?

32-4 Calculation of Electric Potential for Various Charge Distributions

There are two way of calculating the potential at some point due to a given charge distribution. If the electric field **E** is already known, the potential function can be found from Equation 32-3. Consider, for example, the case of an infinite-plane-charge distribution of surface charge density σ. Let the charge be in the yz plane. In Chapter 30 we found that the electric field for an infinite plane charge has magnitude $\sigma/2\epsilon_0$, independent of the distance from the plane. For positive x,

$$E_x = \frac{\sigma}{2\epsilon_0}$$

Equation 32-3 then gives for the potential difference,

$$dV = -\mathbf{E} \cdot d\mathbf{s} = -E_x\, dx = -\frac{\sigma}{2\epsilon_0}\, dx$$

Integrating, we obtain

$$V = -\frac{\sigma}{2\epsilon_0} x + C \qquad x > 0 \qquad\qquad \text{32-13}$$

where again C is an arbitrary constant of integration. The potential decreases linearly with distance from the plane. Since it does not approach any limiting value as x approaches infinity, we cannot choose the potential to be zero at $x = \infty$. We can, however, choose V to be zero at any convenient finite value of x. For example, if we choose V to be zero at $x = 0$, the constant C is zero and $V = -(\sigma/2\epsilon_0)x$. Alternatively, we could choose the potential to have some other value at $x = 0$, say 100 V, in which case the constant C would be 100 V. Since only differences in potential are important, it does not matter what value is chosen for the constant C.

The second method of finding the potential due to a given charge distribution is to use Equation 32-10 or 32-11 for the potential due to a point charge. Consider, for example, two point charges. As discussed in Example 32-2, the work that must be done against the electric field in order to bring a test charge q_0 from an infinite distance away to some point is just the sum of the work that must be done against the field due to each charge separately. The potential at some point relative to zero potential at infinity is therefore the sum of the potentials due to each charge. In general, for a system of point charges the potential at point P is given by

$$V = \sum_i \frac{kq_i}{r_{i0}} \qquad\qquad \text{32-14}$$

where r_{i0} is the distance between the field point P and the ith charge q_i. If we have a continuous charge distribution of finite size, the potential at some point P can be found by treating a small element of the charge dq as a point charge and integrating:

$$V = \int \frac{k\, dq}{r} \qquad\qquad \text{32-15}$$

Potential for continuous distribution of charge

where r is the distance from the charge element dq to point P. This result is based on the assumption that the potential is zero at infinite distance from the charge distribution. Equation 32-15 can be used only if the charge distribution is of finite extent, so that the potential approaches a limiting value at infinity which can be chosen to be zero. Sufficiently far from any finite charge distribution, the charge distribution looks like a point charge. Then the electric field decreases as $1/r^2$, and the potential decreases as $1/r$, which approaches zero as r approaches infinity. However, if the charge distribution is not finite, the electric field does not decrease as $1/r^2$ and in general the potential does not approach any limiting value at infinity. We saw an example of this for the field of an infinite plane charge with electric field independent of the distance from the plane. Another example, which we shall consider below, is the infinite line charge. For such charge distributions of infinite extent we cannot use Equation 32-15 to find the potential V. Instead we find the electric field $\mathbf{E}$ first and then find V from $dV = -\mathbf{E} \cdot d\mathbf{s}$, choosing V to be zero at any convenient point.

Example 32-3 Find the potential on the axis of a ring of radius a and charge Q.

An element of charge dq is shown in Figure 32-3. The distance from this charge element to the field point on the axis of the ring is $s = \sqrt{x^2 + a^2}$, which is the same for all elements of charge on the ring. The potential due to the ring is thus

$$V - \int \frac{k\,dq}{\sqrt{x^2 + a^2}} = \frac{k}{\sqrt{x^2 + a^2}} \int dq = \frac{kQ}{\sqrt{x^2 + a^2}} \qquad \text{32-16}$$

The electric field can be calculated from Equation 32-7:

$$F_x = -\frac{dV}{dx} = -kQ(-\tfrac{1}{2})(x^2 + a^2)^{-3/2}2x = \frac{kQx}{(x^2 + a^2)^{3/2}}$$

which agrees with the direct calculation from Coulomb's law in Chapter 30.

Example 32-4 Find the potential on the axis of a disk of uniform charge density σ.

Let the disk axis be the x axis and consider positive x only. We can treat the disk as a set of ring charges and use our result from Example 32-3. Consider a ring of radius r and width dr (Figure 32-4). The area of this ring is $2\pi r\,dr$, and its charge is $dq = \sigma\,dA = \sigma 2\pi r\,dr$. The potential on the axis of the disk is found by summing from $r = 0$ to $r = R$:

$$V = \int_{r=0}^{r=R} \frac{k\,dq}{\sqrt{x^2 + r^2}} = \int_0^R \frac{k\sigma 2\pi r\,dr}{\sqrt{x^2 + r^2}} = k\sigma\pi \int_0^R (x^2 + r^2)^{-1/2}2r\,dr$$

This integral is of the form $\int u^n\,du$, with $u = x^2 + r^2$ and $n = -\tfrac{1}{2}$.

$$V = k\sigma\pi \left. \frac{(x^2 + r^2)^{+1/2}}{\tfrac{1}{2}} \right|_0^R = 2\pi k\sigma[(x^2 + R^2)^{1/2} - x] \qquad \text{32-17}$$

Again, the electric field can be obtained from Equation 32-7:

$$E_x = -\frac{dV}{dx} = 2\pi k\sigma \left(1 - \frac{x}{\sqrt{x^2 + R^2}} \right) \qquad \text{32-18}$$

which agrees with the result obtained directly from Coulomb's law in Chapter 30. The potential function (Equation 32-17) becomes infinite as R becomes infinite. As discussed, we cannot use Equation 32-15 with the potential at infinity defined to be zero when the charge distribution is of infinite extent. The potential function for an infinite-plane-charge distribution is found from the electric field, as before.

Figure 32-3
Geometry for the calculation of electric potential at a point on the axis of a uniformly charged ring.

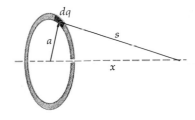

Figure 32-4
Geometry for the calculation of the electric potential at a point on the axis of a uniformly charged disk. The disk is divided into rings of area $2\pi r\,dr$.

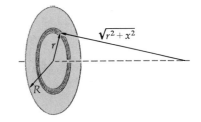

Example 32-5 Find the potential due to a spherical shell of charge Q.

We consider a spherical shell of radius R having a uniform surface charge density. We are interested in the potential at all points inside and outside the shell. Since this charge distribution is of finite extent, we can use Equation 32-15. The calculation is nearly identical to that for the gravitational potential in Section 16-7. We shall not repeat it here, performing instead the simpler calculation using Equation 32-3 and the known result for the electric field.

In the region outside the sphere, the electric field is the same as if all the charge were at the origin,

$$E_r = \frac{kQ}{r^2} \qquad r > R$$

where $Q = 4\pi R^2 \sigma$ is the total charge on the shell. Then the potential is found from

$$dV = -E_r \, dr = -\frac{kQ}{r^2} \, dr$$

$$V = \frac{kQ}{r} + C$$

Again, we choose the arbitrary constant of integration to be zero so that the potential is zero at infinite r. Then

$$V = \frac{kQ}{r}$$

Inside the shell the electric field is zero. Then

$$dV = -E_r \, dr = 0$$

$$V = \text{constant inside}$$

As r approaches R from outside the shell, the potential approaches kQ/R. Hence the constant value of V inside must be kQ/R to make $V(r)$ continuous. Thus

$$V(r) = \begin{cases} \dfrac{kQ}{R} & r \leq R \\[2mm] \dfrac{kQ}{r} & r \geq R \end{cases}$$

32-19

Potential for spherical shell of charge

This potential function is sketched in Figure 32-5.

A common mistake is to think that the potential must be zero inside a spherical shell because the electric field is zero. Actually zero electric field merely implies that the potential does not change. Consider a

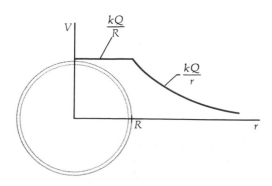

Figure 32-5
Electric potential of a uniformly charged spherical shell of radius R as a function of distance r from the center of the shell. Inside the shell the potential has the constant value kQ/R. Outside the shell the potential is the same as that due to a point charge at the center of the sphere.

spherical shell with a small hole so that we can move a test charge in and out of the shell. If we move the test charge from an infinite distance to the shell, the work per charge we must do against the electric force is kQ/R. Inside the shell there is no electric field, and so it takes no work to move the test charge inside the shell. The total amount of work per charge it takes to bring the test charge from infinity to any point inside the shell is just the work it takes to bring it up to the shell radius R, which is kQ/R. The potential is therefore kQ/R everywhere inside the shell.

Example 32-6 Find the potential due to a spherical ball of uniform charge density.

Let the radius of the ball be R and the total charge be $Q = \frac{4}{3}\pi R^3 \rho$, where ρ is the charge per unit volume. We found the electric field for this charge distribution in Example 30-6. Outside the ball the electric field is kQ/r^2, the same as that due to a point charge. Thus the potential outside is kQ/r. Inside the charged ball the electric field is

$$E_r = \frac{kQ}{R^3}\, r$$

Since the electric field inside the ball is not zero, the potential will not be constant, and since the electric field points radially outward inside the ball, the potential must increase as we move inward. That is, work must be done on a test charge to move it from the outside edge of the ball inward. The potential therefore has its greatest value at the origin. The potential function inside is found from

$$dV = -E_r\, dr = -\frac{kQ}{R^3}\, r\, dr$$

Integrating, we obtain

$$V = -\frac{kQ}{R^3}\frac{r^2}{2} + C$$

The constant of integration is not arbitrary because we have already chosen the potential to be zero at $r = \infty$. The constant C is found by requiring the potential to be continuous at $r = R$. At this point at the edge of the ball the potential is kQ/R. Thus

$$-\frac{kQ}{R^3}\frac{R^2}{2} + C = \frac{kQ}{R}$$

$$C = \frac{3kQ}{2R}$$

The potential inside the ball is then

$$V = \frac{3kQ}{2R} - \frac{kQr^2}{2R^3} = \frac{kQ}{2R}\left(3 - \frac{r^2}{R^2}\right) \qquad \text{32-20}$$

The complete potential function is sketched in Figure 32-6.

Figure 32-6
Plot of the electric potential V of a ball of charge of uniform density versus distance r from the center of the ball. V has its maximum value at the center of the ball. Outside the ball V is the same as that due to a point charge at the center of the sphere.

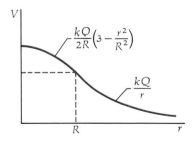

Example 32-7 Find the potential due to a uniform infinite line charge.

Let the charge per unit length be λ. Since this charge distribution extends to infinity, we cannot use Equation 32-15 to find the potential. The potential does not approach a limiting value at infinity and therefore cannot be chosen to be zero there. We use Equation 32-3. Let E_r be the component of the electric field vector which points away from the line charge. E_r is most easily found using Gauss' law, as in Chapter

30. The result is $E_r = 2k\lambda/r$. Then Equation 32-3 gives

$$dV = -\mathbf{E} \cdot d\mathbf{s} = -E_r\, dr = -\frac{2k\lambda}{r}\, dr$$

Integrating gives

$$V = -2k\lambda \ln r + C \qquad\qquad\qquad \text{32-21}$$

The potential decreases as the logarithm of the distance from the line charge. Note that the magnitude increases without limit as r approaches infinity. The constant C can be determined by choosing V to be zero at some distance $r = a$. Then

$$V = 0 = -2k\lambda \ln a + C \qquad C = 2k\lambda \ln a$$

Substituting this expression for the constant C in Equation 32-21, we have

$$V = -2k\lambda \ln r + 2k\lambda \ln a = -2k\lambda \ln \frac{r}{a} \qquad\qquad \text{32-22}$$

Questions

9. In Example 32-3, does it matter whether the charge Q is uniformly distributed around the ring? Would either V or E_x be different if it were not?

10. If the electric potential is constant throughout a region of space, what can you say about the electric field in that region?

32-5 Equipotential Surfaces

A surface on which the electric potential is constant is called an *equipotential surface*. For example, for the potential $V = kQ/r$ produced by a point charge Q at the origin, the surfaces defined by $r = $ constant are equipotential surfaces. These surfaces are concentric spherical surfaces (Figure 32-7).

   If we give a test charge a small displacement $d\mathbf{s}$ on a equipotential surface, the potential change $dV = -\mathbf{E} \cdot d\mathbf{s}$ is zero, implying that the displacement is perpendicular to the electric field $\mathbf{E}$. The lines of force are always perpendicular to an equipotential surface. For a point charge at the origin, the lines of force are radial lines, and the equipotential surfaces are spheres. If we have a uniform electric field in the x direction, e.g., that produced by an infinite plane charge in the yz

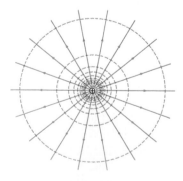

Figure 32-7
Lines of force and equipotential surfaces of a point charge. The equipotentials are spherical surfaces. The lines of force are everywhere perpendicular to the equipotentials.

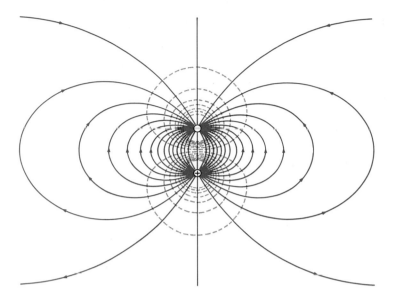

Figure 32-8
Lines of force and equipotential surfaces of a dipole. The
lines of force are always perpendicular to the equipotential surfaces.

plane, the lines of force are parallel lines in the x direction and the equipotential surfaces are planes parallel to the yz plane. Figure 32-8 shows equipotential surfaces for the field produced by an electric dipole. In a two-dimensional drawing of lines of force, the curve representing the intersection of an equipotential surface with the plane of the paper is shown by sketching a continuous curve everywhere perpendicular to the lines of force.

Since the electric field at the surface of a conductor in equilibrium is perpendicular to the surface, the conductor surface is itself an equipotential surface; since the electric field is zero inside a conductor (in electrostatic equilibrium), the potential is constant everywhere on and inside a conductor.

*Conductor surface is an
equipotential surface*

32-6   Charge Sharing

In general, two conductors which are separated in space will not be at the same potential. The potential difference between the conductors depends on the geometrical shapes of the conductors, their separation, and the net charge on each conductor. When two conductors are brought into contact, the charge on the conductors distributes itself so that in electrostatic equilibrium the electric field is zero inside both conductors. In this situation the two conductors in contact may be considered a single conductor. In equilibrium each conductor has the same potential. The transfer of charge from one conductor to another is called *charge sharing*. Coulomb used the method of charge sharing to produce various charges of known ratios to some original charge in his experiment to find the force law between two small (point) charges.

Consider a spherical conductor carrying a charge $+Q$. The lines of force outside the conductor point radially outward, and the potential of the conductor relative to infinity is kQ/R. If we bring up a second identical but uncharged conductor, the potential and field lines will change: negative electrons on the uncharged conductor will be attracted to the positive charge Q, leaving the near side of the uncharged conductor with a negative charge and the far side with a positive charge (Figure 32-9). This charge separation on the neutral conductor will affect the originally uniform charge distribution on the positive

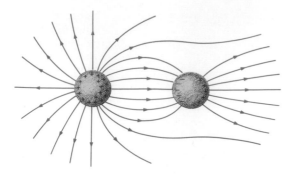

Figure 32-9
Electric field lines for a charged spherical conductor near an uncharged spherical conductor.

conductor. Although the detailed calculation of the charge distributions and potential in this case is quite complicated, we can see that some of the field lines leaving the positive conductor will end on the negative charge on the near side of the neutral conductor and an equal number of lines will leave the far side of that conductor. Since the potential decreases as we move along a field line, the positively charged conductor is at a greater potential than the neutral conductor. If we put the two conductors in contact, positive charge will flow to the neutral conductor until both conductors are at the same potential. (Actually, negative electrons flow from the neutral conductor to the positive conductor. It is slightly more convenient to think of this as a flow of positive charge in the opposite direction.) By symmetry, since the conductors are identical, they will share the original charge equally. If the conductors are now separated, each will carry charge $\frac{1}{2}Q$ and both will be at the same potential.

In Figure 32-10 a small conductor carrying a positive charge q is inside a cavity of a second larger conductor. In equilibrium, the electric field is zero inside the conducting material. The lines of force that leave the positive charge q must end on the inner surface of the large conductor. A negative charge $-q$ must therefore be induced on the inner surface of this conductor. This must occur no matter what the charge is on the outside surface of this conductor. Regardless of the charge on the larger conductor, the small conductor in the cavity is at a greater potential because the lines of force go from this conductor to the larger conductor. If the conductors are now connected, say with a fine conducting wire, *all* the charge originally on the smaller conductor will flow to the larger one. When the connection is broken, there is no charge on the small conductor in the cavity and there are no field lines anywhere within the outer surface of the larger conductor. The positive charge transferred from the smaller conductor to the larger one resides completely on the outside surface. If we bring up a second small positively charged conductor into the cavity (through the small opening shown in the figure) and touch it to the inner surface, we again transfer all the charge to the outer conductor. This procedure can be repeated indefinitely. This method is used to produce large potentials in the Van de Graaff generator, where the charge is brought to the inner surface of a larger spherical conductor by a continuous belt (Figure 32-11). The greater the net charge on the outer conductor the greater its potential. For example, if the conductor is a spherical shell, its potential is kQ/R. The maximum potential obtainable in this way is limited only by the fact that air molecules become ionized in very high electric fields and the air becomes a conductor. This phenomenon, called *dielectric breakdown,* occurs in air at electric field strengths of about $E_{max} \approx 3 \times 10^6$ N/C $= 3 \times 10^6$ V/m. The resulting discharge

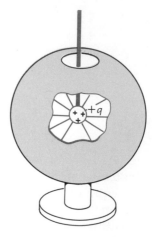

Figure 32-10
Small conductor carrying a positive charge inside a cavity in a larger conductor.

William Vandivert

Corona discharge from overloaded power line at General Electric test facilities in Pittsfield, Massachusetts. Because of the intense electric field near the wire, the air molecules are ionized and the air becomes conducting. The light is given off when the ions and electrons recombine.

Figure 32-11
Small demonstration Van de
Graaff generator. The top is
removed in the photograph
on the right to show the
charging belt. (*Courtesy of
Larry Langrill.*)

through the conducting air is called *corona discharge*. Since the potential of a spherical conductor is kQ/R and the electric field just outside
such a conductor is kQ/R^2, the maximum potential is related to the
field by

Corona discharge

$$V_{max} = RE_{max}$$

A sphere of radius 1 m can therefore be raised to a potential of about
3×10^6 V in air before breakdown.

Our argument that all the charge on the inner conductor in Figure
32-10 will flow to the outer conductor when the two are connected
made use of lines of force to describe the electric field between the two
conductors. As we have shown, the use of lines of force to describe the
electric field is based on the fact that the electric field due to a point
charge varies exactly as $1/r^2$. If the force law were not exactly an
inverse-square law, these arguments based on the lines of force would
not be correct. If a small charged conductor is placed in a cavity inside
a larger conductor and connected to it by a fine wire, all the charge
will flow to the outer conductor only if the force law is exactly an
inverse-square law. If it is not exactly an inverse-square law, some of
the charge will remain on the smaller inner conductor when the two
conductors are at equal potentials. In principle, if the electric field of a
point charge varies as $1/r^n$, the value of n can be calculated from a
measurement of the amount of charge remaining on the smaller inner
conductor. In 1936 Plimpton and Lawton found that the exponent n
has the value 2 ± 0.000000002. In their experiment two concentric
spherical conductors were connected by a sensitive electrometer which
could measure any charge that passed between the conductors. A large
charge was placed on the outer shell. The fact that no charge flowed
through the electrometer to the inner shell indicated that the exponent
of n is 2, with an accuracy of 1 part in 10^9.

When a charge is placed on a conductor of nonspherical shape like
that shown in Figure 32-12, the conductor will be an equipotential surface but the charge density and the electric field just outside the conductor $E = \sigma/\epsilon_0$ will vary from point to point. Near a point where the
radius of curvature is small (A in Figure 32-12) the charge density and
electric field will be large, whereas near a point where the radius of
curvature is large (B in Figure 32-12) the charge density and electric
field will be small. We can understand this as follows. Consider two

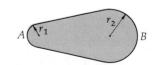

Figure 32-12
Nonspherical conductor. The
charge density σ is greatest at
points of small radius of curvature (near A) and least at
points of large radius of curvature (near B). The electric
field at the conductor surface
$E = \sigma/\epsilon_0$ is greatest where the
radius of curvature is smallest.

spherical conductors of radius R_1 and R_2 isolated from each other and carrying charges q_1 and q_2 such that they have the same potential V. The charges and radii are then related by

$$V = \frac{kq_1}{R_1} = \frac{kq_2}{R_2}$$

In terms of the charge densities $\sigma_1 = q_1/4\pi R_1^2$ and $\sigma_2 = q_2/4\pi R_2^2$ we have

$$V = 4\pi k \sigma_1 R_1 = 4\pi k \sigma_2 R_2$$

If we now touch the spheres together or connect them by a fine conducting wire, no charge will flow because they are already at the same potential. We can now consider the two spheres as a single conductor. We therefore have the important result that the charge density on a conductor varies inversely as the local radius of curvature:

$$\sigma \propto \frac{1}{R} \qquad\qquad 32\text{-}23$$

Since the electric field just outside a conductor is proportional to the charge density, E also varies inversely with the local radius of curvature. For an arbitrarily shaped conductor, the potential at which dielectric breakdown occurs depends on the smallest radius of curvature of any part of the conductor. If the conductor has sharp points of very small radius of curvature, dielectric breakdown will occur at relatively low potentials.

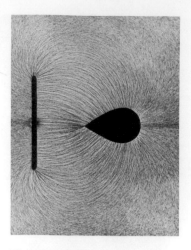

Electric field lines near a nonspherical conductor and plate carrying equal and opposite charges. The lines are shown by small bits of thread suspended in oil. The electric field is strongest near points of small radius of curvature, such as at the ends of the plate and at the pointed left side of the conductor. (*Courtesy of Harold M. Waage, Princeton University.*)

Review

A. Define, explain, or otherwise identify:

Potential difference, 749
Volt, 749
Electric potential, 749
Gradient, 750
Electrostatic potential energy, 752

Electron volt, 753
Equipotential surface, 758
Dielectric breakdown, 760
Corona discharge, 761

B. True or false:

1. If the electric field is zero in some region of space, the electric potential must also be zero in that region.

2. If the electric potential is zero in some region of space, the electric field must also be zero in that region.

3. If the electric field is zero at a point, the potential must also be zero at that point.

4. Lines of electric field point toward regions of lower potential.

5. The value of the electric potential can be chosen to be zero at any convenient point in space.

6. If a conductor is completely surrounded by an outer conductor and the two conductors are put into electrical contact, all the charge originally on the inner conductor will flow to the outer conductor.

Exercises

Section 32-1, Potential Difference

1. A uniform electric field of magnitude 200 N/C is in the x direction. A point charge $Q = 3 \ \mu C$ initially at rest at the origin is released. (a) What is the

kinetic energy of the charge when it is at $x = 4$ m? (b) What is the change in potential energy of the charge from $x = 0$ to $x = 4$ m? (c) What is the potential difference $V(4 \text{ m}) - V(0)$?

2. An infinite plane sheet of charge density $\sigma = +1 \ \mu\text{C/m}^2$ is in the yz plane. (a) What is the magnitude of the electric field in newtons per coulomb? In volts per meter? What is the direction of $\mathbf{E}$ for positive x? (b) What is the potential difference $V_b - V_a$ when point b is at $x = 20$ cm and point a is at $x = 50$ cm? (c) How much work is required by an outside agent to move a test charge $q_0 = +1$ nC from point a to point b without acceleration?

3. A uniform electric field is in the negative x direction. Points a and b are on the x axis, a at $x = 2$ m and b at $x = 6$ m. (a) Is the potential difference $V_b - V_a$ positive or negative? (b) If the magnitude of $V_b - V_a$ is 10^5 V, what is the magnitude of the electric field E?

4. Two parallel conducting plates carry equal and opposite charge densities so that the electric field between them is approximately uniform. The difference in potential between the places is 500 V, and they are separated by 10 cm. An electron is released from rest at the negative plate. (a) What is the magnitude of the electric field between the plates? Is the positive or negative plate at the higher potential? (b) Find the work done by the electric field as the electron moves from the negative plate to the positive plate. Express your answers in both electron volts and joules. (c) What is the change in potential energy of the electron when it moves from the negative plate to the positive plate? What is its kinetic energy when it reaches the positive plate?

Section 32-2, Electric Potential

5. A uniform electric field of magnitude 200 N/C is in the x direction. Find the potential $V(x)$ if $V(x)$ is chosen to be (a) zero at $x = 0$, (b) 4000 V at $x = 0$, (c) zero at $x = 1$ m.

6. In the following, V is in volts and x is in meters. Find E_x when (a) $V(x) = 2000 + 3000x$; (b) $V(x) = 4000 + 3000x$; (c) $V(x) = 2000 - 3000x$; (d) $V(x) = -2000$ independent of x.

7. The electric potential in some region of space is given by $V(x) = C_1 + C_2 x^2$, where V is in volts, x is in meters, and C_1 and C_2 are positive constants. (a) What is the significance of the constant C_1? (b) Find the electric field $\mathbf{E}$ in this region. In what direction is $\mathbf{E}$?

8. An electric field is given by $\mathbf{E} = ax\mathbf{i}$, where E is in newtons per coulomb, x is in meters, and a is a positive constant. (a) What are the SI units of a? (b) How much work is done by this field on a positive point charge q_0 when the charge moves from the origin to some point x? (c) Find the potential function $V(x)$ such that $V = 0$ at $x = 0$.

Section 32-3, Potential Due to a Point Charge and Electrostatic Potential Energy

9. A positive charge of magnitude 2 μC is at the origin. (a) What is the electric potential V at a point 4 m from the origin relative to $V = 0$ at infinity? (b) What is the potential energy when a $+3$-μC charge is placed at $r = 4$ m? (c) How much work must be done by an outside agent to bring the 3-μC charge from infinity to $r = 4$ m assuming the 2-μC charge is held fixed at the origin? (d) How much work must be done by an outside agent to bring the 2-μC charge from infinity to the origin if the 3-μC charge is first placed at $r = 4$ m and held fixed?

10. Use Equations 32-8 and 32-10 to find the electric field at a distance r from a point charge q at the origin.

11. A point charge $q = +3.00\ \mu C$ is at the origin. (a) Find the potential V on the x axis at $x = 3.00$ m and at $x = 3.01$ m, (b) Does the potential increase or decrease as x increases? Compute $-\Delta V/\Delta x$, where ΔV is the change in potential from $x = 3.00$ m to $x = 3.01$ m and $\Delta x = 0.01$ m. (c) Find the electric field at $x = 3.00$ m and compare its magnitude with $-\Delta V/\Delta x$ found in part (b). (d) Find the potential (to three significant figures) at the point $x = 3.00$ m and $y = 0.01$ m and compare your result with the potential on the x axis at $x = 3.00$ m. Discuss the significance of this result.

12. Three point charges are at the corners of an equilateral triangle of side 3 m. Find the electrostatic potential energy of this system if (a) the charges are all positive and equal to $2\ \mu C$, (b) two of the charges are $+2\ \mu C$ and the third charge is $-2\ \mu C$.

13. Three point charges are on the x axis, q_1 at the origin, q_2 at $x = 3$ m, and q_3 at $x = 6$ m. Find the electrostatic potential energy if (a) $q_1 = q_2 = q_3 = 2\ \mu C$, (b) $q_1 = q_2 = 2\ \mu C$ and $q_3 = -2\ \mu C$, (c) $q_1 = q_3 = 2\ \mu C$ and $q_2 = -2\ \mu C$.

Section 32-4, Calculation of Electric Potential for Various Charge Distributions

14. Points A, B, and C are at the corners of an equilateral triangle of side 3 m. Equal positive charges of $2\ \mu C$ are at A and B. (a) What is the potential at point C? (b) How much work is required to bring a positive charge of $5\ \mu C$ from infinity to point C if the other charges are held fixed? (c) Answer parts (a) and (b) if the charge at B is replaced by a charge of $-2\ \mu C$.

15. A charge of $+3.00\ \mu C$ is at the origin, and a charge $-3.00\ \mu C$ is on the x axis at $x = 6.00$ m. (a) Find the potential on the x axis at $x = 3.00$ m. (b) Find the electric field on the x axis at $x = 3.00$ m. (c) Find the potential on the x axis at $x = 3.01$ m and compute $-\Delta V/\Delta x$, where ΔV is the change in potential from $x = 3.00$ m to $x = 3.01$ m and $\Delta x = 0.01$ m, and compare your result with your answer to part (b).

16. Sketch $V(x)$ versus x for points on the x axis for the charge distribution in Exercise 15.

17. Two positive charges $+q$ are on the x axis at $x = +a$ and $x = -a$. (a) Find the potential $V(x)$ as a function of x for points on the x axis. (b) Sketch $V(x)$ versus x. (c) What is the significance of the minimum in your curve halfway between the charges?

18. Sketch $V(x)$ versus x for the uniformly charged ring in the yz plane of Example 32-3. At what point is $V(x)$ a maximum? What is E_x at this point?

19. A charge of $q = +10^{-8}$ C is uniformly distributed on a spherical shell of radius 10 cm. (a) What is the magnitude of the electric field just outside the shell and just inside the shell? (b) What is the magnitude of the electric potential just outside and just inside the shell? (c) What is the electric potential at the center of the shell? What is the electric field at that point?

20. Consider a ball of charge of uniform charge density with radius $R = 10^{-15}$ m and total charge $+e$. (This is a possible model of a proton.) If the center of the charge is at the origin, find the electric field and the potential at (a) $r = R$ and (b) $r = 0$.

Section 32-5, Equipotential Surfaces

21. An infinite sheet of charge has surface charge density $1\ \mu C/m^2$. How far apart are the equipotential planes whose potentials differ by 100 V?

22. A point charge $q = +\frac{1}{9} \times 10^{-8}$ C is at the origin. Taking the potential to be zero at $r = \infty$, locate the equipotential surfaces at 20-V intervals from 20 to 100 V and sketch to scale. Are these surfaces equally spaced?

23. Two equal positive charges are separated by a small distance. Sketch the lines of force and the equipotential surfaces for this system.

Section 32-6, Charge Sharing

24. Sketch lines of force and equipotential surfaces both near to and far from the conductor shown in Figure 32-12, assuming the conductor carries some charge q.

25. Two isolated conducting spheres of radii 4.0 and 8.0 cm are each given a charge of 3×10^{-8} C. The spheres are brought into contact and then separated. (a) What is the ratio of the charge on the larger sphere to that on the smaller sphere? What is the sum of the charges on the two spheres? (b) Find the charge on each sphere. (c) Find the potential of each sphere.

26. Suppose that a Van de Graaff generator has a potential difference of 10^6 V between the belt and the outer shell and that charge is supplied at the rate of 200 μC/sec. What minimum power is needed to drive the moving belt?

27. A conducting sphere of radius 20 cm is raised to a potential of 10^4 V. It is brought into contact with a neutral sphere of radius 30 cm, and then the two spheres are separated. (a) What is the original charge on the 20-cm-radius sphere? (b) What is the charge on each sphere after they are brought into contact? (c) What is the potential of each sphere after they are separated? (d) Find the charge density on each sphere (after they are brought into contact) and show that they are in the ratio $\sigma_1/\sigma_2 = R_2/R_1$.

28. To what potential can a small Van de Graaff sphere of radius 10.0 cm be raised before dielectric breakdown?

Problems

1. In the Bohr model of the hydrogen atom (Example 32-1) the electron moves in a circular orbit of radius r around the proton. (a) Find an expression for the kinetic energy of the electron as a function of r by setting the force on the electron (given by Coulomb's law) equal to ma, where a is the centripetal acceleration. Show that at any distance r, the kinetic energy is half the magnitude of the potential energy. (b) Evaluate $\frac{1}{2}mv^2$ and the total energy $E = \frac{1}{2}mv^2 + U$ for $r = 0.529 \times 10^{-10}$ m. (c) How much energy (in electron volts) must be supplied to the hydrogen atom to ionize it, i.e., to remove the electron to infinity with zero energy?

2. Show that Equation 32-17 for the potential on the axis of a disk charge gives the expected limit $V \rightarrow kq/x$ for x much greater than R, where $q = \pi R^2 \sigma$ is the total charge. *Hint:* Use the binomial expansion on the expression $(x^2 + R^2)^{1/2} = x(1 + R^2/x^2)^{1/2}$ and keep the first two terms.

3. Radioactive ^{210}Po emits alpha particles with energy 5.30 MeV. Assume that just after the alpha particle is formed and escapes from the nucleus, the alpha particle with charge $+2e$ is a distance R from the center of the daughter nucleus ^{206}Pb with charge $+82e$. Calculate R by setting the electrostatic potential energy of the two particles at this separation equal to 5.30 MeV.

4. When uranium ^{235}U captures a neutron, it splits into two nuclei (and emits several neutrons which can cause other uranium nuclei to split). Assume that the fission products are equally charged nuclei with charge $+46e$ and that these nuclei are at rest just after fission and separated by twice their radius $2R \approx 1.3 \times 10^{-14}$ m. (a) Using $U = kq_1q_2/2R$, calculate the electrostatic potential energy of the fission fragments. This is approximately the energy released per fission. (b) About how many fissions per second are needed to produce 1 MW of power in a reactor?

5. An electric dipole consists of a negative charge $-q$ at the origin and a posi-

tive charge $+q$ on the z axis at $z = a$ (Figure 32-13). (a) Find the potential on the z axis for $z > a$ and show that when z is much greater than a, $V \approx kp/z^2$, where $p = qa$ is the dipole moment. (b) Show that this result can be obtained by considering a single charge $+q$ at the origin and finding $V(z) - V(z + \Delta z) \approx -dV/dz \, \Delta z$, where $\Delta z = a$.

6. For the dipole of Problem 5, show that the potential at a point off axis a great distance r from the origin is given approximately by

$$V = \frac{kaq \cos \theta}{r^2} = \frac{kp \cos \theta}{r^2} = \frac{kpz}{r^3}$$

Hint: If r_1 is the distance from the positive charge to the field point, as in Figure 32-13, use the law of cosines to show that $r_1 \approx r[1 - (2a \cos \theta)/r]^{1/2}$ for $r \gg a$. Then use the binomial expansion to approximate $kq/r_1 = kqr_1^{-1}$.

7. Consider two infinite parallel sheets of charge, one in the yz plane and the other at distance $x = a$. (a) Find the potential everywhere in space with $V = 0$ at $x = 0$ if the sheets carry equal positive charge density $+\sigma$. (b) Do the same if the charge densities are equal and opposite with the sheet in the yz plane positive.

8. Two concentric spherical shell conductors carry equal and opposite charges. The inner shell has radius a and charge $+q$; the outer shell has radius b and charge $-q$. Find the potential difference between the shells $V_a - V_b$. *Hint:* First find the electric field between the shells using Gauss' law.

9. Two very long coaxial cylindrical shell conductors carry equal and opposite charges. The inner cylinder has radius a and charge $+q$; the outer shell has radius b and charge $-q$. The length of each cylinder is L. Find the potential difference between the shells. *Hint:* First use Gauss' law to find the electric field between the shells.

10. A hollow spherical conductor has inner radius a and outer radius b. A positive point charge $+q$ is at the center of the sphere, and the conductor is uncharged. Find the potential $V(r)$ everywhere, assuming that $V = 0$ at $r = \infty$, and sketch $V(r)$ versus r. *Hint:* Find the charges induced on the inner and outer surfaces of the conductor and use Gauss' law to find the electric field everywhere.

11. Two large parallel metal sheets carry equal and opposite charge densities of magnitude σ. They have area A and are separated by a small distance d. (a) Show that the potential difference between the sheets can be written $V = (q/\epsilon_0 A)d$, where q is the magnitude of the total charge on either sheet. (b) A third metal sheet of the same area and thickness a less than d is inserted between the original two sheets. (The third sheet has no net charge.) Show that the potential difference between the original two sheets is now $V = (q/\epsilon_0 A)(d - a)$.

12. In a charged spherical droplet of a nonconducting liquid, the charge will tend to be pushed toward the outer surface at radius R. A possible distribution of charge is $\rho = \rho_0 \, r/R$, where ρ_0 is the maximum charge density at the outer surface. (a) Show that the total charge in a spherical shell of radius r and thickness dr inside the drop is $dq = \rho_0 \, 4\pi r^3/R \, dr$. (b) Show that the total charge inside a sphere of radius r is $q = \rho_0 \, \pi r^4/R$ if $r < R$ and $Q = \pi R^3 \rho_0$ if $r > R$. (c) Use Gauss' law to find the electric field E_r everywhere, and use $dV = -E_r \, dr$ to find the potential $V(r)$ assuming that $V = 0$ at $r = \infty$. *Hint:* Remember to include the constant of integration when finding $V(r)$ inside the drop because V is not zero at $r = 0$. The constant can be found by matching the solution inside to the solution outside at $r = R$.

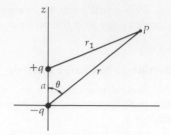

Figure 32-13
Electric dipole on the z axis for Problems 5 and 6. The law of cosines can be used to relate the distance r_1 to r and θ.

CHAPTER 33   Capacitance, Electrostatic Energy, and Dielectrics

The potential (relative to zero potential at infinity) of a single isolated conductor carrying a charge Q is proportional to the charge Q and depends on the size and shape of the conductor. In general, the larger the conductor, the greater the amount of charge it can carry for a given potential. For example, the potential of a spherical conductor or a spherical shell of radius R carrying a charge Q is

$$V = \frac{kQ}{R} = \frac{1}{4\pi\epsilon_0}\frac{Q}{R}$$

33-1

As this equation shows, the greater the radius, the greater the charge Q can be for a given potential. The ratio of the charge to the potential of an isolated conductor is called the *capacitance* of the conductor,

$$C = \frac{Q}{V}$$

33-2   *Capacitance defined*

Since the potential is always proportional to the charge, this ratio does not depend on either V or Q but only on the size and shape of the conductor. The capacitance of a spherical conductor is

$$C = \frac{Q}{V} = \frac{Q}{kQ/R} = \frac{R}{k} = 4\pi\epsilon_0 R$$

33-3

The capacitance of a spherical conductor is just proportional to its radius. The SI unit of capacitance is called the *farad* (F).

33-1 Capacitors

A system of two conductors carrying equal but opposite charges, called a *capacitor*, is a device for storing charge. In doing so it also stores energy, as we shall see later. The capacitance of a capacitor is defined to be the ratio of the magnitude of the charge (on either conductor) to the magnitude of the potential difference between the conductors. We can apply Equation 33-2 to a capacitor if we interpret V as the potential difference between the conductors and Q as the magnitude of the charge on either conductor. Since the potential difference between the conductors is always proportional to the charge in this case, the capacitance depends only on the size, shape, and geometrical arrangement of the conductors. Equation 33-3 illustrates a general result that capacitance can always be expressed as the product of ϵ_0 and a characteristic length. The SI unit of ϵ_0 can therefore be written as a farad per meter:

$$\epsilon_0 = \frac{1}{4\pi k} = 8.85 \times 10^{-12} \text{ F/m} \qquad\qquad 33\text{-}4$$

The smallness of ϵ_0 in SI units indicates that the farad is a very large unit. Because of this, submultiples are commonly used, e.g., the microfarad (1 μF $= 10^{-6}$ F) and the picofarad (1 pF $= 10^{-12}$ F). For example, the capacitance of a sphere of radius 1 m is only

$$C = 4\pi\epsilon_0 R = 4\pi(8.85 \times 10^{-12} \text{ F/m})(1 \text{ m}) = 1.11 \times 10^{-10} \text{ F}$$
$$= 111 \text{ pF}$$

Common capacitors consist of two closely spaced parallel plates or two concentric cylinders.

33-2   Calculation of Capacitance

The calculation of the capacitance of any capacitor is not difficult in principle. Given any two conductors, we place $+Q$ on one and $-Q$ on the other and find the potential difference between them. For simple geometries, e.g., the parallel-plate capacitor or the cylindrical capacitor, we can find the potential difference by first finding the electric field using either Gauss' or Coulomb's law, whichever is more convenient. The potential difference is then found by integrating the electric field along any path connecting the conductors, according to Equation 32-4. We shall illustrate this calculation for three simple arrangements of practical importance. The problem of finding the potential difference between two conductors in more complicated arrangements, such as two nearby (nonconcentric) spherical conductors, is a problem in intermediate or advanced electrostatics and will not be considered here.

Example 33-1   The Parallel-Plate Capacitor   A parallel-plate capacitor consists of two parallel conducting plates very close together. Let each plate have area A with charge $+Q$ on one plate and $-Q$ on the other. Each plate then has a charge density $\sigma = Q/A$, and the field will be nearly uniform between the plates, as indicated by the equally spaced field lines in Figure 33-1. The electric field in the region of space between the plates and far from the edges will essentially be that due

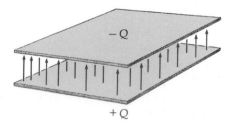

Figure 33-1
Parallel-plate capacitor. If the plates are closely spaced, the electric field is approximately uniform between the plates.

to two infinite plane sheets of charge. The field due to each sheet has the magnitude $\sigma/2\epsilon_0$. Outside the plates these fields cancel, but in the space between the plates they add and the electric field between the plates is

$$E = \frac{\sigma}{\epsilon_0} = \frac{Q}{\epsilon_0 A} \qquad\qquad 33\text{-}5$$

This result could also be obtained from Gauss' law applied to a pillbox gaussian surface with one face between the plates and the other inside one of the conductors. This result is only approximate because effects near the edges of the plates have been neglected. However, in practice, this approximation is quite good because the spacing between the plates is often very much less than the diameter of the plates (if plates are circular) or the length of the shortest edge (if they are rectangular). Since this field is constant in the region between the plates, the potential difference between the plates is just Ed, where d is the separation distance between the plates. Thus

$$V = Ed = \frac{Qd}{\epsilon_0 A}$$

The capacitance is

$$C = \frac{Q}{V} = \frac{Q}{Qd/\epsilon_0 A} = \frac{\epsilon_0 A}{d} \qquad\qquad 33\text{-}6$$

The capacitance is proportional to the area of the plates and inversely proportional to the separation distance.

Example 33-2 The Cylindrical Capacitor A cylindrical capacitor consists of a small cylinder or wire of radius a and a larger concentric cylindrical shell of radius b. We shall assume that the length L of these cylinders is much larger than a or b, so that we can neglect edge effects. Let $+Q$ be the charge on the inner conductor and $-Q$ be on the outer conductor. The electric field between the conductors is most easily found from Gauss' law. Let us consider a cylindrical gaussian surface of radius r and length L_1, as shown in Figure 33-2. According to Gauss' law, the flux through this surface equals q/ϵ_0, where q is the net charge inside the surface. If we assume that L is very large so that we can

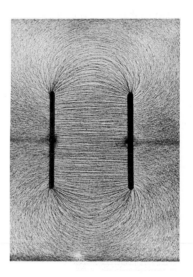

Electric field lines between plates of a parallel-plate capacitor shown by small bits of thread suspended in oil. When the plates are very close together, the fringing of the field near the edges can be neglected. (*Courtesy of Harold M. Waage, Princeton University.*)

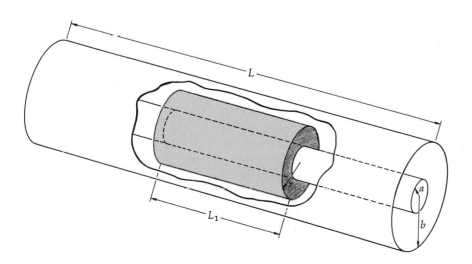

Figure 33-2
Cylindrical capacitor consisting of an inner wire of radius a and a concentric outer shell of radius b. The electric field is calculated from Gauss' law using a cylindrical gaussian surface of radius r and length L_1 between the conductors.

neglect edge effects, the electric field will be perpendicular to the axis of the conductors and depend only on the perpendicular distance from the axis of the conductors. Then there will be flux through only the cylindrical part of the gaussian surface; i.e., there will be no flux through the faces at each end of the surface. Since the electric field is constant in magnitude on the gaussian surface and perpendicular to this surface, the flux is just the magnitude E_r times the area, which is $2\pi r L_1$. The net charge inside the gaussian surface is $q = \lambda L_1 = QL_1/L$, where $\lambda = Q/L$ is the charge per unit length of the inner conductor. Thus Gauss' law gives for the electric field,

$$\phi_{\text{net}} = E_r 2\pi r L_1 = \frac{q}{\epsilon_0} = \frac{\lambda L_1}{\epsilon_0}$$

$$E_r = \frac{\lambda}{2\pi\epsilon_0 r} = \frac{Q}{2\pi\epsilon_0 L}\frac{1}{r} \qquad\qquad 33\text{-}7$$

This result is the same as that for an infinite line charge. The potential difference between the conductors is found from Equation 32-4. Let V_a be the potential of the inner conductor and V_b be that of the outer conductor. Then

$$V_b - V_a = -\int_a^b E_r\,dr = -\frac{Q}{2\pi\epsilon_0 L}\int_a^b \frac{dr}{r} = -\frac{Q}{2\pi\epsilon_0 L}\ln\frac{b}{a}$$

The potential is of course greater on the inner conductor, which carries the positive charge, since the electric field lines point from this conductor to the outer conductor. The magnitude of this potential difference is

$$V = V_a - V_b = \frac{Q\ln(b/a)}{2\pi\epsilon_0 L} \qquad\qquad 33\text{-}8$$

and the capacitance is

$$C = \frac{Q}{V} = \frac{2\pi\epsilon_0 L}{\ln(b/a)} \qquad\qquad 33\text{-}9$$

As expected, the capacitance is proportional to the length of the cylinders. The greater the length the greater the amount of charge that can be put on the conductors for a given potential difference since the electric field, and therefore the potential difference, depends only on the charge per unit length.

Figure 33-3
Spherical capacitor consisting of an inner sphere of radius R_1 and a concentric outer spherical shell of radius R_2. The electric field between the conductors is radial.

Example 33-3 The Spherical Capacitor A spherical capacitor consists of a small inner conducting sphere of radius R_1 and a larger concentric spherical shell of radius R_2 (Figure 33-3). The inner sphere can be supported on an insulator. A small opening is made in the outer shell so that charge can be placed on the inner sphere. If this opening is small enough, it will have a negligible effect on the spherical symmetry of the capacitor. To find the capacitance we place a charge $+Q$ on the inner sphere and $-Q$ on the outer sphere and find the potential difference. The electric field between the conductors is the same as that due to a point charge Q at the origin. As we have seen, the charge on the outer shell does not contribute to the electric field inside it. (We do not need to know the electric field in the regions $r < R_1$ or outside the outer shell $r > R_2$, but it is easy to show from Gauss' law that $\mathbf{E} = 0$ in both these regions.) The electric field between the conductors is thus

$$E_r = \frac{kQ}{r^2}$$

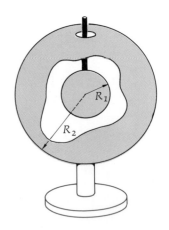

If V_1 is the potential on the inner sphere and V_2 on the outer sphere, the potential difference is

$$V_2 - V_1 = -\int_{R_1}^{R_2} E_r \, dr = -\int_{R_1}^{R_2} \frac{kQ}{r^2} \, dr = \frac{kQ}{R_2} - \frac{kQ}{R_1} = -\frac{kQ(R_2 - R_1)}{R_1 R_2}$$

The potential of the inner sphere V_1 is greater than that of the outer sphere, as evidenced by the fact that the electric lines point radially outward from the inner to outer sphere. The magnitude of the potential difference is

$$V = V_1 - V_2 = \frac{kQ(R_2 - R_1)}{R_1 R_2} \qquad \text{33-10}$$

The capacitance is

$$C = \frac{Q}{V} = \frac{R_1 R_2}{k(R_2 - R_1)} = 4\pi\epsilon_0 R_1 \frac{R_2}{R_2 - R_1} \qquad \text{33-11}$$

As the radius of the outer sphere approaches infinity, the capacitance approaches $4\pi\epsilon_0 R_1$, the capacitance of a single isolated spherical conductor. We can thus think of such a single conductor as a spherical capacitor with the second conductor at $R_2 = \infty$.

A variable parallel-plate capacitor. (*Courtesy of Larry Langrill, Oakland University.*)

33-3 Parallel and Series Combinations of Capacitors

Two or more capacitors are often used in combination. In electric circuits a capacitor is indicated by the symbol ⊣⊢. Figure 33-4 shows two capacitors which are connected in parallel. The upper plates of the two capacitors are connected together by a conducting wire and are therefore at the same potential in electrostatics. (We shall see in Chapter 34 that even when there is a current in a conducting connecting wire, the electric field in the wire is usually negligible, so that all points along the wire are essentially at the same potential.) The lower plates are also connected together and are at a common potential. It is clear that the effect of adding a second capacitor connected in this way is to increase the capacitance; i.e., the area is essentially increased, allowing more charge to be stored for the same potential difference $V = V_a - V_b$. If the capacitances are C_1 and C_2, the charges Q_1 and Q_2 stored on the plates are given by

$$Q_1 = C_1 V \qquad \text{and} \qquad Q_2 = C_2 V$$

where V is the potential difference across either capacitor. The total charge stored is thus

$$Q = Q_1 + Q_2 = C_1 V + C_2 V = (C_1 + C_2)V \qquad \text{33-12}$$

The effective capacitance of two capacitors in parallel is defined to be the ratio of the total charge stored to the potential Q/V. Thus

$$C_{\text{eff}} = \frac{Q}{V} = C_1 + C_2 \qquad \text{33-13}$$

The effective capacitance is that of a single capacitor which could replace the parallel combination and store the same amount of charge for a given potential difference V. This reasoning can be extended to three or more capacitors connected in parallel, as in Figure 33-5. The

Figure 33-4
Two capacitors in parallel. The potential difference across the capacitors is the same, $V_a - V_b$, for each capacitor.

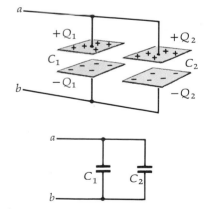

Parallel capacitors

effective capacitance is just the sum of the individual capacitances.

In Figure 33-6 two capacitors C_1 and C_2 are connected in series with a potential difference $V = V_a - V_b$ between the upper plate of the first capacitor and the lower plate of the second. Such a situation can be realized in practice by connecting points a and b to the terminals of a battery. If a charge $+Q$ is placed on the upper plate of the first capacitor, there will be an equal negative charge $-Q$ induced on its lower plate. This charge comes from electrons drawn from the upper plate of the second capacitor. Thus there will be an equal charge $+Q$ on the upper plate of the second capacitor and $-Q$ on its lower plate. The potential difference across the upper capacitor is $V_a - V_c = Q/C_1$. Similarly the potential difference across the second capacitor is $V_c - V_b = Q/C_2$. The potential difference across the two capacitors in series is just the sum of these potential differences:

$$V_a - V_b = (V_a - V_c) + (V_c - V_b) = \frac{Q}{C_1} + \frac{Q}{C_2}$$

Calling this potential difference V, we have

$$V = Q\left(\frac{1}{C_1} + \frac{1}{C_2}\right) = \frac{Q}{C_{\text{eff}}} \qquad \text{33-14}$$

where $C_{\text{eff}} = Q/V$ is the ratio of the charge to the total potential difference across the two capacitors connected in series. From Equation 33-14 we have

$$\frac{1}{C_{\text{eff}}} = \frac{1}{C_1} + \frac{1}{C_2} \qquad \text{33-15}$$

Series capacitors

or

$$C_{\text{eff}} = \frac{C_1 C_2}{C_1 + C_2} \qquad \text{33-16}$$

The effective capacitance of two capacitors in series is less than that of either of the individual capacitors. For example, in Equation 33-16, C_{eff} can be considered the product of C_1 and $C_2/(C_1 + C_2)$, which is less than 1. Equation 33-15 can be generalized to three or more capacitors in series:

$$\frac{1}{C_{\text{eff}}} = \frac{1}{C_1} + \frac{1}{C_2} + \frac{1}{C_3} + \cdots \qquad \text{33-17}$$

33-4   Electrostatic Energy in a Capacitor

We can charge a capacitor by transferring charge from one conductor to the other. In this process the potential of the transferred charge is increased. Thus work must be done to charge a capacitor. Some of this work (or all of it, depending on the charging process) is stored as potential energy. (If the capacitor is charged by connecting it to a battery, for example, the battery does twice as much work as is stored as potential energy. Half the work done by the battery is wasted as heat in the connecting wires and the battery itself. We shall discuss this in detail in Chapter 35.)

Let us consider the charging of a parallel-plate capacitor. Since only the potential *difference* between the plates is important, we are free to choose the potential to be zero at any point. It is convenient to choose

Figure 33-5
Three capacitors in parallel. The effect of adding a parallel capacitor is to increase the effective capacitance.

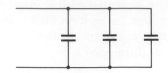

Figure 33-6
Two capacitors in series. The potential difference $V_a - V_b$ is the sum of the potential differences across the capacitors. The charge on each capacitor is the same.

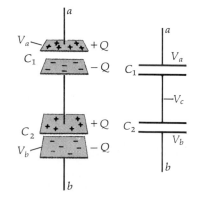

the potential of the negative plate to be zero. At the beginning of the charging process, neither plate is charged. There is no electric field, and both plates are at the same potential. After the charging process a charge Q_0 has been transferred from one plate to the other, and the potential difference is $V_0 = Q_0/C$, where C is the capacitance. We than have a negative charge $-Q_0$ on one plate, which we have chosen to be at zero potential, and $+Q_0$ on the other plate at potential V_0. One might expect that the work needed to accomplish this would be just the charge Q_0 times the potential energy per unit charge V_0, but only the last bit of charge must be raised by the full potential difference V_0. The potential difference between the plates increases from 0 originally to its final value V_0. The average value of the potential difference during the charging process is just $\frac{1}{2}V_0$, and the work needed is $\frac{1}{2}Q_0V_0$. We can see this as follows.

Let q be the charge which has been transferred at some time during the process. The potential difference is then $V = q/C$. If a small amount of charge dq is now transferred from the plate with charge $-q$ at zero potential to the plate with charge q at potential V, its potential energy is increased by

$$dU = V\, dq = \frac{q}{C}\, dq$$

The total increase in potential energy in charging from $q = 0$ to $q = Q_0$ is the energy stored in the capacitor (Figure 33-7).

$$U = \int dU = \int_0^{Q_0} \frac{q}{C}\, dq = \frac{1}{C} \int_0^{Q_0} q\, dq = \frac{1}{2} \frac{Q_0^2}{C}$$

Using $C = Q_0/V_0$, we can write this in a variety of other ways:

$$U = \frac{1}{2} \frac{Q_0^2}{C} = \tfrac{1}{2}Q_0V_0 = \tfrac{1}{2}CV_0^2 \qquad\qquad 33\text{-}18$$

Energy of a charged capacitor

The derivation of Equation 33-18 was carried out for the parallel-plate capacitor, but a review of the steps will show that the geometry of the capacitor played no role in the argument. The expression $C = \epsilon_0 A/d$ was never used. The argument is applicable to any capacitor, and Equation 33-18 is a general expression for the energy stored in a charged capacitor as electrostatic potential energy.

Questions

1. The potential difference of a capacitor is doubled. By what factor does its stored electric energy change?

2. Half the charge is removed from a capacitor. What fraction of its stored energy was removed along with the charge?

33-5   Electrostatic Field Energy

In the process of charging a capacitor, an electric field is created. For example, in a parallel-plate capacitor, there is no electric field when the plates are uncharged. When they have their final charge Q_0, there is a field $E_0 = \sigma/\epsilon_0 = Q_0/\epsilon_0 A$, where A is the area of the plates. We can think of the work done in charging the capacitor in an alternative way

Figure 33-7
The work needed to charge a capacitor is the integral of $V\, dq$ from the original charge $q = 0$ to the final charge $q = Q_0$. This work is the area $\frac{1}{2}Q_0V_0$ under the curve.

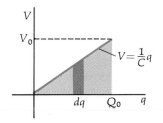

as being the work needed to create an electric field. Let us consider again the charging of a parallel-plate capacitor. When one plate has charge $+q$ and the other $-q$, the field between the plates is

$$E = \frac{\sigma}{\epsilon_0} = \frac{q}{\epsilon_0 A} \qquad\qquad 33\text{-}19$$

The work needed to transfer an additional charge dq is force $dq\,E$ times the distance d, since the field (and therefore the force) is constant. This is the work done to push dq against the field from $-q$ to $+q$,

$$dW = dq\,Ed \qquad\qquad 33\text{-}20$$

From Equation 33-19 we have

$$dE = \frac{dq}{\epsilon_0 A}$$

Substituting $dq = \epsilon_0 A\,dE$ into Equation 33-20, we have

$$dW = (\epsilon_0 Ad)E\,dE \qquad\qquad 33\text{-}21$$

The work done to increase the field from $E = 0$ to $E = E_0 = Q_0/\epsilon_0 A$ is thus

$$W = \epsilon_0 Ad \int_0^{E_0} E\,dE = \tfrac{1}{2}\epsilon_0 E_0{}^2 Ad$$

This work appears as electrostatic potential energy U:

$$U = \tfrac{1}{2}\epsilon_0 E_0{}^2 Ad \qquad\qquad 33\text{-}22$$

Electrostatic potential energy

This result could be obtained directly from Equation 33-18 using $Q_0 = \epsilon_0 AE_0$ and $C = \epsilon_0 A/d$. It is convenient to think of the energy as being stored in the electric field. Note that, neglecting edge effects, the field is constant in the region between the plates and zero outside. The volume $\mathscr{V}$ between the plates is Ad. The energy per unit volume in the field, called the energy density η, is thus

$$\eta = \frac{U}{\mathscr{V}} = \tfrac{1}{2}\epsilon_0 E^2 \qquad\qquad 33\text{-}23$$

Energy density in an electric field

Although we have been considering the simple case of a parallel-plate capacitor, the result (Equation 33-23) for the energy per unit volume in our electrostatic field is generally valid. We shall illustrate the generality of Equation 33-23 by using it to calculate the energy in an electrostatic field which is not constant in space.

Example 33-4 Energy of a Single Isolated Spherical Conductor of Radius R Carrying a Charge Q We noted in Example 33-3 that a single spherical conductor can be thought of as a spherical capacitor with the outer shell at infinity, where the potential is chosen to be zero. The potential of this conductor is kQ/R, and its capacitance is $C = 4\pi\epsilon_0 R = R/k$, where k is the coulomb constant. According to Equation 33-18, the electrostatic energy is

$$U = \tfrac{1}{2}QV = \tfrac{1}{2}Q\,\frac{kQ}{R} = \frac{1}{2}\frac{kQ^2}{R} \qquad\qquad 33\text{-}24$$

A direct derivation of this result by considering the work done in bringing charge from infinity and placing it on the conductor is given as an exercise. We shall give here a derivation using the result that the energy per unit volume in an electric field is $\tfrac{1}{2}\epsilon_0 E^2$, as given in Equation 33-23. At distance r from the center of the sphere the electric field is

$$E_r = \frac{kQ}{r^2}$$

The energy per unit volume is then

$$\eta = \tfrac{1}{2}\epsilon_0 E^2 = \tfrac{1}{2}\epsilon_0 \frac{k^2 Q^2}{r^4} \qquad\qquad 33\text{-}25$$

Consider a spherical shell volume element of radius r, thickness dr, and volume $d\mathcal{V} = 4\pi r^2\, dr$ (Figure 33-8). The energy in this volume element is

$$dU = \eta\, d\mathcal{V} = \tfrac{1}{2}\epsilon_0 k^2 \frac{Q^2}{r^4}\, 4\pi r^2\, dr$$

$$= 2\pi\epsilon_0 k^2 \frac{Q^2}{r^2}\, dr = \tfrac{1}{2}k \frac{Q^2}{r^2}\, dr \qquad\qquad 33\text{-}26$$

using $k = 1/4\pi\epsilon_0$. We calculate the total energy by integrating from $r = R$ to $r = \infty$ since $E = 0$ for $r < R$. Thus

$$U = \int_R^\infty \tfrac{1}{2}k \frac{Q^2}{r^2}\, dr = \tfrac{1}{2}k \frac{Q^2}{R}$$

in agreement with Equation 33-24.

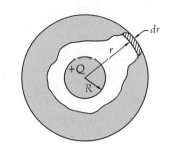

Figure 33-8
Calculation of the electrostatic energy of a spherical conductor with charge Q.

33-6 Dielectrics

A nonconducting material, e.g., glass or wood, is called a *dielectric*. Faraday discovered that when the space between the two conductors of a capacitor is occupied by a dielectric, the capacitance is increased. If the space (between the plates of a parallel-plate capacitor, for example) is completely filled by the dielectric, the capacitance increases by a factor K which is characteristic of the dielectric and called the *dielectric constant*.

Dielectric constant

   Suppose a capacitor of capacitance C_0 is connected to a battery which charges it to a potential difference V_0 by placing a charge $Q_0 = C_0 V_0$ on the plates. If the battery is now disconnected and a dielectric is inserted, filling the space between the plates, the potential difference decreases to a new value,

$$V = \frac{V_0}{K} \qquad\qquad 33\text{-}27$$

Since the original charge Q_0 is still on the plates, the new capacitance is

$$C = \frac{Q_0}{V} = \frac{K Q_0}{V_0} = K C_0 \qquad\qquad 33\text{-}28$$

   If, on the other hand, the dielectric is inserted while the battery is still connected, the battery must supply more charge to maintain the original potential difference. The total charge on the plates is then $Q = K Q_0$. In either case, the capacitance is increased by the factor K.

   Since the potential difference between the plates of a parallel-plate capacitor is just the electric field between the plates times the separation d, the effect of the dielectric (with the battery disconnected) is to

 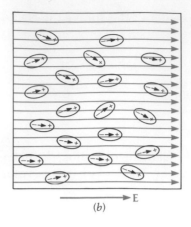

(a)   (b)

Figure 33-9
(a) Randomly oriented electric
dipoles in the absence of an
external electric field. (b) In
the presence of an external
field the dipoles are partially
aligned parallel to the field.

decrease the electric field by the factor K. If E_0 is the original field
without the dielectric, the new field E is

$$E = \frac{E_0}{K} \qquad\qquad 33\text{-}29$$

*Dielectric reduces electric
field*

We can understand this result in terms of molecular polarization of
the dielectric. If the molecules of the dielectric are polar molecules, i.e.,
have permanent dipole moments, these moments are originally randomly oriented. In the presence of the field between the capacitor
plates, these dipole moments experience a torque, which tends to align
them in the direction of the field (Figure 33-9). The amount of alignment depends on the strength of the field and the temperature. At
high temperatures, the random thermal motion of the molecules tends
to counteract the alignment. In any case, the alignment of the molecular dipoles produces an additional electric field due to the dipoles
which is in the direction opposite the original field. The original field
is thus weakened. Even if the molecules of the dielectric are nonpolar,
they will have induced dipole moments in the presence of the electric
field between the plates. The induced dipole moments are in the direction of the original field. Again, the additional electric field due to
these induced moments weakens the original field.

A dielectric which has electric dipole moments predominently in
the direction of the external field is said to be *polarized* by the field,
whether the polarization is due to alignment of permanent dipole
moments of polar molecules or to the creation of induced dipole
moments in nonpolar molecules. The net effect of the polarization of a
homogeneous dielectric is the creation of a surface charge on the
dielectric faces near the plates. Figure 33-10 shows a rectangular slab of
homogeneous dielectric which is in a uniform electric field to the
right. The molecular dipole moments are indicated schematically. As
can be seen from the figure, the net effect of the polarization is to
produce a positive surface charge density on the right face and a negative surface charge density on the left face.[1] The charge densities on
the dielectric faces are due to the displacement of positive and negative molecular charges near the faces. This displacement is due to the
external electric field of the capacitor. The charge on the dielectric,

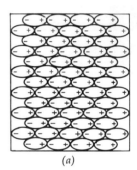

(a)

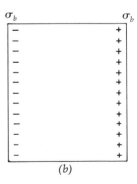

(b)

Figure 33-10
The net effect of alignment of
electric dipole moments distributed uniformly in a volume (a) is a positive bound-
surface-charge density on one
side of the dielectric and a
negative bound-surface-
charge density on the other
side, as shown in (b). The result is to weaken the electric
field in the dielectric.

[1] If the dielectric is not homogeneous or the field is not uniform, there may be a volume-
charge density created inside the dielectric because the amount of positive charge
pushed out the right side of some volume element may not equal the negative charge
pushed out the left side of the element. We shall neglect such complications.

called *bound charge,* is not free to move about like the ordinary free charge on the conducting capacitor plates. Although it disappears when the external electric field disappears, it produces an electric field just like any other charge. We shall now relate the bound-charge density σ_b to the dielectric constant K and to the free-charge density σ_f on the capacitor plates.

Bound charge on dielectric

The electric field inside the dielectric slab due to the bound-charge densities $+\sigma_b$ on the right and $-\sigma_b$ on the left is just the field due to two infinite-plane-charge densities (assuming that the slab is very thin; i.e., the plates of the capacitor are close together). The field E' thus has the magnitude

$$E' = \frac{\sigma_b}{\epsilon_0} \qquad\qquad 33\text{-}30$$

This field is to the left and subtracts from the electric field due to the ordinary free-charge density on the capacitor plates. The original field E_0 has the magnitude

$$E_0 = \frac{\sigma_f}{\epsilon_0} \qquad\qquad 33\text{-}31$$

The magnitude of the resultant field E is just the difference of these magnitudes. It also equals E_0/K

$$E = E_0 - E' = \frac{E_0}{K} \qquad\qquad 33\text{-}32$$

or

$$E' = E_0 \left(1 - \frac{1}{K}\right) = \frac{K-1}{K} E_0 \qquad\qquad 33\text{-}33$$

Writing σ_b/ϵ_0 for E' and σ_f/ϵ_0 for E_0, we have

$$\sigma_b = \frac{K-1}{K}\, \sigma_f \qquad\qquad 33\text{-}34$$

Bound-charge density related to free-charge density

The bound-charge density σ_b is always less than the free-charge density σ_f on the capacitor plates and is zero if $K = 1$, which is the case of no dielectric. Figure 33-11 shows the lines of force for a parallel-plate capacitor with a dielectric of constant $K = 2$ which does not completely fill the space between the plates. According to Equation 33-34, the bound-charge density on the dielectric is half the free-charge density on the plates. Half the lines that leave the positive charge on the lower plate end on the negative bound charge on the dielectric. Similarly, half of the lines which end on the negative charge on the upper plate begin on the positive bound charge on the dielectric. The density of field lines in this diagram shows that the electric field inside the dielectric is just half that outside. The effect of the dielectric on the capacitance depends on how much space is filled.

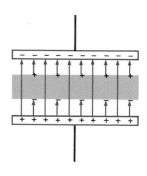

Figure 33-11
Electric field lines for a dielectric of constant $K = 2$ between the plates of a parallel-plate capacitor. The electric field is reduced to half its original strength, as indicated by the fact that half the original field lines end on one side of the dielectric and begin again on the other side.

Example 33-5 A slab of material of dielectric constant K has the same area as the plates of a parallel-plate capacitor but has a thickness $\frac{3}{4}d$, where d is the separation of the plates. How is the capacitance changed when the slab is inserted between the plates?

Let $E_0 = V_0/d$ be the electric field between the plates when there is no dielectric and the potential difference is V_0. If the dielectric is now inserted, the electric field in the dielectric will be $E = E_0/K$. The potential difference will then be

$$V = E_0(\tfrac{1}{4}d) + \frac{E_0}{K}\left(\frac{3}{4}\,d\right) = E_0 d\left(\frac{1}{4} + \frac{3}{4K}\right) = V_0\,\frac{K+3}{4K}$$

The potential difference decreases by the factor $(K+3)/4K$ while the free charge Q_0 on the plates remains unchanged. The capacitance thus increases:

$$C = \frac{Q_0}{V} = \frac{4K}{K+3}\,\frac{Q_0}{V_0} = \frac{4K}{K+3}\,C_0$$

In addition to increasing the capacitance, a dielectric has two other functions in a capacitor: (1) it provides a mechanical means of separating the two conductors, which must be very close together in order to obtain a large capacitance; (2) the dielectric strength is increased because the dielectric strength of a dielectric is usually greater than that of air. We have already mentioned that the dielectric strength of air is 3×10^6 V/m = 3 kV/mm. Fields greater than this magnitude cannot be maintained in air because of dielectric breakdown; i.e., the air becomes ionized and conducts. Many materials have dielectric strengths greater than that of air, allowing greater potential differences between the conductors of a capacitor.

Three functions of a dielectric

An example of these three dielectric functions is a parallel-plate capacitor made from two sheets of metal foil of large area (to increase the capacitance) separated by a sheet of paper. The paper increases the capacitance because of its polarization; that is, K is greater than 1. It also provides a mechanical separation so that the sheets can be very close together without being in electrical contact. (A small separation is important because the capacitance varies inversely with separation.) Finally, the dielectric strength of paper is greater than that of air, so that greater potential differences can be attained without breakdown. Table 33-1 lists the dielectric constant and dielectric strength of some dielectrics.

Table 33-1
Dielectric constant and strength of various materials

Material	Dielectric constant K	Dielectric strength, kV/mm
Air	1.00059	3
Bakelite	4.9	24
Glass (Pyrex)	5.6	14
Mica	5.4	10–100
Neoprene	6.9	12
Paper	3.7	16
Paraffin	2.1–2.5	10
Plexiglas	3.4	40
Polystyrene	2.55	24
Porcelain	7	5.7
Transformer oil	2.24	12
Water (20°C)	80	—

Electrostatics and Xerography

Richard Zallen
Xerox Research Laboratories, Webster, N.Y.

There are many important and beneficial technological applications which could be included in a discussion of uses of electrostatic phenomena. For example, a powerful air-pollution preventor is the electrostatic precipitator, which years ago made life livable near cement mills and ore-processing plants and which is currently credited with extracting better than 99 percent of the ash and dust from the gases about to issue from chimneys of coal-burning power plants. The basic idea of this very effective antipollution technique is shown in Figure 1. The outer wall of a vertical metal duct is grounded, while a wire running down the center of the duct is kept at a very large negative voltage. In this concentric geometry (which corresponds to the cylindrical capacitor in Example 33-2) a very nonuniform electric field is set up, with lines of force directed radially inward toward the negative wire electrode. Close to the wire the field attains enormous values, large enough to produce an electrical breakdown of air, and the normal placid mixture of neutral gas molecules is replaced by a turmoil of free electrons and positive ions. The electrons from this corona discharge are driven outward from the wire by the electric field. Most of them quickly become attached to oxygen molecules to produce negative O_2^- ions, which are also accelerated outward. As this stream of ions passes across the hot waste gas rising in the duct, small particles carried by the gas become charged by capturing ions and are pulled by the field to the outer wall. If the noxious particles are solid, they are periodically shaken down off the duct into a hopper; if they are liquid, the residue simply runs down the wall and is collected below.

Besides electrostatic precipitation, other technological examples include electrocoating with spray paints and the electrostatic separation of granular mixtures used for the removal of rock particles from minerals, garlic seeds from wheat, even rodent excreta from rice. However, the application which is

Figure 1
Schematic diagram of the use of a corona discharge in an electrostatic precipitator.

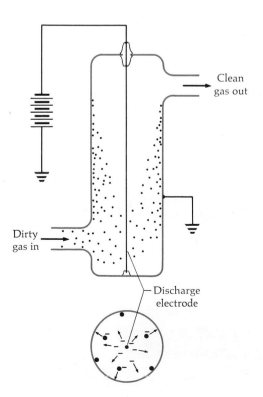

Clean
gas out

Dirty
gas in

Discharge
electrode

the main focus of this essay is xerography, the most widely used form of electrostatic imaging, or electrophotography. This is the most familiar use of electrostatics in terms of the number of people who have occasion to use plain-paper copying machines in offices, libraries, and schools, and it also provides a fine example of a process utilizing a sequence of distinct electrostatic events.

The xerographic process was invented in 1937 by Chester Carlson. The term xerography, literally "dry writing," was actually adopted a bit later to emphasize the distinction from wet chemical processes. Carlson's innovative concept did not find early acceptance, and a practical realization of his idea became available only after a small company (in a famous entrepreneurial success story) risked its future in its intensive efforts to develop the process.

Four of the main steps involved in xerography are illustrated in Figure 2. In the interest of clarity the process has been oversimplified, and several subtleties (as well as gaps in our understanding) have been suppressed. Electrostatic imaging takes place on a large thin plate of a photoconducting material supported by a grounded metal backing. A photoconductor is a solid which is a good insulator *in the dark* but which becomes capable of conducting electric current when exposed to light. The unilluminated, insulating state is indicated by shading in Figure 2. In the dark, a uniform electrostatic charge is laid down on the surface of the photoconductor. This charging step (Figure 2a) is accomplished by means of a positive corona discharge surrounding a fine wire held at about +5000 V. This corona (a miniature version of, and opposite in sign to, the intense precipitator corona of Figure 1) is passed over the photoconductor surface, spraying positive ions onto it and charging it to a potential of the order of +1000 V. Since charge is free to flow within the grounded metal backing, an equal and opposite induced charge (see Chapter 31) develops at the metal-photoconductor interface. In the dark the photoconductor contains no mobile charge, and the large potential difference persists across this dielectric layer, which is only 0.005 cm thick.

The photoconductor plate is next exposed to light in the form of an image reflected from the document being copied. What happens now is indicated in Figure 2b. Where light strikes the photoconductor, light quanta (photons) are absorbed, and pairs of mobile charges are created. Each photogenerated pair consists of a negative charge (an electron) and a positive charge (a hole; crudely, a missing electron). Photogeneration of this free charge depends not only on the photoconductor used and on the wavelength and intensity of the incident light but also on the electric field present. This large field (1000 V/0.005 cm $= 2 \times 10^5$ V/cm $= 2 \times 10^7$ V/m) helps to pull apart the mutually attracting electron-hole pairs so that they are free to move separately. The electrons then move under the influence of the field to the surface, where they neutralize positive charges, while the holes move to the photo-

Figure 2
Steps in the xerographic process: (a) charging, (b) exposure, (c) development, and (d) transfer.

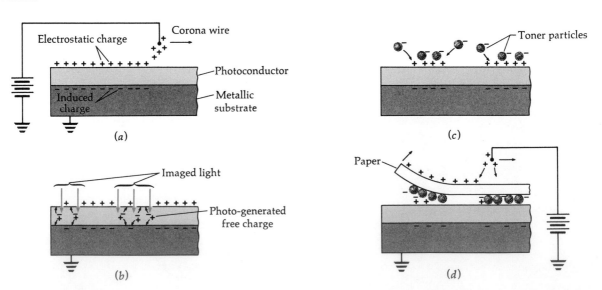

conductor-substrate interface and neutralize negative charges there. Where intense light strikes the photoconductor, the charging step is totally undone; where weak light strikes it, the charge is partially reduced; and where no light strikes it, the original electrostatic charge remains on the surface. The critical task of converting an optical image into an electrostatic image, which is now recorded on the plate, has been completed. This latent image consists of an electrostatic potential distribution which replicates the light and dark pattern of the original document.

To develop the electrostatic image, fine negatively charged pigmented particles are brought into contact with the plate. These *toner particles* are attracted to positively charged surface regions, as shown in Figure 2c, and a visible image appears. The toner is then transferred (Figure 2d) to a sheet of paper which has been positively charged in order to attract them. Brief heating of the paper fuses the toner to it and produces a permanent photocopy ready for use.

Finally, to prepare the photoconductor plate for a repetition of the process, any toner particles remaining on its surface are mechanically cleaned off, and the residual electrostatic image is erased, i.e., discharged, by flooding with light. The photoconductor is now ready for a new cycle, starting with the charging step. In high-speed duplicators the photoconductor layer is often in the form of a moving continuous drum or belt, around the perimeter of which are located stations for performing the various functions of Figure 2. The speed of xerographic printing technology is presently on the order of a few copies per second.[1]

---

[1] For further information on electrostatics in xerography, consult J. H. Dessauer and H. E. Clark (eds.), *Xerography and Related Processes,* Focal Press, New York, 1965; R. M. Schaffert, *Electrophotography,* rev. ed., Focal Press, New York, 1973; and recent technical articles appearing in the *1974 IEEE Conference Record of the 9th Annual Meeting of the IEEE-Industry Applications Society.* Other modern applications of electrostatics are discussed in A. D. Moore, *Scientific American,* March 1972.

Optional

Figure 33-12
Small cylindrical volume of area A and length L containing N molecular dipoles each of moment p_m. The total dipole moment is Np_m, which also equals $\sigma_b AL$. The bound-charge density is therefore equal to the dipole moment per unit volume.

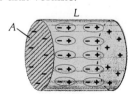

Bound-charge density related to molecular dipole moment

33-7 Molecular Polarizability

The dielectric constant of a material can be related to a molecular property called the *polarizability*. Thus a simple macroscopic measurement of the dielectric constant can be used to gain some information about the microscopic molecular structure of the material.

Consider a small volume of area A and length L (Figure 33-12) containing N molecules. Let p_m be the average dipole moment of a molecule. The total dipole moment in this volume is then

$$p = Np_m$$

We can think of this volume as containing a bound charge $\sigma_b A$ on each face separated by distance L. Then the total dipole moment is

$$p = (\sigma_b A)L = Np_m \qquad 33\text{-}35$$

and the bound-charge density is related to the average molecular dipole moment by

$$\sigma_b = \frac{Np_m}{AL} = \frac{N}{\mathscr{V}} p_m = np_m \qquad 33\text{-}36$$

where $n = N/\mathscr{V}$ is the number of molecules per unit volume. (The dipole moment per unit volume $n\mathbf{p}_m$ is often called the *polarization* $\mathbf{P}$ of the dielectric.)

The average dipole moment of a molecule is related to the electric field which produces the polarization of the dielectric. In many cases it is just proportional to this field:

$$p_m = \alpha E \qquad \text{33-37}$$

Polarizability defined

The proportionality constant α is called the *molecular polarizability.* Then the bound-charge density can be written

$$\sigma_b = n\alpha E \qquad \text{33-38}$$

But according to Equation 33-33 we have

$$E' = \frac{K-1}{K} E_0 = \frac{\sigma_b}{\epsilon_0}$$

or

$$\sigma_b = \epsilon_0 (K-1) \frac{E_0}{K} = \epsilon_0 (K-1)E \qquad \text{33-39}$$

since $E_0/K = E$. Combining Equations 33-39 and 33-38, we obtain

$$n\,\alpha E = \epsilon_0 (K-1)E$$

or

$$\alpha = \frac{\epsilon_0 (K-1)}{n} \qquad \text{33-40}$$

Equation 33-40 relates a microscopic property of molecules, the polarizability, to a macroscopic measurable quantity, the dielectric constant K.[1]

Bohr-model calculation of polarizability

We can get an idea of the order of magnitude of the polarizability for nonpolar molecules by using some simple atomic models. Figure 33-13a illustrates the Bohr model of the hydrogen atom, in which an electron moves about the proton in a circular orbit of radius $r = 0.53$ Å. The centripetal force necessary to keep the electron moving in a circle is provided by the electrostatic attraction of the proton and electron given by Coulomb's law:

$$F_r = \frac{1}{4\pi\epsilon_0} \frac{(+e)(-e)}{r^2} = -m\frac{v^2}{r}$$

In the presence of an external electric field, the orbit is changed. Figure 33-13b shows an electric field perpendicular to the original plane of motion of the electron. The electron now moves in a circle of the same radius but about a center which is displaced a distance x from the proton. Since the center of positive charge (the proton) and the center of negative charge (the center of the electron's orbit) are now displaced, the atom now has a dipole moment of magnitude ex. We can calculate x by setting the external force on the electron eE equal to the component of the force of attraction of the proton perpendicular to the plane

---

[1] In a more precise treatment we should distinguish between the macroscopic electric field **E** and the microscopic field seen by a single molecule $\mathbf{E}_m$, which is somewhat larger than **E**. The polarizability is rigorously defined by $\mathbf{p}_m = \alpha\mathbf{E}_m$. This leads to a slightly different result:

$$\alpha = \frac{\epsilon_0}{n} (K-1) \frac{3}{K+2}$$

This relation is used chiefly for gases, for which $K + 2 \approx 3$, so that Equation 33-40 is a good approximation relating K to α.

Figure 33-13
(a) In the Bohr model of the hydrogen atom the electron orbits the positive proton at radius r with the electrostatic force of attraction providing the centripetal force. (b) In the presence of an external electric field perpendicular to the plane of the orbit, the plane is displaced a small distance x, and the atom has an induced dipole moment parallel to the field of magnitude ex.

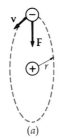

(a)

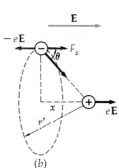

(b)

of the orbit. This component is

$$F_x = \frac{1}{4\pi\epsilon_0} \frac{e^2}{r'^2} \cos\theta = \frac{1}{4\pi\epsilon_0} \frac{e^2}{r'^2} \frac{x}{r'}$$

For typical external electric fields the distance x is much smaller than the orbit radius r. Then the new separation distance r' is nearly equal to the original orbit radius r. Using this approximation and equating these forces on the electron gives

$$\frac{1}{4\pi\epsilon_0} \frac{e^2 x}{r^3} = eE$$

or

$$ex = (4\pi\epsilon_0 r^3)E \qquad\qquad 33\text{-}41$$

The left side of this equation, ex, is the dipole moment of the hydrogen atom. The factor multiplying E on the right side is the polarizability. Thus

$$p = (4\pi\epsilon_0 r^3)E = \alpha E$$

$$\alpha = 4\pi\epsilon_0 r^3 \qquad\qquad 33\text{-}42$$

Example 33-6 A hydrogen atom is in an external field of $E = 10^5$ N/C. What is the magnitude of the induced dipole moment?

The radius of the electron orbit is $r = 0.53$ Å $= 5.3 \times 10^{-11}$ m. The polarizability, according to Equation 33-42, is

$$\alpha = 4\pi\epsilon_0 r^3 = \frac{(5.3 \times 10^{-11} \text{ m})^3}{9 \times 10^9 \text{ N-m}^2/\text{C}^2} = 1.65 \times 10^{-41} \text{ m-C}^2/\text{N}$$

For $E = 10^5$ N/C, the induced dipole moment is

$$p = \alpha E = (1.65 \times 10^{-41} \text{ m-C}^2/\text{N})(10^5 \text{ N/C})$$

$$= 1.65 \times 10^{-36} \text{ C-m}$$

This is more conveniently expressed in electron angstroms:

$$p = 1.65 \times 10^{-36} \text{ C-m} \frac{e}{1.6 \times 10^{-19} \text{ C}} \frac{10^{10} \text{ Å}}{1 \text{ m}} = 1.0 \times 10^{-7} \text{ eÅ}$$

We note that the induced dipole moment is small compared with the permanent dipole moment of the polar NaCl molecule, which has the magnitude of about 2 eÅ.

Review

A. Define, explain, or otherwise identify:

B. True or false:

1. The capacitance of a conductor is defined to be the total amount of charge it can hold.

2. The capacitance of a parallel-plate capacitor depends on the voltage difference between the plates.

3. The capacitance of a parallel-plate capacitor is proportional to the charge on the plates.

4. The effective capacitance of two capacitors in parallel equals the sum of the individual capacitances.

5. The effective capacitance of two capacitors in series is less than that of either capacitor.

6. Since V is energy per charge, the total energy in a capacitor is QV.

7. The electrostatic energy per unit volume at some point is proportional to the square of the electric field at that point.

Exercises

Section 33-1, Capacitors, and Section 33-2, Calculation of Capacitance

1. What is the capacitance of an isolated spherical conductor of radius (a) 1.8 cm, (b) 1.8 m, (c) 1.8 km?

2. Considering the earth to be a spherical conductor of radius 6400 km, calculate its capacitance.

3. (a) Find the radius of an isolated spherical conductor which has a capacitance of 1 F. (b) What is the ratio of the radius of this sphere to the radius of the earth?

4. Calculate the capacitance of a parallel-plate capacitor of plate area 1.0 m² and separation 1.0 cm.

5. If a parallel-plate capacitor has 0.1-mm separation, what must its area be to have a capacitance of 1 F? If the plates are square, what is the length of their sides?

6. A parallel-plate capacitor has capacitance of 2.0 μF and plate separation of 1.0 mm. (a) How much potential difference can be placed across the capacitor before dielectric breakdown of air occurs ($E_{max} = 3 \times 10^6$ V/m)? (b) What is the magnitude of the greatest charge the capacitor can store before breakdown?

7. A coaxial cable between two cities has inner radius 1.0 cm and outer radius 1.1 cm. Its length is 8×10^5 m (about 500 mi). Treat this cable as a cylindrical capacitor and calculate its capacitance.

8. A Geiger tube consists of a wire of radius 0.2 mm and length 12 cm with a coaxial cylindrical conductor of the same length and radius 1.5 cm. (a) Find the capacitance assuming that the gas in the tube has a dielectric constant of 1. (b) Find the charge per unit length on the wire when the capacitor is charged to 1200 V.

Section 33-3, Parallel and Series Combinations of Capacitors

9. A 10.0-μF capacitor is connected in series with a 20.0-μF capacitor across a 6.0-V battery. (a) What is the equivalent capacitance of this combination? (b) Find the charge on each capacitor. (c) Find the potential difference across each capacitor.

10. A 10.0-μF capacitor and a 20.0-μF capacitor are connected in parallel across a 6.0-V battery. (a) What is the equivalent capacitance of this combination? (b) What is the potential difference across each capacitor? (c) Find the charge on each capacitor.

11. Three capacitors have capacitance 2.0, 4.0, and 8.0 μF. Find the equivalent capacitance (a) if the capacitors are in parallel and (b) if they are in series.

12. A 2.0-μF capacitor is charged to a potential difference of 12.0 V and disconnected from the battery. (*a*) How much charge is on the plates? (*b*) When a second capacitor (initially uncharged) is connected in parallel across this capacitor, the potential difference drops to 4.0 V. What is the capacitance of the second capacitor?

13. (*a*) How many 1.0-μF capacitors would have to be connected in parallel to store 10^{-3} C of charge with a potential difference of 10 V across each? (*b*) What would be the potential difference across the combination? (*c*) If these capacitors are connected in series and the potential difference across each is 10 V, find the charge on each and the potential difference across the combination.

14. A 1.0-μF capacitor is connected in parallel with a 2.0-μF capacitor, and the combination is connected in series with a 6.0-μF capacitor. What is the equivalent capacitance of this combination?

15. A 3.0-μF capacitor and a 6.0-μF capacitor are connected in series and the combination is connected in parallel with an 8.0-μF capacitor. What is the equivalent capacitance of this combination?

Section 33-4, Electrostatic Energy in a Capacitor

16. (*a*) A 3-μF capacitor is charged to 100 V. How much energy is stored in the capacitor? (*b*) How much additional energy is required to charge the capacitor from 100 to 200 V?

17. How much work is needed to charge an isolated spherical conductor of radius 10 cm to 3000 V?

18. (*a*) A 10-μF capacitor is charged to $Q = 4\ \mu$C. How much energy is stored? (*b*) If half the charge is removed, how much energy remains?

19. (*a*) Find the energy stored in a 20-pF capacitor when it is charged to 5 μC. (*b*) How much additional energy is required to increase the charge from 5 to 10 μC?

20. A conducting sphere of radius R carries charge q. (*a*) Show that the work needed to bring additional charge dq from infinity to the sphere is $(kq/R)\ dq$. (*b*) Use this result to show that the work needed to increase the charge from 0 to Q is $\frac{1}{2}(kQ^2/R)$.

21. A parallel-plate capacitor of area A and separation d is charged to a potential difference V and then disconnected from the charging source. The plates are then pulled apart until the separation is $2d$. Find expressions in terms of A, d, and V for (*a*) the new capacitance, (*b*) the new potential difference, and (*c*) the new stored energy. (*d*) How much work was required to change the plate separation from d to $2d$?

Section 33-5, Electrostatic Field Energy

22. Find the energy per unit volume in an electric field equal to the breakdown field in air 3×10^6 V/m.

23. A parallel-plate capacitor has plate area 2 m² and separation 1.0 mm. It is charged to 100 V. (*a*) What is the electric field between the plates? (*b*) What is the energy per unit volume in the space between the plates? (*c*) Find the total energy by multiplying your answer to part (*b*) by the total volume between the plates. (*d*) Find the capacitance C, and calculate the total energy from $U = \frac{1}{2}CV^2$, and compare with your answer to part (*c*).

Section 33-6, Dielectrics

24. A parallel-plate capacitor is made by placing polyethylene $(K = 2.3)$

between sheets of aluminum foil. The area of each sheet is 400 cm², and the spacing is 0.3 mm. Find the capacitance.

25. A parallel-plate capacitor has plates of area 600 cm² and separation 4 mm. It is charged to 100 V and disconnected from the battery. (a) Find the electric field E, the charge density σ, and the energy U. A dielectric of constant $K = 4$ is inserted, completely filling the space between the plates. (b) Find the new electric field E and the potential difference V. (c) Find the new energy. (d) Find the bound-charge density.

26. What is the dielectric constant of a dielectric on which the induced bound-charge density is (a) 80 percent of the free-charge density, (b) 20 percent of the free-charge density, and (c) 98 percent of the free-charge density?

Section 33-7, Molecular Polarizability

There are no exercises for this section.

Problems

1. A spherical capacitor consists of two concentric spherical shells of radii R_1 and R_2. Show that when these radii are nearly equal, the capacitance is given approximately by the expression for the capacitance of a parallel-plate capacitor $C = \epsilon_0 A/d$, where A is the area of the sphere and $d = R_2 - R_1$.

2. The effective capacitance of two capacitors in series can be written $C = C_1C_2/(C_1 + C_2)$. Show that the correct expression for the effective capacitance of three capacitors in series is $C = C_1C_2C_3/(C_1C_2 + C_2C_3 + C_1C_3)$.

3. Three identical capacitors are connected so that their maximum equivalent capacitance is 15 μF. Find the three other combinations possible (using all three capacitors) and their equivalent capacitances.

4. A spherical capacitor has an inner sphere of radius R_1 with charge $+Q$ and an outer concentric spherical shell of radius R_2 with charge $-Q$. (a) Find the electric field and the energy density at any point in space. (b) How much energy is in the volume of the spherical shell of radius r, thickness dr, and volume $4\pi r^2\, dr$ between the conductors? (c) Integrate your expression in part (b) to find the total energy stored in the capacitor and compare your result with that obtained from $U = \frac{1}{2}QV$.

5. A cylindrical capacitor consists of a long wire of radius R_1 and length L with positive charge Q and a concentric outer cylindrical shell of radius R_2, length L, and charge $-Q$. (a) Find the electric field and the energy density at any point in space. (b) How much energy is in the cylindrical shell of radius r, thickness dr, and volume $2\pi rL\, dr$ between the conductors? (c) Integrate your expression in part (b) to find the total energy stored in the capacitor and compare your result with that obtained from $U = \frac{1}{2}QV$.

6. Find all the different possible effective capacitances that can be obtained using a 1.0-, a 2.0-, and a 4.0-μF capacitor in any combination which includes all three or any two capacitors.

7. A parallel-plate capacitor has capacitance C_0 and plate separation d. Two dielectric slabs of constants K_1 and K_2, each of thickness $\frac{1}{2}d$ and of the same area as the plates, are inserted between the plates as indicated in Figure 33-14. Show that the capacitance is then given by

$$C = \frac{2K_1K_2}{K_1 + K_2}\, C_0$$

8. A parallel-plate capacitor of area A and separation d is charged to potential difference V and removed from the charging source. A dielectric slab of constant $K = 2$, thickness d, and area $\frac{1}{2}A$ is inserted as shown in Figure 33-15. Let

Figure 33-14
Capacitor with two dielectrics for Problem 7.

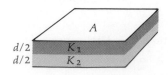

σ_1 be the free-charge density at the conductor-dielectric surface and σ_2 be the charge density at the conductor-vacuum surface. (a) Why must the electric field have the same value inside the dielectric as in the free space between the plates? (b) Show that $\sigma_1 = 2\sigma_2$. (c) Show that the new capacitance is $3\epsilon_0 A/2d$ and the new potential difference is $\frac{2}{3}V$.

9. In Figure 33-16, $C_1 = 2\ \mu F$, $C_2 = 6\ \mu F$, and $C_3 = 3.5\ \mu F$. (a) Find the equivalent capacitance of this combination. (b) If the breakdown voltages of the individual capacitors are $V_1 = 100$ V, $V_2 = 50$ V, and $V_3 = 400$ V, what maximum voltage can be placed across points a and b?

10. A parallel-plate capacitor is filled with two dielectrics of equal size, as shown in Figure 33-17. Show that the capacitance is increased by the factor $(K_1 + K_2)/2$.

11. A parallel-plate capacitor has plate area A and separation d. A metal slab of thickness t and area A is inserted between the plates. Show that the capacitance is given by $C = \epsilon_0 A/(d - t)$ regardless of where the metal plate is placed. Show that this arrangement can be considered to be a capacitor of separation a in series with one of separation b, where $a + b + t = d$.

12. A parallel-plate capacitor of plate area A and separation x is given a charge Q and removed from the charging source. (a) Find the electrostatic energy stored as a function of x. (b) Find the increase in energy dU due to an increase in plate separation dx from $dU = (dU/dx)\ dx$. (c) If F is the force exerted by one plate on the other, the work done to move one plate a distance dx is $F\ dx = dU$. Show that $F = Q^2/2\epsilon_0 A$. (d) Show that the force found in part (c) equals $\frac{1}{2}EQ$, where Q is the charge on one plate and E is the electric field between the plates. Discuss the reason for the factor $\frac{1}{2}$ in this result.

13. Design a circuit of capacitors which has a capacitance of $2\ \mu F$ and breakdown voltage of 400 V using as many $2\text{-}\mu F$ capacitors as needed, each with a breakdown voltage of 100 V.

14. A parallel-plate capacitor of plate area 1.0 m² and plate separation distance 0.5 cm has a glass plate of the same area and thickness between its plates. The glass has dielectric constant 5.0. The capacitor is charged to a potential difference of 12.0 V and removed from its charging source. How much work is required to pull the glass plate out of the capacitor?

15. Early capacitors, called Leyden jars, were actually glass jars coated inside and outside with metal foil. Suppose that the jar is a cylinder 40 cm high with 2.0 mm thick walls of inner diameter 8 cm. Ignore any field fringing. (a) Find the capacitance of this jar. (b) What maximum charge can it take without collapse? The dielectric constant of the glass is 5.0, and its dielectric strength is 15×10^6 V/m.

16. A ball of charge of radius R has a uniform charge density ρ and total charge $Q = \frac{4}{3}\pi R^3\rho$. (a) Find the electrostatic energy density at distance r from the center of the charge for $r < R$ and for $r > R$. (b) Find the energy in a spherical shell of volume $4\pi r^2\ dr$ for both $r < R$ and $r > R$. (c) Compute the total electrostatic energy by integrating your expressions in part (b) and show that your result can be written $U = \frac{3}{5}kQ^2/R$. Explain why this result is greater than that for a spherical conductor of radius R carrying a total charge Q.

17. In a simplified model of the hydrogen atom, the proton is a point charge, but the electron is a uniform ball of charge of radius R, the radius of the atom (about 0.53 Å). When there is no external electric field, the proton is at the center of the negative charge and the dipole moment is zero. In the presence of an external electric field E, the center of the electron charge and the proton are separated by a small distance x, which is less than R. (a) Show that the force exerted by the electron charge on the proton is ke^2x/R^3. (b) Show that the induced dipole moment is $(R^3/k)E = 4\pi\epsilon_0 R^3 E$ and compare this result with that obtained from the Bohr-model calculation in Section 33-7.

Figure 33-15
Problem 8.

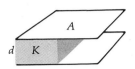

Figure 33-16
Problem 9.

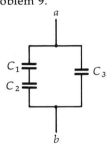

Figure 33-17
Problem 10.

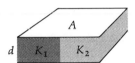

CHAPTER 34   Electric Current

An electric current is a flow of charge past some point. Although the flow of charge often takes place inside a conductor, this is not necessary; e.g., a beam of charged ions in vacuum from an accelerator constitutes a current. We define electric current as follows. Consider a small area element A, as in Figure 34-1, which might be the cross-sectional area of a conducting wire but need not be. The current through the area is defined as the amount of charge flowing through the area per unit time. If ΔQ is the charge that flows through the area in time Δt, the current is

$$I = \frac{\Delta Q}{\Delta t}$$

34-1   *Electric current defined*

The SI unit of current is the ampere (A):

$$1\ \mathrm{A} = 1\ \mathrm{C/sec}$$

34-2

The direction of current is taken to be the direction of flow of positive charge. If the current is due to particles such as electrons which have negative charge, the direction of current is in the direction opposite the flow of the negatively charged particles. This definition of the direction of the current is purely arbitrary. In a conducting wire it is electrons which are free to move and produce the flow of charge. By our definition, the current is in the direction opposite to the motion of the electrons.

There are many examples of electric current outside of conducting wires. For example, a beam of protons from an accelerator produces a current in the direction of the motion of the positively charged protons. In electrolysis, the current is produced by the motion of both electrons and positive ions. Since these particles move in opposite directions, both produce current in the same direction. In nearly all applications, the motion of negative charges to the left is indistinguishable from the motion of positive charges to the right. (An exception is the Hall effect, Chapter 36.) We can always think of current as motion of positive charges in the direction of the current and re-

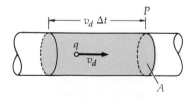

Figure 34-1
The current through area A at point P is the rate of flow of charge through that area. In time Δt all the charge in the shaded volume $Av_d\ \Delta t$ flows past point P, where v_d is the speed of the carriers. If there are n carriers per unit volume each with charge q, the total charge is $\Delta Q = nqv_dA\ \Delta t$ and the current is $I = \Delta Q/\Delta t = nqv_dA$.

member (if we need to) that in conducting wires, for example, the electrons are moving in the direction opposite to the current.

34-1 Current and Motion of Charges

An electric current can be related to the motion of the charged particles responsible for it. Let us first consider the case of current in a conducting wire of cross-sectional area A. Let n be the number of free charge-carrying particles per unit volume. For a conducting wire, these charge carriers are the free electrons. We shall assume that each particle carries a charge q and moves with velocity v_d. In a time Δt all the particles in the volume $Av_d \, \Delta t$, shaded in Figure 34-1, pass through the area element at point P. The number of particles in this volume is $nAv_d \, \Delta t$, and the total charge is

$$\Delta Q = qnAv_d \, \Delta t$$

The current at point P is thus

$$I = \frac{\Delta Q}{\Delta t} = nqv_d A \qquad\qquad 34\text{-}3$$

The current per unit area is the *current density J*:

$$J = \frac{I}{A} = nqv_d \qquad\qquad 34\text{-}4$$

We can generalize the idea of current density to apply to any type of current whether confined to a wire or not. We define the current-density vector $\mathbf{J}$ by

$$\mathbf{J} = nq\mathbf{v}_d \qquad\qquad 34\text{-}5 \qquad \text{\textit{Current density}}$$

The current-density vector is in the direction of $\mathbf{v}_d$ if the charge q is positive and in the opposite direction if q is negative. The quantity $\mathbf{v}_d$ is the average velocity of the charge carriers. For electrons in conductors this velocity is called the *drift velocity*. The current density is a fundamental quantity related by Equation 34-5 to the number density of charge carriers n, the charge q, and the average velocity of the carriers $\mathbf{v}_d$. If the current is due to particles with different densities, charges, or velocities (as in electrolysis), the current density is found by summing $n_i q_i (\mathbf{v}_d)_i$ over the particles:

$$\mathbf{J} = \sum_i n_i q_i (\mathbf{v}_d)_i \qquad\qquad 34\text{-}6$$

If $\mathbf{J}$ is constant over the area A, the current through A is just

$$I = \mathbf{J} \cdot \hat{\mathbf{n}} A = J_n A \qquad\qquad 34\text{-}7$$

where $\hat{\mathbf{n}}$ is the unit vector perpendicular to the plane of the area A and J_n is the component of the current density parallel to $\hat{\mathbf{n}}$. If the current density is not constant, the current through a surface is found by integration:

$$I = \int \mathbf{J} \cdot \hat{\mathbf{n}} \, dA \qquad\qquad 34\text{-}8 \qquad \text{\textit{Current is flux of the current density}}$$

The current is the flux of the current density through the surface.

We can get an idea of the order of magnitude of the drift velocity for electrons in a conducting wire by putting typical magnitudes into Equation 34-4.

Example 34-1 What is the drift velocity of electrons in a typical copper wire (14 gauge) of radius 0.0814 cm carrying a current of 1 A?

If we assume one free electron per copper atom, the density of free electrons is the same as the density of atoms. Then

$$n = \frac{(6.02 \times 10^{23} \text{ atoms/mole})(8.92 \text{ gm/cm}^3)}{63.5 \text{ gm/mole}}$$

$$= 8.46 \times 10^{22} \text{ atoms/cm}^3$$

and

$$v_d = \frac{I}{Ane} = \frac{1 \text{ C/sec}}{\pi (0.0814 \text{ cm})^2 \, (8.46 \times 10^{22}/\text{cm}^3) \, (1.60 \times 10^{-19} \text{ C})}$$

$$\approx 3.55 \times 10^{-3} \text{ cm/sec}$$

We see that typical drift velocities are very small.

The actual instantaneous velocity of an electron in a metal is much larger than the drift velocity, which is the average velocity. The behavior of electrons in a metal is similar to that of gas molecules in air. In still air, the gas molecules move with large instantaneous velocities between collisions, but the average velocity is zero. When there is a breeze, the air molecules have a small drift velocity in the direction of the breeze superimposed on the much larger instantaneous velocity. We shall discuss this point further in Section 34-3 when we consider the classical model of electric conduction in metals.

Question

1. Two wires of the same diameter but different materials are joined together to carry the same electric current. In one wire the density of charge carriers is twice that in the other. How do the drift velocities in the two wires compare?

Georg Simon Ohm (1787–1854).

34-2 Ohm's Law and Resistance

In our study of conductors in electrostatics we argued that the electric field inside a conductor must be zero in electrostatic equilibrium. If this were not so, the free charges inside the conductor would move about. We are now considering nonelectrostatic equilibrium situations, in which the free charge *does* move in a conductor. When a conductor carries a current, there is an electric field inside the conductor. In many conductors, the current density **J** is proportional to the electric field in the conductor which produces the current. The ratio of the magnitude of the current density and the magnitude of the electric field is called the *conductivity* σ of the conductor.[1]

$$\sigma = \frac{J}{E} \quad \text{or} \quad \mathbf{J} = \sigma \mathbf{E} \qquad\qquad 34\text{-}9$$

Conductivity defined

Equation 34-9 defines the conductivity σ for any material. If the conductivity so defined does not depend on the electric field, the material

[1] Care must be taken to distinguish between the conductivity σ and a surface charge density, also designated by σ.

is said to follow *Ohm's law*. Ohm's law is a statement of an experimental result that for many materials

The current density is proportional to the electric field; i.e., the ratio of the magnitudes of the current density and electric field is independent of the field.

This ratio, the conductivity, may depend on the temperature and on the composition of the material, but for materials following Ohm's law, it does not depend on the electric field. This result holds over a wide range of electric fields in many materials, including most metals. Such materials are called *ohmic*. For many other nonohmic materials the conductivity defined by Equation 34-9 does depend on the electric field. For nonohmic materials, the current density is not proportional to the electric field. Ohm's law is not a fundamental law of nature like Newton's laws or the laws of thermodynamics but an empirical description of a property shared by many materials.

For current in a wire, we can write Ohm's law in a more familiar form in terms of the voltage drop across a segment of the wire. Figure 34-2 shows a segment of wire of length L and cross-sectional area A carrying a current I. We choose the segment short enough to ensure that the electric field does not vary appreciably over the distance L. If the electric field is directed from point a to point b, the potential is lower at point b than at point a by the amount

$$V = V_a - V_b = EL$$

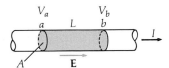

Figure 34-2
Segment of wire carrying current $I = JA = \sigma EA$. The voltage drop from a to b is $V = V_a - V_b = EL = IL/\sigma A$.

where E is the electric field. The current in the wire is the current density times the cross-sectional area:

$$I = JA = \sigma EA = \sigma A\, \frac{V}{L}$$

or

$$V = \frac{L}{\sigma A}\, I \qquad\qquad 34\text{-}10$$

The quantity $V/I = L/\sigma A$ is called the *resistance R* of the wire segment.

Resistance defined

$$R = \frac{V}{I} = \frac{L}{\sigma A} \qquad\qquad 34\text{-}11$$

The unit of resistance is the ohm (Ω), defined by

$$1\ \Omega = 1\ \text{V/A} \qquad\qquad 34\text{-}12$$

An equivalent statement of Ohm's law is:

Resistance is independent of voltage and current.

For ohmic materials, the voltage drop across a segment of wire is proportional to the current in the wire:

$$V = IR \qquad\qquad 34\text{-}13$$

In nonohmic materials, the resistance defined by Equation 34-11 depends on V and I, and V does not vary linearly with I. Figure 34-3 shows plots of V versus I for ohmic and nonohmic materials.

We see from Equation 34-11 that the resistance of a wire is proportional to the length of the wire and inversely proportional to the cross-

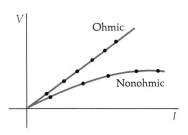

Figure 34-3
V versus I for ohmic and nonohmic materials. The resistance $R = V/I$ is independent of I for ohmic materials, as indicated by the constant slope of the line.

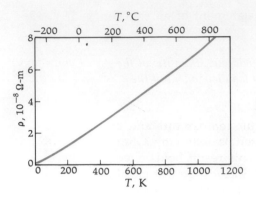

Figure 34-4
Plot of the resistivity ρ versus
absolute temperature T for
copper. (*By permission from
D. Halliday and R. Resnick,
Fundamentals of Physics, p.
512, John Wiley & Sons, Inc.,
New York, 1974.*)

sectional area. The proportionality constant is called *resistivity* ρ.[1] The
resistivity is the reciprocal of the conductivity,

$$\rho = \frac{1}{\sigma} \qquad\qquad 34\text{-}14$$

Resistivity defined

In terms of the resistivity, the resistance is written

$$R = \frac{\rho L}{A} \qquad\qquad 34\text{-}15$$

The unit of resistivity is the ohm-meter (Ω-m). The resistivity of any
given metal depends on the temperature. Figure 34-4 shows this dependence for copper. Except at very low temperatures the resistivity
varies nearly linearly with temperature. The resistivity is often given
in tables in terms of its value ρ_{20} at 20°C and the temperature coefficient of resistivity α, which is the slope of the ρ-versus-T curve. The
resistivity ρ at some other Celsius temperature t is then given by

*Temperature coefficient of
resistivity*

$$\rho = \rho_{20}\,[1 + \alpha\,(t - 20°C)] \qquad\qquad 34\text{-}16$$

(Since the Celsius and absolute temperatures differ only in the choice
of zero, the resistivity has the same slope whether plotted against t or
T.) The resistivity ρ and temperature coefficient α are listed in Table
34-1 for various materials. This table shows that there is a wide range
of values for the resistivity and an enormous difference between conductors and insulators.

Wires used to carry electric current are manufactured in standard
sizes. The diameter of the circular cross section is indicated by a gauge
number; higher numbers correspond to smaller diameters. For example, the diameter of a 10-gauge copper wire is 2.588 mm and that of
14-gauge wire is 1.628 mm. Handbooks give the combination ρ/A in
ohms per centimeter or ohms per foot.

Example 34-2 Calculate ρ/A in ohms per foot for 14-gauge copper wire.
From Table 34-1 we have for the resistivity of copper,

$$\rho = 1.7 \times 10^{-6}\ \Omega\text{-cm}$$

The cross-sectional area of 14-gauge wire is

$$A = \frac{\pi d^2}{4} = \frac{\pi}{4}\,(0.163\ \text{cm})^2 = 2.09 \times 10^{-2}\ \text{cm}^2$$

[1] The use of the same symbol ρ for resistivity, mass density, and charge density can
cause confusion, but this notation is standard usage.

Table 34-1
Resistivities and temperature coefficients

Material	Resisitivity ρ at 20°C, Ω-m	Temperature coefficient α at 20°C, per °C
Silver	1.6×10^{-8}	3.8×10^{-3}
Copper	1.7×10^{-8}	3.9×10^{-3}
Aluminum	2.8×10^{-8}	3.9×10^{-3}
Tungsten	5.5×10^{-8}	4.5×10^{-3}
Iron	10×10^{-8}	5.0×10^{-3}
Lead	22×10^{-8}	4.3×10^{-3}
Mercury	96×10^{-8}	0.9×10^{-3}
Nichrome	100×10^{-8}	0.4×10^{-3}
Carbon	3500×10^{-8}	-0.5×10^{-3}
Germanium	0.45	-48×10^{-3}
Silicon	640	-75×10^{-3}
Wood	$10^{8}-10^{14}$	
Glass	$10^{10}-10^{14}$	
Hard rubber	$10^{13}-10^{16}$	
Amber	5×10^{14}	
Sulfur	10^{15}	

Thus

$$\frac{\rho}{A} = \frac{1.7 \times 10^{-6} \ \Omega\text{-cm}}{2.09 \times 10^{-2} \ \text{cm}^2}$$

$$= (8.13 \times 10^{-5} \ \Omega/\text{cm}) \ \frac{2.54 \times 12 \ \text{cm}}{1 \ \text{ft}} = 2.48 \times 10^{-3} \ \Omega/\text{ft}$$

This example shows that the copper connecting wires used in the laboratory have a very small resistance.

Example 34-3 What is the electric field in a 14-gauge copper wire that carries a current of 1 A?

According to Example 34-2, the resistance of a 1-m length of 14-gauge copper wire is

$$R = \frac{\rho}{A} L = (8.13 \times 10^{-5} \ \Omega/\text{cm}) \ (100 \ \text{cm}) = 8.13 \times 10^{-3} \ \Omega$$

The voltage drop across 1 m of this wire is

$$V = IR = 8.13 \times 10^{-3} \ \text{V}$$

and the electric field is

$$E = \frac{V}{L} = 8.13 \times 10^{-3} \ \text{V/m}$$

Note that the electric field in a conducting wire is very small.

There are many metals for which the resistivity is zero below a certain temperature T_c, called the *critical temperature*. This phenomenon, called *superconductivity*, was discovered in 1911 by the Dutch physicist H. Kamerlingh Onnes. Figure 34-5 shows his plot of the resistance of mercury versus temperature. The critical temperature for mercury is 4.2 K. (Critical temperatures for other superconductors range from 1.2 K for aluminum to 9.2 K for niobium.) The conductivity of a superconductor cannot be defined since there can be a finite current density even when the electric field in the superconductor is zero. Steady currents have been observed to persist in superconducting rings for years with no electric field and no apparent loss. The phenomenon of superconductivity can be understood only with the aid of quantum mechanics, which is beyond the scope of this book. The first successful theory of superconductivity was published by Bardeen, Cooper, and Schrieffer in 1957 and is known as the BCS theory.

Questions

2. Wire a and wire b have the same electric resistance and are made of the same material. Wire a has twice the diameter of wire b. How do the lengths of the wires compare?

3. If wires a and b in Question 2 carry the same currents, how do their current densities compare? The voltage drops across them? The electric fields inside the wires?

4. In our study of electrostatics we concluded that there is no electric field within the material of a conductor. Why do we now find it possible to discuss electric fields inside conducting material?

Superconductivity

Figure 34-5
Plot by Kamerlingh Onnes of resistance of mercury versus temperature showing sudden decrease at the critical temperature $T = 4.2$ K. (*By permission from C. Kittel*, Introduction to Solid State Physics, *3d ed., John Wiley & Sons, Inc., New York, 1966.*)

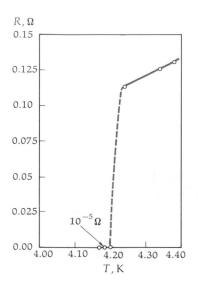

34-3 Classical Model of Electric Conduction

A classical model of electric conduction was first proposed by Drude in 1900 and developed by Lorentz about 1909. This model successfully predicts Ohm's law and relates the conductivity and resistivity to the motion of free electrons in conductors. This classical theory is useful in understanding conduction even though it has been replaced by a modern theory based on quantum mechanics. We present the classical theory of electric conduction in this section and then discuss how some of the parameters are reinterpreted in the modern theory in Section 34-4.

In the classical model of electric conduction, a metal or other conductor is pictured as a regular three-dimensional array of atoms or ions with a large number of electrons free to move about the whole metal. In copper, for example, there is approximately one free electron per copper atom. The number of free electrons per atom can be measured using the Hall effect (Chapter 36). A copper atom with one electron missing is a copper ion with a positive charge $+e$. The regular arrangement of copper ions in metallic copper is called a *lattice*. In the absence of an electric field, the free electrons move about the metal much like gas molecules in a container. The free electrons make collisions with the ions of the lattice and are in thermal equilibrium with it. The rms speed of the electrons can be calculated from the equipartition theorem. The result is the same as that for an ideal-gas molecule

Demonstration of persistent currents. Oppositely directed superconducting current loops are set up in the ring and ball so that the magnetic force between the currents is repulsive. The ball floats above the ring, its weight balanced by the magnetic force of repulsion. (*Courtesy of Cryogenic Technology, Inc.*)

with the electron mass replacing the molecular mass in Equation 17-25:

$$v_{\text{rms}} = \sqrt{\frac{3kT}{m}} \qquad\qquad 34\text{-}17$$

For example, at temperature $T = 300$ K the rms speed is

$$v_{\text{rms}} = \sqrt{\frac{3(1.38 \times 10^{-23} \text{ J/K})(300 \text{ K})}{9.11 \times 10^{-31} \text{ kg}}}$$

$$= 1.17 \times 10^5 \text{ m/sec}$$

This speed is much larger than the drift velocity calculated in Example 34-1.

At first glance it is surprising that any material obeys Ohm's law. In the presence of an electric field, a free electron experiences a force $q\mathbf{E}$ which would cause an acceleration $q\mathbf{E}/m$ if it were the only force acting on the electron. But Ohm's law implies a steady drift velocity (not acceleration) of the electron which is proportional to the electric field. The current density $J = nqv_d$ is proportional to the drift velocity; according to Ohm's law, J is proportional to the electric field E. Of course there are other forces acting on the electrons because they make collisions with the lattice ions. In the classical model of conduction, it is assumed that after an electron makes a collision with a lattice ion, its velocity is completely unrelated to that before the collision. The justification of this assumption is that the drift velocity is very small compared with the random thermal velocity of the electron. We thus picture the electron as being accelerated by the electric field for a short time between collisions with the lattice ions and thereby acquiring a small drift velocity. After each collision, which gives it a random velocity, the electron is again accelerated by the field. The motion of such an electron is shown in Figure 34-6, in which the drift velocity is greatly exaggerated. In Figure 34-7b a small drift velocity $\mathbf{v}_d$ has been added to the electron velocities, which in Figure 34-7a are random. We can relate the drift velocity to the electric field by ignoring the random thermal velocities of the electrons and assuming that the electron starts from rest after a collision. After a time τ the electron will have a velocity of magnitude equal to the acceleration qE/m times the time

$$v = \frac{qE}{m}\tau$$

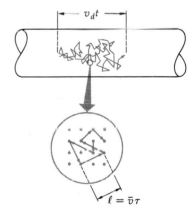

Figure 34-6
Electron path in a wire. Superimposed on the random thermal motion is a slow drift in the direction of the electric force $q\mathbf{E}$. The mean free path ℓ, mean time between collisions τ, and mean speed $\bar{v}$ are related by $\ell = \bar{v}\tau$.

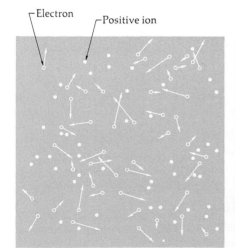
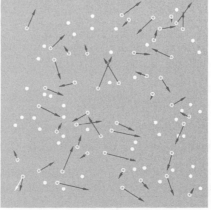

(a) Average electron velocity = 0   (b) Average electron velocity = ➝

Figure 34-7
(a) Stationary positive ions and electrons with a random distribution of velocities. (b) A small drift velocity to the right has been added to each electron velocity, as indicated for the electron in the lower left corner. (*Adapted from* E. M. Purcell, Electricity and Magnetism, *p. 123, Berkeley Physics Course, McGraw-Hill Book Company, New York, 1965. Courtesy of Education Development Center, Inc., Newton, Mass.*)

The average velocity of the electron during this time is just half the final velocity. The distance that the electron drifts in the direction opposite to the electric field is this average velocity times the time between collisions τ. This average velocity is thus the drift velocity

$$v_d = \frac{1}{2} \frac{qE}{m} \tau \tag{34-18}$$

Using this expression for the drift velocity in Equation 34-5 for the current density, we have

$$J = nqv_d = nq \frac{\frac{1}{2}qE}{m} \tau = \frac{nq^2\tau}{2m} E \tag{34-19}$$

Comparing this result with the definition of electric conductivity and its reciprocal, the resistivity, we have

$$\sigma = \frac{nq^2\tau}{2m} \quad \text{and} \quad \rho = \frac{2m}{nq^2\tau} \tag{34-20}$$

The average distance the electron travels between collisions is called the *mean free path* ℓ. It is the product of the mean speed $\bar{v}$ and the mean time between collisions τ:

Mean free path

$$\ell = \bar{v}\tau \tag{34-21}$$

In terms of the mean free path and the mean speed, the conductivity and resistivity are

$$\sigma = \frac{nq^2\ell}{2m\bar{v}} \quad \rho = \frac{2m\bar{v}}{nq^2\ell} \tag{34-22}$$

According to Ohm's law, the conductivity and resistivity are independent of the electric field E. The quantities in Equation 34-22 that might depend on the electric field are the mean speed $\bar{v}$ and the mean free path ℓ. As we have seen, the drift velocity is very much smaller than the rms speed of the electrons in thermal equilibrium with the lattice ions. Thus the electric field has essentially no effect on the mean speed of the electrons. The mean free path of the electron depends on the chance that the electron in traveling some distance will collide with a lattice ion. According to the classical model, this chance depends on the size of the lattice ion and on the density of ions, neither of which depends on the electric field E. Thus this model predicts Ohm's law with the resistivity and conductivity given by Equations 34-22.

Example 34-4 Use Equation 34-18 and the results of Examples 34-1 and 34-3 to estimate the mean time between collisions for electrons in 14-gauge copper wire when the current is 1 A.

From these examples we have $v_d = 3.55 \times 10^{-3}$ cm/sec $= 3.55 \times 10^{-5}$ m/sec, and $E \approx 8.13 \times 10^{-3}$ V/m. Using $q = 1.60 \times 10^{-19}$ C and $m = 9.11 \times 10^{-31}$ kg for the charge and mass of the electron, we have

$$\tau = \frac{2mv_d}{qE} = \frac{2(9.11 \times 10^{-31} \text{ kg}) (3.55 \times 10^{-5} \text{ m/sec})}{(1.60 \times 10^{-19} \text{ C}) (8.13 \times 10^{-3} \text{ V/m})} = 4.97 \times 10^{-14} \text{ sec}$$

The time between collisions is very short.

Example 34-5 Assuming that the mean speed is about 10^5 m/sec, the same order of magnitude as the rms speed calculated in this section, find the order of magnitude of the mean free path of an electron in a copper wire.

Using $\tau \approx 5 \times 10^{-14}$ sec from Example 34-4, we have

$$\ell = \bar{v}\tau = (10^5 \text{ m/sec}) \ (5 \times 10^{-14} \text{ sec}) = 5 \times 10^{-9} \text{ m} = 50 \text{ Å}$$

Although successful in predicting Ohm's law, the classical theory of conduction has several defects. The numerical magnitudes of the conductivity and resistivity calculated from Equation 34-22 using classical methods for calculating the mean free path and the mean speed differ from measured values by a factor of 10 or so, depending on the temperature at which these values are measured. The temperature dependence of resistivity is given completely by the mean speed $\bar{v}$ in Equation 34-22, which is proportional to $\sqrt{T}$. Thus this calculation does not give a linear dependence on temperature. Finally, the classical model says nothing about why some materials are conductors, others insulators, and still others semiconductors, a subject we shall discuss briefly in Section 34-5.

34-4   Corrections to the Classical Theory of Conduction

The results of the modern theory of electric conduction based on quantum mechanics can be discussed briefly and qualitatively. The features of this theory important for our discussion are that rms speed v_{rms} is not given by Equation 34-17, because the equipartition theorem does not hold for electrons in a metal, and that the mean free path must be calculated using the wave nature of propagation of electrons in the metal. The calculation of the mean speed $\bar{v}$ is quite complicated. The results are that $\bar{v}$ is essentially independent of temperature and is about 1.6×10^6 m/sec for copper, roughly 16 times that calculated at $T = 300$ K from the equipartition theorem. The wave analog of electron collisions with the lattice ions is the scattering of electron waves by the ions. A detailed calculation of electron wave scattering by a perfectly periodic lattice of identical ions gives the result that there is no scattering; the mean free path is infinite. Thus for such a perfect crystal, the resistance is zero. Electron waves are scattered only if the lattice is not perfectly periodic. There are two main causes for lattice deviations from perfect periodicity. One is impurities. For example, if some zinc is introduced into pure copper, the previously perfect periodicity is destroyed. At very low temperatures the resistance of a metal is primarily due to impurities. The other cause of deviations is the displacement of the lattice ions due to vibrations. This effect is dominant at ordinary temperatures. Thus the mean free path is not determined by the size of the lattice ions. At very low temperatures the ions effectively look like points to the electron as far as scattering is concerned. At normal temperatures, the effective area of an ion is proportional to the square of the amplitude of vibration. This in turn is proportional to the energy of vibration, which is proportional to T (except at very low temperatures). The modern theory accurately accounts for the temperature dependence of resistance.

As an illustration of these ideas it is instructive to compare the resistivity of copper and brass (an alloy of copper and zinc) at temperatures of 300 and 4 K. The resistivity of copper drops by a factor of about 50 to 100 (depending on the purity) when the temperature drops from 300 to 4 K, indicating that most of the resistivity at 300 K is due to thermal motion. However, the resistivity of brass drops by only a factor of 4 from temperatures of 300 to 4 K. A large part of the resistivity of brass is due to the temperature-independent irregularity of the copper-zinc lattice.

34-5   Conductors, Insulators, and Semiconductors

The classical model for electric conduction with the modifications due to quantum mechanics we have discussed gives a good account of Ohm's law and the temperature dependence of resistivity and conductivity, but it does not deal with the important question of why some materials are good conductors and others are insulators. As we have seen, there is an enormous variation in the conductivity and resistivity from the best conductors to the best insulators. The difference between a conductor and an insulator can be characterized empirically by the density of charge carriers n. For a good conductor the number of charge carriers is about one per atom; for insulators the number of charge carriers is nearly zero.

To discuss why this is so, even qualitatively, we need two important ideas from quantum mechanics. One is that the energy of an electron in an atom cannot take on any value but is quantized. This idea was first put forward by Bohr, who proposed that the possible energy values or levels for the hydrogen atom are given by the simple formula

$$E_n = -\frac{13.6}{n^2}\,\text{eV}$$

where n is any integer, $n = 1, 2, 3. \ldots$. The lowest possible energy for the hydrogen atom corresponds to $n = 1$; the other possible energies are given by this expression with other values of n. No simple formulas exist for the possible energies of electrons in other more complicated atoms, but the important point for this discussion is that for any atom there is a discrete set of possible energies.

The other quantum-mechanical result needed for us to understand why some materials are conductors and others are not is the *Pauli exclusion principle*. Only two electrons (with opposite spins) can occupy a given quantum state. Consider for a moment two identical atoms very far apart. Each atom has the same set of allowed energy levels. Let us focus our attention on one such level (such as $n = 2$ for hydrogen). There is a level of the same energy for each atom. If the atoms are brought close together, the energy of this level changes because of the influence of the other atom. If the atoms are close enough together, we can no longer associate the energy levels with just one of the atoms but must look at the two-atom system. The two previously identical energy levels split into two levels of slightly different energy for the two-atom system. Similarly, if we have N identical atoms, a particular energy level of the isolated atom splits into N different nearly equal energy levels when the atoms are sufficiently close together. The number of copper atoms in a piece of copper is very large, on the order of Avogadro's number. A single atomic energy level identified with an isolated copper atom splits into a *band* of a very large number of energy levels which are closely spaced in energy.

Energy bands

Because the number of levels in the band is so large, they are spaced almost continuously within the band. There is a separate band of levels for each particular energy level of the isolated atom. In the isolated atom the different levels are often far apart. (For example, in hydrogen the energy for $n = 1$ is -13.6 eV, and for $n = 2$ it is $-13.6/4 = -3.4$ eV.) The energy bands corresponding to these individual levels for a large number of atoms in a solid may be widely separated in energy, may be close together, or may even overlap in energy, depending on the kind of atom and the type of bonding in the crystal.

Figure 34-8 shows four possible kinds of band structure for a solid. The band structure for copper is shown in Figure 34-8a. The lower bands are filled with the inner electrons of the atoms; i.e., according to the Pauli exclusion principle, no more electrons can occupy levels in these bands. The uppermost band containing electrons is only about half full. In the normal state, at low temperatures, the lower half of the energy band is filled, and the upper half is empty. At higher temperatures, a few of the electrons are in the higher energy states in that band because of thermal excitation, but there are still many unfilled energy states above the filled ones.

When an electric field is established in the conductor, the electrons in the upper band are accelerated, which means that their energy is increased. This is consistent with the Pauli exclusion principle because there are many empty energy states just above those occupied by electrons in this band. These electrons are thus the conduction electrons, and the band is called the *conduction band.*

Figure 34-8b shows the band structure for a typical insulator. At $T = 0$ K the highest energy band which contains electrons is completely full. The next energy band, containing empty energy states, is separated from the last filled band by an energy gap which is quite large compared with thermal energies (of the order of 0.02 eV). This energy gap is sometimes referred to as a *forbidden energy band.* Very few electrons can be thermally excited to the nearly empty conduction band even at fairly high temperatures. When an electric field is established in the solid, electrons cannot be accelerated because there are no empty energy states at nearby energies. In the classical model, we describe this by saying that there are no free electrons. The small conductivity that is observed is due to the very few electrons that are thermally excited into the upper nearly empty conduction band.

These results—that the conduction band is essentially empty, and the gap between it and the lower energy band which is nearly filled is large—can be derived by the methods of quantum mechanics. In some materials the energy gap between the top filled band and the empty conduction band is very small (Figure 34-8d). At ordinary temperatures there is an appreciable number of electrons in the conduction band due to thermal excitation. Such a material is called a *semiconductor.* In the presence of an electric field, the electrons in the conduction band can be accelerated. Also, for each electron in the conduction band there is a vacancy, or hole, in the nearly filled band (this band is called the *valence band*). In the presence of a field, electrons in this band can be excited to a vacant energy level. This contributes to the electric current and is most easily described as the motion of a hole in the direction of the field and opposite the motion of the electrons. The hole thus acts like a positive charge. (An analogy is a line of cars with a space the size of one car: as the cars move to fill the space, the space moves backward, in the direction opposite the motion of the cars.) An

Figure 34-8
Four possible band structures for a solid. In (a) an allowed band is only partially full, whereas in (c) two allowed bands overlap. In both cases nearby energy states are available for excitation of electrons, and so these materials are conductors. (b) The band structure of an insulator, where there is a forbidden band with a large energy gap between a filled allowed band and the next allowed band. (d) Since the energy gap between the filled band and the next allowed band is small, some electrons are excited to the next allowed band at normal temperatures, making this material a semiconductor.

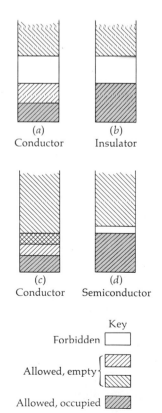

(a)
Conductor

(b)
Insulator

(c)
Conductor

(d)
Semiconductor

Key

Forbidden ☐

Allowed, empty

Allowed, occupied

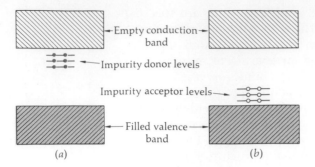

Figure 34-9
(*a*) Energy bands of an *n*-type semiconductor. Impurity atoms provide filled energy levels just below the empty conduction band and donate electrons to the conduction band. (*b*) Energy bands of a *p*-type semiconductor. Impurity atoms provide empty energy levels just above the filled valence band and accept electrons from it.

interesting feature of a semiconductor is that as the temperature increases, the conductivity increases (and the resistivity decreases), contrary to the usual behavior of conductors. The reason is that as the temperature is increased, the number of free electrons n is increased because there are more electrons in the conduction band. (The number of holes of course also increases.) This increase in n outweighs the natural decrease in conductivity due to a smaller mean free path at higher temperatures.

A semiconductor can also be made from an insulator by introducing certain impurities (this procedure is called *doping*) which introduce slight modification of the band structure. These slight modifications have a great effect on the conductivity. In one type of doped semiconductor the impurity atoms provide electrons in energy states which lie in the original forbidden band just below the empty conduction band (Figure 34-9*a*). These impurity energy levels are called *donor levels*. The impurity atoms donate electrons to the conduction band. This is called an *n-type semiconductor* because the charge carriers are negative electrons. In a second kind of doped semiconductor, the impurity atoms have empty energy levels just above the filled band in the original solid, as in Figure 34-9*b*. These levels accept electrons from the filled valence band when these electrons are thermally excited to a higher energy state. Thus a hole is created in the valence band which is free to propagate in the direction of an electric field. This is called a *p-type semiconductor* because the charge carriers are positive holes.

34-6   Conservation of Charge and Approach to Electrostatic Equilibrium

In our brief historical discussion of electric charge, in Chapter 29, we mentioned that implicit in Franklin's one-fluid model of electricity is the idea that electric charge is conserved. This concept is much more extensive than Franklin imagined and is now considered a fundamental law of nature. Even when electrons or other elementary charged particles are created or destroyed, they are always created or destroyed in pairs, so that there is no change in net charge. For example, electrons can be created by the interaction of high-energy electromagnetic radiation with matter. In such an event, a negative electron and a positively charged positron are always created together. Similarly, electrons and positrons can annihilate each other, producing electromagnetic radiation. Again, both a positive and negative charge disappear together, so that the net charge is unchanged.

If the amount of charge in any region in space is increasing or decreasing in time, there must be a corresponding flow of charge into or out of that region. We can write this mathematically as follows. Consider a surface S enclosing some region in space in which there is a net charge Q. The rate at which charge flows out of that region through a small element of area on the surface is $J_n\, dA$, where J_n is the normal component of the current density at that area element. If we integrate this rate of flow of charge over the complete closed surface S, we obtain the total current out of the region. This total current must equal the rate of decrease of the charge Q inside the surface:

$$\oint \mathbf{J} \cdot \hat{\mathbf{n}}\, dA = -\frac{dQ}{dt} \qquad\qquad 34\text{-}23 \qquad \textit{Continuity equation}$$

The net flux of the current density through the closed surface equals the rate of decrease of charge inside the surface. Equation 34-23, called the *continuity equation*, expresses the fact that charge is conserved.

We can use the continuity equation, Gauss' law, and Ohm's law to show that if we have an initial charge within a conductor, the charge will quickly flow out to the conductor surface and electrostatic equilibrium with zero electric field inside the conductor will be approached rapidly. Consider a closed gaussian surface completely inside a conductor. Let there be a net charge Q inside the surface. According to Gauss' law, the net flux of the electric field through the surface equals Q/ϵ_0:

$$\oint \mathbf{E} \cdot \hat{\mathbf{n}}\, dA = \frac{Q}{\epsilon_0}$$

If we multiply each side by the conductivity σ, we have

$$\oint \sigma \mathbf{E} \cdot \hat{\mathbf{n}}\, dA = \oint \mathbf{J} \cdot \hat{\mathbf{n}}\, dA = \frac{\sigma Q}{\epsilon_0}$$

Substituting $-dQ/dt$ for the net flux of the current density from Equation 34-23, we have

$$-\frac{dQ}{dt} = \frac{\sigma Q}{\epsilon_0} \qquad \text{or} \qquad \frac{dQ}{Q} = -\frac{\sigma}{\epsilon_0}\, dt \qquad\qquad 34\text{-}24$$

Integrating gives

$$\ln Q = -\frac{\sigma}{\epsilon_0} t + C$$

where C is a constant of integration. Then

$$Q = e^{C - \sigma t/\epsilon_0} = e^C e^{-\sigma t/\epsilon_0}$$

Let Q_0 be the charge at time $t = 0$. Then

$$Q_0 = e^C$$

The net charge inside the conductor thus obeys the equation

$$Q = Q_0 e^{-\sigma t/\epsilon_0} = Q_0 e^{-t/\rho\epsilon_0} \qquad\qquad 34\text{-}25$$

where $\rho = 1/\sigma$ is the resistivity. We see that the charge decreases exponentially with time. The electric field inside the conductor also decreases exponentially with time (Figure 34-10). For example, if the conductor is a solid sphere with the charge Q at the center, the field at a distance r within the conductor is

$$E = \frac{1}{4\pi\epsilon_0}\frac{Q}{r^2} = \frac{1}{4\pi\epsilon_0}\frac{Q_0}{r^2}\, e^{-t/\rho\epsilon_0} = E_0 e^{-t/\rho\epsilon_0}$$

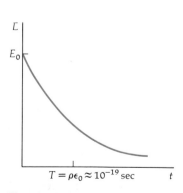

$$T = \rho\epsilon_0 \approx 10^{-19}\,\text{sec}$$

Figure 34-10
Plot of electric field in a conductor versus time after a charge has been introduced inside the conductor.

After a time $T = \rho\epsilon_0$ the charge decreases to $1/e$ of its initial value, as can be seen by substituting this value of t into Equation 34-25. This is called the *decay time*. Using the value $\epsilon_0 = 8.85 \times 10^{-12}$ F/m and $\rho = 1.7 \times 10^{-8}$ Ω-m for copper, we find the decay time to be

$$T = \rho\epsilon_0 = (1.7 \times 10^{-8} \ \Omega\text{-m})(8.85 \times 10^{-12} \ \text{F/m}) = 1.5 \times 10^{-19} \ \text{sec}$$

We see from this calculation that electrostatic equilibrium is approached so rapidly that it is nearly instantaneous. Table 34-1 shows that this holds for any good conductor because of the very small values of ρ.

34-7 Energy in Electric Circuits

When there is electric current in a conductor, electric energy is continuously converted into thermal energy of the conductor. For example, in our simple model of conduction, free electrons are accelerated by the electric field, acquiring an increase in velocity of $qE\tau/m$ and a corresponding additional kinetic energy in the time τ. This kinetic energy is continuously transferred to the conductor by collisions between the electrons and the lattice ions of the conductor. Thus, though the electrons continually gain energy from the electric field, the energy is immediately transferred to thermal energy of the conductor and the electrons maintain a steady drift velocity on the average.

In general, when (positive) charge flows in a conductor, it flows from high potential to low potential in the direction of the electric field. (Negative charge of course flows in the opposite direction.) The charge thus loses potential energy. This potential-energy loss does not appear as kinetic energy of the charge carriers except momentarily, before it is transferred to the lattice ions by collisions. Consider the charge ΔQ which passes point P_1 in Figure 34-11 during time Δt. If the potential at that point is V_1, the charge has potential energy $\Delta Q \ V_1$. During that time interval the same amount of charge passes point P_2, where the potential is V_2. It has potential energy $\Delta Q \ V_2$, which is less than the original potential energy. The energy lost by the charge passing through this segment of conductor is $-\Delta W$, given by

$$-\Delta W = \Delta Q \ (V_1 - V_2) = \Delta Q \ V$$

where $V = V_1 - V_2$ is the potential drop from point 1 to point 2. The rate of energy loss by the charge is

$$-\frac{\Delta W}{\Delta t} = \frac{\Delta Q}{\Delta t} V = IV$$

where I is the current. The power loss in the conductor is thus

$$P = IV \tag{34-26}$$

This expression for electric-power loss can easily be remembered by recalling the definitions of V and I. The voltage drop is the potential-energy decrease per charge, and the current is the charge flowing per unit time. Thus the product of V and I is the energy lost per unit time, or the power lost. As we have seen, this power goes into heating the conductor.

Using the definition of resistance $R = V/I$, we can write Equation

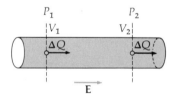

Figure 34-11
At point P_1 the charge ΔQ has potential energy $\Delta Q \ V_1$, and at point P_2 its potential energy is $\Delta Q \ V_2$. The energy loss in the conductor is $\Delta Q \ V$, where $V = V_1 - V_2$, and the power loss is IV.

34-26 in several other useful forms by eliminating either V or I:

$$P = (IR)I = I^2R \qquad \text{or} \qquad P = V\frac{V}{R} = \frac{V^2}{R} \qquad\qquad 34\text{-}27$$

Power dissipated in a conductor

This energy put into a conductor (often called resistor) is called *joule heat.*

In order to have a steady current in a conductor we need to have a supply of electric energy. A device which supplies electric energy is called a seat of electromotive force or simply an *emf.*

emf

An emf is a device which converts chemical, mechanical, or other form of energy into electric energy. It is often a battery, which converts chemical energy into electric energy, or a generator, which converts mechanical energy into electric energy.

Figure 34-12 shows a simple circuit consisting of a resistance R connected to an ideal emf which maintains a constant potential difference $\mathcal{E}$ between the points a and b. This potential difference is the value of the emf. The symbol ⊣⊢ stands for an emf. The longer line indicates the higher-potential side. The resistance is indicated by the symbol ⌇. The straight lines in the circuit diagram indicate connecting wires of negligible resistance. Since the choice of zero potential is always arbitrary, we are interested only in the potential differences between various points in the circuit. By definition, an ideal emf maintains a constant potential difference $\mathcal{E}$ between points b and a. There is no potential difference between points a and c because the connecting wire is assumed to have negligible resistance. Similarly there is no potential difference between points b and d. Thus the potential difference between points c and d is

$$V = V_c - V_d = \mathcal{E}$$

There is an electric field in the resistor from c to d, and charge flows in this direction. The current has the magnitude $I = V/R = \mathcal{E}/R$.

In the emf, this same current is in the direction from b to a. The current must be the same; otherwise charge would accumulate at some point in the circuit. Note that *in the emf, the charge flows from low potential to high potential.* (In some cases, studied in the next chapter, e.g., charging a battery by another emf, the charge flows in the opposite direction through the emf.) When charge ΔQ flows through the emf, its potential energy is *increased* by the amount $\Delta Q\,\mathcal{E}$. It then flows through the conductor, where it loses this potential energy to heat. The value of the emf is thus the *work done per unit charge.* The unit of emf is the volt, the same as the unit of potential difference. The

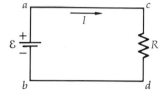

Figure 34-12
Simple circuit with an ideal emf, resistance R, and connecting wires assumed to be resistanceless.

An unusual source of emf is the electric eel (*Electrophorus electricus*), which lives in the Amazon and Orinoco rivers. When attacking, the eel can develop up to 600 V between its head (positive) and tail (negative).

Brown Brothers

The first practical method of maintaining a steady electric current was the battery, invented by Alessandro Volta in 1800. The earliest battery was a series of electric cells containing a copper disk and a zinc disk separated by a moistened pasteboard. Volta is shown here demonstrating his electric cell to Napoleon.

rate at which energy is supplied by the emf is

$$P = \frac{\Delta W}{\Delta t} = \frac{\Delta Q}{\Delta t} \mathcal{E} = \mathcal{E}I \qquad\qquad 34\text{-}28$$

Power supplied by emf

In this simple circuit, the power input by the seat of emf equals that dissipated in the resistor. An emf can be thought of as a sort of charge pump. It pumps the charge from low potential energy to high potential energy, analogous to a water pump pumping water from low to high gravitational potential energy.

Figure 34-13 shows a mechanical analog to the simple electric circuit discussed above. Marbles of mass m roll along an inclined board with many nails in it. The marbles are accelerated by the gravitational field between collisions with the nails. They transfer the kinetic energy obtained between collisions to the nails during the collisions. Because of the many collisions, the marbles move with a small drift velocity toward the ground. When they reach the bottom, a boy picks them up and starts them again. The boy is the analog of the emf. He does work mgh on each marble. The work per mass is gh, analogous to the work per charge done by the emf. The energy source in this case is the internal chemical energy of the boy.

Figure 34-13
Mechanical analog of resistance and emf. As the marbles roll down the incline, their potential energy is converted into kinetic energy, which is quickly converted into heat because of collisions with the nails in the board. A boy lifts the marbles up from low potential energy to high potential energy, converting his internal chemical energy into potential energy of the marbles.

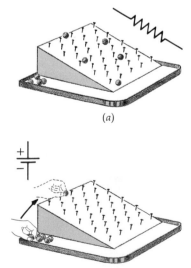

(a)

(b)

Questions

5. What are several common kinds of emfs? What sort of energy is converted into electric energy by the emf in each?

6. In a simple electric circuit like that shown in Figure 34-12 the charge outside the emf flows from positive voltage toward negative voltage, but in the same circuit the current inside the emf flows from minus to plus. Explain how this is possible.

7. Show that 1 W equals 1 V-A.

8. Figure 34-13 illustrates a mechanical analog to a simple electric circuit. Devise another in which the current is a flow of water instead of marbles.

9. An emf is a device which converts chemical, mechanical, or other form of energy into electric energy. Could a device which uses electric energy as its source of energy be considered an emf? Can you think of any examples? What about an electronic power supply which serves the same function as a battery?

Transistors

Reuben E. Alley, Jr.
U.S. Naval Academy

In 1949, John Bardeen, Walter Brattain, and William Shockley, all of Bell Telephone Laboratories, initiated a revolution in electronics with the invention of the transistor. In 1956 they received a Nobel prize for their work.

Transistors, semiconductor diodes, and the integrated circuit, which combines the functions of many transistors and their associated resistors and capacitors into one small package, have almost completely replaced vacuum tubes in all but a few special applications, e.g., television picture tubes and high-power radio transmitters. These solid-state devices can be made very small, very light, and from readily available materials. They are cheap to manufacture, require little power, and operate at room temperatures. A spectacular example of reduction in size and cost is the pocket calculator, available in some models for under $50. The most expensive parts of the modern pocket calculator are said to be the control-button switches. Their size, which is limited by the size of the human finger, prevents further miniaturization by a factor of perhaps 25.

At least one company offers a programmable calculator whose memory capacity and computational abilities equal or perhaps surpass those of the best vacuum-tube computer of the early 1950s. Its cost (a few hundred dollars) is less than one hundredth and its volume is less than one-thirty-thousandth that of the vacuum-tube counterpart. It is portable, operates on tiny batteries, and is highly reliable. In contrast, the vacuum-tube computer required an enormous power supply and air conditioning to remove the great quantities of waste heat resulting from the high operating temperature of the vacuum tubes. This high temperature seriously limited the life and reliability of the individual components.

Manufacture of diodes, transistors, and integrated circuits requires a very pure sample of one element throughout which a carefully controlled, exceedingly small amount (a few parts per million) of another element has been uniformly distributed. The basic material usually is germanium or silicon. Both these elements are semiconductors; i.e., at room temperature they have relatively few free charges (compared with the practically unlimited supply of free charge in a conductor such as copper). The second element (the *impurity* or *doping* material) must lie near silicon or germanium in the periodic table but have one less or one more valence electron than silicon or germanium. Commonly used are boron (valence 3) and arsenic (valence 5). Adding boron produces a material that acts as if it had a large number of free positive charges. Such material is called a *p*-type semiconductor. Similarly, the introduction of arsenic into the pure material produces a sample in which current is carried by flow of negative charges (an *n*-type semiconductor).

A single crystal of silicon made to contain a *p*-type region and an adjacent *n*-type region with a fairly abrupt discontinuity between them is a diode. Within a specified voltage range (below the breakdown voltage indicated in Figure 1) a diode is a good conductor for one polarity of applied voltage and a very poor conductor for applied voltage of opposite polarity.

Semiconductor diodes are used in *rectifiers,* which convert the alternating voltage supplied by all power companies into unidirectional voltage. With *filters* and *regulating circuits* one can construct a *power supply* that provides a direct voltage of constant amplitude.

Radios, television sets, sound-reproducing systems, and many other devices require a circuit that will accept a time-dependent voltage or current of very small amplitude (the *input*) and provide a greatly magnified replica of the input voltage or current as its *output.* Such a circuit, which must be supplied with energy from a battery or a power supply, is called an *amplifier.*

Figure 1
Diode characteristic. Note the different scales in the first and third quadrants.

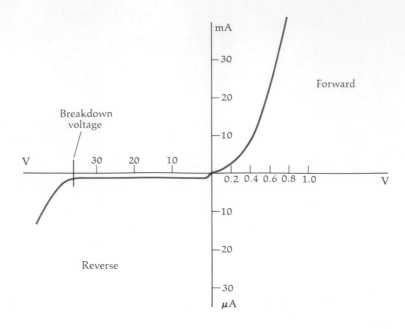

Figure 2

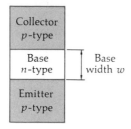

The basic element of most amplifiers is the transistor, which contains three distinct regions of semiconductor material. Figure 2 is an idealized representation of a *pnp* transistor, where the three regions are labeled with their common designations. In a properly designed transistor the level of doping is much greater in the emitter than in the other two regions, and the base width w is very narrow.

The device represented in Figure 2 contains two *pn* junctions. In a practical circuit (Figure 3) voltages are supplied so that the emitter-base (*EB*) junction is forward-biased and the collector-base (*CB*) junction is reverse-biased. The difference in doping levels of the emitter and the base means that practically all the current across the *EB* junction consists of positive charges from the emitter. Because the base width is small and the *CB* junction is reverse-biased, most of the positive charges from the emitter that enter the base diffuse across its narrow width and enter the collector. The few positive charges that are not collected by the *CB* junction leave the base through the external connection. In Figure 3, therefore, I_C is almost but not quite equal to I_E, and I_B is much smaller than either I_C or I_E. It is customary to write

$$I_C = \beta I_B$$

where β, the *current gain,* is an important parameter of the transistor. Transistors can be designed to have values of β as low as 10 or as high as several hundred.

In a useful transistor, β will be constant; i.e., the device will be linear, over a restricted range of base current provided that proper biases are maintained on the two junctions. Figure 4 shows a simple amplifier. A small time-varying voltage, say from the motion of a needle in the groove of a phonograph record, is in series with a bias voltage V_{BB}. The base current is then the sum of a steady current I_B produced by V_{BB} and a varying current i_b produced by v_s, the voltage to be amplified. Because v_s may at any instant be either positive or negative, V_{BB} must be included and must be large enough to ensure that there is always a forward bias on the *EB* junction. Since the device is linear, the collector current will consist of two parts, a steady current $I_C = \beta I_B$ and a time-varying current $i_c = \beta i_b$. We thus have a *current amplifier* where the time-varying output current i_c is β times the input current i_b. In such an amplifier, the steady currents I_C and I_B, while essential to proper operation, are usually not of interest. (If an *npn* transistor is used, the bias voltages must be reversed and the currents will all be in the other direction. Otherwise, operation is identical with that described here.)

Figure 3
A *pnp* transistor biased for normal operation in the common-emitter connection.

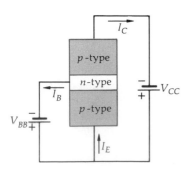

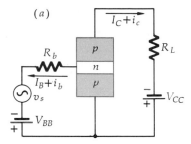

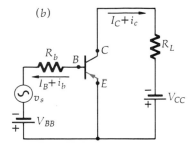

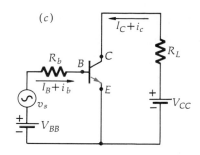

Figure 4
(a) Simple amplifier circuit using *pnp* transistor. (b) The same circuit with the transistor represented by a standard symbol. (c) Amplifier using *npn* transistor. Battery polarities and current directions are opposite those in part (b).

Figure 5
(a) Five-watt audio amplifier using an integrated circuit. The integrated circuit is partly hidden by a piece of metal 3.3 cm long. The five cylinders are capacitors.
(b) Photomicrograph of an integrated circuit with more than 50 components, on a silicon chip approximately 0.25 cm by 0.2 cm.

(a)

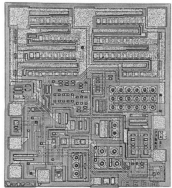

(b)

Ohm's law tells us that v_s and i_b are related by

$$i_b = \frac{v_s}{R_b + r_b}$$

where r_b is the internal resistance of the transistor between base and emitter. Likewise, in the collector circuit the current i_c produces a voltage v_L across the resistor R_L,

$$v_L = i_c R_L$$

We already have

$$i_c = \beta i_b = \beta \frac{v_s}{R_b + r_b}$$

and therefore,

$$v_L = \beta \frac{R_L}{R_b + r_b} v_s$$

and the voltage gain or voltage amplification is

$$\frac{v_L}{v_s} = \beta \frac{R_L}{R_b + r_b}$$

In a practical case β may be 100, and the ratio $R_L/(R_b + r_b)$ may be $\frac{1}{2}$. Then the voltage gain is 50. A more detailed derivation shows that v_L and v_s are exactly *out of phase;* i.e., when v_s has its most positive value, v_L has its most negative value. For a simple amplifier this phase shift is not important because all input voltages, regardless of frequency, are affected identically. The simple voltage amplifier is the basis of many circuits such as oscillators and modulators that constitute communication systems.

The complete amplifier in a record or tape player generally consists of several *stages* similar to Figure 4 connected in *cascade* so that the output of one becomes the input of the next. Thus the very small voltage produced by the motion of the needle or by the passage of magnetized tape through the player head controls the large amounts of power required to drive a system of loudspeakers. The energy delivered by the speakers is supplied by the sources of direct voltage connected to each transistor.

The resistance of a semiconductor material depends upon the impurity concentration. Reverse-biased diodes have capacitance that can be controlled by controlling the bias voltage. These two facts are exploited in the manufacture of *integrated circuits,* in which several transistors along with associated resistors and capacitors are interconnected on a single small piece of silicon. Figure 5a shows a 5-W audiofrequency amplifier that uses one integrated circuit containing 14 transistors, 3 diodes, and 10 resistors. Figure 5b shows a different integrated circuit containing even more circuit elements.

In addition to greater reductions in their size and weight, the introduction of integrated circuits has changed the design of electronics devices radically. For example, one need no longer design an amplifier but can choose from those available a complete amplifier with the desired characteristics. It is then procured and installed as a complete unit. Such possibilities have enabled small companies with little expertise in design to build rather complex electronics equipment and compete favorably in price and quality.

Review

A. Define, explain, or otherwise identify:

B. True or false:

1. The net motion of electrons is in the direction of the current.

2. Current density is proportional to the drift velocity.

3. $R = V/I$ is Ohm's law.

4. The current in a conductor is always proportional to the potential drop across the conductor.

5. $R = V/I$ holds for all materials for which resistance can be defined.

6. The resistance of a metal is due to irregularities in the spacing of the lattice ions.

7. If the net charge in some region decreases, there must be a net flow of charge out of the region.

Exercises

Section 34-1, Current and Motion of Charges

1. A wire carries a steady current of 2.0 A. (*a*) How much charge flows past a point in the wire in 5.0 min? (*b*) If the current is due to flow of electrons, how many electrons flow past a point in this time?

2. A 10-gauge copper wire (diameter 2.59 mm) carries a current of 20 A. Assuming one free electron per copper atom, calculate the electron drift velocity.

3. In a 3.0-cm diameter fluorescent tube, 2.0×10^{18} electrons and 0.5×10^{18} positive ions (charge $+e$) flow past a point each second. (*a*) What is the current in the tube? (*b*) What is the current density?

4. In a certain electron beam, there are 5.0×10^6 electrons per cubic centimeter. The kinetic energy of the electrons is 10.0 keV, and the beam is cylindrical with diameter 1.00 mm. (*a*) What is the velocity of the electrons? (*b*) Find the current density. (*c*) Find the beam current.

5. A charge $+q$ moves in a circle of radius r with speed v. (*a*) Express the frequency f with which the charge passes a point in terms of r and v. (*b*) Show that the average current is qf and express this in terms of v and r.

6. A ring of radius R has charge per unit length λ. The ring rotates with angular velocity ω about its axis. Find an expression for the current at a point on the ring.

7. A 20.0-MeV proton beam of diameter 2.0 mm in a certain accelerator constitutes a current of 1.0 mA (10^{-3} A). The beam strikes a metal target and is absorbed by it. (*a*) What is the density of protons in the beam? (*b*) How many protons strike the target in 1.0 min? (*c*) If the target is initially uncharged, express the charge of the target as a function of time.

8. A 10-gauge wire (diameter 2.59 mm) is welded end to end to a 14-gauge wire (diameter 1.63 mm). The wires carry a current of 15 A. (*a*) What is the cur-

rent density in each wire? (b) If both wires are copper with one free electron per atom, find the drift velocity in each wire.

Section 34-2, Ohm's Law and Resistance

9. A 10-m wire of resistance 0.2 Ω carries a current of 5 A. (a) What is the potential difference across the wire? (b) What is the magnitude of the electric field in the wire?

10. A potential difference of 100 V produces a current of 3 A in a certain resistor. (a) What is its resistance? (b) What is the current when the potential difference is 25 V?

11. A copper wire and an iron wire have the same length and diameter and carry the same current I. (a) Find the potential drop across each wire and the ratio between them. (b) In which wire is the electric field greater?

12. A piece of carbon is 3.0 cm long and has a square cross section 0.5 cm on a side. A potential difference of 8.4 V is maintained across its long dimension. (a) What is the resistance of the block? (b) What is the current in this resistor? (c) What is the current density?

13. A tungsten rod is 50 cm long and has a square cross section 1.0 mm on a side. (a) What is its resistance at 20°C? (b) What is its resistance at 40°C?

14. The third (current-carrying) rail of a subway track is made of steel and has a cross-sectional area of about 8.5 in². What is the resistance of 10 mi of this track?

15. A wire of length 1 m has a resistance of 0.3 Ω. It is uniformly stretched to a length of 2 m. What is its new resistance?

16. What is the potential difference across one wire of a 100-ft extension cord of 16-gauge copper wire (diameter 0.051 in) which carries a current of 5.0 A?

17. At what temperature will the resistance of a copper wire be 10 percent greater than it is at 20°C?

18. How long is a 14-gauge copper wire which has a resistance of 2 Ω?

19. A cube of copper has side 2.0 cm. If it is drawn out into a 14-gauge wire, what will its resistance be?

Section 34-3, Classical Model of Electric Conduction

20. A current of 10^{-8} A exists in a copper wire of cross-sectional area 2.0 mm². (a) What is the current density? (b) What is the drift velocity of the electrons? (c) What is the average time between collisions? (d) What is the mean free path of electrons in this wire (assume $\bar{v} = 10^5$ m/sec)?

21. Repeat Exercise 20 for a current of 5.0 mA in a copper wire of cross-sectional area 1.0 mm².

Section 34-4, Corrections to the Classical Theory of Conduction, and Section 34-5, Conductors, Insulators, and Semiconductors

There are no exercises for these sections.

Section 33-6, Conservation of Charge and Approach to Electrostatic Equilibrium

22. A point charge 2 μC is placed at a point on the surface of a spherical copper conductor of radius 20 cm at time $t = 0$. (a) What is the magnitude of

the electric field at the center of the conductor just after the charge is placed on the surface? (*b*) What is the magnitude of the electric field at the center of the conductor at time $t = 1$ nsec (10^{-9} sec) after the charge is placed on the conductor?

23. Find the decay time for approach to equilibrium for (*a*) silicon and (*b*) rubber, which has resistivity of 10^{14} Ω-m.

Section 34-7, Energy in Electric Circuits

24. What is the power dissipated in a 10.0-Ω resistor if the potential difference across it is 50 V?

25. Find the power dissipated in a resistor of resistance (*a*) 5 Ω and (*b*) 10 Ω connected across a constant potential difference of 110 V.

26. A 200-W heater is used to heat water in a cup. Assume that 90 percent of the energy goes into heating the water. (*a*) How long does it take to heat 0.25 kg of water from 15 to 100°C? (*b*) How long does it take to boil this water away after it reaches 100°C?

27. A 1-kW heater is designed to operate at 220 V. (*a*) What is its resistance, and what current does it draw? (*b*) What is the power of this resistor if it operates at 100 V?

28. A 10.0-Ω resistor is rated to be able to dissipate 5.0 W. (*a*) What maximum current can this resistor tolerate? (*b*) What voltage across this resistor will produce this current?

29. If energy costs 4 cents per kilowatt-hour, (*a*) how much does it cost to operate an electric toaster for 4 min if the toaster has resistance 11.0 Ω and is connected across 100 V? (*b*) How much does it cost to operate a heater of resistance 5.0 Ω across 100 V for 8 h?

30. A battery has an emf of 12.0 V. How much work does it do in 5 sec if it delivers 3 A?

Problems

1. A 16-gauge copper wire (diameter 1.29 mm) can safely carry a maximum current of 6 A (assuming rubber insulation). (*a*) How great a potential difference can be applied across 40 m of such a wire? (*b*) Find the current density and the electric field in the wire when it carries 6 A. (*c*) Find the power dissipated in the wire when it carries 6 A.

2. A Van de Graaff accelerator belt carries a surface charge density of 5 μC/m². The belt is 0.5 m wide and moves at 20 m/sec. (*a*) What current does it carry? (*b*) If this charge is raised to a potential of 100,000 V, what is the least value of power of the motor needed to drive the belt?

3. A linear accelerator produces a beam of electrons in which the current is not constant but consists of a pulsed beam of particles. Suppose that the pulse current is 1.6 A for a 0.1-μsec duration. (*a*) How many electrons are accelerated in each pulse? (*b*) What is the average beam current if there are 1000 pulses per second? (*c*) If the electrons acquire an energy of 400 MeV, what is the average (minimum) power input to the accelerator? (*d*) What is the peak power input? (*e*) What fraction of the time is the accelerator actually accelerating particles? (This is called the *duty factor* of the accelerator.)

4. A 100-W heater resistor is designed to operate across 100 V. (*a*) What is its resistance and what current does it draw? (*b*) Show that if the potential difference across the resistor changes by a small amount ΔV, the power changes by a small amount ΔP, where $\Delta P/P \approx 2\Delta V/V$. *Hint:* Approximate the changes with differentials. (*c*) Find the approximate power dissipated in the resistor if the potential difference increases to 115 V.

5. An 80.0-m copper wire 1.0 mm in diameter is joined end to end with a 49.0-m iron wire of the same diameter. The current in each is 2.0 A. (*a*) Find the current density in each wire. (*b*) Find the electric field in each wire. (*c*) Find the potential difference across each wire. (*d*) Find the equivalent resistance which would carry 2.0 A at a potential difference equal to the sum of that across the two and compare it with the sum of the resistances of the two.

6. Two wires have the same length but different cross-sectional areas and are made from different materials. They are laid side by side and joined at each end so that the potential difference across each wire is the same. Let R_1 be the resistance of one wire and R_2 be that of the other. (*a*) If the potential difference across each is V, find the current in each and find the total current $I = I_1 + I_2$. (*b*) Find an expression for the resistance of a single wire which would carry the same total current when the potential difference is V.

7. A wire of area A, length l_1, resistivity ρ_1, and temperature coefficient α_1 is connected to a second wire of length l_2, resistivity ρ_2, temperature coefficient α_2, and the same area A, so that the wires carry the same current. (*a*) Show that if the current is I, the potential drop across the two wires is IR, where $R = R_1 + R_2$ is the sum of the resistances. (*b*) Show that if $\rho_1 l_1 \alpha_1 + \rho_2 l_2 \alpha_2 = 0$, the total resistance R is independent of temperature for small temperature changes. (*c*) If one wire is made of carbon and the other copper, find the ratio of their lengths such that R is approximately independent of temperature.

8. An expression for the resistivity of pure metals at any temperature t which is more general than Equation 34-16 is

$$\rho = \rho_{20} e^{\alpha(t - 20°C)}$$

where ρ_{20} is the value at $t = 20°C$. (*a*) Show that for small values of $t - 20°C$ this reduces to Equation 34-16. (*b*) Use the expansion $e^x \approx 1 + x + x^2/2 + \cdots$ to show that $\rho - \rho_{20} \approx \rho_{20} x (1 + \frac{1}{2} x)$, where $x = \alpha(t - 20°C)$. (*c*) For what range of x is Equation 34-16 in agreement with the approximation in (*b*) to within 1 percent? (*d*) For copper what range of t corresponds to this range of x?

CHAPTER 35 Direct-Current Circuits

In this chapter we shall analyze some simple circuits consisting of batteries, resistors, and capacitors in various combinations. They are called direct-current (dc) circuits because the direction of the current in any one part of the circuit does not vary. Circuits in which the current alternates in direction (ac circuits) will be discussed in Chapter 40.

35-1 Kirchhoff's Rules

In a typical dc circuit, several batteries and resistors are connected in some way, and the values of the currents in the various parts of the circuit are to be found. Two simple rules, called *Kirchhoff's rules*, are needed for this type of analysis:

1. The sum of the potential drops around any circuit loop must equal the sum of the potential increases.

2. At a junction point in a circuit where the current can divide, the sum of the currents into the junction must equal the sum of the currents out of the junction.

Rule 1 follows from the conservation of energy, and rule 2 from the conservation of charge. A simple circuit consisting of an emf, a resistor, and connecting wires is shown in Figure 35-1. The connecting wires indicated by straight lines in this diagram are assumed to have negligible resistance. Any two points along such a line are therefore always at the same potential. If we start from point a and move around the circuit with the current (clockwise in this case), we first encounter a potential drop of magnitude IR from point b to point c. A charge ΔQ loses energy $IR\,\Delta Q$ flowing from point b to point c. We then encounter a potential increase of magnitude ε as we traverse the emf. The charge thus gains energy $\varepsilon\,\Delta Q$ going from d to a. When the charge is back at

The Granger Collection

Gustav Robert Kirchhoff (1824–1887).

Kirchhoff's rules

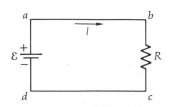

Figure 35-1
Simple circuit with emf ε and resistance R connected by wires of negligible resistance. The potential drop from b to c equals the potential increase from d to a.

point a, it has the same energy as it had before the trip around the circuit (the potential energy $V_a \Delta Q$ is the same, and its kinetic energy is the same because the drift velocity does not change). Thus the energy lost, $IR \Delta Q$, must equal the energy gained, $\varepsilon \Delta Q$. Dividing by ΔQ gives us Kirchhoff's first rule. It should be clear that this rule holds no matter how many resistors or emfs are encountered or what path is followed as long as we arrive back at the starting point.

We shall need Kirchhoff's second rule for multiloop circuits which contain points at which the current can divide, as in Figure 35-2. We have labeled the currents in the three wires I_1, I_2, and I_3. In a time interval Δt, charge $I_1 \Delta t$ flows into the junction from the left. In the same time charges $I_2 \Delta t$ and $I_3 \Delta t$ flow out of the junction to the right. Since there is no way that charge can originate at this point or collect there, conservation of charge implies rule 2, which for this case gives

$$I_1 = I_2 + I_3 \qquad\qquad 35\text{-}1$$

(In Section 35-3 we shall have to modify rule 2 for points in a circuit which contains capacitors, in which charge can accumulate.)

Figure 35-3 shows a simple circuit containing a battery, a resistor, and connecting wires. As before, we ignore any resistance in the connecting wires. A real battery is more complicated than merely a seat of emf. The potential difference across the battery terminals, called the *terminal voltage*, is not simply the emf of the battery. Figure 35-4 shows a typical dependence of the terminal voltage on the current: the terminal voltage decreases slightly as the current increases.[1] As we have seen, an ideal emf maintains a *constant* potential difference between its terminals independent of the current. Because the decrease in terminal voltage is approximately linear, we can represent a real battery by a seat of emf plus a resistor, as shown in Figure 35-5, which is the circuit diagram for the circuit pictured in Figure 35-3. The resistance r is called the *internal resistance of the battery*. Often the internal resistance is quite small, only a few hundredths of an ohm. To simplify calculation in problems, we may either exaggerate the magnitude of r or ignore it completely.

Terminal voltage

Internal resistance

---

[1] This is assuming that the current through the battery is in the same direction as the emf of the battery, which is the case for a circuit with only one emf. When the current is opposite the emf, as when a battery is being charged by another emf, the terminal voltage increases with increasing current.

Figure 35-2
If charge cannot originate or collect at the junction point a, conservation of charge implies that the current I_1 into the point must equal the current $I_2 + I_3$ out of the point.

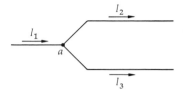

Figure 35-3
A simple circuit consisting of a battery, resistor, and connecting wires. (*Courtesy of Larry Langrill, Oakland University.*)

Figure 35-4
Graph of terminal voltage of battery versus current for the circuit shown in Figure 35-3. The dashed line shows the ideal emf.

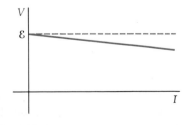

Figure 35-5
Circuit diagram representing the circuit with battery and resistor shown in Figure 35-3. The battery can be represented by an ideal emf ε and a small resistance r.

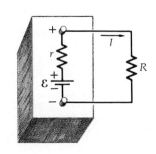

We can find the current in the circuit in Figure 35-5 using rule 1. Let the current I be in the direction shown. If we start at any point and traverse the complete circuit in the direction of the current, we encounter potential drops of Ir across the internal resistance of the battery and IR across the external resistor. These drops must be balanced by the increase through the emf. Thus rule 1 gives

$$\mathcal{E} = Ir + IR$$

or

$$I = \frac{\mathcal{E}}{r + R} \qquad 35\text{-}2$$

Note that the power output $\mathcal{E}I$ of the emf goes partly into joule heating of the external resistor and partly into joule heating of the battery because of its internal resistance. We can see this by multiplying each term in Equation 35-2 by I. We have then

$$\mathcal{E}I = I^2 r + I^2 R \qquad 35\text{-}3$$

Example 35-1 For a battery of given emf and internal resistance r, what value of external resistance R should be placed across the terminals to obtain the greatest joule heating in R?

The resistance R is sometimes called the *load resistance*. The power into R is $I^2 R$, where I is given by Equation 35-2. Thus the power is

$$P = I^2 R = \frac{\mathcal{E}^2}{(r + R)^2} R \qquad 35\text{-}4$$

Figure 35-6 shows a sketch of P versus R. The maximum value of P occurs when $R = r$. Thus the maximum power is obtained when the load resistance equals the internal resistance. A similar result also occurs for more complicated ac circuits and is known as *impedance matching*. It is left as an applied calculus problem for the student to derive the result that P is a maximum when $R = r$.

Example 35-2 Find the current in the circuit shown in Figure 35-7.

In this circuit, we have two seats of emf and three external resistors. We cannot predict the direction of the current unless we know which emf is greater, but we do not have to know the direction of the current before solving the problem. We can assume any direction and solve the problem with that assumption. If the assumption is incorrect, we shall get a negative number for the current, indicating that its direction is opposite to that assumed.

Let us assume I to be clockwise, as indicated in the figure. The potential drops and increases as we traverse the circuit in the assumed direction of the current, starting at point a, are given in Table 35-1. We encounter a potential drop traversing one of the emfs and an increase traversing the other. Kirchhoff's first rule gives

$$\mathcal{E}_1 = IR_1 + IR_2 + \mathcal{E}_2 + Ir_2 + IR_3 + Ir_1 \qquad 35\text{-}5$$

Solving for the current I, we obtain

$$I = \frac{\mathcal{E}_1 - \mathcal{E}_2}{R_1 + R_2 + R_3 + r_1 + r_2} \qquad 35\text{-}6$$

Note that if $\mathcal{E}_2$ is greater than $\mathcal{E}_1$, we get a negative number for the current I, indicating that we have chosen the wrong direction for I. For $\mathcal{E}_2$ greater than $\mathcal{E}_1$ the current is in the counterclockwise direction. On

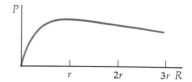

Figure 35-6
Plot of power into resistor $I^2 R$ versus R. The power is a maximum when the load resistance R equals the internal resistance r of the battery.

Impedance matching

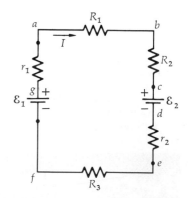

Figure 35-7
Circuit for Example 35-2.

the other hand, if ε_1 is the greater emf, we get a positive number for I, indicating that its assumed direction is correct.

Let us assume for this example that ε_1 is the greater emf and that the current is clockwise, as indicated. A charge ΔQ moving through battery 2 from point c to point d loses energy $\varepsilon_2 \Delta Q$. In this battery, electric energy is converted into chemical energy and stored. Battery 2 is *charging*. We can account for the energy balance in this circuit by multiplying each term in Equation 35-5 by the current I. Then the term on the left, $\varepsilon_1 I$, is the rate at which battery 1 puts energy into the circuit. This energy comes from internal chemical energy of the battery. The first term on the right, I^2R_1, is the rate of production of joule heat in resistor R_1. There are similar terms for each of the other resistors. The term $\varepsilon_2 I$ is the rate at which electric energy is converted into chemical energy in the second battery.

Table 35-1	
$a \to b$	Drop IR_1
$b \to c$	Drop IR_2
$c \to d$	Drop ε_2
$d \to e$	Drop Ir_2
$e \to f$	Drop IR_3
$f \to g$	Increase ε_1
$g \to a$	Drop Ir_1

In Example 35-2, the terminal voltage $V_a - V_f$ for battery 1 is less than its emf by the amount Ir_1, which is the potential drop across the internal resistance of the battery. For battery 2, which is being charged, the terminal voltage $V_c - V_e$ is greater than its emf. Since the current is from point d to point e, the potential at e is lower than that at d by the amount Ir_2. Thus the terminal voltage $V_c - V_e$ is the sum of ε_2 and Ir_2.

We should note from this example that because of its internal resistance, a battery is not completely reversible whereas an ideal seat of emf is. Consider, for example, a battery having an emf of 6 V and internal resistance of 1 Ω and delivering a current of 2 A to some external circuit. The power delivered by the emf is $P = $ (6 V)(2 A) $= 12$ W. The power delivered by the battery to the external circuit is $VI = $ (4 V)(2 A) $= 8$ W, where $V - \varepsilon - Ir = 6 - (2)(1)$ $= 4$ V is the terminal voltage. If we reverse the direction of the current and charge the battery with a current of 2 A (by any means, e.g., putting a larger emf in the external circuit), the terminal voltage of the battery is now $V = \varepsilon + Ir = 6 + 2(1) = 8$ V. The power delivered to the battery is now $VI = 8(2) = 16$ W. However, only part of this power goes into energy storage in the emf. The rate of energy storage is $\varepsilon I = 6(2) = 12$ W. The difference between the 16 W delivered to the battery and the rate of energy storage of 12 W is the power $I^2 r = 2^2(1) = 4$ W dissipated in the internal resistance of the battery.

Our simple model of an ideal emf in series with an internal resistance adequately describes the behavior of some real batteries in many circuit applications, but it is by no means a complete description and should not be taken too seriously. For example, although some batteries, such as the storage batteries used in automobiles, are nearly reversible and can easily be recharged, others (such as dry cells) are not. If you attempt to recharge a dry cell by driving current through it from its positive to negative terminal, most, if not all of the energy expended goes into heat rather than into chemical energy of the dry cell.

Example 35-3 The quantities in the circuit of Example 35-2 have the magnitudes $\varepsilon_1 = 12$ V, $\varepsilon_2 = 4$ V, $r_1 = r_2 = 1$ Ω, $R_1 = R_2 = 5$ Ω, $R_3 = 4$ Ω. Find the potentials at the points indicated in Figure 35-8, assuming that the potential at point f is zero, and discuss the energy balance in the circuit.

Since only potential differences are important, any point in a circuit can be chosen to have zero potential. In this example we have chosen point f in Figure 35-8. The potentials of the other points are then found relative to point f.[1] This procedure often simplifies the analysis of a problem.

We first find the current in the circuit. From Equation 35-6 we have

$$I = \frac{12 - 4}{5 + 5 + 4 + 1 + 1} = \frac{8 \text{ V}}{16 \text{ } \Omega} = 0.5 \text{ A}$$

We can now find the potentials at the indicated points relative to zero potential at point f. Since, by definition, the emf maintains a constant potential difference $\mathcal{E}_1 = 12$ V between point g and point f, the potential at point g is 12 V. The potential at point a is less than at g by the potential drop $Ir_1 = 0.5(1) = 0.5$ V. Thus the potential at point a is $12 - 0.5 = 11.5$ V. Similarly the potential drops across the 5-Ω resistors R_1 and R_2 are each $IR_1 = 0.5(5) = 2.5$ V. The potential at point b is then $11.5 - 2.5 = 9$ V, and that at c is 6.5 V. The potential drop across $\mathcal{E}_2$ is 4 V. Thus point d is at a potential of 2.5 V. Since the drop across the 1-Ω resistance r_2 is 0.5 V, the potential at e is 2 V. The potential drop across the 4-Ω resistance R_3 is $IR_3 = 2$ V. This gives zero for the potential at f, consistent with our original assumption.

The power delivered by the emf $\mathcal{E}_1$ is $\mathcal{E}_1 I = 12(0.5) = 6$ W. The power into the internal resistance of battery 1 is $I^2 r_1 = (0.5)^2(1) = 0.25$ W. Thus the power delivered by the battery to the external circuit is $6 - 0.25 = 5.75$ W. This also equals $V_1 I$, where $V_1 = V_a - V_f = 11.5$ V is the terminal voltage of that battery. The total power into the external resistances in the circuit is $I^2 R_1 + I^2 R_2 + I^2 R_3 = I^2(R_1 + R_2 + R_3) = (0.5)^2(5 + 5 + 4) = 3.5$ W. The power into the battery being charged is $(V_c - V_e)I = (6.5 - 2)(0.5) = 2.25$ W. Part of this, $I^2 r_2 = 0.25$ W, is dissipated in the internal resistance, and part, $\mathcal{E}_2 I = 2$ W, is the rate at which energy is stored in that battery.

We now consider an example of a circuit which contains more than one loop. In such circuits we need to apply Kirchhoff's second rule at junction points where the current splits into two or more parts.

Example 35-4 Find the current in each part of the circuit shown in Figure 35-9.

Let I be the current through the 18-V battery in the direction shown. At point b, this current divides into currents I_1 and I_2, as shown. Until we know the solution for the currents, we cannot be sure of their direction. For example, we need to know whether point b or point c is at the higher potential in order to know the direction of current through the 6-Ω resistor. The current directions indicated are merely guesses. Applying rule 2 to the junction point b, we obtain

$$I = I_1 + I_2$$

There are three possible loops for applying rule 1, loops $abcd$, $befc$, and $abefcd$. We need only two more equations to determine the three unknown currents. Equations for any two of the loops will be sufficient. (The third loop will then give redundant information.) Traversing the

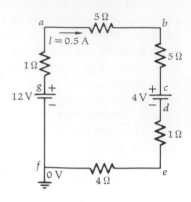

Figure 35-8
Circuit for Example 35-3.

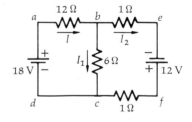

Figure 35-9
Circuit for Example 35-4.

---

[1] In more complicated circuits, we may wish to keep some point such as point f at a fixed potential relative to the earth even though some of the elements of the circuit change. This can be done by connecting that point to the earth, i.e., grounding it. Since the earth is such a large conductor, it has nearly infinite capacitance. Thus its potential remains essentially constant and is usually chosen to be zero.

loop *abcd* in the clockwise direction and applying Kirchhoff's first rule gives

$$(12 \ \Omega)I + (6 \ \Omega)I_1 = 18 \ V$$

$$2I + I_1 - 3 \ A \qquad\qquad 35\text{-}7$$

Applying rule 1 to loop *befc*, we obtain another equation,

$$(1 \ \Omega)I_2 + (1 \ \Omega)I_2 = 12 \ V + (6 \ \Omega)I_1$$

Note that in moving from c to b we encounter a voltage increase because the (assumed) current I_1 is in the opposite direction. Combining terms and substituting $I - I_1$ for I_2 in this equation, we obtain

$$2(I - I_1) = 12 \ A + 6I_1$$

$$2I = 12 \ A + 8I_1 \qquad\qquad 35\text{-}8$$

Equations 35-7 and 35-8 can be solved for the unknown currents I and I_1. The results are

$$I = 2 \ A \qquad I_1 = -1 \ A$$

Then

$$I_2 = I - I_1 = 2 - (-1) = 3 \ A$$

Our original guess about the direction of I_1 was incorrect. The current through the 6-Ω resistor is in the direction from point c to point b.

Question

1. How might you modify Kirchhoff's second rule at a point (such as on the plate of a capacitor) where charge can collect?

35-2 Series and Parallel Resistors

Two or more resistors connected such that the same charge must flow through each are said to be connected in series. Resistors R_1 and R_2 in Figure 35-10a are examples of resistors in series. They must carry the same current I. The potential drop across the two resistors (from point a to point c in Figure 34-10) is

$$V = IR_1 + IR_2 = I(R_1 + R_2)$$

We can often simplify the analysis of a circuit with resistors in series by replacing such resistors with a single equivalent resistance which gives the same total potential drop V when carrying the same current I. The equivalent resistance for series resistances is just the sum of the original resistances:

$$R_{eq} = \frac{V}{I} = R_1 + R_2$$

Figure 35-10
(*a*) Two resistors in series. (*b*) The resistors in part (*a*) can be replaced by a single equivalent resistance $R_{eq} = R_1 + R_2$, which gives the same total potential drop when carrying the same current as in part (*a*).

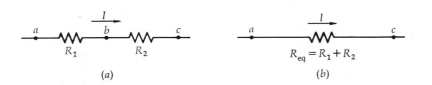

(*a*)   (*b*)

When there are more than two resistances in series, the equivalent resistance is

$$R_{eq} = R_1 + R_2 + R_3 + \cdots \qquad \text{35-9}$$

Two resistors connected as in Figure 35-11a are said to be connected in parallel. Let I be the current from point a to point b. At point a the current splits into two parts, I_1 through resistor R_1 and I_2 through R_2. From Kirchhoff's second rule the total current is the sum of the individual currents:

$$I = I_1 + I_2 \qquad \text{35-10}$$

Let $V = V_a - V_b$ be the potential drop across either resistor. In terms of the currents and resistances, we have

$$V = I_1 R_1 = I_2 R_2$$

The ratio of the two currents is therefore the inverse ratio of the resistances; the smaller resistance carries the larger current and vice versa:

$$\frac{I_1}{I_2} = \frac{R_2}{R_1} \qquad \text{35-11}$$

The equivalent resistance of a combination of parallel resistors is defined to be that resistance R_{eq} for which the same total current I produces the potential drop V:

$$R_{eq} = \frac{V}{I}$$

Equation 35-10 can then be written

$$I = \frac{V}{R_{eq}} = I_1 + I_2 = \frac{V}{R_1} + \frac{V}{R_2}$$

or

$$\frac{1}{R_{eq}} = \frac{1}{R_1} + \frac{1}{R_2} \qquad \text{35-12}$$

This result can be generalized to any number of resistances in parallel:

$$\frac{1}{R_{eq}} = \frac{1}{R_1} + \frac{1}{R_2} + \frac{1}{R_3} + \cdots \qquad \text{35-13}$$

The equivalent resistance of a combination of parallel resistors is less than that of any resistor. Consider, for example, any one resistor such as R_1 carrying current I_1 with potential drop $V = I_1 R_1$. The effect of another resistor or other resistors in parallel with R_1 is to increase the total current for the same potential drop V. The equivalent resistance V/I is thereby decreased.

The equivalent resistance for two parallel resistors can be found from Equation 35-12 by putting the two terms on the right side of the

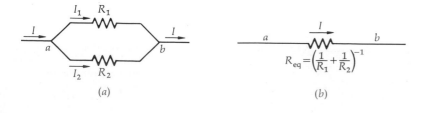

(a)   (b)

Figure 35-11
(a) Two resistors in parallel. (b) The two resistors in part (a) can be replaced by a single equivalent resistance R_{eq} related to R_1 and R_2 by $1/R_{eq} = 1/R_1 + 1/R_2$.

equation over a common denominator and taking the reciprocal of each side. The result is

$$R_{eq} = \frac{R_1 R_2}{R_1 + R_2} \qquad\qquad 35\text{-}14$$

This result explicitly demonstrates that R_{eq} is less than either R_1 or R_2 since the right side of the equation can be considered to be the product of either resistance (such as R_1) and a factor less than 1, such as $R_2/(R_1 + R_2)$. When there are three or more parallel resistors, it is usually easier to solve Equation 35-13 for $1/R_{eq}$ rather than to remember a general result analogous to Equation 35-14. For example, the equivalent resistance of three parallel resistors is given by the somewhat complicated expression (see Problem 14),

$$R_{eq} = \frac{R_1 R_2 R_3}{R_1 R_2 + R_2 R_3 + R_3 R_1} \qquad\qquad 35\text{-}15$$

The current in each of two parallel resistors can be found from the fact that the potential drop across the combination is IR_{eq} and is also equal to $I_1 R_1$ and to $I_2 R_2$:

$$I_1 R_1 = I_2 R_2 = IR_{eq} = I\,\frac{R_1 R_2}{R_1 + R_2}$$

Then

$$I_1 = \frac{R_2}{R_1 + R_2}\,I \quad\text{and}\quad I_2 = \frac{R_1}{R_1 + R_2}\,I \qquad\qquad 35\text{-}16$$

Example 35-5 For the circuit shown in Figure 35-12 find the equivalent resistance of the parallel combination of resistors, the current carried by each, and the total current through the emf.

We first find the equivalent resistance of the 6- and 12-Ω resistors in parallel:

$$\frac{1}{R_{eq}} = \frac{1}{6\ \Omega} + \frac{1}{12\ \Omega} = \frac{3}{12\ \Omega} = \frac{1}{4\ \Omega}$$

$$R_{eq} = 4\ \Omega$$

Note that the equivalent resistance is less than that of either resistor. Figure 35-13 shows the circuit with R_{eq} replacing the parallel combination. The resistances R_{eq} and r are in series. The current in this circuit is

$$I = \frac{\mathcal{E}}{R_{eq} + r} = \frac{6\text{ V}}{4\ \Omega + 2\ \Omega} = 1\text{ A}$$

This is the total current in the seat of emf.

We now find the currents I_1 and I_2 in the original circuit. From Equation 35-16 we have

$$I_1 = \frac{R_2}{R_1 + R_2}\,I = \frac{6\ \Omega}{12\ \Omega + 6\ \Omega}\,I = \tfrac{6}{18}\,(1\text{ A}) = \tfrac{1}{3}\text{ A}$$

$$I_2 = \frac{R_1}{R_1 + R_2}\,I = \frac{12\ \Omega}{18\ \Omega}\,(1\text{ A}) = \tfrac{2}{3}\text{ A}$$

The 6-Ω resistor carries twice as much current as the 12-Ω resistor since the potential drop is the same across each.

Figure 35-12
Circuit for Example 35-5.

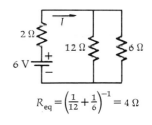

$$R_{eq} = \left(\tfrac{1}{12} + \tfrac{1}{6}\right)^{-1} = 4\ \Omega$$

Figure 35-13
Simplification of the circuit in Figure 35-12 by replacing the parallel resistors by a single equivalent resistance.

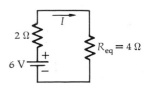

Question

2. A common mistake is to generalize Equation 35-14 to $R_1R_2R_3/(R_1 + R_2 + R_3)$ for the equivalent resistance of three parallel resistors. What simple argument can you give to show that such an expression could not be correct?

35-3   *RC* Circuits

In this section we consider circuits containing capacitors, in which the current is not steady but varies with time. Figure 35-14*a* shows a capacitor with an initial charge $+Q_0$ on the upper plate and $-Q_0$ on the lower plate. It is connected to a resistor R and a switch S, which is open to prevent the charge from flowing through the resistor. The potential difference across the capacitor is initially $V_0 = Q_0/C$, where C is the capacitance. Since there is no current with the switch open, there is no potential drop across the resistor. Thus there is also a potential difference V_0 across the switch.

We close the switch at time $t = 0$. Since there is now a potential difference across the resistor, there must be a current in it. The initial current is $I_0 = V_0/R$. The current is due to the flow of charge from the positive plate to the negative plate through the resistor, and so after a time the charge on the capacitor is reduced. The rate of decrease of the charge is just the current. If Q is the charge on the capacitor at any time, the current at that time is

$$I = -\frac{dQ}{dt} \qquad\qquad 35\text{-}17$$

We can write an equation for the charge and current as functions of time by applying Kirchhoff's first rule to the circuit after the switch is closed. Traversing the circuit in the direction of the current, we encounter a potential drop IR across the resistor and a potential increase Q/C across the capacitor. Kirchhoff's first rule gives

$$IR = \frac{Q}{C} \qquad\qquad 35\text{-}18$$

where both Q and I are functions of time and are related by Equation 35-17. Substituting $-dQ/dt$ for I in this equation, we have

$$-R\frac{dQ}{dt} = \frac{Q}{C} \qquad \text{or} \qquad \frac{dQ}{dt} = -\frac{1}{RC}Q \qquad 35\text{-}19$$

Equation 35-19, which relates the derivative of the charge to the charge, is a *differential equation*. It is solved by standard methods of calculus. We shall discuss its solution without doing the details of the calculus. We can understand how Q varies with time by considering the equation qualitatively. Originally the charge is Q_0, and its rate of decrease is $(1/RC)Q_0$. Later the charge is smaller, and its rate of decrease is correspondingly smaller. Figure 35-15 shows a sketch of Q versus time. After a long time both the value of Q and its slope, which is proportional to Q, approach zero as shown. The analytical form of the function $Q(t)$ is

$$Q = Q_0 e^{-t/RC} = Q_0 e^{-t/t_c} \qquad\qquad 35\text{-}20$$

The time

$$t_c = RC \qquad\qquad 35\text{-}21$$

Figure 35-14
(*a*) Parallel-plate capacitor in series with a switch and a resistor R. (*b*) Circuit diagram representing this circuit.

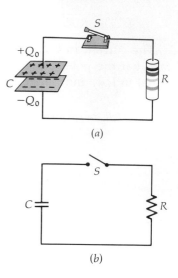

(*a*)

(*b*)

Figure 35-15
Plot of charge Q on capacitor versus time t for the circuit in Figure 35-14. The charge decreases by a factor $1/e$ in the time $t_c = RC$. The time constant t_c is also the time in which the capacitor would be fully discharged if the discharge rate were constant, as indicated by the dashed line.

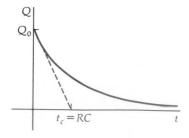

is the time in which the charge decreases to $1/e$ of its original value; it is called the *time constant* of the circuit. After a time t_c the charge is $Q = e^{-1}Q_0 = 0.37\,Q_0$. After a time $2t_c$ the charge is $Q = e^{-2}Q_0 = 0.135\,Q_0$, and so forth. We obtain the current as a function of time by differentiating $Q(t)$:

Time constant

$$\frac{dQ}{dt} = -\frac{1}{RC}\,Q_0 e^{-t/RC} = -\frac{1}{RC}\,Q \qquad 35\text{-}22$$

This function does indeed satisfy Equation 35-19. The current is

$$I = -\frac{dQ}{dt} = \frac{Q_0}{RC}\,e^{-t/RC} = I_0 e^{-t/RC} \qquad 35\text{-}23$$

where $I_0 = Q_0/RC = V_0/R$ is the initial current. The current also decreases exponentially with time and falls to $1/e$ of its initial value after a time $t_c = RC$. We can give a graphical interpretation of the time constant t_c. The slope of the graph of Q versus t at time $t = 0$ is, from Equation 35-22,

$$\frac{dQ}{dt} = \frac{-Q}{RC} = \frac{-Q_0}{t_c} \qquad \text{slope at } t = 0$$

The equation of the line tangent to this curve at time $t = 0$ is

$$Q = Q_0 - \frac{Q_0}{t_c}\,t$$

This straight line intersects the time axis $Q = 0$ at time $t = t_c$. We can thus think of the time t_c as the time after which the capacitor would be completely discharged if the rate were constant and equal to the initial rate.

Figure 35-16 shows a circuit used for charging a capacitor, which we assume to be originally uncharged. The switch, originally open, is closed at time $t = 0$. Charge immediately begins to flow into the capacitor. If the charge on the capacitor at some time is Q and the current in the circuit is I, Kirchhoff's first rule gives

$$\mathcal{E} = IR + \frac{Q}{C} \qquad 35\text{-}24$$

In this circuit, the current equals the rate of increase of charge on the capacitor:

$$I = \frac{dQ}{dt}$$

Substituting this into Equation 35-23, we obtain

$$\mathcal{E} = R\frac{dQ}{dt} + \frac{Q}{C} \qquad 35\text{-}25$$

At time $t = 0$, the charge is zero, and the current is $I_0 = \mathcal{E}/R$. The charge then increases, and the current decreases, as can be seen from Equation 35-24. The charge reaches a maximum value of $C\mathcal{E}$, again from Equation 35-24, when the current I equals zero. Figures 35-17 and 35-18 show the behavior of the charge versus time and the current versus time. The analytical expressions for $Q(t)$ and $I(t)$ are obtained by solving the differential equation 35-25:

$$Q(t) = C\mathcal{E}(1 - e^{-t/RC}) \qquad 35\text{-}26$$

$$I = \frac{dQ}{dt} = \frac{\mathcal{E}}{R}\,e^{-t/RC} \qquad 35\text{-}27$$

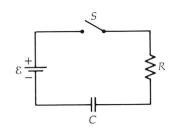

Figure 35-16
Circuit for charging a capacitor to a potential difference $\mathcal{E}$.

Figure 35-17
Plot of charge on capacitor versus time for the circuit in Figure 35-16. After time $t = t_c$ the charge is 63 percent of its final value of $C\mathcal{E}$. If the charging rate were constant and equal to its initial value, the capacitor would be fully charged after time $t = t_c$, as indicated by the dashed line.

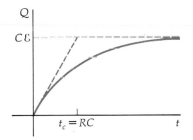

Figure 35-18
Plot of current versus time for the circuit in Figure 35-16. The current decreases by a factor $1/e$ in the time $t = t_c = RC$.

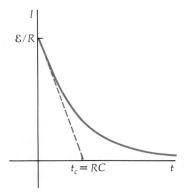

Example 35-6 A 6-V battery of negligible internal resistance is used to charge a 2-μF capacitor through a 100-Ω resistor. Find the initial current, the final charge, and the time required to obtain 90 percent of the final charge.

The initial current is $I_0 = \mathcal{E}/R = (6 \text{ V})/(100 \ \Omega) = 0.06$ A. The final charge is $Q_f = \mathcal{E}C = (6 \text{ V})(2 \ \mu\text{F}) = 12 \ \mu$C. The time constant for this circuit is $RC = 100 \ \Omega \times 2 \ \mu\text{F} = 200 \ \mu$sec. We expect the charge to reach 90 percent of its final value in a time of the order of several time constants. We can find the exact solution from Equation 35-26, using $Q = 0.9 \ \mathcal{E}C$:

$$Q = 0.9 \ \mathcal{E}C = \mathcal{E}C(1 - e^{-t/RC})$$

$$0.9 = 1 - e^{-t/RC}$$

$$e^{-t/RC} = 1 - 0.9 = 0.1$$

$$\ln e^{-t/RC} = -\frac{t}{RC} = \ln 0.1 = \ln 1 - \ln 10 = -\ln 10 = -2.3$$

since $\ln 0.1 = \ln \frac{1}{10} = \ln 1 - \ln 10$ and $\ln 1 = 0$. Thus

$$t = RC \ln 10 = 2.3RC = 2.3 \times 200 \ \mu\text{sec} = 460 \ \mu\text{sec}$$

During the charging process a total charge $Q_f = \mathcal{E}C$ flows through the battery. The battery does work,

$$W = Q_f \mathcal{E} = \mathcal{E}^2 C$$

The energy stored in the capacitor is just half this amount. By Equation 33-18,

$$U = \tfrac{1}{2}QV = \tfrac{1}{2}Q_f\mathcal{E} = \tfrac{1}{2}\mathcal{E}^2 C$$

We shall now show that the other half of the energy provided by the battery goes into joule heat in the resistor. The rate of energy put into the resistor is

$$\frac{dW_R}{dt} = I^2 R$$

Using Equation 35-27 for the current, we have

$$\frac{dW_R}{dt} = \left(\frac{\mathcal{E}}{R} e^{-t/RC}\right)^2 R = \frac{\mathcal{E}^2}{R} e^{-2t/RC}$$

We find the total heat produced by integrating from $t = 0$ to $t = \infty$:

$$W_R = \int_{t=0}^{t=\infty} dW_R = \int_0^\infty \frac{\mathcal{E}^2}{R} e^{-2t/RC} \ dt$$

The integration can be done by substituting $x = 2t/RC$. Then

$$dt = \frac{RC}{2} \ dx$$

$$W_R = \frac{\mathcal{E}^2}{R} \frac{RC}{2} \int_0^\infty e^{-x} \ dx = \tfrac{1}{2}\mathcal{E}^2 C$$

since the integral is 1. This answer is independent of the resistance R. When a capacitor is charged by a constant emf, half the energy provided by the battery (emf) is stored in the capacitor and half goes into heat, independent of the resistance. (This heat energy includes power into the internal resistance of the battery.)

35-4 Ammeters, Voltmeters, and Ohmmeters

We turn now to the consideration of the measurement of electrical quantities in dc circuits. Devices which measure current, potential difference, or resistance are called ammeters, voltmeters, and ohmmeters, respectively. To measure the current through the resistor in the simple circuit in Figure 35-19 we place an ammeter in series with the resistor, as indicated in the figure. Since the ammeter has some resistance, the current in the circuit is changed when the ammeter is inserted. Ideally, the ammeter should have a very small resistance so as to introduce only a small change in the current to be measured. The potential difference across the resistor is measured by placing a voltmeter across the resistor in parallel with it, as shown in Figure 35-20. The voltmeter reduces the resistance between points a and b, thus increasing the total current in the circuit and changing the potential drop across the resistor. An ideal voltmeter has a very large resistance, to minimize its effect on the circuit.

The principal component of an ammeter or voltmeter is a galvanometer, a device which detects a small current through it. The most common type, the *D'Arsonval galvanometer*, consists of a coil of wire free to turn, an indicator of some kind, and a scale. The galvanometer is designed so that the scale reading is proportional to the current in the galvanometer. The galvanometer operates on the principle that a coil carrying a current in a magnetic field experiences a torque which is proportional to the current. This torque rotates the coil until it is balanced by the restoring torque provided by the mechanical suspension of the coil. We shall study the effects of magnetic field on current carrying wires in Chapter 36; here we merely need to know that such a torque exists, that it is proportional to the current in the coil, and that with proper design of the magnet the torque does not depend on the angle of rotation of the coil. Figure 35-21 shows a sketch of a galvanometer coil in a magnetic field.

Since the restoring torque of the suspension is proportional to the angle of rotation of the coil, the equilibrium angle of rotation will be proportional to the current in the coil. The resistance of the galvanometer and the current needed to produce full-scale deflection are the two parameters important for the construction of an ammeter or

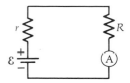

Figure 35-19
To measure the current in the resistor R an ammeter Ⓐ is placed in series with the resistor.

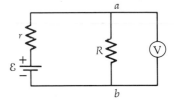

Figure 35-20
To measure the voltage drop across a resistor, a voltmeter Ⓥ is placed in parallel with the resistor.

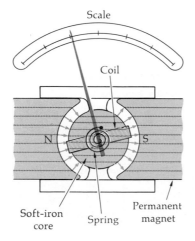

Figure 35-21
D'Arsonval galvanometer. When the coil carries a current, the magnet exerts a torque on the coil proportional to the current, causing the coil to twist. The deflection read on the scale is proportional to the current in the coil.

voltmeter from a galvanometer. Typical values of these parameters for a portable pivoted-coil laboratory galvanometer are $R_g = 20\ \Omega$ and $I_g = 5 \times 10^{-4}$ A. The voltage drop across a galvanometer with these parameters is thus $I_g R_g = 10^{-2}$ V for full-scale deflection. To construct an ammeter from a galvanometer, we place a small resistance, called *a shunt resistor,* in parallel with the galvanometer. Since the shunt resistance is usually much smaller than the resistance of the galvanometer, most of the current is carried by the shunt and the effective resistance of the ammeter is much smaller than the galvanometer resistance.

Resistors are added in series with a galvanometer to construct a voltmeter. Figure 35-22 illustrates the construction of an ammeter and voltmeter from a galvanometer. The choice of the appropriate resistors for the construction of an ammeter or voltmeter from a galvanometer is best illustrated by example.

Example 35-7 Using a galvanometer with a resistance of 20 Ω for which 5×10^{-4} A gives full-scale deflection, design an ammeter which will read full scale when the current is 5 A.

Since the total current through the ammeter must be 5 A when the current through the galvanometer is just 5×10^{-4} A, most of the current must go through the shunt resistor. Let R_p be the shunt resistance and I_p be the current through the shunt. Since the galvanometer and shunt are in parallel, we have $I_g R_g = I_p R_p$ and $I_p + I_g = 5$ A or

$$I_p = 5\ \text{A} - I_g = 5 - 5 \times 10^{-4}\ \text{A} \approx 5\ \text{A}$$

Thus the value of the shunt resistor should be

$$R_p = \frac{I_g}{I_p}\, R_g = \frac{5 \times 10^{-4}}{5} \times 20 = 2 \times 10^{-3}\ \Omega$$

Since the resistance of the shunt is so much smaller than the resistance of the galvanometer, the effective resistance of the parallel combination is approximately equal to the shunt resistance.

Example 35-8 Using the same galvanometer as in Example 35-7, design a voltmeter which will read 10 V full-scale deflection.

Let R_s be the value of a resistor in series with the galvanometer. We want to choose R_s so that a current of $I_g = 5 \times 10^{-4}$ A gives a potential drop of 10 V. Thus

$$I_g(R_s + R_g) = 10\ \text{V}$$

$$R_s + R_g = \frac{10\ \text{V}}{5 \times 10^{-4}\ \text{A}} = 2 \times 10^4\ \Omega$$

$$R_s = 2 \times 10^4 - R_g = 2 \times 10^4 - 20 = 19{,}980\ \Omega \approx 2 \times 10^4\ \Omega$$

Figure 35-23 shows an ohmmeter consisting of a battery, a galvanometer, and a resistor, which can be used to measure resistance. The resistance R_s is chosen so that the galvanometer reads full scale when the terminals a and b are shorted (touched together). Thus full scale on the galvanometer is marked zero resistance. When the terminals are connected across an unknown resistance R, the current is less than I_g and the galvanometer reads less than full scale. The current in this case is

$$I = \frac{\mathcal{E}}{R + R_s + R_g} \tag{35-28}$$

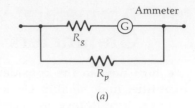

Ammeter

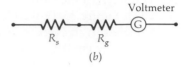

Voltmeter

(a)   (b)

Figure 35-22
(*a*) An ammeter consists of a galvanometer Ⓖ and a small parallel resistance R_p, called a shunt resistance. (*b*) A voltmeter consists of a galvanometer and a large series resistance R_s.

Ammeter

Voltmeter

Ohmmeter

Since this depends on R, the scale can be calibrated in terms of the resistance measured, from zero at full scale to infinite resistance at zero deflection. Since the calibration of the scale is far from linear and depends on the constancy of the emf of the battery, such an ohmmeter is not a high-precision instrument. It is, however, quite useful for making quick, rough determinations of resistance.

Some caution must be exercised in the use of an ohmmeter since it does send a current through the resistance to be measured. For example, consider an ohmmeter with a 1.5-V battery and a galvanometer similar to that in the previous examples. The series resistance needed is

$$I_g (R_s + R_g) = 1.5 \text{ V}$$

or

$$R_s = \frac{1.5}{5 \times 10^{-4}} - R_g = 3000 - 20 = 2980 \ \Omega$$

Suppose we were to use the ohmmeter to measure the resistance of a more sensitive galvanometer which gives a full-scale reading with a current of 10^{-5} A and has a resistance of about 20 Ω. When the terminals a and b are placed across this more sensitive galvanometer, the current will be just slightly less than 5×10^{-4} A because the total resistance is 3020 Ω, which is just slightly more than 3000 Ω. Such a current, about 50 times that needed to produce full-scale deflection, would undoubtedly ruin the sensitive galvanometer.

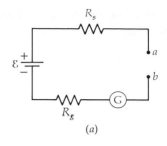

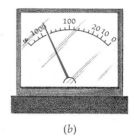

(b)

Figure 35-23
(a) Ohmmeter consisting of a battery in series with a galvanometer and a resistor R_s chosen so that the galvanometer reads full-scale deflection when points a and b are shorted. (b) Calibration of the galvanometer scale.

Questions

3. Under what conditions might it be advantageous to use a galvanometer which is less sensitive, i.e., requires a greater current I_g for full-scale deflection, than the one discussed in Examples 35-7 and 35-8?

4. When the series resistance R_s is properly chosen for a particular emf $\mathcal{E}$ of an ohmmeter, any value of resistance from zero to infinity can be measured. Why are there different scales on a practical ohmmeter for different ranges of resistance to be measured?

35-5   The Wheatstone Bridge

An accurate method of measuring resistance uses a circuit known as a *Wheatstone bridge* (Figure 35-24). In this figure R_x is an unknown resistance, and R_1, R_2, and R_4 are presumed known. The galvanometer is used as a null detector. The resistances R_1 and R_2 are varied until there is no current in the galvanometer; the bridge is then said to be balanced. Under this condition, the potential at point a is the same as that at point b. Thus the potential drop across R_x must equal that across R_1:

$$I_2 R_x = I_1 R_1$$

Similarly, the drop across R_2 must equal that across R_4. Thus

$$I_2 R_4 = I_1 R_2$$

We can eliminate the currents from these two equations by dividing

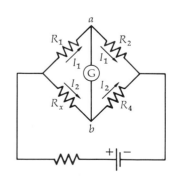

Figure 35-24
Wheatstone-bridge circuit for measuring an unknown resistance R_x in terms of known resistances R_1, R_2, and R_4.

one equation by the other, and then we can solve for the unknown resistance R_x:

$$\frac{R_x}{R_4} = \frac{R_1}{R_2}$$

or

$$R_x = R_4 \frac{R_1}{R_2} \qquad\qquad\text{35-29}$$

Balance condition

In practice, a large variable resistance is placed in series with the galvanometer to reduce its sensitivity and protect it when the bridge is unbalanced. When the bridge is nearly balanced, the series resistance is decreased and eventually shorted out for maximum sensitivity of the galvanometer. Often a meter-long wire is used for the resistors R_1 and R_2 (Figure 35-25). Point a is a sliding contact, which is varied until the bridge is balanced. If the wire is uniform, the ratio R_1/R_2 is just the ratio of the lengths L_1/L_2. The measurement does not depend on the emf used to supply the current.

Figure 35-25
Slide-wire Wheatstone bridge. The resistances R_1 and R_2 are proportional to the lengths L_1 and L_2. Their ratio is varied by moving a sliding contact along the wire until the bridge is balanced.

Question

5. What are the advantages and disadvantages of a Wheatstone bridge compared with an ohmmeter?

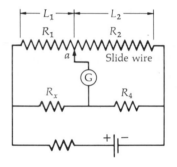

35-6 The Potentiometer

A similar bridge circuit useful in the measurement of emf is called a *potentiometer* (Figure 35-26). A battery supplies a constant current I through the resistances R, R_1, and R_2. Resistances R_1 and R_2 are varied in such a way that the sum $R_1 + R_2$ is kept constant. Then the current I does not change if the emf of the battery is constant. The cell whose emf is to be measured, labeled $\mathcal{E}_x$ in this diagram, is connected in series to a sensitive galvanometer G, a key switch S, and the resistance R_1. (There is also usually a protective resistance in series with the galvanometer which can be reduced to zero as the balance point is approached. This resistance is not shown in the figure.) The resistances R_1 and R_2 are adjusted so that when the switch is closed, there is no current through the galvanometer. The bridge circuit is then said to be balanced. Since there is no current in the unknown emf $\mathcal{E}_x$, the potential difference $V_a - V_b$ equals the emf $\mathcal{E}_x$. This potential difference also equals IR_1. Thus

Figure 35-26
Potentiometer circuit.

$$\mathcal{E}_x = IR_1 \qquad\qquad\text{35-30}$$

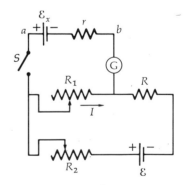

Instead of measuring the current I in the lower circuit, the cell with the unknown emf is replaced by a standard cell of known emf $\mathcal{E}_s$, and the bridge is again balanced with the sum $R_1 + R_2$ still unchanged so that the current I is unchanged. If R_{1s} is the value of R_1 when the circuit is balanced with the standard cell, we have

$$\mathcal{E}_s = IR_{1s}$$

Substituting $I = \mathcal{E}_s/R_{1s}$ in Equation 35-30, we obtain the unknown emf $\mathcal{E}_x$ in terms of the standard $\mathcal{E}_s$:

$$\mathcal{E}_x = \frac{R_1}{R_{1s}} \mathcal{E}_s \qquad\qquad\text{35-31}$$

Question

6. What are the advantages and disadvantages of a potentiometer compared with a voltmeter?

Review

A. Define, explain, or otherwise identify:

Kirchhoff's rules, 812	Galvanometer, 823
Internal resistance, 813	Ammeter, 824
Load resistance, 814	Voltmeter, 824
Impedance matching, 814	Ohmmeter, 824
Series resistors, 818	Wheatstone bridge, 825
Parallel resistors, 818	Potentiometer, 826
Time constant, 821	

B. True or false:

1. The effective resistance of two series resistors is always greater than that of either resistor.

2. The effective resistance of two parallel resistors is always less than that of either resistor.

3. To measure the current through a resistor, an ammeter is placed in series with the resistor.

4. To measure the potential drop across a resistor a voltmeter is placed in parallel with the resistor.

5. The time constant $t_c = RC$ is the time needed to discharge a capacitor completely.

6. A Wheatstone bridge is used to measure emf.

7. A potentiometer can measure the emf of a battery without drawing current from the battery.

Exercises

Section 35-1, Kirchhoff's Rules

1. A battery with 12 V emf has a terminal voltage of 11.4 V when it delivers 20 A to the starter of a car. (*a*) What is the internal resistance *r* of the battery? (*b*) What voltage must be put across the battery terminals to charge the battery at 10 A?

2. (*a*) How much power is delivered by the emf of the battery in Exercise 1 when it delivers 20 A? How much of this power is delivered to the starter? (*b*) By how much does the chemical energy of the battery decrease when it delivers 20 A for 3 min in starting the car? (*c*) How much heat is developed in the battery when it delivers 20 A for 3 min?

3. A 6-V (emf) battery with internal resistance 0.3 Ω is connected to a variable resistance *R*. Find the current and power delivered by the battery when *R* is (*a*) 5 Ω and (*b*) 10 Ω. (*c*) What is the maximum power that can be dissipated in the resistance *R*?

4. A variable resistance *R* is connected across a potential difference *V* which remains constant independent of *R*. At one value $R = R_1$, the current is 6.0 A. When *R* is increased to $R_2 = R_1 + 10.0$ Ω, the current drops to 2.0 A. Find (*a*) R_1 and (*b*) *V*.

5. A battery has emf ε and internal resistance *r*. When a 5.0-Ω resistor is connected across the terminals, the current is 0.5 A. When this resistor is replaced by an 11.0-Ω resistor, the current is 0.25 A. Find the emf ε and internal resistance *r*.

Figure 35-27
Exercises 6 and 8.

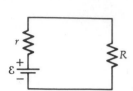

Figure 35-28
Exercise 7.

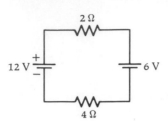

Figure 35-29
Exercise 9.

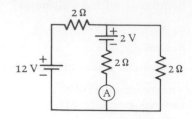

6. In Figure 35-27 the emf is 6 V and $R = 0.5\ \Omega$. The rate of joule heating in R is 8 W. (*a*) What is the current in the circuit? (*b*) What is the potential difference across R? (*c*) What is r?

7. For the circuit shown in Figure 35-28 find (*a*) the current, (*b*) the power delivered or absorbed by each emf, and (*c*) the rate of joule heat production in each resistor. (Assume the batteries have negligible internal resistance.)

8. In the circuit shown in Figure 35-27 let $\varepsilon = 18.0$ V and $r = 3.0\ \Omega$. (*a*) Graph the current through R as a function of R for the range $R = 0$ to $R = 6.0\ \Omega$. (*b*) For the same range of R, graph the power supplied to R as a function of R. (*c*) At what value of R is the power supplied to R a maximum? What is the value of this maximum power?

9. In the circuit shown in Figure 35-29 the batteries have negligible internal resistance, and the ammeter has negligible resistance. (*a*) Find the current through the ammeter. (*b*) Find the energy delivered by the 12-V battery in 3 sec. (*c*) Find the total heat produced in that time. (*d*) Account for the difference in your answers to parts (*b*) and (*c*).

10. In the circuit shown in Figure 35-30 the batteries have negligible internal resistance. (*a*) Find the current in each resistor, (*b*) the potential difference between points a and b, and (*c*) the power supplied by each battery.

11. Repeat Exercise 10 for the circuit shown in Figure 35-31.

Figure 35-30
Exercise 10.

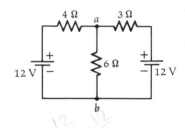

Figure 35-31
Exercise 11.

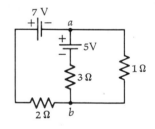

Section 35-2, Series and Parallel Resistors

12. (*a*) Find the equivalent resistance between points a and b in Figure 35-32. (*b*) If the potential drop between a and b is 12 V, find the current in each resistance.

13. Repeat Exercise 12 for the resistors shown in Figure 35-33.

14. Repeat Exercise 12 for the resistors shown in Figure 35-34.

15. Repeat Exercise 12 for the resistors shown in Figure 35-35.

16. In Figure 35-34 the current in the 6-Ω resistor is 3 A. (*a*) What is the potential drop between a and b? (*b*) What is the current in the 2-Ω resistor?

Figure 35-32
Exercise 12.

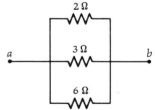

Figure 35-33
Exercise 13.

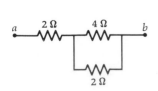

Figure 35-34
Exercises 14 and 16.

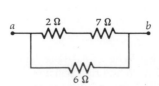

Figure 35-35
Exercise 15.

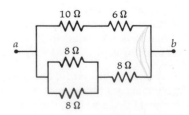

17. Repeat Exercise 12 for the resistors shown in Figure 35-36.

18. (a) Show that the equivalent resistance between a and b in Figure 35-37 is R. (b) What would be the effect of adding a resistance R between points c and d?

19. The battery in Figure 35-38 has negligible resistance. (a) Find the current in each resistor. (b) Find the power delivered by the battery.

20. Consider the equivalent resistance of two parallel resistors R_1 and R_2 as a function of the ratio $R_2/R_1 = x$. (a) Show that $R_{eq} = R_1 x/(1 + x)$. (b) Sketch R_{eq} as a function of x.

Section 35-3, RC Circuits

21. A 6-μF capacitor is initially charged to 100 V and then connected across a 500-Ω resistor. (a) What is the initial charge on the capacitor? (b) What is the initial current just after the capacitor is connected to the resistor? (c) What is the time constant of this circuit? (c) How much charge is on the capacitor after 6 msec?

22. (a) For Exercise 21, find the initial energy stored in the capacitor. (b) Show that the energy stored in the capacitor is given by $U = U_0 e^{-2t/t_c}$, where U_0 is the initial energy and $t_c = RC$ is the time constant. (c) Make a careful sketch of the energy U in the capacitor versus time t.

23. A 0.10-μF capacitor is given a charge Q_0. After 4 sec its charge is observed to be $\frac{1}{2}Q_0$. What is the effective resistance across this capacitor?

24. A 1.0-μF capacitor, initially uncharged, is connected in series with a 10-kΩ resistor and a 5.0-V battery of negligible internal resistance. (a) What is the charge on the capacitor after a very long time? (b) How long does it take for the capacitor to reach 99 percent of its maximum charge?

25. A 1-MΩ resistor is connected in series with a 1.0-μF capacitor and a 6.0-V battery of negligible internal resistance. The capacitor is initially uncharged. After a time $t - t_c - RC$, find (a) the charge on the capacitor, (b) the rate at which the charge is increasing, (c) the current, (d) the power supplied by the battery, (e) the power dissipated in the resistor, and (f) the rate at which the energy stored in the capacitor is increasing.

26. Repeat Exercise 25 for the time $t = 2t_c$.

Section 35-4, Ammeters, Voltmeters, and Ohmmeters

27. A galvanometer has a resistance of 100 Ω. It requires 1 mA to give full-scale deflection. (a) What resistance should be placed in parallel with the galvanometer to make an ammeter that reads 2 A at full-scale deflection? (b) What resistance should be placed in series to make a voltmeter that reads 5 V at full-scale deflection?

28. Sensitive galvanometers can detect currents as small as 10^{-12} A. How many electrons per second does this current represent?

29. A sensitive galvanometer has resistance 100 Ω and requires 1.0 μA of current to produce full-scale deflection. (a) Find the shunt resistance needed to construct an ammeter which reads 1.0 mA full scale. (b) What is the resistance of the ammeter? (c) What resistance would be required to construct a voltmeter reading 3.0 V full scale from this galvanometer?

30. A galvanometer of resistance 100 Ω gives full-scale deflection when its current is 1.0 mA. It is used to construct an ammeter whose full-scale reading is 100 A. (a) Find the shunt resistance needed. (b) What is the resistance of the ammeter? (c) If the shunt resistor consists of a piece of 10-gauge copper wire (diameter 2.59 mm), what should its length be?

Figure 35-36
Exercise 17.

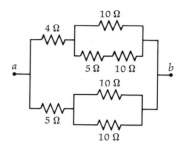

Figure 35-37
Exercise 18.

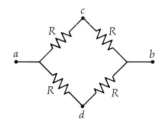

Figure 35-38
Exercise 19.

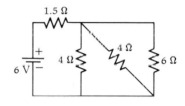

31. The galvanometer in Exercise 30 is used with a 1.5-V battery of negligible internal resistance to make an ohmmeter. (a) What resistance R_s should be placed in series with the galvanometer? (b) What resistance R will give half-scale deflection? (c) What resistance will give a deflection of one-tenth full scale?

Figure 35-39
Exercise 33.

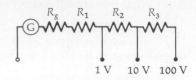

32. For the ohmmeter in Exercise 31, indicate how the galvanometer scale should be calibrated by representing the galvanometer scale on a straight line of some length L where the end of the line ($x = L$) represents full-scale reading at $R = 0$. Divide the line into 10 equal divisions and indicate values of the resistance at each division.

33. A galvanometer of resistance 100 Ω gives full-scale reading when its current is 0.1 mA. It is to be used in a multirange voltmeter as shown in Figure 35-39, where the connections refer to full-scale readings. Determine R_1, R_2, and R_3.

Figure 35-40
Exercise 34.

34. The galvanometer of Exercise 33 is to be used in a multirange ammeter with full-scale readings as indicated in Figure 35-40. Determine R_1, R_2, and R_3.

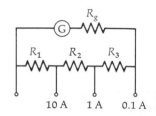

Section 35-5, The Wheatstone Bridge

35. In a slide-wire Wheatstone bridge using a uniform wire of length 1 m, R_1 is proportional to the length of wire from 0 to the balance point and R_2 is proportional to the length from the balance point to 100 cm (the end of the wire). If the fixed resistor is $R_4 = 200$ Ω, find the unknown resistance if (a) the bridge balances at the 18-cm mark, (b) the bridge balances at the 60-cm mark, and (c) the bridge balances at the 95-cm mark.

36. In the Wheatstone bridge of Exercise 35, with $R_4 = 200$ Ω, the bridge balances at the 98-cm mark. (a) What is the unknown resistance? (b) What effect would an error of 2 mm have on the measured value of the unknown resistance? (c) How should R_4 be changed so that this unknown resistor will give a balance point nearer the 50-cm mark?

Figure 35-41
Exercise 37.

Section 35-6, The Potentiometer

37. A basic slide-wire potentiometer circuit is shown in Figure 35-41. The resistor $R_1 + R_2$ is a long uniform wire of length $L_1 + L_2$. Show that if L_{1x} is the value of L_1 at balance for the unknown emf and L_{1s} is the corresponding value for the standard cell, $\mathcal{E}_x$ is given by

$$\mathcal{E}_x = \mathcal{E}_s \frac{L_{1x}}{L_{1s}}$$

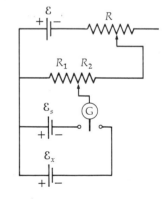

Problems

1. An emf with internal resistance r is connected to an external resistance R. Show that (a) the power P supplied by the emf is given by $\mathcal{E}^2/(r + R)$, (b) the power P_R supplied to the external resistor is $\mathcal{E}^2 R/(r + R)^2$, and (c) the power P_r dissipated in the emf is $\mathcal{E}^2 r/(r + R)^2$. Sketch on the same graph P, P_R, and P_r versus R.

Figure 35-42
Problem 3.

2. Show that as R varies in Problem 1, P_R is maximum at $R = r$, and that the maximum power is $P_R = \mathcal{E}^2/4r$.

3. In the circuit shown in Figure 35-42 find (a) the current in each resistor, (b) the power supplied by each emf, and (c) the power dissipated in each resistor.

4. Two identical batteries with emf $\mathcal{E}$ and internal resistance r may be connected across a resistance R either in series or in parallel. Which method of connection supplies the greater power to R when (a) $R < r$, (b) $R > r$?

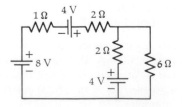

5. In the circuit shown in Figure 35-43, find the potential difference between points a and b.

6. The space between the plates of a parallel-plate capacitor is filled with a dielectric of constant K and resistivity ρ. (a) Show that the charge on the plates decays with time constant $t_c = \epsilon_0 K \rho$. (b) If the dielectric is mica with $K = 5.0$ and $\rho = 9 \times 10^{13}$ Ω-m, find the time for the charge to decrease to $1/e^2 \approx 14$ percent of its initial value.

7. In the circuit shown in Figure 35-44 the battery has internal resistance 0.01 Ω. An ammeter of resistance 0.01 Ω is inserted at point a. (a) What is the reading of the ammeter? (b) By what percentage is the current changed because of the ammeter? (c) The ammeter is removed and a voltmeter of resistance 1000 Ω is connected from a to b. What is the reading of the voltmeter? (d) By what percentage is the voltage drop from a to b changed by the presence of the voltmeter?

8. If you have two batteries, one with $\mathcal{E} = 9.0$ V and $r = 0.8$ Ω and the other with $\mathcal{E} = 3.0$ V and $r = 0.4$ Ω, how would you connect them to give the largest current through a resistor R for (a) $R = 0.2$ Ω, (b) $R = 0.6$ Ω, (c) $R = 1.0$ Ω, and (d) $R = 1.5$ Ω?

9. In the circuit shown in Figure 35-45, the reading of the ammeter is the same with both switches open as with both closed. Find the resistance R.

10. In Figure 35-46 the galvanometer has resistance 10 Ω and is connected across a 90-Ω resistor. The value of the current which causes full-scale deflection may be chosen by using connections ab, ac, ad, or ae. (a) How should the 90-Ω resistor be divided so that the current causing full-scale deflection decreases by a factor of 10 for each successive connection ab, ac, etc.? (b) What should be the full-scale deflection current in the galvanometer I_g so that this ammeter has ranges 1.0 mA, 10 mA, 100 mA, and 1 A?

11. Figure 35-47 shows two possible ways to use a voltmeter and ammeter to measure an unknown resistance. Assume that the internal resistance of the battery is negligible and that the resistance of the voltmeter is 1000 times that of the ammeter. $R_v = 1000 R_a$. The calculated value of R is taken to be $R_c = V/I$, where V and I are the readings of the voltmeter and ammeter. (a) Discuss which circuit is preferred for values of R in the range from $10 R_a$ to $0.9 R_v$. (b) Find R_c for each circuit if $R_a = 0.1$ Ω, $R_v = 100$ Ω, and $R = 0.5$, 3, and 80 Ω.

12. (See Problem 11.) For the circuits shown in Figure 35-47 show that $R_c = V/I$ is related to the true value R by

$$\frac{1}{R_c} = \frac{1}{R} + \frac{1}{R_v}$$

for circuit (a) and $R_c = R + R_a$ for circuit (b). If $\mathcal{E} = 1.5$ V, $R_a = 0.01$ Ω, and $R_v = 10{,}000$ Ω, what range of values of R can be measured such that R_c is within 5 percent of R using circuit (a)? Using circuit (b)?

13. Consider the circuit in Figure 35-48 in which r is the internal resistance of the emf and R_a the resistance of the ammeter. (a) Show that the ammeter reading is given by

$$\mathcal{E}\left(R_2 + R_a + r + \frac{R_2 + R_a}{R_1} r\right)^{-1}$$

(b) Show that if the ammeter and emf are interchanged, the ammeter reading is

$$\mathcal{E}\left(R_2 + R_a + r + \frac{R_2 + r}{R_1} R_a\right)^{-1}$$

Note that if $R_a = r$, or if both are negligible, the reading is the same. (When one can neglect R_a and r, this symmetry can be very useful in analyzing cir-

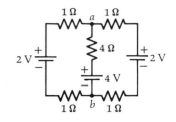

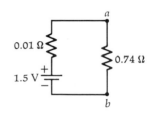

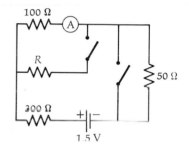

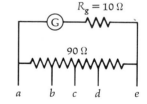

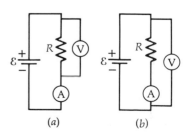

(a)   (b)

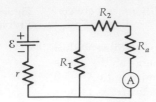

Figure 35-48
Problem 13.

cuits with only one emf; it is not valid for more than one emf.)

14. Derive Equation 35-15 for the equivalent resistance of three parallel resistors.

Figure 35-49
Problem 15.

15. In the circuit shown in Figure 35-49 the capacitor is originally uncharged with the switch open. At $t = 0$ the switch is closed. (*a*) What is the current supplied by the emf just after the switch is closed? (*b*) What is the current a long time after the switch is closed? (*c*) Derive an expression for the current through the emf for any time after the switch is closed. (*d*) After a long time t' the switch is opened. How long does it take for the charge on the capacitor to decrease to 10 percent of its value at $t = t'$ if $R_1 = R_2 = 5$ kΩ and $C = 1.0\ \mu$F?

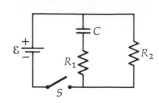

CHAPTER 36   The Magnetic Field

It was known to the early Greeks more than 2000 years ago that certain stones from Magnesia, now called magnetite, attract pieces of iron. Early experiments in magnetism were concerned chiefly with the behavior of permanent magnets. It is not known when a magnet was first used for navigation, but there is written reference to such use as early as the twelfth century.[1] In 1269 an important discovery was made by Pierre de Maricourt, when he laid a needle on a spherical natural magnet at various positions and marked the directions taken by the needle. The lines encircled the magnet just as the meridian lines encircle the earth, passing through two points at opposite ends of the sphere. Because of the analogy with the meridian lines of the earth, he called these points the *poles* of the magnet. It was noted by many experimenters that every magnet of whatever shape had two poles, a north pole and a south pole, where the magnetic force exerted by the magnet is strongest. Like poles of two magnets repel each other, and unlike poles attract.

In 1600 William Gilbert discovered the reason that a compass needle orients itself in definite directions: the earth itself is a permanent magnet. Since the north pole of a compass needle is attracted to the north geographic pole of the earth, this north geographic pole of the earth is actually a south magnetic pole. The attraction and repulsion of magnetic poles was studied quantitatively by John Michell in 1750. Using a torsion balance, Michell showed that the attraction and repulsion of the poles of two magnets are of equal strength and vary inversely as the square of the distance between the poles. These results were confirmed shortly thereafter by Coulomb. The law of force between two magnetic poles is similar to that between two electric charges, but there is an important difference: magnetic poles always occur in pairs. It is impossible to isolate a single magnetic pole. If a

---

[1] Much of this discussion follows the excellent book *A History of the Theories of Aether and Electricity*, by E. Whittaker, Nelson and Sons Ltd., London, 1953; Torchbook edition, Harper & Brothers, New York, 1960.

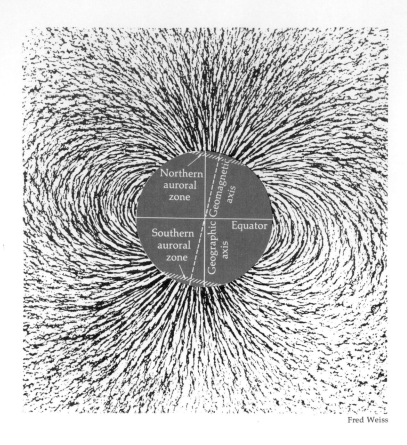

Fred Weiss

Magnetic field lines of the earth indicated by iron filings around a uniformly magnetized sphere.

magnet is broken in half, equal and opposite poles appear at the break point, so that there are two magnets each with equal and opposite poles. Coulomb explained this result by assuming that the magnetism is contained in each molecule of the magnet.

The connection between electricity and magnetism was not known until the nineteenth century, when Oersted discovered that an electric current affects the orientation of a compass needle. Subsequent experiments by Ampère and others showed that electric currents attract bits of iron and that parallel currents attract each other. Ampère proposed the theory that electric currents are the source of all magnetism; in ferromagnets these currents were thought to be "molecular" current loops which are aligned when the material is magnetized. Ampère's model is the basis of the theory of magnetism today, though in many cases the "currents" in magnetic materials are related to the intrinsic electron spin, a quantum property of the electron which could not be envisioned in the nineteenth century. Further connections between magnetism and electricity were demonstrated by experiments of Michael Faraday and Joseph Henry, which showed that a changing magnetic field produces a nonconservative electric field; and by Maxwell's theory, which showed that a changing electric field produces a magnetic field.

The basic magnetic interaction is one between two charges in motion. As with the electrostatic interaction, it is convenient to consider the magnetic force exerted by one moving charge on another to be transmitted by a third agent, the magnetic field. A moving charge produces a magnetic field, and that field in turn exerts a force on a second moving charge. Since a moving charge constitutes an electric current, the magnetic interaction can also be thought of as an interaction between two currents.

In this chapter we consider only the effects of given magnetic fields on moving charges and currents in wires. We postpone consideration of the origin of the magnetic field until Chapter 37.

36-1 Definition of the Magnetic Field **B**

We can define the magnetic field vector **B** at a point in space similar to the way we defined the electric field **E** (recall that we defined **E** at a point in space to be the force per unit charge on a test charge q_0 placed at that point). It is observed experimentally that when a charge has a velocity **v** in the vicinity of a magnet or a current-carrying wire, there is an additional force on it which depends on the magnitude and direction of the velocity. We can easily separate these two forces by measuring the force on the charge when it is at rest and subtracting this *electric* force from the total force acting when the charge is moving. For simplicity, let us assume that there is no electric field at the point in space we are considering. Experiments with various charges moving with various velocities at such a point in space give the following results for the magnetic force:

1. The force is proportional to the magnitude of the charge q.

2. The force is proportional to the speed v.

3. The magnitude and direction of the force depend on the direction of the velocity **v**.

4. If the velocity of the particle is along a certain line in space, the force is zero.

5. If the velocity is not along this line, there is a force which is perpendicular to this line and also perpendicular to the velocity.

6. If the velocity makes an angle θ with this line, the force is proportional to $\sin \theta$.

7. The force on a negative charge is in the opposite direction to that on a positive charge with the same velocity.

We can summarize these experimental results by defining a magnetic field vector **B** to be along the line described in result 4 and by writing the force

$$\mathbf{F} = q\mathbf{v} \times \mathbf{B} \qquad 36\text{-}1$$

For historical reasons, the vector **B** is usually called the *magnetic-induction vector* or the *magnetic flux density*. (A related vector **H**, the magnetic field intensity, will be introduced in Chapter 39 when we study magnetic materials. In free space, the vectors **B** and **H** are proportional to each other. In keeping with custom we shall refer to this **B** field as the *magnetic induction* although the phrase magnetic field **B** may also be used.)

Figure 36-1 shows the force exerted on various moving charges when the magnetic-induction vector **B** is in the vertical direction. Note that the direction of **B** can be found by experiment using Equation 36-1.

Magnetic force on moving charge

Figure 36-1
Direction of the magnetic force on a charged particle with velocity **v** at various orientations in a magnetic field **B**.

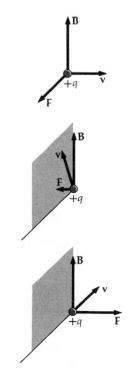

The SI unit of magnetic induction is the tesla (T).[1] A charge of one coulomb moving with a velocity of one meter per second perpendicular to a magnetic field of one tesla experiences a force of one newton.

$$1 \text{ T} = 1 \text{ N-sec/C-m} = 1 \text{ N/A-m}$$

This unit is rather large; a more common unit, derived from the cgs system, is the gauss (G), related to the tesla as follows:

$$1 \text{ T} = 10^4 \text{ G} \qquad\qquad 36\text{-}2$$

Because of the nearly universal use of the gauss as the unit of magnetic field, we often give **B** in gauss and remember to convert to teslas when making calculations.

Example 36-1 The magnetic induction of the earth has a magnitude of 0.6 G and is directed downward and northward, making an angle of about 70° with the horizontal. (These data are approximately correct for the central United States.) A proton moves horizontally in the northward direction with speed $v = 10^7$ m/sec. Calculate the force on the proton.

Figure 36-2 shows the directions of the magnetic induction **B** and the proton's velocity **v**. The angle between them is $\theta = 70°$. The direction of the force is $\mathbf{v} \times \mathbf{B}$, which is west for a proton moving north. The magnitude of the magnetic force is

$$F = qvB \sin \theta = (1.6 \times 10^{-19} \text{ C})(10^7 \text{ m/sec})(0.6 \times 10^{-4} \text{ T})(0.94)$$
$$= 9.02 \times 10^{-17} \text{ N}$$

The acceleration of the proton is

$$a = \frac{F}{m} = \frac{9.02 \times 10^{-17} \text{ N}}{1.67 \times 10^{-27} \text{ kg}} = 5.40 \times 10^{10} \text{ m/sec}^2$$

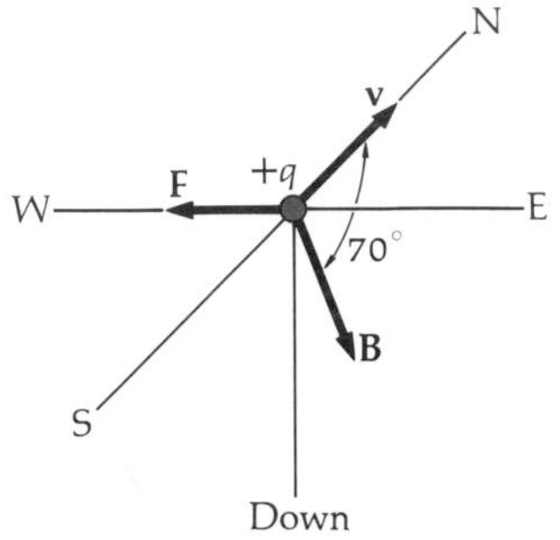

When a wire carries a current in a magnetic field, there is a force on the wire which is merely the sum of the magnetic forces on the charged particles whose motion produces the current. Figure 36-3 shows a short segment of wire of cross-sectional area A and length l carrying a current I. If the wire is in a magnetic field **B**, the magnetic force on each charge is $q\mathbf{v}_d \times \mathbf{B}$, where $\mathbf{v}_d$ is the drift velocity of the charge carriers. The number of charges in the wire segment is the number n per unit volume times the volume Al. Thus the total force **F** on the wire segment is

$$\mathbf{F} = (q\mathbf{v}_d \times \mathbf{B})nAl$$

By Equation 34-3, the current in the wire is

$$I = nqv_dA$$

Thus the force can be written

$$\mathbf{F} = I\mathbf{l} \times \mathbf{B} \qquad\qquad 36\text{-}3$$

where **l** is a vector whose magnitude is the length of the wire and whose direction is parallel to $q\mathbf{v}_d$, the direction of the current I. Equation 36-3 assumes that the wire is straight and that the magnetic induction does not vary over the length of the wire. It is easily generalized for an arbitrarily shaped wire in any magnetic field. We merely

Figure 36-2
Force on a proton moving north in the magnetic field of the earth, which makes an angle of 70° with the horizontal north direction. The force is directed toward the west.

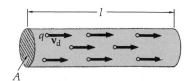

Figure 36-3
Current element of length l. Each charge experiences a magnetic force $q\mathbf{v}_d \times \mathbf{B}$.

---

[1] This unit is also called a weber per square meter, because the unit for magnetic flux is called the weber (Wb).

choose a very small wire segment $d\mathbf{l}$ and write the force on this segment $d\mathbf{F}$:

$$dF = I\, d\mathbf{l} \times \mathbf{B} \qquad\qquad 36\text{-}4$$

where $\mathbf{B}$ is the magnetic-induction vector at the segment. The quantity $I\, d\mathbf{l}$ is called a *current element*. We find the total force on the wire by summing (or integrating) over all the current elements using the appropriate field $\mathbf{B}$ at each element.

Force on current element

Equation 36-4 may be taken as an alternate definition of $\mathbf{B}$ equivalent to Equation 36-1. The magnitude and direction of $\mathbf{B}$ can be found using this definition by measuring the force on a current element $I\, d\mathbf{l}$ for various orientations of the element.

Questions

1. Charge q moves with velocity $\mathbf{v}$ through a magnetic field $\mathbf{B}$. At this instant it experiences a magnetic force $\mathbf{F}$. How would the force differ if the charge had the opposite sign? If the velocity had the opposite direction? If the magnetic field had the opposite direction?

2. For what angle between $\mathbf{B}$ and $\mathbf{v}$ is the magnetic force on q greatest? Least?

3. A moving electric charge may experience both electric and magnetic forces. How could you distinguish whether a force causing a charge to deviate from a straight path is an electric or a magnetic force?

4. How can a charge move through a region of magnetic induction without ever experiencing any magnetic force?

5. Show that the force on a current element is the same in direction and magnitude regardless of whether positive charges, negative charges, or a mixture of positive and negative charges create the current.

6. A current-carrying wire passes through a magnetic field, but the wire does not experience any magnetic force. How is this possible?

36-2 Magnets in Magnetic Fields

When a small permanent magnet such as a compass needle is placed in a magnetic field, it tends to orient itself so that the north pole points in the direction of $\mathbf{B}$. This effect also occurs with previously unmagnetized iron filings, which become magnetized in the presence of a $\mathbf{B}$ field. Figure 36-4 shows a small magnet which makes an arbitrary angle with a magnetic field $\mathbf{B}$. There is a force $\mathbf{F}_1$ on the north pole in the direction of $\mathbf{B}$ and an equal but opposite force $\mathbf{F}_2$ on the south pole. The *pole strength* of the magnet q^* is defined[1] to be the ratio of the magnitude of the force $\mathbf{F}$ on the pole and the magnitude of the magnetic induction $\mathbf{B}$:

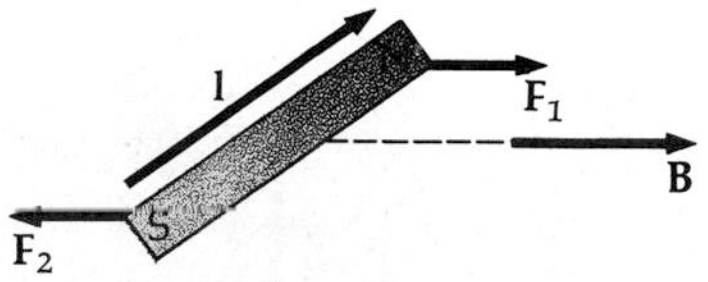

$$q^* = \frac{F}{B} \qquad\qquad 36\text{-}5$$

Figure 36-4
A small magnet experiences a torque in a uniform magnetic field.

Magnetic pole strength defined

---

[1] The notation for magnetic pole strength q^* is used so that the magnetic equations resemble the corresponding equations for electric charges in electric fields. The asterisk reminds us that q^* denotes a magnetic pole and not an electric charge.

Since a tesla is equal to a newton per ampere-meter, the SI unit of magnetic pole strength is the ampere-meter (A-m). If we adopt the sign convention that the north pole is positive and the south pole negative, the force on a pole can be written as a vector equation:

$$\mathbf{F} = q^*\mathbf{B} \qquad\qquad 36\text{-}6$$

From Figure 36-4 we see that there is a torque acting on a magnet in a magnetic field. If $\mathbf{l}$ is a vector pointing from the south pole to the north pole with the magnitude of the distance between the poles, the torque is

$$\boldsymbol{\tau} = \mathbf{l} \times \mathbf{F} = \mathbf{l} \times q^*\mathbf{B} = q^*\mathbf{l} \times \mathbf{B} \qquad\qquad 36\text{-}7$$

The magnetic moment $\mathbf{m}$ of a magnet is defined to be

$$\mathbf{m} = q^*\mathbf{l} \qquad\qquad 36\text{-}8$$

Magnetic moment

The SI unit of magnetic moment is the ampere-meter^2 (A-m^2). In terms of the magnetic moment, the torque on a magnet is

$$\boldsymbol{\tau} = \mathbf{m} \times \mathbf{B} \qquad\qquad 36\text{-}9$$

Torque on magnetic dipole

These equations are completely analogous to those for the torque on an electric dipole in an electric field (Section 29-8).

It is customary to indicate the direction of the magnetic induction $\mathbf{B}$ by drawing lines which are parallel to $\mathbf{B}$ at each point in space and to indicate the magnitude of $\mathbf{B}$ by the density of the lines, just as for the electric field $\mathbf{E}$. Such lines are called *lines of magnetic induction*. For a given magnetic induction $\mathbf{B}$ the lines can be found using a compass needle or iron filings since these small magnets align themselves in the direction of $\mathbf{B}$. Figure 36-5 shows the lines of magnetic induction near a bar magnet and near a long straight current-carrying wire. External to the magnet, the lines of $\mathbf{B}$ leave the north pole and enter the south pole. This is consistent with the fact that the force on a north pole is in the direction of $\mathbf{B}$, and two north poles repel each other. We shall investigate how to calculate $\mathbf{B}$ for these cases in later chapters.

Lines of magnetic induction

Questions

7. Assuming that the magnitude of the magnetic induction $\mathbf{B}$ is known in some region of space, how can the magnetic-pole strength of a small bar magnet be determined experimentally?

Figure 36-5
Magnetic field lines indicated by the alignment of iron filings for (*left*) a magnet and (*right*) a long straight current-carrying wire.

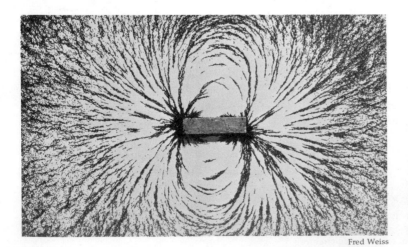
Fred Weiss

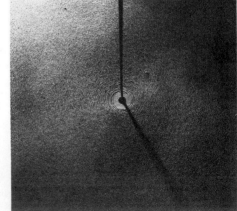
Education Development Center, Inc., Newton, Mass.

8. Although the force on a north pole of a magnet is in the direction of the lines of magnetic induction, the force on an electric charge q moving through a region of magnetic induction **B** is not. Is it possible to draw a set of lines which would indicate the direction of the magnetic force on a moving charge at each point in space?

9. The lines of magnetic induction near the long straight current-carrying wire shown in Figure 36-5b are circles concentric with the wire. If a single isolated magnetic pole did exist, what would happen to it if it were placed near such a wire?

36-3   Torque on a Current Loop in a Uniform Magnetic Field

When a wire carrying a current I is placed in a uniform magnetic field, there are forces on each part of the wire. If the wire is in the shape of a closed loop, there is no net force on the loop because the forces on the different parts of the wire sum to zero (assuming that the magnetic induction **B** is the same at all parts of the wire). However, in general, the magnetic forces produce a torque on the wire which tends to rotate the loop so that the area of the loop is perpendicular to the magnetic induction **B**.

Figure 36-6 shows a rectangular loop carrying a current I. The loop is in a region of uniform magnetic field parallel to the plane of the loop. The forces on each segment of the loop are indicated in the figure. There are no forces on the top or bottom of the loop because for those segments the current is parallel or antiparallel to the magnetic induction **B**, so that $I\,d\mathbf{l} \times \mathbf{B}$ is zero. The forces on the sides of the loop have the magnitude

$$F_1 = F_2 = IaB$$

where a is the length of the side. These forces are equal and opposite. They thus form a couple. The net force is zero, and the torque about any point is independent of the location of the point. Point P_1 is a convenient point about which to compute the torque. The magnitude of the torque is

$$F_1 b = IabB = IAB$$

where b is the width of the loop and $A = ab$ is the area of the loop. The torque is thus the product of the current, the area of the loop, and the

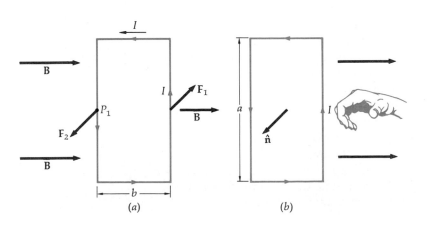

Figure 36-6
(a) Forces exerted on a rectangular current loop in a uniform magnetic field **B** parallel to the plane of the loop. (b) Unit vector $\hat{\mathbf{n}}$ perpendicular to the plane of the loop. The sense of $\hat{\mathbf{n}}$ is defined by the right-hand rule. With the fingers curling in the direction of the current, the thumb points in the direction of $\hat{\mathbf{n}}$. The forces in (a) tend to twist the loop so that $\hat{\mathbf{n}}$ rotates into **B**.

magnetic induction B. This torque tends to twist the loop so that its plane is perpendicular to **B**. The orientation of the loop can be described conveniently by a unit vector $\hat{n}$ perpendicular to the plane of the loop. The sense of $\hat{n}$ is chosen to be that given by the right-hand rule applied to the circulating current, as illustrated in Figure 36-6. The torque thus tends to rotate $\hat{n}$ into the direction of **B**.

Figure 36-7 shows the forces exerted by a uniform magnetic field on a rectangular loop whose normal unit vector $\hat{n}$ makes an angle θ with the magnetic induction **B**. Again, the net force on the loop is zero. The torque about any point (point P, for example) is the product of the force $F_2 = IaB$ and the lever arm $b \sin \theta$. The torque thus has the magnitude

$$\tau = IaBb \sin \theta = IAB \sin \theta$$

where again $A = ab$ is the area of the loop. This torque can be written conveniently in terms of the cross product of $\hat{n}$ and **B**:

$$\tau = IA\hat{n} \times \mathbf{B} \qquad\qquad 36\text{-}10$$

Equation 36-10 is of the same form as Equation 36-9 for the torque on a bar magnet in a uniform magnetic field with the quantity $IA\hat{n}$ for the loop playing the role of the magnetic moment $q^*\mathbf{l}$ of the bar magnet. We thus define the magnetic moment of a current loop to be $IA\hat{n}$. Often a loop contains several turns of wire, each turn having the same area and carrying the same current. The total magnetic moment of such a loop is the product of the number of turns and the magnetic moment of each turn. The total magnetic moment of a loop of N turns is then

$$\mathbf{m} = NIA\hat{n} \qquad\qquad 36\text{-}11$$

With this definition, the torque on the current loop is given by Equation 36-9:

$$\tau = \mathbf{m} \times \mathbf{B}$$

Equation 36-10, which we have derived for a rectangular loop, holds in general for a loop of any shape. The torque on any loop is the cross product of the magnetic moment of the loop and the magnetic induction **B**, where the magnetic moment is defined to be a vector whose magnitude is the current times the area of the loop and whose direction is perpendicular to the area of the loop. A current loop thus behaves like a small bar magnet (Figure 36-8). We shall see in the next chapter that the magnetic field produced by a current loop at points far from the loop is identical to that produced by a bar magnet at points far from the magnet if the magnetic moments are the same. Since such a field is of the same form as the electric field produced by an electric dipole, the analogy of a current loop with a bar magnet is useful for understanding the behavior of the loop in an external magnetic field and for explaining the magnetic field produced by the loop. We say more about the significance of this analogy in Chapter 37.

An important application of the torque exerted on a coil in a magnetic field is the D'Arsonval galvanometer discussed in Section 35-4 (see Figure 35-21). In this case, the magnetic field produced by a permanent magnet is radial at the coil, so that the angle θ between **B** and the normal to the plane of the coil is 90° independent of the orientation of the coil. The torque produced by the magnetic field is proportional to the current in the coil and is balanced by a restoring torque due to

Figure 36-7
Rectangular current loop whose unit normal $\hat{n}$ makes an angle θ with a uniform magnetic field **B**. The torque on the loop is $IA\hat{n} \times \mathbf{B} = \mathbf{m} \times \mathbf{B}$, where $\mathbf{m} = IA\hat{n}$ is the magnetic moment of the loop.

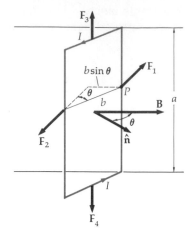

Magnetic moment of a current loop

Torque on current loop

Figure 36-8
A current loop of arbitrary shape experiences a torque $\mathbf{m} \times \mathbf{B}$ in a magnetic field.

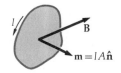

the spring suspension. If this restoring torque is proportional to the angular deflection of the coil, the angular deflection will be proportional to the current I. In any case, the angular deflection of the coil can be calibrated to indicate the current in the coil.

Questions

10. A current loop has its magnetic moment antiparallel to a uniform **B** field. What is the torque on the loop? Is this equilibrium stable or unstable?

11. A charge q moves in a circle of radius r with frequency f. What is the current for this current loop? What is the magnetic moment?

36-4   Motion of a Point Charge in a Magnetic Field

An important characteristic of the magnetic force on a moving charged particle described by Equation 36-1 is that the force is always perpendicular to the velocity of the particle. The magnetic force therefore does no work on the particle, and the kinetic energy of the particle is unaffected by this force. The magnetic force changes the direction of the velocity but not its magnitude.

In the special case where the velocity of a particle is perpendicular to a uniform magnetic field, as shown in Figure 36-9, the particle moves in a circular orbit. The magnetic force provides the centripetal force necessary for circular motion. We can relate the radius of the circle to the magnetic field and the velocity of the particle by setting the resultant force equal to the mass m times the centripetal acceleration v^2/r. The resultant force in this case is just qvB since **v** and **B** are perpendicular. Thus Newton's second law gives

$$qvB = \frac{mv^2}{r}$$

or

$$r = \frac{mv}{qB} \qquad\qquad 36\text{-}12$$

Figure 36-9
Charged particle moving in a plane perpendicular to a uniform magnetic field which is into the plane of the paper (indicated by the crosses). The magnetic force is perpendicular to the velocity of the particle, causing it to move in a circle of radius r, which can be found from $qvB = mv^2/r$.

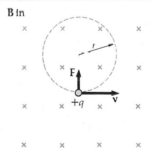

Circular path of electrons moving in a magnetic field produced by two large coils. The electrons ionize the gas in the tube, causing it to give off a bluish glow that indicates the path of the beam. (*Courtesy of Larry Langrill, Oakland University.*)

The angular frequency of the circular motion is

$$\omega = \frac{v}{r} = \frac{qB}{m}$$ 36-13

and the period is

$$T = \frac{2\pi}{\omega} = \frac{2\pi m}{qB}$$ 36-14

Radius, angular frequency, and period of circular orbit

Note that the frequency given by Equation 36-13 does not depend on the radius of the orbit or the velocity of the particle. This frequency is called the *cyclotron* frequency. Two of the many interesting applications of the circular motion of charged particles in a uniform magnetic field, the mass spectrograph and the cyclotron, are discussed in Examples 36-3 and 36-4.

If a charged particle enters a region of uniform magnetic induction with a velocity which is not perpendicular to **B**, the path of the particle is a helix. The component of velocity parallel to **B** is not affected by the magnetic field. Consider, for example, a uniform magnetic field in the z direction, and let v_z be the component of velocity of the particle parallel to the field. In a reference frame moving in the z direction with speed v_z, the particle has a velocity perpendicular to the field and moves in a circle in the xy plane. In the original frame the path of the particle is a helix which winds around the lines of **B**, as shown in Figure 36-10.

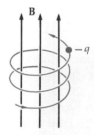

Figure 36-10
When a charged particle has a small component of velocity parallel to a magnetic field **B**, it moves in a helical path.

The magnetic force on a charged particle moving in a uniform magnetic field can be balanced by an electrostatic force if the magnitudes and directions of the magnetic and electric fields are properly chosen. Since the electric force is in the direction of the electric field (for positive particles) and the magnetic force is perpendicular to the magnetic field, the electric and magnetic fields must be perpendicular to each other if the forces are to balance. Figure 36-11 shows a region of space between the plates of a capacitor in which there is an electric field and a perpendicular magnetic field (which can be produced by a magnet not shown). Such an arrangement of perpendicular fields is called *crossed fields*. Consider a particle of charge q entering this space from the left. If q is positive, the electric force qE is down and the magnetic force qv × B is up. If the charge is negative, each of the forces is reversed. Since **v** is perpendicular to **B**, the magnitude of the magnetic force is just qvB. These two forces will balance if qE = qvB or

Crossed electric and magnetic fields

$$v = \frac{E}{B}$$ 36-15

For given magnitudes of the electric and magnetic fields, the forces will balance only for particles with the speed given by Equation 36-15. Any particle, no matter what its mass or charge, will traverse the space

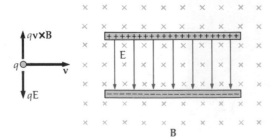

Figure 36-11
Crossed electric and magnetic fields. When a positive particle moves to the right, it experiences a downward electric force qE and an upward magnetic force qv × B, which balance if vB = E.

Cloud chamber photograph of the helical path of an electron moving in a magnetic field. The path of the electron is made visible by the condensation of water droplets in the cloud chamber. (*Courtesy of Carl E. Nielsen, Ohio State University.*)

undeflected if its speed is given by Equation 36-15. A particle of greater speed will be deflected in the direction of the magnetic force; one of less speed will be deflected in the direction of the electric force. Such an arrangement of fields is called a *velocity selector*; its use is illustrated in Examples 36-2 and 36-3.

Velocity selector

The motion of charged particles in nonuniform magnetic fields is quite complicated. We shall discuss qualitatively some cases in which the magnetic field **B** is nearly uniform so that the path of the particle is approximately a helix. Consider a magnetic field indicated by the lines of induction in Figure 36-12, which represents a three-dimensional field axially symmetric about the horizontal axis. The field is strongest at the ends and weakest at the center, as indicated by the line spacing. Such a field is sometimes called a *magnetic bottle* because charged particles can be trapped in it, as we shall show.

Magnetic bottle

Let a positive particle have an initial velocity into the paper at point P_1. If the magnetic induction **B** were horizontal at this point, the particle would move in a circle. However, **B** has a small vertical component, and the force $q\mathbf{v} \times \mathbf{B}$ has a small component to the right. The particle thus accelerates to the right as it moves in a nearly circular path about the axis of symmetry of the field. According to Equation 36-12, the radius of the circular motion is proportional to the velocity component perpendicular to **B** and inversely proportional to **B**. (Since the speed of the particle cannot be changed by the magnetic force, which is perpendicular to the velocity, an increase in the axial component of the velocity is accompanied by a slight decrease in the component perpendicular to the axis.) As the particle moves to the right, it enters a region of weaker magnetic field. The radius of its nearly circular path thus increases. At point P_2 the magnetic field is horizontal. The particle has no axial acceleration at this point but continues to move to the right because of its axial velocity. To the right of point P_2 the magnetic force has a component to the left which decreases the axial velocity until it is zero at point P_3, where the particle begins to move to the left. Since the field is stronger at point P_3, the radius of the circular

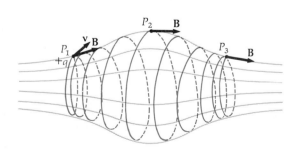

Figure 36-12
Magnetic bottle. The positive particle moves back and forth between P_1 and P_3 along the indicated path.

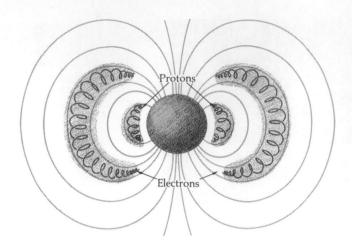

Figure 36-13
Van Allen belts. Protons
(inner belts) and electrons
(outer belts) are trapped in
the earth's magnetic field and
spiral along the field lines
between the north and south
poles, as in a magnetic bottle.

path is smaller than that at point P_2. The particle thus oscillates back and forth between points P_1 and P_3. A motion somewhat similar to this is the oscillation of charged ions back and forth between the earth's magnetic poles in the Van Allen belts (Figure 36-13).

Example 36-2 Thomson's Measurement of q/m for Electrons An example of the use of a velocity selector and the measurement of the deflection of charged particles in an electric field is the famous experiment of J. J. Thomson in 1897, in which he showed that rays in cathode-ray tubes can be deflected by electric and magnetic fields and therefore consist of charged particles. By observing the deflection of these rays with various combinations of electric and magnetic fields, Thomson was able to prove that all the particles had the same charge-to-mass ratio q/m and to determine the ratio. He showed that particles with this charge-to-mass ratio can be obtained using any material for the cathode, which means that these particles, now called electrons, are a fundamental constituent of all matter.

Figure 36-14 shows the cathode-ray tube he used. Electrons are emitted from a cathode C, which is at a negative potential relative to the slits A and B. There is an electric field in the direction from A to C which accelerates the electrons. They pass through the slits A and B into a field-free region and then encounter an electric field between the plates D and F which is perpendicular to the velocity of the electrons. The acceleration produced by this electric field gives the electrons a vertical component of velocity when they leave the region between the plates. They strike the phosphorescent screen S at the far right side of the tube at some displacement Δy from the point at which they strike when there is no field between the plates D and F. When the electrons strike the screen, it glows, indicating the location of the beam.

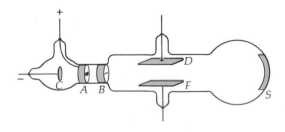

Figure 36-14
Thomson's tube for mea-
suring q/m. Electrons from the
cathode C pass through the
slits at A and B and strike a
phosphorescent screen S. The
beam can be deflected by an
electric field between the
plates D and F or by a magnetic field (not shown). [*From
J. J. Thomson,* Philosophical
Magazine, *Ser. 5, vol. 44
(1897).*]

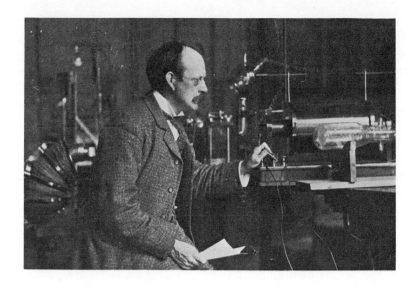

J. J. Thomson in his laboratory. (*Courtesy of the Cavendish Laboratory, University of Cambridge.*)

Let x_1 be the horizontal distance across the deflection plates D and F. If the velocity of the electron is v_0 when it enters the plates, the time spent between the plates will be

$$t_1 = \frac{x_1}{v_0}$$

Assuming that the electron was moving horizontally when it entered the region between the deflection plates, the vertical velocity when it leaves this region will be

$$v_y = at_1 = \frac{qE}{m} \, t_1 = \frac{qE}{m} \frac{x_1}{v_0}$$

where E is the electric field between the plates. The deflection in this region will be

$$\Delta y_1 = \tfrac{1}{2}at_1{}^2 = \frac{1}{2} \frac{qE}{m} \left(\frac{x_1}{v_0}\right)^2$$

The electron then travels an additional horizontal distance x_2 in the field-free region between the deflection plates and the screen. Since there is no force on the electron (except the negligible gravitational force), the velocity of the electron is constant in this region. It therefore takes a time

$$t_2 = \frac{x_2}{v_0}$$

to reach the screen and suffers an additional vertical deflection given by

$$\Delta y_2 = v_y t_2 = \frac{qE}{m} \frac{x_1}{v_0} \frac{x_2}{v_0}$$

The total deflection at the screen is

$$\Delta y = \Delta y_1 + \Delta y_2 = \frac{1}{2} \frac{qE}{m} \left(\frac{x_1}{v_0}\right)^2 + \frac{qE}{m} \frac{x_1 x_2}{v_0{}^2} \qquad \text{36-16}$$

If the initial velocity v_0 is known, a measurement of the deflection for a given electric field E yields a value for q/m. The initial velocity v_0 is de-

termined by introducing a magnetic field **B** between the plates perpendicular to both the electric field and the initial velocity and adjusting its magnitude so that the beam is undeflected. The velocity of
the electrons is then given by Equation 36-15. The magnetic field is
then turned off, and the deflection is measured.

Example 36-3 The Mass Spectrograph The mass spectrograph, first
made by Aston in 1919 and improved by Bainbridge and others, was
designed to measure the masses of isotopes. It measures the mass-to-
charge ratio of ions by determining the velocity of the ions and then
measuring the radius of their circular orbit in a uniform magnetic
field. According to Equation 36-12, the mass-to-charge ratio is given
by

$$\frac{m}{q} = \frac{Br}{v} \qquad \qquad 36\text{-}17$$

where B is the magnetic induction, r the radius of the circular orbit,
and v the speed of the particle. In Figure 36-15, a simple schematic
drawing of a mass spectrograph, ions from the source are accelerated
by an electric field and enter a uniform magnetic field produced by an
electromagnet. If the ions start from rest and move through a potential
drop V, their kinetic energy when they enter the magnet equals the
loss in potential energy qV:

$$\tfrac{1}{2}mv^2 = qV \qquad \qquad 36\text{-}18$$

The ions move in a semicircle of radius r given by Equation 36-17 and
strike a photographic film at point P_2, a distance $2r$ from the point
where they entered the magnet. The speed v can be eliminated from
Equations 36-17 and 36-18 to find q/m in terms of the known quantities V, B, and r. The result is

$$\frac{m}{q} = \frac{B^2r^2}{2V} \qquad \qquad 36\text{-}19$$

Strictly speaking, the mass of an ion can be determined only if the
charge q is known. However this is not important because the charge q
is either one or two electron charges and the spectrograph is used to
compare the masses of several isotopes which have nearly equal mass.
In Aston's original spectrograph this could be done to a precision of
about 1 part in 10,000. The precision is improved by the introduction
of a velocity selector between the ion source and the magnet, making
it possible to determine the velocity of the ions accurately and to limit
the range of velocities of the incoming ions.

Figure 36-15
Schematic drawing of a mass
spectograph. The outward
magnetic field is indicated by
the dots.

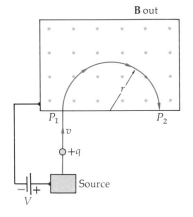

Example 36-4 The Cyclotron The cyclotron was invented by Lawrence
and Livingston in 1934 to accelerate particles such as protons or deuterons to high kinetic energy. The high-energy particles are then used
to bombard nuclei, causing nuclear reactions which are studied to obtain information about the nucleus. High-energy protons or deuterons
are also used to produce radioactive materials and for medical purposes. The operation of the cyclotron is based on the fact that the
period of motion of the charged particle in a uniform magnetic field is
independent of the velocity of the particle.

Figure 36-16 is a schematic drawing of a cyclotron. The particles
move in two semicircular metal containers called *dees* (because of their
shape). The containers lie in a vacuum chamber which is in a uniform

M. S. Livingston and E. O. Lawrence standing in front of their 27-in cyclotron in 1934. Lawrence won the Nobel prize (1939) for the invention of the cyclotron. (*Courtesy of Lawrence Radiation Laboratory, University of California, Berkeley.*)

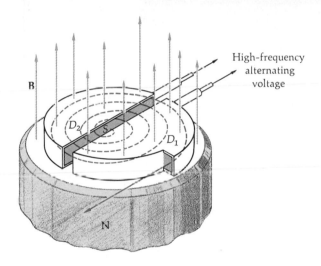

Figure 36-16
Schematic drawing of a cyclotron. The upper (south) pole face of the magnet has been omitted.

magnetic field provided by an electromagnet. (The region in which the particles move must be evacuated so that the particle will not lose energy and be scattered in collisions with air molecules.) The dees are maintained at a potential difference V, which alternates in time with the period T, chosen to be equal to the cyclotron period given by Equation 36-14. The charged particle is initially injected with a small velocity from an ion source S near the center of the magnetic field. It moves in a semicircle in one of the dees and arrives at the gap between the dees after a time $\frac{1}{2}T$, where T is the cyclotron period and also the period of the alternating potential across the dees. Let us call the dee in which the particle traverses this first semicircle dee 1. The alternating potential is phased so that when the particle arrives at the gap between the dees, dee 1 is at a higher potential than dee 2 so that the particle is accelerated across the gap by the electric field across the gap and gains kinetic energy qV. The region inside each dee is shielded from electric fields by the metal dee. Thus the particle moves in a semicircle of larger radius in dee 2 and again arrives at the gap after a time $\frac{1}{2}T$. By this time the potential between the dees has been reversed, and dee 2 is at the higher potential. Once more the particle is

A modern 83-in cyclotron at the University of Michigan. (A deflecting magnet is shown in the foreground.) (*Courtesy of University of Michigan.*)

accelerated across the gap and gains kinetic energy qV. In each half revolution, the particle gains kinetic energy qV and moves into a semicircular orbit of larger and larger radius until it leaves the magnetic field. In a typical cyclotron, a particle may make 50 to 100 revolutions.

Questions

12. By observing the path of a particle, how can you distinguish whether the particle is deflected by a magnetic field or an electric field?

13. A beam of positively charged particles passes undeflected from left to right through a velocity selector in which the electric field is up. The beam is then reversed so that it travels from right to left. Will the beam be deflected in the velocity selector? If so, in which direction?

36-5 The Hall Effect

In Section 36-1 we calculated the force exerted by a magnetic field on a current-carrying wire. This force is actually exerted directly on the charge carriers in the wire, the electrons. The force is transferred to the wire by the forces which bind the electrons to the wire at the surface. Since the charge carriers themselves experience the magnetic force when a current-carrying wire is in a magnetic field, the carriers are accelerated toward one side of the wire. This phenomenon, called the *Hall effect,* allows us to determine the sign of the charge on the carrier and the number density n of charge carriers in a conductor. It also provides a convenient method for measuring magnetic fields.

Figure 36-17a shows a conducting strip carrying a current I to the right in a magnetic field which we have chosen to be into the paper. Let us assume for the moment that the current consists of positively charged particles moving to the right. The magnetic force on these particles will be in the direction $q\mathbf{v}_d \times \mathbf{B}$, which is up in the figure. The positive particles will thus move up to the top of the strip, leaving the bottom of the strip with an excess negative charge. This charge separation causes an electrostatic field in the strip which opposes the magnetic force on the charge carriers. When the electrostatic and magnetic forces cancel, the charge carriers will no longer move upward. In this equilibrium situation, the upper part of the strip will be positively charged and be at a greater potential than the negatively charged lower part. On the other hand, if the current consists of negatively charged particles, they must be moving to the left for a current to the right. The magnetic force $q\mathbf{v}_d \times \mathbf{B}$ will again be up since we have changed the sign of both q and $\mathbf{v}_d$. Again the carriers will be forced to the upper part of the strip, but the upper part of the strip now carries a negative charge (because the charge carriers are negative) and the lower part a positive charge. A measurement of the sign of the potential difference between the upper and lower part of the strip tells us the sign of the charge carriers. Experiments for the current and magnetic field in the directions indicated in the figure show that the upper part of the strip carries a negative charge and is at a lower potential than the lower part. This observation led to the discovery that the charge carriers in metallic conductors are negative.

If we connect the upper and lower portions of the strip with a wire of resistance R, the negative electrons will flow from the upper part of the strip through the wire to the lower part. As soon as some electrons leave the upper part of the strip, more carriers in the strip will be driven up by the magnetic force because the electrostatic force is momentarily weakened by the reduced charge separation. Thus the magnetic force maintains the potential difference across the strip. The strip is thus a seat of emf. The potential difference between the top and bottom of the strip is called the *Hall emf*. Its magnitude is not hard to calculate. The magnetic force on the charge carriers in the strip has the magnitude qv_dB. This magnetic force is balanced by the electrostatic force qE, where E is the electric field due to the charge separation. Thus $E = v_dB$. If the width of the strip is w, the potential difference is Ew. The Hall emf is therefore

$$\mathcal{E}_H = Ew = v_dBw \qquad 36\text{-}20$$

We can see from Equation 36-20 that for ordinary-sized strips and magnetic fields, the Hall emf is very small since the drift velocity for ordinary currents is very small.

From measurements of the size of the Hall emf for a strip of a given size carrying a known current in a known magnetic field we can determine the number of charge carriers per unit volume in the strip. By Equation 34-5 the current density is

$$J = nqv_d \qquad 36\text{-}21$$

The current density is determined by measuring the current I and the cross-sectional area of the strip A. The quantity q is just one electron charge. The drift velocity is determined from Equation 36-20 by measuring the Hall emf. Thus the number density of charge carriers n is given by

$$n = \frac{J}{qv_d} = \frac{I}{Aqv_d} \qquad 36\text{-}22$$

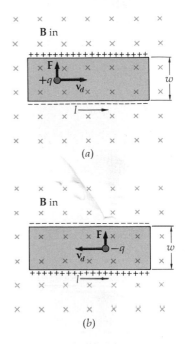

Figure 36-17
The Hall effect. Whether the current to the right is due to positive particles moving to the right as in (*a*) or to negative particles moving to the left as in (*b*), the magnetic force $q\mathbf{v}_d \times \mathbf{B}$ will be upward on the charge carriers. The sign of the carriers can be determined by observing whether the top of the strip is (*a*) positive or (*b*) negative.

Hall emf

(In Example 34-1 we estimated the drift velocity v_d by assuming a value for n, namely one electron per atom for copper. This assumption for n is justified by Hall emf measurements for copper.)

Although the Hall emf is ordinarily very small, it provides a convenient method for measuring magnetic fields. (Even very small emf's can be accurately measured by the potentiometer methods discussed in Chapter 35.) Combining Equations 36-20 and 36-21, we can write for the Hall emf,

$$\mathcal{E}_H = \frac{J}{nq} wB = \frac{Iw}{nqA} B \qquad\qquad 36\text{-}23$$

A given strip can be calibrated by measuring the Hall emf for a given current in a known magnetic field. The strength of the magnetic induction B of an unknown field can then be measured by placing the strip in the unknown field, sending a current through the strip, and measuring $\mathcal{E}_H$.

Review

A. Define, explain, or otherwise identify:

Magnetic induction, 835
Magnetic flux density, 835
Tesla, 836
Gauss, 836
Current element, 837
Pole strength, 837
Magnetic moment of magnet, 838
Lines of magnetic induction, 838
Magnetic moment of current loop, 840

Cyclotron frequency, 842
Crossed fields, 842
Velocity selector, 843
Magnetic bottle, 843
Mass spectrograph, 846
Cyclotron, 846
Hall effect, 848
Hall emf, 849

B. True or false:

1. The magnetic force is always perpendicular to the velocity of a particle.

2. The torque on a magnet tends to align the magnetic moment parallel to **B**.

3. A current loop behaves like a small magnet in a uniform magnetic field.

4. The period of a particle moving in a circle in a magnetic field is proportional to the radius of the circle.

5. The drift velocity of electrons in a wire can be determined from the Hall effect.

Exercises

Section 36-1, Definition of the Magnetic Field B

1. Find the magnetic force on a proton moving with velocity 4×10^6 m/sec in the positive x direction in a magnetic field of 2 T in the positive z direction.

2. A charge $q = -2$ nC moves with velocity $\mathbf{v} = -3.0 \times 10^6\, \mathbf{i}$ m/sec. Find the force on the charge if the magnetic field is (a) $\mathbf{B} = 6000\,\mathbf{j}$ G, (b) $\mathbf{B} = (6000\,\mathbf{i} + 6000\,\mathbf{j})$ G, (c) $\mathbf{B} = 8000\,\mathbf{i}$ G, (d) $\mathbf{B} = (6000\,\mathbf{j} + 6000\,\mathbf{k})$ G.

3. A uniform magnetic field of magnitude 1.5 T is in the positive z direction. Find the force on a proton if its velocity is (a) $\mathbf{v} = 3.0 \times 10^6\,\mathbf{i}$ m/sec, (b) $\mathbf{v} = 2.0 \times 10^6\,\mathbf{j}$ m/sec, (c) $\mathbf{v} = 8 \times 10^6\,\mathbf{k}$ m/sec, (d) $\mathbf{v} = (3.0 \times 10^6\,\mathbf{i} + 4.0 \times 10^6\,\mathbf{j})$ m/sec.

4. An electron moves with velocity 5×10^6 m/sec in the xy plane at an angle of 30° to the x axis and 60° to the y axis. A magnetic field of 15,000 G is in the positive y direction. Find the force on the electron.

5. A straight wire segment 2 m long makes an angle of 30° with a uniform magnetic field of 5000 G. Find the force on the wire if it carries a current of 2 A.

6. A straight wire 10 cm long carrying a current of 3.0 A is in a uniform magnetic field of 1.5 T. The wire makes an angle of 37° with the direction of **B**. What is the magnitude of the force on the wire?

7. A long wire parallel to the x axis carries a current of 10.0 A in the direction of increasing x. There is a uniform magnetic field of magnitude 2.0 T in the positive y direction. Find the force per unit length on the wire.

8. The wire segment in Figure 36-18 carries a current of 2.0 A from a to b. There is a magnetic field **B** $= 1.0$ **k** T. Find the total force on the wire and show that it is the same as if the wire were a straight segment from a to b.

Figure 36-18
Exercise 8.

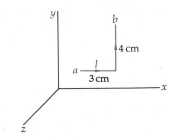

Section 36-2, Magnets in Magnetic Fields

9. A small magnet is placed in a magnetic field of 0.1 T. The maximum torque experienced by the magnet is 0.20 N-m. (*a*) What is the magnetic moment of the magnet? (*b*) If the length of the magnet is 4 cm, what is the pole strength q^*?

10. Show that the SI unit of magnetic moment, A-m², can also be written J/T.

11. The cgs unit for magnetic moment is the dyne-cm/G (1 dyne $= 10^{-5}$ N). (*a*) Find the conversion factor between 1 dyne-cm/G and 1 A-m² $= 1$ J/T (see Exercise 10). (*b*) A useful unit in atomic physics for magnetic moment is eV/G. Find the conversion factor between eV/G and A-m² $=$ J/T.

12. A small magnet of length 5 cm is placed at an angle of 45° to the direction of a uniform magnetic field of magnitude 0.04 T. The observed torque has the magnitude 0.10 N-m. (*a*) Find the magnetic moment of the magnet. (*b*) Find the pole strength q^*.

Section 36-3, Torque on a Current Loop in a Uniform Magnetic Field

13. A small circular coil of 20 turns of wire lies in a uniform magnetic field of 5000 G so that the normal to the plane of the coil makes an angle of 60° with the direction of **B**. The radius of the coil is 4 cm, and it carries a current of 3 A. (*a*) What is the magnitude of the magnetic moment of the coil? (*b*) What torque is exerted on the coil?

14. The SI unit for the magnetic moment of a current loop is A-m². Use this to show that 1 T $= 1$ N/A-m.

15. A rectangular 50-turn coil has sides 6.0 and 8.0 cm long and carries a current of 2.0 A. It is oriented as shown in Figure 36-19 and pivoted about the z axis. (*a*) If the wire in the xy plane makes an angle of 37° with the y axis as shown, what angle does the unit normal $\hat{\mathbf{n}}$ make with the x axis? Write an expression for $\hat{\mathbf{n}}$ in terms of the unit vectors **i** and **j**. (*b*) What is the magnetic moment of the coil? (*c*) Find the torque on the coil when there is a uniform magnetic field **B** $= 1.5$ **j** T.

Figure 36-19
Exercises 15 and 16.

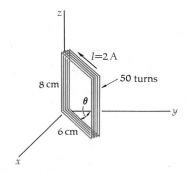

16. The coil of Exercise 15 is pivoted about the z axis and held at various positions in a uniform magnetic field **B** $= 2.0$ **j** T. Make a sketch of the coil position and find the torque exerted when the unit normal is (*a*) $\hat{\mathbf{n}} = \mathbf{i}$, (*b*) $\hat{\mathbf{n}} = \mathbf{j}$, (*c*) $\hat{\mathbf{n}} = -\mathbf{j}$, (*d*) $\hat{\mathbf{n}} = (\mathbf{i} + \mathbf{j})/\sqrt{2}$.

17. What is the maximum torque on a 500-turn circular coil of radius 0.5 cm which carries a current of 1.0 mA and resides in a uniform magnetic field of 1000 G?

18. A particle of charge q and mass M moves in a circle of radius r with angular velocity ω. (*a*) Show that the average current is $I = q\omega/2\pi$ and the magnetic

moment has the magnitude $m = \frac{1}{2}q\omega r^2$. (b) Show that the angular momentum of this particle has the magnitude $L = Mr^2\omega$ and that magnetic-moment and angular-momentum vectors are related by $\mathbf{m} = (q/2M)\mathbf{L}$.

Section 36-4, Motion of a Point Charge in a Magnetic Field

19. A proton moves in a circular orbit of radius 80 cm perpendicular to a uniform magnetic field of magnitude 0.5 T. (a) What is the period for this motion? (b) Find the speed of the proton. (c) Find the kinetic energy of the proton.

20. A particle of charge q and mass m has momentum $p = mv$ and kinetic energy $E_k = \frac{1}{2}mv^2 = p^2/2m$. If it moves in a circular orbit of radius r in a magnetic field B, show that (a) $p = Bqr$ and (b) $E_k = B^2q^2r^2/2m$.

21. An electron of kinetic energy 25 keV moves in a circular orbit in a magnetic field of 2000 G. (a) Find the radius of the orbit. (b) Find the angular frequency and the period of the motion.

22. Protons, deuterons, and alpha particles of the same kinetic energy enter a uniform magnetic field B which is perpendicular to their velocities. Let r_p, r_d, and r_α be the radii of their circular orbits. Find the ratios r_d/r_p and r_α/r_p. Assume $m_\alpha = 2m_d = 4m_p$.

23. An alpha particle travels in a circular path of radius 0.5 m in a magnetic field of 1.0 T. Find (a) the period, (b) speed, and (c) kinetic energy (in electron volts) of the alpha particle. Take $m = 6.65 \times 10^{-27}$ kg for the mass of the alpha particle.

24. Show that the cyclotron frequency is the same for deuterons as for alpha particles and that each is half that of a proton in the same magnetic field.

25. A beam of protons moves along the x axis in the positive x direction with speed 10^4 m/sec through a region of crossed fields. (a) If there is a magnetic field of magnitude 10^4 G in the positive y direction, find the magnitude and direction of the electric field. (b) Would electrons of the same velocity be deflected by these fields? If so, in what direction?

26. A velocity selector has a magnetic field of magnitude 0.1 T perpendicular to an electric field of magnitude 2×10^5 V/m. (a) What must the speed of a particle be to pass through undeflected? What energy must (b) protons and (c) electrons have to pass through undeflected?

27. The plates of a Thomson q/m apparatus are 5.0 cm long and are separated by 1.0 cm. The end of the plates is 25.0 cm from the tube screen. The kinetic energy of the electrons is 2.0 keV. (a) If a potential of 20.0 V is applied across the deflection plates, by how much will the beam deflect? (b) Find the magnitude of a crossed B field which will allow the beam to pass through undeflected.

28. A singly ionized ^{24}Mg ion (mass 24.0 u) is accelerated through 2000 V potential and bent in a magnetic field of 500 G in a mass spectrometer. (a) Find the radius of curvature of the orbit for the ion. (b) What is the difference in radius for ^{26}Mg and ^{24}Mg ions?

29. A cyclotron for accelerating protons has a magnetic field of 1.5 T and a radius of 0.5 m. (a) What is the cyclotron frequency? (b) Find the maximum energy of the protons when they emerge. (c) How do your answers change if deuterons are used instead of protons?

30. A certain cyclotron has a magnetic field of 2.0 T and is designed to accelerate protons to 20 MeV. (a) What is the cyclotron frequency? (b) What must the minimum radius of the magnet be to achieve the 20-MeV emergence energy? (c) If the alternating potential applied to the dees has a maximum value of 50,000 V, how many orbital trips must the protons make before emerging with 20 MeV energy?

Section 36-5, The Hall Effect

31. A metal strip 2 cm wide and 0.1 cm thick carries a current of 20 A in a uniform magnetic field of 2.0 T, as shown in Figure 36-20. The Hall emf is measured to be 4.27 μV. (a) Calculate the drift velocity of the electrons in the strip. (b) Find the number density of charge carriers in the strip.

32. (a) In Figure 36-20 which point (a or b) is at the higher potential? (b) If the metal strip is replaced by a p-type semiconductor in which the charge carriers are positive, which point will be at the higher potential?

33. The number density of free electrons in copper is 8.5×10^{22} electrons per cubic centimeter. If the metal strip in Figure 36-20 is copper and the current is 10 A, find (a) the drift velocity v_d and (b) the Hall emf. (Assume that the magnetic field is 2.0 T, as in Exercise 31.)

34. A copper strip ($n = 8.5 \times 10^{22}$ electrons per cubic centimeter) 2 cm wide and 0.1 cm thick (like that shown in Figure 36-20) is used to measure the magnitudes of unknown magnetic fields which are perpendicular to the strip. Find the magnitude of B in gauss when $I = 20$ A and the Hall emf is (a) 2.00 μV, (b) 5.25 μV, and (c) 8.00 μV.

Figure 36-20
Exercises 31 to 34.

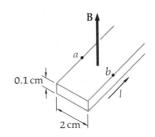

Problems

1. A 10.0-cm length of wire has mass 5.0 gm and is connected to a source of emf by light flexible leads. A magnetic field $B = 0.5$ T is horizontal and perpendicular to the wire. Find the current necessary to float the wire, i.e., such that the magnetic force balances the weight of the wire.

2. A metal crossbar of mass M rides on a pair of long horizontal conducting rails separated by a distance l and connected to a device that supplies constant current I to the circuit, as shown in Figure 36-21. A uniform magnetic field B is established as shown. (a) If there is no friction and the bar starts from rest at $t = 0$, show that at time t the bar has velocity $v = (BIL/M)t$. (b) In which direction will the bar move? (c) If the coefficient of static friction is μ_s, find the minimum field B necessary to start the bar moving.

3. In Figure 36-21 assume the rails are frictionless but tilted upward so that they make an angle θ with the horizontal. (a) What vertical magnetic field B is needed so that the bar will not slide down the rails? (b) What is the acceleration of the bar if B is twice the value found in (a)?

Figure 36-21
Top view of bar on rails for Problems 2 and 3.

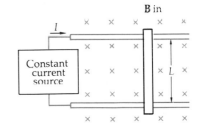

4. A wire bent into some arbitrary shape carries a current I in a uniform magnetic field B. Show that the total force on the part of the wire from some point a to some point b is $\mathbf{F} = I\mathbf{L} \times \mathbf{B}$, where $\mathbf{L}$ is the vector from a to b. *Hint:* Integrate $d\mathbf{F} = I \, d\mathbf{l} \times \mathbf{B}$ for constant $\mathbf{B}$.

5. If you have a wire of fixed length L and make a coil of N turns, the larger the number of turns the smaller the area enclosed by the wire. Show that for a wire of given length carrying current I the maximum magnetic moment is achieved with a coil of just one turn and the magnitude of this magnetic moment is $IL^2/4\pi$. (You need consider only circular coils. Why?)

6. A nonconducting rod of length l has a uniform charge per unit length λ and is rotated with angular velocity ω about an axis through one end and perpendicular to the rod. (a) Consider a small segment of length dx and charge $dq = \lambda \, dx$ at a distance x from the pivot. Show that the magnetic moment of this segment is $\frac{1}{2}\lambda\omega x^2 \, dx$. (b) Integrate your result to show that the total magnetic moment of the rod is $m = \frac{1}{6}\lambda\omega l^3$. (c) Show that the magnetic moment $\mathbf{m}$ and angular momentum $\mathbf{L}$ are related by $\mathbf{m} = (Q/2M)\mathbf{L}$, where M is the mass of the rod and Q is the total charge.

7. A nonconducting disk of radius R has uniform surface charge density σ and rotates with angular velocity ω about its axis. (a) Consider a ring of radius r

and thickness dr. Show that the total current in this ring is $dI = (\omega/2\pi)\, dq = \omega\sigma r\, dr$. (b) Show that the magnetic moment of the ring is $dm = \pi\omega\sigma r^3\, dr$. (c) Integrate your result for part (b) to show that the total magnetic moment of the disk is $m = \frac{1}{4}\pi\omega\sigma R^4$. (d) Show that the magnetic moment $\mathbf{m}$ and angular momentum $\mathbf{L}$ are related by $\mathbf{m} = (Q/2M)\mathbf{L}$, where M is the total mass of the disk and Q is its total charge.

8. Particles of charge q and mass m are accelerated from rest through a potential difference V and enter a region of uniform magnetic field $\mathbf{B}$ perpendicular to the velocity. If r is the radius of curvature of their circular orbit, show that $q/m = 2V/r^2B^2$.

9. A beam of particles enters a region of uniform magnetic field $\mathbf{B}$ with velocity $\mathbf{v}$ which makes a small angle θ with $\mathbf{B}$. Show that after a particle moves a distance $2\pi(m/qB)v\cos\theta$ measured along the direction of $\mathbf{B}$, the velocity of the particle is in the same direction as when it entered the field.

10. A small magnet of moment $\mathbf{m}$ makes an angle θ with a uniform magnetic field $\mathbf{B}$. (a) How much work must be done by an external torque to twist the magnet by a small amount $d\theta$? (b) Show that the work required to rotate the magnet until it is perpendicular to the field is $W = mB\cos\theta$. (c) Use your result for part (b) to show that if the potential energy of the magnet is chosen as zero when the magnet is perpendicular to the field, the potential energy at angle θ is $U(\theta) = -\mathbf{m}\cdot\mathbf{B}$. (d) Would any part of this problem be different if the magnet were replaced by a coil carrying current such that its magnetic moment is $\mathbf{m}$?

CHAPTER 37   Sources of the Magnetic Field

We now turn to a consideration of the origins of the magnetic field **B**. The earliest known sources of magnetism were magnets, but after Oersted described his finding in September 1820 that a compass needle is deflected by an electric current, many scientists investigated the properties of the magnetism associated with electric currents. One month later, Jean Baptiste Biot and Félix Savart announced the results of their measurements of the force on a magnetic pole near a long current-carrying wire and analyzed these results in terms of the magnetic field produced by each element of the current. Ampère extended these experiments and showed that current elements themselves experience a force in the presence of a magnetic field; in particular, he showed that two currents exert forces on each other.

Brown Brothers

André Marie Ampère
(1775–1836).

Ampère also obtained the Biot-Savart result for the magnetic field due to a current element and developed a model which explained the behavior of permanent magnets in terms of microscopic current loops within the magnetic material. In this model, it is proposed that all matter contains microscopic current loops which in nonmagnetic materials are randomly oriented, producing no net effect. In magnetic materials, current loops are aligned and produce a magnetic field with the same characteristics as that produced by aligned current loops of ordinary conduction current in wires. Ampère's model is essentially correct in that we can treat all magnetic fields as arising from an electric current of some kind. However, the microscopic currents (called *amperian currents*) are more complicated than could have been envisioned by Ampère, being related to the motions of atomic electrons which cannot be fully described classically. In particular, ferromagnetism is intimately connected with electron spin, an intrinsic property of the electron.

37-1 The Biot-Savart Law

From their investigations of the force produced by a long wire carrying a current on a magnetic pole, Biot and Savart proposed an expression relating the magnetic induction vector **B** at a point in space to an element of the current that produces it. Let $I\,d\mathbf{l}$ be an element of current (Figure 37-1). The magnetic induction $d\mathbf{B}$ produced by this element at a field point P a distance r away is

$$dB = k_m \frac{I\,d\mathbf{l} \times \hat{\mathbf{r}}}{r^2} \qquad\qquad 37\text{-}1 \qquad \textit{Biot-Savart law}$$

where $\hat{\mathbf{r}}$ is the unit vector pointing from the current element to the field point and k_m is a constant which depends on the units. Equation 37-1, the *Biot-Savart law*, was also deduced by Ampère. The magnetic induction due to the total current in a circuit can be found by using the Biot-Savart law for the field due to each current element and summing over all the current elements in the circuit.

The SI unit of current, the ampere, is defined so that the magnetic constant k_m is exactly 10^{-7}. This definition of the ampere will be discussed in more detail in the next section. The units of k_m will be shown below to be N/A^2. As with the coulomb constant, it is customary to write the magnetic constant k_m in terms of another constant μ_0, called the *permeability of free space*. These constants are related by

$$k_m = \frac{\mu_0}{4\pi} = 10^{-7}\ N/A^2$$

$$\mu_0 = 4\pi k_m = 4\pi \times 10^{-7}\ N/A^2 \qquad\qquad 37\text{-}2$$

In terms of the constant μ_0 the Biot-Savart law is written

$$d\mathbf{B} = \frac{\mu_0}{4\pi} \frac{I\,d\mathbf{l} \times \hat{\mathbf{r}}}{r^2} \qquad\qquad 37\text{-}3$$

The Biot-Savart law is analogous to Coulomb's law. The source of the magnetic field is the current element $I\,d\mathbf{l}$, just as the charge q is the source of the electrostatic field. The magnetic field decreases as the square of the distance from the current element, like the decrease in the electric field with distance from a point charge, but the directional aspects of these fields are quite different. Whereas the electrostatic field points in the radial direction **r** from the point charge to the field point (assuming a positive charge), the magnetic field is perpendicular both to the radial direction and to the direction of the current element $I\,d\mathbf{l}$. At a point along the line of the current element, e.g., point P_2 in Figure 37-1, the magnetic field is zero.

By combining the Biot-Savart law with the expression for the force on a current element in a magnetic field (Equation 36-4) we can write an equation for the force exerted by one current element on another. The force on current element $I_2\,d\mathbf{l}_2$ exerted by element $I_1\,d\mathbf{l}_1$ is

$$d\mathbf{F}_{12} = I_2\,d\mathbf{l}_2 \times \left(k_m \frac{I_1\,d\mathbf{l}_1 \times \hat{\mathbf{r}}}{r^2} \right) \qquad\qquad 37\text{-}4$$

From this equation we can see that if the force is in newtons, the distances dl_1, dl_2, and r in meters, and the currents in amperes, the units of k_m are N/A^2.

This relation is remarkable in that the force exerted by element 1 on element 2 is not equal and opposite to that exerted by element 2 on element 1. That is, these forces do not obey Newton's action-reaction law, as can be demonstrated by considering the special case illustrated in Figure 37-2. Here, the magnetic field at element 2 due to element 1 is into the paper, and the force on element 2 is to the left. However, the magnetic field at element 1 due to element 2 is zero because $d\mathbf{l}_2 \times \hat{\mathbf{r}}$

Figure 37-1
The magnetic field $d\mathbf{B}$ at point P_1 due to the current element $I\,d\mathbf{l}$ is $d\mathbf{B} = k_m\,(I\,d\mathbf{l} \times \hat{\mathbf{r}})/r^2$. At point P_2 the field due to this element is zero because $d\mathbf{l} \times \hat{\mathbf{r}}$ is zero.

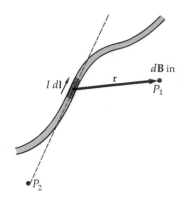

Figure 37-2
The forces exerted by current elements on each other are not equal and opposite. Here **B** at element 2 due to element 1 is into the paper, giving a force on element 2 to the left, but **B** due to element 2 at element 1 is zero.

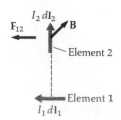

is zero. Thus no force is exerted on element 1. If the force $\mathbf{F}_2$ is the only force acting on the two-current-element system, the system will accelerate in the direction of $\mathbf{F}_2$ and linear momentum of the system will apparently not be conserved. We recall that it was the experimental observation of conservation of momentum in collisions that originally led Newton to the law of action and reaction. In most situations, current elements are but a part of a complete circuit. If the current elements shown are parts of complete circuits, as in Figure 37-3, there will be forces on the elements from other parts of the circuits. A detailed analysis of the total force exerted on one circuit by the other shows that the total forces *do* obey Newton's third law; i.e., the force exerted on circuit 1 by circuit 2 is equal and opposite to the total force exerted on circuit 2 by circuit 1. This analysis can be found in intermediate books on electricity and magnetism.

It is possible to produce the equivalent of isolated current elements by accelerating charges for a short time and then stopping them. Then the problem of the apparent violation of Newton's action-reaction law and of conservation of momentum is real. It is this type of situation for which our discussion of action at a distance in Chapter 4 is pertinent. We need to include in our system the electric and magnetic fields as well as the two current elements. The acceleration of electric charges necessary to produce isolated current elements also produces electromagnetic radiation, which, like all wave motion, carries momentum. A detailed analysis of such a situation as pictured in Figure 37-2 shows that indeed the current element system is accelerated to the left while electromagnetic radiation carries momentum to the right. When we include the field and its momentum in the system, the total momentum of the system is again conserved.

We now illustrate how the Biot-Savart law is used to calculate the magnetic induction for two important current configurations, a long straight wire and a current loop.

Example 37-1 Magnetic Induction of a Long Straight Current Figure 37-4 shows the geometry needed to calculate the magnetic induction $\mathbf{B}$ at point P due to the current in the long wire shown. We choose the wire to be the x axis and the y axis to be perpendicular to the wire through point P. A typical current element $I\,d\mathbf{l}$ at a distance x is shown. The vector $\mathbf{r}$ points from the element to the field point P. The direction of the magnetic field at P due to this element is out of the paper, as determined by the direction of $I\,d\mathbf{l} \times \mathbf{r}$. We note that all such current elements of the wire give contributions in this same direction, and so we need only compute the magnitude of the field. The field due to the current element shown has the magnitude

$$dB = k_m \frac{I\,dx}{r^2}\sin\phi = k_m\frac{I\,dx\cos\theta}{r^2} \qquad \text{37-5}$$

where θ is the angle shown. In order to sum over all the current elements, we need to relate the variables θ, r, and x. It turns out to be easiest to eliminate x and r in favor of θ. We have

$$x = y\tan\theta$$

Then $dx = y\sec^2\theta\,d\theta = y(r^2/y^2)\,d\theta = (r^2/y)\,d\theta$ using $\sec\theta = r/y$. Substituting this expression for dx into Equation 37-5, we have

$$dB = \frac{k_m I}{r^2}\frac{r^2\,d\theta}{y}\cos\theta = \frac{k_m I}{y}\cos\theta\,d\theta$$

Figure 37-3
Complete circuits containing the current elements shown in Figure 37-2. The total force which circuit 1 exerts on circuit 2 is equal and opposite to the total force which circuit 2 exerts on circuit 1.

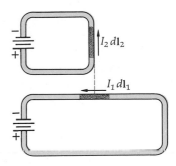

Figure 37-4
(a) Geometry for calculation from the Biot-Savart law of the magnetic field at point P due to a straight current segment. Each element gives a contribution to the magnetic field at point P directed out of the paper. (b) The result is given by Equation 37-6 in terms of the angles θ_1 and θ_2.

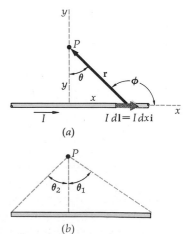

Let us first calculate the contribution from the current elements to the right of the point $x = 0$. We sum over these elements by integrating from $\theta = 0$ to $\theta = \theta_1$, where θ_1 is the angle between the line perpendicular to the wire and the line from P to the right end of the wire, as shown. We have for this contribution

$$B_1 = \int_0^{\theta_1} \frac{k_m I}{y} \cos \theta \, d\theta = \frac{k_m I}{y} \sin \theta_1$$

Similarly, the contribution from elements to the left of the point $x = 0$ is

$$B_2 = \frac{k_m I}{y} \sin \theta_2$$

We thus have for the total magnetic field due to the wire

$$B = \frac{k_m I}{y} (\sin \theta_1 + \sin \theta_2) \tag{37-6}$$

This result gives the magnetic field due to any wire segment in terms of the angles subtended at the field point by the ends of the wire. If the wire is very long, these angles are nearly 90°. The result for an infinitely long wire is obtained from Equation 37-6 by using $\theta_1 = \theta_2 = 90°$:

$$B = \frac{2k_m I}{y} = \frac{\mu_0 I}{2\pi y} \tag{37-7}$$

The direction of **B** is such that the lines of **B** encircle the wire as shown in Figure 37-5. This direction can be remembered by a right-hand rule as shown in that figure. This result for the magnetic field **B** due to a current in a long straight wire was found experimentally by Biot and Savart in 1820. From the analysis of this experimental result they were able to discover the expression given in Equation 37-1 for the magnetic induction due to an element of the current.

Example 37-2 Find the magnetic induction field at the center of a square current loop of side 1 m carrying a current of 1 A.

From Figure 37-6 we see that each side of the loop contributes a field in the direction out of the paper. By the symmetry of the situation we need only calculate the field due to one side and multiply by 4. The distance between one side and the field point $y = \frac{1}{2}l = \frac{1}{2}$ m, and the angles θ_1 and θ_2 are 45°. Thus the field due to one segment is

$$B_1 = \frac{k_m I}{\frac{1}{2}l} (\sin 45° + \sin 45°) = \frac{2k_m I}{l} 2 \frac{1}{\sqrt{2}} = 2\sqrt{2} \frac{k_m I}{l}$$

and the total field is

$$B = 4B_1 = 8\sqrt{2} \frac{k_m I}{l} = 8\sqrt{2} \frac{10^{-7} \times 1}{1} = 11.3 \times 10^{-7} \text{ T}$$

$$= 11.3 \times 10^{-3} \text{ G}$$

Example 37-3 Magnetic Induction on the Axis of a Circular Current Loop The geometry needed for this calculation is shown in Figure 37-7. We first consider the current element at the top of the loop. Here $I \, d\mathbf{l}$ is out of the paper and perpendicular to the vector **r**, which lies in the plane of the paper. The magnetic field due to this element is in the direction shown. The field $d\mathbf{B}$ is perpendicular to **r**, and since it is perpendicular

Figure 37-5
The magnetic field due to a very long straight wire is tangent to a circle around the wire. The direction of **B** is related to the direction of the current I by the right-hand rule illustrated.

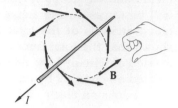

Magnetic field of a straight wire

Field of an infinite straight wire

Figure 37-6
Square current loop for Example 37-2.

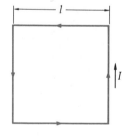

Figure 37-7
Geometry for the calculation of **B** on the axis of a current loop for Example 37-3.

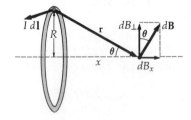

to $I\,d\mathbf{l}$, it must be in the plane of the paper as shown. The magnitude of $d\mathbf{B}$ is

$$|d\mathbf{B}| = k_m \frac{|I\,d\mathbf{l} \times \hat{\mathbf{r}}|}{r^2} = k_m \frac{I\,dl}{x^2 + R^2}$$

since $d\mathbf{l}$ and $\hat{\mathbf{r}}$ are perpendicular and $r^2 = x^2 + R^2$.

When we sum around the current elements in the loop, components of $d\mathbf{B}$ perpendicular to the axis of the loop sum to zero, leaving only the components parallel to the axis. We thus compute only the x component. From the figure we have

$$dB_x = dB \sin \theta = dB \frac{R}{\sqrt{x^2 + R^2}} = \frac{k_m I\,dl}{x^2 + R^2} \frac{R}{\sqrt{x^2 + R^2}}$$

Since neither x nor R varies as we sum over the elements in the loop, this sum merely gives $\oint dl = 2\pi R$.

$$B_x = \frac{k_m I(2\pi R)R}{(x^2 + R^2)^{3/2}} = \frac{\mu_0}{2} \frac{IR^2}{(x^2 + R^2)^{3/2}} \qquad 37\text{-}8 \qquad \textit{B on axis of current loop}$$

An important special case of this result is the magnetic field at the center of the loop which we obtain by setting $x = 0$.

$$B_x = \frac{\mu_0 I}{2R} \qquad 37\text{-}9 \qquad \textit{B at center of current loop}$$

At great distances from the loop, x is much greater than R and we can neglect R^2 compared with x^2 in the denominator of Equation 37-8. Then

$$B_x \rightarrow \frac{2k_m I(\pi R^2)}{x^3} = \frac{2k_m m}{x^3} \qquad 39\text{-}10 \qquad \textit{Dipole field of current loop}$$

where $m = I(\pi R^2)$ is the magnetic moment of the loop. Note the similarity of this expression to that for the electrostatic field on the axis of an electric dipole of moment p:

$$E_x \rightarrow \frac{2kp}{x^3}$$

A current loop behaves like a magnetic dipole both in experiencing a torque $\mathbf{m} \times \mathbf{B}$ when placed in an external magnetic field and in producing a dipole field at great distances from the loop. Figure 37-8 shows the lines of B for a current loop. In the regions far from the loop these lines are identical to those of $\mathbf{E}$ for an electric dipole. Our result that a current loop produces a dipole field is not confined to the axis of the loop.

Figure 37-8
Lines of magnetic induction $\mathbf{B}$ due to a circular current loop indicated by iron filings. (*Avco Research Laboratory.*)

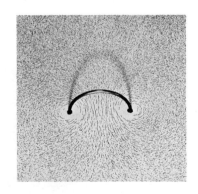

Questions

1. For a given distance r from a current element, in what direction does the contribution of the current element to the total magnetic field have its greatest value?

2. What is the effect of replacing the single loop in Example 36-3 with a coil of N turns of the same radius each carrying the current I?

3. A common method for cutting down on the magnetic fields created by leads carrying current to electrical devices is to twist together the wires carrying the incoming and outgoing currents. Explain.

37-2 The Definition of the Ampere and the Coulomb

One week after Ampère heard of Oersted's discovery of the effect of a current on a compass needle, he showed that two parallel currents attract each other if the currents are in the same direction and repel each other if the currents are in opposite directions. Figure 37-9 shows two long straight wires separated by a distance r and carrying currents I_1 and I_2 in the same direction. We can compute the force exerted by one wire on the other from our result for the magnetic field of a long current-carrying wire and Equation 36-3 for the force exerted by a magnetic field on a current element. Let us consider a segment of the second wire of length l_2. The magnetic field $\mathbf{B}_1$ at this segment is perpendicular to the segment and has the magnitude $2k_mI_1/r$ if the segment is sufficiently close to wire 1 and wire 1 is long enough for the angles in Equation 37-6 to be approximately 90°. The force on this segment has the magnitude

$$F_2 = |I_2\mathbf{l}_2 \times \mathbf{B}_1| = \frac{2k_mI_1I_2l_2}{r}$$

From the direction of the cross product $\mathbf{l}_2 \times \mathbf{B}_1$ we see that this force on the segment of wire 2 is toward wire 1. The force per unit length is

$$\frac{F_2}{l_2} = \frac{2k_mI_1I_2}{r} \qquad\qquad 37\text{-}11$$

In Chapter 29 we deferred the definition of the coulomb as a unit of charge, mentioning that it is defined in terms of the ampere. The ampere is defined as follows:

If two very long parallel wires one meter apart carry equal currents, the current in each is defined to be one ampere if the force per unit length on each wire is 2×10^{-7} newton per meter.

This definition allows the unit of current, and thus of electric charge, to be determined by a mechanical experiment. In practice of course, the currents are chosen to be much closer together than 1 m so that the wires do not need to be so long and the force is large enough to measure accurately. Figure 37-10 shows a current balance which can be used to calibrate an ammeter from the fundamental definition of

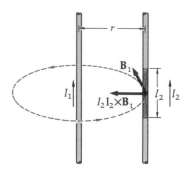

Figure 37-9
Two long straight wires carrying parallel currents. The magnetic field at the current element $I_2\mathbf{l}_2$ due to current I_1 is into the paper, giving an attractive force $\mathbf{F}_2 = I_2\mathbf{l}_2 \times \mathbf{B}_1$.

Ampere defined

Figure 37-10
Current balance used in elementary physics laboratory to calibrate an ammeter. (*Courtesy of Larry Langrill, Oakland University.*)

the ampere. The upper conductor is free to rotate about the knife edges and is balanced so that the wires are a small distance apart. The conductors are wired in series to carry the same current but in opposite directions, so that the wires repel rather than attract. The force of repulsion can be measured by placing weights on the upper conductor until it balances again at the original separation. This definition of the ampere makes the magnetic constant $k_m = \mu_0/4\pi$ exactly equal to 10^{-7}. The ratio of the coulomb constant k and the magnetic constant k_m is

$$\frac{k}{k_m} = \frac{1/4\pi\epsilon_0}{\mu_0/4\pi} = \frac{1}{\epsilon_0\mu_0} = \frac{9 \times 10^9 \text{ N-m}^2/\text{C}^2}{10^{-7} \text{ N-sec}^2/\text{C}^2} = 9 \times 10^{16} \text{ m}^2/\text{sec}^2$$

This ratio equals the square of the speed of light, a fact noted by Maxwell in 1860. He showed that the laws of electricity and magnetism imply that an accelerated charge radiates energy in the form of waves which travel with speed

$$c = \sqrt{\frac{k}{k_m}} = \frac{1}{\sqrt{\epsilon_0\mu_0}} = 3 \times 10^8 \text{ m/sec} \qquad 37\text{-}12$$

Since this speed is the same as that of light, Maxwell speculated that light is an electromagnetic wave produced by the acceleration of atomic charges. In the next section we shall consider a special example which shows that the magnetic field is a necessary consequence of Coulomb's law and the theory of special relativity. Without any reference to the Biot-Savart law we shall show that the magnetic field at a distance r from a long straight wire has the magnitude $2(k/c^2)I/r$, where k is the coulomb constant and c is the speed of light.

37-3 Special Relativity and the Magnetic Field[1]

A simple example illustrates the connection between special relativity and the existence of the magnetic field. Figure 37-11 shows a long wire carrying a current to the right and a point charge q at rest a distance r from the wire. The current is due to the flow of electrons to the left with drift velocity **v**. For simplicity we shall consider the electrons to be a very long line charge with linear charge density of magnitude λ'_- C/m. (We use primes on quantities in this reference frame to distinguish them from corresponding quantities in another reference frame, the unprimed frame, which moves to the left relative to this frame. This notation is in agreement with that in Chapter 28 on special relativity.) In a time $\Delta t'$ the charge passing some point P is that contained

[1] This section may be omitted if special relativity has not been studied; however, the only result of that theory used here is the Lorentz contraction.

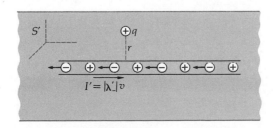

Figure 37-11
Long wire carrying current I' to the right due to motion of negative charges to the left. If the densities of positive and negative charges are equal, the point charge q at rest a distance r from the wire experiences no electric or magnetic force.

in the distance $v\,\Delta t'$, which is $\Delta q' = \lambda'_- v\,\Delta t'$. Thus the current in this frame is

$$I' = \frac{\Delta q'}{\Delta t'} = \lambda'_- v \qquad\qquad 37\text{-}13$$

We shall assume that there is a positive charge density whose magnitude λ'_+ is equal to that of the negative charge density. In this frame the positive charge is at rest. Since there is no net charge density by our assumptions, there is no electric force on point charge q due to the line charges. The point charge q thus does not accelerate but stays at rest a distance r from the wire.

We now look at this same simple situation in a reference frame moving to the left with speed v relative to the primed frame, as shown in Figure 37-12. In this frame, the negative line of charge is at rest, and the positive line moves to the right with speed v. The point charge q also moves to the right with speed v. According to the theory of special relativity, if the total charge is the same in both reference frames, the charge densities will not be the same. (The fact that the total charge is the same in two reference frames is known as *charge invariance.*) For example, if we had N positive charges in a length l' in the previous frame, these charges will now reside in a shorter distance l related to l' by the Lorentz contraction,

$$l = \sqrt{1 - \frac{v^2}{c^2}}\; l'$$

Then the positive charge density will be greater in this frame.

$$\lambda_+ = \frac{\lambda'_+}{\sqrt{1 - \dfrac{v^2}{c^2}}} \qquad\qquad 37\text{-}14$$

Similarly, the negative charge density will be smaller. A distance l containing a given number of negative charges in this frame will be contracted in the original frame in which it moves with speed v. The negative charge density in this frame will be

$$\lambda_- = \sqrt{1 - \frac{v^2}{c^2}}\; \lambda'_- \qquad\qquad 37\text{-}15$$

The ratio of these charge densities is

$$\frac{\lambda_-}{\lambda_+} = \frac{\sqrt{1 - v^2/c^2}\;\lambda'_-}{\lambda'_+/\sqrt{1 - v^2/c^2}} = \left(1 - \frac{v^2}{c^2}\right)\frac{\lambda'_-}{\lambda'_+}$$

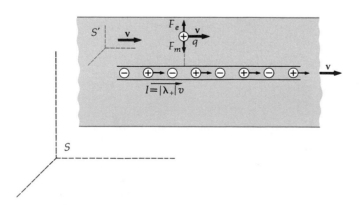

$$I = |\lambda_+| v$$

Figure 37-12
The situation in Figure 37-11 viewed from a reference frame S in which the wire and point charges are moving to the right with speed v. In this frame the negative charges in the wire are at rest, and the current I is due to motion of the positive charges. The positive-charge density is greater and the negative-charge density is less than in frame S' (Figure 37-11) because of the Lorentz contraction; so the positive charge q experiences a repulsive electrostatic force. This force is balanced by the magnetic force on the moving charge.

Since we have assumed that the original charge densities are equal, we can cancel $\lambda'_- = \lambda'_+$, giving

$$\frac{\lambda_-}{\lambda_+} = 1 - \frac{v^2}{c^2} \qquad\qquad 36\text{-}16$$

Because there is a net charge density in this reference frame, there will be an electric field due to the line charges with an electric force on charge q pointing away from the wire (assuming q to be positive). The magnitude of the electric field at a distance r from a very long line charge of density $\lambda_{net} = \lambda_+ - \lambda_-$ is

$$E = \frac{2k\lambda_{net}}{r} = \frac{2k(\lambda_+ - \lambda_-)}{r} = \frac{2k\lambda_+(1 - \lambda_-/\lambda_+)}{r} \qquad\qquad 37\text{-}17$$

where $k = 1/4\pi\epsilon_0$ is the coulomb constant. Using Equation 37-16 for the ratio λ_-/λ_+, we have

$$E = \frac{2k\lambda_+}{r}\left[1 - \left(1 - \frac{v^2}{c^2}\right)\right] = \frac{2k\lambda_+}{r}\frac{v^2}{c^2}$$

The current in this reference frame is

$$I = \lambda_+ v \qquad\qquad 37\text{-}18$$

The electric field in terms of the current is

$$E = \frac{2kI}{r}\frac{v}{c^2}$$

and the electric force on the charge q is

$$F_e = qE = 2\frac{k}{c^2}\frac{I}{r}qv \qquad\qquad 37\text{-}19$$

This electric force points away from the wire since λ_{net} is positive. We see that the Lorentz contraction and Coulomb's law (used to find the electric field due to a line charge) predict that there will be an electric force on the charge q directed away from the wire in this reference frame. If this were the only force on q, the point charge would accelerate away from the wire. Clearly this cannot be the case because in the original reference frame the charge remains at rest a constant distance r from the wire. Thus there must be another force on the charge q in this frame to balance the electric force. This force must be directed toward the wire and have the magnitude given by Equation 37-19. This second force is, in fact, the magnetic force. It is proportional to the charge q, the speed of the charge v, and to the current I. From our definition of the magnetic field B in Section 36-1 the magnetic force is related to the magnetic field by

$$\mathbf{F}_m = q\mathbf{v} \times \mathbf{B}$$

Since $\mathbf{v}$ is to the right in our diagram and the force is toward the wire, the magnetic field at the point charge must be out of the paper and have the magnitude

$$B = \frac{F_m}{qv} = 2\frac{k}{c^2}\frac{I}{r} \qquad\qquad 37\text{-}20$$

This result is the same as that found using the Biot-Savart law with the magnetic constant k_m replaced by k/c^2, where k is the coulomb constant and c the velocity of light.

37-4 Ampère's Law

We have noted that the lines of **B** for a long, straight current-carrying wire encircle the wire. These lines are quite different from any lines of electric field we have studied. The electrostatic field is conservative. The work done on a test charge by the electrostatic field when the charge is moved in a complete circle is always zero. This work per charge is the sum of **E** · $d\mathbf{l}$ around the path. The line integral of the electrostatic field around any closed path is zero because the electrostatic field is conservative.

$$\oint_C \mathbf{E} \cdot d\mathbf{l} = 0 \qquad C \text{ any closed curve}$$

Clearly the sum of **B** · $d\mathbf{l}$ around a closed path is not necessarily zero. If we take this sum along a circular path enclosing a long wire carrying a current, the magnetic-induction vector **B** is everywhere tangent to the path (Figure 37-13). Then **B** · $d\mathbf{l}$ is everywhere positive if we travel in the direction of the field lines. Since **B** is in fact parallel to $d\mathbf{l}$ and has the constant magnitude given by Equation 37-7, we can easily compute this sum:

$$\oint \mathbf{B} \cdot d\mathbf{l} = \frac{2k_m I}{r} \oint dl = \frac{2k_m I}{r} 2\pi r = 4\pi k_m I \qquad \text{37-21}a$$

This integral is independent of the radius chosen for the circular path. This result, known as *Ampère's law*, is more general than indicated by our calculation. It holds for the line integral of **B** · $d\mathbf{l}$ around any closed path which encloses a steady current.

The line integral around any closed path equals $4\pi k_m$ times the total current which crosses any area bounded by the path. In terms of the constant $\mu_0 = 4\pi k_m$, Ampère's law is written

$$\oint_C \mathbf{B} \cdot d\mathbf{l} = \mu_0 I \qquad \text{37-21}b$$

Ampère's law

(It is because of the 4π factor in Equation 37-21a that the magnetic constant k_m is written $\mu_0/4\pi$.)

Ampère's law holds for any curve C as long as the currents are steady. Like Gauss' law, it is useful for the calculation of the magnetic field only in situations with considerable symmetry. If the symmetry is great enough, the line integral can be written as the product of B and some distance, thus relating B to the current enclosed by the curve. Also, like Gauss' law, if the currents are steady, Ampère's law holds even if there is no symmetry; but in those cases it is no help in finding an expression for the magnetic field.

Example 37-4 A long straight wire of radius a carries a current I which is uniformly distributed over the wire with current density $J = I/\pi a^2$ (Figure 37-14). Find the magnetic field both inside and outside the wire.

We can calculate **B** from Ampère's law because of the high degree of symmetry. Consider a circle of radius r concentric with the axis of the wire. We expect **B** to be tangential to the circle, as with a very thin wire. By symmetry, the magnitude of **B** is constant everywhere on the circle. Thus

$$\oint \mathbf{B} \cdot d\mathbf{l} = B \oint dl = 2\pi r B$$

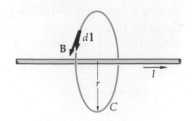

Figure 37-13
Circular path around a long current-carrying wire. The integral of **B** · $d\mathbf{l}$ around this curve is $4\pi k_m I = \mu_0 I$.

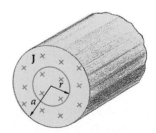

Figure 37-14
Long straight wire of radius a carrying current I uniformly distributed with current density $J = I/\pi a^2$. The magnetic field at distance r inside the wire can be calculated from Ampère's law applied to the circle of radius r.

The current through C depends on whether r is less or greater than the radius of the wire. For r greater than a, the total current I crosses the area bounded by C. Then

$$\oint \mathbf{B} \cdot d\mathbf{l} = B2\pi r = \mu_0 I$$

$$B = \frac{\mu_0 I}{2\pi r} = 2\frac{\mu_0}{4\pi}\frac{I}{r} \qquad r > a$$

This result is the same as obtained in Equation 37-7. To find B inside the wire we choose r less than a. Then the current through C is

$$\pi r^2 J = \frac{r^2}{a^2} I$$

and

$$\oint \mathbf{B} \cdot d\mathbf{l} = 2\pi r B = \mu_0 \frac{r^2}{a^2} I$$

$$B = \frac{\mu_0}{2\pi}\frac{I}{a^2} r \qquad r < a \qquad\qquad\qquad 37\text{-}22$$

Figure 37-15 is a sketch of B versus r.

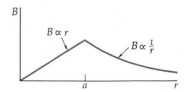

Figure 37-15
Plot of B versus r calculated in Example 37-4. Inside the wire, B is proportional to r, whereas outside, it is inversely proportional to r.

Field inside a wire carrying uniform current

For an example in which Ampère's law is not useful in calculating the magnetic induction due to a steady current, consider a current loop as shown in Figure 37-16. We calculated the magnetic induction on the axis of such a loop from the Biot-Savart law in Example 37-3. According to Ampère's law, the line integral of $\mathbf{B} \cdot d\mathbf{l}$ around a curve such as curve C in Figure 37-16 equals μ_0 times the current I in the loop. Although Ampère's law is true for this curve, the magnetic induction B is not constant along any curve encircling the current nor is it everywhere tangent to any such a curve. Thus there is not enough symmetry in this situation to allow us to calculate B from Ampère's law.

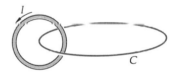

Figure 37-16
Ampère's law holds for curve C encircling the current in the circular loop, but it is not useful in finding $\mathbf{B}$ because $\mathbf{B}$ is neither constant along the curve nor tangent to it.

Example 37-5 Use Ampère's law to find the magnetic field at a point on the perpendicular bisector of a current segment of length l. Let the point be a distance r from the segment.

This situation is shown in Figure 37-17. A direct application of Ampère's law again gives

$$B = \frac{\mu_0}{2\pi}\frac{I}{r}$$

This result is the same as for an infinitely long wire since we have the same symmetry arguments. The result does not agree with that obtained from the Biot-Savart law. That law gives a smaller result which depends on the length of the current segment and agrees with experiment. There are two possibilities for the discrepancy: (1) Ampère's law is correct but was incorrectly applied; (2) Ampère's law is not correct. In order to understand which explanation applies we must ask how we can obtain a current segment as shown in Figure 37-17. One possibility is shown in Figure 37-18, where the current segment is just one segment of a continuous loop carrying a steady current. For this situation, Ampère's law is correct, but our application was not. If there are other current segments in the problem, as in Figure 37-18, we do not have the symmetry we assumed to compute $\oint \mathbf{B} \cdot d\mathbf{l}$. For example, $\mathbf{B}$ is not tangential to the curve shown, and its magnitude is not constant

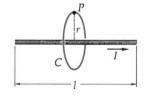

Figure 37-17
Application of Ampère's law to find the magnetic field on the bisector of a finite current segment gives an incorrect result.

Figure 37-18
If the current segment in Figure 37-17 is a part of a complete circuit carrying a steady current, Ampère's law for curve C is correct, but there is not enough symmetry to use it to find the magnetic field at point P.

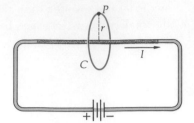

Figure 37-19
If the current segment in Figure 37-17 is due to a momentary flow of charge from a small conductor on the left to one at the right, there is enough symmetry to compute $\oint \mathbf{B} \cdot d\mathbf{l}$ along curve C but Ampère's law does not hold because the current is not steady.

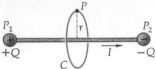

along the curve. Then even though Ampère's law is correct, it is not useful for the calculation of the magnetic field for this situation. Figure 37-19 shows another possibility for obtaining a current segment, i.e., a source of charge at point P_1 and a sink at point P_2. For example, we might have a small spherical conductor with charge $+Q$ at the left, and one with charge $-Q$ at the right. When they are connected, a current $I = -dQ/dt$ exists in the segment for a short time until the spheres are uncharged. For this case we do have the symmetry needed to assume that $\mathbf{B}$ is tangential to the curve and of constant magnitude along the curve, but the current is not a steady current. When the current is not steady, Ampère's law does not hold. We shall see in Section 37-8 how Maxwell generalized Ampère's law to hold for all currents, steady or not. When the generalized form is used, the magnetic field calculated for this situation agrees with that found from the Biot-Savart law.

Figure 37-20
A solenoid produces a strong uniform magnetic field in the space inside it.

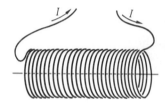

37-5 The Magnetic Field of a Solenoid

A wire wound into a helix, as in Figure 37-20, called a *solenoid*, is used to produce a strong, uniform magnetic field in a small region of space. It plays a role in magnetism analogous to that of the parallel-plate capacitor in providing a strong, uniform electrostatic field between its plates. The magnetic field of a solenoid is essentially that of a set of N identical coils placed side by side. Figure 37-21 shows the lines of magnetic induction for two such coils. In the space between the coils near the axis the fields of the individual coils add; between the coils but at distances from the axis greater than the coil radius, the individual fields tend to cancel. Figure 37-22 shows the field lines for a long, tightly wound solenoid. Inside the solenoid, the field lines are approx-

Figure 37-21
The magnetic fields due to two adjacent loops of a solenoid add in the region inside the solenoid and subtract in the region outside.

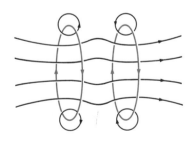

Figure 37-22
Magnetic field lines of a tightly wound solenoid. The field is strong and nearly uniform inside the solenoid. Outside it the field lines are like those of the electric field due to two disk charges (Figure 37-23).

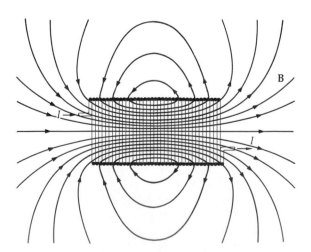

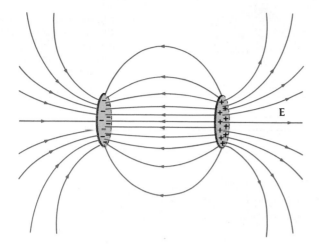

Figure 37-23
Electric field lines due to two
oppositely charged disks. The
lines outside the cylinder
bounded by the disks are like
the magnetic field lines of a
solenoid (Figure 37-22).
Between the disks the **E** lines
are directed opposite to the
corresponding magnetic
field lines of a solenoid.

imately parallel to the axis and closely and uniformly spaced, indicating a strong, uniform magnetic field. Outside the solenoid, the
lines are much less dense. They diverge from one end and converge at
the other end, like the lines of electric field from disks of positive and
negative charge (Figure 37-23). Note that *inside* the solenoid the magnetic field lines are very different from electric field lines from the disk
charges.

Because the magnetic field is very strong and nearly uniform inside
the solenoid, as indicated by the field lines in Figure 37-22, and the
field is very weak outside the solenoid, we can use Ampère's law to
find a useful approximate result for the magnitude of B. Let the
solenoid have radius r, length l, and N turns of wire carrying a current
I, and assume that l is much greater than r. We shall assume that **B** is
uniform and parallel to the axis inside the solenoid and zero outside.
We apply Ampère's law to the curve C (Figure 37-24), which is a
rectangle with sides of length a and b. The only contribution to the
integral $\oint \mathbf{B} \cdot d\mathbf{l}$ around this curve is along side 1 since along the opposite side (outside the solenoid) the magnetic field is assumed to be
zero and along the two short sides **B** is perpendicular to $d\mathbf{l}$. Since **B** is
assumed to be uniform and axial, the total line integral is merely Bb.
The total current through this curve is the current I times the number
of turns of wire enclosed. Since there are N turns in a total length l, the
number of turns in the length b is Nb/l and Ampère's law gives

$$\oint \mathbf{B} \cdot d\mathbf{l} = Bb = \mu_0 \frac{Nbl}{l}$$

or

$$B = \mu_0 \frac{N}{l} I = \mu_0 n I \qquad \qquad 37\text{-}23$$

where $n = N/l$ is the number of turns per unit length. Figure 37-25
shows how the magnetic field on the axis of a solenoid varies with

Magnetic field lines of a solenoid indicated by iron filings.

Fred Weiss

B inside solenoid

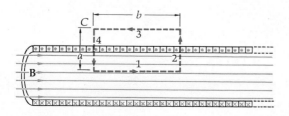

Figure 37-24
The magnetic field inside a
solenoid can be calculated by
applying Ampère's law to the
curve C shown, assuming that
B is uniform inside and zero
outside.

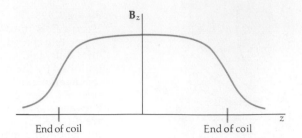

Figure 37-25
Plot of B_z versus z on axis of a solenoid. Note that B_z is nearly constant inside the solenoid except near the ends.

distance from the ends. The approximation that the field is constant independent of position along the axis is quite good except very near the ends. At a point right at the end, the magnitude of **B** is about half that at the center for a long solenoid.

We shall calculate the magnetic field outside the solenoid only for the special case of a point far from the end compared with the radius of the coils. Then each coil gives a dipole field, since we are far from it compared with its radius. We can calculate the resultant field for N coils by summing the fields of the individual coils. We can anticipate the result. Since each loop produces a field like that of a small bar magnet, i.e., a dipole field, we expect a set of adjacent loops to give a field like that of a set of adjacent small bar magnets, i.e., like a bar magnet of length l. We shall now derive this result.

Let us consider a point on the axis a distance x_0 from the center and distances $x_0 - \frac{1}{2}l$ from the close end and $x_0 + \frac{1}{2}l$ from the far end with $x_0 - \frac{1}{2}l$ much greater than the solenoid radius r (Figure 37-26). We first consider an element of the solenoid of length dx at a distance x from the center. If $n = N/l$ is the number of turns per unit length, there are $n\,dx$ turns of wire in this element, each carrying a current I. The element is thus equivalent to a loop carrying a current $di = nI\,dx$, giving a dipole field,

$$dB = 2k_m \frac{di\,A}{x'^3} = 2k_m nIA \frac{dx}{x'^3}$$

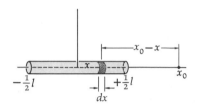

Figure 37-26
Geometry for calculating the magnetic field on the axis of a tightly wound solenoid at a point far from both ends. The element of length dx gives a magnetic dipole field of a current loop carrying current $di = nI\,dx$, where I is the current in the solenoid.

where $x' = x_0 - x$ is the distance from the element to the field point and $A = \pi r^2$ is the area of the coil. We find the total field at x_0 due to the solenoid by summing from $x = -\frac{1}{2}l$ to $x = +\frac{1}{2}l$. Performing the integration between these limits, we obtain

$$B = k_m nIA \left(\frac{1}{r_1{}^2} - \frac{1}{r_2{}^2} \right) \tag{37-24}$$

B on axis outside solenoid

where $r_1 = x_0 - \frac{1}{2}l$ is the distance to the near end and $r_2 = x_0 + \frac{1}{2}l$ is the distance to the far end. We have obtained the expected result. We can put this into a form similar to Coulomb's law by defining a magnetic-pole strength q^*:

$$q^* = nIA$$

Pole strength of solenoid

Then

$$B = k_m \frac{q^*}{r_1{}^2} - k_m \frac{q^*}{r_2{}^2} \tag{37-25}$$

Note that the dipole moment of the solenoid is

$$q^*l = nIAl = NIA \tag{37-26}$$

Magnetic dipole moment of solenoid

which is just the sum of the dipole moments of the individual loops, as expected.

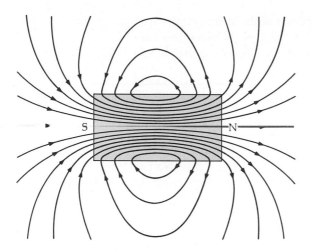

Figure 37-27
Magnetic field lines of a bar magnet. The lines are the same everywhere as those of a solenoid of similar shape.

Figure 37-27 shows the magnetic field lines of a bar magnet of the same shape as the solenoid in Figure 37-22. The lines of **B** are identical for these two cases.

Question

4. The magnetic field inside a long solenoid has magnitude B. How would the field change if the current in the solenoid were doubled? If the diameter were doubled without changing the length or number of turns? If the number of turns were doubled by making the solenoid twice as long? If the number of turns were doubled without changing the length of the solenoid?

37-6 The Magnetic Field of a Bar Magnet

The first quantitative investigations of magnetic fields were made with permanent magnets. Using a torsion balance, Michell found that the force of attraction or repulsion between two poles varies as the product of the pole strengths and inversely with the square of the distance between the poles. Since the force acting on a single pole in a magnetic induction field **B** is $q^*\mathbf{B}$, this result implies that the magnetic field due to a pole is proportional to q^*/r^2 and in the radial direction. According to Ampère's model, all magnetic fields are due to currents of some kind. In permanent magnets or other magnetized material these currents are due to the intrinsic motion of atomic electrons. Although these motions are very complicated, for this model we need only assume that the motions are equivalent to closed-circuit loops (Figure 37-28). The magnetic moments of the current loops shown in this figure are in the direction parallel to the axis. If the material is homogeneous, the net current at any point inside the material is zero because of cancellation of neighboring current loops. However, since there is no cancellation on the surface of the material, the result of these current loops is equivalent to a current on the surface of the material, called an *amperian current*. The surface current is similar to the real conduction current in a tightly wound solenoid. The magnetic field due to the surface current is the same as that due to an equivalent "surface" current in a solenoid. Let M be the amperian current per unit length on the surface of a cylindrical bar magnet. The corre-

Figure 37-28
Model of atomic current loops when all the atomic dipoles are parallel to the axis of the cylinder. The net current at any point inside the material is zero due to cancellation of neighboring atoms. The result is a surface current similar to that of a solenoid.

sponding quantity for a solenoid is nI, where n is the number of turns per unit length and I is the current in each turn. If we replace the current per unit length nI in Equation 37-26 by M, we have for the dipole moment of the magnet,

$$q^*l = M(Al) \qquad 37\text{-}27$$

where A is the cross-sectional area of the magnet and l the length. Since Al is the total volume of the magnet, M equals the magnetic moment per unit volume (Figure 37-29). The vector $\mathbf{M}$, whose magnitude is the magnetic moment per unit volume and whose direction is that of the dipole moment of the magnet, is called the *magnetization vector* of the magnet. The direction of $\mathbf{M}$ is from the south (negative) pole to the north (positive) pole. (Any magnetized material, whether it is a permanent magnet or some material magnetized by an external magnetic field, can be characterized by the magnetic moment per unit volume $\mathbf{M}$, which for nonhomogeneous material may vary from point to point in the material.) The pole strength of a cylindrical bar magnet of cross-sectional area A is related to its magnetization M through Equation 37-27:

$$q^* = MA \qquad 37\text{-}28$$

The magnetic induction B due to a bar magnet is then given by Equation 37-25 with q^* equal to the magnetization times the area. In our derivation of that equation for a solenoid we assumed that the field point was far away compared with the radius of the solenoid. Thus this equation is only an approximation for the field due to a magnet also. This approximation is equivalent to the assumption that the pole strength of the magnet is concentrated at a single point at each end of the magnet. From the observation that the lines of magnetic induction outside a cylindrical bar magnet are identical to the lines of $\mathbf{E}$ for two equally but oppositely charged disks, we see that we can improve this approximation in the region just outside the magnet by replacing these point poles with a surface pole density σ^* on the ends, where σ^* is given by

$$\sigma^* = \frac{q^*}{A} = M \qquad 37\text{-}29$$

In the region inside a solenoid the magnetic induction is approximately $\mu_0 nI$, according to Equation 37-23. This approximation is good as long as the field point is not near the ends, as indicated in Figure 37-25. Replacing the current per unit length of the solenoid nI by the corresponding amperian current per unit length of a bar magnet M, we have for the field inside a magnet far from the ends

$$B = \mu_0 M \qquad 37\text{-}30$$

We emphasize that the magnetic induction $\mathbf{B}$ for a bar magnet of magnetization $\mathbf{M}$ is identical everywhere to that of a solenoid of the same shape and carrying a current per unit length nI equal to M.

37-7 Magnetic Flux

The flux of the magnetic field through a surface is defined similarly to the flux of the electric field. Let dA be an element of area on the surface and $\hat{\mathbf{n}}$ be the unit vector perpendicular to the element. The magnetic

Magnetization vector $\mathbf{M}$

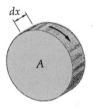

Figure 37-29
Disk for relating the surface current to the dipole moment. If M is the current per unit length on the surface, the total current is $M\,dx$ and the dipole moment is $M\,dx\,A = M\,d\mathcal{V}$, where $d\mathcal{V}$ is the volume.

flux ϕ_m is then defined by

$$\phi_m = \int \mathbf{B} \cdot \hat{\mathbf{n}} \, dA \qquad\qquad 37\text{-}31$$

If the surface is a plane with area A, and $\mathbf{B}$ is constant in magnitude and direction over the surface, making an angle θ with the unit normal vector, the flux is

$$\phi_m = BA \cos \theta \qquad\qquad 37\text{-}32$$

The unit of flux is called the weber (Wb):

$$1 \text{ Wb} = 1 \text{ T-m}^2 \qquad\qquad 37\text{-}33$$

We are often interested in the magnetic flux through a coil containing several turns of wire. Figure 37-30 shows a coil with two turns in a uniform magnetic field $\mathbf{B}$ perpendicular to the plane of the coil. The total area enclosed by the coil is twice the area enclosed by each turn. The magnetic flux through the coil is therefore the product of the field B, the area of each coil A, and the number of turns N, which in this case is two.

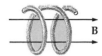

Figure 37-30
The area enclosed by a coil of two turns is twice the area bound by each turn. In general, the area bound by a coil of N turns is N times that of each turn.

Example 37-6 Find the magnetic flux through a long, tightly wound solenoid of length l, cross-sectional area A, and number of turns N, carrying current I.

We shall use the approximation that the magnetic field inside the solenoid is constant, equal to its value at the center:

$$B = \mu_0 nI = \mu_0 \frac{N}{l} I$$

The total area is NA since there are N turns, each of area A. Since the magnetic field is perpendicular to the area of each turn, the flux is

$$\phi_m = BA = \mu_0 \frac{N}{l} INA = \mu_0 \left(\frac{N^2}{l}\right) AI = \mu_0 n^2 (lA)I$$

where $n = N/l$ is the number of turns per unit length and the quantity lA is the volume of the space inside the solenoid containing the magnetic field B.

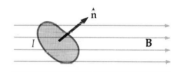

Figure 37-31
Current loop in an external magnetic field. The torque on the loop tends to rotate the loop so as to increase the flux through the loop.

In the next chapter we shall see that there is an emf in a circuit associated with a changing magnetic flux through that circuit. Because of the importance of magnetic flux, the magnetic field is sometimes called the *flux density*. In this section we shall merely express some properties of torques and forces already studied in terms of magnetic flux changes and discuss the magnetic analog of Gauss' law.

In general, when a current loop is in an external field, or if there are two circuits exerting forces on each other, the forces and torques are in such a direction as to *increase the flux* through the circuits. For a current loop in an external field, as shown in Figure 37-31, the torque $\mathbf{m} \times \mathbf{B}$ tends to line up the magnetic moment in the direction of $\mathbf{B}$, which is also the orientation of the loop for which the flux of the external field is maximum, assuming that the direction of magnetic moment is the positive direction for the flux. Figure 37-32 shows two long, parallel wires, each part of a complete circuit. The flux through either circuit is partly caused by the magnetic field due to that circuit and partly caused by the magnetic field due to the other circuit. If the currents in the near wires are parallel, as shown, these two contributions to the flux add. The flux through either circuit will be increased

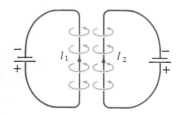

Figure 37-32
The attraction of two parallel currents tends to move the wires together so as to increase the magnetic flux through each circuit.

if the circuits are moved closer together. Since the main force on the long wire is due to the near parallel wire, the force on the wire acts to increase the flux through the circuit.

In Chapter 29 we considered the flux of the electric field through a closed surface and found that the flux, which is proportional to the net number of lines leaving the surface, is proportional to the net electric charge within the surface. The lines of magnetic field differ from those of electric field in that the magnetic lines are continuous and have no beginning or end. Thus for any closed surface, the same number of lines must enter the surface as leave the surface. This means that the net magnetic flux through any *closed* surface is always zero:

$$\oint \mathbf{B} \cdot \hat{\mathbf{n}} \, dA = 0 \qquad\qquad 37\text{-}34$$

The magnetic field differs fundamentally from the electric field in that there are no magnetic charges or poles where the field lines originate or terminate. In Section 37-5 we defined a magnetic-pole strength q^* in order to better understand the magnetic field *outside* a solenoid or bar magnet. Figure 37-33 shows a closed surface surrounding one end of a solenoid or bar magnet (the lines for both are identical). This figure should be compared with Figure 37-34, showing a similar closed surface surrounding the positive charge of an electric dipole. In the region corresponding to that outside the solenoid or bar magnet, the lines of **B** are identical to the lines of **E** for the dipole, and in each case they leave the closed surface indicated. However, inside the solenoid or magnet, the lines of **B** are different from the lines of **E** for the charge distribution. The magnetic field lines enter the closed surface inside the solenoid or magnet and there is no divergence of lines from the region of the pole. The lines which leave the surface also enter it.

Equation 37-34 is a statement of the experimental result that isolated magnetic poles do not exist. There are no regions of space in which lines of magnetic induction originate or terminate.

Figure 37-33
The net magnetic flux through a surface enclosing the north pole of a magnet is zero because each line that leaves the surface outside the magnet enters the surface inside the magnet.

Figure 37-34
Electric flux through a surface enclosing a positive charge for comparison with Figure 37-33. Here the net flux is not zero.

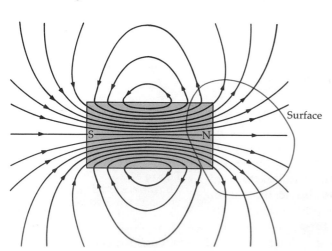

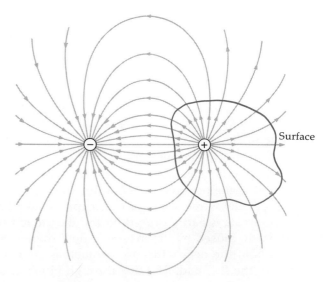

Questions

5. How can a small disk of area A be placed in a magnetic field so that there is no magnetic flux through this area?

6. How should a small loop of wire be oriented in a magnetic field so that the flux of field lines through it is a maximum? A minimum?

7. How does a positive (north) magnetic pole q^* as defined for a solenoid or magnet in Sections 37-5 and 37-6 differ from a true isolated positive pole? Sketch the lines of **B** near the poles for these cases.

37-8 Maxwell's Displacement Current

When the current is not steady, Ampère's law does not hold. We can see why by considering the charging of a capacitor (Figure 37-35). Consider the curve C. According to Ampère's law, the line integral of the magnetic field around this curve equals μ_0 times the total current through any surface bounded by the curve. Such a surface need not be plane. Two surfaces bounded by the curve C are indicated in the figure. The current through surface 1 is I. There is no current through surface 2 since the charge stops on the capacitor plate. There is thus an ambiguity in the phrase "current through any surface bounded by the curve." However, for steady currents, charge does not build up at any point, and we get the same current no matter which surface we choose.

Maxwell recognized this flaw in Ampère's law and showed that the law could be generalized to include all situations if the current I in the equation is replaced by the sum of the true current I and another term I_d, called *Maxwell's displacement current*. The displacement current is defined by

$$I_d = \epsilon_0 \frac{d\phi_e}{dt} \qquad\qquad 37\text{-}35$$

where ϕ_e is the flux of the electric field. The generalized form of Ampère's law is then

$$\oint \mathbf{B} \cdot d\mathbf{l} = \mu_0(I + I_d) = \mu_0 I + \mu_0\epsilon_0 \frac{d\phi_e}{dt} \qquad\qquad 37\text{-}36$$

We can understand the generalization of Ampère's law by considering Figure 37-35 again. Let us call the sum $I + I_d$ the generalized current. According to our arguments above, the same generalized current must cross any area bounded by the curve C. Thus there can be no net generalized current into or out of the closed volume. If there is a net true current I into the volume, there must be an equal net displacement current I_d out of the volume. In the volume pictured there is a net true current I into the volume which increases the charge within the volume:

$$I = \frac{dQ}{dt}$$

The flux of the electric field out of the volume is related to the charge by Gauss' law:

$$\phi_{e\ \text{net out}} = \oint \mathbf{E} \cdot \hat{\mathbf{n}}\ dA = \frac{1}{\epsilon_0} Q$$

Figure 37-35
Two surfaces S_1 and S_2 bounded by the same curve C. The current I passes through surface S_1 but not S_2. Ampère's law, which relates the line integral of B around the curve C to the total current passing through any surface bounded by C, is not valid when the current is not continuous, as here, where it stops at the capacitor plate.

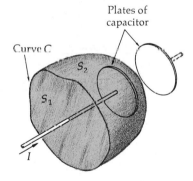

The rate of increase of charge is thus proportional to the rate of increase of the net flux out of the volume:

$$\epsilon_0 \frac{d\phi_{e \text{ net out}}}{dt} = \frac{dQ}{dt}$$

and the net current into the volume equals the net displacement current out of the volume.

Figure 37-36 shows surfaces S_1 and S_2 which enclose one plate of the capacitor. In this situation, all the flux of the electric field goes through S_2. The electric field between the plates is related to the charge by

$$E = \frac{\sigma}{\epsilon_0} = \frac{Q}{\epsilon_0 A}$$

where A is the area of the plates. The flux through S_2 is

$$\phi_e = EA = \frac{Q}{\epsilon_0}$$

The displacement current through S_2 is

$$I_d = \epsilon_0 \frac{d\phi_e}{dt} = \frac{dQ}{dt}$$

which is equal to the true current through S_1.

A significant feature of Maxwell's generalization is that a magnetic field is produced by a changing electric field as well as by true electric currents. Maxwell was undoubtedly led to this generalization by the reciprocal result that an electric field is produced by a changing magnetic flux as well as by electric charges. This latter result, known as *Faraday's law*, preceded Maxwell's generalization. We shall study Faraday's law in the next chapter. These two properties of electric and magnetic field—that a changing flux of one field produces the other—are connected with electromagnetic radiation, of which light is an example.

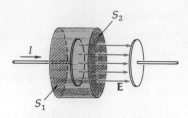

Figure 37-36
The displacement current $I_d = \epsilon_0 \, d\phi_e/dt$, which passes through surface S_2, equals the true current I through surface S_1. If I_d is added to I, the generalized current $I + I_d$ is continuous in the sense that $I + I_d$ is the same through any surface bounded by a given curve.

Review

A. Define, explain, or otherwise identify:

B. True or false:

1. The magnetic field due to a current element decreases as the square of the distance from the element.

2. The magnetic field due to a current element is parallel to the element.

3. The magnetic field due to a very long wire carrying a current decreases as the square of the distance from the wire.

4. A solenoid and a uniformly magnetized magnet of the same shape and same magnetic moment produce the same magnetic field everywhere in space.

5. Lines of **B** never diverge from a point in space.

Exercises

Section 37-1, The Biot-Savart Law

1. A small element of current $d\mathbf{l} = 2\mathbf{k}$ mm has current $I = 2$ A and is centered at the origin. Find the magnetic-induction field $d\mathbf{B}$ at the following points: (a) on the x axis at $x = 3$ m, (b) on the x axis at $x = -6$ m, (c) on the z axis at $z = 3$ m, (d) on the y axis at $y = 3$ m.

2. For the current element of Exercise 1 find the magnitude and indicate the direction of $d\mathbf{B}$ at the point $x = 0$, $y = 3$ m, $z = 4$ m.

3. For the current element of Exercise 1, find the magnitude of $d\mathbf{B}$ and indicate its direction on a diagram for the points (a) $x = 2$ m, $y = 4$ m, $z = 0$; (b) $x = 2$ m, $y = 0$, $z = 4$ m.

4. A long straight wire carries a current of 75 A. Find the magnitude of B at distances (a) 10 cm, (b) 50 cm, and (c) 2 m from the center of the wire.

Exercises 5 to 10 refer to Figure 37-37, which shows two long straight wires in the xy plane parallel to the x axis. One wire is at $y = -6$ cm and the other is at $y = +6$ cm. The current in each wire is 20 A.

5. If the currents in Figure 37-37 are in the negative x direction, find $\mathbf{B}$ at the points on the y axis at (a) $y = -3$ cm, (b) $y = 0$, (c) $y = +3$ cm, (d) $y = +9$ cm.

6. Sketch B_z versus y for points on the y axis when both currents are in the negative x direction.

7. Find $\mathbf{B}$ at points on the y axis as in Exercise 5 but with the current in the wire at $y = -6$ cm in the negative x direction and the current in the wire at $y = +6$ cm in the positive x direction.

8. Sketch B_z versus y for points on the y axis when the currents are in the direction opposite that in Exercise 7.

9. Find $\mathbf{B}$ at the point on the z axis at $z = 8$ cm if (a) the currents are parallel, as in Exercise 5; (b) the currents are antiparallel, as in Exercise 7.

10. Find the magnitude of the force per unit length exerted by one wire on the other.

11. The current in the wire of Figure 37-38 is 8.0 A. Find $\mathbf{B}$ at point P due to each wire segment and sum to find the resultant $\mathbf{B}$.

Figure 37-37
Two long straight parallel wires for Exercises 5 to 10.

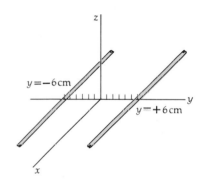

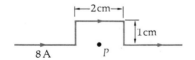

Figure 37-38
Exercise 11.

12. Consider a circular loop of radius R carrying current I. Show directly from the Biot-Savart law that the magnetic field at the center of the loop due to each segment of length dl has the magnitude

$$dB = \frac{\mu_0}{4\pi} \frac{I\, dl}{R^2}$$

and that the resultant magnetic field at the center is $B = \mu_0 I / 2R$.

13. Find the magnetic field at point P in Figure 37-39 if the current is 15 A.

14. A single-turn circular loop of radius 10.0 cm is to produce a field at its center which will just cancel the earth's field at the equator, which is 0.7 G directed north. Find the current in the wire and make a sketch showing the orientation of the loop and current.

Figure 37-39
Exercise 13.

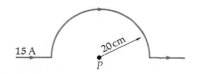

15. A wire of length L carries a current I. Find the magnetic field B at the center when (a) the wire is bent into a square of side $L/4$ and (b) the wire is bent into a circle of circumference L. (c) Which gives the greater value of B?

16. Find the magnetic field at point P in Figure 37-40 which is at the common center of the two semicircular arcs.

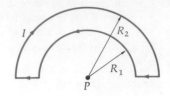

Figure 37-40
Exercise 16.

Section 37-2, The Definition of the Ampere and the Coulomb

17. Two long straight parallel wires 10.0 cm apart carry currents of equal magnitude I. They repel each other with a force per unit length of 4.0×10^{-9} N/m. (a) Are the currents parallel or antiparallel? (b) Find I.

18. A wire of length 12 cm is suspended by flexible leads above a long straight wire. Equal and opposite currents are established in the wires such that the 12-cm wire floats 1.5 mm above the long wire with no tension in its suspension leads. If the mass of the 12-cm wire is 10.0 gm, what is the current?

19. In a student experiment with a current balance, the upper wire of length 30 cm is pivoted such that with no current it balances at 2 mm above a fixed parallel wire also 30 cm long. When the wires carry equal and opposite currents I, the upper wire again balances at its original position when a 2.4-gm mass is placed on it. What is the current I?

20. Three long parallel straight wires pass through the corners of an equilateral triangle of side 10.0 cm, as shown in Figure 37-41, where a dot means that the current is out of the paper and a cross means that it is into the paper. If each current is 15.0 A, find (a) the force per unit length on the upper wire and (b) the magnetic field **B** at the upper wire due to the lower two wires. *Hint:* It is easier to find the force per unit length directly from Equation 37-11 and use your result to find **B** than find **B** first and use it to find the force.

21. Work Exercise 20 with the current in the lower right corner of Figure 37-41 reversed.

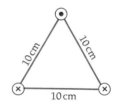

Figure 37-41
Exercises 20 and 21.

Section 37-3, Special Relativity and the Magnetic Field

There are no exercises for this section.

Section 37-4, Ampère's Law

22. A long straight thin-walled cylindrical shell of radius R carries a current I. Find **B** inside and outside the cylinder.

23. In Figure 37-42 one current is 10 A into the paper, the other current is 10 A out of the paper, and each curve is a circular path. (a) Find $\oint \mathbf{B} \cdot d\mathbf{l}$ for each path indicated. (b) Which path, if any, can be used to find **B** at some point due to these currents?

24. A very long coaxial cable has an inner wire and a concentric outer cylindrical conducting shell of radius R. At one end the wire is connected to the shell. At the other end the wire and shell are connected to the opposite terminals of a battery so that there is a current I down the wire and back up the shell. Assume that the cable is straight and find B (a) at points far from the ends and between the wire and shell and (b) outside the cable.

25. A wire of radius 0.5 cm carries a current of 100 A uniformly distributed over the cross section. Find B (a) at 0.1 cm from the center of the wire, (b) at the surface of the wire, and (c) at a point outside the wire 0.2 cm from the surface of the wire. (d) Make a graph of B versus distance from the center of the wire.

26. Show that a uniform magnetic field with no fringing field as shown in Figure 37-43 is impossible because it violates Ampère's law. Do this by applying Ampère's law to the rectangular curve shown by the dashed lines.

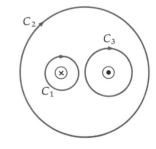

Figure 37-42
Exercise 23.

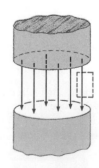

Figure 37-43
Exercise 26.

Section 37-5, The Magnetic Field of a Solenoid

27. A solenoid has a length of 25 cm, radius 1 cm, and 400 turns and carries 3 A of current. Find (a) B on the axis at the center, (b) the magnetic moment of the solenoid, (c) the pole strength q^*, and (d) an approximate value for B on the axis a distance 2 m from one end.

28. Work Exercise 27 for a solenoid of length 30 cm, radius 2 cm, 800 turns, and carrying 2 A.

29. Two tightly wound solenoids, each 10 cm long, have 200 turns, a cross-sectional area of 0.5 cm², and carry current of 4.0 A. Each has its axis along the x axis, and the currents are such that each has its north pole on the right and its south pole on the left. Their centers are 60 cm apart. (a) Find the pole strength of each solenoid. (b) Find the net force of attraction of the solenoids from Coulomb's law of attraction or repulsion between magnetic poles $F = k_m q_1^* q_2^*/r^2$.

Section 37-6, The Magnetic Field of a Bar Magnet

30. What is the magnitude of the magnetization vector **M** that gives a magnetic induction field of 5000 G at the center of a long cylindrical bar magnet?

31. A bar magnet has the same shape as the solenoid in Exercise 27. Find the magnetization M such that the magnetic field produced by the bar magnet is the same as that produced by the solenoid.

32. A long thin rod of iron is uniformly magnetized with magnetization $M = 2 \times 10^5$ A/m along its axis. Its length is 15 cm, and its cross-sectional area is 0.3 cm². (a) What is the magnitude of B at the center? (b) Find the magnetic pole strength q^* at each end. (c) Find the magnetic moment of the magnet.

33. An iron atom is like a small bar magnet of magnetic moment 1.8×10^{-23} A-m². The number of atoms per unit volume in iron is 8.55×10^{28} atoms per cubic meter. (a) If all the iron atoms in an iron bar are aligned with their magnetic moment along the axis, what is the magnitude of the magnetization M? (b) What is B at the center of the magnet, assuming it to be a long rod?

Section 37-7, Magnetic Flux

34. A uniform magnetic field of magnitude 2000 G is parallel to the x axis. A square coil of side 5 cm has a single turn and makes an angle θ with the z axis, as shown in Figure 37-44. Find the magnetic flux through the coil when (a) $\theta = 0$, (b) $\theta = 30°$, (c) $\theta = 60°$, (d) $\theta = 90°$.

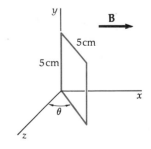

Figure 37-44
Exercise 34.

35. A circular coil has 25 turns and a radius of 5 cm. It is at the equator, where the earth's magnetic field is 0.7 G north. Find the magnetic flux through the coil when (a) its plane is horizontal, (b) its plane is vertical with its axis pointing north, (c) its plane is vertical with its axis pointing east, and (d) its plane is vertical with its axis making an angle of 30° with the north.

36. Find the magnetic flux through the solenoid in Exercise 27. (Assume the magnetic field to be constant inside the solenoid.)

37. Find the magnetic flux through the solenoid in Exercise 28.

38. A circular coil of 15 turns of radius 4 cm is in a uniform magnetic field of 4000 G in the positive x direction. Find the flux through the coil when the unit normal vector to the plane of the coil is (a) $\hat{n} = i$, (b) $\hat{n} = j$, (c) $\hat{n} = (i + j)/\sqrt{2}$, (d) $\hat{n} = k$, and (e) $\hat{n} = 0.6\ i + 0.8\ j$.

Section 37-8, Maxwell's Displacement Current

39. A parallel-plate capacitor has circular plates of radius 1.0 cm separated by 1.0 mm in air. Charge is flowing onto the upper plate and off the lower plate at

a rate of 5 A. (*a*) Find the time rate of change of the electric field between the plates. *Hint:* Write E in terms of the charge Q on the upper plate and find dE/dt. (*b*) Compute the displacement current between the plates and show that it equals 5 A. (*c*) Find the displacement current density J_d between the plates and show that $J_d = \epsilon_0\, dE/dt$.

40. For Exercise 39, show that at distance r from the axis of the plates the magnetic field between the plates is given by $B = \frac{1}{2}\mu_0 J_d r$ if r is less than the radius of the plates and J_d is the displacement current density between the plates.

41. Show that for a parallel-plate capacitor the displacement current is given by $I_d = C\, dV/dt$, where C is the capacitance and V the voltage across the capacitor.

Problems

1. A power cable carrying 50.0 A is located 2.0 m below the earth's surface, but its direction and precise position are unknown. Show how you could locate the cable using a compass. Assume that you are at the equator, where the earth's magnetic field is 0.7 G north.

2. A very long straight wire carries a current of 20.0 A. An electron is 1.0 cm from the center of the wire and is moving with speed 5.0×10^6 m/sec. Find the force on the electron when it moves (*a*) directly away from the wire, (*b*) parallel to the wire in the direction of the current, and (*c*) perpendicular to the wire and tangent to a circle around the wire.

3. A particle of positive charge q moves in a helical path around a long straight wire carrying current I. Its velocity has components $v_\parallel$ parallel to the current and v_t tangent to a circle around the current. Find the relationship between $v_\parallel$ and v_t such that the magnetic force provides the necessary centripetal force.

4. Four long straight parallel wires each carry current I. In a plane perpendicular to the wires, the wires are at the corners of a square of side a. Find the force per unit length on one of the wires if (*a*) all the currents are in the same direction and (*b*) the currents in the wires at adjacent corners are oppositely directed.

5. A relatively inexpensive ammeter, called a *tangent galvanometer,* can be made using the earth's field. A plane circular coil of N turns and radius R is oriented such that the field B_c it produces in the center of the coil is either east or west. A compass is placed at the center of the coil. When there is no current in the coil, the compass points north. When there is a current I, the compass points in the direction of the resultant magnetic field B at an angle θ to the north. Show that the current I is related to θ and the horizontal component of the earth's field B_e by

$$I = \frac{2RB_e}{\mu_0 N}\, \tan\theta$$

6. A very long bar magnet of pole strength q^* lies along the axis of a circular loop with its north pole at the center of the loop, as in Figure 37-45. The loop has radius r and carries current I. (*a*) From the definition $F = q^*B$, find the magnitude and direction of the force exerted by the loop on the magnet. Assume the south pole of the magnet is far enough away to be neglected. (*b*) Let B be the magnetic field at the loop due to the north pole of the magnet. Assuming that B points directly away from the pole, find the magnitude and direction of the total force exerted by the pole on the loop (in terms of B, r, and I). (*c*) Using Newton's third law, set the magnitudes of the forces obtained in (*a*) and (*b*) equal to obtain the relation

$$B = \frac{\mu_0}{4\pi}\frac{q^*}{r^2}$$

Figure 37-45
Problem 6.

for the magnetic field due to a magnetic pole.

7. A disk of radius R carries a fixed charge density σ and rotates with angular velocity ω. (a) Consider a circular strip of radius r and width dr. Show that the current produced by this strip is $dI = (\omega/2\pi) \, dq = \sigma\omega r \, dr$. (b) Use your result for part (a) to show that the magnetic field at the center of the disk is $B = \frac{1}{2}\mu_0\sigma\omega R$. (c) Use your result for part (a) to find the magnetic field at a point on the axis of the disk a distance x from the center.

8. A circular loop of radius R carrying a current I is centered at the origin with its axis along the x axis. Its current is such that it produces a magnetic field in the positive x direction. (a) Sketch B_x versus x for points on the x axis. Include both positive and negative x. Compare this sketch with that for E_x due to a charged ring of the same size. (b) A second identical loop carrying the same current in the same sense is in a plane parallel to the xy plane with its center at $x = d$. Sketch the magnetic field on the x axis due to each coil separately and the resultant field due to the two coils. Show from your sketch that dB_x/dx is zero midway between the coils.

9. Two coils which are separated by a distance equal to their radius and carrying equal currents such that their axial fields add are called *Helmholtz coils*. A feature of Helmholtz coils is that the resultant magnetic field between the coils is very uniform. Let $R = 10$ cm, $I = 20$ A, and $N = 300$ turns for each coil. Place one coil with center at the origin and the other at $x = 10$ cm (as in Problem 8 with $d = R = 10$ cm). (a) Calculate the resultant field B_x at points $x = 5$ cm, $x = 7$ cm, $x = 9$ cm, and $x = 11$ cm and from this sketch B_x versus x. (Note that B_x is symmetric about the midpoint of the coils.) (b) Show that $dB_x/dx = 0$ and $d^2B_x/dx^2 = 0$ at $x = \frac{1}{2}R = 5$ cm. (c) Show that $d^3B_x/dx^3 = 0$ at $x = \frac{1}{2}R$.

10. The magnetic field of a small current-carrying coil can be found at points off axis and a great distance from the coil by considering the coil to be a small magnet of pole strength q^* and length l with $q^*l = IA$, where A is the area of the coil and I is its current. Consider a small coil at the origin with its magnetic moment in the positive z direction. Show that the field at a point on the x axis a great distance away is given by

$$\mathbf{B} = -\frac{\mu_0}{4\pi}\frac{IA}{x^3}\,\mathbf{k}$$

Hint: Calculate $\mathbf{B}$ from Coulomb's law, part (c) of Problem 6, assuming a positive pole on the z axis at $z = +\frac{1}{2}l$ and a negative pole at $z = -\frac{1}{2}l$.

These Helmholtz coils at the Kettering Magnetics Laboratory at Oakland University are used to cancel the earth's magnetic field and to provide a uniform magnetic field in a small region.

11. A long straight wire carries a current of 20 A, as shown in Figure 37-46. A rectangular coil with two sides parallel to the straight wire has sides 5 and 10 cm with its near side a distance 2 cm from the wire. (a) Compute the magnetic flux through the rectangular coil. Hint: Calculate the flux through a strip of area $dA = (10$ cm$)$ dx and integrate from $x = 2$ cm to $x = 7$ cm. (b) If the rectangle carries a counterclockwise current of 5 A, find the net force on it.

Figure 37-46
Problem 11.

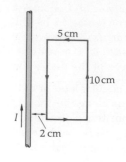

12. A solenoid has n turns per unit length and radius R and carries current I. Its axis is at the x axis with one end at $x = -x_1$ and the other at $x = +x_2$. Consider a part of the solenoid of length dx at distance x from the origin to be a circular loop carrying current nI dx. (a) Use Equation 37-8 for the magnetic field at the origin due to this loop and integrate from $x = -x_1$ to $x = +x_2$ to find the total magnetic field at the origin due to the solenoid. Hint: You will need to use the indefinite integral

$$\int \frac{dx}{r^3} = \frac{x}{R^2 r} \qquad \text{where } r = \sqrt{x^2 + R^2}$$

(b) Show that the result can be written as $B = \frac{1}{2}\mu_0 nI$ (cos θ_1 + cos θ_2), where cos $\theta_1 = x_1/r_1$ and cos $\theta_2 = x_2/r_2$. (c) Use this result to show that $B \approx \mu_0 nI$ at points far from the ends compared with R and $B \approx \frac{1}{2}\mu_0 nI$ at each end.

13. A large 50-turn circular loop of radius 10.0 cm carries a current 4.0 A. At the center of the loop is a small 20-turn coil of radius 0.5 cm carrying current 1.0 A. The planes of the two coils are perpendicular. Find the torque exerted by the large coil on the small coil. (Neglect any variation in **B** due to the large coil over the region occupied by the small coil.)

14. A solenoid has n turns per unit length, radius R_1, and carries current I. (a) A large circular loop of radius $R_2 > R_1$ and N turns encircles the solenoid at a point far away from the ends of the solenoid. Find the magnetic flux through the loop. (b) A small circular loop of radius $R_3 < R_1$ is completely inside the solenoid far from its ends with its axis parallel to that of the solenoid. Find the magnetic flux through the loop.

15. A long cylindrical conductor of radius R carries current I of uniform density $J = I/\pi R^2$. Find the magnetic flux per unit length through the area indicated in Figure 37-47.

Figure 37-47
Problem 15.

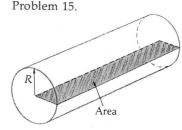

16. In this problem you are to show that the generalized form of Ampère's law (Equation 37-36) and the Biot-Savart law give the same result in a situation in which they both can be used. Figure 37-48 shows two charges $+Q$ and $-Q$ on the x axis at $x = -a$ and $x = +a$ with a current $I = -dQ/dt$ along the line between them. Point P is on the y axis at $y = R$. (a) Use Equation 37-6, obtained from the Biot-Savart law, to show that the magnitude of B at point P is

$$B = \frac{\mu_0}{2\pi R} \frac{Ia}{\sqrt{a^2 + R^2}}$$

Figure 37-48
Problem 16.

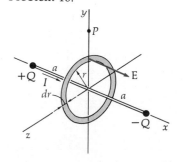

(b) Show that the electric field at any point y on the y axis is in the x direction with magnitude

$$E_x = \frac{1}{2\pi\epsilon_0} \frac{Qa}{(y^2 + a^2)^{3/2}}$$

(c) Consider a circular strip of radius r and width dr in the yz plane with center at the origin. Show that the flux of the electric field through this strip is E_x $dA = (Q/\epsilon_0)a(r^2 + a^2)^{-3/2}r$ dr. (d) Use your result in part (c) to find the total flux ϕ_e through a circular area of radius R. Show that $\epsilon_0\phi_e = Q(1 - a/\sqrt{a^2 + R^2})$. (e) Find the displacement current I_d and show that $I + I_d = Ia/\sqrt{a^2 + R^2}$. Then show that Equation 37-36 gives the same result for B as that found in part (a).

CHAPTER 38   Faraday's Law

The fact that electric currents can be induced by changing magnetic fields was discovered in the early 1830s by Michael Faraday and simultaneously by the American physicist Joseph Henry. The results of many experiments by these men and others can be expressed by a single relation known as Faraday's law. Let ϕ_m be the flux of the magnetic field through a circuit (Figure 38-1). If this flux is changed in any way, there is an emf in the circuit given by

$$\mathcal{E} = -\frac{d\phi_m}{dt}$$
38-1

Faraday's law

The emf is usually detected by observing a current in the circuit. The emf in a circuit is the work done per unit charge by a nonelectrostatic electric field as the charge moves around the complete circuit. (The work done by an *electrostatic* field is always zero around a complete circuit because electrostatic fields are conservative.) When the source of emf is a battery, this nonelectrostatic electric field is localized between the poles of the battery and is due to nonconservative electrochemical forces acting on the charges in the battery. When the emf is due to a changing magnetic field, the nonelectrostatic field is not localized but exists throughout the circuit. This electric field is present even when there is no wire or other material. Let $\mathbf{E}$ be this nonelectrostatic electric field. Since the integral of $\mathbf{E} \cdot d\mathbf{l}$ around a closed curve or circuit is the work done per unit charge by this field, this integral equals the emf in the circuit:

$$\mathcal{E} = \oint \mathbf{E} \cdot d\mathbf{l} = -\frac{d\phi_m}{dt}$$
38-2

The magnetic flux through a circuit can be changed in many different ways: the current producing the flux may be increased or decreased; permanent magnets may be moved toward or away from the circuit; the circuit itself may be moved toward or away from the source of the flux; or the area of the circuit may be increased or decreased in a fixed magnetic field. In every case, an emf is induced in the circuit

Figure 38-1
Flux of **B** through a circuit. If the flux changes, an emf is induced in the circuit.

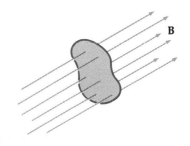

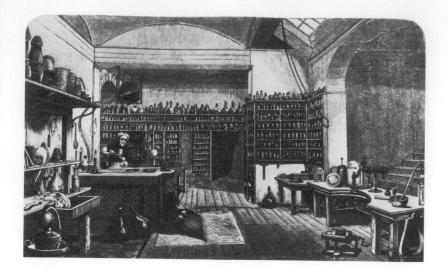

An engraving of Michael Faraday (1791–1867) in his laboratory in the basement of the Royal Institution in London, where he carried out his research on electricity and magnetism. (*Courtesy of the Niels Bohr Library, American Institute of Physics.*)

equal in magnitude to the rate of change of the magnetic flux. We shall discuss the significance of the negative sign in Faraday's law in Section 38-2.

38-1 Motional EMF

Figure 38-2 shows a conducting rod sliding along conducting rails which are connected to a switch and a resistor. A uniform magnetic field is directed into the paper. This example illustrates an induced emf for which we can understand the origin of the forces on the moving charges in the circuit from principles we have already studied. For convenience, assume that the moving rod contains positive charges free to move in the rod. If the rod moves with velocity **v**, there will be a magnetic force $q\mathbf{v} \times \mathbf{B}$ on the charges parallel to the rod, as shown. As with the Hall effect, there will be a charge separation, i.e., an excess positive charge at the top of the rod and an excess negative charge at the bottom (assuming the switch to be open so that there will be no current in the circuit). The positive charges will continue to move to the top of the rod until the electrostatic field produced by the charge separation balances the magnetic force. The magnitude of the electrostatic field will thus be

$$E = |\mathbf{v} \times \mathbf{B}| = vB$$

since **v** and **B** are perpendicular. Point a will then be at a greater potential than point b. The magnitude of this potential difference is

$$V = El = vBl$$

where l is the distance between these two points.

In a reference frame in which the rod is at rest (and the rails and magnetic field are moving), there is no magnetic force on the charges in the rod because they are at rest. The electrostatic force on a charge q due to the charge separation is balanced by a nonconservative force $q\mathbf{E}_{non}$, where $\mathbf{E}_{non}$ is a nonelectrostatic electric field in this frame which has magnitude vB. When the switch is closed, there will be a coun-

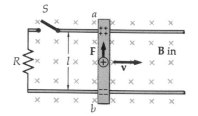

Figure 38-2
Rod sliding on rails in magnetic field.

Faraday and his wife in the 1850s.

terclockwise current in the circuit but the charge separation will be maintained by this nonconservative force. The work done by this non-conservative force when a charge q moves a distance l is $qE_{non}l = qvBl$. The work per unit charge is the emf in the circuit:

$$\mathcal{E} = vBl \qquad\qquad 38\text{-}3 \qquad \textit{Motional emf}$$

This emf is the same as given by Faraday's law in Equation 38-1. The flux through the circuit when the rod is a distance x from the left end is

$$\phi_m = Blx$$

The rate of change of the flux is

$$\frac{d\phi_m}{dt} = Bl\frac{dx}{dt} = Blv$$

Questions

1. A rod moves in a plane perpendicular to a uniform magnetic field **B**. Is a force required to move the rod with constant velocity?

2. The work required to accelerate a rod from rest to velocity **v** in a plane perpendicular to a uniform magnetic field is greater than $\frac{1}{2}mv^2$. Why?

38-2 Lenz's Law

The direction of the induced emf and current can be found from a general statement known as *Lenz's law*.

The emf and induced current are in such a direction as to tend to oppose the change which produced them.

In our statement of Lenz's law we did not specify just what kind of change it is that causes the induced emf. This was left purposefully vague in order to allow for a variety of interpretations all of which are valid and give the same result for the direction of the induced emf and current. A few illustrations will clarify this point.

In the example of the rod moving on the rails, above, the magnetic field is into the paper. Let us call this direction positive. The movement of the rod then tends to increase the flux in this direction. According to Lenz's law, the induced current is in the direction that will oppose this change. As we have seen, the induced emf and current are counterclockwise. The flux produced *by the induced current* is out of the paper, opposing the increase in flux through the circuit which results from the motion of the rod. If the current induced were in the clockwise direction, the induced flux would add to the increase produced by the moving rod, contrary to Lenz's law. Lenz's law is thus a statement that the induced current tends to maintain the status quo. If we try to increase the flux through a circuit, a current will be induced which tends to decrease the flux. If we move the rod to the left in our example, decreasing the flux, Lenz's law tells us that the induced current will be clockwise to help maintain the original flux through the circuit.

We can also interpret Lenz's law in terms of forces and the conservation of energy. In our example of the rod moving to the right, there is a magnetic force on the rod due to the induced current. Since the current is upward in the rod and the magnetic field is inward, the force is to the left,

$$\mathbf{F}_m = I\mathbf{l} \times \mathbf{B}$$

This force tends to oppose the motion of the rod (see Figure 38-3). If the rod is to move to the right with constant velocity, there must be an external force to the right of magnitude IlB. Lenz's law then tells us that there will be a current induced in such a direction that the magnetic force opposes this external force.

When the rod is moving with constant velocity, the power input by the external force on the rod goes into heating the resistor. The power input is Fv, where F is the external force. Thus

$$P = Fv = IlBv = I^2R \qquad\qquad 38\text{-}4$$

or

$$IR = Blv \qquad\qquad 38\text{-}5$$

By Kirchhoff's first rule the potential drop IR must equal the emf in the circuit. Thus we find by means of energy balance that the emf in the circuit is

$$\mathcal{E} = Blv \qquad\qquad 38\text{-}6$$

If we assume the opposite of Lenz's law, energy cannot be conserved. Suppose we start with the rod at rest on the rails and give it a slight push to the right. Assuming the opposite of Lenz's law, we would have a current in the clockwise direction. Then there would be a magnetic force on the rod to the right which would accelerate the rod, increasing its velocity. The emf and induced current would thus be increased. That would further increase the force on the rod to the right. Hence, the opposite of Lenz's law gives a situation in which the rod continues to accelerate to the right and the current continues to increase without limit. With no power input to this circuit, the power I^2R into the resistor would increase without limit.

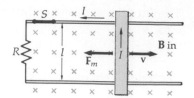

Figure 38-3
The magnetic force $\mathbf{F}_m$ on the rod opposes the motion of the rod. Here the rod is moving to the right, and $\mathbf{F}_m$ is to the left.

38-3   Applications of Faraday's Law

In most applications of Faraday's law it is not possible to understand the origin of the induced emf from our previous knowledge of magnetic forces. That is, we cannot derive Equation 38-1 by considering magnetic or electric forces on charges, as we did for the special case of the rod moving on rails. Instead, we must take Faraday's law as expressed in Equation 38-1 to be an independent law of nature expressing a wide range of experimental results. In this section we consider more examples to illustrate the use of Faraday's law and Lenz's law to find the direction of the induced current.

Figure 38-4 shows a bar magnet moving toward a loop which has a resistance R. Since the magnetic field from the bar magnet is to the right, out of the north pole of the magnet, the movement of the magnet toward the loop tends to increase the flux through the loop to the right. The induced current in the loop is in the direction shown, to create a flux which opposes the change.

Demonstration of induced emf. When the magnet is moved away from the coil, an emf is induced in the coil, as shown by the galvanometer deflection. No deflection is observed when the magnet is stationary. (*Courtesy of Larry Langrill, Oakland University.*)

Figure 38-4
When the bar magnet moves toward the loop, the induced emf in the loop produces a current in the direction shown.

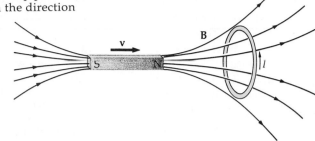

Figure 38-5 shows the induced magnetic moment of the current loop in Figure 38-4. The loop acts like a small magnet with north pole to the left and south pole to the right. Since opposite poles attract and like poles repel, the induced magnetic moment of the loop exerts a force on the bar magnet to the left to oppose its motion toward the loop. Again, we could use Lenz's law in terms of forces rather than flux. If the bar magnet is moved toward the loop, the induced current must produce a magnetic moment to oppose this change.

In Figure 38-6 the bar magnet is at rest, and the loop is moved away from it. The induced current and magnetic moment are shown in the figure. In this case the magnetic moment of the loop attracts the bar magnet, as required by Lenz's law.

In Figure 38-7 when the current in circuit 1 is changed, there is a change in the flux through circuit 2. Suppose the switch in circuit 1 is open with no current in the circuit. When the switch is closed (Figure 38-7b), the current in circuit 1 does not reach its steady value ε_1/R_1 instantaneously but takes a short time to change from zero to this value. During this short time, while the current is increasing, the flux in circuit 2 is changing and there is an induced current in that circuit in the direction shown. When the current in the first circuit reaches its steady value, the flux is no longer changing and there is no induced current in circuit 2. An induced current in circuit 2 in the opposite direction appears momentarily when the switch in circuit 1 is opened (Figure 38-7c) and the current is decreasing to zero.

For our next example we consider a single isolated circuit. When there is a current in the circuit, there is magnetic flux through the circuit due to its own current. When the current is changing, the flux is changing and there is an induced emf in the circuit. This *self-induced* emf opposes the change in the current. It is because of this self-induced emf that the current in a circuit cannot jump instantaneously from zero to some finite value or from some value to zero. Henry first noticed this effect when he was experimenting with a circuit con-

Figure 38-5
The magnetic moment of the loop (indicated by the outlined magnet) due to the induced current is such as to oppose the motion of the bar magnet. Here the bar magnet is moving toward the loop, and so the induced magnetic moment repels the bar magnet.

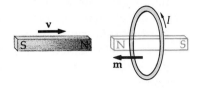

Figure 38-6
When the loop is moved away from a stationary bar magnet, the induced magnetic moment attracts the bar magnet, again opposing the relative motion.

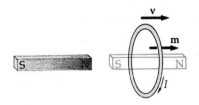

Figure 38-7
(a) Two adjacent circuits. (b) Just after the switch is closed, I_1 is increasing and B is increasing in the direction shown. The changing flux in circuit 2 induces current I_2. The flux due to I_2 opposes the increase in flux due to I_1. (c) As the switch is opened, I_1 decreases and B decreases. The induced current I_2 tends to maintain the flux in the circuit, opposing the decrease.

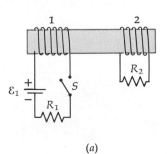

(a)

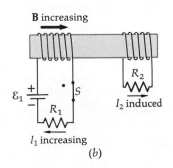

(b)

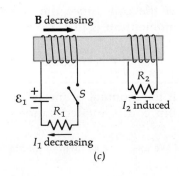

(c)

sisting of many turns of a wire (Figure 38-8), an arrangement that gives a large flux through the circuit for even a small current. Henry noticed a spark across the switch when he tried to break the circuit. This spark is due to the large induced emf which occurs when the current varies rapidly, as with opening the switch. In this case the induced emf tries to maintain the original current. The large induced emf produces a large voltage drop across the switch as it is opened. The electric field between the poles of the switch is large enough to tear the electrons from the air molecules, causing dielectric breakdown. When the molecules in the air dielectric are ionized, the air conducts electric current in the form of a spark.

For our final example, we consider a device for the measurement of magnetic fields called a *flip coil*. Figure 38-9 shows a coil whose plane is perpendicular to a uniform magnetic induction **B** which is to be measured. The coil is connected to a *ballistic galvanometer*, a device designed to measure the total charge passing through it in a short time. The flux through the flip coil is

$$\phi_m = NBA$$

where N is the number of turns and A is the area of the coil. If the coil is suddenly rotated about a diameter through 90°, the flux decreases to zero. While the flux is decreasing, there is an emf in the coil and a current in the coil-galvanometer circuit. The current is

$$I = \frac{\mathcal{E}}{R} = \frac{1}{R}\frac{d\phi_m}{dt}$$

where R is the total resistance of the coil and galvanometer. The total charge that passes through the galvanometer is

$$Q = \int I\,dt = \frac{1}{R}\int d\phi_m = \frac{\phi_m}{R} = \frac{NAB}{R}$$

If the total charge is measured, the magnetic induction B can be found from

$$B = \frac{RQ}{NA} \qquad\qquad 38\text{-}7$$

A ballistic galvanometer is similar to an ordinary galvanometer except that it is constructed to have a large period of oscillation and is only very slightly damped, so that it oscillates nearly freely. The torque on the galvanometer coil is proportional to the current in it:

$$\tau = CI$$

The constant C is just the product of the number of turns, the area, and the magnetic field through the galvanometer coil. The current I lasts only a short time. The integral of the torque over the time equals the angular impulse, which equals the change in the angular momentum of the coil. Since the coil is initially at rest, the change in angular momentum is just $\mathcal{I}\omega$[1]:

$$\int \tau\,dt = \int CI\,dt = CQ = \Delta(\mathcal{I}\omega) = \mathcal{I}\omega \qquad\qquad 38\text{-}8$$

If the moment of inertia of the galvanometer coil is very large and the flip coil is rotated suddenly, the galvanometer coil moves very little

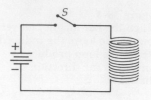

Figure 38-8
Circuit with many turns of wire, giving a large flux for a given current in the circuit. The emf induced in the circuit when the current changes opposes the change, trying to maintain the original current.

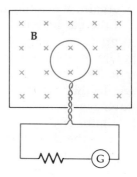

Figure 38-9
Flip-coil circuit for measuring the magnetic field **B**. When the coil is flipped over, the total charge flowing through the galvanometer is proportional to B.

[1] We use $\mathcal{I}$ for moment of inertia here to avoid confusion with current I.

while the charge passes through it. The coil thus receives an angular impulse and begins to rotate with initial angular velocity ω. The coil then rotates to a maximum deflection angle which depends on the restoring torque of the suspension and is proportional to the initial angular velocity. A measurement of this maximum deflection determines ω which is proportional to the total charge Q.

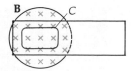

Figure 38-10
Eddy currents. If the magnetic field is changing, an emf is induced in any closed path in the metal such as the curve C shown. The induced emf causes a current in the circuit.

38-4 Eddy Currents

In the examples we have discussed, the currents produced by a changing flux were set up in definite circuits. Often a changing flux sets up circulating currents, called *eddy currents,* in a piece of bulk metal like the core of a transformer. Consider a conducting slab between the pole faces of an electromagnet (Figure 38-10). If the magnetic induction B between the pole faces is changing with time (as it will if the current in the magnet windings is alternating current), the flux through any closed loop in the slab will be changing. For example, the flux through the curve C indicated in the figure is just the magnetic induction B times the area enclosed by the curve. If B varies, the flux will vary and there will be an induced emf around curve C. Since path C is in a conductor, there will be a current given by the emf divided by the resistance of the path. In this figure we have indicated just one of the many closed paths which will contain currents if the magnetic field between the pole faces varies. Circulating or eddy currents are usually unwanted because the heat produced is not only a power loss but must be dissipated. The power loss can be reduced by increasing the resistance of the possible paths for the eddy currents, as in Figure 38-11. Here the conducting slab is laminated, i.e., made up of small strips glued together. Because of the resistance between the strips, the eddy currents are essentially confined to the strips. The large eddy current loops are broken up, and the power loss is greatly reduced.

Figure 38-11
The eddy currents in a metal slab can be reduced by constructing the slab from small strips. The resistance of the path indicated by C is now large because of the glue between the strips.

The existence of eddy currents can be demonstrated by pulling a copper or aluminum sheet between the poles of a strong permanent magnet (Figure 38-12). Part of the area enclosed by curve C in this figure is in the magnetic field, and part is outside the field. As the sheet is pulled to the right, the flux through this curve decreases (assuming that the flux into the paper is positive). According to Faraday's law and Lenz's law, there will be a clockwise current induced around this curve. Since this current is directed upward in the region between the pole faces, the magnetic field exerts a force on the current to the left opposing motion of the sheet. You can feel this force on the sheet if you try to pull a sheet suddenly through a strong magnetic field. If the sheet has cuts in it, as in Figure 38-13, to reduce the eddy currents, the force is greatly reduced.

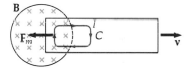

Figure 38-12
Demonstration of eddy currents. When the metal slab is pulled to the right, there is a magnetic force to the left on the induced current, opposing the motion.

Questions

3. A bar magnet is dropped inside a long, vertical copper pipe. The pipe is evacuated so that there is no air resistance, but the falling magnet still reaches a terminal velocity. Explain why.

4. A sheet of metal fixed to the end of a pivoted rod will swing like a pendulum about the pivot, but if the sheet is made to swing through the gap between two poles of a magnet, the oscillation will rapidly damp out. Why?

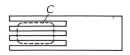

Figure 38-13
If the metal slab has cuts as shown, the eddy currents are reduced because of the lack of good conducting paths.

Figure 38-14
D. W. Kerst with the first betatron in 1941. This betatron
was capable of accelerating
electrons to an energy of 2.3
MeV. The electron beam
strikes a target inside the
doughnut, producing x-rays
used for research in nuclear
physics. (*Courtesy of the Niels
Bohr Library, American Institute of Physics.*)

38-5   The Betatron

An interesting use of Faraday's law is the betatron (Figure 38-14), invented in 1941 by Donald Kerst at the University of Illinois to accelerate electrons (also called beta particles) to high energies. In the betatron, electrons move in a circular orbit at a constant radius in an
evacuated doughnut-shaped tube. They are accelerated by a tangential, nonconservative electric field, which is produced by a changing
magnetic flux through the orbit generated by an electromagnet. The
centripetal force needed to keep the electrons moving in a circle is
provided by the same magnet. This can be done if the magnetic induction B_0 at the orbit radius is just half the average value of the magnetic
induction over the complete area of the orbit. To accomplish this the
pole pieces of the magnet are shaped as shown in Figure 38-15, so that
the magnetic induction decreases with distance from the center of the
circle.

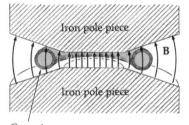

Ceramic
doughnut

Figure 38-15
The pole faces in a betatron
magnet are shaped as shown
so that the magnetic field at
the orbit radius of the electrons has a magnitude half
that of the average magnetic
field through the area enclosed by the orbit.

We define the average magnetic induction B_{av} to be the flux divided
by the area

$$B_{av} = \frac{1}{\pi R^2} \int \mathbf{B} \cdot \hat{\mathbf{n}} \, dA = \frac{\phi_m}{\pi R^2} \qquad \text{38-9}$$

The emf around the orbit is

$$\varepsilon = \frac{d\phi_m}{dt} = \pi R^2 \frac{dB_{av}}{dt} \qquad \text{38-10}$$

This emf is equivalent to a tangential, nonconservative electric field
given by Equation 38-2:

$$\varepsilon = \oint \mathbf{E} \cdot d\mathbf{l} = E 2\pi R = \pi R^2 \frac{dB_{av}}{dt}$$

or

$$E = \tfrac{1}{2} R \frac{dB_{av}}{dt} \qquad \text{38-11}$$

The tangential force $q\mathbf{E}$ increases the magnitude of the momentum of the electron according to Newton's second law:

$$\frac{dp}{dt} = qE = \tfrac{1}{2}qR\,\frac{dB_{av}}{dt} \qquad\qquad 38\text{-}12$$

The centripetal force which keeps the electrons moving in a circle is provided by the radial force qvB_o, where B_o is the value of the magnetic induction at the orbit radius. The condition for circular motion is Equation 36-12, which we rewrite as

$$p = B_o qR \qquad\qquad 38\text{-}13$$

Both Equations 38-12 and 38-13 written in terms of the momentum p are valid relativistically if p is interpreted as relativistic momentum. These equations are consistent with each other if the magnetic induction at the orbit is exactly half the average value over the area:

$$B_o = \tfrac{1}{2}B_{av} \qquad\qquad 38\text{-}14$$

Betatron condition

The magnetic field is produced by coils which carry alternating current. The magnetic induction B then varies as in Figure 38-16. The electrons are injected into orbit just as the magnetic field crosses zero and begins to increase. (The electrons are injected from an electron gun just outside the stable orbit and are quickly brought into the stable orbit by a magnetic field from a current pulse through some auxiliary coils.) Only the first quarter of the cycle can be used. As the flux through the orbit changes, the electrons gain energy and their momentum increases, according to Equation 38-12. They stay in their orbit because the field at the orbit B_o changes at the same rate as the momentum. Typically, the electrons make several hundred thousand revolutions during the quarter cycle in which they are accelerated. When the magnetic field is near its maximum value, the electrons are ejected from their orbit by a current pulse through a set of auxiliary coils which momentarily weakens the induction at the orbit. Usually the electron beam is moved into a larger orbit and hits the injector gun, producing x-rays. The x-rays are used for nuclear research or for medical purposes.

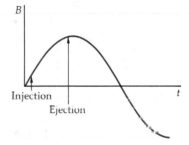

Figure 38-16
B versus t in the accelerating part of the betatron cycle. The electrons are injected just after B crosses zero and ejected just before maximum B.

Questions

5. Why can't the second quarter of the cycle in Figure 38-16 be used to accelerate the electrons?

6. Why can't the fourth quarter of the cycle be used? Is there any other part of the cycle that can be used?

7. The iron core of the betatron magnet-pole face is made not from one piece of iron but from many pie-shaped wedges. Why?

38-6 Inductance

The flux through a circuit can be related to the current in that circuit and the currents in other nearby circuits. (We shall assume that there are no permanent magnets around.) Consider the two circuits in Figure 38-17. The magnetic field at some point P consists of a part due to I_1 and a part due to I_2. These fields are proportional to the currents producing them and could, in principle, be calculated from the Biot-

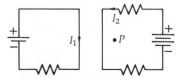

Figure 38-17
Two adjacent circuits. The magnetic field at point P is partly due to current I_1 and partly due to I_2. The flux through either circuit is the sum of two terms, one proportional to I_1 and the other to I_2.

Savart law. We can therefore write the flux through circuit 2 as the sum of two parts; one part is proportional to the current I_1 and the other to the current I_2:

$$\phi_{m2} = L_2 I_2 + M_{12} I_1 \qquad\qquad 38\text{-}15$$

where L_2 and M_{12} are constants. The constant L_2, called the *self-inductance* of circuit 2, depends on the geometrical arrangement of that circuit. The constant M_{12}, called the *mutual inductance* of the two circuits, depends on the geometrical arrangement of both circuits. In particular, we can see that if the circuits are far apart, the flux through circuit 2 due to the current I_1 will be small and the mutual inductance will be small. An equation similar to Equation 38-15 can be written for the flux through circuit 1:

Self-inductance and mutual inductance

$$\phi_{m1} = L_1 I_1 + M_{21} I_2 \qquad\qquad 38\text{-}16$$

The self-inductance L_1 depends only on the geometry of circuit 1, whereas the mutual inductance M_{21} depends on the arrangement of both circuits. Though it is certainly not obvious, it can be shown in general that these two mutual inductances are equal:

$$M_{12} = M_{21} \qquad\qquad 38\text{-}17$$

When the circuits are fixed and only the currents change, their induced emfs are, from Faraday's law,

$$\mathcal{E}_1 = -\frac{d\phi_{m1}}{dt} = -L_1 \frac{dI_1}{dt} - M \frac{dI_2}{dt}$$

$$38\text{-}18$$

$$\mathcal{E}_2 = -L_2 \frac{dI_2}{dt} - M \frac{dI_1}{dt}$$

where we have dropped the subscripts on the mutual inductance.

The SI unit of inductance is the henry (H). From Equations 38-16 and 38-18 we see that the henry is related to other SI units by

$$1\,\text{H} = 1\,\frac{\text{T-m}^2}{\text{A}} = 1\,\frac{\text{V-sec}}{\text{A}}$$

Unit of inductance

We shall see that inductance can always be written as the product of a μ_0 and some characteristic length. The SI unit of the constant μ_0 can therefore be conveniently expressed as henrys per meter:

$$\mu_0 = 4\pi \times 10^{-7}\,\text{H/m}$$

Units of μ_0

In principle, the calculation of the self- and mutual inductances is straightforward for any given circuits and arrangement. For example, to calculate L_1 we need only assume some current I_1 and calculate the magnetic field from the Biot-Savart law at every point on some surface bounded by the circuit. This magnetic field will of course be proportional to the current I_1. We then calculate the flux of this field by integrating over the area. Since B is proportional to I_1, the flux will also be proportional to I_1. The proportionality constant is the self-inductance L_1. Similarly, we calculate the mutual inductance by assuming some current I_1 in circuit 1 and calculating the magnetic field everywhere on a surface bounded by circuit 2 and from it the flux through that circuit. Because of the difficulty in calculating the magnetic field in general from the Biot-Savart law and the difficulty in integrating this field over any surface, it is not surprising that the calculations of self-inductance and mutual inductance are extremely difficult except for a few very simple geometrical arrangements. Consider, for example, the calculation of the self-inductance for a single circular loop of radius R. The magnetic field at the center of the loop is $B = \frac{1}{2}\mu_0 I/R$. But in order to find the flux through the loop, we need to know the value of B everywhere on the circle enclosed by the loop, not just at the center. Such a calculation is too difficult to do here. The self-inductance of a single loop depends not only on the radius of the loop but also on the radius of the conducting wire.

We now present some examples in which inductance can be calculated, at least approximately.

Example 38-1 Self-Inductance of a Solenoid This calculation is not difficult because, to a good approximation, the magnetic field inside the solenoid is uniform and given by

$$B = \mu_0 \frac{N}{l} I = \mu_0 n I$$

where $n = N/l$ is the number of turns per unit length, N the total number of turns, l the total length, and I the current in the solenoid. The area enclosed by this circuit is the area of one turn times the number of turns, and so the flux is

$$\phi_m = BNA = \mu_0 n I N A = \mu_0 n \frac{N}{l} l A I = \mu_0 n^2 (lA) I \qquad \text{38-19}$$

where A is the area of a single loop. As expected, the flux is proportional to the current I. The proportionality constant is the self-inductance:

$$L = \frac{\phi_m}{I} = \mu_0 n^2 A l = \mu_0 N^2 \frac{A}{l} \qquad \text{38-20}$$

The self-inductance is proportional to the square of the number of turns per unit length and to the volume Al. It can also be thought of as the product of μ_0, N^2, and the characteristic length A/l.

Example 38-2 Find the self-inductance of a solenoid of length 10 cm, area 5 cm², and 100 turns.

We can calculate the self-inductance from Equation 38-20 in henrys if we put all quantities in SI units:

$$n = \frac{N}{l} = \frac{100 \text{ turns}}{0.1 \text{ m}} = 10^3 \text{ turns/m}$$

$$Al = (5 \times 10^{-4} \text{ m}^2)(0.1 \text{ m}) = 5 \times 10^{-5} \text{ m}^3$$

$$\mu_0 = 4\pi \times 10^{-7} \text{ H/m}$$

$$L = (4\pi \times 10^{-7})(10^3)^2(5 \times 10^{-5}) = 2\pi \times 10^{-5} \text{ H}$$

Example 38-3 At what rate must the current in the solenoid of Example 38-2 change to induce an emf of 20 V?

From Equation 38-18 with $M = 0$ we have

$$\varepsilon = -L \frac{dI}{dt} = 20 \text{ V}$$

Then

$$\frac{dI}{dt} = -\frac{\varepsilon}{L} = -\frac{20 \text{ V}}{2\pi \times 10^{-5} \text{ H}} = -3.18 \times 10^5 \text{ A/sec}$$

Example 38-4 Mutual Inductance between a Long Wire and a Rectangular Loop Two circuits for which the mutual inductance can be calculated are shown in Figure 38-18, where the wire on the left is assumed to be very long so that the remaining part of its circuit is far away from the rectangular loop and contributes negligible flux. To find the flux through the rectangular loop we must integrate. Since the magnetic field due to a long wire decreases as $1/r$ with the distance from the wire, we choose the strip-shaped area element shown in the figure. The area of this strip is $c\ dx$, and the flux through the strip is

$$d\phi_m = \frac{\mu_0 I}{2\pi x}\ c\ dx \qquad\qquad 38\text{-}21$$

where I is the current in the long wire. The total flux is found by integrating from $x = a$ to $x = b$:

$$\phi_m = \frac{\mu_0}{2\pi}\ cI \int_a^b \frac{dx}{x} = \frac{\mu_0 Ic}{2\pi}\ \ln\frac{b}{a} \qquad\qquad 38\text{-}22$$

The flux through the rectangular loop due to the current I in the long wire is proportional to I, as expected. The proportionality constant is the mutual inductance:

$$M = \frac{\mu_0 c}{2\pi}\ \ln\frac{b}{a}$$

The calculation of the flux through the circuit on the left due to a current I_2 in the rectangular loop is more difficult because of the complicated expression for the magnetic field due to the loop at an arbitrary point to the left of the wire.

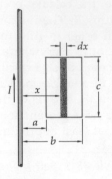

Figure 38-18
Long wire and circuit for calculation of mutual inductance in Example 38-4.

Questions

8. How would the inductance of a solenoid be changed if the same length of wire were wound onto a cylinder of the same diameter but twice as long? If twice as much wire were wound onto the same cylinder?

9. Two solenoidal inductances are wound on cylinders of the same diameter. Is it possible for them to have the same inductance even though one has twice the total length of wire as the other?

38-7 *LR* Circuits

As we have seen, self-inductance in a circuit prevents the current from rising or falling instantaneously. Circuits which contain coils or solenoids of many turns have a large self-inductance. Such a coil or solenoid is called an *inductor*. The symbol for an inductor is ⌇⌇⌇ . Often we can neglect the self-inductance of the rest of the circuit compared with that of an inductor.

A circuit containing batteries, resistors, and inductors is called an *LR* circuit. Since all circuits contain resistance and self-inductance, the analysis can be applied to some extent to all circuits. All circuits also have some capacitance between parts of the circuit at different potentials. We shall include the effects of capacitance in Section 38-9, when we study *LC* and *LCR* circuits. We neglect capacitance here in order to simplify the analysis and to bring out the special features of inductance in a simple situation.

Figure 38-19 shows a typical circuit. We neglect the internal resistance of the battery and treat it as an ideal emf. The current in the circuit is originally zero. At $t = 0$ we close the switch. As we have discussed previously, while the current is increasing there is an induced emf in the inductance in the direction opposing the current increase. The directions of the current and emfs are indicated in the figure. We can write an equation relating the current and its rate of change by applying Kirchhoff's first rule to the circuit after the switch is closed. Let us start at point a. From point a to point b there is a potential drop IR across the resistor, assuming the current to be clockwise as indicated. From b to c there is a potential decrease across the inductor of magnitude $L\, dI/dt$ because the induced emf is from c to b, assuming that dI/dt is positive. From points c to a there is a potential increase $\mathcal{E}_0$ due to the battery. Kirchhoff's first rule gives

$$\mathcal{E}_0 = IR + L\frac{dI}{dt} \qquad\qquad 38\text{-}23$$

This differential equation is similar to the one we obtained for an RC circuit. We can understand the general behavior of the current I versus time without going through the mathematical details of the solution of the equation.

Just after the switch is closed, at $t = 0$, the current is zero and the rate of change of the current is, from Equation 38-23,

$$\left(\frac{dI}{dt}\right)_0 = \frac{\mathcal{E}_0}{L} \qquad\qquad 38\text{-}24$$

The current thus increases as expected. After a short time, the current has reached some positive value, and the rate of change is given by

$$\frac{dI}{dt} = \frac{\mathcal{E}_0}{L} - \frac{IR}{L}$$

At this time the current is still increasing, but its rate of increase is less than at $t = 0$. The final value of the current can be obtained by setting dI/dt equal to zero. From Equation 38-23 we see that the final value of the current is

$$I_f = \frac{\mathcal{E}_0}{R} \qquad \text{when} \qquad \frac{dI}{dt} = 0 \qquad\qquad 38\text{-}25$$

Figure 38-20 shows I versus t. The time for the current to reach an appreciable fraction of its final value depends on the resistance and inductance. The mathematical expression for the current as a function of time obtained by solving Equation 38-23 is

$$I = \frac{\mathcal{E}_0}{R}(1 - e^{-RT/L}) = I_f(1 - e^{-t/t_c}) \qquad\qquad 38\text{-}26$$

where

$$t_c = \frac{L}{R} \qquad\qquad 38\text{-}27$$

is the *time constant* of the circuit. The larger the self-inductance L or the smaller the resistance R, the longer it takes for the current to build up. We can show that I given by Equation 38-26 is indeed the solution of Equation 38-23 by computing the derivative dI/dt. We first note that

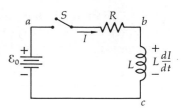

Figure 38-19
Typical LR circuit.

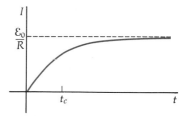

Figure 38-20
Graph of current I versus time t for the circuit in Figure 38-19, in which the switch is closed at $t = 0$. The current increases to its maximum value $\mathcal{E}_0/R$, reaching 63 percent of this value after time $t_c = L/R$.

Time constant

at $t = 0$ Equation 38-26 gives $I = 0$ as required by our assumption that the switch was open until $t = 0$. Taking the derivative of the current with respect to time from Equation 37-26, we obtain

$$\frac{dI}{dt} = \frac{\mathcal{E}_0}{R}\frac{R}{L} e^{-Rt/L} = \frac{\mathcal{E}_0}{L} e^{-t/t_c}$$

Using this result for dI/dt and Equation 38-26 for I, we see that Equation 38-23 is indeed satisfied by our solution. Figure 38-21 graphs dI/dt versus time. In agreement with our qualitative analysis, the rate of increase of the current is maximum at $t = 0$ and falls to zero as the current increases.

We note that the product of the initial slope $\mathcal{E}_0/L$ and the time constant $t_c = L/R$ equals the final current $\mathcal{E}_0/R$:

$$\left(\frac{dI}{dt}\right)_0 t_c = \frac{\mathcal{E}_0}{L}\frac{L}{R} = \frac{\mathcal{E}_0}{R} = I_f$$

If the current continued to increase at its initial rate, it would reach its final value after one time constant.

Figure 38-22 shows a slightly different arrangement with an additional switch that allows us to remove the battery and an additional resistor r to protect the battery so that is is not shorted when both switches are momentarily closed. If both switches are open and we close switch S_1 at some time, the current builds up in the circuit as discussed above except that the total resistance is now $r + R$ and the final current is $\mathcal{E}/(r + R)$. Let us suppose that this switch has been closed for a time long compared with the time constant, which is now $L/(R + r)$, so that the current is approximately steady at its final value, which we shall call I_0. The switch S_2 is then closed, and switch S_1 is opened to remove the battery from consideration completely. Let us choose the time $t = 0$ when switch S_2 is closed. We now have a circuit with just a resistor and inductor (loop $abcd$) with an initial current I_0. Again we analyze this circuit by applying Kirchhoff's law. The potential drop from point a to b across the resistor is IR. Across the inductor, the potential drop from b to c is $L\,dI/dt$. (Since dI/dt will turn out to be negative, this is really a potential increase. Less confusion with signs arises if we always assume that the quantities I and dI/dt are positive when writing the equations. The solutions of the equations will then tell us the correct signs of the quantities.) We are now back to our original point. The sum of the potential drops must be zero. Thus

$$IR + L\frac{dI}{dt} = 0 \qquad\qquad 38\text{-}28$$

Since the current I is positive, the rate of change dI/dt is negative:

$$\frac{dI}{dt} = -\frac{R}{L} I \qquad\qquad 38\text{-}29$$

The induced emf is the direction of the current and opposite to that indicated in Figure 38-22. In agreement with Lenz's law, the induced emf is in the direction that maintains the current and tends to prevent its decrease. Equation 38-29 is similar to Equation 35-19 for the discharge of a capacitor.

Figure 38-23 shows a sketch of the current I versus time t. The original value of the current is I_0, and the original value of the slope is $dI/dt = -RI_0/L$. As the current decreases, the slope decreases in magnitude until both the current and its slope are zero. The mathematical expres-

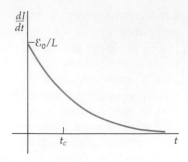

Figure 38-21
Graph of dI/dt versus time t for circuit of Figure 38-19. The maximum rate of change of the current is $\mathcal{E}_0/L$, which occurs just after the switch is closed at time $t = 0$. The rate of increase of the current decreases with time as the current increases.

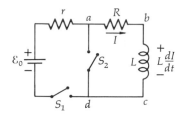

Figure 38-22
LR circuit with two switches so that the battery can be removed from the circuit. After the current in the inductor reaches its maximum value with S_1 closed, S_2 is closed and S_1 opened. The current then decreases in time as shown in Figure 38-23.

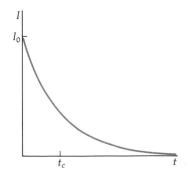

Figure 38-23
Current versus time after switch S_2 is closed in Figure 38-22. The current decreases as e^{-t/t_c}, where $t_c = L/R$ is the time constant.

sion for the current is

$$I = I_0 e^{-Rt/L} = I_0 e^{-t/t_c}$$ 38-30

where again $t_c = L/R$ is the time constant.

Question

10. A current is set up in an inductor with inductance L and resistance R by connecting it to a battery. Later, a resistor r is switched into the circuit parallel to the inductor. Can the voltage across this resistor ever be greater than the battery voltage? How?

38-8   Magnetic Energy

When a current is set up in a circuit like the one in Figure 38-19, only part of the energy supplied by the battery goes into joule heat in the resistor; the rest of the energy is stored in the inductor. We can see this by multiplying each term in Equation 38-23 by the current. We have then

$$\mathcal{E}_0 I = I^2 R + LI \frac{dI}{dt}$$ 38-31

The term on the left side of this equation represents the rate at which energy is supplied by the emf. The first term on the right, $I^2 R$, is the rate at which energy is put into the resistor. This energy appears as heat. The second term, $LI\,dI/dt$, is the rate at which energy is put into the inductor and is the term we wish to discuss. Let U_m be the energy in the inductor. According to Equation 38-31, the rate of increase of this energy is

$$\frac{dU_m}{dt} = LI \frac{dI}{dt}$$ 38-32

We can find the total energy in the inductor by integrating from the time $t = 0$, when the current is zero, to the time $t = \infty$, when the current has reached its final value I_f:

$$U_m = \int dU_m = \int LI\,dI = \tfrac{1}{2}LI_f^2$$ 38-33   *Energy in an inductor*

This energy is stored in the self-inductance of the circuit, as we can see by considering what happens when the emf is removed from the circuit, as in Figure 38-22 with S_2 closed and S_1 open. Though there is now no power input into the circuit by the emf, the current does not immediately fall to zero; so there is still some power put into the resistor. The rate at which energy is put into the resistor in this case equals the rate at which the energy in the inductor is decreased. We can see this by multiplying each term in Equation 38-28 by the current I. We then have

$$I^2 R = -LI \frac{dI}{dt}$$ 38-34

   Whenever a current is set up in a circuit, a magnetic field is set up. We can consider the energy stored in a circuit by virtue of its self-inductance and current as energy stored in the magnetic field. When the current is decreased, decreasing the energy $\frac{1}{2}LI^2$ in the inductor, the magnetic field is also decreased. The idea that energy is stored in a magnetic field is similar to the idea that energy is stored in an electric field when a capacitor is charged. In Chapter 33 we showed that the electrostatic energy stored in a parallel-plate capacitor can be written

$$U_e = \tfrac{1}{2}QV = \tfrac{1}{2}\epsilon_0 E^2 Ad$$

where the area of the plates A times their separation d is the volume in which there is an electrostatic field E. We then argued that this relation is more general than indicated by that example, namely, that whenever an electrostatic field is set up in some volume, there is energy stored $\frac{1}{2}\epsilon_0 E^2$ per unit volume. We now give an analogous argument for magnetic energy. Again, for simplicity, we consider a special case, that of a solenoid.

Let n be the number of turns per unit length of a solenoid and lA be the volume, where A is the cross-sectional area and l the length. The magnetic field inside the solenoid is

$$B = \mu_0 nI$$

and the self-inductance is

$$L = \mu_0 n^2 lA$$

Substituting this expression for L and $I = B/\mu_0 n$ for the current into Equation 38-33 for the energy of an inductor, we have

$$U_m = \tfrac{1}{2}LI^2 = \tfrac{1}{2}\mu_0 n^2 lA \left(\frac{B}{\mu_0 n}\right)^2 = \frac{B^2}{2\mu_0} lA$$

The energy stored in a solenoid can be written as the product of the volume lA and the term $\frac{1}{2}B^2/\mu_0$. Again, the result is more general than is indicated by this discussion. Whenever a magnetic field is set up in some volume $\mathscr{V}$ in space, energy is stored. The energy per unit volume η_m is

$$\eta_m = \frac{U_m}{\mathscr{V}} = \frac{B^2}{2\mu_0} \qquad\qquad\qquad 38\text{-}35 \qquad \textit{Magnetic energy density}$$

We see that there is energy stored in both the electric and magnetic fields. In general, if there is an electric field E and magnetic field B in some volume of space, the energy per unit volume associated with these fields is

Total energy density

$$\eta = \frac{U}{\mathscr{V}} = \tfrac{1}{2}\epsilon_0 E^2 + \frac{B^2}{2\mu_0} \qquad\qquad\qquad 38\text{-}36$$

38-9 *LC* and *LCR* Circuits

Figure 38-24 shows a capacitor connected to an inductor and a switch. Let us assume that the capacitor carries an initial charge Q_0 and the switch is initially open. At $t = 0$ we close the switch, and the charge flows through the inductor. For simplicity, we shall neglect any resistance in the circuit.

We have chosen the direction for the current in the circuit so that when the charge on the bottom capacitor plate is $+Q$, the current is

$$I = \frac{dQ}{dt}$$

Figure 38-24
LC circuit. The capacitor is initially charged. The charge on the plates and the current I are related by $I = dQ/dt$ for the choice of current direction indicated. Just after the switch is closed, I is negative.

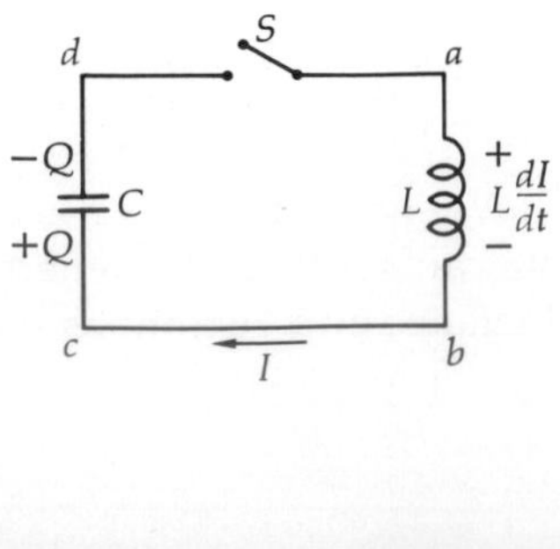

This choice of direction is purely arbitrary. With this choice, the current will be negative just after the switch is closed. With this choice of positive sense for the current, the potential drop across the inductor from point a to b is $L\, dI/dt$. Across the capacitor from c to d there is a potential drop Q/C. Thus Kirchhoff's rule for this circuit gives

$$L\frac{dI}{dt} + \frac{Q}{C} = 0 \qquad\qquad 38\text{-}37$$

Substituting dQ/dt for I in this equation, we obtain for the charge Q in the capacitor

$$L\frac{d^2Q}{dt^2} + \frac{Q}{C} = 0$$

or

$$\frac{d^2Q}{dt^2} = -\frac{1}{LC}\,Q \qquad\qquad 38\text{-}38$$

Equation 38-38 is of the same form as the equation for the acceleration of a mass on a spring,

$$\frac{d^2x}{dt^2} = -\omega^2 x$$

where $\omega^2 = k/m$, k is the spring constant, and m is the mass.

In Chapter 14 we studied the solutions of this equation for simple harmonic motion and found that we could always write the solution in the form

$$x = A\cos(\omega t + \delta)$$

where $\omega = \sqrt{k/m}$ is the angular frequency, A is the maximum value of x (called the amplitude), and δ is the phase constant, which depends on the initial conditions. We can put Equation 38-38 in the same form by writing ω^2 for the quantity $1/LC$. Then

$$\frac{d^2Q}{dt^2} = -\omega^2 Q \qquad\qquad 38\text{-}39$$

where $\omega = 1/\sqrt{LC}$. The solution of Equation 38-39 is

Frequency of LC circuit

$$Q = A\cos(\omega t + \delta) \qquad\qquad 38\text{-}40$$

The current is obtained by differentiating this solution:

$$I = \frac{dQ}{dt} = -\omega A \sin(\omega t + \delta) \qquad\qquad 38\text{-}41$$

For our initial conditions, the phase constant δ must be zero because the current is zero at time $t = 0$. Setting $t = 0$ in Equation 38-41, we have

$$I = -\omega A \sin \delta = 0$$

and hence $\delta = 0$. Then our solution for the charge is

$$Q = A\cos \omega t$$

The constant A is just the value of the charge at $t = 0$ since $\cos 0 = 1$. Writing Q_0 for this initial charge, we have

$$Q = Q_0 \cos \omega t \qquad\qquad 38\text{-}42$$

and

Charge and current in LC circuit

$$I = -\omega Q_0 \sin \omega t \qquad\qquad 38\text{-}43$$

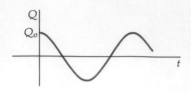
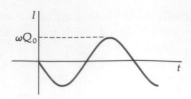

Figure 38-25
Graphs of Q versus t and I
versus t for LC circuit in Figure 38-24.

Figure 38-25 shows graphs of Q and I versus time. The charge oscillates between the values $+Q_0$ and $-Q_0$ with the angular frequency $\omega = 1/\sqrt{LC}$. The current also oscillates with this frequency and is 90° out of phase with the charge. The current is maximum when the charge is zero and zero when the charge is maximum.

In our study of the oscillation of a mass on a spring we found that the total energy is constant but oscillates between potential and kinetic energy. In our LC circuit we also have two kinds of energy, electrostatic energy and magnetic energy. The electrostatic energy of the capacitor is

$$U_e = \tfrac{1}{2}QV = \frac{Q^2}{2C}$$

where $V = Q/C$ is the voltage drop across the capacitor. Substituting $Q_0 \cos \omega t$ for Q in this equation, we have for the electrostatic energy of the capacitor,

$$U_e = \frac{Q_0^2}{2C} \cos^2 \omega t \qquad\qquad 38\text{-}44$$

The electrostatic energy stored in the capacitor oscillates between its maximum value $Q_0^2/2C$ and zero. When the current in the circuit is I, the magnetic energy stored in the inductor is

$$U_m = \tfrac{1}{2}LI^2$$

Substituting the value for the current from Equation 38-43 into this expression, we obtain

$$U_m = \tfrac{1}{2}L\omega^2 Q_0^2 \sin^2 \omega t = \frac{Q_0^2}{2C} \sin^2 \omega t \qquad\qquad 38\text{-}45$$

where we have used the fact that ω^2 is equal to $1/LC$. The magnetic energy also oscillates between its maximum value of $Q_0^2/2C$ and zero. The sum of the electrostatic and magnetic energies is the total energy, which is constant in time:

$$U_{\text{total}} = U_e + U_m = \frac{Q_0^2}{2C} \cos^2 \omega t + \frac{Q_0^2}{2C} \sin^2 \omega t = \frac{Q_0^2}{2C}$$

Energy in LC circuit

Example 38-5 A 2-μF capacitor is initially charged to 20 V and then shorted across a 6-μH inductor. What are (*a*) the frequency of oscillation and (*b*) the maximum value of the current?

The frequency of oscillation is independent of the initial charge and depends only on the values of the capacitance and inductance. The frequency is

$$f = \frac{\omega}{2\pi} = \frac{1}{2\pi}\sqrt{\frac{1}{LC}} = \frac{1}{2\pi\ \sqrt{LC}}$$

$$= \frac{1}{2\pi\ \sqrt{(6 \times 10^{-6} \times 2 \times 10^{-6})}} = 4.59 \times 10^4\ \text{Hz}$$

According to Equation 38-43, the maximum value of the current is related to the maximum value of the charge by

$$I_m = \omega Q_0 = \frac{Q_0}{\sqrt{LC}}$$

The initial charge on the capacitor is

$$Q_0 = CV_0 = (2 \ \mu\text{F})(20 \ \text{V}) = 40 \ \mu\text{C}$$

Thus

$$I_m = \frac{40 \ \mu\text{C}}{\sqrt{(6 \ \mu\text{H})(2 \ \mu\text{F})}} = 11.5 \ \text{A}$$

In Figure 38-26 we include a resistor in series with the capacitor and inductor. Again we assume that the switch is initially open, with the capacitor carrying an initial charge Q_0, and we close the switch at $t = 0$. We need only modify Equation 38-37 by including the potential drop IR across the resistor. We then have from Kirchhoff's rule

$$L \frac{dI}{dt} + \frac{Q}{C} + IR = 0$$

or

$$L \frac{d^2Q}{dt^2} + \frac{Q}{C} + R \frac{dQ}{dt} = 0 \qquad \text{38-46}$$

using $I = dQ/dt$ as before. Equation 38-46 is analogous to Equation 15-3 for a damped harmonic oscillator. If the resistance is small, the charge and current still oscillate with very nearly the same frequency $1/\sqrt{LC}$ but the oscillations are damped; i.e. the maximum values of the charge and current decrease with each oscillation. We can understand this qualitatively from energy considerations. Let us multiply each term in Equation 38-46 by the current I. We then have

$$IL \frac{dI}{dt} + I \frac{Q}{C} + I^2 R = 0 \qquad \text{38-47}$$

The first term in this equation is the current times the voltage across the inductor. This is the rate at which energy is put into the inductor or taken out of it, i.e., the rate of change of magnetic energy, which is positive or negative depending on whether I and dI/dt have the same or different signs. Similarly, the second term is the current times the voltage across the capacitor. This is the rate of change of the energy of the capacitor. Again, it may be positive or negative. The last term, I^2R, the rate at which energy is dissipated in the resistor as joule heat, is always positive regardless of the sign of the current since it depends only on I^2. The sum of the electric and magnetic energies is not constant for this circuit because energy is continually dissipated in the resistor. Figure 38-27 shows graphs of Q versus t and I versus t for small

Figure 38-26
LCR circuit.

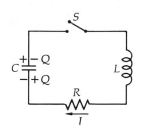

Figure 38-27
Graphs of Q versus t and I versus t for *LCR* circuit of Figure 38-26 for small resistance R.

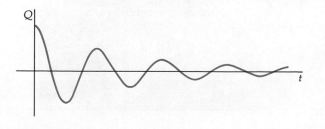

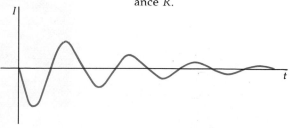

resistance. If we increase R, the oscillations are more heavily damped until a critical value of R is reached for which there is not even one oscillation. Figure 38-28 shows the charge versus t for R greater than the critical damping value. The detailed solution of Equation 38-46 is exactly analogous to the solution of Equation 15-3 for the damped harmonic oscillator and will not be discussed here. (See Problems 17 and 18 for further details of the solution of this equation.)

Figure 38-28
Graph of Q versus t for very large resistance R in LCR circuit of Figure 38-26.

Questions

11. It is not difficult to produce LC circuits with frequencies of oscillation of thousands of hertz or more, but it is difficult to produce LC circuits with small frequencies. Why?

12. How will the frequency of an LCR circuit be altered if the inductance in the circuit is doubled? The capacitance doubled? The voltage with which the capacitor is initially charged is doubled? The resistance in the circuit is doubled?

Electric Motors

Reuben E. Alley, Jr.
U.S. Naval Academy

In Thomas Edison's machine shop the various tools were driven from a single steam engine by means of shafts, pulleys, and belts, as shown in Figure 1. Such a power-distribution system was typical of most industrial establishments around 1900. The installation of electric-power distribution facilities and the development of practical electric motors made it possible to replace the complicated and hazardous system of belts and pulleys with individual energy sources for each machine. The growth of a nationwide electric-power network and the concomitant invention of several types of electric motors resulted in a revolution in our society as profound as the electronics revolution associated with the invention and application of transistors and integrated circuits.

Figure 1
Interior of Thomas Edison's second machine shop, Menlo Park, New Jersey, as it appeared during a New Year's Eve lighting demonstration, 1879. (*Collections of Greenfield Village and the Henry Ford Museum, Dearborn, Michigan.*)

A single electric motor may be supplied from a battery or from a small generator, but a nationwide power network is required to supply the millions of horsepower of electric energy that are converted into mechanical energy in such varied applications as steel rolling mills, paper mills, subway trains, electric locomotives, automobile assembly lines, and mining machinery. Although automobiles, ships, diesel locomotives, trucks, and airplanes do not depend upon the power network, they all use electric motors. In the home, electric motors are essential parts of furnaces, air conditioners, refrigerators, and washing machines. Imagine what your everyday life would be like if there were no electric motors.

Figure 2 shows a simple *dc motor.* Current from the battery magnetizes the soft-iron *armature,* which is free to rotate about axis *AA'* and turns to align itself with the *field* produced by the poles labeled N and S. As the armature rotates, it carries with it the *commutator,* whose two segments provide means for reversing current direction just as the armature reaches its equilibrium position. The mass of the armature ensures that it will rotate past the equilibrium position; then, because of its reversed polarity, there is a further rotation of a half revolution. Because the commutator reverses the current direction every 180°, continuous rotation is achieved. Useful work can be done by the shaft of the motor.

Figure 2
A simple dc motor. The field magnet may be a permanent magnet or a soft-iron core wrapped with many turns of wire, through which a direct current flows.

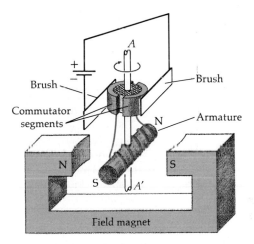

The motor shown in Figure 2 has several disadvantages: (1) when power is disconnected, the motor tends to stop in the equilibrium position, and so there may be no starting torque, and (2) the torque is zero twice during each revolution. If the armature is provided with additional poles and windings, and if the commutator is divided into more segments, the torque is more nearly uniform and the motor always starts.

If the field is supplied from a permanent magnet, as in model trains and most battery-powered toys, the speed depends upon supply voltage and changes with load. For applications requiring appreciable amounts of power, electromagnets supply the field. Adjustment of field current permits control of speed independent of armature current. In general, if the load on a dc motor increases, speed decreases and more armature current is required. Field current can be adjusted to maintain constant speed. Automatic controls are frequently used with dc motors.

Dc motors with field windings are very flexible. They may be connected in *series,* so that the armature current is also the field current; in *parallel* (or *shunt*), so that the field current is independent of armature current; or in a *compound* arrangement using two field windings, so that one is connected in series and the other is connected in parallel. Dc motors are widely used for traction (subways and electric railways) and in applications where speed control is critical, e.g., steel rolling mills.

Figure 3
A simple synchronous motor.
The field magnet is like that
in Figure 2. Only the pole
pieces are shown here.

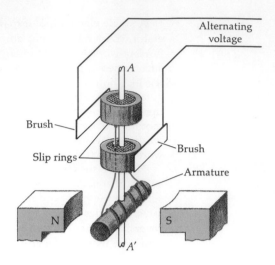

The dc motor in Figure 2 becomes a *synchronous motor* if the commutator is replaced by *slip rings* (Figure 3) and alternating current is supplied. Although this simple device has no starting torque, it will "lock in" if it is accelerated by external means to *synchronous speed* (determined by power-line frequency). As the load is increased, armature current increases but the speed stays constant unless the load is great enough to make the machine stall. Synchronous motors designed for use with three-phase power (see below) can be self-starting. When the horsepower rating is high, field current is much less than armature current. In order that the smaller current can be supplied through slip rings, most such machines have a stationary armature while the field rotates on the shaft that drives the load.

The *universal series motor* has many fractional-horsepower applications such as hand drills and small saws. A simple form of such a motor results if the current, before entering the commutator (Figure 2), first goes through the field winding. This motor will operate on either alternating or direct current. No-load speed depends upon the number of field poles, the number of armature poles, and the supply voltage. Many hand drills now are equipped with solid-state circuits that control speed by controlling applied voltage. These machines slow down under load, but they develop reasonably high torque and so are especially satisfactory for driving hand tools.

Practically all commercially generated electric power in the United States is three-phase, 60-Hz sinusoidal voltage. The term three-phase means that there are three sinusoidal voltages of equal amplitude and frequency whose successive peak values are separated in time by one-third of a cycle (see Figure 4). These three voltages are usually carried in the three-wire transmission lines that are a common sight around the countryside. One phase of the power appears between each pair of wires. Residences, almost without exception, are

Figure 4
Voltages in a three-phase
power system. At any instant
the sum of the three voltages
is zero.

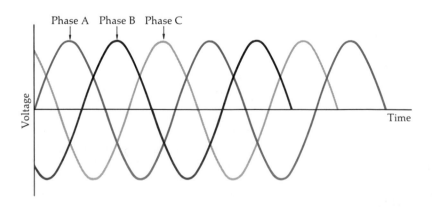

Figure 5
Schematic cross section of a simple three-phase induction motor.

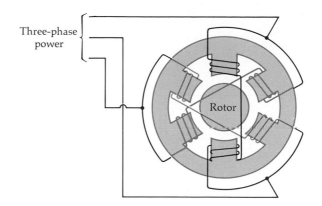

supplied with only a single phase, but all three phases are commonly supplied to industries.

Figure 5 shows a simple three-phase *induction motor*. The winding of each of the three pairs of poles is connected to a different phase. The currents in the windings therefore reach their maximum values successively, and the result is a rotating magnetic field in the region between the poles. A stationary rotor of conducting material in this field will have currents set up in it. The magnetic fields of the *induced currents* interact with the rotating field to produce a torque. If there is no load on the motor, the torque accelerates the rotor to a speed almost equal to the synchronous speed of rotation of the field. (If the rotor actually attained synchronous speed, there would be no relative motion between field and rotor and so no induced currents or torque.) As the motor is loaded, rotor speed decreases; this results in larger induced currents and correspondingly greater torques. Speed may decrease to 75 percent of synchronous speed before the motor stalls. There are no moving electrical contacts in the three-phase induction motor, which can be built to develop hundreds of horsepower. The elimination of any possibility of electric sparks makes these motors especially attractive for applications in explosive atmospheres, e.g., mines or flour mills.

A practical induction motor usually has a cylindrical iron rotor with insulated copper bars embedded in its surface and interconnected to form good conducting paths for induced currents. In some motors the rotor windings, instead of being short-circuited, are connected to slip rings so that external resistors can be used for speed control, but the resulting moving electrical contacts remove one of the important advantages of induction motors.

If only one phase is connected in Figure 5, there is no rotating magnetic field and no starting torque. If a second winding in Figure 5 is connected in series with a large (hundreds of microfarads) capacitor and supplied from the same phase as the first winding, the currents in the two windings are sufficiently out of phase to provide a rotating field and therefore a starting torque. This fact is employed in the design of *single-phase* induction motors, used to drive refrigerators, washing machines, air conditioners, and furnaces.

Since 1965, especially in Great Britain, Germany, and Japan, serious efforts have been made to develop a high-speed mass-transit system based upon the *linear induction motor*. It is possible to achieve both acceleration and levitation (once speeds of about 20 mi/h are reached) by electrical means. Such support for the vehicle eliminates most of the friction losses associated with the rolling of flanged wheels on steel rails.

Suppose the cylindrical shell that supports the poles of an induction motor is cut along an axial line and flattened (just as a tin can may be cut and unrolled). Then the poles will lie in a line and the magnetic field will move across the poles periodically. Now imagine a long strip of magnetic material with many poles equally spaced along its length. Let the first pole and every third succeeding pole be supplied from one phase of the power line; let the second pole and every third succeeding pole be supplied from the second phase; and let the third pole and every third succeeding pole be supplied from

Figure 6
Schematic diagram of a
simple linear induction
motor.

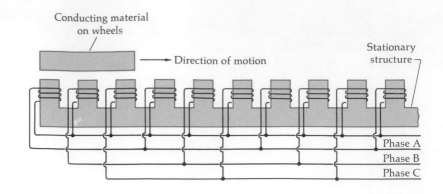

the third phase. The result will be a magnetic field that travels along the strip with a speed relative to the ground that depends upon the distance between poles and the frequency of the power. Now let a mass of conducting material be supported (by wheels running on rails, for example) so that it is free to move in the direction of the strip. When the mass is stationary, currents will be induced whose magnetic fields will interact with the moving field to produce an accelerating force along the direction of the strip (Figure 6). The mass will acquire a speed that is somewhat less than synchronous speed, just like the rotor of the induction motor. What has been described is a simple linear induction motor. Such a device was proposed in 1946 as a means of catapulting aircraft from a ship.

An alternative design of the linear induction motor involves mounting a number of poles on the vehicle and supplying them through flexible collectors in contact with a stationary three-phase distribution system. The interaction of the fields of the poles in the vehicle with currents induced in a stationary metal strip (aluminum backed by a soft magnetic material to provide a path for magnetic flux) produces support and accelerating force for the vehicle (Figure 7).

Figure 7
Linear induction motor with
moving poles.

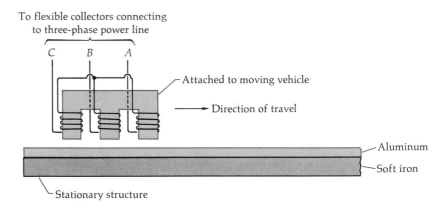

In the United States some engineers are investigating the use of a single-phase *linear synchronous motor*. Speed variation is achieved through change in frequency of the alternating voltage supplied to the vehicle. The required steady magnetic field is stationary. One proposal is to operate the field windings at superconducting temperatures, thus greatly reducing the winding resistance and thereby reducing the power loss from the direct current supplying these windings.

Although no system has yet reached the commercial stage, some prototypes have been tested. There seems to be ample evidence that it will be possible to design and build a practical transportation system that will operate at speeds of 200 mi/h or more based upon the concept of a linear motor.

Review

A. Define, explain, or otherwise identify:

Faraday's law, 881
Motional emf, 883
Lenz's law, 883
Self-induced emf, 885
Flip coil, 886
Ballistic galvanometer, 886

Eddy currents, 887
Betatron, 888
Self-inductance, 890
Mutual inductance, 890
Inductor, 892
Magnetic energy, 895

B. True or false:

1. The induced emf in a circuit is proportional to the magnetic flux through the circuit.

2. There can be an induced emf at an instant when the flux through the circuit is zero.

3. The induced emf in a circuit always decreases the magnetic flux through the circuit.

4. Faraday's law can be derived from the Biot-Savart Law.

5. The energy in a magnetic field is proportional to B^2.

Exercises

Section 38-1, Motional EMF

1. A rod 30 cm long moves at 8 m/sec in a plane perpendicular to a magnetic field of 500 G. Its velocity is perpendicular to the length of the rod. Find (a) the magnetic force on an electron in the rod, (b) the electrostatic field E in the rod, and (c) the potential difference V between the ends of the rod.

2. Find the speed of the rod in Exercise 1 if the potential difference between its ends is 6 V.

3. In Figure 38-3 let B be 8000 G, $v = 10.0$ m/sec, $l = 20$ cm, and $R = 2$ Ω. Find (a) the induced emf in the circuit, (b) the current in the circuit, and (c) the force needed to move the rod with constant velocity assuming negligible friction. (d) Find the power input by the force found in part (c) and the rate of heat production I^2R.

4. Work Exercise 3 for $B = 1.5$ T, $v = 6$ m/sec, $l = 40$ cm, and $R = 1.2$ Ω.

Sections 38-2, Lenz's Law, and Section 38-3, Applications of Faraday's Law

5. The two loops in Figure 38-29 have their planes parallel to one another. As viewed from A toward B there is a counterclockwise current in A. Give the direction of the current in loop B and state whether the loops attract or repel each other if the current in loop A is (a) increasing; (b) decreasing.

6. A bar magnet moves with constant velocity along the axis of a loop as shown in Figure 38-30. (a) Make a qualitative sketch of the flux ϕ_m through the loop as a function of time t. Indicate the time t_1 when the magnet is halfway through the loop. (c) Sketch the current I in the loop versus time choosing I positive when it is counterclockwise as viewed from the left.

7. Give the direction of the induced current in the circuit on the right in Figure 38-31 when the resistance in the circuit on the left is suddenly (a) increased; (b) decreased.

8. A uniform magnetic field **B** is established perpendicular to the plane of a loop of radius 5.0 cm, resistance 0.4 Ω, and negligible self-inductance. The magnitude of B is increasing at a rate of 400 G/sec. Find (a) the induced emf in

Figure 38-29
Exercise 5.

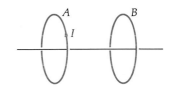

Figure 38-30
Exercise 6.

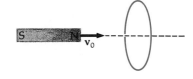

Figure 38-31
Exercise 7.

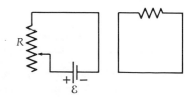

the loop, (b) the induced current in the loop, and (c) the rate of joule heating in the loop.

9. A 100-turn coil has radius of 4.0 cm and resistance 25 Ω. At what rate must a perpendicular magnetic field change to produce a current of 4.0 A in the coil?

10. Show that if the flux through each turn of an N-turn coil of resistance R changes from ϕ_{m1} to ϕ_{m2} in any manner, the total charge that passes through the coil is given by $Q = N(\phi_{m2} - \phi_{m1})/R$.

11. The flux through a loop is given by $\phi_m = (t^2 - 4t) \times 10^{-1}$ T-m^2, where t is in seconds. (a) Find the induced emf $\mathcal{E}$ as a function of time. (b) Find both ϕ_m and $\mathcal{E}$ at $t = 0$, $t = 2$ sec, $t = 4$ sec, and $t = 6$ sec.

12. For the flux given in Exercise 11, sketch ϕ_m versus t and $\mathcal{E}$ versus t. (a) At what time is the flux maximum? What is the emf at this time? (b) At what times is the flux zero? What is the emf at these times?

13. A 100-turn circular coil has a diameter of 2.0 cm and resistance of 50 Ω. The plane of the coil is perpendicular to a uniform magnetic field of magnitude 1.0 T. The field is suddenly reversed in direction. (a) Find the total charge passing through the coil. If the reversal takes 0.1 sec, find (b) the average current in the circuit and (c) the average emf in the circuit.

14. A 1000-turn coil of cross-sectional area 300 cm^2 and resistance 15.0 Ω is aligned with its plane perpendicular to the earth's magnetic field of 0.7 G (at the equator). If the coil is flipped over, how much charge flows through it?

15. A circular coil of 300 turns and radius 5.0 cm is connected to a ballistic galvanometer. The total resistance of the circuit is 20 Ω. The plane of the coil is originally aligned perpendicular to the earth's magnetic field at some point. When the coil is rotated through 90°, the charge passing through the galvanometer is measured to be 9.4 μC. Calculate the magnitude of the earth's magnetic field at that point.

Section 38-4, Eddy Currents

There are no exercises for this section.

Section 38-5, The Betatron

16. Consider a plane perpendicular to a magnetic field **B** which is uniform over a circular region of radius R and is essentially zero outside the circle ($r > R$). Show that if the magnetic field is changing at the rate dB/dt, the induced electric field at distance r from the center of the circle is tangent to the circle of radius r and has the magnitude

$$E = \begin{cases} \frac{1}{2}r \dfrac{dB}{dt} & \text{for } r < R \\[2ex] \dfrac{R^2}{2r} \dfrac{dB}{dt} & \text{for } r > R \end{cases}$$

Section 38-6, Inductance

17. A coil with a self-inductance of 8.0 H carries a current of 3 A, which is changing at a rate of 200 A/sec. (a) Find the magnetic flux through the coil. (b) Find the induced emf in the coil.

18. A coil with self-inductance L carries a current I, given by $I = I_0 \sin 2\pi f t$. Find and graph the flux ϕ_m and the self-induced emf as functions of time.

19. A solenoid has length 25 cm, radius 1 cm, and 400 turns and carries 3 A current. Find (a) B on the axis at the center, (b) the flux through the solenoid

assuming B to be uniform, (c) the self-inductance of the solenoid, and (d) the induced emf in the solenoid when the current changes at 150 A/sec.

20. A circular coil of N_1 turns and area A_1 encircles a long, tightly wound solenoid of N_2 turns, area A_2, and length l_2, as shown in Figure 38-32. The solenoid carries a current I_2. (a) Show that the flux through the circular coil due to the current in the solenoid is given by $\phi_{m1} = \mu_0 N_1 (N_2/l_2) A_2 I_2$. (b) Explain why the flux depends on the area A_2 of the solenoid and not on that of the coil. (c) Find the mutual inductance of the solenoid and coil. (d) Explain why it is much more difficult to calculate the mutual inductance by assuming current I_1 in the coil and then finding the flux through the solenoid due to I_1.

Figure 38-32
Exercise 20.

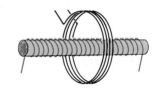

21. A very small circular coil of N_1 turns and area A_1 is completely inside a long, tightly wound solenoid with its plane perpendicular to the axis of the solenoid. The solenoid has N_2 turns, area A_2, and length l_2 and carries current I_2. (a) Show that the flux through the circular coil due to the current in the solenoid is given by $\phi_{m1} = \mu_0 N_1 (N_2/l_2) A_1 I_2$. (b) Find the mutual inductance of the coil and solenoid. (c) Explain why it is much more difficult to calculate the mutual inductance by assuming current I_1 in the coil and then finding the flux through the solenoid due to I_1.

Section 38-7, *LR* Circuits

22. The current in an *LR* circuit is zero at time $t = 0$ and increases to half its final value in 4.0 sec. (a) What is the time constant of this circuit? (b) If the total resistance is 5 Ω, what is the self-inductance?

23. A coil of resistance 8.0 Ω and self-inductance 4.0 H is suddenly connected across a constant potential difference of 100 V. Let $t = 0$ be the time of connection at which the current is zero. Find the current I and its rate of change dI/dt at times (a) $t = 0$, (b) $t = 0.1$ sec, (c) $t = 0.5$ sec, and (d) $t = 1.0$ sec.

24. How many time constants must elapse before the current in an *LR* circuit which is originally zero reaches (a) 90 percent, (b) 99 percent, and (c) 99.9 percent of its final value?

25. The current in a coil of self-inductance 1 mH is 2.0 A at $t = 0$, when the coil is shorted through a resistor. The total resistance of the coil plus resistor is 10.0 Ω. Find the current after (a) 0.5 msec, (b) 1 msec, (c) 10 msec.

26. Compute the initial slope dI/dt at $t = 0$ from Equation 38-30 and show that if the current decreases steadily at this rate, it would be zero after one time constant.

27. In Exercise 25, find the time for the current to decrease to one electron per second.

Section 38-8, Magnetic Energy

28. A solenoid of 2000 turns, area 4 cm², and length 30 cm carries 4.0 A current. (a) Calculate the magnetic energy stored from $\frac{1}{2}LI^2$. (b) Divide your answer in part (a) by the volume of the solenoid to find the magnetic energy per unit volume in the solenoid. (c) Find B in the solenoid. (d) Compute the magnetic energy density from $\eta_m = B^2/2\mu_0$ and compare with part (b).

29. In the circuit in Figure 38-19 let $\mathcal{E}_0 = 12.0$ V, $R = 3.0$ Ω, and $L = 0.6$ H. The switch is closed at time $t = 0$ sec. At time $t = 0.5$ sec, find (a) the rate at which the battery supplies power, (b) the rate of joule heating, and (c) the rate at which energy is being stored in the inductor.

30. Do Exercise 29 for the times $t = 1$ sec and $t = 100$ sec.

31. A coil with self-inductance 2.0 H and resistance 12.0 Ω is connected across a 24-V battery of negligible internal resistance. (a) What is the final current?

(b) How much energy is stored in the inductor when the final current is attained?

32. In a plane electromagnetic wave such as a light wave, the magnitudes of the electric and magnetic fields are related by $E = cB$, where $c = 1/\sqrt{\epsilon_0\mu_0}$ is the speed of light. Show that in this case the electric and magnetic energy densities are equal.

33. Find (a) the magnetic energy, (b) the electric energy, and (c) the total energy in a volume of 1.0 m³ in which there is an electric field of 10^4 V/m and a magnetic field of 10^4 G.

Section 38-9, LC and LCR Circuits

34. Show from the definitions of the henry and farad that $1/\sqrt{LC}$ has units of $\sec^{-1}$.

35. What is the period of oscillation of an LC circuit consisting of a 2-mH coil and a 20-μF capacitor?

36. What inductance is needed with an 80-μF capacitor to construct an LC circuit which oscillates with frequency 60 Hz?

37. An LC circuit has capacitance C_1 and inductance L_1. A second circuit has $C_2 = \frac{1}{2}C_1$ and $L_2 = 2L_1$, and a third circuit has $C_3 = 2C_1$ and $L_3 = \frac{1}{2}L_1$. (a) Show that each of these circuits oscillates with the same frequency. (b) In which circuit would the maximum current be the greatest if in each case the capacitor were charged to potential V?

Figure 38-33
Problem 1.

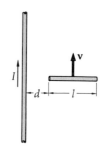

B in

Problems

1. A conducting rod of length l rotates at constant angular velocity ω about one end in a plane perpendicular to a uniform magnetic field **B** (Figure 38-33). (a) Show that the magnetic force on a charge q at distance r from the pivot is $Bqr\omega$. (b) Show that the potential difference between the ends of the rod is $V = \frac{1}{2}B\omega l^2$. (c) Draw any radial line in the plane from which to measure $\theta = \omega t$. Show that the area of the pie-shaped region between the reference line and the rod is $A = \frac{1}{2}l^2\theta$. Compute the flux through this area and show that $\mathcal{E} = \frac{1}{2}B\omega l^2$ follows from Faraday's law applied to this area.

2. A rod of length l lies with its length perpendicular to a long wire carrying current I, as shown in Figure 38-34. The near end of the rod is a distance d away from the wire. The rod moves with speed v in the direction of the current I. (a) Show that the potential difference between the ends of the rod is given by

$$V = \frac{\mu_0 I}{2\pi} v \ln \frac{d+l}{d}$$

(b) Use Faraday's law to obtain this result by considering the flux through a rectangular area $A = lvt$ swept out by the rod.

Figure 38-34
Problem 2.

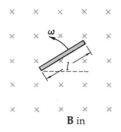

3. A rectangular loop 10 by 5.0 cm with resistance 1 Ω is pulled through a region of uniform magnetic field, $B = 1.0$ T (Figure 38-35), with constant speed $v = 2.0$ cm/sec. The front end of the loop enters the region of magnetic field at time $t = 0$. (a) Find and graph the flux through the loop as a function of time. (b) Find and graph the induced emf and the current in the loop as a function of time. Neglect any self-inductance of the loop and extend your graphs from $t = 0$ to $t = 16$ sec.

4. Show that for two inductors L_1 and L_2 connected in series such that none of the flux of either passes through the other, the effective inductance is given by $L = L_1 + L_2$.

Figure 38-35
Problem 3.

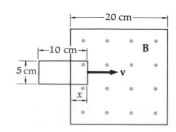

5. Two concentric solenoids have the same length l. The inner solenoid has n_1 turns per unit length and radius r_1, whereas the outer solenoid has n_2 turns per unit length and radius r_2 ($r_2 > r_1$). Find the mutual inductance of these solenoids.

6. Show that for two inductors L_1 and L_2 connected in parallel such that none of the flux of one passes through the other, the effective inductance is given by $1/L = 1/L_1 + 1/L_2$.

7. When the current in a certain coil is 5.0 A and is increasing at the rate of 10.0 A/sec, the potential difference across the coil is 140 V. When the current is 5.0 A and decreasing at the rate of 10.0 A/sec, the potential difference is 60 V. Find the resistance and self-inductance of the coil.

8. A coaxial cable consists of two very thin walled conducting cylinders of radii r_1 and r_2 (Figure 38-36). Current I goes in one direction down the inner cylinder and back up the outer cylinder. (a) Use Ampère's law to find B and show that $B = 0$ except in the region between the conductors. (b) Show that the magnetic energy density in the region between the cylinders is

$$\eta_m = \frac{\mu_0 I^2}{8\pi^2 r^2}$$

(c) Find the magnetic energy in a cylindrical shell volume element of length l and volume $d\mathscr{V} = l2\pi r\, dr$ and integrate your result to show that the total magnetic energy in the volume of length l between the cylinders is

$$U_m = \frac{\mu_0}{4\pi} I^2 l \ln \frac{r_2}{r_1}$$

(d) Use the result of part (c) and $U_m = \frac{1}{2}LI^2$ to show that the self-inductance per unit length is

$$\frac{L}{l} = \frac{\mu_0}{2\pi} \ln \frac{r_2}{r_1}$$

9. In the circuit of Figure 38-19 let $\mathcal{E} = 12.0$ V, $R = 3.0\ \Omega$, and $L = 0.6$ H (as in Exercise 29). After a time $t = t_c$, find (a) the total energy supplied by the battery, (b) the total energy dissipated as joule heat, and (c) the energy stored in the inductor. Hint: Find the rates as functions of time and integrate from $t = 0$ to $t = t_c = L/R$.

10. In Figure 38-36 compute the flux through a rectangular area of sides l and $r_2 - r_1$ between the conductors. Show that the self-inductance per unit length can be found from $\phi_m = LI$ [see part (d) of Problem 8].

11. Electrons are injected into a betatron with initial energy 2.5 keV. They move in an orbit of radius 0.5 m. (a) Find the initial value of the orbit field B_o and the initial average field B_{av}. (b) If the electrons gain 125 eV during each revolution, what must dB_{av}/dt be? (c) The electrons emerge with kinetic energy 10 MeV. How many revolutions do they make? (d) What is the final value of B_o? Hint: $p = mv$ can be used for the initial momentum because the electrons are nonrelativistic, and the final momentum can be found from $p = E/c$, where $E = 10.5$ MeV is the total final energy.

12. An ac generator rectangular loop of dimensions a and b has N turns. The loop is connected to slip rings (Figure 38-37) and rotates with angular velocity ω in a uniform magnetic field **B**. (a) Show that the potential difference between the two slip rings is $\mathcal{E} = NBab\omega \sin \omega t$. (b) If $a = 1.0$ cm, $b = 2.0$ cm, $N = 1000$, and $B = 2$ T, at what angular frequency ω must the coil be rotated to generate an emf whose maximum value is 110 V?

13. A conducting rod of length l, resistance R, and mass m rides on a pair of horizontal frictionless rails of negligible resistance (Figure 38-38). A uniform magnetic field **B** is perpendicular to the plane of the rails. The rod is moving with initial speed v_0 when points a and b are connected with a conductor of

Figure 38-36
Problems 8 and 10.

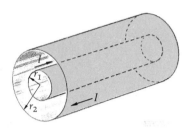

Figure 38-37
Problem 12.

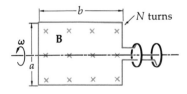

Figure 38-38
Problems 13 and 14.

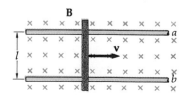

negligible resistance. (a) Find the retarding force on the rod. (b) Show that the speed is given by $v = v_0 e^{-t/T}$, where $T = mR/(Bl)^2$, assuming the connection across ab is made at $t = 0$. (c) Find the rate of joule-heat production I^2R and show that the total heat produced equals $\frac{1}{2}mv_0^2$.

14. A battery of emf $\mathcal{E}$ and negligible internal resistance is connected across ab in Figure 38-38, and the rod is placed at rest across the rails at $t = 0$. (a) Find the force on the rod as a function of speed v and write Newton's second law for the rod when it has speed v. (b) Show that the rod reaches a terminal velocity and find an expression for it. (c) What is the current when the rod reaches its terminal velocity?

15. In the circuit in Figure 38-22 the current is $I_0 = 4.0$ A when switch S_2 is closed and S_1 opened. The inductance is 50 mH, and the resistance is 150 Ω. (a) Find the rate of joule-heat production I^2R as a function of time and (b) calculate the total heat produced. (c) Calculate the initial energy stored in the inductor and show that this equals your answer to part (b).

16. Show that the electric field in a charged parallel-plate capacitor cannot go abruptly to zero at the edge by evaluating $\oint \mathbf{E} \cdot d\mathbf{l}$ for the rectangular curve shown in Figure 38-39 and applying Faraday's law to this curve.

Figure 38-39
Problem 16.

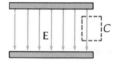

17. Show by direct substitution that Equation 38-46 is satisfied by $Q = Q_0 e^{-Rt/2L} \cos \omega' t$, where $\omega' = \sqrt{(1/LC) - (R/2L)^2}$ and Q_0 is the charge on the capacitor at $t = 0$.

18. (a) Compute the current $I = dQ/dt$ from the solution of Equation 38-46 given in Problem 17 and show that

$$I = -I_0 (\sin \omega' t + [R/2L\omega'] \cos \omega' t)e^{-Rt/2L}$$

where $I_0 = \omega' Q_0$. (b) Show that this can be written

$$I = -\frac{I_0}{\cos \delta} (\cos \delta \sin \omega' t + \sin \delta \cos \omega' t)e^{-Rt/2L}$$

$$= -\frac{I_0}{\cos \delta} \sin (\omega' t + \delta)e^{-Rt/2L}$$

where $\tan \delta = R/2L\omega'$. When $R/2L\omega'$ is small, $\cos \delta \approx 1$ and $I \approx -I_0 \sin (\omega' t + \delta)e^{-Rt/2L}$.

CHAPTER 39   Magnetism in Matter

In studying electric fields in matter we found that the electric field is affected by the presence of electric dipoles. For polar molecules, which have a permanent electric dipole moment, the dipoles are aligned by the electric field; whereas for nonpolar molecules, electric dipoles are induced by the external field. In both cases, the dipoles are aligned parallel to the external electric field, and the alignment tends to weaken the external field.

Somewhat similar but more complicated effects occur in magnetism. Atoms have magnetic moments due to the motion of their electrons. In addition, each electron has an intrinsic magnetic moment associated with its spin. The net magnetic moment of an atom depends on the arrangement of the electrons in the atom. Unlike the situation with electric dipoles, the alignment of magnetic dipoles parallel to an external magnetic field tends to increase the field. We can see this difference by comparing the lines of $\mathbf{E}$ for an electric dipole with the lines of $\mathbf{B}$ for a magnetic dipole, e.g., a small current loop as shown in Figure 39-1. Far from the dipoles, the lines are identical, but in the region inside the dipole the lines of $\mathbf{B}$ and $\mathbf{E}$ are oppositely directed. Thus in an electrically polarized material, the dipoles create an electric field *antiparallel* to their dipole-moment vector, whereas in a magnetically polarized material, the dipoles create a magnetic field *parallel* to the magnetic-dipole-moment vectors.

We can classify materials into three categories, paramagnetic, diamagnetic, and ferromagnetic. Paramagnetic and ferromagnetic materials have molecules with permanent magnetic dipole moments. In *paramagnetism*, these moments do not interact strongly with each other and are normally randomly oriented. In the presence of an external magnetic field, the dipoles are partially aligned in the direction of the field, thereby increasing the field. However, at ordinary temperatures and ordinary external fields, only a very small fraction of the molecules are aligned because thermal motion tends to randomize their orientation. The increase in the total magnetic field is therefore very small. *Ferromagnetism* is much more complicated. Because of a strong interac-

Three types of magnetic materials

tion between neighboring magnetic dipoles, a high degree of alignment can be achieved even with weak external magnetic fields, thereby causing a very large increase in the total field. Even when there is no external magnetic field, ferromagnetic material may have magnetic dipoles aligned, as in permanent magnets. *Diamagnetism* is the result of an induced magnetic moment opposite in direction to the external field. The induced dipoles thus weaken the resultant magnetic field. This effect occurs in all materials but is very small and often masked by the paramagnetic or ferromagnetic effects if the individual molecules have permanent magnetic dipole moments.

In Section 37-6 we discussed the magnetic field produced by a permanent magnet, characterizing the magnet by the magnetic moment per unit volume **M**, known as the magnetization vector. We found that the field due to the magnetized material is the same as that due to a current per unit length M on the surface of the magnet, and we argued that this amperian current can be understood in terms of atomic current loops which do not completely cancel on the surface. The discussion and results of that section apply to any magnetized material, whether it is paramagnetic, diamagnetic, or ferromagnetic. We describe the magnetization of any such material by the magnetic dipole moment per unit volume **M**. Except for permanent magnets, this magnetization is caused by some external magnetic field and disappears when the external field is removed. The magnetization vector **M** is parallel to the external field which causes it for paramagnetism and ferromagnetism and antiparallel for diamagnetism. The magnetic induction **B** due to the magnetization of the material is very small compared with the external field for paramagnetism and diamagnetism and very large for ferromagnetism.

The calculation of the resultant magnetic-induction field **B** is complicated when magnetized material is in the presence of an external field. Although the field due to the material can be calculated from Ampère's law or the Biot-Savart law using the amperian current associated with the magnetization vector **M**, the magnetization itself depends on the resultant magnetic-induction field **B**. Because of this complication, a new vector, the magnetic intensity **H**, is defined in an attempt to separate out the external field from that due to the material. In many cases of interest, **H** can be found from the external conduction currents in wires alone without reference to any material present. The magnetization **M** can then be related to **H**, and the resultant magnetic-induction field **B** can be found from **H** and **M**. We define the magnetic intensity **H** in Section 39-1 and discuss some of its properties. We then estimate the order of magnitude of the magnetic moment of an atom and use this to estimate the possible magnitude of the magnetization **M** for paramagnetic and ferromagnetic materials. After a brief look at paramagnetism and diamagnetism we shall discuss qualitatively the complicated phenomenon of ferromagnetism, which has many important practical applications.

39-1 Magnetic Intensity **H**

Let us consider a long solenoid with n turns per unit length carrying a current I. The magnetic induction at the center of the solenoid is

$$B_0 = \mu_0 n I \qquad\qquad 39\text{-}1$$

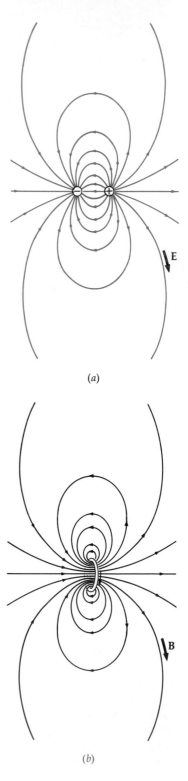

Figure 39-1
(*a*) Electric field lines of an electric dipole. (*b*) Magnetic field lines of a magnetic dipole. Far from the dipoles, the field patterns are identical. In the region inside the dipoles, the lines of **B** and **E** are oppositely directed.

(*a*)

(*b*)

This expression is good if we are far from the ends of the solenoid. Since we wish to avoid any complications from end effects, we shall assume that the solenoid is very long. (End effects can be completely eliminated in practice, using a doughnut-shaped solenoid called a *toroid,* illustrated in Figure 39-2, but because we have studied a solenoid in some detail, we shall continue to use it.) If we now insert some material inside the solenoid, there will be an additional contribution to the magnetic-induction field due to the magnetization of the material. If **M** is the magnetic moment per unit volume, the induction field $\mathbf{B}_m$ produced by this magnetization is given by an expression similar to Equation 39-1, with the amperian current per unit length M replacing the current per unit length in the solenoid nI:

$$\mathbf{B}_m = \mu_0 \mathbf{M} \qquad\qquad 39\text{-}2$$

The total magnetic induction at the center of the solenoid is the vector sum of these separate fields:

$$\mathbf{B} = \mathbf{B}_0 + \mu_0 \mathbf{M} \qquad\qquad 39\text{-}3$$

(We have written this as a vector sum because **M** is in the opposite direction to $\mathbf{B}_0$ for diamagnetic materials.) As mentioned above, it is convenient to separate out the external magnetic field from that due to the material. The *magnetic-intensity* vector **H** is defined by

$$\mathbf{H} = \frac{\mathbf{B}}{\mu_0} - \mathbf{M} \qquad\qquad 39\text{-}4a$$

Then

$$\mathbf{B} = \mu_0 \mathbf{H} + \mu_0 \mathbf{M} \qquad\qquad 39\text{-}4b$$

For a toroid or very long solenoid, for which we can neglect the effects at the ends of the material, **H** is simply the original external magnetic induction $\mathbf{B}_0$ divided by μ_0. The magnitude of **H** at the center of a long solenoid is

$$H = nI \qquad\qquad 39\text{-}5$$

From Equation 39-4 we see that the dimensions of **H** are the same as those for **M**, which are amperes per meter. (Because the number of turns of wire often comes into the equation for **H**, as in Equation 39-5, the units of **H** are often stated as ampere-turns per meter.)

In situations like this where we can neglect end effects of the magnetized material, the magnetic intensity **H** is determined by the external conduction currents and is not affected by the magnetization. In these cases **H** obeys Ampère's law and the Biot-Savart law in the forms

$$\oint_C \mathbf{H} \cdot d\mathbf{l} = I \qquad\qquad 39\text{-}6$$

and

$$d\mathbf{H} = \frac{1}{4\pi} \frac{I \, d\mathbf{l} \times \hat{\mathbf{r}}}{r^2} \qquad\qquad 39\text{-}7$$

It is important to realize that these equations differ from the corresponding equations for the magnetic induction **B** not only in the absence of the constant μ_0 but also because the current I is the macroscopic conduction current whereas in the equations for **B** the current I represents any kind of current, including the atomic or amperian current associated with the magnetization of the material.

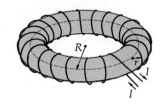

Figure 39-2
Toroid. The magnetic field is similar to that of a solenoid, but there are no end effects.

Magnetic intensity **H**
defined.

Example 39-1 Use Equations 39-6 and 37-21 to find **B** and **H** for a long cylinder which has a magnetization **M** and is wound with n turns per unit length carrying a current I.

Figure 39-3 shows a rectangular path for computing the line integrals in these equations. As in Section 37-5, we assume that the tangential component of the magnetic field is constant along path ab and zero along the other legs of the path. The conduction current through the rectangle is nIl, where l is the length from a to b. Thus Equation 39-6 gives

$$\oint \mathbf{H} \cdot d\mathbf{l} = Hl = nIl \quad \text{or} \quad H = nI$$

To calculate **B** we need to include the atomic current, which is Ml since M is the atomic current per unit length. Thus Equation 37-21 gives

$$\oint \mathbf{B} \cdot d\mathbf{l} = \mu_0 I_{\text{total}} = \mu_0 (I_{\text{conduction}} + I_{\text{magnetization}})$$

or

$$Bl = \mu_0 (nIl + Ml)$$

$$B = \mu_0 (nI + M)$$

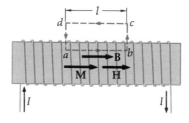

Figure 39-3
Path for calculation of $\oint \mathbf{H} \cdot d\mathbf{l}$ and $\oint \mathbf{B} \cdot d\mathbf{l}$ in Example 39-1. The only contribution is from a to b.

At points near the ends, or poles, of a magnetized material, the magnetic intensity **H** is not completely determined by the external conduction currents because of a contribution from the poles of the material. We can see this by considering the flux of **H** through any closed surface. Using the defining Equation 39-4, we have

$$\oint_S \mathbf{H} \cdot \hat{\mathbf{n}} \, dA = \frac{1}{\mu_0} \oint_S \mathbf{B} \cdot \hat{\mathbf{n}} \, dA - \oint_S \mathbf{M} \cdot \hat{\mathbf{n}} \, dA$$

Since the net flux of the magnetic induction **B** is necessarily zero, we have

$$\oint_S \mathbf{H} \cdot \hat{\mathbf{n}} \, dA = -\oint_S \mathbf{M} \cdot \hat{\mathbf{n}} \, dA \qquad \text{39-8}$$

In a homogeneous material, the magnetization **M** is uniform. The flux of **M** through any closed surface within the material is then zero. There is a net flux of **M** only at the ends of the material. Let us consider the uniformly magnetized cylinder shown in Figure 39-4. The flux of **M** through the surface S_1 enclosing the left end of the cylinder is positive and has the magnitude MA, where A is the cross-sectional area of the cylinder. Similarly, the net flux of **M** through surface S_2 enclosing the right end is $-MA$. From our discussion in Section 37-6 we note that MA is just the pole strength of the material and that the pole at the left is negative and the pole at the right is positive. In general, we can define the pole strength of a magnetized material q^* by

$$q^* = -\oint_S \mathbf{M} \cdot \hat{\mathbf{n}} \quad dA \qquad \qquad \text{39-9} \qquad \textit{Pole strength}$$

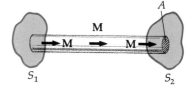

Figure 39-4
Surfaces S_1 and S_2 enclosing the ends of a uniformly magnetized cylinder. The net flux of **M** out of S_1 is $+MA$, and that out of S_2 is $-MA$, where A is the area of the cylinder.

Then the flux of the magnetization vector **H** is

$$\oint_S \mathbf{H} \cdot \hat{\mathbf{n}} = q^*$$

39-10 *Gauss' law for* **H**

This equation is similar to Gauss' law for the flux of the electric field **E** with the magnetic pole strength q^* replacing the electric charge q. It follows from this equation that **H** varies inversely as the square of the distance from a magnetic pole according to the equation

$$\mathbf{H} = \frac{1}{4\pi} \frac{q^*}{r^2} \hat{\mathbf{r}}$$

39-11 *Coulomb's law for* **H**

If we wish to calculate **H** at a point near a magnetic pole, we must add the contribution due to the pole given by Equation 39-11 to that due to the conduction currents, which can be calculated from Ampère's law or the Biot-Savart law. In many cases we are interested in the magnetic intensity **H** only in regions far from any poles, e.g., at the center of a long solenoid, and we can ignore the contribution given by Equation 39-11.

Example 39-2 A permanent cylindrical bar magnet of radius a and length $L = 10a$ has uniform magnetization **M** to the right parallel to its axis. Find **H** and **B** on the axis at the center of the magnet and just outside the right end.

In this example there are no conduction currents, and so **H** is due solely to the magnetic poles. If we consider the center of the magnet to be far from the ends so that the poles can be neglected, $\mathbf{H} = 0$ and $\mathbf{B} = \mu_0\mathbf{M}$ at that point. We can find a correction of this approximation by using Coulomb's law for **H** at the center due to the two poles. At the center, both the positive pole on the right and the negative pole on the left contribute equally to give **H** to the left in the direction opposite **M**. Using $q^* = \pi a^2 M$ and $r = \frac{1}{2}L$ in Equation 39-11, we have for the magnitude of **H** due to either pole,

$$H = \frac{1}{4\pi} \frac{q^*}{r^2} = \frac{1}{4\pi} \frac{\pi a^2 M}{(\frac{1}{2}L)^2} = \frac{a^2}{L^2} M$$

Adding the contributions due to each pole, we obtain for **H** at the center

$$\mathbf{H} = -\frac{2a^2}{L^2}\mathbf{M}$$

The magnetic induction **B** at the center is obtained from Equation 39-4*b*:

$$\mathbf{B} = \mu_0(\mathbf{H} + \mathbf{M}) = \mu_0\mathbf{M}\left(1 - \frac{2a^2}{L^2}\right)$$

We note that for $a = L/10$ the correction due to the poles is 2 percent.

Just outside the right end, the near pole can be treated as a disk of pole density $\sigma^* = M$, and the far pole can be treated as a point pole a distance L away. We found in Chapter 30 that a disk charge produces an electric field at a nearby point given by $\sigma/2\epsilon_0$. Since Equations 39-10 and 39-11 for **H** are the same as the corresponding equations for **E** except for the absence of ϵ_0, the magnetic intensity H produced by a disk of pole density σ^* is $\sigma^*/2$. The total H just outside the right end of the magnet is then

$$H = +\frac{\sigma^*}{2} - \frac{1}{4\pi}\frac{q^*}{L^2} = +\frac{M}{2} - \frac{1}{4\pi}\frac{\pi a^2 M}{L^2}$$

or

$$\mathbf{H} = \frac{\mathbf{M}}{2}\left(1 - \frac{a^2}{2L^2}\right)$$

Since there is no magnetization $\mathbf{M}$ outside the magnet, $\mathbf{B}$ is just $\mu_0\mathbf{H}$,

$$\mathbf{B} = \frac{\mu_0\mathbf{M}}{2}\left(1 - \frac{a^2}{2L^2}\right)$$

If we neglect the term $a^2/2L^2$ due to the far pole, the magnetic induction $\mathbf{B}$ is $\frac{1}{2}\mu_0\mathbf{M}$, approximately half its value at the center of the magnet, a result which also holds for a long solenoid.

Questions

1. How are $\mathbf{B}$ and $\mathbf{H}$ related in a vacuum?

2. Must $\mathbf{B}$ and $\mathbf{H}$ necessarily be parallel?

39-2   Magnetic Susceptibility and Permeability

For paramagnetic and diamagnetic materials the magnetization $\mathbf{M}$ is proportional to the magnetic intensity $\mathbf{H}$:

$$\mathbf{M} = \chi_m\mathbf{H} \qquad\qquad 39\text{-}12$$

The proportionality constant χ_m is called the *magnetic susceptibility*. Since $\mathbf{M}$ and $\mathbf{H}$ have the same dimensions, the susceptibility constant is dimensionless. It is positive for paramagnetic materials and negative for diamagnetic materials. From Table 39-1, listing typical values

---

Table 39-1
Magnetic susceptibility of various materials at 20°C

Material	χ_m
Aluminum	2.3×10^{-5}
Bismuth	-1.66×10^{-5}
Copper	-0.98×10^{-5}
Diamond	-2.2×10^{-5}
Gold	-3.6×10^{-5}
Magnesium	1.2×10^{-5}
Mercury	-3.2×10^{-5}
Silver	-2.6×10^{-5}
Sodium	-0.24×10^{-5}
Titanium	7.06×10^{-5}
Tungsten	6.8×10^{-5}
Hydrogen (1 atm)	-9.9×10^{-9}
Carbon dioxide (1 atm)	-2.3×10^{-9}
Nitrogen (1 atm)	-5.0×10^{-9}
Oxygen (1 atm)	2090×10^{-9}

of the susceptibility, we see that χ_m is much less than 1. If **M** is proportional to **H**, then **B** is also proportional to **H**. Substituting χ_m**H** for **M** in Equation 39-4*b*, we can write

$$\mathbf{B} = \mu_0(\mathbf{H} + \mathbf{M}) = \mu_0(\mathbf{H} + \chi_m\mathbf{H}) = \mu_0(1 + \chi_m)\mathbf{H} \qquad \text{39-13}$$

or

$$\mathbf{B} = \mu\mathbf{H} \qquad \text{39-14}$$

where

$$\mu = (1 + \chi_m)\mu_0 \qquad \text{39-15}$$

*Magnetic permeability
defined*

is called the *permeability* of the material. Because the susceptibility is so small, the permeability of all paramagnetic and diamagnetic material is very nearly equal to the permeability of free space μ_0.

Equation 39-14 is also written for ferromagnetic materials, but it is difficult to interpret because the magnetization **M** is not a linear function of **H**. The permeability defined by Equation 39-14 depends not only on the value of **H** but also on the previous state of magnetization of a ferromagnetic material. In such cases, **B** and **M** are multivalued functions of **H**. The maximum value of μ for ferromagnetic materials is typically several thousand times greater than μ_0. We shall discuss the relation between **B**, **H**, and **M** for ferromagnetic materials in more detail in Section 39-6.

Questions

3. Why are some values of χ_m in Table 39-1 positive and others negative?

4. Why is it more convenient to list χ_m than μ for the materials in Table 39-1?

39-3 Atomic Magnetic Moments

The magnetization **M** of a paramagnetic or ferromagnetic material can be related to the permanent magnetic moments of the individual atoms of the material. In this section we shall calculate the order of magnitude of these magnetic moments from the Bohr theory of the atom, in which the electrons move in circular orbits around the nucleus. Although this theory has severe defects, its results are in agreement with those from the correct quantum-mechanical theory of the atom in many cases, and often much insight can be obtained from the simpler Bohr picture. We can use the simple circular motion of a charged particle to show that there is a general connection between the angular momentum of the particle and the magnetic moment (Figure 39-5). Consider a charge q moving in a circle of radius r with speed v. The time for each revolution, the period T, is related to the speed and radius by

$$vT = 2\pi r$$

This motion results in a current loop and thus produces a magnetic moment. The current is the charge q divided by the period T:

$$I = \frac{q}{T} = \frac{qv}{2\pi r}$$

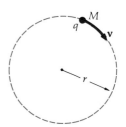

Figure 39-5
Particle of charge q and mass M moving in a circle of radius r. The angular momentum is into the paper with magnitude Mvr, and the magnetic moment is in (if q is positive) with magnitude $\frac{1}{2}qvr$.

The magnetic moment is the product of the current and the area πr^2:

$$m = IA = \frac{qv}{2\pi r}\,\pi r^2 = \tfrac{1}{2}qvr \qquad\qquad 39\text{-}16$$

If the mass of the particle is M (we use M here for the mass to avoid confusion with m for the magnetic moment), the angular momentum is

$$L = Mvr = M\,\frac{2m}{q}$$

If the charge q is positive, the angular momentum and magnetic moment are in the same direction. We can therefore write

$$\mathbf{m} = \frac{q}{2M}\,\mathbf{L} \qquad\qquad 39\text{-}17$$

Magnetic moment related to angular momentum

For the electron, with negative charge, the magnetic moment points in the direction opposite to the angular momentum. Equation 39-17 is the general classical relation between magnetic moment and angular momentum. It also holds in the quantum-mechanical theory of the atom for orbital angular momentum but not for the intrinsic spin angular momentum of the electron. For electron spin, the magnetic moment is twice that predicted by this equation. This extra factor of 2 is a quantum-mechanical result which has no analog in classical mechanics.

In both the Bohr theory of the atom and in the rigorous quantum-mechanical theory, the orbital angular momentum is quantized and takes on only values which are integral multiples of Planck's constant h divided by 2π. Thus

$$L = N\,\frac{h}{2\pi} = N\hbar \qquad\qquad 39\text{-}18$$

where N is an integer and

$$\hbar = \frac{h}{2\pi} = \frac{6.63 \times 10^{-34}}{2\pi}\ \text{J-sec} = 1.05 \times 10^{-34}\ \text{J-sec}$$

The spin angular momentum of the electron is $\tfrac{1}{2}\hbar$; it is therefore convenient to write the magnetic moment as

$$\mathbf{m} = \frac{qh}{2M}\,\frac{\mathbf{L}}{\hbar} \qquad\qquad 39\text{-}19a$$

For an electron $q = -e$ and $M = m_e$. The magnetic moment and angular momentum are then oppositely directed, and their magnitudes are related by

$$m = \frac{e\hbar}{2m_e}\,\frac{L}{\hbar} = m_B\,\frac{L}{\hbar} \qquad\qquad 39\text{-}19b$$

where

$$m_B = \frac{e\hbar}{2m_e} \qquad\qquad 39\text{-}20$$

Bohr magneton

is called a *Bohr magneton*. It is a convenient unit with which to measure atomic magnetic moments. For example, the magnetic moment associated with the spin of an electron is 1 Bohr magneton. Although the calculation of the magnetic moment of any atom is a complicated quantum-mechanical problem, the result of theory and experiment for all atoms is that the magnetic moment is of the order of a few Bohr magnetons (or zero for atoms with closed-shell electron structure). The magnitude of the Bohr magneton in SI units is

$$m_B = \frac{(1.60 \times 10^{-19}\ \text{C})(1.05 \times 10^{-34}\ \text{J-sec})}{2(9.11 \times 10^{-31}\ \text{kg})} = 9.27 \times 10^{-24}\ \text{A-m}^2$$

The SI unit for magnetic moment, the ampere-meter², can also be expressed as a joule per tesla (see Exercise 10 in Chapter 36). The value of the Bohr magneton is therefore

$$m_B = 9.27 \times 10^{-24}\ \text{A-m}^2 = 9.27 \times 10^{-24}\ \text{J/T} \qquad \text{39-21}$$

Bohr magneton in SI units

If all the atoms or molecules in some material have their magnetic moments aligned, the magnetic moment per unit volume of the material is just the product of the number of molecules per unit volume n and the magnetic moment of each molecule. For this extreme case, the saturation magnetization vector $\mathbf{M}_s$ is

$$\mathbf{M}_s = n\mathbf{m} \qquad \text{39-22}$$

The number of molecules per unit volume is related to the molecular weight, the density, and Avogadro's number. For example, the number of iron molecules (atoms) per unit volume is

$$n = \frac{6.02 \times 10^{23}\ \text{atoms/mole}}{55.8\ \text{gm/mole}} \frac{7.9\ \text{gm}}{1\ \text{cm}^3} \frac{10^6\ \text{cm}^3}{1\ \text{m}^3} = 8.52 \times 10^{28}\ \text{atoms/m}^3$$

If we assume that the magnetic moment of each iron atom is 1 Bohr magneton, the maximum value of the magnetization vector in iron has the magnitude

$$M_s = (8.52 \times 10^{28}\ \text{atoms/m}^3)(9.27 \times 10^{-24}\ \text{A-m}^2/\text{atom})$$
$$= 7.90 \times 10^5\ \text{A/m}$$

The magnetic induction on the axis inside a long iron cylinder with maximum magnetization is then

$$B = \mu_0 M_s = (4\pi \times 10^{-7}\ \text{N/A}^2)(7.90 \times 10^5\ \text{A/m}) = 0.993\ \text{T} \approx 1\ \text{T}$$

The measured saturation magnetic induction of annealed iron is about 2.16 T, indicating that the magnetic moment of an iron atom is slightly greater than 2 Bohr magnetons. This magnetic moment is due mainly to the spins of two unpaired electrons in the iron atom.

39-4   Paramagnetism

Paramagnetism occurs in materials whose atoms have permanent magnetic moments interacting with each other only very weakly. When there is no external magnetic field, these magnetic moments are randomly oriented. In the presence of an external magnetic field they tend to line up parallel to the field, but this is counteracted by the tendency for the moments to be randomly oriented due to thermal motion. The fraction of moments that line up with the field depends on the strength of the field and the temperature. At very low temperatures and high external fields nearly all the moments are aligned with the field. In this situation the contribution to the total magnetic field due to the material is very large, as indicated in the numerical estimates in the previous section. Even with the largest magnetic field obtainable in the laboratory, the temperature must be as low as a few degrees absolute in order to obtain a high degree of alignment. At higher temperatures and weaker external fields only a small fraction of the moments are aligned with the field, and the contribution of the material to the total magnetic field is very small. We can state this more quantitatively by comparing the energy of a magnetic moment in an external magnetic field with the thermal energy, which is of the order of kT, where k is Boltzmann's constant and T is the absolute temperature. The potential energy of a magnetic moment in an external magnetic field is least when the moment is parallel to the field and greatest when it is antiparallel to the field. We can find the expression for this potential energy by computing the work that must be done to rotate a magnetic dipole in an external field. Let a dipole of moment **m** make an angle θ with the magnetic induction **B** (Figure 39-6a). The torque exerted by the field on the dipole has the magnitude

$$\tau = mB \sin \theta$$

If we wish to rotate the dipole to increase the angle by $d\theta$, the work we must do is

$$dW = \tau \, d\theta = mB \sin \theta \, d\theta$$

The work we must do to rotate the dipole from $\theta = 0$ to $\theta = 180°$ (Figure 39-6b) equals the change in potential energy:

$$W = \int_0^{180°} mB \sin \theta \, d\theta = -mB \cos \theta \Big]_0^{180°} = 2mB = \Delta U \qquad 39\text{-}23$$

The potential energy when $\theta = 180°$ is thus greater than at $\theta = 0°$ by the amount $2mB$. For a typical magnetic moment of 1 Bohr magneton and a typical strong field of 1 T, this difference in potential energy is

$$2mB = 2(9.27 \times 10^{-24} \text{ J/T})(1 \text{ T}) = 1.85 \times 10^{-23} \text{ J}$$

At a normal temperature of $T = 300$ K, the typical thermal energy kT is 4.14×10^{-21} J, about 200 times greater than the difference in potential energy when the dipole is aligned or antialigned with the field. Thus even in a strong field of 1 T, most of the moments will be essentially randomly oriented because of thermal motions.

Figure 39-7 shows a plot of the ratio of the magnetization M to its saturation value M_s versus a dimensionless parameter $x = mB/kT$. At $x = 0$, corresponding to no external field, the magnetization is zero because the dipoles are randomly aligned. At very large values of x, which can occur only for high fields and very low temperatures, nearly all the dipoles are aligned with the field and $M/M_s = 1$. In the region of small x the curve is approximately linear, with a slope of $\frac{1}{3}$. In this region we have

$$M = \frac{1}{3} \frac{mB}{kT} M_s \qquad 39\text{-}24$$

The result that the magnetization varies inversely with the absolute temperature was discovered experimentally by Pierre Curie and is known as *Curie's law*.

39-5 Diamagnetism

Diamagnetism was discovered by Faraday in 1846 when he found that a piece of bismuth is repelled by either pole of a magnet, indicating that the external field of the magnet induces a magnetic dipole in the bismuth in the direction opposite the field. We can understand this ef-

Figure 39-6
(a) Magnetic dipole whose moment makes an angle θ with a magnetic field **B**. (b) When the dipole is antiparallel to **B**, the energy is greater by the amount $2mB$ than when it is parallel to **B**.

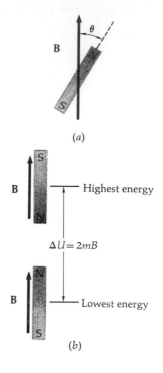

(a)

$\Delta U = 2mB$

(b)

Curie's law

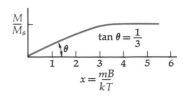

Figure 39-7
M/M_s versus mB/kT. At weak external fields or high temperatures, M/M_s increases linearly with mB/kT. When mB is much greater than kT, M reaches its saturation value M_s.

fect qualitatively using Lenz's law. Figure 39-8 shows two positive charges moving in circular orbits with the same speed but in opposite directions. Their magnetic moments are in opposite directions and cancel. (It is simpler to consider positive charges even though it is the negatively charged electrons which provide the magnetic moments in matter.) Consider now what happens when an external magnetic field **B** is turned on in the direction into the page. According to Lenz's law, currents will be induced to oppose the change in flux. If we assume that the radius of the circle is not changed, the charge on the left will be speeded up to increase its flux out of the page and the charge on the right will be slowed down to decrease its flux into the page. In each case, the *change* in magnetic moment of the charges will be in the direction out of the page, opposite to that of the external field.

We can estimate the magnitude of this effect by relating the change in the speeds of the particles to the change in the centripetal force due to the external magnetic field. We assume that the radius of the orbit does not change and that the change in speed is small compared with the original speed. Both these assumptions can be justified. The original centripetal force is provided by the electrostatic force of attraction of the electron to the nucleus. Calling this force F and setting it equal to the mass times the acceleration, we have

$$F = \frac{m_e v^2}{r} \qquad\qquad 39\text{-}25$$

where m_e is the electron mass. In the presence of an external magnetic field there is an additional force on each particle $q\mathbf{v} \times \mathbf{B}$. Note that on the particle on the left in Figure 39-8 this force is inward, thus increasing the centripetal force. This is necessary because, according to our argument from Lenz's law, this particle must speed up to oppose the change in flux. Similarly, the force is outward on the particle on the right, decreasing the centripetal force. Again, this is in the correct direction since this particle slows down to oppose the change in flux when the field is turned on. In each case the magnitude of the change in force is qvB since **v** and **B** are perpendicular. Since this change is very small, we use the differential approximation. From Equation 39-25 we have

$$\Delta F \approx dF = \frac{2m_e v}{r}\, dv \approx \frac{2m_e v}{r}\, \Delta v$$

Substituting qvB for ΔF and solving for Δv gives

$$\Delta v = \frac{qrB}{2m_e} \qquad\qquad 39\text{-}26$$

This change in speed causes a change in the magnetic moment which is in the outward direction for both particles. The magnetic moment of a charge moving in a circle is related to its speed by Equation 39-16,

$$m = \tfrac{1}{2}qvr$$

Then

$$\Delta m = \tfrac{1}{2}qr\,\Delta v = \tfrac{1}{2}qr\,\frac{qrB}{2m_e} = \frac{q^2 r^2}{4m_e}\,B \qquad\qquad 39\text{-}27$$

When we use $R = 10^{-10}$ m for a typical radius, the values of q and m_e for the electron, and a strong field of $B = 1$ T, the order of magnitude

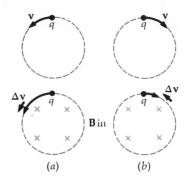

Figure 39-8
(*a*) Charge moving in circle with magnetic moment out (assuming q positive). In the presence of an external magnetic field the speed is increased. (*b*) Charge moving in a circle with its magnetic moment in. An external magnetic field causes a decrease in its speed. In both cases the *change* in the magnetic moment is out.

of this induced magnetic moment is

$$\Delta m = \frac{(1.60 \times 10^{-19}\ \text{C})^2 (10^{-10}\ \text{m})^2}{4(9.11 \times 10^{-31}\ \text{kg})}\ 1\ \text{T} \approx 7 \times 10^{-29}\ \text{A-m}^2$$

which is about 10^5 times smaller than a Bohr magneton. Since m is of the order of a Bohr magneton, $\Delta m/m$ is about 10^{-5}. This justifies our assumption that the change in speed is very small.

Since the induced magnetic moment is antiparallel to the external field whether the original moment is parallel or antiparallel to the field, there is a contribution from each electron in the atom to the induced magnetic moment of the atom. The magnetization vector **M** is the product of the number of *electrons* per unit volume times the induced magnetic moment of each electron. If n is the number of atoms per unit volume and Z is the atomic number, the number of electrons per unit volume is nZ. (Of course, an accurate calculation of the induced dipole moment of an atom must take into account the different values of r^2 for the different electrons and the fact that the orbits are not perpendicular to the external field.)

Atoms which have closed-shell electron structures have zero angular momentum and therefore no permanent magnetic dipole moment. They are diamagnetic. Atoms which have permanent magnetic dipole moments (and are not ferromagnetic) are either paramagnetic or diamagnetic, depending on which effect is stronger. Since the diamagnetic effect does not depend on temperature, and alignment of permanent moments decreases with temperature for both paramagnetic and ferromagnetic substances, all materials are diamagnetic at sufficiently high temperatures.

39-6 Ferromagnetism

Ferromagnetism occurs in pure iron, cobalt, and nickel, in alloys of these metals with each other and with some other elements, and in a few other substances (gadolinium, dysprosium, and a few compounds). In these substances a small external magnetic field can produce a very large degree of alignment of the atomic magnetic dipole moments, which, in some cases, can persist even when there is no external magnetizing field. This occurs because the magnetic dipole moments of atoms of these substances exert strong forces on their neighbors so that over a small region of space the moments are aligned with each other even with no external field. These dipole forces are predicted for these substances by quantum mechanics but cannot be explained with classical physics. At temperatures above a critical temperature, called the *Curie temperature*, these forces disappear and ferromagnetic materials become paramagnetic.

The region of space over which the magnetic dipole moments are aligned is called a *domain*. The size of a domain is usually microscopic. Within the domain, all the magnetic moments are aligned, but the direction of alignment varies from domain to domain so that the net magnetic moment of a macroscopic piece of material is zero in the normal state. Figure 39-9 illustrates this situation. When an external magnetic field is applied, the boundaries of the domains (called *domain walls*) shift and the direction of alignment within a domain may change so that there is a net magnetic moment in the direction of the applied field. Since the degree of alignment is great even for a

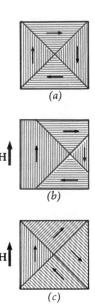

Figure 39-9
Schematic illustration of magnetization by domain changes: (*a*) unmagnetized; (*b*) magnetization by domain boundary changes; (*c*) magnetization by rotation of domains.

Curie temperature

Magnetic domains

Photomicrograph showing magnetic domains in a thin single crystal of yttrium iron garnet magnified 330 times. The photograph was made with transmitted light through the crystal between crossed Polaroids. The different directions of magnetization of the domains give rise to different colors in the transmitted light. The magnetization is pointing down into the paper in the light striped regions and out of the paper in the dark striped regions. In the large gray areas, the magnetization lies in the plane of the paper.

small external field, the magnetic field produced in the material by the dipoles is often much greater than the external field.

Let us consider magnetizing a long iron rod inside a solenoid by gradually increasing the current in the solenoid windings. We shall assume that the rod and solenoid are long enough to permit us to neglect end effects. The magnetic intensity **H** at the center of the solenoid is then simply related to the current by

$$H = nI = \frac{N}{l} I \qquad\qquad 39\text{-}28$$

where N is the number of turns and l is the length. The magnetic induction **B** is given by

$$\mathbf{B} = \mu_0\mathbf{H} + \mu_0\mathbf{M} \qquad\qquad 39\text{-}29$$

where **M** is the magnetization. For iron and other ferromagnetic materials, the magnetization **M** is often much greater than the magnetic intensity **H** by a factor of several thousand or more.

Figure 39-10 shows a plot of B versus H. As H is gradually increased from zero, B increases from zero along the part of the curve from the origin O to point P_1. The flattening of this curve near point P_1 indicates that the magnetization M is approaching its saturation value M_s when all the atomic dipoles are aligned. The external field needed to produce saturation in a given ferromagnetic material is called the saturation intensity H_s. A further increase in H above H_s increases B only through the term μ_0H in Equation 39-29. When H is gradually decreased from point P_1, there is no corresponding decrease in the magnetization. The shifting of the domains in a ferromagnetic material is not completely reversible, and some magnetization remains even when H is reduced to zero, as indicated in Figure 39-10. This effect is called *hysteresis*, from a word which means to lag. The value of the magnetic induction at point r when H is zero is called the *remanent field* B_r. If the current in the solenoid is now reversed so that **H** is in the opposite direction, the magnetic induction B is gradually brought to zero at point c. The value of H needed to reduce B to zero is called the *coercive force* H_c. The remaining part of the hysteresis curve is obtained by further increasing the current in the opposite direction until

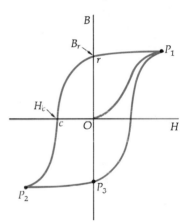

Figure 39-10
B versus H for hard ferromagnetic material. The outer curve is called a hysteresis curve.

Hysteresis

point P_2 is reached, corresponding to saturation in the opposite direction, decreasing the current to zero at point P_3, and increasing the current again to produce a magnetic intensity **H** in the original direction. We can formally define the magnetic susceptibility χ_m just as we did for paramagnetic or diamagnetic materials:

$$\mathbf{M} = \chi_m \mathbf{H}$$

Then

$$\mathbf{B} = \mu_0(\mathbf{H} + \mathbf{M}) = \mu_0(1 + \chi_m)\mathbf{H} = K_m \mu_0 \mathbf{H} = \mu \mathbf{H} \qquad 39\text{-}30$$

where

$$K_m = 1 + \chi_m = \frac{\mu}{\mu_0} \qquad 39\text{-}31$$

is called the *relative permeability*. Since, according to Figure 39-10, **B** and **H** are sometimes parallel and sometimes antiparallel and either can be zero when the other is not, μ and K_m may have any value from zero to infinity and may be positive or negative depending on **H** and the history of the material. If we confine the definitions of μ and K_m to that part of the magnetization curve from the origin O to point P_1, these quantities are always positive and have maximum values characteristic of the material. Table 39-2 lists the saturation magnetization M_s and the maximum value of K_m for some ferromagnetic materials. Note that the maximum values of K_m are much greater than 1. In a ferromagnetic material the quantity $\mu_0 \mathbf{H}$ can usually be neglected compared with $\mu_0 \mathbf{M}$.

The area enclosed by the hysteresis curve represents loss in energy because of the irreversibility of the process. The energy appears in the material as heat. We can see this as follows. The emf in the solenoid due to the changing flux is given by Faraday's law:

$$\varepsilon = \frac{d\phi}{dt} = NA \frac{dB}{dt} \qquad 39\text{-}32$$

where A is the cross-sectional area of the coil and rod. (We neglect the minus sign in Faraday's law because we are interested in magnitudes only.) The rate at which work is done against this emf is

$$\frac{dW}{dt} = \varepsilon I \qquad 39\text{-}33$$

Using Equation 39-32 for the emf and Equation 39-28 for the current, we have

$$\frac{dW}{dt} = \left(NA \frac{dB}{dt}\right)\frac{Hl}{N} = lAH \frac{dB}{dt}$$

Table 39-2
Maximum values of $\mu_0 M$ and K_m for some ferromagnetic materials

Material	$\mu_0 M_s$, T	K_m
Iron (annealed)	2.16	5,500
Iron-silicon (96% Fe, 4% Si)	1.95	7,000
Permalloy (55% Fe, 45% Ni)	1.60	25,000
Mu-metal (77% Ni, 16% Fe, 5% Cu, 2% Cr)	0.65	100,000

or

$$dW = lAH\,dB \qquad\qquad 39\text{-}34$$

The sum of this work for a complete hysteresis cycle is thus the area enclosed in the B-versus-H curve times the volume lA of the material. If the hysteresis effect is small, so that the area is small, indicating a small energy loss, the material is called *magnetically soft* (soft iron is an example). The hysteresis curve for a magnetically soft material is shown in Figure 39-11. Here the remanent field B_r and the coercive force H_c are nearly zero, and the energy loss per cycle is small. Magnetically soft materials are used for transformer cores to increase the induction B without incurring a large energy loss when the field alternates many times per second.

On the other hand, it is desirable to have a large remanent field B_r and a large coercive force H_c in a permanent magnet (the large coercive force is important so that the magnetization will not be destroyed by small stray fields). Magnetically hard materials, e.g., carbon steel and the alloy Alnico 5, are used for permanent magnets.

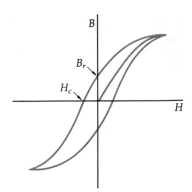

Figure 39-11
Hysteresis curve for magnetically soft material. The remanent field B_r and the coercive force H_c are small compared with those for a magnetically hard material (Figure 39-10).

Review

A. Define, explain, or otherwise identify:

Paramagnetism, 911
Diamagnetism, 911
Ferromagnetism, 911
Magnetization, 912
Magnetic intensity, 912
Magnetic pole strength, 914
Gauss' law for **H**, 915
Coulomb's law for **H**, 915
Magnetic susceptibility, 916
Permeability, 917

Permeability of free space, 917
Bohr magneton, 918
Curie's law, 920
Curie temperature, 922
Magnetic domain, 922
Hysteresis, 923
Remanent field, 923
Coercive force, 923
Magnetically soft, 925
Magnetically hard, 925

B. True or false:

1. Diamagnetism occurs in all materials.

2. Diamagnetism is the result of induced magnetic dipole moments.

3. In free space, **H** is always zero.

4. Paramagnetism is the result of partial alignment of permanent magnetic dipole moments.

5. Hysteresis is associated with a loss in electromagnetic energy.

6. Magnetically hard materials are good for transformer cores.

Exercises

Section 39-1, Magnetic Intensity H

1. A tightly wound solenoid 20 cm long has 400 turns carrying a current of 4 A such that its axial field is in the z direction. Neglecting end effects, find **H** and **B** at the center when (a) there is no core and (b) there is an iron core in the solenoid with magnetization $\mathbf{M} = 1.2 \times 10^6\,\mathbf{k}$ A/m.

2. Consider a single magnetic pole of strength q^* far away from any other pole. Use Gauss' law for **H** (Equation 39-10) to derive Coulomb's law for **H** (Equation 39-11).

3. In Exercise 1 the cross-sectional area of the solenoid core is 4 cm². Assuming the same current and magnetization as in Exercise 1, find (a) the magnetic pole strength q^* of the core, (b) **H** at the center of the solenoid due to the magnetic poles, (c) the resultant **H** at the center due to both the magnetic poles and the current in the windings, and (d) **B** at the center.

4. For the permanent magnet in Example 39-2, show that just inside the right end of the magnet (a) $H = -\frac{1}{2}M(1 + a^2/2L^2)$ and (b) $B = \frac{1}{2}\mu_0 M(1 - a^2/2L^2)$. (c) How do these results compare with those in Example 39-2 for the point just outside the magnet?

Section 39-2, Magnetic Susceptibility and Permeability

5. Which of the four gases listed in Table 39-1 are diamagnetic, and which are paramagnetic?

6. If the solenoid in Exercise 1 has an aluminum core, find **H**, **M**, and **B** at the center, neglecting end effects.

7. Repeat Exercise 6 for a tungsten core.

Section 39-3, Atomic Magnetic Moments

8. Nickel has density 8.7 gm/cm³ and molecular weight 58.7 gm/mole. Its saturation magnetization is $\mu_0 M_s = 0.61$ T. Calculate the magnetic moment in Bohr magnetons of a nickel atom.

9. Repeat Exercise 9 for cobalt, which has density 8.9 gm/cm³, molecular weight 58.9 gm/mole, and saturation magnetization $\mu_0 M_s = 1.79$ T.

Section 39-4, Paramagnetism

10. Show that Curie's law predicts that the magnetic susceptibility for a paramagnetic substance is given by $\chi_m = m\mu_0 M_s/3kT$.

11. Assume that the magnetic moment of an aluminum atom is 1 Bohr magneton. The density of aluminum is 2.7 gm/cm³, and its molecular weight is 27 gm/mole. (a) Calculate M_s and $\mu_0 M_s$ for aluminum. (b) Use the result of Exercise 10 to calculate χ_m at $T = 300$ K. (c) Explain why you expect this result to be larger than that listed in Table 39-1.

12. In a simple model of paramagnetism we can consider that some fraction f of the molecules have their magnetic moment aligned with the external magnetic field and the rest of the molecules are randomly oriented to produce no contribution to the magnetic moment. (a) Use this model and Curie's law to show that at temperature T and external field B this fraction aligned is $f = mB/3kT$. (b) Calculate this fraction for $T = 300$ K, $B = 1$ T assuming m to be 1 Bohr magneton.

Section 39-5, Diamagnetism

There are no exercises for this section.

Section 39-6, Ferromagnetism

13. The saturation magnetic intensity for annealed iron is $H_s = 1.6 \times 10^5$ A/m. Find the permeability μ and the relative permeability K_m at saturation (see Table 39-2).

14. For annealed iron the relative permeability K_m has its maximum value of about 5500 at $H = 125$ A/m. (This is well below the saturation intensity.) Find M and B when K_m is maximum.

15. The coercive force for a certain permanent bar magnet is $H_c = 4.4 \times 10^4$ A/m. The bar magnet is to be demagnetized by placing it inside a long solenoid 15 cm long with 600 turns. What is the minimum current needed in the solenoid to demagnetize the magnet?

16. A long solenoid has 50 turns/cm and carries a current of 2 A. The solenoid is filled with iron, and B is measured to be 1.72 T. (a) What is H (neglecting end effects)? (b) What is M? (c) What is the relative permeability K_m for this case?

17. When the current in the solenoid in Exercise 16 is 0.2 A, the magnetic induction is measured to be 1.58 T. (a) Neglecting end effects, what is H? (b) What is M? (c) What is the relative permeability K_m?

Problems

1. A toroid has mean radius R and cross-sectional radius r (Figure 39-2), where $r \ll R$. (a) Use Equation 39-6 to show that $H = nI = IN/2\pi R$, where n is the number of turns per unit length and I is the current. (b) When the toroid is filled with material, it is called a *Rowland ring*. Find $\mathbf{H}$ and $\mathbf{B}$ in such a ring. Assume a magnetization $\mathbf{M}$ everywhere parallel to $\mathbf{H}$.

2. Show that the normal component of the magnetic induction $\mathbf{B}$ is continuous across any surface. Do this by applying Gauss' law for $\mathbf{B}$ ($\oint \mathbf{B} \cdot \hat{\mathbf{n}} \, dA = 0$) to a pillbox gaussian surface which has a face on each side of the surface.

3. Use Equation 39-10 to show that the normal component of $\mathbf{H}$ on one side of a surface differs from that on the other side of the surface by the amount σ^*, where σ^* is the magnetic pole density on the surface (see Problem 2).

4. A Rowland ring has windings with 5 turns/cm carrying current $I = 1$ A. The magnetic induction in the ring is 1.40 T. (a) Find $\mathbf{H}$ and $\mathbf{M}$ in the ring. (b) What is K_m? (c) A small gap is cut in the ring. Find $\mathbf{B}$ and $\mathbf{H}$ in the gap assuming that they are unchanged in the bulk of the material (see Problems 2 and 3).

5. In our derivation of the magnetic moment induced in an atom we used the Bohr model and assumed that the radius of the orbit did not change in the presence of an external magnetic field. In this problem you are to show that the assumption of constant radius is justified by showing that when $\mathbf{B}$ is applied, there is an impulse which increases or decreases the speed of the electron by just the correct amount given by Equation 39-26. (a) Use Faraday's law and the expression $\mathcal{E} = \oint \mathbf{E} \cdot d\mathbf{l}$ to show that the induced electric field is related to the rate of change of the magnetic field by $E = \frac{1}{2}r \, dB/dt$ assuming r to be constant. (b) Use Newton's second law to show that the change in speed of the electron dv is related to the change in B by $dv = (qr/2m_e) \, dB$. Integrate to obtain Equation 39-26.

6. Equation 39-27 gives the induced magnetic moment for a single electron in an orbit which has its plane perpendicular to $\mathbf{B}$. If an atom has Z electrons, a reasonable simplifying assumption is that on the average one-third have their planes perpendicular to $\mathbf{B}$. Show that the diamagnetic susceptibility obtained from Equation 39-27 is then

$$\chi_m = \frac{-nZq^2r^2}{12m_e}\mu_0$$

where n is the number of atoms per unit volume. Use $n \approx 6 \times 10^{28}$ atoms/m^3 and $r \approx 5 \times 10^{-11}$ m to estimate χ_m for $Z \approx 50$.

7. Use the values in the table on the right to plot B versus H and K_m versus H.

8. A long solenoid of length l and cross-sectional area A has n turns per unit length and carries a current I. It is filled with iron of relative permeability K_m. Find the self-inductance of the solenoid.

H, A/m	B, T
0	0
50	0.04
100	0.67
150	1.00
200	1.2
500	1.4
1000	1.6
10,000	1.7

CHAPTER 40   Alternating-Current Circuits

In Chapter 38 we looked briefly at circuits with inductance, capacitance, and resistance and saw that in an *LC* circuit with no resistance, the energy oscillates between electrostatic energy in the capacitor and magnetic energy in the inductor. With resistance in the circuit, the oscillations are damped as energy is dissipated as heat. In this chapter we shall look at similar circuits driven by a generator of sinusoidal alternating current. Much of the mathematics involved in the detailed analysis of such circuits is studied in a course in differential equations. As in the last sections of Chapter 38 and in the study of forced oscillators in Chapter 15, we shall content ourselves with a qualitative analysis of these circuits which brings out the main features but avoids the need for differential-equation theory.

We shall begin by studying a simple ac generator which puts out a sinusoidal emf and then consider the results of placing a resistor, capacitor, or inductor separately across its terminals. We shall then look at several combinations of these circuit elements in series with such a generator. There are several reasons for studying sinusoidal currents and voltages. Although practical generators are considerably different from the simple device we shall study in Section 40-1, they are designed to put out a sinusoidal emf. We shall see that when the generator output is sinusoidal, the current in an inductor, capacitor, or resistor is also sinusoidal though not necessarily in phase with the generator emf. When the emf and current are both sinusoidal, their maximum values can be simply related. Finally, an important reason for studying sinusoidal currents is that even when the current in some circuit is not sinusoidal, it can be analyzed in terms of sinusoidal components using Fourier analysis. The results for sinusoidal currents can therefore be applied to any kind of alternating current.

40-1 An AC Generator

A simple generator of alternating current is a coil rotating in a uniform magnetic field, as shown in Figure 40-1, where the unit vector $\hat{n}$ normal to the plane of the coil makes an angle θ with a uniform magnetic field $\mathbf{B}$. The magnetic flux through the coil is

$$\phi_m = NBA \cos \theta \qquad\qquad 40\text{-}1$$

where N is the number of turns and A the area of the coil. If the coil is mechanically rotated, the flux through the coil will change and, according to Faraday's law, there will be an emf in the coil. Let ω be the angular velocity of the coil. Then

$$\theta = \omega t \qquad\qquad 40\text{-}2$$

where we have chosen $\theta = 0$ at time $t = 0$ for convenience. Substituting this expression for θ into Equation 40-1, we have for the flux,

$$\phi_m = NBA \cos \omega t \qquad\qquad 40\text{-}3$$

The emf in the coil will then be

$$\mathcal{E} = -\frac{d\phi_m}{dt} = -NBA\,(-\omega \sin \omega t) = +NBA\omega \sin \omega t$$

or

$$\mathcal{E} = \mathcal{E}_{max} \sin \omega t \qquad\qquad 40\text{-}4$$

where

$$\mathcal{E}_{max} = NBA\omega \qquad\qquad 40\text{-}5$$

is the maximum value of the emf. We can thus produce sinusoidal emf in a coil by rotating it with constant angular velocity in a magnetic field. In this type of emf source, mechanical energy is converted into electric energy. In circuit diagrams, an ac generator is represented by the symbol $\bigotimes$. Although practical generators are considerably more complicated, they work on the principle that there is an alternating emf in a coil rotating in a magnetic field and they are designed so that the emf produced is sinusoidal.

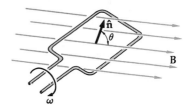

Figure 40-1
A coil rotating in a magnetic field with constant angular velocity generates a sinusoidal emf.

Questions

1. Does the sinusoidal nature of the emf depend on the size or shape of the coil?

2. How could such a coil be used to generate a nonsinusoidal emf?

3. If such a generator delivers electric energy to a circuit, where does the energy come from?

40-2 Alternating Current in a Resistor

The simplest ac circuit consists of a generator and a resistor, as shown in Figure 40-2. We find the current in this circuit as usual by Kirchhoff's rule. Let the emf of the generator be given by

$$\mathcal{E} = \mathcal{E}_{max} \sin \omega t \qquad\qquad 40\text{-}6$$

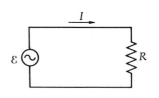

Figure 40-2
An ac generator in series with a resistor R.

Then Kirchhoff's rule gives

$$\mathcal{E} = IR = \mathcal{E}_{max} \sin \omega t \qquad\qquad 40\text{-}7$$

The current is proportional to the generator voltage and is given by

$$I = \frac{\mathcal{E}_{max}}{R} \sin \omega t = I_{max} \sin \omega t \qquad\qquad 40\text{-}8$$

where the maximum current is

$$I_{max} = \frac{\mathcal{E}_{max}}{R} \qquad\qquad 40\text{-}9$$

40-3 Alternating Current in a Capacitor

Figure 40-3 shows a capacitor connected across the terminals of a generator. Assuming I positive in the clockwise direction as indicated, we have

$$I = \frac{dQ}{dt}$$

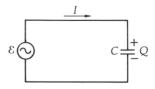

Figure 40-3
An ac generator in series with a capacitor C.

where Q is the charge on the capacitor. Kirchhoff's rule for this circuit gives

$$\mathcal{E} = \frac{Q}{C} = \mathcal{E}_{max} \sin \omega t \qquad \text{or} \qquad Q = \mathcal{E}_{max} C \sin \omega t$$

The current is

$$I = \frac{dQ}{dt} = \omega \mathcal{E}_{max} C \cos \omega t = I_{max} \cos \omega t \qquad\qquad 40\text{-}10$$

where

$$I_{max} = \mathcal{E}_{max} \omega C \qquad\qquad 40\text{-}11$$

The current in this circuit is not in phase with the generator voltage, which is also the potential drop across the capacitor. Using the trigonometric identity

$$\cos \omega t = \sin (\omega t + 90°)$$

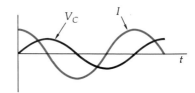

Figure 40-4
Plots of current I and voltage V_C across the capacitor versus time. The voltage lags the current by 90°.

we have

$$I = I_{max} \sin (\omega t + 90°)$$

The current and voltage across the capacitor are plotted in Figure 40-4, from which we see that the maximum value of the voltage occurs 90° after the maximum value of the current. The voltage drop across the capacitor is said to *lag* the current by 90°.

The relation between the maximum current and the maximum voltage can be written in a form similar to Ohm's law. We have

Voltage across capacitor lags the current

$$I_{max} = \omega C \mathcal{E}_{max} = \frac{\mathcal{E}_{max}}{1/\omega C} = \frac{\mathcal{E}_{max}}{X_C} \qquad\qquad 40\text{-}12$$

where

$$X_C = \frac{1}{\omega C} \qquad\qquad 40\text{-}13$$

Capacitive reactance

is called the *capacitive reactance* of the circuit. Like resistance, capacitive reactance has units of ohms, and the larger the reactance for a given emf the smaller the maximum value of the current. Unlike resistance, the capacitive reactance depends on the frequency of the current: the larger the frequency the smaller the reactance.

Example 40-1 A 20-μF capacitor is placed across a generator which has a maximum emf of 100 V. Find the reactance and maximum current if the frequency is 60 Hz and if the frequency is 5000 Hz.

These frequencies correspond to angular frequencies ω_1 and ω_2 given by $\omega_1 = 2\pi f_1 = 2\pi(60 \text{ sec}^{-1}) = 377 \text{ sec}^{-1}$ and $\omega_2 = 2\pi(5000 \text{ sec}^{-1}) = 3.14 \times 10^4 \text{ sec}^{-1}$. The reactances at these frequencies are $X_1 = 1/\omega_1 C = [(377)(20 \times 10^{-6})]^{-1} = 133 \ \Omega$ and $X_2 = 1/\omega_2 C = [(3.14 \times 10^4)(20 \times 10^{-6})]^{-1} = 1.59 \ \Omega$. The maximum currents are then $I_{1,\text{max}} = \mathcal{E}_{\text{max}}/X_1 = (100 \text{ V})/(133 \ \Omega) = 0.754 \text{ A}$ and $I_{2,\text{max}} = (100 \text{ V})/(1.59 \ \Omega) = 62.8 \text{ A}$.

40-4 Alternating Current in an Inductor

Figure 40-5 shows an inductor across the terminals of an ac generator. Kirchhoff's rule for this circuit gives

$$\mathcal{E} = L\frac{dI}{dt} = \mathcal{E}_{\text{max}} \sin \omega t \qquad \text{40-14}$$

Solving for the current I, we obtain

$$I = -\frac{\mathcal{E}_{\text{max}}}{\omega L} \cos \omega t = \frac{\mathcal{E}_{\text{max}}}{\omega L} \sin (\omega t - 90°) \qquad \text{40-15}$$

(We have neglected the constant of the integration because it depends on the initial conditions, which are of no interest for this discussion.) From the plot in Figure 40-6 of the current and voltage across the inductor we see that the maximum value of the voltage occurs 90° before the corresponding maximum value of the current. The voltage across an inductor is said to *lead* the current by 90°.

The maximum current is related to the maximum voltage across the inductor by

$$I_{\text{max}} = \frac{\mathcal{E}_{\text{max}}}{\omega L} = \frac{\mathcal{E}_{\text{max}}}{X_L} \qquad \text{40-16}$$

where

$$X_L = \omega L \qquad \text{40-17}$$

The quantity X_L is called the *inductive reactance*. As the frequency increases, the inductive reactance increases and the maximum current decreases.

40-5 *LCR* Circuit with Generator

An important circuit containing many of the features of most ac circuits is a series *LCR* circuit with a generator (Figure 40-7). We discussed a similar circuit without the generator in Chapter 38, finding that the current oscillates with frequency nearly equal to $\omega_0 = 1/\sqrt{LC}$

Figure 40-5
An ac generator in series with an inductor *L*.

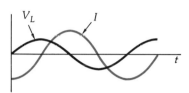

Figure 40-6
Plots of current *I* and voltage V_L across the inductor as functions of time. The voltage leads the current by 90°.

Voltage across an inductor leads the current

Inductive reactance

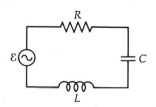

Figure 40-7
A series *LCR* circuit with ac generator.

(assuming that the resistance is small) and that the maximum value of the current decreases exponentially with time. The mathematical solution of the differential equation obtained by applying Kirchhoff's law to this circuit contains a term exactly like that with no generator present, called the *transient current,* plus a term describing a current which does not decrease exponentially in time, called the *steady-state current.* At a sufficient time after the switch is closed, the transient current is negligible compared with the steady-state current. We shall ignore the transient term in discussing this circuit.

Transient and steady-state currents

We can understand the behavior of the circuit in Figure 40-7 best if we first consider the case with no resistance (Figure 40-8). Kirchhoff's law for this circuit gives

$$\varepsilon = L\frac{dI}{dt} + \frac{Q}{C} = \varepsilon_{max} \sin \omega t$$

Substituting dQ/dt for I in this equation gives

$$L\frac{d^2Q}{dt^2} + \frac{Q}{C} = \varepsilon_{max} \sin \omega t \qquad \text{40-18}$$

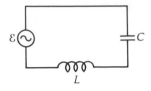

Figure 40-8
A series LC circuit with generator.

We can show by substitution that Equation 40-18 is satisfied by the function

$$Q = Q_{max} \sin \omega t$$

Then

$$\frac{d^2Q}{dt^2} = -\omega^2 Q_{max} \sin \omega t$$

Substituting these into Equation 40-18, we obtain

$$-L\omega^2 Q_{max} \sin \omega t + \frac{Q_{max}}{C} \sin \omega t = \varepsilon_{max} \sin \omega t$$

which is satisfied for any time if Q_{max} is given by

$$Q_{max} = \frac{\varepsilon_{max}}{1/C - L\omega^2} = \frac{\varepsilon_{max}}{L(1/LC - \omega^2)} = \frac{\varepsilon_{max}}{L(\omega_0{}^2 - \omega^2)}$$

where $\omega_0 = 1/\sqrt{LC}$ is the natural frequency. It is customary to write the charge in terms of an amplitude Q_{max} which is intrinsically positive. We can do this by introducing a phase angle δ. We write

$$Q = Q_{max} \sin (\omega t - \delta) \qquad \text{40-19}$$

where

$$Q_{max} = \frac{\varepsilon_{max}}{L|\omega_0{}^2 - \omega^2|} \qquad \text{40-20}$$

and the phase angle δ is zero if $\omega < \omega_0$ or 180° if $\omega > \omega_0$.

The current is obtained from Equation 40-19 by differentiation:

$$I = \frac{dQ}{dt} = \omega Q_{max} \cos (\omega t - \delta) = I_{max} \sin (\omega t - \delta + 90°) \qquad \text{40-21}$$

where

$$I_{max} = \omega Q_{max} = \frac{\omega \varepsilon_{max}}{L|\omega_0{}^2 - \omega^2|} \qquad \text{40-22}$$

The maximum current can also be written in terms of the capacitive and inductive reactances. We have

$$X_C = \frac{1}{\omega C} = \frac{L}{\omega}\frac{1}{LC} = \frac{L}{\omega}\,\omega_0{}^2 \qquad \text{and} \qquad X_L = \omega L = \frac{L}{\omega}\,\omega^2$$

so that

$$X_C - X_L = \frac{L}{\omega}\,(\omega_0{}^2 - \omega^2) \qquad\qquad 40\text{-}23$$

Comparing with Equation 40-22, we have

$$I_{max} = \frac{\mathcal{E}_{max}}{|X_C - X_L|} \qquad\qquad 40\text{-}24$$

When the generator frequency ω equals the natural frequency ω_0, the capacitive and inductive reactances are equal and the current is infinite. This is the condition of *resonance:*

$$\omega = \omega_0 \qquad X_C = X_L \qquad\qquad 40\text{-}25 \qquad \textit{Resonance condition}$$

In real circuits there is always some resistance, and the current is maximum but not infinite at resonance. At frequencies below the resonance frequency, X_C is greater than X_L, the phase δ in Equation 40-21 is zero, and the voltage lags the current by 90°, as in a purely capacitive circuit. At frequencies above the resonance frequency, X_C is less than X_L, the phase δ is 180°, and the voltage leads the current, as in a purely inductive circuit.

In the original circuit with resistance (Figure 40-7) the application of Kirchhoff's rule gives

$$\mathcal{E}_{max} \sin \omega t = L\frac{dI}{dt} + \frac{Q}{C} + IR = L\frac{d^2Q}{dt^2} + \frac{Q}{C} + R\frac{dQ}{dt} \qquad 40\text{-}26$$

This equation has the same form as Equation 15-21 for the forced, damped harmonic oscillator. The current obtained by solving Equation 40-26 is

$$I = I_{max} \sin (\omega t + \phi) \qquad\qquad 40\text{-}27$$

where

$$\tan \phi = \frac{X_C - X_L}{R} \qquad\qquad 40\text{-}28$$

and

$$I_{max} = \frac{\mathcal{E}_{max}}{\sqrt{(X_C - X_L)^2 + R^2}} = \frac{\mathcal{E}_{max}}{Z} \qquad\qquad 40\text{-}29$$

The quantity

$$Z = \sqrt{(X_C - X_L)^2 + R^2} \qquad\qquad 40\text{-}30 \qquad \textit{Impedance}$$

is called the *impedance.* In terms of the impedance, the current is given by

$$I = \frac{\mathcal{E}_{max}}{Z} \sin (\omega t + \phi) \qquad\qquad 40\text{-}31$$

We can obtain these results from a simple vector diagram called a *phasor diagram.* At any instant, the current is the same in each element of the series circuit. We represent this current by a vector (called a phasor) which has magnitude equal to I_{max} and which makes an angle $\omega t + \phi$ with the x axis. This vector rotates counterclockwise with angular frequency ω. The instantaneous current is the y component of this vector. If we multiply this vector by the resistance R, we obtain a vector $\mathbf{V}_R$ representing the voltage drop across the resistor. These vectors are parallel because the voltage drop across the resistor is in phase with the current. The voltage drop across the capacitor lags the current by 90°. This voltage is therefore represented by a vector $\mathbf{V}_C$ which has magnitude $I_{max}X_C$ and lags the vector $\mathbf{V}_R$ by 90°, as shown in Figure 40-9. The y component of this vector equals the instantaneous voltage drop across the capacitor. Similarly, we represent the voltage drop across the inductor by a vector $\mathbf{V}_L$ which has magnitude $I_{max}X_L$ and which leads $\mathbf{V}_R$ by 90°. The sum of the y components of these three vectors equals the instantaneous sum of the voltage drops across the resistor, capacitor, and inductor. By Kirchhoff's law, this sum equals the instantaneous emf. The sum of these y components is the y component of the resultant vector,

$$\mathcal{E} = \mathbf{V}_R + \mathbf{V}_C + \mathbf{V}_L$$

The vector representing the emf makes an angle ϕ with the current vector and the vector $\mathbf{V}_R$ and makes an angle ωt with the x axis as shown. From the right triangle in Figure 40-9(a) we obtain for the magnitude of the resultant vector,

$$\mathcal{E}_{max} = |\mathbf{V}_R + \mathbf{V}_C + \mathbf{V}_L| = \sqrt{V_R{}^2 + (V_C - V_L)^2}$$
$$= I_{max}\sqrt{R^2 + (X_C - X_L)^2} = I_{max}Z$$

This result is Equation 40-29. We see also from the diagram that the phase angle ϕ is given by Equation 40-28.

At low frequencies, $\omega < \omega_0$, the capacitive reactance is greater than the inductive reactance, and so ϕ is positive, indicating that the current leads the generator voltage. (This is the situation shown in Figure 40-9.) At high frequencies, $\omega > \omega_0$, the phase ϕ is negative; then the current lags the generator voltage. At the resonance frequency, $\omega = \omega_0$, the total impedance is just the resistance R because the capacitive and inductive reactances add to zero. The current at resonance is in phase with the generator voltage and has its maximum value of $I_{max} = \mathcal{E}_{max}/R$.

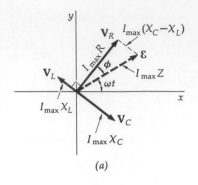

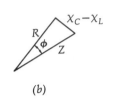

(b)

Figure 40-9
(a) Phasor diagram for finding I_{max} and ϕ in a series LCR circuit with generator. The vector $\mathbf{V}_R$ has magnitude $I_{max}R$ and is in phase with the current. $\mathbf{V}_L$ leads the current by 90°, and $\mathbf{V}_C$ lags the current by 90°. The vector $\mathcal{E}$ is the vector sum $\mathbf{V}_R + \mathbf{V}_L + \mathbf{V}_C$. It has magnitude $I_{max}Z$. (b) Impedance diagram, showing that $Z^2 = R^2 + (X_C - X_L)^2$ and $\tan \phi = (X_C - X_L)/R$.

Example 40-2 A series LCR circuit with $L = 2$ H, $C = 2$ μF, and $R = 20$ Ω is driven by a generator of maximum emf 100 V and variable frequency. Find the resonance frequency ω_0, and the phase ϕ and maximum current I_{max}, when the generator angular frequency is $\omega = 400$ rad/sec.

The resonance frequency is $\omega_0 = 1/\sqrt{LC} = 1/\sqrt{(2\text{ H})(2 \times 10^{-6}\text{ F})} = 500$ rad/sec. When the generator frequency is 400 rad/sec, it is well below the resonance frequency. The capacitive and inductive reactances at 400 rad/sec are

$$X_C = \frac{1}{\omega C} = \frac{1}{(400)(2 \times 10^{-6})} = 1250 \ \Omega$$

and

$$X_L = \omega L = (400)(2) = 800 \ \Omega$$

The total reactance is $X_C - X_L = 1250 \ \Omega - 800 \ \Omega = 450 \ \Omega$. This is much greater than the resistance, a result which always holds far from resonance. The total impedance is

$$Z = \sqrt{(X_C - X_L)^2 + R^2} \approx 450 \ \Omega$$

since 20^2 is negligible compared with 450^2. The maximum current is then

$$I_{max} = \frac{\mathcal{E}_{max}}{Z} = \frac{100 \text{ V}}{450 \text{ }\Omega} = 0.222 \text{ A}$$

This is small compared with I_{max} at resonance, which is $(100 \text{ V})/(20 \text{ }\Omega) = 5 \text{ A}$. Since X_C is greater than X_L, this current leads the generator voltage by ϕ, given by

$$\tan \phi = \frac{X_C - X_L}{R} = \frac{450 \text{ }\Omega}{20 \text{ }\Omega} = 22.5 \qquad \phi = 87°$$

40-6 Power in AC Circuits

The instantaneous power supplied by a generator to any circuit is the product of the generator voltage and the current. For the circuit in Figure 40-7, in which the current is given by Equation 40-27, the instantaneous power is

$$\begin{aligned} P &= \mathcal{E}I = (\mathcal{E}_{max} \sin \omega t) \left[I_{max} \sin (\omega t + \phi) \right] \\ &= \mathcal{E}_{max} I_{max} \sin \omega t \sin (\omega t + \phi) \end{aligned} \qquad 40\text{-}32$$

The instantaneous power is a complicated function of the time. We are usually interested in the average power over one or more cycles. It is easier to compute this average if we first expand the function $\sin (\omega t + \phi)$ using the trigonometric identity for the sine of the sum of two angles:

$$\sin (\omega t + \phi) = \sin \omega t \cos \phi + \cos \omega t \sin \phi$$

Then

$$P = \mathcal{E}_{max} I_{max} (\cos \phi \sin^2 \omega t + \sin \phi \sin \omega t \cos \omega t)$$

The time average of the second term in parentheses is zero. We can see this by writing $\sin \omega t \cos \omega t = \frac{1}{2} \cos 2 \omega t$. This term oscillates twice during each cycle and is negative as often as it is positive. The average value of $\sin^2 \omega t$ over one or more cycles is $\frac{1}{2}$.[1] The average power is therefore

$$P_{av} = \tfrac{1}{2} \mathcal{E}_{max} I_{max} \cos \phi \qquad 40\text{-}33$$

Most ac ammeters and voltmeters measure rms values of current and voltage rather than maximum or peak values. For any sinusoidal quantity, the rms value equals the maximum value divided by $\sqrt{2}$. We can see this by considering the current. The instantaneous value of I^2 is

$$I^2 = I_{max}^2 \sin^2 (\omega t + \phi)$$

Averaging over one or more cycles gives

$$(I^2)_{av} = I_{max}^2 \left[\sin^2 (\omega t + \phi) \right]_{av} = \tfrac{1}{2} I_{max}^2$$

Then

$$I_{rms} = \sqrt{(I^2)_{av}} = \frac{I_{max}}{\sqrt{2}} \qquad 40\text{-}34 \qquad \textit{rms current}$$

[1] This can be seen from the identity $\sin^2 \omega t + \cos^2 \omega t = 1$ and the fact that the average values of $\sin^2 \omega t$ and $\cos^2 \omega t$ are equal.

Similarly $\mathcal{E}_{rms} = \mathcal{E}_{max}/\sqrt{2}$. The average power can then be written

$$P_{av} = \mathcal{E}_{rms} I_{rms} \cos \phi \qquad \qquad 40\text{-}35$$

As in a dc circuit, the power supplied depends on the product of the emf and the current. However, Equation 40-35 also includes the factor $\cos \phi$, called the *power factor*. From the triangle in Figure 40-9*b* we have

$$\cos \phi = \frac{R}{Z} = \frac{R}{\sqrt{(X_C - X_L)^2 + R^2}} \qquad \qquad 40\text{-}36$$

At resonance, $X_C = X_L$, the impedance is just R, and the power factor is 1. Off resonance, the power factor is less than 1. Combining Equations 40-35 and 40-36 and writing $\mathcal{E}_{rms} = I_{rms}Z$, we find for the average power,

$$P_{av} = \mathcal{E}_{rms} I_{rms} \frac{R}{Z} = I_{rms}^2 R = \mathcal{E}_{rms}^2 \frac{R}{Z^2} \qquad \qquad 40\text{-}37$$

Let us now consider changing the generator frequency ω while keeping $\mathcal{E}_{max}$ and $\mathcal{E}_{rms}$ constant. The average power is maximum at resonance, as can be seen from the last expression in Equation 40-37, since Z then has its minimum value R. We have in general,

$$Z^2 = (X_C - X_L)^2 + R^2 = \left(\frac{1}{\omega C} - \omega L\right)^2 + R^2$$

$$= \frac{L^2}{\omega^2}(\omega_0^2 - \omega^2)^2 + R^2$$

where we have used $1/LC = \omega_0^2$. Using this expression for Z^2 in Equation 40-37, we obtain

$$P_{av} = \frac{\mathcal{E}_{rms}^2 R\omega^2}{L^2(\omega_0^2 - \omega^2)^2 + \omega^2 R^2} \qquad \qquad 40\text{-}38$$

Figure 40-10 shows a plot of the average power versus generator frequency ω for two values of the resistance R. The smaller the resistance the more sharply peaked the resonance. Let $\Delta\omega = \omega_2 - \omega_1$ be the *width* of the resonance, where ω_1 and ω_2 are the two values of ω for which P_{av} is half its maximum value. For a sharply peaked resonance we can find a useful approximation for the width as follows. At resonance, the denominator of the expression on the right in Equation 40-38 is just $\omega^2 R^2$. The power will be half its maximum value when this denominator is approximately twice this value (neglecting the variation of the numerator with ω). Then

$$L^2(\omega_0^2 - \omega^2)^2 \approx \omega^2 R^2$$

or

$$L(\omega_0^2 - \omega^2) = L(\omega_0 - \omega)(\omega_0 + \omega) \approx \pm\omega R$$

Since ω and ω_0 are nearly equal for a sharply peaked resonance, we can replace ωR with $\omega_0 R$ and $\omega_0 + \omega$ with $2\omega_0$. Then we have approximately

$$\omega_0 - \omega \approx \pm\frac{R}{2L}$$

$$\omega \approx \omega_0 \pm \frac{R}{2L}$$

that is, $\omega_2 \approx \omega_0 + R/2L$ and $\omega_1 \approx \omega_0 - R/2L$. The width is then

$$\Delta\omega \approx \frac{R}{L} \qquad \qquad 40\text{-}39$$

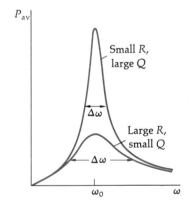

Figure 40-10
Plot of average power P_{av} given by Equation 40-38 versus ω for small and large R. The Q value is the ratio of the resonance frequency ω_0 to the resonance width $\Delta\omega$. A large Q value indicates a narrow resonance.

It is customary to describe the power resonance by a dimensionless parameter Q called the Q *value* (for quality value). In general, Q is defined by $Q = L\omega_0/R$. When Q is greater than about 2 or 3, the resonance is sharply peaked and our approximations for the width are valid. Then Q is simply the ratio of the resonance frequency to the width:

$$Q = \frac{L\omega_0}{R} = \frac{\omega_0}{\Delta\omega} \qquad \text{40-40} \qquad Q \; value$$

A common application of series resonance circuits is a radio receiver, where the resonance frequency of the circuit is varied by varying the capacitance. Resonance occurs when the natural frequency of the circuit equals the frequency of the radio waves picked up at the antenna. At resonance, there is a relatively large current in the antenna circuit. If the Q value of the circuit is sufficiently high, currents due to other stations off resonance will be negligible compared with those due to the station to which the circuit is tuned.

Example 40-3 Find the resonance width and Q value of the circuit of Example 40-2.

In that example we found the resonance frequency to be $\omega_0 = 500$ rad/sec. The Q value is then

$$Q = \frac{L\omega_0}{R} = \frac{2(500)}{20} = 50$$

and the width of the resonance is

$$\Delta\omega = \frac{\omega_0}{Q} = \frac{500}{50} = 10 \text{ rad/sec}$$

Questions

4. In general, does the power factor depend on frequency?

5. Can the instantaneous power delivered by the generator ever be negative?

6. What is the power factor for a circuit which has inductance or capacitance or both but zero resistance?

7. What is the power factor for a circuit with resistance but no inductance or capacitance? Does this power factor depend on frequency?

40-7 The Transformer

In order to transport power with a minimum I^2R heat loss in transmission lines it is economical to use a high voltage and low current. On the other hand, safety and other considerations, e.g., insulation, make it convenient to use power to run motors and other electrical appliances at lower voltage and higher current. This is accomplished by using a transformer, a device for changing ac voltage and current without appreciable loss in power. If V is the voltage and I the current, the instantaneous power is VI. If the voltage is to be changed without change in power, the current must also be changed.

Figure 40-11 diagrams a simple transformer consisting of two coils of wire around a common core of soft iron. The coil carrying the input power is called the *primary*, and the other coil is called the *secondary*. Either coil of a transformer can be used for the primary or secondary. The function of the iron core is to greatly increase the flux for a given current and to guide the flux so that nearly all the flux through one turn of one coil goes through each turn of both coils. The iron core is laminated to reduce eddy-current losses. Other possible losses are the I^2R losses in the coils, which can be reduced by using a low-resistance wire for the coils, and hysteresis losses in the core, which can be reduced by using soft iron. It is relatively easy to design a transformer for which power is transferred from the primary to the secondary with efficiency of 90 to 99 percent. We shall discuss only an ideal transformer, which has no losses.

Consider an ac generator of emf ε across the primary of N_1 turns with the secondary coil of N_2 turns open. Because of the iron core, there is a large flux through each coil even with a very small magnetizing current I_m in the primary circuit. We can neglect the resistance of the coil compared with the inductive reactance. The primary is then a simple circuit consisting of an ac generator and a pure inductance, as discussed in Section 40-4. The current and voltage in the primary are out of phase by 90°, and the average power in the primary circuit is zero. We can see from Faraday's law that the voltage across the secondary is just the turns ratio N_2/N_1 times the voltage across the primary. The induced emf V_1 in the primary circuit is

$$V_1 = -\frac{d\phi}{dt}$$

The total flux ϕ is the flux through each turn ϕ_{turn} times the number of turns N_1:

$$V_1 = -N_1 \frac{d\phi_{\text{turn}}}{dt} \qquad \text{40-41}$$

Assuming no flux leakage out of the iron core, the flux through each turn is the same for both coils. Thus the total flux through the secondary coil is $N_2\phi_{\text{turn}}$, and the voltage across the secondary coil is

$$V_2 = -N_2 \frac{d\phi_{\text{turn}}}{dt} \qquad \text{40-42}$$

Comparing these equations, we see that

$$V_2 = \frac{N_2}{N_1} V_1 \qquad \text{40-43}$$

According to Kirchhoff's law for the primary circuit, the applied emf ε is the negative of the induced emf V_1:

$$\varepsilon - \frac{d\phi}{dt} = 0$$

$$\varepsilon = \frac{d\phi}{dt} = -V_1 \qquad \text{40-44}$$

In terms of the applied emf, the voltage across the secondary coil is

$$V_2 = -\frac{N_2}{N_1} \varepsilon \qquad \text{40-45}$$

The transformer is called a *step-up transformer* if N_2 is greater than N_1 so that the output voltage is greater than the input voltage. If N_2 is

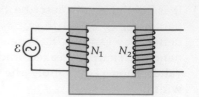

Figure 40-11
Transformer with N_1 turns in the primary and N_2 turns in the secondary.

less than N_1, it is a *step-down transformer.* There is no current in the secondary because that circuit is open. The very small current I_m in the primary coil is 90° out of phase with the emf in that coil.

Consider now what happens when we put a resistance R, called a *load resistance,* across the secondary coil. There will then be a current I_2 in the secondary circuit which is in phase with the voltage V_2 across the resistance. This current will set up an additional flux through each turn ϕ'_{turn} proportional to $N_2 I_2$. This flux adds to the original flux ϕ_{turn} set up by the original magnetizing current in the primary I_m. However, the voltage across the primary coil is determined by the generator emf, which is unaffected by the secondary circuit. According to Equation 40-44, the flux in the iron core must therefore change at the original rate; i.e., the total flux in the iron core must be the same as with no load across the secondary. The primary coil thus draws an additional current I_1 to maintain the original flux ϕ_{turn}. The flux through each turn produced by this additional current is proportional to $N_1 I_1$. Thus the additional current I_1 in the primary is related to the current I_2 in the secondary by

$$N_1 I_1 = -N_2 I_2 \qquad 40\text{-}46$$

The negative sign indicates that these currents are out of phase because they produce counteracting fluxes. Since I_2 is in phase with V_2, the additional current in the primary I_1 is in phase with the applied emf.

Figure 40-12 is a phasor diagram of the phase relationships between the voltages and currents. The total current in the primary I is the "vector" sum of the original magnetizing current I_m and the additional current I_1, which is usually very much greater than I_m. The power delivered by the generator is the product of the rms applied emf, the rms total current in the primary I_{rms}, and the power factor $\cos \phi$, where ϕ is the phase angle between the applied emf and the total current I. Since I_1 is in phase with the applied emf, ϕ is the angle between I_1 and I shown in Figure 40-12.

We note from this figure that $I \cos \phi$ is just the additional current I_1, and so the power input by the primary is

$$P = \mathcal{E}_{rms} I_{rms} \cos \phi = \mathcal{E}_{rms} I_{1,rms}$$

Using Equations 40-45 and 40-46 to relate the applied emf and the additional current I_1 in the primary to the voltage V_2 and the current I_2 in the secondary, we have

$$\mathcal{E}I_1 = \left(-\frac{N_1}{N_2} V_2\right)\left(-\frac{N_2}{N_1} I_1\right) = V_2 I_2 \qquad 40\text{-}47$$

Thus

$$\mathcal{E}_{rms} I_{1,rms} = V_{2,rms} I_{2,rms} \qquad 40\text{-}48$$

The power input to the primary equals the power output in the secondary, as assumed for an ideal transformer with no losses.

In most cases the additional current in the primary I_1 is much greater than the original magnetizing current I_m with no load. This can be demonstrated by putting a light bulb in series with the primary coil: it is much brighter when there is a load across the secondary than when the secondary circuit is open. If I_m can be neglected, Equation 40-46 relates the total currents in the primary and secondary circuit.

The currents I_1 and I_2 can be related to the resistance R across the

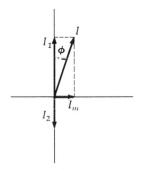

Figure 40-12
Phase relations between the magnetizing current I_m, the secondary current I_2, and the additional primary current I_1. The total primary current I is nearly equal to I_1 because I_m is usually small. Then the power factor, $\cos \phi$, is nearly 1.

secondary. For the secondary current we have simply

$$I_2 = \frac{V_2}{R}$$

Using Equations 40-45 and 40-46 to write I_2 and V_2 in terms of the primary current I_1 and the primary emf $\mathcal{E}$, we have

$$I_1 = \frac{\mathcal{E}}{(N_1/N_2)^2 R} \qquad\qquad 40\text{-}49$$

The current I_1 is the same as if we connected a resistance $(N_1/N_2)^2 R$ across the generator. This effect is called *impedance transformation* since in general some combination of capacitance, inductance, and resistance with impedance Z is connected across the secondary coil as the load.

Review

A. Define, explain, or otherwise identify:

Capacitive reactance, 930

Inductive reactance, 931

Resonance condition, 933

Impedance, 933

Phasor, 933

rms current, 935

Power factor, 936

Resonance width, 936

Q value, 937

Transformer, 937

Primary, 938

Secondary, 938

Step-up transformer, 938

Step-down transformer, 939

Impedance transformation, 940

B. True or false:

1. At very high frequencies a capacitor acts like a short circuit.

2. A circuit with a high Q value has a narrow resonance curve.

3. At resonance the impedance equals the resistance.

4. Off resonance, the current always lags the generator voltage.

5. At resonance, the current and generator voltage are in phase.

6. If a transformer increases the current, it must decrease the voltage.

Exercises

Section 40-1, An AC Generator

1. A coil has an area of 4 cm², has 200 turns, and rotates in a magnetic field of 5000 G. (*a*) What must its angular frequency be to generate an emf of 10 V maximum? (*b*) If it rotates at 60 Hz, what is $\mathcal{E}_{max}$?

2. In what magnetic field should the coil of Exercise 1 rotate to generate an emf of 10 V maximum at 60 Hz?

3. (*a*) Show that the rms emf of a generator is related to the maximum emf by $\mathcal{E}_{rms} = 0.707\mathcal{E}_{max}$. (*b*) The rms value of standard house voltage is 110 V. What is the maximum voltage?

Section 40-2, Alternating Current in a Resistor

4. A resistor R is placed in series with an ac generator of angular frequency ω. Find an expression for the instantaneous power I^2R as a function of time, and draw a sketch. Show that the average power is $I_{rms}^2 R$.

5. A 3-Ω resistor is placed in series with a 12.0-V (maximum) generator of frequency 60 Hz. (a) What is the angular frequency of the current? (b) Find I_{max} and I_{rms}. What is (c) the maximum power into the resistor, (d) the minimum power, and (e) the average power?

Section 40-3, Alternating Current in a Capacitor

6. What is the reactance of a 1.0-nF capacitor at (a) 60 Hz, (b) 6000 Hz, and (c) 6 MHz?

7. Find the reactance of a 10.0-μF capacitor at (a) 60 Hz, (b) 600 Hz, and (c) 6000 Hz.

8. Sketch a graph of X_C versus frequency f for $C = 100$ μF.

9. An emf of 10.0 V maximum and frequency 20 Hz is applied to a 20-μF capacitor. What is the maximum value of the current?

10. Find an expression for the instantaneous power $P = \mathcal{E}I$ delivered by a generator to a capacitor and sketch as a function of time. Show that the average power over one cycle is zero.

11. At what frequency is the reactance of a 10-μF capacitor (a) 1 Ω, (b) 100 Ω, and (c) 0.01 Ω?

Section 40-4, Alternating Current in an Inductor

12. Find an expression for the instantaneous power $P = \mathcal{E}I$ delivered by a generator to an inductor and sketch as a function of time. Show that the average power over one cycle is zero.

13. What is the reactance of a 1.0-mH inductor at (a) 60 Hz, (b) 600 Hz, and (c) 6000 Hz?

14. An inductor has a reactance of 100 Ω at 80 Hz. (a) What is its inductance? (b) What is its reactance at 160 Hz?

15. At what frequency would the reactance of a 10.0-μF capacitor equal that of a 1.0-mH inductor?

Section 40-5, LCR Circuit with Generator

16. Show that Equation 40-29 can be written $I_{max} = \omega \mathcal{E}_{max} / \sqrt{L^2(\omega_0{}^2 - \omega^2)^2 + \omega^2 R^2}$.

17. A series LCR circuit with $L = 10$ mH, $C = 2$ μF, and $R = 5$ Ω is driven by a generator of maximum emf 100 V and variable angular frequency ω. Find (a) the resonance frequency ω_0 and (b) I_{max} at resonance. When $\omega = 8000$ rad/sec, find (c) X_C and X_L, (d) Z and I_{max}, and (e) the phase angle ϕ.

18. (a) Show that Equation 40-28 can be written $\tan \phi = L(\omega_0{}^2 - \omega^2)/\omega R$. Find ϕ approximately (b) at very low frequencies and (c) at very high frequencies.

19. For the circuit of Exercise 17, let the generator frequency be $f = \omega/2\pi = 1000$ Hz. Find (a) the resonance frequency $f_0 = \omega_0/2\pi$, (b) X_C and X_L, (c) the total impedance Z and I_{max}, and (d) the phase angle ϕ.

20. A series LCR circuit in a radio receiver is tuned by a variable capacitor so that it can resonate at frequencies from 500 to 1600 kHz. If $L = 1.0$ μH, find the range of C necessary to cover this range of frequencies.

Section 40-6, Power in AC Circuits

21. FM radio stations have carrier frequencies which are separated by 0.20 MHz. When the radio is tuned to a station such as 100.1 MHz, the resonance width of the receiver circuit should be much smaller than 0.2 MHz so that ad-

jacent stations are not received. If $f_0 = 100.1$ MHz and $\Delta f = 0.01$ MHz, what is the Q value of the circuit?

22. (a) Find the power factor for the circuit in Example 40-2 when $\omega = 400$ rad/sec. (b) At what angular frequency is the power factor $\frac{1}{2}$?

23. Find (a) the Q value and (b) the resonance width for the circuit of Exercise 17. (c) What is the power factor when $\omega = 8000$ rad/sec?

24. Show that when there is resistance and capacitance but no inductance, the power factor is given by $\cos \phi = RC\omega / \sqrt{1 + (RC\omega)^2}$. Sketch the power factor versus ω.

25. An ac generator of maximum emf 20 V is connected in series with a 20-μF capacitor and an 80-Ω resistor. There is no inductance in the circuit. Find (a) the power factor, (b) the rms current, and (c) the average power if the angular frequency of the generator is 400 rad/sec.

26. Express the power factor as a function of ω and sketch for a circuit with resistance and inductance in series with a generator.

27. A coil can be considered to be a resistance and inductance in series. Assume that $R = 100$ Ω and $L = 0.4$ H. The coil is connected across a 110-V, 60-Hz line. Find (a) the power factor, (b) the rms current, and (c) the average power supplied.

28. Find the power factor and the phase ϕ for the circuit of Exercise 17 when the generator frequency is (a) 900 Hz, (b) 1100 Hz, and (c) 1300 Hz.

Section 40-7, The Transformer

29. A transformer has 400 turns on the primary and 8 turns on the secondary. (a) Is this a step-up or step-down transformer? (b) If the primary is connected across 110 V, what is the open-circuit voltage across the secondary? (c) If the primary current is 0.1 A, what is the secondary current assuming negligible magnetizing current and no power loss?

Problems

1. Show by direct substitution that the current given by Equation 40-27 with I_{max} and ϕ given by Equations 40-28 and 40-29 satisfies Equation 40-26. *Hint:* Use trigonometric identities for the sine and cosine of the sum of two angles and write the equation in the form $A \sin \omega t + B \cos \omega t = 0$. Since this equation must hold for all time, $A = 0$ and $B = 0$.

2. A coil with resistance and inductance is connected to a 120-V rms, 60-Hz line. The average power supplied to the coil is 60 W, and the rms current is 1.5 A. Find (a) the power factor, (b) the resistance of the coil, and (c) the inductance of the coil. (d) Does the current lead or lag the voltage? What is the phase angle ϕ?

3. In a certain LCR circuit $X_C = 16$ Ω and $X_L = 4$ Ω at some frequency. The resonance frequency is $\omega_0 = 10^4$ rad/sec. (a) Find L and C. If $R = 5$ Ω and $\mathcal{E}_{max} = 26$ V, find (b) the Q value and (c) the maximum current.

4. In a series LCR circuit connected to an ac generator whose maximum emf is 200 V, the resistance is 60 Ω and the capacitance is 8.0 μF. The inductor can be varied from 8.0 to 40.0 mH by insertion of an iron core into a solenoid. The angular frequency is 2500 rad/sec. If the capacitor voltage is not to exceed 150 V, find (a) the maximum current and (b) the range of L that is safe to use.

5. Sketch the impedance Z versus ω for (a) a series LR circuit, (b) a series RC circuit, and (c) a series LCR circuit.

6. (*a*) Show that Equation 40-28 can be written $\tan \phi = Q(\omega_0{}^2 - \omega^2)/\omega\omega_0$. (*b*) Show that near resonance, $\tan \phi \approx 2Q(\omega_0 - \omega)/\omega$. (*c*) Sketch ϕ versus x where $x = \omega/\omega_0$ for a circuit with high Q and for one with low Q.

7. A resistor and capacitor are connected *in parallel* across a sinusoidal emf $\mathcal{E} = \mathcal{E}_{max} \sin \omega t$, as in Figure 40-13. (*a*) Show that the current in the resistor is $I_R = (\mathcal{E}_{max}/R) \sin \omega t$. (*b*) Show that the current through the capacitor branch is $I_C = (\mathcal{E}_{max}/X_C) \sin (\omega t + 90°)$. (*c*) Show that the total current is given by $I = I_R + I_C = I_{max} \sin (\omega t + \phi)$, where $\tan \phi = R/X_C$ and $I_{max} = \mathcal{E}_{max}/Z$ with $Z^{-2} = R^{-2} + X_C{}^{-2}$.

8. A resistor and inductor are *in parallel* across an emf $\mathcal{E} = \mathcal{E}_{max} \sin \omega t$ as in Figure 40-14. Show that (*a*) $I_R = (\mathcal{E}_{max}/R) \sin \omega t$, (*b*) $I_L = (\mathcal{E}_{max}/X_L) \sin (\omega t - 90°)$, and (*c*) $I = I_R + I_L = I_{max} \sin (\omega t - \phi)$, where $\tan \phi = R/X_L$ and $I_{max} = \mathcal{E}_{max}/Z$ with $Z^{-2} = R^{-2} + X_L{}^{-2}$.

9. A resistor, inductor, and capacitor are *in parallel* across an ac generator $\mathcal{E} = \mathcal{E}_{max} \sin \omega t$ as shown in Figure 40-15. Show that the total current in the generator is given by $I = I_{max} \sin (\omega t + \phi)$, where $I_{max} = \mathcal{E}_{max}/Z$ with $Z^{-2} = R^{-2} + (1/X_C - 1/X_L)^2$ and $\tan \phi = R(1/X_C - 1/X_L)$.

10. Find the power factor for the parallel circuit of Problem 9 and show that $\phi = 0$ when $X_C = X_L$.

11. When an *LRC* series circuit is connected to a 110-V rms, 60-Hz line the current is $I_{rms} = 11.0$ A and the current leads the emf by 45°. (*a*) Find the power supplied to the circuit. (*b*) What is the resistance? (*c*) If the inductance $L = 0.05$ H, find the capacitance C. (*d*) What capacitance or inductance would you add to make the power factor 1?

12. A method for measuring inductance is to connect the inductor in series with a known capacitance, a resistor, an ac ammeter, and a variable-frequency signal generator. The frequency of the signal generator is varied and the emf kept constant until the current is maximum. If $C = 10$ μF, $\mathcal{E}_{max} = 10$ V, $R = 100$ Ω, and I is maximum at $\omega = 5000$ rad/sec, what is L? What is I_{max}?

13. A certain electrical device draws 10 A rms and has an average power of 720 W when connected to a 120-V rms, 60-Hz power line. (*a*) What is the impedance of the device? (*b*) What series combination of resistance and reactance is this device equivalent to? (*c*) If the current leads the emf, is the reactance inductive or capacitive?

Figure 40-13
Parallel *RC* circuit and ac generator for Problem 7.

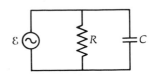

Figure 40-14
Parallel *RL* circuit and ac generator for Problem 8.

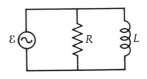

Figure 40-15
Parallel *LCR* circuit and ac generator for Problem 9.

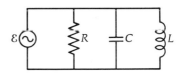

CHAPTER 41
Maxwell's Equations and Electromagnetic Waves

The experimental laws of electricity and magnetism studied in Chapters 29 to 40 are conveniently summarized in a set of equations known as *Maxwell's equations* (after James Clerk Maxwell). These equations relate the electric and magnetic field vectors **E** and **B** to their sources, which are electric charges, currents, and changing fields. Maxwell's equations play a role in classical electromagnetism analogous to that of Newton's laws in classical mechanics. In principle, all problems in electricity and magnetism can be solved from Maxwell's equations, just as all problems in classical mechanics can be solved from Newton's laws. However, Maxwell's equations are considerably more complicated than Newton's laws, and their application to the solution of most problems is beyond the scope of this book. In Section 41-1 we shall state Maxwell's equations and relate each one to the laws of electricity and magnetism already studied. We shall then show in Section 41-2 that Maxwell's equations imply that the electric and magnetic field vectors obey a wave equation which describes waves propagating through free space with speed $c = 1/\sqrt{\mu_0 \epsilon_0}$.

James Clerk Maxwell (1831–1879). (*Courtesy of Trinity College Library, Cambridge University.*)

41-1 Maxwell's Equations

Maxwell's equations are

$$\oint_S \mathbf{E} \cdot \hat{\mathbf{n}} \, dA = \frac{1}{\epsilon_0} Q \qquad\qquad 41\text{-}1$$

$$\oint_S \mathbf{B} \cdot \hat{\mathbf{n}} \, dA = 0 \qquad\qquad 41\text{-}2$$

$$\oint_C \mathbf{E} \cdot d\mathbf{l} = -\frac{d}{dt} \int_S \mathbf{B} \cdot \hat{\mathbf{n}} \, dA \qquad\qquad 41\text{-}3$$

$$\oint_C \mathbf{B} \cdot d\mathbf{l} = \mu_0 I + \mu_0 \epsilon_0 \frac{d}{dt} \int_S \mathbf{E} \cdot \hat{\mathbf{n}} \, dA \qquad\qquad 41\text{-}4$$

Equation 41-1 is Gauss' law; it states that the flux of the electric field through any closed surface equals $1/\epsilon_0$ times the net charge inside the surface. As discussed in Chapter 29, Gauss' law implies that the electric field due to a point charge varies inversely as the square of the distance from the charge. This law describes how the lines of **E** diverge from a positive charge and converge on a negative charge. Its experimental basis is Coulomb's law.

Equation 41-2, sometimes called Gauss' law for magnetism, states that the flux of the magnetic induction vector **B** is zero through any closed surface. This equation describes the experimental observation that the lines of **B** do not diverge from any point in space or converge to any point; i.e., isolated magnetic poles do not exist.

Equation 41-3 is Faraday's law. The integral of the electric field around any closed curve C is the emf. This emf equals the rate of change of the magnetic flux through any surface S bounded by the curve. (This is not a closed surface, and so the magnetic flux through S is not necessarily zero.) Faraday's law describes how the lines of **E** encircle an area through which the magnetic flux is changing. Faraday's law relates the electric field vector **E** to the magnetic field vector **B**.

Equation 41-4, Ampère's law with Maxwell's displacement-current modification, states that the line integral of the magnetic induction **B** around any closed curve C equals μ_0 times the current through any surface bounded by the curve plus $\mu_0\epsilon_0$ times the rate of change of the electric flux through such a surface. This law describes how the magnetic-induction lines encircle an area through which a current is passing or the electric flux is changing.

The charges and currents in these equations include the bound charges which occur on dielectrics and the magnetization or atomic currents which occur in magnetized materials. To be useful when such materials are present, Equations 41-4 and 41-1 are rewritten in terms of free charge and conduction currents (plus displacement currents) only. This is done by substituting **H** for **B** in Equation 41-4 and omitting the μ_0. Similarly, Equation 41-1 can be written in terms of free charge only by defining a new vector **D**, called the *displacement vector*, which is related to **E** and the electric polarization of dielectrics analogously to the way **B**, **H**, and the magnetic polarization **M** are related. We shall not consider the complications introduced by polarized materials and need not be concerned with such refinements.

41-2 The Wave Equation for Electromagnetic Waves

In Section 21-8 we showed that the harmonic wave functions for waves on a string and for sound waves obey a partial differential equation called the *wave equation:*

$$\frac{\partial^2 y(x,t)}{\partial x^2} = \frac{1}{v^2}\frac{\partial^2 y(x,t)}{\partial t^2} \qquad\qquad 41\text{-}5$$

In this equation, $y(x,t)$ is the wave function, which for string waves is the displacement of the string and for sound waves can be either the pressure change or the displacement of air particles from equilibrium. The derivatives are partial derivatives because the wave function depends on both x and t. The quantity v is the velocity of the wave, which depends on the medium and on the frequency if the medium is dispersive. We showed in Section 21-8 that the wave equation for string waves can be derived by applying Newton's laws of motion to a string under tension, and we found that the velocity of the wave is $\sqrt{T/\mu}$, where T is the tension and μ the linear mass density.

The solutions of this equation, studied in Chapter 21, were harmonic wave functions of the form

$$y = y_0 \sin (kx - \omega t)$$

where $k = 2\pi/\lambda$ is the wave number and $\omega = 2\pi f$ is the angular frequency.

In this section we shall use Maxwell's equations to derive the wave equation for electromagnetic waves. We shall not consider how such waves arise from the motion of charges but merely show that the laws of electricity and magnetism imply the existence of a wave equation which implies the existence of electric and magnetic fields $\mathbf{E}$ and $\mathbf{B}$ propagating through space with the velocity of light c. We shall consider only free space, in which there are no charges or currents. We shall also assume that the electric and magnetic fields $\mathbf{E}$ and $\mathbf{B}$ are functions of time and one space coordinate only, which we shall take to be the x coordinate. This assumption is equivalent to the assumption of plane waves.

We can use Equations 41-1 and 41-2 to show that the electric and magnetic fields associated with electromagnetic waves must be transverse to the direction of propagation. For the situation we are considering, i.e., free space with no charge density and no current, Equations 41-1 and 41-2 state that the net flux of either $\mathbf{E}$ or $\mathbf{B}$ through any closed surface is zero. Let us consider the flux of $\mathbf{E}$ through a cube of sides Δx, Δy, and Δz, as shown in Figure 41-1. We are assuming that $\mathbf{E}$ depends only on x, the direction of propagation. Consider first the top and bottom faces of the cube of area $\Delta x\, \Delta z$. Flux through their faces depends on E_y. Since, by our assumption, E_y does not depend on y, the flux of $\mathbf{E}$ into the cube through the bottom face must equal the flux out through the top face. Similar reasoning leads to the result that there can be no contribution to the net flux through the faces of area $\Delta x\, \Delta y$ because E_z does not depend on z. The only possible contribution is through the faces of area $\Delta y\, \Delta z$. The flux out of the right face is $E_{xR}\, \Delta y\, \Delta z$, where E_{xR} is the value of E_x at the right face. The flux in through the left face is $E_{xL}\, \Delta y\, \Delta z$, where E_{xL} is E_x at the left face. The net flux out of the entire cube is thus

$$\Delta E_x\, \Delta y\, \Delta z$$

where ΔE_x is the change in E_x in the distance Δx. Since this flux is zero by Equation 41-1, we find that ΔE_x must be zero. That is, E_x does not

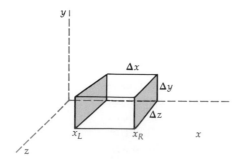

Figure 41-1
Volume element with sides parallel to the coordinate planes. If $\mathbf{E}$ depends only on x and t, there is no contribution to the net flux of $\mathbf{E}$ from the top and bottom or from the front and back sides. If the net flux of $\mathbf{E}$ out of the volume is zero, E_x must have the same value at x_R as at x_L.

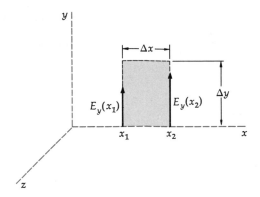

Figure 41-2
Rectangular curve in the xy
plane for derivation of Equa-
tion 41-6. Evaluating $\oint \mathbf{E} \cdot d\mathbf{l}$
for this path gives

$$E_y(x_2) \, \Delta y - E_y(x_1) \, \Delta y$$
$$\approx \frac{\partial E_y}{\partial x} \, \Delta x \, \Delta y$$

The rate of change of the
magnetic flux through this
curve is $\partial B_z/\partial t \, \Delta x \, \Delta y$.

depend on x. If the electric field has any component at all in the x
direction, it must be constant in space. Such a constant field is unrelated to electromagnetic waves and can be ignored for this discussion.
Similar reasoning applied to the flux of $\mathbf{B}$ through such a cube leads to
the conclusion that B_x must be constant in space. Thus the part of the
electric or magnetic fields which varies in time and space must be perpendicular to the x direction, the direction of propagation.

We next consider Faraday's law (Equation 41-3) applied to the curve
of sides Δx and Δy lying in the xy plane, as shown in Figure 41-2.
Assuming that Δx and Δy are very small, the line integral of $\mathbf{E}$ around
this curve is approximately

$$\oint \mathbf{E} \cdot d\mathbf{l} = E_y(x_2) \, \Delta y - E_y(x_1) \, \Delta y = [E_y(x_2) - E_y(x_1)] \, \Delta y$$

The contributions of the type $E_x \, \Delta x$ from the top and bottom of this
curve cancel because we have assumed that $\mathbf{E}$ does not depend on y
(or z). Assuming that Δx is very small, we can approximate the difference in E_y at the points x_1 and x_2 by

$$E_y(x_2) - E_y(x_1) = \Delta E_y \approx \frac{\partial E_y}{\partial x} \, \Delta x$$

Then

$$\oint \mathbf{E} \cdot d\mathbf{l} \approx \frac{\partial E_y}{\partial x} \, \Delta x \, \Delta y$$

The flux of the magnetic induction through this curve is approximately

$$\oint \mathbf{B} \cdot \hat{\mathbf{n}} \, dA = B_z \, \Delta x \, \Delta y$$

Faraday's law then gives

$$\frac{\partial E_y}{\partial x} \, \Delta x \, \Delta y = - \frac{\partial B_z}{\partial t} \, \Delta x \, \Delta y$$

or

$$\frac{\partial E_y}{\partial x} = - \frac{\partial B_z}{\partial t} \qquad\qquad 41\text{-}6$$

Equation 41-6 implies that if there is a component of electric field E_y
which depends on x, there must be a component of magnetic induction B_z which depends on time or, conversely, if there exists a magnetic induction field B_z which depends on time, there must be an electric field E_y which depends on x. We can get a similar equation
relating E_y to B_z by applying Equation 41-4 to the curve of sides Δx
and Δz in the xz plane, as shown in Figure 41-3. For the case with no

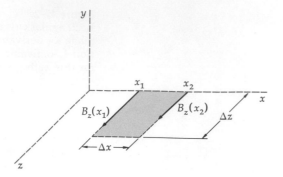

Figure 41-3
Rectangular curve in the xz plane for the derivation of Equation 41-7. Evaluating $\oint \mathbf{B} \cdot d\mathbf{l}$ for this path gives

$$B_z(x_1)\, \Delta z - B_z(x_2)\, \Delta z$$
$$\approx -\frac{\partial B_z}{\partial x}\, \Delta x\, \Delta z$$

The rate of change of the electric flux through this curve is $\partial E_y/\partial t\, \Delta x\, \Delta z$.

currents Equation 41-4 is

$$\oint_C \mathbf{B} \cdot d\mathbf{l} = \mu_0 \epsilon_0 \frac{d}{dt} \int_S \mathbf{E} \cdot \hat{\mathbf{n}}\, dA$$

We omit the details of this calculation, which is similar to that already done; the result is

$$\frac{\partial B_z}{\partial x} = -\mu_0 \epsilon_0 \frac{\partial E_y}{\partial t} \tag{41-7}$$

which relates the space variation of the magnetic induction B_z and the time variation of the electric field E_y.

We can eliminate either B_z or E_y from Equations 41-6 and 41-7 by differentiating either equation with respect to x or t. If we differentiate both sides of Equation 41-6 with respect to x, we obtain

$$\frac{\partial}{\partial x}\left(\frac{\partial E_y}{\partial x}\right) = -\frac{\partial}{\partial x}\left(\frac{\partial B_t}{\partial t}\right) \quad \text{or} \quad \frac{\partial^2 E_y}{\partial x^2} = -\frac{\partial}{\partial t}\left(\frac{\partial B_z}{\partial x}\right)$$

where we have interchanged the order of the time and space derivatives on the right side. We now use Equation 41-7 for $\partial B_z/\partial x$:

$$\frac{\partial^2 E_y}{\partial x^2} = -\frac{\partial}{\partial t}\left(-\mu_0 \epsilon_0 \frac{\partial E_y}{\partial t}\right) = +\mu_0 \epsilon_0 \frac{\partial^2 E_y}{\partial t^2} \tag{41-8}$$

Wave equation for E_y

Comparing this equation with Equation 41-5, we see that E_y obeys a wave equation for waves with speed

$$v = \frac{1}{\sqrt{\mu_0 \epsilon_0}} \tag{41-9}$$

If we had instead chosen to eliminate E_y from Equations 41-6 and 41-7 (by differentiating Equation 41-6 with respect to t, for example), we would have obtained an equation identical to Equation 41-8 with B_z replacing E_y. We have thus shown that both the electric field E_y and the magnetic field B_z obey a wave equation for waves traveling with the velocity $1/\sqrt{\mu_0 \epsilon_0}$, which is the velocity of light. As we noted in discussing harmonic waves, a particularly important solution to Equation 41-8 is the harmonic wave function of the form

$$E_y = E_{y0} \sin(kx - \omega t) \tag{41-10}$$

E and B are in phase

If we substitute this solution into either Equation 41-6 or 41-7, we can show that the magnetic induction B_z is in phase with the electric field E_y. From Equation 41-6 we have

$$\frac{\partial E_y}{\partial x} = kE_{y0} \cos(kx - \omega t) = -\frac{\partial B_z}{\partial t}$$

Solving for B_z gives

$$B_z = \frac{k}{\omega} E_{y0} \sin (kx - \omega t) = B_{z0} \sin (kx - \omega t) \qquad 41\text{-}11$$

where

$$B_{z0} = \frac{k}{\omega} E_{y0} = \frac{E_{y0}}{c} \qquad 41\text{-}12$$

and $c = \omega/k$ is the velocity of the wave.[1]

We have shown that Maxwell's equations imply the wave equation 41-8 for the electric field component E_y and the magnetic induction component B_z and that if E_y varies harmonically, as in Equation 41-10, the magnetic induction B_z is in phase with E_y and has an amplitude related to the amplitude of E_y by Equation 41-12. The electric and magnetic fields are perpendicular to each other and to the direction of the wave propagation, the x direction here. We would have obtained similar results for the components E_z and B_y by applying Faraday's law (Equation 41-3) to the rectangular curve in the xz plane. In either case, the electric field and magnetic induction are perpendicular to the x axis, the direction of propagation. As we have shown, electromagnetic waves are transverse waves. For simplicity, we shall assume that the electric field is in the y direction. This amounts to a choice of one particular polarization for the transverse wave. Having made this choice, we are *not* free to choose the direction for the magnetic induction. If the electric field is in the y direction, the magnetic induction must be in the z direction, according to Equations 41-6 and 41-7, which were derived from Maxwell's equations. Also, if E_y is the harmonic wave function given by Equation 41-10, we have shown that B_z is also a harmonic wave function in phase with E_y and related in amplitude by Equation 41-12 (Figure 41-4).

From Figure 41-4 we note that the direction of propagation (the x direction in this case) is parallel to the vector $\mathbf{E} \times \mathbf{B}$. The vector $\mathbf{E} \times \mathbf{B}/\mu_0 = \mathbf{E} \times \mathbf{H}$ is known as the *Poynting vector* (after John H. Poynting):

$$\mathbf{S} = \frac{\mathbf{E} \times \mathbf{B}}{\mu_0} = \mathbf{E} \times \mathbf{H} \qquad 41\text{-}13$$

We shall now show that its average magnitude equals the intensity of

---

[1] In obtaining Equation 41-11 by integration from the previous equation an arbitrary constant of integration arises, but we have omitted this constant magnetic induction field from Equation 41-11 because it plays no part in the electromagnetic waves we are interested in. Note that if any constant electric field is added to Equation 41-10, the new function still satisfies the wave equation.

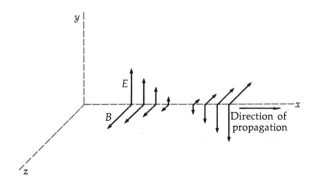

Figure 41-4
Electric and magnetic field vectors in a plane electromagnetic wave at one instant in time. $\mathbf{E}$ and $\mathbf{B}$ are in phase. The direction of propagation of the wave is the direction of $\mathbf{E} \times \mathbf{B}$, which in this case is the x direction.

the electromagnetic wave. In our discussion of the transport of energy by waves of any kind, we showed that the intensity (average energy per unit time per unit area) is in general equal to the product of the average energy density (energy per unit volume) and the speed of the wave. The energy per unit volume associated with an electric field $\mathbf{E}$ is

$$\eta_e = \tfrac{1}{2}\epsilon_0 E^2$$

For the harmonic wave given by Equation 41-10 we have

$$\eta_e = \tfrac{1}{2}\epsilon_0 E_0^2 \sin^2 (kx - \omega t)$$

For simplicity in notation, we shall drop the y- and z-component subscripts. The energy per unit volume associated with a magnetic induction field $\mathbf{B}$ is

$$\eta_m = \frac{B^2}{2\mu_0}$$

From Equations 41-11 and 41-12 we have

$$\eta_m = \frac{B_0^2}{2\mu_0} \sin^2 (kx - \omega t) = \frac{E_0^2}{2\mu_0 c^2} \sin^2 (kx - \omega t)$$

But since $c^2 = 1/\epsilon_0\mu_0$,

$$\eta_m = \tfrac{1}{2}\epsilon_0 E_0^2 \sin^2 (kx - \omega t)$$

and we see that the magnetic and electric energy densities are equal. The total electromagnetic energy per unit volume is

$$\eta = \eta_e + \eta_m = \epsilon_0 E_0^2 \sin^2 (kx - \omega t) = \frac{E_0^2}{\mu_0 c^2} \sin^2 (kx - \omega t) \qquad 41\text{-}14$$

The energy per unit time per unit area is the product of this energy density and the speed c:

$$\eta c = \frac{E_0^2}{\mu_0 c} \sin^2 (kx - \omega t)$$

$$= [E_0 \sin (kx - \omega t)] \left[\frac{E_0}{\mu_0 c} \sin (kx - \omega t) \right]$$

$$= E \frac{B}{\mu_0} = EH = |\mathbf{S}|$$

The intensity at any point x is the time average of the energy per unit time per unit area. From Equation 41-14 we see that the energy density is proportional to $\sin^2 (kx - \omega t)$. The average of this quantity over one or more cycles is $\tfrac{1}{2}$. The average energy density is then

$$\eta_{av} = \frac{E_0^2}{2\mu_0 c^2}$$

and the intensity is

$$I = c\eta_{av} = \frac{E_0^2}{2\mu_0 c} = \frac{E_0}{\sqrt{2}} \frac{B_0}{\mu_0 \sqrt{2}} = E_{rms} H_{rms} = |\mathbf{S}|_{av} \qquad 41\text{-}15$$

where $E_{rms} = E_0/\sqrt{2}$ is the rms value of E and $H_{rms} = B_{rms}/\mu_0 = B_0/\mu_0 \sqrt{2}$ is the rms value of H.

The magnitude of the Poynting vector is the instantaneous power per unit area. The rate at which electromagnetic energy flows through any area is the flux of the Poynting vector through that area:

$$P = \int \mathbf{S} \cdot \hat{n} \, dA = \int (\mathbf{E} \times \mathbf{H}) \cdot \hat{n} \, dA$$

where P is the power through the area and $\mathbf{S} = \mathbf{E} \times \mathbf{H}$ is the Poynting vector. The net power radiated out of any volume is the net flux of the Poynting vector through the surface enclosing the volume.

Review

Write the Maxwell equation which corresponds most closely to the law stated:

1. Faraday's law
2. Ampere's law
3. Coulomb's law
4. Gauss' law
5. Gauss' law for magnetism

Exercises

Section 41-1, Maxwell's Equations

There are no exercises for this section.

Section 41-2, The Wave Equation for Electromagnetic Waves

1. (a) Show that Equation 41-10 can be written $E_y = E_0 \sin k(x - ct)$, where $c = \omega/k$. In which direction is this wave traveling? (b) Substitute this function into Equation 41-8 and show that this equation is satisfied if $c = 1/\sqrt{\mu_0 \epsilon_0}$. (c) Use the known values of μ_0 and ϵ_0 in SI units to show that $1/\sqrt{\mu_0 \epsilon_0}$ is approximately 3×10^8 m/sec.

2. (a) Show that the rms electric field in a wave is related to the intensity I by $E_{rms} = \sqrt{\mu_0 c I}$. (b) Show that $B_{rms} = E_{rms}/c$.

3. An electromagnetic wave has intensity $I = 100$ W/m². Find E_{rms} and B_{rms} (see Exercise 2).

4. Show that the Poynting vector $\mathbf{S} = \mathbf{E} \times \mathbf{H}$ has units of watts per square meter.

5. The amplitude of an electromagnetic wave is $E_0 = 400$ V/m. Find (a) B_0, (b) the average total electromagnetic energy density, and (c) the intensity.

Problems

1. A laser beam has a diameter of 1.0 mm and average power of 1.5 mW. Find the intensity of the beam and E_{rms} and B_{rms}.

2. (a) Derive Equation 41-7. (b) Eliminate E_y from Equations 41-6 and 41-7 to derive the wave equation for B_z.

3. A point source emits radiation uniformly in all directions. (a) Show that if P_{av} is its average power, the intensity at a distance r is $I = P_{av}/4\pi r^2$. (b) If the intensity of sunlight striking the upper atmosphere of the earth is 1400 W/m², find the average power output of the sun. (c) Find E_{rms} and B_{rms} due to the sun at the upper atmosphere of the earth.

4. (a) Using arguments similar to those given in the text, show that

$$\frac{\partial E_z}{\partial x} = \frac{\partial B_y}{\partial t} \quad \text{and} \quad \frac{\partial B_y}{\partial x} = \mu_0 \epsilon_0 \frac{\partial E_z}{\partial t}$$

(b) Show that E_z and B_y also satisfy the wave equation.

5. Assume that a 100-W light bulb is a point source radiating uniformly in all directions (see Problem 3). (a) Find the intensity at 1.0 m from the bulb. (b) Assuming plane waves, find E_{rms} and B_{rms} at this distance.

6. A long cylindrical conductor of radius a and resistivity ρ carries a steady current I uniformly distributed over its cross-sectional area. (*a*) Use Ohm's law to relate the electric field $\mathbf{E}$ in the conductor to I, ρ, and a. (*b*) Find the magnetic field $\mathbf{B}$ just outside the conductor. (*c*) At $r = a$ (the edge of the conductor) use the results of parts (*a*) and (*b*) to compute the Poynting vector $\mathbf{S} = \mathbf{E} \times \mathbf{H} = (1/\mu_0)\, \mathbf{E} \times \mathbf{B}$. In what direction is $\mathbf{S}$? (*d*) Find the flux of $\mathbf{S}$ through the surface of the wire of length L and area $2\pi a L$ and show that the rate of energy flow into the wire equals $I^2 R$, where R is the resistance.

7. Discuss how Maxwell's equations might have to be modified in a region of space containing magnetic monopoles. (A magnetic monopole is a particle carrying a single "magnetic charge" q^*, i.e., an isolated magnetic pole. Its existence was predicted by P.A.M. Dirac in 1931. The discovery of one such particle was reported in 1975, but the evidence is not conclusive.)

CHAPTER 42 Quantization

In Chapter 41 we showed that a wave equation for electromagnetic waves traveling at the speed of light can be derived from Maxwell's equations, which summarize the experimental laws of electricity and magnetism. This unification of the previously separate subjects of optics and electromagnetism through Maxwell's equations was one of the great triumphs of classical physics in the nineteenth century. Some other predictions of electromagnetism which we did not study are also important for understanding the modifications of classical physics in the twentieth century:

1. Light and other electromagnetic waves carry momentum as well as energy. The momentum p and the energy E are related by $p = E/c$, where c is the speed of the waves.

2. If a charged particle has acceleration $\mathbf{a}$, it radiates electromagnetic energy at a rate proportional to a^2.

3. If the acceleration of the charged particle is associated with periodic motion, e.g., simple harmonic oscillation or circular motion, the particle emits electromagnetic radiation with a frequency equal to that of the motion.

All these predictions have been verified in experiments involving macroscopic phenomena. Near the end of the nineteenth century it was generally believed that all natural phenomena could be described by Newton's laws, the laws of thermodynamics, and the laws of electromagnetism (as expressed in Maxwell's equations). There was nothing left for scientists to do but apply these laws to various phenomena and to measure the next decimal in the fundamental constants (the speed of light, the electron charge, the gas constant, Avogadro's number, etc.). However, this optimism (or pessimism if you were a scientist hoping to make important discoveries) proved premature. We have already described in Chapter 28 how newtonian mechanics must be replaced with the special theory of relativity when the speed of a particle is not small compared with the speed of light, and we showed that when the particle's speed is small, the special theory of relativity reduces to Newton's laws. Many startling discoveries, both experimental and theoretical, in the last 20 years of the nineteenth century and first 30 years of the twentieth century showed that the laws of classical physics also break down when applied to microscopic systems such as the particles within an atom. This failure is even more drastic than the failure of newtonian mechanics at high speeds. The interior of the atom can be described only in terms of quantum theory, which requires the modification of some of our fundamental ideas about the relationships between physical theory and the physical world. As with special relativity, quantum theory reduces to classical physics when applied to macroscopic systems. Table 42-1 lists the approximate dates of some of the important experiments and theories from 1881 to 1932.

A proper study of quantum theory is beyond the scope of this book. In this chapter we shall concentrate on the fundamental ideas of energy quantization and the wave-particle duality of nature.

Table 42-1
Approximate dates of some important experiments and theories, 1881–1932

1881	Michelson obtains null result for absolute velocity of earth
1884	Balmer finds empirical formula for spectral lines of hydrogen
1887	Hertz produces electromagnetic waves, verifying Maxwell's theory and accidently discovering photoelectric effect
1887	Michelson repeats his experiment with Morley, again obtaining null result
1895	Röntgen discovers x-rays
1896	Becquerel discovers nuclear radioactivity
1897	J. J. Thomson measures e/m for cathode rays, showing that electrons are fundamental constituents of atoms
1900	Planck explains blackbody radiation using energy quantization involving new constant h
1900	Lenard investigates photoelectric effect and finds energy of electrons independent of light intensity
1905	Einstein proposes the special theory of relativity
1905	Einstein explains photoelectric effect by suggesting quantization of radiation
1907	Einstein applies energy quantization to explain temperature dependence of heat capacities of solids
1908	Rydberg and Ritz generalize Balmer's formula to fit spectra of many elements
1909	Millikan's oil-drop experiment shows quantization of electric charge
1911	Rutherford proposes nuclear model of atom based on alpha-particle scattering experiments of Geiger and Marsden
1912	Friedrich and Knipping and von Laue demonstrate diffraction of x-rays by crystals showing that x-rays are waves and crystals are regular arrays

42-1 The Origin of the Quantum Constant: Blackbody Radiation

One of the most puzzling phenomena studied near the end of the nineteenth century was the spectral distribution of blackbody radiation. A blackbody is an ideal system which absorbs all the radiation incident on it; it can be approximated by a cavity with a very small opening. The characteristics of the radiation in such a cavity of a body in thermal equilibrium depend only on the temperature of the walls. The fraction of the radiant energy density of wavelength λ in the interval $d\lambda$ is called the *spectral distribution* $f(\lambda,T)\ d\lambda$. The function $f(\lambda,T)$ was calculated from classical physics in a straightforward way and the result compared with experiment. The calculation involved finding the number of standing waves in a three-dimensional cavity in the wavelength interval $d\lambda$ and multiplying by the average energy per wave, which the equipartition theorem predicts to be kT. The result, known

Table 42-1
(*continued*)

1913	Bohr proposes model of hydrogen atom
1914	Moseley analyzes x-ray spectra using Bohr model to explain periodic table in terms of atomic number
1914	Franck and Hertz demonstrate atomic energy quantization
1915	Duane and Hunt show that the short-wavelength limit of x-rays is determined from quantum theory
1916	Wilson and Sommerfeld propose rules for quantization of periodic systems
1916	Millikan verifies Einstein's photoelectric equation
1923	Compton explains x-ray scattering by electrons as collision of photon and electron and verifies results experimentally
1924	De Broglie proposes electron waves of wavelength h/p
1925	Schrödinger develops mathematics of electron wave mechanics
1925	Heisenberg invents matrix mechanics
1925	Pauli states exclusion principle
1927	Heisenberg formulates uncertainty principle
1927	Davisson and Germer observe electron wave diffraction by single crystal
1927	G. P. Thomson observes electron wave diffraction in metal foil
1928	Gamow and Condon and Gurney apply quantum mechanics to explain alpha-decay lifetimes
1928	Dirac develops relativistic quantum mechanics and predicts existence of positron
1932	Chadwick discovers neutron
1932	Anderson discovers positron

as the *Rayleigh-Jeans law*, is

$$f(\lambda,T) = 8\pi kT\lambda^{-4}$$

42-1   *Rayleigh-Jeans law*

This result agrees with experiment in the region of long wavelengths but disagrees violently at short wavelengths. As λ approaches zero, the experimentally determined $f(\lambda,T)$ also approaches zero, but the calculated function becomes infinite as λ^{-4}. This result was known as the *ultraviolet catastrophe*. In 1900 the German physicist Max Planck announced that by making a somewhat peculiar modification in the classical calculation he could derive a function $f(\lambda,T)$ which agreed with the experimental data at all wavelengths. He first found an empirical function that fitted the data and then searched for a way to modify the usual calculation. His empirical function is

Ultraviolet catastrophe

$$f(\lambda,T) = \frac{8\pi hc\lambda^{-5}}{e^{hc/\lambda kT} - 1}$$

42-2   *Planck's law*

where h is an adjustable constant. Figure 42-1 compares this function with the Rayleigh-Jeans law and with experimental data. At large wavelengths, the exponential in the denominator of Equation 42-2 is small, and we can use the approximation

$$e^{hc/\lambda kT} \approx 1 + \frac{hc}{\lambda kT}$$

Then

$$f(\lambda,T) \to \frac{8\pi hc\lambda^{-5}}{hc/\lambda kT} = 8\pi kT\lambda^{-4}$$

which is the Rayleigh-Jeans formula. For small wavelengths, the exponential is very large, and we can neglect the 1 in the denominator of Equation 42-2, giving

$$f(\lambda,T) \to 8\pi hc\lambda^{-5}e^{-hc/\lambda kT}$$

42-3

This approaches zero as λ approaches zero. The constant h can be found by fitting this formula to the data points at small λ. Its presently accepted value is

$$h = 6.626 \times 10^{-34} \text{ J-sec} = 4.136 \times 10^{-15} \text{ eV-sec}$$

42-4   *Planck's constant*

Planck found that he could derive Equation 42-2 by modifying the calculation of the average energy per wave in the cavity. Instead of dealing directly with the radiation in the cavity, Planck considered the

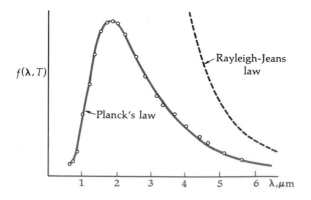

Figure 42-1
Spectral distribution of black-body radiation versus wavelength at $T = 1600$ K. (*Adapted from F. K. Richtmyer, E. H. Kennard, and J. N. C. Lauritsen,* Introduction to Modern Physics, *5th ed., McGraw-Hill Book Company, New York, 1955; by permission.*)

emission and absorption of radiation by the cavity walls, which he represented by a set of oscillators of all frequencies. He reasoned that, in equilibrium, the average energy of an oscillator of frequency f would be associated with the average energy of the electromagnetic radiation of that frequency in the cavity. In the usual calculation of the average energy of an oscillator in thermal equilibrium with its surroundings the energy E is a continuous variable, and integration is used. The result for a one-dimensional oscillator is kT. Planck modified this calculation by assuming that the energy of an oscillator is discrete; i.e., it can take on only certain values E_n, given by

$$E_n = nhf \qquad \text{42-5}$$

where n is an integer, f is the frequency, and h is a constant. With this assumption Planck obtained for the average energy of an oscillator

$$E_{av} = \frac{hc/\lambda}{e^{hc/\lambda kT} - 1} \qquad \text{42-6}$$

Replacing kT in Equation 42-1 by this expression gives Equation 42-2. Although Planck tried to fit the constant h into the framework of classical physics, he was unable to do so. The fundamental importance of his assumption of energy quantization, implied by Equation 42-5, was not generally appreciated until Einstein applied similar ideas to explain the photoelectric effect and suggested that quantization is a fundamental property of electromagnetic radiation.

42-2 Quantization of Electromagnetic Radiation: Photons

As mentioned in Chapter 26, the photoelectric effect was discovered by Hertz in 1887 and studied by Lenard in 1900. Figure 42-2 shows a schematic diagram of the basic apparatus. When light is incident on a clean metal surface, the cathode C, electrons are emitted. If some of these electrons strike the anode A, there is a current in the external circuit. The number of the emitted electrons reaching the anode can be increased or decreased by making the anode positive or negative with respect to the cathode. Let V be the increase in potential from the cathode to anode. Figure 42-3 shows the current versus V for two values of the intensity of light incident on the cathode. When V is positive, the electrons are attracted to the anode. At sufficiently large V all the emitted electrons reach the anode and the current reaches its maximum value. A further increase in V does not affect the current. Lenard observed that the maximum current is proportional to the light intensity, an expected result since doubling the energy per unit time incident on the cathode should double the number of electrons emitted. When V is negative, the electrons are repelled from the anode. Only electrons with initial kinetic energy $\frac{1}{2}mv^2$ greater than $|eV|$ can then reach the anode. From Figure 42-3 we see that if V is less than $-V_0$, no electrons reach the anode. The potential V_0 is called the *stopping potential*. It is related to the maximum kinetic energy of the emitted electrons by

$$(\tfrac{1}{2}mv^2)_{max} = eV_0$$

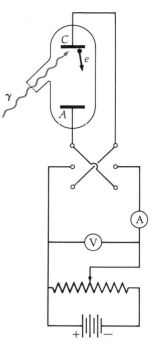

Figure 42-2
Schematic drawing of photoelectric-effect apparatus. Light strikes the cathode C and ejects electrons. The number of electrons which reach the anode A is measured by the current in the ammeter. The anode can be made positive or negative with respect to the cathode to attract or repel the electrons.

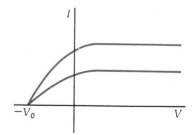

Figure 42-3
Photoelectric current versus voltage V for two values of light intensity. There is no current when V is less than $-V_0$. The saturation current observed for large V is proportional to the light intensity.

Stopping potential

The experimental result that V_0 is independent of the incident light intensity was surprising. Apparently increasing the rate of energy falling
on the cathode does not increase the maximum kinetic energy of the
electrons emitted. In 1905 Einstein demonstrated that this result can be
understood if light energy is not distributed continuously in space but
quantized in small bundles called *photons*. The energy of each photon
is hf, where f is the frequency and h is Planck's constant. An electron
emitted from a metal surface exposed to light receives its energy from
a single photon. When the intensity of the light of a given frequency is
increased, more photons fall on the surface in a unit time but the
energy absorbed by each electron is unchanged. If ϕ is the energy necessary to remove an electron from the surface of a metal, the maximum
kinetic energy of the electrons emitted will be

$$(\tfrac{1}{2}mv^2)_{\max} = eV_0 = hf - \phi \qquad\qquad 42\text{-}7$$

Einstein's photoelectric equation

The quantity ϕ, called the *work function*, is a characteristic of the metal.
Some electrons will have kinetic energy less than $hf - \phi$ because of
energy loss traversing the metal. Equation 42-7 is known as *Einstein's
photoelectric equation*.

Work function

> If the derived formula is correct, then V_0 when represented in car-
> tesian coordinates as a function of the frequency of the incident light
> must be a straight line whose slope is independent of the nature of
> the emitting substance.[1]

From Equation 42-7 we see that the slope of V_0 versus f should equal
h/e. Einstein's equation was a bold prediction, for at that time there
was no evidence that Planck's constant had any applicability outside
of blackbody radiation and there were no experimental data on the
stopping potential V_0 as a function of frequency. The experimental verification of Einstein's theory was quite difficult. Careful experiments
by Millikan reported in 1914 and in more detail in 1916 showed that
the Einstein equation is correct, and measurements of h agreed with
the value found by Planck. Figure 42-4 shows a plot of Millikan's data.

The threshold frequency f_t and the corresponding threshold
wavelength λ_t are related to the work function ϕ by setting V_0 equal to
zero in Equation 42-7. Then

$$\phi = hf_t = \frac{hc}{\lambda_t} \qquad\qquad 42\text{-}8$$

Threshold frequency and wavelength

[1] From Einstein's original paper translated by A. B. Arons and M. B. Peppard in *American Journal of Physics*, vol. 33, p. 367, 1965.

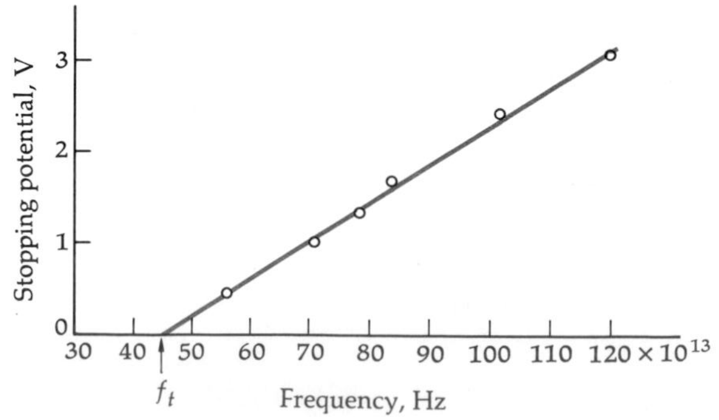

Figure 42-4
Millikan's data for the stopping potential versus
frequency for the photoelectric effect. The data fall on a
straight line which has slope
h/e as predicted by Einstein a
decade before the experiment.
(*From R. C. Millikan,* Physical
Review, *vol. 7, p. 362, 1916.*)

Photons of frequency less than f_t (and therefore wavelengths greater than λ_t) do not have enough energy to eject an electron from the metal. Work functions for metals are typically a few electron volts. Since wavelengths are usually given in nanometers or angstroms and energies in electron volts, it is useful to have the value of hc in electron volt–angstroms or electron volt–nanometers. We have

$$hc = (4.14 \times 10^{-15} \text{ eV sec})(3 \times 10^8 \text{ m/sec}) = 1.24 \times 10^{-6} \text{ eV-m}$$

or

$$hc = 1240 \text{ eV-nm} = 12,400 \text{ eV-Å} \qquad\qquad 42\text{-}9$$

Example 42-1 The threshold wavelength for potassium is 564 nm. What is the work function for potassium? What is the stopping potential when light of wavelength 400 nm is used?

$$\phi = hf_t = \frac{hc}{\lambda_t} = \frac{1240 \text{ eV-nm}}{564 \text{ nm}} = 2.20 \text{ eV}$$

The energy of a photon of wavelength 400 nm is

$$E = \frac{hc}{\lambda} = \frac{1240 \text{ eV-nm}}{400 \text{ nm}} = 3.10 \text{ eV}$$

The maximum kinetic energy of the emitted electrons is then

$$(\tfrac{1}{2}mv^2)_{\text{max}} = hf - \phi = 3.10 \text{ eV} - 2.20 \text{ eV} = 0.90 \text{ eV}$$

The stopping potential is therefore 0.90 V.

Another interesting feature of the photoelectric effect is the absence of time lag between the time light is turned on and electrons appear. In the classical theory, given the intensity (power per unit area) the time can be calculated for enough energy to fall on the area of an atom to eject an electron. However, even when the intensity is so small that such a calculation gives time lags of hours, essentially no time lag is observed. The explanation of this result is that although the number of photons hitting the metal per unit time is very small when the intensity is low, each photon has enough energy to eject an electron and there is a chance that one photon will be absorbed immediately. The classical calculation gives the correct average number of photons absorbed per unit time.

Further evidence of the correctness of the photon concept was furnished by Arthur H. Compton, who measured the scattering of x-rays by free electrons. According to classical theory, when an electromagnetic wave of frequency f_1 is incident on material containing charges, the charges will oscillate with this frequency and reradiate electromagnetic waves of the same frequency. Compton pointed out that if the scattering process were considered to be a collision between a photon and electron, the electron would absorb energy due to recoil and the scattered photon would have less energy and therefore a lower frequency than the incident photon. According to classical theory, the energy and momentum of an electromagnetic wave are related by $E = pc$. This result is also consistent with the relativistic expression relating the energy and momentum of a particle (Equation 28-29),

$$E^2 = p^2c^2 + (mc^2)^2 \qquad\qquad 42\text{-}10$$

if the mass of the photon is assumed to be zero. Figure 42-5 shows the geometry of a collision between a photon of wavelength λ_1 and an electron at rest. Compton related the scattering angle θ to the incident and scattered wavelengths λ_1 and λ_2 by treating the scattering as a relativistic-mechanics problem using conservation of energy and momentum. Let $\mathbf{p}_1$ be the momentum of the incident photon, $\mathbf{p}_2$ that of the scattered photon, and $\mathbf{p}_e$ that of the recoiling electron. Conservation of momentum gives

$$\mathbf{p}_1 = \mathbf{p}_2 + \mathbf{p}_e$$

or

$$p_e{}^2 = (\mathbf{p}_1 - \mathbf{p}_2)^2 = p_1{}^2 + p_2{}^2 - 2\mathbf{p}_1 \cdot \mathbf{p}_2 = p_1{}^2 + p_2{}^2 - 2p_1 p_2 \cos \theta \qquad 42\text{-}11$$

The energy before the collision is $p_1 c + mc^2$, where mc^2 is the rest energy of the electron. After the collision the electron has energy $\sqrt{(mc^2)^2 + p_e{}^2 c^2}$. Conservation of energy then gives

$$p_1 c + mc^2 = p_2 c + \sqrt{(mc^2)^2 + p_e{}^2 c^2} \qquad 42\text{-}12$$

Compton eliminated the electron momentum p_e from Equations 42-11 and 42-12 and expressed the photon momenta in terms of the wavelengths to obtain an equation relating the incident and scattered wavelengths λ_1 and λ_2 and the angle θ. The algebraic details are left as a problem (Problem 4). Compton's result is

$$\lambda_2 - \lambda_1 = \frac{h}{mc} (1 - \cos \theta) \qquad 42\text{-}13$$

The change in wavelength is independent of the original wavelength. The quantity h/mc depends only on the mass of the electron. It has dimensions of length and is called the *Compton wavelength*. Its value is

$$\lambda_c = \frac{h}{mc} = \frac{hc}{mc^2} = \frac{1240 \text{ eV-nm}}{5.11 \times 10^5 \text{ eV}} = 2.43 \text{ pm} \qquad 42\text{-}14$$

Because $\lambda_2 - \lambda_1$ is small, it is difficult to observe unless λ_1 is so small that the fractional change $(\lambda_2 - \lambda_1)/\lambda_1$ is appreciable. Compton used x-rays of wavelength 71.1 pm. The energy of a photon of this wavelength is $E = hc/\lambda = (1240 \text{ eV-nm})/(0.0711 \text{ nm}) = 17.4 \text{ keV}$. Since this is much greater than the binding energy of the valence electrons in carbon, these electrons can be considered to be essentially free. Compton's experimental results for $\lambda_2 - \lambda_1$ as a function of scattering angle θ agreed with Equation 42-13, thereby confirming the correctness of the photon concept.

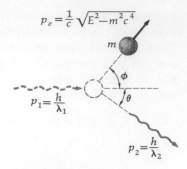

Figure 42-5
Compton scattering of an x-ray by an electron. The scattered photon has less energy and therefore greater wavelength than the incident photon because of the recoil energy of the electron. The change in wavelength $\lambda_2 - \lambda_1$ is found from conservation of energy and momentum.

Compton wavelength

42-3 Quantization of Atomic Energies: The Bohr Model

In Planck's treatment of blackbody radiation he represented the walls of the body by a set of oscillators of frequency f whose energy is quantized in units of hf. In 1907, Einstein applied similar reasoning to explain the temperature dependence of the heat capacity of a solid. The most famous application of energy quantization to microscopic systems was that of Niels Bohr. In 1913, Bohr proposed a model of the hydrogen atom which had spectacular successes in calculating the wavelengths of lines in the known hydrogen spectrum and in predicting new lines (later found experimentally) in the infrared and ultraviolet spectrum.

Many data were collected near the turn of the century on the emission of light by atoms in a gas when excited by an electric discharge. Viewed through a spectroscope with a narrow-slit aperture, this light appears as a discrete set of lines of different color or wavelength; the spacing and intensities of the lines are characteristic of the element. It was possible to determine the wavelengths of these lines accurately, and much effort went into finding regularities in the spectra. In 1884 a Swiss schoolteacher, Johann Balmer, found that the wavelengths of some of the lines in the spectrum of hydrogen can be represented by the formula

$$\lambda = 364.6 \, \frac{m^2}{m^2 - 4} \quad \text{nm} \qquad\qquad 42\text{-}15 \qquad \textit{Balmer formula}$$

where m is a variable integer which takes on the values $m = 3$, 4, 5 . . . . Figure 42-6 shows the set of spectral lines of hydrogen, now known as the *Balmer series,* whose wavelengths are given by Equation 42-15. Balmer suggested that his formula might be a special case of a more general expression applicable to the spectra of other elements. Such an expression, found by Rydberg and Ritz, gives the reciprocal wavelength as

$$\frac{1}{\lambda} = R\left[\frac{1}{(m+a)^2} - \frac{1}{(n+b)^2}\right], \qquad n > m \qquad 42\text{-}16 \qquad \textit{Rydberg-Ritz formula}$$

where a and b are constants, which are different for different series, and R, called the *Rydberg constant* or the *Rydberg,* is the same for all series of the same element and varies only slightly in a regular way from element to element. For hydrogen, $a = b = 0$, and the value of R is $R_H = 109{,}677.8 \text{ cm}^{-1}$. For very massive elements, R approaches the value $R_\infty = 109{,}737.5 \text{ cm}^{-1}$. (The quantity $1/\lambda$ was used instead of the frequency $f = c/\lambda$ because λ can be measured much more accurately than the speed of light c.) Such empirical expressions were successful in predicting other spectra; e.g., other hydrogen lines outside the visible spectrum were predicted and found.

Many attempts were made to construct a model of the atom that would yield these formulas for its radiation spectrum. The most popular model, due to J. J. Thomson, considered various arrangements of electrons embedded in some kind of fluid that contained most of the mass of the atom and had enough positive charge to make the atom electrically neutral. Since classical electromagnetic theory predicted that a charge oscillating with frequency f would radiate light of that frequency, Thomson searched for configurations that were stable and had normal modes of vibration of frequencies equal to those of the spectrum of the atom. A difficulty of this model and all others was that

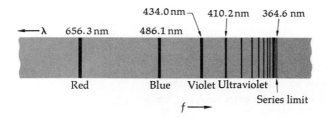

Figure 42-6
Balmer series in hydrogen. The wavelengths of these lines are given by Equation 42-15 for different values of the integer m.

electric forces alone cannot produce stable equilibrium. Thomson was unsuccessful in finding a model which predicted the observed frequencies for any atom.

The Thomson model was essentially ruled out by a set of experiments by Geiger and Marsden under the supervision of Rutherford about 1911, in which alpha particles from radioactive radium were scattered by atoms in a gold foil. Rutherford showed that the number of alpha particles scattered at large angles could not be accounted for by an atom in which the positive charge was distributed throughout the atomic size (known to be about 1 Å in diameter) but required that the positive charge and most of the mass of the atom be concentrated in a very small region, now called the nucleus, of diameter of the order of 10^{-5} Å. Niels Bohr, working in the Rutherford laboratory at the time, proposed a model of the hydrogen atom which combined the work of Planck, Einstein, and Rutherford and successfully predicted the observed spectra. Bohr assumed that the electron in the hydrogen atom moved under the influence of the coulomb attraction to the positive nucleus according to classical mechanics, which predicts circular or elliptical orbits with the force center at one focus, as in the motion of the planets about the sun. For simplicity he chose a circular orbit. Although mechanical stability is achieved because the coulomb attractive force provides the centripetal force necessary for the electron to remain in orbit, such an atom is unstable electrically according to classical theory because the electron must accelerate when moving in a circle and therefore radiate electromagnetic energy of frequency equal to that of its motion. According to classical electromagnetic theory, such an atom would quickly collapse, the electron spiraling into the nucleus as it radiates away its energy. Bohr "solved" this difficulty, modifying the laws of electromagnetism by *postulating* that the electron could move in certain orbits without radiating. He called these stable orbits *stationary states*. The atom radiates when the electron somehow makes a transition from one stationary state to another. The frequency of radiation is not the frequency of motion in either stable orbit but is related to the energies of the orbits by

First Bohr postulate: nonradiating orbits

$$f = \frac{W_i - W_f}{h} \qquad \text{42-17}$$

Second Bohr postulate: photon frequency from energy conservation

where h is Planck's constant and W_i and W_f are the total energies in the initial and final orbits. This assumption, which is equivalent to that of energy conservation with the emission of a photon, is a key one in the Bohr theory because it deviates from the classical theory, which requires the frequency of radiation to be that of the motion of the charged particle. Planck made a similar assumption in his blackbody-radiation theory, but in that case the energy of the oscillators was nhf and the energy difference between two adjacent states hf, so that when the oscillator emitted or absorbed radiation, the frequency of the radiation was the same as that of the oscillator.

If the nuclear charge is $+Ze$ and the electron charge $-e$, the potential energy at a distance r is

$$U = -\frac{kZe^2}{r}$$

where $k = 1/4\pi\epsilon_0$ is the Coulomb constant. (For hydrogen, $Z = 1$, but it is convenient to not specify Z at this time so that the results can be applied to other atoms.) The total energy of the electron moving in a circular orbit with speed v is then

$$W = \tfrac{1}{2}mv^2 + U = \tfrac{1}{2}mv^2 - \frac{kZe^2}{r}$$

The kinetic energy can be obtained as a function of r by using Newton's law $\mathbf{F} = m\mathbf{a}$. Setting the coulomb attractive force equal to the mass times the centripetal acceleration, we have

$$\frac{kZe^2}{r^2} = m\frac{v^2}{r}$$

or

$$\tfrac{1}{2}mv^2 = \frac{1}{2}\frac{kZe^2}{r} \qquad\qquad 42\text{-}18$$

For circular orbits the kinetic energy equals half the magnitude of the potential energy, a result which holds for circular motion in any inverse-square-law force field. The total energy is then

$$W = -\frac{1}{2}\frac{kZe^2}{r} \qquad\qquad 42\text{-}19$$

Using Equation 42-17 for the frequency of radiation when the electron changes from orbit 1 of radius r_1 to orbit 2 of radius r_2, we obtain

$$f = \frac{W_1 - W_2}{h} = \frac{1}{2}\frac{kZe^2}{h}\left(\frac{1}{r_2} - \frac{1}{r_1}\right) \qquad\qquad 42\text{-}20$$

To obtain the Balmer-Ritz formula $f = c/\lambda = cR(1/m^2 - 1/n^2)$ it is evident that the radii of stable orbits must be proportional to the squares of integers. Bohr searched for a quantum condition for the radii of the stable orbits which would yield this result. After much trial and error, he found that he could obtain the correct results if he postulated that in a stable orbit the angular momentum of the electron equals an integer times Planck's constant divided by 2π. Since the angular momentum of a circular orbit is just mvr, this postulate is

$$mvr = \frac{nh}{2\pi} = n\hbar \qquad\qquad 42\text{-}21$$

Third Bohr postulate: quantized angular momentum

where $\hbar = h/2\pi$. (The constant $\hbar = h/2\pi$ is often more convenient than h itself, just as angular frequency $\omega = 2\pi f$ is often more convenient than the frequency f.) We can determine r by eliminating v between Equations 42-18 and 42-21. We have

$$v^2 = n^2\frac{\hbar^2}{m^2r^2} = \frac{kZe^2}{mr}$$

or

$$r = n^2\frac{\hbar^2}{mkZe^2} = n^2\frac{r_0}{Z} \qquad\qquad 42\text{-}22$$

where

$$r_0 = \frac{\hbar^2}{mke^2} \approx 0.529\ \text{Å} \qquad\qquad 42\text{-}23$$

is called the *first Bohr radius*. Combining Equations 42-22 and 42-20, we obtain

$$f = Z^2\frac{mk^2e^4}{4\pi\hbar^3}\left(\frac{1}{n_2{}^2} - \frac{1}{n_1{}^2}\right) \qquad\qquad 42\text{-}24$$

Comparing this for $Z = 1$ with the empirical Balmer-Ritz formula (Equation 42-16 with $a = b = 0$), we have for the Rydberg constant

$$R = \frac{mk^2e^4}{4\pi c\hbar^3} \qquad \text{42-25}$$

Using the values of m, e, and $\hbar$ known in 1913, Bohr calculated R and found his result to agree (within the limits of the uncertainties of the constants) with the value obtained from spectroscopy.

The possible values of the energy of the hydrogen atom predicted by the Bohr model are given by Equation 42-19, with r given by Equation 42-22. They are

$$W_n = -\frac{k^2e^4m}{2\hbar^2}\frac{Z^2}{n^2} = -Z^2\frac{E_1}{n^2} \qquad \text{42-26}$$

where $E_1 = k^2e^4m/2\hbar^2 \approx 13.6$ eV. It is convenient to represent these energies in an energy-level diagram, as in Figure 42-7. Various series of transitions are indicated in this diagram by vertical arrows between the levels. The frequency of the light emitted in one of these transitions is the energy difference divided by h, according to Equation 42-17. At the time of Bohr's paper (1913) the Balmer series corresponding to $n_2 = 2$, $n_1 = 3$, 4, $5 \ldots$, and the Paschen series corresponding to $n_2 = 3$, $n_1 = 4, 5, 6 \ldots$, were known. In 1916 Lyman found the series corresponding to $n_2 = 1$, and in 1922 and 1924 Brackett and Pfund, respectively, found series corresponding to $n_2 = 4$ and $n_2 = 5$. As can be determined by computing the wavelengths of these series, only the Balmer series lies in the visible portion of the electromagnetic spectrum.

In our derivations we have assumed the electron to revolve around a stationary nucleus. This is equivalent to assuming the nucleus to have infinite mass. Since the mass of the hydrogen nucleus is not infinite but only about 2000 times that of the electron, a correction must be made for the motion of the nucleus. In the center-of-mass frame of the atom, the nucleus and electron have equal and opposite momenta, giving zero total momentum to the atom. If p is the magnitude of the momentum of either, the kinetic energy of the nucleus is $p^2/2M$, where M is its mass and the kinetic energy of the electron is $p^2/2m$. The total kinetic energy is

$$E_k = \frac{p^2}{2M} + \frac{p^2}{2m} = \frac{M+m}{mM}\frac{p^2}{2} = \frac{p^2}{2\mu}$$

where μ, called the *reduced mass*, is given by

$$\mu = \frac{mM}{m+M} = \frac{m}{1+m/M} \qquad \text{42-27} \qquad \textit{Reduced mass}$$

This is slightly different from the kinetic energy of just the electron because the reduced mass μ differs slightly from the electron mass m. We can correct any of our results for the motion of the nucleus by replacing the electron mass m with the reduced mass μ. (The validity of this procedure is shown in most intermediate and advanced mechanics books.) This correction leads to a very slight dependence of the Rydberg constant as given in Equation 42-25 on the nuclear mass, in precise agreement with the observed variation.

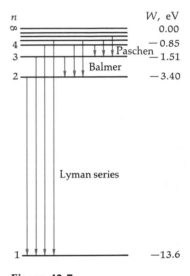

n	W, eV
∞	0.00
4	−0.85
3	−1.51
2	−3.40
1	−13.6

Paschen
Balmer
Lyman series

Figure 42-7
Energy-level diagram for hydrogen showing a few transitions in each of the Lyman, Balmer, and Paschen series. The energies of the levels are given by Equation 42-26.

Example 42-2 Find the energy and wavelength of the longest-wavelength line in the Lyman series.

From Figure 42-7 we see that the Lyman series corresponds to transitions ending at the ground state of energy $W_1 = -13.6$ eV. Since λ varies inversely with energy, the longest-wavelength transition is the lowest-energy transition from the first excited state $n = 2$ to the ground state. The first excited state has energy $W_2 = -13.6$ eV/4 = -3.40 eV. Since this is 10.2 eV above the ground-state energy, the energy of the photon emitted is 10.2 eV. The wavelength of this photon is

$$\lambda = \frac{hc}{E} = \frac{1240 \text{ eV-nm}}{10.2 \text{ eV}} = 121.6 \text{ nm}$$

This photon is outside the visible spectrum toward the ultraviolet. Since all the other lines in the Lyman series have even greater energies and shorter wavelengths, the Lyman series is completely in the ultraviolet region.

Questions

1. If an electron moves to a larger orbit, does its total energy increase or decrease? Does its kinetic energy increase or decrease?

2. How does the spacing of adjacent energy levels change as n increases?

3. What is the energy of the shortest-wavelength photon emitted by the hydrogen atom?

42-4 Electron Waves

In 1924 a French student, L. de Broglie, suggested in his dissertation that since light is known to have both wave and particle properties, perhaps matter—especially electrons—may also have wave as well as particle characteristics. This suggestion was highly speculative; there was no evidence at that time for any wave aspects of electrons. For the frequency and wavelength of electron waves de Broglie chose the equations

$$f = \frac{E}{h} \qquad \qquad \text{42-28}$$

De Broglie relations

$$\lambda = \frac{h}{p} \qquad \qquad \text{42-29}$$

where p is the momentum and E the energy of the electron. These equations also hold for photons, as we have seen. De Broglie pointed out that with these relations, the Bohr quantum condition (Equation 42-21) is equivalent to a standing-wave condition. We have

$$mvr = n\frac{h}{2\pi}$$

Substituting h/λ for the momentum mv gives

$$\frac{h}{\lambda}r = n\frac{h}{2\pi}$$

or

$$n\lambda = 2\pi r = C \qquad \qquad \text{42-30}$$

Figure 42-8
Standing waves around the
circumference of a circle.

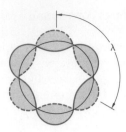

where C is the circumference of the circular Bohr orbit. Thus Bohr's quantum condition is equivalent to saying that an integral number of electron waves must fit into the circumference of the circular orbit (Figure 42-8). The idea of explaining the discrete energy states of matter by standing waves seemed promising. In classical wave theory applied to waves on a string, for example, application of the boundary conditions that the string be fixed at both ends leads to a discrete set of allowed frequencies—those for which an integral number of half wavelengths fit into the length of the string. If energy is associated with the frequency of a standing wave, as in Equation 42-28, standing waves imply quantized energies. These ideas were later developed into a detailed mathematical theory by Erwin Schrödinger in 1925. Schrödinger found a wave equation for electron waves and solved the standing-wave problem for the hydrogen atom, the simple harmonic oscillator, and other systems of interest. He found that the allowed frequencies, combined with the de Broglie relation $E = hf$, lead to the same set of energy levels for the hydrogen atom as found by Bohr (Equation 42-26) and to almost the same set of energies for the simple harmonic oscillator as assumed by Planck and Einstein.[1] Schrödinger's *wave mechanics* therefore gave a general method of finding the quantization condition for a given system.

The first direct measurements of the wavelengths of electrons were made in 1927 by C. J. Davisson and L. H. Germer, who were studying electron scattering from a nickel target at Bell Telephone Laboratories. After heating the target to remove an oxide coating that had accumulated during an accidental break in the vacuum system, they found that the scattered-electron intensity as a function of the scattering angle showed maxima and minima. Their target had crystallized, and by accident they had observed electron diffraction. They then prepared a target consisting of a single crystal of nickel and investigated this phenomenon extensively. From measurements of the diffraction maxima and minima they observed that the wavelength of the electrons could be calculated, and they found agreement with de Broglie's relation (Equation 42-29) in all cases. In the same year G. P. Thomson (son of J. J. Thomson) also observed electron diffraction in transmission of electrons through thin metallic foils. Since then, diffraction has been observed for neutrons, protons, and other particles of small mass. Figure 42-9 shows a diffraction pattern produced by electrons incident on two narrow slits. This pattern is identical to that observed with photons of the same wavelength. Figures 42-10 and 42-11 compare the diffraction patterns of x-rays, electrons, and neutrons of similar wavelength in transmission through thin foils.

Figure 42-9
Two-slit electron diffraction
pattern. This pattern is the
same as that usually obtained
with photons. (*Courtesy of
Claus Jönsson.*)

---

[1] Schrödinger found $E_n = (n + \frac{1}{2})hf$ for the energy levels of the simple harmonic oscillator which differs slightly from $E_n = nhf$ assumed by Planck and Einstein. This difference, which amounts to a shift of each energy level by $\frac{1}{2}hf$, has no effect on their results, which depend only on the difference between energy levels.

(a)   (b)

Figure 42-10
Diffraction pattern produced
by (a) x-rays and (b) electrons
on an aluminum foil target.
(*Courtesy of Film Studio, Education Development Center, Newton, Mass.*)

We can understand why the wave properties of matter are not observed in macroscopic experiments by calculating the de Broglie wavelength for a macroscopic object.

Example 42-3 Find the de Broglie wavelength of a particle of mass 10^{-6} gm moving with speed 10^{-6} m/sec.

From Equation 42-29 we have

$$\lambda = \frac{h}{p} = \frac{h}{mv} - \frac{6.63 \times 10^{-34} \text{ J-sec}}{(10^{-9} \text{ kg})(10^{-6} \text{ m/sec})} - 6.63 \times 10^{-19} \text{ m}$$

Since this wavelength is much smaller than any possible apertures or obstacles (the diameter of the nucleus of the atom is about 10^{-15} m, roughly 10,000 times this wavelength), diffraction or interference of such waves cannot be observed. As discussed in Chapter 25, the propagation of waves of very small wavelength is indistinguishable from the propagation of particles. Note that we have chosen an extremely small value for the momentum in Example 42-3. Any other macroscopic particle with greater momentum will have an even smaller de Broglie wavelength. However, the situation is different for low-energy electrons. Consider an electron that has been accelerated through a potential difference V. Its energy is then

$$E = \frac{p^2}{2m} = eV$$

and its wavelength is

$$\lambda = \frac{h}{p} = \frac{h}{\sqrt{2mE}} = \frac{h}{\sqrt{2meV}}$$

Putting in the values for m, e, and h, we obtain an expression for λ in angstroms when V is in volts:

$$\lambda = \frac{12.26}{\sqrt{V}} \text{ Å}$$ 42-31

Figure 42-11
Diffraction pattern produced
by neutrons on a target of
polycrystalline copper. Note
the similarity in the patterns
produced by x-rays, electrons,
and neutrons. (*Courtesy of C. G. Shull.*)

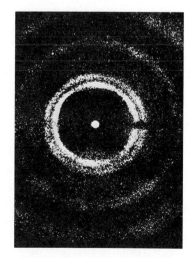

Example 42-4 Find the de Broglie wavelength of an electron whose kinetic energy is 13.6 eV.

From Equation 42-31 we have for $V = 13.6$ V,

$$\lambda = \frac{12.26 \text{ Å}}{\sqrt{13.6}} = 3.32 \text{ Å} = 2\pi(0.529 \text{ Å})$$

which is the circumference of the first Bohr orbit in the hydrogen atom.

From Equation 42-31 we see that electrons with energies of the order of tens of electron volts have de Broglie wavelengths of the order of angstroms, which is the order of magnitude of the size of the atom and of the spacing of atoms in a crystal.

42-5   Wave-Particle Duality

We have seen that light, which we ordinarily think of as a wave motion, exhibits particle properties when it interacts with matter, as in the photoelectric effect or in Compton scattering, whereas electrons, which we usually think of as particles, exhibit the wave properties of interference and diffraction. All phenomena—electrons, atoms, light, sound, etc.—have both particle and wave characteristics. It might be tempting to say that an electron, for example, is both a wave and a particle, but the meaning of such a statement is not clear. In classical physics the concepts of waves and particles are mutually exclusive. A *classical particle* behaves like a piece of shot; it can be localized and scattered; it exchanges energy suddenly in lumps, and it obeys the laws of conservation of energy and momentum in collisions. It does *not* exhibit interference or diffraction. A *classical wave,* on the other hand, behaves like a water wave; it exhibits diffraction and interference, and its energy is spread out continuously in space and time. Nothing can be both a classical particle and a classical wave at the same time.

Until the twentieth century, it was thought that light was a classical wave and electrons were classical particles. We now see that these concepts of classical waves and particles do not adequately describe the complete behavior of any phenomenon. Everything propagates like a classical wave and exchanges energy like a classical particle. Of course, there are times when the classical-particle and classical-wave concepts give the same results. When the wavelength is very small, the propagation of a classical wave cannot be distinguished from that of a classical particle. Similarly, the time averages of energy and momentum exchanges are the same for classical waves as for classical particles. For example, the wave theory of light correctly predicts that the total electron current in the photoelectric effect is proportional to the intensity of light.

42-6   The Uncertainty Principle

The wave-particle duality of nature has many important consequences. In newtonian mechanics the rate of change of momentum of a particle is related to the resultant force acting on the particle by Newton's second law. To determine the position of the particle at any time we must know the forces acting and the position and velocity of the particle at some time; i.e., we must know the initial conditions. Although there are always experimental uncertainties in any measurement of the initial position and velocity, it is assumed in classical mechanics that such uncertainties can in principle be made as small as desired. Because of the wave-particle duality of both radiation and matter, we now understand that it is impossible in principle to measure both the position and velocity of a particle simultaneously with infinite precision. This result is known as the *uncertainty principle*, first enunciated by Werner Heisenberg in 1927. For one-dimensional motion it is expressed in terms of the momentum p and the position x of a particle. Let Δx and Δp be the uncertainties in the position and momentum, respectively. According to the uncertainty principle, the product of Δx and Δp can never be less than $h/4\pi = \frac{1}{2}\hbar$:

Uncertainty principle

$$\Delta x \, \Delta p \geq \tfrac{1}{2}\hbar \qquad\qquad 42\text{-}32$$

Usually the uncertainty product is much greater than $\frac{1}{2}\hbar$. The equality holds only if the measurements of both x and p have ideal gaussian distributions and the experiments are ideal.

We can get a qualitative understanding of the uncertainty principle by considering the measurement of the position and momentum of a particle. If we know the mass of the particle, we can determine its momentum by measuring its position at two nearby times and computing its velocity. A common way to measure the position of an object is to look at it with light. When we do this, we scatter light from the object and determine the position by the direction of the scattered light. If we use light of wavelength λ, we can measure the position only to an uncertainty of the order of λ because of diffraction effects. To reduce the uncertainty in position we therefore use light of very short wavelength, perhaps even x-rays. In principle, there is no limit to the accuracy of such a position measurement because there is no limit on how short a wavelength of electromagnetic radiation can be used. However, since all electromagnetic radiation carries momentum, the scattering of the radiation by the particle will deflect the radiation and change its original momentum in an uncontrollable way. By momentum conservation, the momentum of the particle also changes in an uncontrollable way. According to classical wave theory, this effect on the momentum of the particle could be reduced by reducing the intensity of the radiation. However, the energy and momentum of the radiation are quantized; each photon has momentum h/λ. When the wavelength of the radiation is small, the momentum of each photon will be large and the momentum measurement will have a large uncertainty. This uncertainty cannot be eliminated by reducing the intensity of light; such a reduction merely reduces the number of photons in the beam. To "see" the particle we must scatter at least one photon. Therefore, the uncertainty in the momentum measurement of the particle will be large if λ is small, and the uncertainty in the position measurement will be large if λ is large. A detailed analysis shows that the product of these uncertainties will always be at least of the order of Planck's constant h. Of course we could "look at" the particle by scattering electrons instead of photons, but we have the same difficulty. If we use low-momentum electrons to reduce the uncertainty in the momentum measurement, we have a large uncertainty in the position measurement because of diffraction of the electrons. The relation between the wavelength and momentum $\lambda = h/p$ is the same for electrons as for photons.

Black Holes

Alan P. Lightman
Cornell University

Imagine a region of space where attractive gravitational force is so intense that light rays venturing too near can be bent into circular orbit—a region from which no matter, radiation, or communication of any kind can ever escape. Such is a black hole, one of the most exciting creatures of theoretical physics and possibly the most bizarre object of space.

Although they were implicitly predicted by Einstein's 1915 theory of gravity, general relativity, black holes were first theoretically "discovered" by Oppenheimer and Snyder in 1939. Because of their highly nonintuitive properties, however, black holes were not taken seriously by most physicists and astronomers until the mid-1960s. We are now on the verge of confirming the discovery in space of the first black hole.

The concept of a black hole stretches our ideas of time and space to their limits. The surface of a black hole, called its *horizon,* is a closed boundary within which the escape velocity exceeds the velocity of light. The prediction of such a surface for sufficiently compact bodies can be made just on the basis of Newton's theory of gravity together with special relativity: the escape velocity of a particle launched from the surface of a spherical mass M of radius R is $v_e = \sqrt{2GM/R}$ (see Chapter 9). When M/R satisfies $2GM/R > c^2$, v_e exceeds the speed of light and no particle or photon may escape, as required by special relativity. As a remarkable result, the interior of a black hole is causally disconnected from the rest of the universe; no physical process occurring inside the horizon can communicate its existence or effects to the outside. For a spherical black hole of mass M, the horizon is a sphere of circumference equal to 2π times the *Schwarzschild radius* of the hole R_G, where $R_G = 2GM/c^2$ (exact numerical coincidence of this radius with the newtonian analog above is accidental). A black hole with a mass equal to that of our sun would have a Schwarzschild radius of 2.95 km.

According to general relativity, time and space are warped by the gravitational field of massive bodies, with the warping most severe near a black hole. Gravity affects all physical systems in a universal way so that all clocks (whether transitions of an ammonia molecule or heartbeats of a human being) and all rulers would indicate that time is slowed down and space stretched out near a black hole (see Figure 1). Alternatively, one can describe the effects of the hole's gravitational field on a local measurement of time and space intervals as an acceleration of the local Lorentz frame (in which special relativity is valid) with respect to other local Lorentz frames at different locations.

Black holes are formed when massive stars undergo total gravitational collapse. While emitting heat and light into space, stars balance themselves against their own gravity with the outward thermal pressure force generated by heat from nuclear reactions in their deep interiors. But every star must die. When its nuclear fuel is exhausted, a star will contract. If its mass is less than about three times the mass of our sun, the shrinking star will stabilize at a smaller size when the inward pull of gravity cannot force the star's constituent particles any closer together. Such a star will spend eternity as a *white dwarf* or *neutron star.* But if the star's mass exceeds about 3 solar masses, theory indicates that no amount of outward pressure force can stave off the overwhelming crush of gravity and the star will plunge in upon itself, disappearing forever from sight and giving birth to a black hole.

Recent mathematical proofs, using Einstein's general relativity theory, demonstrate that a black hole is one of the simplest objects of nature and can be completely characterized by only three quantities: its mass, angular momentum, and net electrical charge. Other than these three quantities, all information of the progenitor star, e.g., whether it was composed of particles or antiparticles, whether it was pancake-shaped or spherical, is lost via gravita-

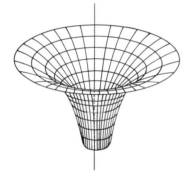

Figure 1
How the geometry of space is warped by a spherical (nonrotating) black hole. The radial distance from any point to the center of the black hole can be measured in two ways. One way is to measure the circumference of a circle passing through the point and centered on the hole, and divide by 2π. A second way is to measure the distance from the point to the center along a radial line on the curved surface. These two measures will be nearly equal far from the hole where space is "flat," but highly discrepant near the black-hole horizon where the funnel narrows to a vertical tube.

tional and electromagnetic waves shortly after formation of the hole. Rotating black holes, called *Kerr holes* (nonrotating holes are called *Schwarzschild holes*), surround themselves with a region called the *ergosphere* in which space and time are so strongly twisted that all particles, photons, and even local Lorentz frames are forced to rotate about the hole.

At great distances from a black hole, its gravitational field behaves as if generated by an ordinary star of mass M, obeying the newtonian inverse-square law. Close to a black hole, the gravitational field is far stronger than would be predicted by newtonian theory. A man being sucked into a solar-mass black hole would be ripped to shreds by the differential force of gravity across his body long before he reached the horizon.

Once the hole's horizon has formed we cannot receive any information about the ultimate fate of the collapsing star inside. Calculations of the collapse (idealized to be spherical) indicate that the star is crushed to zero volume and infinite density at the center of the black hole, forming there a point of infinite gravitational force called a *singularity*. For a collapsing star of a few solar masses, the final stellar death throes would be over in a few hundred-thousandths of a second (as measured locally). Quantum effects, omitted in the classical theory of general relativity, could possibly halt the stellar collapse at the unimaginable density $\rho \sim c^5/hG^2 \sim 5 \times 10^{93}$ gm/cm^3, where h is Planck's constant, and thus prevent creation of singularities; but such effects could not prevent the formation of black holes.

Current theories of stellar evolution suggest that there could be as many as 100 million black holes in our galaxy. The search for these black holes is not easy. We could never detect a tiny black speck a few miles across against the night sky. Instead, we must search for signs of black holes interacting with their astrophysical neighborhoods. Most vulnerable to discovery would be a black hole orbiting a normal star in a binary system. Using Kepler's laws, analysis of the magnitude and period of the doppler shift of the visible normal star allows us to compute whether the unseen companion is massive enough to be a black hole. In addition we might observe intense, flickering x-rays produced when gas from the normal star is sucked toward the black hole and heated to a billion degrees in its spiraling path to destruction in the hole. Such are the telltale signs of the binary x-ray source Cygnus X-1, an excellent candidate for a black hole about 8000 light-years from earth in the constellation Cygnus.

Black holes are a fundamental phenomenon of nature. Space may be littered with black holes, tempting us with their secrets of time and space behind a cloak of impenetrable darkness, a challenge and a prize for the persistence of astronomers and physicists.

Review

A. Define, explain, or otherwise identify:

Blackbody radiation, 955
Ultraviolet catastrophe, 956
Photoelectric effect, 957
Work function, 958
Einstein's photoelectric equation, 958
Compton wavelength, 960
Balmer series, 961
The Rydberg, 961
Stationary state, 962

Bohr quantization condition, 963
Energy-level diagram, 964
Reduced mass, 964
De Broglie wavelength, 965
Classical particle, 968
Classical wave, 968
Wave-particle duality, 968
Uncertainty principle, 968

B. True or false:

1. The spectral distribution of radiation in a blackbody depends only on the temperature of the body.

2. In the photoelectric effect, the maximum current is proportional to the intensity of the incident light.

3. The work function of a metal depends on the frequency of the incident light.

4. The maximum kinetic energy of electrons emitted in the photoelectric effect varies linearly with the frequency of the incident light.

5. In the Bohr model of the hydrogen atom the electron moves in circular orbits.

6. One of Bohr's assumptions is that atoms never radiate light.

7. In a stationary state in the Bohr model the total energy is negative and has the same magnitude as the kinetic energy.

8. In the ground state of the hydrogen atom the potential energy is -27.2 eV.

9. The de Broglie wavelength of an electron varies inversely with its momentum.

10. Electrons can be diffracted.

11. Neutrons can be diffracted.

12. Classical particles can be diffracted.

13. The position and momentum of a particle cannot be measured simultaneously.

14. The uncertainty principle could be violated in principle if the experimental measurements were ideal.

Exercises

Section 42-1, The Origin of the Quantum Constant: Blackbody Radiation

There are no exercises for this section.

Section 42-2, The Quantization of Electromagnetic Radiation: Photons

1. Find the range of photon energies in the visible spectrum from about 400 to 700 nm wavelength.

2. Find the photon energy if the wavelength is (*a*) 1 Å (about 1 atomic diameter) and (*b*) 1 fm (10^{-15} m, about 1 nuclear diameter).

3. Find the photon energy for an electromagnetic wave in the FM band of frequency 100 MHz.

4. The work function for tungsten is 4.58 eV. (*a*) Find the threshold frequency and wavelength for the photoelectric effect. Find the stopping potential if the wavelength of the incident light is (*b*) 200 nm; (*c*) 250 nm.

5. When light of wavelength 300 nm is incident on potassium, the emitted electrons have maximum kinetic energy of 2.03 eV. (*a*) What is the energy of the incident photon? (*b*) What is the work function for potassium? (*c*) What would be the stopping potential if the incident light had a wavelength of 400 nm?

6. The threshold wavelength for the photoelectric effect for silver is 262 nm. (*a*) Find the work function for silver. (*b*) Find the stopping potential if the incident radiation has a wavelength of 200 nm.

7. The work function for cesium is 1.9 eV. (*a*) Find the threshold frequency and wavelength for the photoelectric effect. Find the stopping potential if the wavelength of the incident light is (*b*) 300 nm; (*c*) 400 nm.

8. The wavelength of Compton-scattered photons is measured at $\theta = 90°$. If $\Delta\lambda/\lambda$ is to be 1 percent, what should the wavelength of the incident photons be?

9. Compton used photons of wavelength 0.0711 nm. (a) What is the energy of these photons? (b) What is the wavelength of the photon scattered at $\theta = 180°$? (c) What is the energy of the photon scattered at $\theta = 180°$? (d) What is the recoil energy of the electron for $\theta = 180°$?

Section 42-3, Quantization of Atomic Energies: The Bohr Model

10. Use the known values of the constants in Equation 42-23 to show that r_0 is approximately 0.529 Å.

11. The wavelength of the longest wavelength in the Lyman series was calculated in Example 42-2. Find the wavelengths for the transitions (a) $n = 3$ to $n = 1$ and (b) $n = 4$ to $n = 1$. (c) Find the shortest wavelength in the Lyman series.

12. (a) Use the Balmer formula (Equation 42-15) to calculate the three longest wavelengths in the Balmer series. (b) Find the photon energies corresponding to each of these wavelengths.

13. Calculate the longest three wavelengths and the series limit (shortest wavelength) for the Paschen series ($n_2 = 3$) and indicate their positions on a horizontal linear scale.

14. Repeat Exercise 13 for the Brackett series ($n_2 = 4$).

Section 42-4, Electron Waves

15. Use Equation 42-31 to calculate the de Broglie wavelength for an electron of kinetic energy (a) 1 eV, (b) 100 eV, (c) 1 keV, and (d) 10 keV.

16. When the kinetic energy of an electron is much greater than its rest energy, the relativistic approximation $E \approx pc$ is good. (a) Show that in this case photons and electrons of the same energy have the same wavelength. (b) What is the de Broglie wavelength of an electron of energy 100 MeV?

17. A thermal neutron in a reactor has kinetic energy of about 0.02 eV. Calculate the de Broglie wavelength of this neutron.

18. Find the de Broglie wavelength of a proton of energy 1 MeV.

Section 42-5, Wave-Particle Duality

There are no exercises for this section.

Section 42-6, The Uncertainty Principle

19. A particle of mass 10^{-6} gm moves with speed 1 cm/sec. If its speed is uncertain by 0.01 percent, what is the minimum uncertainty in its position?

20. The uncertainty in the position of an electron in an atom cannot be greater than the diameter of the atom. (a) Calculate the minimum uncertainty in momentum associated with an uncertainty in position of 1.0 Å. (b) If an electron has momentum p equal in magnitude to the uncertainty Δp found in (a), what is its kinetic energy? Express your answer in electron-volts.

21. An electron has kinetic energy 25 eV. If its momentum is uncertain by 10 percent, what is the minimum uncertainty in its position?

Problems

1. Data for stopping potential versus wavelength for the photoelectric effect using sodium are

λ, nm	200	300	400	500	600
V_0, volts	4.20	2.06	1.05	0.41	0.03

Plot these data so as to obtain a straight line and from your plot find (a) the work function, (b) the threshold frequency, and (c) the ratio h/e.

2. A light bulb radiates 100 W uniformly in all directions. (a) Find the intensity at a distance of 1 m. (b) Find the number of photons per second that strike a 1-cm^2 area oriented so that its normal is along the line to the bulb if the wavelength of the light is 600 nm.

3. This problem is one of estimating the time lag (expected classically but not observed) in the photoelectric effect. Let the intensity of the incident radiation be 0.01 W/m^2. (a) Assuming that the area of an atom is 1 Å^2, find the energy per second falling on an atom. (b) If the work function is 2 eV, how long would it take classically for this much energy to fall on one atom?

4. (a) Solve Equation 42-12 for p_e^2 to obtain $p_e^2 = p_1^2 + p_2^2 - 2p_1p_2 + 2mc(p_1 - p_2)$. (b) Eliminate p_e^2 from your result in part (a) and Equation 42-11 to obtain $mc(p_1 - p_2) = p_1p_2(1 - \cos\theta)$. (c) Multiply both sides of your result in part (b) by h/mcp_1p_2 and use the de Broglie relation $h/p = \lambda$ to obtain the Compton formula (Equation 42-13).

5. The photoelectric effect can occur only with electrons that are bound (in atoms); i.e., a photon cannot be absorbed by a single free electron. Prove this by considering the problem of conservation of energy and momentum in the reference frame in which the total momentum of the initial photon and electron is zero.

6. The total energy density of radiation in a blackbody is given by $\eta = \int_0^\infty f(\lambda,T)\, d\lambda$, where $f(\lambda,T)$ is given by the Planck formula (Equation 42-2). Change the variable to $x = hc/\lambda kT$ and show that the total energy density can be written

$$\eta = \left(\frac{kT}{hc}\right)^4 8\pi hc \int_0^\infty \frac{x^3}{e^x - 1}\, dx = \alpha T^4$$

where α is a constant independent of T. This result, that the total energy density is proportional to T^4, is known as the *Stefan-Boltzmann law*.

7. Find the ratio of the reduced mass for the hydrogen atom to the electron mass μ/m. Show that this ratio equals the ratio of the Rydberg constant for hydrogen to that for an infinitely massive nucleus R_H/R_∞, and compare your results with the values of R_H and R_∞ given in the text.

8. The frequency of revolution of an electron in a circular orbit of radius r is $f_{rev} = v/2\pi r$, where v is the speed. (a) Show that in the nth stationary state

$$f_{rev} = \frac{k^2Z^2e^4m}{2\pi h^3} \frac{1}{n^3}$$

(b) Show that when $n_1 = n$ and $n_2 = n - 1$ and n is much greater than 1,

$$\frac{1}{n_2^2} - \frac{1}{n_1^2} \approx \frac{2}{n^3}$$

(c) Use your result of part (b) in Equation 42-24 to show that in this case the frequency of radiation emitted equals the frequency of motion. This result is an example of Bohr's correspondence principle: when n is large, so that the energy difference between adjacent states is a small fraction of the total energy, classical and quantum physics must give the same result.

9. The kinetic energy of rotation of a diatomic molecule can be written $E_k = L^2/2I$, where L is its angular momentum and I the moment of inertia. (a) Assuming that the angular momentum is quantized, as in the Bohr model of the hydrogen atom, show that the energy is given by $E_n = n^2 E_1$, where $E_1 = \hbar^2/2I$, and make an energy-level diagram for such a molecule. (b) Estimate E_1 for the hydrogen molecule assuming the separation of the atoms to be 1 Å and considering rotation about an axis through the center of mass and perpendicular to the line joining the atoms. Express your estimate in electron volts. (c) When E_1 is greater than kT, molecular collisions do not excite rotation and rotational energy does not contribute to the internal energy of the gas. Use your result of part (b) to find the critical temperature T_c, defined by $kT_c = E_1$. (d) Estimate E_1 for rotation about the line joining the atoms. In this case, the moment of inertia I is of the order $m_e r_a^2$, where m_e is the electron mass and r_a is the radius of the atom. Find T_c for rotation about this axis.

10. A particle moves about in a one-dimensional box of length L. Its energy is quantized by the condition $n(\lambda/2) = L$, where λ is the de Broglie wavelength of the particle. (a) Show that the allowed energies are given by $E_n = n^2 E_1$, where $E_1 = h^2/8mL^2$ and m is the mass of the particle. Evaluate E_1 for (b) an electron in a box of size $L = 1$ Å and (c) a proton in a box of size $L = 10^{-15}$ m. (Take the potential energy of the particle in the box to be zero so that its total energy is its kinetic energy $p^2/2m$.)

11. An important consequence of the uncertainty principle is that a particle confined to some region of space has a minimum kinetic energy greater than zero. If Δp is taken to be the standard deviation, it is given by $(\Delta p)^2 = [(p - p_{av})^2]_{av}$. (a) Show that this is equivalent to $(\Delta p)^2 = (p^2)_{av} - (p_{av})^2$. (b) Consider a particle of mass m moving in a box of length L such that $p_{av} = 0$. Its average kinetic energy is then $E_{k,av} = (p^2)_{av}/2m = (\Delta p)^2/2m$. Show that its average kinetic energy must be at least as large as $\hbar^2/8mL^2$ by using the fact that Δx cannot be larger than L.

APPENDIX A   SI Units

Basic units

Length	The *meter* (m) is the length equal to 1,650,763.73 wavelengths in vacuum of the radiation corresponding to the transition between the $2p_{10}$ and $5d_5$ levels of the ^{86}Kr atom
Time	The *second* (sec) is the duration of 9,192,631,770 periods of the radiation corresponding to the transition between the two hyperfine levels of the ground state of the ^{133}Cs atom
Mass	The *kilogram* (kg) is the mass of the international standard body preserved at Sèvres, France
Current	The *ampere* (A) is that current in two very long parallel wires 1 m apart which gives rise to a magnetic force per unit length of 2×10^{-7} N/m
Temperature	The *kelvin* (K) is 1/273.16 of the thermodynamic temperature of the triple point of water
Luminous Intensity	The *candela* (cd) is the luminous intensity, in the perpendicular direction, of a surface of area 1/600,000 m² of a blackbody at the temperature of freezing platinum at pressure of 1 atm

Derived units

Force	Newton (N)	1 N = 1 kg-m/sec²
Work, energy	Joule (J)	1 J = 1 N-m
Power	Watt (W)	1 W = 1 J/sec
Frequency	Hertz (Hz)	1 Hz = 1/sec
Charge	Coulomb (C)	1 C = 1 A-sec
Potential	Volt (V)	1 V = 1 J/C
Resistance	Ohm (Ω)	1 Ω = 1 V/A
Capacitance	Farad (F)	1 F = 1 C/V
Magnetic induction	Tesla (T)	1 T = 1 N/A-m
Magnetic flux	Weber (Wb)	1 Wb = 1 T-m²
Inductance	Henry (H)	1 H = 1 J/A²

APPENDIX B   Numerical Data

For additional data see the front and back endpapers and the following tables in the text:

Physical constants

Gravitational constant	G	6.672×10^{-11} N-m^2/kg^2
Speed of light	c	2.997925×10^8 m/sec
Electron charge	e	1.60219×10^{-19} C
Avogadro's number	N_A	6.0220×10^{23} particles/mole
Gas constant	R	8.314 J/mole-K
		1.9872 cal/mole-K
		8.206×10^{-2} ℓ-atm/mole-K
Boltzmann's constant	$k = R/N_A$	1.3807×10^{-23} J/K
		8.617×10^{-5} eV/K
Unified mass unit	$u = 1/N_A$ gm	1.6606×10^{-24} gm
Coulomb constant	$k = 1/4\pi\epsilon_0$	8.98755×10^9 N-m^2/C^2
Permittivity of free space	ϵ_0	8.85419×10^{-12} C^2/N-m^2
Magnetic constant	$k_m = \mu_0/4\pi$	10^{-7} N/A^2 (exact)

Physical constants
(continued)

Permeability of free space	μ_0	$4\pi \times 10^{-7}$ N/A^2
		1.256637×10^{-6} N/A^2
Planck's constant	h	6.6262×10^{-34} J-sec
		4.1357×10^{-15} eV-sec
	$\hbar = h/2\pi$	1.05459×10^{-34} J-sec
		6.5822×10^{-16} eV-sec
Mass of electron	m_e	9.1095×10^{-31} kg
		511.0 keV/c^2
Mass of proton	m_p	1.67265×10^{-27} kg
		938.28 MeV/c^2
Mass of neutron	m_n	1.67495×10^{-27} kg
		939.57 MeV/c^2
Bohr magneton	$m_B = e\hbar/2m_e$	9.2741×10^{-24} A-m^2
		5.7884×10^{-5} eV/T

Terrestrial data

Acceleration of gravity g	9.80665 m/sec^2 standard value
	32.1740 ft/sec^2
At sea level, at equator*	9.7804 m/sec^2
At sea level, at poles*	9.8322 m/sec^2
Mass of earth M_E	5.98×10^{24} kg
Radius of earth R_E, mean	6.37×10^6 m
	3960 mi
Escape velocity $\sqrt{2R_E g}$	1.12×10^4 m/sec
	6.95 mi/sec
Solar constant†	1.35 kW/m^2
Standard temperature and pressure (STP):	
T	273.15 K
P	1.00 atm
	1.01325×10^5 N/m^2
Molecular weight of air	28.97 gm/mole
Density of air (STP), ρ_{air}	1.293 kg/m^3
Speed of sound (STP)	331 m/sec
Heat of fusion of H$_2$O (0°C, 1 atm)	79.7 cal/gm
Heat of vaporization of H$_2$O (100°C, 1 atm)	539.6 cal/gm

* Measured relative to the earth's surface.
† Average power incident normally on 1 m^2 outside the earth's atmosphere at the mean distance from the earth to the sun.

Astronomical data

Earth, distance to moon*	3.844×10^8 m
	2.389×10^5 mi
Distance (mean) to sun*	1.496×10^{11} m
	9.30×10^7 mi
	1.00 AU
Orbital speed (mean)	2.98×10^4 m/sec
Moon, mass	7.35×10^{22} kg
Radius	1.738×10^6 m
Period	27.32 d
Acceleration of gravity at surface	1.62 m/sec^2
Sun, mass	1.99×10^{30} kg
Radius	6.96×10^8 m

* Center to center.

Densities of selected substances (STP)

Substance	Density, gm/cm^3*
Hydrogen	8.994×10^{-5}
Helium	1.786×10^{-4}
Air	1.293×10^{-3}
Alcohol (ethanol)	0.806
Water	1.00
Mercury	13.60
Aluminum	2.70
Iron	7.86
Copper	8.96
Lead	11.3
Gold	19.6

* 1 gm/cm^3 = 1000 kg/m^3.

APPENDIX C Conversion Factors

Conversion factors are written as equations for simplicity; relations marked with an asterisk are exact.

Length

*1 in = 2.54 cm

*1 ft = 12 in = 30.48 cm

*1 yd = 3 ft = 91.44 cm

1 m = 1.0936 yd = 3.281 ft = 39.37 in

*1 mi = 5280 ft = 1760 yd

1 mi = 1.609 km

1 light-year = 9.461×10^{15} m

*1 Å = 10^{-10} m

Area

*1 in² = 6.4516 cm²

1 ft² = 9.29×10^{-2} m²

*1 cm² = 10^{-4} m²

*1 acre = 43,560 ft²

1 mi² = 640 acres = 2.590 km²

Volume

1 in³ = 16.39 cm³

1 ft³ = 1728 in³ = 28.32 ℓ = 2.832×10^4 cm³

1 ℓ = 1000 cm³ = 10^{-3} m³ = 3.531×10^{-2} ft³

1 gal = 4 qt = 8 pt = 128 oz = 231 in³

1 gal = 3.786 ℓ

Time

*1 h = 60 min = 3600 sec

*1 d = 24 h = 1440 min = 8.64×10^4 sec

1 y = 365.24 d = 3.156×10^7 sec

Speed

*1 ft/sec = 0.3048 m/sec

1 mi/h = 0.4470 m/sec

1 mi/h = 1.467 ft/sec

Angle and angular velocity

*π rad = 180°

1 rad = 57.30°

1° = 1.745×10^{-2} rad

1 rev/min = 0.1047 rad/sec

1 rad/sec = 9.549 rev/min

Mass

*1 kg = 10^3 gm

1 slug = 14.59 kg

1 kg = 6.852×10^{-2} slug

1 kg = 6.022×10^{26} u

1 u = 1.6606×10^{-27} kg

Density

1 gm/cm³ = 10^3 kg/m³

(1 gm/cm³)g = 62.4 lb/ft³

Force

1 lb = 4.4482 N

1 N = 0.2248 lb = 10^5 dyne

(1 kg)g = 2.2046 lb

*1 ton = 2000 lb

Pressure

1 lb/in² = 6.895×10^3 N/m

*1 atm = 1.01325 bar = 1.01325×10^5 N/m²

1 mm Hg = 133.32 N/m²

1 in H₂O = 1.868 mm Hg = 249.1 N/m²

Energy

1 ft-lb = 1.356 J = 3.766×10^{-7} kW-h = 1.286×10^{-3} Btu

*1 cal = 4.1840 J

1 Btu = 778 ft-lb = 252 cal = 1054 J

1 eV = 1.602×10^{-19} J

1 ℓ-atm = 101.325 J = 24.217 cal

*1 kW-h = 3.6×10^6 J

Power

1 horsepower = 550 ft-lb/sec = 745.7 W

1 Btu/min = 0.02357 horsepower = 17.58 W = 4.200 cal/sec

1 W = 1.341×10^{-3} horsepower = 0.7376 ft-lb/sec

Magnetic induction

*1 G = 10^{-4} T

*1 T = 10^4 G

APPENDIX D — Mathematical Symbols and Formulas

Mathematical Symbols and Abbreviations

$=$	is equal to
$\neq$	is not equal to
$\approx$	is approximately equal to
$\propto$	is proportional to
$>$	is greater than
$\gg$	is much greater than
$<$	is less than
$\ll$	is much less than
Δx	change in x
$n!$	$n(n-1)(n-2) \cdots 1$
Σ	sum
lim	limit
$\Delta t \rightarrow 0$	Δt approaches zero
$\int$	integral
$\dfrac{dx}{dt}$	derivative of x with respect to t
$\dfrac{\partial y}{\partial x}$	partial derivative of y with respect to x

Roots of the Quadratic Equation

If

$$ax^2 + bx + c = 0$$

then

$$x = \frac{-b}{2a} \pm \frac{1}{2a} \sqrt{b^2 - 4ac}$$

Binomial Expansion

$$(1 + x)^n = 1 + nx + \frac{n(n-1)}{2!} x^2 + \frac{n(n-1)(n-2)}{3!} x^3 + \cdots$$

To compute $(a + b)^n$ write

$$(a + b)^n = a^n(1 + x)^n \qquad \text{with } x = \frac{b}{a}$$

or

$$(a + b)^n = b^n(1 + y)^n \qquad \text{with } y = \frac{a}{b}$$

Trigonometric Formulas

See also Sec. 1-7.

$$\sin^2 \theta + \cos^2 \theta = 1 \qquad \sec^2 \theta - \tan^2 \theta = 1 \qquad \csc^2 \theta - \text{ctn}^2 \theta = 1$$

$$\sin 2\theta = 2 \sin \theta \cos \theta$$

$$\cos 2\theta = \cos^2 \theta - \sin^2 \theta = 2 \cos^2 \theta - 1 = 1 - 2 \sin^2 \theta$$

$$\tan 2\theta = \frac{2 \tan \theta}{1 - \tan^2 \theta}$$

$$\sin \tfrac{1}{2}\theta = \sqrt{\frac{1 - \cos \theta}{2}} \qquad \cos \tfrac{1}{2}\theta = \sqrt{\frac{1 + \cos \theta}{2}} \qquad \tan \tfrac{1}{2}\theta = \sqrt{\frac{1 - \cos \theta}{1 + \cos \theta}}$$

$$\sin (A \pm B) = \sin A \cos B \pm \cos A \sin B$$

$$\cos (A \pm B) = \cos A \cos B \mp \sin A \sin B$$

$$\tan (A \pm B) = \frac{\tan A \pm \tan B}{1 \mp \tan A \tan B}$$

$$\sin A \pm \sin B = 2 \sin \left[\tfrac{1}{2}(A \pm B)\right] \cos \left[\tfrac{1}{2}(A \mp B)\right]$$

$$\cos A + \cos B = 2 \cos \left[\tfrac{1}{2}(A + B)\right] \cos \left[\tfrac{1}{2}(A - B)\right]$$

$$\cos A - \cos B = 2 \sin \left[\tfrac{1}{2}(A + B)\right] \sin \left[\tfrac{1}{2}(B - A)\right]$$

$$\tan A \pm \tan B = \frac{\sin (A \pm B)}{\cos A \cos B}$$

Exponential and Logarithmic Functions

$$e = 2.71828 \qquad e^0 = 1$$

If $y = e^x$, then $x = \ln y$.

$$e^{\ln x} = x$$

$$e^x e^y = e^{(x+y)}$$

$$(e^x)^y = e^{xy} = (e^y)^x$$

$$\ln e = 1 \qquad \ln 1 = 0$$

$$\ln xy = \ln x + \ln y$$

$$\ln \frac{x}{y} = \ln x - \ln y$$

$$\ln e^x = x \qquad \ln a^x = x \ln a$$

$$\ln x = (\ln 10) \log x = 2.3026 \log x$$

$$\log x = \log e \ln x = 0.43429 \ln x$$

$$e^x = 1 + x + \frac{x^2}{2!} + \frac{x^3}{3!} + \cdots$$

$$\ln (1 + x) = x - \frac{x^2}{2} + \frac{x^3}{3} - \frac{x^4}{4} + \cdots$$

APPENDIX E Derivatives and Integrals

Table E-1 lists some important properties of derivatives and the derivatives of some particular functions which often occur in physics. It is followed by comments aimed at making these properties and rules plausible. More detailed and rigorous discussions can be found in most calculus books. Table E-2 lists some important integration formulas. More extensive lists of differentiation and integration formulas can be found in handbooks, e.g., Herbert Dwight, *Tables of Integrals and Other Mathematical Data*, 4th ed., Macmillan Publishing Co. Inc., New York, 1961.

Table E-1
Properties of derivatives and derivatives of particular functions

---

Linearity

---

1. The derivative of a constant times a function equals the constant times the derivative of the function:

$$\frac{d}{dt}\left[Cf(t)\right] = C\frac{df(t)}{dt}$$

2. The derivative of a sum of functions equals the sum of the derivatives of the functions:

$$\frac{d}{dt}\left[f(t) + g(t)\right] = \frac{df(t)}{dt} + \frac{dg(t)}{dt}$$

Chain rule

---

3. If f is a function of x and x is in turn a function of t, the derivative of f with respect to t equals the product of the derivative of f with respect to x and the derivative of x with respect to t:

$$\frac{d}{dt}f(x) = \frac{df}{dx}\frac{dx}{dt}$$

Derivative of a product

---

4. The derivative of a product of functions $f(t)g(t)$ equals the first function times the derivative of the second plus the second function times the derivative of the first:

$$\frac{d}{dt}\,[f(t)g(t)] = f(t)\,\frac{dg(t)}{dt} + \frac{df(t)}{dt}\,g(t)$$

Reciprocal derivative

---

5. The derivative of t with respect to x is the reciprocal of the derivative of x with respect to t assuming that neither derivative is zero:

$$\frac{dx}{dt} = \left(\frac{dt}{dx}\right)^{-1} \quad \text{if} \quad \frac{dt}{dx} \neq 0$$

Derivatives of particular functions

---

6. $\dfrac{dC}{dt} = 0 \qquad$ where C is a constant

7. $\dfrac{d(t^n)}{dt} = nt^{n-1}$

8. $\dfrac{d}{dt}\,\sin \omega t = \omega \cos \omega t$

9. $\dfrac{d}{dt}\,\cos \omega t = -\omega \sin \omega t$

10. $\dfrac{d}{dt}\,e^{bt} = be^{bt}$

11. $\dfrac{d}{dt}\,\ln bt = \dfrac{1}{t}$

Comment on Rules 1 to 5

Rules 1 and 2 follow from the fact that the limiting process is linear. We can understand rule 3, the chain rule, by multiplying $\Delta f/\Delta t$ by $\Delta x/\Delta x$ and noting that since x is a function of t, both Δx and Δf approach zero as Δt approaches zero. Since the limit of a product of two functions equals the product of their limits, we have

$$\lim_{\Delta t \to 0} \frac{\Delta f}{\Delta t} = \lim_{\Delta t \to 0} \frac{\Delta f}{\Delta x}\frac{\Delta x}{\Delta t} = \left(\lim_{\Delta x \to 0} \frac{\Delta f}{\Delta x}\right)\left(\lim_{\Delta t \to 0} \frac{\Delta x}{\Delta t}\right) = \frac{df}{dx}\frac{dx}{dt}$$

Rule 4 is not immediately apparent. The derivative of a product of functions is the limit of the ratio

$$\frac{f(t + \Delta t)g(t + \Delta t) - f(t)g(t)}{\Delta t}$$

If we add and subtract the quantity $f(t + \Delta t)g(t)$ in the numerator, we can write this ratio as

$$\frac{f(t + \Delta t)g(t + \Delta t) - f(t + \Delta t)g(t) + f(t + \Delta t)g(t) - f(t)g(t)}{\Delta t}$$

$$= f(t + \Delta t)\left[\frac{g(t + \Delta t) - g(t)}{\Delta t}\right] + g(t)\left[\frac{f(t + \Delta t) - f(t)}{\Delta t}\right]$$

As Δt approaches zero, the terms in brackets become $dg(t)/dt$ and $df(t)/dt$, respectively, and the limit of the expression is

$$f(t)\frac{dg(t)}{dt} + g(t)\frac{df(t)}{dt}$$

Rule 5 follows directly from the definition:

$$\frac{dx}{dt} = \lim_{\Delta t \to 0}\frac{\Delta x}{\Delta t} = \lim_{\Delta t \to 0}\left(\frac{\Delta t}{\Delta x}\right)^{-1} = \lim_{\Delta x \to 0}\left(\frac{\Delta t}{\Delta x}\right)^{-1} = \left(\lim_{\Delta x \to 0}\frac{\Delta t}{\Delta x}\right)^{-1}$$

Comment on Rule 7

We can obtain this important result using the binomial expansion (Appendix D). We have

$$f(t) = t^n$$

$$f(t + \Delta t) = (t + \Delta t)^n = t^n\left(1 + \frac{\Delta t}{t}\right)^n$$

$$= t^n\left[1 + n\frac{\Delta t}{t} + \frac{n(n-1)}{2!}\left(\frac{\Delta t}{t}\right)^2 + \frac{n(n-1)(n-2)}{3!}\left(\frac{\Delta t}{t}\right)^3 + \cdots\right]$$

Then

$$f(t-\Delta t) - f(t) = t^n\left[n\frac{\Delta t}{t} + \frac{n(n-1)}{2!}\left(\frac{\Delta t}{t}\right)^2 + \cdots\right]$$

and

$$\frac{f(t-\Delta t) - f(t)}{\Delta t} = nt^{n-1} + \frac{n(n-1)}{2!}t^{n-2}\Delta t + \cdots$$

The next term omitted from the last sum is proportional to $(\Delta t)^2$, the following to $(\Delta t)^3$, and so on. Each term except the first approaches zero as Δt approaches zero. Thus

$$\frac{df}{dt} = \lim_{\Delta x \to 0}\frac{f(t+\Delta t) + f(t)}{\Delta t} = nt^{n-1}$$

Comment on Rules 8 and 9

We first write $\sin \omega t = \sin \theta$ with $\theta = \omega t$ and use the chain rule,

$$\frac{d \sin \theta}{dt} = \frac{d \sin \theta}{d\theta}\frac{d\theta}{dt} = \omega\frac{d \sin \theta}{d\theta}$$

We then use the trigonometric formula for the sine of the sum of two angles θ and $\Delta\theta$:

$$\sin(\theta + \Delta\theta) = \sin \Delta\theta \cos \theta + \cos \Delta\theta \sin \theta$$

Since $\Delta\theta$ is to approach zero, we can use the small-angle approximations

$$\sin \Delta\theta \approx \Delta\theta \quad \text{and} \quad \cos \Delta\theta \approx 1$$

Then

$$\sin(\theta + \Delta\theta) \approx \Delta\theta \cos \theta + \sin \theta$$

and

$$\frac{\sin(\theta + \Delta\theta) - \sin \theta}{\Delta\theta} \approx \cos \theta$$

Similar reasoning can be applied to the cosine function to obtain rule 9.

Comment on Rule 10

Again we use the chain rule,

$$\frac{de^\theta}{dt} = b\frac{de^\theta}{d\theta} \quad \text{with } \theta = bt$$

and use the series expansion for the exponential function (Appendix D):

$$e^{\theta + \Delta\theta} = e^{\theta} e^{\Delta\theta} = e^{\theta} \left[1 + \Delta\theta + \frac{(\Delta\theta)^2}{2!} + \frac{(\Delta\theta)^3}{3!} + \cdots \right]$$

Then

$$\frac{e^{\theta + \Delta\theta} - e^{\theta}}{\Delta\theta} = e^{\theta} + e^{\theta} \frac{\Delta\theta}{2!} + e^{\theta} \frac{(\Delta\theta)^2}{3!} + \cdots$$

As $\Delta\theta$ approaches zero, the right side of the equation above approaches e^{θ}.

Comment on Rule 11

Let

$$y = \ln bt$$

Then

$$e^{y} = bt \qquad \text{and} \qquad \frac{dt}{dy} = \frac{1}{b} e^{y} = t$$

Then using rule 5, we obtain

$$\frac{dy}{dt} = \left(\frac{dt}{dy} \right)^{-1} = \frac{1}{t}$$

Table E-2
Integration formulas*

1. $\displaystyle\int A\, dt = At$	5. $\displaystyle\int e^{bt}\, dt = \frac{1}{b} e^{bt}$
2. $\displaystyle\int At\, dt = \tfrac{1}{2} At^2$	6. $\displaystyle\int \cos \omega t\, dt = \frac{1}{\omega} \sin \omega t$
3. $\displaystyle\int At^n\, dt = A \frac{t^{n+1}}{n+1} \qquad n \neq -1$	7. $\displaystyle\int \sin \omega t\, dt = -\frac{1}{\omega} \cos \omega t$
4. $\displaystyle\int At^{-1}\, dt = A \ln t$	

---

* In these formulas A, b, and ω are constants. An arbitrary constant C can be added to the the right side of each equation.

APPENDIX F — Trigonometric Tables

Degree	Radian	sin θ	cos θ	tan θ	Degree	Radian	sin θ	cos θ	tan θ
0	0.0000	0.0000	1.0000	0.0000					
1	0.0175	0.0175	0.9998	0.0175	46	0.8029	0.7193	0.6947	1.0355
2	0.0349	0.0349	0.9994	0.0349	47	0.8203	0.7314	0.6820	1.0724
3	0.0524	0.0523	0.9986	0.0524	48	0.8378	0.7431	0.6691	1.1106
4	0.0698	0.0698	0.9976	0.0699	49	0.8552	0.7547	0.6561	1.1504
5	0.0873	0.0872	0.9962	0.0875	50	0.8727	0.7660	0.6428	1.1918
6	0.1047	0.1045	0.9945	0.1051	51	0.8901	0.7771	0.6293	1.2349
7	0.1222	0.1219	0.9925	0.1228	52	0.9076	0.7880	0.6157	1.2799
8	0.1396	0.1392	0.9903	0.1405	53	0.9250	0.7986	0.6018	1.3270
9	0.1571	0.1564	0.9877	0.1584	54	0.9425	0.8090	0.5878	1.3764
10	0.1745	0.1736	0.9848	0.1763	55	0.9599	0.8192	0.5736	1.4281
11	0.1920	0.1908	0.9816	0.1944	56	0.9774	0.8290	0.5592	1.4826
12	0.2094	0.2079	0.9781	0.2126	57	0.9948	0.8387	0.5446	1.5399
13	0.2269	0.2250	0.9744	0.2309	58	1.0123	0.8480	0.5299	1.6003
14	0.2443	0.2419	0.9703	0.2493	59	1.0297	0.8572	0.5150	1.6643
15	0.2618	0.2588	0.9659	0.2679	60	1.0472	0.8660	0.5000	1.7321
16	0.2793	0.2756	0.9613	0.2867	61	1.0647	0.8746	0.4848	1.8040
17	0.2967	0.2924	0.9563	0.3057	62	1.0821	0.8829	0.4695	1.8807
18	0.3142	0.3090	0.9511	0.3249	63	1.0996	0.8910	0.4540	1.9626
19	0.3316	0.3256	0.9455	0.3443	64	1.1170	0.8988	0.4384	2.0503
20	0.3491	0.3420	0.9397	0.3640	65	1.1345	0.9063	0.4226	2.1445
21	0.3665	0.3584	0.9336	0.3839	66	1.1519	0.9135	0.4067	2.2460
22	0.3840	0.3746	0.9272	0.4040	67	1.1694	0.9205	0.3907	2.3559
23	0.4014	0.3907	0.9205	0.4245	68	1.1868	0.9272	0.3746	2.4751
24	0.4189	0.4067	0.9135	0.4452	69	1.2043	0.9336	0.3584	2.6051
25	0.4363	0.4226	0.9063	0.4663	70	1.2217	0.9397	0.3420	2.7475
26	0.4538	0.4384	0.8988	0.4877	71	1.2392	0.9455	0.3256	2.9042
27	0.4712	0.4540	0.8910	0.5095	72	1.2566	0.9511	0.3090	3.0777
28	0.4887	0.4695	0.8829	0.5317	73	1.2741	0.9563	0.2924	3.2709
29	0.5061	0.4848	0.8746	0.5543	74	1.2915	0.9613	0.2756	3.4874
30	0.5236	0.5000	0.8660	0.5774	75	1.3090	0.9659	0.2588	3.7321
31	0.5411	0.5150	0.8572	0.6009	76	1.3265	0.9703	0.2419	4.0108
32	0.5585	0.5299	0.8480	0.6249	77	1.3439	0.9744	0.2250	4.3315
33	0.5760	0.5446	0.8387	0.6494	78	1.3614	0.9781	0.2079	4.7046
34	0.5934	0.5592	0.8290	0.6745	79	1.3788	0.9816	0.1908	5.1446
35	0.6109	0.5736	0.8192	0.7002	80	1.3963	0.9848	0.1736	5.6713
36	0.6283	0.5878	0.8090	0.7265	81	1.4137	0.9877	0.1564	6.314
37	0.6458	0.6018	0.7986	0.7536	82	1.4312	0.9903	0.1392	7.115
38	0.6632	0.6157	0.7880	0.7813	83	1.4486	0.9925	0.1219	8.144
39	0.6807	0.6293	0.7771	0.8098	84	1.4661	0.9945	0.1045	9.514
40	0.6981	0.6428	0.7660	0.8391	85	1.4835	0.9962	0.0872	11.430
41	0.7156	0.6561	0.7547	0.8693	86	1.5010	0.9976	0.0698	14.301
42	0.7330	0.6691	0.7431	0.9004	87	1.5184	0.9986	0.0523	19.081
43	0.7505	0.6820	0.7314	0.9325	88	1.5359	0.9994	0.0349	28.636
44	0.7679	0.6947	0.7193	0.9657	89	1.5533	0.9998	0.0175	57.290
45	0.7854	0.7071	0.7071	1.0000	90	1.5708	1.0000	0.0000	∞

APPENDIX G

Periodic Table of Elements

The values listed are based on $^{12}_{6}C = 12$ u exactly. For artificially produced elements, the approximate atomic weight of the most stable isotope is given in brackets.

PERIOD	SERIES	GROUP I	II	III	IV	V	VI	VII	VIII			0
1	1	1 H 1.00797										2 He 4.003
2	2	3 Li 6.942	4 Be 9.012	5 B 10.81	6 C 12.011	7 N 14.007	8 O 15.9994	9 F 19.00				10 Ne 20.183
3	3	11 Na 22.990	12 Mg 24.31	13 Al 26.98	14 Si 28.09	15 P 30.974	16 S 32.064	17 Cl 35.453				18 Ar 39.948
4	4	19 K 39.102	20 Ca 40.08	21 Sc 44.96	22 Ti 47.90	23 V 50.94	24 Cr 52.00	25 Mn 54.94	26 Fe 55.85	27 Co 58.93	28 Ni 58.71	
4	5	29 Cu 63.54	30 Zn 65.37	31 Ga 69.72	32 Ge 72.59	33 As 74.92	34 Se 78.96	35 Br 70.909				36 Kr 83.80
5	6	37 Rb 85.47	38 Sr 87.62	39 Y 88.905	40 Zr 91.22	41 Nb 92.91	42 Mo 95.94	43 Tc [98]	44 Ru 101.1	45 Rh 102.905	46 Pd 106.4	
5	7	47 Ag 107.870	48 Cd 112.40	49 In 114.82	50 Sn 118.69	51 Sb 121.75	52 Te 127.60	53 I 126.90				54 Xe 131.30
6	8	55 Cs 132.905	56 Ba 137.34	57-71 Lanthanide series*	72 Hf 178.49	73 Ta 180.95	74 W 183.85	75 Re 186.2	76 Os 190.2	77 Ir 192.2	78 Pt 195.09	
6	9	79 Au 196.97	80 Hg 200.59	81 Tl 204.37	82 Pb 207.19	83 Bi 208.98	84 Po [210]	85 At [210]				86 Rn [222]
7	10	87 Fr [223]	88 Ra [226]	89-103 Actinide series†								

*Lanthanide Series	57 La 138.91	58 Ce 140.12	59 Pr 140.91	60 Nd 144.24	61 Pm [147]	62 Sm 150.35	63 Eu 152.0	64 Gd 157.25	65 Tb 158.92	66 Dy 162.50	67 Ho 164.93	68 Er 167.26	69 Tm 168.93	70 Yb 173.04	71 Lu 174.97
†Actinide Series	89 Ac [227]	90 Th 232.04	91 Pa [231]	92 U 238.03	93 Np [237]	94 Pu [242]	95 Am [243]	96 Cm [247]	97 Bk [247]	98 Cf [251]	99 Es [254]	100 Fm [253]	101 Md [256]	102 No [254]	103 Lw [257]

Answers

These answers are calculated using $g = 32$ ft/sec^2 = 9.8 m/sec^2 unless otherwise specified in the exercise or problem. The results are usually rounded to three significant figures. Differences in the third figure which can result from differences in rounding the input data are not important.

Chapter 1

True or False

1. True 2. False; e.g., $x = vt$, where the speed v and time t have different dimensions. 3. True 4. True 5. False; $\cos \theta \approx 1$ for small θ.

Exercises

1. (a) C_1 in meters; C_2 in meters per second; C_3 in meters per second per second (b) C_1 in meters; C_2 in sec^{-1} (c) C_1 in meters; C_2 in sec^{-1}

3. (a) length/(time)^2 (b) m/sec^2

5. C_1 dimensions: length/time; units: meters per second
C_2 dimensions: time^{-1}; units: sec^{-1}

7. Circumference 2.48×10^4 mi; radius 3.95×10^3 mi

9. (a) 3.16×10^7 sec/year (b) 31.7 years (c) 1.90×10^{16} years

11. (a) 1.61 km/mi (b) 2.20 lb

13. (a) 1.00 (b) 0.00 (c) 0.00 (d) 1.00 (e) 0.707 (f) −0.707

15. (a) $\pi/3$ rad (b) $\pi/2$ rad (c) $\pi/6$ rad (d) $\pi/4$ rad (e) π rad (f) 0.643 rad (g) 4π rad

17. (a) 3.49 rad = 200°

19. $\sin \theta_1 = 0.6$; $\cos \theta_1 = 0.8$; $\tan \theta_1 = 0.75$
$\sin \theta_2 = 0.8$; $\cos \theta_2 = 0.6$; $\tan \theta_2 = 1.33$
$\theta_1 = 37°$; $\theta_2 = 53°$

21. (a) 0.14 (b) 0.087

23. $1 + 3x + 3x^2 + x^3$

25. (a) 9.95 (b) 0.99 (c) 4.99 (rounded off from 4.986667)

Problems

1. (a) $n = 0.5$; $C = 1.8$ sec-kg$^{-1/2}$ (b) $m = 0.4$ kg and $m = 1.5$ kg

3. (a) $\sqrt{l/g}$ (c) $T = 2\pi \sqrt{l/g}$

5. (a) increase with increasing H; increase with increasing v (b) R proportional to $v \sqrt{H/g}$

7. $\tan \theta = (\tan \theta_1 + \tan \theta_2)/(1 - \tan \theta_1 \tan \theta_2)$

9. (a) 1.25×10^{-5}, 0.001% correction; 1.0×10^{-4}, 0.01% correction; -3.55×10^{-5}, 7×10^{-4}% correction (b) 0.001%; 0.01%, 7×10^{-4}%

Chapter 2

True or False

1. False; holds only for constant acceleration. 2. False; e.g., motion with constant velocity. 3. True 4. False; only for constant acceleration.

Exercises

1. (a) −2 m/sec (b) 2.25 m/sec (c) −0.3 m/sec

3. (a) 195 mi (b) 48.8 mi/h

5. (a) velocity and speed at t_2 less than at t_1 (b) both equal (c) velocity greater at t_2 but speed less at t_2 (d) velocity less at t_2 but speed greater at t_2

7. about 2 m/sec; 2.7 m/sec; 3.2 m/sec; 4.0 m/sec
$v \approx 4.2$ m/sec from measurement of slope at 0.75 sec

9. (a) 2 m, 2 m/sec (b) $\Delta x = 2t\,\Delta t - 5\Delta t + (\Delta t)^2$ (c) $v = (2t - 5)$ m/sec

11. 12 mi/h-sec = 17.6 ft/sec² = 0.55g

13. −2 m/sec²

15. (a) $a = 0$ (b) $a > 0$ (c) $a < 0$ (d) $a = 0$

17. $v = (8t - 8)$ ft/sec; $a = 8$ ft/sec²

19. (a) $v = A$ at $t = 0$ so A is initial velocity; sec⁻¹ (b) $a = dv/dt = -bAe^{-bt} = -bv$

21. $x(t) = 3.5t^2 + 5t + x_0$

23. $x(t) = (v_0/\omega)\sin \omega t + x_0$

25. $x = 124$ m

27. (a) −36 m (b) $\Delta x = -36$ m (c) −9 m/sec

29. (a) 0.25 m/sec (b) about 0.95 m/sec; 3.2 m/sec; 6.2 m/sec (c) about 7 m

31. (a) 80 m/sec (b) 400 m (c) 40 m/sec

33. $15\frac{5}{8}$ m/sec²

35. 4.47 sec; 44.7 m/sec; 22.4 m/sec

37. (a) −13.9 ft/sec (b) 11.3 ft/sec (c) 1.26×10^3 ft/sec²

39. (a) 193.44 ft (b) $v = 640$ ft/sec, $dx = 192$ ft

41. 2%

Problems

1. (b) 15 sec (c) 300 m from the intersection

7. (b) 20 sec (c) 100 mi/h

9. (b) 58.7 ft/sec; 5.11 sec

13. (a) yes (c) 10 ft/sec; 1 ft/sec; 50 ft

15. (b) $v_{av} = 0.72$ cm/sec; 0.83 cm/sec; 0.85 cm/sec; 0.865 cm/sec; 0.87 cm/sec; 0.875 cm/sec (c) 0.875 cm/sec

17. (b) distance run during time T is $x_1 = \frac{1}{2}aT^2 = \frac{1}{2}v_0T$; from time T to t distance is $x_2 = v_0(t - T)$; total distance is $x_1 + x_2 = v_0(t - \frac{1}{2}T)$ (c) $T \approx 2.4$ sec; $v_0 \approx 11.5$ m/sec

Chapter 3

True or False

1. True 2. False; e.g., projectile moving up, **a** down. 3. False; e.g., circular motion. 4. True 5. False 6. False; e.g., magnitude of **A** + (−**A**) is zero. 7. True 8. True unless speed is zero. 9. True

Exercises

1. resultant displacement vector is 18.6 m, 22° north of east

3. (a) 2 mi at 0°; 1 mi at 135°; 3 mi at 90° (b) $A_x = 2$ mi, $A_y = 0$ mi; $A_x = -0.707$ mi, $A_y = +0.707$ mi; $A_x = 0$ mi, $A_y = 3$ mi (c) 3.93 mi, 70.8° north of east (d) no; no: possibly but not necessarily

5. (−5 m, 5 m); **A** = −5**i** + 5**j** m

7. (a) $A = 5.83$; $\phi = 31°$ (b) $B = 12.2$; $\phi = 325°$ or −35° (c) $C = 5.39$; $\phi = 235°$; $\theta = 42°$. The angles θ and ϕ are those shown in Fig. 3-2.

9. $g_x = 4.9$ m/sec², $g_y = -8.49$ m/sec²

11. (a) **v** = 5**i** + 8.66**j** m/sec (b) **A** = −3.54**i** − 3.54**j** m (c) **r** = 14**i** − 6**j** m

15. 5**A** = 10**i** − 30**j**; − 7**A** = −14**i** + 42**j**

17. **B** = −1.5**A**

19. $\Delta\mathbf{A}$ and $\mathbf{A}$ are approximately perpendicular.

21. (*a*) $2\mathbf{i} + 2\mathbf{j}$ m/sec (*b*) $\frac{11}{5}\mathbf{i} + \frac{11}{5}\mathbf{j}$ m/sec

23. (*b*) $\mathbf{v} = 5\mathbf{i} + 10\mathbf{j}$ m/sec; $v = 11.2$ m/sec

25. (*a*) 90°, 45°, 0°, −45° and −90° with x axis (*b*) BC toward center of arc; DE toward center of arc

27. (*a*) $\mathbf{v}_{av} = 33.3\mathbf{i} + 26.7\mathbf{j}$ m/sec (*b*) $\mathbf{a}_{av} = -3\mathbf{i} - 1.77\mathbf{j}$ m/sec²

29. 0.5 sec

31. (*a*) 2.30×10^3 m (*b*) 43.3 sec (*c*) 9.18×10^3 m

33. $H = (v_0{}^2 \sin^2 \theta_0)/2g$

35. 45 m/sec²

37. $v^2/r = 2.47$ in/sec²

39. 194 ft/sec² up

41. (*a*) 31.6 m/sec² (*b*) 26.2 m/sec² (*c*) 23.9 m/sec². *Note:* It is easiest to express $\mathbf{v}_2$ and $\mathbf{v}_2 - \mathbf{v}_1$ in rectangular components and then find $|\Delta\mathbf{v}/\Delta t|$.

43. (*a*) $v = 10$ m/sec; $dv/dt = 0$ (*b*) $v = 11.4$ m/sec; $dv/dt = 15$ m/sec² (*c*) $v = 13.3$ m/sec; $dv/dt = -35.4$ m/sec²

45. (*a*) circular with constant speed (*b*) $\mathbf{v} = 20\pi\hat{\boldsymbol{\theta}}$ (*c*) 20π; yes (*d*) $\mathbf{a} = -40\pi^2\hat{\mathbf{r}}$

Problems

1. 76°

3. (*a*) 39 ft (*b*) $v_x - 32$ ft/sec, $v_y = 48$ ft/sec (*c*) 57.7 ft/sec (*d*) 56.3°

5. 3.19 m/sec toward ball

7. (*a*) $80\mathbf{i} + 21.2\mathbf{j} + 21.2\mathbf{k}$ ft/sec (*b*) it is in the air 1.49 sec; it lands at a distance 31.7 ft from the road at the point $119\mathbf{i} + 31.7\mathbf{j}$ ft from the car's original position, or $0\mathbf{i} + 31.7\mathbf{j}$ ft from the car's present position

9. (*b*) $\mathbf{v} = 8\pi \cos 2\pi t \, \mathbf{i} - 8\pi \sin 2\pi t \, \mathbf{j}$ (*c*) $\mathbf{a} = -4\pi^2\mathbf{R}$

11. (*b*) $\mathbf{v} = 6\pi \cos 2\pi t \, \mathbf{i} - 4\pi \sin 2\pi t \, \mathbf{j}$ (*c*) $\mathbf{a} = -4\pi^2\mathbf{R}$ (*d*) v is maximum at $t = 0$ sec; $\frac{1}{2}$ sec, 1 sec; $\frac{3}{2}$ sec, 2 sec, . . . , and minimum at $t = \frac{1}{4}$ sec; $\frac{3}{4}$ sec; $\frac{5}{4}$ sec . . . ; show this by writing $v = 2\pi \sqrt{9 \cos^2 \theta + 4 \sin^2 \theta} = 2\pi \sqrt{4 + 5 \cos^2 \theta}$, where $\theta = 2\pi t$; then v is maximum when $\cos \theta = \pm 1$ and minimum when $\cos \theta = 0$

13. 1360 ft

15. 114 ft/sec

Chapter 4

True or False

1. True 2. False; could be balancing forces. 3. False; e.g., circular motion. 4. True 5. False 6. True 7. False; this is a common misconception. 8. True

Exercises

1. (*a*) 8 m/sec² (*b*) second object has half the mass of first (*c*) $\frac{8}{3}$ m/sec²

3. AB: toward B; BC: toward center of arc; $CD: a = 0$; DE: toward center of arc; EF: toward E

5. from 2 sec to about 5 sec (−); from about 6 sec to about 8 sec (+)

7. (*a*) 7.07 m/sec² (*b*) 14.0 m/sec²

9. $\mathbf{F}_2 = 2\mathbf{F}_1$

11. 3 kg

13. 10 kg

15. (*a*) 8.77 m/sec² (*b*) 26.3 N

19. (a) 6i N (b) 2i + 3j N (c) 6ti + 3j N (d) $-18e^{-3t}$i N

21. (a) 547 slugs using $g = 32.0$ (544 slugs using $g = 32.174$) (b) 79.5 kg using 2.20 lb = (1 kg)g (c) 7.95×10^4 gm

23. 4.9×10^4 dynes = 0.49 N

25. (a) 126 lb (b) 5 slugs, same as at the earth's surface

27. (a) 1.67 m/sec^2 (b) 1.67 m/sec^2; 1.67 N; contact force exerted by 2-kg mass (c) 3.33 N

29. (a) momentum changes are equal in magnitude, opposite in direction (b) velocity change of 1-kg mass is twice that of 2-kg mass and opposite in direction

31. 33.3 mi/h at $\theta = 31°$ north of east

33. $\mathbf{v} = 10\mathbf{j}$ in water frame; $\mathbf{v} = 4\mathbf{i} + 10\mathbf{j}$ in land frame

Problems

1. (a) 7.2 N (b) 9.3 cm

3. (a) 5.5 sec (b) 2000 lb (c) 60.5 ft; 430 ft

5. (a) 400 N (b) 625 N

7. (a) 80 N (b) net force is 600 N; $F = 680$ N (c) 6.8 m/sec^2

9. (a) 3×10^6 m/sec^2; 6.71×10^6 mi/h (b) $5 \times 10^{-3}\%$

11. (b) apparent weight is less than force of gravity (c) in inertial frame $g = 981.4$ cm/sec^2 at equator and $g = 983.4$ cm/sec^2 at $\theta = 45°$ *Note:* The angle between the force of gravity (radial line) and the apparent free-fall acceleration at $\theta = 45°$ is small and can be neglected here.

13. (a) deflection direction opposite to $\mathbf{a}$ (c) $\theta = 8.6°$ forward (acceleration is 4.84 ft/sec^2 = 0.15g backward)

Chapter 5

True or False

1. False 2. False; true under conditions stated on page 119. 3. False; $T = mg + mv^2/R$. 4. True 5. True

Exercises

1. (a) 2.60i + 0.500j (b) -26.0i − 5.00j

3. (a) 1.5i + 2j m/sec^2 (b) 4.5i + 6j m/sec (c) 6.75i + 9j m

5. (a) 2 lb in each wire (b) $W/(2 \sin \theta)$ (c) T least at $\theta = 90°$, $T \to \infty$ as $\theta \to 0$

7. *Hint:* Consider the components of forces perpendicular to the wires.

9. (a) 8.8 m/sec^2 down (b) 7.8 m/sec^2 down (c) 10.2 m/sec^2 up

11. $a = 2.45$ m/sec^2, m_1 up the incline; $T = 36.8$ N

13. (a) 0.5 m/sec^2 (b) 1 N (c) 2 N (d) 2 N

15. (a) 19.6 N, the weight of the mass (b) 19.6 N (c) 39.6 N (d) 19.6 N from $t = 0$ to $t = 2$ sec; 9.6 N from $t = 2$ sec to $t = 4$ sec

17. $a = 17.0$ m/sec^2; for greater a, block would slide up incline

19. (a) 1.73 lb toward center of circle (b) 3.46 lb (c) 6.08 ft/sec

21. 5.56×10^{-3} N; factor of 1.83

23. $\tan \theta = v^2/Rg = 0.605$; $\theta = 31°$

25. (a) 1750 lb (b) 299 ft

27. (a) 8.17 N (b) copper moves up relative to jar as spring returns to unstretched position

29. 0.44

31. To the right; the air pressure must be greater on the left side of car to accelerate air to the right

33. force-time/length = mass/time; N-sec/m = kg/sec

35. 150 lb; 1.70 lb-sec/ft

Problems

1. (*a*) $T = F/(2 \sin \theta) = 955$ lb (*b*) 1433 lb

3. $F = 202$ N; $T = 100$ N at A, 101 N at B, and 201 N at C

5. for acceleration 0.6g stopping in 4.6 sec and 202 ft, force is 120 lb for 200-lb driver and 72 lb for 120-lb driver

7. (*a*) 2 kg (*b*) 13 kg (*c*) 0.2 kg (*d*) $a = 2g/7$

9. (*a*) $a = F/(m_1 + m_2)$; $F_c = m_2 a = m_2 F/(m_1 + m_2)$

11. $m_{3,\text{max}} = \dfrac{(m_1 + m_2)\mu}{1 - \mu}$

13. (*a*) $a = \dfrac{F_0 - 3mg}{3m}$; $T = \dfrac{F_0}{3}$ (*b*) $a_1 = \dfrac{3F_0}{11m} - g$; $a_2 = \dfrac{4F_0}{11m} - g$; $a_3 = \dfrac{2F_0}{11m} - g$; $T = \dfrac{4F_0}{11}$

15. (*a*) except at the top and bottom the weight has a tangential component which is unbalanced (*b*) resultant force is mv^2/R; force exerted by track is $mv^2/R - mg$ (*c*) $v = \sqrt{Rg}$; block leaves the track and moves in parabolic path as projectile until hitting track

17. 13.9% by volume; 53.5% by weight, or 2.67 lb of gold

19. 1.26 lb upper scale; 3.74 lb lower scale

21. (*b*) $dm = \rho A\, dy = \rho_0 A e^{-Kgy}\, dy$ (*c*) $M = A\rho_0/Kg = AP_0/g$

23. (*a*) 58.8 N (*b*) about twice the weight of Al (26.5 N) (*c*) no; it would take 6 mg = 159 N to accelerate aluminum upward with $a = 5g$

Chapter 6

True or False

1. True 2. True 3. False 4. False 5. True 6. True

Exercises

1. 5.96×10^{24} kg

3. 1.40×10^{-7} lb

5. magnitude is 2.89×10^{-5} N

7. 0.13% decrease

9. 2.04 N

11. magnitude is 2.69×10^{-2} N

13. (*a*) 2.82×10^{22} atoms; 8.18×10^{23} electrons; 1.31×10^5 C (*b*) 1.31×10^5 sec = 36.4 h (*c*) 3.86×10^{19} N *Note:* It is impossible to produce such a large charge on an object as small as a penny.

15. 4.10×10^{-9} N, same order of magnitude as atomic forces, much smaller than nuclear forces

17. 167 N/cm

19. $x_1/x_2 = k_2/k_1$

21. (*a*) static friction is greater than sliding friction (*b*) 0.229

23. 280 ft

25. (*a*) lifting easier than pushing down because the decrease in normal force also decreases frictional force (*b*) 60 lb horizontal; 106 lb at 30° down; 51.4 lb at 30° up

27. (a) $x = x_0 + 10t - 2.5t^2$; $v = 10 - 5t$ relative to car (b) stops at $t = 2$ sec and reaches original position at $t = 4$ sec

29. $x = x_0 + 10t - 4t^2$; $v = 10 - 8t$ for $0 < t < 1.25$ sec; object stops at $t = 1.25$ sec and remains at rest relative to car

31. 6.7 cm (vertical force 60 N; horizontal force 30 N)

Problems

1. (a) 179 (c) 0.46

3. (a) $C = 4\pi^2/gR_E^2$ (b) 84.4 min (c) 2.23×10^4 mi

5. for spherical mountain of radius r, deflection is approximately $\theta \approx r/R_E$; for $r = 6$ km, $\theta \approx 10^{-3}$ rad $= 0.06°$

7. (a) 0.6 g (b) 22.5 kg (c) 0.7g for m_2 and m_3; 0.4g for m_1; $T = 88.2$ N

9. (a) $F = \mu mg/(\mu \sin \theta + \cos \theta)$

11. $x = x_0 + \dfrac{m}{k}\left(1 - \dfrac{\rho_w}{\rho}\right) a$ for a positive upward; spring extends when elevator starts up, has length x_0 when velocity is constant, and contracts as elevator slows down

13. 0.50 gm/cm³

15. $\mu_s = 0.577$; $\mu_k = 0.400$

17. (a) 11.76 N (b) $a = 0.5g$ for 4-kg block; $a = 0.2g$ for 2-kg block (c) $a = 0.1g$ for both blocks; $f = 1.96$ N

19. 27 rev/min

21. (a) circular motion with angular velocity ω and linear velocity $r\omega$; resultant force is inward $mr\omega^2$ (b) centrifugal force is outward $mr\omega^2$; coriolis force is inward $2mv = 2mr\omega^2$

Chapter 7

True or False

1. True 2. False 3. True 4. True 5. False; unit of energy. 6. False 7. True

Exercises

1. (a) 490 J (b) −490 J

3. (a) 7.5 J (b) −3.0 J (c) 4.5 J

5. (a) 296 J (b) −196 J

7. 1 J = 10^7 ergs

9. (a) 5000 J (b) 1250 J

11. 400 J for 2-kg mass; 4 J for 200-kg mass

13. (a) 320 J; −196 J (b) 124 J

15. (a) 9 J (b) 19.5 J (c) 4.42 m/sec

17. (a) 74 N (b) 222 J (c) −147 J (d) 75 J

19. (a) 104 J (b) 104 J (c) 6.45 m/sec

21. (a) 1.5 N (b) tension; weight; normal force; each does zero work because each is perpendicular to motion

23. (a) weight 98 J; normal force 0; friction −50.9 J (b) 51.1 J

25. (a) 3920 J (b) 3920 J (c) he could do more work and increase his kinetic energy

27. (a) $W_{man} = 375$ ft-lb; $W_{earth} = -375$ ft-lb (b) 0 (c) 375 ft-lb (d) $W_{man} = 375$ ft-lb; $W_{earth} = 0$

29. (a) 392 J (b) 4.9 m/sec² (c) 9.8 m down incline or height of 15.1 m; $v = 9.8$

m/sec (d) at $t = 2$ sec potential energy 296 J, kinetic energy 96 J, total energy 392 J (e) 392 J

31. (a) $\frac{5}{3}$ m/sec (b) 15 J

33. (a) normal force = 17.6 N; $\mu = 0.197$ (b) 1.73 W (c) -5.20 J

35. (a) 140 lb (b) 22.3 horsepower (c) work done against air resistance and internal friction

37. 83°

39. (a) 142° (b) 101° (c) 90°

41. $B_z = 0$, $B_y = 2B_x/3$

Problems

3. (a) Assume weight of 700 N (157 lb) is raised 3 cm for each step; then $mgh = 21$ J per step. Using 2000 steps per mile gives 4.2×10^4 J (b) 4.2×10^4 J/1200 sec = 35 W. Alternative estimation assumes work is done to accelerate leg to speed v and back to rest in each step. Taking $m = 10$ kg for mass of leg and $v = 1.61$ km/1200 sec gives $2(\frac{1}{2}mv^2) = 18$ J per step, same order of magnitude.

5. (a) $-mg \sin \theta$ (b) $-mgL(1 - \cos \theta_0) = -mgH$

7. (a) $-\frac{3}{8}mv_0^2$ (b) $\mu = \dfrac{3}{16\pi} \dfrac{v_0^2}{rg}$ (c) $\frac{1}{3}$ rev

Chapter 8

True or False

1. False 2. False; only with conservative forces. 3. False 4. True 5. True 6. False

Exercises

1. (a) $W_{AB} = 0$; $W_{BC} = 78.4$ J; $W_{total} = 78.4$ J (b) $W_{AD} = 117.6$ J; $W_{DE} = 0$; $W_{EC} = -39.2$ J; $W_{total} = 78.4$ J (c) $W_{AC} = 78.4$ J

3. (a) from (0,0) to (2,0) $W = 0$; from (2,0) to (2,2) $W = 10$ J; $W_{total} = 10$ J (b) from (0,0) to (0,2) $W = 10$ J; from (0,2) to (2,2) $W = 16$ J; $W_{total} = 26$ J

5. (a) -10 J (b) 0

7. (a) $E_k = 25$ J $= E_{total}$ (b) $E_k = 0$, $U = 25$ J, $E_{total} = 25$ J (c) 5.10 m (d) no

9. (a) -25 J (b) -25 J (c) 100 m/sec

11. (a) -200 J (b) $+200$ J

13. 121 ft

15. (a) $U = -4x + U_0$ (b) $U = -4(x - 6) = 24 - 4x$ (c) $U = -4(x - 6) + 12 = 36 - 4x$

17. (a) 14.1 cm (b) 10 cm

19. conservative: (b), (c), and (e); nonconservative: (a), (d), (f), and (g)

21. (a) $U = -F_0z$ (b) $U = -F_0(z + 15)$ (c) $U = -F_0(z - 15)$

23. (a) $W = U = 3.12 \times 10^7$ J (b) $mgh = 6.24 \times 10^7$ J (c) force varies from 9.8 N at $r = R_E$ to 2.45 N at $r = 2R_E$

25. (a) 1.88 eV (b) 625 eV (c) 9.38×10^5 MeV (d) 510 keV = 0.510 MeV

27. (a) 3.20×10^{-11} J (b) 8.20×10^{10} J

29. (a) 0 at B and E; $+$ at C and D; $-$ at A and F (b) D (c) stable at E; unstable at B

31. (a) $F_x = -U_0/x$ (b) no

33. (a) C/r^2 (b) away; F_r is positive (c) U decreases as r increases (d) if C is negative, force is toward origin and U increases as r increases

35. 203 MeV

37. (a) conservative (b) conservative (c) nonconservative (d) nonconservative

Problems

1. (a) −11 J, −10 J, −7 J, −3 J, 0, +1 J, 0, −2 J, −3 J (b) F_x is given as a function of x only (c) graph should be the negative of that in (a)

3. $U = ax^3/3$

5. (a) minimum of $U(r)$ curve is point of stable equilibrium

7. (a) Along direct path from (4,1) to (4,4) on line x = 4 m the work is 36 J. Along path from (4,1) to (0,1) to (0,4) to (4,4) the work is zero. Other similar paths give different results because work along paths parallel to x axis adds to zero while work along vertical path depends on value of x. (b) second path given in (a)

9. (a) $F_r = -dU/dr = 12(U_0/r_0)[(r_0/r)^{13} - (r_0/r)^7]$ (b) $U = 0$ at $r = 2^{-1/6}r_0$; $F_r = 0$ at $r = r_0$ (c) U has minimum value $U = -U_0$ at $r = r_0$

Chapter 9

True or False   1. False 2. True 3. False 4. False 5. True

Exercises

1. (a) 75 J (b) −29.4 J (c) 45.6 J (d) 4.27 m/sec

3. (a) 4.43 m/sec (b) −19.6 J (c) 0.167

5. 2.55 m

7. (a) 294 J (b) 147 J (c) decrease by 147 J (d) −147 J

9. (a) 50 ft-lb (b) $\frac{1}{2}mv_x^2 + \frac{1}{2}mv_y^2 = 18.1$ ft-lb + 31.9 ft-lb; minimum E_k is $\frac{1}{2}mv_x^2 = 18.1$ ft-lb at top of flight (c) 31.9 ft-lb (d) 63.8 ft

11. 3.61 m/sec

15. (a) -6.27×10^9 J (b) -3.14×10^9 J (c) 7.92 km/sec

17. 2.38 km/sec

19. 51.2 km/sec

21. If r increases, U increases, E_k decreases, and E_{total} increases because $\Delta U = -2\Delta E_k$.

23. (a) 128 ft/sec (b) 8.51 atm

25. (a) 12 m/sec (b) 1.33 N/m² = 1.31 atm (c) same

Problems

1. (a) $mg(h - 2R)$ (b) $2g(h - 2R)/R$ (c) $h = 2.5R$

3. (a) $v = \sqrt{2g(2x - L)}$ (b) $a \geq g$

5. (a) 0.989 m   Note: The mass falls a distance $h = (4 + x) \sin 30°$, where x is the compression of the spring. If you neglect x in calculating the potential-energy loss, you get 0.885 for x. (b) 0.783 m (c) 1.54 m above spring (d) It will slide up and down a distance which decreases each time and eventually come to rest with the spring compressed about 8.6 cm.

7. (a) $\sqrt{2m_2gy/(m_1 + m_2)}$ (b) $a = m_2g/(m_1 + m_2)$ Note: $dy/dt = v$.

9. (b) When r decreases, E_k increases but U and E_{total} both decrease.

11. (b) about 1.6×10^{-10} m and 5×10^{-10} m

13. 6.6

15. (a) 463 m/sec (b) 10.74 km/sec (c) about 8%

Chapter 10

True or False   1. True 2. True 3. True 4. True 5. True 6. True

Exercises

1. $x_{CM} = 5$ m

3. (2 m, 1 m)

5. (*a*) $3\mathbf{i} - 1.5\mathbf{j}$ m/sec (*b*) $18\mathbf{i} - 9\mathbf{j}$ kg-m/sec

7. (*b*) 10.7 ft/sec²

9. (*a*) 32.7 ft from shore (*b*) CM 32.7 ft from shore; near end of log 2.7 ft from shore (*c*) CM 32.7 ft from shore; man and end of log 5.45 ft from shore

11. (*a*) $x = 0$ (*b*) $x = -8$ cm, $v = -2$ cm/sec

13. (*a*) 50 kg-m/sec (*b*) 50 kg-m/sec (*c*) 3.57 m/sec

15. 4.55 ft/sec

17. 3.33 m/sec

19. (*a*) 0.8 m/sec (*b*) before: heavier car 0.2 m/sec right, lighter car 0.3 m/sec left; after: both cars at rest (*c*) before: heavier car 1200 J, lighter car 1800 J; after: 0 for both cars

21. (*a*) 39 J (*b*) 3 m/sec (*c*) 2 m/sec to the right and 2 m/sec to the left (*d*) 12 J (6 J each) (*e*) $\frac{1}{2}Mv_{CM}^2 = 27$ J $= 39$ J $- 12$ J

23. (*a*) Both have same velocity, and so there is no rotation. (*b*) Masses have equal but opposite velocities, and so CM is at rest.

25. (*a*) Both have same velocity, so that spring remains unstretched and there is no rotation. (*b*) Masses have equal but opposite velocities, so that CM is at rest. Motion can be oscillation along line of spring, rotation about CM, or some combination.

27. 0.0176 J

31. (*a*) $3L\lambda_0/2$ (*b*) $5L/9$

Problems

1. (*a*) about 2950 mi from center of earth or 1050 mi beneath earth's surface (*b*) gravitational attraction of sun and other planets; neglecting other planets, $\mathbf{a}_{CM}$ is toward sun (*c*) about 5900 mi

3. (*a*) 2 m/sec (*b*) 8 m/sec (*c*) $2\frac{2}{3}$ m/sec (*d*) frictional force exerted by ground on man (*e*) 1080 J is total increase in kinetic energy

5. $r/6$ from center of large disk along common diameter and away from hole

7. (*a*) 8 kg-m/sec (*b*) 1 m/sec for 8-kg mass, 4 m/sec for 2-kg mass (*c*) 4.12 m/sec and 5.66 m/sec (*d*) 80 J; 20 J (*e*) 5 m/sec for 8-kg mass, 2-kg mass is at rest; 80 J; 20 J

9. $x_{CM} = 6.67$ in, $y_{CM} = 3.33$ in

Chapter 11

True or False

1. False 2. False; true only in center-of-mass frame. 3. False; true only in center-of-mass frame. 4. True

Exercises

1. (*a*) 4 N-sec (*b*) 2 m/sec

3. (*a*) 8 N-sec (*b*) 0.16 sec

5. (*a*) 15.6 lb-sec (*b*) 0.31 sec

7. about 2 N

9. (*a*) 1.4 lb-sec (*b*) 1400 lb

11. $\sqrt{8mE_k}$

13. 8.33×10^{-4} N; $P - 2.60 \times 10^{-4}$ N/m²

15. 5.65×10^{-21} J $= 0.0353$ eV

17. (a) about 5×10^{-4} cm (b) about 0.025% of distance traveled due to club

19. (a) 2.4 lb-sec (b) 2400 lb, which is 7500 times weight of ball; yes

25. (b) and (c)

27. 15-gm mass; 3 m/sec to the left; 6-gm mass; 7.5 m/sec to the right

29. (a) 2 m/sec (b) 10-kg mass: 3 m/sec, 6-kg mass: -5 m/sec (c) 10-kg mass: -3 m/sec, 6-kg mass: 5 m/sec (d) 10-kg mass: -1 m/sec, 6-kg mass: 7 m/sec

31. before: $\mathbf{v}_{1i} = 2u\mathbf{i}$, $\mathbf{v}_{2i} = 0$; after: $\mathbf{v}_{1f} = u\mathbf{i} + u\mathbf{j}$, $\mathbf{v}_{2f} = u\mathbf{i} - u\mathbf{j}$, where $u =$ speed of center of mass and ball on right has been assumed initially at rest

33. 1.76 ft/sec

35. 15 gm

37. 447 m/sec

39. (a) 94.9 m/sec (b) 98%

41. 6.16 eV

43. (a) 20% (b) heat (internal energy of ball and surroundings) (c) 0.89

Problems

3. (c) 2.08 N (d) 1.5 N, 2.67 N

5. $a = (100 \text{ ft}/x)g$, where x is distance moved in hand; for $x = 4$ ft, $a = 25g = 800$ ft/sec^2; $I = 0.8$ lb-sec; $F_{av} = 8$ lb neglecting weight (8.32 lb including weight), $t = 0.1$ sec; neglecting weight results in about 4% error

7. (a) 4.8 N-sec (b) 1600 N (c) 2.4 N-sec (d) 19.2 N

9. 48 ft/sec

11. (a) 7 m/sec (b) $p_1 = +12$ N-sec, $p_2 = -12$ N-sec (c) $p_1 = -6$ N-sec, $p_2 = +6$ N-sec (d) $v_1 = 5.5$ m/sec, $v_2 = 8$ m/sec (e) 22.5 J

13. $v = [2(m_1 + m_2)/m_1]\sqrt{gL}$

15. (a) 0.72 (b) 42 collisions

Chapter 12   **True or False**   1. False 2. True 3. True 4. False; e.g., angular displacement. 5. True 6. False 7. True 8. False; zero resultant torque is also necessary. 9. True 10. False

Exercises

1. 1 rad/sec $= 9.55$ rev/min

3. (a) 0.2 rad/sec (b) 0.955 rev

5. (a) 2.91×10^{-2} rad/sec^2 (b) 1.75 rad/sec (c) 33.3 rev

7. 0.625 sec

9. (a) 40 cm/sec for 3-kg masses; 80 cm/sec for 1-kg masses; 1.12 J (b) $I = 0.56$ kg-m^2

11. (a) 52 kg-m^2 (b) 25.4 rev/min

13. 28 kg-m^2

15. (a) 26 kg-m^2 (b) $I_{x'} = 18$ kg-m^2, $I_y = 8$ kg-m^2

17. $MR^2/4$

19. $3MR^2/2$

21. (a) 2.4 N-m (b) 66.7 rad/sec^2 (c) 200 rad/sec (d) 720 J (e) 300 rad (f) (300 rad) (2.4 N-m) $= 720$ J

23. (a) 0.126 rad/sec^2 (b) 400 N-m (c) 3.18×10^3 kg-m^2

25. (a) $\tau = 1.5$ N-m, $I = 1.5 \times 10^{-2}$ kg-m^2, $\alpha = 100$ rad/sec^2 (b) $\omega = 400$ rad/sec, $P = 600$ W

27. 267 horsepower

29. 1.93 m/sec (1.98 m/sec if I is neglected)

31. 26 lb at left and 74 lb at right

33. 354 lb

Problems

1. (a) π rad/sec^2 (b) $3\pi/2$ rad/sec^2

3. (a) 1.97×10^5 J (b) 26.2 N-m (c) 1200 rev

5. (a) $(2m_2 g \sin\theta)/(2m_2 + m_1)$ (b) $(m_1 m_2 g \sin\theta)/(2m_2 + m_1)$ (c) $m_2 gh$ (d) $m_2 gh$ (e) $\sqrt{2gh/(1 + m_1/2m_2)}$ (f) for $\theta = 0$: $a = 0$, $T = 0$, $E = 0$; for $\theta = 90°$ $a = 2m_2 gh/(2m_2 + m_1)$, $T = m_1 m_2 g/(2m_2 + m_1)$, $E = m_2 gh$; for $m_1 = 0$: $a = g \sin\theta$, $T = 0$, $E = m_2 gh$

7. (b) $\omega = \sqrt{2gh(\cos\theta_0 - \cos\theta)/k^2}$ (c) $F_r = mg(7 \cos\theta - 6 \cos\theta_0)$; $F_t = 2mg \sin\theta$

9. $I = ML^2/6$

13. $\tan\theta = 0.5$, $\theta = 27°$

15. $F = mg \sqrt{2hR - h^2}/(R - h)$

Chapter 13

True or False

1. False; $\mathbf{A} \times \mathbf{B} = -\mathbf{B} \times \mathbf{A}$. 2. True 3. False 4. False; $\mathbf{L}$ must be constant. 5. True 6. True 7. False; f is less than $\mu_s N$ except at limiting angle.

Exercises

1. $\mathbf{F} = -F\mathbf{i}$, $\mathbf{r} = R\mathbf{j}$, $\boldsymbol{\tau} = RF\mathbf{k}$

3. $\boldsymbol{\tau} = mgx\mathbf{k}$

5. (a) $25\mathbf{k}$ (b) $25\mathbf{j}$ (c) $8\mathbf{k}$

9. (a) 60 kg-m^2/sec (b) 75 kg-m^2 (c) 0.8 rad/sec

11. (a) doubled (b) doubled

13. (a) 4 N-m (b) $\frac{2}{3}t$ rad/sec

15. $L_{earth} = 7.15 \times 10^{33}$ kg-m^2/sec; $L_{sun} = 2.69 \times 10^{40}$ kg-m^2/sec

17. (a) 132 gm-cm^2/sec (b) 132 gm-cm^2/sec

19. (a) 0.784 N-m (b) $0.26v$ (c) 3.02 m/sec^2

21. (a) $\frac{1}{3}Ml^2\omega$ (b) $\frac{1}{2}l\omega$ (c) $\frac{1}{12}Ml^2\omega$

23. (a) 600 rev/min (b) since $E = L^2/2I$ and $I_f = 3I_i$, $E_f = E_i/3$

25. (a) 6.47 rad/sec (b) $E_{ki} = 41.9$ ft-lb; $E_{kf} = 86.4$ ft-lb (c) 44.5 ft-lb

27. Without smaller rotor, any variation in the angular speed of the main rotor would produce an opposite rotation in the helicopter body because of conservation of angular momentum.

29. (b) no

31. 527 ft-lb

33. (a) 2.88 rad/sec (b) 0.144 m/sec (c) 0.417 m/sec^2 toward pivot (d) $F_{vert} = 19.6$ N; $F_{horiz} = 0.834$ N

35. *Hint:* First find speed v from $v^2 = rg \tan 30°$, where $r = 0.5$ m, as in Example 5-7, page 122. (a) $\mathbf{L} = -2.91\hat{\mathbf{r}} - 1.68\mathbf{k}$ kg-m^2/sec (b) 9.8 N-m

Problems

3. $E_k = \frac{3}{4}MV^2$

5. 2.7r; 2.5r

7. 1.36 mi

Chapter 14

True or False

1. False 2. True 3. False; period is independent of amplitude. 4. True 5. True 6. False; it is periodic but not simple harmonic for large amplitudes. 7. True 8. True 9. True 10. True

Exercises

1. (a) 2.5 Hz; 0.4 sec (b) 3 m (c) $x(0) = -3$ m; $x(\frac{1}{2}) = 0$

3. $x = 6 \cos \pi t$ in; $v = -6\pi \sin \pi t$ in/sec; $a = -6\pi^2 \cos \pi t$ in/sec^2

5. (a) 4 ft at $t = \pi/4$ sec (b) $v = 8 \cos 2t$ ft/sec; at $t = 0$, $v = 8$ ft/sec (c) $a = -16 \sin 2t$ ft/sec^2; $a = 0$ at $t = 0$; $a_{max} = 16$ ft/sec^2

7. (a) 2 rad/sec (b) 0.318 Hz; 3.14 sec (c) $x = 40 \cos (2t + \delta)$ cm/sec; $y = 40 \sin (2t + \delta)$ cm/sec, where δ depends on initial position

9. (a) 474 N/m (b) 2.37 J (c) $x(t) = 10 \cos (4\pi t + \delta)$ cm; no

11. (a) 1580 N/m (b) 79.0 m/sec^2 (c) 0.0197 J

13. $E_k = 0.75E$; $E_k = U$ at $x = A/\sqrt{2}$

15. 24.8 cm

17. 9.79 m/sec^2

21. 2.0 sec

23. (a) 2 in; $\frac{1}{3}$ ft-lb (b) $\Delta U_{spring} = +\frac{7}{4}$ ft-lb; $\Delta U_{gravity} = -1$ ft-lb; $\Delta U_{total} = \frac{3}{4}$ ft-lb = $\frac{1}{2}(24 \text{ lb/ft})(\frac{1}{4}\text{ ft})^2$ (c) 0.45 sec; 2.2 Hz; 3 in

25. (a) π; 8 cm (b) 0.628 sec

Problems

1. (a) equal (b) equal (c) 2.4 times as far during first second

3. 6.2 cm

5. (b) They meet at the bottom. The period of motion is independent of the amplitude, and so the time to reach the bottom $T/4$ is the same for each mass.

7. (a) $kA/(m_1 + m_2)g$ (b) A and E remain unchanged; ω decreases, $\omega_f = \sqrt{m_1/(m_1 + m_2)}\, \omega_i$; T increases, $T_f = \sqrt{(m_1 + m_2)/m_1}\, T_i$

9. (a) 13 cm (b) 6.5 cm (c) 0.51 sec (d) 0.32 J (e) 0.8 m/sec; 0.13 sec (f) 0.12 sec, 1.13 m/sec

11. (a) $I = ML^2 + \frac{2}{5}Mr^2$ (c) 0.008% error ($= 1.6 \times 10^{-4}$ sec for $T_0 = 2.01$ sec); about 22 cm

13. (a) $d^2x/dt^2 = -(k/m)x$ (b) $E = \frac{1}{2}mv^2 + \frac{1}{2}kx^2 = \frac{1}{2}kA^2$; $v = \sqrt{2E/m - kx^2/m} = \sqrt{k/m}\,\sqrt{A^2 - x^2}$

15. (a) $v_{SHM} = \theta_0 g L$ (b) $v = 2gL(1 - \cos \theta_0)$ (c) $v_{SHM} = 0.5464$ m/sec; $v = 0.5457$ m/sec; difference is 0.07 cm/sec, about 0.13%

19. (a) $dU/dx = -F_x = 0$ at $x = x_0$ (b) $-\frac{1}{2}U_0$ (c) $\omega = \frac{6}{x_0}\sqrt{\frac{U_0}{m}}$

21. 1.28 sec. *Hint:* Use $T = 2\pi \sqrt{L/g'}$, where g' is free-fall acceleration observed in accelerating frame.

Chapter 15

True or False

1. True 2. False 3. True 4. True 5. True 6. False 7. True 8. False 9. True 10. True

Exercises

3. (*a*) 0.44 sec; 0.18 J (*b*) 0.045 kg/sec

5. (*a*) $\frac{3}{4}$

7. 1.0 sec; 2.17 cm

9. (*a*) $a(t) = (F_0/m) \cos \omega t$; $v(t) = (F_0/m\omega) \sin \omega t$ (*b*) $x(t) = -(F_0/m\omega^2)(\cos \omega t - 1)$ (*c*) frequency of oscillation is same but amplitude is different; in steady state, $x(t)$ is given by Equation 15-25

11. for $b = 0$, $A_1/A_2 = \frac{7}{5}$

15. (*a*) 4.98 cm (5.00 cm neglecting the damping term) (*b*) $\omega = 14.1$ rad/sec; $f = 2.25$ Hz (*c*) 35.4 cm

Problems

1. After 14 cycles energy is about 49% of original energy, and amplitude is $0.70A_0$. *Hint:* Easiest to use $E_n = (0.95)^n E_0 = 0.5E_0$ and solve for n.

5. (*a*) at $t = 3.54$ sec (*b*) 6 J

7. (*a*) 219 sec (*b*) $\omega' \approx \omega_0 = 24.5$ rad/sec; $f' = 3.90$ Hz

9. $A(t + T) = e^{-3.05}A(t) = 4.74 \times 10^{-2}A(t)$; $E(t + T) = e^{-6.1}E(t) = 2.24 \times 10^{-3}E(t)$

Chapter 16

True or False

1. False 2. True 3. True 4. True 5. True 6. False 7. True 8. True 9. True 10. False

Exercises

1. 4.37×10^{11} m

3. 1.50×10^{15} m

5. 84.6 y

7. 6.02×10^{24} kg

9. (*a*) 0 (*b*) $\mathbf{g} = -Gm \left| \dfrac{1}{(x - a)^2} + \dfrac{1}{(x + a)^2} \right| \mathbf{i}$

11. (*a*) 10.8 N (*b*) at 56° above the x axis (*c*) $\mathbf{F} = 10\mathbf{i} + 15\mathbf{j}$ N

13. 2.5×10^{-3} N/kg toward the center of the circle

15. (*a*) $V(x) = -2Gm/(x^2 + a^2)^{1/2}$ (*c*) $g_x = -Gmx/(x^2 + a^2)^{3/2}$; $x = 0$

17. (*a*) $V(x) = -GM/(x^2 + r^2)^{1/2}$ (*c*) at $x = 0$, V is maximum and g is zero

21. (*a*) Lines cannot cross because direction of $\mathbf{g}$ is ambiguous at intersection. (*b*) Lines cannot have discontinuous change in direction (kinks) because direction of $\mathbf{g}$ is ambiguous at such a point. (*c*) Lines diverging from a point imply negative mass at the point, which does not exist.

23. (*a*) $\mathbf{g} = -2.0 \times 10^{-10}\ \hat{\mathbf{r}}$ N/kg (*b*) $\mathbf{g} = 0$

25. (*c*) 6.24×10^7 J/kg (*d*) 4.99×10^9 J $= 1.39 \times 10^3$ kW-h; about \$41.60 (in practice the cost is somewhat greater)

Problems

3. (*a*) 3.31 y (the semimajor axis is 1.4×10^8 km) (*b*) 1.49×10^{29} kg (*c*) m_2 has greater speed and greater total energy (*d*) $v_P = 1.8v_A$

5. (*b*) $m_{s_2} = (r_2/r_1)m_{s_1}$; m_{s_1}; toward center of s_1 (*c*) toward center of s_1 (*d*) 0 (*e*) for spherical shell, $m_{s_2} = (r_2^2/r_1^2)m_{s_1}$; the elements would produce equal and opposite fields at point P

7. 198 m/sec $= 444$ mi/h

9. $g_x = -GM/(x_0^2 - L^2/4)$

11. $\omega = \sqrt{4\pi G\rho_0/3}$

Chapter 17   **True or False**   1. True 2. True 3. False 4. False 5. False 6. True

Exercises   1. The system is not in equilibrium when the temperature and pressure readings are taken.

3. (a) adiabatic (b) diathermic (c) adiabatic. *Note:* Both the increase in P_A and the decrease in V_A tend to increase the temperature of A.

5. $t_A \neq t_B$ and $t_B \neq t_C$; no; t_A and t_C may or may not be equal

7. (a) 8.4 cm (b) 107°C

9. (a) 53.9 mm (b) 371 K

11. (a) 33.9 ft (b) 0.89 atm = 673 mm Hg

13. 29.9 in Hg

15. (a) 23.37 lb/in² = 1.59 atm (b) 8.67 lb/in² = 0.59 atm

17. 3.81×10^4 lb

19. (a) 6173 K (b) 10,650°F

21. 37.0°C

23. (a) 10^7 °C (b) about 1.8×10^7 °R (c) about 1.8×10^7 °F

25. (a) 12.3 ℓ (b) 24.6 ℓ (c) 375 K

27. 1.15

29. 5.40×10^3 N due to air inside box; if box is in air at pressure of 1 atm, there will be an opposing force of 4.05×10^3 N due to air outside, giving a net force of 1.35×10^3 N outward.

31. 1.79 moles; 1.08×10^{24} molecules

33. 275 m/sec for argon and 870 m/sec for helium

35. 8.84 K

37. 33×10^{-6} per Celsius degree

39. $1/T = 0.003661$, which differs from 0.003673 by about 0.3%

Problems   1. (a) 6.37 atm (b) 21.4% escapes (c) 3.93 atm

3. (a) 32.0 lb/in² (b) 27.8 lb/in². *Hint:* You must convert gauge pressure to absolute pressure by adding 14.7 lb/in² before using $PV = nRT$.

5. (a) 6.07×10^{-21} J (b) 1.9×10^3 m/sec (c) 7.31×10^3 J (d) 1.9×10^3 J

7. (a) 1.6×10^5 K (b) 1.0×10^4 K (c) yes (see essay in Chapter 9 for complete discussion) (d) about 7.3×10^3 K for O_2 and 4.6×10^2 K for H_2; yes

9. 145°C

11. 5.09 m/sec

Chapter 18   **True or False**   1. False 2. False 3. False 4. False 5. True 6. True 7. True 8. False 9. True 10. False

Exercises   1. 200 kcal

3. (a) 1 Btu = 252 cal (b) 1 Btu = 1054 J (c) 1440 Btu

5. 0.41 cal/gm-°C

7. (a) 0°C (b) 125.6 gm, or 62.8%

9. 49.5 gm

11. (*a*) 4.18×10^3 J/kg-K (*b*) 3.33×10^5 J/kg

13. $t_f = 21.54°C$

15. (*a*) 0.12°C (*b*) 1.73°C

17. 1 ℓ-atm $= 101.3$ J $= 24.22$ cal

19. (*a*) 405 J (*b*) 861 J

21. (*a*) 507 J (*b*) 963 J

23. *T* increases slightly; since *U* decreases when the atoms are farther apart and E_{total} is constant, E_k and therefore *T* must increase.

25. (*a*) 3.00 moles (*b*) $C_v = 8.96$ cal/K; $C_p = 14.9$ cal/K (*c*) $C_v = 14.9$ cal/K; $C_p = 20.9$ cal/K

27. (*a*) $C_v = \frac{3}{2}R$, $C_p = \frac{5}{2}R$, $\gamma = \frac{5}{3}$
    (*b*) $C_v = \frac{5}{2}R$; $C_p = \frac{7}{2}R$; $\gamma = \frac{7}{5}$
    (*c*) $C_v = \frac{7}{2}R$, $C_p = \frac{9}{2}R$, $\gamma = \frac{9}{7}$

29. (*a*) $\Delta U = 894$ cal; $W = 0$; $Q = 894$ cal (*b*) $\Delta U = 894$ cal; $W = 596$ cal; $Q = 1490$ cal (*c*) $W = 596$ cal

31. 3.98 gm/mole

35. (*a*) $V_i = 2.24$ ℓ; $V_f = 5.88$ ℓ (*b*) $T_f = 143$ K

Problems

1. 34 km; variation in *g* can be neglected since H/R_E is only about 0.5%; if most of the energy dissipated because of air resistance were absorbed by the ice, air resistance could be neglected. Since much of the energy is probably absorbed by the surrounding air, *H* would be larger if air resistance were taken into account.

3. Dulong-Petit law gives 6 cal/mole-K, whereas measured value is about 0.5 cal/gm-K = 9 cal/mole-K. This could be explained by three degrees of freedom associated with internal energy in addition to the six degrees associated with center-of-mass vibration.

5. $t_f = 28.5°C$; t_i should be about 16°C

9. $C_p' - C_v' = 1.7 \times 10^{-7}$ cal/mole-K

11. (*b*) calculation gives -65.1 ℓ-atm $= -1.58 \times 10^3$ cal done by gas, that is, 65.1 ℓ-atm done on gas (*c*) 1.58×10^3 cal removed from gas during cycle (*d*) Work done by gas for each step is $+22.2$ ℓ-atm during adiabatic expansion, $+8.9$ ℓ-atm during expansion at 1 atm, 0.0 during heating at constant volume, and -96.2 ℓ-atm during compression at 5 atm.

Chapter 19 **True or False**

1. False 2. False; e.g., isothermal expansion of ideal gas. 3. False 4. False 5. False 6. True 7. False 8. True 9. False 10. True

Exercises

1. (*a*) 500 J (*b*) 400 J

3. (*a*) 40% (*b*) 335 W

5. 66.7 J absorbed and 46.7 J rejected per cycle

7. *Hint:* Use a perfect refrigerator to transfer 140 cal from the cold to hot reservoir.

9. The reverse process violates the Kelvin-Planck statement in (*a*) and (*b*) and the Clausius statement in (*c*).

13. (*a*) $33\frac{1}{3}$% (*b*) $W = 33.3$ cal; $Q_c = 66.7$ cal (*c*) 2

15. a 5-K decrease in temperature of cold reservoir

17. 224 K

19. (a) 2.75 cal/K = 11.5 J/K (b) 0

21. (a) $\Delta U = +50$ cal (b) $\Delta S = 0.167$ cal/K (c) $\Delta S_u = 0$ (d) parts (a) and (b) have same answers but for part (c) $\Delta S_u > 0$

23. (a) 2.75 cal/K (b) 2.75 cal/K

25. 0.417 cal/K

27. +1.45 cal/K

29. (a) $W = 400$ cal (b) $+0.5$ cal/K (c) $Q_{out} = 400$ cal; $W_c = 200$ cal, which is 200 cal less than in (a) (d) 200 cal = (0.5 cal/K)(400 K) (e) since Carnot efficiency is 50% here, 100 cal additional work could be done extracting 200 cal from hot reservoir; 100 cal = (0.5 cal/K)(200 K)

31. (a) 1.38 cal/K (b) 413 cal

33. Still air at 25°C has about 102 J more energy, but only air with center-of-mass motion can drive a windmill.

Problems

1. (a) (1) $W = 0$; $Q = 894$ cal; $\Delta U = 894$ cal
   (2) $W = 49.2$ ℓ-atm $= 1190$ cal; $Q = 2980$ cal; $\Delta U = 1790$ cal
   (3) $W = 0$; $Q = -1790$ cal; $\Delta U = -1790$ cal
   (4) $W = 24.6$ ℓ-atm $= 596$ cal; $Q = -1490$ cal; $\Delta U = -894$ cal
   (b) 15.3%

3. (c) 65% (d) The expansion and compression are not adiabatic, and none of the processes are quasi-static.

5. (a) greatest possible efficiency = 41% (b) 1.68×10^9 J

9. 100 W = 0.135 horsepower

13. (a) 10.2°C (10.0°C if 80 cal/gm is taken for heat of fusion of water) (b) +5.3 cal/K

Chapter 20

True or False

1. True 2. True 3. False 4. False; speed of sound increases by factor $\sqrt{2}$. 5. False

Exercises

3. Segments from $x = 2$ cm to $x = 3$ cm are moving up. Segments from $x = 1$ cm to $x = 2$ cm are moving down. Segment at $x = 2$ cm is at rest.

11. (b) T in pounds; μ in slugs per foot

13. 0.24 sec

15. 924 ft/sec (using $g = 32$ ft/sec²)

17. 776 mi/h

19. 5.06×10^3 m/sec

21. 2.86×10^{10} N/m²

23. $T = 2L/v$; no

25. $T = 4L/v$; no

Problems

1. at 300 K, $v = 0.216$ mi/sec $\approx \frac{1}{5}$mi/sec ($1/v = 4.64$ sec/mi); error is about 8%; correction for travel time of light is insignificant

3. (a) 78.4 m (b) 69.8 m (using 347 m/sec for v, which gives 0.226 sec for sound travel time) (c) 70.6 m

5. Approximation gives 347.4 m/sec; exact calculation gives 347.0, assuming that $v = 331$ m/sec at 0°C. Approximation is good to about 0.1%.

7. 2.21 sec

Chapter 21 **True or False** 1. True 2. True 3. True 4. False; only transverse waves can be polarized. 5. False; 90° out of phase. 6. True 7. True 8. True 9. True 10. False; it has 1000 times the intensity.

Exercises 1. (a) $y(x,t) = y_0 \sin k(x - vt)$ (b) $y(x,t) = y_0 \sin 2\pi\left(\dfrac{x}{\lambda} - ft\right)$

(c) $y(x,t) = y_0 \sin 2\pi\left(\dfrac{x}{\lambda} - \dfrac{t}{T}\right)$ (d) $y(x,t) = y_0 \sin \dfrac{2\pi}{\lambda}(x - vt)$

(e) $y(x,t) = y_0 \sin 2\pi f\left(\dfrac{x}{v} - t\right)$

3. (a) 1.30 m (b) 0.65 m

5. 3000 m for AM; 3 m for FM

7. (a) $y = 0.02 \sin (37.7x - 377t)$ (b) no; any phase constant can be added to the phase

9. (a) linearly polarized (b) circularly polarized if $y_0 = z_0$; elliptically polarized if $y_0 \neq z_0$

11. (a) 0 (b) 3.7×10^{-6} m

13. 8.27×10^{-2} N/m²

15. (a) 2.1×10^{11} m (b) 1.1×10^{11} m

17. 0.035 m

19. (a) 90° (b) $\sqrt{2}\,A$

21. sum is zero

23. (a) 4.74 J (b) 30.6 W

25. (a) 138 N/m² (b) 21.6 W/m² (c) 0.216 W

27. (a) 10^{-11} W/m² (b) 2×10^{-12} W/m² (c) 9.2×10^{-5} N/m²; 4.1×10^{-5} N/m²

29. 99% (power would have to be reduced by factor of 100)

Problems 3. (a) 0.0023 sec; 440 Hz (b) 316 m/sec (c) 0.719 m; 8.74 m⁻¹ (d) $y(x,t) = 5 \times 10^{-4} \sin 2\pi(1.39x - 440t)$ (e) 1.38 m/sec; 3.82×10^3 m/sec² (f) 3.02 W

5. (a) one-ninth reflected; eight-ninths transmitted (b) $\frac{1}{3}A$ and $\frac{2}{3}A$

7. (a) 5.46 cm (b) 120° (c) $9I_0$; $2\pi n$, where $n = 1, 2, \ldots$ (e) 240°

11. 26%. This would imply that the noise energy doubles every 3 years, which is probably too great an increase to be reasonable over a long period of time.

Chapter 22 **True or False** 1. True 2. True 3. True 4. False 5. True

Exercises 1. (a) 440 m/sec (b) 2 m; 220 Hz (c) 440 Hz; 660 Hz

3. 141 Hz

5. (a) 296 m/sec (b) 87.8 N

7. (a) 2 m; 25 Hz (b) $y(x,t) = 0.004 \cos 50\pi t \sin \pi x$

9. (a) 200 m/sec; 0.25 cm (b) 1.26 m (c) 1.26 m

11. (a) 16 m; 5.33 m; 3.2 m (b) 6.25 Hz; 18.75 Hz; 31.25 Hz

13. (a) 4 m; 1.57 m^{-1} (b) 800π rad/sec (c) $y(x,t) = 0.03 \sin \frac{1}{2}\pi x \cos 800\pi t$

15. 17.4 Hz for open; 8.71 Hz for closed

17. (a) 2230 Hz (b) eighth or ninth ($f_9 = 20{,}070$ Hz)

Problems

1. 3.27 cm for A; 6.19 cm for B; 7.56 cm for C; 10 cm for D

3. (a) The powder moves about due to the motion of the gas and collects at the displacement nodes, where the gas is at rest. (b) $v = 2fx$, where x is distance between piles

5. (a) eighth and ninth harmonics (b) 2.16 m (c) 4.32 m

7. 338 m/sec (not accurate because of end correction)

9. (a) 345 m/sec (b) 1.25 cm

11. (a) $v_y = -7.54 \sin 2.36x \sin 377t$ (b) free end at $x = 0.666$ m; maximum speed of this point is 7.54 m/sec (c) $a_y = -2.84 \times 10^3 \sin 2.36x \cos 377t$. The free end at $x = 0.666$ m has the greatest acceleration, which has a maximum value of 2.84×10^3 m/sec^2.

13. this result applies to all harmonics; 1.54%

Chapter 23   **True or False**

1. True 2. True 3. False 4. True 5. False

Exercises

1. (a) 0.4 cm (b) $y(x,t) = 0.4$ cm $\cos (0.1x - 10t) \cos (5.9x - 590t)$; $v_p = 100$ m/sec (c) $v_g = 100$ m/sec (d) 62.8 m (e) dispersionless

3. (a) $s_1 = s_0 \cos (9.24x - 1000\pi t)$; $s_2 = s_0 \cos (9.33x - 1010\pi t)$; $s = 2s_0 \cos (.046x - 5\pi t) \cos (9.29x - 1005\pi t)$

5. 437 Hz

7. (a) 10 μsec (b) $\Delta f \approx 10^5$ Hz

11. $v_p = v_g = \sqrt{\gamma RT/M}$; no, nondispersive

Problems

3. When the zero of your vernier scale is aligned with the 0-cm mark on the meterstick, the first line on the vernier is 0.1 cm before the 1-cm mark, the second vernier mark is 0.2 cm before the 2-cm mark, etc. When the vernier moves 0.1 cm past the 0-cm mark, the first vernier mark is aligned with the 1-cm mark. The frequency of centimeter marks on the meterstick is 1 per centimeter and the frequency of vernier marks is 1.1 per centimeter. The beat frequency is 0.1 per 1 cm or 1 per 10 cm. The marks are therefore in phase once every 10 cm.

Chapter 24   **True or False**

1. True 2. False; it also depends on the "absolute" velocity of source or receiver relative to the still air. 3. True 4. False 5. False 6. True

Exercises

1. (a) 8.84×10^{-5} W/m^2 (b) about 80 dB

3. (a) I varies as $1/r$ (b) A varies as $1/\sqrt{r}$

5. (a) $\lambda = v/f_0$; $f = v/\lambda = f_0$ (b) $\lambda' = (v - u_S)/f_0$; $f' = v/\lambda' = vf_0/(v - u_S)$ (c) $\lambda = v/f_0$; $f' = (v + u_R)/\lambda = (v + u_R)f_0/v$

7. (a) 260 m/sec (b) 1.3 m (c) 261.5 Hz

9. (a) 1.7 m (b) 247 Hz

11. 153 Hz

15. (a) 8.3° (b) 16.5° (c) three maxima beyond the $\theta = 0$ maximum

19. (a) $\sin \theta = m/3$, $m = 0, 1, 2, 3$ (b) $\sin \theta = m/9$, $m = 1, 2, 4, 5, 7, 8$

21. 72°

23. (a) 0 (b) $2I_0$ (c) $4I_0$

25. (a) $r_1 - r_2 = (n + \frac{1}{4})\lambda$ (b) $r_1 - r_2 = (n - \frac{1}{4})\lambda$, where $n = 0, 1, 2, \ldots$

Problems

1. (a) 5.3 ft; 207.6 Hz (b) 215 Hz (c) 15 Hz

3. (a) 8.84×10^{-6} W/m² (b) about 69 dB (c) 892 m

5. 23.8 mi/h

7. 3.33 mm

11. (a) $I_1 = 1.99 \times 10^{-5}$ W/m²; $I_2 = 8.84 \times 10^{-6}$ W/m² (b) $I = 5.53 \times 10^{-5}$ W/m² (c) $I = 2.2 \times 10^{-6}$ W/m² (d) $I = 2.87 \times 10^{-5}$ W/m²

13. Speakers are connected properly if they move in phase when driven by a monophonic source. A stereo source is not used because the speakers would not be coherent (they would be driven by different sources). Bass is used so that listener's position is not critical for detection of constructive interference from in-phase speakers and destructive interference from 180°-out-of-phase speakers. For example, if $f = 100$ Hz, $\lambda \approx 11$ ft, and listener need be equidistant from speakers only to within a few feet. If $f = 1000$ Hz, $\lambda \approx 1.1$ ft, and listener's position is critical.

Chapter 25

True or False

1. True 2. False 3. False; it is caused by limiting the wavefront. 4. True 5. True 6. False; the ray approximation is valid for any type of wave if the wavelength is very small. 7. False 8. False; it occurs where the waves from the top and bottom of the slit are in phase. 9. True 10. True

Exercises

5. (a) $\theta_1 = 19.5°$ (b) $\theta_2 = 41.8°$

7. (b) 12 mm

13. $\lambda = 6.7 \times 10^{-29}$ m; no

15. away from the normal because the speed of sound is greater in water than in air

17. (a) 1.46 (b) $\lambda = 2.05$ cm

19. 13.3°

21. 0, 22.0, and 48.6°

Problems

3. (a) 62.7° (b) No; the critical angle for glass-air refraction is 41.8°. For angles greater than 41.8° but less than 62.7° the light rays will enter the water, but they cannot get into the air.

Chapter 26

Exercises

3. 12.4 keV

5. 3×10^{18} Hz

7. 1240 nm; infrared

9. (a) $\theta = 30°$ or $\theta = 150°$ (b) 7.07 m

11. 2.1×10^6 years

13. 10.7 min

15. 92%

17. The angle of refraction is about 27.5° for $\lambda = 400$ nm and about 28.1° for $\lambda = 700$ nm, a difference of about 0.6°.

19. 526 nm (using $n = 1.33$); the swimmer observes the same color

21. 1.68 cm

23. (a) 8.54×10^{-3} rad (b) 6.83 cm

25. (a) 56.4 km = 35.0 mi (b) 56.4 m = 185 ft

27. (a) 0.15 mm (b) 33.3 bands per centimeter

29. 39

31. (a) dark, the ray reflecting from air-glass surface suffers a phase change of 180° (b) 1.75×10^{-4} rad $= 0.01°$

33. minimum thickness 1.15×10^{-4} mm

35. $\lambda_1 = 486$ nm; $\lambda_2 = 660$ nm

37. (a) $\theta_{579} = 0.1158$ rad, $\theta_{576} = 0.1154$ rad, $\Delta\theta = 4 \times 10^{-4}$ rad (b) 1.44 mm

39. (a) $I_0/8$ (b) $3I_0/32$

41. (a) 30° (b) 1.73

43. 56.3°

Problems

5. (b) 491 (c) 0.99 mm

7. about 6.5×10^6 km using red light ($\lambda = 700$ nm)

9. (c) A transmission interference pattern is produced by the interference of the light directly transmitted with that twice reflected in the air gap. The pattern is the reverse of the reflected pattern. Since only a small fraction of the light is twice reflected, the transmitted light is nearly uniformly bright with weak interference bands. (d) 68 (e) 1.14 cm (f) Since λ is smaller in water than in air, the fringes would be closer together with 0.75 times the original spacing.

11. (a) minimum thickness 6×10^{-7} m (b) 400, 514, and 720 nm (c) 400, 514, and 720 nm

13. (c) $n_{air} = 1.0003$ (d) not necessary

17. (a) 7.94×10^{-4} mm (b) 1.59×10^{-3} mm (c) no; not unless the indexes of refraction varied in just such a way that $\lambda/(n_O - n_E)$ was constant

Chapter 27

True or False

1. True 2. True 3. False 4. True 5. False 6. True 7. False 8. True 9. False; it is smaller than the object. 10. True

Exercises

7. (a) $s' = 33.3$ cm; real, inverted, reduced by $y' = -y/3$
   (b) $s' = 50$ cm; real, inverted, same size
   (c) $s' = \infty$, no image
   (d) $s' = -16.7$ cm, virtual, erect, enlarged $y' = 5y/3$

9. (a) $s' = -20$ cm, virtual, erect, reduced $y' = y/5$
   (b) $s' = -16.7$ cm, virtual, erect, reduced $y' = y/3$
   (c) $s' = -12.5$ cm, virtual, erect, reduced $y' = y/2$
   (d) $s' = -7.14$ cm, virtual, erect, reduced $y' = 0.714\,y$

13. 1.33 cm

15. $s = 10$ cm

17. (a) $s' = -46.15$ cm, virtual (b) $s' = -6.47$ cm, virtual (c) $s' = 44.1$ cm, real

19. (a) 13.6 cm (b) 20 cm (c) −10 cm (d) −40 cm

21. (a) −40 cm (b) +40 cm (c) −40 cm

23. A negative object distance occurs when a surface interrupts light which was converging to a point beyond the surface. The image point is then a virtual object, and the object distance is negative. (a) $s' = +10$ cm, real, erect (b) $s' = +15$ cm, real, erect

25. $s' = +10$ cm; $y' = -1$ cm

27. There are two solutions: (1) $s = 15$ cm produces a real, inverted image at $s' = 30$ cm; (2) $s = 5$ cm produces a virtual, erect image at $s' = -10$ cm.

31. Image is 30 cm to right of second lens. It is erect and twice as large as object.

33. The computed distances from the vertex of the mirror are (to four significant figures) for 0.5 cm, $s' = 3.000$ cm; for 1.0 cm, $s' = 2.998$ cm; for 2.0 cm, $s' = 2.975$ cm; and for 4.0 cm, $s' = 2.674$ cm.

35. Lens should move 2.78 mm forward.

37. $\theta = 4 \times 10^{-5}$ rad; $\Delta x = 0.8$ mm

41. $M_{25} = 2.5$; $M_{50} = 5.0$; for images at infinity, the retinal images are the same size

43. (a) 30 cm (b) −6 (c) −30 (d) 5.83 cm

45. 17.6 cm

Problems

3. (a) $s' = -128$ cm (b) $s' = 14.7$ cm (c) real

7. (b) 2.55 m

9. $f_{\text{blue}} = 17.4$ cm

Chapter 28

True or False

1. True 2. True 3. False 4. True 5. True 6. False 7. False 8. False 9. True 10. True

Exercises

1. (a) 118 μsec (b) 1.18×10^{-12} sec (c) no

3. *Hint:* Use the binomial expansion.

5. (a) 5.96×10^{-8} sec (b) 16.1 m (c) 7.02 m

7. (a) 0.8 m (b) 4.44 nsec

9. 80 c-min

13. 60 min, which is the result of Exercise 12

15. about $0.21c = 39,000$ mi/sec

17. 0.6c

19. (a) $x_2' - x_1' = \gamma(x_2 - x_1)$ (b) Times t_1 and t_2 in Equations 28-19 are equal to t_0, but times t_1' and t_2' in Equations 28-18 are not equal and not known.

21. (a) $u_x = v$; $u_y = c/\gamma$

27. (a) 4.974 MeV/c (b) 0.9948

29. (a) $938\sqrt{3}$ MeV/$c = 1625$ MeV/c (b) $v/c = \frac{1}{2}\sqrt{3} = 0.866$; $v = 2.60 \times 10^8$ m/sec

31. 6.26 MeV

33. (a) 4.5×10^{-9} u (b) about 8×10^{-9}%, a very small error

Problems   3. (c) 0.866 c and 0.9988 c

5. (a) 60 m (b) 0.9756 c (c) 21.95 m (d) 0.25 μsec

11. (a) 120 min (b) 240 min (c) 60 min; 120 min; 240 min; observer in S' makes same table

15. (c) $P' = Mu/\sqrt{1 - u^2/c^2}$ (f) in S: $E_i = E_k + 2mc^2 = 2mc^2\gamma$; $E_f = Mc^2 = 2m\gamma c^2$ since $M = 2m\gamma$; in S': $E_i = mc^2/\sqrt{1 - u'^2/c^2} + mc^2 = 2mc^2\lambda^2$; $E_f = Mc^2\gamma = 2mc^2\lambda^2$

Chapter 29   **True or False**   1. False; only if the charge is positive. 2. True 3. False; they diverge from a positive point charge. 4. True 5. False; the *net* flux through the surface is zero.

Exercises   1. (a) 6.24×10^{12} (b) 6.24×10^6

3. 230 N

5. $1.5 \times 10^{-2}\mathbf{i}$ N

7. 0.914 kq^2/L^2 away from negative charge

9. (a) $8 \times 10^3\mathbf{i}$ N/C at $x = 2$ m; $9.36 \times 10^3\mathbf{i}$ N/C at $x = 10$ m (b) $x = 4$ m (c) E points in the positive x direction just to the right of the origin and in the negative x direction just to the left of the origin.

11. (a) $-2.59 \times 10^{-4}\mathbf{j}$ N/C (b) $-5.18 \times 10^{-5}\mathbf{j}$ N

13. (a) $|q_{\text{left}}| = 4|q_{\text{right}}|$ (b) left positive, right negative (c) Field is strong close to charges and between them where lines are most dense. Field is weak away from charges, particularly to the right of the negative charge, where the lines are less dense.

17. 0

19. (a) yes (b) 0 (c) 0

21. (a) 0.053 μC (b) no, only that the net charge inside is zero

23. (a) 1.76×10^{11} C/kg (b) 1.76×10^{13} m/sec^2 opposite the direction of E (c) 0.17 μsec (d) 25.6 cm

25. (a) $-7.04 \times 10^{13}\mathbf{j}$ m/sec^2 (b) 50 nsec (c) 8.8 cm

27. electric force about 1.46×10^9 times mg

29. $8 \times 10^{-18}\mathbf{i}$ C-m

Problems   1. (a) 0.914 kq^2/L^2 toward other positive charge

3. (a) 4 and 2 μC (b) $3 + \sqrt{17}$ μC $= +7.12$ μC and $3 - \sqrt{17}$ μC $= -1.12$ μC

5. (c) For $x \gg a$, the separation of charges is not important, and so you expect the field to be like that of a point charge $+2q$.

7. (a) stable for displacements of q_0 along y axis; unstable for displacements along x axis assuming other charges held fixed (b) stable for displacements of q_0 along x axis and unstable for displacements along y axis assuming other charges held fixed (c) $q_0 = -\frac{1}{4}q$. If none of the charges is held fixed, the equilibrium is unstable for displacement of any of the charges. A stable equilibrium distribution of electric charges is not possible under electric forces only.

Chapter 30   **True or False**   1. False 2. True 3. True 4. False

Exercises   1. (a) in the negative x direction (b) $E_y = 0$, $E_x = -2kqa/(y^2 + a^2)^{3/2}$

3. (a) $1.42 \times 10^6\mathbf{i}$ N/C (b) $1.92 \times 10^6\mathbf{i}$ N/C (c) $1.77 \times 10^6\mathbf{i}$ N/C (d) 125$\mathbf{i}$ N/C

5. (a) $\dfrac{k\lambda}{y}\left(\dfrac{b}{\sqrt{y^2+b^2}}+\dfrac{a}{\sqrt{y^2+a^2}}\right)$

7. (a) $k\pi a^2\sigma/x^2$

11. (a) 9×10^5 N/C (b) 1.798×10^6 N/C (c) 1.796×10^6 N/C (d) 4.50×10^5 N/C (e) 1.12×10^5 N/C. These results for parts (c), (d), and (e) are the same as in Exercise 10, but for parts (a) and (b), $E=0$ in Exercise 10. [Answers to parts (b) and (c) are given to four significant figures, using $k=9\times10^9$ N-m²/C² to bring out the slight variation in E from 9.99 to 10.01 cm.]

13. (a) $\mathbf{E}=\dfrac{kQr}{R^3}\,\hat{\mathbf{r}}$ (b) $\mathbf{E}=\dfrac{kQ}{r^2}\,\hat{\mathbf{r}}$, where $Q=\frac{4}{3}\pi R^3\rho$

Problems

1. *Hint:* Set dE_x/dx equal to zero and solve for x.

3. $E_x=kQx(x^2+\frac{1}{4}L^2)^{-1}(x^2+\frac{1}{2}L^2)^{-1/2}$, where $Q=4L\lambda$. For a ring of radius $r=\frac{1}{2}L$ but carrying the same charge, $E_x=kQx(x^2+\frac{1}{4}L^2)^{-3/2}$. *Hint:* Use Equation 30-9 for the magnitude of the field of each segment, replacing y with $r=(x^2+\frac{1}{4}L^2)^{1/2}$. Then take 4 times the x component of this field.

5. (a) left: $-\sigma/\epsilon_0$; right: $+\sigma/\epsilon_0$; between: zero, where the $+$ direction is to the right (b) left: 0; right: 0; between: $+\sigma/\epsilon_0$

7. (a) $Q=\pi AR^4$ (b) inside: $E_r=Ar^2/4\epsilon_0$; outside: $E_r=AR^4/4\epsilon_0r^2=kQ/r^2$

Chapter 31

True or False

1. False; it is zero in electrostatic equilibrium but not in general. 2. False 3. False 4. False; e.g., a neutral conducting sphere in an external electric field has positive charge on one side and negative charge on the other. 5. True

Exercises

1. (a) 2.85×10^{22} electrons (b) 4.56×10^3 C

3. 1.13×10^4 N/C

5. (a) $7.07\ \mu C/m^2$ (b) 8.0×10^5 N/C (c) 0 (d) 4.0×10^5 N/C outward both just inside and just outside the hole (if the portion with its charge were replaced, its field would cancel the field inside and add to the field outside)

Problems

3. (c) 14.3 N/m²

Chapter 32

True or False

1. False; V must be constant in the region but need not be zero 2. True; but if $V=0$ at a *point*, $\mathbf{E}$ need *not* be zero at that point; if V is zero or constant in a region of space, it cannot be changing from point to point, and so $\mathbf{E}$ must also be zero. 3. False 4. True 5. True 6. True

Exercises

1. (a) 2.4×10^{-2} J (b) -2.4×10^{-2} J (c) -8000 V

3. (a) positive (b) 2.5×10^4 V/m

5. (a) $V(x)=-200x$ (b) $V(x)=4000-200x$ (c) $V(x)=200-200x$, where x is in meters and V in volts

7. (a) C_1 is the potential at $x=0$ (b) $\mathbf{E}=-2C_2x\mathbf{i}$; in negative x direction

9. (a) 4.5 kV (b) 1.35×10^{-2} J (c) 1.35×10^{-2} J (d) 1.35×10^{-2} J

11. (a) 9000 and 8970 V (b) decrease, $-\Delta V/\Delta x=2990$ V/m (c) $E_x=3000$ V/m, same to 0.3% (d) $V(3.00,0.01)-9000$ V, the same as at $x-3.00$ m, $y=0$ m to this accuracy. The change in V for displacements in the y direction is related to E_y, which is zero at the point $x=3.00$ m, $y=0$.

13. (a) 3×10^{-2} J (b) -6×10^{-3} J (c) -1.8×10^{-2} J

15. (a) 0 (b) $\mathbf{E} = 6000\mathbf{i}$ V/m (c) $V(3.01) = -60$ V; $-\Delta V/\Delta x = 6000$ V/m, the same as E_x to this accuracy

17. (a) $V(x) = \dfrac{kq}{|x-a|} + \dfrac{kq}{|x+a|}$ (c) A positive test charge placed at this point will be in stable equilibrium.

19. (a) 9×10^3 V/m just outside; 0 just inside (b) 900 V both just outside and just inside (c) $V = 900$ V, $E = 0$ at center

21. 1.77 mm

25. (a) $Q_2/Q_1 = 2$; $Q_1 + Q_2 = 6 \times 10^{-8}$ C, where Q_1 is charge on 4-cm sphere and Q_2 is charge on 8-cm sphere (b) $Q_1 = 2 \times 10^{-8}$ C; $Q_2 = 4 \times 10^{-8}$ C (c) both at 4.5 kV

27. (a) 0.222 μC (b) $Q_1 = 0.0889$ μC (20-cm sphere); $Q_2 = 0.133$ μC (30-cm sphere) (c) 4 kV (d) $\sigma_1 = 0.177$ μC/m^2; $\sigma_2 = 0.118$ μC/m^2

Problems

1. (a) $E_k = +ke^2/2r$ (b) $\frac{1}{2}mv^2 = 2.18 \times 10^{-18}$ J $= 13.6$ eV; $E = \frac{1}{2}mv^2 + U = -13.6$ eV (c) 13.6 eV

3. 4.46×10^{-14} m

5. $V = kqa/z(z-a)$

7. (a) $V = \begin{cases} \sigma x/\epsilon_0 & \text{for } x < 0 \\ 0 & \text{for } 0 \leqslant x \leqslant 0 \\ -\sigma(x-a)/\epsilon_0 & \text{for } x > a \end{cases}$ (b) $V = \begin{cases} 0 & \text{for } x < 0 \\ -\sigma x/\epsilon_0 & \text{for } 0 \leqslant x \leqslant a \\ -\sigma a/\epsilon_0 & \text{for } x > a \end{cases}$

9. $V_a - V_b = \dfrac{2kq}{L} \ln \dfrac{b}{a}$

Chapter 33

True or False

1. False 2. False 3. False 4. True 5. True 6. False 7. True

Exercises

1. (a) 2 pF (b) 200 pF (c) 0.2 μF

3. (a) 9×10^9 m (b) 1410

5. 1.13×10^7 m^2; 3.36 km $= 2.09$ mi

7. 466 μF

9. (a) 6.67 μF (b) 40 μC (c) 4 V across 10-μF capacitor and 2 V across 20-μF capacitor

11. (a) 14 μF (b) $\frac{8}{7}$ μF

13. (a) 100 (b) 10 V (c) $q = 10$ μC; $V = 1$ kV

15. 10 μF

17. 5×10^{-5} J

19. (a) 0.625 J (b) 1.875 J

21. (a) $\epsilon_0 A/2d$ (b) 2V (c) $\epsilon_0 AV^2/d$ (d) $\epsilon_0 AV^2/2d$

23. (a) 10^5 V/m (b) 4.42×10^{-2} J/m^3 (c) 8.85×10^{-5} J (d) same, 8.85×10^{-5} J

25. (a) $E = 2.5 \times 10^4$ V/m; $\sigma = 2.21 \times 10^{-7}$ C/m^2; $U = 6.64 \times 10^{-7}$ J (b) 6.25 kV/m; 25 V (c) 1.66×10^{-7} J (d) 1.66×10^{-7} C/m^2

Problems

3. each capacitor is 5 μF; $C_{eff} = \frac{5}{3}$ μF in series; $C_{eff} = 7.5$ μF one in parallel with two in series; $C_{eff} = \frac{10}{3}$ μF, two in parallel with one in series

5. (a) $E_r = 2Q/Lr$ for $R_1 < r < R_2$; $E = 0$ elsewhere (b) $dU = (kQ^2/L)$ dr/r (c) $U = (kQ^2/L) \ln (R_2/R_1)$, same as $\frac{1}{2}QV$ since $V = (2kQ/L) \ln (R_2/R_1)$

9. (a) 5.0 μF (b) 66.7 V

13. Use 16 capacitors with 4 parallel sets of 4 in series

15. (a) 2.28×10^{-9} F. *Note:* The approximation that this is a parallel-plate capacitor with area $2\pi r L$ and separation $d = 2$ mm gives 2.22×10^{-9} F. (b) about 67 μC

Chapter 34

True or False   1. False 2. True 3. False; this is *definition* of resistance. 4. False; true only for conductors obeying Ohm's law 5. True 6. True 7. True

Exercises   1. (a) 600 C (b) 3.75×10^{21} electrons

3. (a) 0.40 A (since ions and electrons flow in opposite directions, their currents add) (b) 566 A/m²

5. (a) $f = v/2\pi r$ (b) $I = qv/2\pi r$

7. (a) 3.2×10^{13} protons per cubic meter (the proton speed is 6.2×10^7 m/sec obtained from $E_k = \frac{1}{2}mv^2$) (b) 3.75×10^{17} protons (c) $Q = 1.0 \times 10^{-3}t$ C since current is 1.0 mA

9. (a) 1.0 V (b) 0.10 V/m

11. (a) $V_{Cu}/V_{Fe} = 0.17$ (b) E greater in iron

13. (a) 2.75×10^{-2} Ω (b) 3.0×10^{-2} Ω

15. 1.2 Ω

17. 46°C

19. 0.031 Ω

21. (a) 5000 A/m² (b) 3.7×10^{-7} m/sec (c) 5.0×10^{-14} sec (d) about 50 Å

23. (a) 5.7 nsec (b) about 15 min

25. (a) 2.42 kW (b) 1.21 kW

27. (a) $R = 48.4$ Ω; $I = 4.55$ A (b) 207 W

29. (a) 0.24 cent (b) 64 cents

Problems   1. (a) 3.12 V (b) $J = 4.59 \times 10^6$ A/m; $E = 7.8 \times 10^{-2}$ V/m (c) 18.7 W

3. (a) 10^{12} electrons (b) 1.6×10^{-4} A (c) 64 kW (d) 640 MW (e) 1/10,000

5. (a) 2.55×10^6 A/m² (b) $E_{Cu} = 0.0433$ V/m; $E_{Fe} = 0.255$ V/m (c) $V_{Cu} = 3.46$ V; $V_{Fe} = 12.48$ V (d) 7.97 Ω $= R_{Cu} + R_{Fe}$, where $R_{Cu} = 1.73$ Ω and $R_{Fe} = 6.24$ Ω

7. $l_{Cu}/l_C = 264$

Chapter 35

True or False   1. True 2. True 3. True 4. True 5. False 6. False 7. True

Exercises   1. (a) 0.03 Ω (b) 12.3 V

3. (a) $I = 1.13$ A; $P = 6.79$ W (6.41 W delivered to R)

5. $\mathcal{E} = 3$ V; $r = 1$ Ω

7. (a) 1 A (b) 12 W delivered by 12-V battery; 6 W absorbed by 6-V battery (c) 2 W in the 2-Ω resistor, 4 W in the 4-Ω resistor

9. (a) $\frac{4}{3}$ A (b) 132 J (c) 124 J (d) the 2-V battery absorbs 8 J in this time

11. (a) $I_{2\Omega} = 3$ A; $I_{3\Omega} = 2$ A; $I_{1\Omega} = 1$ A (b) $V_a - V_b = 1$ V (c) $P_{7V} = 21$ W; $P_{5V} = 10$ W

13. (a) $\frac{10}{3}$ Ω (b) 3.6 A in first 2-Ω resistor; 1.2 A in 4-Ω resistor and 2.4 A in second 2-Ω resistor

15. (a) $\frac{48}{7}$ Ω (b) 0.75 A in upper branch; 1.0 A in lower branch with 0.5 A in each of the parallel 8-Ω resistors

17. (a) 5 Ω (b) upper branch: $I_{4\Omega} = 1.2$ A with 0.72 A in the upper subbranch and 0.48 A in the lower subbranch (5- and 10-Ω resistors); lower branch: $I_{5\Omega} = 1.2$ A with 0.6 A in each 10-Ω resistor

19. (a) 2 A through 1.5-Ω resistor; 0.75 A through each 4-Ω resistor, 0.5 A through 6-Ω resistor (b) 12 W

21. (a) 600 μC (b) 0.2 A (c) 3 msec (d) 81.2 μC

23. 57.7 Ω

25. (a) 3.79 μC (b) 2.21 μA (c) 2.21 μA (d) 13.2 μW (e) 4.87 μW (f) 8.37 μW

27. (a) 0.050 Ω (b) 4.9 kΩ

29. (a) 0.100 Ω (b) 0.100 Ω (c) 3.00 MΩ

31. (a) 1.49 kΩ (b) 1.50 kΩ (c) 13.5 kΩ

33. (a) 9.90 kΩ (b) 90.0 kΩ (c) 900 kΩ

35. (a) 43.9 Ω (b) 300 Ω (c) 3800 Ω

Problems

3. (a) 2 A to the right through the 1-Ω and first 2-Ω resistor; 1 A down through the second 2-Ω resistor; 1 A down through the 6-Ω resistor (b) 16 W by the 8-V battery; 8 W by the upper 4-V battery and 4 W to the lower 4-V battery

5. $V_a - V_b = 2.4$ V

7. (a) 1.974 A (b) 1.3% (c) 1.48 V (d) 0.001%

9. 600 Ω

11. (b) method a gives $R_c = 0.498$ Ω; 2.91 Ω; 44.4 Ω; method b gives 0.6 Ω; 3.1 Ω; 80.1 Ω

15. (a) $I = \dfrac{R_1 + R_2}{R_1 R_2}\, \mathcal{E}$ (b) $I = \mathcal{E}/R_2$ (c) $I(t) = \mathcal{E}\left(\dfrac{1}{R_2} + \dfrac{1}{R_1}\, e^{-tR_1 C}\right)$ (d) 2.3×10^{-2} sec

Chapter 36

True or False

1. True 2. True 3. True 4. False; the period is independent of the radius. 5. True

Exercises

1. $-1.28 \times 10^{-12}\mathbf{j}$ N

3. (a) $-7.2 \times 10^{-13}\mathbf{j}$ N (b) $4.8 \times 10^{-13}\mathbf{i}$ N (c) $9.6 \times 10^{-13}\mathbf{i} - 7.2 \times 10^{-13}\mathbf{j}$ N

5. 1.0 N

7. 20**k** N/m

9. (a) 2 A-m² (b) 50 A-m

11. (a) 1 dyne-cm/G $= 10^{-3}$ A-m² (b) 1 eV/G $= 1.6 \times 10^{-15}$ A-m²

13. (a) 0.30 A-m² (b) 0.13 N-m

15. (a) 37°; $\hat{\mathbf{n}} = 0.8\mathbf{i} - 0.6\mathbf{j}$ (b) 0.48**n** A-m² $= 0.38\mathbf{i} - 0.29\mathbf{j}$ A-m² (c) 0.58**k** N-m

17. 3.93×10^{-6} N-m

19. (a) 0.131 μsec (b) 3.83×10^7 m/sec (c) 7.66 MeV

21. (a) 2.7 mm (b) 3.5×10^{10} rad/sec (c) 0.18 nsec

23. (a) 0.13 μsec (b) 2.41×10^7 m/sec (c) 12.0 MeV

25. (a) $-10^4\mathbf{k}$ V/m (b) no

27. (a) 6.9 mm (b) 0.75 G

29. (a) 1.44×10^8 rad/sec (b) 27.1 MeV (c) assuming $m_d = 2m_p$, both answers are halved

31. (a) 1.07×10^{-2} cm/sec (b) 5.84×10^{22} electrons/cm^3

33. (a) 3.68×10^{-3} cm/sec (b) 1.47 μV

Problems

1. 0.98 A

3. (a) $B = (Mg/IL) \tan \theta$ (b) $g \sin \theta$ up the rails

Chapter 37

True or False

1. True 2. False 3. False 4. True 5. True

Exercises

1. (a) $4.44 \times 10^{-11}\mathbf{j}$ T (b) $-1.11 \times 10^{-11}\mathbf{j}$ T (c) 0 (d) $-4.44 \times 10^{-11}\mathbf{i}$ T

3. (a) 2×10^{-11} T in direction of the unit vector $\mathbf{n} = -(2/\sqrt{5})\mathbf{i} + (1/\sqrt{5})\mathbf{j}$ (b) $2/\sqrt{5} \times 10^{-11}\mathbf{j}$ T

5. (a) $-8.89 \times 10^{-5}\mathbf{k}$ T (b) 0 (c) $8.89 \times 10^{-5}\mathbf{k}$ T (d) $-1.6 \times 10^{-4}\mathbf{k}$ T

7. (a) $-1.78 \times 10^{-4}\mathbf{k}$ T (b) $-1.33 \times 10^{-4}\mathbf{k}$ T (c) $-1.78 \times 10^{-4}\mathbf{k}$ T (d) $+1.07 \times 10^{-4}\mathbf{k}$ T

9. (a) $6.4 \times 10^{-5}\mathbf{j}$ T (b) $-4.8 \times 10^{-5}\mathbf{k}$ T

11. 0 due to segments along line through P; 5.66×10^{-5} T inward due to each vertical segment, 1.13×10^{-4} T inward due to horizontal segment above P; resultant $\mathbf{B}$ 2.26×10^{-4} T inward

13. 2.36×10^{-5} T

15. (a) $32\sqrt{2}\, k_m I/L = 45.3 k_m I/L$ (b) $4\pi^2 k_m I/L = 39.5 k_m I/L$ (c) the square

17. (a) antiparallel (b) 44.7 mA

19. 28 A

21. (a) 4.5×10^{-4} N/m to right (b) 3×10^{-5} T down

23. (a) $+10\mu_0$ for C_1; 0 for C_2; $-10\mu_0$ for C_3 (b) none of these paths can be used because of lack of symmetry

25. (a) $8\ G = 8 \times 10^{-4}$ T (b) 40 G (c) 28.6 G

27. (a) 6.03×10^{-3} T (b) 0.377 A-m^2 (c) 1.51 A-m (d) 7.9×10^{-9} T using the point pole approximation

29. (a) 0.40 A-m (b) 7.76×10^{-9} N

31. 4800 A/m

33. (a) 1.54×10^6 A/m (b) 1.93 T

35. (a) 0 (b) 1.37×10^{-5} T-m$^2 = 1.37 \times 10^{-5}$ Wb (c) 0 (d) 1.19×10^{-5} Wb

37. 6.74×10^{-3} Wb

39. (a) 1.80×10^{15} N/C-sec (b) $I_d = \epsilon_0 A\, dE/dt = 5$ A (c) 1.60×10^4 A/m^2

Problems

1. The magnetic field due to the cable is 0.05 G at the earth's surface, which is about 7% of magnitude of the earth's field.

3. $v_{||} = (2\pi m/\mu_0 q I)v_t^2$

9. (a) 5.40×10^{-2} T at $x = 5$ cm; 5.39×10^{-2} T at $x = 7$ cm; 5.26×10^{-2} T at $x = 9$ cm; 4.86×10^{-2} T at $x = 11$ cm

11. (a) 5.01×10^{-7} Wb (b) 7.14×10^{-5} N away from wire

13. 1.97×10^{-6} N-m

15. $k_m I$

Chapter 38

True or False

1. False; it is proportional to the rate of change of the flux. 2. True 3. False 4. False; it is an independent law. 5. True

Exercises

1. (a) 6.4×10^{-20} N (b) 0.40 V/m (c) 0.12 V

3. (a) 1.6 V (b) 0.8 A counterclockwise (c) 0.128 N (d) 1.28 W; 1.28 W

5. (a) clockwise; repel (b) counterclockwise; attract

7. (a) counterclockwise (b) clockwise

9. 199 T/sec

11. (a) $\mathcal{E} = 0.4 - 0.2t$ V (b) at $t = 0$, $\phi_m = 0$, $\mathcal{E} = 0.4$ V; at $t = 2$ sec, $\phi_m = -0.4$ T-m^2, $\mathcal{E} = 0$; at $t = 4$ sec, $\phi_m = 0$, $\mathcal{E} = -0.4$ V; at $t = 6$ sec, $\phi_m = 1.2$ T-m^2, $\mathcal{E} = -0.8$ V

13. (a) 1.26×10^{-3} C (b) 12.6 mA (c) 0.628 V

15. 0.80 G

17. (a) 24 Wb (b) (−)1.60 kV

19. (a) 60.3 G (b) 7.58×10^{-4} Wb (c) 25.3 mH (d) 37.9 mV

21. (b) $M = \mu_0 N_1 N_2 A_1 / l_2$ (c) The flux through the solenoid depends on B due to the coil which is not uniform, especially outside the coil.

23. (a) $I = 0$; $dI/dt = 25$ A/sec (b) $I = 2.26$ A; $dI/dt = 20.5$ A/sec (c) $I = 7.90$ A; $dI/dt = 9.20$ A/sec (d) $I = 10.8$ A; $dI/dt = 3.38$ A/sec

25. (a) 13.5 mA (b) 90.8 μA (c) 7.44×10^{-44} A

27. 4.4 msec

29. (a) 44.06 W (b) 40.44 W (c) 3.62 W

31. (a) 2.0 A (b) 4.0 J

33. (a) $W_m = 3.98 \times 10^5$ J (b) $W_e = 4.43 \times 10^{-4}$ J (c) 3.98×10^5 J

35. 1.26 msec

37. (b) third circuit since $I_{max} = \omega Q_0 = \omega CV$ and ω and V are the same for all three circuits

Problems

3. (a) for $0 \le t \le 5$ sec, $\phi_m = 10^{-3}t$ Wb; for 5 sec $\le t \le 10$ sec, $\phi_m = 5 \times 10^{-3}$ Wb; for 10 sec $\le t \le 15$ sec, $\phi_m = (15 - t) \times 10^{-3}$ Wb; for 15 sec $\le t$, $\phi_m = 0$ (b) for $0 \le t < 5$ sec, $\mathcal{E} = -1$ mV; for 5 sec $< t < 10$ sec, $\mathcal{E} = 0$; for 10 sec $< t < 15$ sec, $\mathcal{E} = +1$ mV; for 15 sec $< t$, $\mathcal{E} = 0$

5. $\mu_0 n_1 n_2 \pi r_1^2 l$

7. $R = 10 \, \Omega$; $L = 4$ H

9. (a) 3.53 J (b) 1.61 J (c) 1.92 J

11. (a) $B_0 = 3.4$ G; $B_{av} = 6.8$ G (b) 159 T/sec (c) about 80,000 rev (d) 700 G

13. (a) $F = B^2 l^2 v / R$ (c) $I^2 R = (B^2 l^2 v_0^2 / R) e^{-2t/T}$

15. (a) $I^2 R = 2400 e^{-6000t}$ (b) 0.40 J (c) 0.40 J

Chapter 39

True or False

1. True; but it is often masked by paramagnetism. 2. True 3. False 4. True 5. True 6. False

Exercises

1. (a) $\mathbf{H} = 8.00 \times 10^3 \mathbf{k}$ A/m; $\mathbf{B} = 101\mathbf{k}$ G (b) $\mathbf{H} = 8.00 \times 10^3 \mathbf{k}$ A/m; $\mathbf{B} = 729\mathbf{k}$ G

3. (a) 75 A-m (b) $\mathbf{H}_p = -1.19 \times 10^3 \mathbf{k}$ A/m (c) $\mathbf{H} = +6.81 \times 10^3 \mathbf{k}$ A/m (d) $\mathbf{B} = 714\mathbf{k}$ G

5. H_2, CO_2, and N_2 diamagnetic; O_2 paramagnetic

7. $\mathbf{H} = 8.00 \times 10^3\mathbf{k}$ A/m; $\mathbf{M} = 0.544\mathbf{k}$ A/m; $\mathbf{B} = 101\mathbf{k}$ G

9. $1.69 m_B$

11. (a) $M_s = 5.6 \times 10^5$ A/m; $\mu_0 M_s = 0.70$ T (b) 5.2×10^{-4} (c) The measured magnetization includes that due to induced moments (diamagnetism), which reduces the net magnetization.

13. $\mu = 1.48 \times 10^{-5}$ H/m; $K_m = 11.7$

15. 11 A

17. (a) 1000 A/m (b) 1.26×10^6 A/m (c) 1.26×10^3

Problems

1. (b) $H = NI/2\pi R$, the same as without core since there are no end effects: $B = \mu_0 NI/2\pi R + \mu_0 M$

Chapter 40

True or False   1. True 2. True 3. True 4. False 5. True 6. True

Exercises

1. (a) 250 rad/sec (b) 4.8π V $= 15.1$ V

3. (b) 156 V

5. (a) 120π rad/sec $= 377$ rad/sec (b) $I_{max} = 4.0$ A; $I_{rms} = I_{max}/\sqrt{2} = 2.83$ A (c) 48 W (d) 0 (e) $P_{av} = I_{rms}^2 R = P_{max}/2 = 24$ W

7. (a) 265 Ω (b) 26.5 Ω (c) 2.65 Ω

9. 25.1 mA

11. (a) 15.9 kHz; angular frequency 10^5 rad/sec (b) 159 Hz; angular frequency 10^3 rad/sec (c) 1.59 MHz; angular frequency 10^7 rad/sec

13. (a) 0.377 Ω (b) 3.77 Ω (c) 37.7 Ω

15. 1.59 kHz

17. (a) 7.07×10^3 rad/sec (b) 20 A (c) $X_C = 62.5$ Ω; $X_L = 80$ Ω (d) $Z = 18.2$ Ω; $I_{max} = 5.49$ A (e) $\phi = -74°$

19. (a) 1.13 kHz (b) $X_C = 19.9$ Ω; $X_L = 251\Omega$ (c) $Z = 231$ Ω; $I_{max} = 0.432$ A (d) $\phi = -89°$

21. 10^4

23. (a) $Q = 10\sqrt{2} = 14.1$ (b) 500 rad/sec (c) 0.275

25. (a) 0.539 (b) 95.3 mA (c) 0.726 W

27. (a) 0.553 (b) 608 mA (c) 37.0 W

29. (a) step down (b) 2.2 V (c) $(-)5$ A

Problems

3. (a) $L = 0.8$ mH; $C = 12.5$ μF (b) 1.6 (c) for the given values of X_L and X_C, $I_{max} = 2$ A; at resonance $I_{max} = 5.2$ A

11. (a) 856 W (b) 7.07 Ω (c) 102 μF (d) add a 38.4-μF capacitor in parallel or a 19-mH inductor in series

13. (a) 12 Ω (b) $R = 7.2$ Ω, $X = 9.6$ Ω (c) capacitive

Chapter 41

Exercises

1. (a) wave is traveling in $+x$ direction

3. $E_{rms} = 194$ V/m; $B_{rms} = 6.47$ mG

5. (a) 13.3 mG (b) 7.08×10^{-7} J/m^3 (c) 212 W/m^2

Problems   1. $I = 1.91$ kW/m²; $E_{rms} = 849$ V/m; $B_{rms} = 28$ mG

3. (b) 3.9×10^{26} W   (c) $E_{rms} = 726$ V/m; $B_{rms} = 24.2$ mG

Chapter 42   **True or False**   1. True 2. True 3. False; it is a property of the metal. 4. True 5. True 6. False 7. True 8. True 9. True 10. True 11. True 12. False 13. True 14. False

Exercises   1. 1.77 to 3.10 eV

3. 4.14×10^{-7} eV

5. (a) 4.13 eV (b) 2.10 eV (c) 1.00 V

7. (a) $f_t = 4.59 \times 10^{14}$ Hz; $\lambda_t = 653$ nm (b) 2.23 V (c) 1.20 V

9. (a) 17.44 keV (b) 0.0760 nm (c) 16.32 keV (d) 1.12 keV

11. (a) 103 nm (b) 97.2 nm (c) 91.2 nm

13. $\lambda_{4 \to 3} = 1876$ nm; $\lambda_{5 \to 3} = 1282$ nm; $\lambda_{6 \to 3} = 1094$ nm; $\lambda_{\infty \to 3} = 820.5$ nm

15. (a) 12.26 Å (b) 1.23 Å (c) 0.388 Å (d) 0.123 Å

17. 2.02 Å

19. about 5×10^{-20} m $= 5 \times 10^{-5}$ fm, where 1 fm $= 10^{-15}$ m is of the order of the diameter of the nucleus

21. 1.95 Å

Problems   1. Plot c/λ versus V_0. (a) 2.05 eV (b) about 5×10^{14} Hz (c) 4.14×10^{-15} V-sec

3. (a) 10^{-22} W (b) about 53 min

7. $\mu/m = 0.99946 = R_H/R_\infty$

9. (b) $E_1 = h^2/m_p r^2 = 6.25 \times 10^{-3}$ eV (c) $T_c \approx 73$ K (d) about 13 eV using $r = 0.53$ Å; $T_c \approx 1.5 \times 10^5$ K

Index

Some Conversion Factors

1 m = 39.37 in = 3.281 ft = 1.094 yd

1 m = 10^{15} fm = 10^{10} Å = 10^9 nm

1 mi = 5280 ft = 1.609 km

1 in = 2.540 cm

1 ℓ = 10^3 cm^3 = 10^{-3} m^3 = 1.057 qt

1 y = 365.24 d = 3.156×10^7 sec

1 ft/sec = 0.3048 m/sec = 0.6818 mi/h

1 rev = 2π rad = 360°

1 rad = 57.30°

1 rev/min = 0.1047 rad/sec

1 slug = 14.59 kg

1 atm = 1.013×10^5 N/m^2 = 1.013 bars = 76.00 cm Hg = 14.70 lb/in^2 (psi)

1 N = 10^5 dynes = 0.2248 lb

1 lb = 4.448 N

1 J = 10^7 ergs = 0.7373 ft-lb = 9.869×10^{-3} ℓ-atm

1 cal = 4.184 J = 4.129×10^{-2} ℓ-atm

1 ℓ-atm = 101.3 J = 24.22 cal

1 eV = 1.602×10^{-19} J

1 Btu = 778 ft-lb = 252 cal = 1054 J

1 horsepower = 550 ft-lb/sec = 746 W

1 T = 10^4 G

1 kg weighs about 2.205 lb